Numerical Mathematics and Advanced Applications

Alfredo Bermúdez de Castro   Dolores Gómez
Peregrina Quintela   Pilar Salgado   *(Editors)*

Numerical Mathematics and Advanced Applications

Proceedings of ENUMATH 2005,
the 6th European Conference on Numerical
Mathematics and Advanced Applications

Santiago de Compostela, Spain, July 2005

Editors

Alfredo Bermúdez de Castro
Dolores Gómez
Peregrina Quintela

Departamento de Matemática Aplicada
Facultade de Matemáticas
Universidade de Santiago de Compostela
15706 Santiago de Compostela, Spain
E-mail: mabermud@usc.es
       malola@usc.es
       mapere@usc.es

Pilar Salgado

Departamento de Matemática Aplicada
Escola Politécnica Superior
27002 Lugo, Spain
E-mail: mpilar@usc.es

Library of Congress Control Number: 2006930397

Mathematics Subject Classification (2000): 65-XX, 74-XX, 76-XX, 78-XX

ISBN-10 3-540-34287-7 Springer Berlin Heidelberg New York
ISBN-13 978-3-540-34287-8 Springer Berlin Heidelberg New York

This work is subject to copyright. All rights are reserved, whether the whole or part of the material is concerned, specifically the rights of translation, reprinting, reuse of illustrations, recitation, broadcasting, reproduction on microfilm or in any other way, and storage in data banks. Duplication of this publication or parts thereof is permitted only under the provisions of the German Copyright Law of September 9, 1965, in its current version, and permission for use must always be obtained from Springer. Violations are liable for prosecution under the German Copyright Law.

Springer is a part of Springer Science+Business Media
springer.com
© Springer-Verlag Berlin Heidelberg 2006

The use of general descriptive names, registered names, trademarks, etc. in this publication does not imply, even in the absence of a specific statement, that such names are exempt from the relevant protective laws and regulations and therefore free for general use.

Typesetting: by the author and techbooks using a Springer LATEX macro package
Cover design: *design & production* GmbH, Heidelberg

Printed on acid-free paper   SPIN: 11755623   46/techbooks   5 4 3 2 1 0

Preface

The European Conference on Numerical Mathematics and Advanced Applications (ENUMATH) is a series of meetings held every two years to provide a forum for discussion on recent aspects of numerical mathematics and their applications. They seek to convene leading experts and young scientists with special emphasis on contributions from Europe. The first ENUMATH meeting held in Paris (1995), and the series continued by the ones in Heidelberg (1997), Jÿvaskÿla (1999), Ischia (2001) and Prague (2003).

This book collects the major part of the lectures given at ENUMATH 2005, that took place in Santiago de Compostela, Spain, from July 18 to 22, 2005. It contents texts of invited speakers, and a selection of papers presented in minisymposia and works communicated within the sessions.

The importance of numerical methods has increased dramatically in science and engineering, reflecting today's unprecedented use of computers. The increasing importance of modeling in addition to numerical simulation was again evident in ENUMATH 2005. Indeed, nodaways mathematics is generally accepted as a technology, playing a crucial role in many branches of industrial activity. Recent results and new trends in the analysis of numerical algorithms as well as their application to challenging scientific and industrial problems were discussed during the meeting. Apart from the theoretical aspects, a major part of the conference was devoted to numerical methods for interdisciplinary applications, with emphasis on showing the potential of new computational methods for solving practical multidisciplinary problems.

We are happy that so many people have shown their interest in this meeting. In addition to the ten invited presentations, we had more than 192 talks during the five-day meeting and about 215 participants from thirty four countries, specially from Europe. A total of 123 contributions appear in these proceedings. The contents range over several of the most active research fields, and survey many of the latest developments in scientific computing. Topics include applications such as atmosphere and ocean, water pollution, electromagnetism, interface problems, waves, finance, heat transfer, unbounded domains, numerical linear algebra, convection-diffusion, fluid-structure, plates, solids, hyperbolic equations, multiphase flow, Navier-Stokes, singular perturbation problems, non-linear PDE, control, parabolic equations, as well as methodologies such as a posteriori error estimates, discontinuous Galerkin methods, multiscale methods, optimization, adaptive methods, domain decomposition techniques, exponential integrators, hp-finite elements, level set methods, fractional step methods, penalty procedures, and finite volumes.

We would like to thank all the participants for the attendance and for their valuable contributions to discussions during the meeting. Special thanks to minisymposium organizers, who made a large contribution to the conference, the chairpersons, the speakers, and, in particular, to the contributors of this volume.

We would like to address our warmest thanks to the invited speakers: A. Buffa (Italy), R. Codina (Spain), W. Dahmen (Germany), Z. Dostál (Czech Republic), A. Ern (France), A. Iserles (United Kingdom), K. Kunisch (Austria), P. Monk (USA), S. Repin (St.Petersburg), E. Zuazua (Spain), for coming to Santiago de Compostela and contributing to the success of the conference with the high quality of their presentations.

A big share of the success of this conference should be given to the members of the Programme Committee (F. Brezzi, M. Feistauer, R. Glowinski, R. Jeltsch, Yu. Kuznetsov, J. Periaux, R. Rannacher) who contribute with their time and energy to produce this series of meetings.

We are greatly indebted to the Scientific Committee (O. Axelsson, C. Bernardi, C. Canuto, E. Fernández-Cara, M. Griebel, R. Hoppe, G. Kobelkov, M. Krizek, P. Hansbo, P. Neittaanmäki, O. Pironneau, A. Quarteroni, J. Sanz-Serna, C. Schwab, E. Süli, W. Wendland) and the external anonymous reviewers who performed the invaluable task of reviewing and selecting the contributed material.

We gratefully acknowledge the financial support provided by the the Spanish Ministerio de Educación y Ciencia, the Xunta de Galicia and the Universidade de Santiago de Compostela, which, in particular, allowed us to grant the participation of many young researchers. We also thank to Springer-Verlag for its cooperation in publishing these proceedings.

Finally, we would like to thank Manuel Porto for his administrative help, Tono Lago for the computer support, and the research students Marta Benítez, Ana María Ferreiro, Laura Saavedra, Inés Santos, Rafael Vázquez, and, particularly, to M$^{\underline{a}}$ Cristina Naya for their help during the meeting.

We think that this book presents a valuable state of the art of the most recent research in scientific computing, providing to the reader the latest developments concerning the mathematical issues and the applications of this active field of science.

Santiago de Compostela, Spain *Alfredo Bermúdez*
July 2006 *Dolores Gómez*
Peregrina Quintela
Pilar Salgado

Contents

PLENARY LECTURES

Compatible Discretizations in Two Dimensions
Annalisa Buffa 3

Finite Element Approximation of the Three Field Formulation of the Elasticity Problem Using Stabilization
Ramon Codina 21

Convergence of Adaptive Wavelet Methods for Goal–Oriented Error Estimation
Wolfgang Dahmen, Angela Kunoth, Jürgen Vorloeper 39

Quadratic Programming and Scalable Algorithms for Variational Inequalities
Zdeněk Dostál, David Horák, Dan Stefanica 62

Discontinuous Galerkin Methods for Friedrichs' Systems
Alexandre Ern, Jean-Luc Guermond 79

Highly Oscillatory Quadrature: The Story so Far
A. Iserles, S.P. Nørsett, S. Olver 97

The 3D Inverse Electromagnetic Scattering Problem for a Coated Dielectric
Fioralba Cakoni, Peter Monk 119

Functional Approach to Locally Based A Posteriori Error Estimates for Elliptic and Parabolic Problems
Sergey Repin 135

Finite Element Approximation of 2D Parabolic Optimal Design Problems
Miguel Cea, Enrique Zuazua 151

CONTRIBUTED LECTURES

ACCURATE FINITE VOLUME SOLVERS FOR FLUID FLOWS

3D Free Surface Flows Simulations Using a Multilayer Saint-Venant Model. Comparisons with Navier-Stokes Solutions
E. Audusse, M.O. Bristeau, A. Decoene 181

Some Well-Balanced Shallow Water-Sediment Transport Models
M. J. Castro-Díaz, E. D. Fernández-Nieto, A. M. Ferreiro 190

Highly Accurate Conservative Finite Difference Schemes and Adaptive Mesh Refinement Techniques for Hyperbolic Systems of Conservation Laws
Pep Mulet, Antonio Baeza 198

Finite Volume Solvers for the Shallow Water Equations Using Matrix Radial Basis Function Reconstruction
L. Bonaventura, E. Miglio, F. Saleri 207

ADAPTIVE METHODS

On Numerical Schemes for a Hierarchy of Kinetic Equations
Hans Babovsky, Laek S. Andallah 217

Computational Aspects of the Mesh Adaptation for the Time Marching Procedure
Jiří Felcman, Petr Kubera 225

On the Use of Slope Limiters for the Design of Recovery Based Error Indicators
M. Möller, D. Kuzmin 233

A POSTERIORI ERROR ESTIMATION

On a Superconvergence Result for Mixed Approximation of Eigenvalue Problems
Francesca Gardini 243

Comparative Study of the a Posteriori Error Estimators
for the Stokes Problem
Elena Gorshkova, Pekka Neittaanmäki, Sergey Repin 252

Error Control for Discretizations of Electromagnetic-
Mechanical Multifield Problem
Marcus Stiemer 260

A Safeguarded Zienkiewicz-Zhu Estimator
Francesca Fierro, Andreas Veeser 269

ATMOSPHERIC AND OCEAN

Some Remarks on a Model for the Atmospheric Pressure
in Ocean Dynamics
T. Chacón Rebollo, E. D. Fernández Nieto, M. Gómez Mármol 279

Computational Time Improvement for Some Shallow
Water Finite Volume Models Applying Parallelization and
Optimized Small Matrix Computations.
M. J. Castro, J. A. García, J. M. González, C. Parés 288

CONTROL

Discretization Error Estimates for an Optimal Control
Problem in a Nonconvex Domain
Th. Apel, A. Rösch, G. Winkler 299

A Posteriori Estimates for Cost Functionals of Optimal
Control Problems
Alexandra Gaevskaya, Ronald H.W. Hoppe, Sergey Repin 308

CONVECTION-DIFUSSION

Optimization of a Duality Method for the Compressible
Reynolds Equation
Iñigo Arregui, J. Jesús Cendán, Carlos Parés, Carlos Vázquez 319

Time-Space & Space-Time Elements for Unsteady
Advection-Dominated Problems
Maria Isabel Asensio, Blanca Ayuso, Giancarlo Sangalli 328

On Discontinuity–Capturing Methods for Convection–
Diffusion Equations
Volker John, Petr Knobloch 336

Algebraic Flux Correction for Finite Element Approximation of Transport Equations
Dmitri Kuzmin 345

A Parallel Multiparametric Gauss-Seidel Method
N. M. Missirlis, F. I. Tzaferis 354

A Numerical Scheme for the Micro Scale Dissolution and Precipitation in Porous Media
I.S. Pop, V.M. Devigne, C.J. van Duijn, T. Clopeau 362

Degenerate Parabolic Equations: Theory, Numerics and Applications

---

Discrete Kinetic Methods for a Degenerate Parabolic Equation in Dimension Two.
Denise Aregba-Driollet 373

Anisotropic Doubly Nonlinear Degenerate Parabolic Equations
Mostafa Bendahmane, Kenneth H. Karlsen 381

A Multiresolution Method for the Simulation of Sedimentation-Consolidation Processes
Raimund Bürger, Alice Kozakevicius 387

Diffusive Relaxation Limit for Hyperbolic Systems
Corrado Lattanzio 396

Parallel Algorithms for Nonlinear Diffusion by Using Relaxation Approximation
Fausto Cavalli, Giovanni Naldi, Matteo Semplice 404

On a Degenerated Parabolic-Hyperbolic Problem Arising From Stratigraphy
Guy Vallet 412

Discontinuous Galerkin

Schwarz Domain Decomposition Preconditioners for Interior Penalty Approximations of Elliptic Problems
Paola F. Antonietti, Blanca Ayuso, Luca Heltai 423

Higher Order Semi-Implicit Discontinuous Galerkin Finite Element Schemes for Nonlinear Convection-Diffusion Problems
Vít Dolejší 432

On Some Aspects of the Discontinuous Galerkin Method
Miloslav Feistauer 440

Mixed Discontinuous Galerkin Methods with Minimal Stabilization
Daniele Marazzina 448

Discontinuous Galerkin Finite Element Method for a Fourth-Order Nonlinear Elliptic Equation Related to the Two-Dimensional Navier–Stokes Equations
Igor Mozolevski, Endre Süli, Paulo Rafael Bösing 457

DOMAIN DECOMPOSITION

Fourier Method with Nitsche-Mortaring for the Poisson Equation in 3D
Bernd Heinrich, Beate Jung 467

Substructuring Preconditioners for the Bidomain Extracellular Potential Problem
Micol Pennacchio, Valeria Simoncini 475

EDGE ORIENTED INTERIOR PENALTY PROCEDURES IN COMPUTATIONAL FLUID DYNAMICS

A Face Penalty Method for the Three Fields Stokes Equation Arising from Oldroyd-B Viscoelastic Flows
Andrea Bonito, Erik Burman 487

Anisotropic H^1-Stable Projections on Quadrilateral Meshes
Malte Braack 495

Continuous Interior Penalty hp-Finite Element Methods for Transport Operators
Erik Burman, Alexandre Ern 504

A Nonconforming Finite Element Method with Face Penalty for Advection–Diffusion Equations
L. El Alaoui, A. Ern, E. Burman 512

Efficient Multigrid and Data Structures for Edge-Oriented FEM Stabilization
Abderrahim Ouazzi, Stefan Turek 520

Electromagnetism

Adaptive Methods for Dynamical Micromagnetics
Ľubomír Baňas 531

Stability for Walls in Ferromagnetic Nanowire
G. Carbou, S. Labbé 539

Continuous Galerkin Methods for Solving Maxwell Equations in 3D Geometries
Patrick Ciarlet, Jr, Erell Jamelot 547

Exponential Integrators

On the Use of the Gautschi-Type Exponential Integrator for Wave Equations
Volker Grimm 557

Positivity of Exponential Multistep Methods
Alexander Ostermann, Mechthild Thalhammer 564

Fluid-Structure Interaction

Stability Results and Algorithmic Strategies for the Finite Element Approach to the Immersed Boundary Method
Daniele Boffi, Lucia Gastaldi, Luca Heltai 575

Heat Transfer and Combustion

A Comparison of Enthalpy and Temperature Methods for Melting Problems on Composite Domains
J.H. Brusche, A. Segal, C. Vuik, H.P. Urbach 585

Qualitative Properties of a Numerical Scheme for the Heat Equation
Liviu I. Ignat 593

Modeling Radiation and Moisture Content in Fire Spread
L. Ferragut, M.I. Asensio, S. Monedero 601

Fast Multipole Method for Solving the Radiosity Equation
J. Morice, K. Mer-Nkonga, A. Bachelot 609

Numerical Modelling of Kinetic Equations
J. Banasiak, N. Parumasur, J.M. Kozakiewicz 618

HYPERBOLIC EQUATIONS

On a Subclass of Hölder Continuous Functions with Applications to Signal Processing
Sergio Amat, Sonia Busquier, Antonio Escudero, J. Carlos Trillo 629

Modelisation and Simulation of Static Grain Deep-Bed Drying
Aworou-Waste Aregba, Denise Aregba-Driollet 638

Hybrid Godunov-Glimm Method for a Nonconservative Hyperbolic System with Kinetic Relations
Bruno Audebert, Frédéric Coquel 646

Cell-Average Multiwavelets Based on Hermite Interpolation
F. Aràndiga, A. Baeza, R. Donat 654

On a General Definition of the Godunov Method for Nonconservative Hyperbolic Systems. Application to Linear Balance Laws
M.J. Castro, J.M. Gallardo, M.L. Muñoz, C. Parés 662

Sequential Flux-Corrected Remapping for ALE Methods
Pavel Váchal, Richard Liska 671

---

hp-FINITE ELEMENTS

Orthogonal *hp*-FEM for Elliptic Problems Based on a Non-Affine Concept
Pavel Šolín, Tomáš Vejchodský, Martin Zítka 683

On Some Aspects of the *hp*-FEM for Time-Harmonic Maxwell's Equations
Tomáš Vejchodský, Pavel Šolín, Martin Zítka 691

INTERFACE PROBLEMS

Numerical Simulation of Phase-Transition Front Propagation in Thermoelastic Solids
A. Berezovski, G.A. Maugin 703

The Level Set Method for Solid-Solid Phase Transformations
E. Javierre, C. Vuik, F. Vermolen, A. Segal, S. van der Zwaag 712

LEVEL-SET METHODS, HAMILTON-JACOBI EQUATIONS AND APPLICATIONS

A Non-Monotone Fast Marching Scheme for a Hamilton-Jacobi Equation Modelling Dislocation Dynamics
Elisabetta Carlini, Emiliano Cristiani, Nicolas Forcadel 723

A Time–Adaptive Semi–Lagrangian Approximation to Mean Curvature Motion
Elisabetta Carlini, Maurizio Falcone, Roberto Ferretti................. 732

MULTISCALE METHODS

Heterogeneous Multiscale Methods with Quadrilateral Finite Elements
Assyr Abdulle 743

Stabilizing the $\mathbb{P}^1/\mathbb{P}^0$ Element for the Stokes Problem via Multiscale Enrichment
Rodolfo Araya, Gabriel R. Barrenechea, Frédéric Valentin............. 752

Adaptive Multiresolution Methods for the Simulation of Shocks/Shear Layer Interaction in Confined Flows
L. Bentaleb, O. Roussel, C. Tenaud 761

Local Projection Stabilization for the Stokes System on Anisotropic Quadrilateral Meshes
Malte Braack, Thomas Richter 770

An Interior Penalty Variational Multiscale Method for High Reynolds Number Flows
Erik Burman... 779

Variational Multiscale Large Eddy Simulation of Turbulent Flows Using a Two-Grid Finite Element or Finite Volume Method
Volker Gravemeier 788

Issues for a Mathematical Definition of LES
Jean-Luc Guermond, Serge Prudhomme 796

Stabilized FEM with Anisotropic Mesh Refinement for the Oseen Problem
Gert Lube, Tobias Knopp, Ralf Gritzki............................... 805

Semi-Implicit Multiresolution for Multiphase Flows
N. Andrianov, F. Coquel, M. Postel, Q. H. Tran 814

Numerical Simulation of Vortex-Dipole Wall Interactions
Using an Adaptive Wavelet Discretization with Volume
Penalisation
Kai Schneider, Marie Farge 822

Inviscid Flow on Moving Grids with Multiscale Space
and Time Adaptivity
Philipp Lamby, Ralf Massjung, Siegfried Müller, Youssef Stiriba 831

Multiphase Flow

A Relaxation Method for a Two Phase Flow with Surface
Tension
C. Berthon, B. Braconnier, J. Claudel, B. Nkonga 843

Extension of Interface Coupling to General
Lagrangian Systems
*A. Ambroso, C. Chalons, F. Coquel, E. Godlewski, F. Lagoutière,
P.-A. Raviart, N. Seguin* 852

A Numerical Scheme for the Modeling of Condensation
and Flash Vaporization in Compressible Multi-Phase Flows
Vincent Perrier, Rémi Abgrall, Ludovic Hallo 861

Navier-Stokes

An Adaptive Operator Splitting of Higher Order for the
Navier-Stokes Equations
Jörg Frochte, Wilhelm Heinrichs 871

The POD Technique for Computing Bifurcation Diagrams:
A Comparison among Different Models in Fluids
Pedro Galán del Sastre, Rodolfo Bermejo 880

Filtering of Singularities in a Marangoni Convection Problem
Henar Herrero, Ana M. Mancho, Sergio Hoyas 889

On Application of Stabilized Higher Order Finite Element
Method on Unsteady Incompressible Flow Problems
Petr Sváček, Jaromír Horáček 897

Numerical Simulation of Coupled Fluid-Solid Systems by Fictitious Boundary and Grid Deformation Methods
Decheng Wan, Stefan Turek 906

NON-LINEAR PDE

An Iterative Method for Solving Non-Linear Hydromagnetic Equations
C. Boulbe, T.Z. Boulmezaoud, T. Amari 917

Mathematical and Numerical Analysis of a Class of Non-linear Elliptic Equations in the Two Dimensional Case
Nour Eddine Alaa, Abderrahim Cheggour, Jean R. Roche 926

NUMERICAL LINEAR ALGEBRA AND APPROXIMATION METHODS

A s-step Variant of the Double Orthogonal Series Algorithm
J.A. Alvarez-Dios, J.C. Cabaleiro, G. Casal 937

Linear Equations in Quaternions
Drahoslava Janovská, Gerhard Opfer 945

Computing the Analytic Singular Value Decomposition via a Pathfollowing
Vladimír Janovský, Drahoslava Janovská, Kunio Tanabe 954

A Jacobi-Davidson Method for Computing Partial Generalized Real Schur Forms
Tycho van Noorden, Joost Rommes 963

NUMERICAL METHODS IN FINANCE

Pricing Multi-Asset Options with Sparse Grids and Fourth Order Finite Differences
C.C.W. Leentvaar, C.W. Oosterlee 975

ODE AND FRACTIONAL STEP METHODS

A Third Order Linearly Implicit Fractional Step Method for Semilinear Parabolic Problems
Blanca Bujanda, Juan Carlos Jorge 987

Numerical Solution of Optimal Control Problems with Sparse SQP-Methods
Georg Wimmer, Thorsten Steinmetz, Markus Clemens 996

Optimization

Semi-Deterministic Recursive Optimization Methods for Multichannel Optical Filters
Benjamin Ivorra, Bijan Mohammadi, Laurent Dumas, Olivier Durand . 1007

A Multigrid Method for Coupled Optimal Topology and Shape Design in Nonlinear Magnetostatics
Dalibor Lukáš 1015

Nonsmooth Optimization of Eigenvalues in Topology Optimization
K. Moritzen 1023

Derivative Free Optimization of Stirrer Configurations
M. Schäfer, B. Karasözen, Ö. Uğur, K. Yapıcı 1031

Mathematical Modelling and Numerical Optimization in the Process of River Pollution Control
L.J. Alvarez-Vázquez, A. Martínez, M.E. Vázquez-Méndez, M. Vilar .. 1040

Plates

A Family of C^0 Finite Elements for Kirchhoff Plates with Free Boundary Conditions
L. Beirão da Veiga, J. Niiranen, R. Stenberg 1051

A Postprocessing Method for the MITC Plate Elements
Mikko Lyly, Jarkko Niiranen, Rolf Stenberg 1059

A Uniformly Stable Finite Difference Space Semi-Discretization for the Internal Stabilization of the Plate Equation in a Square
Karim Ramdani, Takéo Takahashi, Marius Tucsnak 1068

Singular Perturbation

An ε-Uniform Hybrid Scheme for Singularly Perturbed 1-D Reaction-Diffusion Problems
S. Natesan, R.K. Bawa, C. Clavero 1079

Solids

A Dynamic Frictional Contact Problem of a Viscoelastic Beam
M. Campo, J.R. Fernández, G.E. Stavroulakis, J.M. Viaño 1091

Numerical Analysis of a Frictional Contact Problem for Viscoelastic Materials with Long-Term Memory
A. Rodríguez-Arós, M. Sofonea, J. Viaño 1099

A Suitable Numerical Algorithm for the Simulation of the Butt Curl Deformation of an Aluminium Slab
P. Barral, M.T. Sánchez 1108

An Efficient Solution Algorithm for Elastoplasticity and its First Implementation Towards Uniform h- and p- Mesh Refinements
Johanna Kienesberger, Jan Valdman 1117

Unbounded Domains

A LDG-BEM Coupling for a Class of Nonlinear Exterior Transmission Problems
Rommel Bustinza, Gabriel N. Gatica, Francisco-Javier Sayas 1129

High Order Boundary Integral Methods for Maxwell's Equations: Coupling of Microlocal Discretization and Fast Multipole Methods
L. Gatard, A. Bachelot, K. Mer-Nkonga 1137

Indirect Methods with Brakhage–Werner Potentials for Helmholtz Transmission Problems
María–Luisa Rapún, Francisco–Javier Sayas 1146

A FEM–BEM Formulation for a Time–Dependent Eddy Current Problem
S. Meddahi, V. Selgas 1155

Mixed Boundary Element–Finite Volume Methods for Thermohydrodynamic Lubrication Problems
J. Durany, J. Pereira, F. Varas 1164

WATER POLLUTION

Numerical Modelling for Leaching of Pesticides in Soils Modified by a Cationic Surfactant
M.I. Asensio, L. Ferragut, S. Monedero, M.S. Rodríguez-Cruz, M.J. Sánchez-Martín 1175

Formulation of Mixed-Hybrid FE Model of Flow in Fractured Porous Medium
Jiřina Královcová, Jiří Maryška, Otto Severýn, Jan Šembera 1184

Newton–Type Methods for the Mixed Finite Element Discretization of Some Degenerate Parabolic Equations
Florin A. Radu, Iuliu Sorin Pop, Peter Knabner 1192

WAVES

Domain Decomposition Methods for Wave Propagation in Heterogeneous Media
R. Glowinski, S. Lapin, J. Periaux, P.M. Jacquart, H.Q. Chen 1203

Galbrun's Equation Solved by a First Order Characteristics Method
Rodolfo Rodríguez, Duarte Santamarina 1212

Open Subsystems of Conservative Systems
Alexander Figotin, Stephen P. Shipman 1220

Author Index 1229

PLENARY LECTURES

Compatible Discretizations in Two Dimensions

Annalisa Buffa

Istituto di Matematica Applicata e Tecnologie Informatiche del CNR,
Via Ferrata 1, 27100 Pavia, ITALY
annalisa@imati.cnr.it

Summary. In this paper we recall the construction of the dual finite element complex introduced in [11] and we investigate some applications. More precisely, we propose and analyze fully compatible discretizations for the magnetostatics and the Darcy flow equations in two dimensions, and we introduce an optimal matching condition for domain decomposition methods for Maxwell equations in three dimensions.

1 Introduction

The use of differential complexes has become increasingly popular in the numerical analysis for partial differential equations. As shown in [2] (see also [3, 4]), they provide a framework for the understanding of the properties of numerical schemes for systems of first order equations such as magnetostatics, Darcy flow, the Stokes problem and so on. The use of differential complexes, or finite elements which form suitable differential complexes, allow to construct stable discretizations which also enjoy some (not all) local conservation properties. In what follows we say that a discrete method is "compatible" when conformity and *all* conservations are preserved locally. These ideas in the field of electromagnetics has been put forward by Bossavit in [9] and then used by many authors (see [17, 18] and the references there in). Moreover, finite element techniques bases on differential complexes are strictly related with finite difference techniques like the Finite Integration Technique (see [14] and the reference therein), or the Mimetic Finite Differences Technique (see [8] and the references there in). We believe that the deep relation existing among these ideas is still not completely understood.

A missing step for the use of differential complexes to provide compatible discretizations is the construction of discrete stable Hodge-$\star$ operators (see. e.g. [16] for a first attempt in this direction). Without entering into the details of differential forms, we explain the concept through an example borrowed by physics: magnetostatics. The magnetic induction **B** and the magnetic field

H coexists, they verify div**B** $= 0$ and curl**H** $=$ **J**, where **J** is a fixed current density. Measurements of **B** are fluxes and measurements of **H** are circulations. In the modeling of constitutive relation **B** $= \mu(\mathbf{H})$ we need an operator μ which maps circulation into fluxes, this is an Hodge-$\star$ operator. A compatible numerical method should be able to reproduce this action. On the other hand, this concept can be useful in the numerical analysis of PDEs only if it is combined with a metric and, in particular, with Sobolev spaces. Let us then introduce some notation. Given a Lipschitz bounded domain $\Omega \subset \mathbb{R}^2$ and any $s \in [-1, 1]$ we denote by $\mathrm{H}^s(\Omega)$ the standard Sobolev space of regularity s (see [15, p. 16]), $\widetilde{\mathrm{H}}^s(\Omega) := \{u \in \mathrm{H}^s(\Omega) \; : \; \widetilde{u} \in \mathrm{H}^s(\mathbb{R}^2)\}$, where $\widetilde{u}$ is the extension by zero of u (see [15, p. 18]). Note that it holds $\widetilde{\mathrm{H}}^{-s}(\Omega) := \left(\mathrm{H}^s(\Omega)\right)'$ and $\mathrm{H}^{-s}(\Omega) := \left(\widetilde{\mathrm{H}}^s(\Omega)\right)'$. In a similar way, we introduce, for $d =$ div and $d =$ curl:[1]

$$\mathrm{H}^s(d, \Omega) = \{\mathbf{u} \in \mathrm{H}^s(\Omega)^2 \; : \; d\mathbf{u} \in \mathrm{H}^s(\Omega)\}; \tag{1}$$

$$\widetilde{\mathrm{H}}^s(d, \Omega) = \{\mathbf{u} \in \mathrm{H}^s(d, \Omega) \; : \; \widetilde{\mathbf{u}} \in \mathrm{H}^s(d, \mathbb{R}^2)\}. \tag{2}$$

The following duality relation holds true (see [12, 13]):

$$\left(\mathrm{H}^{-s}(\mathrm{curl}, \Omega)\right)' = \widetilde{\mathrm{H}}^{s-1}(\mathrm{div}, \Omega) \quad s \in [0, 1].$$

We can draw the following diagram, $s \in (-\tfrac{1}{2}, \tfrac{1}{2})$:

$$\begin{array}{ccccc}
\widetilde{\mathrm{H}}^s(\Omega) & \stackrel{\mathrm{curl}}{\Longrightarrow} & \widetilde{\mathrm{H}}^{s-1}(\mathrm{div}, \Omega) & \stackrel{\mathrm{div}}{\longrightarrow} & \widetilde{\mathrm{H}}^{s-1}(\Omega) \\
\star_0 \downarrow & & \star_1 \downarrow & & \star_2 \downarrow \\
\mathrm{H}^{-s}(\Omega) & \stackrel{\mathrm{curl}}{\Longleftarrow} & \mathrm{H}^{-s}(\mathrm{curl}, \Omega) & \stackrel{-\mathrm{grad}}{\longleftarrow} & \mathrm{H}^{1-s}(\Omega)
\end{array}$$

It is apparent that the Hodge-$\star$ operators are the (isomorphic) identifications between spaces on the first line and their duals on the second line: the vertical arrows. In the paper [11], a discrete analog of this diagram is provided when the first line is discretized by the complex centered around low order Raviart-Thomas (RT) finite elements on a given simplicial mesh $\mathcal{T}_h$ (see (3) for the definition of spaces, and [10] for details). More precisely, another discrete complex $(\mathrm{Y}_h^0, \mathrm{Y}_h^1, \mathrm{Y}_h^2)$ is built as discretization of the second line in order to ensure that the vertical lines remains uniformly stable isomorphisms in natural norms. This is way we call it "dual complex". In other words, the authors built couples of finite dimensional spaces $(\mathrm{X}_h^i, \mathrm{Y}_h^i)$ which are linked by discrete, uniformly stable Hodge-$\star$ operators, i.e., which are inf-sup stable in natural norms, with respect to the L^2 duality pairing.

In this paper, after recalling the construction and main properties of the new family of finite elements, we analyze its applications. In Section 3 we first provide a compatible discretization scheme for a general div $-$ curl problem. The scheme is then adapted in Section 3.1 to magnetostatics and Darcy flow

---

[1] We recall: curl$u = \partial_x u_2 - \partial_y u_1$ and $\underline{\mathrm{curl}}\, u = (\partial_y u, -\partial_x u)$.

equation. For both problems, two different compatible discretizations are proposed and their stability properties are analyzed. Finally, in Section 3.2, we use the dual complex to provide an optimal matching conditions for domain decomposition methods on non-matching grids for Maxwell equations in three dimensions. For all these examples only the stability properties are analyzed and the corresponding error estimates are object of on-going research.

2 Construction of the dual complex

2.1 Definition and algebraic properties

Let Ω be a bounded Lipschitz polygon in $\mathbb{R}^2$, Γ be its boundary and **n** be the outer unit normal. We equip Ω with a simplicial mesh denoted $\mathcal{T}_h$, and we denote by $\mathcal{T}_h^i$, $i = 0, 1, 2$ the set of vertices, (closed) edges and (closed) triangles of $\mathcal{T}_h$. For further use, we also introduce the barycentric refinement of $\mathcal{T}_h$ whish is constructed by dividing each triangle $s \in \mathcal{T}_h^2$, into six triangles by drawing the six edges joining the barycentre of s with the vertexes of s as well as the midpoints of its edges. The barycentric refinement of $\mathcal{T}_h$ is denoted $\mathcal{T}_h'$. In the figures 1, 3 and 4 the edges of $\mathcal{T}_h$ are drawn in bold, whereas non-bold segments are edges of $\mathcal{T}_h'$ (all bold segments are also edges of $\mathcal{T}_h'$).

On $\mathcal{T}_h$ we consider the lowest order finite-element complex (X_h^0, X_h^1, X_h^2) based on Raviart-Thomas divergence conforming vector fields RT_0. It is defined by:

$$X_h^0 = \{u \in \widetilde{\mathrm{H}}^1(\Omega) \quad : \forall\, t \in \mathcal{T}_h^2 \quad u|_t \in \mathcal{P}_1\}, \tag{3a}$$

$$X_h^1 = \{u \in \widetilde{\mathrm{H}}(\mathrm{div}, \Omega) : \forall\, t \in \mathcal{T}_h^2 \quad u|_t \in \mathrm{RT}_0\}, \tag{3b}$$

$$X_h^2 = \{u \in \mathrm{L}_0^2(\Omega) \quad : \forall\, t \in \mathcal{T}_h^2 \quad u|_t \in \mathcal{P}_0\}, \tag{3c}$$

where $\mathrm{L}_0^2(\Omega)$ denotes the space of L^2 functions with zero mean value. For generalities about mixed finite elements and Raviart-Thomas vector fields in particular, we refer to [10]. These spaces satisfy $\underline{\mathrm{curl}} X_h^0 \subset X_h^1$ and $\mathrm{div} X_h^1 \subset X_h^2$, so that the spaces do indeed form a complex:

$$X_h^0 \xRightarrow{\underline{\mathrm{curl}}} X_h^1 \xrightarrow{\mathrm{div}} X_h^2. \tag{4}$$

We denote by $\lambda^i = (\lambda_s^i)$ indexed by $s \in \mathcal{T}_h^i$ the standard basis of X_h^i. For each i, the usual family of degrees on freedom relative to X_h^i will be denoted $l^i = (l_s^i)$ indexed by $s \in \mathcal{T}_h^i$. Then l_v^0 is evaluation at the vertex v, l_e^1 is integration of the normal component along the edge e in some orientation, and l_t^2 is integration on the triangle t. In a sense, for each i and each s, l_s^i can be represented as integration on the simplex s. The basis λ^i of X_h^i is characterized by the property that $l_s^i(\lambda_t^i) = \delta_{st}$.

On $\mathcal{T}_h'$ we consider the slightly different finite-element complex $(X_h'^0, X_h'^1, X_h'^2)$ defined by:

$$X_h'^0 = \{u \in H^1(\Omega) \quad : \forall t \in \mathcal{T}_h'^2 \quad u|_t \in \mathcal{P}_1\}, \tag{5a}$$

$$X_h'^1 = \{u \in H(\text{curl}, \Omega) : \forall t \in \mathcal{T}_h'^2 \quad u|_t \in \text{RT}_0 \times n\}, \tag{5b}$$

$$X_h'^2 = \{u \in L^2(\Omega) \quad : \forall t \in \mathcal{T}_h'^2 \quad u|_t \in \mathcal{P}_0\}. \tag{5c}$$

The only two differences from the spaces corresponding to X_h^i on the refined mesh $\mathcal{T}_h'$ is that we rotate the middle one by the operation $u \to u \times n$, and that we remove the boundary conditions. These spaces satisfy $\text{grad} X_h'^0 \subset X_h'^1$ and $\text{curl} X_h'^1 \subset X_h'^2$ so that we have the complex:

$$X_h'^0 \xrightarrow{\text{grad}} X_h'^1 \xrightarrow{\text{curl}} X_h'^2. \tag{6}$$

Basis are constructed for the spaces $X_h'^i$ associated with $\mathcal{T}_h'$ as for the spaces associated with $\mathcal{T}_h$, and denoted $(\lambda_s'^i : s \in \mathcal{T}_h'^i)$ (the corresponding degrees of freedom will not be needed).

In the paper [11], the authors construct subspaces $Y_h^i \subset X_h'^i$ such that on the one hand Y_h^i is L^2-dual to X_h^{2-i} (in the sense of satisfying a Babuska-Brezzi Inf-Sup condition uniformly in h, in appropriate norms), and on the other hand they should form a complex:

$$Y_h^0 \xrightarrow{\text{grad}} Y_h^1 \xrightarrow{\text{curl}} Y_h^2. \tag{7}$$

We recall here the construction of these spaces, by means of explicit definition of their basis functions. To this aim we fix some notation. For each triangle $t \in \mathcal{T}_h^2$, let t' denote its barycentre. For each edge $e \in \mathcal{T}_h^1$, let e' be union of (the geometric realizations of) the two edges of $\mathcal{T}_h'$ joining the barycentre of e to the barycentres of the two neighboring triangles. The oriented tangent vector along e' is denoted $\tau_{e'}$, orientation being chosen such that $\tau_{e'} \cdot \tau_e \times n < 0$. For each vertex $v \in \mathcal{T}_h^0$, denote by v' the union of (the geometric realizations of) the triangles of $\mathcal{T}_h'$ containing v.

For each $i \in \{0, 1, 2\}$ and each simplex $s \in \mathcal{T}_h^{2-i}$, let $\mu_s^i \in X_h'^i$ be the field attached to s constructed as a linear combination of the functions $\lambda_s'^i$ with the following coefficients:

- For $i = 0$, let $t \in \mathcal{T}_h^2$. We need to distinguish three cases:
  (i) $t \cap \partial\Omega = \emptyset$, the coefficients are shown in Figure 1; t is the triangle of $\mathcal{T}_h$ whose barycentre carries the coefficient 1. Thus μ_t^0 is the continuous piecewise affine function on $\mathcal{T}_h'$ with non-zero values at the vertices shown in that figure.
  (ii) $t \cap \partial\Omega \in \mathcal{T}_h^1$, i.e., t shares an edge with the boundary, then the basis function associated with t is $\mu_t^0 \in \widetilde{X}_h'^0$ having the coefficients shown in Figure 2(a), where t is the triangle of $\mathcal{T}_h$ whose barycentre carries the coefficient 1.
  (iii) $t \cap \partial\Omega \in \mathcal{T}_h^0$, i.e., t shares a vertex with the boundary, then $\mu_t^0 \in \widetilde{X}_h'^0$ has the coefficients shown in Figure 2(b)

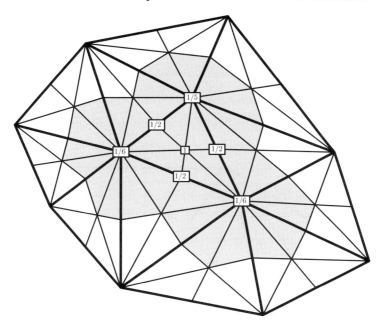

Fig. 1. A basis element for Y_h^0 expressed in the basis of $X_h'^0$.

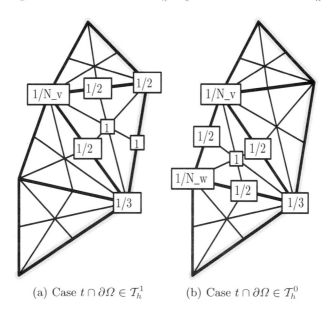

(a) Case $t \cap \partial \Omega \in \mathcal{T}_h^1$ (b) Case $t \cap \partial \Omega \in \mathcal{T}_h^0$

Fig. 2. The two types of boundary basis element for Y_h^0 expressed in the basis of $X_h'^0$. The shaded gray region corresponds to the boundary $\partial \Omega$.

- For $i = 1$, let $s \in \mathcal{T}_h^1$. We need to distinguish two cases:

  (i) $s \cap \partial\Omega = \emptyset$. The coefficients are shown in Figure 3; s is the central edge and we have oriented the edges as pointing away from it. Thus μ_s^1 is the Nédélec vector field on $\mathcal{T}_h'$ such that the integrals of the tangent component on edges is the coefficient shown in the figure. The coefficient of each edge should be multiplied by the one indicated at its origin e.g. to the left we have coefficients ranging from $5/12$ to $-5/12$ when ordered counterclockwise.

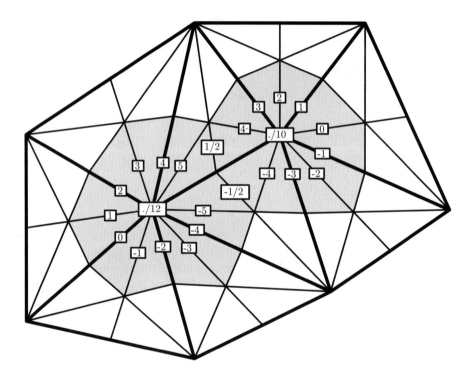

Fig. 3. A basis element for Y_h^1 expressed in the basis of $X_h'^1$.

(ii) $s \cap \partial\Omega \neq \emptyset$. We associate a basis function only with those edges $e \in \mathcal{T}_h^1$ such that $e \cap \partial\Omega \in \mathcal{T}_h^0$, i.e., which share a vertex with the boundary. Let $v \in \mathcal{T}_h^0$ be on $\partial\Omega$ and $m_v + 1$ the number of triangles $t \in \mathcal{T}_h^2$ sharing v as a vertex. We number these triangles as $t_0, \ldots, t_{m_v}$ turning around v in a counterclockwise sense. Accordingly we number edges e such that $e \cap \partial\Omega = v$ as $e_1, e_2, \ldots e_{m_v - 1}$ and denote by w_i the other vertex of e_i. We suppose each e_i to be oriented from w_i to v, and we denote by $\mu^\star_{e_i}$ the basis function associated with e_i built as in (i) here above. The basis function $\mu_{e_i}^1$ associated with e_i is then defined as:

$$\mu_{e_i}^1 = \begin{cases} \mu_{e_i}^\star & \text{on } w_i' \\ \sum_{j=0}^{i-1} \operatorname{grad} \mu_{t_j}^0 & \text{on } v'. \end{cases} \tag{8}$$

- For $i = 2$ and $s \in T_h^0$. The coefficients are shown in Figure 4; s is the central vertex. All 12 triangles of T_h' in the shaded region should carry the same weight $1/12$. Thus μ_s^2 is the piecewise constant field on T_h' whose integral is $1/12$ on each shaded triangle. We associate a basis function only to those vertices s such that $s \in \Omega$.

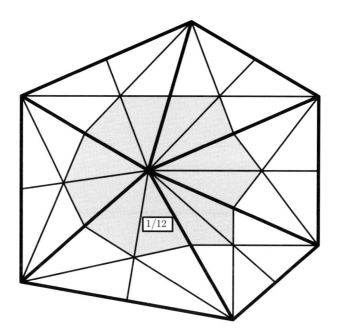

Fig. 4. A basis element for Y_h^2 expressed in the basis of $X_h'^2$.

In each figure the shaded region is the support of the corresponding field We define Y_h^i by:

$$Y_h^i = \operatorname{span}\{\mu_s^i \;:\; s \in T_h^{2-i}, (s \setminus \partial s) \cap \partial \Omega = \emptyset\}. \tag{9}$$

For each integer $i \in \{0, 1, 2\}$, we now construct families of linear forms on fields (scalar or vector according to the case) whose restrictions to Y_h^i are linearly independent. These linear forms are the degrees of freedom (dof).

We now define three families of degrees of freedom:

$$\mathrm{M}_h^0 = (m_t^0 : u \mapsto u(t') \quad : t \in T_h^2), \tag{10a}$$
$$\mathrm{M}_h^1 = (m_e^1 : u \mapsto \int_{e'} u \cdot \tau_{e'} \;:\; e \in T_h^1), \tag{10b}$$
$$\mathrm{M}_h^2 = (m_v^2 : u \mapsto \int_{v'} u \quad : v \in T_h^0). \tag{10c}$$

It is perhaps worth remarking that the first family of linear forms M_h^0 can also be written as integrals (with respect to the trivial measure on points). In this sense the three preceding definitions may be written:

$$M_h^i = (m_s^i : u \mapsto \int_{s'} u \;:\; s \in \mathcal{T}_h^{2-i}), \tag{11}$$

where we integrate on certain dual geometric objects s' relative to $\mathcal{T}_h'$ and attached to simplexes $s \in \mathcal{T}_h$, defined above.

Proposition 1. *For each $i \in \{0,1,2\}$ and each i-dimensional simplexes $s, s' \in \mathcal{T}_h^i$ we have:*

$$m_s^i(\mu_{s'}^i) = \delta_{ss'}. \tag{12}$$

In particular, for each i the family $\mu^i = (\mu_s^i)$ indexed by $s \in \mathcal{T}_h^{2-i}$ is a basis for Y_h^i, and an element $u \in Y_h^i$, $i = 0, 1, 2$ is uniquely determined the values $m_s^i(u)$ for $s \in \mathcal{T}_h^{2-i}$.

Proof. This is a matter of straightforward checking. ∎

We also remark that:

Proposition 2. *The family of functions $(\mu_s^i \;:\; s \in \mathcal{T}_h^2)$ is a partition of unity.*

Proof. It is enough to remark that for each $s \in \mathcal{T}_h^2$ the nonzero values of μ_s^i at the vertexes v of the barycentric refinement $\mathcal{T}_h'$ are the inverses of the number of triangles $t \in \mathcal{T}_h^2$ such that $v \in t$. Therefore the sum of the functions μ_s^i evaluated at any such vertex v is 1. ∎

Proposition 3. *We have* $\mathrm{grad} Y_h^0 \subset Y_h^1$ *and* $\mathrm{curl} Y_h^1 \subset Y_h^2$. *Moreover the matrix of* $\mathrm{grad} : Y_h^0 \to Y_h^1$ *in the basis* $\mu^0 \to \mu^1$ *is minus the transpose of the matrix of* $\mathrm{div} : X_h^1 \to X_h^2$ *in the standard basis, and similarly the matrix of* $\mathrm{curl} : Y_h^1 \to Y_h^2$ *in the basis* $\mu^1 \to \mu^2$ *is the transpose of the matrix of* $\mathrm{curl} : X_h^0 \to X_h^1$ *in the standard basis.*

Proof. Concerning the grad operator one checks that for each triangle $t \in \mathcal{T}_h$, $\mathrm{grad}\mu_t^0$ is a linear combination of the three vector-fields μ_e^1 where e is an edge of t. The coefficients are 1 or -1 according to orientations of the edges. Checking this is a matter of elementary but tedious computations using only the definitions of basis functions. The matrix thus formed is known as an incidence matrix and its transpose is also known to be the matrix of $-\mathrm{div} : X_h^1 \to X_h^2$ in the standard basis. The case of the curl operator is similar. ∎

For each $i \in \{0, 1, 2\}$ we denote by I_h^i the interpolation operator associated with the d.o.f M_h^i. Explicitly I_h^i associates with a field u (scalar or vector according to i) the element u_h of Y_h^i such that :

$$\forall s \in \mathcal{T}_h^{2-i} \quad m_s^i(u_h) = m_s^i(u). \tag{13}$$

Let $\Delta^0 \subset \tilde{H}^1(\Omega)$, $\Delta^1 \subset (\mathrm{curl}, \Omega)$ and $\Delta^2 \subset L^2(\Omega)$ be the subspaces consisting of piecewise smooth fields. We then have:

Proposition 4. *The interpolators satisfy the following commuting diagram:*

$$
\begin{array}{ccccc}
\Delta^0 & \xrightarrow{\mathrm{grad}} & \Delta^1 & \xrightarrow{\mathrm{curl}} & \Delta^2 \\
I_h^0 \downarrow & & I_h^1 \downarrow & & I_h^2 \downarrow \\
Y_h^0 & \xrightarrow{\mathrm{grad}} & Y_h^1 & \xrightarrow{\mathrm{curl}} & Y_h^2
\end{array}
\qquad (14)
$$

Proof. This follows from an application of Stokes theorem on the geometric elements s' we associated with the simplexes $s \in \mathcal{T}_h$ in order to define the degrees of freedom. ∎

Moreover,

Proposition 5. *In the following complex, the cohomology groups have the "right" dimension.*

$$ 0 \longrightarrow Y_h^0 \xrightarrow{\mathrm{grad}} Y_h^1 \xrightarrow{\mathrm{curl}} Y_h^2 \longrightarrow 0. \qquad (15) $$

Specifically, for a connected domain, for the first cohomology group an element of Y_h^0 has gradient 0 iff it is constant, whereas for the last cohomology group an element of Y_h^2 is the curl of an element of Y_h^1.

Proof. By Proposition 3, we already know that $\mathrm{grad} Y_h^0 \subseteq Y_h^1$, and $\mathrm{curl} Y_h^1 \subset Y_h^2$. We need now to prove that $\{u \in Y_h^1 : \mathrm{curl} u = 0\} = \mathrm{grad} Y_h^0$. Take $u \in Y_h^1 : \mathrm{curl} u = 0$. Since $Y_h^1 \subset X_h'^1$, there exists a $p \in X_h'^0$ such that $u = \mathrm{grad} p$. Given a edge e, let v_1 and v_2 its end point. Note that $\int_{e'} u \cdot \tau_{e'} = \int_{e'} \mathrm{grad} p \cdot \tau_{e'} = \pm(p(v_1') - p(v_2'))$. Let $q \in X_h^0$ be such that $q(v') = p(v'), \forall v \in \mathcal{T}_h^0$. We have $\mathrm{grad} q \subset Y_h^1$. On the other hand, by construction $\int_{e'} u \cdot \tau_{e'} = \int_{e'} \mathrm{grad} q \cdot \tau_{e'}$. Since the d.o.fs M_h^1 are uni-solvent, $u = \mathrm{grad} q$. The last statement is an application of the Euler identity. ∎

Proposition 4 and 5 are the main tool to prove the validity of a uniform discrete Friedrichs inequality (see [11, Section 3.2]): for all $u \in Y_h^1$ it holds

$$ \int_\Omega u \cdot \mathrm{grad} q = 0 \quad \forall q \in Y_h^0 \Rightarrow \|u\|_0 \le C_F \|\mathrm{curl} u\|_0; \qquad (16) $$

where C_F is a constant which depends only upon the domain Ω. The same result holds true in a slightly more general situation: let **a** be a positive definite 2×2 matrix with piecewise regular coefficients, then for all $u \in Y_h^1$ it holds

$$ \int_\Omega \mathbf{a}\, u \cdot \mathrm{grad} q = 0 \quad \forall q \in Y_h^0 \Rightarrow \|u\|_0 \le C_F' \|\mathrm{curl} u\|_0; \qquad (17) $$

where C_F' is a constant which depends only upon the domain Ω and $\|\mathbf{a}\|$. The same type of result holds true for Raviart-Thomas finite elements (see [10] and [1] for details).

2.2 LBB inf-sup conditions

In the work [11], the authors prove that the couples $X_h^i - Y_h^{2-i}$, $i = 0, 1, 2$ are inf-sup stable for the $L^2(\Omega)$ scalar product for a range of Sobolev indices. Here we report only the ones we need in the applications we propose in Section 3. Under the assumption that the mesh $\mathcal{T}_h$ is quasi-uniform there hold:

(i) The couple $X_h^0 - Y_h^2$ is inf-sup stable: i.e., there exists $\alpha > 0$ independent of the mesh size s.t.

$$\inf_{u \in X_h^2} \sup_{v \in Y_h^0} \frac{\int uv}{\|u\|_0 \|v\|_0} \geq \alpha \quad \text{and} \quad \inf_{v \in Y_h^0} \sup_{u \in X_h^2} \frac{\int uv}{\|u\|_0 \|v\|_0} \geq \alpha. \quad (18)$$

(ii) The couple $X_h^2 - Y_h^0$ is inf-sup stable: i.e., there exists $\alpha > 0$ independent of the mesh size s.t.

$$\inf_{u \in X_h^0} \sup_{v \in Y_h^2} \frac{\int uv}{\|u\|_0 \|v\|_0} \geq \alpha \quad \text{and} \quad \inf_{v \in Y_h^2} \sup_{u \in X_h^0} \frac{\int uv}{\|u\|_0 \|v\|_0} \geq \alpha. \quad (19)$$

(iii) The question whether the couple $X_h^1 - Y_h^1$ is L^2 inf-sup stable is still open. In [11], the authors prove that this couple is stable with respect to other norms but L^2 which are relevant for the application proposed later in Section 3.2. We consider the spaces $\widetilde{H}^{-\frac{1}{2}}(\text{div}, \Omega)$ and $H^{-\frac{1}{2}}(\text{curl}, \Omega)$ as defined in the Introduction, formulae (1) endowed with their graph norms $\|\cdot\|_{-\frac{1}{2}, \text{div}}$ and $\|\cdot\|_{-\frac{1}{2}, \text{curl}}$. Following [13], we know it holds $\widetilde{H}^{-\frac{1}{2}}(\text{div}, \Omega) = \left(H^{-\frac{1}{2}}(\text{curl}, \Omega)\right)'$, i.e. these spaces are in duality with L^2 as a pivot space: with a little abuse of notation (because one should make explicit the meaning of the numerator in the next fraction), this can be expressed in formulae as:

$$\inf_{u \in \widetilde{H}^{-\frac{1}{2}}(\text{div}, \Omega)} \sup_{v \in H^{-\frac{1}{2}}(\text{curl}, \Omega)} \frac{\int u \cdot v}{\|u\|_{-\frac{1}{2}, \text{div}} \|v\|_{-\frac{1}{2}, \text{curl}}} > \beta > 0. \quad (20)$$

We have $X_h^1 \subset \widetilde{H}^{-\frac{1}{2}}(\text{div}, \Omega)$, $Y_h^1 \subset H^{-\frac{1}{2}}(\text{curl}, \Omega)$ and the following discrete counterpart of (20) is proved in [11]: there exists $\alpha > 0$ independent of the mesh size such that

$$\inf_{u \in X_h^1} \sup_{v \in Y_h^1} \frac{\int u \cdot v}{\|u\|_{-\frac{1}{2}, \text{div}} \|v\|_{-\frac{1}{2}, \text{curl}}} \geq \alpha > 0. \quad (21)$$

3 Applications

This section is devoted to some applications of the finite element complex introduced in Section 2. In [11], the complex $\{Y_h^i\}$ is used to provide an optimal preconditioner for integral equations in electromagnetics, and more precisely for the Electric Field Integral Equation. Here, we devote our attention to other applications where a suitable modeling of the Hodge-$\star$ operator allow for "compatible" discretization of some partial differential equations in the sense of [2]. The meaning we give to the word "compatible" is made clear through in the following Section 3.1. Finally, in the Section 3.2, we show how to apply this theory to the construction of an optimal mortar method for Maxwell equation in three dimension. For all examples, the attention is devoted only to the wellposedness and stability of the discrete problems we propose and no error analysis is provided. A complete error analysis for the schemes proposed in this section will be the object of a future work.

In this section, $\|\cdot\|_s$, $s \in [-1, 1]$ will denote the standard Sobolev norm of $H^s(\Omega)$.

3.1 The div-curl problem and some applications

We formulate the div-curl problem in two dimensions in the following way: given $f \in L^2(\Omega)$, $g \in L^2_0(\Omega)$, find

$$\begin{cases} \text{div}\mathbf{x} = g & \Omega \\ \text{curl}\mathbf{y} = f & \Omega \\ \mathbf{x} \cdot \mathbf{n} = 0 & \partial\Omega. \end{cases} \quad (22)$$

together with the constitutive law $\mathbf{x} = \mathbf{a}\mathbf{y}$ where $\mathbf{a}$ is assumed here to be a positive smooth function or a positive definite 2×2 matrix with smooth coefficients. Equations (22) can be reformulated in terms of conservation and continuity in the following way: for any subset $T \subset \Omega$ it holds

$$[\![\mathbf{x}]\!]_{\nu,\partial T} = 0 \quad \text{and} \quad \int_{\partial T} \mathbf{x} \cdot \boldsymbol{\nu} = \int_T g \quad (23a)$$
$$[\![\mathbf{y}]\!]_{\tau,\partial T} = 0 \quad \text{and} \quad \int_{\partial T} \mathbf{y} \cdot \boldsymbol{\tau} = \int_T f \quad (23b)$$

where $\boldsymbol{\tau}$ and $\boldsymbol{\nu}$ are the tangential and normal unit vectors at ∂T, respectively; $[\![\cdot]\!]_{\nu,\partial T}$ and $[\![\cdot]\!]_{\tau,\partial T}$ denote the jumps of the normal and tangential component, respectively. We say that a discretization of (22) is "compatible" when (23) are verified on each element of the mesh (or of a dual mesh), or, in order words, when the discretization spaces are conforming and the local conservations are preserved.

A well known discretization of this problem is by means of RT finite elements. In this case, the discretization reads: Find $\mathbf{x}_h \in X_h^1$ such that:

$$\int_\Omega \text{div}\mathbf{x}_h q = \int_\Omega g q \quad \forall q \in X_h^2 \qquad \int_\Omega \mathbf{a}\mathbf{x}_h \cdot \text{curl}v = \int_\Omega f v \quad \forall v \in X_h^0. \quad (24)$$

This discretization provides a wellposed problem [10], which is not compatible: the conservation and continuity (23b) are not satisfied on any subset T of Ω. This can be seen as a lack of modeling for the constitutive relation $\mathbf{y} = \mathbf{a}\mathbf{x}$.

Thanks to the new finite element complex (Y_h^0, Y_h^1, Y_h^2), we can propose a compatible discretization of (22) as follows: Find $\mathbf{x}_h \in X_h^1$ and $\mathbf{y}_h \in Y_h^1$ such that:

$$\int_\Omega \mathrm{div}\mathbf{x}_h \, q = \int_\Omega g q \quad \forall q \in X_h^2 \qquad \int_\Omega \mathrm{curl}\mathbf{y}_h \, q' = \int_\Omega f q' \quad \forall q' \in Y_h^2 \qquad (25a)$$

$$\int_\Omega (\mathbf{y}_h - \mathbf{a}^{-1}\mathbf{x}_h)\mathbf{x}_h^t = 0 \quad \forall \mathbf{x}_h^t \in X_h^1. \qquad (25b)$$

In alternative, (25b) can be replaced by its analog:

$$\int_\Omega (\mathbf{a}\mathbf{y}_h - \mathbf{x}_h)\mathbf{y}_h^t = 0 \quad \forall \mathbf{y}_h^t \in Y_h^1. \qquad (26)$$

All next theorems and remarks will apply with no change to the discrete problem (25a)-(26).

A few remarks are due:

1. conformity steams directly from the choice of the spaces;
2. conservation is achieved on all $T \in \mathcal{T}_h^2$ for the divergence equation, and on all dual cells v', $v \in \mathcal{T}_h^0$ for the curl equation;
3. apparently we multiply by two the number of unknowns. Indeed, both projections (25b) or (26) provide a mapping from one set of unknowns (dof for $\mathbf{x}_h$) to the other (dof for $\mathbf{y}_h$) which is, in general, cheap to compute once that the basis functions for both X_h^1 and Y_h^1 have been built.

Thus, the discretization (25) is compatible. It remains to prove that the discrete problem is wellposed and this is the object of the following theorem.

Theorem 1. *The problem (25) admits a unique solution $(\mathbf{x}_h, \mathbf{y}_h) \in X_h^1 \times Y_h^1$ which verifies:*

$$\|\mathbf{x}_h\|_0 + \|\mathbf{y}_h\|_{-1} \leq C(\|f\|_0 + \|g\|_0) \qquad (27)$$

where the constant C does not depend upon the data and the mesh size.

Proof. Since the problem is finite dimensional, uniqueness implies existence. We prove uniqueness. Let $(\mathbf{x}_h, \mathbf{y}_h) \in X_h^1 \times Y_h^1$ be a solution of (25). Using the properties of the complexes (4) and (15), we decompose $\mathbf{x}_h$ and $\mathbf{y}_h$ as follows:

$$\mathbf{x}_h = \underline{\mathrm{curl}}q + \boldsymbol{\xi}, \; q \in X_h^0, \quad \boldsymbol{\xi} \in X_h^1 \; : \; \int_\Omega \mathbf{a}^{-1}\boldsymbol{\xi} \cdot \underline{\mathrm{curl}}\chi = 0 \; \forall \chi \in X_h^0;$$

$$\mathbf{y}_h = \mathrm{grad}p + \boldsymbol{\psi}, \; p \in Y_h^0, \quad \boldsymbol{\psi} \in Y_h^1 \; : \; \int_\Omega \boldsymbol{\psi} \cdot \mathrm{grad}p^t = 0 \; \forall p^t \in Y_h^0.$$

$$(28)$$

By means of the discrete Friedrichs inequalities for the space X_h^1 (see [10]) and for the space Y_h^1 (see (16)), we have $\|\boldsymbol{\xi}\|_0 \leq C\|\mathrm{div}\boldsymbol{\xi}\|_0$, and $\|\boldsymbol{\psi}\|_0 \leq C\|\mathrm{curl}\boldsymbol{\psi}\|_0$. Thus, the quantities $\boldsymbol{\psi}$ and $\boldsymbol{\xi}$ are determined by the equations (25a) only, and it holds:

$$\|\boldsymbol{\xi}\|_0 \leq C\|g\|_0 \qquad \|\boldsymbol{\psi}\|_0 \leq C\|f\|_0. \qquad (29)$$

Using the decomposition (28) for the test function $\mathbf{x}_h^t \in X_h^1$, (25b) splits in two:

$$\int_\Omega (\mathbf{a}^{-1}\underline{\mathrm{curl}q} - \boldsymbol{\psi})\underline{\mathrm{curl}q}^t = 0 \ \forall q^t \qquad \int_\Omega (\mathbf{a}^{-1}\boldsymbol{\xi} - \boldsymbol{\psi} - \mathrm{grad}p)\boldsymbol{\xi}^t = 0 \ \forall \boldsymbol{\xi}^t. \quad (30)$$

Now, from the first equation, we deduce $\|\underline{\mathrm{curl}q}\|_0 \leq C\|\boldsymbol{\psi}\|_0$, and from the second one, integrating by parts:

$$\int_\Omega p \, \mathrm{div}\boldsymbol{\xi}^t = \int_\Omega (\boldsymbol{\psi} - \mathbf{a}^{-1}\boldsymbol{\xi})\boldsymbol{\xi}^t.$$

Using now (18) and the discrete Friedrichs inequality for X_h^1, we obtain $\|p\|_0 \leq C\|\mathbf{a}^{-1}\boldsymbol{\xi} - \boldsymbol{\psi}\|_0$. This concludes the proof. ∎

Corollary 1. *The problem (25a)-(26) admits a unique solution which verifies:*

$$\|\mathbf{x}_h\|_{-1} + \|\mathbf{y}_h\|_0 \leq C(\|f\|_0 + \|g\|_0)$$

where the constant C does not depend upon the data and the mesh size.

Remark 1. Note that the Hodge decompositions (28) are just a tool to prove wellposedness and they are not part of the numerical scheme. Thus, we never need to compute them explicitly.

A new compatible discretization of magnetostatics

Magnetostatics corresponds to (22) with $g = 0$: $\mathbf{x}$ is the magnetic induction $\mathbf{B}$, $\mathbf{y}$ the magnetic field $\mathbf{H}$ and $\mathbf{a}$ is the inverse of the magnetic permeability. Thus, the schemes (25) or (25a)-(26) are a compatible discretization of the magnetostatics. Moreover, following the steps in the proof of Theorem 1 and looking in particular to (29), we realize that the computed magnetic induction $\mathbf{B}_h := \mathbf{x}_h$ takes the form of $\mathbf{B}_h = \underline{\mathrm{curl}}q_h$, for some $q_h \in X_h^0$. This suggests a way to simplify the discrete problem by using this information explicitly in the scheme.

Performing this simplification on the discretization (25a)-(26), we obtain the following discrete problem: Find $q_h \in X_h^0$, $\mathbf{H}_h \in Y_h^1$ such that:

$$\int_\Omega \underline{\mathrm{curl}}\mathbf{H}_h q' = \int_\Omega f \, q' \ \forall q' \in Y_h^2 \qquad \int_\Omega (\underline{\mathrm{curl}}q_h - \mathbf{a}\mathbf{H}_h)\mathbf{H}^t = 0 \ \forall \mathbf{H}^t \in Y_h^1. \quad (31)$$

Proposition 6. *The problem (31) admits a unique solution which verifies:*

$$\|q_h\|_0 + \|\mathbf{H}_h\|_0 \leq C\|f\|_0. \qquad (32)$$

Proof. We decompose any element $\mathbf{y} \in Y_h^1$ as follows:

$$\mathbf{y} = \mathrm{grad}p + \boldsymbol{\psi}, \ p \in Y_h^0, \text{ and } \boldsymbol{\psi} \in Y_h^1 \ : \ \int_\Omega \mathbf{a}\boldsymbol{\psi} \cdot \mathrm{grad}p^t = 0 \ \forall p^t \in Y_h^0.$$

The equation (31-left) together with the discrete Friedrichs inequality (17) implies that $\|\psi\|_0 \leq C\|\mathrm{curl}\psi\|_0 \leq C\|f\|_0$. The equation (31-right) reads, after rearrangement and integration by parts,

$$\int_\Omega q_h\, \mathrm{curl}\psi^t = \int_\Omega \mathbf{a}\,\psi\cdot\psi^t\ \forall \psi^t \qquad \int_\Omega \mathbf{a}\,\mathrm{grad}p\cdot\mathrm{grad}p^t = 0\ \forall p^t.$$

Thus, $\mathrm{grad}p = 0$, and using (19) together with the discrete Friedrichs inequality (17), we obtain:

$$\|q_h\|_0 \leq C\|\psi\|_0$$

which concludes the proof. ∎

Remark 2. The problem (31) is simpler, and "smaller" (less unknowns) than (25). On the other hand, it has lost symmetry.

The estimate (32) is not completely satisfactory, since we would like to provide a stability for q_h is $\widetilde{H}^1$, i.e., $\|q_h\|_1 \leq C\|f\|_0$. Such an estimate would rely on an inf-sup condition for the pair (X_h^0, Y_h^2) with respect to the norms $\widetilde{H}^1 - H^{-1}$, and on an optimal discrete Friedrichs inequality $\|\psi\|_0 \leq C\|\mathrm{curl}\psi\|_{-1}$ for all discrete $\psi \in Y_h^1$ orthogonal to gradients. The inf-sup condition has been proved in [11, Section 3.3], whereas the validity of such a discrete Friedrichs inequality is an open problem. In [11], it is proved that $\|\psi\|_0 \leq C\|\mathrm{curl}\psi\|_{-1+s}$, for $s \in (0,1]$ and for all $\psi \in Y_h^1$ orthogonal to gradients. This estimate ensures only the following:

$$\|q_h\|_{1-s} \leq C(s)\|f\|_0 \qquad s \in (0,1],$$

where the constant $C(s)$ may be unbounded when $s \to 0$.

A new compatible discretization of the Darcy flow equation

The Darcy flow equation corresponds to (22) with $f = 0$: $\mathbf{x}$ is the flow usually denoted by $\boldsymbol{\sigma}$, $\mathbf{y}$ is the gradient of a potential p and $\mathbf{a}$ is the diffusion coefficient. As before, we can use the information that $\mathbf{y} = \mathrm{grad}p$ to simplify the discretization (22) and we will see that we obtain a scheme which is very similar to the standard discretization of the Darcy flow equation by means of RT element (see [10] for details). Performing this simplification on the scheme (25), we obtain the following discrete problem: Find $\boldsymbol{\sigma}_h \in X_h^1$ and $p_h \in Y_h^0 \setminus \mathbb{R}$ such that:

$$\int_\Omega \mathrm{div}\boldsymbol{\sigma}_h q = \int_\Omega gq\ \forall q \in X_h^2 \quad \int_\Omega (\mathrm{grad}p_h - \mathbf{a}^{-1}\boldsymbol{\sigma}_h)\boldsymbol{\sigma}^t = 0\ \forall \boldsymbol{\sigma}^t \in X_h^1. \quad (33)$$

Note that, an integration by parts can be performed in the second equation and we obtain:

$$\int_{\Omega} (a^{-1}\boldsymbol{\sigma}_h \cdot \boldsymbol{\sigma}^t + p_h \mathrm{div}\boldsymbol{\sigma}^t) = 0 \quad \forall \boldsymbol{\sigma}^t \in X_h^1$$

The following Proposition holds.

Proposition 7. *The problem (33) admits a unique solution which verifies:*

$$\|p_h\|_0 + \|\boldsymbol{\sigma}_h\|_0 \leq C\|g\|_0 \tag{34}$$

Proof. It is enough to proceed as for the proof of Proposition 6. ∎

As for the magnetostatics, the estimate on the discrete scalar potential p_h is not optimal, we would like a stability estimate of the type $\|p_h\|_1 \leq C\|g\|_0$. Such an estimate would rely again on an inf-sup condition which has been proved in [11, Section 3.3] and on an optimal discrete Friedrichs inequality for RT elements which is not known (to the author's knowledge). The inf-sup condition and the discrete Friedrichs inequality provided in [11] ensure only the following:

$$\|p_h\|_{1-s} \leq C(s)\|g\|_0 \quad s \in (0,1],$$

where, as before, the constant $C(s)$ may be unbounded when $s \to 0$.

3.2 An optimal mortar method for Maxwell equations

In this Section we will show how the space Y_h^1 can be used as Lagrange multiplier space in the definition of domain decomposition methods with non-matching grids for Maxwell equations. We restrict ourselves to simplified situation and leave the generality for future investigation. In this section we suggest a way to reformulate the mortar method proposed in [6] in an optimal way. Let $\Omega \subset \mathbb{R}^3$ be a bounded Lipschitz polyhedron, $\mathbf{n}$ be unit outer normal at the boundary $\partial\Omega$ and

$$\mathrm{H}_0(\mathbf{curl}, \Omega) := \{\mathbf{u} \in L^2(\Omega)^3 \ : \ \mathbf{curl}\,\mathbf{u} \in L^2(\Omega)^3, \ (\mathbf{u} \times \mathbf{n})|_{\partial\Omega} = 0\}.$$

This is the energy space for the following problem: Given $\mathbf{f} \in L^2(\Omega)^3$, find $\mathbf{u} \in \mathrm{H}_0(\mathbf{curl}, \Omega)$ such that:

$$\mathbf{curl}\,\mathbf{curl}\,\mathbf{u} + \mathbf{u} = \mathbf{f}. \tag{35}$$

This is a simplified and coercive version of Maxwell equations in three space dimensions. Let Γ be a flat interface which split the domain Ω into two non-empty subsets Ω^+ and Ω^-: in particular $\partial\Gamma \subset \partial\Omega$. We denote by $\boldsymbol{\nu}$ the unit normal on Γ pointing into Ω^+. Given two triangulations $\mathcal{T}_h^+$ and $\mathcal{T}_h^-$ of Ω^+ and Ω^-, respectively, we define the continuous and discrete broken spaces:

$$\begin{aligned}
\mathbf{V}^b &:= \{\mathbf{v} \in L^2(\Omega)^3 \ : \quad \mathbf{v}^{\pm} = \mathbf{v}_h|_{\Omega^{\pm}} \in \mathrm{H}(\mathbf{curl}, \Omega^{\pm}), \, (\mathbf{v} \times \mathbf{n})|_{\partial\Omega} = 0\} \\
\mathbf{V}_h^b &:= \{\mathbf{v}_h \in L^2(\Omega)^3 \ : \quad \mathbf{v}_h^{\pm} = \mathbf{v}_h|_{\Omega^{\pm}} \in N_0(\mathcal{T}_h^{\pm}), \, (\mathbf{v}_h \times \mathbf{n})|_{\partial\Omega} = 0\}
\end{aligned}$$

where $N_0(\mathcal{T}_h^\pm)$ stands for low order edge elements of the first family (see [19] for the definition).

It is well known that

$$H_0(\mathbf{curl}, \Omega) = \{\mathbf{v} \in \mathbf{V}^b : (\mathbf{v}^+ - \mathbf{v}^-) \times \boldsymbol{\nu} = 0 \text{ on } \Gamma\} \tag{36}$$

and we want to reproduce the matching condition $(\mathbf{v}^+ - \mathbf{v}^-) \times \boldsymbol{\nu} = 0$ in a suitable way at the discrete level. Let us first analyze its consequences:

(i) if $\mathbf{v}$ happens to be a gradient $\mathbf{v} = \text{grad} p$, then $p \in \widetilde{H}^1(\Omega)$ and $p^+ - p^- = 0$ on Γ;

(ii) it also holds that $\text{div}((\mathbf{v}^+ - \mathbf{v}^-) \times \boldsymbol{\nu}) = 0$ on Γ, which means

$$(\mathbf{curl}\,\mathbf{v}^+ - \mathbf{curl}\,\mathbf{v}^-) \cdot \boldsymbol{\nu} = 0 \text{ on } \Gamma.$$

Indeed, it is easy to see that *(i)* and *(ii)* provide a characterization of the matching condition in (36).

The discretization we propose here will reproduce *(i)* and *(ii)* at the discrete level in a stable way.

The interface Γ is triangulated by two possibly different grids. We choose one of them, say the trace of $\mathcal{T}_h^+$ and we introduce the spaces $Y_h^i(\Gamma)$, $i = 0, 1, 2$ as Y_h^i on this given triangulation on Γ, i.e., $\mathcal{T}_h^+|_\Gamma$. We define the constrained space by mimicking (36):

$$\mathbf{V}_h := \{\mathbf{v}_h \in \mathbf{V}_h^b : \int_\Gamma \left((\mathbf{v}_h^+ - \mathbf{v}_h^-) \times \boldsymbol{\nu}\right) \cdot \mathbf{y}_h = 0 \,\forall \mathbf{y}_h \in Y_h^1(\Gamma)\}$$

and solve the following discrete problem: Find $\mathbf{u}_h \in \mathbf{V}_h$ such that

$$\int_\Omega (\mathbf{curl}\,\mathbf{u}_h \cdot \mathbf{curl}\,\mathbf{v}_h + \mathbf{u}_h \cdot \mathbf{v}_h) = \int_\Omega \mathbf{f} \cdot \mathbf{v}_h \quad \forall \mathbf{v}_h \in \mathbf{V}_h. \tag{37}$$

It is now enough to realize that $\mathbf{v}^+ \times \boldsymbol{\nu}$ belongs to the space $X_h^1(\Gamma)$ defined on Γ and on the triangulation $\mathcal{T}_h^+$ of Γ. Thus, the inf-sup condition (21), together with the standard theory of mortar method (see [7] or [5]), implies the wellposedness of (37):

Theorem 2. *The problem (37) admits a unique solution which verifies:*

$$\|\mathbf{u}_h^\pm\|_{0,\Omega^\pm} + \|\mathbf{curl}\,\mathbf{u}_h^\pm\|_{0,\Omega^\pm} \leq C\|\mathbf{f}\|_{0,\Omega}.$$

For this theorem, only the inf-sup condition (21) is needed. In general, Maxwell equations do not correspond to a positive definite coercive bilinear form, and then the structure of the interface condition should matter in order to ensure wellposedness, spectral correctness and so on. We analyze then the discrete counterpart of the characterization *(i)* and *(ii)* above. We have:

$(i)_d$ Suppose that $\mathbf{v}_h^\pm = \mathrm{grad}\, p_h^\pm$. Then, the interface condition reads

$$\int_\Gamma ((\mathrm{grad}\, p_h^+ - \mathrm{grad}\, p_h^-) \times \boldsymbol{\nu})\mathbf{y}_h = 0 \quad \forall \mathbf{y}_h \in Y_h^1(\Gamma). \tag{38}$$

Note that the potential $p_h = p_h^\pm$ on $\Omega^\pm$ verifies $p_h|_{\partial\Omega} = 0$. We rewrite (38) and by integration by parts we obtain: for all $\mathbf{y}_h \in Y_h^1(\Gamma)$

$$\int_\Gamma ((\mathrm{grad}\, p_h^+ - \mathrm{grad}\, p_h^-) \times \boldsymbol{\nu}) = \int_\Gamma (\mathrm{curl}(p_h^+ - p_h^-) \cdot \mathbf{y}_h =$$
$$= \int_\Gamma (p_h^+ - p_h^-)\mathrm{curl}\, \mathbf{y}_h = 0.$$

Thus the jump $(p_h^+ - p_h^-)$ is orthogonal to the space $\mathrm{curl} Y_h^1(\Gamma) = Y_h^2(\Gamma)$. Note that $p_h^+|_\Gamma \in X_h^0(\Gamma)$ and that the pair $(X_h^0(\Gamma), Y_h^2(\Gamma))$ verifies an inf-sup condition. Thus the potential p_h matches in an optimal way on the interface Γ;

$(ii)_d$ Choosing only Lagrange multipliers of the type $\mathrm{grad}\, q_h$, $q_h \in Y_h^0(\Gamma)$, we perform integration by parts:

$$\int_\Gamma ((\mathbf{v}_h^+ - \mathbf{v}_h^-) \times \boldsymbol{\nu}) \cdot \mathrm{grad}\, q_h = \int_\Gamma \mathrm{div}((\mathbf{v}_h^+ - \mathbf{v}_h^-) \times \boldsymbol{\nu})q_h$$
$$= \int_\Gamma (\mathbf{curl}\, \mathbf{v}_h^+ - \mathbf{curl}\, \mathbf{v}_h^-) \cdot \boldsymbol{\nu}\, q_h = 0.$$

This means that the quantity $(\mathbf{curl}\, \mathbf{v}_h^+ - \mathbf{curl}\, \mathbf{v}_h^-) \cdot \boldsymbol{\nu}$ is orthogonal to all $q_h \in Y_h^0(\Gamma)$. Noting that $\mathbf{curl}\, \mathbf{v}_h^+ \cdot \boldsymbol{\nu} \in X_h^2(\Gamma)$ and recalling that the pair $(X_h^2(\Gamma), Y_h^0(\Gamma))$ is inf-sup stable, we can argue that also the condition $(\mathbf{curl}\, \mathbf{v}_h^+ - \mathbf{curl}\, \mathbf{v}_h^-) \cdot \boldsymbol{\nu} = 0$ is reproduced at the discrete level in an optimal way.

References

1. Amrouche, C., Bernardi, C., Dauge, M., Girault, V.: Vector potentials in three-dimensional non-smooth domains. Math. Meth. Appl. Sci. **21**, 823–864 (1998)
2. Douglas N. Arnold: Differential complexes and numerical stability. Proceedings of the International Congress of Mathematicians, Vol. I (2002) (Beijing), Higher Ed. Press, 2002, pp. 137–157.
3. Arnold, Douglas N. , Falk, Richard S., Winther, R.: Differential complexes and stability of finite element methods. i. the de rham complex. Proceedings of the IMA workshop on Compatible Spatial Discretizations for PDE, (2006) (to appear)
4. _____: Differential complexes and stability of finite element methods. ii. the elasticity complex. Proceedings of the IMA workshop on Compatible Spatial Discretizations for PDE, (2006) (to appear)

5. Ben Belgacem, F.: The mortar finite element method with lagrange multipliers. Numer. Math. **84, Issue 2**, 173–197 (2000)
6. Ben Belgacem, F., Buffa, A., Maday, Y.: The mortar element method for Maxwell equations: first results. SIAM J. Num. Anal. **39**, no. 3, 880–901 (2001)
7. Bernardi, C., Maday, Y., Patera, A. T.: A new nonconforming approach to domain decomposition: The mortar elements method. Nonlinear partial differential equations and their applications (H. Brezis and J.L. Lions, eds.), Pitman, pp. 13–51 (1994)
8. Bochev, P.B., Hyman, J.M.: Principles of mimetic discretizations of differential operators. Proceedings of the IMA "Hot Topic conference" on Compatible spatial discretizations, (2004)
9. Bossavit, A.: The mathematics of finite elements and applications, vol. **VI** (Uxbridge, 1987), ch. Mixed finite elements and the complex of Whitney forms, pp. 1377–144, Academic Press, London (1988)
10. Brezzi, F., Fortin, M.: Mixed and hybrid finite element methods, vol. **15**, Springer-Verlag, Berlin (1991)
11. Buffa, A., Christiansen, S. H.: A dual finite element complex on the barycentric refinement. Tech. Report PV-18, IMATI-CNR (2005)
12. Buffa, A., Ciarlet, Jr.,P.: On traces for functional spaces related to Maxwell's equations. Part I: An integration by parts formula in Lipschitz polyhedra. Math. Meth. Appl. Sci. **21**, no. 1, 9–30 (2001)
13. _____: On traces for functional spaces related to Maxwell's equations. Part II: Hodge decompositions on the boundary of Lipschitz polyhedra and applications, Math. Meth. Appl. Sci. **21**, no. 1, 31–48 (2001)
14. Clemens, M., Weiland, T.: Discrete Electromagnetism with the Finite Integration Technique. Progress in Electromagnetic research **PIER 32**, 65–87 (2001)
15. Grisvard, P.: Elliptic problems in nonsmooth domains. Monographs and studies in Mathematics, vol. **24**, Pitman, London (1985)
16. Hiptmair, R.: Discrete Hodge operators. Numer. Math. **90**, no. 2, 265–289 (2001)
17. Hiptmair, R.: Finite elements in computational electromagnetism. Acta Numerica, 237–339 (2002)
18. Monk, P.: Finite Element Methods for Maxwell's Equations. Numerical Mathematics and Scientific Computation, Oxford University Press, (2003)
19. Nédélec, J.C.: Mixed finite element in $\mathbb{R}^3$. Numer. Math. **35**, 315–341 (1980)

Finite Element Approximation of the Three Field Formulation of the Elasticity Problem Using Stabilization

Ramon Codina

Universitat Politècnica de Catalunya
Jordi Girona 1-3, Edifici C1, 08034 Barcelona, Spain
ramon.codina@upc.edu

Summary. The stress-displacement-pressure formulation of the elasticity problem may suffer from two types of numerical instabilities related to the finite element interpolation of the unknowns. The first is the classical pressure instability that occurs when the solid is incompressible, whereas the second is the lack of stability in the stresses. To overcome these instabilities, there are two options. The first is to use different interpolation for all the unknowns satisfying two inf-sup conditions. Whereas there are several displacement-pressure interpolations that render the pressure stable, less possibilities are known for the stress interpolation. The second option is to use a stabilized finite element formulation instead of the plain Galerkin approach. If this formulation is properly designed, it is possible to use equal interpolation for all the unknowns. The purpose of this paper is precisely to present one of such formulations. In particular, it is based on the decomposition of the unknowns into their finite element component and a subscale, that will be approximated and whose goal is to yield a stable formulation. A singular feature of the method to be presented is that the subscales will be considered orthogonal to the finite element space. We describe in detail the original formulation and a simplified variant and present the results of their numerical analysis.

1 Introduction

The analysis of the three field formulation of the linear elastic problem is probably not a goal by itself, but rather a simple model to study problems in which it is important to interpolate the stresses independently from the displacements and, in the case we will consider, also the pressure. Perhaps the most salient problem that requires the interpolation of the (deviatoric) stresses is the viscoelastic one. In this case, the algebraic constitutive equation (linear or nonlinear) that relates stresses and strains has to be replaced by an evolution equation (see [1] for a review).

The problem we will study in this paper is the simple Stokes problem arising in linear elasticity or creeping flows, taking as unknowns the displacement field (or velocity field, in a fluid problem), the pressure and the deviatoric part of the stresses. In particular, we shall consider that the material is *incompressible*.

When the finite element approximation of the problem is undertaken, it is well known that incompressibility poses a stringent requirement in the way the pressure is interpolated with respect to the displacement field. The displacement and pressure finite element spaces have to satisfy the classical inf-sup condition [4]. Several interpolations are known that satisfy this condition and yield a stable displacement-pressure numerical solution. However, less is known about another inf-sup condition that needs to be satisfied when the stresses are interpolated independently from the displacement. This inf-sup condition is trivially satisfied for the continuous problem, but only a few interpolations are known that verify it for the discrete case.

The inf-sup conditions for the displacement-pressure and stresses-displacement interpolations are needed if the standard Galerkin method is used for the space discretization. However, there is also the possibility to resort to a *stabilized* finite element method, in which the discrete variational form of the Galerkin formulation is modified in order to enhance its stability. The purpose of this paper is precisely to present one of such formulations. In particular, the one proposed here is based on the decomposition of the unknowns into their finite element component and a subscale, that is, the component of the continuous unknown that can not be captured by the finite element mesh. Obviously, this subscale needs to be approximated in one way or another. This idea was proposed in the finite element context in [11, 12], although there are similar concepts developed in different situations (both in physical and numerical modeling).

The important property of the formulation to be presented here is that the subscale will be considered *orthogonal* to the appropriate finite element space. This idea was first applied to the Stokes problem in displacement-pressure form in [5], and subsequently applied to general incompressible flows in [6].

Different stabilized formulations for the three-field Stokes problem can be found in the literature. The GLS (Galerkin/least-squares) method is used for example in [2, 9]. In [10, 8] the authors propose what they call EVSS (elastic-viscous-split-stress), that is related to the formulation proposed in this paper in what concerns the way to stabilize the stress interpolation. An analysis of both approaches, GLS and EVSS, is presented in [3].

The paper is organized as follows. In the following section we present the problem to be solved and its Galerkin finite element approximation, explaining the sources of numerical instability. Then we present the stabilized finite element formulation we propose, for which we present a complete numerical analysis in Section 4. Section 5 is concerned with a modified formulation, slightly simpler but that in fact allows us to obtain stability and error estimates in natural norms (H^1 for the displacement and L^2 for the pressure and the stresses). The paper concludes with some final remarks.

2 Problem statement and Galerkin finite element discretization

2.1 Boundary value problem

Let Ω be the computational domain of $\mathbb{R}^d$ ($d=2$ or 3) occupied by the solid (or fluid) and $\partial\Omega$ its boundary. If $\boldsymbol{u}$ is the displacement field, p the pressure (taken as positive in compression) and $\boldsymbol{\sigma}$ the deviatoric component of the stress field, the field equations to be solved in the domain Ω are

$$-\nabla \cdot \boldsymbol{\sigma} + \nabla p = \boldsymbol{f}, \tag{1}$$
$$\nabla \cdot \boldsymbol{u} = 0, \tag{2}$$
$$\boldsymbol{\sigma} - 2\mu \nabla^S \boldsymbol{u} = 0, \tag{3}$$

where $\boldsymbol{f}$ is the vector of body forces, μ the shear modulus and $\nabla^S \boldsymbol{u}$ the symmetrical part of $\nabla \boldsymbol{u}$. For simplicity, we shall consider the simplest boundary condition $\boldsymbol{u} = \boldsymbol{0}$ on $\partial\Omega$.

2.2 Variational form

To write the weak form of problem (1)-(3) we need to introduce some functional spaces. Let $\mathcal{V} = (H_0^1(\Omega))^d$, $\mathcal{Q} = L^2(\Omega)/\mathbb{R}$ and $\mathcal{T} = (L^2(\Omega))^{d \times d}$. If we call $U = (\boldsymbol{u}, p, \boldsymbol{\sigma})$, $\mathcal{X} = \mathcal{V} \times \mathcal{Q} \times \mathcal{T}$, the weak form of the problem consists in finding $U \in \mathcal{X}$ such that

$$B(U, V) = L(V), \tag{4}$$

for all $V = (\boldsymbol{v}, q, \boldsymbol{\tau}) \in \mathcal{X}$, where

$$B(U, V) = (\nabla^S \boldsymbol{v}, \boldsymbol{\sigma}) - (p, \nabla \cdot \boldsymbol{v}) + (q, \nabla \cdot \boldsymbol{u}) + \frac{1}{2\mu}(\boldsymbol{\sigma}, \boldsymbol{\tau}) - (\nabla^S \boldsymbol{u}, \boldsymbol{\tau}), \tag{5}$$
$$L(V) = \langle \boldsymbol{f}, \boldsymbol{v} \rangle, \tag{6}$$

where $(\cdot, \cdot)$ is the L^2 inner product and $\langle \cdot, \cdot \rangle$ is the duality pairing betwen $\mathcal{V}$ and its dual, $(H^{-1}(\Omega))^d$, where $\boldsymbol{f}$ is assumed to belong.

2.3 Stability of the Galerkin finite element discretization

let us consider a finite element partition of the domain Ω of diameter h. For simplicity, we will consider quasi uniform refinements, and thus all the element diameters can be bounded above and below by constants multiplying h.

From the finite element partition we may build up conforming finite element spaces $\mathcal{V}_h \subset \mathcal{V}$, $\mathcal{Q}_h \subset \mathcal{Q}$ and $\mathcal{T}_h \subset \mathcal{T}$ in the usual manner. If $\mathcal{X}_h = \mathcal{V}_h \times \mathcal{Q}_h \times \mathcal{T}_h$ and $U_h = (\boldsymbol{u}_h, p_h, \boldsymbol{\sigma}_h)$, the Galerkin finite element approximation consists in finding $U_h \in \mathcal{X}_h$ such that

$$B(U_h, V_h) = L(V_h), \tag{7}$$

for all $V_h = (\boldsymbol{v}_h, q_h, \boldsymbol{\tau}_h) \in \mathcal{X}_h$.

In principle, we have posed no restrictions on the choice of the finite element spaces. However, let us analyze the numerical stability of problem (6). If we take $V_h = U_h$, it is found that

$$B(U_h, U_h) = \frac{1}{2\mu}\|\boldsymbol{\sigma}_h\|^2, \tag{8}$$

where $\|\cdot\|$ is the $L^2(\Omega)$ norm. It is seen from (7) that B_h is not coercive in $\mathcal{X}_h$, the displacement and the pressure being out of control. Moreover, the inf-sup condition

$$\inf_{U_h \in \mathcal{X}_h} \sup_{V_h \in \mathcal{X}_h} \frac{B(U_h, V_h)}{\|U_h\|_\mathcal{X} \|V_h\|_\mathcal{X}} \geq \beta$$

is *not* satisfied for any positive constant β unless the two conditions

$$\inf_{q_h \in \mathcal{Q}_h} \sup_{\boldsymbol{v}_h \in \mathcal{V}_h} \frac{(q_h, \nabla \cdot \boldsymbol{v}_h)}{\|q_h\|_{\mathcal{Q}_h} \|\boldsymbol{v}_h\|_{\mathcal{V}_h}} \geq C_1 > 0, \tag{9}$$

$$\inf_{\boldsymbol{v}_h \in \mathcal{V}_h} \sup_{\boldsymbol{\tau}_h \in \mathcal{T}_h} \frac{(\boldsymbol{\tau}_h, \nabla^S \boldsymbol{v}_h)}{\|\boldsymbol{\tau}_h\|_{\mathcal{T}_h} \|\boldsymbol{v}_h\|_{\mathcal{V}_h}} \geq C_2 > 0, \tag{10}$$

hold for positive constants C_1 and C_2. In all the expressions above, $\|\cdot\|_\mathcal{Y}$ stands for the appropriate norm in space $\mathcal{Y}$.

Conditions (9) and (10) pose stringent requirements on the choice of the finite element spaces. Our intention in this paper is to present a stabilized finite element formulation that avoids the need for such conditions and, in particular, allows *equal interpolation for all the unknowns*. Although this is only a particular choice for the finite element spaces, we will concentrate on this. Therefore, in what follows we will assume that $\mathcal{V}_h$, $\mathcal{Q}_h$ and $\mathcal{T}_h$ are all constructed from *continuous* finite element interpolations of degree k.

3 Finite element approximation using subscales

3.1 Decomposition of the unknowns

Let us start by explaining the basic idea of the multiscale formulation proposed in [11] and applying it to our problem. If we split $U = U_h + U'$, where U_h belongs to the finite element space $\mathcal{X}_h$ and U' to any space $\mathcal{X}'$ to complement $\mathcal{X}_h$ in $\mathcal{X}$, problem (4) is exactly equivalent to

$$B(U_h, V_h) + B(U', V_h) = L(V_h) \quad \forall V_h \in \mathcal{X}_h, \tag{11}$$
$$B(U_h, V') + B(U', V') = L(V') \quad \forall V' \in \mathcal{X}'. \tag{12}$$

Integrating some terms by parts and using the fact that $\boldsymbol{u}_h = \boldsymbol{u}' = \boldsymbol{0}$ on $\partial\Omega$, it is easy to see that (11) in our case can be written as the system

$$(\nabla^S \boldsymbol{v}_h, \boldsymbol{\sigma}_h) + (\nabla^S \boldsymbol{v}_h, \boldsymbol{\sigma}') - (p_h, \nabla \cdot \boldsymbol{v}_h) - (p', \nabla \cdot \boldsymbol{v}_h) = \langle \boldsymbol{f}, \boldsymbol{v} \rangle, \quad (13)$$
$$(q_h, \nabla \cdot \boldsymbol{u}_h) - (\nabla q_h, \boldsymbol{u}') = 0, \quad (14)$$
$$(\boldsymbol{\sigma}_h, \boldsymbol{\tau}_h) + (\boldsymbol{\sigma}', \boldsymbol{\tau}_h) - 2\mu(\nabla^S \boldsymbol{u}_h, \boldsymbol{\tau}_h) + 2\mu(\boldsymbol{u}', \nabla \cdot \boldsymbol{\tau}_h) = 0, \quad (15)$$

which must hold for all test functions $\boldsymbol{v}_h$, q_h and $\boldsymbol{\tau}_h$. On the other hand, (12) implies that

$$-\nabla \cdot \boldsymbol{\sigma}' + \nabla p' = \boldsymbol{r}_1 := \boldsymbol{f} + \nabla \cdot \boldsymbol{\sigma}_h - \nabla p_h + \xi_1, \quad (16)$$
$$\nabla \cdot \boldsymbol{u}' = r_2 := -\nabla \cdot \boldsymbol{u}_h + \xi_2, \quad (17)$$
$$\boldsymbol{\sigma}' - 2\mu \nabla^S \boldsymbol{u}' = \boldsymbol{r}_3 := -\boldsymbol{\sigma}_h + 2\mu \nabla^S \boldsymbol{u}_h + \xi_3, \quad (18)$$

where ξ_1, ξ_2 and ξ_3 are responsible to enforce that the previous equations hold in the space for the subscales, that still needs to be approximated (see [6] for more details). The way to approximate the solution of problem (16)-(18) and to choose the space for the subscales is the objective of the following section.

3.2 Approximation of the subscales

The subscales, solution of problem (16)-(18), need now to be approximated. Once this is done, inserting them in (13)-(14) will lead to a problem for the finite element unknowns which will hopefully have better stability properties than the standard Galerkin method.

The are several possibilities to deal with problem (16)-(18). As in [6], we will approximate $\boldsymbol{\sigma}'$, p' and $\boldsymbol{u}'$ by using an (approximate) Fourier analysis of the problem. We will omit the details, for which we refer to the above reference, and sketch only the idea.

Denoting by $\widehat{}$ the Fourier transform, and assuming that the values of the subscales are negligible on the element boundaries (which is reasonable for highly fluctuating subscales), the transformed problem (16)-(18) is, *within each element of the finite element partition*,

$$-\mathrm{i}\frac{\boldsymbol{k}}{h} \cdot \widehat{\boldsymbol{\sigma}'} + \mathrm{i}\frac{\boldsymbol{k}}{h}\widehat{p'} = \widehat{\boldsymbol{r}}_1,$$
$$\mathrm{i}\frac{\boldsymbol{k}}{h} \cdot \widehat{\boldsymbol{u}'} = \widehat{r}_2,$$
$$\widehat{\boldsymbol{\sigma}'} - \mu\mathrm{i}\left(\frac{\boldsymbol{k}}{h} \otimes \widehat{\boldsymbol{u}'} + \widehat{\boldsymbol{u}'} \otimes \frac{\boldsymbol{k}}{h}\right) = \widehat{\boldsymbol{r}}_3,$$

where $\boldsymbol{k}$ is the dimensionless wave number. From these equations it is possible to obtain the Fourier transform of the subscales and to compute its L^2 norm, which will have the form $\|\widehat{U'}(\boldsymbol{k})\| = \|\boldsymbol{\alpha}(\boldsymbol{k})\widehat{R}(\boldsymbol{k})\|$ for a certain matrix $\boldsymbol{\alpha}(\boldsymbol{k})$ depending on $\boldsymbol{k}$ and $\widehat{R}$ being the vector that contains $(\widehat{r}_1, \widehat{r}_2, \widehat{r}_3)$. The mean value theorem guarantees that there is a wave number $\boldsymbol{k}_0$ for which $\|\widehat{U'}(\boldsymbol{k})\| \leq \|\boldsymbol{\alpha}(\boldsymbol{k}_0)\widehat{R}(\boldsymbol{k})\|$. Parseval's formula allows now to state that $\|U'\| \leq \|\boldsymbol{\alpha}(\boldsymbol{k}_0)R\|$. Thus, if we take $U' = \boldsymbol{\alpha}_0 R$, where now $\boldsymbol{\alpha}_0$ is a matrix of constants, there will be values of these constants for which the approximated subscales U' will have the correct L^2 norm over each element.

In principle, $\boldsymbol{\alpha}_0$ is a full (symmetric) matrix. However, assuming that the components of $\boldsymbol{k}_0$ are high and neglecting tensors of rang lower that d, it can be heuristically argued that $\boldsymbol{\alpha}_0$ can be approximated by a diagonal matrix, and therefore the subscales approximated by

$$\boldsymbol{u}' = \alpha_1 \frac{h^2}{\mu} \boldsymbol{r}_1, \qquad (19)$$

$$p' = \alpha_2 2\mu r_2, \qquad (20)$$

$$\boldsymbol{\sigma}' = \alpha_3 \boldsymbol{r}_3. \qquad (21)$$

These are the expressions we were looking for. Here, α_1, α_2 and α_3 are constants that play the role of the algorithmic parameters of the formulation. The possibility of using the full matrix $\boldsymbol{\alpha}_0$ needs to be further explored.

It only remains to determine which is the space of the subscales, that is, to choose the functions ξ_i, $i = 1, 2, 3$. Our particular choice is *to take the space for the subscales L^2 orthogonal to the finite element space*. In view of (19)-(15), this implies that r_1, r_2 and r_3 must be orthogonal to $\mathcal{V}_h$, $\mathcal{Q}_h$ and $\mathcal{T}_h$, respectively. Denoting by P_u, P_p and P_σ the L^2 projections onto these spaces and by $P_u^\perp$, $P_p^\perp$ and $P_\sigma^\perp$ the orthogonal projections, we will have that

$$\xi_1 = -P_u(\boldsymbol{f} + \nabla \cdot \boldsymbol{\sigma}_h - \nabla p_h) \quad \text{and} \quad \boldsymbol{u}' = \alpha_1 \frac{h^2}{\mu} P_u^\perp(\boldsymbol{f} + \nabla \cdot \boldsymbol{\sigma}_h - \nabla p_h),$$

$$\xi_2 = -P_p(-\nabla \cdot \boldsymbol{u}_h) \qquad \text{and} \quad p' = \alpha_2 2\mu P_p^\perp(-\nabla \cdot \boldsymbol{u}_h),$$

$$\xi_3 = -P_\sigma(-\boldsymbol{\sigma}_h + 2\mu \nabla^S \boldsymbol{u}_h) \quad \text{and} \quad \boldsymbol{\sigma}' = \alpha_3 P_\sigma^\perp(-\boldsymbol{\sigma}_h + 2\mu \nabla^S \boldsymbol{u}_h).$$

Clearly, we have that $P_\sigma^\perp(-\boldsymbol{\sigma}_h) = 0$. We may also assume for simplicity that the body force belongs to the finite element space, and thus $P_u^\perp(\boldsymbol{f}) = 0$. Hence, the expression for the subscales we finally propose is

$$\boldsymbol{u}' = \alpha_1 \frac{h^2}{\mu} P_u^\perp(\nabla \cdot \boldsymbol{\sigma}_h - \nabla p_h), \qquad (22)$$

$$p' = -\alpha_2 2\mu P_p^\perp(\nabla \cdot \boldsymbol{u}_h), \qquad (23)$$

$$\boldsymbol{\sigma}' = \alpha_3 2\mu P_\sigma^\perp(\nabla^S \boldsymbol{u}_h). \qquad (24)$$

3.3 Stabilized finite element problem

Once arrived to (22)-(17), the stabilized finite element problem is obtained by inserting these approximations for the subscales into (13)-(14). Noting that $(\boldsymbol{\sigma}', \boldsymbol{\tau}_h) = 0$, the result is the following:

$$(\nabla^S \boldsymbol{v}_h, \boldsymbol{\sigma}_h) + \alpha_3 2\mu(\nabla^S \boldsymbol{v}_h, P_\sigma^\perp(\nabla^S \boldsymbol{u}_h)) - (p_h, \nabla \cdot \boldsymbol{v}_h)$$
$$+ \alpha_2 2\mu(P_p^\perp(\nabla \cdot \boldsymbol{u}_h), \nabla \cdot \boldsymbol{v}_h) = \langle \boldsymbol{f}, \boldsymbol{v} \rangle, \tag{25}$$

$$(q_h, \nabla \cdot \boldsymbol{u}_h) + \alpha_1 \frac{h^2}{\mu}(\nabla q_h, P_u^\perp(\nabla p_h - \nabla \cdot \boldsymbol{\sigma}_h)) = 0, \tag{26}$$

$$\frac{1}{2\mu}(\boldsymbol{\sigma}_h, \boldsymbol{\tau}_h) - (\nabla^S \boldsymbol{u}_h, \boldsymbol{\tau}_h) + \alpha_1 \frac{h^2}{\mu}(P_u^\perp(\nabla \cdot \boldsymbol{\sigma}_h - \nabla p_h), \nabla \cdot \boldsymbol{\tau}_h) = 0. \tag{27}$$

Introducing the bilinear form

$$B_{\text{stab}}(U_h, V_h) := B(U_h, V_h) + \alpha_3 2\mu(\nabla^S \boldsymbol{v}_h, P_\sigma^\perp(\nabla^S \boldsymbol{u}_h))$$
$$+ \alpha_2 2\mu(P_p^\perp(\nabla \cdot \boldsymbol{u}_h), \nabla \cdot \boldsymbol{v}_h)$$
$$+ \alpha_1 \frac{h^2}{\mu}(\nabla q_h - \nabla \cdot \boldsymbol{\tau}_h, P_u^\perp(\nabla p_h - \nabla \cdot \boldsymbol{\sigma}_h)), \tag{28}$$

problem (18)-(27) can be written as follows: find $U_h \in \mathcal{X}_h$ such that

$$B_{\text{stab}}(U_h, V_h) = L(V_h), \tag{29}$$

for all $V_h \in \mathcal{X}_h$. This is the stabilized finite element method we propose and whose stability and convergence properties are established in the following section. In Section 5 we will also present and analyze a modified formulation.

4 Numerical analysis of the original formulation

In this section we present the results of the numerical analysis of the method proposed in the previous section. The norm in which the results will be presented is

$$\|V_h\|^2 := \frac{1}{2\mu}\|\boldsymbol{\tau}_h\|^2 + \alpha_3 2\mu\|\nabla^S \boldsymbol{v}_h\|^2 + \alpha_2 2\mu\|\nabla \cdot \boldsymbol{v}_h\|^2$$
$$+ \alpha_1 \frac{h^2}{\mu}\|\nabla q_h - \nabla \cdot \boldsymbol{\tau}_h\|^2. \tag{30}$$

In fact, the term multiplied by α_2 is unnecessary, since it already appears in the term multiplied by α_3. However, we will keep it for generality, to see the effect of the subscale associated to the pressure introduced in the previous section. In all what follows we will assume that $\alpha_i > 0$, $i = 1, 2, 3$.

As it has been mentioned in Section 2, we will consider for the sake of conciseness quasi-uniform finite element partitions. Therefore, we assume that there is a constant C_{inv}, independent of the mesh size h (the maximum of all the element diameters), such that

$$\|\nabla v_h\| \leq \frac{C_{\text{inv}}}{h}\|v_h\|, \tag{31}$$

for all finite element functions v_h. Since, again for the sake of simplicity, we have assumed equal interpolation for all the unknowns, this inequality can be used for scalars, vectors or tensors.

In all what follows, C will denote a positive constant, independent of the discretization and the physical coefficient μ, and possibly different at different occurrences.

We start proving what is in fact the key result, which states that the formulation presented is stable in the norm (30). This stability is presented in the form of an inf-sup condition:

Theorem 1 (Stability I). *There is a constant $C > 0$ such that*

$$\inf_{U_h \in \mathcal{X}_h} \sup_{V_h \in \mathcal{X}_h} \frac{B_{\text{stab}}(U_h, V_h)}{\|U_h\| \|V_h\|} \geq C. \tag{32}$$

Proof. Let us start noting that, for any function $U_h \in \mathcal{X}_h$, we have

$$B_{\text{stab}}(U_h, U_h) = \frac{1}{2\mu} \|\boldsymbol{\sigma}_h\|^2 + \alpha_3 2\mu \|P_\sigma^\perp(\nabla^S \boldsymbol{u}_h)\|^2$$
$$+ \alpha_2 2\mu \|P_p^\perp(\nabla \cdot \boldsymbol{u}_h)\|^2 + \alpha_1 \frac{h^2}{\mu} \|P_u^\perp(\nabla p_h - \nabla \cdot \boldsymbol{\sigma}_h)\|^2. \tag{33}$$

The basic idea is to obtain control on the components on the finite element space for the terms whose orthogonal components appear in this expression. The key point is that this control comes from the Galerkin terms in the bilinear form B_{stab}.

Let us consider $V_{h1} := \alpha_1 \frac{h^2}{\mu}(P_u(\nabla p_h - \nabla \cdot \boldsymbol{\sigma}_h), 0, 0)$. A straightforward application of Schwarz's inequality, Young's inequality and the inverse estimate (31) leads to

$$B_{\text{stab}}(U_h, V_{h1}) \geq \alpha_1 \frac{h^2}{2\mu} \|P_u(\nabla p_h - \nabla \cdot \boldsymbol{\sigma}_h)\|^2$$
$$- 4\alpha_1 \alpha_3^2 \mu C_{\text{inv}}^2 \|P_\sigma^\perp(\nabla^S \boldsymbol{u}_h)\|^2$$
$$- 4\alpha_1 \alpha_2^2 \mu C_{\text{inv}}^2 \|P_p^\perp(\nabla \cdot \boldsymbol{u}_h)\|^2. \tag{34}$$

Consider now $V_{h2} := (\mathbf{0}, \alpha_2 2\mu P_p(\nabla \cdot \boldsymbol{u}_h), \mathbf{0})$. The same strategy as before now leads to

$$B_{\text{stab}}(U_h, V_{h2}) \geq \alpha_2 \mu \|P_p(\nabla \cdot \boldsymbol{u}_h)\|^2$$
$$- \alpha_1^2 \alpha_2 C_{\text{inv}}^2 \frac{h^2}{\mu} \|P_u^\perp(\nabla p_h - \nabla \cdot \boldsymbol{\sigma}_h)\|^2. \tag{35}$$

Finally, taking $V_{h3} := (\mathbf{0}, 0, -\alpha_3 2\mu P_\sigma(\nabla^S \boldsymbol{u}_h))$ what we obtain is

$$B_{\text{stab}}(U_h, V_{h3}) \geq \alpha_3 \mu \|P_\sigma(\nabla^S \boldsymbol{u}_h)\|^2$$
$$- \alpha_3 \frac{1}{\mu} \|\boldsymbol{\sigma}_h\|^2$$
$$- 2\alpha_1^2 \alpha_3 C_{\text{inv}}^2 \frac{h^2}{\mu} \|P_u^\perp(\nabla p_h - \nabla \cdot \boldsymbol{\sigma}_h)\|^2. \tag{36}$$

Let $V_h = \beta_1 V_{h1} + \beta_2 V_{h2} + \beta_3 V_{h3}$, with V_{hi}, $i=1,2,3$, introduced above. Adding up inequalities (34)-(35)-(36) multiplied by β_1, β_2 and β_3, respectively, and adding also (33), it is trivially verified that there exist values of the coefficients β_i, $i = 1,2,3$, for which

$$B_{\text{stab}}(U_h, V_h) \geq C\|U_h\|^2. \tag{37}$$

On the other hand, we have that

$$\|V_{h1}\|^2 \leq 2\alpha_1^2(\alpha_2+\alpha_3)C_{\text{inv}}^2 \frac{h^2}{\mu}\|\nabla p_h - \nabla\cdot\boldsymbol{\sigma}_h\|^2 \leq C\|U_h\|^2,$$
$$\|V_{h2}\|^2 \leq 4\alpha_1\alpha_2\mu C_{\text{inv}}^2\|\nabla\cdot\boldsymbol{u}_h\|^2 \leq C\|U_h\|^2,$$
$$\|V_{h3}\|^2 \leq 2\alpha_3^2\mu(1+2\alpha_1 C_{\text{inv}}^2)\|\nabla^S\boldsymbol{u}_h\|^2 \leq C\|U_h\|^2,$$

from where it follows that $\|V_h\| \leq C\|U_h\|$. Using this fact in (37) we have shown that for each $U_h \in \mathcal{X}_h$ there exists $V_h \in \mathcal{X}_h$ such that $B_{\text{stab}}(U_h, V_h) \geq C\|U_h\|\|V_h\|$, from where the theorem follows. ∎

Once stability is established, a more or less standard procedure leads to convergence. In this case, we will assume that all the components of the continuous solution $U = (\boldsymbol{u}, p, \boldsymbol{\sigma}) \in \mathcal{X}$ belong to $H^{k+1}(\Omega)$, where k is the order of the finite element interpolation. A remark on this requirement will be made after the final convergence result.

Let $\mathcal{W}_h \subset H^1(\Omega)$ be a finite element space of degree k, constructed as any of the spaces for the displacement, the pressure or the deviatoric stress. For any function $v \in H^{k+1}(\Omega)$ and for $i = 0, 1$, we define the interpolation errors

$$\inf_{v_h \in \mathcal{W}_h} \|v - v_h\|_{H^i(\Omega)} \leq Ch^{k+1-i}\|v\|_{H^{k+1}(\Omega)} =: \varepsilon_i(v). \tag{38}$$

We will denote by $\tilde{v}_h$ the best approximation of v in $\mathcal{W}_h$. Clearly, we have that $\varepsilon_0(v) = h\varepsilon_1(v)$. This will allow us to prove that the error function of the method is

$$E(h) := h^k\left(\sqrt{\mu}\|\boldsymbol{u}\|_{H^{k+1}(\Omega)} + \frac{1}{\sqrt{\mu}}h\|\boldsymbol{\sigma}\|_{H^{k+1}(\Omega)} + \frac{1}{\sqrt{\mu}}h\|p\|_{H^{k+1}(\Omega)}\right). \tag{39}$$

To prove convergence, we need to preliminary lemmas. The first concerns the consistency of the formulation:

Lemma 1 (Consistency I). *Let $U \in \mathcal{X}$ be the solution of the continuous problem and $U_h \in \mathcal{X}_h$ the finite element solution of (29). Then, if $\boldsymbol{f} \in \mathcal{V}_h$,*

$$B_{\text{stab}}(U - U_h, V_h) = 0 \quad \forall V_h \in \mathcal{X}_h. \tag{40}$$

Proof. This lemma is a trivial consequence of the consistency of the finite element method proposed (considering the force term $\boldsymbol{f}$ in the finite element space). Note that all the terms added to B in the definition (20) of B_{stab} vanish if U_h is replaced by U (recall that $\boldsymbol{\sigma}_h$ could have been added to $\nabla^S \boldsymbol{u}_h$, since $P_\sigma^\perp(\boldsymbol{\sigma}_h) = \boldsymbol{0}$). ∎

Remark 1. If $P_u^\perp(\boldsymbol{f}) \neq \boldsymbol{0}$ there are two options. The first is to include this orthogonal projection in the definition of the method, and therefore to modify the right-hand-side of (29). All the analysis carries over to this case. The second is to take into account the consistency error coming from $\boldsymbol{f}$ in (40). It is easy to see that in this case this equation can be replaced by $B_{\text{stab}}(U - U_h, V_h) \leq CE(h)\|V_h\|$ and the following results can be immediately adapted.

The next step is to express the interpolation error in terms of the norm $\|\cdot\|$ and the bilinear form B_{stab}. We do this in the following:

Lemma 2 (Interpolation error I). *Let $U \in \mathcal{X}$ be the continuous solution and $\widetilde{U}_h \in \mathcal{X}_h$ its best finite element approximation. Then, the following inequalities hold:*

$$B_{\text{stab}}(U - \widetilde{U}_h, V_h) \leq CE(h)\|V_h\|, \tag{41}$$

$$\|U - \widetilde{U}_h\| \leq CE(h), \tag{42}$$

where $E(h)$ is given in (39).

Proof. Let us start proving (42). By the definition (30) of the norm $\|\cdot\|$ it is immediately checked that

$$\|U - \widetilde{U}_h\| \leq \frac{1}{2\mu}\varepsilon_0^2(\boldsymbol{\sigma}) + \alpha_3 2\mu\varepsilon_1^2(\boldsymbol{u}) + \alpha_2 2\mu\varepsilon_1^2(\boldsymbol{u})$$
$$+ \alpha_1 \frac{h^2}{\mu}\varepsilon_1^2(p) + \alpha_1 \frac{h^2}{\mu}\varepsilon_1^2(\boldsymbol{\sigma}),$$

and (42) follows from the fact that $\varepsilon_0(v) = h\varepsilon_1(v)$ for any function $v \in H^1(\Omega)$.

The proof of (41) is as follows:

$$B_{\text{stab}}(U - \widetilde{U}_h, V_h) \leq \sqrt{\mu}\|\nabla^S \boldsymbol{v}_h\|\frac{1}{\sqrt{\mu}}\varepsilon_0(\boldsymbol{\sigma}) + \sqrt{\mu}\|\nabla \cdot \boldsymbol{v}_h\|\frac{1}{\sqrt{\mu}}\varepsilon_0(p)$$
$$+ \frac{1}{2\sqrt{\mu}}\|\boldsymbol{\tau}_h\|\frac{1}{\sqrt{\mu}}\varepsilon_0(\boldsymbol{\sigma}) + \frac{1}{\sqrt{\mu}}\|\boldsymbol{\tau}_h\|\sqrt{\mu}\varepsilon_1(\boldsymbol{u})$$
$$+ 2\alpha_3\sqrt{\mu}\|\nabla^S \boldsymbol{v}_h\|\sqrt{\mu}\varepsilon_1(\boldsymbol{u}) + 2\alpha_2\sqrt{\mu}\|\nabla \cdot \boldsymbol{v}_h\|\sqrt{\mu}\varepsilon_1(\boldsymbol{u})$$
$$+ \alpha_1\frac{h^2}{\mu}\|\nabla q_h - \nabla \cdot \boldsymbol{\tau}_h\|(\varepsilon_1(p) + \varepsilon_1(\boldsymbol{\sigma})).$$

All the terms have been organized to see that they are all bounded by $CE(h)\|V_h\|$, from where (41) follows. ∎

We are finally in a position to prove convergence. The proof is standard, but we include it for completeness.

Theorem 2 (Convergence I). *Let $U = (\boldsymbol{u}, p, \boldsymbol{\sigma}) \in \mathcal{X}$ be the solution of the continuous problem. Assume that all the components of this solution belong to $H^{k+1}(\Omega)$, where k is the order of the finite element interpolation. Then, there is a constant $C > 0$ such that*

$$\|U - U_h\| \leq CE(h),$$

where $E(h)$ is given in (39).

Proof. Consider the finite element function $\widetilde{U}_h - U_h \in \mathcal{X}_h$ where, as in Lemma 2, $\widetilde{U}_h \in \mathcal{X}_h$ is the best finite element approximation to U. Starting from the inf-sup condition (16) it follows that there exists $V_h \in \mathcal{X}_h$ such that

$$\begin{aligned}
C\|\widetilde{U}_h - U_h\| \|V_h\| &\leq B_{\text{stab}}(\widetilde{U}_h - U_h, V_h) \\
&= B_{\text{stab}}(\widetilde{U}_h - U, V_h) \qquad \text{(from the consistency (40))} \\
&\leq CE(h)\|V_h\| \qquad \text{(from (41))},
\end{aligned}$$

from where $\|\widetilde{U}_h - U_h\| \leq CE(h)$. The theorem follows now from the triangle inequality $\|U - U_h\| \leq \|U - \widetilde{U}_h\| + \|\widetilde{U}_h - U_h\|$ and the interpolation error estimate (42). ■

Clearly, this convergence result is optimal.

Remark 2. In the error estimate obtained with the standard Galerkin method and using finite element interpolations satisfying the inf-sup conditions (9)-(10), the error function would involve $\|\boldsymbol{\sigma}\|_{H^k(\Omega)}$ and $\|p\|_{H^k(\Omega)}$ instead of $h\|\boldsymbol{\sigma}\|_{H^{k+1}(\Omega)}$ and $h\|p\|_{H^{k+1}(\Omega)}$, respectively. Therefore, the stabilized finite element method requires more regularity for the continuous solution than what would be needed using the Galerkin method. This is a common feature of all stabilized methods of the type presented in this paper.

5 A modified stabilized problem

The problem presented in Section 3 and analyzed in Section 4 comes directly from the variational multiscale concept. However, once arrived to the stabilized problem (29) we may *a posteriori* modify it. We do this here. As we shall see, the modified method has both *improved convergence behavior* and *smaller computational cost*. The only price to be paid is a consistency error that has to be taken into account in the convergence analysis, which otherwise follows exactly the same lines as in the one presented in the previous section.

The starting observation is that it would be computationally convenient to drop the last term in the left-hand-side of (27), that is to say, to replace it by the equation that would come from the standard Galerkin method. In this case we would simply have $\boldsymbol{\sigma}_h = P_\sigma(2\mu\nabla^S \boldsymbol{u}_h)$. But then the discrete equations would be non-symmetric, due to the presence of $\nabla \cdot \boldsymbol{\sigma}_h$ in (19). The next idea is thus to drop also this term. The final discrete system of equations to be solved, instead of (18)-(27), is

$$(\nabla^S \boldsymbol{v}_h, \boldsymbol{\sigma}_h) + \alpha_3 2\mu(\nabla^S \boldsymbol{v}_h, P_\sigma^\perp(\nabla^S \boldsymbol{u}_h)) - (p_h, \nabla \cdot \boldsymbol{v}_h)$$
$$+ \alpha_2 2\mu(P_p^\perp(\nabla \cdot \boldsymbol{u}_h), \nabla \cdot \boldsymbol{v}_h) = \langle \boldsymbol{f}, \boldsymbol{v} \rangle, \qquad (43)$$

$$(q_h, \nabla \cdot \boldsymbol{u}_h) + \alpha_1 \frac{h^2}{\mu}(\nabla q_h, P_u^\perp(\nabla p_h)) = 0, \qquad (44)$$

$$\frac{1}{2\mu}(\boldsymbol{\sigma}_h, \boldsymbol{\tau}_h) - (\nabla^S \boldsymbol{u}_h, \boldsymbol{\tau}_h) = 0, \qquad (45)$$

that must hold for all $V_h = (\boldsymbol{v}_h, q_h, \boldsymbol{\tau}_h) \in \mathcal{X}_h$. This problem can now be written as: find $U_h \in \mathcal{X}_h$ such that

$$B_{\text{stab},*}(U_h, V_h) = L(V_h) \qquad \forall V_h \in \mathcal{X}_h, \qquad (46)$$

where the bilinear form $B_{\text{stab},*}$ is now defined as

$$B_{\text{stab},*}(U_h, V_h) := B(U_h, V_h) + \alpha_3 2\mu(\nabla^S \boldsymbol{v}_h, P_\sigma^\perp(\nabla^S \boldsymbol{u}_h))$$
$$+ \alpha_2 2\mu(P_p^\perp(\nabla \cdot \boldsymbol{u}_h), \nabla \cdot \boldsymbol{v}_h) + \alpha_1 \frac{h^2}{\mu}(\nabla q_h, P_u^\perp(\nabla p_h)). \qquad (47)$$

Remark 3. Even though we will not discuss here the extension of the present formulation to nonlinear problems, let us briefly discuss some of its implications in a nonlinear situation. Suppose for example that the constitutive law is of the form $\boldsymbol{\sigma} = \boldsymbol{F}(\nabla^S \boldsymbol{u})$, with $\boldsymbol{F}$ a nonlinear function. Equation (45) has to be replaced by $(\boldsymbol{\sigma}_h, \boldsymbol{\tau}_h) - (\boldsymbol{F}(\nabla^S \boldsymbol{u}_h), \boldsymbol{\tau}_h) = 0$, that is to say, $\boldsymbol{\sigma}_h = P_\sigma(\boldsymbol{F}(\nabla^S \boldsymbol{u}_h))$. A straightforward application of the variational multi-scale concept would lead us to replace the second term in the left-hand-side of (43) by $\alpha_3(\nabla^S \boldsymbol{v}_h, P_\sigma^\perp(\boldsymbol{F}(\nabla^S \boldsymbol{u})))$ and therefore the first two terms of this equation would add up to $(\nabla^S \boldsymbol{v}_h, (1-\alpha_3)P_\sigma(\boldsymbol{F}(\nabla^S \boldsymbol{u}_h)) + \alpha_3 \boldsymbol{F}(\nabla^S \boldsymbol{u}_h))$. For $\alpha_3 = 0$ *the formulation would be unstable, whereas for $\alpha_3 = 1$ we would recover an irreductibe formulation, without the stress as unknwon.* Nothing is gained for $0 < \alpha_3 < 1$. However, *there is no need to take $\alpha_3(\nabla^S \boldsymbol{v}_h, P_\sigma^\perp(\boldsymbol{F}(\nabla^S \boldsymbol{u})))$ in the second term of (43)*. We could for example take $\alpha_3 \mu_0 (\nabla^S \boldsymbol{v}_h, P_\sigma^\perp(\nabla^S \boldsymbol{u}_h))$, with μ_0 a constant. Once more, the only price to be paid is *an optimal consistency error*, and the gain is that the constitutive law only appears in (45) which, as it has been said, implies $\boldsymbol{\sigma}_h = P_\sigma(\boldsymbol{F}(\nabla^S \boldsymbol{u}_h))$.

Let us proceed to analyze now problem (46) with the bilinear form $B_{\text{stab},*}$ given by (47). The analysis now is based on the norm $\|\cdot\|_*$, defined by

$$\|V_h\|_*^2 := \frac{1}{2\mu}\|\boldsymbol{\tau}_h\|^2 + \alpha_3 2\mu\|\nabla^S \boldsymbol{v}_h\|^2 + \alpha_2 2\mu\|\nabla \cdot \boldsymbol{v}_h\|^2 + \alpha_1 \frac{h^2}{\mu}\|\nabla q_h\|^2. \qquad (48)$$

Clearly, the first point to be noticed is that this norm *is finer than* $\|\cdot\|$, since it involves the norm of the pressure gradient directly, and not a combination of the pressure gradient and the stress divergence. The same stability and convergence estimate in this norm gives, in principle, more information than in the norm $\|\cdot\|$.

The results to be presented follow the same scheme as in Section 4. Let us start proving stability:

Theorem 3 (Stability II). *There is a constant $C > 0$ such that*

$$\inf_{U_h \in \mathcal{X}_h} \sup_{V_h \in \mathcal{X}_h} \frac{B_{\mathrm{stab},*}(U_h, V_h)}{\|U_h\|_* \|V_h\|_*} \geq C. \tag{49}$$

Proof. Taking $V_h = U_h$ in the definition of $B_{\mathrm{stab},*}$ yields

$$B_{\mathrm{stab},*}(U_h, U_h) = \frac{1}{2\mu}\|\boldsymbol{\sigma}_h\|^2 + \alpha_3 2\mu \|P_\sigma^\perp(\nabla^S \boldsymbol{u}_h)\|^2$$
$$+ \alpha_2 2\mu \|P_p^\perp(\nabla \cdot \boldsymbol{u}_h)\|^2 + \alpha_1 \frac{h^2}{\mu}\mu\|P_u^\perp(\nabla p_h)\|^2. \tag{50}$$

The control on the components on the finite element space for the terms whose orthogonal components appear in this expression is obtained in a manner completely analogous to that of Theorem 1. Some of the details will be omitted.

Taking $V_{h1} := \alpha_1 \frac{h^2}{\mu}(P_u(\nabla p_h), 0, \boldsymbol{0})$ it is now found that

$$B_{\mathrm{stab}}(U_h, V_{h1}) \geq \alpha_1 \frac{h^2}{4\mu}\|P_u(\nabla p_h)\|^2$$
$$- 4\alpha_1 \alpha_3^2 \mu C_{\mathrm{inv}}^2 \|P_\sigma^\perp(\nabla^S \boldsymbol{u}_h)\|^2$$
$$- 4\alpha_1 \alpha_2^2 \mu C_{\mathrm{inv}}^2 \|P_p^\perp(\nabla \cdot \boldsymbol{u}_h)\|^2$$
$$- \alpha_1 C_{\mathrm{inv}}^2 \frac{1}{\mu}\|\boldsymbol{\sigma}_h\|^2. \tag{51}$$

Considering $V_{h2} := (\boldsymbol{0}, \alpha_2 2\mu P_p(\nabla \cdot \boldsymbol{u}_h), \boldsymbol{0})$, $V_{h3} := (\boldsymbol{0}, 0, -\alpha_3 2\mu P_\sigma(\nabla^S \boldsymbol{u}_h))$, as in Theorem 1, yields:

$$B_{\mathrm{stab}}(U_h, V_{h2}) \geq \alpha_2 \mu \|P_p(\nabla \cdot \boldsymbol{u}_h)\|^2 - \alpha_1^2 \alpha_2 C_{\mathrm{inv}}^2 \frac{h^2}{\mu}\|P_u^\perp(\nabla p_h)\|^2, \tag{52}$$

$$B_{\mathrm{stab}}(U_h, V_{h3}) \geq \alpha_3 \mu \|P_\sigma(\nabla^S \boldsymbol{u}_h)\|^2 - \alpha_3 \frac{1}{4\mu}\|\boldsymbol{\sigma}_h\|^2. \tag{53}$$

Let $V_h = \beta_1 V_{h1} + \beta_2 V_{h2} + \beta_3 V_{h3}$, with V_{hi}, $i = 1, 2, 3$, introduced above. Adding up inequalities (51)-(52)-(53) multiplied by β_1, β_2 and β_3, respectively, and adding also (50), it is easily shown that there exist β_i, $i = 1, 2, 3$, for which

$$B_{\mathrm{stab},*}(U_h, V_h) \geq C\|U_h\|_*^2. \tag{54}$$

On the other hand, it can be shown that $\|V_h\|_* \leq C\|U_h\|_*$, which, together with (54) completes the proof of the theorem. ∎

We shall prove convergence under the same assumptions as in Section 4. We will see that *the error function in this case is again (39)*. The main difference in the analysis is in fact the consistency established in the following:

Lemma 3 (Consistency II). *Let $U \in \mathcal{X}$ be the solution of the continuous problem and $U_h \in \mathcal{X}_h$ the finite element solution of (46). Then*

$$B_{\text{stab},*}(U - U_h, V_h) \leq CE(h)\|V_h\|_* \qquad \forall V_h \in \mathcal{X}_h, \tag{55}$$

where $E(h)$ is given in (39).

Proof. It is readily checked that

$$\begin{aligned} B_{\text{stab},*}(U - U_h, V_h) &= \alpha_3 2\mu(\nabla^S v_h, P_\sigma^\perp(\nabla^S u)) \\ &\quad + \alpha_1 \frac{h^2}{\mu}(\nabla q_h, P_u^\perp(\nabla p)). \end{aligned} \tag{56}$$

We could have neglected the first term in the right-hand-side of this expression assuming that $P_u^\perp(\boldsymbol{f}) = \boldsymbol{0}$ and noting that $P_\sigma^\perp(\boldsymbol{\sigma}_h) = \boldsymbol{0}$. However, we have in any case a consistency error due to the last term, and therefore there is no need to assume that $\boldsymbol{f}$ is a finite element function (see Remark 1).

To prove (55) from (56) it is enough to recall the best approximation property of the $L^2(\Omega)$-projection onto the finite element spaces, which implies $\|P_\sigma^\perp(\nabla^S \boldsymbol{u})\| \leq C\varepsilon_1(\boldsymbol{u})$ and $\|P_u^\perp(\nabla p)\| \leq Ch\varepsilon_0(p)$, with $\varepsilon_i(\cdot)$, $i = 0, 1$, defined in (38). ∎

Now we need to express the interpolation error in terms of the norm $\|\cdot\|_*$ and the bilinear form $B_{\text{stab},*}$. The result is

Lemma 4 (Interpolation error II). *Let $U \in \mathcal{X}$ be the continuous solution and $\tilde{U}_h \in \mathcal{X}_h$ its best finite element approximation. Then, the following inequalities hold:*

$$B_{\text{stab},*}(U - \tilde{U}_h, V_h) \leq CE(h)\|V_h\|_*, \tag{57}$$
$$\|U - \tilde{U}_h\|_* \leq CE(h), \tag{58}$$

where $E(h)$ is given in (39).

Proof. It follows the same steps as that of Lemma 2. ∎

We finally give the convergence result. The modification of the standard proof due to the consistency error is trivial:

Theorem 4 (Convergence II). *Let $U = (\boldsymbol{u}, p, \boldsymbol{\sigma}) \in \mathcal{X}$ be the solution of the continuous problem. Assume that all the components of this solution belong* to $H^{k+1}(\Omega)$, where k is the order of the finite element interpolation. Then, there is a constant $C > 0$ such that

$$\|U - U_h\|_* \leq CE(h),$$

where $E(h)$ is given in (39).

The direct control on the pressure gradient provided by Theorem 3 and Theorem 4 (instead of a combination of pressure gradient and stress divergence, as in the previous formulation) allows us to obtain stability and error estimates for the pressure in its natural norm, namely, $L^2(\Omega)$. We do this next, extending the strategy employed for example in [7] for the classical displacement-pressure formulation of the Stokes problem:

Theorem 5 (Stability and convergence in natural norms). *The solution of the discrete problem (46), $U_h = (\boldsymbol{u}_h, p_h, \boldsymbol{\sigma}_h) \in \mathcal{X}_h$, is bounded as*

$$\sqrt{\mu}\|\boldsymbol{u}_h\|_{H^1(\Omega)} + \frac{1}{\sqrt{\mu}}\|\boldsymbol{\sigma}_h\| + \frac{1}{\sqrt{\mu}}\|p_h\| \leq \frac{C}{\sqrt{\mu}}\|\boldsymbol{f}\|_{H^{-1}(\Omega)}. \tag{59}$$

Moreover, under the assumptions of Theorem 4 it follows that

$$\sqrt{\mu}\|\boldsymbol{u} - \boldsymbol{u}_h\|_{H^1(\Omega)} + \frac{1}{\sqrt{\mu}}\|\boldsymbol{\sigma} - \boldsymbol{\sigma}_h\| + \frac{1}{\sqrt{\mu}}\|p - p_h\| \leq CE(h), \tag{60}$$

where $U = (\boldsymbol{u}, p, \boldsymbol{\sigma}) \in \mathcal{X}$ is the solution of the continuous problem.

Proof. Let us first recall that Korn's inequality implies that $\|\nabla^S \boldsymbol{v}\|$ is a norm in $\mathcal{V}$ equivalent to $\|\boldsymbol{v}\|_{H^1(\Omega)}$. On the other hand, it is clear that

$$\langle \boldsymbol{f}, \boldsymbol{v}_h \rangle \leq \frac{C}{\sqrt{\mu}}\|\boldsymbol{f}\|_{H^{-1}(\Omega)}\sqrt{\mu}\|\boldsymbol{v}_h\|_{H^1(\Omega)} \leq \frac{C}{\sqrt{\mu}}\|\boldsymbol{f}\|_{H^{-1}(\Omega)}\|V_h\|_*,$$

where $V_h = (\boldsymbol{v}_h, q_h, \boldsymbol{\tau}_h) \in \mathcal{X}_h$ is arbitrary. Therefore the inf-sup condition proved in Theorem 3 implies that $\|U_h\|_* \leq \frac{C}{\sqrt{\mu}}\|\boldsymbol{f}\|_{H^{-1}(\Omega)}$, which, together with the definition of $\|\cdot\|_*$ in (48) yields the bound (59) for the first two terms in the left-hand-side of this inequality. Likewise, Theorem 4 implies the error estimate (60) for the displacement and the stresses.

The point is thus to prove the stability and the error estimate for the pressure stated in (59) and (60), respectively. We do this using a duality argument. Let $(\boldsymbol{\omega}, \pi, \boldsymbol{S}) \in \mathcal{X}$ be the solution of the following problem:

$$-\nabla \cdot \boldsymbol{S} + \nabla \pi = \boldsymbol{0} \quad \text{in } \Omega, \tag{61}$$

$$\nabla \cdot \boldsymbol{\omega} = \gamma p - p_h \quad \text{in } \Omega, \tag{62}$$

$$\boldsymbol{S} - 2\mu\nabla^S \boldsymbol{\omega} = \boldsymbol{0} \quad \text{in } \Omega, \tag{63}$$

$$\boldsymbol{\omega} = \boldsymbol{0} \quad \text{on } \partial\Omega,$$

where γ can only take the values 0 and 1. Testing (61) by $\boldsymbol{\omega}$, (62) by π, (63) by $\boldsymbol{S}$ and adding the results, it follows that

$$\frac{1}{2\mu}\|\boldsymbol{S}\|^2 \leq \|\gamma p - p_h\| \|\pi\|. \tag{64}$$

On the other hand, the continuous inf-sup condition for $\mathcal{V}$ and $\mathcal{Q}$ implies that there exists $\boldsymbol{\xi} \in \mathcal{V}$ such that $\|\pi\| \|\boldsymbol{\xi}\|_{H^1(\Omega)} \leq C(\pi, \nabla \cdot \boldsymbol{\xi})$, and, from (61), $(\pi, \nabla \cdot \boldsymbol{\xi}) \leq C \|\boldsymbol{\xi}\|_{H^1(\Omega)} \|\boldsymbol{S}\|$, from where $\|\pi\| \leq C\|\boldsymbol{S}\|$. The continuous equation (63) yields also $2\mu \|\boldsymbol{\omega}\|_{H^1(\Omega)} \leq C\|\boldsymbol{S}\|$. Using this in (64) we have the stability bound

$$\|\boldsymbol{\omega}\|_{H^1(\Omega)} \leq C\|\gamma p - p_h\|. \tag{65}$$

Let $\widetilde{\boldsymbol{\omega}}_h \in \mathcal{V}_h$ be an approximation to $\boldsymbol{\omega}$ such that

$$\|\boldsymbol{\omega} - \widetilde{\boldsymbol{\omega}}_h\|_{H^m(\Omega)} \leq Ch^{1-m}\|\boldsymbol{\omega}\|_{H^1(\Omega)}, \quad m = 0, 1. \tag{66}$$

If now we test (62) by $\gamma p - p_h$, we obtain:

$$\|\gamma p - p_h\|^2 = (\gamma p - p_h, \gamma p - p_h)$$
$$= (\nabla \cdot \boldsymbol{\omega}, \gamma p - p_h)$$
$$= (\nabla \cdot (\boldsymbol{\omega} - \widetilde{\boldsymbol{\omega}}_h), \gamma p - p_h) - (\widetilde{\boldsymbol{\omega}}_h, \nabla(\gamma p - p_h))$$
$$= -(\boldsymbol{\omega} - \widetilde{\boldsymbol{\omega}}_h, \nabla(\gamma p - p_h)) + (\nabla^S \widetilde{\boldsymbol{\omega}}_h, \gamma\boldsymbol{\sigma} - \boldsymbol{\sigma}_h) - (\gamma - 1)\langle \widetilde{\boldsymbol{\omega}}_h, \boldsymbol{f} \rangle$$
$$\leq \|\boldsymbol{\omega} - \widetilde{\boldsymbol{\omega}}_h\| \|\gamma \nabla p - \nabla p_h\|$$
$$\quad + C\|\boldsymbol{\omega}\|_{H^1(\Omega)} \left(\|\gamma\boldsymbol{\sigma} - \boldsymbol{\sigma}_h\| + (1-\gamma)\|\boldsymbol{f}\|_{H^{-1}(\Omega)} \right)$$
$$\leq C\|\boldsymbol{\omega}\|_{H^1(\Omega)} h \|\gamma \nabla p - \nabla p_h\|$$
$$\quad + C\|\boldsymbol{\omega}\|_{H^1(\Omega)} \left(\|\gamma\boldsymbol{\sigma} - \boldsymbol{\sigma}_h\| + (1-\gamma)\|\boldsymbol{f}\|_{H^{-1}(\Omega)} \right)$$
$$\leq C \left(h\|\gamma \nabla p - \nabla p_h\| + \|\gamma\boldsymbol{\sigma} - \boldsymbol{\sigma}_h\| + (1-\gamma)\|\boldsymbol{f}\|_{H^{-1}(\Omega)} \right) \|\gamma p - p_h\|.$$

The stability and error estimate for the pressure we wished to prove follow taking $\gamma = 0$ and $\gamma = 1$, respectively, and using the stability and convergence provided by Theorems 3 and 4. ∎

6 Concluding remarks

Let us conclude with some remarks concerning the numerical formulations presented in this paper. From the point of view of the numerical analysis, which has been our main concern, the two methods presented are stable and optimally accurate *using equal interpolation for the displacement, the pressure and the stresses*. Therefore, the main goal has been achieved.

Let us comment on two aspects that have been not treated in the paper and that refer to the pontential of these formulations. The first remark is the implementation of the orthogonal projections, say $P^\perp$. In practice, this projection applied to any derivative of a finite element function, v_h, can be expressed as $P^\perp(\nabla v_h) = \nabla v_h - P(\nabla v_h)$. In an iterative scheme, the term $P(\nabla v_h)$ can be evaluated in a previous iteration. This allows us to maintain the stencil of the Galerkin formulation in the matrix of the final discrete system. Of course for a linear problem, as the one analyzed here, this iterative procedure implies an additional cost, but for a nonlinear problem this iterative treatment can be coupled with the iterations due to the nonlinearity. Our experience indicates that this causes no significant deterioration of the nonlinear convergence of the scheme.

As it has been mentioned in the Introduction, the problem analyzed here is nothing but a model for more complex situations. Typically, viscoelastic flows are often posed as example of a problem that requires the interpolation of the stresses, but this can also be done for nonlinear models such as damage or plasticity in solid mechanics, and non-Newtonian fluids or even turbulence models in fluid mechanics. When designing an extension of the formulations presented here to these more complex situations, the most important idea to bear in mind is which is the stabilization mechanism introduced by the formulations proposed. The analysis dictates that pressure is stabilized by the term proportional to $P_u^\perp(\nabla p_h)$ introduced in the continuity equation (see (19) and (44)) and the displacement gradient is stabilized by the term proportional to $P_\sigma^\perp(\nabla^S \boldsymbol{u}_h)$ introduced in the momentum equation (see (18) and (43)). This is the essential point. The only condition on the factors that multiply these terms is that they have to yield an adequate scaling and order of convergence.

References

1. Baaijens, F.P.T., Hulsen, M.A., Anderson, P.D.: The use of mixed finite element methods for viscoelastic fluid flow analysis, chapter 14 in encyclopedia of computational mechanics, e. stein, r. de borst and t.j.r. hughes (eds.), pp. 481–498, John Wiley & Sons (2004)
2. Behr, M., Franca, L.P., Tezduyar, T.E.: Stabilized finite element methods for the velocity-pressure-stress formulation of incompressible flows, **104**, 31–48 (1993)
3. Bonvin, J., Picasso, M., Stenberg, R.: GLS and EVSS methods for a three fields Stokes problems arising from viscoelastic flows, **190**, 3893–3914 (2001)
4. Brezzi, F., Fortin, M.: Mixed and hybrid finite element methods, Springer Verlag (1991)
5. Codina, R.: Stabilization of incompressibility and convection through orthogonal sub-scales in finite element methods, **190**, 1579–1599 (2000)
6. _____: Stabilized finite element approximation of transient incompressible flows using orthogonal subscales, **191**, 4295–4321 (2002)
7. Codina, R., Blasco, J.: Analysis of a pressure-stabilized finite element approximation of the stationary Navier-Stokes equations, Numerische Mathematik**87**, 59–81 (2000)

8. Fortin, M., Guénette, R., Pierrer.: Numerical analysis of the modified EVSS method, **143**, 79–95 (1997)
9. Franca, L., Stenberg, R.: Error analysis of some Galerkin least-squares methods for the elasticity equations, SIAM Journal on Numerical Analysis**28** 1680–1697 (1991)
10. Guénette, R., Fortin, M.: A new mixed finite element method for computing viscoelastic flows, Journal of Non-Newtonian Fluid Mechanics **60**, 27–52 (1995)
11. Hughes, T.J.R.: Multiscale phenomena: Green's function, the Dirichlet-to-Neumann formulation, subgrid scale models, bubbles and the origins of stabilized formulations, **127**, 387–401 (1995)
12. Hughes, T.J.R., Feijóo, G.R., Mazzei, L., Quincy, J.B.: The variational multiscale method–a paradigm for computational mechanics, **166**, 3–24 (1998)

Convergence of Adaptive Wavelet Methods for Goal–Oriented Error Estimation*

Wolfgang Dahmen[1], Angela Kunoth[2] and Jürgen Vorloeper[1]

[1] Institut für Geometrie und Praktische Mathematik
RWTH Aachen, Templergraben 55, 52056 Aachen, Germany
{dahmen,jvor}@igpm.rwth-aachen.de

[2] Institut für Angewandte Mathematik und Institut für Numerische Simulation
Universität Bonn, Wegelerstr. 6, 53115 Bonn, Germany
kunoth@iam.uni-bonn.de

Summary. We investigate adaptive wavelet methods which are *goal–oriented* in the sense that a *functional* of the solution of a linear elliptic PDE is computed up to arbitrary accuracy at possibly low computational cost measured in terms of degrees of freedom. In particular, we propose a scheme that can be shown to exhibit convergence to the target value without insisting on energy norm convergence of the primal solution. The theoretical findings are complemented by first numerical experiments.

1 Introduction

The importance of *adaptive* solution concepts for large scale computational tasks arising in Numerical Simulation based on PDEs or integral equations is nowadays well accepted. The evidence provided by numerical experience is, however, nor quite in par with the theoretical foundation of such schemes. A thorough analytical understanding, in turn, has recently proven to lead to new algorithmic paradigms in connection with wavelet based schemes. Rigorous complexity and convergence estimates were obtained for adaptive wavelet methods for a wide class of linear and nonlinear variational problems, see, e.g., [8, 9, 12, 14]. These estimates relate for the first time the computational work and the adaptively generated number of degrees of freedom to the *target accuracy* of the approximate solution. This accuracy refers to the approximation in some (energy) norm, i.e., the whole unknown solution is recovered. These

---

* This research was supported in part by the EEC Human Potential Programme under contract HPRN-CT-2002-00286, "Breaking Complexity", the SFB 401, "Flow Modulation and Fluid-Structure Interaction at Airplane Wings", and SFB 611, "Singular Phenomena and Scaling in Mathematical Models", funded by the German Research Foundation.

developments have meanwhile spilled over to the Finite Element setting where analogous results could be obtained for a much more restricted problem class, though, see, e.g., [3, 24].

However, in many applications one is only interested in some *functional* of the solution which, in particular, might be local such as point values or integrals on some lower dimensional manifold. In such a case one might expect to obtain the desired information at a much lower expense than computing the whole solution. This is exactly the objective of *goal-oriented* error estimation which gives rise to the so called *dual weighted residual method* (DWR), see, e.g., [7] and the references cited therein.

Many striking examples indicate that one may indeed reach the goal with the aid of this paradigm at the expense of much less computational work in comparison with schemes driven by *norm approximation*. On the other hand, a rigorous analysis of the DWR faces a number of severe obstructions related, in particular, to the fact that the central error representation involves the (unknown) solution to the dual problem. Thus, the dual solution has to be estimated along the way. Although this problem arises, in principle, already when dealing with linear problems, it becomes more delicate in the nonlinear case since the dual solution depends then on the primal one. It is fair to say that the mutual intertwinement of the accuracies of dual and primal solutions, especially with regard to the spatial distribution of degrees of freedom, is far from a rigorous understanding. It is not even clear in the linear case that adaptive refinements based on the practiced versions of the DWR paradigm actually converge in the sense that the searched value is actually approached better and better by the computed one as the refinement goes on. It is this issue that will be the primary concern of this paper.

To appreciate this issue, it is helpful to keep a few principal facts in mind. Approximability of a function in some norm can always be understood in terms of the *regularity* of that function (with respect to some nonclassical regularity measure). In a typical application of the DWR, adaptivity is not driven by the regularity of the searched for object, but primarily by the *locality* of the targeted information, conveyed by the dual solution which is often termed *generalized Green's function*, see, e.g., [19]. This generalized Green's function indicates the influence of parts of the primal solution away from the spatial location of the target functional. Thus, the experience gained with adaptive wavelet schemes for energy norm approximation is not immediately seen to be helpful in the context of the DWR.

Nevertheless, the primary goal of this paper is to contribute to the understanding of the DWR by looking at this paradigm from a wavelet point of view. Here is a rough indication why this might indeed be a promising perspective: The key to the above mentioned results from [8, 9] is to formulate an iteration (e.g., a gradient or a Newton scheme) for the full infinite dimensional problem formulated in wavelet coordinates. This idealized iteration is then mimicked by the adaptive evaluation of the involved operators within any desired error tolerance. Staying in that sense controllably close to the infinite dimensional problem may therefore be expected to help also in the context of the DWR.

In this note we wish to explore this aspect for an admittedly simple class of model problems, namely, linear elliptic boundary value problems. Moreover, we shall consider only linear evaluation functionals that belong to the dual of the energy space. Further linearization and/or regularization can be, of course, performed as explained in many foregoing investigations. The main point is to identify the key mechanisms so as to draw also conclusions for more complex problems.

We shall occasionally use the following convention for estimates containing generic constants. The relation $a \sim b$ always stands for $a \lesssim b$ and $a \gtrsim b$, i.e., a can be estimated from above and below by a constant multiple of b independent of all parameters on which a or b may depend.

2 Goal–oriented error estimation

2.1 Problem formulation

Let V denote a Hilbert space living on some bounded Lipschitz domain $\Omega \subset \mathbb{R}^d$ and let V' be its topological dual. Its associated dual form will be denoted as $\langle \cdot, \cdot \rangle_{V \times V'}$, or shortly as $\langle \cdot, \cdot \rangle$.

Moreover, let $a(\cdot, \cdot)$ be a symmetric bilinear form which will here always supposed to be continuous and elliptic on V, i.e., there exist constants c_A, C_A such that

$$\sqrt{c_A}\|v\|_V \leq a(v,v)^{1/2} \leq \sqrt{C_A}\|v\|_V, \qquad v \in V. \tag{2.1}$$

In this case the variational problem: given any $f \in V'$, find $u \in V$ such that

$$a(v, u) = \langle v, f \rangle, \qquad v \in V, \tag{2.2}$$

is well posed. It will be convenient to introduce the induced operator $A : V \to V'$ given by $\langle v, Aw \rangle := a(v, w)$ for all $v, w \in V$.

Instead of approximating the whole solution u we are interested in evaluating only a *functional* of the unknown solution. Specifically, we consider the following problem: Given a fixed linear functional $J \in V'$, compute

$$J(u) := \langle u, J \rangle, \tag{2.3}$$

where u is the solution of (12). $J(u)$ may be a very local quantity, such as the point evaluation of u at some point $x_* \in \Omega$, if the Dirac functional is in V' (as in the case of Plateau's equation on an interval), or a local quantity like the mean of u over some small domain $\Omega_\delta \subset \Omega$, i.e., $J(u) = |\Omega_\delta|^{-1} \int_{\Omega_\delta} u(x)\,dx$, or a weighted integral of u over some lower dimensional manifold in Ω. We shall exclude first more general situations such as *nonlinear* functionals J which would require an additional linearization process as shown in [7], as well as functionals that are not contained in V' but require additional regularity of the solution.

Of course, one might approximate the quantity $J(u)$ by determining first some approximation u_Λ to u sitting in some finite dimensional trial space indicated by the subscript Λ, and take then $J(u_\Lambda)$ as an approximation to the desired value $J(u)$. Moreover, in the above framework it is natural to take u_Λ as a *Galerkin solution* with respect to some subspace $V_\Lambda \subset V$, i.e.,

$$a(v, u_\Lambda) = \langle v, f \rangle, \qquad v \in V_\Lambda. \tag{2.4}$$

Under the circumstances (2.1), (12), u_Λ is uniquely determined for any $V_\Lambda \subset V$. (For conceptual reasons that will become clear later, we deliberately do not even insist at this point on V_Λ being finite dimensional.) We shall frequently use the shorthand notation

$$e_\Lambda := u - u_\Lambda.$$

Our goal now is to determine u_Λ such that for a given *target accuracy* $\varepsilon > 0$

$$|J(u) - J(u_\Lambda)| = |J(u - u_\Lambda)| = |J(e_\Lambda)| \leq \varepsilon, \tag{2.5}$$

while the computational cost needed to determine u_Λ is to be kept as low as possible. Since, by assumption, $J \in V'$, we have

$$|J(e_\Lambda)| \leq \|J\|_{V \to \mathbb{R}} \|e_\Lambda\|_V, \tag{2.6}$$

where, as usual, $\|J\|_{V \to \mathbb{R}} := \sup_{v \in V, \|v\|_V \leq 1} \langle v, J \rangle$.

Remark 1. When $J \notin V'$ but $J \in (V^+)'$ where $V^+ \hookrightarrow V$ and $u, u_\Lambda \in V^+$, we obtain an analogous estimate of the form $|J(e_\Lambda)| \leq \|J\|_{V^+ \to \mathbb{R}} \|e_\Lambda\|_{V^+}$.

Staying with the simpler former situation, a principal gain is that the target accuracy ε can be achieved by solving two problems, namely, the primal (12) and the dual one (2.8) with accuracies of the order $\sqrt{\varepsilon}$. Thus, choosing some subspace V_Λ, based on some a-priori estimates, such that the Galerkin error satisfies

$$\|u - u_\Lambda\|_V < \varepsilon / \|J\|_{V \to \mathbb{R}}, \tag{2.7}$$

this, together with (2.6), would yield (2.5). In general, such an a-priori choice would require a too large V_Λ. In any case, an adaptive choice of V_Λ with respect to the energy norm may lead to an overestimation since such a norm approximation does not take the locality of J into account.

2.2 The dual weighted residual method: error representation

It is the very purpose of the *dual weighted residual method* (DWR) to take the locality of J into account when refining a given discretization so as to improve on the accuracy of the approximate value, possibly without approximating the whole solution everywhere in the domain with a comparable accuracy. In order to motivate the subsequent development, we briefly review some basic facts concerning this methodology from [7, 19]. The key is to obtain an *error representation* comprised of local quantities that reflect residual terms which *can* be evaluated. The derivation of such representations relies on *duality arguments* to be explained next.

Let $z \in V$ be the solution of the *dual problem*

$$a(z,w) = \langle w, J \rangle, \qquad w \in V, \tag{2.8}$$

with $J \in V'$ serving as right hand side. Inserting $w = u - u_\Lambda = e_\Lambda$ yields the error representation

$$J(e_\Lambda) = \langle e_\Lambda, J \rangle = a(z, e_\Lambda) = a(z - y_\Lambda, e_\Lambda), \qquad \text{for any } y_\Lambda \in V_\Lambda, \tag{2.9}$$

where we have used Galerkin orthogonality in the last step. This suggests several options for bounding these residuals. First, we obtain the estimate

$$|J(u - u_\Lambda)| = |a(z - y_\Lambda, u - u_\Lambda)| \lesssim \|u - u_\Lambda\|_V \inf_{y_\Lambda \in V_\Lambda} \|z - y_\Lambda\|_V. \tag{2.10}$$

Thus, if the computational work (measured in terms of problem size expressed as the number of degrees of freedom N) needed to compute such approximations for the primal and dual solution with accuracy ε scales like $N(\varepsilon) = \varepsilon^{-\alpha}$ for some $\alpha > 0$, the error in (2.10) can be bounded by ε^2. So the computational work needed to determine the value $J(u)$ within a tolerance ε scales like $2\varepsilon^{-\alpha/2}$. This is asymptotically better than just computing the primal solution with tolerance ε in the energy norm (2.7).

This still does not exploit the locality of the functional J of interest. In the framework of Finite Element discretizations, one usually treats this latter objective by bounding the error representation $a(z - y_\Lambda, u - u_\Lambda)$ by a sum of local computable quantities. To specify this, let Λ denote then a current triangulation of the domain Ω. Such estimates have then the form

$$|a(z - y_\Lambda, u - u_\Lambda)| \lesssim \sum_{T \in \Lambda} w_T(y_\Lambda) \, r_T(u_\Lambda), \tag{2.11}$$

where the $r_T(u_\Lambda)$ are *local residuals* of the approximate solution u_Λ and the $w_T(y_\Lambda)$ are *weights* computed in terms of the dual solution. For the simple case $a(v,w) = \int_\Omega (\nabla y)^T \nabla w \, dx$, they look like

$$r_T(u_\Lambda) = \|f + \Delta u_\Lambda\|_{L_2(T)} + \frac{1}{2} h_T^{-1/2} \left\| \left[\frac{\partial u_\Lambda}{\partial n} \right] \right\|_{L_2(\partial T)}. \tag{2.12}$$

The weights or *stability factors* are of the form

$$w_T(y_\Lambda) = \|z - y_\Lambda\|_{L_2(T)} + h_T^{1/2} \|z - y_\Lambda\|_{L_2(\partial T)}, \tag{2.13}$$

see, e.g., [7, 19].

Note that, while the $r_T(u_\Lambda)$ are computable, the weights $w_T(y_\Lambda)$ depend on the unknown dual solution z. One can argue that, in practical applications it suffices to know only the "trend" of these weights to see the influence of the local residual $r_T(u_\Lambda)$ and, consequently, of the local error caused by u_Λ. There are several ways of obtaining approximations to these weights:

(i) One can compute an approximate solution $\bar{z}$ of z on some finer mesh than the one used for the primal solution and substitute $\bar{z}$ for z.
(ii) One can compute a *higher order Galerkin approximation* as a substitute for z in (2.13).
(iii) Instead of computing the difference $z - y_\Lambda$, one determines a higher order Galerkin approximation $\bar{z}$ to z, computes its second order derivatives and replaces $w_T(y_\Lambda)$ by some constant multiple of $h_T^2 \|\bar{z}\|_{H^2(T)}$.
(iv) A lower order Galerkin approximation is postprocessed to provide second order approximations that can then be used as in (iii).

In simple cases, all these strategies are expected to work fine. Nevertheless, even in the simple linear model case, none of them give rigorous bounds for the actual error resulting from any refinement strategy and from corresponding decisions on how accurately the dual solution needs to be approximated. The amount of confidence one can put in either of them may vary considerably: Neither is it clear that any fixed mesh refinement or a higher order approximation is sufficiently closer to the true solution to provide a reliable trend (in particular, near singularities), nor is it clear that the second order derivatives behave as those of the true dual solution (again, especially, when singularities interfere).

Thus, already at a rather basic level, one faces the essential question as to how accurately should the dual solution be computed and how localized the distribution of degrees of freedom can be chosen without loosing essential information.

The subsequent discussion attempts to shed some further light on these issues exploiting some concepts that have been developed in connection with adaptive wavelet schemes, see, e.g., [7, 8, 9].

2.3 Wavelet coordinates

Let $\Psi := \{\psi_\lambda : \lambda \in \mathbb{I}\} \subset V$ be a wavelet basis for V. By this we mean that every $v \in V$ has a unique expansion $v = \sum_{\lambda \in \mathbb{I}} v_\lambda \psi_\lambda$ with coefficient array $\mathbf{v} = (v_\lambda)_{\lambda \in \mathbb{I}}$ such that for fixed constants c_Ψ, C_Ψ one has

$$c_\Psi \|\mathbf{v}\| \leq \|v\|_V \leq C_\Psi \|\mathbf{v}\|, \qquad (2.14)$$

where $\|\mathbf{v}\|^2 := \sum_{\lambda \in \mathbb{I}} |v_\lambda|^2 = \mathbf{v}^T \mathbf{v}$ denotes the ℓ_2-norm. Only when the ℓ_2-norm with respect to a specific subset $\Lambda \subset \mathbb{I}$ is meant we write for clarity $\|\mathbf{v}\|^2_{\ell_2(\Lambda)} := \sum_{\lambda \in \Lambda} |v_\lambda|^2$. Recall that, by a simple duality argument (see, e.g., [13]), one has

$$C_\Psi^{-1}\|\langle\psi_\lambda,w\rangle\| \leq \|w\|_{V'} \leq c_\Psi^{-1}\|\langle\psi_\lambda,w\rangle\|, \quad w \in V'. \tag{2.15}$$

For typical constructions of wavelet bases that are suitable, e.g., for $V = H_0^1(\Omega)$, we refer to [5, 6, 15, 16, 11, 17]. Here it suffices to add a few remarks on the structure of the index set $I\!\!I$. Each index λ comprises information on the scale, denoted by $|\lambda|$, and on the spatial location of the associated basis function $k(\lambda)$. There is usually a finite number of "scaling function type" basis functions on some coarsest level of resolution j_0. This subset will be denoted by $I\!\!I_\phi$. All remaining indices refer to "true" wavelets gathered in $I\!\!I_\psi$. These wavelets are always of compact support whose diameter scale like $2^{-|\lambda|}$. Moreover, these true wavelets have cancellation properties of some specified order $\tilde{m}$ usually derived from a corresponding order of vanishing moments $\langle\psi_\lambda,P\rangle = 0$ for all $\lambda \in I\!\!I_\psi$ and any polynomial P of total order at most $\tilde{m}$. Furthermore, it follows from (2.14) that the wavelets are normalized such that $\|\psi_\lambda\|_V \sim 1$.

Testing (12) by $v = \psi_\lambda$, $\lambda \in I\!\!I$, we obtain an equivalent formulation in wavelet coordinates

$$\mathbf{Au} = \boldsymbol{f}, \tag{2.16}$$

where

$$\mathbf{A} = \bigl(a(\psi_\lambda,\psi_\nu)\bigr)_{\lambda,\nu \in I\!\!I} \tag{2.17}$$

is the wavelet representation of the operator $A : V \to V'$ induced by $a(v,w) = \langle v, Aw\rangle$ for all $v,w \in V$. Likewise the dual problem (2.8) is equivalent to

$$\mathbf{A}^T \mathbf{z} = \mathbf{J}, \tag{2.18}$$

where $\mathbf{J} := \bigl(\langle\psi_\lambda,E\rangle\bigr)_{\lambda \in I\!\!I}$. Combining (2.14), (2.15) with (2.1) yields

$$c_\Psi^2 c_A \|\mathbf{v}\| \leq \|\mathbf{A}\mathbf{v}\| \leq C_\Psi^2 C_A \|\mathbf{v}\|, \quad \mathbf{v} \in \ell_2, \tag{2.19}$$

i.e., the wavelet representation is well conditioned in the Euclidean metric ℓ_2, see e.g. [9].

For any subset $\Lambda \subset I\!\!I$ we let $\Psi_\Lambda := \{\psi_\lambda : \lambda \in I\!\!I\} \subset V$ be the corresponding subset of wavelets and denote by V_Λ the closure in V of the linear span of Ψ_Λ. We continue denoting by u_Λ the Galerkin solution, now with respect to the subspace V_Λ, and by $\mathbf{u}_\Lambda$ the corresponding array of wavelet coefficients supported in Λ.

Note that for any $w = \sum_{\lambda \in I\!\!I} w_\lambda \psi_\lambda =: \mathbf{w}^T \Psi$

$$J(w) = \sum_{\lambda \in I\!\!I} w_\lambda J(\psi_\lambda) = \mathbf{J}^T \mathbf{w}. \tag{2.20}$$

Thus, abbreviating $\mathbf{e}_\Lambda := \mathbf{u} - \mathbf{u}_\Lambda$, $e_\Lambda := (\mathbf{u}-\mathbf{u}_\Lambda)^T \Psi$, the representation (2.9) then takes on the form

$$J(u) - J(u_\Lambda) = \mathbf{J}^T \mathbf{e}_\Lambda = (\mathbf{z}-\mathbf{y}_\Lambda)^T(\boldsymbol{f}-\mathbf{A}\mathbf{u}_\Lambda) = (\mathbf{A}^T(\mathbf{z}-\mathbf{y}_\Lambda))^T (\mathbf{u}-\mathbf{u}_\Lambda), \tag{2.21}$$

where $\mathbf{y}_\Lambda$ is any vector supported in Λ and the primal residual is given by

$$\mathbf{r}_\Lambda(\mathbf{u}) := \mathbf{f} - \mathbf{A}\mathbf{u}_\Lambda = \mathbf{A}\mathbf{e}_\Lambda. \tag{2.22}$$

It is important to note here that (2.22) is the *true* residual for the infinite dimensional operator $\mathbf{A}$.

We shall frequently exploit that, by definition, one has

$$\mathbf{r}_\Lambda(\mathbf{u})|_\Lambda = \mathbf{0}. \tag{2.23}$$

Moreover, it immediately follows from (2.19) that

$$c_A c_\Psi^2 \|\mathbf{u} - \mathbf{u}_\Lambda\| \leq \|\mathbf{r}_\Lambda(\mathbf{u})\| \leq C_A C_\Psi^2 \|\mathbf{u} - \mathbf{u}_\Lambda\|. \tag{2.24}$$

Hence, approximations in V and V' on the function side reduce to approximation in ℓ_2 for the primal and dual wavelet coefficient arrays.

Of course, the problem that the representation (2.21) involves the unknown dual solution remains the same as in conventional discretization settings. However, while the terms in (2.11) reflect primarily spatial localization, the summands in (2.21) convey spatial and frequency information in terms of (dual) wavelet coefficients (of the residual) and of the error. We shall explore next whether this can be exploited for a reliable error estimation.

3 Adaptive error estimation

Our objective is to develop a-posteriori refinement strategies that aim at computing $J(u)$ within some error tolerance at possibly low computational cost. This amounts to a DWR method in wavelet coordinates. (2.20) suggests to take (the computable quantity)

$$J(u_\Lambda) = \mathbf{J}(\mathbf{u}_\Lambda) = \sum_{\lambda \in \Lambda} \mathbf{J}^T \mathbf{u}_\Lambda \tag{3.25}$$

as an approximate value of the target functional, where Λ is a suitable finite index set. Concerning the incurred error, since, by (2.23), one has $\mathbf{r}_\Lambda(\mathbf{u})|_\Lambda = 0$, we infer from (2.21)

$$\mathbf{J}^T \mathbf{e}_\Lambda = \sum_{\lambda \in \mathbb{I} \setminus \Lambda} z_\lambda (\mathbf{r}_\Lambda(\mathbf{u}))_\lambda. \tag{3.26}$$

As a natural heuristics this suggests an analog to option (i) in the Finite Element context, namely, to select some larger index set $\hat{\Lambda} \supset \Lambda$ and replace $\mathbf{z}$ in (3.26) by the Galerkin solution $\mathbf{z}_{\hat{\Lambda}}$ in $V_{\hat{\Lambda}}$. But again the question remains, how large has $\hat{\Lambda}$ to be chosen in order to provide a reliable estimate. The following simple observations suggest how to deal with this question. By (2.21) we have

$$|\mathbf{J}^T(\mathbf{u} - \mathbf{u}_\Lambda)| \leq \Big| \sum_{\lambda \in \Lambda_\delta \setminus \Lambda} z_{\hat{\Lambda},\lambda}\, r_{\Lambda,\lambda}(\mathbf{u}) \Big| + \sum_{\lambda \in \mathit{II} \setminus \Lambda} |(z_\lambda - z_{\hat{\Lambda},\lambda})\, r_{\Lambda,\lambda}(\mathbf{u})|. \quad (3.27)$$

The first part is a finite sum that is computable through the primal residual on a finite set and the computed $\mathbf{z}_{\hat{\Lambda}}$. The second part can be estimated as follows:

$$|\mathbf{J}^T(\mathbf{u} - \mathbf{u}_\Lambda)| \leq \Big| \sum_{\lambda \in \hat{\Lambda} \setminus \Lambda} z_{\hat{\Lambda},\lambda}\, r_{\Lambda,\lambda}(\mathbf{u}) \Big| + \inf_{\substack{1 \leq p, p' \leq \infty \\ \frac{1}{p} + \frac{1}{p'} = 1}} \|\mathbf{z} - \mathbf{z}_{\hat{\Lambda}}\|_{\ell_p} \|\mathbf{r}_\Lambda(\mathbf{u})\|_{\ell_{p'}}. \quad (3.28)$$

Specifically, $p = p' = 1/2$ yields

$$|\mathbf{J}^T(\mathbf{u} - \mathbf{u}_\Lambda)| \leq \Big| \sum_{\lambda \in \hat{\Lambda} \setminus \Lambda} z_{\hat{\Lambda},\lambda}\, r_{\Lambda,\lambda}(\mathbf{u}) \Big| + \|\mathbf{z} - \mathbf{z}_{\hat{\Lambda}}\| \, \|\mathbf{r}_\Lambda(\mathbf{u})\|. \quad (3.29)$$

Thus, due to the norm equivalences (2.24), (2.15), (2.14) the second term on the right hand side is the product of the primal and dual energy norm error. Thus, whenever the dual solution is approximated in the energy norm and the growth of Λ depends on the energy norm approximation of $\mathbf{z}$ the target value is approximated with increasing accuracy even though the global primal residual does not tend to zero at all in ℓ_2. It may tend to zero in some weaker norm which, according to (3.28), could give a better estimate.

Led by the above considerations, we formulate now in precise terms an algorithm which, for any given target accuracy ε, computes $J(u_\Lambda) = \mathbf{J}^T(\mathbf{u}_\Lambda)$ such that $|J(e_\Lambda)| = |\mathbf{J}^T(\mathbf{e}_\Lambda)| \leq \varepsilon$. A central ingredient is the adaptive wavelet scheme from [9] that will be formulated next. The resulting well-posedness in ℓ_2 (2.19) allows one to contrive an (idealized) iteration

$$\mathbf{u}^{n+1} = \mathbf{u}^n - \mathbf{B}(\mathbf{A}\mathbf{u}^n - \boldsymbol{f}), \quad n = 0, 1, 2, \ldots, \quad (3.30)$$

where $\mathbf{B}$ is (a possibly stage dependent) preconditioner, such that for some $\rho < 1$

$$\|\mathbf{u} - \mathbf{u}^{n+1}\| \leq \rho \|\mathbf{u} - \mathbf{u}^n\|, \quad n \in \mathbb{N}_0, \quad (3.31)$$

see [8, 9] for various examples covering also noncoercive problems.

The idea is now to mimic (3) numerically by evaluating the weighted residual $\mathbf{B}(\mathbf{A}\mathbf{u}^n - \boldsymbol{f})$ within a stage dependent dynamical accuracy tolerance. This, in turn, hinges on the *adaptive* evaluation of the involved (at this stage still infinite dimensional) operators when applied to a finitely supported array. We refer to [9, 10, 2] for the precise description of such evaluation schemes for a range of (linear and nonlinear) operators. Therefore we may assume at this point to have a routine of the following form at hand:

RES$[\eta, \mathbf{B}, \mathbf{A}, \boldsymbol{f}, \mathbf{v}] \to \mathbf{r}_\eta$ COMPUTES FOR ANY FINITELY SUPPORTED INPUT $\mathbf{v}$ AND ANY POSITIVE TOLERANCE η AN APPROXIMATE FINITELY SUPPORTED RESIDUAL $\mathbf{r}_\eta$ SUCH THAT

$$\|\mathbf{B}(\mathbf{A}\mathbf{v} - \boldsymbol{f}) - \mathbf{r}_\eta\| \leq \eta. \quad (3.32)$$

We further need the routine

COARSE$[\eta, \mathbf{v}] \to \mathbf{w}_\eta$ DETERMINES FOR ANY FINITELY SUPPORTED INPUT $\mathbf{v}$ AN OUTPUT $\mathbf{w}_\eta$ WITH POSSIBLY SMALL SUPPORT SUCH THAT STILL

$$\|\mathbf{v} - \mathbf{w}_\eta\| \leq \eta. \tag{3.33}$$

Following [9] the announced adaptive solution scheme can now be described as follows.

SOLVE$[\varepsilon, \mathbf{A}, \boldsymbol{f}, \bar{\mathbf{u}}^0] \to (\bar{\mathbf{u}}_\varepsilon, \Lambda_\varepsilon)$ COMPUTES FOR ANY GIVEN TARGET ACCURACY $\varepsilon > 0$ AND ANY INITIAL GUESS $\bar{\mathbf{u}}^0$, SATISFYING $\|\mathbf{u} - \bar{\mathbf{u}}^0\| \leq \delta$, AN APPROXIMATION $\bar{\mathbf{u}}_\varepsilon$ TO (12), SUPPORTED IN SOME FINITE (TREE LIKE) INDEX SET Λ_ε, SUCH THAT

$$\|\mathbf{u} - \bar{\mathbf{u}}_\varepsilon\| \leq \varepsilon, \tag{3.34}$$

ACCORDING TO THE FOLLOWING STEPS:

(I) CHOOSE SOME $C^* > 1$, $\bar{\rho} \in (0,1)$. SET $\varepsilon_0 := \delta$ ACCORDING TO THE ABOVE INITIALIZATION, AND $j = 0$;

(II) IF $\varepsilon_j \leq \varepsilon$ STOP AND OUTPUT $\bar{\mathbf{u}}_\varepsilon := \bar{\mathbf{u}}^j$; ELSE SET $\mathbf{v}^0 := \bar{\mathbf{u}}^j$ AND $k = 0$

(II.1) SET $\eta_k := \omega_k \bar{\rho}^k \varepsilon_j$ AND COMPUTE

$$\mathbf{r}^k = \mathbf{RES}\,[\eta_k, \mathbf{B}, \mathbf{A}, \boldsymbol{f}, \mathbf{v}^k], \quad \mathbf{v}^{k+1} = \mathbf{v}^k - \mathbf{r}^k.$$

(II.2) IF

$$\beta(\eta_k + \|\mathbf{r}^k\|) \leq \varepsilon_j/(2(1+C^*)), \tag{3.35}$$

SET $\tilde{\mathbf{v}} := \mathbf{v}^k$ AND GO TO (III). ELSE SET $k+1 \to k$ AND GO TO (II.1).

(III) COARSE$[\frac{C^*\varepsilon_j}{2(1+C^*)}, \tilde{\mathbf{v}}] \to \bar{\mathbf{u}}^{j+1}$, $\varepsilon_{j+1} = \varepsilon_j/2$, $j+1 \to j$, GO TO (II).

Step (ii) is a block of perturbed iterations of the form (3). As soon as the approximate residual is small enough, the iteration is interrupted by a coarsening step. The constant β in step (ii.2) depends on the constants in (2.19). It can be shown that the number of perturbed iterations between two coarsening steps remains uniformly bounded. Things are arranged such that after an iteration block and a coarsening step the error in the energy norm is at least halved. Thus, under the above conditions the scheme SOLVE terminates always after finitely many steps. Moreover, its computational complexity is in some sense asymptotically optimal in that the number of adaptively generated degrees of freedom and the respective computational work grow at the rate of the best N-term approximation, see [9]. For more general problem classes, the coarsening step ensures optimal complexity rates. It has recently been shown in [20], however, that coarsening can be avoided for the current class of problems.

We shall use (variants of) this algorithm as ingredients in the present weighted dual residual scheme. The routine RES is based on the following ingredients. Suppose for simplicity that $\boldsymbol{f}$ is a finitely supported array, possibly as a result of a preprocessing step. In addition, one needs an approximate application of $\mathbf{A}$:

APPLY$[\eta, \mathbf{A}, \mathbf{v}] \to \mathbf{w}$ COMPUTES FOR ANY FINITELY SUPPORTED INPUT $\mathbf{v}$ AND ANY TOLERANCE $\eta > 0$ A FINITELY SUPPORTED OUTPUT $\mathbf{w}$ SUCH THAT

$$\|\mathbf{A}\mathbf{v} - \mathbf{w}\| \leq \eta. \tag{3.36}$$

Realizations of such a routine satisfying all requirements that render SOLVE having optimal complexity can be found in [1]. For the current type of elliptic problems we can, in principle, choose the preconditioner $\mathbf{B} = \alpha \mathbf{I}$ as a stage independent damped identity which gives rise to a Richardson iteration. In this case the residual approximation scheme takes the form

$$\text{RES}[\eta, \mathbf{A}, \boldsymbol{f}, \mathbf{v}] := \alpha \left(\text{APPLY}[\eta/2\alpha, \mathbf{A}, \mathbf{v}] - \text{COARSE}[\eta/2\alpha, \boldsymbol{f}] \right). \tag{3.37}$$

The quantitative performance of this choice is usually rather poor and we refer to [18] for more efficient versions that are actually used in our experiments here as well.

Since SOLVE produces energy norm approximants, a few preparatory comments on its use in the present context are in order. Let again $\Lambda \subset \mathbb{I}$ be any (possibly infinite) subset of $\mathbb{I}$. For any two such subsets Λ, Λ' let

$$\mathbf{A}_{\Lambda, \Lambda'} := \bigl(a(\psi_\lambda, \psi_\nu) \bigr)_{\lambda \in \Lambda, \nu \in \Lambda'}$$

be the section of $\mathbf{A}$ determined by Λ and Λ'. For simplicity we set $\mathbf{A}_\Lambda := \mathbf{A}_{\Lambda, \Lambda}$. Clearly, (2.4) is then equivalent to

$$\mathbf{A}_\Lambda \mathbf{u}_\Lambda = \boldsymbol{f}_\Lambda := \boldsymbol{f}|_\Lambda. \tag{3.38}$$

Of course, (2.19) remains valid when replacing ℓ_2 by $\ell_2(\Lambda)$ and $\mathbf{A}$ by $\mathbf{A}_\Lambda$ uniformly in Λ. Solving the original problem in V_Λ can therefore be done by running the scheme SOLVE while restricting all arrays to Λ. An adaptive application of the operator $\mathbf{A}$ in this constrained setting can be thought of for the moment as employing the usual (unconstrained) scheme to the constrained input and cutting the result back to Λ. (There may be even better ways taking the special circumstances into account but this satisfies all the properties needed in [9] to establish corresponding error and complexity estimates for the restricted case.) We identify this version of SOLVE by writing SOLVE$_\Lambda[\eta, \mathbf{A}, \boldsymbol{f}, \overline{\mathbf{u}}^0]$ (and accordingly RES$_\Lambda[\eta, \mathbf{A}, \boldsymbol{f}, \mathbf{v}]$). As before, the subscript Λ is omitted when $\Lambda = \mathbb{I}$. All arrays generated by this scheme are then by definition supported in Λ.

It will be important to distinguish between the residual $\alpha(\mathbf{A}_\Lambda \mathbf{v} - \boldsymbol{f}_\Lambda)$ in $\ell_2(\Lambda)$ which is approximated by RES$_\Lambda[\eta, \mathbf{A}, \boldsymbol{f}, \mathbf{v}]$ and the *full* residual $\mathbf{A}\mathbf{v} - \boldsymbol{f}$ which appears in (2.21). The latter one reflects the global deviation of $\mathbf{v}$ from the exact solution $\mathbf{u}$. In fact, for the *exact* solution $\mathbf{u}_\Lambda$ of the restricted problem (3.38) one has $\mathbf{A}\mathbf{u}_\Lambda = \mathbf{A}_{\mathbb{I},\Lambda} \mathbf{u}_\Lambda$ and therefore

$$\mathbf{r}_\Lambda(\mathbf{u}) = \mathbf{A}_{\mathbb{I},\Lambda} \mathbf{u}_\Lambda - \boldsymbol{f} = \begin{pmatrix} \mathbf{A}_\Lambda \mathbf{u}_\Lambda - \boldsymbol{f}_\Lambda \\ \mathbf{A}_{\mathbb{I} \setminus \Lambda, \Lambda} \mathbf{u}_\Lambda - \boldsymbol{f}_{\mathbb{I} \setminus \Lambda} \end{pmatrix} = \begin{pmatrix} 0 \\ \mathbf{A}_{\mathbb{I} \setminus \Lambda, \Lambda} \mathbf{u}_\Lambda - \boldsymbol{f}_{\mathbb{I} \setminus \Lambda} \end{pmatrix}, \tag{3.39}$$

reflecting the pollution caused by the restricted wavelet coordinate domain. A more careful analysis of this aspect will be given in a forthcoming paper. We have collected now the main ingredients for the following scheme:

ALGORITHM I$[\varepsilon, \mathbf{A}, \mathbf{J}, \boldsymbol{f}] \to \bar{J}$ COMPUTES FOR ANY TARGET ACCURACY $\varepsilon > 0$ A VALUE $\bar{J}$ SUCH THAT

$$|\bar{J} - J(u)| \leq \varepsilon, \tag{3.40}$$

WHERE u IS THE SOLUTION TO (12), AS FOLLOWS:

(I) FIX PARAMETERS $c_u, c_z, c_r \in (0,1)$, $m_0 \geq 2$ AND SET $j = 0$, $\delta_u := c_A^{-1}\|\boldsymbol{f}\|$, $\delta_z := c_A^{-1}\|\mathbf{J}\|$ AND CHOOSE $\varepsilon_0 := \min\{\delta_u/2, \delta_z/2\}$.
APPLY SOLVE$[\varepsilon_0, \mathbf{A}, \boldsymbol{f}, \mathbf{0}] \to (\bar{\mathbf{u}}^0, \hat{\Lambda}_0)$;
APPLY SOLVE$[\varepsilon_0, \mathbf{A}^T, \mathbf{J}, \mathbf{0}] \to (\bar{\mathbf{z}}^0, \hat{\Upsilon}_0)$;
SET $\Lambda_0 := \hat{\Lambda}_0 \cup \hat{\Upsilon}_0$.

(II) APPLY SOLVE$[c_z\varepsilon_j, \mathbf{A}^T, \mathbf{J}, \bar{\mathbf{z}}^j] \to (\hat{\mathbf{z}}_j, \hat{\Lambda}_j)$;
APPLY SOLVE$_{\Lambda_j}[c_u\varepsilon_j, \mathbf{A}, \boldsymbol{f}, \bar{\mathbf{u}}^j] \to \bar{\mathbf{u}}_{\Lambda_j}$;
APPLY RES$[c_r\varepsilon_j, \mathbf{A}, \boldsymbol{f}, \bar{\mathbf{u}}_{\Lambda_j}]|_{\mathbf{I}\setminus\Lambda_j} \to \mathbf{r}$;
SET $\widetilde{\mathbf{w}} := \hat{\mathbf{z}}_j|_{\hat{\Lambda}_j \setminus \Lambda_j}$ AND COMPUTE

$$e_j := \Big|\sum_{\lambda \in \hat{\Lambda}_j \setminus \Lambda_j} \widetilde{w}_\lambda r_\lambda\Big|. \tag{3.41}$$

IF

$$e_j + \varepsilon_j\Big\{(C_A c_u + c_r)(\|\widetilde{\mathbf{w}}|_{\hat{\Lambda}_j \setminus \Lambda_j}\| + c_z\varepsilon_j) + c_z\|\mathbf{r}\|\Big\} \leq \varepsilon \tag{3.42}$$

STOP AND ACCEPT

$$\bar{J} = \mathbf{J}^T \bar{\mathbf{u}}_{\Lambda_j} := \sum_{\lambda \in \Lambda_j} \bar{u}_{\Lambda_j,\lambda} J_\lambda \tag{3.43}$$

AS TARGET VALUE.
OTHERWISE

(III) SET

$$\bar{\mathbf{u}}^{j+1} := \bar{\mathbf{u}}_{\Lambda_j}, \quad \bar{\mathbf{z}}^{j+1} := \hat{\mathbf{z}}_j, \quad \Lambda_{j+1} := \Lambda_j \cup \hat{\Lambda}_j, \quad \varepsilon_{j+1} = \varepsilon_j/m_0, \quad j+1 \to j, \tag{3.44}$$

AND GO TO (II).

A few comments on this scheme are in order. Step (i) should be viewed as an initialization where ε_0 is a crude initial tolerance whose square is typically still larger than the target accuracy ε. The initial approximate solutions for the primal and dual problem are energy norm approximations. Because of the crude target accuracy, one expects that the degrees of freedom generated in Λ_0 are necessary anyway.

Note that the approximations $\bar{\mathbf{u}}_{\Lambda_j}$ are then generated through the *restricted scheme* SOLVE_{Λ_j} while the corresponding residual approximations are unrestricted. Moreover, the application of SOLVE for the dual problem in step (ii) is unconstrained. We have explained the rationale of this step above. It essentially enforces the approximation of $\mathbf{z}$ in the norm but is expected to draw in only the relevant degrees of freedom concentrated near the support of J. It presumably requires only a few iterations with the initial guess $\bar{\mathbf{z}}_{\Lambda_j}$ which already is a good norm approximation for a somewhat larger tolerance.

In summary, in the above version the primal problem is *always* solved in a *constrained* subspace determined by the *norm approximation* of the dual solution.

Theorem 1. *For any target accuracy $\varepsilon > 0$ the above scheme terminates after a finite number of steps and outputs a result $\overline{J}$ satisfying $|J(u) - \overline{J}| \leq \varepsilon$.*

Proof: First note that at the jth stage we have, according to (3.26),

$$J(e_{\Lambda_j}) = \mathbf{z}^T \mathbf{r}_{\Lambda_j}(\mathbf{u}) = \sum_{\lambda \in \hat{\Lambda}_j \setminus \Lambda_j} \widetilde{w}_\lambda r_\lambda + \sum_{\lambda \in \hat{\Lambda}_j \setminus \Lambda_j} \widetilde{w}_\lambda (r_{\Lambda_j,\lambda}(\mathbf{u}) - r_\lambda)$$
$$+ \sum_{\lambda \in I \setminus \Lambda_j} (z_\lambda - \widetilde{w}_\lambda) r_\lambda + \sum_{\lambda \in I \setminus \Lambda_j} (z_\lambda - \widetilde{w}_\lambda)(r_{\Lambda_j,\lambda}(\mathbf{u}) - r_\lambda)$$
$$= \left(\widetilde{\mathbf{w}}|_{\hat{\Lambda}_j \setminus \Lambda_j}\right)^T \mathbf{r} + \left(\widetilde{\mathbf{w}}|_{\hat{\Lambda}_j \setminus \Lambda_j}\right)^T (\mathbf{r}_{\Lambda_j}(\mathbf{u}) - \mathbf{r})$$
$$+ \left((\mathbf{z} - \widetilde{\mathbf{w}})|_{I \setminus \Lambda_j}\right)^T \mathbf{r} + \left((\mathbf{z} - \widetilde{\mathbf{w}})|_{I \setminus \Lambda_j}\right)^T (\mathbf{r}_{\Lambda_j}(\mathbf{u}) - \mathbf{r}),$$

so that

$$|J(e_{\Lambda_j})| \leq e_j + \|\widetilde{\mathbf{w}}|_{\hat{\Lambda}_j \setminus \Lambda_j}\| \|\mathbf{r}_{\Lambda_j}(\mathbf{u}) - \mathbf{r}\| + \|(\mathbf{z} - \widetilde{\mathbf{w}})|_{I \setminus \Lambda_j}\| \|\mathbf{r}\|$$
$$+ \|(\mathbf{z} - \widetilde{\mathbf{w}})|_{I \setminus \Lambda_j}\| \|\mathbf{r} - \mathbf{r}_{\Lambda_j}(\mathbf{u})\|. \qquad (3.45)$$

We collect now several auxiliary estimates for the various terms in (1). By definition of $\widetilde{\mathbf{w}}$ we have

$$\|(\mathbf{z} - \widetilde{\mathbf{w}})|_{I \setminus \Lambda_j}\| \leq \|\mathbf{z} - \widetilde{\mathbf{w}}\| \leq c_z \varepsilon_j. \qquad (3.46)$$

As for the exact residual of the exact Galerkin solution $\mathbf{u}_{\Lambda_j}$, we have, on account of (3.38), the very rough estimate

$$\|\mathbf{r}_\Lambda(\mathbf{u})\| \leq \|\mathbf{f}\| + \|\mathbf{A}\mathbf{u}_\Lambda\| = \|\mathbf{f}\| + \|\mathbf{A}\mathbf{A}_\Lambda^{-1}\mathbf{f}_\Lambda\|. \qquad (3.47)$$

Alternatively, because the exact Galerkin solution $\mathbf{u}_\Lambda$ is a best approximation to $\mathbf{u}$ from $\ell_2(\Lambda)$ in the norm $\|\|\mathbf{v}\|\|^2 := \mathbf{v}^T \mathbf{A} \mathbf{v}$, one could argue that

$$\|\mathbf{r}_\Lambda(\mathbf{u})\| \leq C_A^{1/2} \|\mathbf{A}^{1/2}(\mathbf{u} - \mathbf{u}_\Lambda)\| \leq C_A^{1/2} \|\mathbf{A}^{1/2}(\mathbf{u} - \bar{\mathbf{u}}^0)\| \leq C_A \varepsilon_0, \qquad (3.48)$$

which would allow us to use the initial norm approximation to $\mathbf{u}$ in step (i) of ALGORITHM I to influence the constant.

Moreover, the approximate residual $\mathbf{r}$ deviates from the exact one for the exact Galerkin solution $\mathbf{u}_{\Lambda_j}$ by

$$\|\mathbf{r}_{\Lambda_j}(\mathbf{u}) - \mathbf{r}\| \leq \|\mathbf{A}(\mathbf{u}_{\Lambda_j} - \overline{\mathbf{u}}_{\Lambda_j})\| + \|\mathbf{A}\overline{\mathbf{u}}_{\Lambda_j} - \boldsymbol{f} - \mathbf{r}\|$$
$$\leq \|\mathbf{A}(\mathbf{u}_{\Lambda_j} - \overline{\mathbf{u}}_{\Lambda_j})\| + c_r \varepsilon_j \leq (C_A c_u + c_r)\varepsilon_j. \quad (3.49)$$

Inserting (3.46) and (3) into (1), yields

$$|J(e_{\Lambda_j})| \leq e_j + \|\widetilde{\mathbf{w}}|_{\hat{\Lambda}_j \setminus \Lambda_j}\|(C_A c_u + c_r)\varepsilon_j + c_z \varepsilon_j \Big(\|\mathbf{r}\| + (C_A c_u + c_r)\varepsilon_j\Big), \quad (3.50)$$

which is the computable error bound (3.42). Thus the termination criterion ensures that the asserted target tolerance is met.

In order to prove convergence it remains to estimate the terms $\|\widetilde{\mathbf{w}}|_{\hat{\Lambda}_j \setminus \Lambda_j}\|$, $\|\mathbf{r}\|$ and e_j. Clearly

$$\|\widetilde{\mathbf{w}}|_{\hat{\Lambda}_j \setminus \Lambda_j}\| \leq \|(\mathbf{z} - \widetilde{\mathbf{w}})|_{\mathbb{I} \setminus \Lambda_j}\| + \|\mathbf{z}|_{\hat{\Lambda}_j \setminus \Lambda_j}\|$$
$$\leq c_z \varepsilon_j + \|\mathbf{z} - \hat{\mathbf{z}}_{j-1}\| \leq c_z(\varepsilon_j + \varepsilon_{j-1})$$
$$= c_z(1 + m_0)\varepsilon_j. \quad (3.51)$$

Furthermore, by (3.47) and (3),

$$\|\mathbf{r}\| \leq \|\mathbf{r} - \mathbf{r}_{\Lambda_j}(\mathbf{u})\| + \|\mathbf{r}_{\Lambda_j}(\mathbf{u})\| \leq (C_A c_u + c_r)\varepsilon_j + C_A \varepsilon_0. \quad (3.52)$$

Finally, by (3.51) and (3.52), we obtain

$$e_j \leq \|\widetilde{\mathbf{w}}|_{\hat{\Lambda}_j \setminus \Lambda_j}\| \|\mathbf{r}\| \leq c_z(1 + m_0)\varepsilon_j\big((C_A c_u + c_r)\varepsilon_j + C_A \varepsilon_0\big), \quad (3.53)$$

which also tends to zero as j grows. This finishes the proof. ∎

To prepare for the numerical experiments in the subsequent section, we address next several further issues concerning the scheme ALGORITHM I.

We have not specified yet the choice of the parameters c_u, c_z, c_r. Of course, the smaller these parameters are chosen, the more will the computed error terms e_j dominate the true error. It is also clear that one should take $c_z < c_u$. The numerical experiments in the subsequent section will shed some more light on the quantitative behavior of ALGORITHM I regarding this point.

Concerning the progressive improvement of accuracy, let

$$\bar{e}_j(\widetilde{\mathbf{w}}, \mathbf{r}) := e_j + \varepsilon_j\Big\{(C_A c_u + c_r)(\|\widetilde{\mathbf{w}}|_{\hat{\Lambda}_j \setminus \Lambda_j}\| + c_z \varepsilon_j) + c_z \|\mathbf{r}\|\Big\}, \quad (3.54)$$

see step (ii) in ALGORITHM I. An alternative choice of the tolerances ε_j might be

$$\varepsilon_{j+1} := \frac{1}{m_0} \min\{\varepsilon_j, \bar{e}_j(\widetilde{\mathbf{w}}, \mathbf{r})\}, \quad (3.55)$$

in order to exploit the fact that the error decay is superlinear. In fact, in view of (3.50) and (3.51), the estimate (3.42) says that

$$|J(e_{\Lambda_j})| \leq c\varepsilon_j(\|\mathbf{r}\| + \varepsilon_j).$$

Thus, up to the approximate residual $\|\mathbf{r}\|$, the error decay is quadratic in the refinement tolerances ε_j. If instead of using the constraint scheme SOLVE$_{\Lambda_j}$ for the primal problem in step (ii) of ALGORITHM I, one applies the unconstraint SOLVE also to the primal problem, the term $\|\mathbf{r}\|$ would decay like ε_j as well. In this case, an overall quadratic error decay would result which is the point of view taken in [22]. In fact, during the final stage of this work, we became aware of recent results by M. S. Mommer and R. P. Stevenson [22] who derive convergence *rates* for a goal oriented scheme in the Finite Element framework. There, however, they combine adaptive energy norm approximations to the primal *and* dual solution to arrive at concrete rates. Of course, this may increase the number of degrees of freedom required for the primal solution even in regions where they may only weakly contribute to the accuracy of the target functional. We shall address this issue in the experiments in the subsequent section.

Even though in the present scheme the primal problem is solved only in a constrained way, one expects that the third term on the right hand side of (1) is too crude an estimate. In fact, as shown in later experiments, $\|\mathbf{r}\|$ may not tend to zero at all but $\mathbf{r}$ may be "locally" small where $\mathbf{z}$ has its most significant terms and large contributions may be damped by negligible components of $\mathbf{z}$. Therefore, the Cauchy Schwarz inequality produces a significant overestimation. Better estimates would require some a-priori knowledge about the decay of the coefficients in the dual solution $\mathbf{z}$ which will be discussed in a forthcoming paper.

As another practical variant, one could tame the increase of degrees of freedom by modifying step (ii) in ALGORITHM I as follows. When (3.42) is not satisfied, for $g_\lambda := |\widetilde{w}_\lambda r_\lambda|$, $\lambda \in \hat{\Lambda}_j \setminus \Lambda_j$, let $\mathbf{g} := (g_\lambda)_{\lambda \in \hat{\Lambda}_j \setminus \Lambda_j}$ and determine the smallest subset $\Gamma \subset \hat{\Lambda}_j \setminus \Lambda_j$ such that

$$\|\mathbf{g}|_\Gamma\|_{\ell_1(\Gamma)} \geq \frac{1}{2}\|\mathbf{g}\|_{\ell_1(\hat{\Lambda}_j \setminus \Lambda_j)}. \tag{3.56}$$

In the subsequent step (iii), one would then set

$$\bar{\mathbf{u}}^{j+1} := \bar{\mathbf{u}}_{\Lambda_j},\ \bar{\mathbf{z}}^{j+1} := \hat{\mathbf{z}}_j,\quad \Lambda_{j+1} := \Lambda_j \cup \Gamma,\quad \varepsilon_{j+1} = \varepsilon_j/m_0,\quad j+1 \to j, \tag{3.57}$$

and go to (ii). This may be viewed as a coarsening based on the error representation. To ensure convergence, one could add in (3.57), in addition, the support of a norm approximation to $\mathbf{z}$ with respect to the coarser tolerance $c'_z \varepsilon_j$, $c'_z > c_z$. The reasoning remains then the same while the constants change somewhat.

As for the computational complexity of any of these versions, most of the applications of SOLVE are actually just tightenings of already good initial guesses where the current accuracy is improved only by a constant factor. So the corresponding computational work remains, in principle, proportional to the current number of degrees of freedom.

4 Numerical experiments

We complement next the above findings by some first numerical experiments that are to shed some light on the quantitative behavior of the various error components.

Our test case is the Poisson equation on the L-shaped domain $\Omega = (-1,1)^2 \setminus ((-1,0] \times [0,1))$ so that

$$a(u,v) = \int_\Omega (\nabla u)^T \nabla v \, dx \qquad (4.58)$$

and $V = H_0^1(\Omega)$ in (12). This problem is interesting since the solution may exhibit a singularity caused by the shape of the domain even for smooth right hand sides, see, e.g., [21]. Thus, we can monitor the quantitative influence of such a singularity on the growth of the sets Λ_j. For the discretization, we use a globally continuous and piecewise linear wavelet basis.

The linear functional in our experiments is given by

$$J(u) = \frac{1}{|\Omega_{v,\delta}|} \int_{\Omega_{v,\delta}} u(x) dx \qquad (4.59)$$

with

$$\Omega_{v,\delta} := \{x \in \mathbb{R}^2 : \|v - x\|_\infty \leq \delta\} \subset \Omega.$$

We choose $v = (0.5, 0.5)^T$ and $\delta = 0.1$. The right hand side is scaled such that $J(u) \approx 1$. Hence $J(e_\Lambda)$ is close to the relative error $|J(e_\Lambda)|/|J(u)|$. Using approximations to u of very high accuracy, we use the resulting value of J for the validation of the results.

In the experiments below, e_j is defined as before by (3.41) while the second summand on the right hand side of (3.42) is denoted by f_j, so that $e_j + f_j$ is the computed error bound at the jth stage of ALGORITHM I.

4.1 Example 1: Smooth right hand side

In the first example, we choose $f := 10$ so that the solution u of (12) exhibits only a singularity at the reentrant corner.

Table 1 shows that the "true" error $J(e_\Lambda)$ decays at least as fast as the parameter ε_j. The component e_j is much smaller than the true error and the computed error bound $e_j + f_j$ exceeds the true error only by a factor around 2.

Table 1. Convergence history of ALGORITHM I in Example 1.

j	ε_j	$e_j + f_j$	e_j	f_j	$J(e_\Lambda)$
1	2.07e+00	8.11e-01	3.10e-01	5.00e-01	1.02e+00
2	1.03e+00	8.91e-01	5.77e-01	3.14e-01	7.47e-01
3	5.17e-01	3.82e-01	2.20e-01	1.61e-01	2.55e-01
4	2.58e-01	1.21e-01	3.98e-02	8.08e-02	1.32e-01
5	1.29e-01	3.45e-02	3.72e-03	3.07e-02	4.21e-02
6	6.46e-02	2.03e-02	5.05e-03	1.53e-02	2.35e-02
7	3.23e-02	9.03e-03	1.70e-03	7.34e-03	7.30e-03
8	1.61e-02	4.24e-03	6.84e-04	3.56e-03	3.63e-03
9	8.07e-03	1.93e-03	2.24e-04	1.71e-03	8.77e-04

This is illustrated in Figure 3 which displays the computed dual error and the computed primal residual. While the dual energy norm error is halved within each iteration, the primal residual shows very poor convergence in accordance with the spirit of the scheme. As mentioned earlier, the slight overestimation is probably due to the crude estimate in the third term of the right hand side of (1). This is substantiated by Figure 1 which depicts the computed primal and dual solution $\mathbf{u}_{\Lambda_j}$ and $\mathbf{z}_{\Lambda_j}$ for $j = 1, \ldots, 5$. The strong concentration of the generalized Green's function around the support of J indicates that the primal residual, being large far away from the support of J, would hardly influence accuracy.

Moreover, the actual behavior of the primal approximate solutions is illustrated in Figures 2 and 4. With each wavelet ψ_λ, we associate a reference point $\kappa_\lambda \in \mathbb{R}^2$ which is located in the 'center' of its support. Locations where wavelets on many scales overlap therefore appear darker. Therefore, plotting the reference points $(\kappa_\lambda)_{\lambda \in \Lambda}$ gives an impression of the distribution of active indices in $u = \sum_{\lambda \in \Lambda} \bar{u}_\lambda$. Specifically, in Figure 2 the distribution of the elements of Λ_9 is displayed. As expected, most wavelets are located near the support of J and near the reentrant corner.

To see where the largest coefficients of the primal residual $\mathbf{r}$ are located, we plot the reference points of the largest (in modulus) 5% of the coefficients r_λ. The result is displayed in Figure 4. It can be seen that, near the support of J, the residual is small, reflecting a 'local' (in the wavelet coordinate domain) convergence behavior of $\bar{\mathbf{u}}_{\Lambda_j}$.

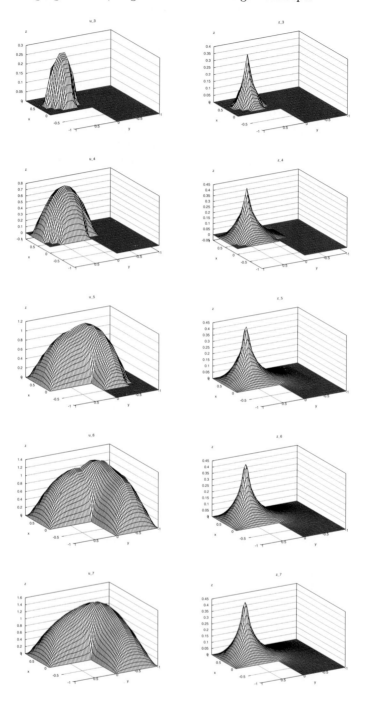

Fig. 1. Computed primal and dual solution in Example 1.

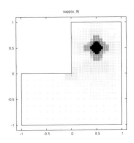

Fig. 2. Set of active coefficients Λ_9 used to evaluate $J(u_\Lambda)$ in Example 1.

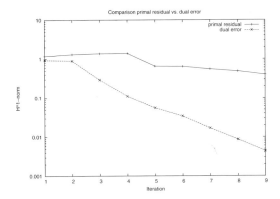

Fig. 3. Convergence of primal residual and dual solution in Example 1.

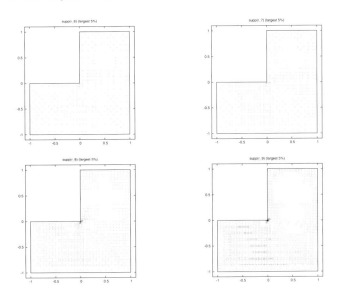

Fig. 4. Largest (in modulus) 5% of coefficients appearing in the primal residual vector in Example 1.

4.2 Example 2: Singular Right Hand Side

Next we wish to test the influence of a strong singularity of the primal solution u located far away from the support of J. This is realized by constructing a corresponding right hand side as follows. All (dual) wavelet coefficients of f are set equal to zero except the ones that overlap a fixed given point in the domain. These coefficients are chosen as $\langle \psi_\lambda, f \rangle := 1/(|\lambda| + 1)$. Since on each dyadic level only a uniformly bounded finite number of indices contributes and since the sequence $(\langle \psi_\lambda, f \rangle)_{\lambda \in \mathbb{I}}$ therefore belongs to ℓ_2, the resulting functional f is not contained in $L_2(\Omega)$, but certainly in $H^{-1}(\Omega)$. We finally add to f the constant function from Example 1. We expect that the singularity of the right hand side causes a strong concentration of relevant coefficients in the solution u that are spatially close to the singularity of f and comprise a wide range of relevant scales.

As we see from Table 2, the overestimation of the true error is slightly stronger than in Example 1. The reason is that, according to Figure 5, the primal residual is in this case larger (away from the support of J) due to the unresolved singularity caused by the right hand side f, so that the third term on the right hand side of (1) is overly pessimistic.

Table 3 sheds some more light on the local behavior of the primal residual. It shows that in the lower left patch where the singularity of f is located it

Table 2. Convergence history of ALGORITHM I in Example 2.

j	ε_j	$e_j + f_j$	e_j	f_j	$J(e_\Lambda)$	$\#\Lambda_j$
1	1.03e+00	1.31e+00	5.77e-01	7.33e-01	7.5092e-01	16
2	5.17e-01	5.85e-01	2.20e-01	3.65e-01	2.5913e-01	53
3	2.58e-01	2.27e-01	4.47e-02	1.82e-01	1.3628e-01	139
4	1.29e-01	9.21e-02	3.72e-03	8.84e-02	5.7297e-02	279
5	6.46e-02	4.90e-02	4.87e-03	4.41e-02	2.7194e-02	570
6	3.23e-02	2.37e-02	1.68e-03	2.20e-02	6.8861e-03	1752
7	1.61e-02	1.17e-02	6.95e-04	1.10e-02	2.7267e-03	5726

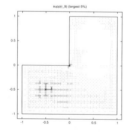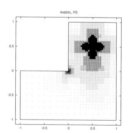

Fig. 5. Largest (in modulus) 5% of coefficients appearing in the primal residual vector and index set Λ_{10} generated in Example 2.

Table 3. Convergence of dual error, primal residual, primal residual restricted to upper right patch $P1$ and lower left patch $P3$ in Example 2.

j	$\|\widetilde{\mathbf{w}}\|$	$\|\mathbf{r}\|$	$\|\mathbf{r}|_{P1}\|$	$\|\mathbf{r}|_{P3}\|$
1	6.92e-01	5.35e+00	8.60e-01	5.23e+00
2	2.23e-01	5.37e+00	6.80e-01	5.23e+00
3	1.02e-01	5.40e+00	3.22e-01	5.29e+00
4	5.37e-02	5.20e+00	3.23e-01	5.19e+00
5	3.30e-02	5.20e+00	2.85e-01	5.18e+00
6	1.63e-02	5.19e+00	1.94e-01	5.18e+00
7	8.46e-03	5.19e+00	1.13e-01	5.18e+00
8	4.34e-03	5.18e+00	5.66e-02	5.17e+00

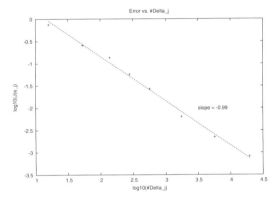

Fig. 6. Error $J(e_\Lambda)$ vs. number of degrees of freedom in Example 2.

does not converge to zero at all which, however, does not appear to affect the accuracy in a strong way.

The complexity of the scheme is indicated in Figure 6 which shows that the true error actually decays like N^{-1}, where N is the size of the index set needed to compute the approximate target value. Note that the rate for the energy norm error would be $N^{-1/2}$ at best.

References

1. Barinka, A., Barsch, T., Charton, Ph., Cohen, A., Dahlke, S., Dahmen, W., Urban, K.: Adaptive wavelet schemes for elliptic problems — Implementation and numerical experiments, SIAM J. Sci. Comp., **23**, 910–939 (2001)
2. Barinka, A., Dahmen, W., Schneider, R.: Fast Computation of Adaptive Wavelet Expansions, IGPM Report 244, RWTH Aachen, August (2004)
3. Binev, P., Dahmen, W., DeVore, R.: Adaptive Finite Element Methods with Convergence Rates, Numer. Math., **97**, 219–268 (2004)

4. Becker, R., Rannacher, R.: An optimal error control approach to a–posteriori error estimation, Acta Numerica, 1–102 (2001)
5. Canuto, C., Tabacco, A., Urban, K.: The wavelet element method, part I: Construction and analysis, Appl. Comput. Harm. Anal., **6**, 1–52 (1999)
6. Canuto, C., Tabacco, A., Urban, K.: The Wavelet Element Method, Part II: Realization and additional features in 2D and 3D, Appl. Comp. Harm. Anal. **8**, 123–165 (2000)
7. Cohen, A., Dahmen, W., DeVore, R.: Adaptive wavelet methods for elliptic operator equations – Convergence rates, Math. Comp. **70**, 27–75 (2001)
8. Cohen, A., Dahmen, W., DeVore, R.: Adaptive wavelet methods II – Beyond the elliptic case, Found. Computat. Math. **2**, 203–245 (2002)
9. Cohen, A., Dahmen, W., DeVore, R.: Adaptive wavelet scheme for nonlinear variational problems, SIAM J. Numer. Anal. **41** (5), 1785–1823 (2003)
10. Cohen, A., Dahmen, W., DeVore, R.: Sparse evaluation of compositions of functions using multiscale expansions, SIAM J. Math. Anal., **35**, 279–303 (2003)
11. Cohen, A., Masson, R.: Wavelet adaptive methods for second order elliptic problems, boundary conditions and domain decomposition, Numer. Math., **86**, 193–238 (2000)
12. Dahlke, S., Dahmen, W., Urban, K.: Adaptive wavelet methods for saddle point problems – Convergence rates, SIAM J. Numer. Anal., **40** (No. 4), 1230–1262 (2002)
13. Dahmen, W.: Multiscale and Wavelet Methods for Operator Equations, C.I.M.E. Lecture Notes in Mathematics, *Multiscale Problems and Methods in Numerical Simulation*, Springer Lecture Notes in Mathematics, Vol. **1825**, Springer-Verlag, Heidelberg, 31–96 (2003)
14. Dahmen, W., Kunoth, A.: Adaptive wavelet methods for linear–quadratic elliptic control problems: Convergence rates, SIAM J. Contr. Optim. **43**(5), 1640–1675 (2005)
15. Dahmen, W., Schneider, R.: Composite wavelet bases for operator equations, Math. Comp., **68**, 1533–1567 (1999)
16. Dahmen, W., Schneider, R.: Wavelets on manifolds I: Construction and domain decomposition, SIAM J. Math. Anal., **31**, 184–230 (1999)
17. Dahmen, W., Stevenson, R. P. Element-by-element construction of wavelets – stability and moment conditions, SIAM J. Numer. Anal., **37**, 319–325 (1999)
18. Dahmen, W., Vorloeper, J.:, Adaptive Application of Operators and Newton's Method, in preparation.
19. Eriksson, K., Estep, D., Hansbo, P., Johnson, C.: Introduction to adaptive methods for differential equations, Acta Numerica, 105–158 (1995)
20. Gantumur, T., Harbrecht, H., Stevenson, R. P.: An optimal adaptive wavelet method without coarsening of the iterands, Preprint No. 1325, Department of Mathematics, Utrecht University, March 2005, submitted, revised version, June 2005
21. Grisvard, P.: Elliptic Problems in Nonsmooth Domains, Pitman, Boston (1985)
22. Mommer, M. S. Stevenson, R. P.: A goal oriented adaptive finite element method with a guaranteed convergence rate, December 21, 2005, lecture at the "Utrecht Workshop on Fast Numerical Solutions of PDEs", (December 20–22, 2005, Utrecht, The Netherlands)

23. Stevenson, R. P.: On the compressibility of operators in wavelet coordinates, SIAM J. Math. Anal. 35(5), 1110–1132 (2004)
24. Stevenson, R. P.: Optimality of a standard adaptive finite element method, Preprint No. 1329, Department of Mathematics, Utrecht University, May 2005, submitted, revised version, January 2006

Quadratic Programming and Scalable Algorithms for Variational Inequalities

Zdeněk Dostál[1], David Horák[1] and Dan Stefanica[2]

[1] FEI VŠB-Technical University Ostrava, CZ-70833 Ostrava, Czech Republic
zdenek.dostal@vsb.cz, david.horak@vsb.cz
[2] Baruch College, City University of New York, NY 10010, USA
dstefan@math.mit.edu

Summary. We first review our recent results concerning optimal algorithms for the solution of bound and/or equality constrained quadratic programming problems. The unique feature of these algorithms is the rate of convergence in terms of bounds on the spectrum of the Hessian of the cost function. Then we combine these estimates with some results on the FETI method (FETI-DP, FETI and Total FETI) to get the convergence bounds that guarantee the scalability of the algorithms. i.e. asymptotically linear complexity and the time of solution inverse proportional to the number of processors. The results are confirmed by numerical experiments.

1 Introduction

One of the most impressive results in numerical analysis of the twentieth century was discovery that the systems of linear equations arising from the discretization of an elliptic partial differential equation may be solved by the multigrid or domain decomposition methods with asymptotically linear complexity. In this paper, we show how to extend these results to get scalable algorithms for variational inequalities. Our basic tool is the FETI method, which was proposed by Farhat and Roux [28] for parallel solution of problems described by elliptic partial differential equations. Its key ingredient is the decomposition of the spatial domain into non-overlapping subdomains that are "glued" by Lagrange multipliers, so that, after eliminating the primal variables, the original problem is reduced to a small, relatively well conditioned, typically equality constrained quadratic programming problem that is solved iteratively. Observing that the equality constraints may be used to define so called "natural coarse grid", Farhat, Mandel and Roux [27] modified the basic FETI algorithm so that they were able to prove its numerical scalability. A similar results were achieved by the Dual-Primal FETI method (FETI–DP) introduced by Farhat et al. [26]; see also [32].

If the FETI procedure is applied to the contact problems, the resulting quadratic programming problem has not only the equality constraints, but also the non-negativity constraints. Even though the latter is a considerable complication as compared with the linear problem, the resulting problem is still easier to solve than the contact problem in displacements as it is smaller, better conditioned having constraints with simpler structure. Promising experimental results by Dureisseix and Farhat [24] support this claim and even indicate numerical scalability of their metod. Similar results were achieved also with the FETI–DP method by Avery, Rebel, Lesoinne and Farhat [1]. A different approach based on the augmented Lagrangian method was used by Dostál, Friedlander, Gomes and Santos [12, 13].

In this paper we review our recent improvements that resulted in development of theoretically supported scalable algorithms for variational inequalities that combine various FETI based domain decomposition methods with our optimal quadratic programming algorithms [6, 23, 7]. We present optimal algorithms based on scalable variant of FETI [27] or on its easier implementable variant called TFETI [19], on FETI–DP [26] and on optimal dual penalty [17]. Let us point out that the effort to develop scalable solvers for variational inequalities was not restricted to FETI. For example, developing ideas of Mandel [35], Kornhuber, Krause and Wohlmuth [33, 34, 40] gave an experimental evidence of numerical scalability of the algorithm based on monotone multigrid. Nice results concerning development of scalable algorithms were proved by Schöberl [37].

We start our exposition by presenting our MPRGP (Modified Proportioning with Reduced Gradient Projection) and SMALBE (Semimonotonic Augmented Lagrangians for Bound and Equality constrained problems) algorithms with in a sense optimal rates of convergence. Then we present a simple model problem and the FETI methodology [12] that turns the variational inequality into the quadratic programming problem with bound and possibly equality constraints. Combining these ingredients, we shall get new algorithms for numerical solution of boundary elliptic variational inequalities. A unique feature of these algorithms is theoretically guaranteed numerical scalability. We report results of numerical experiments that are in agreement with the theory and indicate high parallel and numerical scalability of the algorithm presented.

2 Bound constrained problems

Let us consider the problem

$$\text{minimize} \quad q(\mathbf{x}) \quad \text{subject to} \quad \mathbf{x} \in \Omega_B \tag{1}$$

with $q(\mathbf{x}) = \frac{1}{2}\mathbf{x}^T \mathsf{A}\mathbf{x} - \mathbf{b}^T \mathbf{x}$, A a symmetric positive definite matrix, $\mathbf{b} \in I\!\!R^n$, $\Omega_B = \{\mathbf{x} : \mathbf{x} \geq \boldsymbol{\ell}\}$ and $\boldsymbol{\ell} \in I\!\!R^n$. The unique solution $\overline{\mathbf{x}}$ of (1) is fully determined by the Karush-Kuhn-Tucker optimality conditions [3] so that for $i = 1, \ldots, n$,

$$\overline{x}_i = \ell_i \text{ implies } \overline{g}_i \geq 0 \text{ and } \overline{x}_i > \ell_i \text{ implies } \overline{g}_i = 0, \tag{2}$$

where $\mathbf{g} = \mathbf{g}(\mathbf{x})$ denotes the gradient of q defined by

$$\mathbf{g} = \mathbf{g}(\mathbf{x}) = \mathbf{Ax} - \mathbf{b}. \tag{3}$$

The conditions (2) can be described alternatively by the *free gradient* φ and the *chopped gradient* β that are defined by

$$\varphi_i(\mathbf{x}) = g_i(\mathbf{x}) \text{ for } x_i > \ell_i, \varphi_i(\mathbf{x}) = 0 \text{ for } x_i = \ell_i,$$

$$\beta_i(\mathbf{x}) = 0 \text{ for } x_i > \ell_i, \beta_i(\mathbf{x}) = g_i^-(\mathbf{x}) \text{ for } x_i = \ell_i,$$

where we have used the notation $g_i^- = \min\{g_i, 0\}$. Thus the conditions (2) are satisfied iff the *projected gradient* $\mathbf{g}^P(\mathbf{x}) = \varphi(\mathbf{x}) + \beta(\mathbf{x})$ is equal to the zero. The algorithm for the solution of (1) that we describe here exploits a given constant $\Gamma > 0$, a test to decide about leaving the face and three types of steps to generate a sequence of the iterates $\{\mathbf{x}^k\}$ that approximate the solution of (1). The *expansion step* may expand the current active set and is defined by

$$\mathbf{x}^{k+1} = \mathbf{x}^k - \overline{\alpha}\widetilde{\varphi}(\mathbf{x}^k) \tag{4}$$

with the fixed steplength $\overline{\alpha} \in (0, \|\mathbf{A}\|^{-1}]$ and the *reduced free gradient* $\widetilde{\varphi}(\mathbf{x})$ with the entries $\widetilde{\varphi}_i = \widetilde{\varphi}_i(\mathbf{x}) = \min\{(x_i - \ell_i)/\overline{\alpha}, \varphi_i\}$. If the inequality

$$\|\beta(\mathbf{x}^k)\|^2 \leq \Gamma^2 \widetilde{\varphi}(\mathbf{x}^k)^\top \varphi(\mathbf{x}^k) \tag{5}$$

holds then we call the iterate $\mathbf{x}^k$ *strictly proportional*. The test (5) is used to decide which component of the projected gradient $\mathbf{g}^P(\mathbf{x}^k)$ will be reduced in the next step. The *proportioning step* may remove indices from the active set and is defined by

$$\mathbf{x}^{k+1} = \mathbf{x}^k - \alpha_{cg}\beta(\mathbf{x}^k) \tag{6}$$

with the steplength α_{cg} that minimizes $q\left(\mathbf{x}^k - \alpha\beta(\mathbf{x}^k)\right)$. It is easy to check [3] that α_{cg} that minimizes $q(\mathbf{x} - \alpha\mathbf{d})$ for a given $\mathbf{d}$ and $\mathbf{x}$ may be evaluated by the formula

$$\alpha_{cg} = \alpha_{cg}(\mathbf{d}) = \frac{\mathbf{d}^\top \mathbf{g}(\mathbf{x})}{\mathbf{d}^\top \mathbf{Ad}}. \tag{7}$$

The *conjugate gradient step* is defined by

$$\mathbf{x}^{k+1} = \mathbf{x}^k - \alpha_{cg}\mathbf{p}^k \tag{8}$$

where $\mathbf{p}^k$ is the conjugate gradient direction [3] which is constructed recurrently. The recurrence starts (or restarts) from $\mathbf{p}^s = \varphi(\mathbf{x}^s)$ whenever $\mathbf{x}^s$ is generated by the expansion or proportioning step. If $\mathbf{p}^k$ is known, then $\mathbf{p}^{k+1}$ is given [3] by

$$\mathbf{p}^{k+1} = \varphi(\mathbf{x}^{k+1}) - \gamma\mathbf{p}^k, \quad \gamma = \frac{\varphi(\mathbf{x}^{k+1})^\top \mathbf{Ap}^k}{(\mathbf{p}^k)^\top \mathbf{Ap}^k}. \tag{9}$$

Algorithm 1. Modified proportioning with reduced gradient projections (MPRGP).
Let $\mathbf{x}^0 \in \Omega$, $\overline{\alpha} \in (0, \|A\|^{-1}]$, and $\Gamma > 0$ be given. For $k \geq 0$ and $\mathbf{x}^k$ known, choose $\mathbf{x}^{k+1}$ by the following rules:
Step 1. If $\mathbf{g}^P(\mathbf{x}^k) = \mathbf{o}$, set $\mathbf{x}^{k+1} = \mathbf{x}^k$.
Step 2. If $\mathbf{x}^k$ is strictly proportional and $\mathbf{g}^P(\mathbf{x}^k) \neq \mathbf{o}$, try to generate $\mathbf{x}^{k+1}$ by the conjugate gradient step. If $\mathbf{x}^{k+1} \in \Omega$, then accept it, else use the expansion step.
Step 3. If $\mathbf{x}^k$ is not strictly proportional, define $\mathbf{x}^{k+1}$ by proportioning.

Algorithm 1 has been proved to enjoy the R-linear rate of convergence in terms of the spectral condition number [23].

To formulate the optimality results, let $\mathcal{T}$ denote any set of indices and assume that for any $t \in \mathcal{T}$ there is defined the problem

$$\text{minimize } q_t(\mathbf{x}) \text{ s.t. } \mathbf{x} \in \Omega_B^t \tag{10}$$

with $\Omega_B^t = \{\mathbf{x} \in \mathbb{R}^{n_t} : \mathbf{x} \geq \boldsymbol{\ell}\}$, $q_t(\mathbf{x}) = \frac{1}{2}\mathbf{x}^\top \mathsf{A}_t \mathbf{x} - \mathbf{b}_t^\top \mathbf{x}$, $\mathsf{A}_t \in \mathbb{R}^{n_t \times n_t}$ symmetric positive definite, and $\mathbf{b}_t, \mathbf{x}, \boldsymbol{\ell}_t \in \mathbb{R}^{n_t}$. Our optimality result then reads as follows.

Theorem 1. *Let the Hessian matrices $\mathsf{A}_t = \nabla^2 q_t$ of (10) satisfy*

$$0 < a_{\min} \leq \lambda_{\min}(\mathsf{A}_t) \leq \lambda_{\max}(\mathsf{A}_t) \leq a_{\max},$$

let $\{\mathbf{x}_t^k\}$ be generated by Algorithm 1 for (10) with a given $\mathbf{x}_t^0 \in \Omega_B^t$, $\overline{\alpha} \in (0, a_{\max}^{-1}]$, and let $\Gamma > 0$. Let there be a constant a_b such that $\|\mathbf{x}_t^0\| \leq a_b \|\mathbf{b}_t\|$ for any $t \in \mathcal{T}$.
(i) If $\epsilon > 0$ is given, then the approximate solution $\overline{\mathbf{x}}_t$ of (10) which satisfies

$$\|\mathbf{x}_t^k - \overline{\mathbf{x}}_t\| \leq \epsilon \|\mathbf{b}_t\|$$

may be obtained at $O(1)$ matrix-vector multiplications by A_t.
(ii) If $\epsilon > 0$ is given, then the approximate solution $\mathbf{x}_t^k$ of (10) which satisfies

$$\|\mathbf{g}_t^P(\mathbf{x}_t^k)\| \leq \epsilon \|\mathbf{b}_t\|$$

may be obtained at $O(1)$ matrix-vector multiplications by A_t.

Proof. See [23]. ∎

Numerical experiments and implementation details may be found in [23].

3 Bound and equality constrained problems

We shall now be concerned with the problem of finding the minimizer of the strictly convex quadratic function $q(\mathbf{x})$ subject to the bound and linear equality constraints, that is

$$\text{minimize } q(\mathbf{x}) \quad \text{subject to} \quad \mathbf{x} \in \Omega_{BE} \tag{11}$$

with $\Omega_{BE} = \{\mathbf{x} \in \mathbb{R}^n : \mathbf{x} \geq \boldsymbol{\ell} \text{ and } \mathsf{C}\mathbf{x} = \mathbf{o}\}$ and $\mathsf{C} \in \mathbb{R}^{m \times n}$. We do not require that C is a full row rank matrix, but we shall assume that Ω_{BE} is not empty. Let us point out that confining ourselves to the homogeneous equality constraints does not mean any loss of generality, as we can use a simple transform to reduce any non-homogeneous equality constraints to our case. The algorithm that we describe here combines in a natural way the augmented Lagrangians and MPRGP described above. It is related to the earlier work of Friedlander and Santos with the present author [11]. Let us recall that the basic scheme that we use was proposed by Conn, Gould and Toint [4] who adapted the augmented Lagrangian method to the solution of the problems with a general cost function subject to general equality constraints and simple bounds.

Algorithm 2. (Semi-monotonic augmented Lagrangians for bound and equality constraints (SMALBE))
Given $\eta > 0$, $\beta > 1$, $M > 0$, $\rho_0 > 0$, and $\boldsymbol{\mu}^0 \in \mathbb{R}^m$, set $k = 0$.
Step 1. {Inner iteration with adaptive precision control.}
  Find $\mathbf{x}^k$ such that

$$\|\mathbf{g}^P(\mathbf{x}^k, \boldsymbol{\mu}^k, \rho_k)\| \leq \min\{M\|\mathsf{C}\mathbf{x}^k\|, \eta\}. \tag{12}$$

Step 2. {Update $\boldsymbol{\mu}$.}
$$\boldsymbol{\mu}^{k+1} = \boldsymbol{\mu}^k + \rho_k \mathsf{C}\mathbf{x}^k. \tag{13}$$

Step 3. {Update ρ provided the increase of the Lagrangian is not sufficient.}
  If $k > 0$ and

$$L(\mathbf{x}^k, \boldsymbol{\mu}^k, \rho^k) < L(\mathbf{x}^{k-1}, \boldsymbol{\mu}^{k-1}, \rho_{k-1}) + \frac{\rho_k}{2}\|\mathsf{C}\mathbf{x}^k\|^2 \tag{14}$$

  then
$$\rho_{k+1} = \beta \rho_k, \tag{15}$$

  else
$$\rho_{k+1} = \rho_k. \tag{16}$$

Step 4. Set $k = k + 1$ and return to *Step 1*.

In (12), we use the augmented Lagrangian defined by

$$L(\mathbf{x}, \boldsymbol{\mu}, \rho) = q(\mathbf{x}) + \boldsymbol{\mu}^\top \mathsf{C}\mathbf{x} + \frac{\rho_k}{2}\|\mathsf{C}\mathbf{x}\|^2. \tag{17}$$

Algorithm 2 has been shown to be well defined [11], that is, any convergent algorithm for the solution of the auxiliary problem required in Step 1 which guarantees convergence of the projected gradient to zero will generate either $\mathbf{x}^k$ that satisfies (12) in a finite number of steps or a sequence of approximations that converges to the solution of (11). To present explicitly the optimality of Algorithm 2 with Step 1 implemented by Algorithm 1, let $\mathcal{T}$ denote any set of indices and let for any $t \in \mathcal{T}$ be defined the problem

$$\text{minimize } q_t(\mathbf{x}) \text{ s.t. } \mathbf{x} \in \Omega_{BE}^t \tag{18}$$

with $\Omega_{BE}^t = \{\mathbf{x} \in \mathbb{R}^{n_t} : \mathsf{C}_t \mathbf{x} = \mathbf{o} \text{ and } \mathbf{x} \geq \boldsymbol{\ell}_t\}$, $q_t(\mathbf{x}) = \frac{1}{2}\mathbf{x}^\top \mathsf{A}_t \mathbf{x} - \mathbf{b}_t^\top \mathbf{x}$, $\mathsf{A}_t \in \mathbb{R}^{n_t \times n_t}$ symmetric positive definite, $\mathsf{C}_t \in \mathbb{R}^{m_t \times n_t}$, and $\mathbf{b}_t, \boldsymbol{\ell}_t \in \mathbb{R}^{n_t}$. Our optimality result reads as follows.

Theorem 2. *Let $\{\mathbf{x}_t^k\}$, $\{\boldsymbol{\mu}_t^k\}$ and $\{\rho_{t,k}\}$ be generated by Algorithm 2 for (18) with $\|\mathbf{b}_t\| \geq \eta_t > 0$, $\beta > 1$, $M > 0$, $\rho_{t,0} = \rho_0 > 0$, $\boldsymbol{\mu}_t^0 = \mathbf{o}$. Let Step 1 of Algorithm 2 be implemented by Algorithm 1 (MPRGP) which generates the iterates $\mathbf{x}_t^{k,0}, \mathbf{x}_t^{k,1}, \ldots, \mathbf{x}_t^{k,l} = \mathbf{x}_t^k$ for the solution of (18) starting from $\mathbf{x}_t^{k,0} = \mathbf{x}_t^{k-1}$ with $\mathbf{x}_t^{-1} = \mathbf{o}$, where $l = l_{k_t}$ is the first index satisfying*

$$\|\mathbf{g}^P(\mathbf{x}_t^{k,l}, \boldsymbol{\mu}_t^k, \rho_k)\| \leq M\|\mathsf{C}_t \mathbf{x}_t^{k,l}\| \tag{19}$$

or

$$\|\mathbf{g}^P(\mathbf{x}_t^{k,l}, \boldsymbol{\mu}_t^k, \rho_k)\| \leq \epsilon \|\mathbf{b}_t\| \min\{1, M^{-1}\}. \tag{20}$$

Let $0 < a_{\min} < a_{\max}$ and $0 < c_{\max}$ be given and let the class of problems (18) satisfy

$$a_{\min} \leq \lambda_{\min}(\mathsf{A}_t) \leq \lambda_{\max}(\mathsf{A}_t) \leq a_{\max} \text{ and } \|\mathsf{C}_t\| \leq c_{\max}. \tag{21}$$

Then Algorithm 2 generates an approximate solution $\mathbf{x}_t^{k_t}$ of any problem (18) which satisfies

$$\|\mathbf{g}^P(\mathbf{x}_t^{k_t}, \boldsymbol{\mu}_t^{k_t}, \rho_{t,k_t})\| \leq \epsilon \|\mathbf{b}_t\| \text{ and } \|\mathsf{C}_t \mathbf{x}_t^{k_t}\| \leq \epsilon \|\mathbf{b}_t\| \tag{22}$$

at $O(1)$ matrix-vector multiplications by the Hessian of the augmented Lagrangian L_t.

Proof. See [7, 8]. ∎

4 Model problem

To simplify our exposition, we restrict our attention to a simple scalar variational inequality. The computational domain is $\Omega = \Omega^1 \cup \Omega^2$, where $\Omega^1 = (0,1) \times (0,1)$ and $\Omega^2 = (1,2) \times (0,1)$, with boundaries Γ^1 and Γ^2, respectively. We denote by Γ_u^i, Γ_f^i, and Γ_c^i the fixed, free, and potential contact parts of Γ^i, $i = 1, 2$. We assume that Γ_u^1 has non-zero measure, i.e., $\Gamma_u^1 \neq \emptyset$. For a coercive model problem, $\Gamma_u^2 \neq \emptyset$, while for a semicoercive model problem, $\Gamma_u^2 = \emptyset$; see Figure 1. Let $H^1(\Omega^i), i = 1, 2$ denote the Sobolev space of the first order in the space $L^2(\Omega^i)$ of functions on Ω^i whose squares are integrable in the Lebesgue sense. Let

$$V^i = \{v^i \in H^1(\Omega^i) : v^i = 0 \text{ on } \Gamma_u^i\}$$

denote the closed subspaces of $H^1(\Omega^i), i = 1, 2$, and let

$$V = V^1 \times V^2 \quad \text{and} \quad \mathcal{K} = \{(v^1, v^2) \in V : v^2 - v^1 \geq 0 \text{ on } \Gamma_c\}$$

denote the closed subspace and the closed convex subset of $\mathcal{H} = H^1(\Omega^1) \times H^1(\Omega^2)$, respectively. The relations on the boundaries are in terms of traces. On $\mathcal{H}$ we shall define a symmetric bilinear form

$$a(u,v) = \sum_{i=1}^{2} \int_{\Omega^i} \left(\frac{\partial u^i}{\partial x} \frac{\partial v^i}{\partial x} + \frac{\partial u^i}{\partial y} \frac{\partial v^i}{\partial y} \right) d\Omega$$

and a linear form

$$\ell(v) = \sum_{i=1}^{2} \int_{\Omega^i} f^i v^i d\Omega,$$

where $f^i \in L^2(\Omega^i), i = 1, 2$ are the restrictions of

$$f(x,y) = \begin{cases} -1 | -3 & \text{for } (x,y) \in (0,1) \times [0.75, 1), \\ 0 | \ 0 & \text{for } (x,y) \in (0,1) \times [0, 0.75) \text{ and } (x,y) \in (1,2) \times [0.25, 1), \\ -3 | -1 & \text{for } (x,y) \in (1,2) \times [0, 0.25), \end{cases}$$

for coercive | semicoercive model problem. Thus we can define a problem to find

$$\min \ q(u) = \frac{1}{2} a(u,u) - \ell(u) \text{ subject to } u \in \mathcal{K}. \tag{23}$$

The solution of the model problem may be interpreted as the displacement of two membranes under the traction f. The membranes are fixed as in Fig. 1 and the left edge of the right membrane is not allowed to penetrate below the right edge of the left membrane. In the first case, when the Dirichlet conditions are prescribed on the parts $\Gamma_u^i, i = 1, 2$ of the boundaries with a positive measure, the quadratic form a is coercive which guarantees the existence and uniqueness

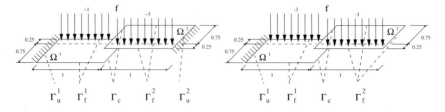

Fig. 1. The coercive (left) and semicoercive (right) model problem

of the solution [31]. In the second case, only the left membrane is fixed on the outer edge and the right membrane has no prescribed displacement as in Fig. 1 (right), so that

$$\Gamma_u^1 = \{(0,y) \in I\!R^2 : y \in [0,1]\}, \quad \Gamma_u^2 = \emptyset.$$

Even though a is in this case only semidefinite, the form q is still coercive due to the choice of f so that it has again the unique solution [31].

5 FETI and total FETI domain decomposition

To enable efficient application of the domain decomposition methods, we can optionally decompose each Ω^i into square subdomains $\Omega^{i1}, \ldots, \Omega^{ip}, p = s^2 > 1, i = 1, 2$. The outer subdomains Ω^{ij} can either inherit the Dirichlet boundary conditions from Γ_u^i as in the original FETI [28], or they can be treated as floating with the Dirichlet conditions enforced by the Lagrange multipliers. The latter approach was coined Total FETI (TFETI) [19]. The continuity in Ω^1 and Ω^2 of the global solution assembled from the local solutions u^{ij} will be enforced by the "gluing" conditions $u^{ij}(x) = u^{ik}(x)$ that should be satisfied for any x in the interface $\Gamma^{ij,ik}$ of Ω^{ij} and Ω^{ik}. After modifying appropriately the definition of problem (23), introducing regular grids in the subdomains Ω^{ij} that match across the interfaces $\Gamma^{ij,kl}$, indexing contiguously the nodes and entries of corresponding vectors in the subdomains, and using the finite element discretization, we get the discretized version of problem (23) with the auxiliary domain decomposition that reads

$$\min \frac{1}{2}\mathbf{u}^\top \mathsf{K}\mathbf{u} - \mathbf{f}^\top \mathbf{u} \quad \text{s.t.} \quad \mathsf{B}^I \mathbf{u} \leq \mathbf{o} \quad \text{and} \quad \mathsf{B}^E \mathbf{u} = \mathbf{o}. \tag{24}$$

In (24), $\mathsf{K} = \text{diag}[\mathsf{K}_1, \ldots, \mathsf{K}_{2p}]$ denotes a positive semidefinite stiffness matrix, the full rank matrices B^I and B^E describe the discretized inequality and gluing conditions, respectively, and $\mathbf{f}$ represents the discrete analog of the linear term $\ell(u)$. Denoting

$$\boldsymbol{\lambda} = \begin{bmatrix} \boldsymbol{\lambda}^I \\ \boldsymbol{\lambda}^E \end{bmatrix} \quad \text{and} \quad \mathsf{B} = \begin{bmatrix} \mathsf{B}^I \\ \mathsf{B}^E \end{bmatrix},$$

we can write the Lagrangian associated with problem (30) briefly as

$$L(\mathbf{u}, \boldsymbol{\lambda}) = \frac{1}{2}\mathbf{u}^\top \mathsf{K}\mathbf{u} - \mathbf{f}^\top \mathbf{u} + \boldsymbol{\lambda}^\top \mathsf{B}\mathbf{u}.$$

It is well known that (24) is equivalent to the saddle point problem

$$\text{Find} \quad (\overline{\mathbf{u}}, \overline{\boldsymbol{\lambda}}) \quad \text{s.t.} \quad L(\overline{\mathbf{u}}, \overline{\boldsymbol{\lambda}}) = \sup_{\boldsymbol{\lambda}^I \geq \mathbf{o}} \inf_{\mathbf{u}} L(\mathbf{u}, \boldsymbol{\lambda}). \tag{25}$$

After eliminating the primal variables $\mathbf{u}$ from (25), we shall get the minimization problem

$$\min \Theta(\boldsymbol{\lambda}) \quad \text{s.t.} \quad \boldsymbol{\lambda}^I \geq \mathbf{o} \quad \text{and} \quad \mathsf{R}^\top(\mathbf{f} - \mathsf{B}^\top\boldsymbol{\lambda}) = \mathbf{o}, \tag{26}$$

where

$$\Theta(\boldsymbol{\lambda}) = \frac{1}{2}\boldsymbol{\lambda}^\top \mathsf{B}\mathsf{K}^\dagger\mathsf{B}^\top\boldsymbol{\lambda} - \boldsymbol{\lambda}^\top\mathsf{B}\mathsf{K}^\dagger\mathbf{f}, \tag{27}$$

$\mathsf{K}^\dagger$ denotes a generalized inverse that satisfies $\mathsf{K}\mathsf{K}^\dagger\mathsf{K} = \mathsf{K}$, and R denotes the full rank matrix whose columns span the kernel of K. We shall choose R so that its entries belong to $\{0, 1\}$ and each column corresponds to some floating auxiliary subdomain Ω^{ij} with the nonzero entries in the positions corresponding to the indices of nodes belonging to Ω^{ij}. The action of $\mathsf{K}^\dagger = \text{diag}[\mathsf{K}_1^\dagger, \ldots, \mathsf{K}_{2p}^\dagger]$ can be evaluated in parallel at the cost comparable with the action of the inverse of the regular matrix with the same sparsity pattern [25]. When TFETI method is used, the implementation is easy as the kernels of K_i are known a priori. Even though problem (26) is much more suitable for computations than (24), further improvement may be achieved by adapting some simple observations and the results of Farhat, Mandel and Roux [27]. Let us denote

$$\mathsf{F} = \mathsf{B}\mathsf{K}^\dagger\mathsf{B}^\top, \quad \widetilde{\mathsf{G}} = \mathsf{R}^\top\mathsf{B}^\top, \quad \widetilde{\mathbf{e}} = \mathsf{R}^\top\mathbf{f}, \quad \widetilde{\mathbf{d}} = \mathsf{B}\mathsf{K}^\dagger\mathbf{f},$$

and let $\widetilde{\boldsymbol{\lambda}}$ solve $\widetilde{\mathsf{G}}\widetilde{\boldsymbol{\lambda}} = \widetilde{\mathbf{e}}$, so that we can transform the problem (26) to minimization on the subset of the vector space by looking for the solution in the form $\boldsymbol{\lambda} = \boldsymbol{\mu} + \widetilde{\boldsymbol{\lambda}}$. Since

$$\frac{1}{2}\boldsymbol{\lambda}^\top \mathsf{F}\boldsymbol{\lambda} - \boldsymbol{\lambda}^\top\widetilde{\mathbf{d}} = \frac{1}{2}\boldsymbol{\mu}^\top\mathsf{F}\boldsymbol{\mu} - \boldsymbol{\mu}^\top(\widetilde{\mathbf{d}} - \mathsf{F}\widetilde{\boldsymbol{\lambda}}) + \frac{1}{2}\widetilde{\boldsymbol{\lambda}}^\top\mathsf{F}\widetilde{\boldsymbol{\lambda}} - \widetilde{\boldsymbol{\lambda}}^\top\widetilde{\mathbf{d}},$$

problem (26) is, after returning to the old notation, equivalent to

$$\min \frac{1}{2}\boldsymbol{\lambda}^\top\mathsf{F}\boldsymbol{\lambda} - \boldsymbol{\lambda}^\top\mathbf{d} \quad \text{s.t} \quad \mathsf{G}\boldsymbol{\lambda} = \mathbf{o} \quad \text{and} \quad \boldsymbol{\lambda}^I \geq -\widetilde{\boldsymbol{\lambda}}^I, \tag{28}$$

where $\mathbf{d} = \widetilde{\mathbf{d}} - \mathsf{F}\widetilde{\boldsymbol{\lambda}}$ and G denotes a matrix arising from the orthonormalization of the rows of $\widetilde{\mathsf{G}}$. Our final step is based on observation that the problem (28) is equivalent to

$$\min \frac{1}{2}\boldsymbol{\lambda}^\top\mathsf{P}\mathsf{F}\mathsf{P}\boldsymbol{\lambda} - \boldsymbol{\lambda}^\top\mathsf{P}\mathbf{d} \quad \text{s.t} \quad \mathsf{G}\boldsymbol{\lambda} = \mathbf{o} \quad \text{and} \quad \boldsymbol{\lambda}^I \geq -\widetilde{\boldsymbol{\lambda}}^I \tag{29}$$

where

$$\mathsf{Q} = \mathsf{G}^\top\mathsf{G} \quad \text{and} \quad \mathsf{P} = \mathsf{I} - \mathsf{Q}$$

denote the orthogonal projectors on the image space of $\mathsf{G}^\top$ and on the kernel of G.

Theorem 3. *If* F *and* P *denote the matrices of the problem (29) (generated either by FETI or TFETI), then the following spectral bounds hold:*

$$\lambda_{\max}(\mathsf{PFP}) \leq \|\mathsf{F}\| \leq C\frac{H}{h}; \quad \lambda_{\min}(\mathsf{PFP}|\text{Im}\mathsf{P}) \geq C.$$

Proof. See [27, 9]. ∎

6 FETI–DP domain decomposition and discretization

We shall now assume that the subdomains are not completely separated, but joined in the joint corners that we shall call crosspoints. We call a crosspoint either a corner that belongs to four subdomains, or a corner that belongs to two subdomains and is located on $\partial\Omega^1 \setminus \Gamma_u^1$ or on $\partial\Omega^2 \setminus \Gamma_u^2$. An important feature for developing FETI–DP type algorithms is that a single degree of freedom is considered at each crosspoint, while two degrees of freedom are introduced at all the other matching nodes across subdomain edges as in FETI or TFETI. Using the finite element discretization, we get again the discretized version of problem (23) with the auxiliary domain decomposition

$$\min \frac{1}{2}\mathbf{u}^\top \mathsf{K}\mathbf{u} - \mathbf{f}^\top \mathbf{u} \quad \text{s.t.} \quad \mathsf{B}^I \mathbf{u} \leq \mathbf{o} \quad \text{and} \quad \mathsf{B}^E \mathbf{u} = \mathbf{o}, \tag{30}$$

where the full rank matrices B^I and B^E describe the non-penetration (inequality) conditions and the gluing (equality) conditions, respectively, and $\mathbf{f}$ represents the discrete analog of the linear form $\ell(\cdot)$. In (30), using suitable numbering, $\mathsf{K} = \mathrm{diag}(\mathsf{K}^1, \mathsf{K}^2)$ is the block diagonal stiffness matrix with the nonzero blocks

$$\mathsf{K}^i = \begin{bmatrix} \mathsf{K}^i_{11} & & & \mathsf{K}^i_{1,p+1} \\ & \ddots & & \vdots \\ & & \mathsf{K}^i_{p,p} & \mathsf{K}^i_{p,p+1} \\ \mathsf{K}^i_{p+1,1} & \cdots & \mathsf{K}^i_{p+1,p} & \mathsf{K}^i_{p+1,p+1} \end{bmatrix}.$$

The block K^1 corresponding to Ω^1 is nonsingular due to the Dirichlet boundary conditions on Γ_u^1. The block K^2 corresponding to Ω^2 is nonsingular for a coercive problem, and is singular, with the kernel made of a vector $\mathbf{e}$ with all the entries equal to 1, for a semicoercive problem. In the latter case, the kernel of K is spanned by the matrix $\mathsf{R} = \begin{bmatrix} \mathbf{o}^\top, \mathbf{e}^\top \end{bmatrix}^\top$. Using the duality theory [3], we can again transform (30) to the dual problem. For a coercive problem, K is nonsingular and we obtain the problem of finding

$$\min \frac{1}{2}\boldsymbol{\lambda}^\top \mathsf{F}\boldsymbol{\lambda} - \boldsymbol{\lambda}^\top \mathbf{d} \quad \text{s.t.} \quad \boldsymbol{\lambda}^I \geq \mathbf{o}, \tag{31}$$

with $\mathsf{F} = \mathsf{B}\,\mathsf{K}^{-1}\mathsf{B}^\top$ and $\mathbf{d} = \mathsf{B}\,\mathsf{K}^{-1}\mathbf{f}$. For an efficient implementation of F, it is important to exploit the structure of K; see [21] for more details. For a semicoercive problem, we obtain the problem of finding

$$\min \frac{1}{2}\boldsymbol{\lambda}^\top \mathsf{F}\boldsymbol{\lambda} - \mathbf{d}^\top \boldsymbol{\lambda} \quad \text{s.t} \quad \mathsf{G}\boldsymbol{\lambda} = \mathbf{o} \quad \text{and} \quad \boldsymbol{\lambda}^I \geq -\widetilde{\boldsymbol{\lambda}}^I, \tag{32}$$

with $\mathbf{d} = \widetilde{\mathbf{d}} - \mathsf{F}\widetilde{\boldsymbol{\lambda}}$ and G and $\widetilde{\boldsymbol{\lambda}}$ defined similarly as in FETI. Our final step is again based on the observation that the Hessian of the augmented Lagrangian for problem (32) may be decomposed by the orthogonal projectors

$$\mathsf{Q} = \mathsf{G}^\top \mathsf{G} \quad \text{and} \quad \mathsf{P} = \mathsf{I} - \mathsf{Q}$$

on the image space of $\mathsf{G}^\top$ and on the kernel of G, respectively. Since $\mathsf{P}\boldsymbol{\lambda} = \boldsymbol{\lambda}$ for any feasible $\boldsymbol{\lambda}$, problem (32) is equivalent to

$$\min \ \frac{1}{2}\boldsymbol{\lambda}^\top \mathsf{PFP}\boldsymbol{\lambda} - \boldsymbol{\lambda}^\top \mathsf{Pd} \quad \text{s.t} \quad \mathsf{G}\boldsymbol{\lambda} = \mathbf{o} \quad \text{and} \quad \boldsymbol{\lambda}^I \geq -\widetilde{\boldsymbol{\lambda}}^I. \qquad (33)$$

The optimality follows from the following theorem.

Theorem 4. *If F denotes the matrix of the problem (32) generated by FETI–DP for the coercive problem, then the following spectral bounds hold:*

$$\lambda_{\max}(\mathsf{F}) = \|\mathsf{F}\| \leq C\left(\frac{H}{h}\right)^2; \quad \lambda_{\min}(\mathsf{F}) \geq C.$$

If F and P denote the matrices of the problem (33) generated by FETI–DP for the semicoercive problem, then the following spectral bounds hold:

$$\lambda_{\max}(\mathsf{PFP}|\mathrm{Im}\mathsf{P}) \leq \|\mathsf{F}\| \leq C\left(\frac{H}{h}\right)^2; \quad \lambda_{\min}(\mathsf{PFP}|\mathrm{Im}\mathsf{P}) \geq C.$$

Proof. See [21, 22]. ■

7 Numerical scalability

To show that Algorithm 2 with the inner loop implemented by Algorithm 1 is optimal for the solution of our model problems (or a class of problems) discretized by means of FETI, TFETI and FETI–DP, we shall use

$$\mathcal{T} = \{(H, h) \in \mathbb{R}^2 : H \leq 1, \ 2h \leq H \ \text{and} \ H/h \in \mathbb{N}\}$$

as the set of indices. Given a constant $C \geq 2$, we shall define a subset $\mathcal{T}_C$ of $\mathcal{T}$ by

$$\mathcal{T}_C = \{(H, h) \in \mathbb{R}^2 : H \leq 1, \ 2h \leq H, \ H/h \in \mathbb{N} \ \text{and} \ H/h \leq C\}.$$

For any $t \in \mathcal{T}$, and a given $\bar{\rho} > 0$, we shall define

$$\mathsf{A}_t = \mathsf{PFP} + \bar{\rho}\mathsf{Q}, \ \mathbf{b}_t = \mathsf{Pd}$$
$$\mathsf{C}_t = \mathsf{G}, \qquad \boldsymbol{\ell}_t^I = -\widetilde{\boldsymbol{\lambda}}^I \ \text{and} \ \boldsymbol{\ell}_t^E = -\infty$$

with the vectors and matrices generated with the discretization and decomposition parameters H and h, respectively, so that the problem (29) is equivalent to the problem

$$\text{minimize} \ \Theta_t(\boldsymbol{\lambda}_t) \quad \text{s.t.} \quad \mathsf{C}_t\boldsymbol{\lambda}_t = 0 \ \text{and} \ \boldsymbol{\lambda}_t \geq \boldsymbol{\ell}_t \qquad (34)$$

with $\Theta_t(\boldsymbol{\lambda}) = \frac{1}{2}\boldsymbol{\lambda}^\top \mathsf{A}_t \boldsymbol{\lambda} - \mathbf{b}_t^\top \boldsymbol{\lambda}$. Using these definitions and $\mathsf{GG}^\top = \mathsf{I}$, we obtain

$$\|\mathsf{C}_t\| \leq 1 \quad \text{and} \quad \|\boldsymbol{\ell}_t^+\| = 0, \tag{35}$$

where for any vector $\mathbf{v}$ with the entries v_i, $\mathbf{v}^+$ denotes the vector with the entries $v_i^+ = \max\{v_i, 0\}$. Moreover, it follows by Theorem 4 that for any $C \geq 2$ there are constants $a_{\max}^C > a_{\min}^C > 0$ such that

$$a_{\min}^C \leq \alpha_{\min}(\mathsf{A}_t) \leq \alpha_{\max}(\mathsf{A}_t) \leq a_{\max}^C \tag{36}$$

for any $t \in \mathcal{T}_C$. Moreover, there are positive constants C_1 and C_2 such that $a_{\min}^C \geq C_1$ and $a_{\max}^C \leq C_2 C$. In particular, it follows that the assumptions of Theorem 5 (i.e. the inequalities (35) and (36)) of [8] are satisfied for any set of indices $\mathcal{T}_C$, $C \geq 2$, and we have the following result:

Theorem 5. *Let $C \geq 2$ denote a given constant, let $\{\boldsymbol{\lambda}_t^k\}, \{\boldsymbol{\mu}_t^k\}$ and $\{\rho_{t,k}\}$ be generated by Algorithm 2 (SMALBE) for (34) with $\|\mathbf{b}_t\| \geq \eta_t > 0$, $\beta > 1$, $M > 0$, $\rho_{t,0} = \rho_0 > 0$, and $\boldsymbol{\mu}_t^0 = \mathbf{o}$. Let $s \geq 0$ denote the smallest integer such that $\beta^s \rho_0 \geq M^2/a_{\min}$ and assume that Step 1 of Algorithm 2 is implemented by means of Algorithm 1 (MPRGP) with parameters $\Gamma > 0$ and $\overline{\alpha} \in (0, (a_{\max} + \beta^s \rho_0)^{-1}]$, so that it generates the iterates $\boldsymbol{\lambda}_t^{k,0}, \boldsymbol{\lambda}_t^{k,1}, \ldots, \boldsymbol{\lambda}_t^{k,l} = \boldsymbol{\lambda}_t^k$ for the solution of (34) starting from $\boldsymbol{\lambda}_t^{k,0} = \boldsymbol{\lambda}_t^{k-1}$ with $\boldsymbol{\lambda}_t^{-1} = \mathbf{o}$, where $l = l_{t,k}$ is the first index satisfying*

$$\|\mathbf{g}^P(\boldsymbol{\lambda}_t^{k,l}, \boldsymbol{\mu}_t^k, \rho_{t,k})\| \leq M \|\mathsf{C}_t \boldsymbol{\lambda}_t^{k,l}\| \tag{37}$$

or

$$\|\mathbf{g}^P(\boldsymbol{\lambda}_t^{k,l}, \boldsymbol{\mu}_t^k, \rho_{t,k})\| \leq \epsilon \|\mathbf{b}_t\| \min\{1, M^{-1}\}. \tag{38}$$

Then for any $t \in \mathcal{T}_C$ and problem (34), Algorithm 2 generates an approximate solution $\boldsymbol{\lambda}_t^{k_t}$ which satisfies

$$M^{-1} \|\mathbf{g}^P(\boldsymbol{\lambda}_t^{k_t}, \boldsymbol{\mu}_t^{k_t}, \rho_{t,k_t})\| \leq \|\mathsf{C}_t \boldsymbol{\lambda}_t^{k_t}\| \leq \epsilon \|\mathbf{b}_t\| \tag{39}$$

at $O(1)$ matrix-vector multiplications by the Hessian of the augmented Lagrangian L_t for (34) and $\rho_{t,k} \leq \beta^s \rho_0$.

Proof. See [9]. ∎

8 Numerical experiments

We have implemented all three domain decomposition methods described above to the solution of both variants of the model problems of Fig. 1. The solution of both problems is in Fig. 2. For the solution of the quadratic programming problems generated by FETI1 and TFETI, we used the SMALBE

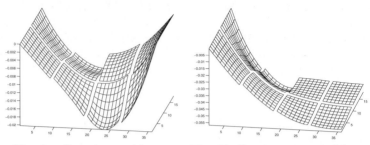

Fig. 3a: Coercive problem   Fig. 3b: Semicoercive problem

Fig. 2. Solution of model problems

algorithm of Section 3 with the inner loop generated by the MPRGP algorithm of Section 2. We have implemented the solver in C exploiting PETSc to solve the semicoercive model problem with varying decomposition and discretization parameters. The results of computations which were carried out to the relative precision 1e-4 are in Table 1.

Table 1. Numerical scalability of FETI and TFETI for H/h=const and ρ=1e3

primal dim.	2312	9248	36992	133128	532512	2130048
FETI/TFETI dual dim.	167/201	863/931	3839/3975	1287/-	6687/-	29823/-
subdomains	8	32	128	8	32	128
FETI iterations	47	58	64	59	36	47
TFETI iterations	39	54	45	-	-	-

Since the algorithms are closely related to the original FETI method, it is not surprising that they enjoy good parallel scalability as documented in Table 2. The experiments with semicoercive problem were run on the Lomond 52-processor Sun Ultra SPARC-III based system with 900 MHz, 52 GB of shared memory, nominal peak performance 93.6 GFlops, 64 kB level 1 and 8 MB level 2 cache in EPCC Edinburgh, to the relative precision 1e-4.

Table 2. Parallel scalability for semicoer.problem with prim.dim 540800, dual dim.14975, 2 outer iters, 43 cg iters, 128 subdomains using Lomond, ρ=1e3

processors	1	2	4	8	16	32
time [sec]	879	290	138	50	27	15

We have implemented also the basic FETI-DP algorithms for the solution of both coercive and semicoercive problems in MATLAB. We have used MPRGP of Section 2 for the solution of the coercive problems and the SMALBE algorithm of Section 3 with the inner loop generated by the MPRGP algorithm to the solution of the semicoercive problem to the relative precision 1e-6. The results are in Table 3.

Table 3. Numerical scalability of the basic FETI-DP for coer. and semi-coer.problem, ρ=1e3

prim./dual/corner dim.	2312/153/10	9248/785/42	36992/3489/154
subdomains	8	32	128
cg iters for coer.problem	27	48	51
cg iters for semicoer.problem	41	57	63

9 Comments and conclusions

We have reviewed our recent results related to application of the augmented Lagrangians with the FETI based domain decomposition method to the solution of variational inequalities using recently developed algorithms for the solution of special QP problems. In particular, we have shown that the solution of the discretized problem to a prescribed precision may be found in a number of iterations bounded independently of the discretization parameter. Numerical experiments with the model variational inequality are in agreement with the theory and indicate that the algorithms presented are efficient. The research in progress includes implementation of preconditioners, the mortar discretization and the generalization to the contact problems with friction.

Acknowledgements

The first two authors were supported by Grant 101/04/1145 of the GA CR and by Project 1ET400300415 of the Ministry of Education of the Czech Republic and HPC-EUROPA project (RII3-CT-2003-506079) with the support of the European Community- Research Infrastructure Action under FP6 "Structuring the European Research Area" Programme. The third author was supported by the National Science Foundation Grant NSF-DMS-0103588 and by the Research Foundation of the City University of New York Awards PSC-CUNY 665463-00 34 and 66529-00 35.

References

1. Avery, P., Rebel, G., Lesoinne, M., Farhat, C.: A umerically scalable dual-primal substructuring method for the solution of contact problems. I: the frictionless case. Comput. Methods Appl. Mech. Eng., **193**, 2403–2426 (2004)

2. Axelsson, O.: Iterative Solution Methods. Cambridge University Press, Cambridge (1994)
3. Bertsekas, D.P.: Nonlinear Optimization. Athena Scientific, Belmont (1999)
4. Conn, A.R., Gould, N.I.M., Toint, Ph.L.: A globally convergent augmented Lagrangian algorithm for optimization with general constraints and simple bounds. SIAM J. Num. Anal. **28**, 545–572 (1991)
5. Dostál, Z.: Box constrained quadratic programming with proportioning and projections. SIAM Journal on Optimization **7**, 871–887 (1997)
6. Dostál, Z.: A proportioning based algorithm for bound constrained quadratic programming with the rate of convergence. Numerical Algorithms, **34**, 293–302 (2003)
7. Dostál, Z.: Inexact semi-monotonic Augmented Lagrangians with optimal feasibility convergence for quadratic programming with simple bounds and equality constraints. SIAM Journal on Numerical Analysis, **43**, 96–115 (2005)
8. Dostál, Z.: An optimal algorithm for bound and equality constrained quadratic programming problems with bounded spectrum. Submitted
9. Dostál, Z.: Scalable FETI for numerical solution of variational inequalities. Submitted
10. Dostál, Z., Friedlander, A., Santos, S.A.: Solution of contact problems of elasticity by FETI domain decomposition. Contemporary Mathematics, **218**, 82–93 (1998)
11. Dostál, Z., Friedlander, A., Santos, S.A.: Augmented Lagrangians with adaptive precision control for quadratic programming with simple bounds and equality constraints. SIAM Journal on Optimization, **13**, 1120–1140 (2003)
12. Dostál, Z., Gomes Neto, F.A.M., Santos, S. A.: Duality based domain decomposition with natural coarse space for variational inequalities. Journal of Computational and Applied Mathematics, **126**, 397–415 (2000)
13. Dostál, Z., Gomes Neto, F.A.M., Santos, S.A.: Solution of contact problems by FETI domain decomposition with natural coarse space projection. Computer Methods in Applied Mechanics and Engineering, **190**, 1611–1627 (2000)
14. Dostál, Z., Haslinger, J., Kučera, R.: Implementation of fixed point method for duality based solution of contact problems with friction. Journal of Computational and Applied Mathematics, **140**, 245–256 (2002)
15. Dostál, Z., Horák, D.: Scalability and FETI based algorithm for large discretized variational inequalities. Mathematics and Computers in Simulation, **61**, 347–357 (2003)
16. Dostál, Z., Horák, D.: Scalable FETI with Optimal Dual Penalty for Semicoercive Variational Inequalities. Contemporary Mathematics **329**, 79–88 (2003)
17. Dostál, Z., Horák, D.: Scalable FETI with Optimal Dual Penalty for a Variational Inequality. Numerical Linear Algebra and Applications, **11**, 455 - 472 (2004)
18. Dostál, Z., Horák, D.: On scalable algorithms for numerical solution of variational inequalities based on FETI and semi-monotonic augmented Lagrangians. In: Kornhuber, R., Hoppe, R.H.W., Périaux, J., Pironneau, O., Widlund, O.B., Xu., J. (eds) Proceedings of DDM15. Springer, Berlin (2003)
19. Dostál, Z., Horák, D., Kučera, R.: Total FETI - an easier implementable variant of the FETI method for numerical solution of elliptic PDE. Submitted

20. Dostál, Z., Horák, D., Stefanica, D.: A Scalable FETI–DP Algorithm with Non-penetration Mortar Conditions on Contact Interface. Submitted
21. Dostál, Z., Horák, D., Stefanica, D.: A scalable FETI–DP algorithm for coercive variational inequalities. IMACS Journal Applied Numerical Mathematics, **54**, 378–390 (2005)
22. Dostál, Z., Horák, D., Stefanica, D.: Scalable FETI–DP Algorithm for Semi-coercive Variational Inequalities. Submitted
23. Dostál, Z., Schöberl, J.: Minimizing quadratic functions over non-negative cone with the rate of convergence and finite termination. Computational Optimization and Applications, **30**, 23–44 (2005)
24. Dureisseix, D., Farhat, C.: A numerically scalable domain decomposition method for solution of frictionless contact problems. International Journal for Numerical Methods in Engineering, **50**, 2643–2666 (2001)
25. Farhat, C., Gérardin, M.: On the general solution by a direct method of a large scale singular system of linear equations: application to the analysis of floating structures. International Journal for Numerical Methods in Engineering, **41**, 675–696 (1998)
26. Farhat, C., Lesoinne, M., LeTallec, P., Pierson, K., Rixen, D.: FETI-DP: A dual-primal unified FETI method. I: A faster alternative to the two-level FETI method. Int. J. Numer. Methods Eng. **50**, 1523–1544 (2001)
27. Farhat, C., Mandel, J., Roux, F.X.: Optimal convergence properties of the FETI domain decomposition method. Computer Methods in Applied Mechanics and Engineering, **115**, 365–385 (1994)
28. Farhat, C., Roux, F.X.: An unconventional domain decomposition method for an efficient parallel solution of large-scale finite element systems. SIAM Journal on Scientific Computing, **13**, 379–396 (1992)
29. Friedlander, A., Martínez, J.M.: On the maximization of a concave quadratic function with box constraints. SIAM J. Optimiz., **4**, 177–192 (1994)
30. Friedlander, A., Martínez, J.M., Santos, S.A.: A new trust region algorithm for bound constrained minimization. Applied Math. & Optimiz., **30**, 235–266 (1994)
31. Hlaváček, I., Haslinger, J., Nečas, J., Lovíšek J.: Solution of Variational Inequalities in Mechanics. Springer, Berlin (1988)
32. Klawonn, A., Widlund, O.B., Dryja, M.: Dual-Primal FETI Methods for Three-dimensional Elliptic Problems with Heterogeneous Coefficients. SIAM J. Numer. Anal. **40**, 159–179 (2002)
33. Kornhuber, R.: Adaptive Monotone Multigrid Methods for Nonlinear Variational Problems. Teubner-Verlag, Stuttgart (1997)
34. Kornhuber, R., Krause, R.: Adaptive multigrid methods for Signorini's problem in linear elasticity. Computer Visualization in Science **4**, 9–20 (2001)
35. Mandel, J.: Étude algébrique d'une méthode multigrille pour quelques problèmes de frontière libre (French). Comptes Rendus de l'Academie des Sciences **I 298**, 469–472 (1984)
36. Mandel, J., Tezaur, R.: On the Convergence of a Dual-Primal Substructuring Method. Numerische Mathematik **88**, 543–558 (2001)
37. Schöberl, J.: Solving the Signorini problem on the basis of domain decomposition techniques. Computing, **60**, 323–344 (1998)

38. Schöberl, J.: Efficient contact solvers based on domain decomposition techniques. Comput. Math. Appl. **42**, 1217–1228 (1998)
39. Wohlmuth, B.: Discretization Methods and Iterative Solvers Based on Domain Decomposition. Springer, Berlin (2001)
40. Wohlmuth, B., Krause, R.: Monotone methods on nonmatching grids for nonlinear contact problems. SIAM J. Sci. Comput. **25** 324–347 (2003)

Discontinuous Galerkin Methods for Friedrichs' Systems

Alexandre Ern[1] and Jean-Luc Guermond[2]

[1] CERMICS, Ecole nationale des ponts et chaussées, Champs sur Marne, 77455 Marne la Vallée Cedex 2, France
ern@cermics.enpc.fr
[2] Dept. Math, Texas A&M, College Station, TX 77843-3368, USA (on leave from LIMSI-CNRS, BP 133, 91403, Orsay, France)
guermond@math.tamu.edu

Summary. This work presents a unified analysis of Discontinuous Galerkin methods to approximate Friedrichs' systems. A general set of boundary conditions is identified to guarantee existence and uniqueness of solutions to these systems. A formulation enforcing the boundary conditions weakly is proposed. This formulation is the starting point for the construction of Discontinuous Galerkin methods formulated in terms of boundary operators and of interface operators that mildly penalize interface jumps. A general convergence analysis is presented. The setting is subsequently specialized to two-field Friedrichs' systems endowed with a particular 2×2 structure in which some of the unknowns can be eliminated to yield a system of second-order elliptic-like PDE's for the remaining unknowns. A general Discontinuous Galerkin method where the above elimination can be performed in each mesh cell is proposed and analyzed. Finally, details are given for four examples, namely advection–reaction equations, advection–diffusion–reaction equations, the linear elasticity equations in the mixed stress–pressure–displacement form, and the Maxwell equations in the so-called elliptic regime.

1 Introduction

Since their introduction in 1973 by Reed and Hill [19] to simulate neutron transport, Discontinuous Galerkin (DG) methods have sparked extensive interest owing to their flexibility in handling non-matching grids, heterogeneous data, and high-order hp-adaptivity. However, the development and analysis of DG methods has followed two somewhat parallel routes depending on whether the PDE is hyperbolic or elliptic.

For hyperbolic PDE's, the first analysis of DG methods in an already rather abstract form was performed by Lesaint and Raviart in 1974 [16, 17] and subsequently improved by Johnson et al. [15] in 1984. More recently, DG methods for hyperbolic and nearly hyperbolic equations experienced a significant development based on the ideas of numerical fluxes, approximate

Riemann solvers, and slope limiters; see, e.g., Cockburn et al. [8] and the references therein.

For elliptic PDE's, DG methods originated from the early work of Nitsche on boundary-penalty methods [18] and the use of Interior Penalties (IP) to weakly enforce continuity on the solution or its derivatives across the interfaces between adjoining elements; see, e.g., Babuška [3], Babuška and Zlámal [4], Baker [5], Wheeler [20], and Arnold [1]. DG methods for elliptic problems in mixed form were introduced more recently (see, e.g., Bassi and Rebay [6]) and further extended by Cockburn and Shu [9] leading to the so-called Local Discontinuous Galerkin (LDG) method. The fact that several of the above DG methods (including IP methods) share common features and can be tackled by similar analysis tools called for a unified analysis. A first important step in that direction has been recently accomplished in Arnold et al. [2], where it is shown that it is possible to cast many DG methods for the Poisson equation with homogeneous Dirichlet boundary conditions into a single framework amenable to a unified error analysis.

The goal of the present work is to propose a unified analysis of DG methods that goes beyond the traditional hyperbolic/elliptic classification of PDE's. To this purpose, we make systematic use of the theory of Friedrichs' systems [14], i.e., systems of first-order PDE's endowed with a symmetry and a positivity property, to formulate DG methods and to perform the convergence analysis. For brevity, the main theoretical results are stated without proof; see [11, 12, 13] for full detail.[3]

This paper is organized as follows. In mS2 we revisit Friedrichs' theory and formulate a set of abstract conditions ensuring well–posedness of the continuous problem while avoiding to invoke traces at the boundary. In mS3 we formulate and analyze a general DG method to approximate Friedrichs' systems. The design of the method is based on an operator enforcing boundary conditions weakly and an operator penalizing the jumps of the solution across the mesh interfaces. All the design constraints to be fulfilled by the boundary and the interface operators for the error analysis to hold are stated. Moreover, using integration by parts, the DG method is re-interpreted locally by introducing the concept of element fluxes, thus providing a direct link with engineering practice where approximation schemes are often designed by specifying such fluxes. In mS4 we specialize the setting to a particular class of Friedrichs' systems with a 2×2 structure in which some of the unknowns can be eliminated to yield a system of second-order elliptic-like PDE's for the remaining unknowns. For such systems, a general Discontinuous Galerkin method is proposed and analyzed. The key feature of the method is that the unknowns that can be eliminated at the continuous level can also be eliminated at the discrete level by solving local problems. In mS5, we apply the theoretical results to advection–reaction equations, advection–diffusion–reaction equations, the linear elasticity equations in the mixed stress–pressure–displacement form,

---

[3] Internal reports available at `cermics.enpc.fr/reports/CERMICS-2005`

and the Maxwell equations in the so-called elliptic regime. Concluding remarks are reported in mS6.

2 Friedrichs' systems

Let Ω be a bounded, open, and connected Lipschitz domain in $\mathbb{R}^d$. We denote by $\mathfrak{D}(\Omega)$ the space of $\mathfrak{C}^\infty$ functions that are compactly supported in Ω. Let m be a positive integer. Let $\mathcal{K}$ and $\{\mathcal{A}^k\}_{1\leq k\leq d}$ be $(d+1)$ functions on Ω with values in $\mathbb{R}^{m,m}$ such that

$$\mathcal{K} \in [L^\infty(\Omega)]^{m,m}, \tag{A1}$$

$$\forall k \in \{1,\ldots,d\},\ \mathcal{A}^k \in [L^\infty(\Omega)]^{m,m} \ \text{and} \ \sum_{k=1}^d \partial_k \mathcal{A}^k \in [L^\infty(\Omega)]^{m,m}, \tag{A2}$$

$$\forall k \in \{1,\ldots,d\},\ \mathcal{A}^k = (\mathcal{A}^k)^t \ \text{a.e. in}\ \Omega, \tag{A3}$$

$$\mathcal{K} + \mathcal{K}^t - \sum_{k=1}^d \partial_k \mathcal{A}^k \geq 2\mu_0 \mathcal{I}_m \ \text{a.e. on}\ \Omega, \tag{A4}$$

where $\mathcal{I}_m$ is the identity matrix in $\mathbb{R}^{m,m}$. Assumptions (A3) and (A4) are, respectively, the symmetry and the positivity property referred to above.

Set $L = [L^2(\Omega)]^m$. A function z in L is said to have an A-weak derivative in L if the linear form $[\mathfrak{D}(\Omega)]^m \ni \phi \longmapsto -\int_\Omega \sum_{k=1}^d z^t \partial_k(\mathcal{A}^k \phi) \in \mathbb{R}$ is bounded on L. In this case, the function in L that can be associated with the above linear form by means of the Riesz representation theorem is denoted by Az. Clearly, if z is smooth, e.g., $z \in [\mathfrak{C}^1(\overline{\Omega})]^m$, $Az = \sum_{k=1}^d \mathcal{A}^k \partial_k z$. Define the so-called graph space $W = \{z \in L;\ Az \in L\}$ equipped with the graph norm $\|z\|_W = \|Az\|_L + \|z\|_L$. The space W is endowed with a Hilbert structure when equipped with the scalar product $(z,y)_L + (Az,Ay)_L$. Define the operators $T \in \mathcal{L}(W;L)$ and $\widetilde{T} \in \mathcal{L}(W;L)$ as

$$Tz = \mathcal{K}z + Az, \qquad \widetilde{T}z = \mathcal{K}^t z + \widetilde{A}z, \tag{1}$$

with $\widetilde{A}z = -\sum_{k=1}^d \partial_k(\mathcal{A}^k z)$. Assumption (A4) implies that $T+\widetilde{T}$ is L-coercive on L.

Let $f \in L$ and consider the problem of seeking $z \in W$ such that $Tz = f$ in L. In general, boundary conditions must be enforced for this problem to be well–posed. In other words, one must find a closed subspace V of W such that $T : V \to L$ is an isomorphism. Let $D \in \mathcal{L}(W;W')$ be the operator defined by

$$\forall (z,y) \in W \times W,\quad \langle Dz,y \rangle_{W',W} = (Az,y)_L - (z,\widetilde{A}y)_L. \tag{2}$$

Let W_0 be the closure of $[\mathfrak{D}(\Omega)]^m$ in W. For every subspace $Z \subset W$, let $Z^\perp$ denote the polar set of Z, i.e., the set of linear forms on W that vanish on Z and use a similar notation for the polar sets of subspaces of W'. A key result concerning the operator D is the following

Lemma 1. *The operator D is self-adjoint. Moreover, the following holds:*

$$\mathrm{Ker}(D) = W_0 \quad \text{and} \quad \mathrm{Im}(D) = W_0^\perp. \tag{3}$$

To enforce boundary conditions, a simple approach inspired from Friedrichs' work consists of assuming that there is an operator $M \in \mathcal{L}(W; W')$ such that

$$M \text{ is positive, i.e., } \langle Mz, z \rangle_{W', W} \geq 0 \text{ for all } z \text{ in } W, \tag{M1}$$
$$W = \mathrm{Ker}(D - M) + \mathrm{Ker}(D + M). \tag{M2}$$

Then by setting $V = \mathrm{Ker}(D - M)$ and $V^* = \mathrm{Ker}(D + M^*)$ where $M^* \in \mathcal{L}(W; W')$ is the adjoint of M and equipping V and V^* with the graph norm, the following theorem can be proved:

Theorem 1. *Assume* (A1)–(A4) *and* (M1)–(M2). *Then, the restricted operators $T : V \to L$ and $\widetilde{T} : V^* \to L$ are isomorphisms.*

The proof of Theorem 1 relies on the following fundamental result, the so-called Banach–Nečas–Babuška (BNB) Theorem, that is restated below for completeness (see, e.g., [5]).

Theorem 2 (BNB). *Let V and L be two Banach spaces, and denote by $\langle \cdot, \cdot \rangle_{L', L}$ the duality pairing between L' and L. Then, $T \in \mathcal{L}(V; L)$ is bijective if and only if*

$$\exists \alpha > 0, \quad \forall w \in V, \quad \sup_{y \in L' \setminus \{0\}} \frac{\langle y, Tw \rangle_{L', L}}{\|y\|_{L'}} \geq \alpha \|w\|_V, \tag{4}$$

$$\forall y \in L', \quad (\langle y, Tw \rangle_{L', L} = 0, \ \forall w \in V) \implies (y = 0). \tag{5}$$

Remark 1. It is possible to formulate an intrinsic criterion for the bijectivity of the operators T and $\widetilde{T}$ that circumvents the somewhat *ad hoc* operator M by introducing the concept of maximal boundary conditions. To this purpose, introduce the cones $C^\pm = \{w \in W; \ \pm \langle Dw, w \rangle_{W', W} \geq 0\}$. Let V and V^* be two subspaces of W such that

$$V \subset C^+ \text{ and } V^* \subset C^-, \tag{V1}$$
$$V = D(V^*)^\perp \text{ and } V^* = D(V)^\perp. \tag{V2}$$

Then, under the assumptions (A1)–(A4) and (V1)–(V2), the conclusions of Theorem 1 still hold. Furthermore, one can prove that if V and V^* are two subspaces of W satisfying (V1)–(V2), then V is maximal in C^+ (there is no $x \in W$ such that $V_x := V + \mathrm{span}(x) \subset C^+$ and V is a proper subspace of V_x) and V^* is maximal in C^- (there is no $y \in W$ such that $V_y^* := V^* + \mathrm{span}(y) \subset C^-$ and V^* is a proper subspace of V_y^*). In this sense, the boundary conditions embodied in V and V^* are maximal.

Owing to Theorem 1, the following problems are well-posed:

$$\text{Seek } z \in V \text{ such that } Tz = f, \tag{6}$$

$$\text{Seek } z^* \in V^* \text{ such that } \tilde{T}z^* = f. \tag{7}$$

The boundary conditions in (6) and (7) are enforced strongly by seeking the solutions in V and V^*, respectively. A key feature of Friedrichs' systems is that it is possible to enforce boundary conditions naturally, thus leading to a suitable framework for developing a DG theory. To see this, introduce the following bilinear forms on $W \times W$,

$$a(z,y) = (Tz,y)_L + \tfrac{1}{2}\langle (M-D)z,y\rangle_{W',W}, \tag{8}$$

$$a^*(z,y) = (\tilde{T}z,y)_L + \tfrac{1}{2}\langle (M^*+D)z,y\rangle_{W',W}. \tag{9}$$

It is clear that a and a^* are in $\mathcal{L}(W \times W; \mathbb{R})$. Consider the following problems:

$$\text{Seek } z \in W \text{ such that } a(z,y) = (f,y)_L, \forall y \in W, \tag{10}$$

$$\text{Seek } z^* \in W \text{ such that } a^*(z^*,y) = (f,y)_L, \forall y \in W. \tag{11}$$

Contrary to (6) and (7), the boundary conditions in (10) and (11) are weakly enforced. For this reason, problem (10) will constitute our working basis for designing DG methods. The key result of this section is the following

Theorem 3. *Assume* (A1)–(A4) *and* (M1)–(M2). *Then, there is a unique solution to* (10) *(resp.,* (11)*) and this solution solves* (6) *(resp.,* (7)*).*

3 Design and analysis of DG methods

The purpose of this section is to design and analyze a general DG method to approximate the unique solution to (10).

3.1 The discrete setting

Let $\{\mathcal{T}_h\}_{h>0}$ be a family of meshes of Ω. The meshes are assumed to be affine to avoid unnecessary technicalities, i.e., Ω is assumed to be a polyhedron. However, we do not make any assumption on the matching of element interfaces. Let p be a non-negative integer and set

$$P_{h,p} = \{v_h \in L^2(\Omega); \forall K \in \mathcal{T}_h, v_h|_K \in \mathbb{P}_p\}, \tag{12}$$

where $\mathbb{P}_p$ denotes the vector space of polynomials with real coefficients and total degree less than or equal to p. Define

$$W_h = [P_{h,p}]^m, \qquad W(h) = [H^1(\Omega)]^m + W_h. \tag{13}$$

We denote by $\mathcal{F}_h^i$ the set of interior faces (or interfaces), i.e., $F \in \mathcal{F}_h^i$ if F is a $(d-1)$-manifold and there are $K_1(F), K_2(F) \in \mathcal{T}_h$ such that $F = K_1(F) \cap K_2(F)$. We denote by $\mathcal{F}_h^\partial$ the set of the faces that separate the mesh from the exterior of Ω, i.e., $F \in \mathcal{F}_h^\partial$ if F is a $(d-1)$-manifold and there is $K(F) \in \mathcal{T}_h$ such that $F = K(F) \cap \partial\Omega$. Finally, we set $\mathcal{F}_h = \mathcal{F}_h^i \cup \mathcal{F}_h^\partial$. Since every function v in $W(h)$ has a (possibly two-valued) trace almost everywhere on $F \in \mathcal{F}_h^i$, it is meaningful to set $v^n(x) = \lim_{\substack{y \to x \\ y \in K_n(F)}} v(y)$, $n \in \{1, 2\}$, for a.e. $x \in F$ and

$$[\![v]\!] = v^1 - v^2, \qquad \{v\} = \tfrac{1}{2}(v^1 + v^2), \qquad \text{a.e. on } F. \tag{14}$$

Nothing that is said hereafter depends on the arbitrariness in the sign of $[\![v]\!]$.

For any measurable subset of Ω, say E, $(\cdot, \cdot)_{L,E}$ denotes the usual L^2-scalar product on E. The same notation is used for scalar- and vector-valued functions. For $K \in \mathcal{T}_h$ (resp., $F \in \mathcal{F}_h$), h_K (resp., h_F) denotes the diameter of K (resp., F). The mesh family $\{\mathcal{T}_h\}_{h>0}$ is assumed to be shape-regular so that the usual inverse and trace inverse inequalities hold on W_h. Henceforth, we use the notation $A \lesssim B$ to represent the inequality $A \leq cB$ where c is independent of h.

3.2 The design of the DG bilinear form

Set $\mathcal{D} = \sum_{k=1}^d n_k \mathcal{A}^k$, where $n = (n_1, \ldots, n_d)^t$ is the outward unit normal to Ω, and assume that there is a matrix-valued field $\mathcal{M} : \partial\Omega \longrightarrow \mathbb{R}^{m,m}$ such that for all functions y, w smooth enough (e.g., $y, w \in [H^1(\Omega)]^m$),

$$\langle \mathcal{D} y, w \rangle_{W',W} = \int_{\partial\Omega} w^t \mathcal{D} y, \qquad \langle \mathcal{M} y, w \rangle_{W',W} = \int_{\partial\Omega} w^t \mathcal{M} y. \tag{15}$$

To enforce boundary conditions weakly, we introduce for all $F \in \mathcal{F}_h^\partial$ a linear operator $M_F \in \mathcal{L}([L^2(F)]^m; [L^2(F)]^m)$ such that for all $y, w \in [L^2(F)]^m$,

$$(M_F(y), y)_{L,F} \geq 0, \tag{DG1}$$
$$(\mathcal{M} y = \mathcal{D} y) \implies (M_F(y) = \mathcal{D} y), \tag{DG2}$$
$$|(M_F(y) - \mathcal{D} y, w)_{L,F}| \lesssim |y|_{M,F} \|w\|_{L,F}, \tag{DG3}$$
$$|(M_F(y) + \mathcal{D} y, w)_{L,F}| \lesssim \|y\|_{L,F} |w|_{M,F}, \tag{DG4}$$

where for all $y \in W(h)$, $|y|_M^2 = \sum_{F \in \mathcal{F}_h^\partial} |y|_{M,F}^2$ with $|y|_{M,F}^2 = (M_F(y), y)_{L,F}$.

For $K \in \mathcal{T}_h$, define the matrix-valued field $\mathcal{D}_{\partial K} : \partial K \to \mathbb{R}^{m,m}$ as

$$\mathcal{D}_{\partial K}(x) = \sum_{k=1}^d n_{K,k} \mathcal{A}^k(x) \quad \text{a.e. on } \partial K, \tag{16}$$

where $n_K = (n_{K,1}, \ldots, n_{K,d})^t$ is the unit outward normal to K on ∂K. We extend the matrix-valued field $\mathcal{D}$ on $\mathcal{F}_h = \mathcal{F}_h^i \cup \mathcal{F}_h^\partial$ as follows. On $\mathcal{F}_h^\partial$, $\mathcal{D}$ is defined as above. On $\mathcal{F}_h^i$, $\mathcal{D}$ is two-valued and for all $F \in \mathcal{F}_h^i$, its two values are $\mathcal{D}_{\partial K_1(F)}$ and $\mathcal{D}_{\partial K_2(F)}$. Note that $\{\mathcal{D}\} = 0$ a.e. on $\mathcal{F}_h^i$. To control the jumps of functions in W_h across mesh interfaces, we introduce for all $F \in \mathcal{F}_h^i$ a linear operator $S_F \in \mathcal{L}([L^2(F)]^m; [L^2(F)]^m)$ such that for all $y, w \in [L^2(F)]^m$,

$$(S_F(y), y)_{L,F} \geq 0, \tag{DG5}$$

$$|(S_F(y), w)_{L,F}| \lesssim |y|_{S,F} |w|_{S,F}, \tag{DG6}$$

$$|(\mathcal{D}_{\partial K(F)} y, w)_{L,F}| \lesssim |y|_{S,F} \|w\|_{L,F}, \tag{DG7}$$

where $F \subset \partial K(F)$ and where for all $y \in W(h)$, $|y|_S^2 = \sum_{F \in \mathcal{F}_h^i} |y|_{S,F}^2$ with $|y|_{S,F}^2 = (S_F(y), y)_{L,F}$. A simple way of enforcing (DG5)–(DG7) consists of setting $S_F(y) = |\mathcal{D}_{\partial K(F)}| y$.

Introduce the bilinear form a_h such that for all z, y in $W(h)$,

$$\begin{aligned} a_h(z, y) = &\sum_{K \in \mathcal{T}_h} (Tz, y)_{L,K} + \sum_{F \in \mathcal{F}_h^\partial} \tfrac{1}{2}(M_F(z) - \mathcal{D}z, y)_{L,F} \\ &- \sum_{F \in \mathcal{F}_h^i} 2(\{\mathcal{D}z\}, \{y\})_{L,F} + \sum_{F \in \mathcal{F}_h^i} (S_F(\llbracket z \rrbracket), \llbracket y \rrbracket)_{L,F}. \end{aligned} \tag{17}$$

Observe that owing to (DG2), the second term in the definition of a_h weakly enforces the boundary conditions in a way which is consistent with (8). The purpose of the third term is to ensure that an L-coercivity property holds on W_h. The last term controls the jump of the discrete solution across interfaces. Some user-dependent arbitrariness appears in the second and fourth term through the definition of the operators M_F and S_F. An equivalent definition of the DG bilinear form obtained by integration by parts is the following:

$$\begin{aligned} a_h(z, y) = &\sum_{K \in \mathcal{T}_h} (z, \tilde{T}y)_{L,K} + \sum_{F \in \mathcal{F}_h^\partial} \tfrac{1}{2}(M_F(z) + \mathcal{D}z, y)_{L,F} \\ &+ \sum_{F \in \mathcal{F}_h^i} \tfrac{1}{2}(\llbracket \mathcal{D}z \rrbracket, \llbracket y \rrbracket)_{L,F} + \sum_{F \in \mathcal{F}_h^i} (S_F(\llbracket z \rrbracket), \llbracket y \rrbracket)_{L,F}. \end{aligned} \tag{18}$$

3.3 Convergence analysis

An approximation to the solution of (10) is constructed as follows: For $f \in L$,

$$\begin{cases} \text{Seek } z_h \in W_h \text{ such that} \\ a_h(z_h, y_h) = (f, y_h)_L, \quad \forall y_h \in W_h. \end{cases} \tag{19}$$

The error analysis uses the following discrete norms on $W(h)$,

$$\|y\|_{h,A}^2 = \|y\|_L^2 + |y|_J^2 + |y|_M^2 + \sum_{K \in \mathcal{T}_h} h_K \|Ay\|_{L,K}^2, \tag{20}$$

$$\|y\|_{h,\frac{1}{2}}^2 = \|y\|_{h,A}^2 + \sum_{K \in \mathcal{T}_h} [h_K^{-1} \|y\|_{L,K}^2 + \|y\|_{L,\partial K}^2]. \tag{21}$$

where for all $y \in W(h)$, $|y|_J^2 = \sum_{F \in \mathcal{F}_h^i} |y|_{J,F}^2$ with $|y|_{J,F} = |[\![y]\!]|_{S,F}$. The convergence analysis is performed in the spirit of Strang's Second Lemma. The main result is the following

Theorem 4. *Let z solve* (10) *and let z_h solve* (19). *Assume that for all $k \in \{1,\ldots,d\}$, $\mathcal{A}^k \in [\mathcal{C}^{0,\frac{1}{2}}(\overline{\Omega})]^{m,m}$. Then,*

$$\|z - z_h\|_{h,A} \lesssim \inf_{y_h \in W_h} \|z - y_h\|_{h,\frac{1}{2}}, \tag{22}$$

if $z \in [H^1(\Omega)]^m$, and $\lim_{h \to 0} \|z - z_h\|_L = 0$ if $z \in V$ only, assuming there is $\gamma > 0$ such that $[H^{1+\gamma}(\Omega)]^m \cap V$ is dense in V.

Using standard interpolation results on W_h, the above result implies that

$$\|z - z_h\|_{h,A} \lesssim h^{p+\frac{1}{2}} \|z\|_{[H^{p+1}(\Omega)]^m} \tag{23}$$

whenever z is in $[H^{p+1}(\Omega)]^m$. In particular, $\|z - z_h\|_L$ converges to order $h^{p+\frac{1}{2}}$, and if the mesh family $\{\mathcal{T}_h\}_{h>0}$ is quasi-uniform, $(\sum_{K \in \mathcal{T}_h} \|A(z - z_h)\|_{L,K}^2)^{\frac{1}{2}}$ converges to order h^p. These estimates are identical to those that can be obtained by other stabilization methods like Galerkin/Least-Squares, subgrid viscosity, etc.

3.4 Localization and the notion of fluxes

The purpose of this section is to discuss briefly some equivalent formulations of the discrete problem (19) in order to emphasize the link with other formalisms derived previously for DG methods based on the notion of fluxes (see, e.g., Arnold et al. [2]). Let $K \in \mathcal{T}_h$. For $v \in W(h)$ and $x \in \partial K$, set $v^i(x) = \lim_{\substack{y \to x \\ y \in K}} v(y)$, $v^e(x) = \lim_{\substack{y \to x \\ y \notin K}} v(y)$ (with $v^e(x) = 0$ if $x \in \partial \Omega$), and

$$[\![v]\!]_{\partial K}(x) = v^i(x) - v^e(x), \quad \{v\}_{\partial K}(x) = \tfrac{1}{2}(v^i(x) + v^e(x)). \tag{24}$$

The *element flux* of a function v on ∂K, say $\phi_{\partial K}(v) \in [L^2(\partial K)]^m$, is defined on a face $F \subset \partial K$ by

$$\phi_{\partial K}(v)|_F = \begin{cases} \tfrac{1}{2} M_F(v|_F) + \tfrac{1}{2} \mathcal{D}v, & \text{if } F \subset \partial K^\partial, \\ S_F([\![v]\!]_{\partial K}|_F) + \mathcal{D}_{\partial K}\{v\}_{\partial K}, & \text{if } F \subset \partial K^i, \end{cases} \tag{25}$$

where ∂K^i denotes that part of ∂K that lies in Ω and ∂K^∂ that part of ∂K that lies on $\partial \Omega$. The relevance of the notion of flux is clarified by the following

Proposition 1. *The discrete problem* (19) *is equivalent to each of the following two local formulations:*

$$\begin{cases} \text{Seek } z_h \in W_h \text{ such that } \forall K \in \mathcal{T}_h \text{ and } \forall y_h \in [\mathbb{P}_p(K)]^m, \\ (z_h, \widetilde{T} y_h)_{L,K} + (\phi_{\partial K}(z_h), y_h)_{L,\partial K} = (f, y_h)_{L,K}, \end{cases} \tag{26}$$

$$\begin{cases} \text{Seek } z_h \in W_h \text{ such that } \forall K \in \mathcal{T}_h \text{ and } \forall y_h \in [\mathbb{P}_p(K)]^m, \\ (T z_h, y_h)_{L,K} + (\phi_{\partial K}(z_h) - \mathcal{D}_{\partial K} z_h^i, y_h)_{L,\partial K} = (f, y_h)_{L,K}. \end{cases} \tag{27}$$

In engineering practice, approximation schemes such as (26) are often designed by specifying the element fluxes. The above analysis then provides a practical means to assess the properties of the scheme. Indeed, once the element fluxes are given, the boundary operators M_F and the interface operators S_F can be directly retrieved from (25). Then, properties (DG1)–(DG7) provide sufficient conditions for convergence.

Remark 2. The element fluxes are conservative in the sense that for all $F = K_1(F) \cap K_2(F) \in \mathcal{F}_h^i$, $\phi_{\partial K_1(F)}(v) + \phi_{\partial K_2(F)}(v) = 0$ on F. The concept of conservativity as such does not play any role in the present analysis of DG methods. It plays a role when deriving improved L^2-error estimates by using the Aubin–Nitsche lemma; see, e.g., Arnold et al. [2] and mS4.3.

4 DG approximation of two-field Friedrichs' systems

In this section the setting is specialized to Friedrichs' systems endowed with a 2×2 block structure in which some of the unknowns can be eliminated to yield a system of elliptic-like PDE's for the remaining unknowns. A general DG method to approximate such systems is proposed and analyzed. The key feature is that the unknowns that can be eliminated at the continuous level can be also eliminated at the discrete level by solving local problems. To achieve this goal we will see that at variance with the DG method formulated in mS3, where jumps and boundary values are equally controlled among the unknowns, the boundary values and jumps of the discrete unknowns to be eliminated must no longer be controlled whereas the boundary values and jumps of the remaining discrete unknowns must be controlled with an $\mathcal{O}(h^{-1})$ weight.

4.1 The setting

We now assume that there are two positive integers m_σ and m_u with $m = m_\sigma + m_u$ such that the $(d+1)$ $\mathbb{R}^{m,m}$-valued fields $\mathcal{K}$ and $\{\mathcal{A}^k\}_{1 \le k \le d}$ have the following 2×2 block structure:

$$\mathcal{K} = \begin{bmatrix} \mathcal{K}^{\sigma\sigma} & \mathcal{K}^{\sigma u} \\ \mathcal{K}^{u\sigma} & \mathcal{K}^{uu} \end{bmatrix}, \qquad \mathcal{A}^k = \begin{bmatrix} 0 & \mathcal{B}^k \\ [\mathcal{B}^k]^t & \mathcal{C}^k \end{bmatrix}, \qquad (28)$$

with obvious notation for the blocks of $\mathcal{K}$ and where for all $k \in \{1, \ldots, d\}$, $\mathcal{B}^k$ is an $m_\sigma \times m_u$ matrix field and $\mathcal{C}^k$ is a symmetric $m_u \times m_u$ matrix field. Define the operators $B = \sum_{k=1}^d \mathcal{B}^k \partial_k$, $B^\dagger = \sum_{k=1}^d [\mathcal{B}^k]^t \partial_k$, and $C = \sum_{k=1}^d \mathcal{C}^k \partial_k$. The two key hypotheses on which the present work is based are the following:

$$\exists k_0 > 0, \ \forall \xi \in \mathbb{R}^{m_\sigma}, \ \xi^t \mathcal{K}^{\sigma\sigma} \xi \ge k_0 \|\xi\|_{\mathbb{R}^{m_\sigma}}^2 \quad \text{a.e. on } \Omega, \qquad \text{(A5)}$$

$$\forall k \in \{1, \ldots, d\}, \ \text{the } m_\sigma \times m_\sigma \ \text{upper-left block of } \mathcal{A}^k \ \text{is zero.} \qquad \text{(A6)}$$

Set $L_\sigma = [L^2(\Omega)]^{m_\sigma}$ and $L_u = [L^2(\Omega)]^{m_u}$. Consider the PDE system $Tz = f$ with $f \in L = L_\sigma \times L_u$ and partition z and f into (z^σ, z^u) and (f^σ, f^u), respectively. Assumption (A5) (which implies that the matrix $\mathcal{K}^{\sigma\sigma}$ is invertible) together with assumption (A6) allow for the elimination of z^σ from the PDE system, yielding $z^\sigma = [\mathcal{K}^{\sigma\sigma}]^{-1}(f^\sigma - \mathcal{K}^{\sigma u}z^u - Bz^u)$, and it comes that z^u solves the following second-order PDE:

$$-B^\dagger[\mathcal{K}^{\sigma\sigma}]^{-1}Bz^u + (C - B^\dagger[\mathcal{K}^{\sigma\sigma}]^{-1}\mathcal{K}^{\sigma u} - \mathcal{K}^{u\sigma}[\mathcal{K}^{\sigma\sigma}]^{-1}B)z^u$$
$$+ (\mathcal{K}^{uu} - \mathcal{K}^{u\sigma}[\mathcal{K}^{\sigma\sigma}]^{-1}\mathcal{K}^{\sigma u})z^u = f^u - (\mathcal{K}^{u\sigma} + B^\dagger)[\mathcal{K}^{\sigma\sigma}]^{-1}f^\sigma. \quad (29)$$

The leading order term in this PDE has a very particular structure since the matrices $(\mathcal{B}^k)^t[\mathcal{K}^{\sigma\sigma}]^{-1}\mathcal{B}^k$ are positive semi-definite. Hence, the PDE's covered hereafter are elliptic-like.

4.2 The design of the DG bilinear form

Let p and p_σ be two non-negative integers such that $p - 1 \leq p_\sigma \leq p$. Define the vector spaces

$$U_h = [\mathbb{P}_{h,p}]^{m_u}, \qquad \Sigma_h = [\mathbb{P}_{h,p_\sigma}]^{m_\sigma}, \qquad W_h = U_h \times \Sigma_h. \quad (30)$$

Consider the DG bilinear form defined in (17) and the discrete problem (19). Partition the discrete unknown into $z_h = (z_h^\sigma, z_h^u)$. We now want to design a DG method in which z_h^σ can be eliminated by solving local problems. It is then readily seen from (26) that this is possible only if the σ-component of the flux $\phi_{\partial K}(z_h)$ solely depends on z_h^u. Owing to (25), it is inferred that the boundary operators M_F and the interface operators S_F must be such that

$$M_F^{\sigma\sigma} = 0 \quad \text{and} \quad S_F^{\sigma\sigma} = 0. \quad (31)$$

Let $U(h) = [H^1(\Omega)]^{m_u} + U_h$. We define the mapping $\theta_h^1 : U(h) \longrightarrow \Sigma_h$ such that for all $z^u \in U(h)$ and for all $K \in \mathcal{T}_h$, $\theta_h^1(z^u)|_K$ solves the following problem: For all $q^\sigma \in [\mathbb{P}_{p_\sigma}(K)]^{m_\sigma}$,

$$(\mathcal{K}^{\sigma\sigma}\theta_h^1(z^u), q^\sigma)_{L_\sigma,K} = -(\mathcal{K}^{\sigma u}z^u + Bz^u, q^\sigma)_{L_\sigma,K}$$
$$- (\phi_{\partial K}^\sigma(z^u) - \mathcal{D}_{\partial K}^{\sigma u}(z^u)^i, q^\sigma)_{L_\sigma,\partial K}. \quad (32)$$

Owing to (A5), this problem is well-posed. Similarly, we define the mapping $\theta_h^2 : L_\sigma \longrightarrow \Sigma_h$ such that for all $f^\sigma \in L_\sigma$ and for all $K \in \mathcal{T}_h$, $\theta_h^2(f^\sigma)|_K$ solves the following local problem: For all $q^\sigma \in [\mathbb{P}_{p_\sigma}(K)]^{m_\sigma}$,

$$(\mathcal{K}^{\sigma\sigma}\theta_h^2(f^\sigma), q^\sigma)_{L_\sigma,K} = (f^\sigma, q^\sigma)_{L_\sigma,K}. \quad (33)$$

Finally, define the bilinear form ϕ_h on $U(h) \times U(h)$ by

$$\phi_h(z^u, y^u) = a_h((\theta_h^1(z^u), z^u), (0, y^u)), \quad (34)$$

and the linear form ψ_h on $U(h)$ by $\psi_h(y^u) = a_h((\theta_h^2(f^\sigma), 0), (0, y^u))$. This readily leads to the following

Proposition 2. If the pair (z_h^σ, z_h^u) solves (19), then,

$$z_h^\sigma = \theta_h^1(z_h^u) + \theta_h^2(f^\sigma), \tag{35}$$

and z_h^u solves the following problem:

$$\begin{cases} \text{Seek } z_h^u \in U_h \text{ such that} \\ \phi_h(z_h^u, y_h^u) = (f^u, y_h^u)_{L_u} - \psi_h(y_h^u), \quad \forall y_h^u \in U_h. \end{cases} \tag{36}$$

Conversely, if z_h^u solves (36) and if z_h^σ is defined by (35), then the pair (z_h^σ, z_h^u) solves (19).

For the convergence analysis of mS4.3 to hold, the boundary operators M_F and the interface operators S_F must comply with certain design criteria that are formulated in [12]. This set of conditions simplifies into the following whenever Dirichlet-type boundary conditions are enforced on z^u: For all $F \in \mathcal{F}_h^\partial$ and for all $y = (y^\sigma, y^u) \in [L^2(F)]^m$, we assume that

$$(\mathcal{M}y - \mathcal{D}y = 0) \implies (M_F(y) - \mathcal{D}y = 0), \tag{LDG1}$$

$$(\mathcal{M}^t y + \mathcal{D}y = 0) \implies (M_F^*(y) + \mathcal{D}y = 0), \tag{LDG2}$$

$$M_F^{\sigma\sigma} = 0, \quad M_F^{\sigma u}(y^u) = -\mathcal{D}^{\sigma u} y^u, \quad M_F^{u\sigma}(y^\sigma) = \mathcal{D}^{u\sigma} y^\sigma, \tag{LDG3}$$

$$M_F^{uu} \text{ is self-adjoint}, \tag{LDG4}$$

$$h_F^{-1}(\mathcal{D}^{u\sigma}\mathcal{D}^{\sigma u})^{\frac{1}{2}} + h_F|\mathcal{D}^{uu}| \lesssim M_F^{uu} \lesssim h_F^{-1}\mathcal{I}_{m_u}, \tag{LDG5}$$

where M_F^* denotes the adjoint operator of M_F and $\mathcal{I}_{m_u}$ the identity matrix in $\mathbb{R}^{m_u,m_u}$. Similarly, for all $F \in \mathcal{F}_h^i$, we assume that

$$S_F^{\sigma\sigma} = 0, \quad S_F^{\sigma u} = 0, \quad S_F^{u\sigma} = 0, \tag{LDG6}$$

$$S_F^{uu} \text{ is self-adjoint}, \tag{LDG7}$$

$$h_F^{-1}(\mathcal{D}^{u\sigma}\mathcal{D}^{\sigma u})^{\frac{1}{2}} + h_F|\mathcal{D}^{uu}| \lesssim S_F^{uu} \lesssim h_F^{-1}\mathcal{I}_{m_u}. \tag{LDG8}$$

Remark 3. Assumption (LDG1) is a consistency assumption similar to (DG2). Assumption (LDG2) is an adjoint-consistency assumption needed to obtain an improved error estimate for z_h^u in the L_u-norm. Assumption (LDG3) is suitable to enforce Dirichlet boundary conditions and must be modified if other boundary conditions are considered. In this case, assumption (LDG5) must also be modified: M_F^{uu} no longer scales as h_F^{-1}, but is of order 1.

4.3 Convergence analysis

The error analysis uses the following discrete norms on $W(h)$,

$$\|y\|_{h,A'}^2 = \|y^\sigma\|_{L_\sigma}^2 + \|y^u\|_{L_u}^2 + |y^u|_J^2 + |y^u|_M^2 + \sum_{K \in \mathcal{T}_h} \|By^u\|_{L_\sigma,K}^2, \tag{37}$$

$$\|y\|_{h,1}^2 = \|y\|_{h,A'}^2 + \sum_{K \in \mathcal{T}_h}[h_K^{-2}\|y^u\|_{L_u,K}^2 + h_K^{-1}\|y^u\|_{L_u,\partial K}^2 + h_K\|y^\sigma\|_{L_\sigma,\partial K}^2], \tag{38}$$

where for all $y^u \in U(h)$, $|y^u|_M^2 = \sum_{F \in \mathcal{F}_h^\partial}(M_F^{uu}(y^u), y^u)_{L_u, F}$ and $|y^u|_J^2 = \sum_{F \in \mathcal{F}_h^i}(S_F^{uu}(\llbracket y^u \rrbracket), \llbracket y^u \rrbracket)_{L_u, F}$. The main result is the following

Theorem 5. *Let z solve (10) and let z_h solve (19). Assume that for all $k \in \{1, \ldots, d\}$, $\mathcal{B}^k \in [\mathcal{C}^{0,1}(\overline{\Omega})]^{m,m}$. Then*

$$\|z - z_h\|_{h, A'} \lesssim \inf_{y_h \in W_h} \|z - y_h\|_{h, 1}, \tag{39}$$

if $z \in [H^1(\Omega)]^m$, and $\lim_{h \to 0}(\|z - z_h\|_L^2 + \sum_{K \in \mathcal{T}_h} \|B(z^u - z_h^u)\|_{L_\sigma, L}^2) = 0$ if $z \in V$ only, assuming there is $\gamma > 0$ s.t. $[H^\gamma(\Omega)]^{m_\sigma} \times [H^{1+\gamma}(\Omega)]^{m_u} \cap V$ is dense in V.

Using standard interpolation results on W_h and since $p - 1 \leq p_\sigma \leq p$, the above result implies

$$\|z - z_h\|_{h, A'} \lesssim h^p(\|z^\sigma\|_{[H^{p_\sigma + 1}(\Omega)]^{m_\sigma}} + \|z^u\|_{[H^{p+1}(\Omega)]^{m_u}}). \tag{40}$$

whenever z is in $[H^{p_\sigma + 1}(\Omega)]^{m_\sigma} \times [H^{p+1}(\Omega)]^{m_u}$. In particular, $\|z - z_h\|_L$ converges to order h^p and if the mesh family $\{\mathcal{T}_h\}_{h>0}$ is quasi-uniform, $(\sum_{K \in \mathcal{T}_h} \|B(z^u - z_h^u)\|_{L_\sigma, K}^2)^{\frac{1}{2}}$ also converges to order h^p. If $p_\sigma = p$, the L-norm error estimate is suboptimal when compared with that obtained using the DG method analyzed in mS3. The reason for this optimality loss is that the interface jumps of the σ-component are no longer controlled to allow for this component to be locally eliminated, and the jumps of the u-component are penalized with an $\mathcal{O}(h^{-1})$ weight. If $p_\sigma = p - 1$, the L-norm error estimate is still suboptimal for the u-component, but is optimal for the σ-component.

To derive an optimal error estimate for the u-component in the L_u-norm, we use a duality argument. Let $\psi \in V^*$ solve

$$\widetilde{T}\psi = (0, z^u - z_h^u). \tag{41}$$

Assuming the above problem yields elliptic regularity, i.e., $\|\psi^u\|_{[H^2(\Omega)]^{m_u}} + \|\psi^\sigma\|_{[H^1(\Omega)]^{m_\sigma}} \lesssim \|z^u - z_h^u\|_{L_u}$, the main result is the following

Theorem 6. *The following holds:*

$$\|z^u - z_h^u\|_{L_u} \lesssim h \inf_{y_h \in W_h} \|z - y_h\|_{h, 1+}, \tag{42}$$

where $\|y\|_{h,1+}^2 = \|y\|_{h,1}^2 + \sum_{K \in \mathcal{T}_h}[h_K^2 \|y^\sigma\|_{[H^1(K)]^{m_\sigma}}^2 + h_K \|y^\sigma\|_{L_\sigma, \partial K}^2]$. In particular, if $z \in [H^{p_\sigma + 1}(\Omega)]^{m_\sigma} \times [H^{p+1}(\Omega)]^{m_u}$, then

$$\|z^u - z_h^u\|_{L_u} \lesssim h^{p+1}(\|z^\sigma\|_{[H^{p_\sigma + 1}(\Omega)]^{m_\sigma}} + \|z^u\|_{[H^{p+1}(\Omega)]^{m_u}}). \tag{43}$$

5 Examples

In this section we apply the methods formulated in mS3 and mS4 to various Friedrichs' systems encountered in engineering applications. To alleviate notation, an index h indicates that the norm is broken on the mesh elements and $\mathfrak{h}$ denotes the piecewise constant function equal to h_K on each $K \in \mathcal{T}_h$.

5.1 Advection–reaction

Let $\mu \in L^\infty(\Omega)$, let $\beta \in [L^\infty(\Omega)]^d$ with $\nabla \cdot \beta \in L^\infty(\Omega)$, and assume that there is $\mu_0 > 0$ such that $\mu(x) - \frac{1}{2}\nabla \cdot \beta(x) \geq \mu_0$ a.e. in Ω. Let $f \in L^2(\Omega)$. The PDE

$$\mu u + \beta \cdot \nabla u = f \qquad (44)$$

is recast as a Friedrichs' system by setting $m = 1$, $\mathcal{K} = \mu$, and $\mathcal{A}^k = \beta^k$ for $k \in \{1, \ldots, d\}$. The graph space is $W = \{w \in L^2(\Omega);\ \beta \cdot \nabla w \in L^2(\Omega)\}$. To enforce boundary conditions, define $\partial\Omega^\pm = \{x \in \partial\Omega;\ \pm\beta(x) \cdot n(x) > 0\}$, and assume that $\partial\Omega^-$ and $\partial\Omega^+$ are well-separated, i.e., $\mathrm{dist}(\partial\Omega^-, \partial\Omega^+) > 0$. Then, the boundary operator D has the following representation: For all $v, w \in W$,

$$\langle Dv, w \rangle_{W', W} = \int_{\partial\Omega} vw(\beta \cdot n). \qquad (45)$$

Letting $\langle Mv, w \rangle_{W', W} = \int_{\partial\Omega} vw|\beta \cdot n|$, then (M1)–(M2) hold and $V = \{v \in W;\ v|_{\partial\Omega^-} = 0\}$, i.e., homogeneous Dirichlet boundary conditions are enforced at the inflow boundary.

Let $\alpha > 0$ (α can vary from face to face) and for all $F \in \mathcal{F}_h$, set

$$\mathcal{M}_F = |\beta \cdot n| \quad \text{and} \quad \mathcal{S}_F = \alpha|\beta \cdot n_F|, \qquad (46)$$

where n_F is a unit normal vector to F (the orientation is irrelevant). Then, letting $M_F(v) = \mathcal{M}_F v$ and $S_F(v) = \mathcal{S}_F v$, assumptions (DG1)–(DG7) hold. Hence, if $\beta \in [\mathcal{C}^{0,\frac{1}{2}}(\Omega)]^d$ and the exact solution is smooth enough,

$$\|u - u_h\|_{L^2(\Omega)} + \|\mathfrak{h}^{\frac{1}{2}} \beta \cdot \nabla(u - u_h)\|_{h, L^2(\Omega)} \lesssim h^{p+\frac{1}{2}} \|u\|_{H^{p+1}(\Omega)}. \qquad (47)$$

Remark 4. The specific value $\alpha = \frac{1}{2}$ leads to the so-called upwind scheme. This coincidence has led many authors to believe that DG methods are methods of choice to solve hyperbolic problems. Actually DG methods, as presented herein, are merely stabilization techniques tailored to solve symmetric positive systems of first-order PDE's.

5.2 Advection–diffusion–reaction

Let μ, β, and f be as in mS5.1. Then, the PDE $-\Delta u + \beta \cdot \nabla u + \mu u = f$ written in the mixed form

$$\begin{cases} \sigma + \nabla u = 0, \\ \mu u + \nabla \cdot \sigma + \beta \cdot \nabla u = f, \end{cases} \qquad (48)$$

falls into the category of Friedrichs' systems by setting $m = d + 1$ and

$$\mathcal{K} = \begin{bmatrix} \mathcal{I}_d & 0 \\ \hline 0 & \mu \end{bmatrix}, \quad \mathcal{A}^k = \begin{bmatrix} 0 & e^k \\ \hline (e^k)^t & \beta^k \end{bmatrix}, \qquad (49)$$

where $\mathcal{I}_d$ is the identity matrix in $\mathbb{R}^{d,d}$ and e^k is the k-th vector in the canonical basis of $\mathbb{R}^d$. The graph space is $W = H(\text{div}; \Omega) \times H^1(\Omega)$.

The boundary operator D is such that for all $(\sigma, u), (\tau, v) \in W$,

$$\langle D(\sigma, u), (\tau, v) \rangle_{W', W} = \langle \sigma \cdot n, v \rangle_{-\frac{1}{2}, \frac{1}{2}} + \langle \tau \cdot n, u \rangle_{-\frac{1}{2}, \frac{1}{2}} + \int_{\partial \Omega} (\beta \cdot n) uv, \quad (50)$$

where $\langle, \rangle_{-\frac{1}{2}, \frac{1}{2}}$ denotes the duality pairing between $H^{-\frac{1}{2}}(\partial \Omega)$ and $H^{\frac{1}{2}}(\partial \Omega)$. Dirichlet boundary conditions are enforced by setting $\langle M(\sigma, u), (\tau, v) \rangle_{W', W} = \langle \sigma \cdot n, v \rangle_{-\frac{1}{2}, \frac{1}{2}} - \langle \tau \cdot n, u \rangle_{-\frac{1}{2}, \frac{1}{2}}$, yielding $V = H(\text{div}; \Omega) \times H_0^1(\Omega)$. Neumann and Robin boundary conditions can be treated similarly.

Let $\alpha_1 > 0$, $\alpha_2 > 0$, and $\eta > 0$ (these design parameters can vary from face to face) and for all $F \in \mathcal{F}_h$, set

$$\mathcal{M}_F = \begin{bmatrix} 0 & -n \\ \hdashline n^t & \eta \end{bmatrix}, \quad \mathcal{S}_F = \begin{bmatrix} \alpha_1 n_F \otimes n_F & 0 \\ \hdashline 0 & \alpha_2 \end{bmatrix}. \quad (51)$$

Then, letting $M_F(\sigma, u) = \mathcal{M}_F(\sigma, u)$ and $S_F(\sigma, u) = \mathcal{S}_F(\sigma, u)$, assumptions (DG1)–(DG7) hold. Hence, if $\beta \in [C^{0, \frac{1}{2}}(\Omega)]^d$ and the exact solution is smooth enough,

$$\|u - u_h\|_{L^2(\Omega)} + \|\sigma - \sigma_h\|_{[L^2(\Omega)]^d} + \|\mathfrak{h}^{\frac{1}{2}} \nabla(u - u_h)\|_{h, [L^2(\Omega)]^d}$$
$$+ \|\mathfrak{h}^{\frac{1}{2}} \nabla \cdot (\sigma - \sigma_h)\|_{h, L^2(\Omega)} \lesssim h^{p+\frac{1}{2}} \|(\sigma, u)\|_{[H^{p+1}(\Omega)]^{d+1}}. \quad (52)$$

The above Friedrichs' system can be equipped with the 2×2 block structure analyzed in mS4 by setting $z^\sigma := \sigma$ and $z^u := u$. Take

$$\mathcal{M}_F = \begin{bmatrix} 0 & -n \\ \hdashline n^t & \eta h_F^{-1} \end{bmatrix}, \quad \mathcal{S}_F = \begin{bmatrix} 0 & 0 \\ \hdashline 0 & \alpha_2 h_F^{-1} \end{bmatrix}. \quad (53)$$

Then, letting $M_F(\sigma, u) = \mathcal{M}_F(\sigma, u)$ and $S_F(\sigma, u) = \mathcal{S}_F(\sigma, u)$, assumptions (LDG1)–(LDG8) hold. If the exact solution is smooth enough,

$$\|u - u_h\|_{L^2(\Omega)} + h\|\sigma - \sigma_h\|_{[L^2(\Omega)]^d} + h\|\nabla(u - u_h)\|_{h, [L^2(\Omega)]^d}$$
$$\lesssim h^{p+1} \|(\sigma, u)\|_{[H^p(\Omega)]^d \times H^{p+1}(\Omega)}. \quad (54)$$

Remark 5. Other choices for the operators M_F and S_F are possible. In particular, one can show that the DG method of Brezzi et al. [7], that of Bassi and Rebay [6], the IP method of Baker [5] and Arnold [1] (provided (LDG5) and (LDG8) are slightly weakened), and the LDG method of Cockburn and Shu [9] fit into the present framework.

5.3 Linear elasticity

Let ς and γ be two positive functions in $L^\infty(\Omega)$ uniformly bounded away from zero. Let $f \in [L^2(\Omega)]^d$. Let u be the $\mathbb{R}^d$-valued displacement field and let σ

be the $\mathbb{R}^{d,d}$-valued stress tensor. The PDE's $\sigma = \frac{1}{2}(\nabla u + (\nabla u)^t) + \frac{1}{\gamma}(\nabla \cdot u)\mathcal{I}_d$ and $-\nabla \cdot \sigma + \varsigma u = f$ can be written in the following mixed stress–pressure–displacement form

$$\begin{cases} \sigma + \pi \mathcal{I}_d - \frac{1}{2}(\nabla u + (\nabla u)^t) = 0, \\ \operatorname{tr}(\sigma) + (d+\gamma)\pi = 0, \\ -\frac{1}{2}\nabla \cdot (\sigma + \sigma^t) + \varsigma u = f. \end{cases} \quad (55)$$

The tensor σ in $\mathbb{R}^{d,d}$ can be identified with the vector $\bar{\sigma} \in \mathbb{R}^{d^2}$ by setting $\bar{\sigma}_{[ij]} = \sigma_{ij}$ with $1 \le i,j \le d$ and $[ij] = d(j-1)+i$. Then, (55) falls into the category of Friedrichs' systems by setting $m = d^2 + 1 + d$ and

$$\mathcal{K} = \begin{bmatrix} \mathcal{I}_{d^2} & \mathcal{Z} & 0 \\ (\mathcal{Z})^t & (d+\gamma) & 0 \\ 0 & 0 & \varsigma \mathcal{I}_d \end{bmatrix}, \quad \mathcal{A}^k = \begin{bmatrix} 0 & 0 & \mathcal{E}^k \\ 0 & 0 & 0 \\ (\mathcal{E}^k)^t & 0 & 0 \end{bmatrix}, \quad (56)$$

where $\mathcal{Z} \in \mathbb{R}^{d^2}$ has components given by $\mathcal{Z}_{[ij]} = \delta_{ij}$, and for all $k \in \{1,\ldots,d\}$, $\mathcal{E}^k \in \mathbb{R}^{d^2,d}$ has components given by $\mathcal{E}^k_{[ij],l} = -\frac{1}{2}(\delta_{ik}\delta_{jl} + \delta_{il}\delta_{jk})$; here, $i,j,l \in \{1,\ldots,d\}$ and the δ's denote Kronecker symbols. The graph space is $W = H_{\bar{\sigma}} \times L^2(\Omega) \times [H^1(\Omega)]^d$ with $H_{\bar{\sigma}} = \{\bar{\sigma} \in [L^2(\Omega)]^{d^2}; \nabla \cdot (\sigma + \sigma^t) \in [L^2(\Omega)]^d\}$. The boundary operator D is s.t. for all $(z := (\bar{\sigma}, \pi, u), y := (\bar{\tau}, \rho, v)) \in W$,

$$\langle Dz, y \rangle_{W',W} = -\langle \tfrac{1}{2}(\tau + \tau^t)\cdot n, u \rangle_{-\frac{1}{2},\frac{1}{2}} - \langle \tfrac{1}{2}(\sigma + \sigma^t)\cdot n, v \rangle_{-\frac{1}{2},\frac{1}{2}}. \quad (57)$$

Letting $\langle Mz, y \rangle_{W',W} = \langle \tfrac{1}{2}(\tau + \tau^t)\cdot n, u \rangle_{-\frac{1}{2},\frac{1}{2}} - \langle \tfrac{1}{2}(\sigma + \sigma^t)\cdot n, v \rangle_{-\frac{1}{2},\frac{1}{2}}$, then (M1)–(M2) hold and $V = H_{\bar{\sigma}} \times L^2(\Omega) \times [H_0^1(\Omega)]^d$, i.e., homogeneous Dirichlet boundary conditions are enforced on the displacement.

Let $\alpha_1 > 0$, $\alpha_2 > 0$, and $\eta > 0$ (these design parameters can vary from face to face) and for all $F \in \mathcal{F}_h$, set

$$\mathcal{M}_F = \begin{bmatrix} 0 & 0 & -\mathcal{H} \\ 0 & 0 & 0 \\ \mathcal{H}^t & 0 & \eta \mathcal{I}_d \end{bmatrix}, \quad \mathcal{S}_F = \begin{bmatrix} \alpha_1 \mathcal{H}_F \cdot \mathcal{H}_F^t & 0 & 0 \\ 0 & 0 & 0 \\ 0 & 0 & \alpha_2 \mathcal{I}_d \end{bmatrix}, \quad (58)$$

where $\mathcal{H} = \sum_{k=1}^d n_k \mathcal{E}^k \in \mathbb{R}^{d^2,d}$ and $\mathcal{H}_F$ is defined similarly by substituting n_F to n. Then, letting $M_F(\bar{\sigma}, \pi, u) = \mathcal{M}_F(\bar{\sigma}, \pi, u)$ and $S_F(\bar{\sigma}, \pi, u) = \mathcal{S}_F(\bar{\sigma}, \pi, u)$, assumptions (DG1)–(DG7) hold. Hence, if the exact solution is smooth enough,

$$\|u - u_h\|_{[L^2(\Omega)]^d} + \|\pi - \pi_h\|_{L^2(\Omega)} + \|\sigma - \sigma_h\|_{[L^2(\Omega)]^{d,d}}$$
$$+ \|\mathfrak{h}^{\frac{1}{2}}\nabla(u - u_h)\|_{h,[L^2(\Omega)]^{d,d}} + \|\mathfrak{h}^{\frac{1}{2}}\nabla \cdot ((\sigma + \sigma^t) - (\sigma_h + \sigma_h^t))\|_{h,[L^2(\Omega)]^d}$$
$$\lesssim h^{p+\frac{1}{2}}\|(\bar{\sigma}, \pi, u)\|_{[H^{p+1}(\Omega)]^{d^2+1+d}}. \quad (59)$$

The above Friedrichs' system can be equipped with the 2×2 block structure analyzed in mS4 by setting $z^\sigma := (\bar{\sigma}, \pi)$ and $z^u := u$. Take

$$\mathcal{M}_F = \begin{bmatrix} 0 & 0 & -\mathcal{H} \\ 0 & 0 & 0 \\ \mathcal{H}^t & 0 & \eta h_F^{-1} \mathcal{I}_d \end{bmatrix}, \quad \mathcal{S}_F = \begin{bmatrix} 0 & 0 & 0 \\ 0 & 0 & 0 \\ 0 & 0 & \alpha_2 h_F^{-1} \mathcal{I}_d \end{bmatrix}. \tag{60}$$

Then, letting $M_F(\overline{\sigma}, \pi, u) = \mathcal{M}_F(\overline{\sigma}, \pi, u)$ and $S_F(\overline{\sigma}, \pi, u) = \mathcal{S}_F(\overline{\sigma}, \pi, u)$, assumptions (LDG1)–(LDG8) hold. Hence, if the exact solution is smooth enough,

$$\|u - u_h\|_{[L^2(\Omega)]^d} + h\|\pi - \pi_h\|_{L^2(\Omega)} + h\|\sigma - \sigma_h\|_{[L^2(\Omega)]^{d,d}}$$
$$+ h\|\nabla(u - u_h)\|_{h,[L^2(\Omega)]^{d,d}} \lesssim h^{p+1} \|(\overline{\sigma}, \pi, u)\|_{[H^p(\Omega)]^{d^2+1} \times [H^{p+1}(\Omega)]^d}. \tag{61}$$

5.4 Maxwell's equations in the elliptic regime

Let σ and μ be two positive functions in $L^\infty(\Omega)$ uniformly bounded away from zero. A simplified form of Maxwell's equations in $\mathbb{R}^3$ in the elliptic regime, i.e., when displacement currents are negligible, consists of the PDE's

$$\mu H + \nabla \times E = f, \qquad \sigma E - \nabla \times H = g, \tag{62}$$

with data $f, g \in [L^2(\Omega)]^3$. The above PDE's fall into the category of Friedrichs' systems by setting $m = 6$ and

$$\mathcal{K} = \begin{bmatrix} \mu \mathcal{I}_3 & 0 \\ 0 & \sigma \mathcal{I}_3 \end{bmatrix}, \quad \mathcal{A}^k = \begin{bmatrix} 0 & \mathcal{R}^k \\ (\mathcal{R}^k)^t & 0 \end{bmatrix}, \tag{63}$$

with $\mathcal{R}_{ij}^k = \epsilon_{ikj}$ for $i, j, k \in \{1, 2, 3\}$, ϵ_{ikj} being the Levi–Civita permutation tensor. The graph space is $W = H(\text{curl}; \Omega) \times H(\text{curl}; \Omega)$. The boundary operator D is such that for all $(H, E), (h, e) \in W$,

$$\langle D(H, E), (h, e) \rangle_{W', W} = (\nabla \times E, h)_{[L^2(\Omega)]^3} - (E, \nabla \times h)_{[L^2(\Omega)]^3} \\ + (H, \nabla \times e)_{[L^2(\Omega)]^3} - (\nabla \times H, e)_{[L^2(\Omega)]^3}. \tag{64}$$

Letting $\langle M(H, E), (h, e) \rangle_{W', W} = -(\nabla \times E, h)_{[L^2(\Omega)]^3} + (E, \nabla \times h)_{[L^2(\Omega)]^3} + (H, \nabla \times e)_{[L^2(\Omega)]^3} - (\nabla \times H, e)_{[L^2(\Omega)]^3}$, then (M1)–(M2) hold and $V = \{(H, E) \in W; (E \times n)|_{\partial \Omega} = 0\}$, i.e., homogeneous Dirichlet boundary conditions are enforced on the tangential component of the electric field.

Let $\alpha_1 > 0$, $\alpha_2 > 0$, and $\eta > 0$ (these design parameters can vary from face to face) and for all $F \in \mathcal{F}_h$, set

$$\mathcal{M}_F = \begin{bmatrix} 0 & -\mathcal{N} \\ \mathcal{N}^t & \eta \mathcal{N}^t \mathcal{N} \end{bmatrix}, \quad \mathcal{S}_F = \begin{bmatrix} \alpha_1 \mathcal{N}_F^t \mathcal{N}_F & 0 \\ 0 & \alpha_2 \mathcal{N}_F^t \mathcal{N}_F \end{bmatrix}, \tag{65}$$

where $\mathcal{N} = \sum_{k=1}^d n_k \mathcal{R}^k \in \mathbb{R}^{3,3}$ and $\mathcal{N}_F$ is defined similarly by substituting n_F to n. Then, letting $M_F(H, E) = \mathcal{M}_F(H, E)$ and $S_F(H, E) = \mathcal{S}_F(H, E)$, assumptions (DG1)–(DG7) hold. Hence, if the exact solution is smooth enough,

$$\|E - E_h\|_{[L^2(\Omega)]^3} + \|H - H_h\|_{[L^2(\Omega)]^3} + \|\mathfrak{h}^{\frac{1}{2}}\nabla\times(E - E_h)\|_{h,[L^2(\Omega)]^3}$$
$$+ \|\mathfrak{h}^{\frac{1}{2}}\nabla\times(H - H_h)\|_{h,[L^2(\Omega)]^3} \lesssim h^{p+\frac{1}{2}}\|(H,E)\|_{[H^{p+1}(\Omega)]^6}. \quad (66)$$

The above Friedrichs' system can also be equipped with the 2×2 block structure analyzed in mS4 by setting $z^\sigma := H$ and $z^u := E$. Take

$$\mathcal{M}_F = \begin{bmatrix} 0 & -\mathcal{N} \\ \hline \mathcal{N}^t & \eta h_F^{-1}\mathcal{N}^t\mathcal{N} \end{bmatrix}, \quad \mathcal{S}_F = \begin{bmatrix} 0 & 0 \\ \hline 0 & \alpha_2 h_F^{-1}\mathcal{N}_F^t\mathcal{N}_F \end{bmatrix}. \quad (67)$$

Then, letting $M_F(H,E) = \mathcal{M}_F(H,E)$ and $S_F(H,E) = \mathcal{S}_F(H,E)$, assumptions (LDG1)–(LDG8) hold. Hence, if the exact solution is smooth enough,

$$\|E - E_h\|_{[L^2(\Omega)]^3} + h\|H - H_h\|_{[L^2(\Omega)]^3} + h\|\nabla\times(E - E_h)\|_{h,[L^2(\Omega)]^3}$$
$$\lesssim h^{p+1}\|(H,E)\|_{[H^p(\Omega)]^3 \times [H^{p+1}(\Omega)]^3}. \quad (68)$$

6 Concluding remarks

In this paper we have presented a unified analysis of DG methods by making systematic use of Friedrichs' systems. As already pointed out by Friedrichs, such systems go beyond the traditional hyperbolic/elliptic classification of PDE's. Furthermore, DG methods as presented herein appear to be merely stabilization methods where the boundary operators M_F and the interface operators S_F have to be set (tuned) by the user so as to comply with the design criteria (DG1)–(DG7) or (LDG1)–(LDG8).

References

1. Arnold, D.N.: An interior penalty finite element method with discontinuous elements. SIAM J. Numer. Anal., **19**, 742–760 (1982)
2. Arnold, D.N., Brezzi, F., Cockburn, B., Marini, L.D.: Unified analysis of discontinuous Galerkin methods for elliptic problems. SIAM J. Numer. Anal., **39**(5), 1749–1779 (2001/02)
3. Babuška, I.: The finite element method with penalty. Math. Comp., **27**, 221–228 (1973)
4. Babuška, I., Zlámal, M.: Nonconforming elements in the finite element method with penalty. SIAM J. Numer. Anal., **10**(5), 863–875 (1973)
5. Baker, G.A.: Finite element methods for elliptic equations using nonconforming elements. Math. Comp., **31**(137), 45–59 (1977)
6. Bassi, F., Rebay, S.: A high-order accurate discontinuous finite element method for the numerical solution of the compressible Navier-Stokes equations. J. Comput. Phys., **131**(2), 267–279 (1997)
7. Brezzi, F., Manzini, M., Marini, L.D., Pietra, P., Russo, A.: Discontinous Finite Elements for Diffusion Problems. In Francesco Brioschi (1824-1897). Convegno di studi matematici, volume **16** of Incontri di Studio, 197–217. Istituto Lombardo Accademia di Scienze e Lettere (1999)

8. Cockburn, B., Karniadakis, G.E., Shu, C. W.: Discontinuous Galerkin Methods – Theory, Computation and Applications, volume **11** of Lecture Notes in Computer Science and Engineering. Springer (2000)
9. Cockburn, B., Shu, C. W.: The local discontinuous Galerkin method for time-dependent convection-diffusion systems. SIAM J. Numer. Anal., **35**, 2440–2463 (1998)
10. Ern, E., Guermond, J.-L.: Theory and Practice of Finite Elements, volume **159** of Applied Mathematical Sciences. Springer-Verlag, New York, NY (2004)
11. Ern, E., Guermond, J.-L.: Discontinuous Galerkin methods for Friedrichs' systems. I. General theory. SIAM J. Numer. Anal. (2005) In press
12. Ern, E., Guermond, J.-L.: Discontinuous Galerkin methods for Friedrichs' systems. II. Second-order PDEs. SIAM J. Numer. Anal. (2005) submitted
13. Ern, E., Guermond, J.-L., Caplain, G.: An intrinsic criterion for the bijectivity of Hilbert operators related to Friedrichs' systems. Comm. Partial. Differ. Equ. (2005) In press
14. Friedrichs, K.O.: Symmetric positive linear differential equations. Comm. Pure Appl. Math., **11**, 333–418 (1958)
15. Johnson, C., Nävert, U., Pitkäranta, J.: Finite element methods for linear hyperbolic equations. Comput. Methods Appl. Mech. Engrg., **45**, 285–312 (1984)
16. Lesaint, P.: Sur la résolution des systèmes hyperboliques du premier ordre par des méthodes d'éléments finis. PhD thesis, University of Paris VI (1975)
17. Lesaint, P., Raviart, P.-A.: On a finite element method for solving the neutron transport equation. In Mathematical Aspects of Finite Elements in Partial Differential Equations, pages 89–123. Publication No. 33. Math. Res. Center, Univ. of Wisconsin-Madison, Academic Press, New York (1974)
18. Nitsche, J.: Über ein Variationsprinzip zur Lösung von Dirichlet-Problemen bei Verwendung von Teilräumen, die keinen Randbedingungen unterworfen sind. Abh. Math. Sem. Univ. Hamburg, **36**, 9–15 (1971)
19. Reed, W.H., Hill, T.R.: Triangular mesh methods for the neutron transport equation. Technical Report LA-UR-73-479, Los Alamos Scientific Laboratory, Los Alamos, NM (1973)
20. Wheeler, M.F.: An elliptic collocation-finite element method with interior penalties. SIAM J. Numer. Anal., **15**, 152–161 (1978)

Highly Oscillatory Quadrature: The Story so Far

A. Iserles[1], S.P. Nørsett[2] and S. Olver[3]

[1] Department of Applied Mathematics and Theoretical Physics, Centre for Mathematical Sciences, University of Cambridge, Wilberforce Road, Cambridge CB3 0WA, United Kingdom
`A.Iserles@damtp.cam.ac.uk`
[2] Department of Mathematical Sciences, Norwegian University of Science and Technology, 7491 Trondheim, Norway
`S.P.Norsett@math.ntnu.no`
[3] Department of Applied Mathematics and Theoretical Physics, Centre for Mathematical Sciences, University of Cambridge, Wilberforce Road, Cambridge CB3 0WA, United Kingdom
`S.Olver@damtp.cam.ac.uk`

Summary. The last few years have witnessed substantive developments in the computation of highly oscillatory integrals in one or more dimensions. The availability of new asymptotic expansions and a Stokes-type theorem allow for a comprehensive analysis of a number of old (although enhanced) and new quadrature techniques: the asymptotic, Filon-type and Levin-type methods. All these methods share the surprising property that their accuracy increases with growing oscillation. These developments are described in a unified fashion, taking the multivariate integral $\int_\Omega f(\boldsymbol{x}) \mathrm{e}^{\mathrm{i}\omega g(\boldsymbol{x})} \mathrm{d}V$ as our point of departure.

1 The challenge of high oscillation

Rapid oscillation is ubiquitous in applications and is, by common consent, considered a 'difficult' problem. Indeed, the standard technique of dealing with high oscillation is to make it disappear by sampling the signal sufficiently frequently, and this typically leads to prohibitive cost.

The subject of this article is a review of recent work on the computation of integrals of the form

$$I[f,\Omega] = \int_\Omega f(\boldsymbol{x}) \mathrm{e}^{\mathrm{i}\omega g(\boldsymbol{x})} \mathrm{d}V, \tag{1}$$

where $\Omega \subset \mathbb{R}^n$ is a bounded open domain with piecewise-smooth boundary, while f and the *oscillator* g are smooth. We assume in (1) that $\omega \in \mathbb{R}$ is large in modulus, hence $I[f,\Omega]$ oscillates rapidly as a function of ω.

A natural technique to compute (1) in a univariate setting is *Gaussian quadrature*. Yet, a moment's reflection clarifies that it is likely to be absolutely useless unless $|\omega|$ is small. Classical quadrature (with a trivial weight function) is just an exact integration of a polynomial interpolation of the integrand. However, if the integrand oscillates rapidly, and unless we use an astronomical number of function evaluations, polynomial interpolation is useless! This is vividly demonstrated in Fig. 1. We have computed

$$\int_{-1}^{1} \cos x e^{i\omega x^2} \, dx = -\frac{\pi^{\frac{1}{2}}}{2(-i\omega)^{\frac{1}{2}}} \exp\left(\frac{1}{4}\frac{1}{i\omega}\right) \left[\operatorname{erf}\left(i\frac{\omega+\frac{1}{2}}{(-i\omega)^{\frac{1}{2}}}\right) + \operatorname{erf}\left(i\frac{\omega-\frac{1}{2}}{(-i\omega)^{\frac{1}{2}}}\right)\right]$$

by Gaussian quadrature with different number of points. The figure displays the absolute value of the error as a function of $\omega \in [0, 100]$. Note that, as long as ω is small, everything is fine, but as soon as ω is large in comparison with the number of quadrature points and high oscillation sets in, the error becomes $\mathcal{O}(1)$. As a matter of fact, given that $I[f] \sim \mathcal{O}\left(\omega^{-\frac{1}{2}}\right)$, the trivial approximation $I[f] \approx 0$ is far superior to Gaussian quadrature with 30 points!

Yet, efficient and cheap quadrature of (1) is perfectly possible. Indeed, once we understand the mathematical mechanism underlying (1), we can compute it to high precision with minimal effort and, perhaps paradoxically, *the quality of approximation increases with* ω.

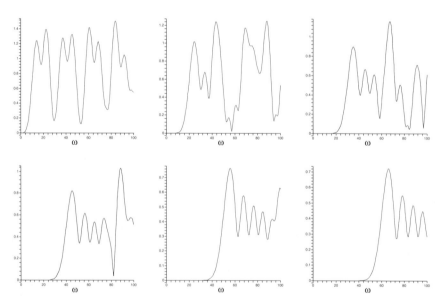

Fig. 1. Error in Gaussian quadrature with $\Omega = (-1, 1)$, $f(x) = \cos x$, $g(x)$ and ν points. Here ν increases by increments of five, from 5 to 30.

This article collates a sequence of papers by the authors into unified narrative. In particular, we revisit here the work of [8, 10, 16] and [17], to which the reader is referred for technical details, more comprehensive exposition and a wealth of further numerical examples.

The conventional organising principle of quadrature is a Taylor expansion. Once the integrand oscillates rapidly, a Taylor expansion converges very slowly indeed and is, to all intents and purposes, useless. Instead, we need to exploit *an asymptotic expansion in negative powers of* ω. In Section 2 we present an asymptotic expansion of (1) in the case when the oscillator g has no *critical points*: $\nabla g(\boldsymbol{x}) \neq \boldsymbol{0}$ for all $\boldsymbol{x} \in \mathrm{cl}\,\Omega$ and subject to the *nonresonance condition*: $\nabla g(\boldsymbol{x})$ is not allowed to be normal to the boundary $\partial\Omega$ for any $\boldsymbol{x} \in \partial\Omega$.

The availability of an asymptotic expansion allows us to design and analyse effective quadrature methods, and this is the subject of Section 3. We single out for consideration three general techniques: *asymptotic methods*, consisting of a truncation of the asymptotic expansion of Section 2, *Filon-type methods*, which interpolate just $f(\boldsymbol{x})$, rather than the entire integral [3], and *Levin-type methods*, which collocate the integrand [11].

In Section 4 we consider the case when critical points are allowed. A comprehensive theory exists, as things stand, only in one dimension, hence we focus on $g : [a, b] \to \mathbb{R}$ and study the case of $g'(\xi) = 0$ for some $\xi \in [a, b]$, $g' \neq 0$ for $[a, b] \setminus \{\xi\}$. (Obviously, we are allowed, without loss of generality, to assume the existence of just one critical point: otherwise we integrate in a finite number of subintervals.) An asymptotic expansion in the presence of a critical point presents us with new challenges. In principle, we could have used here the standard technique of *stationary phase* [15, 18], except that it is not equal to our task. We present an alternative that leads to an explicit and workable expansion. It is subsequently used to design asymptotic and Filon-type methods: unfortunately, Levin-type methods are not available in this setting.

The purpose of the final section is the sketch gaps in the theory and comment on ongoing challenges and developments. Moreover, we describe there briefly the recent method of [5], as well as the work in progress in Cambridge and Trondheim.

Quadrature of (1) represents but one problem in the wide range of issues originating in high oscillation. Quite clearly, a more significant challenge is to solve highly oscillatory differential equations. It is thus of interest to mention that the availability of efficient highly oscillatory quadrature is critical to a number of contemporary methods for ordinary differential equations that exhibit rapid oscillation [2, 6, 7, 12].

2 Asymptotic expansion in the absence of critical points

We restrict our analysis to $\mathbb{R}^2$, directing the reader to [10] for the general case. Let first $\Omega = \mathcal{S}_2$, the triangle with vertices at $(0,0)$, $(1,0)$ and $(0,1)$.

The nonresonance condition is thus

$$g_y(x,0) \neq 0, \quad x \in [0,1], \qquad g_x(0,y) \neq 0, \quad y \in [0,1],$$
$$g_x(x,y) - g_y(x,y) \neq 0, \quad x,y \geq 0, \ x+y \in [0,1].$$

Integrating by parts in the inner integral,

$$\begin{aligned}
I[g_x^2 f, \mathcal{S}_2] &= \int_0^1 \int_0^{1-y} g_x^2(x,y) f(x,y) e^{i\omega g(x,y)} \, dx \, dy \\
&= \frac{1}{i\omega} \int_0^1 g_x(1-y, y) f(1-y, y) e^{i\omega g(1-y, y)} \, dy \\
&\quad - \frac{1}{i\omega} \int_0^1 g_x(0,y) f(0,y) e^{i\omega g(0,y)} \, dy - \frac{1}{i\omega} I\left[\frac{\partial}{\partial x}(g_x f), \mathcal{S}_2\right] \\
&= \frac{1}{i\omega} \int_0^1 g_x(x, 1-x) f(x, 1-x) e^{i\omega g(x, 1-x)} \, dx \\
&\quad - \frac{1}{i\omega} \int_0^1 g_x(0,y) f(0,y) e^{i\omega g(0,y)} \, dy - \frac{1}{i\omega} I\left[\frac{\partial}{\partial x}(g_x f), \mathcal{S}_2\right].
\end{aligned}$$

By the same token,

$$\begin{aligned}
I[g_y^2 f, \mathcal{S}_2] &= \frac{1}{i\omega} \int_0^1 g_y(x, 1-x) f(x, 1-x) e^{i\omega g(x, 1-x)} \, dx \\
&\quad - \frac{1}{i\omega} \int_0^1 g_y(x,0) f(x,0) e^{i\omega g(x,0)} \, dy - \frac{1}{i\omega} I\left[\frac{\partial}{\partial y}(g_y f), \mathcal{S}_2\right].
\end{aligned}$$

Adding up, we have

$$I[\|\nabla g\|^2 f, \mathcal{S}_2] = \frac{1}{i\omega}(M_1 + M_2 + M_3) - \frac{1}{i\omega} I[\nabla^\top (f \nabla g), \mathcal{S}_2],$$

where

$$M_1 = \int_0^1 f(x,0) [\boldsymbol{n}_1^\top \nabla g(x,0)] e^{i\omega g(x,0)} \, dx,$$

$$M_2 = \sqrt{2} \int_0^1 f(x, 1-x) [\boldsymbol{n}_2^\top \nabla g(x, 1-x)] e^{i\omega g(x, 1-x)} \, dx,$$

$$M_3 = \int_0^1 f(0,y) [\boldsymbol{n}_3^\top \boldsymbol{g}(0,y)] e^{i\omega g(0,y)} \, dy.$$

Here

$$\boldsymbol{n}_1 = \begin{bmatrix} 0 \\ -1 \end{bmatrix}, \qquad \boldsymbol{n}_2 = \begin{bmatrix} \frac{\sqrt{2}}{2} \\ \frac{\sqrt{2}}{2} \end{bmatrix}, \qquad \boldsymbol{n}_3 = \begin{bmatrix} -1 \\ 0 \end{bmatrix}$$

are outward unit normals at the edges of $\mathcal{S}_2$.

Since $\nabla g(x,y) \neq \boldsymbol{0}$ in $\operatorname{cl} \mathcal{S}_2$, we may replace above f by $f/\|\nabla g\|^2$ without any danger of dividing by zero. The outcome is

$$I[f, \mathcal{S}_2] = \frac{1}{i\omega} \int_{\partial \mathcal{S}_2} \boldsymbol{n}^\top(\boldsymbol{x}) \nabla g(\boldsymbol{x}) \frac{f(\boldsymbol{x})}{\|\nabla g(\boldsymbol{x})\|^2} e^{i\omega g(\boldsymbol{x})} dS \qquad (2)$$
$$- \frac{1}{i\omega} I\left[\nabla^\top \left(\frac{f}{\|\nabla g\|^2} \nabla g \right), \mathcal{S}_2 \right].$$

Extending this technique to $\mathbb{R}^n$, it is possible to prove that (2) remains true once we replace $\mathcal{S}_2$ by $\mathcal{S}_n \subset \mathbb{R}^n$, the *regular simplex* with vertices at $\boldsymbol{0}$ and $\boldsymbol{e}_1, \ldots, \boldsymbol{e}_n$.

Let
$$f_0(\boldsymbol{x}) = f(\boldsymbol{x}), \qquad f_m = \nabla^\top \left[\frac{f_{m-1}(\boldsymbol{x})}{\|\nabla g(\boldsymbol{x})\|^2} \nabla g(\boldsymbol{x}) \right], \qquad m \in \mathbb{N}.$$

We deduce from (2) (extended to $\mathbb{R}^n$) that

$$I[f_m, \mathcal{S}_n] = \frac{1}{i\omega} \int_{\partial \mathcal{S}_n} \boldsymbol{n}^\top \nabla g(\boldsymbol{x}) \frac{f_m(\boldsymbol{x})}{\|\nabla g(\boldsymbol{x})\|^2} e^{i\omega g(\boldsymbol{x})} dS - \frac{1}{i\omega} I[f_{m+1}, \mathcal{S}_n], \qquad m \in \mathbb{Z}_+.$$

Finally, we iterate the above expression to obtain a *Stokes-type formula*, expressing $I[f, \mathcal{S}_n]$ as an asymptotic expansion on the boundary of the simplex,

$$I[f, \mathcal{S}_n] \sim -\sum_{m=0}^{\infty} \frac{1}{(-i\omega)^{m+1}} \int_{\partial \mathcal{S}_n} \boldsymbol{n}^\top \nabla g(\boldsymbol{x}) \frac{f_m(\boldsymbol{x})}{\|\nabla g(\boldsymbol{x})\|^2} e^{i\omega g(\boldsymbol{x})} dS. \qquad (3)$$

We wish to highlight four important issues. Firstly, a trivial inductive proof confirms that each f_m can be expressed as a linear combination of f and the first m directional derivatives (altogether, $\binom{n+m+1}{m}$ quantities), with coefficients that depend on the oscillator g and its derivatives.

Secondly, the simplest (and most useful) special case is $n = 1$, whence (3) reduces to

$$I[f, (0,1)] \sim -\sum_{m=0}^{\infty} \frac{1}{(-i\omega)^{m+1}} \left[\frac{e^{i\omega g(1)}}{g'(1)} f_m(1) - \frac{e^{i\omega g(0)}}{g'(0)} f_m(0) \right]. \qquad (4)$$

Thirdly, using an affine transformation, we can map $\mathcal{S}_n$ to an arbitrary simplex in $\mathbb{R}^n$. Applying an identical transformation to (3), we deduce that it is valid for $I[f, \mathcal{S}]$, where $\mathcal{S} \subset \mathbb{R}^n$ is any simplex.

Fourthly, the boundary of $\mathcal{S}$ is itself composed of $n+1$ simplices in $\mathbb{R}^{n-1}$. Because of the nonresonance condition, the gradient of the oscillator does not vanish in any of these simplices and we can apply (3) therein: this expresses $I[f, \mathcal{S}]$ as an asymptotic expansion over $(n-2)$-dimensional simplices. We continue with this procedure until we reach 0-dimensional simplices: the $n+1$ vertices of the original simplex. Bearing in mind our first observation, we thus deduce that

$$I[f, \mathcal{S}] \sim \sum_{m=0}^{\infty} \frac{1}{(-i\omega)^{m+n}} \Theta_m[f], \qquad (5)$$

where each Θ_m is a linear functional which depends on $\partial^{|i|} f/\partial \boldsymbol{x}^i$, $|i| \leq m$, at the $n+1$ vertices of $\mathcal{S}$. Note that $I[f, \mathcal{S}] = \mathcal{O}(\omega^{-n})$.

In general, the functionals Θ_m are fairly complicated, the univariate case (4) being an exception. However, it is the *existence* of (5), rather than its exact form, which render possible the design of efficient quadrature methods in the next section.

Let $\Omega \subset \mathbb{R}^n$ be a *polytope*, a bounded (open) domain with piecewise-linear boundary. (Note that Ω need be neither convex nor even simply connected.) We may then tessellate Ω with simplices $\Omega_1, \Omega_2, \ldots, \Omega_r \in \mathbb{R}^n$, therefore

$$I[f, \Omega] = \sum_{k=1}^{r} I[f, \Omega_r]. \tag{6}$$

A *simplicial complex* is a collection $\mathcal{C}$ of simplices in $\mathbb{R}^n$ such that every face of $\Phi \in \mathcal{C}$ is also in $\mathcal{C}$ and if $\Phi_1 \cap \Phi_2 \neq \emptyset$ for $\Phi_1, \Phi_2 \in \mathcal{C}$ then $\Phi_1 \cap \Phi_2$ is a face of both Φ_1 and Φ_2 [14]. We may always choose a tessellation composed of all n-dimensional simplices in a simplicial complex. In finite-element terminology, this corresponds to a tessellation without 'hanging nodes'.

Assume that the nonresonance condition condition holds for the oscillator g. We may always choose a simplicial complex so that the nonresonance condition is valid in each Ω_k, otherwise we vary the internal nodes. Clearly, once we can expand asymptotically each $I[f, \Omega_k]$, we may use (6) to expand $I[f, \Omega]$. Bearing in mind (5), this means that the entire information needed to construct such an expansion is the values of f and its derivatives at the vertices of the Ω_ks. However, a moment's reflection clarifies that *only* the original vertices of Ω may influence the expansion: the internal vertices are arbitrary, since there is an infinity of simplicial complexes consistent with the nonresonance condition. In other words, because of our construction of the tessellation via a simplicial complex, the contributions from neighbouring simplices cancel at internal vertices and each Θ_m depends on f and its derivatives at the original vertices of Ω.

3 Asymptotic, filon and levin methods

3.1 Asymptotic methods

The simplest and most natural means of approximating (1) consists of a truncation of the asymptotic expansion (5) (replacing $\mathcal{S}$ by a polytope Ω). This results in the *asymptotic method*

$$Q_s^{\mathrm{A}}[f, \Omega] = \sum_{m=0}^{s-1} \frac{1}{(-\mathrm{i}\omega)^{m+n}} \Theta_m[f], \tag{7}$$

bearing an asymptotic error of

$$Q_s^A[f, \Omega] - I[f, \Omega] \sim \mathcal{O}(\omega^{-n-s}), \qquad |\omega| \gg 1.$$

We say that Q_s^A is of an *asymptotic order* $s + n$.

Asymptotic quadrature is particularly straightforward in a single dimension, since then its coefficients are readily provided explicitly by an affine mapping of (4) from $(0, 1)$ to an arbitrary bounded real interval.

In Fig. 2 we have plotted the absolute value of the error once

$$\int_{-1}^{1} x \sin x e^{i\omega(x + \frac{1}{4}x^2)} \mathrm{d}x$$

is approximated by Q_s^A with $s = 1$ and $s = 2$. The error (here and in the sequel) is scaled by ω^p, where p is the asymptotic order, otherwise the rate of decay at the plot would have been so rapid as to prevent much useful insight. It is clear that, exactly as predicted by our theory, the error indeed decays as $\psi(\omega)/\omega^p$, where ψ is a bounded function.

The coefficients of an asymptotic method are becoming fairly elaborate in $n \geq 2$ dimensions. Thus, for example, for the linear oscillator $g(x, y) = \kappa_1 x + \kappa_2 y$ we have

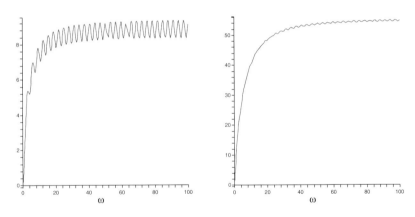

Fig. 2. Error, scaled by ω^p, in asymptotic quadrature of asymptotic order p with $\Omega = (-1, 1)$, $f(x) = x \sin x$, $g(x) = x + \frac{1}{4}x^2$ and $s = 1$, $p = 2$ (on the left) and $s = 2$, $p = 3$ (on the right).

$$Q_2^A[f, S_2] = \frac{1}{(-i\omega)^2} \left[\frac{1}{\kappa_1 \kappa_2} f(0,0) + \frac{e^{i\kappa_1 \omega}}{\kappa_1(\kappa_1 - \kappa_2)} f(1,0) - \frac{e^{i\kappa_2 \omega}}{\kappa_2(\kappa_1 - \kappa_2)} f(0,1) \right]$$
$$+ \frac{1}{(-i\omega)^3} \left\{ \left[\frac{1}{\kappa_1^2 \kappa_2} f_x(0,0) + \frac{1}{\kappa_1 \kappa_2^2} f_y(0,0) \right] \right.$$
$$+ e^{i\kappa_1 \omega} \left[\frac{2\kappa_1 - \kappa_2}{\kappa_1^2(\kappa_1 - \kappa_2)^2} f_x(1,0) - \frac{1}{\kappa_1(\kappa_1 - \kappa_2)^2} f_y(1,0) \right]$$
$$\left. + e^{i\kappa_2 \omega} \left[-\frac{1}{\kappa_2(\kappa_1 - \kappa_2)^2} f_x(0,1) + \frac{-\kappa_1 + 2\kappa_2}{\kappa_2^2(\kappa_1 - \kappa_2)^2} f_y(0,1) \right] \right\}.$$

Note that all the coefficients are well defined, because of the nonresonance condition.

Figure 3 exhibits the scaled error of two asymptotic methods, of asymptotic orders 3 and 4, respectively, in S_2. Yet, it is fair to comment that the sheer complexity of the coefficients for general oscillators and polytopes limits the application of (1) mainly to the univariate case. Another important shortcoming of an asymptotic method is that, given ω and the number of derivatives that we may use, its accuracy, although high, is predetermined. Often we may increase accuracy by using higher derivatives, but even this is not assured, since asymptotic expansions do not converge in the usual sense. Once ω is fixed, it is entirely possible that Q_s^A for some $s \geq 1$ is superior to Q_r^A for all $r > s$.

3.2 Filon-type methods

Although an asymptotic method (1) is the most obvious consequence of the asymptotic expansion (5), it is by no means the most effective. A more

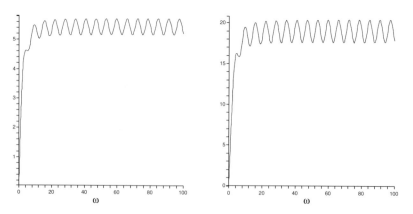

Fig. 3. Scaled error for Q_1^A (on the left) and Q_2^A (on the right) for $\Omega = S_2$, $f(x,y) = e^{x-2y}$ and $g(x,y) = x + 2y$.

sophisticated use of the asymptotic expansion rapidly leads to far superior, accurate and versatile quadrature schemes.

Let φ be an arbitrary smooth function in the closure of the polytope $\Omega \subset \mathbb{R}^n$ and suppose that at every vertex $\boldsymbol{v} \in \mathbb{R}^n$ of Ω it is true that

$$\frac{\partial^{|\boldsymbol{i}|}}{\partial \boldsymbol{x}^{\boldsymbol{i}}}\varphi(\boldsymbol{v}) = \frac{\partial^{|\boldsymbol{i}|}}{\partial \boldsymbol{x}^{\boldsymbol{i}}}f(\boldsymbol{v}), \qquad 0 \leq |\boldsymbol{i}| \leq s-1.$$

It then follows at once from (5) (where, again, we have replaced $\mathcal{S}$ with Ω) that

$$I[\varphi, \Omega] - I[f, \Omega] = I[\varphi - f, \Omega] \sim \mathcal{O}(\omega^{-s-n}), \qquad |\omega| \gg 1.$$

This motivates the *Filon-type method*

$$Q_s^{\mathsf{F}}[f, \Omega] = I[\varphi, \Omega] = \int_\Omega \varphi(\boldsymbol{x}) \mathrm{e}^{\mathrm{i}\omega g(\boldsymbol{x})} \mathrm{d}S. \qquad (8)$$

Needless to say, the above is a 'method' only if $I[\varphi, \Omega]$ can be evaluated exactly. In the most obvious case when φ is a polynomial, this is equivalent to the explicit computability of relevant *moments* of the oscillator g,

$$\mu_{\boldsymbol{i}}(\omega) = \int_\Omega \boldsymbol{x}^{\boldsymbol{i}} \mathrm{e}^{\mathrm{i}\omega g(\boldsymbol{x})} \mathrm{d}S, \qquad \boldsymbol{x}^{\boldsymbol{i}} = x_1^{i_1} \cdots x_n^{i_n}, \qquad \boldsymbol{i} \in \mathbb{Z}_+^n.$$

We will return to this restriction upon the applicability of (2) in the sequel.

It is important to observe that in the 'minimalist' case, when φ interpolates only at the vertices of Ω, (1) and (2) use *exactly the same information*. The difference in their performance, which is often substantive, is due solely to the different way this information is processed. While the error in (1) is determined by the asymptotic expansion (5) of f, the error of (2) follows from an asymptotic expansion of the interpolation error $\varphi - f$. The latter is likely to be smaller.

Historically, Louis Napoleon George [3] was the first to contemplate this approach in a single dimension, replacing f by a quadratic approximation at the endpoints and the midpoint. This was generalized by [13] and [4], who have considered general univariate interpolatory quadrature in which $\mathrm{e}^{\mathrm{i}\omega g(x)}$ plays the role of a complex-valued weight function. Yet, a thorough qualitative understanding of such methods and an analysis of their asymptotic order (indeed, the very observation that this concept is germane to their understanding) has been presented only recently: in the univariate case in [8] and in a multivariate setting in [10].

In one dimension we construct Filon-type methods similarly to the familiar *interpolatory quadrature rules*. Thus, we choose *nodes* $c_1 < c_2 < \cdots < c_\nu$, where c_1 and c_ν are the endpoints of Ω, as well as *multiplicities* $\boldsymbol{m} \in \mathbb{N}^\nu$. The function φ is the unique *Hermite interpolating polynomial* of degree $\mathbf{1}^\top \boldsymbol{m} - 1$ such that

$$\varphi^{(i)}(c_k) = f^{(i)}(c_k), \qquad i = 0, \ldots, m_k, \quad k = 1, 2, \ldots, \nu.$$

This is consistent with (2) with $s = \min\{m_1, m_\nu\}$.

Note that, although asymptotic order is assured by interpolation at the endpoints, it is often useful to interpolate also at internal points, since this usually decreases the error. This is demonstrated in Fig. 4, where we revisit the calculation of Fig. 2 using three Filon-type methods.

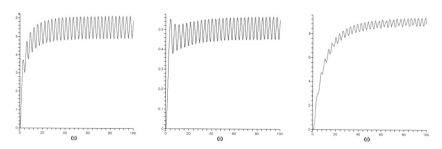

Fig. 4. Scaled error for Q_1^{F} with $\boldsymbol{c} = [-1, 1]$, $\boldsymbol{m} = [1, 1]$ (on the left), Q_1^{F} with $\boldsymbol{c} = [-1, -\frac{3}{4}, \frac{3}{4}, 1]$, $\boldsymbol{m} = [1, 1, 1, 1]$ (at the centre) and Q_2^{A} with $\boldsymbol{c} = [-1, 1]$, $\boldsymbol{m} = [2, 2]$ (on the right) for $\Omega = (-1, 1)$, $f(x) = x \sin x$ and $g(x) = x + \frac{1}{4}x^2$.

Unlike (1), it is fairly straightforward to implement Filon-type methods in a multivariate setting, using standard multivariate approximation theory. The most natural approach is to take a leaf off finite-element theory, tessellate a polytope with simplices (taking care to respect nonresonance) and interpolate in each simplex with suitable polynomials. Note that there is no need to force continuity across edges. In general, the computation of the moments might be problematic, but it is trivial for linear oscillators $g(\boldsymbol{x}) = \boldsymbol{\kappa}^\top \boldsymbol{x}$.

Figure 5 displays a bivariate Filon-type quadrature of the integral of Fig. 3. On the left we have used a standard linear interpolation at the vertices. On the right the ten degrees of freedom of a bivariate cubic were quenched by imposing function and first-derivative interpolation at the vertices and simple interpolation at the centroid $(\frac{1}{3}, \frac{1}{3})$.

We mention that it is possible to implement Filon-type methods without the computation of derivatives, using instead finite differences with spacing of $\mathcal{O}(\omega^{-1})$ [9].

Filon-type methods are highly accurate, affordable and very simple to construct. Yet, there is no escaping their main shortcoming: we must be able to evaluate the moments μ_i of the underlying oscillator. In the next subsection we describe another kind of quadrature methods that use identical information and attain identical asymptotic order without any need to calculate moments.

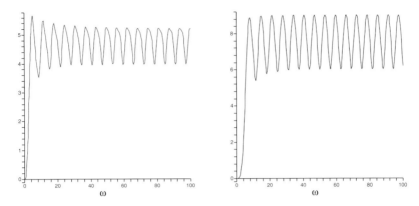

Fig. 5. Scaled error for Q_1^F (on the left) and Q_2^F (on the right) for $\Omega = \mathcal{S}_2$, $f(x,y) = e^{x-2y}$ and $g(x,y) = x+2y$.

3.3 Levin-type methods

Levin-type methods are quadrature techniques which do not require the computation of moments. Indeed, if Ω satisfies the nonresonance condition, a Levin-type method can be used to approximate $I[f, \Omega]$ even if Ω is not a polytope. We begin with an overview of the method described in [11]. If we have a function F such that

$$\frac{\mathrm{d}}{\mathrm{d}x}\left[F(x)\mathrm{e}^{\mathrm{i}\omega g(x)}\right] = [F'(x) + \mathrm{i}\omega g'(x)F(x)]\mathrm{e}^{\mathrm{i}\omega g(x)} = f(x)\mathrm{e}^{\mathrm{i}\omega g(x)},$$

then we can compute $I[f, (a,b)]$ trivially. Defining the differential operator $L[F] = F' + \mathrm{i}\omega g'F$ and rewriting the above equation as $L[F] = f$, we can now approximate F by a function v that is a linear combination of ν basis functions $\psi_1, \psi_2, \ldots, \psi_\nu$, using collocation with the operator L. In other words, we choose *nodes* $c_1 < c_2 < \cdots < c_\nu$, where c_1 and c_ν are the endpoints of the interval Ω, and solve for v using the system

$$L[v](c_k) = f(c_k), \qquad k = 1, 2, \ldots, \nu.$$

[16] generalized this method in a manner similar to a Filon-type method, equipping collocation points with *multiplicities* $\boldsymbol{m} \in \mathbb{N}^\nu$. Now v is a linear combination of $\tau = \mathbf{1}^\top \boldsymbol{m} - 1$ functions. This results in a new system,

$$\frac{\mathrm{d}^i}{\mathrm{d}x^i}L[v](c_k) = \frac{\mathrm{d}^i}{\mathrm{d}x^i}f(c_k), \qquad i = 1, 2, \ldots, m_k, \quad k = 1, 2, \ldots, \nu.$$

We then define

$$Q_s^\mathsf{L}[f, (a,b)] = v(b)\mathrm{e}^{\mathrm{i}\omega g(b)} - v(a)\mathrm{e}^{\mathrm{i}\omega g(a)},$$

which is equivalent to $I[L[v]]$.

One huge benefit of Levin-type methods is that they work easily on complicated domains and complicated oscillators for which Filon-type methods utterly fail. We demonstrate the method on the quarter-circle $H = \{(x,y) : x^2 + y^2 < 1,\ x, y > 0\}$, however it works equally well on other domains that satisfy the nonresonance condition, including those in higher dimensions. In the univariate version we approximated F, where $L[F] = f$, which enabled us to 'push' the integral to the boundary of the interval, namely its endpoints. We use this idea as an inspiration for the multivariate case: we begin by determining an operator L that will allow us to 'push' the integral to the boundary. To do so, we use differential forms along with the Stokes theorem. Suppose we have a function F such that

$$\int_{\partial H} F(x,y) e^{i\omega g(x,y)} (\mathrm{d}x + \mathrm{d}y) = \int_H f(x,y) e^{i\omega g(x,y)} \mathrm{d}V.$$

Stokes' theorem tells us that

$$I[f] = \int_{\partial H} F e^{i\omega g} (\mathrm{d}x + \mathrm{d}y) = \int_H \mathrm{d}\left[F e^{i\omega g} (\mathrm{d}x + \mathrm{d}y) \right]$$
$$= \int_H (F_y + i\omega g_y F) e^{i\omega g} \mathrm{d}y \wedge \mathrm{d}x + (F_x + i\omega g_x F) e^{i\omega g} \mathrm{d}x \wedge \mathrm{d}y$$
$$= I[F_x + i\omega g_x F - F_y - i\omega g_y F].$$

Hence we use the collocation operator $L[F] = F_x + i\omega g_x F - F_y - i\omega g_y F$. For simplicity, we write both the univariate and multivariate operator as $L[F] = J[F] + i\omega J[g]F$, where in two dimensions $J[F] = F_x - F_y$, and in one dimension $J[F] = F'$. Thus we determine a linear combination of basis functions v by solving the system

$$\frac{\partial^{|\boldsymbol{i}|}}{\partial x^{\boldsymbol{i}}} L[v](\boldsymbol{c}_k) = \frac{\partial^{|\boldsymbol{i}|}}{\partial x^{\boldsymbol{i}}} f(\boldsymbol{c}_k), \qquad 0 \le |\boldsymbol{i}| \le m_k - 1, \qquad k = 1, 2, \ldots, \nu, \quad (9)$$

where $\boldsymbol{c}_1, \ldots, \boldsymbol{c}_\nu$ is a sequence of nodes. Consequently,

$$I[f, H] \approx I[L[v], H] = \int_{\partial H} v e^{i\omega g} (\mathrm{d}x + \mathrm{d}y)$$
$$= \int_0^{\frac{\pi}{2}} (\cos t - \sin t) v(\cos t, \sin t) e^{i\omega g(\cos t, \sin t)} \mathrm{d}t \quad (10)$$
$$- \int_0^1 v(0, 1-t) e^{i\omega g(0, 1-t)} \mathrm{d}t + \int_0^1 v(t, 0) e^{i\omega g(t, 0)} \mathrm{d}t.$$

We thus define $Q^L[f, H]$ by approximating each of these univariate integrals using univariate Levin-type methods. For the proof of the asymptotic order we assume that the endpoints of each of these integrals have the same multiplicity as the associated vertex. For example, the multiplicity at $t = 0$ of the first integral is the same as the multiplicity at $(\cos 0, \sin 0) = (1, 0)$.

We will show that, as in a Filon-type method, $I[f,H] - Q^L[f,H] = \mathcal{O}(\omega^{-s-n}) = \mathcal{O}(\omega^{-s-2})$, where s is again the smallest vertex multiplicity. We begin by showing that $I[f,\Omega] - I[L[v],\Omega] = \mathcal{O}(\omega^{-s-n})$, where $\Omega = H$ or a univariate interval. One might be tempted to prove this by considering it as a Filon-type method with $\phi = L[v]$. Indeed, it satisfies all the conditions of a Filon-type method, except for the fact that $L[v]$ depends on ω. Hence, in order to prove the error, we also need to show that $f - L[v]$ and its derivatives are bounded for increasing ω. To do so, we impose the *regularity condition*, which requires that the vectors $g_1, g_2, \ldots, g_\tau$, where $\tau = \mathbf{1}^\top \mathbf{m} - 1$, are linearly independent. Here

$$g_j = \begin{bmatrix} \rho_{j,1} \\ \vdots \\ \rho_{j,\nu} \end{bmatrix},$$

where

$$\rho_{j,k} = \begin{bmatrix} \frac{\partial^{|p_{k,1}|}}{\partial x^{p_{k,1}}}(J[g]\psi_j)(c_k) \\ \vdots \\ \frac{\partial^{|p_{k,n_k}|}}{\partial x^{p_{k,n_k}}}(J[g]\psi_j)(c_k) \end{bmatrix},$$

while $p_{k,1}, \ldots, p_{k,n_k} \in \mathbb{N}^n$, $n_k = \frac{1}{2}m_k(m_k + 1)$, are all the vectors such that $|p_{k,i}| \le m_k - 1$, lexicographically ordered.

Note that we can rewrite the system (9) in the form $(P + i\omega G)\mathbf{d} = \mathbf{f}$, where G is the matrix whose jth column is g_j, P is a matrix independent of ω, $\mathbf{d}$ is the vector of unknown coefficients in v, and $\mathbf{f}$ is defined as

$$\mathbf{f} = \begin{bmatrix} \sigma_1 \\ \vdots \\ \sigma_\tau \end{bmatrix}, \quad \sigma_k = \begin{bmatrix} \frac{\partial^{|p_{k,1}|}}{\partial x^{p_{k,1}}} f(c_k) \\ \vdots \\ \frac{\partial^{|p_{k,n_k}|}}{\partial x^{p_{k,n_k}}} f(c_k) \end{bmatrix}, \quad k = 1, \ldots, \nu.$$

From Cramer's rule we know that $d_k = \det D_k / \det(P + i\omega G)$, where D_k is the matrix $P + i\omega G$ with the kth column replaced by $\mathbf{f}$. Due to the regularity condition, G is nonsingular, hence $[\det(P + i\omega G)]^{-1} = \mathcal{O}(\omega^{-\tau})$, where τ is equal to the number of rows in G. Moreover, it is clear that $\det D_k = \mathcal{O}(\omega^{\tau-1})$. Hence $d_k = \mathcal{O}(\omega^{-1})$, and $L[v] = \mathcal{O}(1)$ for increasing ω. Thus, as in a Filon-type method, $I[f,\Omega] - I[L[v],\Omega] = \mathcal{O}(\omega^{-s-n})$.

If Ω is a univariate interval then we have just demonstrated that $I[f,\Omega] - Q^L[f,\Omega] = \mathcal{O}(\omega^{-s-1})$. In the multivariate case (and sticking to our example of a quarter-circle: the general case is similar) we need to prove that $I[L[v],H] - Q^L[f,H] = \mathcal{O}(\omega^{-s-n})$. Each of the integrands in (10) is of order $\mathcal{O}(\omega^{-1})$. It follows that the approximations by Q^L are of order $\mathcal{O}(\omega^{-s-2})$. Hence we have demonstrated that $I[f, H] - Q^L[f, H] = \mathcal{O}(\omega^{-s-2})$. It is clear that this proof can be generalized to other domains, with an asymptotic order $n + s$.

It should be emphasized that a Levin-type method attains exactly the same asymptotic order as a Filon-type method, using the same information about f. In fact, if Ω is a simplex and g is a linear oscillator then the two methods are equivalent, assuming that the subintegrals in a Levin-type method have a sufficient number of data points [17]. However, the latter requires significantly more operations, assuming that the computation of moments is efficient, since a system must be solved for each dimension. Moreover, [16] presents experimental evidence that suggests that Levin-type methods are typically less accurate than Filon-type methods, though this depends on the choice of oscillator g, on interpolation nodes, the closeness of f to a polynomial and the choice of interpolation basis for the Levin-type method.

In Fig. 6, we approximate the same univariate integral as in Fig. 4, now with Levin-type methods in place of Filon-type methods. As can be seen, in conformity with the theory, the two methods share the same asymptotic order, while the Levin-type method exhibits somewhat lesser accuracy.

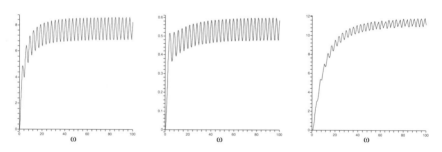

Fig. 6. Scaled error for Q_1^L with $\boldsymbol{c} = [-1, 1]$, $\boldsymbol{m} = [1, 1]$ (on the left), Q_1^L with $\boldsymbol{c} = [-1, -\frac{3}{4}, \frac{3}{4}, 1]$, $\boldsymbol{m} = [1, 1, 1, 1]$ (at the centre) and Q_2^L with $\boldsymbol{c} = [-1, 1]$, $\boldsymbol{m} = [2, 2]$ (on the right) for $\Omega = (-1, 1)$, $f(x) = x \sin x$ and $g(x) = x + \frac{1}{4}x^2$.

In Fig. 7 we see how well can a Levin-type method handle two-dimensional domains with nonlinear g. Specifically, we consider the quarter-circle $H = \{(x, y) : x^2 + y^2 < 1,\ x, y > 0\}$. In the first figure we collate at each vertex with multiplicity one for the bivariate system, and at the endpoints with multiplicity one for each univariate integral in (10). The second figure collocates with multiplicity two at each vertices and with multiplicity one at $\left(\frac{1}{3}, \frac{1}{3}\right)$, for the bivariate system, and collocates with just the endpoints with multiplicities two for each univariate system. Note that H, not being a polytope, represents a domain for which no viable theory exists for Filon-type methods.

In the univariate case it is possible to identify basis functions ψ_k which lead to the highest-possible asymptotic order. Specifically, $\psi_k = f_{k+1}/g'$, where the

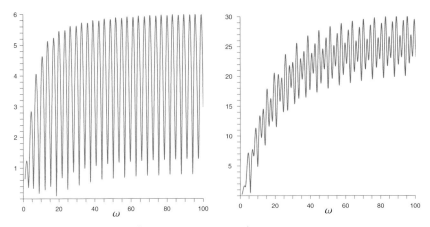

Fig. 7. Scaled error for Q_1^{L} (on the left) and Q_2^{L} (on the right) for $\Omega = H$, $f(x,y) = \mathrm{e}^{x-2y}$ and $g(x,y) = x^3 + x - y$.

functions f_k have already featured in the asymptotic expansion (4). We dwell no further on this issue, referring the reader to [16].

4 Critical points

Once ∇g is allowed to vanish in cl Ω, the asymptotic formula (5) is no longer valid. Worse, in a multivariate setting surprisingly little is known about asymptotic expansions in the presence of high oscillation and critical points [18]. The situation is much clearer and better understood in a single dimension.[4] This is due to the *van der Corput theorem,* which allows us to determine the asymptotic order of magnitude of (1) [18]. Moreover, the classical *method of stationary phase* provides an avenue of sorts, once we have taken care of the behaviour at the endpoints, toward an asymptotic expansion [15, 18]. Unfortunately, this technique falls short of providing the entire information required to construct an asymptotic expansion, while being complicated and cumbersome.

In this section we describe an alternative to the method of stationary phase which has been introduced in [8]. We revisit the method of proof of Section 2, taking full advantage of the considerable simplification due to univariate setting. Let us suppose for simplicity that $\Omega = (a, b)$ and there exists a unique $\xi \in (a, b)$ such that $g'(\xi) = 0$, $g''(\xi) \neq 0$ and $g'(x) \neq 0$ for $x \in [a, b] \setminus \{\xi\}$. Clearly, the assumption that there is just one critical point hardly represents loss of generality, since we can always partition (a, b) into such subintervals. We will comment later on the case when also higher derivatives of g vanish at

---

[4] In the univariate case critical points are often termed "stationary points", but for consistency's sake we employ 'multivariate' terminology.

ξ. Finally, the case of $\xi = a$ or $\xi = b$ can be obtained by fairly straightforward generalization of our technique and is left to the reader.

A single step of our expansion technique *in the absence of critical points* in a single dimension is

$$I[f,(a,b)] = \frac{1}{i\omega} \int_a^b \frac{f(x)}{g'(x)} \frac{d}{dx} e^{i\omega g(x)} dx$$

$$= \frac{1}{i\omega} \left[\frac{e^{i\omega g(b)}}{g'(b)} f(b) - \frac{e^{i\omega g(a)}}{g'(a)} f(a) \right] - \frac{1}{i\omega} I\left[\left(\frac{f}{g'}\right)', (a,b) \right]$$

and it does not generalize to our setting since division by g' introduces polar singularity at ξ. Instead, we add and subtract $f(\xi)$ in the integrand,

$$I[f,(a,b)] = f(\xi) \int_a^b e^{i\omega g(x)} dx + \frac{1}{i\omega} \int_a^b \frac{f(x) - f(\xi)}{g'(x)} \frac{d}{dx} e^{i\omega g(x)} dx \quad (11)$$

$$= f(\xi)\mu_0(\omega) + \frac{1}{i\omega} \left\{ \frac{e^{i\omega g(b)}}{g'(b)}[f(b) - f(\xi)] - \frac{e^{i\omega g(a)}}{g'(a)}[f(a) - f(\xi)] \right\}$$

$$- \frac{1}{i\omega} I\left[\left(\frac{f - f(\xi)}{g'}\right)', (a,b) \right].$$

Note that $[f(x) - f(\xi)]/g'(x)$ is a smooth function, since the singularity at ξ is removable.

Iterating the last identity leads to an asymptotic expansion in the presence of a simple critical point. Thus, we define

$$f_0(x) = f(x), \qquad f_m(x) = \frac{d}{dx} \frac{f_{m-1}(x) - f_{m-1}(y)}{g'(x)}, \qquad m \in \mathbb{N},$$

whence

$$I[f,(a,b)] \sim \mu_0(\omega) \sum_{m=0}^{\infty} \frac{1}{(-i\omega)^m} f_m(y) \quad (12)$$

$$- \sum_{m=0}^{\infty} \frac{1}{(-i\omega)^{m+1}} \left\{ \frac{e^{i\omega g(b)}}{g'(b)}[f_m(b) - f_m(y)] - \frac{e^{i\omega g(a)}}{g'(a)}[f_m(a) - f_m(y)] \right\}.$$

For $x \neq \xi$ each f_m is a linear combination of $f, f', \ldots, f^{(m)}$, but at $x = \xi$ we have

$$f_0(\xi) = f(\xi),$$

$$f_1(\xi) = \frac{1}{2} \frac{1}{g''(\xi)} f''(\xi) - \frac{1}{2} \frac{g'''(\xi)}{g''^2(\xi)} f'(\xi),$$

$$f_2(\xi) = \frac{1}{8} \frac{1}{g''^2(\xi)} f^{(iv)}(\xi) - \frac{5}{12} \frac{g'''(\xi)}{g''^3(\xi)} f'''(\xi) + \left[\frac{5}{8} \frac{g'''^2(\xi)}{g''^4(\xi)} - \frac{1}{4} \frac{g^{(iv)}(\xi)}{g''^3(\xi)} \right] f''(\xi)$$

$$+ \left[-\frac{5}{8} \frac{g'''^2(\xi)}{g''^5(\xi)} + \frac{2}{3} \frac{g^{(iv)}(\xi)}{g''^4(\xi)} - \frac{1}{8} \frac{g^{(v)}(\xi)}{g''^3(\xi)} \right] f'(\xi).$$

and so on: in general, each $f_m(\xi)$ is a linear combination of $f^{(i)}(\xi)$, $i = 0, 1, \ldots, 2m$. The price tag of quadrature in the presence of critical point is the imperative to evaluate more derivatives there.

Note that (12) is not a 'proper' asymptotic expansion, because of the presence of the function $\mu_0(\omega)$. In principle, it might have been possible to replace μ_0 by its asymptotic expansion, e.g. using the method of stationary phase. This, however, is neither necessary nor, indeed, advisable. Assuming that μ_0 can be computed – and we need this anyway for Filon-type methods! – it is best to leave it in place. According to the van der Corput theorem, $\mu_0(\omega) \sim \mathcal{O}\left(\omega^{-\frac{1}{2}}\right)$.

It is straightforward to generalize our method of analysis to higher-order critical points. Thus, if $g^{(i)}(\xi) = 0$, $i = 1, 2, \ldots, r$, $g^{(r+1)}(\xi) \neq 0$, in place of (11) we integrate by parts on the right in

$$I[f,(a,b)] = \sum_{k=0}^{r-1} \tfrac{1}{k!} f^{(k)}(\xi) \int_a^b (x-\xi)^k e^{i\omega g(x)} dx$$
$$+ \frac{1}{i\omega} \int_a^b \frac{f(x) - \sum_{k=0}^{r-1} \tfrac{1}{k!} f^{(k)}(\xi)(x-\xi)^k}{g'(x)} \frac{d}{dx} e^{i\omega g(x)} dx.$$

Again, we obtain removable singularity inside the integral. Note that by the van der Corput theorem $I[f,(a,b)] = \mathcal{O}(\omega^{-1/(r+1)})$.

Truncation of (12) results in an asymptotic method, a generalization of (1). Specifically,

$$Q_s^{\mathrm{A}}[f,(a,b)] = \mu_0(\omega) \sum_{m=0}^{s-1} \frac{1}{(-i\omega)^m} f_m(y)$$
$$- \sum_{m=0}^{s-1} \frac{1}{(-i\omega)^{m+1}} \left\{ \frac{e^{i\omega g(b)}}{g'(b)} [f_m(b) - f_m(y)] - \frac{e^{i\omega g(a)}}{g'(a)} [f_m(a) - f_m(y)] \right\}$$

bears asymptotic error of $s + \tfrac{1}{2}$.

Figure 8 revisits the calculation from Section 1 that persuaded us in the inadequacy of Gaussian quadrature in the present setting: the calculation of $\int_{-1}^{1} \cos x \, e^{i\omega x^2} dx$. Note that Q_1^{A} requires just the values of f at $-1, 0, 1$, while Q_2^{A} needs f and f' at the endpoints and f, f', f'' at the critical point.

It is easy to generalize Filon-type methods to this setting. Nothing of essence changes. Thus, we choose nodes $a = c_1 < c_2 < \cdots < c_\nu = b$, taking care to include ξ: thus, $c_r = \xi$ for some $r \in \{2, \ldots, \nu - 1\}$. We interpolate to f and its first $m_k - 1$ derivatives at c_k, $k = 1, 2, \ldots, \nu$, with a polynomial φ of degree $m^\top 1 - 1$ and set

$$Q_s^{\mathrm{F}}[f,(a,b)] = I[\varphi,(a,b)]. \tag{13}$$

Here $s = \min\{m_1, \lfloor (m_r - 1)/2 \rfloor, m_\mu\}$. It follows at once from the asymptotic expansion that

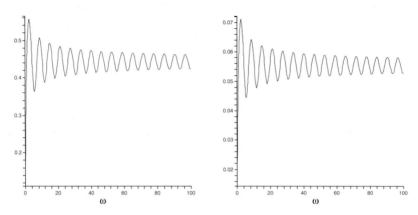

Fig. 8. The error for Q_1^A (on the left) and Q_2^A (on the right), scaled by $\omega^{\frac{3}{2}}$ and $\omega^{\frac{5}{2}}$ respectively, for $\Omega = (-1, 1)$, $f(x) = \cos x$ and $g(x) = x^2$, with a stationary point at the origin.

$$Q_s^F[f,(a,b)] = I[f,(a,b)] + \mathcal{O}\left(\omega^{-s-\frac{1}{2}}\right), \qquad |\omega| \gg 1,$$

and the method is of asymptotic order $s + \frac{1}{2}$. As a matter of fact, a generalization of Filon in the presence of critical points is much more flexible than that of the asymptotic method. We can easily cater for any number of critical points, possibly of different degrees, once we include them among the nodes and choose sufficiently large multiplicities.

Figure 9 shows the scaled error for three different Filon-type methods for the same problem as in Figs 1 and 8. Note how the accuracy greatly improves upon the addition of extra internal nodes. It is at present unclear why the addition of extra internal nodes has a much more dramatic effect in the presence of critical points.

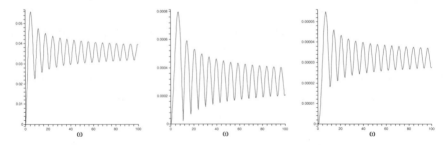

Fig. 9. Scaled error for Q_1^F with $\mathbf{c} = [-1, 0, 1]$, $\mathbf{m} = [1, 3, 1]$ (on the left), Q_1^F with $\mathbf{c} = [-1, -\frac{1}{2}, 0, \frac{1}{2}, 1]$, $\mathbf{m} = [1, 1, 3, 1, 1]$ (at the centre) and Q_2^F with $\mathbf{c} = [-1, 0, 1]$, $\mathbf{m} = [2, 5, 2]$ (on the right) for $\Omega = (-1, 1)$, $f(x) = \cos x$ and $g(x) = x^2$.

5 Conclusions and pointers for further research

The first and foremost lesson to be drawn from our analysis is that, once we can understand the *mathematics* of high oscillation, we gain access to a wide variety of effective and affordable algorithms. This, of course, is a truism that we might apply to just about every issue in mathematical computation, yet it is of particular importance in the current framework. The overwhelming wisdom in much of classical treatment of rapidly oscillating phenomena is to find means to make high oscillation go away. Thus, the 'rule of a thumb', ubiquitous in signal processing, that a function should be sampled sufficiently often within each period: in the current setting this translates to an approximation of $I[f,(a,b)]$, say, by partitioning (a,b) into a very large number of subintervals of length $\mathcal{O}(\omega^{-1})$ and using Gaussian quadrature within each 'panel'. However, the conclusion of this paper, and also of much contemporary work in the discretization of highly oscillatory ordinary differential equations, is that high oscillation renders solution *easier!*

Another reason why it is important to emphasize the role of mathematical understanding in our endeavour is that so little is known about the asymptotics of $I[f, \Omega]$ in general domains Ω. A fairly complete theory exists for $\Omega = \mathbb{R}^n$ and for $\Omega = \mathbb{S}^{n-1}$ (the $(n-1)$-sphere), at least as long as there are no critical points [18]. Yet, once we concern ourselves with bounded domains with boundary and allow for the presence of critical points, a great deal remains to be done. It is a sobering thought that the asymptotic behaviour of $I[f, \Omega]$, where $\Omega \subset \mathbb{R}^2$ is bounded and with piecewise-smooth boundary, is unknown in general even if there are no critical points! Clearly, it depends on the geometry of $\partial \Omega$, an issue to which we will return, but it is presently unclear how.

A thread running through our entire analysis is the centrality of an asymptotic expansion of $I[f, \Omega]$. Once (5) is available, its truncation presents us with an immediate means to compute the integral. Moreover, even if the explicit form of (5) is unavailable, the very existence and known structure of an asymptotic formula allow us to analyse better and more flexible quadrature methods.

The assumption that Ω is a polytope is very restrictive. A naive means of a generalization to arbitrary bounded domains Ω with piecewise-smooth boundary is to approximate it from within by a convergent sequence of polytopes and use the dominating convergence theorem. This, however, might fall foul of the nonresonance condition. Consider, for example, the linear oscillator $g(\boldsymbol{x}) = \boldsymbol{\kappa}^\top \boldsymbol{x}$, $\boldsymbol{x} \in \mathbb{R}^2$ and a circular wedge Ω with angle α,

$$\Omega = \left\{ \boldsymbol{x} \in \mathbb{R}^2 \,:\, x_1^2 + x_2^2 < 1,\, \arctan \frac{x_2}{x_1} < \alpha \right\}.$$

As long as

$$\pm \frac{1}{\sqrt{\kappa_1^2 + \kappa_2^2}} \begin{bmatrix} -\kappa_2 \\ \kappa_1 \end{bmatrix} \notin \partial \Omega,$$

we can approximate Ω with narrow wedges, pass to a limit and obtain an asymptotic expansion, expressing $I[f,\Omega]$ in terms of f and its derivatives at $(0,0)$ and $(\cos\alpha, \pm\sin\alpha)$. Yet, if the above condition fails, the nonresonance condition *must* be breached upon passage to the limit. It is important to make it clear that the fault is definitely *not* in our method of proof. Once the resonance condition fails, an (5)-like expansion is no longer true! For simplicity, consider the bivariate unit disc $\Omega = \mathbb{S}^1$ and, again, a linear oscillator. We have

$$I[f,\mathbb{S}^1] = \int_{-1}^{1}\int_{-(1-x^2)^{\frac{1}{2}}}^{(1-x^2)^{\frac{1}{2}}} f(x,y)e^{i\omega(\kappa_1 x + \kappa_2 y)}dydx.$$

Expanding asymptotically in the inner interval similarly to (4), we thus have (assuming for simplicity that $\kappa_2 \neq 0$)

$$I[f,\mathbb{S}^1] \sim -\frac{1}{\kappa_2}\int_{-1}^{1}\sum_{m=0}^{\infty}\frac{1}{(-i\omega)^{m+1}}\left[e^{i\omega\kappa_2(1-x^2)^{\frac{1}{2}}}\frac{d^m}{dy^m}f(x,(1-x^2)^{\frac{1}{2}})\right.$$
$$\left. - e^{-i\omega\kappa_2(1-x^2)^{\frac{1}{2}}}\frac{d^m}{dy^m}f(x,-(1-x^2)^{\frac{1}{2}})\right]e^{i\omega\kappa_1 x}dx$$
$$= -\frac{1}{\kappa_2}\sum_{m=0}^{\infty}\frac{1}{(-i\omega)^{m+1}}\int_{-1}^{1}\frac{d^m}{dy^m}f(x,(1-x^2)^{\frac{1}{2}})e^{i\omega[\kappa_1 x+\kappa_2(1-x^2)^{\frac{1}{2}}]}dx$$
$$+ \frac{1}{\kappa_2}\sum_{m=0}^{\infty}\frac{1}{(-i\omega)^{m+1}}\int_{-1}^{1}\frac{d^m}{dy^m}f(x,-(1-x^2)^{\frac{1}{2}})e^{i\omega[\kappa_1 x-\kappa_2(1-x^2)^{\frac{1}{2}}]}dx,$$

an infinite sum of univariate integrals. However, before we rush to expand them asymptotically, we observe that the new oscillators have critical points at $\pm\kappa_1/(\kappa_1^2+\kappa_2^2)^{\frac{1}{2}}$. Our immediate conclusion is that $I[f,\mathbb{S}^1] \sim \mathcal{O}\left(\omega^{-\frac{3}{2}}\right)$, rather than the $\mathcal{O}(\omega^{-2})$ which we might have expected. Worse, all our three approaches fail. The moments of $g(x) = \kappa_1 x \pm \kappa_2(1-x^2)^{\frac{1}{2}}$ are unknown, hence we have neither an asymptotic expansion *á la* (12) nor a Filon-type method. Moreover, a Levin-type method fails because of the presence of critical points.

As long as the nonresonance condition is maintained throughout the approximation of Ω by polytopes, our methods can be extended to this setting. This has been already done for Levin-type methods in [17]: cf. the discussion leading to Fig. 7 in Section 3.

Our narrative underlies the importance of further research into quadrature methods for highly oscillatory integrals, in particular in the presence of critical points and when exact moments are unavailable. There are a few natural ways forward, in particular Filon-type methods with suitable approximate moments and Levin-type methods with special treatment of small neighbourhoods surrounding critical points (where the integral does not oscillate rapidly). Both approaches are under active consideration. Another option is quadrature methods based on altogether new principles, e.g. the recent technique of [5], who approximate (1) in a single dimension using a complex-valued path along which $e^{i\omega g(x)}$ does not oscillate. The underlying idea there, assuming that both f and g can be analytically extended to the complex plane, is to find a path from each endpoint of $\Omega = (a,b)$ to infinity alongside which $g(z) - g(a)$ and $g(z) - g(b)$, respectively, are real and negative. In place of (1) it is then possible to integrate from b to $z = \infty$ and then from ∞ to a. Because of exponential decay of the integrand, each integration can be accomplished by familiar Gauss–Laguerre quadrature and the outcome matches Filon-type and Levin-type methods in its asymptotic behaviour. We further note that in the presence of critical points there is a need to integrate also along paths joining them with $z = \infty$ in a fairly nontrivial manner.

Other challenges in highly oscillatory quadrature abound. One obvious generalization of (1) is

$$\int_\Omega f(\boldsymbol{x}) K(\omega, \boldsymbol{x}) \mathrm{d}V,$$

where K oscillates rapidly for $|\omega| \gg 1$. Filon-type methods have been generalized to this setting in the important special case of the *Bessel oscillator*, when $\Omega = (a,b)$ and $K(\omega, x) = J_\nu(\omega x)$ [20] but, by and large, this is an uncharted territory. Another *terra incognita* is (1) where Ω is a general bounded manifold with boundary, immersed in $\mathbb{R}^n$.

We have already touched upon applications of highly oscillatory quadrature to numerical methods for rapidly oscillating differential equations. Even more ambitious goal is the analysis of highly oscillatory *Fredholm equations of the second kind*

$$\int_a^b f(x,\omega) K(x,y,\omega) \mathrm{d}x = \lambda(\omega) f(y,\omega) - g(y), \qquad y \in [a,b], \qquad (14)$$

where $\lambda(\omega) \in \mathbb{C}$ is not an eigenvalue of the underlying operator, and of the corresponding spectral problem

$$\int_{-1}^1 \varphi(x,\omega) K(x,y,\omega) = \lambda(\omega) \varphi(y,\omega), \qquad y \in [a,b]. \qquad (15)$$

Both (14) and (15) are highly interesting because of their applications in electromagnetics and in laser theory, but their treatment by 'our' methods is hampered by the fact that the function f in (14) and the eigenfunction φ in (15) themselves oscillate. This renders integration by parts, along the lines of Section 2, fairly useless.

The spectral problem (15) has been solved for the kernel $K(x,y,\omega) = e^{i\omega xy}$, by demonstrating that φ obeys a specific Sturm–Liouville problem [1]. The asymptotic behaviour of the spectrum for $K(x,y,\omega) = e^{i\omega|x-y|}$ has been investigated by [19]. A detailed investigation of this kernel, inclusive of an asymptotic expansion of both eigenvalues and the solution of (14) in negative powers of ω will feature in a forthcoming paper by Brunner, Iserles and Nørsett. Yet, in their full generality, highly oscillatory integral equations of this kind represent an enduring and difficult challenge.

References

1. Cochran, J. A., Hinds, E. W.: Eigensystems associated with the complex-symmetric kernels of laser theory. SIAM J. Appld Maths, **26**, 776-786 (1974).
2. Degani, I., Schiff, J.: RCMS: Right correction Magnus series approach for integration of linear ordinary differential equations with highly oscillatory terms, Technical report, Weizmann Institute of Science.(2003)
3. Filon, L. N. G.: On a quadrature formula for trigonometric integrals. Proc. Royal Soc. Edinburgh,**49**, 38-47 (1928).
4. Flinn, E. A.: A modification of Filon's method of numerical integration. J ACM,**7**, 181-184 (1960).
5. Huybrechs, D., Vandewalle, S., On the evalution of highly oscillatory integrals by analytic continuation, Technical report, Katholieke Universiteit Leuven, (2005).
6. Iserles, A.: On the global error of discretization methods for highly-oscillatory ordinary differential equations. BIT, **42**, 561-599 (2002).
7. Iserles, A.: On the method of Neumann series for highly oscillatory equations. BIT, **44**, 473-488 (2004).
8. Iserles, A., Nørsett, S. P.: Efficient quadrature of highly oscillatory integrals using derivatives. Proc. Royal Soc. A, **461**, 1383–1399 (2005a).
9. Iserles, A., Nørsett, S. P.: On quadrature methods for highly oscillatory integrals and their implementation. BIT.(2005 b) To appear.
10. Iserles, A., Nørsett, S. P.: Quadrature methods for multivariate highly oscillatory integrals using derivatives. Maths Comp. (2006) To appear.
11. Levin, D.: Procedure for computing one- and two-dimensional integrals of functions with rapid irregular oscillations. Maths Comp., **38**, 531-538 (1982).
12. Lorenz, K., Jahnke, T., Lubich, C.: Adiabatic integrators for highly oscillatory second order linear differential equations with time-varying eigendecomposition. BIT, **45**, 91–115 (2005).
13. Luke, Y. L.: On the computation of oscillatory integrals. Proc. Cambridge Phil. Soc., **50**, 269-277 (1954).
14. Munkres, J. R., Analysis on Manifolds, Addison-Wesley, Reading, MA (1991).
15. Olver, F. W. J., Asymptotics and Special Functions, Academic Press, New York (1974).
16. Olver, S.: Moment-free numerical integration of highly oscillatory functions. IMA J. Num. Anal. (2005a) To appear.
17. Olver, S., Multivariate Levin-type method, Technical Report TBD, DAMTP, University of Cambridge (2005b).
18. Stein, E.: Harmonic Analysis: Real-Variable Methods, Orthogonality, and Oscillatory Integrals, Princeton University Press, Princeton, NJ (1993).
19. Ursell, F.: Integral equations with a rapidly oscillating kernel. J. London Math. Soc., **44**, 449-459 (1969).
20. Xiang, S.: On quadrature of Bessel transformations. J. Comput. Appld Maths,**177**, 231-239 (2005).

The 3D Inverse Electromagnetic Scattering Problem for a Coated Dielectric

Fioralba Cakoni and Peter Monk

Department of Mathematical Sciences University of Delaware, Newark, Delaware 19716, USA.
cakoni@math.udel.edu, monk@udel.edu

Summary. We use the linear sampling method to determine the shape and surface conductivity of a partially coated dielectric from a knowledge of the far field pattern of the scattered electromagnetic wave at fixed frequency. A mathematical justification of the method is provided for the full 3D vector case based on the use of a complete family of solutions. Numerical examples are given.

1 Introduction

In a previous paper together with D. Colton [6], we have analyzed the use of the linear sampling method to identify the shape of a coated dielectric in the 2D TM-polarized case. In addition we proposed and tested a heuristic formula for calculating the surface conductivity from far field data. In this paper we extend the techniques of [6] to the full three dimensional electromagnetic scattering problem at a fixed frequency. Using approximation arguments we shall provide a mathematical justification of the linear sampling method of finding the shape of a coated dielectric. Such arguments avoid the need to analyze an appropriate interior problem appearing in the theory (the "interior transmission problem"). Assuming that the interior transmission problem is well-posed (currently an open problem), we then derive a formula for the surface conductivity. Computational results for simple model problems show that the linear sampling method can reconstruct the surface conductivity .

The physical relevance and background for the inverse problem in this paper is discussed in [6] and we direct the reader there for further references.

The plan of our paper is as follows. In Section 2 we formulate the direct and inverse scattering problem for a dielectric that is partially coated by a highly conductive layer. In Section 3, we then use the linear sampling method [10] to determine the shape of the scattering object. We also discuss how to additionally recover the surface conductivity from the scattering data. In Section 4, we conclude by providing some numerical examples.

2 Formulation of the direct and inverse scattering problem

Let $D \subset \mathbb{R}^3$ be a bounded region with boundary Γ such that $D_e := \mathbb{R}^3 \setminus \overline{D}$ is connected. Each simply connected piece of D is assumed to be a Lipschitz curvilinear polyhedron. Moreover we assume that the boundary $\Gamma = \Gamma_1 \cup \Pi \cup \Gamma_2$ is split into two disjoint parts Γ_1 and Γ_2 having Π as their possible common boundary in Γ which is assumed to be a union of Lipschitz curves. The domain D is the support of an anisotropic object that is partially coated on a portion Γ_2 of the boundary by a very thin homogeneous layer of a highly conductive material and the incident field is a time-harmonic electromagnetic plane wave with frequency ω (Γ_1 may be the empty set which corresponds to a fully coated obstacle!). The interior electric and magnetic fields $\widetilde{E}^{int}$, $\widetilde{H}^{int}$, and the exterior electric and magnetic fields $\widetilde{E}^{ext}$, $\widetilde{H}^{ext}$, satisfy

$$\begin{cases} \nabla \times \widetilde{E}^{ext} - i\omega\mu_0 \widetilde{H}^{ext} = 0 \\ \nabla \times \widetilde{H}^{ext} + i\omega\epsilon_0 \widetilde{E}^{ext} = 0 \end{cases} \quad \text{in } D_e \qquad (1)$$

$$\begin{cases} \nabla \times \widetilde{E}^{int} - i\omega\mu_0 \widetilde{H}^{int} = 0 \\ \nabla \times \widetilde{H}^{int} + (i\omega\epsilon(x) - \sigma(x))\widetilde{E}^{int} = 0 \end{cases} \quad \text{in } D \qquad (2)$$

and on the boundary Γ

$$\nu \times \widetilde{E}^{ext} - \nu \times \widetilde{E}^{int} = 0 \qquad \text{on } \Gamma \qquad (3)$$

$$\nu \times \widetilde{H}^{ext} - \nu \times \widetilde{H}^{int} = 0 \qquad \text{on } \Gamma_1 \qquad (4)$$

$$\nu \times \widetilde{H}^{ext} - \nu \times \widetilde{H}^{int} = \widetilde{\eta}\,(\nu \times \widetilde{E}^{ext}) \times \nu \qquad \text{on } \Gamma_2. \qquad (5)$$

The electric permittivity ϵ_0 and magnetic permeability μ_0 of the exterior dielectric medium are positive constants whereas the scatterer has the same magnetic permeability μ_0 as the exterior medium but the electric permittivity ϵ and conductivity σ are real 3×3 matrix valued functions. The constant $\widetilde{\eta} > 0$ describes the physical properties of the thin coating layer [1]. If we define $\widetilde{E}^{(ext,int)} = \frac{1}{\sqrt{\epsilon_0}} E^{(ext,int)}$, $\widetilde{H}^{(ext,int)} = \frac{1}{\sqrt{\mu_0}} H^{(ext,int)}$, $k^2 = \epsilon_0 \mu_0 \omega^2$, $N(x) = \frac{1}{\epsilon_0}\left(\epsilon(x) + i\frac{\sigma(x)}{\omega}\right)$, and $\widetilde{\eta} = \sqrt{\frac{\mu_0}{\epsilon_0}}\eta$ we obtain the transmission problem

$$\begin{cases} \nabla \times E^{ext} - ikH^{ext} = 0 \\ \nabla \times H^{ext} + ikE^{ext} = 0 \end{cases} \quad \text{in } D_e \qquad (6)$$

$$\begin{cases} \nabla \times E^{int} - ikH^{int} = 0 \\ \nabla \times H^{int} + ikN(x)E^{int} = 0 \end{cases} \quad \text{in } D \qquad (7)$$

$$\nu \times E^{ext} - \nu \times E^{int} = 0 \quad \text{on} \quad \Gamma \tag{8}$$

$$\nu \times H^{ext} - \nu \times H^{int} = 0 \quad \text{on} \quad \Gamma_1 \tag{9}$$

$$\nu \times H^{ext} - \nu \times H^{int} = \eta \left(\nu \times E^{ext} \right) \times \nu \quad \text{on} \quad \Gamma_2, \tag{10}$$

where the exterior field E^{ext}, H^{ext} is given by

$$E^{ext} = E^i + E^s \tag{11}$$

$$H^{ext} = H^i + H^s, \tag{12}$$

E^s, H^s is the scattered field satisfying the Silver Müller radiation condition

$$\lim_{r \to \infty} (H^s \times x - rE^s) = 0 \tag{13}$$

uniformly in $\hat{x} = x/|x|$, $r = |x|$, the incident field E^i, H^i is given by

$$E^i(x) := \frac{i}{k} \nabla \times \nabla \times p e^{ikx \cdot d} = ik(d \times p) \times d e^{ikx \cdot d} \tag{14}$$

$$H^i(x) := \nabla \times p e^{ikx \cdot d} = ikd \times p e^{ikx \cdot d},$$

where the wave number k is a positive constant, $d \in \Omega := \{x \in \mathbb{R}^3 : |x| = 1\}$ is a unit vector giving the direction of propagation and p is the polarization vector.

In the following we assume that N is a 3×3 symmetric matrix-valued function whose entries are in $C^1(\overline{D})$, and N satisfies $\bar{\xi} \cdot \text{Im}(N) \xi \geq 0$ and $\bar{\xi} \cdot \text{Re}(N) \xi \geq \gamma |\xi|^2$ for all $\xi \in \mathbb{C}^3$ and all $x \in \overline{D}$ where γ is a positive constant. In order to formulate precisely the forward problem we need the following spaces. Letting $(H^s(D))^3$, $(H^s_{loc}(D_e))^3$ and $(H^s(\Gamma))^3$, $s \in \mathbb{R}$, denote the product of the standard Sobolev spaces defined on D, D_e and Γ respectively (with the convention $H^0 = L^2$), and

$$H(\text{curl}, D) := \{u \in (L^2(D))^3 : \nabla \times u \in (L^2(D))^3\}$$

$$L^2_t(\Gamma) := \{u \in (L^2(\Gamma))^3 : \nu \cdot u = 0 \quad \text{on} \quad \Gamma\}$$

$$L^2_t(\Gamma_2) := \{u|_{\Gamma_2} : u \in L^2_t(\Gamma)\},$$

we introduce the space

$$X(D, \Gamma_2) := \{u \in H(\text{curl}, D) : \nu \times u|_{\Gamma_2} \in L^2_t(\Gamma_2)\}$$

equipped with the norm

$$\|u\|^2_{X(D, \Gamma_2)} = \|u\|^2_{H(\text{curl}, D)} + \|\nu \times u\|^2_{L^2(\Gamma_2)}. \tag{15}$$

For the exterior domain D_e we define the above spaces in the same way for every $D_e \cap B_R$, with B_R a ball of radius R containing D and denote these spaces by $H_{loc}(\text{curl}, D_e)$ and $X_{loc}(D_e, \Gamma_2)$, respectively. Finally, we introduce the trace space of $X(D, \Gamma_2)$ on Γ by

$$Y(\Gamma) := \left\{ h \in (H^{-1/2}(\Gamma))^3 \ : \ \exists u \in H_0(\text{curl}, B_R), \begin{array}{l} \nu \times u|_{\Gamma_2} \in L_t^2(\Gamma_2) \\ \text{and} \quad h = \nu \times u|_\Gamma \end{array} \right\}$$

where $H_0(\text{curl}, B_R)$ is the space of functions u in $H(\text{curl}, B_R)$ satisfying $\hat{x} \times u = 0$ on the boundary of B_R. As shown in [7] $Y(\Gamma)$ is a Banach space with respect to the norm

$$\|h\|_{Y(\Gamma)}^2 := \inf \{\|u\|_{H(\text{curl}, B_R)}^2 + \|\nu \times u\|_{L_t^2(\Gamma_2)}^2\} \tag{16}$$

where the infimum is taken over all functions $u \in H_0(\text{curl}, B_R)$ such that $\nu \times u|_{\Gamma_2} \in L_t^2(\Gamma_2)$ and $h = \nu \times u|_\Gamma$. In fact $Y(\Gamma)$ coincides with $H_{\text{div}}^{-\frac{1}{2}}(\Gamma) \cap L_t^2(\Gamma_2)$ where

$$H_{\text{div}}^{-\frac{1}{2}}(\Gamma) := \left(u \in (H^{-\frac{1}{2}}(\Gamma))^3, \quad \nu \cdot u = 0, \quad \text{div}_\Gamma u \in H^{-\frac{1}{2}}(\Gamma) \right)$$

is the trace space of $\nu \times u|_\Gamma$ for $u \in H_0(\text{curl}, B_R)$ (see [4] and [2, 3] for the case of Lipshitz boundaries). We also recall that the trace space of $(\nu \times u) \times \nu|_\Gamma$ for $u \in H(\text{curl}, B_R)$ is defined by

$$H_{\text{curl}}^{-\frac{1}{2}}(\Gamma) := \left(u \in (H^{-\frac{1}{2}}(\Gamma))^3, \quad \nu \cdot u = 0, \quad \text{curl}_\Gamma u \in H^{-\frac{1}{2}}(\Gamma) \right),$$

and a duality relation is defined between $H_{\text{div}}^{-\frac{1}{2}}(\Gamma)$ and $H_{\text{div}}^{-\frac{1}{2}}(\Gamma)$.

Expressing the magnetic fields in terms of the electric fields, the direct scattering problem becomes a particular case of the following problem: Given $f \in Y(\Gamma)$, $h \in Y(\Gamma)$, $h_2 \in L_t^2(\Gamma_2)$ find $E^s \in X_{\text{loc}}(D_e, \Gamma_2)$, $E^{int} \in X(D, \Gamma_2)$ such that

$$\nabla \times \nabla \times E^s - k^2 E^s = 0 \quad \text{in} \quad D_e \tag{17}$$

$$\nabla \times \nabla \times E^{int} - k^2 N(x) E^{int} = 0 \quad \text{in} \quad D \tag{18}$$

$$\nu \times E^s - \nu \times E^{int} = f \quad \text{on} \quad \Gamma \tag{19}$$

$$\nu \times (\nabla \times E^s) - \nu \times (\nabla \times E^{int}) = h + \begin{cases} 0 & \text{on } \Gamma_1 \\ ik\eta E_T^s + h_2 & \text{on } \Gamma_2 \end{cases} \tag{20}$$

$$\lim_{r \to \infty}((\nabla \times E^s) \times x - ikr E^s) = 0 \tag{21}$$

where u_T denotes the tangential component $(\nu \times u) \times \nu$. Note that the direct scattering problem corresponds to $f := -\nu \times E^i|_\Gamma$, $h := -\nu \times (\nabla \times E^i)|_\Gamma$, and $h_2 := ik\eta E_T^i|_{\Gamma_2}$.

The following theorem concerning the well-posedness of the above problem was proved in [9].

Theorem 1. *The transmission problem (17)–(21) has a unique solution $E^{int} \in X(D, \Gamma_2)$, $E^s \in X_{\text{loc}}(D_e, \Gamma_2)$ which satisfies*

$$\|E^{int}\|_{X(D, \Gamma_2)} + \|E^s\|_{X(B_R \setminus \overline{D}, \Gamma_2)} \leq C \left(\|f\|_{Y(\Gamma)} + \|h\|_{Y(\Gamma)} + \|h_2\|_{L_t^2(\Gamma_2)} \right) \tag{22}$$

for some positive constant C depending on R but not on f, h and h_2.

It is known [11] that the radiating solution E^s to (17)–(21) has the asymptotic behavior
$$E^s(x) = \frac{e^{ik|x|}}{|x|}\left\{E_\infty(\hat{x}) + O\left(\frac{1}{|x|}\right)\right\} \tag{23}$$
as $|x| \to \infty$, where E_∞ is defined on the unit sphere Ω and is known as the *electric far field pattern*. In the case of incident plane waves given by (14) the electric far field pattern depends on the incident direction and polarization which will be indicated by $E_\infty(\hat{x}) = E_\infty(\hat{x}, d, p)$.

The *inverse scattering problem* we are concern with is to determine D and η from a knowledge of the electric far field pattern $E_\infty(\hat{x}, d, p)$ of the scattered field E^s for $\hat{x}, -d \in \Omega_0$, where Ω_0 is a subset of the unit sphere Ω, and three linearly independent polarizations p. Note that no a priori knowledge of the amount of coating is required.

3 Analysis of the inverse problem

Now we turn to the *inverse problem* for the vector case. Given the incident plane wave $E^i = ik(d \times p) \times d\, e^{ikx \cdot d}$ and the corresponding electric far field pattern $E_\infty(\hat{x}, d, p)$ for $\hat{x}, d$ in the unit sphere Ω and three linearly independent polarizations p, determine D and η. Uniqueness results for the inverse problem can be found in [9]. The aim of this paper is to show how to reconstruct D and η from the given data.

3.1 Shape reconstruction

The analysis of this inverse problems follows the lines of the analysis of the inverse problem in the scalar case treated in [6]. We define *Maxwell eigenvalues* to be the values of k for which
$$\nabla \times \nabla \times E + k^2 N(x) E = 0 \quad \text{in} \quad D$$
$$\nu \times E = 0 \quad \text{on} \quad \Gamma,$$
has a nontrivial solution, and *transmission eigenvalues* the values of k for which
$$\begin{cases} \nabla \times \nabla \times E_0 - k^2 E_0 = 0 \\ \nabla \times \nabla \times E - k^2 N(x) E = 0 \end{cases} \quad \text{in } D \tag{24}$$
$$\nu \times E - \nu \times E_0 = 0 \quad \text{on } \Gamma \tag{25}$$
$$\nu \times (\nabla \times E) - \nu \times (\nabla \times E_0) = 0 \quad \text{on } \Gamma_1 \tag{26}$$
$$\nu \times (\nabla \times E) - \nu \times (\nabla \times E_0) = -ik\eta(\nu \times E_0) \times \nu \quad \text{on } \Gamma_2. \tag{27}$$
has a nontrivial solution. Note that if $\bar{\xi} \cdot \text{Im}(N)\xi > 0$ at a point $x_0 \in D$ Maxwell eigenvalues and transmission eigenvalues do not exist.

We now define an *electromagnetic Herglotz pair* to be a pair of vector fields of the form

$$E_g(x) = \int_\Omega e^{ikx\cdot d} g(d) ds(d), \qquad H_g(x) = \frac{1}{ik} \nabla \times E_g(x) \qquad (28)$$

where $g \in L_t^2(\Omega)$. It is easy to see that $\nabla \times \nabla \times E_g - k^2 E_g = 0$. Next we consider the vector space

$$E(D) := \{E \in X(D, \Gamma_2), \, \nabla \times E \in X(D, \Gamma_2), \\ \nabla \times \nabla \times E - k^2 N(x) E = 0 \text{ in } D\}$$

and define the subset of $\mathcal{Y}(\Gamma) := H_{div}^{-\frac{1}{2}}(\Gamma) \times Y(\Gamma_1) \times L_t^2(\Gamma_2)$ by

$$\mathcal{E} := \{\nu \times (E_g - E), \, \nu \times \nabla \times (E_g - E)|_{\Gamma_1}, \\ \nu \times \nabla \times (E_g - E) - ik\eta(\nu \times E_g) \times \nu|_{\Gamma_2}\}$$

for all $E \in E(D)$, $g \in L_t^2(\Gamma)$, E_g the electric field of the electromagnetic Herglotz pair with kernel g, where $Y(\Gamma_1) := \{h|_{\Gamma_1} : h \in Y(\Gamma)\}$.

Theorem 2. *Suppose that k is neither a Maxwell eigenvalue nor a transmission eigenvalue. Then $\mathcal{E}$ is dense in $\mathcal{Y}(\Gamma)$.*

Proof. Let $\varphi \in H_{curl}^{-\frac{1}{2}}(\Gamma)$ and $\psi \in H_{curl}^{-\frac{1}{2}}(\Gamma) \cap L_t^2(\Gamma_2)$ such that

$$\int_\Gamma \nu \times (E_g - E) \cdot \varphi \, ds + \int_\Gamma \nu \times \nabla \times (E_g - E) \psi \, ds - \int_{\Gamma_2} ik\eta (E_g)_T \cdot \psi \, ds = 0 \quad (29)$$

for all $g \in L_t^2(\Omega)$ and $E \in E(D)$. Note that $\varphi \in H_{curl}^{-\frac{1}{2}}(\Gamma)'$ and $\psi|_{\Gamma_1} \in Y(\Gamma_1)'$ (see [7], Section 2.2 for the characterization of the dual space $Y(\Gamma_1)'$). The first and the second integral in (29) is understood in the sense of duality paring between $H_{div}^{-\frac{1}{2}}(\Gamma)$ and $H_{div}^{-\frac{1}{2}}(\Gamma)$ while the third integral containing η is understood in the $L_t^2(\Gamma_2)$ sense. Setting first $E = 0$ in (29) and interchanging the order of integrations we obtain

$$0 = \hat{x} \times \left\{ \int_\Gamma (\varphi \times \nu) e^{-iky\cdot \hat{x}} \, ds + ik\, \hat{x} \times \int_\Gamma (\psi \times \nu) e^{-iky\cdot \hat{x}} \, ds \right. \\ \left. - ik\eta \int_{\Gamma_2} [(\nu \times \psi) \times \nu] e^{-iky\cdot \hat{x}} \, ds \right\} \times \hat{x} \qquad (30)$$

The right hand side of the above expression is the far field pattern of the following electric and magnetic dipole distribution defined by

$$P(x) = \frac{1}{k^2} \nabla \times \nabla \times \int_\Gamma (\varphi(y) \times \nu) \Phi(x,y) \, ds_y$$

$$- \nabla \times \int_\Gamma (\psi(y) \times \nu) \Phi(x,y) \, ds_y$$

$$- i \frac{\eta}{k} \nabla \times \nabla \times \int_{\Gamma_2} [(\nu \times \psi(y)) \times \nu] \Phi(x,y) \, ds_y \tag{31}$$

where $\Phi(x,y) := \frac{1}{4\pi} \frac{e^{ik|x-y|}}{|x-y|}$. Therefore we conclude that $P(x) = 0$ in $D_e := \mathbb{R}^3 \setminus \overline{D}$. Hence taking the limit of $P(x)$ as $x \to \Gamma$ from both sides we obtain

$$\nu \times P^- = -\nu \times \psi \qquad \nu \times \nabla \times P^- - ik\widetilde{\eta}(\nu \times P^-) \times \nu = \nu \times \varphi$$

on Γ, where the superscript - indicates the limit obtained by approaching the boundary Γ from D, and $\widetilde{\eta} = 0$ on Γ_1 and $\widetilde{\eta} = \eta$ on Γ_2. We remark that $P(x)$ and $\operatorname{curl} P(x)$ are both square integrable in any compact subset of D and D_e. Furthermore, since $\varphi \times \nu$ and $\psi \times \nu$ are in $H_{div}^{-\frac{1}{2}}(\Gamma)$, the potentials over Γ in (31) and the corresponding jump relations are well defined from potential theory for single layer potentials with $H^{-\frac{1}{2}}$ densities [17], while the jump relations for the potential over Γ_2 with L^2 density is interpreted in the sense of the L^2 limit ([11] p.172). Next, setting $E_g = 0$ in (29), using the expressions for φ and ψ and Green's formula together with a parallel surfaces argument (see [14]) we obtain

$$0 = \int_\Gamma [(\nu \times P^-) \cdot \nabla \times E - (\nu \times E) \cdot \nabla \times P^-$$

$$- ik\widetilde{\eta}(\nu \times E) \cdot (\nu \times P^-)] \, ds$$

$$= k^2 \int_D P^- \cdot (I - N) E \, dx - ik\eta \int_{\Gamma_2} (\nu \times E) \cdot (\nu \times P^-) \, ds. \tag{32}$$

Note that $P \in L^2(D)$. Now let $F \in H(\operatorname{curl}, D)$ be the unique solution (c.f. [18]) of

$$\nabla \times \nabla \times F - k^2 N F = k^2 (I - N) P \quad \text{in} \quad D$$
$$\nu \times F = 0 \quad \text{on} \quad \Gamma.$$

Using the vector Green formula for E and F, from (32) we obtain

$$\int_D (\nu \times E) \cdot \nabla \times F \, ds = -k^2 \int_D P^- \cdot (I - N) E \, dx$$

$$= -ik\eta \int_{\Gamma_2} (\nu \times E) \cdot (\nu \times P^-) \, ds.$$

Hence
$$\int_\Gamma (\nu \times E) \cdot [\nabla \times F + ik\widetilde{\eta}(\nu \times P^-)] \, ds = 0$$
for all $E \in E(D)$ whence
$$\nu \times \nabla \times F + ik\widetilde{\eta}(\nu \times P^-) \times \nu = 0 \quad \text{on} \quad \Gamma$$
since k is not a Maxwell eigenvalue. Now we observe that P and $\widetilde{E} = P + F$ satisfy
$$\begin{cases} \nabla \times \nabla \times P - k^2 P = 0 \\ \nabla \times \nabla \times \widetilde{E} - k^2 N(x)\widetilde{E} = 0 \end{cases} \quad \text{in } D \quad (33)$$
$$\nu \times \widetilde{E} - \nu \times P = 0 \quad \text{on } \Gamma \quad (34)$$
$$\nu \times (\nabla \times \widetilde{E}) - \nu \times (\nabla \times P) = 0 \quad \text{on } \Gamma_1 \quad (35)$$
$$\nu \times (\nabla \times \widetilde{E}) - \nu \times (\nabla \times P) = -ik\eta(\nu \times P) \times \nu \quad \text{on } \Gamma_2 \quad (36)$$

which implies that $P = \widetilde{E} = 0$ in D provided k is not a transmission eigenvalue. Therefore $\varphi = \psi = 0$ which proves the theorem. We remark that, in order to conclude that $P = \widetilde{E} = 0$ in D, the $H(\text{curl}, D_h)$-regularity of P, where $\overline{D_h} \subset D$ allows us to apply the vector Green's formula in any compact subset D_h of D and then take the limit of the surface integrals since the boundary relations in (33)–(36) hold in the L^2-limit sense (see Lemma 2.1 of [6] for a similar proof in the scalar case). ∎

Next we define the *far field operator* $F : L_t^2(\Omega) \to L_t^2(\Omega)$ by
$$(Fg)(\hat{x}) := \int_\Omega E_\infty(\hat{x}, d, g(d)) ds(d), \quad \hat{x} \in \Omega \text{ and } g \in L_t^2(\Omega) \quad (37)$$

and look for solutions $g \in L_t^2(\Omega)$ of the *far field equation*
$$(Fg)(\hat{x}) := E_{e,\infty}(\hat{x}, z, q) \quad (38)$$
where
$$E_{e,\infty}(\hat{x}, z, q) = \frac{ik}{4\pi}(\hat{x} \times q) \times \hat{x}\, e^{-ik\hat{x}\cdot z}$$
is the electric far field pattern of the electric dipole with polarization q given by
$$E_e(x, z, q) := \frac{i}{k} \nabla_x \times \nabla_x \times q\, \Phi(x, z) \quad (39)$$
with $\Phi(x, z) := \frac{1}{4\pi} \frac{e^{ik|x-z|}}{|x-z|}$. We can now prove the following theorem.

Theorem 3. *Assume that k is neither a transmission eigenvalue nor a Maxwell eigenvalue and let F be the far field operator corresponding to (6)–(13). Then we have:*

1. For $z \in D$ and every $\epsilon > 0$ there exists a solution $g_\epsilon^z \in L_t^2(\Omega)$ of the inequality
$$\|Fg_\epsilon^z - E_{e,\infty}(\cdot, z, q)\|_{L^2(\Omega)} < \epsilon$$
such that $\|E_{g_\epsilon^z}\|_{X(D,\Gamma_2)} < \infty$ where $E_{g_\epsilon^z}$ is the electric field of the electromagnetic Herglotz pair with kernel g_ϵ^z. Moreover, for a fixed $\epsilon > 0$
$$\lim_{z \to \Gamma} \|g_\epsilon^z\|_{L_t^2(\Omega)} = \infty \quad \text{and} \quad \lim_{z \to \Gamma} \|E_{g_\epsilon^z}\|_{X(D,\Gamma_2)} = \infty.$$

2. For $z \in \mathbb{R}^3 \setminus \overline{D}$ and every $\epsilon > 0$ and $\delta > 0$ there exists a solution $g_{\epsilon,\delta}^z \in L_t^2(\Omega)$ of the inequality
$$\|Fg_{\epsilon,\delta}^z - E_{e,\infty}(\cdot, z, q)\|_{L^2(\Omega)} < \epsilon + \delta$$
such that
$$\lim_{\delta \to 0} \|g_{\epsilon,\delta}^z\|_{L_t^2(\Omega)} = \infty \quad \text{and} \quad \lim_{\delta \to 0} \|E_{g_{\epsilon,\delta}^z}\|_{X(D,\Gamma_2)} = \infty,$$
where $E_{g_{\epsilon,\delta}^z}$ is the electric field of the electromagnetic Herglotz pair with kernel $g_{\epsilon,\delta}^z$.

Remark 1. Note that, in Theorem 3, $E_{g_\epsilon^z}$ for $z \in D$ is such that $\nu \times E_{g_\epsilon^z}$ and $\nu \times \nabla \times E_{g_\epsilon^z}$ converge with respect to the $Y(\Gamma)$ norm as $\epsilon \to 0$.

Proof. The proof follows the lines of the proof of Theorem 2.6 of [6]. Let $\mathcal{B}$ denotes the linear bounded operator which maps $f := \nu \times E|_\Gamma$, $h := \nu \times (\nabla \times E)|_\Gamma$ and $h_2 := ik\eta(\nu \times E) \times \nu|_{\Gamma_2}$, where $E \in X(D, \Gamma_2)$ satisfies $\nabla \times \nabla \times E - k^2 E = 0$, onto the electric far field pattern of the corresponding solution of (17)–(21). Exactly in the same way as in Lemma 2.5 of [6] by making use of the result of Theorem 2 and using the divergence free vector spherical wave functions [11], one can show that $\mathcal{B} : Y(\Gamma) \times Y(\Gamma) \times L_t^2(\Gamma_2) \to L_t^2(\Omega)$ is compact, injective and has dense range provided that k is neither a Maxwell eigenvalue nor a transmission eigenvalue.

Next consider $z \in D$. Given $\epsilon > 0$ from Theorem 2 there exists $E_{g_\epsilon^z}$ with $g_\epsilon^z \in L_t^2(\Omega)$ and $E_\epsilon^z \in E(D)$ such that $\nu \times (E_{g_\epsilon^z} - E_\epsilon^z)$, $\nu \times \nabla \times (E_{g_\epsilon^z} - E_\epsilon^z) - ik\widetilde{\eta}(\nu \times E_{g_\epsilon^z}) \times \nu$ approximates $\nu \times E_e(\cdot, z, q)$, $\nu \times \nabla \times E_e(\cdot, z, q) - ik\widetilde{\eta}(\nu \times E_e(\cdot, z, q)) \times \nu$ in the $\mathcal{Y}(\Gamma)$ norm with discrepancy ϵ, where $\widetilde{\eta} = \eta$ on Γ_2 and $\widetilde{\eta} = 0$ on Γ_1. Noting that Fg_ϵ^z is the far field pattern of the electric scattered field corresponding to $E_{g_\epsilon^z}$ as the incident field, from the estimate (22) and the fact that the far field pattern depends continuously on the scattered field we obtain that
$$\|Fg_{\epsilon,\delta}^z - E_{e,\infty}(\cdot, z, q)\|_{L^2(\Omega)} < C\epsilon$$
where C is a positive constant independent of ϵ. As $z \to \Gamma$, using (22) for the solution of the direct scattering problem E_ϵ and $E_e(\cdot, z, q)$ together with the fact that $\lim_{z \to \Gamma} \|E_e(\cdot, z, q)\|_{H(\mathrm{curl}, B_R \setminus \overline{D})} = \infty$ one obtain that

$$\lim_{z \to \Gamma} \|g_\epsilon^z\|_{L^2_t(\Omega)} = \infty \quad \text{and} \quad \lim_{z \to \Gamma} \|E_{g_\epsilon^z}\|_{X(D,\Gamma_2)} = \infty.$$

Now let $z \in \mathbb{R}^3 \setminus \overline{D}$. From the theory of the ill-posed problems applied to the compact operator $\mathcal{B}$, we obtain

$$\mathcal{B}(f_z^\alpha, h_z^\alpha, h_{1z}^\alpha) - E_{e,\infty}(\cdot, z, q)\|_{L^2(\Omega)} < \delta$$

for an arbitrary small but fixed δ where $f_z^\alpha = \nu \times E_z^\alpha, h_z^\alpha = \nu \times \nabla \times E_z^\alpha, h_{1z}^\alpha = ik\eta(\nu \times E_z^\alpha \times \nu)$ with $E_z^\alpha \in X(D,\Gamma_2)$ is the regularized solution corresponding to the regularization parameter α chosen by a regular regularization strategy (e.g. the Morozov discrepancy principle). Furthermore, we have that the $Y(\Gamma) \times Y(\Gamma) \times L^2_t(\Gamma_2)$ norm of $(f_z^\alpha, h_z^\alpha, h_{1z}^\alpha)$ goes to infinity as $\alpha \to \infty$. Note that $\alpha \to 0$ as $\delta \to 0$. Now the second part of the theorem follows from the fact that E_z^α can be approximated arbitrarily close with respect to the $X(D,\Gamma_2)$-norm by a Herglotz wave function E_g (see Theorem 2.5 of [7]) and the fact that $Fg = \mathcal{B}(\nu \times E_g, \nu \times \nabla \times E_g, ik\eta(\nu \times E_g \times \nu))$. This ends the proof. ∎

The above result provides a characterization for the boundary Γ of the scattering object D. Unfortunately, since the behavior of $E_{g_\epsilon^z}$ is described in terms of a norm depending on the unknown region D, $E_{g_\epsilon^z}$ can not be used to characterize D. Instead the linear sampling method characterizes the obstacle by the behavior of g_ϵ^z. In particular, given a discrepancy $\epsilon > 0$ and g_ϵ^z the ϵ-approximate solution of the far field equation (38), the boundary of the scatterer is reconstructed as the set of points z where the $L^2_t(\Omega)$ norm of g_ϵ^z becomes large. Alternatively one can try to use $|E_{g_\epsilon^z}|$ as an indicator function of the boundary ∂D of the scattering object D as it will be shown in the numerical examples presented in Section 4 (although this is not justified by the foregoing theory).

3.2 Identification of the surface conductivity

Assuming now that D is known, we want to determine the surface conductivity η by making use of the approximate solution g to the far field equation (38). In [5] a formula for computing η in the 2D TE-polarized case is derived and the mathematical justification is based on the analysis of a appropriate boundary value problem called the *interior transmission problem*. The interior transmission problem corresponding to our scattering problem reads: Find a solution E, E_0 of the following boundary value problem

$$\begin{cases} \nabla \times \nabla \times E_0^z - k^2 E_0^z = 0 \\ \nabla \times \nabla \times E^z - k^2 N(x) E^z = 0 \end{cases} \quad \text{in } D \quad (40)$$

$$\nu \times E^z - \nu \times (E_0^z + E_e(\cdot, z, q)) = 0 \quad \text{on } \Gamma \quad (41)$$

$$\nu \times (\nabla \times E^z) - \nu \times [\nabla \times (E_0^z + E_e(\cdot, z, q))] = 0 \quad \text{on } \Gamma_1 \quad (42)$$

$$\nu \times (\nabla \times E^z) - \nu \times [\nabla \times (E_0^z + E_e(\cdot, z, q))] = \\ -ik\eta \, [\nu \times (E_0^z - E_e(\cdot, z, q))] \times \nu \quad \text{on } \Gamma_2 \quad (43)$$

where $E_e(\cdot, z, q)$ is the electric dipole given by (39), $z \in D$ and $q \in \mathbb{R}^3$.

As noticed in [6] the completeness result given by Theorem 2 does not suffice to proceed further with the reconstruction of η. It is essential in the following analysis to know that the interior transmission problem has a (weak) solution in appropriate Sobolev spaces. Unfortunately, the well posedness of (40)–(43) is not yet established. In the case where $\eta = 0$, Haddar in [13] has shown that, provided k is not a transmission eigenvalue and under some assumptions on N, the interior transmission problem has a unique weak solution $E^z \in L^2(D)$ and $E_0^z \in L^2(D)$ such that $E^z - E_0^z \in H(\text{curl}, D)$ and $\nabla \times (E^z - E_0^z) \in H(\text{curl}, D)$.

Conjecture 1. Assume that k is not a transmission eigenvalue and either $\bar{\xi} \cdot \text{Re}\,(N - I)^{-1} \xi \geq \gamma |\xi|^2$ or $\bar{\xi} \cdot \text{Re}\,(N - I)\xi \geq \gamma |\xi|^2$ for all $\xi \in \mathbb{C}^3$, all $x \in \overline{D}$ and some $\gamma > 0$. Then the interior transmission problem (40)–(43) has a unique solution $E^z \in L^2(D)$, $E_0^z \in L^2(D)$ and $E_0^z|_{\Gamma_2} \in L^2(\Gamma_2)$ such that $E^z - E_0^z \in H(\text{curl}, D)$, $\nabla \times (E^z - E_0^z) \in H(\text{curl}, D)$ and $\nu \times \nabla \times (E^z - E_0^z)|_{\Gamma_2} \in L^2(\Gamma_2)$.

Assuming Conjecture 1, we now use the approximate solution g^z for $z \in D$ of the far field equation (38) to give an approximation for the surface conductivity η. To this end we need the following lemma.

Lemma 1. *Assume that k is neither a Maxwell eigenvalue nor a transmission eigenvalue. For any point z in D we have that*

$$\int_D \overline{E}^z \cdot \text{Im}\,(N) E^z\, dx + k\eta \int_{\Gamma_2} |\nu \times (E_0^z + E_e(\cdot, z, q))|^2\, ds$$
$$= -\frac{k^3}{6\pi} \|q\|^2 + k\text{Re}\,(E_0^z(z)) \tag{44}$$

where E^z and E_0^z is a solution to the interior transmission problem (40)–(43).

Proof. From Theorem 2, for given $\epsilon > 0$, there exists a $E_\epsilon^z \in E(D)$ and a electromagnetic Herglotz pair with electric field $E_{g_\epsilon^z}$ and kernel $g_\epsilon^z \in L_t^2(\Omega)$ such that

$$\begin{cases} \nu \times E_\epsilon^z - E_e(\cdot, z, q)) = E_{g_\epsilon^z} + \alpha_\epsilon \\ \nu \times \nabla \times (E_\epsilon^z - E_e(\cdot, z, q)) + ik\widetilde{\eta}(\nu \times E_e(\cdot, z, q)) \times \nu \\ = \nu \times \nabla \times E_{g_\epsilon^z} - ik\widetilde{\eta}(\nu \times E_{g_\epsilon^z}) \times \nu + \beta_\epsilon \end{cases}$$

on Γ where

$$\|(\alpha_\epsilon, \beta_\epsilon)\|_{\mathcal{Y}(\Gamma)} < \epsilon. \tag{45}$$

Now, let E^z and E_0^z be the unique solution of the interior transmission problem (40)–(43). Obviously, E_ϵ^z and $E_{g_\epsilon^z}$ and converge to E^z and E_0^z, respectively as $\epsilon \to 0$ with respect to the graph norm $L^2(D) \cap L_t^2(\Gamma_2)$. Hence, E_ϵ^z and $E_{g_\epsilon^z}$ are uniformly bounded together with their curl in the $L^2(D)$ norm. Applying the vector Green's formula to E_ϵ^z and $\overline{E}_\epsilon^z$ in D (see [18] for the case of $H(\text{curl}, D)$ functions) we obtain

$$\int_\Gamma (\nu \times E_\epsilon^z \cdot \operatorname{curl} \overline{E_\epsilon^z} - \nu \times \overline{E_\epsilon^z} \cdot \operatorname{curl} E_\epsilon^z)\, ds = 2i \int_D \overline{E_\epsilon^z} \cdot \operatorname{Im}(N) E_\epsilon^z\, dx. \qquad (46)$$

On the other hand, using (45) and defining $W_\epsilon^z := E_{g_\epsilon^z} + E_e(\cdot, z, q)$, we have that

$$\int_\Gamma (\nu \times E_\epsilon^z \cdot \operatorname{curl} \overline{E_\epsilon^z} - \nu \times \overline{E_\epsilon^z} \cdot \operatorname{curl} E_\epsilon^z)\, ds$$
$$= \int_\Gamma (\nu \times W_\epsilon^z \cdot \operatorname{curl} \overline{W_\epsilon^z} - \nu \times \overline{W_\epsilon^z} \cdot \operatorname{curl} W_\epsilon^z)\, ds$$
$$- 2ik\eta \int_{\Gamma_2} |(\nu \times W_\epsilon^z) \times \nu|^2\, ds + R_\epsilon^z \qquad (47)$$

where $|R_\epsilon^z| \leq C\epsilon$ for a positive constant C independent of ϵ. Again using the vector Green's formula, the integral representation formula and connecting the radiating solution $E_e(\cdot, z, q)$ to its far field pattern as in [8] Theorem 3.1, we obtain

$$\int_\Gamma (\nu \times W_\epsilon^z \cdot \operatorname{curl} \overline{W_\epsilon^z} - \nu \times \overline{W_\epsilon^z} \cdot \operatorname{curl} W_\epsilon^z)\, ds \qquad (48)$$
$$= -\frac{ik^3}{3\pi} \|q\|^2 + ikq \cdot [E_{g_\epsilon^z}(z) + \overline{E_{g_\epsilon^z}}(z)].$$

Hence, combining (46), (47) and (48) we have that

$$2i \int_D \overline{E_\epsilon^z} \cdot \operatorname{Im}(N) E_\epsilon^z\, dx + 2ik\eta \int_{\Gamma_2} |(\nu \times W_\epsilon^z) \times \nu|^2\, ds \qquad (49)$$
$$= -\frac{ik^3}{3\pi} \|q\|^2 + ikq \cdot [E_{g_\epsilon^z}(z) + \overline{E_{g_\epsilon^z}}(z)] - R_\epsilon^z.$$

Now letting $\epsilon \to 0$ in (49) we obtain the result. ∎

Theorem 4. *Let z be a fixed point in D, $\operatorname{Im}(N) = 0$ and assume that k is neither a Maxwell eigenvalue nor a transmission eigenvalue. Then for every $\epsilon > 0$ there exists an electromagnetic Herglotz function $E_{g_\epsilon^z}$ with kernel $g_\epsilon^z \in L_t^2(\Omega)$ an approximate solution of the far field equation (38) such that*

$$\left| \eta + \frac{\frac{k^2}{6\pi} \|q\|^2 - \operatorname{Re}(E_{g_\epsilon^z}(z))}{\|\nu \times (E_{g_\epsilon^z} + E_e(\cdot, z, q))\|^2_{L_t^2(\Gamma_2)}} \right| \leq \epsilon. \qquad (50)$$

Proof. From the proof of Theorem 3 we have that the kernel g_ϵ^z of the Herglotz wave function $E_{g_\epsilon^z}$ in the proof of Lemma 1 is the ϵ-approximate solution to the far field equation (38). Hence the result of the theorem follows from Lemma 1. ■

A draw back of (50) is that the extent of the coating Γ_2 is not known. So, in practice this expression only provides a lower bound for η. In addition, due to the accuracies in the determination of Γ by the linear sampling method, the computation of the outward normal ν can be problematic hence the most reliable lower bound for η is the following estimate

$$\eta \geq \frac{-\frac{k^2}{6\pi}\|q\|^2 + \mathrm{Re}\,(E_{g^z}(z))}{\|(E_{g^z} + E_e(\cdot, z, q))\|^2_{L^2_t(\Gamma)}} \qquad (51)$$

where g_z is the regularized solution of the far field equation (38) which is previously computed to determine D.

4 Numerical example

For detailed numerical examples of shape reconstruction for coated objects, and also of estimating the surface conductivity the reader can consult [12]. Here we will give a single numerical example that illustrates our more general experience with the method. The numerical experiment is performed on synthetic far field data computed using the Ultra Weak Variational Formulation of Maxwell's equations as described in [16]. This (already approximate) far field data is further corrupted by noise as described in [12]. The far field data is then used first to reconstruct the shape of the scatterer using the standard Linear Sampling Method. This involves computing an approximate solution to the far field equation (38) for many sampling points z by discretizing g on the unit sphere and applying Tikhonov regularization and Morozov's principle to this ill-posed problem.

Once an approximation to the boundary of the scatterer is determined, the conductivity η can be approximated using (50) or a lower bound estimated using (51).

For this example we choose as test object the cube $[-1,1]^3$. Outside this cube $N = 1$ and within the cube $N = 2$. The entire cube is coated with $\eta = .1$ and $k = 3$ so the wavelength of the radiation is $\lambda = 2.09$. Figure 1 shows the result of reconstructing the cube using the Linear Sampling Method with 96 incoming waves (and 96 measurements) for each of two linearly independent polarizations (the other parameters in the method including the surface chosen for display are as in [12]). It is interesting to see that the Herglotz wave function gives a much better reconstruction of the scatterer than the Herglotz kernel.

Using the reconstructed surface in panel (c) of Fig. 1 we can estimate η. Alternatively we can test the formula (50) using the exact boundary in (a). The exact value is $\eta = 0.1$ using (50) gives $\eta \approx 0.14$ and using (51) gives

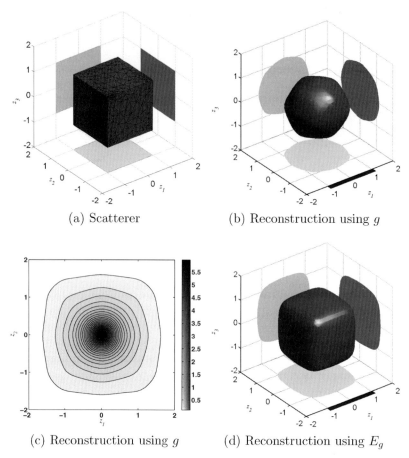

Fig. 1. Reconstructing the cube: (a) the original scatterer showing the surface mesh, (b) the reconstructed surface using the Linear Sampling Method and g, (c) a contour map of $1/\|g_z\|$ in the plane $z_3 = 0$ showing how the surface in (b) is obtained, (d) a reconstruction of the scatterer using $|E_{g_z}(z)|$. Surprisingly, use of the Herglotz wave function E_{g_z} gives a much better reconstruction of the scatterer than use of the kernel.

the same approximation. Of course the reconstructed scatterer is not very accurate and this accounts for the rather poor approximation to η (the lower bound is an overestimate for this reason also).

5 Conclusion

We have given some mathematical theory to substantiate the use of the Linear Sampling Method for reconstructing the shape of coated dielectrics. Assuming a conjecture on the existence of solutions of an interior transmission problem we have also derived a formula for the surface conductivity. Numerical results here and elsewhere show that the method can be applied in practice. We hope that the conjectured existence theory will be proved shortly.

Acknowledgment

We gratefully acknowledge the support of our research by the Air Force Office of Scientific Research under grant FA9550-05-1-0127. The computations were performed on a computer purchased under and NSF SCREMS grant DMS-0322583.

References

1. Angell, T. S., Kirsch, A.: The conductive boundary condition for Maxwell's equations. SIAM J. Appl. Math. **52**, 1597-1610 (1992)
2. Buffa, A., Ciarlet, P. Jr.: On traces for functional spaces related to Maxwell's equations. Part I: An integration by parts formula in Lipschitz polyhedra. Math. Meth. Appl. Sci., **24**, 9-30 (2001)
3. Buffa, A., Ciarlet, P. Jr.: On traces for functional spaces related to Maxwell's equations. Part II: Hodge decompositions on the boundary of Lipschitz polyhedra and applications. Math. Meth. Appl. Sci., **24**, 31-48 (2001)
4. Cessenat, M.: Mathematical Methods in Electromagnetism. World Sciences, Singapore (1996)
5. Cakoni, F., Colton, D., Monk, P.: The determination of the surface conductivity of a partially coated dielectric, SIAM J. Appl. Math. **64**, 709-723 (2004)
6. Cakoni, F., Colton, D., Monk, P.: The Inverse Electromagnetic Scattering Problem for a Partially Coated Dielectric, submitted for publication (2004)
7. Cakoni, F., Colton, D., Monk, P.: The electromagnetic inverse scattering problem for partially coated Lipschitz domains. Proc. Royal Soc. Edinburgh, **134 A**, 661-682 (2004)
8. Cakoni, F., Colton, D.: The determination of the surface impedance of a partially coated obstacle from far field data. SIAM J. Appl. Math., **64**, 709-723 (2004)
9. Cakoni, F., Colton, D.: A uniqueness theorem for an inverse electomagnetic scattering problem in inhomogeneous anisotropic media. Proc. Edinburgh Math. Soc., **46**, 293-314 (2003)
10. Colton, D., Haddar, H.,Piana, M.: The linear sampling method in inverse electromagnetic scattering theory. Inverse Problems, **19**, S105-S137 (2003)
11. Colton, D., Kress, R.: Inverse Acoustic and Electromagnetic Scattering Theory, 2nd ed. Springer Verlag (1998)
12. Colton, D., Monk, P.: Target Identification of Coated Objects, to appear in IEEE Trans. Antennas and Propagation (2006)
13. Haddar, H.: The interior transmission problem for anisotropic Maxwell's equations and its applications to the inverse problem. Math. Methods Appl. Sci., **27**, 2111-2129 (2004)

14. Hähner, P.: An approximation theorem in inverse electromagnetic scattering. Math. Methods Appl. Sci., **17**, 293-303 (1994)
15. Hoppe, D. J., Rahmat-Samii, Y.: Impedance Boundary Conditions in Electromagnetics. Taylor&Francis Publishers (1995)
16. Huttunen, T., Malinen, M., Monk, P.: Solving Maxwell's Equations Using the Ultra Weak Variational Formulation, submitted for publication.
17. McLean, W.: Strongly Elliptic Systems and Boundary Integral Equations. Cambridge University Press (2000)
18. Monk, P.: Finite Element Methods for Maxwell's Equations. Oxford University Press, Oxford (2003)

Functional Approach to Locally Based A Posteriori Error Estimates for Elliptic and Parabolic Problems

Sergey Repin

St. Petersburg Division of the Steklov Mathematical Institute,
Russian Academy of Sciences, Fontanka 27, 191023, St. Petersburg, Russia
repin@pdmi.ras.ru

Summary. The paper is concerned with functional approach to the a posteriori error control for approximate solutions of differential equations. Functional a posteriori estimates are derived by purely functional methods using the analysis of variational problems or integral identities. They are intended to give computable minorants and majorants for various measures of the difference between exact solutions and their conforming approximations. Functional estimates contain no mesh dependent constants and provide guaranteed lower and upper bounds of errors. In this paper, the major attention is paid on a posteriori estimates in terms of local norms or locally based linear functionals. It is shown that for linear elliptic and parabolic problems functional estimates in global (energy) norms imply a posteriori estimates in terms of local quantities.

1 Introduction

A posteriori error estimation methods for partial differential equations started receiving attention in the middle of the 20th century (see [10, 13]). In general, they are intended to solve two problems: (a) find reliable bounds of the overall error encompassed in an approximate solution and (b) give an error indicator for mesh–adaptive procedures. In the last decades, such topics as adaptive methods, reliable computer simulation methods and a posteriori estimates for differential equation were in the focus of numerous researches and are exposed in a vast amount of publications. It is not surprising that finite element methods (FEM) where one of the first were such methods were developed. At present, such methods as "explicit residual", "dual–weighted residual", and "equilibrated residual" (see, e.g., [1, 4, 2, 4, 27]) are widely used by numerical analysts for a posteriori control of the quality of approximate solutions. Methods based on post–processing (e.g., gradient averaging) form another group of cheap and efficient error indicators (see, e.g., [3, 5, 28]). They gained high popularity in engineering computations. Typically, a posteriori methods for FEM exploit specific features of approximations (Galerkin orthogonality, superconvergence, higher regularity etc.).

In this paper, we consider another (functional) approach to the a posteriori error control of approximate solutions of differential equations. These estimates are derived by purely functional methods using the analysis of variational problems or integral identities. Functional a posteriori estimates are intended to give *computable* minorants and majorants for various measures of the difference between exact solution u and *any conforming approximation* v. In general, they have the following form:

$$M_\ominus(\mathcal{D}, v) \leq \Phi(u - v) \leq M_\oplus(\mathcal{D}, v) \qquad \forall v \in V, \qquad (1)$$

where $\mathcal{D}$ is the data set (coefficients, domain, parameters, etc.) and $\Phi : V \to \mathbb{R}_+$ is a given functional. $M_\ominus$ and $M_\ominus$ must be explicitly computable and continuous in the sense that

$$M_\ominus \text{ and } M_\oplus \to 0, \quad \text{if } v \to u$$

Typically, the the functional Φ is presented on one of the following three forms:

$$\Phi(u-v) = \|u-v\|_\Omega \quad (\|\cdot\|_\Omega \text{ is the global (energy) norm});$$
$$\Phi(u-v) = \|u-v\|_\omega \quad (\|\cdot\|_\omega \text{ is a local norm});$$
$$\Phi(u-v) = <\ell, u-v> \quad (\ell \quad \text{is a linear functional}).$$

Estimate (1) provides a computable measure for the deviation from the exact solution u and, therefore, is also called a *deviation estimate*. The latter gives the principal form of the a posteriori bounds for all conforming approximations of a boundary–value problem considered that follows from the theory of partial differential equations. It does not attract specific features of the numerical method, approximations, and the mesh used. This information should be used on the next stage when deviation estimates are applied to a particular approximate solution.

Such type estimates were primarily derived in 1996-99 by means of variational methods in the duality theory of the calculus of variations and convex analysis (see, [14]-[16] and some other papers cited therein). A systematic exposition of the variational approach to a posteriori error estimation is presented in [12, 16]. Functional a posteriori estimates can be also used for approximations that violate boundary conditions (see [24, 25]). Other important areas of their application arise due to a possibility to evaluate *modeling* and *indeterminacy* errors (see [20, 21, 25]).

For elliptic type problems, a non–variational approach to the derivation of functional a posteriori estimates was suggested in [17]. In [7, 18] it was extended to parabolic problems. In [17], it was also shown that for linear elliptic PDE's deviation estimates obtained by variational and non-variational methods are identical. Deviation estimates for the Stokes problem can be found in [19].

All the estimates mentioned above has been derived for energy norms. Deviation estimates in terms of non–energy quantities were considered in [22, 23], where such estimates were obtained for local norms of $u - v$ and for $\langle \ell, u - v \rangle$. In this paper, we present advanced forms of this type a posteriori estimates and discuss their properties.

For the convenience of readers, in Section 2, we first consider functional a posteriori estimates in global norms and present the main ideas of the approach on the paradigm of the problem

$$\operatorname{div} A \nabla u + f = 0 \quad \text{in } \Omega; \tag{2}$$

$$u = u_0 \quad \text{on } \partial_1 \Omega; \tag{3}$$

$$\nu \cdot A \nabla u = F \quad \text{on } \partial_2 \Omega, \tag{4}$$

where $\Omega \in \mathbb{R}^d$ is a bounded connected domain with Lipschitz continuous boundary that consists of two measurable disjoint parts $\partial_1 \Omega$ and $\partial_2 \Omega$. For this problem, we derive two–sided bounds for the energy norm $|\!|\!| u - v |\!|\!|$, where v is an arbitrary conforming approximation of u.

In Section 3, these results are used to derive guaranteed upper bounds for errors measured in terms of the local norm $|\!|\!| u - v |\!|\!|_\omega$ related to a certain domain $\omega \subset \Omega$. We discuss particular forms of such estimates and methods of their practical implementation. Local a posteriori estimates are also derived for the linear elasticity problem.

Section 4 is concerned with estimates in terms of goal–oriented quantities. One example of such a quantity is $\int_\Omega \ell(u - v) dx$. If ℓ is a locally supported function, then such a quantity can be also used to characterize local behavior of approximation errors.

Finally, in Section 5, we consider the parabolic equation

$$u_t - \operatorname{div} A \nabla u - f = 0 \quad \text{in } \Omega.$$

For the respective initial boundary–value problem we obtain a posteriori estimates in terms of local quantities and discuss how to apply them in practice.

2 Functional a posteriori estimates for elliptic problems

2.1 Two–sided estimates in the energy norm

In this section, we shortly recall a non–variational method. In a more general form, it has been presented in [17] where it was shown that functional a posteriori estimates can be obtained from the integral identities of the respective boundary–value problems without using methods of the duality theory in the calculus of variations. For the convenience of readers, we present below this method using (2)–(3) as an example. We assume that $A = \{a_{ij}\}$ is a symmetric positive definite matrix, which has a positive definite inverse matrix A^{-1} and satisfies the usual condition

$$c_1^2|\xi|^2 \leq A\xi \cdot \xi \leq c_2^2|\xi|^2 \qquad \forall \xi, \qquad (5)$$

where $|\xi|^2 = \xi \cdot \xi = \sum_i \xi_i^2$. Here and later on, $\|\cdot\|$ defines the norm in $L^2(\Omega)$. To simplify the notation, we denote the norm in the space of vector–valued functions $Y := L^2(\Omega, R^d)$ by the same symbol. Also, we use the norms

$$\| y \|^2 := \int_\Omega Ay \cdot y \, dx \quad \text{and} \quad \| y \|_*^2 := \int_\Omega A^{-1} y \cdot y \, dx,$$

which are equivalent to the natural norm of Y.

Generalized solution u is an function in $V_0 + u_0$, where

$$V_0 := \{ v \in H^1(\Omega) \mid v = 0 \text{ on } \partial_1 \Omega \}$$

that meets the integral identity

$$\int_\Omega A\nabla u \cdot \nabla w \, dx = \int_\Omega f w \, dx + \int_{\partial_2 \Omega} F w \, ds \qquad \forall w \in V_0(\Omega) \qquad (6)$$

For any $w \in V_0$ and any

$$y \in Y_{\text{div}} := \{ y \in Y := L^2(\Omega, R^n) \mid \text{div} y \in L^2(\Omega), \ y \cdot \nu \in L^2(\partial_2 \Omega) \},$$

we have

$$\int_\Omega ((\text{div} y) w + \nabla w \cdot y) \, dx = \int_{\partial_2 \Omega} (y \cdot \nu) w \, ds.$$

Now, from (6) it follows that

$$\int_\Omega A\nabla(u - v) \cdot \nabla w \, dx = \int_\Omega (f + \text{div} y) w \, dx + \int_\Omega (y - A\nabla v) \cdot \nabla w \, dx + \int_{\partial_2 \Omega} (F - y \cdot \nu) w \, ds. \qquad (7)$$

Let $\lambda_1(\Omega, \partial_2 \Omega)$ be such a constant that

$$\lambda_1^2(\Omega, \partial_2 \Omega) = \inf_{w \in V_0} \frac{\| \nabla w \|^2}{\|w\|^2 + \|w\|_{\partial_2 \Omega}^2}. \qquad (8)$$

Since

$$\int_\Omega (A\nabla v - y) \cdot \nabla w \, dx \leq \| A\nabla v - y \|_* \| \nabla w \|$$

and

$$\left|\int_\Omega (f+\operatorname{div}y)w\,dx + \int_{\partial_2\Omega}(F-y\cdot\nu)w\,ds\right| \le$$
$$\le \left(\|f+\operatorname{div}y\|^2 + \|F-y\cdot\nu\|^2_{\partial_2\Omega}\right)^{1/2} C_1\,\|\,\nabla w\,\|$$

we arrive at the *deviation estimate in the global (energy) norm*

$$\|\nabla(u-v)\|^2 \le M^2_\oplus(v,y,\beta,C_1) := (1+\beta)\,\|\,A\nabla v - y\,\|^2_* +$$
$$+\frac{1+\beta}{\beta}C_1^2\left(\|f+\operatorname{div}y\|^2 + \|F-y\cdot\nu\|^2_{\partial_2\Omega}\right), \quad (9)$$

where β is an arbitrary positive number, and C_1 is any constant greater than $\lambda_1^{-1}(\Omega,\partial_2\Omega)$.

Minimization with respect to β leads to the estimate

$$\|\nabla(u-v)\| \le \|\,A\nabla v - y\,\|_* + C_1\left(\|f+\operatorname{div}y\|^2 + \|F-y\cdot\nu\|^2_{\partial_2\Omega}\right)^{1/2}. \quad (10)$$

It is worth noting that (9) may be less convenient for practical computations than (9) because its right–hand side is given by a non-quadratic functional.

Estimates (9) and (9) are directly computable and give *guaranteed upper bounds* of the energy norm of the difference between the exact solution and an arbitrary conforming approximation v. These bounds are exact in the sense that by choosing proper β and y it is possible to make the right hand side of the estimate arbitrarily close to the left hand one. In other words (see [17]),

$$\|\nabla(u-v)\| =$$
$$= \inf_{y\in Y_{\operatorname{div}}(\Omega)} \|\,A\nabla v - y\,\|_* + C_1\left(\|f+\operatorname{div}y\|^2 + \|F-y\cdot\nu\|^2_{\partial_2\Omega}\right)^{1/2}.$$

A lower bound of $\|\nabla(u-v)\|$ can be derived as follows. Note that

$$\sup_{w\in V_0}\int_\Omega\left(A\nabla(u-v)\cdot\nabla w - \frac{1}{2}A\nabla w\cdot\nabla w\right)dx \le$$
$$\le \sup_{\tau\in L^2(\Omega,R^n)}\int_\Omega\left(A\nabla(u-v)\cdot\tau - \frac{1}{2}A\tau\cdot\tau\right)dx = \frac{1}{2}\,\|\nabla(u-v)\|^2\,.$$

However,

$$\sup_{w\in V_0}\int_\Omega\left(A\nabla(u-v)\cdot\nabla w - \frac{1}{2}A\nabla w\cdot\nabla w\right)dx \ge$$
$$\int_\Omega\left(A\nabla(u-v)\cdot\nabla(u-v) - \frac{1}{2}A\nabla(u-v)\cdot\nabla(u-v)\right)dx = \frac{1}{2}\|\nabla(u-v)\|^2\,.$$

Thus, we conclude that

$$\frac{1}{2}\|\nabla(u-v)\|^2 = \sup_{w\in V_0}\int_\Omega \left(A\nabla(u-v)\cdot\nabla w - \frac{1}{2}A\nabla w\cdot\nabla w\right)dx \geq$$

$$\geq \sup_{w\in V_0}\int_\Omega \left(-\frac{1}{2}A\nabla w\cdot\nabla w - A\nabla v\cdot\nabla w + fw\right)dx + \int_{\partial_2\Omega} Fw\,ds.$$

It is easy to see that this lower bound is sharp (set $w = u - v$). Thus, for any $w \in V_0$

$$\|\nabla(u-v)\|^2 \geq M_\ominus^2(v,w) :=$$
$$= \int_\Omega (-A\nabla w \cdot \nabla w - 2A\nabla v \cdot \nabla w + 2fw)\,dx + 2\int_{\partial_2\Omega} Fw\,ds. \quad (11)$$

2.2 Application to FEM

There are 3 basic ways to measure errors of a finite element approximations by means of the above estimates:

- (a) *Direct* (flux averaging on the mesh $\mathcal{T}_h$);
- (b) *One step retardation* (flux averaging on the mesh h_{ref});
- (c) *Optimization* (minimization (maximization) of the Majorant (Minorant) with respect to y (w)).

Let us discuss these methods on the paradigm of Dirichlét type problem (i.e., for the case $\partial\Omega = \partial_1\Omega$).

(a) Use recovered fluxes on $\mathcal{T}_h$. Let $u_h \in V_h$, then

$$p_h := \nabla u_h \in L_2(\Omega, \mathrm{R}^d), \quad p_h \notin H(\Omega, \mathrm{div}).$$

Use an averaging operator $G_h : L_2(\Omega, \mathrm{R}^d) \to H(\Omega, \mathrm{div})$ and have a directly computable estimate

$$\|\nabla(u - u_h)\| \leq \| A\nabla u_h - G_h p_h \|_* + C_1 \|\mathrm{div} G_h p_h + f\|. \quad (12)$$

(b) Take the recovered fluxes from a refined mesh. Let $u_{h_1}, u_{h_2}, ..., u_{h_k}, ...$ be a sequence of approximations on meshes $\mathcal{T}_{h_k}$. Compute $p_k := \nabla u_k$, average it by G_{h_k} and for $u_{h_{k-1}}$ use the estimate

$$\|\nabla(u - u_{h_{k-1}})\| \leq \| A\nabla u_{h_{k-1}} - G_{h_k} p_{h_k} \|_* + C_1 \|\mathrm{div} G_{h_k} p_{h_k} + f\|. \quad (13)$$

It is worth mentioning, that this estimate gives *a quantitative form of the heuristic Runge's rule* that dates back to the 19th century. This rule reads: *If the difference between approximate solutions computed on two consequent meshes is small, then probably both of them are close to the exact one.*

In other words, it was suggested to use the quantity $\|u_{h_k} - u_{h_{k-1}}\|$ as a heuristic a posteriori error indicator based on the information contained in two consequent approximations.

Estimate (6) presents a functional, which is defined on approximations computed on the two consequent meshes $\mathcal{T}_{h_{k-1}}$ and $\mathcal{T}_{h_k}$. It gives a *cheaply computable* and *guaranteed* upper bound of $\|u - u_{h_{k-1}}\|$.

A computable lower bound is given by (4). By setting $w = u_{h_k} - u_{h_{k-1}}$ we find that the quantity

$$M_\ominus(u_{h_{k-1}}, u_{h_k} - u_{h_{k-1}})$$

gives a guaranteed lower bound for $\|\nabla(u - u_{h_{k-1}})\|$.

(c) *Minimization with respect to y.* Select a certain subspace Y_τ in $H(\Omega, \mathrm{div})$. In the simplest case, this space is constructed with help of the same mesh $\mathcal{T}_h$. However, in general, any another suitable mesh $\mathcal{T}_\tau$ and trial functions can be used.

Then, we have

$$\|\nabla(u - u_h)\| \leq \inf_{y_\tau \in Y_\tau} \{\|\nabla u_h - y_\tau\| + C_\Omega \|\mathrm{div} y_\tau + f\|\} \tag{14}$$

Let us denote the respective minimizer by $\widehat{y}_\tau$. It is clear that the wider $Y_\tau \subset H(\Omega, \mathrm{div})$ we take the sharper upper bound we obtain.

Similarly, if take a subspace $\widehat{V}_h$ wider than V_h, then the quantity

$$\sup_{w_h \in \widehat{V}_h} M_\ominus^2(u_h, w_h)$$

gives a positive lower bound for $\|\nabla(u - u_h)\|$. As we will see, the respective maximizer $\widehat{w}_h$ can be also used in a posteriori estimates of local errors and errors estimated in terms of goal–oriented quantities.

More information on the practical implementation of the functional a posteriori estimates is presented in [6, 8, 11, 12, 16, 24].

2.3 A posteriori error estimates in local norms

Functional a posteriori estimates implies computable upper bounds for the local errors. Let ω be a subdomain of Ω with Lipschitz continuous boundary $\partial \omega$ and $\| \cdot \|_\omega$ denotes the norm in $L_2(\omega)$. Take a function $\varphi \in V_0$. Since $\widetilde{v} = (v + \varphi)$ can be viewed as an approximation of u, we apply (9) and obtain

$$\|\nabla(u - \widetilde{v})\|^2 = \|\nabla(u - \widetilde{v})\|_\omega^2 + \|\nabla(u - \widetilde{v})\|_{\Omega \setminus \omega}^2 \leq$$
$$\leq (1 + \beta) \|A\nabla(v + \varphi) - y\|_*^2 +$$
$$+ \frac{1 + \beta}{\beta} C_1^2 \left(\|f + \mathrm{div} y\|^2 + \|F - y \cdot \nu\|_{\partial_2 \Omega}^2 \right), \tag{15}$$

where β is an arbitrary positive number and y is an function from Y_div. For any $\gamma \in (0, 1)$,

$$\|\nabla(u - \widetilde{v})\|_\omega^2 \geq (1 - \gamma) \|\nabla(u - v)\|_\omega^2 + \left(1 - \frac{1}{\gamma}\right) \|\nabla \varphi\|_\omega^2.$$

Then, we obtain
$$c_1^2 \|\nabla(u-v)\|_{2,\omega}^2 \leq \|\nabla(u-v)\|_\omega^2 \leq M_{\oplus\omega}^2(v), \qquad (16)$$
where
$$M_{\oplus\omega}^2(v) := \inf_{\substack{\beta\in\mathbb{R}_+,\ \varphi\in V_0 \\ \gamma\in(0,1),\ y\in Y_{\text{div}}}} \frac{1}{1-\gamma}\Big\{(1+\beta)\|A\nabla(v+\varphi)-y\|_*^2 +$$
$$+\frac{1+\beta}{\beta}C_1^2\left(\|f+\text{div}\,y\|^2+\|F-y\cdot\nu\|_{\partial_2\Omega}^2\right)\Big\} + \frac{1}{\gamma}\|\nabla\varphi\|_\omega^2.$$

Note that the second inequality in (13) holds as equality (i.e., $M_{\oplus\omega}(v)$ gives the exact bound of the local error). To show this it suffices to set $\phi = u - v$, $y = A\nabla u$ and tend γ to 1.

In particular, if take
$$\varphi \subset V_{0\omega} := \{v \in V_0 \mid v(x) = \text{const}\ \forall x \in \omega\}.$$
then the third term vanishes and we observe that *an upper bound of the local error is obtained by the minimization of the global (energy) majorant with respect to an additional variable φ in the space $V_{0\omega}$ of all the functions from the space V_0 having zero gradients on ω gives a guaranteed upper bound of $\|u-v\|_\omega$ for any conforming approximation v.*

Practical implementation of the above estimate follows the lines of the scheme presented in 2.2. If u_{h_k} and $u_{h_{k-1}}$ are two approximate solutions computed on two consequent meshes, then from (13) we find that
$$\|\nabla(u-u_{h_{k-1}})\|_\omega \leq \|\nabla(u_{h_k}-u_{h_{k-1}})\|_\omega +$$
$$+ \|A\nabla u_{h_k} - G_{h_k}p_{h_k}\| + C_1\left(\|f+\text{div}\,G_{h_k}p_{h_k}\|^2 + \|F-G_{h_k}p_{h_k}\cdot\nu\|^2\right)^{1/2}. \qquad (17)$$

If we have only one approximate solution u_h computed on $\mathcal{T}_h$, then bounds of local errors can be easily found provided that two-sided estimates of the global energy error norm has been accurately determined. In this case, we may use the functions $\widehat{y}_\tau$ and $\widehat{w}_h$ found in the framework of the method (c) and obtain
$$\|\nabla(u-u_h)\|_\omega \leq \|\nabla \widehat{w}_h\|_\omega +$$
$$+ \|A\nabla(u_h+\widehat{w}_h) - \widehat{y}_\tau\| + C_1\left(\|f+\text{div}\,\widehat{y}_\tau\|^2 + \|F-\widehat{y}_\tau\cdot\nu\|^2\right)^{1/2}. \qquad (18)$$

Linear elasticity problem gives another practically interesting elliptic problem. In this problem, we need to find a vector–valued function u (displacement) and a tensor–valued function σ (stress) such that

$$\text{div}\,\sigma + f = 0 \quad \text{in}\,\Omega; \tag{19}$$

$$\sigma = L\varepsilon(u); \quad L = \{L_{ijkm}\},\ L_{ijkm} = L_{kmij} = L_{jikm}, \tag{20}$$

$$u = u_0 \quad \text{on}\ \partial_1\Omega, \qquad \sigma \cdot \nu = F \quad \text{on}\ \partial_2\Omega, \tag{21}$$

where $\varepsilon(u)$ is the symmetric part of the tensor ∇u, ν is the unit outward normal to $\partial\Omega$ and $(\sigma \cdot \nu)_i = \sigma_{ij}\nu_j$. Assume that

$$c_1^2|\xi|^2 \le L\xi : \xi \le c_2^2|\xi|^2 \qquad \forall \xi \in M^{n\times n}, \tag{22}$$

where two dots denote the scalar product in the space of real $n \times n$ matrices and $|\xi| = \sqrt{\xi:\xi}$. Generalized solution u of this problem is a function in $V_0 := \{v \in W_2^1(\Omega, R^n) \mid v = 0 \text{ on } \partial_1\Omega\}$ that meets the integral identity

$$\int_\Omega L\varepsilon(u) \cdot \varepsilon(w)\,dx = \int_\Omega fw\,dx + \int_{\partial_2\Omega} Fw\,ds \qquad \forall w \in V_0(\Omega)$$

For this problem, a posteriori error estimates in the energy norm has been derived in [15, 17] and tested in [11]. It has the form

$$\|\varepsilon(u-v)\|^2 \le 2(1+\beta)D(\varepsilon(v),\tau) + \tag{23}$$
$$+ \frac{1+\beta}{\beta}C_1^2\left(\|f + \text{div}\,\tau\|^2 + \|F - \tau\cdot\nu\|^2\right).$$

Here

$$D(\varepsilon(v),\tau) = \frac{1}{2}\|\varepsilon(v)\|^2 + \frac{1}{2}\|\tau\|_*^2 - \int_\Omega \varepsilon(v):\tau\,dx,$$

$$\|\varepsilon(v)\|^2 := \int_\Omega L\varepsilon(v):\varepsilon(v)\,dx, \quad \|\tau\|_*^2 := \int_\Omega L^{-1}\tau:\tau\,dx,$$

τ is an arbitrary tensor–valued function in the space

$$\Sigma_{\text{div}} := \{\tau \in L_2(\Omega, M^{n\times n}) \mid \text{div}\,\tau \in L_2(\Omega, R^n)\},$$

β is an arbitrary positive number, L^{-1} is the tensor inverse to L, and C_1 is a constant greater than $\lambda_1(\Omega, \partial_2\Omega)$, where

$$\lambda_1^2(\Omega, \partial_2\Omega) = \inf_{w \in V_0} \frac{\|\varepsilon(w)\|^2}{\|w\|^2 + \|w\|_{\partial_2\Omega}^2}. \tag{24}$$

Estimate (23) yields local estimates. To derive them we use the same arguments. Take $\varphi \in V_0$. Then $\tilde{v} = v + \varphi \in V_0 + u_0$. We apply (23) and obtain

$$\|\varepsilon((u-\tilde{v}))\|^2 \le (1+\beta)D(\varepsilon(v+\varphi),\tau) +$$
$$+ \frac{1+\beta}{\beta}C_1^2\left(\|f - \text{div}\,\tau\|^2 + \|F - \tau\cdot\nu\|^2\right), \tag{25}$$

Thus,
$$c_1^2 \|\varepsilon((u-v))\|_{2,\omega}^2 \le \| \varepsilon((u-v)) \|_\omega^2 \le M_\omega^\oplus(v), \qquad (26)$$

where

$$M_{\oplus\omega}^2(v) := \inf_{\substack{\beta\in\mathbb{R}_+,\ \varphi\in V_0 \\ \gamma\in(0,1),\ \tau\in\Sigma_{\mathrm{div}}}} \frac{1}{1-\gamma}\Big\{(1+\beta)D(\varepsilon(v+\varphi),\tau)+ \\ +\frac{1+\beta}{\beta}C_1^2\left(\|f+\mathrm{div}\tau\|^2+\|F-\tau\cdot\nu\|^2\right)\Big\}+\frac{1}{\gamma}\|\nabla\varphi\|_\omega^2 .$$

Similarly to the previous case, we establish that the last inequality in (26) holds as the equality, i.e., the upper bound of the local error is sharp. Also, (26) leads to practically computable a posteriori estimates analogous to (5) and (18).

2.4 Error estimates in terms of linear functionals

Very often error control is performed in terms of the so–called "goal–oriented" functionals (see, e.g., [4]). In this case, a linear functional $\ell \in V_0^*$ is specially selected in order to control some specific properties of the solution. If ℓ is defined by the integral relation with a locally based integrand, then the quantity $|\langle \ell, u-v\rangle|$, is a certain characteristic of the local accuracy. In this section, we discuss how guaranteed upper bounds of such a quantity can be derived with help of the functional type a posteriori estimates derived for the energy norm of the error (see also [22, 23]). Let $\varphi \in V_0$. Then

$$\langle \ell, u-v \rangle = \langle \ell, u-v-\varphi \rangle + \langle \ell, \varphi \rangle$$

and, therefore,

$$|\langle \ell, u-v\rangle| \le |\ell| \inf_{\varphi\in V_0} \|\nabla(u-v-\varphi)\| + |\langle \ell,\varphi\rangle|, \qquad (27)$$

where

$$|\ell| := \sup_{w\in V_0} \frac{|\langle \ell, w\rangle|}{\|\nabla w\|}.$$

This estimate is sharp. Indeed, if $\varphi = u-v$, then (27) holds as equality. Usually, the quantity $|\ell|$ is not difficult to estimate. For example, if

$$\langle \ell, u-v\rangle = \int_\Omega \lambda(u-v)\,dx, \qquad \lambda \in L_2(\Omega)$$

then $|\ell| \le \|\lambda\|\frac{C_\Omega}{c_1}$.

Now, we apply the energy estimate (9) and obtain

$$|\langle \ell, u-v \rangle | \leq |\ell| \inf_{\substack{\varphi \in V_0, \beta > 0 \\ y \in Y_{\text{div}}(\Omega)}} \Big\{ \| A\nabla(v+\varphi) - y \|_* +$$
$$+ C_1 \left(\|f + \text{div} y\|^2 + \|F - y \cdot \nu\|^2_{\partial_2 \Omega} \right)^{1/2} + |\langle \ell, \varphi \rangle | \Big\}. \quad (28)$$

If $\varphi \in V_{0\ell}(\Omega) := \{ \varphi \in V_0(\Omega) \mid \langle \ell, \varphi \rangle = 0 \}$, then a particular form of (28) arises. It reads,

$$|\langle \ell, u-v \rangle | \leq |\ell| \inf_{\substack{\varphi \in V_{0\ell}, \beta > 0 \\ y \in Y_{\text{div}}(\Omega)}} \Big\{ \| A\nabla(v+\varphi) - y \|_* +$$
$$+ C_1 \left(\|f + \text{div} y\|^2 + \|F - y \cdot \nu\|^2_{\partial_2 \Omega} \right)^{1/2} \Big\}. \quad (29)$$

This estimate shows that *minimization of the global (energy) error majorant with respect to an additional variable φ in the space $V_{0\ell}$ of all the functions orthogonal to ℓ gives a guaranteed upper bound of $|\langle \ell, u-v \rangle|$ for any conforming approximation v.*

Note that actually (28) and (29) hold as equalities. For (28) it is easily observed (take $\varphi = u - v$, $y = A\nabla u$). For (29) the respective proof is more complicated (see [23]).

Practical implementation of the above estimate follows the lines of the scheme presented in 2.2. If u_{h_k} and $u_{h_{k-1}}$ are two approximate solutions computed on two consequent meshes, then from

$$|\langle \ell, u - u_{h_{k-1}} \rangle | \leq |\langle \ell, u_{h_k} - u_{h_{k-1}} \rangle | + |\ell| \Big\{ \| A\nabla u_{h_k} - G_{h_k} p_{h_k} \| +$$
$$+ C_1 \left(\|f + \text{div} G_{h_k} p_{h_k}\|^2 + \|F - G_{h_k} p_{h_k} \cdot \nu\|^2 \right)^{1/2} \Big\}. \quad (30)$$

If we have only one approximate solution u_h computed on $\mathcal{T}_h$, then bounds of the goal–oriented quantity can be directly computed provided that we have found the functions $\widehat{y}_\tau$ and $\widehat{w}_h$ that give sufficiently good two-sided estimates of the global energy error norm. In this case,

$$|\langle \ell, u - u_h \rangle | \leq |\langle \ell, \widehat{u}_h - u_h \rangle | +$$
$$+ |\ell| \Big\{ \| A\nabla(u_h + \widehat{w}_h) - \widehat{y}_\tau \| + C_1 \left(\|f + \text{div} \widehat{y}_\tau\|^2 + \|F - \widehat{y}_\tau \cdot \nu\|^2 \right)^{1/2} \Big\}. \quad (31)$$

Another way to compute an upper bound of the goal–oriented error follows from (29): set $y = \widehat{y}_\tau$ and minimize the functional

$$|\ell|^2 \Big\{ (1+\beta) \| A\nabla(v+\varphi) - y \|_*^2 +$$
$$+ \frac{1+\beta}{\beta} C_1^2 \left(\|f + \text{div} y\|^2 + \|F - y \cdot \nu\|^2_{\partial_2 \Omega} \right) \Big\} \quad (32)$$

with respect to $\beta > 0$ and $\varphi \in V_{0\ell h}$, where $V_{0\ell h}$ is a certain finite dimensional subspace of $V_{0\ell}$.

Guaranteed error bounds in terms of goal–oriented quantities for linear elasticity and Stokes problem can be derived in a similar way (see [22, 23]).

3 Functional a posteriori estimates for a model evolutionary problem

Consider the classical linear parabolic problem: find a function $u(x,t)$ such that

$$u_t - \operatorname{div} A\nabla u - f = 0, \quad \text{in } Q_T, \tag{33}$$
$$u(x,0) = \varphi(x), \quad x \in \Omega, \tag{34}$$
$$u(x,t) = 0, \quad (x,t) \in S_T. \tag{35}$$

Here, $Q_T := \Omega \times (0,T)$ is the space–time cylinder, $S_T := \partial\Omega \times [0,T]$, $T > 0$, A is a symmetric matrix that satisfies the conditions

$$\nu_1 |\xi|^2 \leq A(x,t)\xi \cdot \xi \leq \nu_2 |\xi|^2, \quad \forall \xi \in \mathbb{R}^d, \; t \in [0,T],$$

with $0 < \nu_1 \leq \nu_2$.

Weak solution of the problem (33)–(35) (see, e.g., [9]) is a function $u \in \overset{\circ}{W}_2^{1,0}(Q_T) := L_2((0,T); \overset{\circ}{W}_2^1(\Omega))$ such that

$$\int_{Q_T} A\nabla u \cdot \nabla \eta \, dx dt - \int_{Q_T} u\eta_t \, dx dt +$$
$$+ \int_{\Omega} (u(x,T)\eta(x,T) - \varphi(x)\eta(x,0))dx = \int_{Q_T} f\eta \, dx dt \; \forall \eta \in W^1_{2,0}(Q_T), \tag{36}$$

where

$$W^1_{2,0}(Q_T) = \{w \in W^1_2(Q_T) \mid w(x,t) = 0 \text{ on } S_T\}.$$

Hereafter, we assume that $f \in L_2(Q_T)$ and $\varphi(x) \in \overset{\circ}{W}_2^1(\Omega)$. Then, $u \in W^{\Delta,1}_{2,0}(Q_T)$, where $W^{\Delta,1}_{2,0}(Q_T)$ is a space with the norm

$$\| w \|_{2,0}^{\Delta,1} = \left(\int_{Q_T} (w^2 + w_t^2 + |\nabla w|^2 + (\Delta w)^2) dx \, dt \right)^{\frac{1}{2}}.$$

Let $v \in W^1_{2,0}(Q_T)$ be an approximation of u. In [7, 18], an upper bound of the deviation $e := u - v$ was evaluated in terms of the quantity

$$[e]^2_{(\gamma_1,\gamma_2)} := \gamma_1 \|e(x,T)\|^2 + \gamma_2 \|\nabla e\|^2_{Q_T},$$

where

$$\|\nabla e\|^2_{Q_T} := \int_0^T \|\nabla e\|^2 \, dt = \int_{Q_T} A\nabla e \cdot \nabla e \, dx \, dt,$$

and γ_1 and γ_2 are some positive numbers.

An upper bound of the deviation is given by the estimate

$$[e]^2_{(2-\delta,1)} \leq M^I_\oplus(v,y,\beta,\delta) :=$$

$$= \|v(x,0) - \varphi(x)\|^2_{2,\Omega} + \frac{1}{\delta} \int_0^T \Big[(1+\beta) \|A\nabla v - y\|^2_* +$$

$$+ C_\Omega^2 \left(1 + \frac{1}{\beta}\right) \|f - v_t + \mathrm{div}\, y\|^2 \Big] dt,$$

where $\delta \in (0,2]$, C_Ω is the constant in the Friedrichs inequality,

$$y \in Y_{\mathrm{div}}(Q_T) := \{y(x,t) \in L_2(Q_T; \mathrm{R}^d) \,|\, \mathrm{div}\, y \in L_2(Q_T)\},$$

and $\beta = \beta(t)$ is a positive valued function.

As in the elliptic case, the majorant $M^I_\oplus(v,y,\beta,\delta,C_\Omega)$ consists of the terms that can be interpreted as penalties for possible violations of the relations

$$u_t - \mathrm{div}\, y - f = 0, \qquad \text{in } Q_T,$$
$$y = A\nabla u, \qquad \text{in } Q_T,$$
$$u(x,0) = \varphi(x), \qquad x \in \Omega.$$

Since $v(x,t) = 0$ on S_T, we observe that $M^I_\oplus(v,y,\beta,\delta,C_\Omega)$ vanishes if and only if $v = u$ and $y = A\nabla u$.

In [7, 18], it was also derived a sharper upper bound of the error norm. It is as follows:

$$[e]^2_{(2-\delta,1-\frac{1}{\gamma})} \leq M^{II}_\oplus(v,w,y,\beta,\gamma,\delta) := \gamma \|w(x,T)\|^2 +$$

$$+ \frac{1}{\delta} \int_0^T \Big[(1+\beta) \|y - A\nabla v + A\nabla w\|^2_* + \left(1 + \frac{1}{\beta}\right) C_\Omega^2 \|f - v_t - w_t + \mathrm{div}\, y\|^2 \Big] dt +$$

$$+ \int_{Q_T} (A\nabla v \cdot \nabla w + v_t w - fw) \, dx \, dt +$$

$$+ \int_\Omega \Big(|\varphi(x) - v(x,0)|^2 - 2w(x,0)(\varphi(x) - v(x,0))\Big) \, dx, \quad (37)$$

where $y \in Y_{\text{div}}(Q_T)$, $\delta \in (0, 2]$, $\gamma \geq 1$, $\beta = \beta(t)$ is a positive valued function and $w \in W^1_{2,0}(Q_T)$ is an additional free variable.

It is worth noting, that both majorants $M^I_\oplus$ and $M^{II}_\oplus$ give certain quantitative forms of the Runge's rule for the parabolic problem. Indeed, let $U_{\mathcal{T}H}$ be an approximate solution of the problem computed on the mesh with mesh size $\mathcal{T}$ for time variable and H for spatial variables and let $u_{\tau,h}$ be another approximate solution computed on a finer mesh (τ, h) (e.g. $\tau = \mathcal{T}/2$, $h = H/2$). Then the estimates can be applied as follows:

$$[u - U_{\mathcal{T}H}]^2_{(2-\delta,1)} \leq M^I_\oplus(U_{\mathcal{T}H}, G_{\mathcal{T}H}(\nabla u_{\tau h}), \beta, \delta), \tag{38}$$

$$[u - U_{\mathcal{T}H}]^2_{(2-\delta,1-\frac{1}{\gamma})} \leq M^{II}_\oplus(U_{\mathcal{T}H}, u_{\tau h} - U_{\mathcal{T}H}, G_{\mathcal{T}H}(\nabla u_{\tau h}), \beta, \gamma, \delta). \tag{39}$$

Properties of $M^I_\oplus$ and $M^{II}_\oplus$ were investigated in [7, 18] where it was shown that

- for any approximation $v \in W^1_{2,0}(Q_T)$ the majorants gives *guaranteed* upper bounds of the error in terms of the quantity $[e]^2_{(\gamma_1,\gamma_2)}$;
- majorants vanish if and only if v coincides with the exact solution u and $y = \nabla u$;
- majorants does not depend on mesh parameters and contains only global constants;
- to obtain a sharper upper bound one should minimize $M^I_\oplus$ over $y \in Y_{\text{div}}(Q_T)$ and $\beta = \beta(t) > 0$ and $M^{II}_\oplus$ additionally with respect to w;
- majorants are given by certain integrals in Q_t and Ω; therefore in practice they are presented as sums of local quantities distributed in space and time, which can be used as error indicators for time and space adaptation strategies (see [7]).

Recalling the idea used for deriving local estimates in elliptic problems, we observe that functional majorants $M^I_\oplus$ and $M^{II}_\oplus$ also imply certain local a posteriori estimates.

Indeed, let $Q_0 = \omega \times [T - t_0, T]$ be a subdomain of the space–time cylinder Q_T.

Let $\varphi \in W^1_{2,0}(Q_T)$. Denote by $[\cdot]_{Q_0;\gamma_1,\gamma_2}$ the restriction of $[\cdot]_{\gamma_1,\gamma_2}$ on Q_0. We have

$$[e]_{Q_0;\gamma_1,\gamma_2} \leq [e - \varphi]_{\gamma_1,\gamma_2} + [\varphi]_{Q_0;\gamma_1,\gamma_2}$$

From this relation, we obtain guaranteed upper bounds for the local error norm

$$[e]_{Q_0;2-\delta,1} \leq \left(M^I_\oplus(v + \varphi, y, \beta, \delta)\right)^{1/2} + [\varphi]_{Q_0;2-\delta,1}, \tag{40}$$

$$[e]_{Q_0;2-\delta,1-\frac{1}{\gamma}} \leq \left(M^{II}_\oplus(v + \varphi, w, y, \beta, \gamma, \delta)\right)^{1/2} + [\varphi]_{Q_0;2-\delta,1-\frac{1}{\gamma}}. \tag{41}$$

These estimates are valid for any variables φ, y, and w in the respective spaces.

Practical implementation of (40) and (41) follows the scheme discussed for elliptic problems (cf. (5) and (18)). Namely, if $u_{\tau,h}$ and $U_{\mathcal{T}H}$ are two approximate solutions computed on a "coarse" and "refined" mesh, then directly computable bounds are given by the relations

$$[u - U_{\mathcal{T}H}]_{Q_0; 2-\delta, 1} \leq \left(M_\oplus^I(u_{\tau,h}, G_{\mathcal{T}H}(\nabla u_{\tau h}), \beta, \delta)\right)^{1/2} +$$
$$+ [u_{\tau,h} - U_{\mathcal{T}H}]_{Q_0; 2-\delta, 1}, \tag{42}$$

$$[u - U_{\mathcal{T}H}]^2_{Q_0; 2-\delta, 1-\frac{1}{\gamma}} \leq \left(M_\oplus^{II}(u_{\tau,h}, u_{\tau h} - U_{\mathcal{T}H}, G_{\mathcal{T}H}(\nabla u_{\tau h}), \beta, \gamma, \delta)\right)^{1/2} +$$
$$+ [u_{\tau,h} - U_{\mathcal{T}H}]^2_{Q_0; 2-\delta, 1-\frac{1}{\gamma}}. \tag{43}$$

Another option is to define certain finite dimensional subspaces for the variables φ, τ, and w and minimize the majorants over these subspaces.

Acknowledgements.
The research was partially supported by the grant of Civilian Research and Development Foundation RU M1-2596-ST-04 (USA) and DAAD Program (Germany). The author wishes to express his thanks to the organizers of ENUMATH2005.

References

1. Ainsworth, M., Oden, J. T.: A posteriori error estimation in finite element analysis. Wiley, New York, (2000)
2. Babuška, I., Strouboulis, T.: The finite element method and its reliability. Clarendon Press, Oxford (2001)
3. Babuška, I., Rodriguez, R.: The problem of the selection of an a posteriori error indicator based on smoothing techniques. Internat. J. Numer. Meth. Engrg., **36**, 539-567 (1993)
4. Bangerth, W., Rannacher, R.: Adaptive finite element methods for differential equations. Birkhäuser, Berlin (2003)
5. Carstensen, C., Bartels, S.: Each averaging technique yields reliable a posteriori error control in FEM on unstructured grids. I: Low order conforming, nonconforming, and mixed FEM. Mathematics of Computation, **71 (239)**, 945-969 (2002)
6. Frolov, M., Neittaanmäk, P., Repin, S.: On practical implementation of the duality error majorants for boundary–value problems arising in the theory of plates. In *European Congress on Computational Methods in Applied Sciences and Engineering, ECCOMAS 2004, P.Neittaanmäki, T. Rossi, K. Majava and O. Pironeau (eds.), O. Nevanlinna and R. Rannacher (assoc. eds.), Jyväskylä, 24-28 July, 2004* (electronic)
7. Gaevskaya, A., Repin, S.: A posteriori error estimates for approximate solutions of linear parabolic problems. Differential Equations, **41 (7)**, 970–983 (2005)
8. Gorshkova, E., Repin, S.: Error control of the approximate solution to the Stokes equation using a posteriori error estimates of functional type. In *European Congress on Computational Methods in Applied Sciences and Engineering, ECCOMAS 2004, P.Neittaanmäki, T. Rossi, K. Majava and O. Pironeau*

(eds.), O. Nevanlinna and R. Rannacher (assoc. eds.), Jyväskylä, 24-28 July, 2004 (electronic)

9. Ladyzhenskaya, O.: The boundary value problems of mathematical physics. Springer New York, (1985)
10. Mikhlin, S.: Variational methods in mathematical physics. Pergamon, Oxford, (1964)
11. Muzalevsky, A., Repin, S.: On two-sided error estimates for approximate solutions of problems in the linear theory of elasticity. Russian J. Numer. Anal. Math. Modelling, **18** (1), 65–85 (2003)
12. Neittaanmäki, P., Repin, S.: Reliable methods for computer simulation. Error control and a posteriori estimates. Elsevier, New York, London, (2004)
13. Prager, W., Synge, J. L.: Approximation in elasticity based on the concept of function spaces. Quart. Appl. Math., **5**, 241-269, (1947)
14. Repin, S.: A posteriori error estimation for nonlinear variational problems by duality theory. Zapiski Nauchnych Semin. POMI, **243**, 201–214 (1997)
15. Repin, S.: A posteriori error estimation for variational problems with uniformly convex functionals. Mathematics of Computation, **69 (230)**, 481–500, (2000)
16. Repin, S.: A unified approach to a posteriori error estimation based on duality error majorants. Mathematics and Computers in Simulation, **50**, 313–329 (1999)
17. Repin, S.: Two-sided estimates for deviation from an exact solution to uniformly elliptic equation. Proceedings of St.-Petersburg Math. Society, **IX**, 143–171 (2001) (in Russian; English translation in American Mathematical Translations Series 2, **209**, 143–171, (2003))
18. Repin, S.: Estimates of deviation from exact solutions of initial-boundary value problems for the heat equation. Rend. Mat. Acc. Lincei, **13**, 121–133 (2002)
19. Repin, S.: Aposteriori estimates for the Stokes problem. J. Math. Sci. New York, **109 (5)**, 1950–1964 (2002)
20. Repin, S.: Estimates for errors in two-dimensional models of elasticity theory. J. Math. Sci. (New York), **106 (3)**, 3027-3041 (2001)
21. Repin, S.: A posteriori error estimates with account of indeterminacy of the problem data. Russ. J. Numer. Anal. Math. Modelling, **18 (6)**, 507-519 (2003)
22. Repin, S.: Local a posteriori estimates for the Stokes problem. Zap. Nauchn. Sem. S.-Peterburg. Otdel. Mat. Inst. Steklov. (POMI), **318**, 233–245, (2004)
23. Repin, S.: A posteriori estimates in local norms. J. Math. Sci. New York, **124 (3)**, 5026–5035 (2004)
24. Repin, S., Sauter, S., Smolianski, A.: A posteriori error estimation for the Dirichlet problem with account of the error in approximation of boundary conditions. Computing, **70 (3)**, 147-168 (2003)
25. Repin, S., Sauter, S., Smolianski, A.: A posteriori estimation of dimension reduction errors for elliptic problems in thin domains. SIAM J. Numer. Anal., **42 (4)**, 1435–1451 (2004)
26. Repin, S., Sauter, S., Smolianski, A.: A posteriori error estimation for the Poisson equation with mixed Dirichlet/Neumann boundary conditions. J. Comput. Appl. Math., **164/165**, 601–612 (2004)
27. Verfürth, R.: A review of a posteriori error estimation and adaptive mesh-refinement techniques. Wiley, Teubner, New-York (1996)
28. Zienkiewicz, O. C., Zhu, J. Z.: A simple error estimator and adaptive procedure for practical engineering analysis. Internat. J. Numer. Meth. Engrg., **24**, 337-357 (1987)

Finite Element Approximation of 2D Parabolic Optimal Design Problems

Miguel Cea[1] and Enrique Zuazua[2]

[1] Departamento de Matemáticas, Facultad de Ciencias,
Universidad Autónoma de Madrid, 28049 Madrid, Spain
miguel.cea@uam.es

[2] Departamento de Matemáticas, Facultad de Ciencias,
Universidad Autónoma de Madrid, 28049 Madrid, Spain
enrique.zuazua@uam.es

Summary. In this paper we consider a problem of parabolic optimal design in 2D for the heat equation with Dirichlet boundary conditions. We introduce a finite element discrete version of this problem in which the domains under consideration are polygons defined on the numerical mesh. The discrete optimal design problem admits at least one solution. We prove that, as the mesh size tends to zero, any limit in H^c of discrete optimal shapes is an optimal domain for the continuous optimal design problem. We work in the functional and geometric setting introduced by V. Šverák in which the domains under consideration are assumed to have an a priori limited number of holes. We present in detail a numerical algorithm and show the efficiency of the method through various numerical experiments.

1 Introduction

We consider a problem of optimal control in which the control variable is the domain on which a partial differential equation is posed. The function we want to minimize depends on Ω through the solution of the PDE. In the present paper we analyze the heat equation in 2D with Dirichlet boundary conditions extending previous works by D. Chenais and the second author on the elliptic problem in [6] and [7].

We focus on the problem of numerical approximation of optimal shapes. We build a finite element approximation of the optimal design problem and prove that, as the mesh size tends to zero, in the H^c-topology, every limit of discrete optimal shapes is an optimal shape for the continuous equation. We work in the functional setting introduced by Šverák [25] in which the domains under consideration have an a priori limited finite number of holes, later adapted to the finite element setting in [6] and [7].

Let us describe more precisely the problem under the consideration.

- $\mathcal{C}$ is a non-empty bounded Lipschitz open set of $\mathbf{R}^2$.
- $\mathcal{O}$ is the set of all open subsets of $\mathcal{C}$.
- For all $\Omega \in \mathcal{O}$ and $T > 0$, we consider the heat equation in Ω

$$\begin{cases} u_t - \Delta u = f & \Omega \times [0,T], \\ u = 0 & \partial\Omega \times [0,T], \\ u(0) = \psi_0 & \Omega, \end{cases} \quad (1)$$

where $f \in L^2(0,T;L^2(\mathbf{R}^2))$ and $\psi_0 \in L^2(\mathbf{R}^2)$. The variational formulation of (1) is as follows (see [4]):

$$\begin{cases} \text{To find } u \in C([0,T];L^2(\Omega)) \cap L^2(0,T;H_0^1(\Omega)) \text{ such that} \\ \frac{d}{dt}(u,\varphi) + a(u,\varphi) = (f,\varphi), & \forall \varphi \in H_0^1(\Omega), \\ u(0) = \psi_0, \end{cases} \quad (2)$$

where

$$a(u,v) = \int_\Omega \nabla u \cdot \nabla v \, dx,$$

and $(\cdot,\cdot)$ stands for the scalar product in $L^2(\Omega)$.

- We also consider the functional $J : \mathcal{O} \to \mathbf{R}$ to be minimized. Typically in applications J is defined as an integral involving the solution u of (1). Therefore, the continuity of J (with respect to the H^c convergence of domains) requires the continuity of the solutions of (1) with respect to the domain. For that to be the case one often needs to restrict the functional to a suitable subclass of domains.

To be more precise we consider functionals of the form

$$J(\Omega) = \int_0^T \int_\Omega L(t,x,u,\nabla u) \, dx \, dt, \quad (3)$$

where $L(t,x,z,s)$ is assumed to be non-negative, continuous in (t,x,z,s), strictly convex in s and and such that there exists $c > 0$ such that

$$|L(t,x,z,s)| \le c(|z|^2 + |s|^2).$$

In (3) u denotes the solution of (1) in Ω.

These assumptions may be greatly simplified in specific applications. We do not intend to describe the most general framework but only give a few relevant examples in which our developments apply.

Let us give some examples of functionals $J(\Omega)$ which often arise in applications and fulfill the previous requirements:

- The first one concerns the compliance of the system (1). It is defined by

$$J(\Omega) = \int_0^T \int_\Omega fu \, dx \, dt.$$

The assumptions are fulfilled when $f = f(x,t)$ is continuous although our methods apply when $f \in L^2(0,T;L^2(\mathbf{R}^2))$.

- A second important example concerns shape identification problems. Let us consider a subdomain $E \in \mathcal{O}, E \neq \emptyset$. We suppose that a function u_E has been measured on E, which is a known or accesible part of the set Ω which is unknown and has to be identified.

In this case, the functional to be minimized is, for example, of the form

$$J(\Omega) = \frac{1}{2} \int_0^T \int_\Omega |\nabla(u - \widetilde{u}_E)|^2 dx dt.$$

Here and in the sequel we denote by $\widetilde{u}$ the extension by zero of u so that $\widetilde{u} = 0$ in $\mathcal{C} \setminus \Omega$. The assumptions above are satisfied by this functional too.

The continuous optimal design problem we consider is as follows:

To find $\Omega^* \in \mathcal{O}$ such that $J(\Omega^*) = \min_{\Omega \in \mathcal{O}} J(\Omega)$. (4)

In practice, often, this problem is formulated in a suitable subset of $\mathcal{O}$ in order to guarantee the compactness and continuity properties that are needed for the minimum to be achieved. The results by Šveràk [25] guarantee that this occurs when working in the subclass of domains with complementary sets with at most a finite prescribed number of connected components. We shall denote by $\mathcal{O}^N$ that class where $\#(\Omega^c) \leq N$ for all $\Omega \in \mathcal{O}^N$, N being a finite number and $\#(K)$ the number of connected components of K.

In other words, we shall be mainly concerned with the following minimization problem:

To find $\Omega^* \in \mathcal{O}^N$ such that $\mathcal{I} := J(\Omega^*) = \min_{\Omega \in \mathcal{O}^N} J(\Omega)$. (5)

The question we address in this paper is the numerical approximation of the optimal design problem (5). In particular we address the issue of whether the discrete optimal shapes for a suitable discretization of the above problem converge in H^c (see Section 2 for the precise definition), to an optimal shape for the continuous one as the mesh-size tends to zero. This problem was successfully formulated and solved by D. Chenais and the second author in [6] and [7] for the elliptic case and this article is aimed to give an extension to the parabolic one.

In order to do this, we now introduce a discretization of this problem as follows.

- For any $h > 0$, we consider a triangulation $\mathcal{T}_h = \{(\tau_i^h)_{i \in I_h}\}$ of $\mathcal{C}$ made of finite elements τ_i^h so that

$$\mathcal{C} = \overset{\circ}{\overline{\bigcup_{i \in I_h} \tau_i^h}},$$

where $\overset{\circ}{\overline{A}}$ denotes the interior of $A \subset \mathbf{R}^2$. To this end, we suppose that the triangulations are uniformly regular, that is

$$\exists \sigma > 0 \text{ s.t. } \forall h > 0, \quad \tau_i^h \in \mathcal{T}_h, \quad 0 < \frac{h}{\rho_i} \leq \sigma,$$

where the grid size h is defined as the maximum diameter of the elements τ_i^h and ρ_i is the radius of the largest ball contained in τ_i^h.

- $\mathcal{O}_h$ is the set of open subsets of $\mathcal{C}$ constituted by unions of triangles of the triangulation $\mathcal{T}_h$ and $\mathcal{O}_h^N = \mathcal{O}_h \cap \mathcal{O}^N$, the subset of those polygonal domains for which the number of connected components of the complement is a priori bounded by N.

- We use the implicit Euler method with time step $\triangle t = T/M$, for some $M \in \mathbf{N}$, to discretize the heat equation (1) in time and a P1 finite element approximation for the elliptic component. For doing that we consider the P1 finite element space $X_h \subset H_0^1(\Omega_{h,\triangle t})$, and we denote by $u_{h,\triangle t}^k$ the discrete solution in the time step k, $u_{h,\triangle t}^k \sim u(x, t_k)$ where $t_k = k\triangle t$. We also denote by $U_{h,\triangle t} := (u_{h,\triangle t}^k)_{k=1}^M$ the vector-valued solution containing the solution for all time-steps. The discrete solution we consider is characterized by the following system:

$$\begin{cases} \text{To find } u_{h,\triangle t}^k \in X_h \text{ such that} \\ \left(\dfrac{u_{h,\triangle t}^k - u_{h,\triangle t}^{k-1}}{\triangle t}, \varphi_h \right) + a(u_{h,\triangle t}^k, \varphi_h) = (f^k, \varphi_h), \forall \varphi_h \in X_h, k = 1, \ldots, M \\ u_h^0 = \psi_{0,h}, \end{cases} \quad (6)$$

where

$$f^k = \frac{1}{\triangle t} \int_{t_k - \triangle t}^{t_k} f(t) dt,$$

and $\psi_{0,h}$ is the orthogonal projection of ψ_0 over X_h.

- We approximate $J(\Omega)$ by a well-chosen functional $J_h^{\triangle t}(\Omega_{h,\triangle t}) : \mathcal{O}_h^N \to \mathbf{R}$. In practice this is done by keeping the same structure of the functional as in (3) in what concerns its x-dependence and replacing the time-integral by a discrete sum.

Thus, the discrete problem we consider is

To find $\Omega_{h,\triangle t}^* \in \mathcal{O}_h^N$ such that $\mathcal{I}_h^{\triangle t} := J_h^{\triangle t}(\Omega_{h,\triangle t}^*) = \min_{\Omega_{h,\triangle t} \in \mathcal{O}_h^N} J_h^{\triangle t}(\Omega_{h,\triangle t}).$

$$(7)$$

As indicated above, this is a natural extension to the parabolic setting of the elliptic optimal design problem addressed in [6] and [7].

The main result of this paper asserts that, for any fixed N, the discrete optimal design problems (7) converge towards (5) as $h \to 0$ and $\triangle t \to 0$ in the sense that the minima converge and that the limits of $\Omega_{h,\triangle t}^*$ are optimal domains for the continuous optimization problem (5).

The techniques we employ and the results we obtain in this article can be adapted and extended to other discretization schemes. In particular this can be done for the semi-discrete approximation and other time-discretization methods of (1).

This paper is divided in five sections after this introduction. In Section 2 we recall some definitions and properties concerning Hausdorff topology, γ-convergence, Mosco-convergence and some useful results from previous papers. In Section 3 we prove the convergence of the numerical scheme. In Section 4 we prove the convergence of discrete optimal shapes. In Section 5 we develop a classical optimization algorithm to obtain the optimal design in the continuous and the discrete time cases respectively. In particular, we present a fully discrete numerical algorithm allowing to obtain an approximation of the optimal domain. Moreover, we present in detail some numerical experiments that allow checking the efficiency of the method. Finally, Section 6 is devoted to summarize the main results of the paper.

2 Preliminaries

2.1 Hausdorff convergence

In this section we recall some notations and basic results.

The Hausdorff distance between two compact sets K_1 and K_2 of $\mathbf{R}^2$ is defined by

$$d(K_1, K_2) = \max\left(\sup_{x \in K_1} \inf_{y \in K_2} ||x - y||, \sup_{x \in K_2} \inf_{y \in K_1} ||x - y||\right).$$

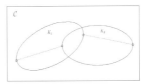

Fig. 1. Hausdorff distance between two compact sets

Definition 1. *The complementary Hausdorff distance between two open subsets Ω_1 and Ω_2 of $\mathcal{C}$ is defined by*

$$d_{H^c}(\Omega_1, \Omega_2) = d_H(\overline{\mathcal{C}} \backslash \Omega_1, \overline{\mathcal{C}} \backslash \Omega_2).$$

We denote by H^c the corresponding convergence of sets, i.e., $\Omega_n \xrightarrow{H^c} \Omega$ if only if $d_{H^c}(\Omega_n, \Omega) \to 0$.

In addition to the set $\mathcal{O}^N$ defined above, for any open non-empty subset ω of $\mathcal{C}$ we define the class $\mathcal{O}_\omega^N$ of domains of $\mathcal{O}^N$ containing ω, i.e.

$$\mathcal{O}_\omega^N = \{\Omega \in \mathcal{O}^N : \omega \subset \Omega\}.$$

The following result on the H^c-compactness of the sets $\mathcal{O}^N$ and $\mathcal{O}_\omega^N$ will be useful for addressing the optimal design problems above.

Lemma 1. *([25, 12]) For any finite N, and ω open subset of $\mathcal{C}$, the sets $\mathcal{O}^N$ and $\mathcal{O}_\omega^N$ are H^c-compact.*

2.2 Dependence of the dirichlet problem with respect to the domain

For each function $\varphi \in H_0^1(\Omega)$, we define $\widetilde{\varphi}$ its extension by zero to $\mathcal{C}$ so that $\widetilde{\varphi} \in H_0^1(\mathcal{C})$ (see [4]).

We recall the definition of γ-convergence and Mosco-convergence.

Definition 2. *([12]) Given a sequence $(\Omega_n)_n \subset \mathcal{O}$ and a domain $\Omega \in \mathcal{O}$, Ω_n γ-converges to Ω, and we denote it as $\Omega_n \xrightarrow{\gamma} \Omega$, if*

$$\forall f \in H^{-1}(\mathcal{C}), \quad \widetilde{u}_{\Omega_n} \to \widetilde{u}_\Omega \text{ strongly in } H_0^1(\mathcal{C}),$$

where $u_{\Omega_n} \in H_0^1(\Omega_n)$ is defined as the solution of the Dirichlet elliptic problem in Ω_n:

$$a(u_{\Omega_n}, \varphi) = <f, \varphi>_{H^{-1}(\Omega_n) \times H_0^1(\Omega_n)}, \quad \forall \varphi \in H_0^1(\Omega_n).$$

Definition 3. *([18]) Ω_n Mosco-converges to Ω and we denote it as $\Omega_n \xrightarrow{\text{Mosco}} \Omega$, if*

1. *For all $\varphi \in H_0^1(\Omega)$, there exists $\varphi_n \in H_0^1(\Omega_n)$ such that $\widetilde{\varphi}_n \to \widetilde{\varphi}$ strongly in $H_0^1(\mathcal{C})$.*
2. *For all subsequence of domains $(\Omega_{n_k})_k$, and for all $\varphi_{n_k} \in H_0^1(\Omega_{n_k})$, one has*

$$\{\widetilde{\varphi}_{n_k} \rightharpoonup w \text{ weakly in } H_0^1(\mathcal{C})\} \Rightarrow \{\exists \varphi \in H_0^1(\Omega) \text{ such that } w = \widetilde{\varphi}\}.$$

It is by now well known that these two notions coincide (see [12]), i.e. $\Omega_n \xrightarrow{\gamma} \Omega$ if and only if $\Omega_n \xrightarrow{\text{Mosco}} \Omega$.

Now, let us recall some relations between H^c-convergence and γ-convergence.

Lemma 2. *([5]) If a sequence H^c-converges, then the first point of the definition of the Mosco convergence is satisfied. In other words, if Ω_n converges to Ω in H^c, then, for all $\varphi \in H_0^1(\Omega)$, there exists $\varphi_n \in H_0^1(\Omega_n)$ such that $\widetilde{\varphi}_n \to \widetilde{\varphi}$ strongly in $H_0^1(\mathcal{C})$.*

In general, H^c-convergence does not imply γ-convergence, nevertheless, several situations are known where this implication holds true. In [5], a list of subsets $\mathcal{U}$ of $\mathcal{O}$ on which H^c-convergence implies γ-convergence is given. The following one is due to V. Šveràk [25]:

Theorem 1. *In two space dimensions, for any finite N, H^c-convergence and γ-convergence are equivalent properties on $\mathcal{O}^N$.*

In order to deal with the time-dependent continuous and discrete heat equations we have to work with functions depending on the time variable. The following technical result is a natural consequence of γ-convergence for sequences of functions depending both on x and t.

Lemma 3. *Assume that $\Omega_j \xrightarrow{\gamma} \Omega$ and consider a sequence of functions u_j in $L^\infty(0,T;L^2(\Omega_j)) \cap L^2(0,T;H_0^1(\Omega_j))$ satisfying*

$$\widetilde{u}_j \rightharpoonup w \quad \text{weakly } * \text{ in } L^\infty(0,T;L^2(\mathcal{C})) \cap L^2(0,T;H_0^1(\mathcal{C})). \tag{8}$$

Then $w = \widetilde{y}$, with $y \in L^\infty(0,T;L^2(\Omega)) \cap L^2(0,T;H_0^1(\Omega))$

Proof (of Lemma 3). As $\Omega_j \xrightarrow{\gamma} \Omega$ we know that $\Omega_j \xrightarrow{\text{Mosco}} \Omega$.
Let $\theta \in L^2(0,T)$ be given. We obtain

$$\widetilde{u}_j^\theta(x,t) = \int_0^T \theta(t)\widetilde{u}_j(x,t)dt \rightharpoonup w^\theta(x,t) = \int_0^T \theta(t)w(x,t)dt \text{ in } H_0^1(\mathcal{C}).$$

Since $\widetilde{u}_j^\theta \in H_0^1(\Omega_j)$, by the γ-convergence of the sets Ω_j, we get $w^\theta \in H_0^1(\Omega)$.
By the Lebesgue Differentiation Theorem we have

$$w(x,t_0) = \lim_{j \to 0} \int_0^T \frac{1}{2j}\chi_{[t_0-j,t_0+j]}(t)w(x,t)dt \quad \text{a.e. } t_0 \in [0,T].$$

Therefore, $w(t) \in H_0^1(\Omega)$ a.e. $t \in [0,T]$.
Now, we have to prove that the function $w : [0,T] \mapsto H_0^1(\Omega)$ is measurable.
Since $H_0^1(\Omega)$ is separable, it is sufficient to prove (see [8]) that w is weakly measurable, i.e., that for any $\varphi \in C_c^\infty(\Omega)$, the function $t \mapsto \int_\Omega w(x,t)\varphi(x)dx$ is measurable. According to (8), $\int_\Omega w(x,t)\varphi(x)dx$ is the weak $*$ limit in $L^\infty(0,T)$ of $\int_\Omega u_j(x,t)\varphi(x)dx$, and, in particular, it is measurable with respect to t.
This completes the proof of the Lemma. ∎

3 Preliminaries on the convergence of the numerical scheme

We first define the set of discrete admissible domains. This set is independent of $\triangle t$.

Definition 4. *For each $h > 0$, we consider the set $\mathcal{O}_h$ of subdomains of $\mathcal{C}$ constituted by elements of the triangulations $\mathcal{T}_h$. Then we set*

$$\mathcal{O}_h^N = \{\Omega_h \in \mathcal{O}_h : \#(\Omega_h^c) \leq N\}.$$

For all $\Omega_h \in \mathcal{O}_h^N$, we consider the P1 finite element space $X_h \subset H_0^1(\Omega_h)$. We use the implicit Euler method to discretize in time and we get the discrete system (6). At each time step k, it consist on solving a linear system of the form

$$(\mathcal{M} + \Delta t \mathcal{A})\xi^k = \eta^{k-1},$$

where $\eta^{k-1} = \mathcal{M}\xi^{k-1} + F^k$ is known, with $F^k = (f^k, \varphi_j)$, $\mathcal{M} = (\varphi_i, \varphi_j)$ is the mass matrix, $\mathcal{A} = a(\varphi_i, \varphi_j)$ is the stiffness matrix for $i, j = 1, \ldots, S$, and $\xi^k = (\xi_j^k)_{j=1}^S$ is the vector of the coefficients of the solution on the finite-elements basis, i.e.

$$u_{h,\Delta t}^k = \sum_{j=1}^S \xi_j^k \varphi_j,$$

$(\varphi_j)_{j=1}^S$ being the basis functions for X_h. Obviously $\mathcal{M} + \Delta t \mathcal{A}$ is symmetric and positive definite so that the system above is solvable.

We recall that, for a fixed bounded domain Ω with Lipschitz boundary, the fully discrete solutions $u_{h,\Delta t}^k$ converge to the solution u of the continuous heat equation (1) as $h \to 0$ and $\Delta t \to 0$. The proof of this result is based on the classical consistency plus stability analysis. In particular, the implicit method (6) is unconditionally stable with respect to the $L^2(0, T; H_0^1(\Omega))$-norm.

The following estimate on the rate of convergence is also well known. Given $\psi_0 \in H^2(\Omega)$, for each $k = 1, \ldots, M$ it follows that (see Section 11.3, pp. 394, Corollary 11.3.1, [21]):

$$\|u_{h,\Delta t}^k - u(t_k)\|_{L^2(\Omega)} \leq \|\psi_{0,h} - \psi_0\|_{L^2(\Omega)}$$
$$+ Ch^2 \Big(|\psi_0|_{H^2(\Omega)} + \int_0^{t_k} |\partial_t u(s)|_{H^2(\Omega)} \Big) + \Delta t \Big(\int_0^{t_k} \|\partial_t^2 u(s)\|_{L^2(\Omega)} \Big), \quad (9)$$

where the seminorm in $H^2(\Omega)$ is denoted by $|\cdot|_{H^2(\Omega)}$ and the norm in $L^2(\Omega)$ is denoted by $\|\cdot\|_{L^2(\Omega)}$.

Remark 1. We choose the implicit method for the time-discretization because it is unconditionally stable, so that the choice of Δt is dictated from accuracy requirements only. Recall that, by the contrary, explicit methods are conditionally stable, and, therefore, they require the time-step Δt to be sufficiently small with respect to the spatial mesh size h.

In the analysis of the convergence of the optimal design problems we will need to pass to the limit in the solution of the discrete problems towards those of the continuous heat equation when the domain varies. The following Proposition provides the needed convergence result:

Proposition 1. *Let $\Omega \in \mathcal{O}^N$ be given. Let $\Omega_{h,\triangle t} \in \mathcal{O}_h^N$ be a sequence such that $\Omega_{h,\triangle t} \xrightarrow{H^c} \Omega$.*
Then $\widetilde{u}_{h,\triangle t}^k \to \widetilde{u}$ strongly in $L^2(0,T; H_0^1(\mathcal{C}))$ when $h \to 0$ and $\triangle t \to 0$.

Remark 2. This convergence property holds for the piecewise constant or linear extension of $\widetilde{u}_{h,\triangle t}^k$ to all $t \in [0,T]$. For the sake of simplicity we denote it simply as $\widetilde{u}_{h,\triangle t}^k$.

Proof (of Proposition 1). Let us denote by $\widetilde{X}_h$ the vector space of all functions of X_h extended by zero to $\mathcal{C}$.

Let $\varphi(x,t) = \sigma(t)w(x) \in C_c^\infty(\Omega \times [0,T])$ be given. We define the time discrete test function:

$$\varphi^k = \sigma^k w(x) = \sigma(t_k)w(x), \quad k = 1,...,M$$

and $\widetilde{\varphi}^k = \sigma^k \widetilde{w}(x)$ its extension by zero to $\mathcal{C}$.

The equation (6) can be rewritten as follows,

$$\int_{\mathcal{C}} \frac{\widetilde{u}_{h,\triangle t}^k - \widetilde{u}_{h,\triangle t}^{k-1}}{\triangle t} \widetilde{\varphi}^k \, dx + \int_{\mathcal{C}} \nabla \widetilde{u}_{h,\triangle t}^k \cdot \nabla \widetilde{\varphi}^k \, dx = \int_{\mathcal{C}} f^k \widetilde{\varphi}^k \, dx, \quad k = 1,...,M.$$

Taking the test function $\widetilde{\varphi}^k = \widetilde{u}_{h,\triangle t}^k$ and rewriting

$$\widetilde{u}_{h,\triangle t}^k = \frac{\triangle t}{2} \frac{\widetilde{u}_{h,\triangle t}^k - \widetilde{u}_{h,\triangle t}^{k-1}}{\triangle t} + \frac{\widetilde{u}_{h,\triangle t}^k + \widetilde{u}_{h,\triangle t}^{k-1}}{2},$$

we get

$$\frac{\|\widetilde{u}_{h,\triangle t}^k\|_{L^2(\mathcal{C})}^2 - \|\widetilde{u}_{h,\triangle t}^{k-1}\|_{L^2(\mathcal{C})}^2}{2\triangle t} + \|\widetilde{u}_{h,\triangle t}^k\|_{H_0^1(\mathcal{C})}^2 \le \frac{1}{2}\|f^k\|_{L^2(\mathcal{C})}^2 + \frac{1}{2}\|\widetilde{u}_{h,\triangle t}^k\|_{L^2(\mathcal{C})}^2.$$

Therefore, we conclude that

$$\|\widetilde{u}_{h,\triangle t}^k\|_{L^2(0,T;H_0^1(\mathcal{C}))} \le c_1 [\|\widetilde{f}\|_{L^2(0,T;L^2(\mathcal{C}))} + \|\widetilde{\psi}_{0,h}\|_{L^2(\mathcal{C})}]$$

for any h and $\triangle t$.

Thus, up to the extraction of subsequences, $\widetilde{u}_{h,\triangle t}^k$ weakly converges in $L^2(0,T; H_0^1(\mathcal{C}))$ to w. We have to show that its limit coincides with $\widetilde{u}$, u being the solution of (1), to later prove strong convergence. By Lemma 3, we know that there exists $y \in L^2(0,T; H_0^1(\Omega))$ such that $w = \widetilde{y}$.

Now, let us prove that $y = u$ and that the convergence holds in the strong topology.

First we prove that $y = u$. Observe that the solution u of (1) is characterized by the fact that $u \in C([0,T]; L^2(\Omega)) \cap L^2(0,T; H_0^1(\Omega))$ and

$$\begin{cases} -\int_0^T \int_\Omega u\sigma_t w \, dxdt + \int_\Omega \psi_0 \sigma(0) w(x) dx - \int_0^T \int_\Omega \nabla u \cdot \sigma \nabla w \, dxdt \\ \qquad = \int_0^T \int_\Omega f\sigma w \, dxdt, \\ \forall \sigma(t)w(x) \in H^1(\Omega \times (0,T)) \text{ s.t. } \sigma(t)w(x)|_{\partial\Omega \times [0,T]} = 0 \text{ and } \sigma(T) \equiv 0. \end{cases}$$

We have to prove that y is the solution of the previous equation.

We have

$$\sum_{k=1}^M \int_\mathcal{C} \frac{\widetilde{u}^k_{h,\Delta t} - \widetilde{u}^{k-1}_{h,\Delta t}}{\Delta t} \sigma^k \widetilde{w}_h \, dx + \sum_{k=1}^M \int_\mathcal{C} \nabla \widetilde{u}^k_{h,\Delta t} \cdot \sigma^k \nabla \widetilde{w}_h \, dx = \sum_{k=1}^M \int_\mathcal{C} f^k \sigma^k \widetilde{w}_h \, dx.$$

By Lemma 3.1 pp. 17, [7], there exists $\widetilde{w}_h \in X_h$ such that $\widetilde{w}_h \to \widetilde{w}$ strongly in $H^1_0(\mathcal{C})$ as $h \to 0$.

Adding by parts we get

$$-\int_0^T \int_\mathcal{C} \widetilde{u}^k_{h,\Delta t} \frac{\sigma^{k+1} - \sigma^k}{\Delta t} \widetilde{w}_h \, dxdt - \int_\mathcal{C} \widetilde{\psi}_{0,h} \sigma^0 \widetilde{w}_h \, dx$$
$$+ \int_0^T \int_\mathcal{C} \nabla \widetilde{u}^k_{h,\Delta t} \cdot \nabla \widetilde{w}_h \sigma^k \, dxdt = \int_0^T \int_\mathcal{C} f^k \sigma^k \widetilde{w}_h \, dxdt. \quad (10)$$

On the other hand

$$\left| \frac{\sigma^{k+1} - \sigma^k}{\Delta t} - \sigma_t(t) \right| \leq \left| \frac{\sigma^{k+1} - \sigma^k}{\Delta t} - \sigma_t(t_k) \right| + \left| \sigma_t(t) - \sigma_t(t_k) \right|$$
$$\leq C(\Delta t) \|\sigma_{tt}\|_{L^\infty(0,T)}.$$

Furthermore, we know that $\widetilde{u}^k_{h,\Delta t} \rightharpoonup \widetilde{y}$ weakly in $L^2(0,T;H^1_0(\mathcal{C}))$. Thus, we can pass to the limit in equation (10) and get

$$-\int_0^T \int_\mathcal{C} \widetilde{y} \sigma_t \widetilde{w} \, dxdt - \int_\mathcal{C} \widetilde{\psi}_0 \sigma(0) \widetilde{w} \, dx + \int_0^T \int_\mathcal{C} \nabla \widetilde{y} \cdot \sigma \nabla \widetilde{w} \, dxdt = \int_0^T \int_\mathcal{C} f\sigma \widetilde{w} \, dxdt.$$

Using that $\widetilde{y}$ is vanishes on $\mathcal{C}\setminus\Omega$ and $\widetilde{y} = y$ on Ω we have that y satisfies the same equation on Ω. So $y = u$.

Now, we prove the strong convergence in $L^2(0,T;H^1_0(\mathcal{C}))$. For $\widetilde{u}$ the energy estimate yields

$$\int_0^T \int_\mathcal{C} |\nabla \widetilde{u}|^2 \, dxdt + \frac{1}{2}\|\widetilde{u}(T)\|^2_{L^2(\mathcal{C})} = \frac{1}{2}\|\widetilde{u}(0)\|^2_{L^2(\mathcal{C})} + \int_0^T \int_\mathcal{C} f\widetilde{u} \, dxdt. \quad (11)$$

For $\widetilde{u}^k_{h,\Delta t}$ taking as test function $\widetilde{\varphi}^k_h = \widetilde{u}^k_{h,\Delta t}$ we get

$$\Delta t \sum_{k=1}^M \int_\mathcal{C} |\nabla \widetilde{u}^k_{h,\Delta t}|^2 \, dx + \frac{1}{2} \int_\mathcal{C} |\widetilde{u}^M_{h,\Delta t}|^2 \, dx = \frac{1}{2} \int_\mathcal{C} |\widetilde{u}^0_{h,\Delta t}|^2 \, dx$$
$$+ \Delta t \sum_{k=1}^M \int_\mathcal{C} f^k \widetilde{u}^k_{h,\Delta t} \, dx. \quad (12)$$

Under the assumptions on the initial data and the weak convergence in $L^2(0,T; H_0^1(\mathcal{C}))$ we can easily pass to the limit in the right hand side term of (14). On the other hand, by weak convergence of the solutions and the weak lower semi-continuity of norms, we have

$$\|\nabla \widetilde{u}\|_{L^2(0,T;L^2(\mathcal{C}))}^2 + \frac{1}{2}\|\widetilde{u}(T)\|_{L^2(\mathcal{C})}^2$$
$$\leq \liminf_{h} \left[\|\nabla \widetilde{u}_{h,\triangle t}^k\|_{L^2(0,T;L^2(\mathcal{C}))}^2 + \frac{1}{2}\|\widetilde{u}_{h,\triangle t}^M\|_{L^2(\mathcal{C})}^2\right]$$
$$= \frac{1}{2}\|\widetilde{u}(0)\|_{L^2(\mathcal{C})}^2 + \int_0^T \int_\mathcal{C} f\widetilde{u}\,dxdt.$$

On the other, by the energy identity (13) for the heat equation (1) we deduce that

$$\|\nabla \widetilde{u}_{h,\triangle t}^k\|_{L^2(0,T;L^2(\mathcal{C}))}^2 + \frac{1}{2}\|\widetilde{u}_{h,\triangle t}^M\|_{L^2(\mathcal{C})}^2 \to \|\nabla \widetilde{u}\|_{L^2(0,T;L^2(\mathcal{C}))}^2 + \frac{1}{2}\|\widetilde{u}(T)\|_{L^2(\mathcal{C})}^2.$$

This, together with weak convergence, implies the strong convergences:

$$\widetilde{u}_{h,\triangle t}^k \to \widetilde{u} \quad \text{in } L^2(0,T; H_0^1(\mathcal{C})).$$
$$\widetilde{u}_{h,\triangle t}^M \to \widetilde{u}(T) \quad \text{in } L^2(\mathcal{C}).$$

Note that, in the proof above, we have used the weak convergence of $\widetilde{u}_{h,\triangle t}^M$ towards $\widetilde{u}(T)$ in $L^2(\mathcal{C})$. This is due to the uniform bounds on (14), the weak convergence in $L^2(0,T; H_0^1(\mathcal{C}))$ and a classical compactness argument which uses the Aubin-Lions Lemma and the equation satisfied by $\widetilde{u}_{h,\triangle t}^M$ which allows getting uniform bounds on the time-derivative of its piecewise linear and continuous extension in time in $L^2(0,T; H^{-1}(\mathcal{C}))$. ∎

4 Convergence of discrete optimal shapes

The question we address here is the numerical approximation of the optimal design problem (5). In particular, we address the issue of whether the discrete optimal shapes for a suitable discretization of the above problem converge in H^c to a continuous optimal shape. As we shall see, the answer to this question is positive if the discrete optimization problem is conveniently built, as above, in the context of finite element approximations.

The triangulation $\mathcal{T}_h$ being fixed, for any $h > 0$, the number of triangular domains in $\mathcal{O}_h^N$ under consideration for the discrete optimal design problem (7) is finite. Thus, the existence of discrete optimal shapes is obvious, and we denote them by $\Omega_{h,\triangle t}^*$.

Now, we prove that any limit in H^c of discrete optimal shapes is an optimal domain for the continuous optimal design problem.

Theorem 2. Let J be the functional as in (3). Suppose that the discretization $J_h^{\triangle t}$ of J has been chosen such that:

1. If $\Omega, \Omega_{h,\triangle t} \in \mathcal{O}^N$ are such that $\Omega_{h,\triangle t} \xrightarrow{H^c} \Omega$, then $J_h^{\triangle t}(\Omega_{h,\triangle t}) \to J(\Omega)$ when $h \to 0$ and $\triangle t \to 0$.

Then, the discrete optimal design problems (7) converge as $h \to 0$ and $\triangle t \to 0$ to the continuous one (5) in the sense that

(a) $J_h^{\triangle t}$ reaches its minimum on $\mathcal{O}_h^N$ for all $h > 0$ and $\triangle t > 0$.
(b) Any accumulation point as $h \to 0$, $\triangle t \to 0$ in the topology H^c of any sequence $\left(\Omega^*_{h,\triangle t}\right)_{h,\triangle t}$ of discrete minimizers is a continuous minimizer.
(c) The whole sequence $\left(\mathcal{I}_h^{\triangle t}\right)_{h,\triangle t}$ converges to $\mathcal{I}$.

Remark 3. Similar results hold in the class $\mathcal{O}_\omega^N$ of domains.

Proof (of Theorem 2). Let $(\Omega^*_{h,\triangle t})_{h,\triangle t}$ be a sequence of discrete minimizers for problem (6). Any $\Omega^*_{h,\triangle t}$ belongs to $\mathcal{O}^N$ which is H^c-compact. Let Ω_{ap} be an accumulation point of this sequence. By Lemma 1 $\Omega_{ap} \in \mathcal{O}^N$. From Proposition 1 we have

$$\widetilde{u}^k_{h,\triangle t} \to \widetilde{u^{\Omega_{ap}}} \text{ strongly in } L^2(0,T;H^1_0(\mathcal{C})),$$

where $u^{\Omega_{ap}}$ is the solution of the continuous problem (1) in Ω_{ap}. Due to the assumption of the Theorem, we obtain

$$\mathcal{I}_h^{\triangle t} = J(\Omega^*_{h,\triangle t}) \to J(\Omega_{ap}) \quad \text{when } h \to 0 \text{ and } \triangle t \to 0. \tag{13}$$

Let us now check that Ω_{ap} is a minimizer for J.

Given $\Omega \in \mathcal{O}^N$, there exist $\Omega_h \in \mathcal{O}_h^N$ such that $\Omega_h \xrightarrow{H^c} \Omega$ (see Section 4.2.1, [7]). For each h and $\triangle t$, we have

$$\mathcal{I}_h^{\triangle t} \leq J_h^{\triangle t}(\Omega_h).$$

Passing to the limit in this inequality and using (13) and hypothesis 1, we obtain $J(\Omega_{ap}) \leq J(\Omega)$ for all $\Omega \in \mathcal{O}^N$. This proves points a) and b) of the theorem.

Also, we have seen that the only accumulation point of the sequence $\left(\mathcal{I}_h^{\triangle t}\right)_{h,\triangle t}$ is nothing but $\mathcal{I}$. ∎

Remark 4. We have proved that any limit in H^c of discrete optimal shapes is an optimal domain for the continuous optimal design problem. The obtention of convergence rates would be of interest, but this subject is completely open.

5 Gradient calculations: A numerical approach

5.1 Preliminaries

We have proven that the discrete optimal shapes converge in H^c to an optimal shape for the continuous problem. Now we address the problem of efficiently computing the discrete optimal shapes. Despite of the fact that, for $h > 0$ and $\triangle t > 0$ given, the existence of the discrete optimal shapes is trivial, its computation may be rather complex because of the very large number of existing admissible domains.

The search of the discrete optimal shapes is usually performed by gradient type methods. The main idea of these methods is to iterate in the discrete domain using the information provided by the gradient of the functional with respect to perturbations of the domain in the continuous framework. This gradient can be calculated using classical methods of differentiation with respect to the domain (see [9, 16, 19, 20]).

As far as we know, the convergence of an iterative method based on these ideas is not proved so far. In fact, in principle, taking into account that the information we are using to iterate on the discrete domains comes from the continuous framework, it is not even clear that the discrete functional decreases along the iteration. We refer to [9] and [19] for an analysis of the comparison between discrete and continuous gradients. As we shall see, however, the method turns out to be efficient in practice.

The second drawback of this procedure is that it is based on tools coming from the differentiation with respect to the shape of the domain. This requires a minimal amount of regularity of the domains under consideration and, consequently, can not be applied in the general geometric setting in which our convergence result in Theorem 2 has been established.

The use of differentiation with respect to domain deformations can be fully justified by restricting the class of admissible domains to consider only sufficiently smooth ones (see [22, 23, 24]). In that setting the existence of optimal domains can be proved by classical regularity and compactness results for the solutions of the PDE under consideration both in the elliptic and the parabolic case (see [16, 17]). However, as far as we know, the convergence of these iterative numerical methods is still to be proved in this context too.

Let us now describe how to use differentiation with respect to the domain to build an iterative method for searching optimal shapes.

5.2 A example for the continuous problem

Consider the functional

$$J(\Omega) = \frac{1}{2} \int_0^T \int_\Omega |\nabla(u - \widetilde{u}_E)|^2 dx dt, \qquad (14)$$

where u_E is the solution of the problem (1) in the domain E that we want to recover. Obviously the solution of this minimization problem is $\Omega = E$. We use it as a test of the efficiency of our method.

The aim of this section is to obtain an expression for the variation of the functional (14). The main tool is the so-called shape differentiation ([16, 20, 22]). To do this, we consider normal variations of the domain and the new domains of the form

$$\Omega + \alpha = \{x + \alpha(x) : x \in \Omega\},$$

where α represents the variations of Ω, with $\alpha \in C^2$. These variations α are assumed to be small enough and oriented along the normal direction over the boundary $\partial\Omega$. This induces a variation on the solution: $\delta u = u(\Omega + \alpha) - u(\Omega)$. Differentiating in (14) we obtain

$$\langle \delta J(\Omega), \alpha \rangle = \frac{1}{2} \int_0^T \int_\Gamma \alpha |\partial_n (u - \tilde{u}_E)|^2 d\sigma dt + \int_0^T \int_\Omega \nabla (u - \tilde{u}_E) \cdot \nabla (\delta u) dx dt \tag{15}$$

where $\Gamma = \partial\Omega$.

On the other hand, differentiating the state equation (1) we have (see [16, 20, 22])

$$\begin{cases} \delta u_t - \Delta(\delta u) = 0 & \Omega \times [0, T], \\ \delta u = -\alpha(\partial_n u) & \Gamma \times [0, T], \\ \delta u(0) = 0 & \Omega. \end{cases} \tag{16}$$

Let $\phi \in H^1(0, T; L^2(\Omega)) \cap L^2(0, T; H^2(\Omega) \cap H_0^1(\Omega))$. Multiplying the previous equation by ϕ and integrating by parts, we get

$$0 = \int_0^T \int_\Omega [-\phi_t - \Delta\phi] \delta u \, dx dt + \int_\Omega \phi(T) \delta u(T) dx - \int_0^T \int_\Gamma \partial_n(\delta u) \phi \, d\sigma dt$$
$$+ \int_0^T \int_\Gamma (\partial_n \phi) \delta u \, d\sigma dt.$$

Let us choose ϕ as the solution of the adjoint problem

$$\begin{cases} -\phi_t - \Delta\phi = -\Delta(u - \tilde{u}_E) & \Omega \times [0, T], \\ \phi = 0 & \Gamma \times [0, T], \\ \phi(T) = 0 & \Omega. \end{cases} \tag{17}$$

Multiplying this equation (17) by δu and integrating by parts we get

$$\int_0^T \int_\Omega \nabla(u - \tilde{u}_E) \cdot \nabla(\delta u) dx dt - \int_0^T \int_\Gamma \partial_n(u - \tilde{u}_E) \delta u \, d\sigma dt$$
$$= -\int_0^T \int_\Omega \phi_t \delta u \, dx dt - \int_0^T \int_\Omega \Delta\phi \delta u \, dx dt$$
$$= \int_0^T \int_\Omega \phi[\delta u_t - \Delta(\delta u)] dx dt - \int_0^T \int_\Gamma (\partial_n \phi) \delta u \, d\sigma dt.$$

Therefore,

$$\int_0^T \int_\Omega \nabla(u - \tilde{u}_E) \cdot \nabla(\delta u) dx dt = \int_0^T \int_\Gamma \alpha(\partial_n u)(\partial_n \phi) d\sigma dt$$
$$- \int_0^T \int_\Gamma \alpha \partial_n (u - \tilde{u}_E) \partial_n u d\sigma dt. \quad (18)$$

Taking (18) in (15) we obtain

$$\langle \delta J(\Omega), \alpha \rangle = \frac{1}{2} \int_0^T \int_\Gamma \alpha |\partial_n (u - \tilde{u}_E)|^2 d\sigma dt + \int_0^T \int_\Gamma \alpha(\partial_n u)(\partial_n \phi) d\sigma dt$$
$$- \int_0^T \int_\Gamma \alpha \partial_n (u - \tilde{u}_E) \partial_n u d\sigma dt$$
$$= \int_0^T \int_\Gamma \alpha \Big(\partial_n (u - \tilde{u}_E) \Big(\frac{1}{2} \partial_n (u - \tilde{u}_E) - \partial_n u \Big) + \partial_n u \partial_n \phi \Big) d\sigma dt.$$

In this way we obtain the following expression for the variation of J:

$$\langle \delta J(\Omega), \alpha \rangle = \int_0^T \int_\Gamma \alpha \Big(-\frac{1}{2} \big((\partial_n u)^2 - (\partial_n \tilde{u}_E)^2 \big) + \partial_n u \partial_n \phi \Big) d\sigma dt. \quad (19)$$

Note that using the adjoint state, the expression of $\langle \delta J(\Omega), \alpha \rangle$ in (15) has been simplified. Indeed, in the final one (19), the variation of the state δu does not enter. This is a significant improvement since, according to (16), computing δu would require solving an initial boundary value problem for each α. In view of (19), it is sufficient to compute the adjoint solution ϕ and then an integral for each α.

5.3 Optimization algorithm

We introduce a full-discretization of the functional (14):

$$J_h^{\Delta t}(\Omega_{h,\Delta t}) = \frac{\Delta t}{2} \sum_{k=1}^M \int_{\Omega_{h,\Delta t}} |\nabla(u_{h,\Delta t}^k - \tilde{u}_E)|^2 dx. \quad (20)$$

We discretize the adjoint problem (17) in the same way as the state equation, i.e. let $\phi_{h,\Delta t}^k \in X_h$ be the solution of

$$\begin{cases} \int_{\Omega_{h,\Delta t}} \frac{\phi_{h,\Delta t}^k - \phi_{h,\Delta t}^{k+1}}{\Delta t} \varphi dx + \int_{\Omega_{h,\Delta t}} (\nabla \phi_{h,\Delta t}^k) \cdot (\nabla \varphi) dx \\ \qquad = \int_{\Omega_{h,\Delta t}} \nabla(u_{h,\Delta t}^k - \tilde{u}_E) \cdot \nabla \varphi dx \quad \forall \varphi \in X_h, k = 1, \ldots, M, \\ \phi_{h,\Delta t}^{M+1} = 0. \end{cases}$$
$$(21)$$

We discretize (19) to get an approximate estimate of the variation of the discrete functional (20):

$$\langle \delta J_h^{\Delta t}(\Omega_{h,\Delta t}), \alpha \rangle \sim \Delta t \sum_{k=1}^{M} \int_{\Gamma_{h,\Delta t}} \alpha \Big(-\frac{1}{2}\big((\partial_n u_{h,\Delta t}^k)^2 - (\partial_n \widetilde{u}_E)^2\big)$$
$$+ (\partial_n u_{h,\Delta t}^k)(\partial_n \phi_{h,\Delta t}^k)\Big) d\sigma. \quad (22)$$

Note that both in the continuous and the discrete case $\partial_n u = \nabla u \cdot \mathbf{n}$.

We denote by Γ_{int} the interior boundary of the domain, and Γ_{out} the outer one, and by τ_i^{h-} the triangles belonging to in Γ_{int} and τ_i^{h+} those in Γ_{out}. The inner τ_i^{h-} and outer triangles τ_i^{h+} are linked by the fact that they have a common edge on the boundary Γ_h, and $\mathcal{F}_h$ is the set of nodes of the boundary (see Fig. 2).

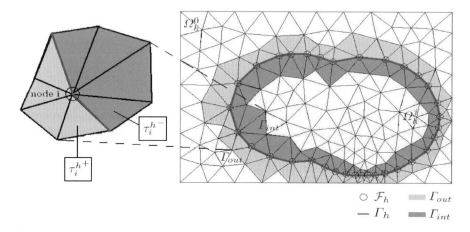

Fig. 2. Outer and inner boundaries of the domain, Γ_{out} and Γ_{int} respectively.

As we mentioned above, in the continuous case the deformations α considered are oriented in the normal direction along the boundary. In the discrete setting it is natural to interpret this fact by considering perturbations in which one adds triangles τ_i^{h+} or drops τ_i^{h-} depending how they contribute to the decrease of the functional.

To do this we compute the contribution of each edge of the boundary to the gradient of the discrete functional as follows:

$$\delta J_h^{\Delta t}(\Omega_{h,\Delta t}^j)\Big|_{\Gamma_{h,\Delta t}^j \cap \tau_i^{h-}} := \Delta t \sum_{k=1}^{M} \int_{\Gamma_{h,\Delta t}^j \cap \tau_i^{h-}} \Big(-\frac{1}{2}\big((\partial_n u_{h,\Delta t}^k)^2 - (\partial_n \widetilde{u}_E)^2\big)$$
$$+ (\partial_n u_{h,\Delta t}^k)(\partial_n \phi_{h,\Delta t}^k)\Big) d\sigma, \quad (23)$$

the contribution of the edge $\Gamma^j_{h,\triangle t} \cap \tau_i^{h^-}$ to this approximation of the variation of the functional $J_h^{\triangle t}$.

The functional $J_h^{\triangle t}(\Omega_{h,\triangle t})$ being defined on a finite number of polygonal domains its continuous derivative is not well defined. But (23) provides an approximation to its change rate locally on each edge of the boundary. However one has to interprete the estimated variation in (23) in the context of the given triangulation and the possible polygonal configurations.

To do this, given a discrete domain, in view of (23), we analyze the contribution of each one of its boundary triangles, both inner and outer ones, and we obtain the new domain adding or cutting triangles based on their contribution to decreasing the value of $J_h^{\triangle t}$ (see [9, 16, 17, 19, 20]).

To compute $\delta J_h^{\triangle t}(\Omega^j_{h,\triangle t})\big|_{\Gamma^j_{h,\triangle t}\cap\tau_i^{h^-}}$, according to (23), we need to solve the discrete state equation (6) and the discretization of the adjoint problem (21) with $\Omega_{h,\triangle t} = \Omega^j_{h,\triangle t}$.

For each node of the boundary, $\ell \in \mathcal{F}_h$, we compute the variation of the functional at this node as the average of the variation of the funcional in the edges $\Gamma^j_{h,\triangle t} \cap \tau_i^{h^-}$ containing the node ℓ. We denote by $\delta J_h^{\triangle t}(\Omega^j_{h,\triangle t})\big|_\ell$ the variation of the functional at the node ℓ.

Following this procedure, the new domain $\Omega^j_{h,\triangle t}$ is obtained from the previous one $\Omega^{j-1}_{h,\triangle t}$ adding the triangles containing the node ℓ where the contribution of $\delta J_h^{\triangle t}(\Omega^j_{h,\triangle t})\big|_\ell$ is negative, and cutting ones where its contribution is positive.

We explain this procedure in more detail below.

Let us now describe the algorithm. We fix a tolerance $TOL > 0$.

1. We choose $h > 0$ and $\triangle t > 0$ and construct the mesh $\mathcal{T}_h$ of $\mathcal{C}$.
2. Consider the initial guess $\Omega^0_{h,\triangle t} = \mathcal{C}$.
3. Iteration scheme, $j \geq 0$. It is applied while $|J_h^{\triangle t}(\Omega^j_{h,\triangle t})| > TOL$:
   a) Solve the discrete state problem (6) with $\Omega_{h,\triangle t} = \Omega^j_{h,\triangle t}$.
   b) Solve the adjoint discrete problem (21) with $\Omega_{h,\triangle t} = \Omega^j_{h,\triangle t}$.
   c) Compute $\delta J_h^{\triangle t}(\Omega^j_{h,\triangle t})\big|_{\Gamma^j_{h,\triangle t}\cap\tau_i^{h^-}}$ as in (23).
   d) Compute $\delta J_h^{\triangle t}(\Omega^j_{h,\triangle t})\big|_\ell$.
   e) Deformation of the domain.
      We build the new domain $\Omega^{j+1}_{h,\triangle t}$ as follows:

$$\Omega^{j+1}_{h,\triangle t} = \Omega^j_{h,\triangle t} \bigcup \{\tau_\ell^h : \delta J_h^{\triangle t}(\Omega^j_{h,\triangle t})\big|_\ell < 0\} \setminus \{\tau_\ell^h : \delta J_h^{\triangle t}(\Omega^j_{h,\triangle t})\big|_\ell > 0\},$$

where τ_ℓ^h are the triangles that contain the node ℓ.
   f) Compute the functional (20) in the new domain $\Omega^{j+1}_{h,\triangle t}$.

g) We take $\Omega_{h,\triangle t}^j = \Omega_{h,\triangle t}^{j+1}$ and go back to the beginning of this iteration scheme.

5.4 Numerical results

All the numerical experiments we present here have been performed with a Pentium M 715 processor and 512 MB RAM.

Let the set $\mathcal{C}$ be the rectangle $(-1.5, 1.5) \times (-1, 1)$. We consider problems (1) and (5), with force term $f = 1$, initial data $\psi_0(x, y) = \sin(2\pi x)$ and where the functional to be minimized is

$$J(\Omega) = \frac{1}{2} \int_0^T \int_\Omega |\nabla (u - \tilde{u}_E)|^2 dx dt, \qquad (24)$$

where u_E is the solution of the problem (1) in the domain E that we want to recover, and u is the solution in Ω.

Numerical experiment # 1

We take $T = 20$, $\triangle t = 0.1$ and $h = 0.19197$. This time the computation is done over a mesh of 206 nodes and 372 triangles. Our goal is to recover the circle E (see Fig 3).

Fig. 3. The unknown body

In order to do this, we compute u_E, the solution of the problem (6) in the domain E, then we minimize the functional (24) by the algorithm that we have described in the previous section.

Figure 4 depicts the evolution of the domain with respect to the iteration j. We find the circle E in 6 steps and 3621 seconds (CPU time). Figure 5 depicts the evolution of the cost function. As expected, the limit of the sequence Ω_j is close to the circle E.

Numerical experiment # 2

We take $T = 20$, $\triangle t = 0.1$ and $h = 0.31062$. This time the computation is done over a mesh of 118 nodes and 213 triangles. Now, our goal is to recover E as in Fig. 6 bellow:

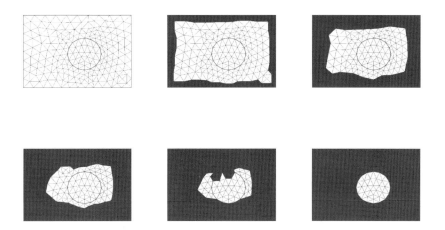

Fig. 4. Evolution of the domain converging to the circle E

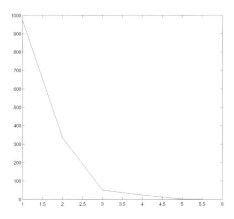

Fig. 5. Evolution of the functional (24)

Fig. 6. The unknown body

Figure 7 depicts the evolution of the domain with respect to the iteration j. In this case, we find E in 5 steps and 1768 seconds (CPU time). Figure 8 depicts the evolution of the cost function.

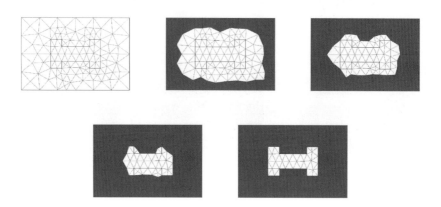

Fig. 7. Evolution of the domain converging to E

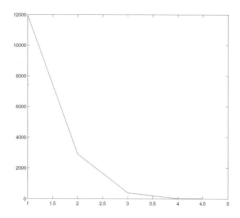

Fig. 8. Evolution of the functional (24)

Numerical experiment # 3

We take $T = 20$, $\triangle t = 0.1$ and $h = 0.14509$. This time the computation is done over a mesh of 281 nodes and 506 triangles. Now, our goal is to recover E as in Fig. 9 bellow:

Fig. 9. The unknown body

Figure 10 depicts the evolution of the domain with respect to the iteration j. In this case, we find E in 6 steps and 6866 seconds (CPU time). Figure 11 depicts the evolution of the cost function.

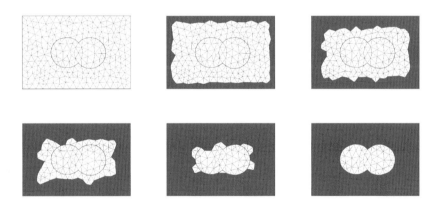

Fig. 10. Evolution of the domain converging to E

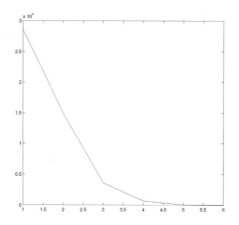

Fig. 11. Evolution of the functional (24)

Numerical experiment # 4

We take $T = 20$, $\triangle t = 0.1$ and $h = 0.13798$. This time the computation is done over a mesh of 284 nodes and 516 triangles. Now, our goal is to recover E as in Fig. 12 bellow:

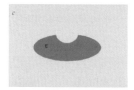

Fig. 12. The unknown body

Figure 13 depicts the evolution of the domain with respect to the iteration j. In this case, we find E in 6 steps and 10372 seconds (CPU time). Figure 14 depicts the evolution of the cost function.

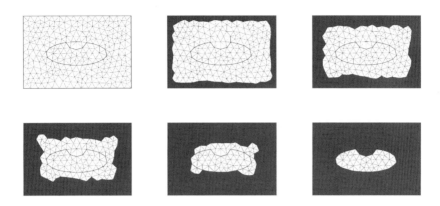

Fig. 13. Evolution of the domain converging to E

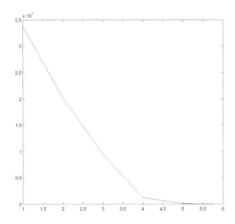

Fig. 14. Evolution of the functional (24)

6 Conclusions

We have considered the problem of numerically approximating optimal shapes in the context of the 2D linear heat equation with Dirichlet boundary conditions. We have addressed the issue of whether discrete optimal shapes for a suitable discretization of the original continuous optimal design problem provide an approximation of the continuous optimal shape.

We have developed a P1 finite-element approximation in space and an implicit discretization in time for which this convergence result holds in the 2D case, in the class of domains with an a priori bounded number of holes, introduced by V. Šverák ([25]). According to our results convergence holds in the complementary-Hausdorff topology.

Our results can be extended to a more general framework of evolution problems provided a number of properties are guaranteed: (a) the continuous dependence of the solution of the PDE with respect to the domain on which it is posed, and (b) the H^c-compactness of the set of admissible continuous domains. These continuity properties, and the convergence properties of the numerical scheme under consideration, allow proving sufficient continuity conditions of numerical schemes with respect to the numerical mesh, to guarantee the convergence of the optimal shapes.

These results extend to the evolution framework those previously developed in [6] and [7] in the elliptic case.

Then, we use a classical iterative optimization algorithm to obtain a numerical approximation of the discrete optimal domains. Using differentiation with respect to the domain, we can find explicit formulas of the approximate variation of the discrete functional to build numerical methods for the search of the discrete optimal shape, by means of the solution of the discrete adjoint problem. The convergence of the iterative numerical methods we obtain by this procedure is not proved but its efficiency is illustrated by various experiments.

Acknowledgements

The second author is grateful to D. Chenais for fruitful discussions. Partially supported by grant MTM2005-00714 of the Spanish MEC, the DOMINO Project CIT-370200-2005-10 in the PROFIT program of the MEC (Spain), the SIMUMAT projet of the CAM (Spain) and the European network "Smart Systems".

References

1. Arendt, W.: Approximation of degenerate semigroups, Taiwanese J. Math. 5, 279–295, (2001).
2. Arendt, W., Daners, D.: Uniform Convergence for Elliptic Problems on Varying Domains. Mathematische Nachrichten, to appear.
3. Attouch, H.: Variational convergence for functions and operators. Applicable Math. series, Pitman, Longon, (1984).
4. Brezis, H.: Analyse fonctionnelle. Théorie et applications. Collection Mathématiques Appliquées pour la Maîtrise. Masson, Paris, (1983).

5. Bucur, D., Buttazzo, G.: Variational methods in shape optimization problems. Progress in Nonlinear Differential Equations and their Applications, 65. Birkhšuser Boston, Inc., Boston, (2005).
6. Chenais, D., Zuazua, E.: Controllability of an Elliptic Equation and its Finite Difference, Numer. Math., 95, no 1, 63–99, (2003).
7. Chenais, D., Zuazua, E.: Finite element approximation for elliptic shape-optimization problems. C. R. Math. Acad. Sci. Paris, 338, no. 9, 729–734, (2004).
8. Cioranescu, D., Donato, P., Murat, F., Zuazua, E.: Homogenization and corrector for the wave equation in domains with small holes. Ann. Scuola Norm. Sup. Pisa Cl. Sci., (4), 18, no. 2, 251–293, (1991).
9. Giles, M., Pierce, N., Süli, E.: Progress in adjoint error correction for integral functionals. Comput. Vis. Sci. 6, no. 2-3, 113–121, (2004).
10. Grisvard, P.: Elliptic problems in nonsmooth domains. Monographs and Studies in Mathematics, 24. Pitman (Advanced Publishing Program), Boston, MA., (1985).
11. Hayouni, M., Henrot, A., Samouh, N.: On the Bernoulli free boundary problem and related shape optimization problems. Interfaces & Free Bound. 3, no. 1, 1–13, (2001).
12. Henrot, A., Pierre, M.: Variation et optimisation de formes. Une analyse géométrique. Collection: Mathématiques et Applications, vol. 48, (2005).
13. Kawohl, B., Pironneau, O., Tartar, L., Zolésio, J.-P.: Optimal shape design. Lectures given at the Joint C.I.M./C.I.M.E. Summer School held in Tróia, June 1–6, 1998. Edited by A. Cellina and A. Ornelas. Lecture Notes in Mathematics, 1740. Fondazione C.I.M.E.. Springer-Verlag, Berlin, (2000).
14. Lions, J.L.: Contrôlabilité exacte, stabilisation et perturbations de systèmes distribués. Tome 1. Contrôlabilité exacte, Masson, RMA 8, Paris, (1988).
15. Liu, W. B., Rubio, J. E.: Optimal shape design for systems governed by variational inequalities. Part 1: Existence theory for the elliptic case, Part 2: Existence theory for the evolution case. J. Optim. Theory Appl. 69, 351–371, 373–396, (1991).
16. Mohammadi, B., Pironneau, O.: Applied Shape Optimization for Fluids. Oxford science publications, (2001).
17. Mohammadi, B., Pironneau, O.: Shape optimization in fluid mechanics. Annual review of fluid mechanics. vol. 36, 255–279, Annu. Rev. Fluid Mech., 36, Annual Reviews, Palo Alto, CA, (2004).
18. Mosco, U.: Convergence of convex sets and solutions of variationnal inequalities, Adv. in Math., 3, 510–585, (1969).
19. Pierce, N., Giles, M.: Adjoint and defect error bounding and correction for functional estimates. J. Comput. Phys. 200, no. 2, 769–794, (2004).
20. Pironneau, O.: Optimal Shape Design for Elliptic Systems. Springer-Verlag, (1984).
21. Quarteroni, A., Valli, A.: Numerical approximation of partial differential equations. Springer Series in Computational Mathematics, 23. Springer-Verlag, Berlin, (1994).
22. Simon, J., Murat, F.: Sur le contrôle par un domaine géométrique. Preprint no. 76015, University of Paris 6, 725–734 (1976).
23. Simon, J.: Differentiation with respect to the Domain in Boundary Value Problems. Numer. Funct. and Optimiz., 2, no. 7-8, 649–687, (1980).

24. Simon, J.: Diferenciación de problemas de contorno respecto del dominio. Lecture notes. Universidad de Sevilla (1991).
    http://math.univ-bpclermont.fr/~simon/
25. Šverák, V.: On optimal shape design. J. Math. Pures. Appl., 72, 537–551 (1993).

CONTRIBUTED LECTURES

Accurate Finite Volume Solvers for Fluid Flows

3D Free Surface Flows Simulations Using a Multilayer Saint-Venant Model. Comparisons with Navier-Stokes Solutions

E. Audusse[1], M.O. Bristeau[1] and A. Decoene[2]

[1] INRIA, Domaine de Voluceau, B.P. 105, 78153 Le Chesnay, France
Emmanuel.Audusse@inria.fr, Marie-Odile.Bristeau@inria.fr
[2] INRIA and LNHE, EDF, 6 Quai Watier, BP 49, 78401 Chatou, France
Astrid.Decoene@inria.fr

Summary. We present a multilayer Saint-Venant system for the simulation of 3D free surface flows. A precise analysis of the shallow water assumption leads to a set of coupled Saint-Venant type systems. For each time dependent layer, a Saint-Venant type system is solved on the same 2D mesh by a kinetic solver using a finite volume framework. We validate the model by comparisons with Navier-Stokes solutions.

1 Introduction

In this paper, we present a multilayer Saint-Venant system for the simulation of 3D free surface flows. The idea is to introduce, when the hydrostatic assumption is valid, an alternative to the solution of the free surface Navier-Stokes system, leading to a precise description of the vertical profile of the horizontal velocity while preserving the robustness and the computational efficiency of the usual Saint-Venant system. This study generalizes the work of Audusse [1] to the 3D problem with slow varying bottom (see also [4]).

In Section 2, we recall the incompressible Navier-Stokes equations, the boundary conditions and the hydrostatic approximation. The multilayer Saint-Venant system is described in Section 3 and the associated numerical method in Section 4. A numerical example is presented in Section 5.

2 Navier-Stokes equations and hydrostatic approximation

We consider the classical incompressible Navier-Stokes system

$$\nabla . \mathbf{U} = 0, \qquad (1)$$

$$\frac{\partial \mathbf{U}}{\partial t} + \nabla . (\mathbf{U} \otimes \mathbf{U}) = \nabla . \sigma + \mathbf{g}, \qquad (2)$$

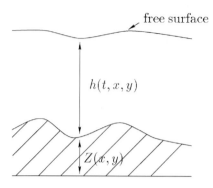

Fig. 1. Flow domain

with the stress tensor σ given by

$$\sigma = -p\mathbb{I}d + \mu[\nabla \mathbf{U} + (\nabla \mathbf{U})^T] \qquad (3)$$

and where $\mathbf{U} = (u, v, w)$ is the velocity, $\mathbf{u} = (u, v)$ is the horizontal velocity, p is the pressure, $\mathbf{g}$ represents the gravity forces, $\mathbf{g} = \begin{pmatrix} 0 \\ 0 \\ -g \end{pmatrix}$ and μ is the viscosity coefficient.

We consider a free surface flow (see Fig. 1), so we assume

$$Z(x, y) \leq z \leq H(t, x, y) = h(t, x, y) + Z(x, y)$$

with $Z(x, y)$ the bottom elevation and $h(t, x, y)$ the water depth.

On the bottom we prescribe an impermeability condition

$$\mathbf{U}.\mathbf{n} = 0 \qquad (4)$$

and a friction condition given by a Navier law

$$(\sigma.\mathbf{n}).\mathbf{t} = -\kappa\ \mathbf{U}.\mathbf{t} \qquad (5)$$

with κ a Navier coefficient, $\mathbf{n}$ a unit outward normal and $\mathbf{t}$ a tangential vector. For applications, we use also the Strickler friction.

On the free surface, the kinematic boundary condition is satisfied

$$\frac{\partial H}{\partial t} + \mathbf{u}(t, x, y, H).\nabla H - w(t, x, y, H) = 0 \qquad (6)$$

and also the no stress condition

$$\sigma.\mathbf{n} = 0. \qquad (7)$$

Then we introduce the shallow water assumption. We consider two characteristic dimensions $\mathcal{H}$ and $\mathcal{L}$ in the vertical and horizontal directions respectively and we assume that $\mathcal{H}$ is small compared to $\mathcal{L}$, so we can write $\epsilon = \frac{\mathcal{H}}{\mathcal{L}}$ with ϵ a small parameter. We assume also slow varying bottom ([4]). We introduce dimensionless variables and we obtain a dimensionless Navier-Stokes system (see [1, 4]). By an asymptotic analysis, we deduce the approximation at zero order in ϵ of the system (1)–(7) which gives the horizontally inviscid hydrostatic model

$$\nabla.\mathbf{U} = 0, \tag{8}$$

$$\frac{\partial \mathbf{u}}{\partial t} + \nabla(\mathbf{u} \otimes \mathbf{u}) + \frac{\partial \mathbf{u}w}{\partial z} + \nabla p = \mu \frac{\partial^2 \mathbf{u}}{\partial z^2}, \tag{9}$$

$$\frac{\partial p}{\partial z} = -g, \tag{10}$$

with the boundary conditions

$$w(t, x, y, Z(x, y)) = 0, \tag{11}$$

$$\mu \frac{\partial \mathbf{u}}{\partial z}(t, x, y, Z(x, y)) = \kappa \; \mathbf{u}(t, x, y, Z(x, y)), \tag{12}$$

$$\frac{\partial \mathbf{u}}{\partial z}(t, x, y, H(t, x, y)) = 0, \tag{13}$$

$$p(t, x, y, H(t, x, y)) = 0. \tag{14}$$

The system is still associated with the kinematic boundary condition (6).

Taking into account the pressure boundary condition on the free surface (14), the equation (10) is equivalent to

$$p(t, x, y, z) = g(H(t, x, y) - z). \tag{15}$$

3   A Multilayer saint-venant system

In order to define a vertical discretization of the system (8)–(14), we introduce a discretization of the water domain in the z direction (see Fig.2). For some $M \in \mathbb{N}$ we define M intermediate water heights $H_\alpha(t, x, y)$ such that

$$0 = H_0(t, x, y) \leq H_1(t, x, y) \leq H_2(t, x, y) \leq \ldots$$
$$\ldots \leq H_{M-1}(t, x, y) \leq H_M(t, x, y) = h(t, x, y).$$

Then for each layer we define its water height $h_\alpha(t, x, y)$ by

$$\forall \alpha \in \{1, M\}, \quad h_\alpha(t, x, y) = H_\alpha(t, x, y) - H_{\alpha-1}(t, x, y),$$

and so

$$\sum_{\alpha=1}^{M} h_\alpha(t, x, y) = h(t, x, y).$$

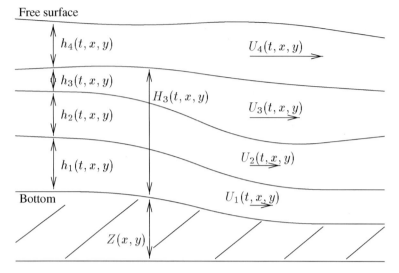

Fig. 2. Water domain discretization in the z direction

We assume that the interfaces are advected by the flow.
We also define an average velocity $\mathbf{U}_\alpha(t, x, y)$ by

$$\forall \alpha \in \{1, M\}, \quad \mathbf{U}_\alpha(t, x, y) = \frac{1}{h_\alpha(t, x, y)} \int_{H_{\alpha-1}}^{H_\alpha} \mathbf{u}(t, x, y, z) dz. \qquad (16)$$

An extension of the arguments in [1] leads to the following result:

The multilayer Saint-Venant system with friction defined by

$$\frac{\partial h_\alpha}{\partial t} + \nabla.(h_\alpha \mathbf{U}_\alpha) = 0, \qquad (17)$$

$$\frac{\partial h_\alpha \mathbf{U}_\alpha}{\partial t} + \nabla (h_\alpha \mathbf{U}_\alpha \otimes \mathbf{U}_\alpha) + g h_\alpha \nabla h = -g h_\alpha \nabla Z$$

$$-\kappa_\alpha \mathbf{U}_\alpha + 2\mu_\alpha \frac{\mathbf{U}_{\alpha+1} - \mathbf{U}_\alpha}{h_{\alpha+1} + h_\alpha} - 2\mu_{\alpha-1} \frac{\mathbf{U}_\alpha - \mathbf{U}_{\alpha-1}}{h_\alpha + h_{\alpha-1}}, \quad \text{for} \quad \alpha = 1, ..., M \qquad (18)$$

with

$$\kappa_\alpha = \begin{cases} \kappa & \text{if } \alpha = 1, \\ 0 & \text{if } \alpha \neq 1, \end{cases} \quad \mu_\alpha = \begin{cases} 0 & \text{if } \alpha = 0, \\ \mu & \text{if } \alpha = 1, ..., M-1, \\ 0 & \text{if } \alpha = M. \end{cases}$$

results from a formal asymptotic approximation in $O(\epsilon)$, coupled with a vertical discretization, of the hydrostatic model and therefore of the Navier-Stokes equations.

The multilayer Saint-Venant system satisfies some fundamental properties (see [1]), we just mention here that the multilayer system (17)–(18) preserves the positivity of the water height in each layer. It preserves also the steady-state of still water.

However the formulation of the multilayer system (17)–(18) has two main drawbacks. The pressure terms are not in a conservative form and thus their definition is not obvious when shocks occur. And if we consider a two layers system satisfying

$$\mathbf{U}_\alpha(t,x,y) = \mathbf{U}(t,x,y) + O(\epsilon) \qquad \forall \alpha = 1,2, \qquad (19)$$

we verify (see [1]) that the two layers system is not hyperbolic.

In order to define a stable approximation of the multilayer system, it is shown in [1] that the following new set-up of the same system but with a conservative form of the left hand side is better

$$\frac{\partial h_\alpha}{\partial t} + \nabla.(h_\alpha \mathbf{U}_\alpha) = 0, \qquad (20)$$

$$\frac{\partial h_\alpha \mathbf{U}_\alpha}{\partial t} + \nabla(h_\alpha \mathbf{U}_\alpha \otimes \mathbf{U}_\alpha) + \frac{g}{2}\nabla(h_\alpha h) = \frac{g}{2}h^2\nabla(\frac{h_\alpha}{h}) - gh_\alpha \nabla Z$$

$$-\kappa_\alpha \mathbf{U}_\alpha + 2\mu_\alpha \frac{\mathbf{U}_{\alpha+1} - \mathbf{U}_\alpha}{h_{\alpha+1} + h_\alpha} - 2\mu_{\alpha-1}\frac{\mathbf{U}_\alpha - \mathbf{U}_{\alpha-1}}{h_\alpha + h_{\alpha-1}}, \quad \text{for} \quad \alpha = 1,...,M \quad (21)$$

4 Numerical method

In this section we give some short information concerning space and time discretization of the system (20)–(21).

With Δt the time step, knowing the solution $(h_\alpha^n, \mathbf{U}_\alpha^n)$ at time t^n, we compute the solution at time t^{n+1} with an explicit treatment of the hyperbolic part (left hand side), of the non conservative pressure source term and of the bottom topography term, and an implicit treatment of the viscous and friction terms, so the scheme is written:

$$\frac{h_\alpha^{n+1} - h_\alpha^n}{\Delta t} + \nabla.(h_\alpha^n \mathbf{U}_\alpha^n) = 0, \qquad (22)$$

$$\frac{h_\alpha^{n+1}\mathbf{U}_\alpha^{n+1} - h_\alpha^n \mathbf{U}_\alpha^n}{\Delta t} + \nabla(h_\alpha^n \mathbf{U}_\alpha^n \otimes \mathbf{U}_\alpha^n) + \frac{g}{2}\nabla(h_\alpha^n h^n)$$

$$= \frac{g}{2}(h^n)^2 \nabla(\frac{h_\alpha^n}{h^n}) - gh_\alpha^n \nabla Z \qquad (23)$$

$$-\kappa_\alpha \mathbf{U}_\alpha^{n+1} + 2\mu_\alpha \frac{\mathbf{U}_{\alpha+1}^{n+1} - \mathbf{U}_\alpha^{n+1}}{h_{\alpha+1}^{n+1} + h_\alpha^{n+1}} - 2\mu_{\alpha-1}\frac{\mathbf{U}_\alpha^{n+1} - \mathbf{U}_{\alpha-1}^{n+1}}{h_\alpha^{n+1} + h_{\alpha-1}^{n+1}}, \quad \text{for } \alpha=1,...,M$$

We notice that h_α^{n+1} is obtained explicitly and $\mathbf{U}_\alpha^{n+1}$ is the solution of a tridiagonal $M \times M$ linear system.

Concerning the space discretization, we consider finite volumes defined on an unstructured mesh. For the hyperbolic part, the fluxes at the interfaces are computed by a kinetic solver analogous to the one explained in details in [2] for the Saint-Venant system. Here the Gibbs equilibrium for the layer α is

$$M_\alpha(t,x,y,\xi) = \frac{h_\alpha(t,x,y)}{c^2(t,x,y)} \chi(\frac{\xi - \mathbf{U}_\alpha(t,x,y)}{c(t,x,y)}), \tag{24}$$

with $c(t,x,y) = \sqrt{\frac{gh(t,x,y)}{2}}$, and the notations defined in [2].

To discretize the bottom topography term we generalize the *hydrostatic reconstruction* [3] in order to preserve steady state of still water.

We define:
- a piecewise constant approximation of the bottom topography $Z(x,y)$

$$Z_i = \frac{1}{|C_i|} \int_{C_i} Z(x,y) dx dy, \tag{25}$$

with $|C_i|$ the area of the cell C_i surrounding the node P_i,
- an interface topography (we denote Z_{ij}, Z_{ji} the values at the interface between nodes P_i and P_j)

$$Z_{ij} = Z_{ji} = max(Z_i, Z_j), \tag{26}$$

- an hydrostatic reconstructed total water depth

$$h_{ij} = (h_i + Z_i - Z_{ij})_+, \tag{27}$$

- a proportional reconstructed water depth for each layer

$$h_{\alpha,ij} = h_{\alpha,i} \frac{h_{ij}}{h_i}. \tag{28}$$

For the non conservative source term of the right hand side $h^2 \nabla(\frac{h_\alpha}{h})$ we use the following approximation $S_{\alpha,i}$ which has proved to be robust and gives stable results

$$S_{\alpha,i} = \min_j(h_{ij}^2) \text{ minmod } (\nabla_{T_k}(\frac{h_\alpha}{h})) \tag{29}$$

where ∇_{T_k} denotes the constant gradient on each triangle surrounding the node P_i.

The vertical velocity is an output variable and is deduced from the impermeability condition at the bottom and the integration in z of the incompressibility condition (8).

5 Numerical results

We compare the results obtained with the multilayer system described above and the hydrostatic Navier-Stokes solver "Telemac" presented in [5]. The main ingredients of the Telemac solver are finite elements, operator splitting, semi-implicit scheme, σ transformation (A.L.E. type transformation) along vertical axis.

We consider a classical test, a stationary transcritical flow over a parabolic bump and the geometric data are the following: channel length ≈ 21 m, channel width ≈ 2 m, bump length ≈ 5.75 m, bump height ≈ 0.2 m. At the inflow boundary, the given discharge is 2.m^3/s and at the outflow the prescribed water depth is 0.6 m. The vertical viscosity is 10^{-2}m^2/s and the Strickler coefficient is 30. The results shown in Figures (3)–(6) have been obtained with 6 layers along the vertical axis (1452 nodes, 2620 triangles for the 2D mesh). We can see the good agreement of the results obtained with the two different models though the approximation of the velocities are different (constant by layer or piecewise linear). The CPU times are 10 minutes for the multilayer and 33 minutes for the hydrostatic Navier-Stokes solver.

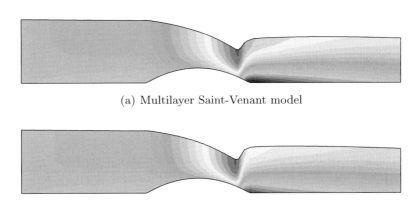

(a) Multilayer Saint-Venant model

(b) Hydrostatic Navier-Stokes model

Fig. 3. Horizontal velocities

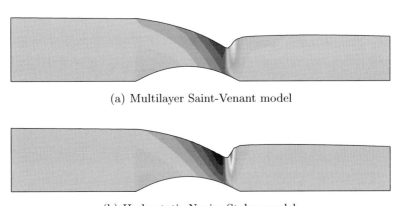

(a) Multilayer Saint-Venant model

(b) Hydrostatic Navier-Stokes model

Fig. 4. Vertical velocities

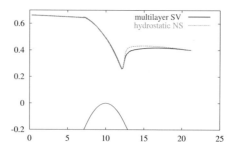

Fig. 5. Free surface comparisons. Multilayer Saint-Venant model (red line) and hydrostatic Navier-Stokes model (green dotted line).

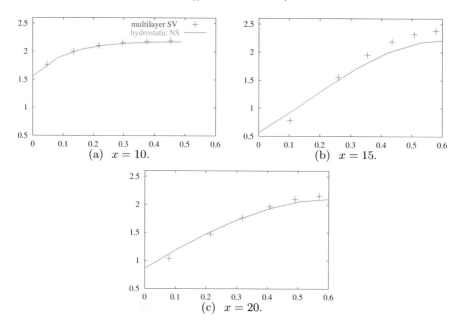

Fig. 6. Vertical profiles of horizontal velocity . Comparison of the multilayer (red crosses) and of the hydrostatic Navier-Stokes (green lines) solutions.

References

1. Audusse, E.: A multilayer Saint-Venant model. Discrete Cont. Dyn. Syst. Ser. B, **5**, No 2, pp 189–214 (2005)
2. Audusse, E., Bristeau, M.O.: A well-balanced positivity preserving second-order scheme for shallow water flows on unstructured meshes. J. Comp. Phys., **206**, pp 311-333 (2005)

3. Audusse, E., Bouchut, F., Bristeau, M.O., Klein, R., Perthame B.: A fast and stable well-balanced scheme with hydrostatic reconstruction for shallow water flows. SIAM J. Sc. Comp., **25**(6), pp 2050-2065 (2004)
4. Ferrari, S., Saleri, F.: A new two dimensionnal shallow water model including pressure term and slow varying bottom topography. M2AN **38**, No 2, pp 211-234 (2004)
5. Hervouet, J.M.: Hydrodynamique des écoulements à surface libre; Modélisation numérique avec la méthode des éléments finis (in french). Presses des Ponts et Chaussées, Paris (2003)

Some Well-Balanced Shallow Water-Sediment Transport Models

M. J. Castro-Díaz[1], E. D. Fernández-Nieto[2] and A. M. Ferreiro[3]

[1] Dpto. Análisis Matemático, Universidad de Málaga. Campus de Teatinos s/n, 29071 Málaga, Spain
   castro@anamat.cie.uma.es
[2] Dpto. Matemática Aplicada I, Universidad de Sevilla. E.T.E. Arquitectura. Avda. de Reina Mercedes s/n. 41012 Sevilla. Spain
   edofer@us.es
[3] Dpto. de Ecuaciones Diferenciales y Análisis Numérico. Universidad de Sevilla. C/Tarfia, s/n. 41080 Sevilla. Spain
   anafefe@us.es

Summary. This paper is concerned with the numerical approximation of bed-load sediment transport due to water evolution. We introduce an unified formulation for several bed-load models. Some numerical simulations are presented.[4]

1 Sediment transport model

In order to understand and predict geomorphological evolutions in coastal seas and estuaries a model, which describes the dynamics of the water motion and bed-load sediment transport movement, is needed.

In this paper, the hydrodynamical model is given by shallow water equations, and the morphological model is modelized using a bed evolution equation. Both systems can be written as a coupled system of conservations laws, with non-conservative products and source terms. The model equations are described in Section 1.4.

1.1 Hydrodynamical model: shallow water equations

The system of equations governing a flow of a shallow layer of fluid through a straight channel with a constant rectangular cross-section is given by the well known shallow water model,

---

[4] This research was partially supported by Spanish Government Research Projects BFM2003-07530-C02-01 and BFM2003-07530-C02-02.

$$\begin{cases} \dfrac{\partial h}{\partial t} + \dfrac{\partial q}{\partial x} = 0, \\ \dfrac{\partial q}{\partial t} + \dfrac{\partial}{\partial x}\left(\dfrac{q^2}{h} + \dfrac{1}{2}gh^2\right) = gh\dfrac{dH}{dx} - ghS_f. \end{cases} \quad (1)$$

In this system, it is supposed that the fluid is homogeneous and inviscid; coordinate x refers to the axis of the channel, t is the time; $h(x,t)$ is the thickness of the fluid layer and $q(x,t)$ represents the mass-flow, being $q(x,t) = h(x,t)u(x,t)$ where $u(x,t)$ is the velocity of the fluid; g is gravity and $H(x)$ the depth function measured from a fixed level of reference (A_R).

The term S_f models bottom friction, that it is supposed given by a Manning's law,

$$S_f = \dfrac{g\eta^2 u^2}{R_h^{4/3}}, \quad (2)$$

being η the Manning coefficient. R_h is the hydraulic ratio, that can be aproximated by h.

To study bed-load sediment transport it is necessary to consider a sediment layer of thickness z_b, and a fixed layer (without sediments), with thickness given by $z_f = -H + A_R$. In this case, system (1) can be rewritten as,

$$\begin{cases} \dfrac{\partial h}{\partial t} + \dfrac{\partial q}{\partial x} = 0, \\ \dfrac{\partial q}{\partial t} + \dfrac{\partial}{\partial x}\left(\dfrac{q^2}{h} + \dfrac{1}{2}gh^2\right) = -gh\dfrac{\partial z_b}{\partial x} + gh\dfrac{dH}{dx} - ghS_f. \end{cases} \quad (3)$$

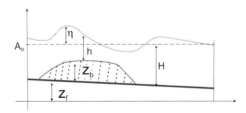

Fig. 1. Sediment layer over a fixed bed

1.2 Morphological model

The continuity sediment equation models bed-load sediment transport. The temporal variation of sediment layer must be equal to the total variation of the solid transport.

The expression of the conservation law of sediment volume is given by,

$$\frac{\partial z_b}{\partial t} + \xi \frac{\partial q_b}{\partial x} = 0. \qquad (4)$$

$z_b(x,t)$ represent the sediment layer; $\xi = 1/(1-\rho_s)$ where ρ_s is the sediment porosity. q_b denotes the solid transport flux, that depends on fluid velocity $q_b = q_b(h,q)$.

1.3 Flux of bed-load sediment transport equations

In the literature there are several formulae for q_b, which have been obtained using different empirical methods, studying hydrodynamical problems in rivers, currents in coastal areas, etc. Some of the most popular equations are (see [4, 5, 6]),

- Grass equation uses the hypothesis that sediment movement begins at the same time as fluid movement,

$$q_b = A_g \frac{q}{h} u |u|^{m_g-1}, \quad 1 \leq m_g \leq 4 \qquad (5)$$

where, the constant A_g (s^2/m) is determined using experimental data, and its value is between 0 and 1. Usually the constant m_g is set to $m_g = 3$.

- Meyer-Peter&Muller equation is a flux formula based on the size of the grain of a porous media. The expression of q_b is obtained from the following identity:

$$\frac{q_b}{\sqrt{(G-1)gd_i^3}} = sgn(u)8\left(\tau_* - \tau_{*c}\right)^{3/2} ; \qquad (6)$$

where, $\tau_{*c} = 0.047$ is the critical stress shear; d_i denotes grain size. $G = \frac{\gamma_s}{\gamma}$ is the rate between the specific weight of the fluid, γ, and the specific weight of the sediment, γ_s. Finally, u is fluid velocity and, τ_* denotes the shear stress whose expression is:

$$\tau_* = \frac{\tau}{(\gamma_s - \gamma)d_i} \quad \text{where } \tau = \frac{\gamma \eta^2 u^2}{R_h^{1/3}}. \qquad (7)$$

Note that bed-load transport occurs when the bed shear stress τ_* exceed the critical value τ_{*c}.

Other models that can be found in literature are: Nielsen, Van Rijn, FL&Van Beek, etc.

Unified formulation

The different formulae of q_b can be written under an unified formulation in this way,

$$q_b = c_1 g_2(h,q)(c_2 + c_3 g_1(h,q))^m, \qquad (8)$$

where c_1, c_2, c_3 and m are constants and g_1, g_2 are functions of h and q.

1.4 One-dimensional coupled model

The one-dimensional model used to modelize the bed-load sediment transport due to water movement, is obtained coupling the shallow water system (3) and the conservation law of the sediment volume (4), resulting the following coupled system of conservations laws with non-conservative products and source terms:

$$\begin{cases} \dfrac{\partial h}{\partial t} + \dfrac{\partial q}{\partial x} = 0, \\[1ex] \dfrac{\partial q}{\partial t} + \dfrac{\partial}{\partial x}\left(\dfrac{q^2}{h} + \dfrac{1}{2}gh^2\right) = -gh\dfrac{\partial z_b}{\partial x} + gh\dfrac{dH}{dx} - ghS_f, \\[1ex] \dfrac{\partial z_b}{\partial t} + \xi\dfrac{\partial q_b}{\partial x} = 0. \end{cases} \quad (9)$$

If a new variable S is defined by $S = H - z_b$, and taking into account that $\dfrac{\partial S}{\partial t} = -\dfrac{\partial z_b}{\partial t}$, is possible to re-write the system (9) as a coupled system of conservation laws with a non-conservative product and a source term S_F:

$$\dfrac{\partial W}{\partial t} + \dfrac{\partial F(W)}{\partial x} = B(W)\dfrac{\partial W}{\partial x} + S_F, \quad (10)$$

where,

$$W = \begin{bmatrix} h \\ q \\ S \end{bmatrix},\ F = \begin{bmatrix} q \\ \dfrac{q^2}{h} + \dfrac{1}{2}gh^2 \\ -\xi q_b \end{bmatrix},\ B(W) = \begin{bmatrix} 0 & 0 & 0 \\ 0 & 0 & gh \\ 0 & 0 & 0 \end{bmatrix},\ S_F = \begin{bmatrix} 0 \\ -ghS_f \\ 0 \end{bmatrix}. \quad (11)$$

2 Finite volume method for non conservative hyperbolic systems

The friction term S_F will be discretized in a semi-implicit way. So, in what follows, to simplify notation, we can forget the friction term and work with the system:

$$\dfrac{\partial W}{\partial t} + \dfrac{\partial F(W)}{\partial x} = B(W)\dfrac{\partial W}{\partial x}, \quad (12)$$

Note that, (12) can be written as a non-conservative hyperbolic system,

$$\dfrac{\partial W}{\partial t} + \mathcal{A}(W)\dfrac{\partial W}{\partial x} = 0, \quad (13)$$

where,

$$\mathcal{A}(W) = A(W) - B(W) = \begin{bmatrix} 0 & 1 & 0 \\ -\dfrac{q^2}{h^2} + gh & 2\dfrac{q}{h} & -gh \\ -\xi\dfrac{\partial q_b}{\partial h} & -\xi\dfrac{\partial q_b}{\partial q} & 0 \end{bmatrix}, \qquad (14)$$

being $A(W)$ the jacobian matrix of $F(W)$. Let us assume that the system is hyperbolic, that is, the eigenvalues of matrix $\mathcal{A}(W)$, $\{\lambda_j, j = 1, 2, 3\}$ are real and distinct.

In order to construct a numerical scheme for solving (13), computing cells $I_i = [x_{i-1/2}, x_{i+1/2}]$ are considered. Let us suppose for simplicity that the cells have constant size, Δx, and that $x_{i+1/2} = i\Delta x$. We will note $x_i = (i-1/2)\Delta x$, the center of the cell I_i. Let Δt be the constant time step and define $t^n = n\Delta t$. Noting by W_i^n the approximation of the cell averages of the exact solution provided by the numerical scheme, that is,

$$W_i^n \cong \frac{1}{\Delta x} \int_{x_{i-1/2}}^{x_{i+1/2}} W(x, t^n)\, dx. \qquad (15)$$

Then, the numerical scheme advances in time by solving Linear Riemann problems at each intercell at time t^n and taking the averages of their solutions on the cells at time t^{n+1}. Under usual CFL conditions, the resulting scheme can be written:

$$W_i^{n+1} = W_i^n + (G_{i+1/2} - G_{i-1/2}) \\ + \frac{\Delta t}{2\Delta x} \left(B_{i-1/2}(W_i^n - W_{i-1}^n) + B_{i+1/2}(W_{i+1}^n - W_i^n) \right), \qquad (16)$$

where,

$$G_{i+1/2} = \frac{1}{2}\left(F(W_i^n) + F(W_{i+1}^n)\right) - \frac{1}{2}\mathcal{D}_{i+1/2}\left(W_{i+1}^n - W_i^n\right); \qquad (17)$$

being $\mathcal{D}_{i+1/2}$ the viscosity matrix of the scheme. Depending on the choice of this matrix, different schemes are obtained. Roe method is obtained by

$$\mathcal{D}_{i+1/2} = |\mathcal{A}_{i+1/2}| = \mathcal{K}\left(W_i^n, W_{i+1}^n\right) |\mathcal{L}\left(W_i^n, W_{i+1}^n\right)| \mathcal{K}^{-1}\left(W_i^n, W_{i+1}^n\right), \qquad (18)$$

where $\mathcal{A}_{i+1/2}$ is the Roe matrix for the states W_i^n and W_{i+1}^n. $\mathcal{L}_{i+1/2}$ is a diagonal matrix whose coefficients are the eigenvalues of $\mathcal{A}_{i+1/2}$:

$$\lambda_1^{i+1/2} < \lambda_2^{i+1/2} < \lambda_3^{i+1/2},$$

and $\mathcal{K}_{i+1/2}$ is a 3×3 matrix whose columns are the associated eigenvectors.

Other definitions of matriz $\mathcal{D}_{i+1/2}$ can be obtained including flux limiters. The basic idea is to use Lax-Wendroff method in the regular parts of the solution.

3 High order schemes based on state reconstruction

Methods based on state reconstruction are built using the following procedure: given a first order scheme with numerical flux function $G(U,V)$, a reconstruction operator of order p is considered, that is, an operator that associates to each given sequence of states, W_i, two new sequences $W^+_{i+1/2}$, $W^-_{i+1/2}$ in a way that, whenever

$$W_i = \frac{1}{\Delta x} \int_{I_i} W(x)dx,$$

for some smooth function W, then:

$$W^\pm_{i+1/2} = W(x_{i+1/2}) + O(\Delta x^p).$$

In [3], a high order numerical scheme based on the state reconstruction for coupled systems of conservation laws with non-conservative products like (12) is obtained. The space discretization is provided by:

$$\begin{aligned} W'_i &= \frac{\Delta t}{\Delta x}\left(\widetilde{G}_{i-1/2} - \widetilde{G}_{i+1/2}\right) \\ &+ \frac{\Delta t}{2\Delta x}\left(B_{i-1/2} \cdot \left(W^+_{i-1/2} - W^-_{i-1/2}\right) + B_{i+1/2} \cdot \left(W^+_{i+1/2} - W^-_{i+1/2}\right)\right) \\ &+ \frac{\Delta t}{\Delta x}\mathcal{I}_{B,i} \end{aligned} \tag{19}$$

where,

$$\widetilde{G}_{i-1/2} = \frac{1}{2}\left(F\left(W^-_{i+1/2}\right) + F\left(W^+_{i+1/2}\right)\right) - \frac{1}{2}\mathcal{D}_{i+1/2} \cdot \left(W^+_{i+1/2} - W^-_{i+1/2}\right), \tag{20}$$

being $\mathcal{D}_{i+1/2}$ the viscosity matrix given in (18).
Moreover,

$$\mathcal{I}_{B,i} = \int_{x_{i-1/2}}^{x_{i+1/2}} B\left[P^t_i\right] \frac{d}{dx}P^t_i(x)dx, \tag{21}$$

being P^t_i a regular function provided by the reconstruction operator and defined at every cell I_i, verifying (see [3] to details),

$$\lim_{x \to x^+_{i-1/2}} P^t_i(x) = W^+_{i-1/2}(t); \quad \lim_{x \to x^-_{i+1/2}} P^t_i(x) = W^-_{i+1/2}(t). \tag{22}$$

The time discretization is provided by a TVD method such RK2 or RK3 proposed in [1].

4 Numerical test: comparison with an analytical solution

Hudson and Sweby in [4] propose a family of asymptotic analytical solutions for the Grass model, when the constant $A_g < 10^{-2}$, supposing that the sediment layer z_b is present over all the domain, and the fluid movement is slow, with a constant flow q. We have considered an analytical solution of this family given by:

$$h = 10 - z_b(x, t), \quad q = 10,$$

where,

$$z_b(x, t) = \begin{cases} 0.1 + \sin^2\left(\dfrac{\pi(x_o - 300)}{200}\right) & \text{if } 300 \leq x_o \leq 500, \\ 0.1 & \text{otherwise,} \end{cases} \quad (23)$$

where x_o is the solution of the equation,

$$\begin{cases} x = x_o + A_g\, \xi\, m_g\, 10^{m_g}\, t \left(10 - \sin^2\left(\dfrac{\pi(x_o - 300)}{200}\right)\right)^{-(m_g+1)} \\ \hspace{20em} \text{if } 300 \leq x_o \leq 500, \\ x = x_o + A_g\, \xi\, m_g\, t/10 \hspace{10em} \text{otherwise.} \end{cases}$$

We consider a rectangular channel with length $L = 1000$ meters, discretized with 250 cells. The CFL parameter is set to 0.8. The sediment porosity is set by $\rho_0 = 0.4$. The constant A_g of Grass formula (5) is set to $A_g = 0.001$ (weak interaction) and $m_g = 3$. Free boundary conditions are considered and the initial condition is given by:

$$h(x, 0) = 10 - z_b(x, 0), \quad q(x, 0) = 10,$$

$$z_b(x, 0) = \begin{cases} 0.1 + \sin^2\left(\dfrac{\pi(x - 300)}{200}\right) & \text{if } 300 \leq x \leq 500; \\ 0.1 & \text{otherwise.} \end{cases} \quad (24)$$

In Figures 2, 3, 4 the analytical solution (continued line) is compared with the approximation provided by different numerical schemes. In Figure 2 the first order Roe method is compared with the second order Roe-flux limiter scheme. In Figure 3 the second order Roe-Flux limiter scheme is compared with the second order Roe-Weno2-RK2 scheme. Finally, in Figure 4 the comparison between the third order Roe-Weno3-RK3 scheme and the second order Roe-Weno2-Rk2 scheme is provided. The best approximation is obtained, as expected, using the third order Roe-Weno3-RK3 scheme.

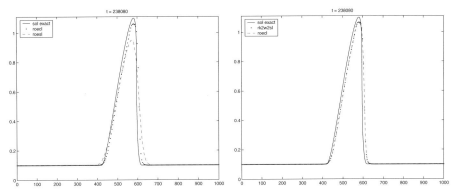

Fig. 2. Roe-Flux limeter (dotted line) and Roe-Euler (dash line)

Fig. 3. Roe-Flux limiters (dash line). Reconstruction Weno2-Rk2 (dotted line)

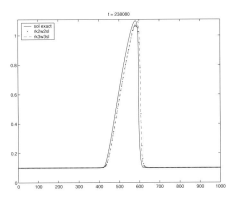

Fig. 4. Reconstruction: Weno2-Rk2 (dotted line), Weno3-Rk3 (dash line)

References

1. Shu, C.W., Osher, S.: Efficient implementation of essentially non-oscilatory shock capturing schemes. J. Comput. Phys. **77**, 439-471, (1998)
2. Parés, C., Castro, M.J.: On the well-balance property of Roe's method for nonconservative hyperbolic systems. Applications to shallow-water systems. ESAIM: Math. Mod. Numer. Anal. **38**(5), 821–852, (2004)
3. Castro, M., Gallardo, J.M., Parés, C.: Finite volume schemes based on weno reconstruction of states for solving nonconservaative hyperbolic systems. Applications to shallow water systems. Accepted on Mathematics of Computation.
4. Hudson, J.: Numerical technics for morphodynamic modelling. Ph. D. Thesis, University of Whiteknights (2001)
5. Peña González, E.: Estudio numérico y experimental del transporte de sedimentos en cauces aluviales. Ph. D. Thesis, Universidade da Coruña. Grupo de Ingeniería del agua y del medio ambiente (2002)
6. Julien, P.Y.: Erosion and Sedimentation. Cambridge Unilversity Press (1998)

Highly Accurate Conservative Finite Difference Schemes and Adaptive Mesh Refinement Techniques for Hyperbolic Systems of Conservation Laws*

Pep Mulet and Antonio Baeza

Departament de Matemàtica Aplicada. Universitat de València
C/Doctor Moliner, 50, 46100 Burjassot (València), Spain
{mulet, Antonio.Baeza}@uv.es

Summary. We review a conservative finite difference shock capturing scheme that has been used by our research team over the last years for the numerical simulations of complex flows [3, 6]. This scheme is based on Shu and Osher's technique [9] for the design of highly accurate finite difference schemes obtained by flux reconstruction procedures (ENO, WENO) on Cartesian meshes and Donat-Marquina's flux splitting [4]. We then motivate the need for mesh adaptivity to tackle realistic hydrodynamic simulations on two and three dimensions and describe some details of our *Adaptive Mesh Refinement* (AMR) ([2, 7]) implementation of the former finite difference scheme [1]. We finish the work with some numerical experiments that show the benefits of our scheme.

1 Introduction

This work is concerned with the numerical solution of hyperbolic systems of conservation laws of the form:

$$\begin{cases} U_t + \sum_{i=1}^{d} F_i(U)_{x_i} = 0 \\ U(x,0) = U_0(x), \end{cases} \tag{1}$$

where $U = (U_1, \ldots, U_m)^T$, $x = (x_1, \ldots, x_d)$, $U_i \colon \mathbb{R}^d \longrightarrow \mathbb{R}$ and $F_i \colon \mathbb{R}^m \longrightarrow \mathbb{R}^m$, by means of a numerical scheme built from Shu-Osher's conservative finite-difference formulation [9], a fifth order weighted essentially non-oscillatory (WENO) interpolatory technique, Donat-Marquina's flux-splitting [4], and a third order TVD Runge-Kutta ODE solver [9], merged with the adaptive mesh refinement (AMR) algorithm [2].

---

* Research supported by EUCO Projects HPRN-CT-2002-00282 and HPRN-CT-2002-00286

The paper is organized as follows: in Sect. 2 we review the numerical method used to solve (1) on a fixed grid. In Sect. 3 we review the AMR technique and we explain our implementation of the method described in Sect. 2 within the AMR framework. In Sect. 5 we experimentally validate our algorithm. Finally, the conclusions are pointed out in Sect. 5.

2 Finite-difference Shu-Osher schemes

Shu and Osher [9] proposed a finite-difference scheme to solve (1) based on highly accurate conservative approximations of the fluxes $F_i(U)$, to ease multi-dimensional extensions. This dimensional-splitting facility allows us to restrict the exposition to the one dimensional case, i.e., $d = 1$ in (1). In this case system (1) is written as:

$$\begin{cases} U_t + F(U)_x = 0 \\ U(x,0) = U_0(x). \end{cases} \quad (2)$$

We will denote by $U^n = \{U_j^n\}_j$ the vector of the numerical approximations to the exact solution $U(x,t)$ of (2) at the points (x_j, t^n), , where $x_j = (j+\frac{1}{2})\Delta x$ and $t^n = n\Delta t$. We start with the case of a scalar conservation law, i.e., $m = 1$ in (2).

The key idea of Shu-Osher's formulation is to express the derivative of the flux as a finite difference. For an (unknown) function ϕ such that

$$F(U(x,t)) = \frac{1}{\Delta x} \int_{x-\frac{\Delta x}{2}}^{x+\frac{\Delta x}{2}} \phi(s)ds,$$

we have

$$F(U(x,t))_x = \frac{\phi(x+\frac{\Delta x}{2}) - \phi(x-\frac{\Delta x}{2})}{\Delta x}.$$

The conservation law (2) is thus equivalent to

$$U_t + \frac{\phi(x+\frac{\Delta x}{2}) - \phi(x-\frac{\Delta x}{2})}{\Delta x} = 0. \quad (3)$$

We can apply a method of lines to solve (3) to obtain a conservative scheme if we approximate the values $\phi(x + \frac{\Delta x}{2})$ using the known values of the cell-averages of the function ϕ (i.e. the values $F(U(x_j,t))$) in the mesh. We denote by $\hat{F}_{j+\frac{1}{2}}$ such a reconstruction in the cell interface $x_{j+\frac{1}{2}}$.

This reconstruction can be performed with the same methods used in the classical finite-volume formulation, in which the point-values of the conserved variables are reconstructed from its cell-averages. The time accuracy is obtained by a high order ODE solver.

The extension to systems is performed by local characteristic decompositions and Donat-Marquina's flux splitting [4]. The idea is that the numerical flux $\hat{F}_{j+\frac{1}{2}}$ is the sum of contributions of the characteristic fluxes corresponding to cells $[x_{j-\frac{1}{2}}, x_{j+\frac{1}{2}}]$ and $[x_{j+\frac{1}{2}}, x_{j+\frac{3}{2}}]$. The contribution corresponding to the p-th field at each contributor cell takes into account the sign of the corresponding eigenvalue $\lambda_p(U)$ and is consistent with the characteristic structure of the Jacobian Matrix at the cell.

At a given cell interface $x_{j+\frac{1}{2}}$, we compute two sided interpolations $U_{j+\frac{1}{2}}^{L,R}$ of the conserved variables. The values coming from the left, $U_{j+\frac{1}{2}}^{L}$, and from the right, $U_{j+\frac{1}{2}}^{R}$, are computed using high order essentially non-oscillatory interpolation procedures with upwind biased stencils that contain the points x_j and x_{j+1}, respectively.

At each point x_k belonging to some upwind biased stencil that contains the given cell interface $x_{j+\frac{1}{2}}$, these interpolated quantities are used to define two sets of characteristic variables $w_{p,k}^{L,R} = \mathbf{l}_p(U_{j+\frac{1}{2}}^{L,R}) \cdot U_k$ and characteristic fluxes $F_{p,k}^{L,R} = \mathbf{l}_p(U_{j+\frac{1}{2}}^{L,R}) \cdot F(U_k)$, where $\mathbf{l}_p(U)$, $\mathbf{r}_p(U)$ stand for normalized left and right eigenvectors of the Jacobian matrix $F'(U)$ corresponding to the eigenvalue $\lambda_p(U)$. Upwind characteristic fluxes are then computed according to the characteristic speeds at both sides, except at sonic points, at which a local Lax-Friedrichs splitting (see [9]) is applied.

Let $R(g_{-s_1}, \ldots, g_{s_2}, x)$ denote the evaluation at x of a reconstruction based on cell-averages g_j of a function at some $s_1 + s_2 + 1$ adjacent cells (we use here the WENO5 procedure [5]). The algorithm to compute the numerical fluxes at $x_{j+\frac{1}{2}}$ is as follows:

---

if $\lambda_p(U)$ does not change sign in a path in phase space connecting U_j and U_{j+1}
  if $\lambda_p(U_j) > 0$
    $\psi_{p,j}^L = R(F_{p,j-s_1}^L, \ldots F_{p,j+s_2}^L, x_{j+\frac{1}{2}})$
    $\psi_{p,j}^R = 0$
  else
    $\psi_{p,j}^L = 0$
    $\psi_{p,j}^R = R(F_{p,j-s_1+1}^R, \ldots F_{p,j+s_2+1}^R, x_{j+\frac{1}{2}})$
else
  $\psi_{p,j}^L = R(\frac{1}{2}(F_{p,j-s_1}^L + \alpha_{j+\frac{1}{2}} w_{p,j-s_1}^L), \ldots \frac{1}{2}(F_{p,j+s_2}^L + \alpha_{j+\frac{1}{2}} w_{p,j+s_2}^L), x_{j+\frac{1}{2}})$
  $\psi_{p,j}^R = R(\frac{1}{2}(F_{p,j-s_1+1}^R - \alpha_{j+\frac{1}{2}} w_{p,j-s_1+1}^R), \ldots \frac{1}{2}(F_{p,j+s_2+1}^R - \alpha_{j+\frac{1}{2}} w_{p,j+s_2+1}^R), x_{j+\frac{1}{2}})$

With the numerical flux defined as

$$\hat{F}_{j+\frac{1}{2}} = \sum_p \psi_{p,j}^L \mathbf{r}_p(U_{j+\frac{1}{2}}^L) + \psi_{p,j}^R \mathbf{r}_p(U_{j+\frac{1}{2}}^R), \tag{4}$$

the spatial semi-discretization of (2)

$$\frac{\partial U_j}{\partial t} + \frac{\hat{F}_{j+\frac{1}{2}} - \hat{F}_{j-\frac{1}{2}}}{\Delta x} = 0 \tag{5}$$

is then solved with a high order ODE solver. We have used a third order TVD Runge-Kutta ODE solver [9].

For higher space dimensions, this scheme admits a straightforward tensorial (dimension by dimension) extension. Each flux function $F_i(U)_{x_i}$ in (1) is discretized using the one-dimensional algorithm in the $i-th$ coordinate, and the ODE solver is then applied to the semi-discrete system

$$\frac{\partial U_j}{\partial t} + \sum_{i=1}^{d} \frac{\hat{F}_{i,j+\frac{1}{2}} - \hat{F}_{i,j-\frac{1}{2}}}{\Delta x} = 0$$

A more detailed description of the overall algorithm can be found in [6].

3 Adaptive mesh refinement for Shu-Osher schemes

The mesh size Δx imposes a limit in the features of the numerical solution that can be resolved by a numerical scheme. To be able to resolve phenomena whose physical scale is small we need very fine computational grids, thus increasing the computational cost of the calculation.

Since fine grids are usually needed only in a part of the computational domain (where the solution has non-smooth structure) the resolution of the computational grid can be increased locally. We have adopted the adaptive mesh refinement technique of Berger et al. [2].

The AMR algorithm uses a grid hierarchy $G_0, \ldots, G_{L-1}$, where G_l is formed by the union of Cartesian patches $G_{l,k}$ of uniform mesh size. The grids at different levels are *nested*, i.e. $G_{l+1} \subseteq G_l$ (our description implies that G_{l+1} is finer than G_l).

As singularities move as time advances the grid system has to be adapted in a way such that singularities cannot move from a given grid to a coarser one before the grid is adapted

Given a grid G_l the adaption process obtains a grid at level $l+1$ that will substitute the existing grid G_{l+1}. This new grid will take into account the features of the recently computed solution at G_l. The adaption process consists of three main building blocks: first a procedure decides which cells of the grid G_l have to be refined to form the grid G_{l+1}. Since the main three kind of singularities that appear in the solutions of hyperbolic systems of conservation laws constitute variations in the solution or in the gradient of the solution it could be enough to use the gradient of the solution as an indicator of the presence of such discontinuities. If the change in some seminorm of the gradient (e.g., the absolute value of the density component of the gradient for Euler equations or some norm of the gradient for general equations) of the solution between two adjacent cells is above a given tolerance R_{tol}, then both cells are flagged for refinement.

Once the coarse grid has been flagged we add a certain number of safety flags to ensure that the cells adjacent to a singularity are refined. The safety flags will avoid singularities to escape from the fine grid during one coarse time step.

The cells corresponding to the fine grid will be composed by the subdivision of the coarse cells flagged for refinement. The second process obtains, from a given set of flagged cells, a set of Cartesian mesh patches containing all the flagged cells, and possibly some non-flagged cells. A parameter C_{tol} controls the percentage of non-flagged cells that can be admitted into a patch. Finally, the third process transfers a numerical solution to the newly created grid. The numerical solution can come from three sources: interpolation from the coarser grid, copy from the grid at level $l+1$ that existed before the adaption, or the application of boundary conditions.

A leading principle of the AMR algorithm is that each Cartesian mesh patch can be integrated by the basic numerical scheme independently of any other patch. To this aim each mesh patch is augmented by some ghost cells that are filled with a numerical solution prior to the integration of the patch. This feature allows the AMR algorithm to integrate each grid with a time step coherent with its mesh size, so that the Courant number $\frac{\Delta t_l}{\Delta x_l}$ remains constant independently of l. This time refinement is expected to reduce the number of cell updates needed to integrate the whole grid hierarchy from time t to time $t + \Delta t_0$, with respect to a fixed grid of size Δx_{L-1}.

Once the grid hierarchy has been evolved from time t to time $t + \Delta t_l$, the numerical solution in a grid G_{l+1} is more precise than in a coarser grid G_l so the coarse solution is modified conservatively using information coming from the fine grid. This process is performed by modifying the numerical fluxes at the interfaces of cells in G_l with the numerical fluxes computed in G_{l+1}, in the regions in which both grids overlap. Then the numerical solution at G_l is modified according to the new fluxes. This projection from fine fluxes to coarse fluxes entails communication among grids and is fundamental for the efficiency of the algorithm.

As stated above, the main issue to be taken into account is to ensure that singularities cannot escape from fine to coarse grids. In [7] it is shown that, for the linear advection equation it is enough to adapt a grid G_l after all grids $G_{l+1}, \ldots, G_{L-1}$ have been evolved until time $t + \Delta t_l$, provided that the Courant numbers do not depend on l, the typical CFL condition $\frac{\Delta t_0}{\Delta x_0}$ is satisfied for the coarsest grid and at least one safety flag is added in the adaption process. The integration and adaption processes can be organized in a way such that the adaption process follows the integration process in the correct order, see [7, 1].

4 Numerical examples

To validate our algorithm we take two particular instances of the Riemann problems for gas dynamics described in [8], which have been used as test problems in other adaptive schemes (see e.g. [3]). Our setup is identical to the one in [3] for both problems.

We consider the Euler equations in two dimensions, $U_t + F(U)_x + G(U)_y = 0$, where

$$U = \begin{pmatrix} \rho \\ \rho u \\ \rho v \\ E \end{pmatrix}, \quad F = \begin{pmatrix} \rho u \\ \rho u^2 + p \\ \rho uv \\ u(E+p) \end{pmatrix}, \quad G = \begin{pmatrix} \rho v \\ \rho uv \\ \rho v^2 + p \\ v(E+p) \end{pmatrix}, \quad (6)$$

where ρ is the density, u and v are the velocity components of the fluid in the x and y directions respectively, $E = \frac{p}{\gamma(\phi)} + \frac{1}{2}\rho(u^2 + v^2)$ is the total energy The internal energy is given by the equation of state

$$p = (\gamma - 1)\rho\epsilon. \quad (7)$$

The computational domain consists of the unit square. Four (constant) different states, initially separated by simple one-dimensional waves are evolved in time. In the first test case the four states are separated by shock waves. The initial data is as follows:

$$U(x,0) = U_0(x) = \begin{cases} U_A & \text{if } 0.75 \leq x \leq 1 \text{ and } 0.75 \leq y \leq 1 \\ U_B & \text{if } 0 \leq x < 0.75 \text{ and } 0.75 \leq y \leq 1 \\ U_C & \text{if } 0 \leq x < 0.75 \text{ and } 0 \leq y < 0.75 \\ U_D & \text{if } 0.75 \leq x \leq 1 \text{ and } 0 \leq y < 0.75 \end{cases}, \quad (8)$$

where the respective values of U_A, U_B and U_C and U_D are taken from the initial states:

$$\begin{array}{llll}
\rho_A = 1.5, & u_A = 0, & v_A = 0, & p_A = 1.5, \\
\rho_B = 0.5323, & u_B = 1.206, & v_B = 0, & p_B = 0.3, \\
\rho_C = 0.138, & u_C = 1.206, & v_C = 1.206, & p_C = 0.029, \\
\rho_D = 0.5323, & u_D = 0, & v_D = 1.206, & p_D = 0.3.
\end{array}$$

We have used a coarse mesh of 100×100 cells to discretize the computational domain. Three levels of refinement with all refinement factors set to 2 have been used to obtain a resolution equivalent to a fixed grid of 400×400 cells. In this experiment we have used the following parameters: the CFL condition has been set to 0.25, the refinement parameter is $R_{tol} = 3.0$ and the clustering parameter is $C_{tol} = 0.8$. In Fig. 1 we display a contour plot of the numerical solution at time $t = 0.8$ as computed with a fixed grid of 400×400 cells and with the AMR algorithm.

The second test corresponds to a 4-contact configuration, with initial data

$$U(x,0) = U_0(x) = \begin{cases} U_A \text{ if } 0.5 \leq x \leq 1 \text{ and } 0.5 \leq y \leq 1 \\ U_B \text{ if } 0 \leq x < 0.5 \text{ and } 0.5 \leq y \leq 1 \\ U_C \text{ if } 0 \leq x < 0.5 \text{ and } 0 \leq y < 0.5 \\ U_D \text{ if } 0.5 \leq x \leq 1 \text{ and } 0 \leq y < 0.5 \end{cases} \quad (9)$$

where

$$\begin{aligned} \rho_A &= 1, \ u_A = 0.75, \ v_A = -0.5, \ p_A = 1, \\ \rho_B &= 2, \ u_B = 0.75, \ v_B = 0.5, \ p_B = 1, \\ \rho_C &= 1, \ u_C = -0.75, \ v_C = 0.5, \ p_C = 1, \\ \rho_D &= 3, \ u_D = -0.75, \ v_D = -0.5, \ p_D = 1. \end{aligned}$$

We compute the numerical solution with the same grids as in the first experiment, and with parameters $CFL = 0.3$, $R_{tol} = 5.0$ and $C_{tol} = 0.8$ (see Fig. 2). In both results we can see that the AMR algorithm has been able to resolve all the structure of the solution with the same quality as with the fixed grid algorithm. With this setup at time $t = 0.8$ the AMR algorithm has computed a 24.71% of integrations with respect to the fixed grid algorithm and has required a 30.20% of computational time for the 4-shock problem, corresponding to the initial data (8), and a 39.59% of integrations with a 39.65% of computational time in the 4-contact problem, corresponding to the initial data (9).

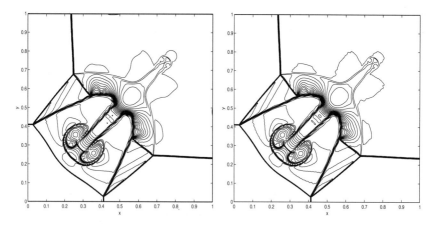

Fig. 1. Contour plot (30 lines) of the density computed with the AMR algorithm (left) and with a fixed grid algorithm (right) for the initial data (8)

5 Conclusions

We have presented a numerical method for the solution of hyperbolic systems of conservation laws, obtained by the combination of a fifth order high resolution shock capturing scheme, built from Shu-Osher's conservative formulation

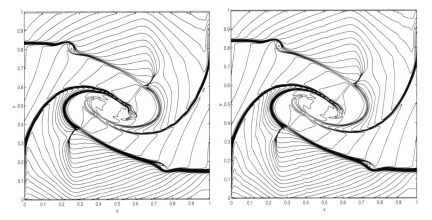

Fig. 2. Contour plot (30 lines) of the density computed with the AMR algorithm (left) and with a fixed grid algorithm (right) for the initial data (9)

[9], a fifth order weighted essentially non-oscillatory (WENO) interpolatory technique [5] and Donat-Marquina's flux-splitting method [4] , with the adaptive mesh refinement technique of Berger et al. [2], in the simplified form proposed by Quirk [7]. The scheme inherits the robustness of Donat-Marquina's basic scheme and has shown to be able to resolve the structure of the numerical solution with an accuracy comparable to the computations made with fixed grids, with a significant reduction of the computational cost.

References

1. Baeza, A., Mulet, P.: Adaptive mesh refinement techniques for high order shock capturing schemes for hyperbolic systems of conservation laws, GrAN report 05-01, Departament de Matemàtica Aplicada, Universitat de València, Spain (2005). http://gata.uv.es/cat/reports.html
2. Berger, M. J., Oliger, J.: Adaptive mesh refinement for hyperbolic partial differential equations, J. Comput. Phys, **53**, 484–512 (1984)
3. Chiavassa, G., Donat, R., Marquina, A.: Fine-mesh numerical simulations for 2D Riemann problems with a multilevel scheme, International Series of Numerical Mathematics, **140**, 247–256 (2001)
4. Donat, R., Marquina, A.: Capturing shock reflections: an improved flux formula, J. Comput. Phys., **125**, 42–58 (1996)
5. Jiang, G.-S., Shu, C.-W.: Efficient implementation of weighted essentially non-oscillatory schemes J. Comp. Phys., **126**, 202–228 (1996)
6. Marquina, A. Mulet, P.: A Flux-Split Algorithm Applied to Conservative Models for Multicomponent Compressible Flows, J. Comput. Phys., **185**, 120–138 (2003)
7. Quirk, J. J.: An adaptive grid algorithm for computational shock hydrodynamics, PhD Thesis, Cranfield Institute of Technology, United Kingdom (1991)

8. Schult-Rinne, C. W.: Classification of the Riemann problem for two-dimensional gas dynamics, SIAM J. Math. Anal., **24**, 76–88 (1993)
9. Shu, C.-W., Osher, S.: Efficient implementation of essentially non-oscillatory shock-capturing schemes, II, J. Comput. Phys., **83**, 32–78 (1989)

Finite Volume Solvers for the Shallow Water Equations Using Matrix Radial Basis Function Reconstruction

L. Bonaventura, E. Miglio and F. Saleri

MOX - Dipartimento di Matematica
Politecnico di Milano
P.zza Leonardo da Vinci 32
20133 Milano Italy
luca.bonaventura@polimi.it
edie.miglio@polimi.it
fausto.saleri@polimi.it

The accuracy of low order numerical methods for the shallow water equations is improved by using vector reconstruction techniques based on matrix valued radial basis functions. Applications to geophysical fluid dynamics problems show that these reconstruction techniques allow to maintain important discrete conservation properties while greatly reducing the error with respect to low order discretizations.

1 Finite volume methods for shallow water models

The shallow water equations result from the Navier-Stokes equations when the hydrostatic assumption holds and only barotropic and adiabatic motions are considered. They can be written as

$$\frac{\partial h}{\partial t} + \nabla \cdot \left(H\mathbf{v} \right) = 0, \tag{1}$$

$$\frac{\partial \mathbf{v}}{\partial t} + (\mathbf{v} \cdot \nabla)\mathbf{v} = -f\mathbf{k} \times \mathbf{v} - g\nabla h. \tag{2}$$

Here, $\mathbf{v}$ denotes the two-dimensional velocity vector, $\mathbf{k}$ is the radial unit vector perpendicular to the plane on which $\mathbf{v}$ is defined (or to the local tangent plane, in case of applications in spherical geometry), h is the height of the fluid layer above a reference level, $H = h - h_s$ is the thickness of the fluid layer, h_s is the orographic or bathymetric profile, g is the gravitational constant and f is the Coriolis parameter. Eulerian-Lagrangian discretizations for the shallow water equations using formulation (1), (2) have been proposed in [5],

[6], which couple a mass conservative, semi-implicit discretization on unstructured Delaunay meshes to an Eulerian-Lagrangian treatment of momentum advection. The resulting methods are highly efficient because of their mild stability restrictions, while mass conservation allows for their practical (and successful) application to a number of pollutant and sediment transport problems. A key step of the Eulerian-Lagrangian method is the interpolation at the foot of characteristic lines, which in the papers quoted above is performed by RT0 elements (see e.g. [9]) or by low order interpolation procedures based on area weighted averaging. These interpolators have at most first order convergence rate and can introduce large amounts of numerical diffusion, thus making their application questionable especially for long term simulations. Another widely used formulation for applications to large scale atmospheric dynamics is the so called *vector invariant form* (see e.g. [11]), in which the momentum equation is rewritten as:

$$\frac{\partial \mathbf{v}}{\partial t} = -(\zeta + f)\mathbf{k} \times \mathbf{v} - \nabla\left(gh + K\right). \tag{3}$$

Here, ζ is the component of relative vorticity in the direction of $\mathbf{k}$ and K denotes the kinetic energy. This formulation is usually the starting point for the derivation of energy, potential enstrophy and potential vorticity preserving discretizations (see e.g. [1]). Eulerian discretizations of equations (1),(3) have been proposed in [3, 4], which preserve discrete approximations of mass, vorticity and potential enstrophy. These properties are important for numerical models of general atmospheric circulation, especially for applications to climate modelling. The two time level, semi-implicit scheme proposed in these papers used RT reconstruction to compute the nonlinear terms in the discretization of (3).

In this paper, vector interpolators based on the technique of matrix valued Radial Basis Functions (RBF) proposed in [7] are applied to improve the accuracy of the above mentioned Eulerian or Eulerian-Lagrangian finite volume solvers. The use of matrix RBF interpolators allows to achieve this goal without having to resort e.g. to higher order RT elements, which would make more difficult or impossible to preserve the important discrete conservation properties of the methods reviewed above. For simplicity, in this paper we restrict ourselves to the two dimensional case, although all the results and the methods can be generalized to 3D. Although in general this is not sufficient to raise the convergence order of the overall methods, models employing RBF reconstructions display significantly smaller errors and have in general less numerical dissipation, making their use attractive for a number of applications. More extensive tests of the accuracy of matrix valued RBF reconstructions have been reported in [2].

2 Matrix valued Radial Basis Functions for vector field reconstruction

In this section, the vector reconstruction based on RBF proposed in [7] is briefly summarized in a context that is appropriate for the applications to hydrodynamic models. Similar applications of scalar RBF reconstructions have been presented e.g. in [10].

Consider a set of N distinct points in the plane $x_i, i = 1, \ldots, N$, $x_i \in \mathbf{R}^2$, and assume that for each x_i a two dimensional unit vector $\mathbf{n}_i$ is given. Consider then a smooth vector field $\mathbf{u} : \mathbf{R}^2 \to \mathbf{R}^2$. The interpolation data are the values $u_i = \mathbf{u}(x_i) \cdot \mathbf{n}_i$. The interpolation problem consists of the reconstruction of the field $\mathbf{u}(x)$ at an arbitrary point $x \in \mathbf{R}^2$, given the values u_i. This problem can be reformulated as follows: consider the vector valued distribution denoted formally by $\boldsymbol{\lambda}_i = \delta(x - x_i) \cdot \mathbf{n}_i$, whose action on a vector valued function $\mathbf{f}(x)$ is such that $(\boldsymbol{\lambda}_i, \mathbf{f}) = \mathbf{f}(x_i) \cdot \mathbf{n}_i$. Given a matrix valued radial basis function

$$\boldsymbol{\Phi}(x) = \begin{bmatrix} \phi_{11}(x) & \phi_{12}(x) \\ \phi_{21}(x) & \phi_{22}(x) \end{bmatrix},$$

where the functions ϕ_{ij} are e.g. Gaussian or multiquadric kernels, the convolution $\boldsymbol{\Phi} * \boldsymbol{\lambda}_i$ is defined according to [7] as

$$\boldsymbol{\Phi} * \boldsymbol{\lambda}_i(x) = \begin{bmatrix} \phi_{11}(x - x_i) n_i^1 + \phi_{12}(x - x_i) n_i^2 \\ \phi_{21}(x - x_i) n_i^1 + \phi_{22}(x - x_i) n_i^2 \end{bmatrix}$$

where $\mathbf{n}_i = [n_i^1, n_i^2]^T$. The interpolation problem consists then of finding coefficients $c_j, j = 1, \ldots, N$ and a vector valued polynomial $\mathbf{p}(x)$ such that the vector valued distribution denoted formally by $\boldsymbol{\lambda} = \sum_{j=1}^{N} c_j \boldsymbol{\lambda}_j$ and the polynomyal $\mathbf{p}$ satisfy the conditions $(\boldsymbol{\lambda}_i, \boldsymbol{\Phi} * \boldsymbol{\lambda} + \mathbf{p}) = u_i$, $i = 1, \ldots, N$. Furthermore, if $\mathbf{u}$ is actually a polynomial, the polynomial $\mathbf{p}$ determined by this procedure should coincide with $\mathbf{u}$. These conditions can be rewritten as

$$\sum_{j=1}^{N} c_j (\boldsymbol{\lambda}_i, \boldsymbol{\Phi} * \boldsymbol{\lambda}_j + \mathbf{p}) = u_i, \quad i = 1, \ldots, N.$$

It can be seen that determination of the coefficients $c_j, j = 1, \ldots, N$ requires the inversion of the interpolation matrix $\mathbf{A} = (a_{i,j})_{i,j=1,\ldots,N}$ whose entries are given by $a_{i,j} = (\boldsymbol{\lambda}_i, \boldsymbol{\Phi} * \boldsymbol{\lambda}_j)$. Conditions under which this matrix is symmetric and positive definite are given in [7]. In the simple case in which it is assumed that no polynomial constraint is imposed and that $\boldsymbol{\Phi} = \phi \mathbf{I}$, where $\mathbf{I}$ is the identity matrix and ϕ is a single scalar radial basis function, the problem reduces for example to $\sum_{j=1}^{N} c_j \phi(x_i - x_j) \mathbf{n}_i \cdot \mathbf{n}_j = u_i$, $i = 1, \ldots, N$.

3 Applications to environmental modelling

Numerical results obtained with the models described in section 1 will now be presented. In all cases, it will be shown that RBF vector reconstruction leads to a substantial improvement of the accuracy of the considered methods. For all the tests, we employed a reconstruction using the Gaussian or multiquadric RBF. The polynomial reproduction constraint, when applied, imposed exact reproduction of constant vectors. A simple 9 point stencil has been adopted, using the normal velocity components to the edges of the triangle on which the reconstruction is being carried out and to the edges of its nearest neighbours (i.e. of the triangles which have common edges with it).

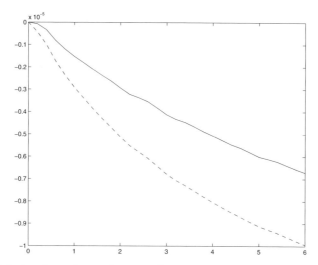

Fig. 1. Relative decay in total energy for Eulerian-Lagrangian model, computed using RT reconstruction (full line) and RBF reconstruction with 9 points stencil (dotted line) in a free oscillations test.

Firstly, the results obtained with the Eulerian-Lagrangian method of [6] will be discussed. In this context, the RBF reconstruction can be used as opposed to RT0 elements when performing the interpolation at the foot of the characteristics. In a first test, a square domain of width 20 m was considered, which was discretized by an unstructured triangular mesh with 3984 elements and 2073 nodes. A constant basin depth of 2 m was assumed. At the initial time, still water was assumed and the free surface profile was taken to be a gaussian hill centered at the center of the domain, with amplitude 0.1 m and standard deviation 2 m. In absence of any explicit dissipative term, the total energy of the system should be conserved. The free oscillations of the fluid were simulated for a total of 6 s with a time step $\Delta t = 0.01$ s. The time evolution of total energy is shown in figure 1, while the height field computed at various timesteps is shown in figures 2, 3. It can be observed that the energy dissipation caused by the interpolation of the Eulerian-Lagrangian method is reduced by 40% if the RBF reconstruction is used, while the values of the maxima and minima in the height field are improved by approximately 20%.

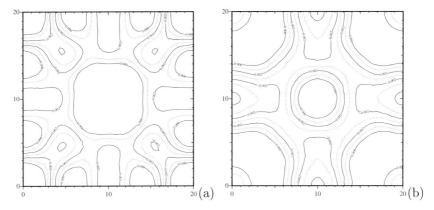

Fig. 2. Height field computed using RT reconstruction for the Eulerian-Lagrangian method in free oscillations test at time (a) $t = 4$ s and (b) $t = 6$ s.

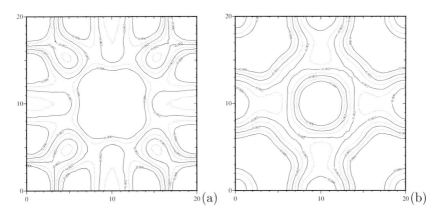

Fig. 3. Height field computed using RBF reconstruction with 9 points stencil for Eulerian-Lagrangian method in a free oscillations test at time (a) $t = 4$ s and (b) $t = 6$ s.

Similar experiments have also been carried out with an Eulerian discretization of the shallow water equations in spherical geometry. In this particular case, a three-time level, semi-implicit time discretization was coupled to the potential enstrophy preserving spatial discretization of [4], using either the Raviart Thomas algorithm or a vector RBF reconstruction of the velocity field necessary for the solution of equation (3). The algorithm performance was studied when applied to test case 3 of the standard shallow water suite [11], which consists of a steady-state, zonal geostrophic flow with a narrow jet at midlatitudes. For this test case, an analytic solution is available, so that errors can be computed by applying the numerical method at different resolutions (denoted by the refinement level in a dyadic refinement procedure starting from the regular icosahedron, see [4] for a complete description of the grid construction). The values of the relative error in various norms as computed at day 2 with different spatial resolutions and with time step $\Delta t = 1800$ s is displayed in Tables 1, 2 for Raviart Thomas algorithm and vector RBF reconstruction, respectively. It can be observed that, although the convergence rates remain approximately unchanged (due to the fact that the approximately second order discretization of the geopotential gradient was the same in both tests), the errors both in the height and velocity fields have decreased by an amount that ranges between 30% and 50% approximately.

Table 1. Relative errors in shallow water test case 3 with RT reconstruction for nonlinear terms.

Level	l_2 error, h	l_2 error, **v**	l_∞ error, h	l_∞ error, **v**
3	7.42e-3	0.25	2.53e-2	0.33
4	1.94e-3	5.9e-2	8.1e-3	9.1e-2
5	6.05e-4	1.27e-2	2.9e-3	1.87e-2
6	2.54e-4	3.19e-3	1.24e-3	4.17e-3

Table 2. Relative errors in shallow water test case 3 with 9 points RBF reconstruction for nonlinear terms.

Level	l_2 error, h	l_2 error, **v**	l_∞ error, h	l_∞ error, **v**
3	7.27e-3	0.16	2.08e-2	0.17
4	1.52e-3	3.38e-2	6.74e-3	5.77e-2
5	4.05e-4	7.7e-3	1.7e-3	1.22e-2
6	1.45e-4	2.11e-3	4.8e-4	2.89e-3

We have then considered the nonstationary test case 6 of [11], for which the inital datum consists of a Rossby - Haurwitz wave of wavenumber 4. This type of wave is an analytic solution for the barotropic vorticity equation and can also be used to test shallow water models on a time scale of up to 10-15 days. Plots of the meridional velocity component at simulation day 5 are

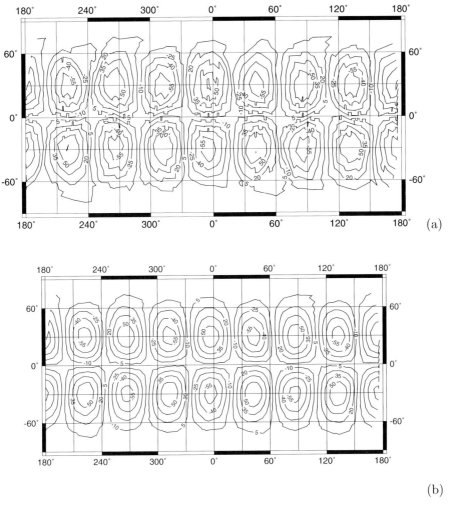

Fig. 4. Meridional velocity in shallow water test case 5, computed using (a) RT0 reconstruction (b) RBF reconstruction with 9 points stencil. Contour lines spacing is 15 m s^{-1}.

shown in figure 4, as computed with a timestep of $\Delta t = 600$ s on a spherical quasi-uniform triangular mesh with a spatial resolution of approximately 400 km. It can be observed for example that, when using RT0 reconstruction, the meridional velocity field obtained is much less regular than in the case of matrix RBF reconstruction, which compares better with results obtained in reference high resolution simulations. Furthermore, the total energy loss is reduced in the RBF computation by approximately 30%, thus improving the energy conservation properties of the model, which conserves potential enstrophy but not energy as discussed in [4].

Concerning the computational cost of RBF reconstructions, it should be observed that, in the case of Eulerian models, it is possible to carry out most of the RBF computations at startup, so that the extra computational cost due to the use of RBFs is approximately 20% of the cost of a model run using simple RT0 reconstruction. On the other hand, in the case of Eulerian-Lagrangian models, the extra computational cost is higher than in the Eulerian case, since the RBF coefficients have to be recomputed at each time step for each of the trajectory departure points.

References

1. Arakawa, A., Lamb, V.: A potential enstrophy and energy conserving scheme for the shallow water equations. Monthly Weather Review, **109**, 18–136 (1981)
2. Baudisch, J.: Reconstruction of Vector Fields Using Radial Basis Functions. MA Thesis, Munich University of Technology, Munich (2005)
3. Bonaventura, L., Kornblueh, L., Heinze, T., Ripodas, P.: A semi-implicit method conserving mass and potential vorticity for the shallow water equations on the sphere, Int. J. Num. Methods in Fluids, **47**, 863–869 (2005)
4. Bonaventura, L., Ringler, T.: Analysis of discrete shallow water models on geodesic Delaunay grids with C-type staggering, Monthly Weather Review, **133**, 2351–2373 (2005)
5. Casulli, V., Walters, R.A.: An unstructured grid, three-dimensional model based on the shallow water equations, Int. J. Num. Methods in Fluids, **32**, 331–348 (2000)
6. Miglio, E., Quarteroni, A., Saleri, F.: Finite element approximation of quasi-3d shallow water equations, Comp. Methods in Appl. Mech. and Eng., **174**, 355–369 (1999)
7. Narcowich, F.J., Ward, J.D.: Generalized Hermite interpolation via matrix-valued conditionally positive definite functions, Math. Comp., **63**, 661–687 (1994)
8. Pedlosky, J.: Geophysical Fluid Dynamics. Springer Verlag, New York - Berlin (1987)
9. Quarteroni, A., Valli, A.: Numerical approximation of partial differential equations. Springer Verlag, New York - Berlin (1994)
10. Rosatti, G., Bonaventura, L., Cesari, D.: Semi-implicit, semi-Lagrangian environmental modelling on cartesian grids with cut cells, J. Comp. Phys., **204**, 353–377 (2005)
11. Williamson, D.L., Drake, J.B., Hack, J.J., Jakob, R., Swarztrauber, P.N.: A Standard Test Set for Numerical Approximations to the Shallow Water Equations in Spherical Geometry, J. Comp. Phys., **102**, 211–224 (1992)

Adaptive Methods

On Numerical Schemes for a Hierarchy of Kinetic Equations

Hans Babovsky[1] and Laek S. Andallah[2]

[1] Institute of Mathematics, Ilmenau Technical University, D-98693 Ilmenau, Germany
hans.babovsky@tu-ilmenau.de
[2] Departement of Mathematics, Jahangirnagar University, Savar, Dhaka-1342, Bangladesh

Abstract. We investigate the hierarchical structure of hexagonal kinetic models as a tool for the numerical simulation of the Boltzmann equation. This is of use for a number of applications, e.g. in the context of domain decomposition and of multigrid techniques.

1 Introduction

Due to the development of hexagonal kinetic models [3] there is an efficient tool available for the numerical simulation of rarefied gas flows [5]. This tool has been proven to be numerically efficient, is reliable from a theoretical point of view and covers a broad spectrum of physical situations, ranging from slow interior to fast exterior 2D flows. Examples are given in [5]. As one particular example, Fig. 1 shows the gas kinetic analogon of the well-known (1D) shock tube problem for the Euler equations (see, e.g. [6]). It is calculated with a kinetic 54-velocity model (2-layer model, cf. Section 2). We easily identify the four plateaus of the density and those interfaces which in the fluid dynamic case correspond to the rarefaction zone, the contact discontinuity, and the shock interface. Of course, in rarefied gas kinetics discontinuities through the shock turn into continuous profiles with steep gradient. If we want to come closer to the fluid dynamic situation described by the Euler equations, we have to cut down viscosity; for this it is necessary to refine the discretization of the velocity space, i.e. to use a larger discrete velocity model.

An interesting aspect is that the hexagonal model is accessible to a multigrid approach, since it can be refined giving rise to a hierarchy of hexagonal models. This hierarchy is the subject of the paper. Since more refined models require essentially more time and memory resources, it seems favorable to supplement them with calculations on coarser grids. One might think e.g. of one of the following three scenarios. *First*, one might try to establish a

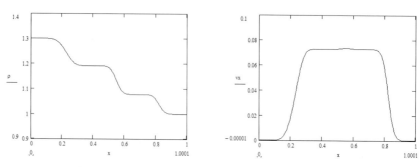

Fig. 1. Shock tube problem. (a) density, (b) velocity.

multigrid scheme as it is applied in different context for partial differential equations, where the calculations switch in an appropriate manner between the different grid levels. *Second*, numerical results on the coarse grid may be seen as a predictor which is corrected on the refined level. e.g., if steady solutions are calculated via a time marching algorithm, the initial phase could be calculated by a rough scheme and followed by a refinement procedure. *Third*, the refined model could be restricted to sensitive areas of the computational domain and coupled to the coarse system in the complementary region.

The scope of the paper is as follows. First, we give a short review on hexagonal collision models (Section 2). Then we introduce a refinement strategy which leads to a hierarchy of grid levels. Given densities on one of the levels, we investigate coarsening and refinement procedures to switch between the levels (Section 3). Finally, we discuss the coupling of different models in a domain decomposition approach (spatially 1D) and work out the problems to be attacked for a successful implementation.

2 Hexagonal kinetic models

Hexagonal kinetic models are founded on two basic features. First, on the hexagonal discretization $\mathcal{G}$ of $\mathbb{R}^2$ (in 2D velocity space; a similar discretization for $\mathbb{R}^3$ has been worked out in [2] but will not be considered here) as given in Fig. 2(a). Second, it is based on a collision operator on each regular hexagon with nodes in $\mathcal{G}$. The number of regular hexagons with nodes in $\mathcal{G}$ is large. As can be shown, for any two points $g_1 \neq g_2 \in \mathcal{G}$ there are two hexagons containing (g_1, g_2) as an edge. Given a hexagon $H = (g_0, \ldots, g_5) \in \mathcal{G}^6$ (g_i numbered in consecutive order as appearing in the hexagon), a collision operator for a density $f = (f_0, \ldots, f_5)$ on H is given as $J = J_{\text{bin}} + J_{\text{ter}}$ with the *binary* collision operator J_{bin} and the ternary J_{ter} being defined via $S[f] := f_0 f_3 + f_1 f_4 + f_2 f_5$ and $T[f] := f_1 f_3 f_5 - f_0 f_2 f_4$ as

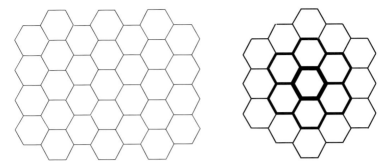

Fig. 2. Hexagonal discretization. (a) the grid, (b) n-layer models.

$$J_{\text{bin}}[f] = S[f] \cdot (1,1,1,1,1,1)^T - 3(f_0 f_3, f_1 f_4, f_2 f_5, f_0 f_3, f_1 f_4, f_2 f_5)^T \quad (1)$$

$$J_{\text{ter}}[f] = T[f] \cdot (1,-1,1,-1,1,-1)^T. \quad (2)$$

The corresponding kinetic theory has been worked out in [3, 2, 1].

For numerical purposes one has to restrict to finite grids. Convenient are so-called n-layer grids which are grouped in a symmetric fashion around a central hexagon as shown in Fig. 2(b). The six nodes of the bold hexagon form the 0-layer system. Including the hexagons with lines of medium thickness yields the 1-layer model with 24 velocities. All lines together form the 2-layer system with 54 velocities. The total number of regular hexagons in the grid and with this the required numerical effort for the calculation of the collision operator increase significantly with n as can be read off from Table 1. (Here, $|\mathcal{G}|$ denotes the number of grid points; *basic* hexagons are the small hexagons producing the grid; $|\mathcal{H}|$ is the number of all regular hexagons.) Thus for the design of efficient algorithms it is necessary to choose n as small as possible.

Table 1. n-**layer models**

#(layers)	0	1	2	3	4	5	6	7	8	9	10
$|\mathcal{G}|$	6	24	54	96	150	216	294	384	486	600	726
#(basic hexagons)	1	7	19	37	61	91	127	169	217	271	331
$|\mathcal{H}|$	1	16	81	256	625	1296	2401	4096	6561	10000	14641

On the other hand, consider a physical flow situation with a typical temperature T_0 and bulk velocities ranging in a certain bounded domain V. The choice of the grid size h and the number n of layers depends significantly on T_0 and V. The lowest temperature to be resolved is certainly restricted by the size of the basic hexagons. If T_0 is too large, then we find distorting boundary effects. Furthermore, bulk velocities close to the boundary of the n-layer grid are not well reproduced. Detailed investigations are found in [2]. In Fig. 3(a),

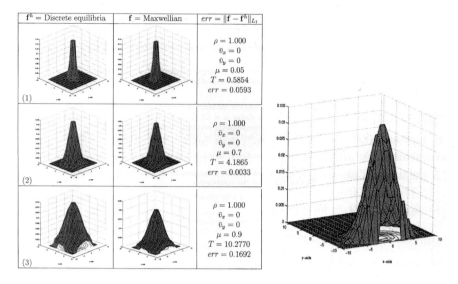

Fig. 3. Discrete equilibria. (a) centered equilibria, (b) boundary distortion.

discrete equilibrium distributions of the 2-layer model with zero bulk velocities are compared to the equilibria of the continuous Boltzmann equation ("Maxwellians"). It turns out, that there is best agreement (with an error of 0.33%) if the temperature is not too small (restriction of the grid length) and not too large (restriction of the bounded grid size). Fig. 3(b) shows distortion effects when the bulk velocity comes into the vicinity of the grid boundary. In this case we find an error of 17% for the discrete equilibrium due to boundary effects.

3 Hexagonal hierarchy

Given an (infinite) grid $\mathcal{G}^{(h)}$ with grid length h, a coarser grid $\mathcal{G}^{(2h)} \subset \mathcal{G}^{(h)}$ with double grid size can be generated by replacing 4-point stencils of the form given in Fig. 4(a) by their center points. In a similar manner, $\mathcal{G}_h$ can be refined to a grid $\mathcal{G}^{(h/2)} \supset \mathcal{G}^{(h)}$ by blowing up every point to a 4-point stencil. This refinement procedure can be continued *ad infinitum* ending with a continuous kinetic model [5]. In this section we restrict to the three-layer model $\mathcal{G}_{96}$ with 96 grid points as given in Fig. 4(b) and its one-layer restriction $\mathcal{G}_{24} \subset \mathcal{G}_{96}$ with 24 grid points (edges of the solid lines in Fig. 4(b)). Switching between the grids requires a coarsening and a refinement procedure.

Coarsening: Suppose given a distribution function f_{96} on $\mathcal{G}_{96}$ to be coarsened to a distribution function f_{24} on $\mathcal{G}_{24}$ in such a manner that the physical conservation laws (mass, momenta and kinetic energy) are satisfied. A first attempt consists in moving the masses of the outer points of each stencil

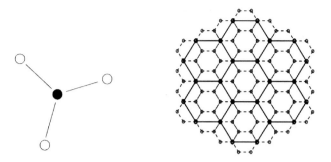

Fig. 4. Switching between grids: (a) 4-point stencil, (b) 3- and 1-layer grid.

to their center points (call the resulting distribution $\widetilde{f}_{24}$). This procedure is mass conserving but momenta and energy are (slightly) perturbed. One way to remedy the situation is to choose methods from optimization theory and find the vector f_{24} closest to $\widetilde{f}_{24}$ (in some appropriate norm) satisfying the conservation laws and the restriction $f_{24} \geq 0$. It should be pointed out that in fluid dynamics there are further quantities which are of interest; among them are the momenta flows and the heat flow which are given by second and third momenta of the distribution. These should be considered to be conserved quantities in the coarsening procedure and included into the optimization described above.

Refinement: The simplest procedure to refine a distribution function f_{24} on $\mathcal{G}_{24}$ to a function on $\mathcal{G}_{96}$ is to imbed f_{24} into $\mathcal{G}_{96}$, i.e. to let the center points of the stencils take over the values of f_{24} and put $f_{96} = 0$ on the outer stencil points. In this case, all quantities mentioned above are conserved. However, there is a severe drawback, since an equilibrium function f_{24} on $\mathcal{G}_{24}$ is mapped onto a function which is quite far apart from the corresponding equilibrium function on $\mathcal{G}_{96}$. Thus for calculations close to the fluid dynamic limit, a better choice is to find among all distributions on $\mathcal{G}_{96}$ satisfying the conservation laws that one which minimizes the H-functional $H[f] = \sum f_i \ln(f_i)$. Again optimization theory supplies the tools.

4 Coupling of two kinetic models

4.1 General remarks

When coupling two different kinetic models within one computational domain, two modeling aspects have to be considered. First, at least close to the fluid dynamic limit it is essential that both models have to be *compatible* in the sense that they exhibit the same macroscopic behavior. An example may illustrate this.

Apart from the (physical) boundaries, the temperature profile $T(x)$ of the classical *1D steady heat layer problem* turns out to be (close to) a straight line. The temperature gradient in this area is ruled by the heat flux q and the heat conduction coefficient κ. A well accepted assumption is $q = \kappa \cdot \partial_x T$. It is an easy exercise to prove that the heat flux has to be constant along the whole line. Thus matching two kinetic models with different heat coefficients leads to the matching of two linear temperature profiles with different slopes which leads to a discontinuity of the first derivative of T and is unphysical.

Besides the adaptation of the heat coefficient, consideration of the viscosity coefficient is important as well. In [5, Section 3.4.2] an example is presented how to vary the viscosity coefficient of a hexagonal kinetic model by changing the collision frequencies. Matching of different kinetic models is in some sense straightforward and is not the subject of the present paper.

A second aspect playing a crucial role in the coupling of kinetic models are *interface* conditions mapping the flux leaving one of the computational domains into one entering the second domain. This will be investigated in some detail.

4.2 Interface conditions

For simplicity we restrict on the spatially 1D case with 2D velocity space.

Suppose given a Discrete Velocity Model (DVM) defined on some finite set (set of admissible velocities) $\mathcal{V} \subset \mathbb{R}^2$. We denote the elements of $\mathcal{V}$ by $v = (v_x, v_y)$, where v_x is the velocity component in x-direction and v_y that in y-direction. Furthermore we define $\mathcal{V}_+ = \{v \in \mathcal{V} : v_x > 0\}$ and $\mathcal{V}_- = \{v \in \mathcal{V} : v_x < 0\}$. The dynamics for the model is given in terms of a kinetic equation. A special role is played by the equilibrium solutions M which we assume to depend – as in standard kinetic theory – uniquely on the density ρ, the mean velocity $\bar{v}$ and temperature T, i.e. $M = M[\rho, \bar{v}, T]$.

Suppose we want to model the transport on the real line by coupling two different kinetic models – say model A on the negative and model B on the positive part with the velocity sets $\mathcal{V}^A$ and $\mathcal{V}^B$. Let f^A and f^B denote the corresponding densities. The most essential question is that of coupling conditions for f^A and f^B at the artificial boundary $x = 0$. There is a flux $v_x f_+^A(0, v)$ entering from the left which serves as a source term for model B; similarly, the flux $v_x f_-^B(0, v)$ from the right serves as a source for A. Here, $f_\pm$ denotes the restriction of f to arguments $v = (v_x, v_y)$ with $v_x > 0$ resp. $v_x < 0$. The most straightforward way to formulate coupling conditions is to choose transmission laws $\mathcal{T}_\pm$ and define the interface conditions

$$|v_x| f^A(0, v) = \sum_{v' \in \mathcal{V}_+^B} \mathcal{T}_-(v|v') |v'_x| f^B(0, v') \quad \text{for all} \quad v \in \mathcal{V}_-^A \qquad (3)$$

$$|v_x| f^B(0, v) = \sum_{v' \in \mathcal{V}_+^A} \mathcal{T}_+(v|v') |v'_x| f^A(0, v') \quad \text{for all} \quad v \in \mathcal{V}_+^B \qquad (4)$$

In order not to introduce artificial sources or sinks we have to require that $\mathcal{T}_\pm(.|v')$ are probability distributions, i.e. they are nonnegative and satisfy

$$\forall v' \in \mathcal{V}_+^A : \sum_{v \in \mathcal{V}_+^B} \mathcal{T}_+(v|v') = 1, \quad \forall v' \in \mathcal{V}_-^B : \sum_{v \in \mathcal{V}_-^A} \mathcal{T}_-(v|v') = 1 \tag{5}$$

A plausible choice for a nonlinear transmission law is (in analogy to diffuse reflection laws) given by

$$\mathcal{T}_\pm(v|v') = c|v_x| M^\pm [\rho^\mp, \overline{v}^\mp, T^\mp](v) \tag{6}$$

with Maxwellians $M^\pm$ and moments $\rho^\mp, \overline{v}^\mp, T^\mp$ depending on the moments of the outgoing flows of the adjacent areas; c is a normalizing constant. Having chosen the moments properly, reproduces correctly global equilibria. For nonequilibrium flows, results from linearized theory tell us that we have to expect interface perturbations which fade away exponentially. This is confirmed in a *first numerical experiment* to produce a constant gradient temperature profile. The result is shown in Fig. 5; at the interface, we find temperature jumps compared to the linear profile of approximately 1.2%. What makes the problem quite complicated is to find the correct interface temperatures which determine the inflow Maxwellians. Between the temperature of the linear profile at $x = 0$ and $T^\mp$ there is a jump of $+5.2\%$ for the left hand side and of -5.1% on the right. At present, there is no possibility to calculate these jumps – they have to be found out experimentally for each single kinetic model.

In a *second series of numerical experiments* we have chosen $\mathcal{T}_\pm$ in such a way that passing from outgoing to ingoing flows, none of the moments of f up to third order is changed, and the H-functional is minimized (cf. Sec. 3). Here, we coupled the 3-layer and the 1-layer hexagonal models and solved the heat layer problem. The results are presented in Fig. 6(a) (refined for $x > 0.5$, coarse for $x > 0.5$) and Fig. 6(b) (refined only in the boundary layers). This interface model produces smaller jumps at the interfaces ($\sim 0.6\%$), is easier to handle and thus should be given preference over the first model.

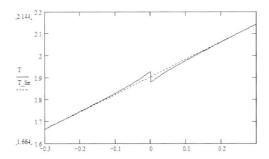

Fig. 5. Temperature profile, artificial interface layers.

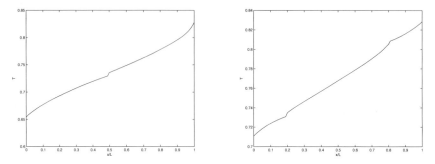

Fig. 6. Temperature profiles of heat layer problem.

4.3 Conclusions

The coupling of different kinetic models in a smooth way is not a straightforward matter as it turned out in the above numerical experiments. Using e.g. the diffuse transmission law requires the matching of the temperatures at the artificial boundary. For hierarchical hexagonal models, an easier to handle alternative is to use refinement and prolongation techniques leaving all up to the third moments invariant and minimizing the H-functional. However, in all cases we experience a (slight) interface perturbation.

As an alternative approach, the coarsening and refinement techniques presented above can be used to establish multigrid schemes.

References

1. Andallah, L.S.: On the generation of a hexagonal collision model for the Boltzmann equation. Comp. Meth. in Appl. Math., **4**, 271–289 (2004)
2. Andallah, L.S.: A hexagonal collision model for the numerical solution of the Boltzmann equation. PhD Thesis, Technical University Ilmenau, Germany (2004)
3. Andallah, L.S., Babovsky, H.: A discrete Boltzmann equation based on hexagons. Math. Models Methods Appl. Sci., **13**, 1537–1563 (2003)
4. Babovsky, H.: Kinetic boundary layers: on the adequate discretization of the Boltzmann collision operator. J. Comp. Appl. Math., **110**, 225–239 (1999)
5. Babovsky, H.: Hexagonal kinetic models and the numerical simulation of kinetic boundary layers. In: G. Warnecke (ed) Analysis and Numerics for Conservation Laws. Springer, Berlin Heidelberg New York (2005)
6. LeVeque, R.J.: Numerical Methods for Conservation Laws. Birkhäuser, Basel (1990)

Computational Aspects of the Mesh Adaptation for the Time Marching Procedure

Jiří Felcman[1,2] and Petr Kubera[2,1]

[1] Charles University in Prague
Faculty of Mathematics and Physics
felcman@karlin.mff.cuni.cz
[2] Jan Evangelista Purkyně, University in Ãšsrã- nad Labem
Faculty of Science
kubera@sci.ujep.cz

Summary. The paper deals with a construction of an adaptive mesh in the framework of the cell-centred finite volume scheme. The adaptive strategy is applied to the numerical solution of problems governed by hyperbolic partial differential equations. Starting from the adaptation techniques for the stationary problems (for a general overview see e.g. [9]), the nonstationary case is studied. The main attention is paid to an adaptive part of a time marching procedure. The *main feature* of the proposed method is to keep the mass conservation of the numerical solution at each adaptation step. We apply an anisotropic mesh adaptation from [1]. This is followed by a recovery of the approximate solution on the new mesh satisfying the geometric conservation law. The adaptation algorithm is formulated in the framework of an N-dimensional numerical solution procedure. A new strategy for moving a vertex of the mesh, based on a gradient method, is presented. The results from [4] are further developed. The *general significance* of the proposed method is the ability to solve problems with moving discontinuities. A numerical example is presented.

1 Euler equations

Let us consider the flow of an inviscid perfect gas in a bounded domain $\Omega \subset \mathbb{R}^N$ and time interval $(0, T)$ with $T > 0$. Here $N = 2$ or 3 for 2D or 3D flow, and we suppose that Ω is polygonal in 2D or polyhedral in 3D, respectively. Further we suppose that the flow is adiabatic and we neglect the outer volume force. Our goal is to solve numerically the Euler equations

$$\frac{\partial \boldsymbol{w}}{\partial t} + \sum_{s=1}^{N} \frac{\partial \boldsymbol{f}_s(\boldsymbol{w})}{\partial x_s} = 0 \quad \text{in } Q_T = \Omega \times (0, T) \tag{1}$$

equipped with the initial condition

$$\boldsymbol{w}(x, 0) = \boldsymbol{w}^0(x), \quad x \in \Omega, \tag{2}$$

with a given vector function $\boldsymbol{w}^0$ and boundary conditions

$$B(\boldsymbol{w}(x,t)) = 0 \quad \text{for } (x,t) \in \partial\Omega \times (0,T). \tag{3}$$

Here B is a suitable boundary operator. The specification of the boundary conditions and their approximation can be found, e.g. in [9, pages 227–233]. For recent results concerning boundary conditions see [3] The state vector $\boldsymbol{w} = (\rho, \rho v_1, \ldots, \rho v_N, E)^T \in \mathbb{R}^m$, $m = N + 2$ (i.e. $m = 4$ or 5 for 2D or 3D flow, respectively). Here ρ, $v_1, \ldots, v_N$ and E denote the density, the velocity components and the total energy, respectively. The fluxes $\boldsymbol{f}_s$, $s = 1, \ldots, N$, are m-dimensional mappings. For their definition see e.g. [9, page 102].

2 Adaptive algorithm

The problem (1)–(3) is solved by an explicit finite volume (FV) method. Its description and the use for a solution of steady 3D problems can be found e.g. in [9]. Here we are dealing with the adaptive time marching procedure in the non-stationary case. Let $0 < t_0 < t_1 < \ldots < t_k < \ldots < T$ be the partition of the time interval $(0,T)$ and $\mathcal{D}^k = \{D_i^k\}_{i \in J^k}$ be a system of N-simplicial (i.e. triangular in 2D, tetrahedral in 3D, respectively) FV meshes of the computational domain Ω, where J^k is an index set. As $\boldsymbol{w}_i^k$ we denote an approximation of an integral average of the vector of conserved quantities on the finite volume D_i^k at the time level t_k:

$$\frac{1}{|D_i^k|} \int_{D_i^k} \boldsymbol{w}(x, t_k)\, dx \approx \boldsymbol{w}_i^k. \tag{4}$$

We define a finite volume approximate solution of (1) as piecewise constant vector-valued functions $\mathbf{w}_{\mathcal{D}^k}^k$, $k = 0, 1, \ldots$, defined a.e. in Ω so that $\mathbf{w}_{\mathcal{D}^k}^k \big|_{\overset{\circ}{D_i^k}} = \boldsymbol{w}_i^k$ for all $i \in J^k$, where $\overset{\circ}{D_i^k}$ is the interior of D_i^k and $\boldsymbol{w}_i^k$ are obtained from the FV formula. The function $\mathbf{w}_{\mathcal{D}^k}^k$ is the approximate solution on the mesh $\mathcal{D}^k$ at time $t = t_k$. The vector $\boldsymbol{w}_i^k$ is the value of the approximate solution on the finite volume D_i^k at time t_k. Analogously we denote by $\mathbf{w}_{\mathcal{D}^k}^{k+1}$ the approximate solution on the mesh $\mathcal{D}^k$ at time $t = t_{k+1}$.

In [4] a time marching FV method for non-stationary problems was worked out. Here we present its further development. The new algorithm consists of three basic sections at each time step: the time evolution of the numerical solution, the mesh adaptation and the recomputing of the numerical solution from the mesh before the adaptation to the mesh after the adaptation. In one time step the finite volume scheme is evaluated twice. Firstly for the *prediction* how to adapt the mesh, further for the *update* of the numerical solution itself.

Prediction part

In the prediction part, we forecast the evolution of the numerical solution and adapt the mesh. The anisotropic mesh adaptation (AMA) is applied. For its description see e.g. [1].

1. Prediction: $\mathbf{w}_{\mathcal{D}^k}^{k+1} := \text{FVsol}\left(\mathbf{w}_{\mathcal{D}^k}^{k}, \mathcal{D}^k\right)$. For the update of the numerical solution $\mathbf{w}_{\mathcal{D}^k}^{k}$ on the mesh $\mathcal{D}^k$ the explicit FV scheme is applied and the new approximation $\mathbf{w}_{\mathcal{D}^k}^{k+1}$ at time level t_{k+1} is constructed.
2. Adaptation: $\mathcal{D}^{k+1} := \text{MeshAdapt}\left(\mathcal{D}^k, \mathbf{w}_{\mathcal{D}^k}^{k+1}\right)$. Using the anisotropic mesh adaptation the new mesh $\mathcal{D}^{k+1}$ is constructed based on the computed prediction $\mathbf{w}_{\mathcal{D}^k}^{k+1}$.
3. Recovery: $\widetilde{\mathbf{w}}_{\mathcal{D}^{k+1}}^{k} := \text{SolRecovery}\left(\mathbf{w}_{\mathcal{D}^k}^{k}, \mathcal{D}^k, \mathcal{D}^{k+1}\right)$. The solution $\mathbf{w}_{\mathcal{D}^k}^{k}$ on the mesh $\mathcal{D}^k$ is recomputed on the mesh $\mathcal{D}^{k+1}$. Such a recovery has to satisfy a geometric mass conservation law (GMCL).

PDE Evolution Part

4. Update: $\mathbf{w}_{\mathcal{D}^{k+1}}^{k+1} := \text{FVsol}\left(\widetilde{\mathbf{w}}_{\mathcal{D}^{k+1}}^{k}, \mathcal{D}^{k+1}\right)$. The numerical solution $\mathbf{w}_{\mathcal{D}^{k+1}}^{k+1}$ at the time level t_{k+1} is computed using the explicit FVM.

3 Anisotropic mesh adaptation

In [1] the necessary condition for the properties of the N-simplicial mesh, on which the discretization error is below the prescribed tolerance, is formulated. It is shown, how to control this necessary condition by the interpolation error and the anisotropic mesh adaptation technique is applied. For 2D and 3D numerical examples see e.g. [9, Section 3.7]. In the AMA technique, the equilateral mesh is constructed in the least squares sense. The length of an edge of an N-simplicial mesh is measured in the numerical solution dependent Riemann norm. For a given mesh $\mathcal{D}^k$ and the solution $\mathbf{w}_{\mathcal{D}^k}^{k+1}$ on it we define the quality parameter of the mesh

$$Q_{\mathcal{D}^k} := \frac{1}{\#\mathcal{D}^k} \sum_{D \in \mathcal{D}^k} \sum_{e = \text{edge of } D} \left(\|e\|_{\mathbf{w}_{\mathcal{D}^k}^{k+1}} - c_N\right)^2 \tag{5}$$

Here $\#\mathcal{D}^k$ is the number of N-simplexes in $\mathcal{D}^k$, $\|\cdot\|_{\mathbf{w}_{\mathcal{D}^k}^{k+1}}$ denotes the energy norm of the vector of the edge e given by a matrix related to the Hesse matrix of the numerical solution and c_N is the dimension dependent constant related to the tolerance for the discretization error. The details can be found in [1]. The equilateral (in the least squares sense) N-simplicial grid with respect to the Riemann norm $\|\cdot\|_{\mathbf{w}_{\mathcal{D}^k}^{k+1}}$ is constructed by minimizing the quality parameter $Q_{\mathcal{D}^k}$.

For a given $\mathcal{D}^k$ the quality parameter $Q_{\mathcal{D}^k}$ is a computable quantity. We adapt the grid $\mathcal{D}^k$ in order to decrease $Q_{\mathcal{D}^k}$ and we want to find a new grid $\mathcal{D}^{k+1}$ such that the quality parameter $Q_{\mathcal{D}^{k+1}}$ of $\mathcal{D}^{k+1}$ is smaller. To this end, the iterative process including the face swappings ($\mathcal{F}$), the edge swappings ($\mathcal{E}$), the edge bisections ($\mathcal{B}$), the removal of edges ($\mathcal{R}$) and the moving of vertices ($\mathcal{M}$) is used. The iterative process reads

$$\mathcal{M} + \mathcal{R} + \mathcal{S} + \mathcal{M} + \mathcal{B} + \mathcal{S} + \mathcal{M} + \mathcal{S} + \mathcal{M}, \tag{6}$$

where $\mathcal{S}$ includes both face and edge swappings and the sequence of operations goes from the left to the right.

The above-mentioned local operations are performed as long as the quality parameter $Q_{\mathcal{D}^k}$ decreases. The process is stopped if no local operation leads to a decrease of the quality parameter $Q_{\mathcal{D}^k}$.

Next we shall deal with a new strategy for moving a vertex. This allows also to compare the anisotropic mesh adaptation with moving mesh adaptation techniques proposed e.g. in [6].

3.1 Moving a vertex

Let us denote by σ^k the set of all vertices of the mesh $\mathcal{D}^k$. Moving a vertex is a local operation on the mesh which moves a vertex $P \in \sigma^k$ towards its new position with the aim to decrease the quality parameter (5) of the adapted mesh. We denote by $\mathcal{K}_P$ the admissible set in which interior the vertex P can move

$$\mathcal{K}_P = \bigcup_{\substack{D \in \mathcal{D}^k \\ D \ni P}} D. \tag{7}$$

Further we define the quality parameter of the vertex P as the function of its coordinates $x = (x_1, \ldots, x_N)$

$$Q(x) = \sum_{e \in \mathcal{E}(x)} \left(\|e\|_{\mathbf{w}_{\mathcal{D}^k}^{k+1}} - c_N \right)^2, \tag{8}$$

where $\mathcal{E}(x)$ is the set of edges connecting the point x with those vertices of $\mathcal{D}^k$ lying on the boundary $\partial \mathcal{K}_P$. See Fig. 1.

For the motion of the vertex P to its new position $\widetilde{P}$ we use the interior point method, which minimizes the following function

$$\Phi(x, \alpha) := \alpha Q(x) + B(x), \tag{9}$$

where $B(x)$ is the so called barrier function and $\alpha > 0$ is a weighted parameter between the quality parameter of the vertex $Q(x)$ and the barrier function $B(x)$. The barrier function $B(x)$ for the vertex P is a non-negative continuous function defined on the interior of the admissible region $\mathcal{K}_P$ which satisfies

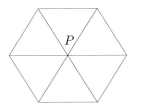
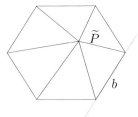

Fig. 1. Admissible set $\mathcal{K}_P$

$$\lim_{x \to y \in \partial \mathcal{K}_P} B(x) \to \infty.$$

We use the following barrier function:

$$B(x) := \sum_{b \subset \partial \mathcal{K}_P} \frac{1}{\text{dist}(x, b)}, \tag{10}$$

where $\text{dist}(x, b)$ is a distance of the point x from the boundary edge $b \subset \mathcal{K}_P$. (See Fig. 1.)

Moving Vertex Algorithm

1. Set the initial position $\widetilde{P} := P$ and set the initial value of the weighted parameter $\alpha := 1$
2. Find $\widetilde{P} = \arg\ \min_{x \in \mathcal{K}_P} \Phi(x; \widetilde{P}, \alpha)$.

The step 2 is repeated with the increasing parameter $\alpha := \alpha\beta, \beta = 2$, until the maximum number of repetition is reached or $Q(\widetilde{P})$ decreases. The BFGS quasi-Newton method from [5] is used for the minimization of Φ. We set the threshold for the number of repetitions to 10.

4 Geometric mass conservation law

After the adaptive mesh is constructed it is necessary to recompute the solution $\mathbf{w}_{\mathcal{D}^k}^k$ on the old mesh $\mathcal{D}^k$ to its recovery $\widetilde{\mathbf{w}}_{\mathcal{D}^{k+1}}^k$ on the newly adapted mesh $\mathcal{D}^{k+1}$. According to [6] the geometric mass conservation law has to be satisfied in this computational step. It reads

$$\sum_{i \in J^k} |D_i^k| \mathbf{w}_i^k = \sum_{i \in J^{k+1}} \left| D_i^{k+1}\ \widetilde{\mathbf{w}}_i^k \right|, \tag{11}$$

where $\widetilde{\mathbf{w}}_i^k = \widetilde{\mathbf{w}}_{\mathcal{D}^{k+1}}^k \big|_{\overset{\circ}{D_i^{k+1}}}$. ($\circ$ denotes the interior of D_i^{k+1}.)

In what follows we shall concentrate on the recomputing of the solution after a moving a vertex. In this case the number of finite volumes $\#\mathcal{D}^k$ of the mesh $\mathcal{D}^k$ is the same as is the number $\#\mathcal{D}^{k+1}$ of the mesh $\mathcal{D}^{k+1}$. The recovery strategy for other local mesh operations $\mathcal{R}, \mathcal{S}$ and $\mathcal{B}$ can be found in [4].

4.1 Perturbation method

The perturbation method from [6] is applied. By the displacement of vertices P_ℓ of the finite volume D_i^k to their new positions $\widetilde{P}_\ell$ the linear mapping $\boldsymbol{c}$ on D_i^k is defined. The point $x \in D_i^k$ is transformed to the point $\widetilde{x} \in D_i^{k+1}$ via the relation

$$\widetilde{x} = x - \boldsymbol{c}(x) \tag{12}$$

with the Jacobian

$$J(x) = \det \frac{D\widetilde{x}}{Dx} = \det \begin{pmatrix} 1 - \frac{\partial c_1}{\partial x_1}, & \frac{\partial c_1}{\partial x_2}, & \frac{\partial c_1}{\partial x_3} \\ \frac{\partial c_2}{\partial x_1}, & 1 - \frac{\partial c_2}{\partial x_2}, & \frac{\partial c_2}{\partial x_3} \\ \frac{\partial c_3}{\partial x_1}, & \frac{\partial c_3}{\partial x_2}, & 1 - \frac{\partial c_3}{\partial x_3} \end{pmatrix}. \tag{13}$$

Supposed that the displacement $\boldsymbol{c}$ is small we can write

$$\begin{aligned}
\int_{D_i^{k+1}} \boldsymbol{w}(\widetilde{x})\, d\widetilde{x} &= \int_{D_i^k} \boldsymbol{w}(x - \boldsymbol{c}(x)) J(x)\, dx \\
&= \int_{D_i^k} \boldsymbol{w}(x - \boldsymbol{c}(x))(1 - \operatorname{div}\boldsymbol{c}(x) + \mathcal{O})\, dx \\
&= \int_{D_i^k} (\boldsymbol{w} - \nabla \boldsymbol{w} \cdot \boldsymbol{c} + \mathcal{O})(1 - \operatorname{div}\boldsymbol{c}(x) + \mathcal{O})\, dx \\
&= \int_{D_i^k} (\boldsymbol{w} - \boldsymbol{w}\operatorname{div}\boldsymbol{c} - \nabla \boldsymbol{w} \cdot \boldsymbol{c} + (\nabla \boldsymbol{w} \cdot \boldsymbol{c})\operatorname{div}\boldsymbol{c} + \mathcal{O})\, dx \\
&\approx \int_{D_i^k} (\boldsymbol{w} - \operatorname{div}(\boldsymbol{w}\boldsymbol{c}))\, dx \\
&= \int_{D_i^k} \boldsymbol{w}\, dx - \int_{\partial D_i^k} \boldsymbol{w} c_n\, dS,
\end{aligned} \tag{14}$$

where $\mathcal{O}$ denotes the generic higher order terms that, together with $(\nabla \boldsymbol{w} \cdot \boldsymbol{c})\operatorname{div}\boldsymbol{c}$, are neglected in (14). In (14) $c_n = \boldsymbol{c} \cdot \boldsymbol{n}$, $\boldsymbol{n}$ being the unit outer normal to ∂D_i^k. The passage to volume averages and the approximation of the surface integral in (14) leads to the following formula for the evaluation of $\widetilde{\boldsymbol{w}}_i^k$ satisfying the geometric conservation law (11)

$$|D_i^{k+1}|\widetilde{\boldsymbol{w}}_i^k = |D_i^k|\boldsymbol{w}_i^k - \sum_{j \in s(i)} |\Gamma_{ij}| \left(c_{n_{ij}}^+ \boldsymbol{w}_i^k + c_{n_{ij}}^- \boldsymbol{w}_j^k \right). \tag{15}$$

Here $s(i)$ is the set of neigbouring finite volumes to D_i^k, $\Gamma_{ij} = \partial D_i^k \cap \partial D_j^k$, $|\Gamma_{ij}|$ is the N-dimensional measure of Γ_{ij}, $\boldsymbol{n}_{ij}$ denotes the unit outer normal to ∂D_i^k on Γ_{ij} and $\pm$ denotes the positive ($\geq$) and negative ($\leq$) part of a scalar quantity, respectively. The constants $c_{n_{ij}}$ in (15) are evaluated at centers of

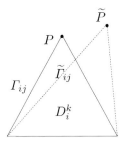

Fig. 2. $P \to \widetilde{P}$, $\Gamma_{ij} \to \widetilde{\Gamma}_{ij}$ displacement.

gravity of Γ_{ij}. Fig. 2 illustrates the choice of $\boldsymbol{w}_i^k$ for Γ_{ij}, where $c_{n_{ij}}$ evaluated at the center of gravity is positive. This means that in the situation in Fig. 2

$$\int_{\Gamma_{ij}} \boldsymbol{w} c_n \, dS \approx |\Gamma_{ij}| c_{n_{ij}} \boldsymbol{w}_i^k. \tag{16}$$

Note that the center of gravity of $\widetilde{\Gamma}_{ij} = \partial D_i^{k+1} \cap \partial D_j^{k+1}$ lies in D_i^k. Equivalently, the approximation used in (14) can be expressed as

$$\int_{\Gamma_{ij}} \boldsymbol{w} c_n \, dS \approx \begin{cases} |\Gamma_{ij}| c_{n_{ij}} \boldsymbol{w}_i^k & \text{if} \quad \widetilde{e}_{ij} \in D_i^k, \\ |\Gamma_{ij}| c_{n_{ij}} \boldsymbol{w}_j^k & \text{if} \quad \widetilde{e}_{ij} \in D_j^k, \end{cases} \tag{17}$$

where $\widetilde{e}_{ij}$ denotes the center of gravity of $\widetilde{\Gamma}_{ij}$.

5 Numerical example

We illustrate the proposed algorithm on a 1D example of Burgers equation:

$$\frac{\partial w}{\partial t} + w \frac{\partial w}{\partial x} = 0 \quad \text{in } (0, 2\pi) \times (0, T),$$

$$w(x, 0) = 0.5 + \sin(x) \quad x \in (0, 2\pi),$$

$$w(0, t) = w(2\pi, t) \quad t \in (0, T).$$

The moving of nodes in an anisotropic mesh adaptation framework is used. The evolution of the exact solution (full line) and numerical solution (rectangles) with the underlying finite volume mesh is presented. The results correspond in left-right-down order to the time instants $t = 0.9, 1.2, 1.6, 2$.

The multidimensional example is the subject -matter of the forthcomming paper [7]. Here, the proposed adaptation strategy is applyed in the framework of the ADER higher order scheme for the Euler equations.

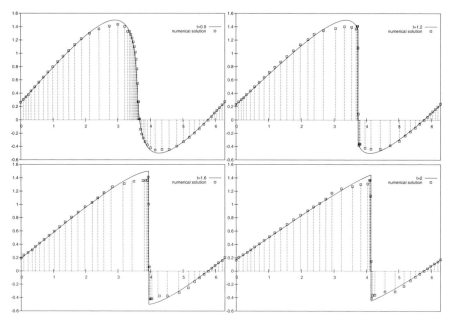

Fig. 3. Evolution of the anisotropic mesh refinement at time $t = 0.9, 1.2, 1.6, 2$.

Acknowledgement

The work is a part of the research project MSM 0021620839 financed by MSMT and partly supported by the grant No. 343/2005/B-MAT/MFF of the Grant Agency of the Charles University.

References

1. Dolejší, V., Felcman, J.: Anisotropic mesh adaptation for numerical solution of boundary value problems. Numerical Methods for Partial Differential Equations, 576–608 (2003)
2. Feistauer, M., Felcman, J., Straškraba, I.: Mathematical and Computational Methods for Compressible Flow. Oxford University Press (2003)
3. Kyncl, M., Felcman, J., Pelant, J.: Numerical boundary conditions for the 3D Euler equations. In: Jonáš, P., Uruba, V. (ed) Proceedings of the Colloquium FLUID DYNAMICS 2004. Institute of Thermomechanics AS CR Prague (2004)
4. Kubera, P., Felcman, J.: Computational aspects of the mesh adaptation. In: Jonáš, P., Uruba, V. (ed) Proceedings of the Colloquium FLUID DYNAMICS 2004. Institute of Thermomechanics AS CR Prague (2004)
5. Štěcha, J.: Optimal decision and control. ČVUT (2003) (in Czech)
6. Tang, H., Tang, T.:Adaptive mesh methods for one- and two-dimensional hyperbolic conservation laws. SIAM J. Appl. Math., 487–515 (2003)
7. Felcman, J., Kubera, P.: Adaptive mesh method for hyperbolic conservative laws. In: Proceedings of the ECT06 Conference. Las Palmas, Spain (2006) (in preparation)

On the Use of Slope Limiters for the Design of Recovery Based Error Indicators

M. Möller[1] and D. Kuzmin[2]

[1] Institute of Applied Mathematics (LS III), University of Dortmund,
Vogelpothsweg 87, D-44227 Dortmund, Germany
matthias.moeller@math.uni-dortmund.de

[2] kuzmin@math.uni-dortmund.de

Summary. A slope limiting approach to the design of recovery based *a posteriori* error indicators for P_1 finite element discretizations is presented. The smoothed gradient field is recovered at edge midpoints by means of limited averaging of adjacent slope values. As an alternative, the constant gradient values may act as upper and lower bounds to be imposed on edge gradients resulting from traditional reconstruction techniques such as averaging projection or discrete patch recovery schemes. In either case, the difference between consistent and reconstructed gradient values measured in the L_2-norm provides a usable indicator for grid adaptivity.

1 Introduction

In a series of recent publications (c.f. [3, 4] and the references therein) an algebraic framework for the construction of high-resolution schemes for convection dominated partial differential equations was developed. The *algebraic flux correction* (AFC) paradigm renders a high-order discretization local extremum diminishing (LED) by applying discrete (anti-)diffusion in a nonlinear conservative fashion. The antidiffusive fluxes are limited node-by-node either by a symmetric FCT limiter or by its upwind-biased counterpart of TVD type.

The adaptive blending of high- and low-order methods prevents us from using error estimators that require an *a priori* knowledge of the order of approximation such as those based on Richardson extrapolation. Gradient recovery techniques [8] seem to be a promising alternative, but their use in error estimation requires that the true solutions be sufficiently smooth.

This paper focuses on hyperbolic problems featuring shocks and discontinuities so that traditional recovery procedures may fail to be reliable. In what follows, limited averaging of consistent slopes is used to compute improved gradient values at midpoints of edges. As an alternative, classical recovery procedures are employed to predict provisional gradient values at edge midpoints to be corrected by means of a slope limiter. The upper and lower bounds to be imposed are given by the constant slopes in two adjacent triangles.

2 A posteriori error indication

As a model problem, consider the weak form of a generic PDE $\mathcal{L}u = f$

$$\int_\Omega w[\mathcal{L}u - f]\,d\mathbf{x} = 0 \tag{1}$$

where the solution is approximated by means of finite elements

$$\mathbf{u} \approx u_h = \sum_j u_j \varphi_j. \tag{2}$$

In this article, we shall concentrate on the numerical error resulting from the *approximation* of spatial derivatives and devise an *a posteriori* indicator for the vector-valued gradient error $\mathbf{e} = \nabla u - \nabla u_h$. In the sequel, the consistent gradient $\nabla u_h = \sum_j u_j \nabla \varphi_j$ will be referred to as low-order gradient.

The aim of recovery based error estimators, introduced by Zienkiewicz and Zhu in [8], is to replace the unknown exact value ∇u by a smoothed gradient field $\hat{\nabla} u_h$, so as to obtain a good approximation to the true error

$$\mathbf{e} \approx \hat{\mathbf{e}} = \hat{\nabla} u_h - \nabla u_h. \tag{3}$$

In general, pointwise error estimates are difficult to obtain, so integral measures are typically employed in the finite element framework. Let Ω_h denote a partition of the domain into a set of non-overlapping elements Ω_e so that the L_2-norm represents a usable measure for the error both globally and locally

$$\|\hat{\mathbf{e}}\|_{L_2}^2 = \sum_{\Omega_e} \|\hat{\mathbf{e}}\|_{L_2(\Omega_e)}^2, \qquad \|\hat{\mathbf{e}}\|_{L_2(\Omega_e)}^2 = \int_{\Omega_e} \hat{\mathbf{e}}^T \hat{\mathbf{e}}\, d\mathbf{x}. \tag{4}$$

We only consider linear (P_1) finite elements for which the consistent gradient ∇u_h is piecewise constant on each triangle. Hence, the improved slopes should be at least piecewise linear so as to provide a better approximation to the exact gradient. It suffices to specify slope values at all midpoints of edges, i.e., $\mathbf{x}_{ij} := \frac{1}{2}(\mathbf{x}_i + \mathbf{x}_j)$, to obtain a smoothed quantity $\hat{\nabla} u_h$ that varies linearly in Ω_e and is allowed to exhibit jumps across interelement boundaries. This approach can be seen as determining the nodal values for a non-conforming approximation of $\hat{\nabla} u_h$ by means of linear Crouzeix-Raviart finite elements for which the local degrees of freedom are located on edge midpoints. For bilinear finite elements used on quadrilateral meshes, the gradient approximation can be based on the nonconforming Rannacher Turek element.

Let (4) be integrated via the second order accurate quadrature rule

$$\int_{\Omega_e} \hat{\mathbf{e}}^T \hat{\mathbf{e}}\, d\mathbf{x} = \frac{|\Omega_e|}{3} \sum_{ij} \hat{\mathbf{e}}_{ij}^T \hat{\mathbf{e}}_{ij}, \qquad \hat{\mathbf{e}}_{ij} = \hat{\nabla} u_{ij} - \nabla u_{ij}, \tag{5}$$

where $|\Omega_e|$ stands for the element area and all quantities are evaluated at the midpoints of surrounding edges indicated by subscript ij. It remains to devise a procedure for constructing an improved gradient value $\hat{\nabla} u_{ij}$ for edge **ij**.

3 Limited gradient averaging

Our first approach to obtaining a smoothed edge gradient is largely inspired by slope limiting techniques employed in the context of high-resolution finite volume schemes and later carried over to discontinuous Galerkin finite element methods. For simplicity, let us illustrate the basic ideas for a one-dimensional finite volume discretization. The task is to define a suitable slope value u'_j on the jth interval $I_j = (x_{j-1/2}, x_{j+1/2})$ so as to recover a piecewise linear approximate solution from the mean value $\bar{u}_j$:

$$u_h(x) = \bar{u}_j + u'_j(x - x_j), \qquad \forall x \in I_j. \tag{6}$$

In the simplest case, one-sided or centered slopes can be utilized to obtain first- and second-order accurate schemes which lead to rather diffusive profiles and are quite likely to produce nonphysical oscillations in the vicinity of steep fronts and discontinuities, respectively. For a numerical scheme to be nonoscillatory, it should possess certain properties [3], e.g., be monotone, positivity preserving, total variation diminishing or satisfy the LED condition.

To this end, Jameson [2] introduced a family of limited average operators $\mathcal{L}(a, b)$ which are characterized by the following properties:

P1. $\mathcal{L}(a, b) = \mathcal{L}(b, a)$.
P2. $\mathcal{L}(ca, cb) = c\mathcal{L}(a, b)$.
P3. $\mathcal{L}(a, a) = a$.
P4. $\mathcal{L}(a, b) = 0$ if $ab \leq 0$.

While conditions P1–P3 are natural properties of an average, P4 is to be enforced by means of a limiter function. It has been demonstrated [2] that a variety of standard TVD limiters can be written in such form. Let the modified sign function be given by $\mathcal{S}(a, b) = \frac{1}{2}(\text{sign}(a) + \text{sign}(b))$ which equals zero for $ab \leq 0$ and returns the common sign of a and b otherwise. Then the most widely used two parameter limiters for TVD schemes can be written as:

1. minmod: $\mathcal{L}(a, b) = \mathcal{S}(a, b) \min\{|a|, |b|\}$
2. maxmod: $\mathcal{L}(a, b) = \mathcal{S}(a, b) \max\{|a|, |b|\}$
3. MC: $\mathcal{L}(a, b) = \mathcal{S}(a, b) \min\{\frac{1}{2}|a + b|, 2|a|, 2|b|\}$
4. superbee: $\mathcal{L}(a, b) = \mathcal{S}(a, b) \max\{\min\{2|a|, |b|\}, \min\{|a|, 2|b|\}\}$

Finally, the limited counterpart of u'_j in (6) can be computed as follows

$$\hat{u}'_j = \mathcal{L}\left(\frac{\bar{u}_{j-1} - \bar{u}_j}{x_{j-1} - x_j}, \frac{\bar{u}_{j+1} - \bar{u}_j}{x_{j+1} - x_j}\right). \tag{7}$$

Let us return to our original task that requires the reconstruction of solution gradients at edge midpoints. This is where the benefits of an edge based formulation come into play. Except at the boundary, *exactly* two elements are adjacent to edge **ij** such that an improved gradient can be determined efficiently from the constant slopes to the left and to the right as follows:

$$\hat{\nabla} u_{ij} = \mathcal{L}(\nabla u_{ij}^+, \nabla u_{ij}^-). \tag{8}$$

For all limiter functions $\mathcal{L}$ presented above, the recovered gradient value equals zero if $\nabla u_{ij}^+ \nabla u_{ij}^- \leq 0$ and satisfies the following inequality otherwise

$$\nabla u_{ij}^{\min} \leq \hat{\nabla} u_{ij} \leq \nabla u_{ij}^{\max}, \quad \text{where} \quad \nabla u_{ij}^{\max}{}_{\min} = \max_{\min} \{\nabla u_{ij}^+, \nabla u_{ij}^-\}. \tag{9}$$

If the upper and lower bounds have different signs, this indicates that the approximate solution attains a local extremum across the edge. Hence, property P4 of limited average operators acts as a discrete analog to the necessary condition in the continuous case which requires the derivative to be zero.

Clearly, the recovered gradient (8) depends on the choice of the limiter function to some extent. In the authors' experience, MC seems to be a safe choice as it tries to select the standard average whenever possible without violating the natural bounds provided by the low-order slopes.

4 Limited gradient reconstruction

As an alternative to the limited averaging approach, traditional recovery procedures can be used to *predict* provisional gradient values at edge midpoints which are *corrected* by edgewise slope limiting so as to satisfy the geometric constraints defined in (9). Since the advent of recovery based schemes [8], a family of *averaging projection* schemes has been proposed in the literature to construct a smoothed gradient from the finite element solution as follows

$$\hat{\nabla} u_h = \sum_j \hat{\nabla} u_j \phi_j, \tag{10}$$

where the coefficients $\hat{\nabla} u_j$ are obtained by solving the discrete problem

$$\int_\Omega \phi_i (\hat{\nabla} u_h - \nabla u_h) \, d\mathbf{x} = 0. \tag{11}$$

Note that the element shape functions used to construct the basis functions ϕ_i may by different from those used in the finite element approximation (2). A detailed analysis by Ainsworth et al. [1] reveals that the corresponding polynomial degrees should satisfy $\deg \phi \geq \deg \varphi$ whereby the original choice $\phi = \varphi$ proposed in [8] 'is not only effective, but also the most economical' [1] one. The substitution of equation (10) into (11) yields a linear algebraic system for each component of the smoothed gradient

$$M_C \hat{\nabla} u_h = \mathbf{C} u. \tag{12}$$

The consistent mass matrix $M_C = \{m_{ij}\}$ and the matrix of discretized spatial derivatives $\mathbf{C} = \{\mathbf{c}_{ij}\}$ are assembled from the following integral terms

$$m_{ij} = \int_\Omega \phi_i \phi_j \, d\mathbf{x}, \qquad \mathbf{c}_{ij} = \int_\Omega \phi_i \nabla \varphi_j \, d\mathbf{x}. \tag{13}$$

For a fixed mesh, the coefficients m_{ij} and $\mathbf{c}_{ij}$ remain unchanged throughout the simulation and, consequently, need to be evaluated just once at the beginning of the simulation and each time the grid has been modified. In case $\phi \equiv \varphi$, the coefficients defined in (13) coincide with the matrix entries of the finite element approximation and, hence, are available at no additional costs.

An edge-by-edge assembly of the right-hand side is also feasible

$$(\mathbf{C} u)_i = \sum_{j \neq i} \mathbf{c}_{ij}(u_j - u_i) \tag{14}$$

since $\mathbf{C}$ features the zero row sum property $\sum_j \mathbf{c}_{ij} = 0$ as long as the sum of basis functions equals one. The solution to system (12) can be computed iteratively by successive approximation preconditioned by the lumped mass matrix $M_L = \text{diag}\{m_i\}$, where $m_i = \sum_j m_{ij}$, as follows:

$$\hat{\nabla} u_h^{(m+1)} = \hat{\nabla} u_h^{(m)} + M_L^{-1}[\mathbf{C} u - M_C \hat{\nabla} u_h^{(m)}], \qquad m = 0, 1, 2, \ldots. \tag{15}$$

If mass lumping is applied directly to equation (12), the values of the projected gradient can be determined at each node from the explicit formula

$$\hat{\nabla} u_i = \frac{1}{m_i} \sum_{j \neq i} \mathbf{c}_{ij}(u_j - u_i). \tag{16}$$

From the nodal values obtained either from (12) or (16), provisional slopes at edge midpoints can be interpolated according to equation (10). For linear finite elements this corresponds to taking the mean values for each edge $\mathbf{ij}$, i.e., $\hat{\nabla} u_h(\mathbf{x}_{ij}) := \frac{1}{2}(\hat{\nabla} u_i + \hat{\nabla} u_j)$. It follows from (10) and (11) that it is also feasible to project the low-order gradient ∇u_h into the space of non-conforming (bi-)linear finite element by letting $\phi_j \in \widetilde{P}_1$ or $\widetilde{Q}_1$, respectively, so as to obtain its smoothed counterpart directly at edge midpoints.

Over the years, a more accurate patch recovery technique (SPR) was introduced [9] which relies on the superconvergence property of the finite element solution at some exceptional, yet *a priori* known, points. Let the smoothed gradient be represented in terms of a polynomial expansion of the form

$$\hat{\nabla} u_h = p(\mathbf{x})\mathbf{a} \tag{17}$$

where the row vector $p(\mathbf{x})$ contains all monomials of degree k at most. Since each vertex, say i, is surrounded by a patch of elements sharing this node, the vector of coefficients $\mathbf{a}$ can be computed from a discrete least square fit to the set $\mathcal{S}_i$ of sampling points $\mathbf{x}_j$ [9]. As a consequence, the multicomponent quantity $\mathbf{a}$ can be determined by solving the linear system

$$M_p \mathbf{a} = \mathbf{f}, \qquad (18)$$

where the local matrix M_p and the right-hand side vector $\mathbf{f}$ are given by

$$M_p = \sum_{j \in \mathcal{S}_i} p^\mathrm{T}(\mathbf{x}_j)\, p(\mathbf{x}_j), \qquad \mathbf{f} = \sum_{j \in \mathcal{S}_i} p^\mathrm{T}(\mathbf{x}_j)\, \nabla u_h(\mathbf{x}_j). \qquad (19)$$

For linear elements, $p(\mathbf{x}) = [1, x, y]$ and the low-order gradient is sampled at the centroids of triangles in the patch. In this case the lumped L_2-projection yields almost the same results on uniform grids but only patch recovery retains its superconvergence property if the grid becomes increasingly distorted.

Regardless of which procedure is employed to predict the high-order gradient values, it may fail if the solution exhibits jumps or the gradient is too steep. This can be attributed to the fact that the averaging process extends over an *unsettled* number of surrounding element gradients which may strongly vary in magnitude and even possess different signs. Thus, it is very difficult the find admissible bounds to be imposed on such *nodal* gradients. The transition to an edge based formulation makes it possible to correct the provisional values according to the constraints (9), set up by the low-order slopes, such that

$$\nabla u_{ij}^{\min} \leq \hat{\nabla} u_{ij} \leq \nabla u_{ij}^{\max}. \qquad (20)$$

It is also advisable to enforce the sign-preserving property (P4) of limited average operators so as to mimic the necessary condition of a local extremum attained across edge $\mathbf{ij}$ in the discrete context. Let $s_{ij} := \mathcal{S}(\nabla u_{ij}^{\min}, \nabla u_{ij}^{\max})$, then the corrected slope values $\hat{\nabla} u_{ij}^*$ can be computed as follows:

$$\hat{\nabla} u_{ij}^* = s_{ij} \left| \max\{ \nabla u_{ij}^{\min}, \min\{ \hat{\nabla} u_{ij}, \nabla u_{ij}^{\max} \}\} \right| \qquad (21)$$

The generality of this predictor-corrector *edgewise limited recovery* (ELR) approach, enables us to use arbitrary reconstruction techniques in the prediction step, e.g., polynomial preserving recovery (PPR) [6] schemes or some recent 'meshless' variants which have been presented by Zhang et al. [7].

5 Adaptation strategy

In adaptive solution procedures for steady state simulations of hyperbolic flows, one typically starts with a moderately coarse grid on which an initial solution can be computed efficiently. Nevertheless, the mesh needs to be fine enough in order to capture all essential flow features in the solution and to enable the error indicator to detect 'imperfect' zones. Next, the grid is locally refined or coarsened according to some adaptation parameter and the whole process is repeated until (ideally) the global relative percentage error

$$\eta := \frac{||\mathbf{e}||_{L_2}}{||\nabla u||_{L_2}} \leq \eta_{\text{tol}} \qquad (22)$$

is below the prescribed tolerance η_{tol}. Replacing the unknown exact quantities by their approximate values and assuming that the relative error is distributed equally between cells the gradient error for each element Ω_e should not exceed

$$||\hat{\mathbf{e}}||_{L_2(\Omega_e)} \leq \eta_{\text{tol}} \left[\frac{||\nabla u_h||_{L_2}^2 + ||\hat{\mathbf{e}}||_{L_2}^2}{|\Omega_h|} \right]^{1/2}, \qquad (23)$$

where $|\Omega_h|$ represents the number of employed elements. Depending on the ratio of estimated and tolerated error, cells are flagged for refinement or coarsening. For a detailed presentation of the grid adaptation procedure including some grid improvement techniques the interested reader is referred to [5].

6 Numerical examples

Let us illustrate the performance of the new algorithm by considering a supersonic flow which enters a converging channel at $M_\infty = 2$. The bottom wall is sloped at 5° which gives rise to the formation of multiple shock reflections. The initial mesh consists of 60×16 quadrilaterals each of which is divided into two triangles. After three sweeps of local mesh refinement ($\eta_{\text{ref}} = 1\%$) and coarsening ($\eta_{\text{crs}} = 0.1\%$) governed by the MC-limited averaging error indicator, the zone of highest grid point concentration confines itself more and more to the vicinity of the shock as depicted in Figure 1. Algebraic flux correction of TVD type [4] was employed to compute the solution, making use of the moderately diffusive CDS-limiter applied to the characteristic variables.

The density distribution for the finest grid (15,664 elements) demonstrates the precise separation into five zones of uniform flow. The crisp resolution of the reflected shock wave can also be observed by considering the density 'cascade' drawn along the straight line $y = 0.6$ for all four grid levels.

7 Conclusions

Slope limiting techniques provide a valuable tool for the construction of high-resolution gradient recovery procedures. Improved slopes can be directly computed at edge midpoints as a limited average of adjacent low-order gradients. Moreover, the consistent slope values serve as natural upper and lower bounds to be imposed on any edge gradient. In addition, traditional (nodal) recovery procedures can be used to predict the high-order gradient which is corrected according to geometric constraints by invoking a slope limiter edge-by-edge.

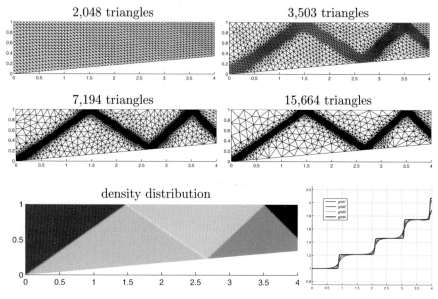

Fig. 1. 5° converging channel at $M_\infty = 2$

References

1. Ainsworth, M., Zhu, J.Z., Craig, A.W., Zienkiewicz, O.C.: Analysis of the Zienkiewicz-Zhu a-posteriori error estimator in the finite element method. Int. J. Numer. Meth. Engng. **28**, 2161–2174 (1989)
2. Jameson, A.: Analysis and design of numerical schemes for gas dynamics 1. Artificial diffusion, upwind biasing, limiters and their effect on accuracy and multigrid convergence. Int. Journal of CFD **4**, 171–218 (1995)
3. Kuzmin, D., Möller, M.: Algebraic flux correction I. Scalar conservation laws. In: D. Kuzmin, R. Löhner, S. Turek (eds.) Flux-Corrected Transport: Principles, Algorithms, and Applications. Springer, 155–206 (2005)
4. Kuzmin, D., Möller, M.: Algebraic flux correction II. Compressible Euler equations. In: D. Kuzmin, R. Löhner, S. Turek (eds.) Flux-Corrected Transport: Principles, Algorithms, and Applications. Springer, 207–250 (2005)
5. Kuzmin, D., Möller, M.: Adaptive mesh refinement for high-resolution finite element schemes. Submitted to: Int. J. Numer. Meth. Fluids.
6. Naga, A., Zhang, Z.: A Posteriori error estimates based on polynomial p reserving recovery. SIAM J. Numer. Anal. **42**, 1780–1800 (2004)
7. Naga, A., Zhang, Z.: A new finite element gradient recovery method: Sup erconvergence property. *SIAM J. Sci. Comput.* **26**, 1192–1213 (2005)
8. Zienkiewicz, O.C., Zhu, J.Z.: A simple error estimator and adaptive procedure for practical engineering analysis. Int. J. Numer. Methods Eng. **24**, 337–357 (1987)
9. Zienkiewicz, O.C., Zhu, J.Z.: The superconvergent patch recovery and a posteriori error estimates. Part 1: The recovery techniques. Int. J. Numer. Methods Eng. **33**, 1331–1364 (1992)

A Posteriori Error Estimation

On a Superconvergence Result for Mixed Approximation of Eigenvalue Problems

Francesca Gardini

Department of Numerical Analysis, Universität Ulm, Helmholtzstr. 18, D-89069 Ulm, Germany,
francesca.gardini@uni-ulm.de

Summary. We state a superconvergence result for the lowest order Raviart-Thomas approximation of eigenvalue problems. Numerical experiments confirm the superconvergence property and suggest that it holds also for the lowest order Brezzi-Douglas-Marini approximation.

1 Introduction

This paper deals with a superconvergence result for mixed approximation of eigenvalue problems. It is well-known that a superconvergence property holds for the mixed approximation of Laplace problem, provided the solution is smooth enough (see [6]). Nevertheless, the proof given by Brezzi and Fortin [6] cannot be generalised in a easy way to eigenvalue problems. Indeed, the key point of the proof strongly relies on the Galerkin orthogonality, which holds for the source problem but not for the eigenvalue one.

In order to prove the superconvergence property we will use the equivalence between the lowest order Raviart-Thomas (RT_0) approximation of Laplace eigenproblem with Neumann boundary condition and the non-conforming piecewise linear Crouziex-Raviart approximation (see [3]). We will also make use of a superconvergence result proved by Durán et al. in [8] for Laplace eigenproblem with Dirichlet boundary condition.

An outline of the paper is as follows. Section 2 is devoted to the mathematical formulation of the model eigenvalue problem and to its mixed lowest order Raviart-Thomas approximation. In Sect. 3 we recall the equivalence with a non-conforming approximation and we state the main results concerning the superconvergence property. Finally, in Sect. 4 we report the results of some numerical experiment, which confirm the supercorvergece property in the case of regular eigenmode. Moreover, we investigate numerically if the superconvergence property holds for the lowest order Brezzi-Douglas-Marini (BDM_1) space as well.

2 Statement of the problem and its discretization

Let $\Omega \subset \mathbb{R}^d$ ($d = 2, 3$) be a simply connected bounded polygonal or polyhedral domain. We consider the following eigenvalue problem:

$$\text{find } \lambda \in \mathbb{R} \text{ s.t. there exists } \varphi \neq 0: \begin{cases} -\Delta \varphi = \lambda \varphi & \text{in } \Omega, \\ \dfrac{\partial \varphi}{\partial n} = 0 & \text{on } \partial \Omega, \end{cases} \qquad (1)$$

where n denotes the outward normal unit vector.

For easy presentation we shall develop the analysis in two dimensions, being the extension to three dimensions straightforward.

We shall use the standard notation for the Sobolev spaces $H^m(\Omega)$, their norms $\|\cdot\|_m$, and seminorms $|\cdot|_m$ (see [1]). As usual we denote by $(\cdot, \cdot)$ the L^2-inner product.

Introducing $\boldsymbol{\sigma} = \nabla \varphi$, we obtain the usual mixed formulation of problem (1) which in weak form is given by

$$\text{find } \lambda \in \mathbb{R} \text{ s.t. there exist } (\boldsymbol{\sigma}, \varphi) \in H_0(\text{div}, \Omega) \times L_0^2(\Omega), \text{ with } \varphi \neq 0: \begin{cases} (\boldsymbol{\sigma}, \boldsymbol{\tau}) + (\text{div } \boldsymbol{\tau}, \varphi) = 0 & \forall \boldsymbol{\tau} \in H_0(\text{div}, \Omega), \\ (\text{div } \boldsymbol{\sigma}, \psi) = -\lambda(\varphi, \psi) & \forall \psi \in L_0^2(\Omega), \end{cases} \qquad (2)$$

where $L_0^2(\Omega)$ is the space consisting of square Lebesgue-integrable functions having zero mean value and

$$H_0(\text{div}, \Omega) = \{ \mathbf{v} \in L^2(\Omega)^2 : \text{div } \mathbf{v} \in L^2(\Omega) \text{ and } \mathbf{v} \cdot \mathbf{n} = \mathbf{0} \text{ on } \partial \Omega \}$$

is endowed with the usual norm $\|\mathbf{v}\|_{\text{div}}^2 = \|\mathbf{v}\|_0^2 + \|\text{div } \mathbf{v}\|_0^2$. Here and thereafter conditions of the type $v = 0$ on $\partial \Omega$ are to be understood in the sense of traces (see [13]).

It is well-known that problem (2) admits a countable set of real and positive eigenvalues, which can be ordered in an increasing divergent sequence and the corresponding eigenfunctions give rise to an orthonormal basis of $L^2(\Omega)^2$. Moreover each eigenspace is finite dimensional. Finally, due to regularity results (see [12]), there exists a constant $s \in (\frac{1}{2}, 1]$ (depending on Ω), such that $(\boldsymbol{\sigma}, \varphi)$ belongs to the space $H^s(\Omega)^2 \times H^{1+s}(\Omega)$. Moreover, the following estimate holds true:

$$\|\mathbf{u}\|_s + \|\text{div } \mathbf{u}\|_{1+s} \leq C \|\mathbf{u}\|_0, \qquad (3)$$

where C is a constant depending on the eigenvalue λ. Here $s = 1$ if Ω is convex and $s = \pi/\omega - \varepsilon$ for a nonconvex domain, $\omega < 2\pi$ being the maximum interior angle of Ω.

Let $\{\mathcal{T}_h\}$ denote a shape-regular family (i.e., satisfying the minimum angle condition, see [7]) of simplicial decomposition of Ω. As usual we require that any two elements in $\mathcal{T}_h$ share at most a common edge or a common vertex, and we denote by h the maximum diameter of the elements K in $\mathcal{T}_h$.

The lowest order Raviart-Thomas space is defined (see [6]) on each element K as
$$RT_0(K) = \mathbb{P}_0(K)^2 + (x_1, x_2)\mathbb{P}_0(K), \tag{4}$$
where $\mathbb{P}_k(K)$ denotes the space of polynomials of degree at most k on K.

Setting
$$\Sigma_h = \{\tau \in H_0(\mathrm{div}, \Omega) : \tau|_K \in RT_0(K) \ \forall K \in \mathcal{T}_h\}$$
and
$$\Phi_h = \{\psi \in L_0^2(\Omega) : \psi|_K \in \mathbb{P}_0(K) \ \forall K \in \mathcal{T}_h\},$$
then the mixed finite element approximation of problem (2) reads

find $\lambda_h \in \mathbb{R}$ s.t. there exist $(\boldsymbol{\sigma_h}, \varphi_h) \in \Sigma_h \times \Phi_h$, with $\varphi_h \neq 0$:
$$\begin{cases} (\boldsymbol{\sigma_h}, \boldsymbol{\tau}) + (\mathrm{div}\,\boldsymbol{\tau}, \varphi_h) = 0 & \forall \boldsymbol{\tau} \in \Sigma_h, \\ (\mathrm{div}\,\boldsymbol{\sigma_h}, \psi) = -\lambda_h(\varphi_h, \psi) & \forall \psi \in \Phi_h, \end{cases} \tag{5}$$

Assume for simplicity that λ is a simple eigenvalue. Then, taking $\|\boldsymbol{\sigma}\|_0 = \|\boldsymbol{\sigma}_h\|_0 = 1$, it follows from the abstract theory (see [4, 5]) and known a priori error estimates that for h small enough (depending on λ)
$$\|\boldsymbol{\sigma} - \boldsymbol{\sigma}_h\|_{\mathrm{div}} = O(h^t), \tag{6}$$
$$|\lambda - \lambda_h| = O(h^{2t}), \tag{7}$$
where $t = \min\{1, s\}$.

3 Main results

Following the arguments given in [3, 14], it can be seen that problem (5) is equivalent to a nonconforming approximation of the standard formulation of (1). Let us introduce the nonconforming space of Crouxiez and Raviart enriched by local bubbles. Denoting by $\mathcal{B}(K)$ the space of cubic polynomials vanishing on ∂K, we define

$$\begin{aligned} X_h &= \{\phi \in L^2(\Omega) : \phi|_K \in \mathbb{P}_1(K) \ \ \forall K \in \mathcal{T}_h, \phi \text{ is continuous at interior midpoints}\}, \\ B_h &= \{b \in H_0^1(\Omega) : b|_K \in \mathcal{B}(K) \ \ \forall K \in \mathcal{T}_h\}, \\ W_h &= X_h \oplus B_h. \end{aligned} \tag{8}$$

Let
$$\Sigma_h^d = \{\mathbf{v} \in L^2(\Omega)^2 : \mathbf{v}|_K \in RT_0(K) \ \ \forall K \in \mathcal{T}_h\}.$$
We also introduce the following L^2-projection operator:
$$\begin{aligned} P_{\Sigma_h^d} &: L^2(\Omega)^2 \longrightarrow \Sigma_h^d \\ P_{\Sigma_h^d}\mathbf{v} &\in \Sigma_h^d \text{ such that} (\mathbf{v} - P_{\Sigma_h^d}\mathbf{v}, \mathbf{v}_h) = 0 \ \ \forall\, \mathbf{v}_h \in \Sigma_h^d. \end{aligned} \tag{9}$$

In the following we denote by $\nabla_h \psi_h$ the elementwise gradient of ψ_h. Then problem (5) is equivalent to the following one:

find $\lambda_h \in \mathbb{R}$ s.t. there exists $\phi_h \in W_h$, with $\phi_h \neq 0$ s.t.
$$(P_{\Sigma_h^d}(\nabla_h \phi_h), \nabla_h \psi_h) = \lambda_h (P_h \phi_h, \psi_h) \quad \forall \psi_h \in W_h, \tag{10}$$

in the sense that they have the same eigenvalues λ_h and the eigenvectors are related by $\sigma_h = P_{\Sigma_h^d}(\nabla_h \phi_h)$ and $\varphi_h = P_h \phi_h$ (see [3, 14]).

Applying the general theory developed in [4], Durán et al. in [8] proved (for Dirichlet boundary condition) the following result, which can be extended to our problem as stated in [2]:
$$\|\varphi - \bar{\phi}_h\|_0 = O(h^{2t}), \tag{11}$$

where $\bar{\phi}_h$ is a multiple of ϕ_h such that $\|\bar{\phi}_h\|_0 = \|\varphi\|_0$ and $t = \min\{1, s\}$.

We now state a result which is useful in the proof of the superconvergence property.

Proposition 1. *If (λ_h, ϕ_h) is an eigensolution of problem (10). Then the elementwise H^1-seminorm $|\phi_h|_{1,h}$ is bounded.*

Then, the following result which generalises to eigenvalue problem the superconvergence property which holds for the source problem is true.

Theorem 1. *Let (λ, φ) be the eigensolution of Laplace eigenproblem (1) and (λ_h, φ_h) be the corresponding discrete eigenpair of problem (5). Then it holds*
$$\|P_h \varphi - \varphi_h\|_0 = O(h^{2t}),$$
where $t = \min\{1, s\}$.

Remark 1. From Theorem 1 together with the a priori error estimate (6), we get that $\|P_h \varphi - \varphi_h\|_0$ is of higher order than $\|\varphi - \varphi_h\|_0$ and $\|\sigma - \sigma_h\|_0$.

For the details of the analysis and for some possible application of this superconvergence result to a posteriori error estimates we refer to [9, 10, 11] and to forthcoming papers.

4 Numerical results

In this section we present the results of some numerical experiments which, in the case of regular eigenmodes, confirm the superconvergence property stated in the previous section. Moreover, we investigate numerically whether the superconvergence property holds for BDM_1 elements as well.

The lowest order BDM space is defined (see [6]) by
$$BDM_1(K) = \mathbb{P}_1(K)^2.$$

Then the BDM_1 mixed approximation of problem (2) is obtained taking

$$\Sigma_h = \{\tau \in H_0(\mathrm{div}, \Omega) : \tau|_K \in BDM_1(K) \ \forall K \in \mathcal{T}_h\}$$

and

$$\Phi_h = \{\psi \in L_0^2(\Omega) : \psi|_K \in \mathbb{P}_0(K) \ \forall K \in \mathcal{T}_h\},$$

in (5).

Let $(\lambda_h, \boldsymbol{\sigma}_h)$ denote the BDM_1 approximation to (λ, φ) and let us assume $\|\boldsymbol{\sigma}\|_0 = \|\boldsymbol{\sigma}_h\|_0 = 1$. Then, it follows from the abstract theory (see [4, 5]) and known a priori error estimates that for h small enough (depending on λ),

$$\|\boldsymbol{\sigma} - \boldsymbol{\sigma}_h\|_{\mathrm{div}} = O(h^t), \qquad (12)$$
$$\|\boldsymbol{\sigma} - \boldsymbol{\sigma}_h\|_0 = O(h^r), \qquad (13)$$
$$|\lambda - \lambda_h| = O(h^{2t}), \qquad (14)$$

where $r = \min\{2, s\}$, and $t = \min\{1, s\}$.

Remark 2. If the eigenfunction $\boldsymbol{\sigma}$ is smooth enough (i.e. belongs to the space $H(\mathrm{div}, \Omega) \cap H^\alpha(\Omega)$ for some $\alpha > 1$) then, contrary to RT_0 elements, BDM_1 ones provide a L^2-approximation of higher order than the $H(\mathrm{div})$-approximation.

The numerical tests have been performed taking $\Omega = (0, \pi) \times (0, \pi)$. In this case in fact the eigensolutions of Laplace eigenproblem with homogeneous Neumann boundary condition are given by eigenvalues

$$\lambda = n^2 + m^2,$$

with the corresponding eigenfunctions

$$\varphi = \cos(nx)\cos(my),$$

where $n, m \in \mathbb{N}$ are not simultaneously vanishing.

We choose as exact eigenpair $(\lambda, \varphi) = (2, \cos x \cos y)$ and we use RT_0 and BDM_1 as approximation spaces.

We test the superconvergence property on two sequences of meshes, both structured and unstructured as shown in Figs. 1–2. The meshes are obtained from an initial triangulation of the square by uniform refinement, namely subdividing each triangle by joining the midpoints of each edge. The first test concerns RT_0 approximation. In Tables 1–2 we report both the L^2 and the $H(\mathrm{div})$-norms of the error in the approximation of the eigenfunction and we compute the numerical rate of convergence, which is 1 as predicted by the a priori error estimate (6) for regular eigenmodes. We also report the L^2-norm of the error between the projection of the continuous eigenfunction and the discrete one. In this case the order of convergence is 2, as predicted by Theorem 1 for regular eigenmodes. Eventually, in Fig. 3 we plot the above errors using a log/log scale.

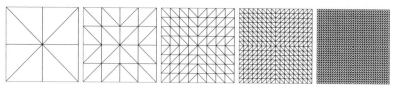

Fig. 1. Structured meshes

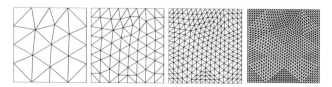

Fig. 2. Unstructured meshes

Table 1. Error table: RT_0 on structured mesh

mesh size	$\|\boldsymbol{\sigma} - \boldsymbol{\sigma}_h\|_0$		$\|\text{div}(\boldsymbol{\sigma} - \boldsymbol{\sigma}_h)\|_0$		$\|P_h\varphi - \varphi_h\|_0$	
	err.	order	err.	order	err.	order
h_0	1.018359		1.894110		0.149686	
$h_0/2$	0.504852	1.01	0.902151	1.07	0.041618	1.84
$h_0/4$	0.251870	1.00	0.450850	1.00	0.010459	1.99
$h_0/8$	0.125915	1.00	0.225469	0.99	0.002625	1.99
$h_0/16$	0.062959	0.99	0.112742	0.99	0.000657	1.99

Table 2. Error table: RT_0 on unstructured mesh

mesh size	$\|\boldsymbol{\sigma} - \boldsymbol{\sigma}_h\|_0$		$\|\text{div}(\boldsymbol{\sigma} - \boldsymbol{\sigma}_h)\|_0$		$\|P_h\varphi - \varphi_h\|_0$	
	err.	order	err.	order	err.	order
h_0	0.583376		0.750504		0.063330	
$h_0/2$	0.294074	0.98	0.383364	0.96	0.016101	1.97
$h_0/4$	0.147403	0.99	0.192606	0.99	0.004053	1.98
$h_0/8$	0.073762	0.99	0.096416	0.99	0.001016	1.99

In the second test we consider BDM_1 finite elements. The results of these experiments are shown in Tables 3–4. As predicted by the a priori error estimates (12) and (13) for regular eigenmodes, the order of convergence is 1 in the $H(\text{div})$-norm and 2 in the L^2-norm. Moreover, the numerical results suggest that the superconvergence property holds as well. Finally, in Fig. 4 we plot the errors in a log/log scale.

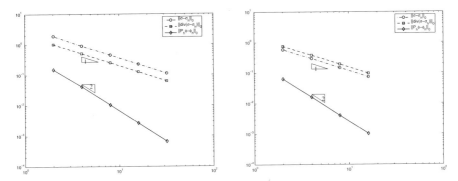

Fig. 3. Errors versus h^{-1} in log/log-scale for RT_0 on structured (left) and unstructured meshes (right)

Table 3. Error table: BDM_1 on structured mesh

mesh size	$\|\sigma - \sigma_h\|_0$		$\|\text{div}(\sigma - \sigma_h)\|_0$		$\|P_h\varphi - \varphi_h\|_0$	
	err.	order	err.	order	err.	order
h_0	0.653615		2.020459		0.057680	
$h_0/2$	0.134918	2.27	0.921033	1.13	0.006028	3.25
$h_0/4$	0.032997	2.03	0.453273	1.02	0.002522	1.25
$h_0/8$	0.008265	1.99	0.225774	1.00	0.000712	1.82
$h_0/16$	0.002130	1.95	0.112780	1.00	0.000183	1.96

Table 4. Error table: BDM_1 on unstructured mesh

mesh size	$\|\sigma - \sigma_h\|_0$		$\|\text{div}(\sigma - \sigma_h)\|_0$		$\|P_h\varphi - \varphi_h\|_0$	
	err.	order.	err.	order	err.	order
h_0	0.083746		0.766087		0.020439	
$h_0/2$	0.022797	1.87	0.385542	0.99	0.006066	1.75
$h_0/4$	0.005837	1.96	0.192885	0.99	0.001578	1.94
$h_0/8$	0.001553	1.91	0.096451	0.99	0.000398	1.98

Possible future developments of the present work go towards the study of the superconvergence property for BDM elements and for RT elements of any order.

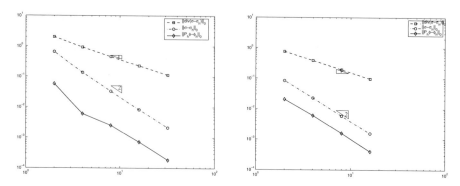

Fig. 4. Errors versus h^{-1} in log/log-scale for BDM_1 on structured (left) and unstructured meshes (right)

References

1. Adams, R. A.: Sobolev spaces. In: Pure and Applied Mathematics, Vol. 65. Academic Press, New York-London (1975)
2. Alonso, A., Dello Russo, A., Vampa, A.: A posteriori error estimates in finite element acoustic analysis. J. Comput. Appl. Math. **117**(2) 105-119 (2000)
3. Arnold, D. N., Brezzi, F.: Mixed and nonconforming finite element methods: implementation, postprocessing and error estimates. RAIRO Modél. Math. Anal. Numér., **19**(1), 7–32 (1985)
4. Babuška, I., Osborn, J.: Eigenvalue Problems. In: Ciarlet, P. G. and Lions, J.L. (eds.) Handbook of Numerical Analysis, Vol. 2. North Holland (1991)
5. Boffi, D., Brezzi, F., Gastaldi, L.: On the convergence of eigenvalues for mixed formulations. Ann. Sc. Norm. Sup. Pisa Cl. Sci., **25** 131–154 (1997)
6. Brezzi, F., Fortin, M.: Mixed and hybrid finite elements methods. In: Springer Series in Computational Mathematics, Vol. 15. Springer-Verlag, New York (1991)
7. Ciarlet, P. G.: The finite element method for elliptic problems. In: Studies in Mathematics and Its Application, Vol. 4. North Holland, Amsterdam (1978)
8. Durán, R., Gastaldi, L., Padra, C.: A posteriori error estimations for mixed approximation of eigenvalue problems. Math. Model. Meth. Appl. Sci. **9**(8), 1165–1178
9. Gardini, F.: A posteriori error estimates for an eigenvalue problem arising from fluid-structure interactions. In: Computational Fluid and Solid Mechanics. Elsevier, Amsterdam (2005)
10. Gardini, F.: A posteriori error estimates for eigenvalue problems in mixed form. Istit. Lombardo Accad. Sci. Lett. Rend. A., **138**, 17–34 (2004)
11. Gardini, F.: A posteriori error estimates for eigenvalue problems in mixed form. PhD Thesis, Università degli Studi di Pavia, Pavia (2005)
12. Grisvard, P.: Elliptic problem in nonsmooth domains. In: Monographs and Studies in Mathematics, Vol. 24. Pitman, Boston (1985)

13. Lions, J.-L., Magenes, E.: Problèmes aux limites non homogènes et applications. In: Travaux et Recherches Mathématiques, Vol. 17. Dunod, Paris (1968)
14. Marini, L. D.: An inexpensive method for the evaluation of the solution of the lowest order Raviart-Thomas mixed method. SIAM J. Numer. Anal., **22**(3), 493–496, (1985)

Comparative Study of the a Posteriori Error Estimators for the Stokes Problem

Elena Gorshkova[1], Pekka Neittaanmäki[1] and Sergey Repin[2]

[1] Department of Mathematical Information Technology, University of Jyväskylä P.O. Box 35, FIN-40014, Finland
egorshko@cc.jyu.fi, pn@mit.jyu.fi

[2] V.A. Steklov Institute of Mathematics, Fontanka 27, 191011, St. Petersburg, Russia
repin@pdmi.ras.ru

Summary. The research presented is focused on a comparative study of a posteriori error estimation methods to various approximations of the Stokes problem. Mainly, we are interested in the performance of functional type a posterior error estimates and their comparison with other methods.

We show that functional type a posteriori error estimators are applicable to various types of approximations (including non-Galerkin ones) and robust with respect to the mesh structure, type of the finite element and computational procedure used. This allows the construction of effective mesh adaptation procedures in all cases considered. Numerical tests justify the approach suggested.

1 Introduction

Reliable methods of numerical modeling are an important and rapidly developing part of modern numerical analysis. In particular, such methods are of utmost significance for the development of the theory of fluids. Most of the a posteriori error estimators are based on a well-known residual method and its modification (see [1, 3, 5, 12]) and averaging techniques (see [3]). These methods use specific features of FEM solution and have certain restrictions in their applicability. First of all, they are valid only for Galerkin approximation of the problem, i.e., for the exact solution of finite dimensional problem. Moreover they depend on the discretization and the type of approximation used. Theoretically they provide an upper bound of the error. However, it requires effective calculation of many local (so-called interpolation) constants. Inaccurate estimation of these constants leads to a major overestimation of the error (see [4] for elliptic equation). Regardless of this fact, they are widely used mainly as error indicators.

In works [9] and [11], a new approach was proposed. Estimators suggested have been derived by the investigation of the respective differential problem on purely functional ground. Therefore, they are applicable for any function from the required functional space.

This paper is devoted to numerical justification of the functional type a posteriori error estimates for the Stokes problem. Some investigation of computational properties of proposed error majorant was made in [7]. The robustness and effectiveness of the numerical approach have been confirmed by series of numerical tests. The results justify not only the effectiveness of a posteriori estimation of the difference between exact and numerical solutions but also give an opportunity to evaluate error distribution in the domain. A posteriori error estimates introduced in this research are valid for a wide class of conforming approximations and do not depend on the method by which a numerical solution was obtained.

2 The stokes problem and its approximation

Let Ω be a bounded domain in $\mathbb{R}^n$, with Lipschitz continuous boundary $\partial\Omega$. Let $\mathbf{f} \in \mathbf{L}^2(\Omega, \mathbb{R}^n)$ be a given vector-valued function. The classical Stokes problem consists in determination a vector-valued function $\mathbf{u}$ (the velocity of the fluid), and a scalar-valued function p (the pressure), which are defined in Ω and satisfy the following equations and boundary conditions:

$$-\nu \triangle \mathbf{u} = \mathbf{f} - \nabla p \quad \text{in } \Omega,$$
$$\text{div } \mathbf{u} = 0 \quad \text{in } \Omega,$$
$$\mathbf{u} = \mathbf{u}_g \quad \text{on } \partial\Omega,$$

where $\nu > 0$ is the kinematic viscosity coefficient.

Two variational formulations of the Stokes problem (see e.g., [6]) can be formulated as follows:

$$\nu \int_\Omega \nabla \mathbf{u} : \nabla \mathbf{v} dx = \int_\Omega (\mathbf{f} - \nabla p) \cdot \mathbf{v}\, dx \quad \forall \mathbf{v} \in \overset{\circ}{\mathbf{W}}{}^1_2(\Omega, \mathbb{R}^n), \tag{1}$$

$$-\int_\Omega q\, \text{div } \mathbf{u} = 0 \quad \forall q \in \overset{\circ}{\mathbf{L}}_2(\Omega) = \{q \in \mathbf{L}_2(\Omega) |\int_\Omega q\, dx = 0\} \tag{2}$$

and

$$\nu \int_\Omega \nabla \mathbf{u} : \nabla \mathbf{v} dx = \int_\Omega \mathbf{f} \cdot \mathbf{v}\, dx \quad \forall \mathbf{v} \in \overset{\circ}{\mathbf{J}}{}^1_2(\Omega, \mathbb{R}^n), \tag{3}$$

where $\overset{\circ}{\mathbf{J}}{}^1_2(\Omega, \mathbb{R}^n)$ - is the closure of the set $\overset{\circ}{\mathbf{J}}{}^\infty(\Omega, \mathbb{R}^n)$ in the norm of space $\mathbf{W}^1_2(\Omega, \mathbb{R}^n)$:

$$\overset{\circ}{\mathbf{J}}{}^\infty(\Omega, \mathbb{R}^n) = \{\mathbf{v} \in C^\infty_0(\Omega, \mathbb{R}^n) \mid \text{div } \mathbf{v} = 0, \text{supp } \mathbf{v} \subset\subset \Omega\}. \tag{4}$$

In the first variational formulation (1), (2), test functions are taken from the space $\overset{\circ}{\mathbf{W}}{}^1_2(\Omega, \mathbb{R}^n)$, in the second (3), from the space of divergence-free function.

These approximations lead to different methods for solving the Stokes problem, including method of stream function, penalty method, Uzawa algorithm etc.

The approximations of the Stokes problem can be divided into three groups: (a) fully-conforming approximation (exactly fulfilling the incompressibility condition); (b) conforming approximation (approximation from energy class, but without divergence-free property, e.g. Taylor-Hood approximations, mini-elements, macro-elements (see [6] for explanation)); (c) non-conforming approximations (approximations which do not belong to the energy space e.g., Crouzeix-Raviart approximation).

It is important to note that the functional type a posteriori error estimates considered in this paper depend only on the type of approximation, but not on the mesh, method used and other properties.

3 Estimation of the deviation from the exact solution

Consider v to be some approximate solution of the Stokes problem obtained by any numerical method. Then (see [11]) the difference between it and the exact solution can be estimated as follows:

$$\nu \parallel \nabla(\mathbf{u}-\mathbf{v}) \parallel \leq \parallel \nu\nabla\mathbf{v}-\tau \parallel + C_\Omega \parallel \operatorname{div} \tau + \mathbf{f} - \nabla q \parallel + \frac{2}{\mathcal{C}_{LBB}}\nu \parallel \operatorname{div} \mathbf{v} \parallel . \quad (5)$$

Here and later on we call the right-hand side of (5) *the functional type error majorant*. This estimator is valid for any tensor-function $\tau \in \{\mathbf{L}_2(\Omega, \mathbb{M}^{n \times n}) \mid \operatorname{div} \tau \in \mathbf{L}_2(\Omega, \mathbb{R}^n)\}$ (by $\mathbb{M}^{n \times n}$ we denote the space of real symmetric $n \times n$ matrices) and scalar-function $q \in \overset{\circ}{L}_2(\Omega) \cap W_2^1(\Omega)$, which can be chosen in order to minimize the right-hand side of (5). The constant C_Ω comes from Friedrichs-Poincaré inequality, $\mathcal{C}_{LBB}$ is the constant that appears in Ladyzhenskaya-Babuška-Brezzi inequality (inf-sup inequality).

$$\inf_{\phi \in \overset{\circ}{L}_2(\Omega); \ \phi \neq 0} \sup_{\mathbf{w} \in \overset{\circ}{W}_2^1(\Omega, \mathbb{R}^n); \ \mathbf{w} \neq 0} \frac{\int_\Omega \phi \operatorname{div} \mathbf{w} \, dx}{\parallel \phi \parallel \parallel \nabla\mathbf{w} \parallel} \geq \mathcal{C}_{LBB}.$$

The functional in the right-hand side of (5) has a clear physical meaning. It represents a linear combination of the error in constitutive law, residual error and error in incompressibility condition. We refer to the first term of (5) as the primary term. It dominates in the whole functional (5) and shows the distribution of the error over the domain. The second term of (5) is the reliability term. When the error majorant is closed to the true error, it is closed to 0. The third term is called the div-term. See [7] and [11] for more details.

For fully-conforming approximations in view of fulfilling the incompressibility condition, the error estimate has a more simple form:

$$\nu \parallel \nabla(\mathbf{u}-\mathbf{v}) \parallel \leq \parallel \nu\nabla\mathbf{v} - \tau \parallel + C_\Omega \parallel \mathrm{div}\ \tau + \mathbf{f} - \nabla q \parallel . \qquad (6)$$

The main advantage of this form is the following: it does not require the value of C_{LBB}, which estimation is a very important, but a separate problem in modern applied mathematics. Both estimates (5) and (6) are valid only for the conforming approximations. Thus, for non-conforming approximation (Crouzeix-Raviart elements) a special algorithm was constructed, that projects the approximate solution obtained to the space of divergence-free functions (see [8]). This projection can be considered as a new approximation. For error estimation, error majorant in the form (6) can be used. Note that each term can be computed directly and that this method does not require the value of C_{LBB}.

The error majorant satisfies the following very important property: it is "exact". This means that it is consistent and allows getting a very sharp estimation of the error. In fact, by substituting in (6) the gradient of the exact solution as τ and exact pressure as q the error is equal to the error estimation. This property does not depend on mesh and characteristics of the approximate solution.

Other a posteriori error estimators are exact only asymptotically for $h \to 0$ and only for the Galerkin approximations. For this reason they cannot provide "exactness property" on any particular mesh and for any approximation.

4 Numerical experiments

In this section, we consider two boundary–value problems whose exact solutions are known. They are solved numerically and the respective approximation errors are estimated by means of the method described above. The error estimation results are compared with exact values of the errors. In this analysis we pay a major attention to two points: (a) the quality of the error estimation in the global (energy) norm and (b) the quality of local error estimation performed either by the error indicator that comes from the global error majorant or by the local error estimation techniques. The latter information is used for the element marking and further mesh refinement. In the experiments we mainly used the following (rather typical) adaptation criterion: *an element is to be refined if the error is bigger than one half of the maximum error.* In this rule, by the "error on a triangle" we mean the contribution of this element to the overall error (or error estimator). Those elements that provide local contributions higher than the mean value are subject to the refinement together with certain neighbor elements, which is necessary to avoid hanging nodes (the so–called red-green-blue refinement, see, e.g., [12]).

Certainly, the best possible adaptive algorithm can be constructed on the basis of the true error distribution obtained by comparing the true and approximate solutions. Since in our examples the true solutions are known, we can compute that distribution and use it as an "etalon". Thus, we construct the refined meshes by using such an etalon and compare them with those obtained by various error indication techniques (the adaptation criteria defined above is one and the same in all the cases). By the results presented below, we can compare the efficiency of different error estimation methods. To compare them not only qualitatively but also quantitatively we introduce a special coefficient p_{eff} that shows the percentage of the elements that has been colored correctly, i.e, colored by the same colors as in the etalon marking.

Another important quantity typically used in the a posteriori error estimation is the effectivity index I_{eff} which is the value of the error estimate divided by the energy norm of the true error. It is always greater than one and characterizes the overestimation of the true error.

4.1 Example 1

First example is on the L-shape domain $\Omega = (-1,1) \times (-1,1) \setminus [0,1] \times [-1,0]$ with $\mathbf{f} = 0$. The boundary values are taken from the exact solution $(\mathbf{u}, p)$ which reads in polar coordinates for $\alpha = 856399/1572864 \approx 0.54448$, $\omega = 3\pi/2$,

$$w(\phi) = (\sin((1+\alpha)\phi)\cos(\alpha\omega))/(1+\alpha) - \cos((1+\alpha)\phi) - \\ - (\sin((1-\alpha)\phi)\cos(\alpha\omega))/(1-\alpha) + \cos((1-\alpha)\phi),$$

$$\mathbf{u}(r,\phi) = r^\alpha((1+\alpha)(\sin(\phi), -\cos(\phi))w(\phi) + (\cos(\phi), \sin(\phi))w_\phi(\phi)),$$

$$p(r,\phi) = -r^{\alpha-1}((1+\alpha)^2 w_\phi(\phi) + w_{\phi\phi\phi}(\phi))/(1-\alpha).$$

We use Taylor-Hood elements and the standard adaptation algorithm described above. Error control is obtained by using projection on the space of divergence-free functions. For guaranteed estimations of the error, we use the error majorant in form (6). We use second order finite elements for approximation of the duality variables τ and q. As an initial guess for τ, we use an averaging of $\nu\nabla\mathbf{v}$, while an initial guess for q is p^h, obtained via the Uzawa algorithm. Then estimation is improved by minimization over τ and q. The plot of the initial (a) and the final (b) mesh generated by using functional type error control depicted in Fig.1. It clearly shows hight refinement of the mesh near the singularity. Similar mesh can be obtained by using some other method of a posteriori error estimation control (see [1, 3, 5, 12]) or by the information about the true error. Obviously, we can expect improvement of the convergence rate in comparison with uniform discretization. Corresponding results are depicted in Fig.1(c). Error and error majorant are plotted against the number of degrees of freedom on a log-scale. This improvement is very similar to the results obtained by other methods of a posteriori control for the error indication and mesh adaptation (see [1, 3, 5, 12]).

Moreover, besides error indication, the functional type error majorant provides guaranteed upper bounds of the error (see Fig.1 (c) and Table 1).

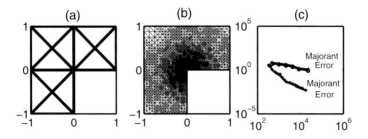

Fig. 1. Example 1. Initial mesh (a), Final mesh (b), error and error majorant for uniform and adaptive meshes (c)

Table 1. Example 1. Error estimation.

iteration	N	error	error majorant	I_{eff}
5	472	0.94	1.288	1.37
9	2174	0.041	0.057	1,41
14	5734	0.013	0.0166	1.28
26	12552	0.008	0.0095	1.19

4.2 Example 2

In the second example data and the exact solution are smooth, so it is not obvious, where the error should be concentrated. Consider an example from [2]. Let $\Omega = (0,1) \times (0,1)$, $\nu = 1$, the exact solution and effective force are defined as follows:

$$\mathbf{u} = (-\sin(\frac{\pi}{2}x)\sin(\frac{\pi}{2}y), -\cos(\frac{\pi}{2}x)\cos(\frac{\pi}{2}y))^T,$$

$$p = \pi\cos(\frac{\pi}{2}x)\sin(\frac{\pi}{2}y), \quad \mathbf{f} = (0, -\pi^2\cos(\frac{\pi}{2}x)\cos(\frac{\pi}{2}y))^T.$$

This problem can be solved by different methods. As an example, we present results obtained by using the Uzawa algorithm, Hesteness-Powel algorithm and Taylor-Hood elements. For the error control a similar procedure to that in Example 1 is used. But error majorant is implemented in the form (5). For the value of C_{LBB} for the rectangular domain we refer to [10].

The procedure for the majorant minimization requires additional computational work. Computational time spent on majorant improvement is determined in compliance with the time spent for obtaining the numerical solution ($1TU$). Table 4.2 demonstrates the dependence of the quality of the error estimation on the computational time spend on majorant improvement. Note that $t = 0\,TU$ denotes the substitution $\tau = \nu \mathbb{G}\nabla\mathbf{v}$, $q = p$ ($\mathbb{G}$ is operator of averaging (see e.g. [13])). This allows one to get guaranteed error estimation almost without additional computational time. Table 4.2 contains information about components of the error majorant and main characteristics, standard

Table 2. Example 2. Dependence of the quality of the error estimation on the computational time spend on majorant improvement

	t=0	$t=0.5TU$	$t=1TU$	$t=2TU$
$\nu \, \|\, \nabla(\mathbf{u}-\mathbf{v}) \,\|$	5.89 e-4	5.89 e-4	5.89 e-4	5.89 e-4
error majorant	0.0159	1.86 e-3	1.0 e-3	6.95 e-4
primary term	1.3 e-4	5.3e-4	5.91 e-4	6.3 e-4
reliability term	0.0157	1.3 e-3	3.8 e-4	6.2 e-5
div term	3.1e-6	3.1e-6	3.1e-6	3.1e-6
I_{eff}	27	3.16	1.71	1.18
p_{eff}	63 %	87 %	96 %	97 %

for a posteriori error estimation. They show overestimation of the error and quality determination local distribution of the error over the domain. Theoretically it is known that the error majorant achieves its minimum when $\tau = \nu \nabla \mathbf{u}$ and $q = p$. By this substitution the second component of (6) turns into 0 and the error estimation turns to be equal to the error. In $t = 0\,TU$ the second component in error majorant prevails and the error indication is not very accurate. But after some time one can see near equality in error estimates and true error, and the local error indication is also very accurate.

Finally, we demonstrate robustness of the functional type estimator in situation, where other error indicator does not work. In Fig. 2 by the dark color we depict the zones in the domain, where cardinal error is concentrated. These elements are need to be refined. On the left there are depicted such zones according to the true error, on the right according to the error majorant.

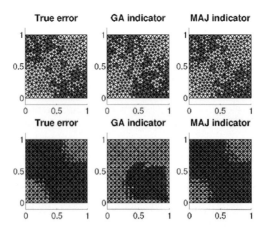

Fig. 2. Example2. Indication of the error by the true error, gradient averaging and the functional type majorant (top: Taylor Hood elements, Uzawa algorithm; bottom: Taylor Hood elements, Hesteness-Powel method)

It is easy to see that they display almost the same zones, what tells about good quality of error indication of error majorant. In the middle there are indications of the error according to the gradient averaging. At some situations (see Fig. 2, top) it also quite closed to the true distribution, but in some cases (see Fig. 2, bottom) it shows totally wrong zones.

In all the examples indication of the error by the error majorant is *always* very close to the true error distribution even in the example without any singularity. By having information about local distribution of the error, one can construct an effective adaptive algorithm. Moreover, guaranteed error estimation allows one to solve important problems with any controllable accuracy.

References

1. Bank, R.E., Welferrt, B.D.: A posteriori error estimates for the Stokes problem. SIAM J. Numer. Anal., **28 (3)**, 591–623 (1991)
2. Braess, D., Sarazin, R.: An efficient smoother for the Stokes problem. Appl. Numer. Math., **23**, 3–19 (1997)
3. Carstensen, C., Funken, A.: A posteriori error control in low-order finite element discretizations of incompressible stationary flow problem. Math. Comp., **70 (236)**, 1352–1381 (2000)
4. Carstensen, C., Funcen, S.A.: Constant in Clement's-interpolation error and residual based a posteriori error estimates in finite element methods. East-West J. Numer. Anal. **8 (3)**, 153–175 (2000)
5. Dari, E., Durán, R., Padra, C.: Error estimators for nonconforming finite element approximation of the Stokes problem. Math. Comp. **64 (211)**, 1017–1033 (1995)
6. Girault V., Raviart P.A.: Finite element approximation of Navier-Stokes equations. Springer, Berlin, (1986).
7. Gorshkova E., Repin S.: On the functional type a posteriori error estimates of the Stokes problem, Proceedings of the 4th European Congress in Applied Sciences and Engineering ECCOMAS (2004), CD-ROM.
8. Gorshkova, E.: A posteriori control of precision of approximate solution of the Stokes problem using projection on solenoidal fields, proceedings of the Polytechnic symposium. Saint-Petersburg State Polytechnic University, 19–23 (2005)(in Russian).
9. Neittaanmäki, P., Repin, S.: Reliable methods for computer simulation. Error control and a posteriori estimates. ELSEVIER, New York, London, (2004)
10. Chizhonkov, E., Olshanskii, M. : On the domain geometry dependence of the LBB condition. Math. Model. Numer. Anal., **34 (5)**, 935–951 (2000)
11. Repin, S.I.: A posteriori estimates for the Stokes problem. J. Math. Sciences, **109 (5)**, 1950–1964 (2002)
12. Verfurth, R.: A posteriori error estimators for the Stokes equations. Numer. Math., **55**, 309–325 (1989)
13. Zienkeiewicz, O.C., Zhu,J.Z.: A simple error estimator and adaptive procedure for practical engineering analysis. Internat. J. Numer. Methods Engrg., **24**, 337–357 (1987)

Error Control for Discretizations of Electromagnetic-Mechanical Multifield Problem

Marcus Stiemer

Chair of Scientific Computation, University of Dortmund, D-44221 Dortmund, Germany,
stiemer@math.uni-dortmund.de

Summary. The modeling of numerous industrial processes leads to multifield problems, which are governed by the coupled interaction of several physical fields. As an example, consider electromagnetic forming, where the evolution of the deformation field of a mechanical structure consisting of well conducting material is coupled with an electromagnetic field, triggering a Lorentz force, which drives the deformation process. The purpose of the work reported on here is to develop techniques for a posteriori error control for the finite element approximation to the solution of certain systems of two boundary value problems that are coupled via their coefficients and their right-hand sides. As a first step, an error estimator for the right-hand side of the mechanical subsystem is presented in the case of a simplified model problem for the electromagnetic system. The particular influence of the mixed character of the evolution equations is discussed for a numerical example.

1 Introduction

Many technological processes are governed by the interaction of different physical fields. As an example consider electromagnetic metal forming: In this process, a pulsed magnetic field induces eddy currents in a work piece consisting of good conducting material like aluminum or copper. The interaction of the eddy currents with the triggering field results in a material body force, the Lorentz force, which drives the deformation of the work piece (see Figure 1). Considering the high strain rates typical of this process, Svendsen and Chanda formulated a material model [14, 15] for a wide class of materials under the influence of strong electromagnetic fields based on the Perzyna model of elasto-viscoplasticity (see [12]). The evolution of the electromagnetic field is determined by Maxwell's equations under the quasistatic hypothesis. The two systems are coupled via the Lorentz force representing the source term in the impulse balance of the mechanical structure and via the distribution of conductivity, which depends on the position of the moving structure.

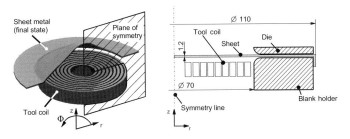

Fig. 1. A typical arrangement for electromagnetic sheet metal forming.

Particularly, a boundary value problem of mixed type arises for the electromagnetic field, which is parabolic in areas of positive conductivity and elliptic elsewhere. In [13], finite element formulations for the electromagnetic and for the mechanical system have been derived and implemented within a staggered strong coupling scheme. See [6, 7, 8, 9, 11, 16] for different approaches to the simulation of coupled electromagnetic mechanical systems.

The present note begins with a brief review of the coupled finite element model. Then, on the electromagnetic side, a method for a posteriori error control of the input quantity to the mechanical subsystem is presented considering a simplified model problem. Its derivation is guided by Eriksson's and Johnson's techniques for parabolic a posteriori error estimation [2, 3, 4, 5]. The methods presented here represents a first step towards a rigorous error control for coupled problems: A more appropriate approach would be the use of dual weighted residual error estimators (see, e.g., [1]), which allow to control the error in the quantity of interest under additional consideration of those errors introduced by the coupling process.

2 Electromagnetic forming

2.1 The mechanical model

The simulation of sheet metal forming requires to account for large deformations. In the following, a suitable mechanical model in Lagrange formulation is described. The applied material model [14, 15] considers viscous effects which become significant at high strain rates. Starting point is a pull back of the weak momentum balance from the current configuration of the work piece Σ to its reference configuration Σ_0, yielding

$$m(\xi) := \int_{\Sigma_0} KF^{-T} : \nabla \Phi \, dx + \int_{\Sigma_0} \left(\rho \ddot{\xi} - Jf(\xi) \right) \Phi \, dx = 0 \qquad (1)$$

for all test functions Φ in the Sobolev space H^1 fulfilling the adequate boundary conditions. Here, $\xi = \xi(x)$ denotes the deformation field, $F = \nabla \xi$ the deformation gradient and $J = \det F$ its determinant. Moreover, $K = J\sigma$ represents the Kirchhoff stress computed from the Cauchy stress σ. The corresponding strain measure results from a multiplicative split of the deformation gradient $F = F^{el} F^{vp}$ in elastic and visco-plastic part, as usual in finite strain plasticity (see [12]). Then the elastic part of the strain is given by $\epsilon^{el} = \log V^{el}$, where V^{el} is the symmetric part of the polar decomposition $F^{el} = V^{el} R^{el}$, and its visco-plastic part by $\epsilon^{vp} = \log \det F^{vp}$. Solving the momentum balance (1) requires a method to compute the stresses for the current thermodynamic state of the structure. They are obtained from the free Helmholtz energy

$$\Psi(\epsilon^{el}, \alpha) = \frac{1}{2}\lambda \left(\operatorname{tr} \epsilon^{el}\right)^2 + \mu \operatorname{tr}\left(\epsilon^{el} : \epsilon^{el}\right) + \psi(\epsilon^{vp}) ,$$

which represents the energy reversibly stored in the material, via the relation

$$K = J\sigma = \frac{\partial \Psi}{\partial \epsilon^{el}} = C : \epsilon^{el} .$$

Here, λ and μ are Lamé's constants and ψ represents energy storage due to hardening (see [13]). The Helmholtz energy depends on a finite set of internal variables, whose values characterize the thermodynamic state of the mechanic structure and represent its memory of the load history. For these, evolution equations have to be constituted. Characteristic for viscoplasticity is an explicit equation

$$\dot{\epsilon}^{vp} = \begin{cases} 0 & , \mathcal{F}(\sigma) \leq 0 , \\ \frac{1}{\eta}\mathcal{F}(\sigma)^m \cdot (\sigma - \prod \sigma) / \|\sigma - \prod \sigma\| & , \mathcal{F}(\sigma) > 0 , \end{cases}$$

for the plastic strain rate. Here, $\mathcal{F}(\sigma) = \|\operatorname{dev} \sigma\|_{L^2(\Omega)} - \sigma_0$ describes the von Mises yield surface and $\prod \sigma = \sigma_0 \operatorname{dev} \sigma / \|\operatorname{dev} \sigma\|_{L^2(\Omega)} + 1/3 \operatorname{tr} \sigma$ denotes a projection on it. For simplicity, the influence of strain hardening is not considered in this notation. The non linear momentum balance is iteratively solved by Newton's method $\xi_{i+1} = \xi_i - (\partial m/\partial \xi)^{-1} m(\xi_i)$. To compute the required linearization, the evolution equations for the internal variables need to be discretized in time, leading to a non-linear system of equations which has to be solved with a further *inner* Newton-iteration.

2.2 Electromagnetic field computation

Since the occurring wave lengths are much longer than the distances relevant for the forming process, the quasistatic approximation to Maxwell's equations applies (see [10, 13]). An Eulerian description of the evolution of the electromagnetic field is given by

$$\operatorname{curl} \frac{1}{\mu} \operatorname{curl} a + \sigma a_t - \sigma(v \times \operatorname{curl} a) = -\sigma \nabla \phi ,$$

$$\operatorname{div} a_t - \operatorname{div}(v \times \operatorname{curl} a) = \Delta \phi , \qquad (2)$$

where $a = a(x,t)$ denotes the magnetic vectorpotential and ϕ the electrostatic potential to be determined. Further, v represents the velocity of the body in which the field is computed, $\mu = \mu(x,t)$ the permeability and $\sigma = \sigma(x,t)$ the conductivity distribution. For the considered geometries, σ is spatially piecewise constant, and μ is entirely constant for the materials considered. The stated differential equations yield in areas, where σ is continuous. At material interfaces, where σ is discontinuous, transition conditions are necessary, stating that tangential components of a remain continuous. Due to the fast decay of the electromagnetic dipol field of order $||a(x)|| = O(||x||^{-2})$, $||x|| \to \infty$, the problem can be tackled in a large bounded open set $\Omega \subset \mathbb{R}^k$ ($k = 2, 3$) with sufficient accuracy. In general, the vector potential is not unique and has to be gauged. However, in two-dimensional or axisymmetric situations the Coulomb gauge $\text{div } a = 0$ is always fulfilled. Since $\text{div}(v \times b) = 0$ is also true in these situations, the system (2) decouples. The weak form for the vector potential reads as follows: Find $a \in H_{\text{curl},0}(\Omega)^3$ such that for all $a^* \in H_{\text{curl},0}(\Omega)^3$

$$-\int_\Omega \frac{1}{\mu} \text{curl } a \cdot \text{curl } a^* \, dx + \int_\Omega \sigma a_t \cdot a^* \, dx = -\int_\Omega \sigma \nabla \phi \cdot a^* \, dx \ .$$

In general, a possesses no H^1 regularity, which has to be considered in the finite element discretization. However, in axisymmetric situations, we are back in the realm of H^1-regularity since $H_{\text{curl},0}(\Omega)^3 \cap H_{\text{div},0}(\Omega)^3 \cong H_0^1(\Omega)^3$, provided that Ω possesses a C^2-boundary.

2.3 Coupling

The simulation is carried out in two meshes, a fixed Eulerian mesh for the electromagnetic field and a Lagrangian mesh for the mechanical structure. At a certain time step, the magnetic vector potential depending on the input amperage and the position of the structure is computed in the electromagnetic mesh and the Lorentz forces $f_L = a_t \times \text{curl } a$ are derived. After that, the forces are transferred into the mechanical mesh and imposed on the structure to determine its corresponding position. The altered position of the work piece is then transferred into the electromagnetic mesh and a corrected force distribution is computed. The two steps *field computation* and *structure simulation* are iterated until both f_L and the position of the structure do not change in the scope of accuracy. After that, the next time step is started. It has turned out that a fine resolution is required at the boundaries of the moving structure to avoid oscillations of the Lorentz forces (see [13]). However, the computed deformations of the strucure are quite accurate even for relatively coarse meshes (see [13]). Nevertheless, the implementation of an ALE-based formulation to improve the accuracy of the computed forces represents work in progress.

3 Error control for coupled and mixed problems

Dual weighted residual error estimators represent an appropriate aproach to error control for problems of the type described above. They allow a goal oriented control of exactly those quantities of interest (see, e.g., [1]). Applied to a staggered solution algorithm, such techniques enable also a control of the error of the quantities that realize the coupling between the two subsystems. Thus, error accumulation due to the coupling procedure can be controlled and, moreover, be reduced by mesh adaption in both subsystems. Here, we only focus on the electromagnetic subsystem and present techniques for the control of the Lorentz force. The incorporation of the strategy outlined before represents work in progress. Further, we neglect here additional errors due to data transfer from one mesh into the other.

Assume that the deformation field u of the mechanical structure is characterized by, e.g., $\mathcal{M}(u,\varphi) = (f_L,\varphi)$ for all suitable test functions φ, where f_L denotes the exact values of the Lorentz force, $(\cdot,\cdot)$ the space-time scalar product in $L^2(L^2(\Sigma))$ w. r. t. a time interval I_n, Σ the current configuration of the structure and $\mathcal{M}$ an operator that is linear in φ, but possibly non linear in u. For simplicity, the Lagrangian formulation of the mechanical mesh has been dismissed here. As an example, consider the error $||u - U||_\infty$, where U represents a Galerkin approximation to u. For brevity, we write here and below $||\cdot||_k$ for the norm in $L^k(L^k(\Sigma))$. To account for the influence of the approximation error w. r. t. f_L, the following estimate on the right-hand side of the equation for u is reasonable:

$$\left| \int_{I_n} \int_\Sigma (f_L - A_t \times \operatorname{curl} A)\,\varphi \right| \leq ||\varphi||_\infty \int_{I_n} \int_\Sigma |a_t \times \operatorname{curl} a - A_t \times \operatorname{curl} A|$$
$$\leq ||\varphi||_\infty \left(||A_t||_2\, ||\operatorname{curl}(a-A)||_2 + ||\operatorname{curl} A||_\infty\, ||a_t - A_t||_1 \right.$$
$$\left. + ||a_t - A_t||_2\, ||\operatorname{curl}(a-A)||_2 \right), \qquad (3)$$

where A is a Galerkin approximation to a and A_t any approximation to a_t derived from A. Here, dx and dt have been dispensed with. Hence, a control on the error in the right-hand side of the mechanical simulation is achieved by controlling $||\operatorname{curl}(a-A)||_2$ and $||a_t - A_t||_1 \leq \sqrt{\operatorname{mes}\Sigma} \left(|||(a-A)(t_n)||_{L^2(\Sigma)} + ||(a-A)(t_{n-1})||_{L^2(\Sigma)} \right) + \mathcal{O}(k_n^2)$, $k_n \to 0$, in the electromagnetic mesh. For brevity, we consider now an analogous problem for a mixed heat- / Laplace-equation and derive an estimator for $||\nabla(a-A)||_{L^2(L^2(\Omega))}$. To estimate $||(a-A)(t_n)||_{L^2(\Sigma)}$, the techniques presented in [2] for parabolic problems can immediately be applied here.

3.1 A mixed elliptic-parabolic model problem

Let $\Omega =]-1,1[\times]-1,1[\subset \mathbb{R}^2$ be the domain we exemplarily use for field computation and $\Sigma \subset \Omega$ the area in which diffusion takes place. As "right-hand side" a source term $s \in L^2(L^2(\Sigma))$ is given and initial values $u_0 \in L^2(\Omega)$. The problem is now to find $u \in L^2(H_0^1(\Omega))$ with $u(t_0) = u_0$ and

$$\int_\Omega \nabla u(x,t)\, \nabla \phi(x,t)\, dx + \int_\Sigma u_t(x,t)\, \phi(x,t)\, dx = \int_\Sigma s(x,t)\, \phi(x,t)\, dx \quad (4)$$

for all $\phi \in H_0^1(\Omega)$ and for $t_0 \leq t \leq T$ a.e.

To discretize (4), a partition $0 = t_0 < \ldots < t_n = T$ of $[0,T]$ in intervals $I_n = (t_{n-1}, t_n)$ of length $k_n = t_n - t_{n-1}$ is introduced. We briefly write $u_n = u(t_n)$. Further, a discrete space $\mathcal{S}_n$ as spatial test and trial space is chosen. This leads to the local test- and trial spaces $V_{q,n} = \left\{ v : v = \sum_{j=0}^{q} t^j \varphi_j, \varphi_j \in \mathcal{S}_n \right\}$, $q \in \mathbb{N}_0$, in the time-space setting. To compute an approximation to (4) the following (temporal) discontinuous Galerkin method is applied: Find a function U with $U|_{\Omega \times I_n} \in V_{q,n}$, such that

$$\int_{I_n} \int_\Omega \nabla U(x,t)\, \nabla v(x,t)\, dx\, dt + \int_{I_n} \int_\Sigma U_t(x,t)\, v(x,t)\, dx\, dt$$
$$+ \int_\Sigma [U]_{n-1}(x)\, v_{n-1}^+(x)\, dx = \int_{I_n} \int_\Sigma s(x,t)\, v(x,t)\, dx\, dt \quad (5)$$

for all $v \in V_{q,n}$ and $n \in \mathbb{N}$, where $[w]_n = w_n^+ - w_n^-$, $w_n^\pm = \lim_{s \to 0^\pm} w(t_n + s)$. To obtain discrete spaces $\mathcal{S}_n$, we consider triangulations $\mathcal{T}_n$ and choose continuous functions whose restriction to an element of the triangulation is linear. Let further h be a positive function in $C^1(\bar{\Omega})$ with bounded gradient. We assume, that for the diameter h_K of each triangle $K \in \mathcal{T}$ the estimates $c_1 h_K^2 \leq \text{mes}\, K$ and $c_2 h_K \leq h(x) \leq h_K$, $x \in K$, are true with constants $c_1, c_2 > 0$.

3.2 On Error Estimation for the Mixed Problem

Let $[U]_n = U_n^+ - U_n^-$ represent the temporal jumps of U and let $D_{h,1}(U_n) = \left(\sum_{\tau \in E_i} h_\tau^2 \left\| \left[\frac{\partial U_n}{\partial n_\tau} \right] \right\|^2 \right)^{1/2}$ dentote the spatial jumps of the gradient, where E_i is the set of all internal edges of the triangulation of Ω. Further, P_n dentotes the L^2-projection on $\mathcal{S}_n$. The qualtity of P_n can be estimated by interpolation. Eriksson and Johnson [2] show

$$|(f, (I - P_n)v)| \leq \alpha \|hf\|_{L^2(\Omega)} \|\nabla v\|_{L^2(\Omega)} \qquad f \in L^2(\Omega), v \in H_0^1(\Omega)$$
$$|(\nabla w, \nabla(I - P_n)v)| \leq \beta D_{h,1}(w) \|\nabla v\|_{L^2(\Omega)} \qquad w \in \mathcal{S}_n, v \in H_0^1(\Omega) \quad (6)$$

with constants $\alpha, \beta > 0$ and with mesh density function h. To facilitate the notation, the sought error estimator is now derived in the particular case $q = 0$ and $f(x,t) = f_n(x)$ for $t \in I_n = (t_{n-1}, t_n)$. Further, $U_0 = P_0 u_0 = u_0$ is assumed.

Theorem 1. *Let C be the constant from the Poincaré-Friedrichs inequality in Σ and α and β as above. Then*

$$\int_{I_n} \|\nabla u(t) - \nabla U(t)\|^2_{L^2(\Omega)}\, dt \leq 2\,\|u_{n-1} - U_{n-1}^-\|_{L^2(\Sigma)} + \mathcal{E}_n(U) + \mathcal{O}(k_n^2)$$

for $k_n \to \infty$, with

$$\mathcal{E}_n(U) = 4\alpha^2 \|h[U]_n\|^2_{L^2(\Sigma)} + 4\beta^2 k_n D_{h,1}(U_n)^2 + 4\alpha^2 k_n \|hs_n\|^2_{L^2(\Sigma)}$$
$$+ 4\alpha^2 C^2 k_n \|h\|^2_{L^\infty(\Sigma)} \|\nabla U_{n-1}^-\|^{2*}_{L^2(\Sigma)}\,.$$

The star indicates that this term only needs to be accounted for in elements that have been coarsened the step before. An estimator for $\|u_{n-1} - U_{n-1}^-\|_{L^2(\Sigma)}$ is presented in [2]. Further, the terms summed up in $\mathcal{O}(k_n^2)$ also admit an a posteriori error control. For brevity, this is not further discussed here.

Proof. To derive the error estimator, one tests with $u - U$ in a time integrated version of (4) and with the projection of $u - U$ on the relevant discrete spaces in (5). The difference of these expressions can then be estimated with the help of (16) and Young's inequality. ∎

As an example, $\Sigma = \Sigma_1 \cup \Sigma_2$ has been chosen with $\Sigma_1 = [-0.6, 0.6] \times [0.4, 0.6]$ and $\Sigma_2 = [-0.2, 0.2] \times [-0.2, 0.2]$. On Σ_1, a constant source $s(x,t) = 1$ is imposed. Outside Σ_1, s vanishes. In (4), the u_t term has additionally been multiplied with a coefficient α with $\alpha(x) = 500$ on Σ_1 and $\alpha(x) = 5000$ on Σ_2. The error estimator has now been used to equilibrate the elementwise error contributions among all elements K of the triangulation of Ω. For spatial mesh adaption, a fixed fraction strategy is applied several times in each time step, refining those $r\%$ elements, that introduce the highest and coarsening those $r\%$ that produce the smallest contribution to the error. The spatial mesh adaption is carried out in each time step several times. In Figure 2 a typical mesh resulting from this algorithm is presented ($r = 20$). The error estimator

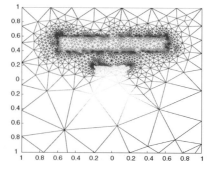

Fig. 2. Typical meshes produced by the autoadaptive algorithm described above

recognizes the transition zones between parabolic and elliptic regions, entailing a huge number of refinement steps there. In Σ_2, the gradient remains small for a quite long time, since convergence to the equilibrium state is retarded due to the small diffusivity of 1/5000 in this area. Consequently, only very few triangulation points lie inside Σ_2. Note that the only geometrical disposition during mesh refinement is that the corner points of Σ_1 and Σ_2 are held fixed. Everything else has been "realized" by the error estimator. While the solution to the above problem possesses locally H^2-regularity away from the transition zones, its overall regularity is reduced. The question, whether adaptive mesh refinement enables to increase the rate of convergence as a function of the numerical efforts to the order of the corresponding purely parabolic problem represents work in progress.

References

1. Becker, R., Rannacher, R.: A feed-back approach to error control in finite element methods: Basic analysis and examples. East-West J. Numer. Math., **4**, 137–164 (1996)
2. Eriksson, K., Johnson, C.: Adaptive finite element methods for parabolic problems I: A linear model problem. SIAM J. Numer. Anal., **28**, 43–77 (1991)
3. Eriksson, K., Johnson, C.: Adaptive finite element methods for parabolic problems II: Optimal error estimates in $L_\infty L_2$ and $L_\infty L_\infty$. SIAM J. Numer. Anal., **32**, 706–740 (1995)
4. Eriksson, K., Johnson, C.: Adaptive finite element methods for parabolic problems IV: Nonlinear problems. SIAM J. Numer. Anal., **32**, 1729–1749 (1995)
5. Eriksson, K., Johnson, C.: Adaptive finite element methods for parabolic problems V: Long-time integration SIAM J. Numer. Anal., **32**, 1750–1763 (1995)
6. Fenton, G., Daehn, G. S.: Modeling of electromagnetically formed sheet metal. J. Mat. Process. Tech. **75**, 6–16 (1998)
7. Gourdin, W. H.: Analysis and assessment of electromagnetic ring expansion as a high-strain rate test. J. Appl. Phys. **65**, 411–422 (1989)
8. Gourdin, W. H., Weinland, S. L., Boling, R. M.: Development of the electromagnetically-launched expanding ring as a high strain-rate test. Rev. Sci. Instrum. **60**, 427–432, (1989)
9. Kleiner, M., Brosius, A., Beerwald, C.: Determination of flow stress at very high strain rates by a combination of magnetic forming and FEM calculation. In: International Workshop on Friction and Flow Stress in Cutting and Forming (ENSAM), Paris, 175–182 (2000)
10. Moon, F.: Magnetic interactions in solids. John-Wiley & Sons (1984)
11. Schinnerl, M., Schöberl, J., Kaltenbacher, M., Lerch, R.: Multigrid methods for three-dimensional simulation of nonlinear magnetomechanical systems. IEEE Trans. Mag. **38**, 1497–1511 (2002)
12. Simo, J. C., Hughes, T. J. R.: Computational inelasticity. Springer Series on Interdisciplinary Applied Mathematics **7**, Springer Verlag (1998)
13. Stiemer, M., Unger, J., Blum, H., Svendsen, B.: Algorithmic formulation and numerical implementation of coupled multifield models for electromagnetic metal forming simulations. Internat. J. Numer. Methods Engng. (accepted) (2006).

14. Svendsen, B., Chanda, T.: Continuum thermodynamic modeling and simulation of electromagnetic forming, Technische Mechanik **23**, 103–112 (2003)
15. Svendsen, B., Chanda, T.: Continuum thermodynamic formulation of models for electromagnetic thermoinlastic materials with application to electromagnetic metal forming, Cont. Mech. Thermodyn., **17**, 1–16 (2005)
16. Takata, N., Kato, M., Sato, K., Tobe, T.: High-speed forming of metal sheets by electromagnetic forces. Japan Soc. Mech. Eng. Int. J. **31**, 142, (1988)

A Safeguarded Zienkiewicz-Zhu Estimator

Francesca Fierro and Andreas Veeser

Dipartimento di Matematica, Università degli Studi di Milano, Via C. Saldini 50, 20133 Milano, Italia
{fierro,veeser}@mat.unimi.it

Summary. For the linear finite element approximation to a linear elliptic model problem, we propose to safeguard the Zienkiewicz-Zhu estimator by an additional estimator for the residual of the averaged gradient. We give a brief account of the theoretical results on reliability, (local) efficiency, and asymptotic exactness of the full estimator and illustrate these properties in numerical tests, incorporating singular solutions and anisotropic ellipticity.

1 Introduction

The gradient averaging and the appertaining a posteriori error estimator introduced in Zienkiewicz/Zhu [9] have striking asymptotic properties. Restricting ourselves to recent references, we mention the superconvergence results in Xu/Zhang [8] and the observed asymptotic exactness in Carstensen/Bartels [3]. On the other hand, since this estimator relies solely on post-processing of the approximate solution, it cannot in general be reliable.

In [4] the authors therefore propose and analyze the Zienkiewicz-Zhu estimator (or, for short, ZZ estimator) with a complementing 'security part', which is based upon the residual of the averaged gradient. The resulting estimator combines the good properties of the ZZ estimator and the standard residual estimator: it is reliable, (locally) efficient, and asymptotically exact whenever the averaged gradient is superconvergent. The latter is a consequence of the fact that the security part is an efficient estimator for the error of the averaged gradient. This proven properties do not hinge on the particular gradient averaging associated with the ZZ estimator and distinguish the approach in [4] from previous ones in Rodríguez [6], Repin [5], Carstensen/Bartels [3], and Carstensen [2].

The purpose of this work is to give an account of the theoretical results of [4] in a simplified setting and to provide further numerical tests.

2 Error estimator and theoretical results

As usual, $L_2(U)$ denotes the space of functions that are Lebesgue measurable and square-integrable in the domain U and $H_0^1(U)$ stands for the space of $L_2(U)$-functions that have first weak derivatives in $L_2(U)$ and zero trace on the boundary ∂U.

2.1 Model problem and discretization

Let $\Omega \subset \mathbb{R}^2$ be a bounded polygonal domain with boundary $\Gamma := \partial \Omega$ that is locally a Lipschitz graph. The load term fulfills $f \in L_2(\Omega)$ and the constant coefficients of the linear operator are given by a symmetric matrix $A \in \mathbb{R}^{2\times 2}$ satisfying

$$\forall \xi \in \mathbb{R}^2 \quad \lambda |\xi|^2 \leq A\xi \cdot \xi \leq \Lambda |\xi|^2$$

with $0 < \lambda \leq \Lambda$.

Let u be the typically unknown weak solution of the elliptic boundary value problem

$$-\mathrm{div}(A\nabla u) = f \text{ in } \Omega, \quad u = 0 \text{ on } \Gamma.$$

In other words:

$$u \in H_0^1(\Omega) \quad \text{and} \quad \forall \varphi \in H_0^1(\Omega) \quad \int_\Omega A\nabla u \cdot \nabla \varphi = \int_\Omega f\varphi. \tag{1}$$

In view of the Riesz representation theorem, u exists and is unique.

In order to approximate the solution u of (9), we shall use linear finite elements that are subordinated to a macro triangulation. Suppose that the macro triangulation $\mathcal{T}_0$ is a conforming (admissibile) triangulation of Ω. The following two quantities of $\mathcal{T}_0$ will be important:

$$\alpha_{\min} := \text{smallest angle occurring in } \mathcal{T}_0 \quad \text{and} \quad \mu := \frac{\max_{T \in \mathcal{T}_0} h_T}{\min_{T \in \mathcal{T}_0} h_T}, \tag{2}$$

where $h_T := \mathrm{diam}\, T$ denotes the diameter of a triangle T.

Let $\mathcal{T}$ be any (not necessarily quasi-uniform) refinement of $\mathcal{T}_0$ that was obtained with the help of the newest-vertex bisection; see e.g. [7]. Hereafter, we suppose that, together with $\mathcal{T}_0$ itself, we are given an appropriate fixed set of refinement edges. The set of the nodes (or vertices) of $\mathcal{T}$ is denoted by $\mathcal{N}$. Let S be the space of continuous piecewise affine finite elements over $\mathcal{T}$:

$$S := \{ w \in C(\bar{\Omega}) \mid \forall T \in \mathcal{T}\ w_{|T} \in P_1(T) \},$$

where $P_k(T)$, $k \in \mathbb{N}$, stands for the space of polynomials over T with degree $\leq k$.

The finite element approximation $\bar{u}_S$ of u in (9) is then characterized by

$$\bar{u}_S \in S_0 \quad \text{and} \quad \forall \chi \in S_0 \quad \int_\Omega A\nabla \bar{u}_S \cdot \nabla \chi = \int_\Omega f\chi, \tag{3}$$

where $S_0 := S \cap H_0^1(\Omega)$. Like u, the finite element approximation $\bar{u}_S$ exists and is unique. In practice, one often does not solve the linear system resulting from (3) exactly. We therefore suppose that $u_S \in S_0$ is an approximation of $\bar{u}_S$ and will provide an a posteriori analysis for the *approximate* finite element solution u_S and its error in the the so-called *energy norm* defined by

$$\|\nabla w\|_A := \left(\int_\Omega A \nabla w \cdot \nabla w \right)^{1/2} \quad \text{for } w \in H_0^1(\Omega). \tag{4}$$

2.2 Estimator and local indicators

We now define the safeguarded ZZ estimator. For linear elements, the ZZ estimator is given by

$$\zeta := \|\nabla u_S - G u_S\|_A, \tag{5}$$

where $G u_S \in S \times S$ is the nodewise averaged gradient of ∇u_S:

$$\forall z \in \mathcal{N} \quad G u_S(z) = \mathcal{L}^2(\omega_z)^{-1} \int_{\omega_z} \nabla u_S \in \mathbb{R}^2, \tag{6}$$

where $\omega_z := \bigcup \{T \in \mathcal{T} : T \ni z\}$ is the star around the node z and $\mathcal{L}^2(\omega_z)$ denotes its area. To split ζ into local contributions, we use the partition of unity $\sum_{z \in \mathcal{N}} \phi_z = 1$ provided by the canonical basis functions of S: for all $z \in \mathcal{N}$, we set

$$\zeta_z^2 := \|\nabla u_S - G u_S\|_{A\phi_z}^2 := \int_{\omega_z} (\nabla u_S - G u_S) \cdot A(\nabla u_S - G u_S) \phi_z. \tag{7}$$

The complementing security part consists of two contributions related to the residual of the averaged gradient $G u_S$, which is continuous. The first part builds upon the 'strong residual' $r := f + \text{div}(A G u_S)$ and is given by

$$\rho := \left[\sum_{z \in \mathcal{N}} \rho_z^2 \right]^{1/2} \quad \text{with} \quad \rho_z^2 := h_z^2 \int_{\omega_z} |r - \bar{r}_z|^2 \phi_z \tag{8}$$

and

$$h_z := \text{diam } \omega_z, \qquad \bar{r}_z := \begin{cases} \left(\int_{\omega_z} \phi_z \right)^{-1} \int_{\omega_z} r \phi_z & \text{if } z \in \mathcal{N} \setminus \Gamma, \\ 0 & \text{if } z \in \mathcal{N} \cap \Gamma. \end{cases}$$

For the second contribution, we shall use a multilevel decomposition of S. The bisections generating $\mathcal{T}$ from $\mathcal{T}_0$ can be recorded by a forest of binary trees $\mathcal{F}$, where each triangle corresponds to a node, the triangles of $\mathcal{T}_0$ are roots, and those of $\mathcal{T}$ are leafs; see e.g. [7] and Figure 1 for an example. Let $\mathcal{F}_\ell$ be the maximal subforest of $\mathcal{F}$ with depth equal or smaller than $\ell \geq 0$ such that its leafs constitute a conforming triangulation, which will be called

 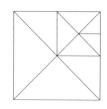 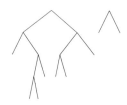

Fig. 1. From left to right: a macro triangulation, a refinement, and its corresponding forest of binary trees, which has maximal depth 4

$\mathcal{T}_\ell$. We denote by $\mathcal{N}_\ell$ the vertices (or nodes) in $\mathcal{T}_\ell$ and by S_ℓ the continuous linear finite elements over $\mathcal{T}_\ell$. Clearly, there holds

$$S_\ell \subset S_{\ell+1}, \quad S = \bigcup_{\ell \geq 0} S_\ell \quad \text{and} \quad \mathcal{N}_\ell \subset \mathcal{N}_{\ell+1}, \quad \mathcal{N} = \bigcup_{\ell \geq 0} \mathcal{N}_\ell. \tag{9}$$

The indicators of the ℓth level are given by

$$\gamma_{\ell z} := \begin{cases} |\int_\Omega AGu_S \cdot \nabla \phi_{\ell z} - f\phi_{\ell z}| & \text{if } z \in \mathcal{N}_\ell \setminus \Gamma, \\ 0 & \text{if } z \in \mathcal{N}_\ell \cap \Gamma, \end{cases} \tag{10}$$

where $(\phi_{\ell z})_{z \in \mathcal{N}_\ell}$ are the canonical basis function in S_ℓ satisfying $\phi_{\ell z}(z) = 1$ and $\phi_{\ell z}(y) = 0$ for all $y \in \mathcal{N}_\ell \setminus \{z\}$. Moreover, we define

$$\widetilde{\mathcal{N}}_\ell := (\mathcal{N}_\ell \setminus \mathcal{N}_{\ell-1}) \cup \{z \in \mathcal{N}_{\ell-1} \mid \phi_{\ell z} \neq \phi_{\ell-1,z}\}$$

for $\ell \geq 1$ and $\widetilde{\mathcal{N}}_0 := \mathcal{N}_0$. To any node $z \in \widetilde{\mathcal{N}}_\ell$ with $\ell \geq 1$, there corresponds a hat function $\phi_{\ell z}$ that is not contained in $S_{\ell-1}$. Consequently only the corresponding indicators $\gamma_{\ell z}$, $z \in \widetilde{\mathcal{N}}_\ell$ provide new information that cannot be seen on the previous level $\ell - 1$. We therefore define the global contribution by

$$\gamma := \left[\sum_{\ell \geq 0} \sum_{z \in \widetilde{\mathcal{N}}_\ell} \gamma_{\ell z}^2 \right]^{1/2}, \tag{11}$$

which measures how much $Gu_S - \nabla u$ misses to mimic the Galerkin orthogonality of $\nabla(\bar{u}_S - u)$.

2.3 Error control

The ZZ estimator ζ alone cannot in general be a reliable. Indeed, consider $A = \text{Id}$ and a load term $f \neq 0$ that is $L_2(\Omega)$-orthogonal to S_0. Then $u \neq 0$ but $u_S := \bar{u}_S = 0$, whence $\|\nabla(u_S - u)\|_A > 0$. However, due to (6), we have $Gu_S = 0$ and thus $\zeta = 0$. For a concrete example and related underestimation of ζ, see [1, mS4.7] or [4, mS6.1 and mS6.2].

The proofs of the following results are given in [4] for more general gradient averaging procedures and more general linear elliptic boundary values problems. The letter C indicates constants that depend only on $\alpha_{\min}$ and μ in (2) of the macro triangulation $\mathcal{T}_0$. We start with the reliability of the safeguarded ZZ estimator.

Theorem 1 (Global upper bound). *The energy norm error of the approximate finite element solution u_S is globally bounded in the following way:*

$$\|\nabla(u_S - u)\|_A \leq \zeta + \frac{C}{\sqrt{\lambda}}(\rho + \gamma).$$

For sake of simplicity, we assume in the following local lower bounds that the load term f is piecewise constant over $\mathcal{T}$. Given a subdomain $\omega \subset \Omega$ and a vector field $W \in L_2(\omega)^2$, define the local norm $\|W\|_{A;\omega} := \left[\int_\omega W \cdot AW\right]^{1/2}$.

Proposition 1 (Lower bounds with averaged gradient). *The indicators ζ_z, ρ_z, $z \in \mathcal{N}$, and $\gamma_{\ell z}$, $\ell \geq 0$, $z \in \mathcal{N}_\ell$, are bounded as follows:*

$$\zeta_z \leq \|\nabla(u_S - u)\|_{A\phi_z} + \|Gu_S - \nabla u\|_{A\phi_z},$$

$$\frac{\rho_z}{\sqrt{\Lambda}} \leq C_1\|Gu_S - \nabla u\|_{A;\omega_z}, \qquad \frac{\gamma_{\ell z}}{\sqrt{\Lambda}} \leq C_2\|Gu_S - \nabla u\|_{A;\omega_{\ell z}},$$

where $\omega_{\ell z} = \operatorname{supp} \phi_{\ell z}$ indicates a star on the level ℓ.

All these local lower bounds have global counterparts, the derivation of which is not trivial for γ. Consequently, the safeguarded ZZ estimator is asymptotically exact whenever the averaged gradient is superconvergent.

For the last two results, let $\mathcal{B}(\omega) := \bigcup\{T \in \mathcal{T} : T \cap \omega \neq \emptyset\}$ be the smallest ball in $\mathcal{T}$ around the set ω.

Proposition 2 (Nondeterioration of averaging). *For any $T \in \mathcal{T}$, the averaged gradient Gu_S satisfies*

$$\|Gu_S - \nabla u\|_{A;T} \leq C\frac{\sqrt{\Lambda}}{\sqrt{\lambda}}\|\nabla(u_S - u)\|_{A;\mathcal{B}(T)}.$$

Combining the two propositions yields the efficiency in any case.

Theorem 2 (Lower bounds). *The indicators ζ_z, ρ_z, γ_z, are bounded by the local energy error. More precisely, for any $z \in \mathcal{N}$,*

$$\zeta_z + \frac{\rho_z}{\sqrt{\lambda}} + \frac{\gamma_z}{\sqrt{\lambda}} \leq C\frac{\Lambda}{\lambda}\|\nabla(u_S - u)\|_{A;\mathcal{B}(\omega_z)}.$$

3 Numerical results

Using the safeguarded estimator $\mathcal{E} = \zeta + \lambda^{-1/2}(C_1\rho + C_2\gamma)$ and the maximum strategy, the authors design in [4] an adaptive algorithm, which is implemented within finite element toolbox **ALBERTA** [7]. It is worth mentioning that this algorithm produces adaptive meshes even in situations when the ZZ estimator vanishes everywhere.

In what follows, we report on two more experiments that enter in the more general setting of [4] and complement the tests therein. As in [4], the estimator constants are $C_1 = 1/5$ and $C_2 = 1/3$ and the various quantities depending on the current finite element space S are indexed by the counter k of the adaptive iteration.

3.1 L-shaped domain

For the L-shaped domain $\Omega =]-1,1[^2 \setminus ([0,1] \times [-1,0])$ and the Laplace operator, $A = \text{Id}$, we approximate the exact solution given in polar coordinates (r, θ) by

$$u(r, \theta) = r^{2/3} \sin(2\theta/3). \tag{12}$$

Due to the $r^{2/3}$-singularity, $u \notin H^2(\Omega)$. The second derivatives of u however exist in $L_1(\Omega)$ and so the error of nonlinear approximation decays with #DOFs$^{-1/2}$.

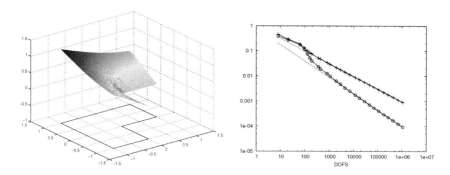

Fig. 2. Example (12): domain and approximate solution of iteration $k = 17$ (left). Log-log plot of error of the untreated ('+') and averaged ('○') gradient versus number of DOFs; the decay rates -0.5 and -0.65 are indicated by dashed lines (right)

Figure 2 and Table 1 reveal that, in spite of the singularity, the averaged gradient is superconvergent, whence the effectivity index of $\mathcal{E}_k$ approaches 1 in accordance with Proposition 1. Notice also that the decay of $\|\nabla(u_k - u)\|_A$ is optimal in that it coincides with the one of nonlinear approximation.

Table 1. Example (12): number of DOFs, error of untreated and averaged gradient, and effectivity indices for selected iterations

k	#DOFs	$\|\nabla(u_k - u)\|_A$	$\|Gu_k - \nabla u\|_A$	$\dfrac{\zeta_k}{\|\nabla(u_k - u)\|_A}$	$\dfrac{\mathcal{E}_k}{\|\nabla(u_k - u)\|_A}$
0	8	4.649e−01	3.917e−01	1.064	1.468
5	161	8.357e−02	5.235e−02	1.004	1.392
10	1 080	2.910e−02	9.808e−03	0.999	1.275
15	11 707	8.639e−03	1.747e−03	0.998	1.190
20	108 993	2.837e−03	4.000e−04	0.999	1.141
25	1 061 834	9.010e−04	9.323e−05	0.999	1.106

3.2 Anisotropic ellipticity

We conclude with an example where the condition number of the coefficient matrix A is large. Let $\Omega :=]0,1[^2$ and let A be the diagonal matrix with diagonal $(0.1, 10)$ and consider two exact solutions

$$u^i(x_1, x_2) = 10^{-(i-1)} \sin(\pi x_i), \qquad i = 1, 2. \tag{13}$$

The solutions u^1 and u^2 are smooth, have the same energy norm and profile, but depend, respectively, only on the direction associated to the minimum eigenvalue 0.1 or the maximum one 10.

To approximate both exact solutions u^1 and u^2, we employ the algorithm of [4] and the standard algorithm of ALBERTA using the maximum strategy and the explicit residual estimator η_k, equilibrated with the same constant $1/5$ as ρ_k; this leads to effectivity indices of $\sqrt{10}\,\eta_k$ close to 1 when approximating on regular meshes the solution u^1 associated with the minimum eigenvalue 0.1. In [4] all four simulation are started from a regular (structured) macro triangulation. Here we start from the irregular macro triangulation in Figure 3 (left) and obtain Table 2.

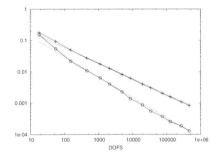

Fig. 3. Example (13): macro triangulation (left). Log-log plot of error of the untreated ('+') and averaged ('o') gradient versus number of DOFs; the decay rates -0.5 and -0.65 are indicated by dashed lines (right)

For the case $i = 1$ corresponding to the minimum eigenvalue 0.1, the safeguarded ZZ estimator $\mathcal{E}_k$ has moderate effectivity indices that appear to decrease to 1, while the residual estimator $\sqrt{10}\,\eta_k$ has quite big effectivity indices, probably due to an error component in the direction of the maximum eigenvalue that is introduced by the irregularity of the macro triangulation.

For the case $i = 2$ corresponding to the maximum eigenvalue 10, $\mathcal{E}_k$ starts with relatively big effectivity indices. However, they improve with refinement, while the ones of the residual estimator $\sqrt{10}\,\eta_k$ are always above 10.

The improving effectivity indices of $\mathcal{E}_k$ in both cases are again a consequence of the superconvergence of the averaged gradient, see Figure 3 (right) for $i = 2$, and Proposition 1.

Table 2. Example (13): number of DOFs, error and effectivity indices of $\mathcal{E}_k$ (left subcolumns) and η_k (right subcolumns) related to minimum ($i=1$) and maximum ($i=2$) eigenvalue of A

k		#DOFs		$\|\nabla(u_S - u)\|_A$		$\dfrac{\mathcal{E}_k \text{ or } \sqrt{10}\,\eta_k}{\|\nabla(u_S - u)\|_A}$	
				$i=1$			
0	0	18	18	5.785e−01	5.785e−01	1.217	4.951
3	4	492	355	1.191e−01	1.384e−01	2.227	7.031
6	9	6513	5893	3.072e−02	3.286e−02	1.753	7.344
9	17	68510	62549	9.486e−03	9.850e−03	1.624	7.381
12	33	354026	338423	4.018e−03	4.204e−03	1.461	7.391
				$i=2$			
0	0	18	18	1.775e−01	1.775e−01	5.795	11.702
3	4	458	440	2.714e−02	2.678e−02	4.821	11.568
6	7	4880	3162	8.148e−03	1.027e−02	4.009	11.642
9	10	34029	25567	3.089e−03	3.393e−03	2.861	11.728
13	14	458924	380611	8.294e−04	8.825e−04	2.728	11.723

References

1. Ainsworth, M., Oden, J.T.: A posteriori error estimation in finite element analysis Pure and Applied Mathematics (New York), Wiley-Interscience, New York (2000)
2. Carstensen, C.: All first-order averaging techniques for a posteriori finite element error control on unstructured grids are efficient and reliable. Math. Comp., **73**, 1153–1165 (2004)
3. Carstensen, C., Bartels, S.: Each averaging technique yields reliable a posteriori error control in FEM on unstructured grids. I. Low order conforming, nonconforming, and mixed FEM. Math. Comp., **71**, 945–969 (2002)
4. Fierro, F., Veeser, A.: A posteriori error estimators, gradient recovery by averaging, and superconvergence. Numer. Math. **103**, 267-298 (2006)
5. Repin, S.I.: A posteriori error estimation for approximate solutions of variational problems by duality theory. In: Bock, H.G. et al. (ed) ENUMATH 97 (Heidelberg). World Sci. Publishing, River Edge, NJ (1998)
6. Rodríguez, R.: A posteriori error analysis in the finite element method. In: Krizek, M. et al. (ed) Finite element methods (Jyväskylä). Marcel Dekker, New York, 389–397 (1994)
7. Schmidt, A., Siebert, K.G.: Design of adaptive finite element software: the finite element toolbox **ALBERTA**. Springer (2004)
8. Xu, J., Zhang, Z.: Analysis of recovery type a posteriori error estimators for mildly structured grids. Math. Comp., **73**, 1139–1152 (2004)
9. Zienkiewicz, O.C., Zhu, J.Z.: A simple error estimator and adaptive procedure for practical engineering analysis. Internat. J. Numer. Methods Engrg., **24**, 337-357 (1987)

Atmosphere and Ocean

Some Remarks on a Model for the Atmospheric Pressure in Ocean Dynamics

T. Chacón Rebollo[1], E. D. Fernández Nieto[2] and M. Gómez Mármol[1]

[1] Departamento de Ecuaciones Diferenciales y Análsis Numérico, Universidad de Sevilla
   chacon@us.es, macarena@us.es
[2] Departamento de Matemática Aplicada I, Universidad de Sevilla
   edofer@us.es.
   Research partially funded by Spanish MEC and EU Feder funds BFM 2003-07530-C02-01 Grant .

Summary. We analyse some questions concerning splitting solution techniques of non-hydrostatic models with atmospheric forcing. We prove that at the free surface the dynamic pressure must exactly vanish. We also analyse a linearised model of free surface and give simple rules to construct stable pairs of (horizontal velocities, free surfaces) for mixed discretizations.

1 Introduction

In this paper we analyse some issues regarding the mathematical modelling of the hydrodynamic forcing of the Ocean by the atmospheric pressure. Hydrodynamic forcing is relevant in ocean areas were important vertical accelerations occur, such as closed and semi-closed seas, straits, flow induced by hurricanes, etc.

Primitive equation models, extensively used in Physical Oceanography, are addressed to large oceanic zones, and only include hydrostatic pressure modelling. Consequently, these models are not suitable to simulate the flows mentioned above (Cf. [6]).

Non-hydrostatic models for ocean flows include an additional pressure (the hydrodynamic pressure) to take into account vertical acceleration effects. These models are also able to drive these flows by horizontal gradients of the atmospheric pressure. The numerical approximations of non-hydrostatic flows usually follow a time splitting into hydrostatic + hydrodynamic steps (Cf. [2, 5, 7]). In this splitting, there is a lack of boundary conditions for the intermediate unknowns, particularly for the hydrodynamic pressure. Thus, the forcing of the flow by the atmospheric pressure is only approximated. However, some of the mentioned flows are quite sensitive to small variations of the forcing conditions. This is the case, for instance, of stratified flows with slight variations of the vertical density gradient. Consequently, to simulate this kind of flows, far from hydrostaticity, it is relevant to correctly impose the atmospheric pressure forcing.

A theory supporting the well-possedness of the models and of their numerical approximations is also lacking. This has been successfully analysed by several authors in the case of the primitive equations (Cf. [8, 6, 3], but it is still an open question for non-hydrostatic models. In fact, this causes the generation of spurious solutions in the numerical solution of the free surface equations. In particular the determination of stable pairs of spaces for velocity and pressure (both hydrodynamic and hydrostatic) is required. (Cf. [7], for instance)

In this paper we address some aspects of the difficulties we have mentioned: We properly impose the atmospheric pressure as a boundary condition at the free surface for the total (hydrodynamic + hydrostatic) pressure, and we derive stable approximations of the free surface equations on a linearised steady model.

On one hand, the main innovation introduced is to give a surface boundary condition for the non-hydrostatic pressure. This is usually set to zero by heuristic reasons, but we prove that it should exactly be set to zero, if the flow is forced by the atmospheric pressure gradient. On another hand, we analyse a linear steady version of the hydrostatic step that yields the free surface. We assume known a solution of the problem, and prove its well-possedness, based upon an inf-sup condition that relates horizontal velocities and free-surfaces. We give a rule to build stable pairs of Finite Element spaces for this steady problem.

2 Modelling of non-hydrostatic free-surface flows

In this section we motivate our work by describing with some detail the main issues related to the modelling of 3D non-hydrostratic free-surface flows.

Let us consider a flow of oceanic water filling at any time $t \in [0,T]$ a domain $\Omega(t)$. We assume that this domain may be described in terms of a 2D domain $\omega(t)$, a continuous depth function $D: \omega(t) \mapsto \mathbb{R}^+$, and a continuous surface function $\eta(\mathbf{x},t) : \omega(t) \mapsto \mathbb{R}$ (we assume $\eta > D$ on $\overline{\omega}$ for simplicity):

$$\Omega(t) = \{(\mathbf{x}, z) \in \mathbb{R}^3 \text{ such that } \mathbf{x} \in \omega(t),\ -D(\mathbf{x}) < z < \eta(\mathbf{x},t)\ \}.$$

We assume that the physical behaviour of the flow is described by the velocity $\mathbf{U}$, the pressure P and the density of the water ρ, and that these variables satisfy the Boussinesq equations, forced by the gravitatory field of the Earth $(-\mathbf{g} = -(0,0,g))$ and the Coriolis forces (that we denote by $\mathbf{C}$):

$$\partial_t \mathbf{U} + (\mathbf{U}\cdot\nabla)\mathbf{U} - \nu\Delta\mathbf{U} + \mathbf{C} + \frac{1}{\rho_0}\nabla P = -\frac{\rho}{\rho_0}\mathbf{g}, \\ \nabla\cdot\mathbf{U} = 0. \qquad (1)$$

Here, ρ_0 is a mean value of density the water, which suffers only of small variations. It depends on salinity and temperature, through the state equation of the water $\rho = \rho(S,T)$. Salinity and temperature satisfy convection-diffusion equations, typically

$$\partial_t S + \mathbf{U}\cdot\nabla S - \mathcal{K}_S \Delta S = f_S.$$

Boussinesq's equations are derived from Navier-Stokes equations with the assumption that density fluctuations are only relevant in the equation describing the conservation of vertical momentum (Cf. [6]).

2.1 An exact boundary condition for the dynamic pressure

The total pressure is the sum of its hydrostatic and hydrodynamic parts,

$$P = P_H + P_D. \qquad (2)$$

P_D is linked to the incompressibility of the flow while P_H is due to the potential nature of the gravity field. As a consequence, it is determined by

$$\partial_z P_H = -\rho g = -(\rho_0 + \rho')g, \qquad (3)$$

Vertical integration of (3) from some depth z to the free surface η yields

$$P_H(z) = P_H(\eta) + \rho_0 g(\eta + b - z), \quad b = \int_z^\eta \frac{\rho'}{\rho_0}, \qquad (4)$$

where $P_H(\eta)$ is the 2D value of the hydrostatic pressure at the free-surface, and b is the hydrostatic pressure due to density fluctuations. It is called the "baroclinic" part of the hydrostatic pressure. In some models, $P_H(\eta)$ is assumed constant, so the forcing by the atmospheric pressure is neglected. To take into consideration this effect, some other models assume that the vertical equilibrium of the free surface requires (Cf. [7]) $P_H(\eta) = P_{atm}$, where P_{atm} denotes the atmospheric pressure. The horizontal gradient of the total pressure is then split into its barotropic ($\nabla_H \eta$), baroclinic ($\nabla_H b$) and hydrodynamic parts, plus a forcing term due to the atmospheric pressure:

$$\nabla_H P = \rho_0 g \nabla_H(\eta + b) + \nabla_H P_D + \nabla_H P_{atm} \qquad (5)$$

If we inject this expression in the horizontal momentum conservation equation of Boussinesq equations, we obtain a model in which apparently only the hydrostatic component of the oceanic flow is driven by the horizontal gradient of the atmospheric pressure.

However, one readily proves that the above expression (5) is exact, if a zero value for the hydrodynamic pressure at the free surface is set. Indeed, let us properly set $P_D(\eta) + P_H(\eta) = P_{atm}$, so that equation (2) is re-written as

$$P(\mathbf{x}, z) = [P_D(\mathbf{x}, z) - P_D(\eta)] + \rho_0\, g\, (\eta + b - z) + P_{atm}, \tag{6}$$

If we now re-define the hydrodynamic pressure as $Q_D = P_D + P_S$, with $P_S(\mathbf{x}) = -P_D(\eta)$, then $Q_D(\eta) = 0$, and (6) reads

$$P(\mathbf{x}, z) = Q_D(\mathbf{x}, z) + \rho_0\, g\, (\eta + b - z) + P_{atm}, \tag{7}$$

where both the hydrodynamic and the hydrostatic contributions to the pressure vanish at the free surface. Injecting this expression in model (1), we obtain the reduced model

$$\left.\begin{aligned}
\partial_t \mathbf{U}_H + (\mathbf{U}\cdot\nabla)\mathbf{U}_H - \nu\Delta\mathbf{U}_H + \mathbf{C}_H + \frac{1}{\rho_0}\nabla_H(Q_D + g\eta) &= \mathbf{f}_H, \\
\partial_t \mathbf{U}_3 + (\mathbf{U}\cdot\nabla)\mathbf{U}_3 - \nu\Delta\mathbf{U}_3 + \mathbf{C}_3 + \frac{1}{\rho_0}\partial_3 Q_D &= \mathbf{f}_3 \\
\nabla\cdot\mathbf{U} &= 0,
\end{aligned}\right\} \tag{8}$$

where

$$\mathbf{f}_H = \frac{1}{\rho_0}\nabla_H\left(-P_{atm} + \rho_0 g b\right), \quad \mathbf{f}_3 = -g\,\partial_3\,(b - z). \tag{9}$$

In this model P_S is the surface value of the hydrodynamic pressure corresponding to a forcing by the atmospheric pressure.

2.2 The hydrostatic + Non-hydrostatic splitting

The equation of the free surface of the flow domain $\Omega(t)$:

$$\partial_t \eta + \mathbf{U}_H \nabla \eta = \mathbf{U}_3 \text{ at } z = \eta(\mathbf{x}, t) \tag{10}$$

when $\nabla \cdot \mathbf{U} = 0$, is equivalent (in a convenient sense) to

$$\partial_t \eta + \nabla_H\cdot\langle\mathbf{U}_H\rangle = 0, \quad \text{on } \omega, \quad \text{where } \langle\mathbf{U}_H\rangle = \int_{-D(x)}^{\eta(\mathbf{x},t)} \mathbf{U}_H(\mathbf{x}, z)\, dz.$$

This allows to decompose the solution of problem (8) by a splitting technique in time by means of hydrostatic and hydrodynamic steps. The procedure maybe sketched as follows:

- First, a convection stage is performed to update a known velocity $\mathbf{U}^n$ into $\hat{\mathbf{U}}^{n+1}$.
- Next, in the *hydrostatic step*, the free boundary is updated:

$$\left.\begin{aligned}
\frac{\tilde{\mathbf{U}}_H^{n+1} - \hat{\mathbf{U}}_H^{n+1}}{\Delta t} - \nu\Delta\tilde{\mathbf{U}}_H^{n+1} + \mathbf{C}_H^{n+1} + \frac{g}{\rho_0}\nabla_H \eta^{n+1} &= \mathbf{f}_H^{n+1}, \\
\frac{\eta^{n+1} - \eta^n}{\Delta t} + \nabla_H\cdot<\tilde{\mathbf{U}}_H^{n+1}> &= 0
\end{aligned}\right\}, \tag{11}$$

- A *diagnostic* vertical velocity $\tilde{\mathbf{U}}_3^{n+1}$ is next computed by solving the continuity equation,
$$\partial_3 \tilde{\mathbf{U}}_3^{n+1} = \nabla_H \cdot \tilde{\mathbf{U}}_H^{n+1}.$$

- Finally, in the *hydrodynamic* step, the computed velocity is corrected to take into account vertical acceleration effects, and continuity:

$$\left. \begin{array}{r} \dfrac{\mathbf{U}_H^{n+1} - \tilde{\mathbf{U}}_H^{n+1}}{\Delta t} + \dfrac{1}{\rho_0} \nabla_H Q_D^{n+1} = 0, \\ \dfrac{\mathbf{U}_3^{n+1} - \tilde{\mathbf{U}}_3^{n+1}}{\Delta t} - \nu \Delta \mathbf{U}_3^{n+1} + \mathbf{C}_3^{n+1} + \dfrac{1}{\rho_0} \partial_3 Q_D^{n+1} = \mathbf{f}_3^{n+1}, \\ \nabla \cdot \mathbf{U}^{n+1} = 0 . \end{array} \right\} \quad (12)$$

At this stage, frequently diffusive and Coriolis forces are neglected as their size is small compared to convective ones.

To compute the hydrodynamic pressure Q_D^{n+1} most often projection techniques are used, as this is computationally less expensive than using mixed methods. This requires to solve an equation of the form

$$-\Delta Q_D^{n+1} = \sigma^{n+1}.$$

So, Dirichlet data at the free surface are needed. These data are usually set to zero, as it is assumed that this pressure is small. This is fully justified by our modelling approach, as we have seen that the exact surface boundary condition is $Q_D = 0$.

3 A linearised model for the free surface equation

We focus now on the building of stable pairs of spaces for the numerical solution of the free surface equation. Our purpose is to find ways to construct stable pairs of (horizontal velocities, free surfaces) to discretize the free surface equation.

We shall specifically focus on the solution of a linearised steady version of problem (8) with some simplifying assumptions: We assume that we already know a steady solution (velocity $\mathbf{U}$, free surface η and hydrodynamic pressure Q). We fix the domain Ω given by η, and consider Q as a data that we integrate into the forcing terms. We also consider $\mathbf{U}$ as a data in the convection term. We finally neglect the Coriolis forces, as these are not relevant for our analysis, and assume $\rho_0 = g = 1$. This yields the linear problem:

Find $\mathbf{U}_H : \Omega \mapsto \mathbb{R}^2, \quad \eta : \overline{\omega} \mapsto \mathbb{R}$ such that

$$\left. \begin{array}{r} \mathbf{U} \cdot \nabla \mathbf{U}_H - \nu \Delta \mathbf{U}_H + \nabla_H \eta = \mathbf{f_H}, \text{ in } \Omega, \\ \nabla_H \cdot <\mathbf{U}_H> = 0 \quad \text{ in } \omega, \end{array} \right\} . \quad (13)$$

We set the following boundary conditions,

$$-\nu\,\partial_n \mathbf{U}_H = \tau_w \text{ on } \Gamma_s, \quad \mathbf{U}_H = 0 \text{ on } \Gamma_b \cup \Gamma_l$$

where Γ_l is a vertical piece of the boundary of Ω formed from a subset γ of $\partial \omega$ as $\Gamma_l = \{(\mathbf{x}, z) \in \mathbb{R}^3 \text{ s.t. } -D(\mathbf{x}) < z < \eta(\mathbf{x}), \mathbf{x} \in \gamma\}$, and Γ_s, Γ_b are the surface and the bottom of the domain, defined as
$\Gamma_s = \{(\mathbf{x}, z) \in \mathbb{R}^3 \text{ s.t. } z = \eta(\mathbf{x}), \mathbf{x} \in \omega\}$, $\Gamma_b = \{(\mathbf{x}, z) \in \mathbb{R}^3 \text{ s.t. } z = -D(\mathbf{x})\}$.
The vertical velocity is recovered by integration of the continuity equation, for this reason is not considered in the PDE system above.

We shall assume the functions η and D to be Lipschitz continuous.

The condition on Γ_s models the wind friction on the surface boundary layer. The no-slip conditions on $\Gamma_b \cup \Gamma_l$ has been set for the sake of simplicity. For the same reason we have not included Coriolis forces in the above model.

Consider the velocity space $H_b^1(\Omega) = \{\mathbf{V} \in H^1(\Omega) \text{ s.t. } \mathbf{V} = 0 \text{ on } \Gamma_b \cup \Gamma_l\}$. We shall look for $\mathbf{U}_H$ in $[H_b^1(\Omega)]^2$.

Our variational formulation is based upon the observation that if $\sigma \in L^2(\Omega)$ and, $\partial_3 \sigma = 0$, then for any $\mathbf{V} \in H_b^1(\Omega)^2$, we have

$$\begin{aligned}(\nabla_H \sigma, \mathbf{V}_H)_\Omega &= (\sigma, \mathbf{V}_H \cdot n_H)_{\partial\Omega} - (\sigma, \nabla_H \cdot \mathbf{V}_H)_\Omega \\ &= (\sigma, \mathbf{V}_H \cdot n_H)_{\partial\Omega} - \int_\omega \sigma \left[\int_{-D(\mathbf{x})}^{\eta(\mathbf{x})} \nabla_H \cdot \mathbf{V}_H \, dz\, d\mathbf{x}\right] \\ &= (\sigma, \mathbf{V}_H \cdot n_H)_{\partial\Omega} - \int_\omega \sigma \left[-\mathbf{V}_H(\eta) \cdot \nabla_H \eta\right] d\mathbf{x} \\ &\quad - \int_\omega \sigma \nabla_H \cdot <\mathbf{V}_H> d\mathbf{x},\end{aligned}$$

where the first integral is understood as a duality pairing.

Using that $z = \eta(x, y)$ is a parameterization of Γ_s, we deduce

$$(\sigma, \mathbf{V}_H \cdot n_H)_{\partial\Omega} = (\sigma, \mathbf{V}_H \cdot n_H)_{\Gamma_s} = \int_\omega \sigma \left[-\mathbf{V}_H(\eta) \cdot \nabla_H \eta\right] d\mathbf{x},$$

so that

$$(\nabla_H \sigma, \mathbf{V}_H)_\Omega = -(\sigma, \nabla_H \cdot <\mathbf{V}_H>)_\omega, \quad \forall \mathbf{V}_H \in [H_b^1(\Omega)]^2. \tag{14}$$

We shall look for η in the the pressure/free-surface space

$$L_{S,0}^2(\Omega) = \{q_s \in L_0^2(\Omega) \text{ s.t. } \partial_z q_s = 0.\}$$

We give the following variational formulation to the steady version of Problem (11):

Obtain $\mathbf{U}_H \in [H_b^1(\Omega)]^2, \eta \in L_{S,0}^2(\Omega)$ such that

$$\left.\begin{aligned}a(\mathbf{U}_H, \mathbf{V}_H) + b(\eta, \mathbf{V}_H) &= L(\mathbf{V}_H), \quad \forall \mathbf{V}_H \in [H_b^1(\Omega)]^2 \\ b(\sigma, \mathbf{U}_H) &= 0, \qquad \forall \sigma \in L_{S,0}^2(\Omega),\end{aligned}\right\}, \tag{15}$$

where

$$a(\mathbf{U}_H, \mathbf{V}_H) = (\mathbf{U} \cdot \nabla \mathbf{U}_H, \mathbf{V}_H)_\Omega + \nu(\nabla \mathbf{U}_H, \nabla \mathbf{V}_H)_\Omega,$$

$$b(\sigma, \mathbf{V}_H) = -(\sigma, \nabla_H \cdot <\mathbf{V}_H>)_\omega, \quad L(\mathbf{V}_H) = <\mathbf{f_H}, \mathbf{V}_H>_\Omega - <\tau_w, \mathbf{V}_H>_{\Gamma_s}.$$

Here, we recall that the convection velocity $\mathbf{U}$ is considered as a data.

To prove that this variational formulation yields a weak form of problem (13), one must at first consider identity (14) to recover the first equation and the boundary conditions on $\partial \Omega$.

Also, as η and D are assumed to be Lipschitz continuous functions, then $\nabla_H \cdot <\mathbf{U}_H> \in L^2(\Omega)$, and the following estimate holds,

$$\|\nabla_H \cdot <\mathbf{U}_H>\|_{0,\Omega} \leq C \left(\|\nabla \mathbf{U}_H\|_{0,\Omega} + \|\mathbf{U}_H\|_{0,\Gamma_s} + \|\mathbf{U}_H\|_{0,\Gamma_b} \right),$$

where the constant C depends on $\|\eta + D\|_{\infty,\omega}$, $\|\nabla_H \eta\|_{\infty,\omega}$, $\|\nabla_H D\|_{\infty,\omega}$. The second identity in (15) then implies

$$\nabla_H \cdot <\mathbf{U}_H> = m, \quad \text{for some constant } m.$$

Now, if we take $\eta = 1$ and $\mathbf{V} = \mathbf{U}$ in (14), we deduce $m = 0$.

The well-possedness of problem (15) lies on the following inf-sup condition, reported in Chacón-Guillén [3]:

Lemma 1. *Assume that Ω is a bounded domain of $\mathbb{R}^d$, ($d = 2$ or 3) with a Lipschitz-continuous boundary. Then there exists a constant $\beta > 0$ depending on Ω and the dimension d such that*

$$\forall \sigma \in L^2_{S,0}(\Omega), \quad \beta \|\sigma\|_{0,\Omega} \leq \sup_{\mathbf{V}_H \in H^1_0(\Omega)^d} \frac{(\nabla_H \cdot <\mathbf{V}_H>, \sigma)_\omega}{\|\nabla \mathbf{V}_H\|_{0,\Omega}}. \tag{16}$$

This result lets us to prove our main result, stated as follows:

Theorem 1. *Under hypotheses of Lemma 1, assume $\mathbf{U} \in H^1(\Omega)^3$, $\tau_w \in H^{-1/2}(\Gamma_s)^2$, $\mathbf{f} \in H^{-1}(\Omega)^2$. Then, Problem (15) is well posed: It admits a unique solution $\mathbf{U}_H \in [H^1_b(\Omega)]^2, \eta \in L^2_{S,0}(\Omega)$ that satisfies the estimates*

$$\nu \|\nabla \mathbf{U}_H\|_{0,\Omega} \leq C M \tag{17}$$

$$\|\eta\|_{0,\Omega} \leq C (1 + \|\mathbf{U}\|_{1,\Omega}) M \tag{18}$$

where $M = \left(\|\tau_w\|_{-1/2,\Gamma_s} + \|\mathbf{f}\|_{-1,\Omega} \right)$, for some constant C depending on the domain Ω and the space dimension d.

Proof. As $H^1_0(\Omega)^d \subset H^1_b(\Omega)^d$, then the inf-sup (16) also holds between $L^2_{S,0}(\Omega)$ and $H^1_b(\Omega)^d$. The result follows using the standard theory for saddle point problems (Cf. [1]). ∎

3.1 Some hints for mixed approximation

The main interest of the above derivation lies on the fact that it may be applied to obtain stable pairs of (Horizontal velocities, free-surface) discrete spaces for mixed approximations of the free-surface problem.

The derivation of the inf-sup condition of [3] allows to construct stable pairs of Finite Elements in a simple way: Given a standard stable 3D pair of finite element spaces, say (Y_h, Q_h), consider the space X_h formed by the horizontal components of the velocities of Y_h and the pressures of M_h that do not depend on z. There exists a constant $\beta > 0$ such that

$$\forall q_h \in Q_h \ \beta \|q_h\|_{0,\Omega} \leq \sup_{\mathbf{V}_h \in Y_h} \frac{(\nabla \cdot \mathbf{V}_h, q_h)_\Omega}{\|\nabla \mathbf{V}\|_{0,\Omega}}.$$

If $\partial_z q_h = 0$, then $(\nabla \cdot \mathbf{V}_h, q_h)_\Omega = (\nabla_H \cdot <\mathbf{V}_{hH}>, q_h)_\omega$, and then

$$\beta \|q_h\|_{0,\Omega} \leq \sup_{\mathbf{V}_h \in X_h} \frac{(\nabla_H \cdot <\mathbf{V}_{hH}>, q_h)_\omega}{\|\nabla \mathbf{V}_{hH}\|_{0,\Omega}}.$$

So, this pair (X_h, M_h) satisfies a discrete inf-sup condition similar to (16). Now, we may build a mixed stable approximation of problem (15): Let us consider the approximated problem

Obtain $\mathbf{U}_h \in X_h$, $\eta_h \in M_h$ such that

$$\left.\begin{aligned} a(\mathbf{U}_h, \mathbf{V}_H) + b(\eta_h, \mathbf{V}_h) &= L(\mathbf{V}_h), \quad \forall \mathbf{V}_h \in X_h, \\ b(\sigma_h, \mathbf{U}_h) &= 0, \quad \forall \sigma_h \in M_h. \end{aligned}\right\} \tag{19}$$

Then, we may conclude the following

Theorem 2. *Assume the family of pairs of spaces $\{(X_h, M_h)\}_{h>0}$ is a convergent internal approximation of $[H_b^1(\Omega)]^2 \times L_{S,0}^2(\Omega)$ satisfying a discrete inf-sup condition for form b. Then problem (19) admits a unique solution that converges strongly in $[H_b^1(\Omega)]^2 \times L_{S,0}^2(\Omega)$ to the solution of problem (15).*

This analysis yields some relevant indications to build stable computations of steady free surfaces. The linear problem (13) is a hydrostatic sub-problem of the general Boussinesq equations (8), and any stable solver of these last equations must be able to solve our simplified problem. This suggests to use pairs of spaces derived as above to solve the transient hydrostatic equations.

References

1. Brezzi, F., Fortin, M.: Mixed and Hybrid finite element methods Springer - Verlag (1991)
2. Casulli, V.: A semi-implicit Finite difference method for non-hydrostatic, free-surface flows. Int. J. Numer. Meth. Fluids, **30**, 425–440 (1999)

3. Chacón Rebollo, T., Guillén González, F.: An intrinsic analysis of Existence of Solutions for the Hydrostatic Approximation of Navier-Stokes Equations. C. R. Acad. Sci. Paris,Série I **330**, 841–846 (2000)
4. http://www.damflow.com Modelling system of surface and ocean water flows.
5. Hervouet, J.M.: Hydrodynamique des Écoulements à Surface Libre. Eds. Ponts et Chaussées, Paris (2003)
6. Lewandowski, R.: Analyse Mathématique et Ocánographie. Masson, Paris (1977)
7. Saleri, F.: Numerical approximation for non hydrostatic 3D shallow water system. In: INRIA Course Support, October 2002. INRIA (2002)
8. Lions,J.-L., Temam, R., Wang, S.: On the equations of the large-scale ocean. Nonlinearity **5**, 1007–10053 (1992)

Computational Time Improvement for Some Shallow Water Finite Volume Models Applying Parallelization and Optimized Small Matrix Computations.

M. J. Castro, J. A. García, J. M. González and C. Parés

Dpt. Análisis Matemático, Facultad de Ciencias, Universidad de Málaga, Campus de Teatinos s/n, 29071 Málaga, Spain
castro,joseanto,vida,pares@anamat.cie.uma.es

Summary. The goal of this paper is to construct efficient finite volume parallel solvers on non-structured grids for 2d hyperbolic systems of conservation laws with source terms and nonconservative products using SIMD registers. Line method is applied: at every intercell a projected Riemann problem along the normal direction is considered (see [2]). The resulting 2d numerical schemes are explicit and first order accurate. The solver is parallelized following a SIMD approach, by means of SSE (*"Streaming SIMD Extensions"*), which are present in common processors. A generic C++ wrapper to small matrices libraries that make use of SIMD instructions has been implemented in an efficient way and an application to IPP small matrix library is presented.

1 Introduction

This article deals with the development of efficient implementations of finite volume solvers on non-structured grids for 2d hyperbolic systems of conservation laws with source terms and nonconservative products. We are concerned in particular with the simulation of one or two layer fluids that can be modelled by the shallow water systems, formulated under the form of a conservation law with source terms or *balance law*. We are mainly interested in the application of these systems to geophysical flows: models based on shallow water systems are useful for the simulation of rivers, channels, dambreak problems, etc... Simulating this phenomena leads to very long lasting simulations in big computational domains, so extremely efficient solvers are needed to solve and analyze these problems in small computational time. In [2] an efficient implementation of the first order well-balanced numerical scheme for general systems of balance laws with nonconservative products was carried out using domain decomposition techniques and MPI in a PC cluster. Very good speed-up results were obtained and the scheme was assessed with numerical and experimental data. In this article we follow a different approach to reduce calculus time. This kind of algorithms essentially consist of performing a huge number of small matrix computations, similar to those carried out in 3d software, CAD, physics computation for games, etc. Modern CPU's are provided with specific SIMD units devoted to these purposes. We introduce a technique to develop a high level C++ small matrix library that takes advantage of SIMD registers, hiding the difficulties related to the use of very low level coding (mostly assembler).

The organization of the article is as follows: in the second section, we present the general formulation of systems of balance laws with nonconservative products and source terms in 2d domains. Next, the numerical scheme is presented for the general case. Section 4 is devoted to SSE description and the description of the high level C++ matrix library implementation.

2 Equations

We consider a general problem consisting of a system of conservation laws with non conservative products and source terms given by:

$$\frac{\partial W}{\partial t} + \frac{\partial F_1}{\partial x_1}(W) + \frac{\partial F_2}{\partial x_2}(W) = B_1(W) \cdot \frac{\partial W}{\partial x_1} + B_2(W) \cdot \frac{\partial W}{\partial x_2} \quad (1)$$
$$+ S_1(W)\frac{\partial H}{\partial x_1} + S_2(W)\frac{\partial H}{\partial x_2},$$

where $W(mx,t): D \times (0,T) \mapsto \Omega \subset \mathbb{R}^N$, being D a bounded domain of $\mathbb{R}^2$; $mx = (x_1, x_2)$ denotes an arbitrary point of D; Ω is an open convex subset of $\mathbb{R}^N$. Finally $F_i: \Omega \mapsto \mathbb{R}^N$, $B_i: \Omega \mapsto \mathcal{M}_N$, $S_i: \Omega \mapsto \mathbb{R}^N$, $i = 1, 2$, are regular functions, and $H: D \mapsto \mathbb{R}$ is a known function. Observe that if $B_1 = B_2 = S_1 = S_2 = 0$, (1) is a system of conservation laws; and if $B_1 = B_2 = 0$, (1) is a system of conservation laws with source term or balance law. The shallow water Systems are particular cases of this general problem (see [2]). Let $J_i(W) = \frac{\partial F_i}{\partial W}(W)$, $i = 1,2$ denote the Jacobians of the fluxes F_i, $i = 1, 2$. Given a unit vector $m\eta = (\eta_1, \eta_2) \in \mathbb{R}^2$, we define the matrix

$$\mathcal{A}(W, m\eta) = J_1(W)\eta_1 + J_2(W)\eta_2 - (B_1(W)\eta_1 + B_2(W)\eta_2).$$

We assume here that (1) is strictly hyperbolic, i.e. for every W in Ω and every unit vector $m\eta \in \mathbb{R}^2$, $\mathcal{A}(W, m\eta)$ has N real distinct eigenvalues so that $\mathcal{A}(W, m\eta) = \mathcal{K}(W, m\eta)\mathcal{D}(W, m\eta)\mathcal{K}^{-1}(W, m\eta)$, where $\mathcal{D}(W, m\eta)$ is the diagonal matrix whose coefficients are the eigenvalues of $\mathcal{A}(W, m\eta)$ and $\mathcal{K}(W, m\eta)$ is a matrix whose columns are associated eigenvectors. Notice that the non-conservative products $B_1(W)\partial_{x_1}W$, $B_2(W)\partial_{x_2}W$, do not make sense in the framework of distributions for discontinuous solutions. The problem of giving a sense to the solution is difficult, and we refer to [4] and [7].

3 Numerical scheme

In this section we present the discretization of System (1) by means of a finite volume scheme. First, the computational domain is divided into discretization cells or finite volumes, $V_i \subset \mathbb{R}^2$, which are supposed to be closed polygons. Let us denote by $\mathcal{T}$ the set of cells. Hereafter we will use the following notation: given a finite volume V_i, $N_i \in \mathbb{R}^2$ is the center of V_i, $\mathcal{N}_i$ is the set of indexes j such that V_j is a neighbor of V_i; Γ_{ij} is the common edge of two neighbor cells V_i and V_j, and $|\Gamma_{ij}|$ its length; $m\eta_{ij} = (\eta_{ij,1}, \eta_{ij,2})$ is the normal unit vector to the edge Γ_{ij} and points toward the cell V_j. The approximations to the cell averages of the exact solution produced by the numerical scheme will be denoted as follows:

$$W_i^n \cong \frac{1}{|V_i|} \int W(x_1, x_2, t^n) dx_1 dx_2, \qquad (2)$$

where $|V_i|$ is the area of the cell and $t^n = n\Delta t$, being Δt the time step which is supposed to be constant for simplicity. Let us suppose that the approximations at time t^n, W_i^n, have been yet calculated. To advance in time, a projected Riemann Problem is considered at every edge Γ_{ij}, obtaining a 1d system of conservation laws with source terms and nonconservative product as those studied in [3]. Following this work, this one-dimensional problem is discretized by means of a generalized Q-scheme of Roe. W_i^{n+1} is then calculated by averaging the approximations obtained at every edge. The resulting scheme is as follows (see [2]):

$$W_i^{n+1} = W_i^n - \frac{\Delta t}{|V_i|} \sum_{j \in \mathcal{N}_i} |\Gamma_{ij}| F_{ij}^-, \qquad (3)$$

where $F_{ij}^- = \mathcal{P}_{ij}^-(\mathcal{A}_{ij}(W_j^n - W_i^n) - \mathcal{S}_{ij}(H_j - H_i))$, with $\mathcal{A}_{ij} = \mathcal{A}(W_{ij}, m\eta_{ij})$; W_{ij}^n an "intermediate state" between W_i^n and W_j^n; and

$$\mathcal{P}_{ij}^- = \frac{1}{2}\mathcal{K}_{ij} \cdot (I - \text{sgn}(\mathcal{D}_{ij})) \cdot \mathcal{K}_{ij}^{-1},$$

$$\mathcal{S}_{ij} = \eta_{ij,1} S_1(W_{ij}^n) + \eta_{ij,2} S_2(W_{ij}^n),$$

being $\mathcal{D}_{ij}$ the diagonal matrix whose coefficients are the eigenvalues of $\mathcal{A}_{ij}$, and $\mathcal{K}_{ij}$ a matrix whose columns are associated eigenvectors. Finally $\text{sgn}(\mathcal{D}_{ij})$ is the diagonal matrix whose coefficients are the sign of the eigenvalues of matrix $\mathcal{A}_{ij}$.

4 Parallel SIMD implementation

In [2] a parallelization of the resulting algorithm based in domain decomposition techniques was carried out. However, if we want to obtain a bigger reduction referring to CPU time with a medium cluster, it is necessary to make a better use of the computational power at each node. As the more demanding operations in this algorithms are matrix computations, our main interest is to have an efficient small matrix library. Performing small matrix operations in commodity processors in a more efficient way can be achieved using SSE instructions. SSE provides a set of eight registers (16 in 64 bit processors) of 128 bits each one, that can store data in 128 bits, 64 bits, 32 bits, 16 bits, etc; and a set of functions providing the elementary algebra. To make use of SIMD registers we must program using assembler or intrinsics, which are not well suited to develop numerical methods due to their lack of portability and because the obtained code is hard to debug. In this section we present a general framework for the development of a generic C++ matrix library with high level characteristics on top of matrix libraries developed making use of SIMD instruccions for common processors. To achieve this goal we make use of the advanced characteristics of C++.

4.1 Application to the intel IPP small matrix library

The *"Intel Performance Primitives"* (IPP) are a set of numerical libraries developed by Intel to help software developers to make use of SSE registers using a higher programming level than rough assembler. These libraries are grouped in several categories: video and audio compression, cryptography, signal analysis and Fourier transform, small matrix and vector operations for physics modeling, etc. The small matrix library (up to size 6×6) is well suited for the kind of problems we are interested in: one layer and two layer flows, as it focuses in the matrix sizes we need and contains all the necessary operations. A detailed description of this library can be seen at [6]. If we want, for example, to add two matrices, A, B of size 6×6 and type 64 bits, and store the result in C, the code would read as follows:

```
ippmAdd_mm_64f_6x6(A,LR,B,LR,C,LR);
```

where LR denotes de distance in bits between the columns of the matrix (in this example LR=48=6×8 bits). So, the name of each function depends on the data type, on the matrix size and on the distance in bites between columns. This implementation would lead us to a difficult to debug code, which is not desirable for implementing numerical methods. To obtain a portable implementation that can benefit from SSE we can build a C++ wrapper.

Overloading usual matrix operations

To develop a more readable code, it is necessary to employ the concept of *operator overloading*, present in object oriented programming languages. So, the main work to develop our matrix library is to implement the most common operator overloading, that afterwards will be used in the finite volume code. To do this in an efficient way, sophisticated C++ techniques must be used.

1. *Templates*

    As we have seen, it is necessary to distinguish the data type which we are using and the size of the matrix to consider, in the development of the C++ matrix library. This decision must be taken in precompile time, avoiding the use of conditionals, so we have used *"templates"* to implement the class `Matrix`, in a generic way for the case of simple and double precision and for different matrix sizes; all these characteristics will be parameters that will be passed as arguments to the class `Matrix`. In this way, `Matrix<TYPE>A,B;` creates a matrix of the given `TYPE`, where `TYPE` refers to matrices of size 3×3 or 6×6 in single or double precision. For each type and size of matrices, we must define the basic operations, using the optima function from IPP library.

2. *Avoiding using temporal variables*

    If we carry out a traditional operator overloading, a great part of the improvement of the calculus time is lost (see Section 4.2, Table 1). This is due to the creation of temporal variables. When we carry out a binary operation in a processor the process in which this operation is carried out is the following: for example, to perform the addition of two matrices (A,B), that is $C = A + B$, in the computer, a temporal variable is created, $\widetilde{C}$, where the result of the addition is stored and then assigned to variable C. As we will see in Section 4.2, creating this temporal variable in memory can nearly double the calculus time. To overcome this difficulty, we use a technique described in [9], that consists of creating a class, ``MMsum'', that does not perform any operation, but saving references to the operands that take place in the operation, as follows:

    ```
    class Matrix;
    class MMsum {
     public:
     const Matrix& m0;
     const Matrix& m1;
     MMsum(const Matrix&mm0,const Matrix& mm1): m0(mm0),m1(mm1) {};
     operator Matrix();
     friend inline MMsum operator+(const Matrix& mm0,const Matrix& mm1)
     {
        return MMsum(mm0,mm1);
     }
    };
    MMsum::operator Matrix() {
       Matrix m;
       ippmAdd_mm_64f_6x6(m0.v,LR,m1.v,LR,m.v,LR);
       return m;
    }
    Matrix& operator=(const MMsum& m){
        ippmAdd_mm_64f_6x6(m.m0.v,LR,m.m1.v,LR,v,LR);
        return (*this);
    }
    ```

With this implementation the compiler does the following: when it finds an expression of the type C = A + B, it begins reading from right to left, and when it finds two matrices separated by the sign "+", the compiler identifies it as an object of the class MMsum, then it saves the corresponding references to the operands A, B and C and the operation type; after that, it continues reading till finding the sign "=" and a matrix on the other side. Then it looks for the operation corresponding to this case (after considering all the operands) and it performs this operation: in this example it should choose the operation defined by Matrix& operator= (ippmAdd_mm_64f_6x6 in this example).

3. *Function inlining*

Finally, another aspect to consider if we want to achieve a efficient implementation is the use of *function inlining*, as we want to call very small functions many times and we do not want the program to go and search for them in execution time.

One of the main advantages of using this matrix library is that this technique can be easily mixed with a domain decomposition based implementation.

4.2 Performance tests for the C++ matrix library

We will present only results for the case of matrices of size 3×3; similar results are obtained for matrices of size 6×6 (see [5]).

Comparison between the wrapper efficiency and the original matrix library

In this section we present comparisons of the performance of the implemented C++ wrapper and the direct use of IPP functions. Times corresponding to the implementation of the overloading using temporal variables (the usual implementation) are also presented. The referred operations are carried out 1.000.000.000 times in order to be able to measure a significant calculus time. Note that the differences in performance between the functions of IPP and our C++ library are neglectible. A fact to consider is that using a temporal variable to define the overloading of operators doubles the calculus time (see Table 1).

Matrix operations test

In this section we present a comparison between the developed matrix library and some usual C++ matrix libraries. To carry out this comparison we have considered Newmat v10.0, which is a C++ matrix library with the usual matrix operations and Gmm++, which is based in Blas and also contains the needed matrix operations. We have performed typical operations used in our finite volume algorithm. In the tables we will use the following notation: V will

Table 1. Efficiency of the C++ wrapper: 3 × 3.

Operations	IPP	Optimized wrapper	Wrapper with temporals
V+V	0m 7.263s	0m 7.266s	0m 10.772s
M+M	0m 18.660s	0m 18.667s	0m 29.312s
M · V	0m 13.130s	0m 13.145s	0m 24.533s
M · V+V	0m 16.387s	0m 16.391	0m 29.852s
M^{-1}	0m 44.735s	0m 44.745s	0m 49.s97s

mean vectors of 3 components, and M matrices of 3 × 3. Again, the referred operations are carried out 1.000.000.000 times in order to be able to measure a significant calculus time. Note that for the reference operation in our case, that is $M \cdot D \cdot M^{-1} \cdot V$, we are able to reduce 48 times the calculus time if compared to Newmat, which is possibly the most used free C++ matrix library.

Table 2. Calculus time: Different matrix libraries performance: 3 × 3.

Operaciones	Newmat v. 10.0	Gmm++	IPP wrapper
M+M	10m 2.675s	6m 4.012s	0m 18.657s
M·V	11m 6.830s	1m 24.316s	0m 13.145s
M·V+V	14m 51.380s	3m 10.810s	0m 16.381s
M^{-1}	35m 22.920s	7m 20.120s	0m 44.745s
$M \cdot D \cdot M^{-1} \cdot V$	85m 41.870s	16m 52.053s	1m 48.260s

4.3 Numerical performance of the matrix library

We consider a rectangular channel of $1\,m$ width and $10\,m$ long with a bump placed at the middle of the domain given by the depth function $H(x_1, x_2) = 1 - 0.2\, e^{-(x_1-5)^2}$. Three meshes of the domain are constructed with 2590, 5162 and 10832 volumes respectively. The initial condition is $mq(x_1, x_2) = m0$, and:

$$h(x_1, x_2) = \begin{cases} H(x_1, x_2) + 0.7 & \text{if } 4 \leq x_1 \leq 6, \\ H(x_1, x_2) + 0.5 & \text{other case.} \end{cases} \quad (4)$$

The numerical scheme is run in the time interval $[0, 10]$ with $CFL = 0.9$. Wall boundary conditions $mq \cdot m\eta = 0$ are considered. Table (3) shows the CPU time for each run. As it can be seen in Figure 1 the linearity of the speed-up of the domain decomposition parallelization noticeably diminishes for meshes 1 and 3 in the one layer case, with respect to the case in which IPP are not used. This phenomena is due to the fact that, due to the great efficiency of the SSE parallelization, the calculus time for each iteration in

Table 3. Calculus time: meshes 1, 2 and 3.

CPUs.	mesh 1		mesh 2		mesh 3	
	SSE	NON-SSE	SSE	NON-SSE	SSE	NON-SSE
1	0m 18.507s	4m 52.201s	0m 51.764s	14m 16.735s	3m 5.985s	50m 21.319s
2	0m 10.685s	2m 32.606s	0m 29.066s	7m 6.800s	1m 38.830s	25m 25.037s
4	0m 6.876s	1m 17.556s	0m 17.078s	3m 38.655s	0m 53.459s	12m 43.717s
8	0m 4.340s	0m 40.120s	0m 10.032s	1m 51.360s	0m 29.315s	6m 26.135s

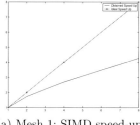
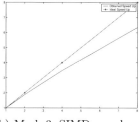

(a) Mesh 1: SIMD speed-up.   (b) Mesh 3: SIMD speed-up.

Fig. 1. Speed-up for meshes 1 and 3: one layer model.

each node is very small, so most of the time is spent in communications. The efficiency of mixing both kinds of parallelism increases with the mesh size. To explain this behaviour, we consider a much finer mesh than mesh number 3 (mesh4, with 244.163 volumes) to compute again test 1 and compare the speed-up (see Table 4).

Table 4. Speed-up: meshes 3 and 4.

N. CPUs.	1	2	4	8
Time for mesh 4	25m 26.436s	12m 53.427s	6m34.203s	3m24.476s
Speed-up for mesh 3	1	1.8818	3.4790	6.3443
Speed-up for mesh 4	1	1.9736	3.8722	7.4651

References

1. Bermúdez, A., Vázquez-Cendón, M.E.: Upwind methods for hyperbolic conservation laws with source terms. Computers and Fluids, **23**(8), 1049–1071 (1994)
2. Castro, M.J., García, J.A. , González, J.M., Parés, C.: A parallel 2D finite volume scheme for solving systems of balance laws with nonconservative products: application to shallow flows. Computer Methods in Applied Mechanics and Engineering, to appear (2006)

3. Castro, M.J., Macías, J., Parés, C.: A Q-Scheme for a class of systems of coupled conservation laws with source term. Application to a two-layer 1-D Shallow Water system. ESAIM: M2AN, **35**(1), 107–127 (2001)
4. Dal Maso, G., LeFloch, P.G., Murat, F.: Definition and weak stability of nonconservative products. J. Math. Pures Appl. **74**, 483–548 (1995)
5. García Rodríguez, J.A.: Paralelización de esquemas de volúmenes finitos: aplicación a la resolución de sistemas de tipo aguas someras. MA Thesis, Universidad de Málaga, (2005)
6. Intel Corporation: Intel Integrated Performance Primitives for Intel Architecture. Reference Manual. Volume 3: Small Matrices. Document Number: A68761-005, (2004)
7. LeFloch, P.G.: Hyperbolic systems of conservation laws, the theory of classical and nonclassical shock waves. ETH Lecture Note Series, Birkauser, (2002)
8. Parés, C., Castro, M.J.: On the well-balance property of Roe's method for nonconservative hyperbolic systems. Applications to shallow-water systems. ESAIM: M2AN, **38**(5), 821–852 (2004)
9. Stroustrup, B.: The C++ Programming Language. Addison Wesley, (1997)

CONTROL

Discretization Error Estimates for an Optimal Control Problem in a Nonconvex Domain

Th. Apel[1], A. Rösch[2] and G. Winkler[1]

[1] Institut für Mathematik und Bauinformatik, Fakultät für Bauingenieur- und Vermessungswesen, Universität der Bundeswehr München, 85577 Neubiberg, Germany
  `thomas.apel@unibw.de`
  `gunter.winkler@unibw.de`
[2] Johann Radon Institute for Computational and Applied Mathematics (RICAM), Austrian Academy of Sciences, Altenbergerstraσe 69, A-4040 Linz, Austria
  `arnd.roesch@oeaw.ac.at`

Summary. An optimal control problem for a 2-d elliptic equation and with pointwise control constraints is investigated. The domain is assumed to be polygonal but non-convex. The corner singularities are treated by a priori mesh grading. A second order approximation of the optimal control is constructed by a projection of the discrete adjoint state. Here we summarize the results from [1] and add further numerical tests.

1 Introduction

This paper is concerned with the a 2-d elliptic optimal control problem with pointwise control constraints. The state and the adjoint state are discretized by continuous, piecewise linear functions on a family of graded finite element meshes. The control is initially discretized with piecewise constants on the same meshes, but this control is used only for solving the system of discretized equations. Finally, an improved control is constructed by postprocessing the adjoint state. This approach was suggested and analysed for sufficiently smooth solutions by Meyer and Rösch [3]. The results of our analysis [1] of the case of non-smooth solutions are summarized in Section 2.

In Section 3, we present some new numerical tests of this method. It can be seen that graded meshes are indeed suited to retain the convergence order of smooth solutions in the non-smooth case. Moreover, we see that the boundary between active and non-active controls is approximated well although the method does not specially target to this aim. The results show that it is not necessary to adapt the mesh to these a priori unknown curves.

2 Theory

In this section, we summarize our results from [1]; and therefore we closely follow that paper. We consider the elliptic optimal control problem

$$J(\bar{u}) = \min_{u \in U_{\mathrm{ad}}} J(u), \quad J(u) := F(Su, u), \tag{1}$$

$$F(y, u) := \frac{1}{2}\|y - y_d\|_{L^2(\Omega)}^2 + \frac{\nu}{2}\|u\|_{L^2(\Omega)}^2, \tag{2}$$

where the associated state $y = Su$ to the control u is the weak solution of the state equation

$$Ly = u \quad \text{in } \Omega, \quad y = 0 \quad \text{on } \Gamma = \partial\Omega, \tag{3}$$

and the control variable is constrained by

$$a \leq u(x) \leq b \quad \text{for a.a. } x \in \Omega. \tag{4}$$

The function $y_d \in L^\infty(\Omega)$ is the desired state, a and b are real numbers, and the regularization parameter $\nu > 0$ is a fixed positive number. Moreover, $\Omega \subset \mathbb{R}^2$ is a bounded polygonal domain with boundary Γ. The set of admissible controls is $U_{\mathrm{ad}} := \{u \in L^2(\Omega) : a \leq u \leq b \text{ a.e. in } \Omega\}$. The second order elliptic operator L is defined by

$$Ly := -\nabla \cdot (A\nabla y) + \mathbf{a} \cdot \nabla y + a_0 y, \tag{5}$$

where the coefficients $A = A^T \in C(\bar{\Omega}, \mathbb{R}^{2\times 2})$, $\mathbf{a} \in C(\bar{\Omega}, \mathbb{R}^2)$, $a_0 \in C(\bar{\Omega})$, satisfy the usual ellipticity and coercivity conditions $\xi^T A \xi \geq m_0 \xi^T \xi$ for all $\xi \in \mathbb{R}^2$ and $a_0 - \frac{1}{2}\nabla \cdot \mathbf{a} \geq 0$.

We focus on state equations with non-smooth solutions. Let us assume that the domain $\Omega \subset \mathbb{R}^2$ has exactly one reentrant corner with interior angle $\omega > \pi$ located at the origin. Due to the local nature of corner singularities in elliptic problems this not a loss of generality. We denote by $r := r(x) = |x|$ the Euclidean distance to this corner. The solution of the elliptic boundary value problem

$$Ly = g \quad \text{in } \Omega, \quad y = 0 \quad \text{on } \Gamma,$$

has typically an r^λ-singularity where $\lambda \in (1/2, 1)$ is a real number which is defined by the coefficient matrix A and the angle ω. In the case of the Dirichlet problem for the Laplace operator, the value of λ is explicitly known, $\lambda = \pi/\omega$. In more general cases this can also be computed.

Via (2), the operator S associates a state $y = Su$ to the control u. We denote by S^* the solution operator of the adjoint problem

$$L^* p = y - y_d \quad \text{in } \Omega, \quad p = 0 \quad \text{on } \Gamma, \tag{6}$$

that means, we have $p = S^*(y - y_d)$. Since we can also write $p = S^*(Su - y_d) = Pu$ with an affine operator P we call the solution $p = Pu$ the associated adjoint state to u.

Fig. 1. Ω with a quasi-uniform mesh ($\mu = 1.0$) and with graded meshes ($\mu = 0.6$)

Introducing the projection
$$\Pi_{[a,b]} f(x) := \max(a, \min(b, f(x))),$$
the condition
$$\bar{u} = \Pi_{[a,b]}\left(-\frac{1}{\nu}\bar{p}\right). \tag{7}$$
is necessary and sufficient for the optimality of $\bar{u}$.

The optimal control problem is now discretized by a finite element method. We analyze a family of graded triangulations $(T_h)_{h>0}$ of $\bar{\Omega}$ with the global mesh size h and a grading parameter $\mu < \lambda$. We assume that the individual element diameter $h_T := \operatorname{diam} T$ of any element $T \in T_h$ is related to the distance $r_T := \inf_{x \in T} |x|$ of the triangle to the corner by the relation

$$\begin{aligned} c_1 h^{1/\mu} \leq h_T \leq c_2 h^{1/\mu} & \quad \text{for } r_T = 0, \\ c_1 h r_T^{1-\mu} \leq h_T \leq c_2 h r_T^{1-\mu} & \quad \text{for } r_T > 0. \end{aligned} \tag{8}$$

For a 2-dimensional domain the number of elements of such a triangulation is of order h^{-2}. Figure 1 shows an example domain with a uniform mesh and graded meshes. Implementational aspects are given in Section 3. On these meshes, we define the finite element spaces

$$U_h := \{u_h \in L^\infty(\Omega) : u_h|_T \in \mathcal{P}_0 \text{ for all } T \in T_h\}, \quad U_h^{\mathrm{ad}} := U_h \cap U_{\mathrm{ad}},$$
$$V_h := \{y_h \in C(\bar{\Omega}) : y_h|_T \in \mathcal{P}_1 \text{ for all } T \in T_h \text{ and } y_h = 0 \text{ on } \Gamma\},$$

where $\mathcal{P}_k$, $k = 0, 1$, is the space of polynomials of degree less than or equal to k.

For each $u \in L^2(\Omega)$, we denote by $S_h u$ the unique element of V_h that satisfies $a(S_h u, v_h) = (u, v_h)_{L^2(\Omega)}$ for all $v_h \in V_h$, where $a : H^1(\Omega) \times H^1(\Omega) \to \mathbb{R}$ is the bilinear form defined by $a(y, v) := \int_\Omega (\nabla y \cdot (A \nabla v) + \mathbf{b} \nabla v + a_0 y v) \, dx$. In other words, $S_h u$ is the approximated state associated with a control u.

The finite dimensional approximation of the optimal control problem is defined by
$$J_h(\bar{u}_h) = \min_{u_h \in U_h^{\mathrm{ad}}} J_h(u_h) \tag{9}$$

with $J_h(u_h) := \frac{1}{2}\|S_h u_h - y_d\|_{L^2(\Omega)}^2 + \frac{\nu}{2}\|u_h\|_{L^2(\Omega)}^2$. The adjoint equation is discretized in the same way: We search $p_h = S_h^*(S_h u_h - y_d) = P_h u_h \in V_h$ such that $a(v_h, p_h) = (S_h u_h - y_d, v_h)_{L^2(\Omega)}$ for all $v_h \in V_h$. The optimal control problem (9) admits a unique solution $\bar{u}_h$, and we denote by $\bar{y}_h = S_h \bar{u}_h$ the optimal discrete state and by $\bar{p}_h = P_h \bar{u}_h$ the optimal discrete adjoint state. In analogy to (6) we define a postprocessed approximate control $\tilde{u}_h$ by a simple projection of the piecewise linear adjoint state $\bar{p}_h$ onto the admissible set U_{ad},

$$\tilde{u}_h := \Pi_{[a,b]}\left(-\frac{1}{\nu}\bar{p}_h\right).$$

Let us now summarize discretization error estimates. Under the assumption that the mesh grading parameter μ satisfies the condition

$$\mu < \lambda, \qquad (10)$$

the optimal, piecewise constant approximate control u_h satisfies

$$\|\bar{u} - \bar{u}_h\|_{L^2(\Omega)} \leq ch\left(\|\bar{u}\|_{L^\infty(\Omega)} + \|y_d\|_{L^\infty(\Omega)}\right) \qquad (11)$$

The first order convergence is also observed in numerical tests. Although the difference $\bar{u} - \bar{u}_h$ is of first order, the associated states and adjoint states differ by second order,

$$\|\bar{y} - \bar{y}_h\|_{L^2(\Omega)} \leq ch^2\left(\|\bar{u}\|_{L^\infty(\Omega)} + \|y_d\|_{L^\infty(\Omega)}\right), \qquad (12)$$

$$\|\bar{p} - \bar{p}_h\|_{L^2(\Omega)} \leq ch^2\left(\|\bar{u}\|_{L^\infty(\Omega)} + \|y_d\|_{L^\infty(\Omega)}\right), \qquad (13)$$

from which one can conclude that the error of the postprocessed control is also of second order,

$$\|\bar{u} - \tilde{u}_h\|_{L^2(\Omega)} \leq ch^2\left(\|\bar{u}\|_{L^\infty(\Omega)} + \|y_d\|_{L^\infty(\Omega)}\right). \qquad (14)$$

These results were first proved by Meyer and Rösch [3] for uniform meshes in the smooth case, where the solution of $Ly = f$ is contained in $W^{2,2}(\Omega) \cap W^{1,\infty}(\Omega)$. The main result of our paper [1] is that the error estimates (11)–(14) are also valid in the case of non-convex domains and appropriately graded meshes, (10). Without local mesh grading ($\mu = 1$), only a reduced convergence order is observed.

For the proof of the superconvergence results, we needed the following assumption. The formula (6) computes the optimal control $\bar{u}$ by a projection of the adjoint state $\bar{p}$. This reduces the smoothness. While $|r^{1-\mu}\bar{p}|_{W^{2,2}(\Omega)} \leq c|r^{1-\mu}\bar{p}|_{W^{2,2}(\Omega)} < \infty$ for $\mu < \lambda < 1$, this is not true for $\bar{u}$ due to kinks at the boundary of the active set. We assume that

$$\sum_{T \in \mathcal{T}_h:\, r^{1-\mu}\bar{u}\notin W^{2,2}(T)} \operatorname{meas} T \leq ch.$$

3 Numerical results

Let Ω be a circular sector as shown in Figure 1. In order to construct meshes that fulfil the conditions (8) we transformed the mesh using the mapping $T(x) = x\|x\|^{\frac{1}{\mu}-1}$ near the corner, see figure 1, middle image. An alternative is to use a partitinong strategy, see figure 1, right image.

We choose the example such that the state and dual state have a singularity near the corner. Consider

$$-\Delta y + y = u + f \quad \text{in } \Omega,$$
$$-\Delta p + p = y - y_d \quad \text{in } \Omega,$$
$$u = \Pi_{[a,b]}\left(-\frac{1}{\nu}p\right)$$

with homogeneous Dirichlet boundary conditions for y and p.

First Example. In order to have an exact solution we choose the data $f = Ly - u = Ly - \Pi\left(-\frac{1}{\nu}p\right)$ and $y_d = y - L^*p$ such that

$$y(r,\varphi) = (r^\lambda - r^\alpha)\sin\lambda\varphi$$
$$p(r,\varphi) = \nu(r^\lambda - r^\beta)\sin\lambda\varphi$$

are the exact solutions of the optimal control problem. We set $\lambda = \frac{2}{3}$, $\alpha = \beta = \frac{5}{2}$, $\nu = 10^{-4}$, $a = -0.3$ and $b = -0.1$. Figure 2 displays a piecewise linear approximation of the corresponding control function $\bar{u}$. Table 1 shows the reduced convergence rate 2λ on a quasi-uniform mesh ($\mu = 1$) and the optimal rate of convergence of the control on a graded mesh ($\mu = 0.6$).

Figure 3 shows that the error near the corner dominates the global error. The picture visualizes the contribution of each triangle to the global L^2-error. Using graded meshes this error diminishs at least as fast as the global error.

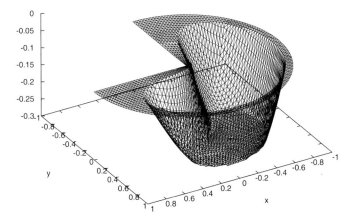

Fig. 2. Example 1. Optimal control function $-0.3 \leq u(x) \leq -0.1$

Table 1. Example 1. L^2-error of the computed control $\tilde{u}_h$, $-0.3 \le u(x) \le -0.1$

	$\mu = 0.6$		$\mu = 1.0$	
ndof	$\|u - \tilde{u}\|_{L^2}$	rate	$\|u - \tilde{u}\|_{L^2}$	rate
18	1.95e-01	0.00	1.95e-01	0.00
55	1.92e-01	0.02	1.92e-01	0.02
189	1.24e-01	0.63	1.31e-01	0.56
697	4.44e-02	1.48	5.87e-02	1.16
2673	1.38e-02	1.69	2.42e-02	1.28
10465	3.79e-03	1.86	9.84e-03	1.30
41409	9.58e-04	1.98	3.93e-03	1.32
164737	2.17e-04	2.14	1.57e-03	1.33

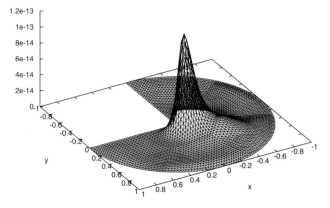

Fig. 3. Example 1. Visualization of the L^2-error of p_h, $-0.3 \le u(x) \le 1$, $\mu = 1$

Second Example. We choose now the data f and y_d such that

$$y(r, \varphi) = (r^\lambda - r^\alpha) \sin 3\lambda\varphi$$
$$p(r, \varphi) = \nu(r^\lambda - r^\beta) \sin 3\lambda\varphi$$

with $\lambda = \frac{2}{3}$ and $\alpha = \beta = \frac{5}{2}$. Further we set $a = -0.2$, $b = 0$ and $\nu = 10^{-4}$. We used a mesh that did not even coincide with the boundary of the upper active set $\{x : u(x) = b\}$ in order to show that the method does not need any apriori information about the active set. Figure 4 shows the piecewise constant approximation of the optimal control $\bar{u}$.

Table 2 shows that the convergence rate of the control $\tilde{u}$ is about 2 which was proven in [1]. Table 3 contains the absolute errors and error reduction rates of the approximated state y_h in both the L^2-norm and the H^1-seminorm.

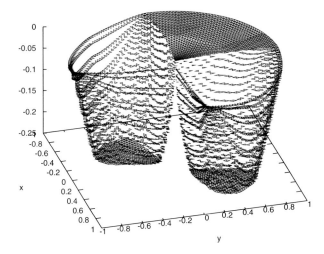

Fig. 4. Example 2. Piecewise constant approximation of optimal control function $\bar{u}$, $-0.2 \leq u(x) \leq 0$. One can see a singularity near the corner.

Table 2. Example 2. L^2-error of the computed control $\tilde{u}_h$, $-0.2 \leq u(x) \leq 0$

	$\mu = 0.6$	
ndof	$\|u - \tilde{u}\|_{L^2}$	rate
18	2.63e-01	0.00
55	2.59e-01	0.02
189	2.33e-01	0.15
697	8.44e-02	1.47
2673	2.36e-02	1.84
10465	6.04e-03	1.96
41409	1.57e-03	1.95
164737	4.31e-04	1.86

Active Sets. The approximation of the boundary of the active sets is very important for the quality of the computed control, see e.g. [2]. The method presented here approximates the active set by a union of triangles. However, after postprocessing the piecewise linear function $\tilde{u}_h$ gives a much better representation of the active sets. Figure 5 shows the active set of Example 1 on different meshes. The active triangles are shaded. The black curve shows the computed boundary of the active set as represented by $\tilde{u}_h$. The second curve displays the exact boundary. Clearly, the approximation improves with decreasing mesh size. Figure 6 shows the same behavior for the second example.

Table 3. Example 2. L^2- and H^1-errors of the computed state y_h, $-0.2 \leq u(x) \leq 0$

ndof	$\mu = 0.6$				$\mu = 1$			
	$\|y - y_h\|_{L^2}$	rate	$\|y - y_h\|_{H^1}$	rate	$\|y - y_h\|_{L^2}$	rate	$\|y - y_h\|_{H^1}$	rate
18	1.55e-01	0.00	1.78e+00	0.00	1.55e-01	0.00	1.78e+00	0.00
55	3.92e-02	1.98	1.04e+00	0.77	4.35e-02	1.83	1.10e+00	0.69
189	7.68e-03	2.35	5.74e-01	0.86	1.10e-02	1.98	6.84e-01	0.69
697	1.99e-03	1.94	3.06e-01	0.91	3.55e-03	1.63	4.24e-01	0.69
2673	6.18e-04	1.69	1.61e-01	0.93	1.23e-03	1.53	2.64e-01	0.68
10465	1.58e-04	1.97	8.38e-02	0.94	3.91e-04	1.66	1.65e-01	0.68
41409	3.97e-05	1.99	4.33e-02	0.95	1.23e-04	1.67	1.04e-01	0.67
164737	1.00e-05	1.99	2.22e-02	0.96	3.87e-05	1.67	6.52e-02	0.67

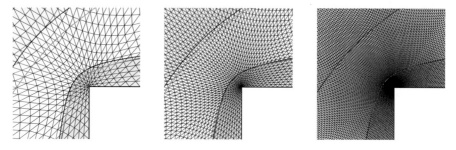

Fig. 5. Example 1: Active triangles and boundary of active sets, (zoom of region near singularity), left: ndof=2673, middle: ndof=10465, right: ndof=41409

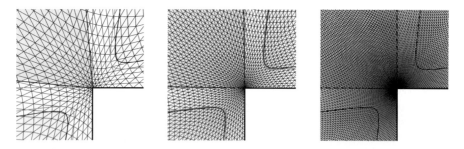

Fig. 6. Example 2: Active triangles and boundary of active sets, $-0.2 \leq u(x) \leq 0$, $\mu = 1.0$, $\alpha = \beta = \frac{5}{2}$, (zoom of region near singularity), left: ndof=2673, middle: ndof=10465, right: ndof=41409

References

1. Apel, T., Rösch, A., Winkler, G.: Optimal control in nonconvex domains. RICAM Report, 17 (2005-17). http://www.ricam.oeaw.ac.at/publications/reports/.
2. Hinze, M.: A variational discretization concept in control constrained optimization: The linear-quadratic case. Computational Optimization and Applications, **30** 45–63 (2005)
3. Meyer, C., Rösch, A.: Superconvergence properties of optimal control problems. SIAM J. Control and Optimization, **43** 970–985 (2004)

A Posteriori Estimates for Cost Functionals of Optimal Control Problems

Alexandra Gaevskaya,[1] Ronald H.W. Hoppe[1,2] and Sergey Repin[3]

[1] Institute of Mathematics, Universität Augsburg, D-86159 Augsburg, Germany
  gaevskaya@math.uni-augsburg.de
[2] Department of Mathematics, University of Houston, Houston,
  TX 77204-3008, USA
  rohop@math.uh.edu
[3] St. Petersburg Division of the Steklov Mathematical Institute,
  Russian Academy of Sciences, 191011 St. Petersburg, Russia
  repin@pdmi.ras.ru

1 Introduction

A posteriori analysis has become an inherent part of numerical mathematics. Methods of a posteriori error estimation for finite element approximations were actively developed in the last two decades (see, e.g., [1, 2, 3, 12] and the references therein). For problems in the theory of optimization, these methods started receiving attention much later. In particular, for optimal control problems governed by PDEs the literature on this matter is rather scarce. In this work, we present a new approach to a class of optimal control problems associated with elliptic type partial differential equations. In the framework of this approach, we obtain directly computable upper bounds for the cost functionals of the respective optimal control problems.

Let $\Omega \in \mathbb{R}^n$ be a Lipschitz domain with boundary $\Gamma := \partial \Omega$.

Problem 1. Given $\psi \in L_\infty(\Omega)$, $y^d \in L_2(\Omega)$, $u^d \in L_2(\Omega)$, $f \in L_2(\Omega)$, and $a \in \mathbb{R}_+$, consider the distributed control problem

$$\text{minimize } J(y(v), v) := \frac{1}{2} \|y - y^d\|^2 + \frac{a}{2} \|v - u^d\|^2 \tag{1a}$$

over $(y, v) \in H_0^1(\Omega) \times L^2(\Omega)$,

$$\text{subject to } -\Delta y = v + f \quad \text{a.e. in } \Omega, \tag{1b}$$

$$v \in K := \{v \in L^2(\Omega) \mid v \leq \psi \text{ a.e. in } \Omega\}. \tag{1c}$$

The function y^d is given and presents the desired shape of the state function y, whereas u^d presents the desired control. It is well-known that under the above assumptions Problem 1 has a unique solution (see, e.g. [9]).

There exist many different approaches to optimal control problems of this type. The numerical solution of optimal control problems is usually based on applying specific iterative schemes to the system of optimality conditions, e.g., active set strategies or interior point methods (cf., e.g., [6, 7] and the references therein). Adaptive techniques for optimal control problems governed by PDEs are presented in [4] and [8].

In this work, we follow another approach which is based on so-called functional type a posteriori error estimates. To explain the meaning of such estimates, as a model problem we consider Poisson's equation with homogeneous boundary conditions

$$-\Delta y = v + f \quad \text{in } \Omega, \tag{2a}$$

$$y = 0 \quad \text{on } \Gamma, \tag{2b}$$

which describes the dependence between the control and the state in the optimal control problem (1a)-(1c). Let $\tilde{y}$ be any function from the admissible set $Y := H_0^1(\Omega)$ which we view as an approximation of the solution of the elliptic problem (2a)-(2b). It was shown (see, e.g., [10] and [11]) that the error of the approximation $\tilde{y}$ satisfies the following estimate:

$$\|\nabla(y(v) - \tilde{y})\| \leq \|\tau - \nabla \tilde{y}\| + C_\Omega \|\mathrm{div}\,\tau + v + f\|. \tag{3}$$

Here, C_Ω is the constant in the Friedrichs inequality

$$\|w\| \leq C_\Omega \|\nabla w\|, \quad w \in H_0^1(\Omega) \tag{4}$$

for the domain Ω and τ is an arbitrary function from the functional class $\Sigma := H_{\mathrm{div}}(\Omega, \mathbb{R}^n)$. Mathematical justifications of functional type a posteriori estimates and their analysis can be found in the above cited literature. Below, we recall the main properties of such estimates:

- For any approximation $\tilde{y} \in Y$, the right–hand side of (3) gives an upper bound of the error in the natural energy norm of the problem considered;
- Its value is equal to zero if and only if $\tilde{y}$ coincides with $y(v)$ and $\tau = \nabla y(v)$;
- The estimate is consistent in the sense that its value tends to zero for any sequences $\{\tilde{y}_k\}$ and $\{\tau_k\}$, converging to the exact solution y and its gradient ∇y, respectively;
- The estimate is exact in the sense that there exists a function τ such that equality holds true;
- The estimate does not depend on the mesh parameters and only contains a global constant.

The function τ in the expression of the error majorant (3) serves as an image of the exact flux $\nabla y(v)$. It is easy to observe that two terms of the majorant represent the respective errors in the *constitutive relation* $\tau = \nabla y(v)$ and in the *equilibrium equation* $\mathrm{div}\,\tau + v + f = 0$.

In this paper, we apply this estimate in order to reformulate the original optimal control problem. As a result, we obtain a directly computable and *guaranteed* majorant for the cost functional. Besides, we prove that the sequences of approximate state and control functions, computed by the minimization of the majorant, converge to the exact state and control functions.

2 Majorants for the cost functional

One of the major difficulties in (1a)-(1c) is that the state and control functions must satisfy the equality constraint presented by the boundary-value problem for an elliptic PDE.

Let $v \in K$ and $y \in Y$ be two functions related by the differential equation (1b). For this pair, the cost functional is as follows:

$$J(y(v), v) := \frac{1}{2}\|y - y^d\|^2 + \frac{a}{2}\|v - u^d\|^2 .$$

Let $\widetilde{y} \in Y$ be some approximation of y so that we may include it in the first term of the cost functional. By the triangle and Friedrichs inequalities, we obtain the estimate

$$J(y(v), v) \leq \frac{1}{2}\left(\|\widetilde{y} - y^d\| + C_\Omega \|\nabla(y - \widetilde{y})\|\right)^2 + \frac{a}{2}\|v - u^d\|^2 . \quad (5)$$

Now, using the error majorant (3) we can estimate the weak norm of the error and substitute it to the estimate of the cost functional (5). By this procedure, we exclude the explicit entry of the exact solution y of (2a)-(2b) from our estimate and arrive at the relation

$$J(y(v), v) \leq \frac{1}{2}\left(\|\widetilde{y} - y^d\| + C_\Omega \|\nabla \widetilde{y} - \tau\| + C_\Omega^2 \|\mathrm{div}\tau + v + f\|\right)^2 + \frac{a}{2}\|v - u^d\|^2 .$$

However, from a computational point of view it is desirable to reformulate this estimate such that the right–hand side is given by a quadratic functional. For this purpose, we introduce parameters $\alpha, \beta > 0$ and obtain the following upper bound (hereafter called *the majorant*):

$$J(y(v), v) \leq J^\oplus(\alpha, \beta; \widetilde{y}, \tau, v) , \quad \forall v \in K . \quad (6)$$

Here,

$$J^\oplus(\alpha, \beta; \widetilde{y}, \tau, v) := \frac{1+\alpha}{2}\|\widetilde{y} - y^d\|^2 + \frac{(1+\alpha)(1+\beta)}{2\alpha}C_\Omega^2\|\tau - \nabla\widetilde{y}\|^2 + \quad (7)$$
$$+ \frac{(1+\alpha)(1+\beta)}{2\alpha\beta}C_\Omega^4\|\mathrm{div}\tau + v + f\|^2 + \frac{a}{2}\|v - u^d\|^2 ,$$

where $\widetilde{y} \in Y$ and τ is an arbitrary function in Σ.

Remark 1. A similar upper estimate can be derived for the optimal control problem with the cost functional

$$J(y,v) = \frac{1}{2}\|\nabla y - \sigma^d\|^2 + \frac{a}{2}\|u - u^d\|^2,$$

where the vector-valued function σ^d is given and presents the desired gradient of the state function.

Let us consider the majorant as a functional that generates a new minimization problem

Problem 1*. Given $\psi \in L_\infty(\Omega)$, $y^d \in L_2(\Omega)$, $u^d \in L_2(\Omega)$, $f \in L_2(\Omega)$, and $a \in \mathbb{R}_+$,

$$\text{minimize } J^\oplus(\alpha, \beta; \widetilde{y}, \tau, v) \tag{8a}$$
$$\text{over } v \in K,\ \widetilde{y} \in Y,\ \tau \in \Sigma,\ \alpha, \beta \in \mathbb{R}_+,$$
$$K := \{v \in L^2(\Omega) \mid v \leq \psi \text{ a.e. in } \Omega\}. \tag{8b}$$

We see that in this problem the differential equation (which in (1a)-(1c) defines the respective admissible set) does not appear explicitly. In (8a)-(8b), the functions τ, $\widetilde{y}$ and v act as independent variables. In the next section, we present properties of the majorant (7) and show that Problem 1* and Problem 1 have one and the same exact lower bound attained on the same state and control functions.

Remark 2. It is worth noting that the majorant $J^\oplus(\alpha, \beta; \widetilde{y}, \tau, v)$ can be used to find *guaranteed upper bounds* for the cost functional when the minimization problem is solved by known methods. Indeed, since the functions $\widetilde{y}$ and v are arbitrary, we can take them as approximate solutions computed by some optimization procedure and minimize the majorant w.r.t. the function τ and the parameters β and α. The respective value $J^\oplus$ will represent the guaranteed upper bound for the value of the cost functional.

3 Properties of majorants

Theorem 1. *The exact lower bound of the majorant (7) coincides with the optimal value of the cost functional of the problem (1a)-(1c), i.e,*

$$\inf_{\substack{\widetilde{y} \in Y, \tau \in \Sigma, \\ v \in K, \alpha, \beta \in \mathbb{R}_+}} J^\oplus(\alpha, \beta; \widetilde{y}, \tau, v) = J(y(u), u).$$

The infimum of $J^\oplus$ is attained for $v = u$, $\widetilde{y} = y(u)$, $\tau = \nabla y(u)$.

This property means that our transformation of the original problem is mathematically correct in the sense that the new problem is solvable and has the same lower bound as the original one.

Let $\{V_k\}_{k=1}^\infty$, $\{Y_k\}_{k=1}^\infty$ and $\{\Sigma_k\}_{k=1}^\infty$ be sequences of finite-dimensional subspaces that are limit dense in $V := L^2(\Omega)$, Y and Σ, respectively. The discrete control constraints are given by $K_k := V_k \cap K$. It is not difficult to show that K_k is limit dense in K.

We define the sequence of numbers

$$J_k^\oplus := J^\oplus(\alpha_k, \beta_k; \widetilde{y}_k, \tau_k, v_k) = \inf_{\substack{\widetilde{y} \in Y_k, \tau \in \Sigma_k, \\ v \in K_k, \alpha, \beta \in \mathbb{R}_+}} J^\oplus(\alpha, \beta; \widetilde{y}, \tau, v), \qquad (9)$$

which is obtained by solving the problem on sequences of the selected finite-dimensional subspaces.

Theorem 2. *If K_k, Y_k and Σ_k are limit dense in K, Y, and Σ, respectively, then*

(i) $J_k^\oplus \to J(y(u), u)$ as $k \to \infty$;

(ii) the sequence $\{y(v_s), v_s\}$ converges to the exact solution of the control problem $\{y(u), u\}$ in $Y \times K$.

The theorem shows that a numerical strategy based upon using the majorant produces sequences of control and state functions which provide a value of the cost functional as close to the value $J(y(u), u)$ as it is required. Moreover, the respective sequences of control and state functions tend to the desired solution of the original problem.

4 Practical implementation

In this section, we briefly discuss the practical implementation of the numerical strategy based on the majorants.

4.1 Discretization of the problem

In the resultes exposed below, we restrict ourselves to the case when the problem is solved by usual finite element approximations on a simplicial mesh which is the same for all functions involved. Let $\mathcal{T}_h(\Omega)$ denote such a shape-regular simplicial triangulation of Ω. For the state function, we use continuous piecewise affine approximations $\widetilde{y}_h \in Y_h$ vanishing on the boundary Γ, whereas for the control $v \in K$ we use piecewise constant approximations $v_h \in K_h$ where K_h is chosen such that $K_h \subset K$. The vector-valued functions $\tau \in \Sigma$ are approximated by piecewise affine functions $\tau_h \in \Sigma_h$.

4.2 Minimization algorithm

To obtain a sharp upper bound of the cost functional, we minimize the majorant $J^\oplus(\alpha, \beta; \widetilde{y}_h, \tau_h, v_h)$ over $(\widetilde{y}_h, \tau_h, v_h) \in Y_h \times \Sigma_h \times K_h$ and $\alpha, \beta \in \mathbb{R}^+$.

The numerical results presented below have been obtained using the following algorithm:

Step 1. Initialization. Set $i = 0$, define α^0, β^0, v_h^0, $\widetilde{y}_h^0$.
Step 2. Minimize $J^\oplus(\alpha^i, \beta^i; \widetilde{y}_h, \tau_h, v_h)$ over $(\widetilde{y}_h, \tau_h, v_h) \in Y_h \times \Sigma_h \times K_h$.
Step 3. Minimize $J^\oplus(\alpha, \beta; \widetilde{y}_h^{i+1}, \tau_h^{i+1}, v_h^{i+1})$ w.r.t. $\beta, \alpha \in \mathbb{R}_+$. Set $i = i + 1$.

Steps 2 and 3 are repeated until

$$\frac{|J_i^\oplus - J_{i-1}^\oplus|}{J_i^\oplus} + \frac{\|v_h^i - v_h^{i-1}\|}{\|v_h^i\|} + \frac{\|\nabla(\widetilde{y}_h^i - \widetilde{y}_h^{i-1})\|}{\|\nabla \widetilde{y}_h^i\|} > \epsilon,$$

where ϵ is a given tolerance and $J_i^\oplus = J^\oplus(\alpha^i, \beta^i; \widetilde{y}_h^i, \tau_h^i, v_h^i)$.

5 Numerical experiments

The method described in the previous sections has been numerically tested on a set of various optimal control problems. In all examples, it has been observed that the sequences of computed upper bounds of the cost functionals rapidly converge to the exact lower bound whose value has been computed at high accuracy. Also, it has been observed that the sequences of the state and control functions converge to the exact ones.

Below, we show these results for the model problem in $\Omega = (0,1)^2$. In this case, $C_\Omega = \frac{1}{\sqrt{2}\pi}$.

The efficiency of the approach is measured by three quantities. The index

$$I = J^\oplus / J(y, u)$$

shows the relation between the value of majorant computed for the control function v and the exact lower bound of the cost functional $J(y, u)$. The quantities

$$\eta_y = (\|y - \widetilde{y}\|_{H^1} / \|y\|_{H^1}) * 100\%, \quad \eta_u = (\|v - u\| / \|u\|) * 100\%,$$

represent the *relative errors* in the state and control functions, respectively.

Example

As an example we take the problem from [6] with the following data:
$a = 0.01$, $\psi(x,y) = 1$, $f(x,y) = 0$, $u^d(x,y) = 0$ and

$$y^d(x,y) = \begin{cases} 200(x-0.5)^2(1-y)yx, & x \leq 0.5, \\ 200(x-0.5)^2(1-y)y(x-1), & \text{else}. \end{cases}$$

The exact solution of this optimal control problem is unknown. Therefore, in order to analyze the efficiency of the method, we have computed a 'reference solution' using a mesh much finer than those used in the actual computations. For this task, we have used the primal-dual active set strategy (cf., e.g., [5]). The reference value of the cost functional in this case is $J(y, u) = 9.5838 \cdot 10^{-2}$.

The discrete problem has been solved for various uniform meshes with N nodes. Table 1 shows the relative errors in the state and control functions and the index I. In Figure 1, we depict values of the majorant with respect to the minimization time ($N = 1089$). In this figure, the horizontal line shows $J(y(u), u)$ (actual value of the cost functional) whereas the rapidly decaying curve reflects the reduction of the computable upper bound given by the majorant. The desired tolerance $\epsilon = 10^{-4}$ was achieved after $i = 16$ iterations. Approximations $(\widetilde{y}_h, v_h)$ obtained by the algorithm and the reference state and control functions are displayed in Figure 2.

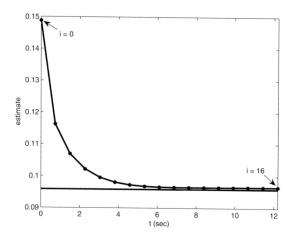

Fig. 1. Reduction of the upper bound of the cost functional w. r. t. CPU time.

Table 1. Index I and relative errors in the state and control.

N	$\eta_y, \%$	$\eta_u, \%$	I
25	67.51	54.39	1.050
81	31.50	25.23	1.029
289	14.59	12.07	1.014
1089	7.55	6.49	1.007
4225	4.67	4.18	1.003
16641	3.65	3.39	1.002

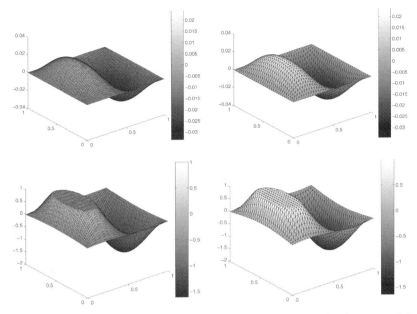

Fig. 2. Exact state y (upper left) and approximate state $\tilde{y}_h$ (upper right), exact control u (lower left) and approximate control v_h (lower right).

References

1. Ainsworth, M., Oden, J.T.: A posteriori error estimation in finite element analysis. Wiley, New York (2000)
2. Babuška, I., Strouboulis, T.: The finite element method and its reliability, Clarendon Press, Oxford (2001)
3. Bangerth, W., Rannacher, R.: Adaptive finite element methods for differential equations. Birkhäuser, Berlin (2003)
4. Becker, R., Rannacher, R.: An optimal control approach to error estimation and mesh adaptation in finite element methods. In: Acta Numerica (A. Iserles, ed.), 10: 1–102, Cambridge University Press, (2001)
5. Bergounioux, M., Haddou, M, Hintermüller, M., Kunisch, K.: A comparison of a Moreau-Yosida-Based Active Set Strategy and Interior Point Methods for Constrained Optimal Control Problems. SIAM J. Optim., **11**, No. 2, 495–521 (2000)
6. Bergounioux, M., Ito, K., Kunisch, K.: Primal-dual strategy for constrained optimal control problems, SIAM J. Control Optim., **37**, 1176–1194 (1999)
7. Bonnans, J.F., Pola, C., Rebaï, R.: Perturbed path following interior point algorithms, Optim. Methods Softw., 11–12, 183–210 (1999)
8. Li, R., Liu, W., Ma, H., Tang, T.: Adaptive finite element approximation for distributed optimal control problems. SIAM J. Control Optim. **41**, 5, 1321–1349 (2002)

9. Lions, J. L.: Optimal Control of Systems Governed by Partial Differential Equations. Springer, Berlin–Heidelberg–New York, (1971)
10. Repin, S.: A posteriori error estimation for nonlinear variational problems by duality theory, Zapiski Nauchn. Semin. POMI, **243**, 201–214 (1997)
11. Repin, S.: A posteriori error estimation for variational problems with uniformly convex functionals, Math. Comp., **69**, 230, 481–500 (2000)
12. Verfürth, R.: A review of a posteriori error estimation and adaptive mesh-refinement techniques, Wiley and Sons, Teubner, New York, (1996)

Convection-Diffusion

Optimization of a Duality Method for the Compressible Reynolds Equation

Iñigo Arregui[1], J. Jesús Cendán[1], Carlos Parés[2] and Carlos Vázquez[1]

[1] Departamento de Matemáticas, Universidad de La Coruña, Facultad de Informática, Campus de Elviña, s/n, 15071 La Coruña, Spain
arregui@udc.es, suceve@udc.es, carlosv@udc.es
[2] Departamento de Análisis Matemático, Universidad de Málaga, Facultad de Ciencias, Campus de Teatinos, 29080 Málaga, Spain
pares@anamat.cie.uma.es

Summary. Mathematical modelling of air lubrication phenomena taking place during read/write processes in magnetic storage devices (hard–disks, for example) can be addressed by using a compressible Reynolds equation for the air pressure. In the present paper, we propose a duality algorithm with optimal functional parameters to numerically solve the nonlinear diffusive term. A theoretical result is stated and some numerical examples are presented to illustrate the performance of the method.

1 The mathematical model

A real hard–disk magnetic recording device consists in a rigid head and a rigid disk. A thin layer of air fills the gap between both elements and acts as a lubricant. Following [4], we will assume in this paper that the device is wide enough (in order to use a 1–D model), and the disk moves with constant velocity V. Moreover, air will be considered a perfect gas in newtonian and laminar regime, inertial forces and stress effects are negligible, and constant viscosity and temperature will be also assumed.

Let be $0 \leq l_1 < l_2$. The thin gap between the head and the disk is given by a function $h \in L^\infty(l_1, l_2)$. Under the previous hypotheses, Burgdorfer [5] proposes the classical compressible Reynolds equation to model the hydrodynamic behaviour of the device:

$$6V\mu \frac{d}{dx}(ph) - 6\lambda p_a \frac{d}{dx}\left(h^2 \frac{dp}{dx}\right) - \frac{d}{dx}\left(h^3 p \frac{dp}{dx}\right) = 0 \quad \text{in } (l_1, l_2) \quad (1)$$

$$p(l_1) = p(l_2) = p_a \quad (2)$$

where p is the air pressure, μ is its viscosity, λ is the molecular mean free path of the air and p_a is the ambient pressure.

If we introduce the dimensionless variables [7]: $X = 100x$, $P = p/p_a$, $H = 10^6 h$, then problem (1)–(2) can be written in the form:

$$\begin{cases} \dfrac{d}{dX}(PH) - \alpha \dfrac{d}{dX}\left(H^2 \dfrac{dP}{dX}\right) - \beta \dfrac{d}{dX}\left(H^3 P \dfrac{dP}{dX}\right) = 0, & X \in (L_1, L_2) \\ P(L_1) = P(L_2) = 1, \end{cases}$$

where $\alpha = 10^{-4} \lambda p_a / (\mu V)$ and $\beta = 10^{-10} p_a / (6 \mu V)$.

Several difficulties arise when addressing the numerical solution of this problem. First, the dimensionless compressible Reynolds equation presents a nonlinear diffusive term; secondly, in real applications $\alpha = O(10^{-2})$ and $\beta = O(10^{-2})$, so that the advection effects are larger than the diffusion ones. In this paper, we focus on the numerical solution of this problem and we propose some specific techniques to overcome these difficulties. In particular, a duality method is used to solve the nonlinearity appearing in the diffusive term, which is optimized in order to reduce the number of iterations needed to attain the convergence.

2 Numerical solution

In order to use a finite element method, let us consider the following variational formulation:

Find $P \in V_1$ such that:

$$\int_{L_1}^{L_2} \frac{d(PH)}{dX} \varphi \, dX + \int_{L_1}^{L_2} \left(\alpha H^2 + \beta H^3 P\right) \frac{dP}{dX} \frac{d\varphi}{dX} \, dX = 0, \quad \forall \varphi \in V_0, \quad (3)$$

where the functional sets are:

$$V_0 = H_0^1(L_1, L_2) \quad \text{and} \quad V_1 = \{\varphi \in H^1(L_1, L_2) \, / \, \varphi(L_1) = \varphi(L_2) = 1\}.$$

Next, taking into account the dominating convection feature in (3), we propose in [1] a characteristics technique for steady state problems. Thus, after a time discretization procedure, an iterative method is deduced in order to reach the stationary solution. More precisely, for $m \geq 0$ and P_m given, we search $P_{m+1} \in V_1$ such that:

$$\int_{L_1}^{L_2} P_{m+1} H \varphi \, dX + k \int_{L_1}^{L_2} \left(\alpha H^2 + \beta H^3 P_{m+1}\right) \frac{dP_{m+1}}{dX} \frac{d\varphi}{dX} \, dX =$$
$$= \int_{L_1}^{L_2} \left((P_m H) \circ \chi^k\right) \varphi \, dX, \quad \forall \varphi \in V_0, \quad (4)$$

where k is the artificial time step, and $\chi^k(X) = X - k$ is related to the characteristics method [1]. Notice that (4) is a nonlinear diffusive problem.

Next, we apply a duality algorithm to solve (4), which is an extension of the one proposed in [3] to solve variational inequalities. For this, we consider the maximal monotone operator f:

$$f(P) = \begin{cases} 0, & \text{if } P < 0 \\ P^2, & \text{if } P \geq 0 \end{cases}$$

so that the variational equation (4) can be written as:

$$\int_{L_1}^{L_2} P_{m+1} H\varphi \, dX + k \int_{L_1}^{L_2} \left(\alpha H^2 \frac{dP_{m+1}}{dX} + \frac{\beta H^3}{2} \frac{d(f(P_{m+1}))}{dX} \right) \frac{d\varphi}{dX} \, dX =$$

$$= \int_{L_1}^{L_2} ((P_m H) \circ \chi^k) \, \varphi \, dX, \qquad \forall \varphi \in V_0. \tag{5}$$

Next, following the version of the algorithm with variable parameters [9], given a function $\omega > 0$, we introduce the new unknown $\theta_{m+1} = f(P_{m+1}) - \omega P_{m+1}$, and search (P_{m+1}, θ_{m+1}) verifying the still nonlinear problem:

$$\int_{L_1}^{L_2} P_{m+1} H \, \varphi \, dX + k \int_{L_1}^{L_2} \left(\alpha H^2 + \frac{\beta \omega}{2} H^3 \right) \frac{dP_{m+1}}{dX} \frac{d\varphi}{dX} \, dX +$$

$$+ \frac{k\beta}{2} \int_{L_1}^{L_2} H^3 \frac{d\omega}{dX} P_{m+1} \frac{d\varphi}{dX} \, dX = \int_{L_1}^{L_2} ((P_m H) \circ \chi^k) \, \varphi \, dX -$$

$$- \frac{k\beta}{2} \int_{L_1}^{L_2} H^3 \frac{d\theta_{m+1}}{dX} \frac{d\varphi}{dX} \, dX, \qquad \forall \varphi \in V_0 \tag{6}$$

$$\theta_{m+1} = f(P_{m+1}) - \omega P_{m+1}. \tag{7}$$

Now, we can use Bermúdez–Moreno lemma for functional parameters [9] and replace (7) by

$$\theta_{m+1} = f^\omega_{1/2\omega}\left(P_{m+1} + \frac{1}{2\omega} \theta_{m+1} \right), \tag{8}$$

where $f^\omega_{1/2\omega}$ is the Yosida approximation of $f - \omega I$ with parameter $1/2\omega$, I being the identity operator. Finally, in order to overcome the nonlinearity (8), we propose the following fixed–point algorithm that iterates between equations (6) and (8):

- θ^ℓ_{m+1} known, find $P^{\ell+1}_{m+1}$ verifying the linear problem:

$$\int_{L_1}^{L_2} P_{m+1}^{\ell+1} H \, \varphi \, dX + k \int_{L_1}^{L_2} \left(\alpha H^2 + \frac{\beta \omega}{2} H^3 \right) \frac{dP_{m+1}^{\ell+1}}{dX} \frac{d\varphi}{dX} dX +$$

$$+ \frac{k\beta}{2} \int_{L_1}^{L_2} H^3 \frac{d\omega}{dX} P_{m+1}^{\ell+1} \frac{d\varphi}{dX} dX = \int_{L_1}^{L_2} \left((P_m H) \circ \chi^k \right) \varphi \, dX -$$

$$- \frac{k\beta}{2} \int_{L_1}^{L_2} H^3 \frac{d\theta_{m+1}^{\ell}}{dX} \frac{d\varphi}{dX} dX \quad , \qquad \forall \varphi \in V_0 \qquad (9)$$

- $P_{m+1}^{\ell+1}$ known, $\theta_{m+1}^{\ell+1}$ is updated by:

$$\theta_{m+1}^{\ell+1} = f_{1/2\omega}^{\omega} \left(P_{m+1}^{\ell+1} + \frac{1}{2\omega} \theta_{m+1}^{\ell} \right).$$

Lagrange P_1 finite elements have been used for the spatial discretization of (9).

3 Optimization of the duality algorithm

In order to analyze the optimal choice of parameter ω, we will consider an abstract mathematical frame. For this, let us introduce the Hilbert spaces $E = L^2(L_1, L_2)$ and $V = H^1(L_1, L_2)$. Let be:

- $\Lambda_E : E \longrightarrow E'$ the canonical isomorphism between E and its dual space;
- $A : V \longrightarrow V'$ the operator given by:

$$A\psi = H\psi - \frac{d}{dX} \left(H^2 \frac{d\psi}{dX} \right), \quad \forall \psi \in V ;$$

- $B : E \longrightarrow V'$ the operator given by:

$$< Bw, \varphi > = \int_{L_1}^{L_2} H^{3/2} w \frac{d\varphi}{dX} dX, \qquad \forall w \in E, \ \forall \varphi \in V ;$$

- $G : V \longrightarrow V$, the maximal monotone operator given by:

$$G(\varphi)(X) = \begin{cases} (\varphi(X))^2, & \text{if } \varphi(X) \geq 0 \\ 0, & \text{if } \varphi(X) \leq 0 ; \end{cases}$$

G is well posed, thanks to inclusion $V \subset L^{\infty}(L_1, L_2)$;
- $f \in V'$, given by:

$$< f, \varphi > = \int_{L_1}^{L_2} \left((P_m H) \circ \chi^k \right) \varphi \, dX, \qquad \forall \varphi \in V .$$

In this abstract frame, the hydrodynamical problem of finding $P_{m+1} \in V_1$ solution of (5) is equivalent to find $y \in V_1$ such that:

$$Ay + c(B\Lambda_E^{-1}B^*)(G(y)) = f , \qquad (10)$$

where $c = k\beta/2$. Next, for $w \in W^{1,\infty}(L_1, L_2)$ such that $1/w \in W^{1,\infty}(L_1, L_2)$, we introduce $\theta = G(y) - wy \in V$ and apply Bermúdez–Moreno lemma to pose the fixed–point algorithm:

$$Ay^{\ell+1} + c\,(B\Lambda_E^{-1}B^*)(wy^{\ell+1}) = f - c(B\Lambda_E^{-1}B^*)(\theta^\ell)$$

$$\theta^{\ell+1} = G_{1/2w}^w\left(y^{\ell+1} + \frac{1}{2w}\theta^\ell\right).$$

First, notice that Yosida approximation is given by:

$$G_{1/2w}^w(z) = 2w\left(I - \left(\frac{1}{2}I + \frac{1}{2w}G\right)^{-1}\right)(z).$$

Next, we define the function $F_w : V \longrightarrow V$, given by:

$$F_w(\theta) = G_{1/2w}^w\left(y(\theta) + \frac{\theta}{2w}\right),$$

where, given $q \in V$, $y(q)$ is such that:

$$Ay(q) + c\,(B\Lambda_E^{-1}B^*)(wy(q)) = f - c\,(B\Lambda_E^{-1}B^*)(q).$$

Let $\bar\theta$ be a fixed point of F_w. Our aim is to accelerate the convergence of the fixed–point algorithm, by choosing w such that:

$$DF_w(\bar\theta) = 0 , \qquad (11)$$

$DF_w(\bar\theta)$ being the Gâteaux–derivative of F_w in $\bar\theta$.

Proposition 1. *A sufficient condition for (11) is $w = 2y(\bar\theta)$.*

Proof. Let $z = y(\theta) + \dfrac{\theta}{2w}$. Some straigthforward calculations show that:

$$\langle DF_w(\theta), q\rangle = \frac{dG_{1/2w}^w}{dz}\left(y(\theta) + \frac{\theta}{2w}\right)\left(y(q) + \frac{q}{2w}\right), \quad \forall q \in V . \qquad (12)$$

If we can choose w so that:

$$\frac{dG_{1/2w}^w}{dz}\left(y(\bar\theta) + \frac{\bar\theta}{2w}\right) = 0 ,$$

then (11) is achieved. An easy application of the inverse function theorem gives:

$$\frac{dG^\omega_{1/2\omega}}{dz}(z) = 2\omega\left(1 - \frac{1}{\frac{1}{2} + \frac{1}{2\omega}G'(t)}\right)(z), \qquad (13)$$

where t verifies the equation:

$$\frac{1}{2}t + \frac{1}{2\omega}G(t) = z, \qquad (14)$$

provided that $z \neq 0$. So, using (12) and (13), a sufficient condition for (11) is:

$$\frac{1}{2} + \frac{1}{2\omega}G'(t) = 1, \qquad \text{with } z = y(\bar{\theta}) + \frac{\bar{\theta}}{2\omega} \text{ in (14)},$$

which is equivalent to $\omega(X) = G'(t)(X)$. Finally, as $\bar{\theta} = G(y(\bar{\theta})) - \omega\bar{\theta}$, it easily follows the equivalence:

$$z = y(\bar{\theta}) + \frac{\bar{\theta}}{2\omega} \quad \Longleftrightarrow \quad t = y(\bar{\theta})$$

and the optimal choice for the parameter is $\omega(X) = 2y(\bar{\theta})(X)$. ∎

Remark 1. Notice that the optimal choice of the parameter depends on solution $y(\bar{\theta})$. So, in our practical implementation $\omega = 2P_m$ is taken, P_m being the approximation of solution in the last step of the characteristics method.

Remark 2. The particular case with constant ω is more classical, although (10) is out of the frame of previous works [3, 2]. In this case, a convergence result is stablished for Lagrange P_1 finite elements discretized problem in [6].

4 Numerical examples

In this section we present several tests that show the behaviour of the previously described numerical techniques.

Test 1. Let us consider the following nonlinear diffusion problem:

$$\begin{cases} -\frac{d}{dx}\left(h^2\frac{dp}{dx} + h^3p\frac{dp}{dx}\right) = f & \text{in } (0,1) \\ p(0) = p(1) = 1 \end{cases}$$

where $h(x) = 2 - x$ and f is such that the solution is $p(x) = 1 + x - x^2$.

Table 1 shows the number of iterations of the duality method (with optimal constant and variable parameters) and the relative quadratic error e_p (between the numerical approximation and the analytical solution) for different mesh sizes and a relative error stopping test equal to 10^{-7}. Table 2

Table 1. Number of iterations (I_d) and quadratic error in Test 1

Δx	$\omega = 2$		$\omega = 2p$	
	I_d	e_p	I_d	e_p
10^{-2}	7	1.2×10^{-6}	4	1.1×10^{-6}
10^{-3}	7	1.2×10^{-8}	4	4.9×10^{-9}
10^{-4}	7	7.7×10^{-10}	4	6.1×10^{-10}

Table 2. Number of iterations (I_d) for different ω in Test 1

ω	0.02	0.2	1	2	3	20	200
I_d	231	41	12	7	8	60	525
ω	$0.02p$	$0.2p$	p	$2p$	$3p$	$20p$	$200p$
I_d	200	34	8	4	8	48	455

illustrates the optimality of parameters in terms of the number of iterations.

Test 2. Let us now consider the convection–diffusion problem which consists in finding the pressure, p, such that:

$$\begin{cases} \dfrac{d}{dx}(h\,p) - \dfrac{d}{dx}\left(h^2 \dfrac{dp}{dx} + h^3 p \dfrac{dp}{dx}\right) = f & \text{in } (0,1) \\ p(0) = p(1) = 1 \end{cases}$$

where $h(x) = 2 - x$ and f is such that the solution is $p(x) = 1 + x - x^2$. We have taken $k = 0.5\,\Delta x$ as time step. The obtained results are shown in Table 3, where I_c is the number of iterations of the characteristics algorithm and $\overline{I_d}$ is the average of iterations of the duality algorithm.

Table 3. Number of iterations for different parameter choices in Test 2

Δx	$\omega = 2$		$\omega = 2p_m$		$\omega = 2p$	
	I_c	$\overline{I_d}$	I_c	$\overline{I_d}$	I_c	$\overline{I_d}$
10^{-2}	90	4.2	90	3.3	90	2.8
10^{-3}	712	3.4	712	2.9	712	2.6
10^{-4}	5922	2.5	5922	2.4	5923	2.3

Test 3. In [8], the following compressible Reynolds equation is proposed to model lubricated rough surfaces:

$$\begin{cases} 300 \dfrac{d}{dx}(h\,p) - \dfrac{d}{dx}\left(0.4 h^2 \dfrac{dp}{dx} + h^3 p \dfrac{dp}{dx}\right) = 0 & \text{in } (0,1) \\ p(0) = p(1) = 1 \end{cases}$$

Table 4. Number of iterations in Test 3

Δx	$\omega = 2$		$\omega = 2p_m$	
	I_c	$\overline{I_d}$	I_c	$\overline{I_d}$
10^{-2}	415	7.0	415	3.5
10^{-3}	4862	7.5	4802	2.9
10^{-4}	45039	4.4	45006	2.8

where $h(x) = 2 - x + 0.6\sin(100\pi x)$. Table 4 shows the number of iterations obtained for constant and functional parameters for different mesh sizes and maximum relative error $\varepsilon = 5 \times 10^{-9}$.

5 Conclusions

In this paper we present an original duality algorithm with functional parameters to solve the nonlinear diffusive term appearing in the first order compressible Reynolds equation. Moreover, the optimal choice of parameters is theoretically proved and illustrated by some numerical experiments. Although the method is presented for 1–D problems, its extension to two dimensions is straightforward (see [6], for example). The algorithm can also be coupled with elastic equations to simulate flexible storage devices.

Aknowledgments

The authors thank Ministerio de Educación y Ciencia (projects MTM2004–05796–C02–01 and BFM2003–07530–C02–02) and Xunta de Galicia (project PGIDIT–02–PXIC–10503–PN) for their financial support.

References

1. Arregui, I., Cendán, J.J., Vázquez, C.: A duality method for the compressible Reynolds equation. Aplication to simulation of read/write processes in magnetic storage devices. J. Comput. Appl. Math., **175**, 31–40 (2005)
2. Bermúdez, A.: Un método numérico para la resolución de ecuaciones con varios términos no lineales. Aplicación a un problema de flujo de gas en un conducto. Rev. Acad. Cienc. Exactas Fís. Nat., **78**, 89–96 (1981)
3. Bermúdez, A., Moreno, C.: Duality methods for solving variational inequalities. Comp. Math. with Appl., **7**, 43–58 (1981)
4. Bhushan, B.: Tribology and Mechanics of Magnetic Storage Devices. Springer, New York (1996)
5. Burgdorfer, A.: The influence of the molecular mean free path on the performance of hydrodynamic gas lubricated bearings. ASME J. Basic Engrg., **81**, 99–100 (1959)

6. Cendán, J.J.: Estudio matemático y numérico del modelo de Reynolds–Koiter y de los modelos tribológicos en lectura magnética. PhD Thesis, University of Vigo (2005)
7. Friedman, A.: Mathematics in Industrial Problems. 7. Springer, New York (1995)
8. Jai, M.: Homogenization and two–scale convergence of the compressible Reynolds lubrification equation modelling the flying characteristics of a rough magnetic head over a rough rigid–disk surface. Math. Modelling Numer. Anal., **29**, 199–233 (1995)
9. Parés, C., Macías, J., Castro, M.: Duality methods with authomatic choice of parameters. Application to shallow water equations in conservative form. Numer. Math., **89**, 161–189 (2001)

Time-Space & Space-Time Elements for Unsteady Advection-Dominated Problems

Maria Isabel Asensio[1], Blanca Ayuso[2] and Giancarlo Sangalli[2]

[1] Departamento de Matemática Aplicada, Plaza de la Merced s/n, Universidad de Salamanca, Salamanca 37008, Spain
`mas@usal.es`

[2] Istituto di Matematica Applicata e Tecnologie Informatiche CNR, Via Ferrata 1, Pavia 27100, Italy
`blanca@imati.cnr.it, sangalli@imati.cnr.it`

Summary. We present some stabilized methods for a nonstationary advection-diffusion problem. The methods are designed by combining of some stabilized finite element methods and Discontinuous Galerkin time integration. Numerical experiments are presented comparing the new schemes with the space time elements of [3].

1 Introduction

The numerical simulation of advection-diffusion problems has been a subject of active research during the last thirty years. In this paper we look at the unsteady problem. Following with the research initiated in [4], our aim is to study the issue of how some of the stabilization techniques proposed for the steady problem could be appropriately combined and used with time integration Discontinuous Galerkin (DG) methods, so that the resulting fully discretized scheme is able to capture and reproduce the small scales into the coarse ones. Our starting point is based on the simple observation that in the non-stationary problem we have two types of partial differentiation which might be considered of different nature: the spatial convection-diffusion-reaction operator and the time derivative which determines the evolution of the convection-diffusion-reaction processes. Therefore, at the very first step of designing the numerical method, two rather different strategies arise:

- discretize at first in time by using a DG method and then apply a stabilized method to approximate the resulting family of stationary problems;
- discretize first in space by means of a stabilized finite element method and then use a DG scheme to integrate the corresponding system of ODE's.

The resulting methods from these two approaches will be described and further compared with the "classical" *space-time elements* introduced in the 80's by Johnson, Nävert and Pitkäranta in [3].

The outline of the paper is as follows. In Sect. 2 we review the stabilization techniques proposed for the stationary problem, that will be further considered. In Sect. 3 we revise the time DG integration and introduce the stabilized methods for the time-dependent problem. Numerical experiments are presented in Sect. 4. For the sake of simplicity we restrict ourselves to the one dimensional problem.

2 Stabilization techniques for the stationary problem

Let $\Omega = (0, L)$ and let $f \in L^2(\Omega)$ be given. Consider the stationary problem

$$\mathcal{L}u = -\epsilon u_{xx} + \beta u_x + \sigma u = f, \quad \text{in } \Omega, \quad u = 0 \quad \text{on } \partial\Omega, \tag{1}$$

where $\epsilon > 0$, $\sigma \geq 0$ and β are assumed to be constants and ϵ will be typically small. Let $\mathcal{V} = H_0^1(\Omega)$. The bilinear form associated to $\mathcal{L}$ is defined by

$$a(u,v) =< \mathcal{L}u, v >\equiv \epsilon \int_\Omega u_x v_x dx + \beta \int_\Omega u_x v dx + \sigma \int_\Omega uv dx, \quad \forall u, v \in \mathcal{V}. \tag{2}$$

We denote by $\mathcal{L}_{Sym}$ and $\mathcal{L}_{Skew}$ the symmetric and skew-symmetric parts of $\mathcal{L}$, respectively. The formal adjoint of $\mathcal{L}$ will be denoted by $\mathcal{L}^* = \mathcal{L}_{Sym} - \mathcal{L}_{Skew}$.

Let $\mathcal{T}_h$ be a partition of Ω into elements (subintervals) K and let $V_h \subset \mathcal{V}$ be the corresponding finite element space of piecewise linear polynomials. The standard Galerkin (SG) approximation of (1) reads:

$$\text{Find } u_h^{SG} \in V_h \quad \text{such that} \quad a(u_h^{SG}, v_h) = (f, v_h), \quad \forall v_h \in V_h.$$

It is well known that the plain Galerkin method on a uniform grid fails to furnish a satisfactory approximation if the diffusion coefficient ϵ is small with respect to the advection or/and reaction coefficients and to the mesh size h. To cope with these difficulties, we consider the next family of strongly consistent methods, which following [1] can be presented in the unified way

$$\begin{cases} \text{Find } u_h^{Stb} \in V_h \text{ such that} \\ a(u_h^{Stb}, v_h) + \sum_{K \in \mathcal{T}_h} \delta_K \left(\mathcal{L} u_h^{Stb}, \mathcal{L}_{Skew} v_h + \rho \mathcal{L}_{Sym} v_h\right)_K \\ = (f, v_h) + \sum_{K \in \mathcal{T}_h} \delta_K \left(f, \mathcal{L}_{Skew} v_h + \rho \mathcal{L}_{Sym} v_h\right)_K, \quad \forall v_h \in V_h, \end{cases} \tag{3}$$

where $\rho = 0$ gives the SUPG (*Stramline Upwind Petrov Galerkin*) method[5]; $\rho = 1$ gives the GLS (*Galerkin/Least Squares*) method [11]; and $\rho = -1$ gives the DWG (*Douglas-Wang Galerkin*) method [7]. These schemes require an appropriate tuning of the problem-dependent parameter δ_K. A straightforward calculation shows that by taking

$$\delta_K := \left((2\sigma) + (2|\beta|)/h + (12\epsilon)/h^2\right)^{-1}, \tag{4}$$

the bilinear form $\mathcal{B}^\rho$ defining these methods,

$$\mathcal{B}^p(w_h, v_h) := a(w_h, v_h) + \sum_{K \in \mathcal{T}_h} \delta_K \left(\mathcal{L} w_h, \mathcal{L}_{Skew} v_h + \rho \mathcal{L}_{Sym} v_h \right)_K, \quad (5)$$

with $w_h, v_h \in V_h$ is coercive in the norm $|||v|||^2 := (\epsilon + h|\beta|)|v|^2_{H^1_0(\Omega)} + \sigma \|v\|^2_{L^2(\Omega)}$ in all possible regimes and consequently the methods are stable.

We consider next the *Link-Cutting Bubble* strategy [4], based on the enrichment of the finite element space V_h. The idea behind is to augment V_h by adding a space of *discrete* bubbles V_B, which is constructed element by element on a suitable subgrid. The LCB method can be regarded from two different standpoints. On the one hand, by considering $V_E = V_h \oplus V_B$ as a space of piecewise linear functions on a suitable refined grid, the LCB-approximation reduces to the plain Galerkin method:

$$\text{Find } u_E^{LCB} \in V_E \text{ such that } a(u_E^{LCB}, v_E) = (f, v_E), \quad \forall v_E \in V_E. \quad (6)$$

On the other hand, by means of *static condensation* of the bubble degrees of freedom, one gets the stabilized method: Find $u_h^{LCB} \in V_h$ s.t.:

$$a(u_h^{LCB}, v_h) + \sum_{K \in \mathcal{T}_h} \left(M_K(f - \mathcal{L} u_h^{LCB}), \mathcal{L}^* v_h \right)_K = (f, v_h), \quad \forall v_h \in V_h. \quad (7)$$

where for each $K \in \mathcal{T}_h$, $M_K : L^2(K) \to V_B|_K$ is the solution operator of the *local bubble problems*: $a(u_B^{LCB}, v_B)_K = (f - \mathcal{L} u_h^{LCB}, v_B)_K \; \forall v_B \in V_B|_K$, $a(\cdot, \cdot)_K$ and $V_B|_K$ being the restrictions to K of $a(\cdot, \cdot)$ and V_B, respectively.

3 Stabilized methods for the non-stationary problem

Given $f \in L^2([0,T]; L^2(\Omega))$ and $u_0 \in L^2(\Omega)$, consider the model problem:

$$\begin{cases} \dfrac{\partial}{\partial t} u + \mathcal{L} u = f & \text{in } Q = \Omega \times (0, T), \\ u|_{t=0} = u_0 & \text{on } \Omega, \quad u = 0 \quad \text{on } \partial\Omega \times (0, T). \end{cases} \quad (8)$$

We next briefly revise the DG method for the time integration of (8). Then, we shall describe the classical space-time elements and the stabilized methods resulting from the two approaches mentioned in the Introduction.

3.1 DG-methods for the time integration

Let $0 = t_0 < t_1 < \ldots < t_N = T$ a subdivision of the time interval $(0, T)$, set $J_n = (t_n, t_{n+1}]$ with $k = t_{n+1} - t_n$, and introduce the strips $\mathcal{S}_n := \{(x, t) \in \Omega \times J_n\}$, for $n = 0, \ldots N - 1$. The DG approximation in time to u, solution of (8), is sought as a piecewise polynomial of degree at most $q \geq 0$

in t on each subinterval J_n, with coefficients in $\mathcal{V}$, i.e., it belongs to the space $\mathcal{W}^q := \{v : [0,T] \vee \mathcal{V}; \ v_{|J_n} = \sum_{j=0}^{q} \psi_j t^j, \ \psi_j \in \mathcal{V}\}$. Note that any $v \in \mathcal{W}^q$ is allowed to be discontinuous at the nodes of the partition. Let $(\cdot, \cdot)$ denote the standard L^2-inner product and for $v, w \in \mathcal{W}^q$ we denote by

$$(v,w)^n := \int_{S_n} vw\, dx\, dt = \int_{J_n} (v,w)\, dt, \qquad v_+(x,t) = \lim_{s \to 0^+} v(x, t+s),$$

$$<v, w>^n := \int_\Omega v(x,t_n) w(x,t_n)\, dx = (v^n, w^n), \qquad v_-(x,t) = \lim_{s \to 0^-} v(x, t+s),$$

and set $[[v]]_n = v_+^n - v_-^n$. The DG time-discretization of (8) is obtained by imposing on S_n the initial value at $t = t_n$ weakly. Thus, the method reduces to find $U \in \mathcal{W}^q$ such that on each J_n (for $n = 0, \ldots N-1$), satisfies

$$(\frac{dU}{dt} + \mathcal{L}U, V)^n + <U_+, V_+>^n = <U_-, V_+>^n + (f,V)^n \quad \forall V \in \mathcal{W}^q. \quad (9)$$

For $q = 0$ (i.e., piecewise constants) one has $dU/dt = 0$ and $U(t) = U^{n+1} = U_+^n$ in J_n, so that the method reduces to the *modified backward Euler*:

$$a(U^{n+1}, \psi) + \frac{1}{k}(U^{n+1}, \psi) = \frac{1}{k}(U^n, \psi) + \left(\frac{1}{k}\int_{J_n} f\, dt, \psi\right) \quad \forall \psi \in \mathcal{V}. \quad (10)$$

For $q=1$, let $U(t) = U_0^{n+1} + \frac{(t-t_n)}{k} U_1^{n+1}$ on J_n so that we have to find $\mathbf{U}^{(n+1)}$ s.t.:

$$\frac{1}{k}(D \cdot \mathbf{U}^{n+1}, \mathbf{V}) + C \cdot a(\mathbf{U}^{n+1}, \mathbf{V}) = \frac{1}{k}(E \cdot \mathbf{U}^n, \mathbf{V}) + (\mathbf{F}^{n+1}, \mathbf{V}) \quad \forall \mathbf{V}, \quad (11)$$

where $\mathbf{U}^{n+1} = [U_0^{n+1}, U_1^{n+1}]^T$ with $U_0^{n+1}, U_1^{n+1} \in \mathcal{V}$, $\mathbf{V} = [\psi, \eta]^T$, $\psi, \eta \in \mathcal{V}$ and

$$D = \begin{bmatrix} 1 & 1 \\ 0 & \frac{1}{2} \end{bmatrix} \quad C = \begin{bmatrix} 1 & \frac{1}{2} \\ \frac{1}{2} & \frac{1}{3} \end{bmatrix} \quad E = \begin{bmatrix} 1 & 1 \\ 0 & 0 \end{bmatrix} \quad \mathbf{F}^{n+1} = \begin{bmatrix} \frac{1}{k}\int_{J_n} f(t)\, dt \\ \frac{1}{k^2}\int_{J_n} (t-t_n) f(t)\, dt \end{bmatrix}.$$

With a small abuse of notation, $\mathbf{a}(\cdot, \cdot)$ should be understood as the matrix

$$\mathbf{a}(\mathbf{U}, \mathbf{V}) = \begin{bmatrix} a(U_0, \psi), a(U_1, \psi) \\ a(U_0, \eta), a(U_1, \eta) \end{bmatrix}.$$

Similarly, in what follows we shall denote by $\mathcal{L}$ the scalar operator acting component-wise; i.e., $\mathcal{L}\mathbf{U} = [\mathcal{L}U_0, \mathcal{L}U_1]^T$.

3.2 Classical Space-Time Elements

We describe briefly the method introduced in [3]. For each n consider a quasi-uniform partition of the strip S_n with elements of size $h > \epsilon$, and let V_h^n be a FE subspace of $H^1(S_n)$ based on such partition, such that for $v \in V_h^n$ it holds $v = 0$ on $\partial \Omega \times J_n$. By applying successively on each strip S_n the stabilized methods of (3) and imposing the initial value at $t = t_n$ weakly and the boundary conditions strongly, one obtains the following method: given u_-^0 an approximation to the initial data u_0, for $n = 0, \ldots, N-1$ find $u^n \in V_h^n$

$$(u_t + \mathcal{L}u, v)^n + \left(u_t + \mathcal{L}u, \hat{\delta} \cdot [v_t + \beta v_x + \rho \mathcal{L}_{sym} v]\right)^n + <u_+, v_+>^n =$$
$$= (f, v + \hat{\delta} \cdot [v_t + \beta v_x + \rho \mathcal{L}_{sym} v])^n + <u_-, v_+>^n, \qquad \forall v \in V_h^n, \quad (12)$$

where the parameter $\hat{\delta}$ is set to $\bar{C}h$ if $\epsilon < h$ and 0 otherwise.

3.3 First Approach: DG in Time + Stabilized Method in Space

To present the fully discretized methods resulting from the first approach, the key point is to observe that on each slab S_n, the solution $U^{n+1} \in \mathcal{V}$ of the DG in time method (10) (and resp. (11)), might be regarded as the solution of a "steady" convection-diffusion-reaction problem with some "added extra reaction" $\frac{1}{k}$, coming from the time discretization. Thus, by discretizing (10) in space with any of the stabilized methods (3), leads to the problem: for each $n = 0, \ldots N - 1$, find $u_h^{n+1} \in V_h$ s.t.

$$\sum_{K \in \mathcal{T}_h} \tilde{\delta}_K^0 \left(\frac{(u_h^{n+1} - u_h^n)}{k} + \mathcal{L} u_h^{n+1} - \frac{1}{k} \int_{J_n} f\,dt, \left[\mathcal{L}_{skew} + \rho \mathcal{L}_{sym} + \frac{\rho}{k}\right] v_h \right)_K +$$
$$a(u_h^{n+1}, v_h) + \frac{1}{k}(u_h^{n+1}, v_h) - \left(\frac{1}{k} \int_{J_n} f\,dt, v_h\right) - \frac{1}{k}(u_h^n, v_h) = 0, \ \forall v_h \in V_h.$$

Similarly for the discretization (11) ($q = 1$), we get: for $n = 0, \ldots, N - 1$ find $\mathbf{U}_h^{n+1} = [U_0^{n+1}, U_1^{n+1}]^T$ that satisfies for all $\mathbf{V}_h = [v_h, w_h]^T$ with $v_h, w_h \in V_h$

$$\sum_{K \in \mathcal{T}_h} \tilde{\delta}_K^1 \left(\frac{(D\mathbf{U}_h^{n+1} - E\mathbf{U}_h^n)}{k} + C\mathcal{L}\mathbf{U}_h^{n+1} - \mathbf{F}^{n+1}, \left[C\mathcal{L}_{skew} + \rho C\mathcal{L}_{sym} + \rho \frac{D}{k}\right] \mathbf{V}_h \right)_K$$
$$+ C\mathbf{a}(\mathbf{U}_h^{n+1}, \mathbf{V}_h) + \frac{D}{k}(\mathbf{U}_h^{n+1}, \mathbf{V}_h) - (\mathbf{F}^{n+1}, \mathbf{V}_h) - \frac{1}{k}(E\mathbf{U}_h^n, \mathbf{V}_h) = 0. \quad (13)$$

where $\mathbf{a}(\cdot, \cdot)$ is defined as in Sect. 3.1. Note that the weighting operators resulting from the stabilization in this approach, $\left(\mathcal{L}_{skew} + \delta\left[\mathcal{L}_{sym} + \frac{1}{k_n}\right]\right)$ for $q = 0$, and $\left(C\mathcal{L}_{skew} + \rho\left[C\mathcal{L}_{sym} + \frac{1}{k}D\right]\right)$ for $q = 1$, contain a term coming from the time derivative, but it acts as a reaction term. To ensure the stability of the method in all possible regimes, it can be shown that it is enough to take $\tilde{\delta}_K^0 := \left((2\sigma + 2/k) + 2|\beta|/h + 12\epsilon/h^2\right)^{-1}$ and $\tilde{\delta}_K^1 := \left(2D/k + 2\sigma C + (2|\beta|C)/h + (12\epsilon C)/h^2\right)^{-1}$, for $q = 0$ and $q = 1$, respectively. In the last case, we have taken into account that (13) is a system.

For the sake of brevity and clarity in the exposition, we only consider the method that results by discretizing (10) in space by means of the LCB strategy. As before, the key observation is that (10) might be regarded as a convection-difussion-reaction stationary problem with the extra reaction $1/k$. Then, the idea is to define a new bilinear form on each strip S_n

$$\tilde{a}(w, v) = a(w, v) + \frac{1}{k}(w, v), \qquad w, v \in W^q \quad (14)$$

and construct the bubble subgrid, and consequently the bubble space $\widetilde{V}_B$, according to this bilinear form rather than (15), the one associated to the stationary problem. Then, for each n one consider either the (SG) approach (6), but with $\widetilde{V}_E = V_h \oplus \widetilde{V}_B$.

3.4 Second Approach: Stabilized Method in Space + DG in Time

We first discretize (8) in space by means of the stabilized methods given in Sect. 2. As for the techniques (3), we are lead to the following system of ODE's:

$$\frac{d}{dt}(u_h, v_h) + a(u_h, v_h) + \sum_{K \in \mathcal{T}_h} \delta_K \left(\frac{\partial u_h}{\partial t} + \mathcal{L}u_h, \mathcal{L}_{Skew}v_h + \rho\mathcal{L}_{Sym}v_h \right)_K =$$

$$= (f(t), v_h) + \sum_{K \in \mathcal{T}_h} \delta_K \left(f(t), \mathcal{L}_{Skew}v_h + \rho\mathcal{L}_{Sym}v_h \right)_K, \quad \forall v_h \in V_h, \quad (15)$$

where $u_h : [0, T] \longrightarrow V_h$ and δ_K is taken as in (4). By Integrating (15) with (10), we get for each $n = 0, \ldots N - 1$

$$\sum_{K \in \mathcal{T}_h} \delta_K \left(\frac{(u_h^{n+1} - u_h^n)}{k} + \mathcal{L}u_h^{n+1} - \frac{1}{k}\int_{J_n} f dt, \left[\mathcal{L}_{skew} + \rho\mathcal{L}_{sym}\right] v_h \right)_K +$$

$$a_h(u_h^{n+1}, v_h) + \frac{1}{k}(u_h^{n+1}, v_h) - \left(\frac{1}{k}\int_{J_n} f dt, v_h\right) - \frac{1}{k}(u_h^n, v_h) = 0, \quad \forall v_h \in V_h.$$

and upon integration in time of (15) with (11) we have: for $n = 0, \ldots, N-1$ find $\mathbf{U}_h^{n+1} = [U_0^{n+1}, U_1^{n+1}]^T$ that satisfies for all $\mathbf{V}_h = [v_h, w_h]^T \; v_h, w_h \in V_h$

$$\sum_{K \in \mathcal{T}_h} \delta_K \left(\frac{(D\mathbf{U}_h^{n+1} - E\mathbf{U}_h^n)}{k} + C\mathcal{L}\mathbf{U}_h^{n+1} - \mathbf{F}^{n+1}, \left[\mathcal{L}_{skew} + \rho\mathcal{L}_{sym}\right] \mathbf{V}_h \right)_K +$$

$$C\mathbf{a_h}(\mathbf{U}_h^{n+1}, \mathbf{V}_h) + \frac{D}{k}(\mathbf{U}_h^{n+1}, \mathbf{V}_h) - (\mathbf{F}^{n+1}, \mathbf{V}_h) - \frac{1}{k}(E\mathbf{U}_h^n, \mathbf{V}_h) = 0. \quad (16)$$

Notice that unlike for methods (12) or (13) no explicit reference to the time integration or time-discretization is contained in the weighting operator for these stabilized methods[3].

For the method resulting by considering the LCB strategy, one starts by constructing the subgrid for the local bubble space V_B, according to the steady operator $\mathcal{L}$, i.e. according to the bilinear form a (is done as for the steady problem). Then, the enriched space $V_E = V_h \oplus V_B$ is built and the LCB approximation is defined by the scheme: Find $u_E : [0, T] \to V_E$ such that

$$\frac{d}{dt}(u_E(t), v_E) + a(u_E(t), v_E) = (f, v_E), \quad \forall v_E \in V_E, \quad (17)$$

Then, one uses either (10) or (11) for the DG integration in time of system (17), noting that now $\mathcal{V}$ is approximated by the enriched space $V_E = V_h \oplus V_B$.

---

[3] For this reason it is enough to take δ_K as in (4) to ensure the stability of the method in all the regimes we will look at; in particular the advection dominated.

4 Numerical Experiments

The next set of experiments is devoted to evaluate the performance of the stabilization methods introduced before. We have considered problem (8) over $\Omega = (0, 1)$ and subject to homogeneous Dirichlet boundary conditions. We have set $\epsilon = 10^{-6}$, $\beta = 1$, $\sigma = 0$, the final time $T = 0.2$ and we assume $f = 0$. The inital data is taken as $u_0 = 1$ if $|x - 0.3| \leq 0.1$ and is set to zero otherwise. We have taken a uniform partition of Ω into subintervals of length $h = |\Omega|/N$, with $N = 20, 40, 80, 160, 320$. For each h, every experiment was carried out with different values of the time step k below which the local time discretizations are desired. k is selected so that the Courant-number $CFL = k|\beta|/h = 1, 1/2, 1/3, 1/4, 1/5, 1/10$. For the three approaches, linear DG integration in time has been used.

To valuate the quality of the approximate solutions obtained by the different methods we have represented them in Fig. 1 at time $t = 0.15$. For the methods obtained with the first and second approaches, we have only represented the approximation obtained with DWG (o joined by a continuous line) and LCB (squares joined with a dotted line). For the classical space-time elements the approximations with all the methods in (3) are represented. It can be observed, that while the classical time-space elements reduce almost completely the spurious oscillations in the numerical approximations, the solution appears to be extremely dissipated. Nevertheless, the approximation with the other approaches while not very diffusive still presents spurious oscillations. We next look to the relative errors in $L^\infty([0, T]; L^1(\Omega))$-norm against N, for the three approaches. They are represented in Fig. 1 for $CFL = 1/3$ and all diagrams are depicted with the same vertical axes to ease the comparation. Among the stabilization techniques of (3) depicted with $-o-$, no significant differences can be observed. The LCB stabilization is represented by squares joined with a dotted line, and in both the first and second approaches, is the method producing the smallest errors. For the first and second approaches, an almost first order of convergence can be observed while for the space-time elements the rate of convergence seems to be close to 0.6. Moreover, the first apprach seems to be the most accurate from the error-diagrams. For space time elements, the errors are substantially higher than for the other

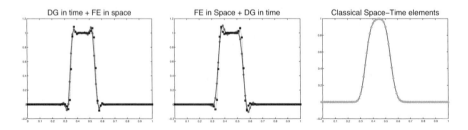

Fig. 1. Approximate solutions with $h = 80$ and $CFL = 1/3$, at time $t = 0.15$.

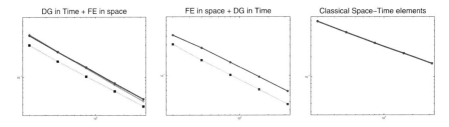

Fig. 2. Convergence Diagrams in $L^\infty([0,T]; L^1(\Omega))$.

two approaches, possibly due to the amount of dissipation that the scheme introduces.

Acknowledgements

This research has been partially supported by the Integrated Action Italo-Spanish HI2004-0383 and project HPRN-CT-2002-00286.

References

1. Baiocchi, C., Brezzi, F.: Stabilization of unstable methods, in: Problemi attuali dell'Analisi e della Fisica Matematica, P.E. Ricci Ed., Universit "La Sapienza", Roma, 59–64 (1993).
2. Brezzi, F., Hauke, G., Marini, L. D. ,Sangalli, G.: Link-cutting bubbles for the stabilization of convection-diffusion-reaction problems. Math. Models Methods Appl. Sci., **13**, 445–461 (2003)
3. Johnson, C., Nävert, U., Pitkäranta, J.: Finite element methods for linear hyperbolic problems. Comput. Methods Appl. Mech. Engrg. **45**, 285–312 (1984)
4. Asensio, M.I., Ayuso, B., Sangalli, G.: Coupling Stabilized Finite Element methods with Finite Difference time integration for the unsteady advection-diffussion-reaction problem, Tech. Report PV-8, (2006)
5. A. N. Brooks, T. J. R. Hughes: Streamline upwind/Petrov-Galerkin formulations for convection dominated flows with particular emphasis on the incompressible Navier-Stokes equations. Comp. Met. Appl. Mech. Engr., **32**, 199–259 (1982)
6. Douglas, J., Wang, Jr., Wang, J. P.: An absolutely stabilized finite element method for the Stokes problem. Math. Comp., **52**, 495–508 (1989)
7. Hughes, T. J. R., Franca, L. P., Hulbert, G. M.: A new finite element formulation for computational fluid dynamics. VIII. The Galerkin/least-squares method for advective-diffusive equations. Comp. Met. Appl. Mech. Engr., **73**, 173–189 (1989)

On Discontinuity–Capturing Methods for Convection–Diffusion Equations

Volker John[1] and Petr Knobloch[2]

[1] Universität des Saarlandes, Fachbereich 6.1 – Mathematik,
  Postfach 15 11 50, 66041 Saarbrücken, Germany
  john@math.uni-sb.de
[2] Charles University, Faculty of Mathematics and Physics,
  Sokolovská 83, 186 75 Praha 8, Czech Republic
  knobloch@karlin.mff.cuni.cz

Summary. This paper is devoted to the numerical solution of two–dimensional steady scalar convection–diffusion equations using the finite element method. If the popular streamline upwind/Petrov–Galerkin (SUPG) method is used, spurious oscillations usually arise in the discrete solution along interior and boundary layers. We review various finite element discretizations designed to diminish these oscillations and we compare them computationally.

1 Introduction

This paper is devoted to the numerical solution of the scalar convection–diffusion equation

$$-\varepsilon \, \Delta u + \boldsymbol{b} \cdot \nabla u = f \quad \text{in } \Omega, \qquad u = u_b \quad \text{on } \partial\Omega, \tag{1}$$

where $\Omega \subset \mathbb{R}^2$ is a bounded domain with a polygonal boundary $\partial\Omega$, $\varepsilon > 0$ is constant and $\boldsymbol{b}$, f and u_b are given functions.

If convection strongly dominates diffusion, the solution of (1) typically contains interior and boundary layers and solutions of Galerkin finite element discretizations are usually globally polluted by spurious oscillations. To enhance the stability and accuracy of these discretizations, various stabilization strategies have been developed during the past three decades. One of the most efficient procedures is the SUPG method developed by Brooks and Hughes [2]. Unfortunately, the SUPG method does not preclude spurious oscillations localized in narrow regions along sharp layers and hence various terms introducing artificial crosswind diffusion in the neighbourhood of layers have been proposed to be added to the SUPG formulation. This procedure is often referred to as discontinuity capturing (or shock capturing). The literature on discontinuity–capturing methods is rather extended and numerical tests published in the literature do not allow to draw conclusions concerning their advantages and drawbacks. Therefore, the aim of this paper is to provide a review of various discontinuity–capturing methods and to compare these methods computationally.

The plan of the paper is as follows. In the next section, we recall the Galerkin discretization of (1) and, in Section 3, we formulate the SUPG method. Section 4 contains a review and a computational comparison of discontinuity–capturing methods and, in Section 5, we present our conclusions.

2 Galerkin's finite element discretization

We introduce a triangulation $\mathcal{T}_h$ of the domain Ω consisting of a finite number of open polygonal elements K. We assume that $\overline{\Omega} = \bigcup_{K \in \mathcal{T}_h} \overline{K}$ and that the elements of $\mathcal{T}_h$ satisfy the usual compatibility conditions. Further, we introduce a finite element space V_h approximating the space $H_0^1(\Omega)$ and satisfying

$$V_h \subset \{v \in L^2(\Omega);\ v|_K \in C^\infty(\overline{K})\ \forall\ K \in \mathcal{T}_h\}.$$

Since the functions from V_h may be discontinuous across edges of the triangulation $\mathcal{T}_h$, we define the 'discrete' operators ∇_h and Δ_h by

$$(\nabla_h v)|_K = \nabla(v|_K), \qquad (\Delta_h v)|_K = \Delta(v|_K) \qquad \forall\ K \in \mathcal{T}_h.$$

Finally, let $u_{bh} \in L^2(\Omega)$ be a piecewise smooth function whose trace on $\partial\Omega$ approximates u_b. Then a discrete solution of (1) can be defined as a function $u_h \in L^2(\Omega)$ satisfying $u_h - u_{bh} \in V_h$ and $a_h(u_h, v_h) = (f, v_h)\ \forall\ v_h \in V_h$, where

$$a_h(u, v) = \varepsilon\,(\nabla_h u, \nabla_h v) + (\boldsymbol{b} \cdot \nabla_h u, v)$$

and $(\cdot, \cdot)$ denotes the inner product in the space $L^2(\Omega)$ or $L^2(\Omega)^2$.

3 The SUPG method

Brooks and Hughes [2] enriched the Galerkin method by a stabilization term yielding the streamline upwind/Petrov–Galerkin (SUPG) method. The discrete solution $u_h \in L^2(\Omega)$ satisfies $u_h - u_{bh} \in V_h$ and

$$a_h(u_h, v_h) + (R_h(u_h), \tau\,\boldsymbol{b} \cdot \nabla_h v_h) = (f, v_h) \qquad \forall\ v_h \in V_h, \qquad (2)$$

where $R_h(u) = -\varepsilon\,\Delta_h u + \boldsymbol{b} \cdot \nabla_h u - f$ is the residual and τ is a nonnegative stabilization parameter. As we see, the SUPG method introduces numerical diffusion along streamlines in a consistent manner. A delicate question is the choice of the parameter τ which may dramatically influence the accuracy of the discrete solution. Here we shall use the formula (cf. Galeão et al. [8])

$$\tau|_K = \frac{h_K}{2\,|\boldsymbol{b}|\,p_K}\left(\coth(\mathrm{Pe}_K) - \frac{1}{\mathrm{Pe}_K}\right) \quad \text{with} \quad \mathrm{Pe}_K = \frac{|\boldsymbol{b}|\,h_K}{2\,\varepsilon\,p_K}, \tag{3}$$

where h_K is the diameter of $K \in \mathcal{T}_h$ in the direction of $\boldsymbol{b}$, p_K is the order of approximation of V_h on K (usually the maximum degree of polynomials in V_h on K), $|\cdot|$ is the Euclidean norm and Pe_K is the local Péclet number.

4 Methods diminishing spurious oscillations in layers

In this section, we present a review and a computational comparison of most of the methods introduced during the last two decades to diminish the oscillations arising in discrete solutions of the problem (1). These methods can be divided into upwinding techniques and into methods adding additional artificial diffusion to the SUPG discretization (2). The artificial diffusion may be either isotropic, or orthogonal to streamlines, or based on edge stabilizations. These four classes of methods will be discussed in the following subsections. It is not possible to describe here thoroughly the ideas on which the design of the methods relies, see [11] for a more comprehesive description. Generally, one can say that the methods are based either on convergence analyses or on investigations of the discrete maximum principle (called DMP in the following) or on heuristic arguments. As we shall see, most of the methods will be nonlinear. The computational comparison of the methods will be performed by means of two test problems specified by the following data of (1):

Example 1. $\Omega = (0,1)^2$, $\varepsilon = 10^{-7}$, $\boldsymbol{b} = (\cos(-\pi/3), \sin(-\pi/3))^T$, $f = 0$, $u_b(x,y) = 0$ for $x = 1$ or $y \leq 0.7$, $u_b(x,y) = 1$ otherwise.

Example 2. $\Omega = (0,1)^2$, $\varepsilon = 10^{-7}$, $\boldsymbol{b} = (1,0)^T$, $f = 1$, $u_b = 0$.

The solution of Ex. 1 possesses an interior layer and exponential boundary layers whereas the solution of Ex. 2 possesses parabolic and exponential boundary layers but no interior layers. All results were computed on uniform $N \times N$ triangulations of the type depicted in Fig. 1. Unless stated otherwise, we used the conforming linear finite element P_1, $N = 20$ for Ex. 1 and $N = 10$ for Ex. 2. The SUPG solutions of Ex. 1 and 2 are shown in Fig. 2 and 5, respectively. It is important that the parameter τ is optimal for the P_1 element in the sense that the SUPG method approximates the boundary layers at $y = 0$ in Ex. 1 and at $x = 1$ in Ex. 2 sharply and without oscillations.

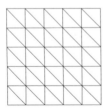

Fig. 1. Type of triangulations ($N = 5$)

Fig. 2. Ex. 1, SUPG

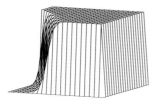 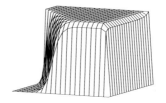

Fig. 3. Ex. 1, IMH [14]   **Fig. 4.** Ex. 1, do Carmo, Galeão [6]

4.1 Upwinding techniques

Initially, stabilizations of the Galerkin discretization of (1) imitated upwind finite difference techniques. However, like in the finite difference method, the upwind finite element discretizations remove the unwanted oscillations but the accuracy attained is often poor since too much numerical diffusion is introduced. According to our experiences, one of the most successful upwinding techniques is the improved Mizukami–Hughes (IMH) method, see Knobloch [14]. It is a nonlinear Petrov–Galerkin method for P_1 elements which satisfies the DMP on weakly acute meshes. In contrast with many other upwinding methods for P_1 elements satisfying the DMP, the IMH method adds much less numerical diffusion and provides rather accurate solutions, cf. Knobloch [15]. The IMH solution for Ex. 1 is depicted in Fig. 3. For Ex. 2, it is even nodally exact.

4.2 Methods adding isotropic artificial diffusion

Hughes *et al* . [10] came with the idea to change the upwind direction in the SUPG term of (2) by adding a multiple of the function $\boldsymbol{b}_h^\|$ which is the projection of $\boldsymbol{b}$ into the direction of ∇u_h. This leads to the additional term

$$(R_h(u_h), \sigma\, \boldsymbol{b}_h^\| \cdot \nabla_h v_h) \qquad (4)$$

on the left–hand side of (2), where σ is a nonnegative stabilization parameter. Since $\boldsymbol{b}_h^\|$ depends on u_h, the resulting method is nonlinear. Hughes *et al* . [10] proposed to set $\sigma = \max\{0, \tau(\boldsymbol{b}_h^\|) - \tau(\boldsymbol{b})\}$ where we use the notation $\tau(\boldsymbol{b}^\star)$ for τ defined by (3) with $\boldsymbol{b}$ replaced by $\boldsymbol{b}^\star$. Other definitions of σ in (4) were proposed by Tezduyar and Park [17]. Since the term (4) equals to

$$(\widetilde{\varepsilon}\, \nabla_h u_h, \nabla_h v_h) \qquad (5)$$

with $\widetilde{\varepsilon} = \sigma\, R_h(u_h)\, \boldsymbol{b}\cdot\nabla u_h / |\nabla u_h|^2$, it introduces an isotropic artificial diffusion.

Another stabilization strategy was introduced by Galeão and do Carmo [9] who proposed to replace the flow velocity $\boldsymbol{b}$ in the SUPG stabilization term by an approximate upwind direction. This gives rise to the additional term

$$(R_h(u_h), \sigma\, \boldsymbol{z}_h \cdot \nabla_h v_h) \qquad (6)$$

on the left-hand side of (2), where $z_h = R_h(u_h)\nabla u_h/|\nabla u_h|^2$ and $\sigma = \max\{0, \tau(z_h) - \tau(\boldsymbol{b})\}$. If $f = 0$ and $\Delta_h u_h = 0$, we have $z_h = \boldsymbol{b}_h^\|$ and hence the method of Galeão and do Carmo [9] is identical with the method of Hughes et al. [10]. Do Carmo and Galeão [6] proposed to simplify σ to

$$\sigma = \tau(\boldsymbol{b}) \max\left\{0, \frac{|\boldsymbol{b}|}{|z_h|} - 1\right\}. \tag{7}$$

Almeida and Silva [1] suggested to replace (7) by

$$\sigma = \tau(\boldsymbol{b}) \max\left\{0, \frac{|\boldsymbol{b}|}{|z_h|} - \zeta_h\right\} \quad \text{with} \quad \zeta_h = \max\left\{1, \frac{\boldsymbol{b}\cdot\nabla_h u_h}{R_h(u_h)}\right\}, \tag{8}$$

which reduces the amount of artificial diffusion along the z_h direction.

Do Carmo and Galeão [6] also introduced a feedback function which should minimize the influence of the term (6) in regions where the solution u of (1) is smooth. Since this approach was rather involved, do Carmo and Alvarez [5] introduced another procedure (still defined using several formulas) suppressing the addition of the artificial diffusion in regions where u is smooth.

Again, the term (6) can be written in the form (5), now with $\widetilde{\varepsilon} = \sigma|R_h(u_h)|^2/|\nabla u_h|^2$. To prove error estimates, Knopp et al. [16] proposed to replace this $\widetilde{\varepsilon}$, on any $K \in \mathcal{T}_h$, by

$$\widetilde{\varepsilon}|_K = \sigma_K(u_h)\,|Q_K(u_h)|^2 \quad \text{with} \quad Q_K(u_h) = \frac{\|R_h(u_h)\|_{0,K}}{S_K + \|u_h\|_{1,K}}, \tag{9}$$

where $\sigma_K(u_h) \geq 0$ and $S_K > 0$ are appropriate constants.

The stabilization term (5) was also used by Johnson [12], who considered

$$\widetilde{\varepsilon}|_K = \max\{0, \alpha\,[\mathrm{diam}(K)]^\nu\,|R_h(u_h)| - \varepsilon\} \quad \forall\,K \in \mathcal{T}_h$$

with some constants α and $\nu \in (3/2, 2)$. He suggested to take $\nu \sim 2$.

If the above methods are applied to Ex. 1, the discrete solution improves in comparison to the SUPG method. However, most of the methods do not remove the spurious oscillations completely and/or lead to an excessive smearing of the layers. The best methods are the methods of do Carmo and Galeão [6] and Almeida and Silva [1] which are identical in this case, see Fig. 5.

4.3 Methods adding artificial diffusion orthogonally to streamlines

Since the streamline diffusion introduced by the SUPG method seems to be enough along the streamlines, an alternative approach to the above methods is to modify the SUPG discretization (2) by adding artificial diffusion in the crosswind direction only as considered by Johnson et al. [13]. A straightforward generalization of their approach leads to the additional term

$$(\widetilde{\varepsilon}\,D\,\nabla_h u_h, \nabla_h v_h) \tag{10}$$

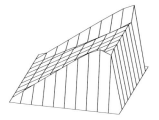
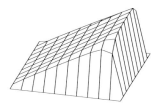

Fig. 5. Ex. 2, SUPG   **Fig. 6.** Ex. 2, MBE

on the left–hand side of (2), where $\tilde{\varepsilon}|_K = \max\{0, |\boldsymbol{b}|\, h_K^{3/2} - \varepsilon\}\ \forall K \in \mathcal{T}_h$ and $D = I - \boldsymbol{b} \otimes \boldsymbol{b}/|\boldsymbol{b}|^2$ is the projection onto the line orthogonal to $\boldsymbol{b}$, I being the identity tensor.

Investigating the validity of the DMP for several model problems, Codina [7] came to the conclusion that the artificial diffusion $\tilde{\varepsilon}$ in (10) should be defined, for any $K \in \mathcal{T}_h$, by

$$\tilde{\varepsilon}|_K = \frac{1}{2} \max\left\{0, C - \frac{2\varepsilon}{|\boldsymbol{b}_h^\parallel|\,\mathrm{diam}(K)}\right\} \mathrm{diam}(K)\, \frac{|R_h(u_h)|}{|\nabla u_h|}, \qquad (11)$$

where C is a suitable constant (we use $C = 0.6$ for linear elements and $C = 0.35$ for quadratic elements). Motivated by assumptions and results of general a priori and a posteriori error analyses, Knopp et al. [16] changed (11) to

$$\tilde{\varepsilon}|_K = \frac{1}{2} \max\left\{0, C - \frac{2\varepsilon}{Q_K(u_h)\,\mathrm{diam}(K)}\right\} \mathrm{diam}(K)\, Q_K(u_h), \qquad (12)$$

where $Q_K(u_h)$ is defined in (9) (the constants S_K equal to 1 in numerical experiments of [16]). Combining the above two definitions of $\tilde{\varepsilon}$, we further propose to use (10) with $\tilde{\varepsilon}$ defined by (12) where $Q_K(u_h) = |R_h(u_h)|/|\nabla u_h|$. This modified method of Codina is called MC method in the following. It is equivalent to (11) if $f = 0$ and $\Delta_h u_h = 0$.

Based on investigations of the DMP for strictly acute meshes and linear simplicial finite elements, Burman and Ern [3] suggested to use (10) with $\tilde{\varepsilon}$ defined, on any $K \in \mathcal{T}_h$, by

$$\tilde{\varepsilon}|_K = \frac{\tau(\boldsymbol{b})\,|\boldsymbol{b}|^2\,|R_h(u_h)|}{|\boldsymbol{b}|\,|\nabla_h u_h| + |R_h(u_h)|}\, \frac{|\boldsymbol{b}|\,|\nabla_h u_h| + |R_h(u_h)| + \tan\alpha_K\,|\boldsymbol{b}|\,|D\nabla_h u_h|}{|R_h(u_h)| + \tan\alpha_K\,|\boldsymbol{b}|\,|D\nabla_h u_h|}.$$

Here, α_K is equal to $\pi/2$ minus the largest angle of K (if K is a triangle). In case of right triangles, it is recommended to set $\alpha_K = \pi/6$.

Our numerical experiments indicate that the above value of $\tilde{\varepsilon}$ is too large and therefore we also consider (10) with $\tilde{\varepsilon}$ defined, on any $K \in \mathcal{T}_h$, by

$$\tilde{\varepsilon}|_K = \frac{\tau(\boldsymbol{b})\,|\boldsymbol{b}|^2\,|R_h(u_h)|}{|\boldsymbol{b}|\,|\nabla_h u_h| + |R_h(u_h)|}. \qquad (13)$$

Fig. 7. Ex. 1, MC, P_2, $N = 10$ **Fig. 8.** Ex. 1, do Carmo, Galeão [6], P_1^{nc}

This modified Burman–Ern method is called MBE method in the following.

If we apply the methods of this subsection to Ex. 2, then only the MC and MBE methods give satisfactory results (and they are comparable), see Fig. 1. For Ex. 1, these methods provide comparable results to the solution in Fig. 5. On the other hand, the two best methods of the previous subsection (do Carmo and Galeão [6], Almeida and Silva [1]) give almost the same results for Ex. 2, which are comparable to the results of the MC and MBE methods.

4.4 Edge stabilizations

Another stabilization strategy for linear simplicial finite elements was introduced by Burman and Hansbo [4]. The term to be added to the left–hand side of (2) is defined by

$$\sum_{K \in \mathcal{T}_h} \int_{\partial K} \Psi_K(u_h) \operatorname{sign}(\boldsymbol{t}_{\partial K} \cdot \nabla(u_h|_K)) \, \boldsymbol{t}_{\partial K} \cdot \nabla(v_h|_K) \, \mathrm{d}\sigma \,,$$

where $\boldsymbol{t}_{\partial K}$ is a unit tangent vector to the boundary ∂K of K, $\Psi_K(u_h) = \operatorname{diam}(K)\,(C_1\,\varepsilon + C_2\operatorname{diam}(K))\,\max_{E \subset \partial K} |\,[[\boldsymbol{n}_E \cdot \nabla u_h]]_E\,|$, $\boldsymbol{n}_E$ are normal vectors to edges E of K, $[[v]]_E$ denotes the jump of a function v across the edge E and C_1, C_2 are appropriate constants. Burman and Hansbo proved that, using an edge stabilization instead of the SUPG term, the DMP is satisfied. Other choices of $\Psi_K(u_h)$ based on investigations of the DMP were recently proposed by Burman and Ern. However, all these edge stabilizations add much more artificial diffusion than the best methods of the previous subsections.

5 Conclusions

Our computations indicate that, among the methods mentioned in this paper, the best ones are: the IMH method [14], the method of do Carmo, Galeão [6] defined by (6), (7), the method of Almeida and Silva [1] defined by (6), (8), the MC method introduced below (12) and the MBE method defined by (10), (13). The IMH method can be used for the P_1 element only but gives best results in this case. The other methods can be successfully also applied to other finite elements as Figs. 2 and 8 show (for the conforming quadratic element P_2 and the nonconforming Crouzeix–Raviart element P_1^{nc}). However, much more comprehensive numerical studies are still necessary to obtain clear conclusions of the advantages and drawbacks of the discontinuity–capturing methods.

Acknowledgements

The work of Petr Knobloch is a part of the research project MSM 0021620839 financed by MSMT and it was partly supported by the Grant Agency of the Charles University in Prague under the grant No. 343/2005/B–MAT/MFF.

References

1. Almeida, R.C., Silva, R.S.: A stable Petrov–Galerkin method for convection–dominated problems. Comput. Methods Appl. Mech. Eng., **140**, 291–304 (1997)
2. Brooks, A.N., Hughes, T.J.R.: Streamline upwind/Petrov–Galerkin formulations for convection dominated flows with particular emphasis on the incompressible Navier–Stokes equations. Comput. Methods Appl. Mech. Eng., **32**, 199–259 (1982)
3. Burman, E., Ern, A.: Nonlinear diffusion and discrete maximum principle for stabilized Galerkin approximations of the convection–diffusion–reaction equation. Comput. Methods Appl. Mech. Eng., **191**, 3833–3855 (2002)
4. Burman, E., Hansbo, P.: Edge stabilization for Galerkin approximations of convection–diffusion–reaction problems. Comput. Methods Appl. Mech. Eng., **193**, 1437–1453 (2004)
5. do Carmo, E.G.D., Alvarez, G.B.: A new stabilized finite element formulation for scalar convection–diffusion problems: The streamline and approximate upwind/Petrov–Galerkin method. Comput. Methods Appl. Mech. Eng., **192**, 3379–3396 (2003)
6. do Carmo, E.G.D., Galeão, A.C.: Feedback Petrov–Galerkin methods for convection–dominated problems. Comput. Methods Appl. Mech. Eng., **88**, 1–16 (1991)
7. Codina, R.: A discontinuity–capturing crosswind–dissipation for the finite element solution of the convection–diffusion equation. Comput. Methods Appl. Mech. Eng., **110**, 325–342 (1993)
8. Galeão, A.C., Almeida, R.C., Malta, S.M.C., Loula, A.F.D.: Finite element analysis of convection dominated reaction–diffusion problems. Appl. Numer. Math., **48**, 205–222 (2004)
9. Galeão, A.C., do Carmo, E.G.D.: A consistent approximate upwind Petrov–Galerkin method for convection–dominated problems. Comput. Methods Appl. Mech. Eng., **68**, 83–95 (1988)
10. Hughes, T.J.R., Mallet, M., Mizukami, A.: A new finite element formulation for computational fluid dynamics: II. Beyond SUPG. Comput. Methods Appl. Mech. Eng., **54**, 341–355 (1986)

11. John, V., Knobloch, P.: A comparison of spurious oscillations at layers diminishing (SOLD) methods for convection–diffusion equations: Part I. Preprint Nr. **156**, FR 6.1 – Mathematik, Universität des Saarlandes, Saarbrücken (2005)
12. Johnson, C.: Adaptive finite element methods for diffusion and convection problems. Comput. Methods Appl. Mech. Eng., **82**, 301–322 (1990)
13. Johnson, C., Schatz, A.H., Wahlbin, L.B.: Crosswind smear and pointwise errors in streamline diffusion finite element methods. Math. Comput., **49**, 25–38 (1987)
14. Knobloch, P.: Improvements of the Mizukami–Hughes method for convection–diffusion equations. Preprint No. MATH–knm–2005/6, Faculty of Mathematics and Physics, Charles University, Prague (2005)
15. Knobloch, P.: Numerical solution of convection–diffusion equations using upwinding techniques satisfying the discrete maximum principle. Submitted to the Proceedings of the Czech–Japanese Seminar in Applied Mathematics 2005
16. Knopp, T., Lube, G., Rapin, G.: Stabilized finite element methods with shock capturing for advection–diffusion problems. Comput. Methods Appl. Mech. Eng., **191**, 2997–3013 (2002)
17. Tezduyar, T.E., Park, Y.J.: Discontinuity–capturing finite element formulations for nonlinear convection–diffusion–reaction equations. Comput. Methods Appl. Mech. Eng., **59**, 307–325 (1986)

Algebraic Flux Correction for Finite Element Approximation of Transport Equations

Dmitri Kuzmin

Institute of Applied Mathematics (LS III), University of Dortmund
Vogelpothsweg 87, D-44227, Dortmund, Germany
kuzmin@math.uni-dortmund.de

Summary. An algebraic approach to the design of high-resolution finite element schemes for convection-dominated flows is pursued. It is explained how to get rid of nonphysical oscillations and remove excessive artificial diffusion in regions where the solution is sufficiently smooth. To this end, the discrete transport operator and the consistent mass matrix are modified so as to enforce the positivity constraint in a mass-conserving fashion. The concept of a *target flux* and a new definition of upper/lower bounds make it possible to design a general-purpose flux limiter which provides an optimal treatment of both stationary and time-dependent problems.

1 Introduction

Algebraic flux correction [3] constitutes a promising approach to the design of high-resolution schemes for convection-dominated transport problems. In the present paper, flux limiting for consistent-mass Galerkin schemes is addressed. Building on the multidimensional limiters of FCT and TVD type [1, 2, 3], we design a general-purpose algorithm which combines their advantages. It represents a simple way to satisfy the discrete maximum principle for both explicit and implicit FEM on structured and unstructured meshes.

2 Flux decomposition

As a representative model problem, consider the continuity equation

$$\frac{\partial u}{\partial t} + \nabla \cdot (\mathbf{v}u) = 0 \qquad (1)$$

discretized in space by a high-order finite element method which yields an ODE system for the vector of time-dependent nodal values

$$M_C \frac{mDu}{mDt} = Ku, \qquad (2)$$

where $M_C = \{m_{ij}\}$ denotes the consistent mass matrix and $K = \{k_{ij}\}$ is the discrete operator resulting from the discretization of the convective term.

A fully discrete scheme proves *positivity-preserving* (PP) if each solution update $u^n \to u^{n+1}$ satisfies an equivalent algebraic system [3]

$$Au^{n+1} = Bu^n, \tag{3}$$

where $A = \{a_{ij}\}$ is an *M-matrix* and $B = \{b_{ij}\}$ has no negative entries. In the linear case, this algebraic criterion can be readily enforced using the 'discrete upwinding' technique which yields a linear PP scheme of the form [3]

$$M_L \frac{mDu}{mDt} = Lu, \qquad L = K + D, \tag{4}$$

where $M_L = \text{diag}\{m_i\}$ is the lumped mass matrix and $D = \{d_{ij}\}$ is the artificial diffusion operator which is supposed to be a symmetric matrix with zero row and column sums. Its off-diagonal coefficients are given by the relation $d_{ij} = \max\{-k_{ij}, 0, -k_{ji}\} = d_{ji}$ so that $l_{ij} := k_{ij} + d_{ij} \geq 0$, as required by the positivity criterion. Without loss of generality, it is assumed that $l_{ij} \leq l_{ji}$, which implies that node i is located 'upwind' of node j [3].

By construction, the difference between the high- and low-order schemes admits decomposition into a sum of antidiffusive *target fluxes* given by

$$f_{ij} = \left[m_{ij} \frac{mD}{mDt} + d_{ij} \right] (u_i - u_j) = f_{ij}^m + f_{ij}^d, \tag{5}$$

where $f_{ij}^m = m_{ij}(\dot{u}_i - \dot{u}_j)$ and $f_{ij}^d = d_{ij}(u_i - u_j)$ offset the error induced by mass lumping and discrete upwinding, respectively. Note that the former contains a time derivative which still needs to be discretized.

3 Algebraic flux correction

In order to prevent the formation of nonphysical local extrema, the raw antidiffusive fluxes are multiplied by suitable correction factors (see below)

$$f_{ij}^* := \alpha_{ij} f_{ij}, \qquad \text{where} \qquad 0 \leq \alpha_{ij} \leq 1. \tag{6}$$

The task of the flux limiter is to determine an optimal value of α_{ij} so as to remove as much artificial diffusion as possible without generating wiggles. Antidiffusive fluxes which violate the positivity constraint (3) and need to be limited are of the form $f_{ij} = p_{ij}(u_j - u_i)$, where $p_{ij} < 0$. On the other hand, edge contributions with nonnegative coefficients resemble diffusive fluxes and are harmless. Hence, some antidiffusion is admissible as long as there exists a set of (solution-dependent) coefficients $c_{ik} \geq 0$, $\forall k \neq i$ such that

$$\sum_{j \neq i} [l_{ij}(u_j - u_i) + f_{ij}^*] = \sum_{k \neq i} c_{ik}(u_k - u_i). \tag{7}$$

In order to enforce this sufficient condition, we resort to a node-based limiting strategy which was largely inspired by Zalesak's FCT algorithm [5] but is even more general. The net antidiffusion received by each node may consist of both positive and negative edge contributions. Assuming the worst-case scenario, let us limit them separately by the following generic algorithm

1. Compute the sums of positive and negative antidiffusive fluxes represented as edge contributions $f_{ij} = p_{ij}(u_j - u_i)$ with negative coefficients $p_{ij} \leq 0$

$$P_i^+ = \sum_{j \neq i} p_{ij} \min\{0, u_j - u_i\}, \qquad P_i^- = \sum_{j \neq i} p_{ij} \max\{0, u_j - u_i\}. \qquad (8)$$

2. Define the upper/lower bounds to be imposed in the course of flux correction as a sum of edge contributions with nonnegative coefficients $q_{ij} \geq 0$

$$Q_i^+ = \sum_{j \neq i} q_{ij} \max\{0, u_j - u_i\}, \qquad Q_i^- = \sum_{j \neq i} q_{ij} \min\{0, u_j - u_i\}. \qquad (9)$$

3. Evaluate the *nodal correction factors* for positive/negative fluxes

$$R_i^+ = \min\{1, Q_i^+/P_i^+\}, \qquad R_i^- = \min\{1, Q_i^-/P_i^-\}. \qquad (10)$$

4. Multiply the target flux f_{ij} by a combination of $R_i^\pm$ and $R_j^\mp$ such that

$$f_{ij}^* = \alpha_{ij} f_{ij}, \qquad \alpha_{ij} = \begin{cases} \alpha(R_i^+, R_j^-), & \text{if } f_{ij} > 0, \\ \alpha(R_i^-, R_j^+), & \text{otherwise.} \end{cases} \qquad (11)$$

The last part calls for further explanation. Recall that the edges of the sparsity graph are oriented so that $0 \leq l_{ij} \leq l_{ji} = k_{ji} + d_{ij}$ and we have

$$l_{ji}(u_i - u_j) - f_{ij}^* = (l_{ji} + \alpha_{ij} p_{ij})(u_i - u_j), \qquad (12)$$

so that the positivity constraint is satisfied if $l_{ji} + \alpha_{ij} p_{ij} \geq 0$. The contribution of the limited antidiffusive flux f_{ij}^* to node i is harmless since

$$Q_i^- \leq R_i^- P_i^- \leq \sum_{j \neq i} \alpha_{ij} f_{ij} \leq R_i^+ P_i^+ \leq Q_i^+. \qquad (13)$$

In light of the above, flux correction can be performed in two different ways:

- Upwind-biased flux correction: 'prelimit' the coefficient $p_{ij} := f_{ij}/(u_j - u_i)$ if it violates the positivity condition (12) for the downwind node j

$$f_{ij}' = \min\{-p_{ij}, l_{ji}\}(u_i - u_j) \qquad (14)$$

and multiply the so-defined antidiffusive flux f_{ij}' by $\alpha_{ij} = R_i^\pm$ to enforce the positivity condition (13) for the upwind node i.

- Symmetric flux correction: multiply f_{ij} by the minimum of nodal correction factors, i.e., $\alpha_{ij} = \min\{R_i^\pm, R_j^\mp\}$ regardless of the edge orientation.

The optimal choice of the limiting strategy depends on the magnitude of the antidiffusion coefficient p_{ij} as compared to that of l_{ji}. The above algorithm leads to a variety of algebraic flux correction schemes which differ in the definition of upper/lower bounds as well as in the type of flux limiting.

3.1 Treatment of convective antidiffusion

For the time being, let us assume that the problem at hand is stationary and neglect the contribution of the consistent mass matrix. The prelimited target flux (14) for a lumped-mass Galerkin discretization is given by

$$f'_{ij} = \min\{d_{ij}, l_{ji}\}(u_i - u_j), \tag{15}$$

where d_{ij} is the artificial diffusion coefficient for discrete upwinding. It is worth mentioning that there is actually no need for prelimiting as long as $l_{ji} - \alpha_{ij}d_{ij} = k_{ji} + (1-\alpha_{ij})d_{ij} \geq 0$. Therefore, the above target flux reduces to f^d_{ij} as defined above, unless both off-diagonal coefficients of the high-order operator K were negative (a rather unusual situation). In this particular case, the upwind-biased limiting strategy is preferable. The total amount of raw antidiffusion received by node i from its downwind neighbors is given by

$$P_i^{\pm} = \sum_{j \in \mathcal{J}_i} {\max \atop \min} \{0, f'_{ij}\}, \quad \text{where} \quad \mathcal{J}_i = \{j \neq i \,|\, 0 = l_{ij} < l_{ji}\}. \tag{16}$$

The off-diagonal coefficients of the low-order operator L can be used to define the upper/lower bounds as in the case of algebraic TVD schemes [3]

$$Q_i^{\pm} = \sum_{j \neq i} l_{ij} {\max \atop \min} (u_j - u_i), \quad l_{ij} \geq 0, \quad \forall j \neq i. \tag{17}$$

Flux limiting is performed using the correction factor for the upwind node:

$$f^*_{ij} = \begin{cases} R_i^+ f'_{ij}, & \text{if } f'_{ij} > 0, \\ R_i^- f'_{ij}, & \text{otherwise}, \end{cases} \qquad f^*_{ji} := -f^*_{ij}. \tag{18}$$

The same approach can be used to construct a family of positivity-preserving schemes based on standard TVD limiters [3]. However, the associated target fluxes are certain to ensure second-order accuracy only for a finite difference approximation in one dimension, whereas the real target fluxes for a finite element scheme are **uniquely** defined by (5). Therefore, such ad hoc extensions of TVD schemes are likely to pollute the solution in smooth regions and are not to be recommended for multidimensional FEM discretizations.

3.2 Treatment of mass antidiffusion

The contribution of the mass matrix to target fluxes of the form (5) may be large enough to render the upwind-biased limiting strategy impractical. Furthermore, the upper and lower bounds based on the coefficients of the low-order operator (17) are independent of the time step and may turn out to be too restrictive. In this subsection, we concentrate on the treatment of mass antidiffusion f^m_{ij} assuming that the convective part f^d_{ij} of the target flux vanishes. In this case, the flow direction is unknown and the antidiffusive flux may violate the positivity condition for both nodes. Therefore, we adopt the symmetric limiting strategy and discuss the choice of constraints to be imposed on the fully discretized target flux f_{ij}^m which corresponds to

$$f_{ij} = \frac{m_{ij}}{\Delta t}(u_i^{n+1} - u_j^{n+1}) - \frac{m_{ij}}{\Delta t}(u_i^n - u_j^n). \qquad (19)$$

Interestingly enough, this flux consists of a truly antidiffusive implicit part and a diffusive explicit part which has a strong damping effect.

If the standard FEM-FCT algorithm is employed, the corresponding upper and lower bounds $Q_i^\pm$ depend on the local extrema $\tilde{u}_i^\pm$ of the low-order solution $\tilde{u} = u^n + \Delta t M_L^{-1} L u^n$ which reduces to u^n in the case $L = 0$ (no convection). In order to avoid the computation of $\tilde{u}$ and accommodate the contribution of the convective term in what follows, we use a weaker constraint and take

$$P_i^\pm = \sum_{j \neq i} \max_{\min}\{0, f_{ij}\}, \qquad Q_i^\pm = \sum_{j \neq i} \frac{m_{ij}}{\Delta t} \max_{\min}\{0, u_j^n - u_i^n\}, \qquad (20)$$

where the coefficients m_{ij} are tacitly assumed to be nonnegative. Note that the nodal correction factors $R_i^\pm = \min\{1, Q_i^\pm/P_i^\pm\}$ are independent of the time step, since both $P_i^\pm$ and $Q_i^\pm$ are inversely proportional to it.

If the coefficient $p_{ij}^n = f_{ij}/(u_j^n - u_i^n)$ is negative, the target flux (19) proves truly antidiffusive and should be limited in a symmetric fashion

$$f_{ij}^* = \begin{cases} \min\{R_i^+, R_j^-\} f_{ij}, & \text{if } f_{ij} > 0, \\ \min\{R_i^-, R_j^+\} f_{ij}, & \text{otherwise}, \end{cases} \qquad f_{ji}^* = -f_{ij}^*. \qquad (21)$$

The above limiting strategy is closely related to Zalesak's multidimensional FCT algorithm [5] but the bounds $Q_i^\pm$ are defined in a different way and there is no need to compute a provisional low-order solution.

3.3 General-purpose flux limiter

Now that we have a stand-alone flux limiter for convective antidiffusion and a stand-alone flux limiter for mass antidiffusion at our disposal, we can proceed to the treatment of antidiffusive fluxes (5) which involve both contributions. Our experience with flux correction of FCT type indicates that it is worthwhile to prelimit f_{ij} so as to prevent it from becoming diffusive and creating numerical artifacts [3]. Therefore, let us adjust the target fluxes thus:

$$f_{ij} := p_{ij}(u_j - u_i), \qquad p_{ij} = \min\{0, (f_{ij}^m + f_{ij}^d)/(u_j - u_i)\}. \qquad (22)$$

It remains to specify the upper/lower bounds $Q_i^\pm$ and choose the flux limiting strategy. Both algorithms considered so far are directly applicable to target fluxes of the form (22) but their performance is highly problem-dependent. It is not unusual that $p_{ij} + l_{ji} < 0$ if mass antidiffusion is strong enough, which means that a significant portion of the target flux cannot be recovered by the upwind-biased flux limiter alone. In other cases, symmetric flux limiting may produce inferior results because taking the minimum of nodal correction factors turns out to be more restrictive than prelimiting based on (14).

A straightforward but inefficient way to combine the two flux limiting techniques is to apply them sequentially. For instance, one can use the upwind-biased algorithm (16)–(18) to predict f_{ij}^* and limit the rejected antidiffusion $\Delta f_{ij} = f_{ij} - f_{ij}^*$ according to (20)–(21) or vice versa. In any event, the effective upper and lower bounds for the sum of limited antidiffusive fluxes $f_{ij}^* + \Delta f_{ij}^*$ consist of the 'stationary' upwind part (17) and the 'time-dependent' symmetric part (20) which complement each other so that

- a certain fraction of admissible antidiffusion is independent of the time step, which prevents a loss of accuracy in steady-state computations;
- solutions to truly time-dependent problems become more accurate as Δt is refined, since a larger portion of the target flux may be retained.

Instead of limiting the target fluxes by the algorithms (16)–(18) and/or (20)–(21) in a segregated way or sequentially, it is worthwhile to combine these special-purpose limiters, which can be accomplished as follows

1. Decompose the target flux $f_{ij} = p_{ij}(u_j - u_i)$ into the prelimited 'upwind' part (14) and the remainder which must be limited in a symmetric fashion

$$f'_{ij} = \min\{-p_{ij}, l_{ji}\}(u_i - u_j), \quad \Delta f_{ij} = f_{ij} - f'_{ij}. \qquad (23)$$

2. Compute the total sums of raw antidiffusive fluxes to be constrained

$$P_i^\pm = \sum_{j \in \mathcal{J}_i} {\max \atop \min} \{0, f'_{ij}\} + \sum_{j \neq i} {\max \atop \min} \{0, \Delta f_{ij}\}. \qquad (24)$$

3. Define the combined upper/lower bounds to be enforced on $P_i^\pm$ as follows

$$Q_i^\pm = \sum_{j \neq i} \left[\frac{m_{ij}}{\Delta t} + l_{ij}\right] {\max \atop \min} (u_j - u_i). \qquad (25)$$

4. Evaluate the nodal correction factors (10) for the flux limiting step

$$R_i^\pm = \min\{1, Q_i^\pm / P_i^\pm\}. \qquad (26)$$

5. In a loop over edges, compute the limited antidiffusive correction

$$f_{ij}^* = R_i^\pm f'_{ij} + \min\{R_i^\pm, R_j^\mp\} \Delta f_{ij}. \qquad (27)$$

Note that the first sum in (24) is evaluated over $j \in \mathcal{J}_i$ (see (16)) while the second one contains antidiffusive fluxes from all neighboring nodes. The resulting nonlinear algebraic system can be solved by an iterative defect correction scheme preconditioned by the 'monotone' low-order operator [3].

4 Numerical example

Let us illustrate the performance of the new algorithm by applying it to the solid body rotation problem proposed by LeVeque [4]. After one full revolution ($t = 2\pi$) the exact solution of the continuity equation (1) coincides with the initial data. The numerical solutions presented in Fig. 1-2 were computed on a uniform mesh of 128×128 Q_1–elements using Crank-Nicolson time-stepping with $\Delta t = 10^{-3}$. The general-purpose (GP) algorithm (24)–(27) produces the results shown in Fig. 1. The cone and hump are reproduced very well and even the narrow bridge of the slotted cylinder is largely preserved. Not surprisingly, this solution is very similar to that computed by an FCT algorithm based on the same target flux [3]. On the other hand, the performance of standard TVD limiters for this time-dependent test problem leaves a lot to be desired. The strongly antidiffusive *superbee* entails a pronounced flattening of smooth peaks [3], while the 'default' MC limiter proves overly diffusive (see Fig. 2).

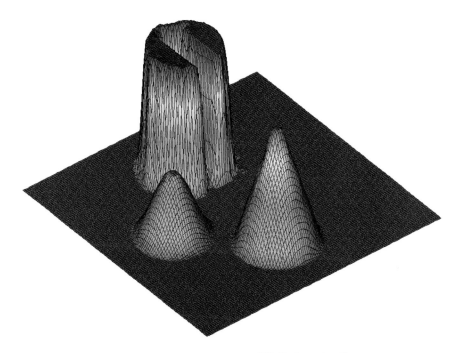

Fig. 1. Solid body rotation: GP limiter, $t = 2\pi$.

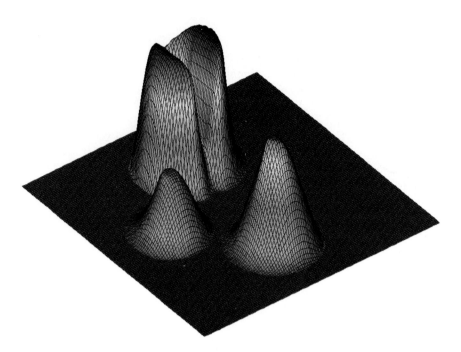

Fig. 2. Solid body rotation: MC limiter, $t = 2\pi$.

5 Conclusions

Algebraic flux correction of the form (8)–(11) provides a very general framework for the derivation of nonoscillatory high-resolution schemes. Unlike other limiting techniques, it is readily applicable to finite element discretizations and unstructured meshes. This paper bridges the gap between algebraic FCT and TVD schemes [3] proposed previously and paves the way to the design of general-purpose flux limiters for implicit FEM including the consistent mass matrix. Of course, there are many other ways to define the upper/lower bounds and perform algebraic flux corection. Moderate improvements can be achieved – at a disproportionately high overhead cost – but our numerical experiments indicate that the accuracy of the target flux rather than the choice of constraints and the type of flux limiting is decisive in many cases. Hence, it is not the limiter but the antidiffusive flux itself that still needs to be optimized.

References

1. Kuzmin, D, Turek, S.: Flux correction tools for finite elements. J. Comput. Phys. **175**, 525-558 (2002)

2. Kuzmin, D, Turek, S.: High-resolution FEM-TVD schemes based on a fully multidimensional flux limiter. J. Comput. Phys. **198** 131-158 (2004)
3. Kuzmin, D., Möller, M.: Algebraic flux correction I. Scalar conservation laws. In: D. Kuzmin, R. Löhner and S. Turek (eds.) Flux-Corrected Transport: Principles, Algorithms, and Applications. Springer, 155-206 (2005)
4. LeVeque, R. J.: High-resolution conservative algorithms for advection in incompressible flow. Siam J. Numer. Anal. **33** 627–665 (1996)
5. Zalesak, S. T.: Fully multidimensional flux-corrected transport algorithms for fluids. J. Comput. Phys. **31** 335–362 (1979)

A Parallel Multiparametric Gauss-Seidel Method*

N. M. Missirlis and F. I. Tzaferis

Department of Informatics and Telecommunications, University of Athens
`nmis@di.uoa.gr, ftzaf@di.uoa.gr`

Summary. In this paper we consider the local Modified Extrapolated Gauss-Seidel($LMEGS$) method combined with Semi-Iterative techniques for the numerical solution of the Convection Diffusion equation and compare it with the local SOR method. *Subject classification* : AMS(MOS), 65F10. *Keywords* : Parallel Iterative methods, linear systems, semi-iterative methods, Fourier analysis, Gauss-Seidel method, convection diffusion equation.

1 Introduction

The model problem considered here is that of solving the second order convection diffusion equation

$$\Delta u - f(x,y)\frac{\partial u}{\partial x} - g(x,y)\frac{\partial u}{\partial y} = 0 \qquad (1)$$

defined on a domain $\Omega = \{ (x,y) | 0 \le x \le \ell_1, \ 0 \le y \le \ell_2 \}$, where Δ is the Laplacian operator and $u = u(x,y)$ is prescribed on the boundary $\partial\Omega$. The discretization of (1) on a rectangular grid $M_1 \times M_2 = N$ unknowns within Ω using the 5–point stencil leads to

$$u_{ij} = l_{ij}u_{i-1,j} + r_{ij}u_{i+1,j} + t_{ij}u_{i,j+1} + b_{ij}u_{i,j-1}, \ i=1,2,\cdots,M_1, \ j=1,2,\cdots M_2 \qquad (2)$$

with

$$l_{ij} = \frac{k^2}{2(k^2+h^2)}\left(1+\frac{1}{2}hf_{ij}\right), \quad r_{ij} = \frac{k^2}{2(k^2+h^2)}\left(1-\frac{1}{2}hf_{ij}\right), \qquad (3)$$

$$t_{ij} = \frac{h^2}{2(k^2+h^2)}\left(1-\frac{1}{2}kg_{ij}\right), \quad b_{ij} = \frac{h^2}{2(k^2+h^2)}\left(1+\frac{1}{2}kg_{ij}\right), \qquad (4)$$

---

* Research is supported by the Action PYTHAGORAS, Program EPEAEK II of the Greek Ministry of Education, grant no. 70/3/7418.

where $h = \ell_1/(M_1+1)$, $k = \ell_2/(M_2+1)$, $f_{ij} = f(ih, jk)$ and $g_{ij} = g(ih, jk)$. For a particular ordering of the grid points, (2) yields a large, sparse linear system of equations of the form

$$Ax = b. \tag{5}$$

For the numerical solution of (5) we consider iterative methods. In the present paper we extend our work in [2] to include the case of complex eigenvalues for the Jacobi iteration matrix. Convergence ranges and good (near the optimum) values for the involved set of parameters of the Local Modified Extrapolated Gauss-Seidel($LMEGS$) method are obtained in case the eigenvalues of the Jacobi iteration matrix are complex of the form $\mu = \alpha + i\beta$, where α, β are real with $\alpha \in [\alpha_m, \alpha_M]$ and $\beta \in [\beta_m, \beta_M]$. Numerical results indicate that $LMEGS$, combined with semi-iterative techniques, posesses the same order of converegence as the local SOR method.

2 The local modified extrapolated Gauss-Seidel (LMEGS) method

The local Jacobi operator J_{ij} is defined as

$$J_{ij} = d_{ij}^{-1}(l_{ij}E_x^{-1} + r_{ij}E_x + t_{ij}E_y + b_{ij}E_y^{-1}) \tag{6}$$

where E_x, E_y are shift operators along the x and y directions defined by $E_x u_{ij} = u_{i+1,j}$, $E_x^{-1}u_{ij} = u_{i-1,j}$, $E_y u_{ij} = u_{i,j+1}$, $E_y^{-1}u_{ij} = u_{i,j-1}$. We can choose to call a grid point (i,j) as red when $i+j$ is even and black when $i+j$ is odd. For the numerical solution of (5) the $LMEGS$ method becomes

$$u_{ij}^{(n+1)} = (1-\tau_{ij})u_{ij}^{(n)} + \tau_{ij}J_{ij}u_{ij}^{(n)}, \quad i+j \text{ even} \tag{7}$$

$$u_{ij}^{(n+1)} = (1-\tau_{ij}')u_{ij}^{(n)} + J_{ij}u_{ij}^{(n+1)} + (\tau_{ij}'-1)J_{ij}u_{ij}^{(n)}, \quad i+j \text{ odd} \tag{8}$$

where τ_{ij}, τ_{ij}' are called the local parameters and correspond to red ($i+j$ even) and black ($i+j$ odd) grid points, respectively. We remark that the LMEGS method generalizes the GS method, allowing the introduction of two sets of parameters (τ_{ij} and τ_{ij}'). The advantage of using the above parameters is (i) the possible increase in the rate of convergence and (ii) that each node in the mesh has its own parameter, thus avoiding global communication when the method is parallelized [1].

3 The eigenvalue relationship

In this section we apply the local Fourier analysis to find an eigenvalue relatioship between the eigenvalues of the LMEGS iteration operator and J_{ij}

the local Jacobi operator. At this point it should be mentioned that Fourier analysis applies exactly only to linear constant coefficient PDEs on an infinite or on a rectangular domain with Dirichlet or periodic boundary conditions. Although this would seem a restriction to our analysis it has been shown that there is a strong correspondence with results for other boundary conditions [6]. Writing (7) and (8) in terms of the error vector $e^{(n)} = u_{ij}^{(n)} - u_{ij}$, we have

$$e_R^{(n+1)} = (1-\tau_{ij})e_R^{(n)} + \tau_{ij}J_{ij}e_B^{(n)}, \quad i+j \text{ even} \tag{9}$$

$$e_B^{(n+1)} = (1-\tau_{ij}')e_B^{(n)} + \tau_{ij}'J_{ij}e_R^{(n+1)} + (\tau_{ij}'-1)J_{ij}e_B^{(n)}, \quad i+j \text{ odd} \tag{10}$$

where $e_R^{(n)}$ and $e_B^{(n)}$ represent the errors at the red and black points, respectively. By eliminating $e_R^{(n+1)}$, (10) is written as

$$e_B^{(n+1)} = (\tau_{ij}' - \tau_{ij})J_{ij}e_R^{(n)} + (1-\tau_{ij}') + \tau_{ij}'J_{ij}^2 e_R^{(n+1)}. \tag{11}$$

Equations (9) and (11) can be written as

$$\begin{pmatrix} e_R^{(n+1)} \\ e_B^{(n+1)} \end{pmatrix} = \mathcal{L}_{\tau_{ij},\tau_{ij}'}(J_{ij}) \begin{pmatrix} e_R^{(n)} \\ e_B^{(n)} \end{pmatrix}, \tag{12}$$

where

$$\mathcal{L}_{\tau_{ij},\tau_{ij}'}(J_{ij}) = \begin{bmatrix} 1-\tau_{ij} & \tau_{ij}J_{ij} \\ (\tau_{ij}'-\tau_{ij})J_{ij} & 1-\tau_{ij}'+\tau_{ij}'J_{ij}^2 \end{bmatrix} \tag{13}$$

is called the LMEGS iteration operator. By assuming that an eigenfunction of $\mathcal{L}_{\tau_{ij},\tau_{ij}'}(J_{ij})$ possesses the form $(c_1 e^{i(k_1 x + k_2 y)}, c_2 e^{i(k_1 x + k_2 y)})^T$ and that the corresponding eigenvalue is λ_{ij}, we have

$$\mathcal{L}_{\tau_{ij},\tau_{ij}'}(J_{ij}) \begin{pmatrix} c_1 e^{i(k_1 x + k_2 y)} \\ c_2 e^{i(k_1 x + k_2 y)} \end{pmatrix} = \lambda_{ij} \begin{pmatrix} c_1 e^{i(k_1 x + k_2 y)} \\ c_2 e^{i(k_1 x + k_2 y)} \end{pmatrix}, \tag{14}$$

yielding

$$\mathcal{L}_{\tau_{ij},\tau_{ij}'}(\mu_{ij}) \begin{pmatrix} c_1 \\ c_2 \end{pmatrix} = \lambda_{ij} \begin{pmatrix} c_1 \\ c_2 \end{pmatrix} \tag{15}$$

since

$$J_{ij} e^{i(k_1 x + k_2 y)} = \mu_{ij}(k_1, k_2) e^{i(k_1 x + k_2 y)}, \tag{16}$$

where

$$\mu_{ij}(k_1, k_2) = l_{ij} e^{-ik_1 h} + r_{ij} e^{ik_1 h} + t_{ij} e^{ik_2 k} + b_{ij} e^{-ik_2 k}. \tag{17}$$

Furthermore, from (15) it follows that $\det(\mathcal{L}_{\tau_{ij},\tau_{ij}'}(\mu_{ij}) - \lambda_{ij} I) = 0$ or because of (13)

$$\lambda_{ij}^2 - (2 - \tau_{ij} - \tau_{ij}' + \tau_{ij}\mu_{ij}^2)\lambda_{ij} + (1-\tau_{ij})(1-\tau_{ij}') + \tau_{ij}(1-\tau_{ij}')\mu_{ij}^2 = 0. \tag{18}$$

4 Determination of good values

Let us assume that the eigenvalues of J_{ij} are complex of the form $\mu_{ij} = \alpha_{ij} + i\beta_{ij}$, where α_{ij}, β_{ij} are real with $\alpha_{ij} \in [\alpha_{mij}, \alpha_{Mij}] = I_1$ and $\beta_{ij} \in [\beta_{mij}, \beta_{Mij}] = I_2$. Solving (18) we find that it has the following two roots

$$\lambda_{ij} = 1 - \tau_{ij} r_{ij} \quad \text{and} \quad \lambda_{ij} = 1 - \tau'_{ij}, \tag{19}$$

where $r_{ij} = 1 - \mu_{ij}^2$ and $r_{ij} = a_{ij} + ib_{ij}$, with $a_{ij} \in [\underline{a}_{ij}, \overline{a}_{ij}]$ and $b_{ij} \in [-\overline{b}_{ij}, \overline{b}_{ij}]$.

Theorem 1. If $S(\mathcal{L}_{\tau,\tau'}) < 1$ [2], then

(i) an upper bound on $S(\mathcal{L}_{\tau,\tau'})$ is given by

$$S(\mathcal{L}_{\tau,\tau'}) \leq \begin{cases} 1 - \tau', & 0 < \tau' \leq 1 - \Lambda^{1/2} \\ \Lambda^{1/2}, & 1 - \Lambda^{1/2} \leq \tau' \leq 1 + \Lambda^{1/2} \\ \tau' - 1, & 1 + \Lambda^{1/2} \leq \tau' < 2, \end{cases} \tag{20}$$

where Λ given by (30),

(ii) if $\underline{a} > 0$, then the bound on $S(\mathcal{L}_{\tau,\tau'})$ is minimised for τ_b given by (39) and any $\tau'_b \in [\tau'_m, \tau'_M]$, where τ'_m and τ'_M are given by (41).

Proof. The spectral radius of $\mathcal{L}_{\tau,\tau'}$ is $S(\mathcal{L}_{\tau,\tau'}) = \max\{|\lambda_1|, |\lambda_2|\}$ where λ_1, λ_2 are the roots of (18). Next, we distinguish the following cases:

$$\textbf{I.} \ |\lambda_1| \geq |\lambda_2| \quad \text{and} \quad \textbf{II.} \ |\lambda_1| < |\lambda_2|. \tag{21}$$

Case I : Suppose that $|\lambda_1| \geq |\lambda_2|$, then

$$S(\mathcal{L}_{\tau,\tau'}) = |\lambda_1| = |1 - \tau'| < 1 \quad \text{or} \quad 0 < \tau' < 2. \tag{22}$$

Furthermore, $|\lambda_1| \geq |\lambda_2|$ yields successively

$$|1 - \tau'| \geq |1 - \tau(1 - \mu^2)| \tag{23}$$

or

$$\phi(\tau') = \tau'^2 - 2\tau' + k(\tau, a, b) \geq 0, \tag{24}$$

where

$$k(\tau, a, b) = 2\tau a - \tau^2(a^2 + b^2). \tag{25}$$

Since for LMEGS to converge $0 < \tau' < 2$, we distinguish the following two subcases:

$$\textbf{I}_1. \ 0 < \tau' \leq 1 \quad \text{or} \quad \textbf{I}_2. \ 1 < \tau' < 2. \tag{26}$$

If $0 < \tau' \leq 1$

$$S(\mathcal{L}_{\tau,\tau'}) = 1 - \tau'. \tag{27}$$

---

[2] In the sequel we drop the subscripts $_{ij}$ for notation simplicity.

Also, if $1 < \tau' < 2$
$$S(\mathcal{L}_{\tau,\tau'}) = \tau' - 1. \tag{28}$$
A similar analysis can be followed for Case **II**. In this case
$$S(\mathcal{L}_{\tau,\tau'}) = \max_{\mu^2} |1 - \tau(1-\mu^2)| \le \Lambda^{1/2}, \tag{29}$$
where
$$\Lambda = \max_{a,b} \lambda(\tau,a,b), \quad \lambda = \min_{a,b} \lambda(\tau,a,b) \quad \text{and} \quad \lambda(\tau,a,b) = 1 - k(\tau,a,b). \tag{30}$$
Summarizing our results so far we have
$$S(\mathcal{L}_{\tau,\tau'}) \le \begin{cases} 1 - \tau', & 0 < \tau' \le 1 - \Lambda^{1/2} \\ \Lambda^{1/2}, & 1 - \lambda^{1/2} < \tau' < 1 + \lambda^{1/2} \\ \tau' - 1, & 1 + \Lambda^{1/2} \le \tau' < 2. \end{cases} \tag{31}$$
By determining that an upper bound on $S(\mathcal{L}_{\tau,\tau'})$, in the intervals for τ'
$$\left[1 - \Lambda^{1/2},\ 1 - \lambda^{1/2}\right] \quad \text{and} \quad \left[1 + \lambda^{1/2},\ 1 + \Lambda^{1/2}\right], \tag{32}$$
is $\Lambda^{1/2}$ we prove the validity of (20). Therefore, (i) holds. In the sequence we will minimize the bound on $S(\mathcal{L}_{\tau,\tau'})$ first with respect to τ' and then with respect to τ. Let us consider the first branch of (20). Then, $S(\mathcal{L}_{\tau,\tau'})$ is minimized for the largest value of τ', say τ'_b which is given by
$$\tau'_b = 1 - \Lambda^{1/2} \tag{33}$$
and its correspoding value is given by
$$S(\mathcal{L}_{\tau,\tau'_b}) \le \Lambda^{1/2}. \tag{34}$$
Similarly, considering the third branch of (20), the bound on $S(\mathcal{L}_{\tau,\tau'})$ is minimized for the smallest value of τ' which now is given by
$$\tau'_b = 1 + \Lambda^{1/2} \tag{35}$$
and its correspoding value is given by (34) also, which coincides with the value of the bound of the second branch of (20). Our conclusion so far is
$$S(\mathcal{L}_{\tau,\tau'_b}) \le \Lambda^{1/2} \tag{36}$$
for
$$\tau'_b \in [\tau'_m,\ \tau'_M], \tag{37}$$
where
$$\tau'_m = 1 - \Lambda^{1/2} \quad \text{and} \quad \tau'_M = 1 + \Lambda^{1/2}, \tag{38}$$

thus (ii) is proved. Next, we have to determine $\Lambda_{\tau_b}^{1/2} = \min_\tau \{\max_{a,b} \Lambda^{1/2}\}$. However, if $\underline{a} > 0$ this is achieved (see Theorem 3.1 of [5]) at

$$\tau_b = \begin{cases} \dfrac{\underline{a}}{\underline{a}^2 + \overline{b}^2}, & \text{if } \overline{a} \leq \overline{b} \\ \dfrac{2}{\overline{a} + \underline{a}}, & \text{if } \overline{b} \leq \underline{a} \end{cases} \tag{39}$$

and

$$\Lambda_{\tau_b}^{1/2} \leq \begin{cases} \dfrac{\overline{b}}{(\underline{a}^2 + \overline{b}^2)^{\frac{1}{2}}}, & \text{if } \overline{a} \leq \overline{b} \\ \dfrac{[4\overline{b}^2 + (\overline{a} - \underline{a})^2]^{\frac{1}{2}}}{\overline{a} + \underline{a}}, & \text{if } \overline{b} \leq \underline{a}. \end{cases} \tag{40}$$

∎

Therefore, (38), because of (40), becomes

$$\tau'_m = 1 - \Lambda_{\tau_b}^{1/2} \quad \text{and} \quad \tau'_M = 1 + \Lambda_{\tau_b}^{1/2}. \tag{41}$$

A similar result holds for $\overline{a} < 0$ (see Theorem 3.1 of [5]).

5 Numerical results and conclusions

In order to predict the performance of the LMEGS method we have to study the eigenvalue spectrum of the local Jacobi operator J_{ij}. From (17) we have that for periodic boundary conditions

$$|\mu_{ij}(k_1, k_2)| = \big([(r_{ij} + \ell_{ij})\cos k_1 h + (t_{ij} + b_{ij})\cos k_2 k]^2 + [(r_{ij} - \ell_{ij})\sin k_1 h + (t_{ij} - b_{ij})\sin k_2 k]^2 \big)^{1/2} \tag{42}$$

ij indicating that $\mu_{ij}(k_1, k_2)$ depends upon the coefficients of the particular PDE, where $k_1, k_2 = \pi, 2\pi, \cdots, (\sqrt{N}-1)\pi$. If the coefficients of the PDE are constant, then $\overline{\mu}_{ij} = \overline{\mu}$, $\tau_{ij} = \tau$, and LMEGS becomes the classic MEGS. Moreover, for $f(x,y) = g(x,y) = 0$ (Poisson equation) on the unit square with Dirichlet boundary conditions (42) yields

$$\mu_{ij}(k_1, k_2) = \frac{1}{2}(\cos k_1 h + \cos k_2 k) \tag{43}$$

hence

$$\overline{\mu}_{ij} = \max_{\pi \leq k_1, k_2 \leq (\sqrt{N}-1)\pi} |\mu_{ij}(k_1, k_2)| = \cos \pi h. \tag{44}$$

If $\sqrt{N}$ is even, then the Jacobi operator has an odd number of eigenvalues which occur in pairs $\pm \mu_{ij}$, therefore zero will be one of its eigenvalues. In this case $\underline{\mu}_{ij} = 0$. If $\sqrt{N}$ is odd, then

$$\underline{\mu}_{ij} = \min_{\pi \leq k_1,\, k_2 \leq (\sqrt{N}-1)\pi} |\mu_{ij}(k_1, k_2)| = \cos\frac{\pi(1-h)}{2} \quad (45)$$

with $M_1 = M_2 = \sqrt{N}$. If the PDE has space-varying coefficients with Dirichlet boundary conditions, then for the 5-point stencil the quantities $\overline{\mu}_{ij}$ and $\underline{\mu}_{ij}$ are determined by [3]

$$\mu_{ij} = 2\left(\sqrt{\ell_{ij}r_{ij}}\cos\frac{k_1\pi}{M_1+1} + \sqrt{t_{ij}b_{ij}}\cos\frac{k_2\pi}{M_2+1}\right), \quad (46)$$

where $k_1 = 1, 2, \ldots, M_1$ and $k_2 = 1, 2, \ldots, M_2$. From (46) we find

$$\overline{\mu}_{ij} = 2\left(\sqrt{\ell_{ij}r_{ij}}\cos\pi h + \sqrt{t_{ij}b_{ij}}\cos\pi k\right) \quad (47)$$

and

$$\underline{\mu}_{ij} = 2\left(\sqrt{\ell_{ij}r_{ij}}\cos\frac{\pi(1-h)}{2} + \sqrt{t_{ij}b_{ij}}\cos\frac{\pi(1-k)}{2}\right). \quad (48)$$

Note that (47) and (48) yield (44) and (45), respectively, for constant coefficient PDEs. From (46) we see that the eigenvalues μ_{ij} may be real, imaginary or complex.

The optimum values for the local relaxation parameters are given by Theorem 1. Next, in an attempt to improve the rate of convergence of $LMEGS$, we apply semi-iterative techniques. In order to test our theoretical results we considered the numerical solution of (1) with $u = 0$ on the boundary of the unit square. The initial vector was chosen as $u^{(0)}(x, y) = xy(1-x)(1-y)$. The solution of the above problem is zero. For comparison purposes we considered the application of the local SOR with red black ordering (LSOR R/B) as described in [4]. The iterative process was terminated when the criterion $||u^{(n)}||_\infty \leq 10^{-6}$ was satisfied. Various functions for the coefficients $f(x, y)$ and $g(x, y)$ were chosen such that the eigenvalue μ_{ij} to be real, imaginary or complex. The coefficients used in each problem are

Real case : 1. $f(x, y) = Re \cdot x^2$, $g(x, y) = 0$
2. $f(x, y) = Re \cdot (10 - 2x)$, $g(x, y) = Re \cdot (10 - 2y)$
Imaginary case : 3. $f(x, y) = \frac{Re}{2}(1 + x^2)$, $g(x, y) = 100$
Complex case : 4. $f(x, y) = Re \cdot (2x - 1)^3$, $g(x, y) = 0$.

The number of iterations for the problems considered are presented in Table 1. These results show that $SI - LMEGS$ has the same convengence behavior as the local SOR method in case the eigenvalues of the Jacobi iteration matrix posesses real or imaginary eigenvalues (see cases 1, 2, 3 of Table 1). This phenomenon was expected since $LMEGS$ has the same order of convergence rate as GS and when one applies semi-iterative techniques to GS, then its rate of convergence is equivalent to that of SOR. However, when the Jacobi iteration matrix posesses complex eigenvalues the corresponding problem was an open one. In this case selecting the involved parameters τ and τ' appropriately (as in Theorem 1) we found that there are cases (case 4

Table 1. Number of iterations of SI-LMEGS and LSOR methods for $h = 1/21, 1/41$.
∗ denotes no convergence after $5 \cdot 10^4$ iterations.

#	Method	$h = 1/21$ $Re = 1$	$h = 1/21$ $Re = 10$	$h = 1/41$ $Re = 1$
1	LSOR	50	69	97
	SI-LMEGS	44	66	86
2	LSOR	29	26	73
	SI-LMEGS	29	24	53
		$Re = 2 \cdot 10^4$	$Re = 10^5$	$Re = 10^4$
3	LSOR	1399	6933	∗
	SI-LMEGS	180	203	878
		$Re = 100$	$Re = 10^5$	$Re = 100$
4	LSOR	173	1673	330
	SI-LMEGS	170	756	326

of Table 1), where $SI - LMEGS$ has significantly better performance than $LSOR$, a fact which needs further investigation. We therefore conclude that $SI - LMEGS$ is a promising method, like local SOR, and has an efficient parallel implementation. As GS is used as a smoother in Multigrid methods [6] it would be interesting to study its replacement by $LMEGS$.

Acknowledgement

The project is co-funded by the European Social Fund and National Resources (EPEAK II) Pythagoras

References

1. Boukas, L.A., Missirlis, N.M.: The Parallel Local Modified SOR for nonsymmetric linear systems. Inter. J. Computer Math., **68**, 153–174 (1998)
2. Consta, A.A, Missirlis, N.M., Tzaferis, F.I.: The local modified extrapolated Gauss-Seidel(LMEGS) method. Computers and Structures, **82**, 2447–2451 (2004)
3. Ehrlich, L.W.: An Ad-Hoc SOR Method. J. Comput. Phys., **42**, 31–45 (1981)
4. Kuo, C.-C.J., Levy, B.C., Musicus, B.R.: A local relaxation method for solving elliptic PDE's on mesh-connected arrays. SIAM J. Sci. Statist. Comput., **8**, 530-573 (1987)
5. Missirlis, N.M.: The extrapolated first order method for solving systems with complex eigenvalues. BIT, **24**, 357-365 (1984)
6. Stuben, K., Trottenberg, U.: Multigrid methods: Fundamental algorithms, model problem analysis and applications in Multigrid Methods. U. Trottenberg ed., Springer Verlag, New York (1982)

A Numerical Scheme for the Micro Scale Dissolution and Precipitation in Porous Media

I.S. Pop[1], V.M. Devigne[2,3], C.J. van Duijn[1] and T. Clopeau[3]

[1] CASA, Technische Universiteit Eindhoven, PO Box 513, 5600 MB Eindhoven, The Netherlands
{I.Pop,C.J.v.Duijn}@tue.nl
[2] Centre SITE, ENS des Mines, 158 cours Fauriel, 42023 Saint-Etienne, France
Vincent.Devigne@emse.fr
[3] Institut Camille Jordan, Université Claude Bernard Lyon I, site de Gerland, Bât. A, bur. 1304, 50, Av. Tony Garnier, 69367 Lyon Cedex 07, France

Summary. In this paper we discuss numerical method for a pore scale model for precipitation and dissolution in porous media. We focus here on the chemistry, which is modeled by a parabolic problem that is coupled through the boundary conditions to an ordinary differential inclusion. A semi-implicit time stepping is combined with a regularization approach to construct a stable and convergent numerical scheme. For dealing with the emerging time discrete nonlinear problems we propose here a simple fixed point iterative procedure.

1 Introduction

In this paper we consider a pore scale model for crystal dissolution and precipitation processes in porous media. This model is proposed in [2] and represents the pore–scale analogue of the one proposed in [7]. We continue here the analysis in [2] by investigating a semi-implicit time discretization of the model. The resulting nonlinear elliptic problems are solved by a simple linear iterative scheme.

Without going into the modeling details, we give here the background of the problem under consideration. A fluid in which cations and anions are dissolved occupies the void region of a porous medium. Under certain conditions, these ions can precipitate and form a crystalline solid, which is attached to the surface of the grains (the porous skeleton) and thus is immobile. The reverse reaction of dissolution is also possible. Therefore the model consists of several components: the Stokes flow in the pores, the transport of dissolved ions by convection and diffusion, and dissolution/precipitation reactions on the surface of the porous skeleton (grains). Here we assume that the flow geometry, as well as the fluid properties are not affected by the chemical processes.

Our main interest is focused on the chemistry, this being the challenging part of the model. To be specific, we denote by $\Omega \subset \mathbb{R}^d$ ($d > 1$) the void space of the porous medium, which is assumed open, connected and bounded. Its boundary is Lipschitz continuous and consists of two disjoint parts: the internal part (Γ_G) represents the surface of the porous skeleton (the grains), and the external part Γ_D, which is the outer boundary of the domain. Further, ν denotes the outer normal to $\partial\Omega$ and $T > 0$ a fixed but arbitrarily chosen value of time. For X being Ω, Γ_G, or Γ_D, we define

$$X^T = (0, T] \times X.$$

Assuming, for the simplicity of presentation, that the boundary and initial data are compatible (see [3], or [7]) we reduce our model to

$$\begin{cases} \partial_t u + \nabla \cdot (\mathbf{q}u - D\nabla u) = 0, & \text{in } \Omega^T, \\ -D\boldsymbol{\nu} \cdot \nabla u = \varepsilon \tilde{n} \partial_t v, & \text{on } \Gamma_G^T, \\ u = 0, & \text{on } \Gamma_D^T, \\ u = u_I, & \text{in } \Omega, \text{ for } t = 0, \end{cases} \quad (1)$$

for the ion transport, and

$$\begin{cases} \partial_t v = D_a(r(u) - w), & \text{on } \Gamma_G^T, \\ w \in H(v), & \text{on } \Gamma_G^T, \\ v = v_I, & \text{on } \Gamma_G, \text{ for } t = 0, \end{cases} \quad (2)$$

for the precipitation and dissolution. Here v denotes the concentration of the precipitate, which is defined only on the interior boundary Γ_G, while u stands for the cation concentration. $\mathbf{q}$ denotes the divergence free fluid velocity. The initial data u_I and v_I are assumed non–negative and essentially bounded. Moreover, for simplicity we assume that $u_I \in H^1_{0,\Gamma_D}(\Omega)$, the space of H^1 functions defined on Ω and having a vanishing trace on Γ_D.

All the quantities and variables in the above are assumed dimensionless. D denotes the diffusion coefficient and $\tilde{n}$ the anion valence. D_a represents the ratio of the characteristic precipitation/dissolution time scale and the characteristic transport time scale - the Damköhler number, which is assumed to be of moderate order. By ε we mean the ratio of the characteristic pore scale and the reference (macroscopic) length scale. Throughout this paper we keep the value of ε fixed, but this can be taken arbitrarily small.

Assuming mass action kinetics, with $[\cdot]+$ denoting the non-negative part, the precipitation rate is defined by

$$r(u) = [u]_+^{\tilde{m}} [(\tilde{m}u - c)/\tilde{n}]_+^{\tilde{n}}, \quad (3)$$

where $\tilde{m}$ is the cation concentration and c the total negative charge, which is assumed here constant in time and space. The analysis here is not restricted to the typical example above, but assumes that r is an increasing, positive, locally Lipschitz continuous function. Further, since c is fixed in (3), there exists a unique $u_* \geq 0$ such that $r(u) = 0$ for all $u \leq u_*$, and r is strictly increasing for $u > u_*$.

By H we mean the Heaviside graph,

$$H(u) = \begin{cases} 0, & \text{if } u < 0, \\ [0,1], & \text{if } u = 0, \\ 1, & \text{if } u > 0, \end{cases}$$

and w is the actual value of the dissolution rate. The dissolution rate is constant (1, by the scaling) in the presence of crystal, i.e. for $v > 0$ somewhere on Γ_G. In the absence of crystals, the overall rate is either zero, if the fluid is not containing sufficient dissolved ions, or positive.

Remark 1. In the setting above, a unique u^* exists for which $r(u^*) = 1$. If $u = u^*$ for all t and x, then the system is in equilibrium: no precipitation or dissolution occurs, since the precipitation rate is balanced by the dissolution rate regardless of the presence of absence of crystals.

The particularity of the model considered here is in the description of the dissolution and precipitation processes taking place on the surface of the grains Γ_G, involving a multi–valued dissolution rate function. In mathematical terms, this translates into a graph–type boundary condition that couples the convection–diffusion equation for the concentration of the ions to an ordinary differential equation defined only on the grain boundary and describing the concentration of the precipitate. Models of similar type are analyzed in a homogenization context in [5] and [6], or [4], where also a numerical scheme is briefly discussed. However, the analysis there does not cover our model.

Due to the occurrence of the multi–valued dissolution rate, classical solutions do not exists, except for some particular cases. For defining a weak solution we consider the following sets

$$\mathcal{U} := \{u \in L^2((0,T); H^1_{0,\Gamma_D}(\Omega)) : \partial_t u \in L^2((0,T); H^{-1}(\Omega))\},$$
$$\mathcal{V} := \{v \in H^1((0,T); L^2(\Gamma_G))\},$$
$$\mathcal{W} := \{w \in L^\infty(\Gamma_G^T), : 0 \leq w \leq 1\}.$$

Here we have used standard notations in the functional analysis.

Definition 1. *A triple $(u,v,w) \in \mathcal{U} \times \mathcal{V} \times \mathcal{W}$ is called a weak solution of (1) and (2) if $(u(0), v(0)) = (u_I, v_I)$ and if*

$$(\partial_t u, \varphi)_{\Omega^T} + D(\nabla u, \nabla \varphi)_{\Omega^T} - (\mathbf{q}u, \nabla \varphi)_{\Omega^T} = -\varepsilon \tilde{n}(\partial_t v, \varphi)_{\Gamma_G^T}, \quad (4)$$

$$(\partial_t v, \theta)_{\Gamma_G^T} = D_a(r(u) - w, \theta)_{\Gamma_G^T},$$
$$w \in H(v) \quad \text{a.e. in } \Gamma_G^T, \quad (5)$$

for all $(\varphi, \theta) \in L^2((0,T); H^1_{0,\Gamma_D}(\Omega)) \times L^2(\Gamma_G^T)$.

By [2], Theorem 2.21, a weak solution exists. The proof is based on regularization arguments and provides a solution for which, in addition, we have

$$w = r(u,c) \quad \text{a.e. in } \{v = 0\} \cap \Gamma_G^T.$$

2 The numerical scheme

In this section we analyze a semi–implicit numerical scheme for the numerical solution of (1) and (2). To overcome the difficulties that are due to the multi–valued dissolution rate we start with approximating it by

$$H_\delta(v) := \begin{cases} 0, & \text{if } v \leq 0, \\ v/\delta, & \text{if } v \in (0, \delta), \\ 1, & \text{if } v \geq \delta, \end{cases} \qquad (6)$$

where $\delta > 0$ is a small regularization parameter. Next we consider a time stepping that is implicit in u and explicit in v. With $N \in \mathbb{N}$, $\tau = T/N$, and $t_n = n\tau$, the approximation (u^n, v^n) of $(u(t_n), v(t_n))$ is the solution of the following problem:

Problem P_δ^n: Given u^{n-1}, v^{n-1}, compute $u^n \in H^1_{0,\Gamma_D}(\Omega)$, and $v^n \in L^2(\Gamma_G)$ such that

$$(u^n - u^{n-1}, \phi)_\Omega + \tau D(\nabla u^n, \nabla \phi)_\Omega + \tau(\nabla.(\mathbf{q}u^n), \phi)_\Omega \\ + \epsilon\tilde{n}(v^n - v^{n-1}, \phi)_{\Gamma_G} = 0, \qquad (7)$$

$$(v^n, \theta)_{\Gamma_G} = (v^{n-1}, \theta)_{\Gamma_G} + \tau D_a(r(u^n) - H_\delta(v^{n-1}), \theta)_{\Gamma_G}, \qquad (8)$$

for all $\phi \in H^1_{0,\Gamma_D}(\Omega)$ and $\theta \in L^2(\Gamma_G)$.

Here $n = 1, \ldots, N$, while $u^0 = u_I$ and $v^0 = v_I$. For completeness we define

$$w^n := H_\delta(v^n). \qquad (9)$$

To simplify the notations, we have given up the subscript δ for the solution triple (u^n, v^n, w^n).

Remark 2. It is worth noticing that the cation concentration u is treated implicitly, whereas for the crystalline concentration v an explicit discretization is considered. A fully implicit discretization is also possible, at the expense of an additional nonlinearity. Further, the Γ_G scalar product in (7) can be replaced by the last term in (8). In this way the two equations can be partially decoupled. Firstly one has to solve an elliptic problem defined in Ω with a nonlinear boundary term. Once u^n has been obtained, v^n can be determined straightforwardly from (8).

In what follows we restrict to announcing the results without proofs. Details will be given in a forthcoming paper. First, since the initial data is positive and essentially bounded, the same holds for the numerical solution at each time step t_n. This can be obtained assuming $\delta \geq \tau D_a$. With

$$M_u := \max\{\|u_I\|_{\infty,\Omega}, u^*\}, \\ M_v := \max\{\|v_I\|_{\infty,\Omega}, 1\}, \text{ and } C_v = \frac{r(M_u)D_a}{M_v}, \qquad (10)$$

the time discrete numerical approximations are essentially bounded for all $0 \leq n \leq N$:

$$0 \leq u^n(x) \leq M_u, \quad \text{and} \quad 0 \leq v^n(x) \leq M_v e^{C_v n \tau}. \tag{11}$$

For a fixed time step t_n, the nonlinear time discrete problem P_δ^n can be solved, for example, by the fixed point iteration proposed in Section 3. This also provides the existence and uniqueness of a solution pair $\{(u^n, v^n),$ whereas w_n is defined in (9). Having the sequence of time discrete triples $\{(u^n, v^n, w^n), n = \overline{1, N}\}$ solving the problems P_δ^n, we can construct an approximation of the solution of (1) and (2) for all times $t \in [0, T]$. To do so, define for $t \in (t_{n-1}, t_n]$

$$Z_\tau(t) := z^n \frac{(t - t_{n-1})}{\tau} + z^{n-1} \frac{(t_n - t)}{\tau}, \tag{12}$$

where Z is either U, V, or W, whereas z is either u, v, or w. Notice that Z_τ does not only depend on τ, but also on the regularization parameter δ. By compactness arguments, for this construction we obtain:

Theorem 1. *Assume $\delta = O(\tau^\alpha)$, with some $\alpha \in (0, 1)$. Then we have*

$$\|\partial_t U^\tau\|^2_{L^2(0,T;H^{-1}(\Omega))} + \|\nabla U^\tau\|^2_{\Omega^T} + \varepsilon \|\partial_t V^\tau\|^2_{\Gamma_G^T} \leq C.$$

Here $C > 0$ does not depend on τ or δ. Further, along a sequence $\tau \searrow 0$, the triple (U_τ, V_τ, W_τ) converges to a weak solution (u, v, w) of (1) and (2).

The estimates stated above are uniform in δ and τ, and are in good agreement with the ones obtained for the solution defined in Definition 1 (see [2]). Further, for a ε–periodic porous medium, the estimates are also ε–independent.

The above convergence should be understood in a weak sense. Unfortunately, no error estimates can be given. Specifically, we have

a) $U_\tau \to u$ weakly in $L^2((0,T); H^1_{0, \Gamma_D}(\Omega))$,
b) $\partial_t U_\tau \to \partial_t u$ weakly in $L^2((0,T); H^{-1}(\Omega))$,
c) $V_\tau \to v$ weakly in $L^2((0,T); L^2(\Gamma_G))$,
d) $\partial_t V_\tau \to \partial_t v$ weakly in $L^2(\Gamma_G^T)$,
e) $W_\tau \to w$ weakly–star in $L^\infty(\Gamma_G^T)$.

3 A fixed point iteration

For each $n \geq 1$, the time discrete problem P_δ^n is nonlinear. Even though the nonlinearity appears only in the boundary term, instabilities in the form of negative concentrations or artificial precipitation can be encountered when applying a straightforward Newton iteration. Moreover, there is no guarantee of convergence, unless the time stepping is not small enough. In this section we discuss a simple iteration scheme for solving the nonlinear elliptic problem. This method is of fixed point type and produces stable results. Moreover, the scheme converges linearly regardless of the parameters ε, τ, or δ.

Assume u^{n-1} and v^{n-1} given. To construct the iteration scheme we recall Remark 2 and decouple the ion transport equation from the dissolution/precipitation equation on the boundary. Using (8), (7) gives

$$(u^n - u^{n-1}, \phi)_\Omega + \tau D(\nabla u^n, \nabla \phi)_\Omega + \tau(\nabla \cdot (\mathbf{q}u^n), \phi)_\Omega \\ + \tau \epsilon \tilde{n} D_a (r(u^n) - H_\delta(v^{n-1}), \theta)_{\Gamma_G} = 0, \tag{13}$$

for all $\phi \in H^1_{0,\Gamma_D}(\Omega)$. This is a scalar elliptic equation with nonlinear boundary conditions on Γ_G. We first construct a sequence $\{u^{n,i}, i \geq 0\}$ approximating the solution u^n of (13). Once this is computed, we use (8) for directly determining v^n.

Let L_r be the Lipschitz constant of the precipitation rate r on the interval $[0, M_u]$. With a given $u^{n,i-1} \in H^1_{0,\Gamma_D}(\Omega)$, the next iteration $u^{n,i}$ is the solution of the linear elliptic equation

$$(u^{n,i} - u^{n-1}, \phi)_\Omega + \tau D(\nabla u^{n,i}, \nabla \phi)_\Omega + \tau(\nabla(\mathbf{q}u^{n,i}), \phi)_\Omega \\ = \tau \epsilon \tilde{n} D_a\, L_r\, (u^{n,i-1} - u^{n,i}, \phi)_{\Gamma_G} \\ - \tau \epsilon \tilde{n} D_a (r(u^{n,i-1}) - H_\delta(v^{n-1}), \theta)_{\Gamma_G}, \tag{14}$$

for all $\phi \in H^1_{0,\Gamma_D}(\Omega)$. The starting point of the iteration can be chosen arbitrarily in $H^1_{0,\Gamma_D}(\Omega)$, but essentially bounded by 0 and M_u. A good initial guess is $u^{n,0} = u^{n-1}$, but this is not a restriction.

Comparing the above to (13) and up to the presence of the superscripts $i-1$ and i, the only difference is in the appearance of the first term on the right in (14). As $u^{n,i}$ approaches u^n, this term will cancel. Before making this sentence more precise we mention that the above construction is common in the analysis of nonlinear elliptic problems, in particular when sub- or supersolutions are sought. In [10] this approach is used in a fixed point context, for approximating the solution of an elliptic problem with a nonlinear and possibly unbounded source term. Following the same ideas, a similar scheme is considered for the implicit discretization of a degenerate (fast diffusion) problem in both conformal and mixed formulation (see [12] and [9]). Since the scheme is of fixed point type, we expect only a linear convergence rate. The advantage is in the stability of the approximation and the guaranteed convergence. For being specific we let $e^{n,i} := u^n - u^{n,i}$ denote the error at iteration i and define the H^1-equivalent norm

$$|||f|||^2 := \|f\|_\Omega^2 + \tau D \|\nabla f\|_\Omega^2 + \frac{\tau}{2} \epsilon \tilde{n} D_a L_r \|f\|_{\Gamma_G}^2. \tag{15}$$

Here f is any function in $H^1_{0,\Gamma_D}(\Omega)$. For this norm, the iteration is a contraction and the iteration converges linearly in H^1 to the solution u^n of (13).

Theorem 2. *With $u^{n,0} \in H^1_{0,\Gamma_D}(\Omega)$ bounded essentially by 0 and M_u, the iteration defined in (14) is stable and convergent. Specifically, if $\tau < 2/(\varepsilon \tilde{n} D_a L_r)$, an i–independent constant $\gamma \in [0,1)$ exists such that for all $i > 0$ we have*

$$0 \leq u^{n,i} \leq M_u$$

almost everywhere in Ω, and

$$|||e^{n,i}|||^2 \leq \gamma^i |||e^{n,0}|||^2.$$

4 A numerical example

We conclude this presentation with a numerical example obtained for the undersaturated regime. In this case we have $u \leq u^*$ almost everywhere in Ω^T, where u^* is the equilibrium concentration mentioned in Remark 1. Extensive numerical results for both dissolution and precipitation, and for high or low Damköhler numbers, will be presented in a forthcoming paper.

The present computations are carried out in a reference cell Ω, where the square $(-1,1)^2$ is including a circular grain of radius $R = 0.5$ centered in the origin. For symmetry reasons, the computations are restricted to the upper half of the domain. The fluid velocity $\mathbf{q}$ is obtained by solving numerically the Stokes model in Ω. To this aim the bubble stabilized finite element method proposed in [11] (see also [8]) has been applied.

We have used the following parameters and rate function:

$$D = 0.25, \ \varepsilon = 1, \ \tilde{m} = \tilde{n} = 1.0, \text{ and } r(u,c) = \frac{10}{9}[u]_+[u - 0.1]_+.$$

Two different regimes are considered, $D_a = 1$ and $D_a = 10$. The Peclet number is moderate. The time discretization discussed in the above is completed by standard piecewise finite elements. No special stabilizing techniques were needed. The initial conditions are $v_I = 0.01$ and $u_I = u^* = 1.0$. On the external boundaries we take $u(t, -1) = u_* = 0.1$, and $\partial_\nu u = 0$ on the remaining part. In this case, only dissolution is possible. This is guaranteed by the L^∞ estimates.

The present computations are implemented in the research software package *SciFEM* (Scilab Finite Element Method, [1]). In Figure 1 we present the cation concentration u. This concentration nis higher at the outflow, since undersaturated fluid is flowing in at the left boundary. Further, notice that for a higher Damköhler number, the cation concentration is higher along the grain boundary. This is due to the enhanced dissolution. As a result, the crystal concentration is bigger if $D_a = 1$, as resulting from Figure 2.

Acknowledgment

The work of I.S. Pop was supported through the BRICKS project MSV1-3 (Modeling, Simulation and Visualization). This research was initiated while V.

Fig. 1. Cation concentration at $t = 0.15$, with $D_a = 1$ (left) and $D_a = 10$ (right)

Fig. 2. Crystal concentration at $t = 0.15$, with $D_a = 1$ (left) and $D_a = 10$ (right)

Devigne spent six months at the Technische Universiteit Eindhoven, supported through an European Community Marie Curie Fellowship (contract number HPMT-CT-2001-00422).

References

1. Devigne, V.M.: Ecoulements et Conditions aux Limites Particulières Appliquées en Hydrogéologie et Théorie Mathématique des Processus de Dissolution/Précipitation en Milieux Poreaux. Ph'D thesis, Université Lyon 1 (2006)
2. van Duijn, C.J., Pop, I.S.: Crystal dissolution and precipitation in porous media: pore scale analysis. J. Reine Angew. Math. **577**, 171–211 (2004)
3. van Duijn, C.J., Knabner, P.: Travelling wave behaviour of crystal dissolution in porous media flow. European J. Appl. Math. **8**, 49–72 (1997)
4. Hornung, U., Jäger, W.: A model for chemical reactions in porous media. In: Warnatz, J., Jäger, W. (eds.) Complex Chemical Reaction Systems. Mathematical Modeling and Simulation. Chemical Physics **47**, Springer, Berlin, (1987)
5. Hornung, U., Jäger, W.: Diffusion, convection, adsorption, and reaction of chemicals in porous media. J. Differ. Equations **92**, 199–225 (2001)
6. Hornung, U., Jäger, W., Mikelić, A: Reactive transport through an array of cells with semipermeable membranes. RAIRO Modél. Math. Anal. Numér. **28**, 59–94 (1994)
7. Knabner, P., van Duijn, C.J., Hengst, S.: An analysis of crystal dissolution fronts in flows through porous media. Part 1: Compatible boundary conditions. Adv. Water Res. **18**, 171–185 (1995)
8. Knobloch, P.: On the application of the p_1^{mod} element to incompressible flow problems. Comput. Vis. Sci. **6**, 185–195 (2004)
9. Pop, I.S., Radu, F., Knabner, P.: Mixed finite elements for the Richards' equation: linearization procedure. J. Comput. Appl. Math. **168**, 365–373 (2004)
10. Pop, I.S., Yong, W.A.: On the existence and uniqueness of a solution for an elliptic problem. Studia Univ. Babeş-Bolyai Math. **45**, 97–107 (2000)

11. Russo, A.: Bubble stabilization of finite element methods for the linearized incompressible Navier-Stokes equations. Comput. Methods appl. Mech. Engrg. **132**, 335–343 (1996)
12. Slodička, M.: A robust and efficient linearization scheme for doubly nonlinear and degenerate parabolic problems arising in flow in porous media. SIAM J. Sci. Comput. **23**, 1593–1614 (2002)

Degenerate Parabolic Equations: Theory, Numerics and Applications

Discrete Kinetic Methods for a Degenerate Parabolic Equation in Dimension Two.

Denise Aregba-Driollet

IMB, MAB, Université Bordeaux 1, 351 cours de la Libération, 33405 Talence cedex, France
aregba@math.u-bordeaux1.fr

Summary. We design finite volume schemes for two dimensional parabolic degenerate systems by using a kinetic, formally BGK, approach. The hyperbolic and parabolic parts are not splitted and the schemes are Riemann solver free. Moreover the spatial discretization can be written analytically, so that the implementation is easy. Some numerical tests are presented.

1 Introduction

Our aim is to construct numerical approximations of the two-dimensional nonlinear degenerate parabolic system:

$$\partial_t u + \partial_x A_1(u) + \partial_y A_2(u) = \Delta B(u). \tag{1}$$

Here $u(x,y,t) \in \Omega$, a domain of $\mathbb{R}^K$, and $\Delta B = \partial_{xx}^2 B + \partial_{yy}^2 B$. A and B are smooth functions satisfying hyperbolic-parabolic conditions for $u \in \Omega$:

(C1) for all $(\xi_1, \xi_2) \in \mathbb{R}^2$ such that $\xi_1^2 + \xi_2^2 = 1$, the matrix $\xi_1 A_1'(u) + \xi_2 A_2'(u)$ has real eigenvalues and is diagonalizable,

(C2) the real parts of the eigenvalues of $B'(u)$ are non negative.

It is possible that $B'(u) = 0$ on a subset of Ω with nonzero measure. As a particular case, the set of equations (15) can be a pure system of conservation laws. Such systems arise in the context of multiphase flows in porous media.

There is a new interest for degenerate parabolic systems for a few years. We refer the reader to the references in [1] for theoritical results and one-dimensional or cartesian discretizations. In [3], a diffusive relaxation approximation is used to design a parallel algorithm. Multi-dimensional finite volume schemes have been studied in [4, 8], with convergence results in some scalar cases, see also references therein.

In [1] we proposed a discrete diffusive kinetic approximation to the system (15). Numerical schemes on cartesian meshes were constructed. Here, we show that our framework also allows the design of finite volume schemes on unstructured meshes and that the spatial discretizations can be computed with explicit formulas. The obtained schemes are robust and easy to implement, even for complicated flux functions: they are Riemann solver free and the formalism for systems is the same as for the scalar case. They are also very flexible. Moreover, the solutions of those kinetic models have been rigorously proved to converge to solutions of (15), in the scalar case as well as for some systems, see [2, 7].

The plan of the paper is the following: in Sect. 2, we describe the kinetic models and related stability conditions. In Sect. 4 we design the numerical schemes. Sect. 4 is devoted to some numerical experiments.

2 Kinetic models

Let us consider the following system:

$$\begin{cases} \partial_t f_l^\varepsilon & +\lambda_{l1}\partial_x f_l^\varepsilon + \lambda_{l2}\partial_y f_l^\varepsilon = \frac{1}{\varepsilon}\left(M_l(u^\varepsilon) - f_l^\varepsilon\right), \ 1 \le l \le N, \\ \partial_t f_{N+m}^\varepsilon & +\gamma^\varepsilon \sigma_{m1}\partial_x f_{N+m}^\varepsilon \\ & +\gamma^\varepsilon \sigma_{m2}\partial_y f_{N+m}^\varepsilon = \frac{1}{\varepsilon}\left(\frac{B(u^\varepsilon)}{\theta^2} - f_{N+m}^\varepsilon\right), \ 1 \le m \le 3, \end{cases} \quad (2)$$

where $u^\varepsilon(x,y,t) = \sum_{l=1}^{N+3} f_l^\varepsilon(x,y,t)$, each f_l^ε and M_l take values in $\mathbb{R}^K$, ε is a positive parameter, the λ_{ld} and σ_{md} are some fixed real constants, $\gamma^\varepsilon = \mu + \frac{\theta}{\sqrt{\varepsilon}}$, $\mu > 0$, $\theta > 0$, and the σ_{md} are such that

$$\sum_{m=1}^{3} \sigma_{m1}\sigma_{m2} = 0, \ \sum_{m=1}^{3} \sigma_{md} = 0, \ \sum_{m=1}^{3} \sigma_{md}^2 = 1, \ d=1,2. \quad (3)$$

Moreover, systems (15) and (2) are linked by the following compatibility conditions, for all $u \in \Omega$:

$$\sum_{l=1}^{N} M_l(u) = u - \frac{3B(u)}{\theta^2}, \ \sum_{l=1}^{N} \lambda_{ld} M_l(u) = A_d(u), \ d=1,2. \quad (4)$$

By analogy with the kinetic theory, M is called a maxwellian function. This approach is based on the ideas of the relaxation approximation of conservation laws [5], as well as on the ones of kinetic (BGK) schemes for compressible fluid flows.

Let us denote

$$v_j^\varepsilon = \sum_{l=1}^{N} \lambda_{lj} f_l^\varepsilon + \gamma^\varepsilon \sum_{m=1}^{3} \sigma_{mj} f_{N+m}^\varepsilon, \ j=1,2. \quad (5)$$

Then it is easy to prove that an equivalent formulation of system (2) is

$$\begin{cases} \partial_t f_l^\varepsilon + \lambda_{l1}\partial_x f_l^\varepsilon + \lambda_{l2}\partial_y f_l^\varepsilon = \dfrac{1}{\varepsilon}\left(M_l(u^\varepsilon) - f_l^\varepsilon\right), \quad 1 \le l \le N, \\ \partial_t u^\varepsilon + \partial_x v_1^\varepsilon + \partial_y v_2^\varepsilon = 0, \\ \partial_t v_j^\varepsilon + \displaystyle\sum_{l=1}^N \lambda_{lj}\lambda_{l1}\partial_x f_l^\varepsilon + \mu^2 \sum_{m=1}^3 \sigma_{mj}\sigma_{m1}\partial_x f_{N+m}^\varepsilon \\ \qquad + \displaystyle\sum_{l=1}^N \lambda_{lj}\lambda_{l2}\partial_y f_l^\varepsilon + \mu^2 \sum_{m=1}^3 \sigma_{mj}\sigma_{m2}\partial_y f_{N+m}^\varepsilon = \dfrac{1}{\varepsilon}(A_j - v_j) \\ \qquad - \dfrac{1}{\varepsilon}(2\mu\theta\sqrt{\varepsilon} + \theta^2) \displaystyle\sum_{m=1}^3 \sigma_{mj}(\sigma_{m1}\partial_x f_{N+m}^\varepsilon + \sigma_{m2}\partial_y f_{N+m}^\varepsilon), \quad j = 1, 2. \end{cases} \quad (6)$$

Suppose that the sequence f^ε converges to some limit function f in a suitable (strong) topology, then the limit function is a maxwellian state: $f_l = M_l(u)$ for $l = 1, \ldots, N$ and $f_{N+m} = B(u)/\theta^2$ for $m = 1, 2, 3$, with $u = \sum_{l=1}^{N+3} f_l$. Consequently, if v^ε converges to v, by the third equation of (6), we obtain:

$$v = A(u) - \nabla B(u),$$

which proves that u is a weak solution of system (15).

Convergence has been proved rigorously in [2] for the scalar case, and in [7] for some systems. A necessary condition for convergence is that M is monotone: for all $u \in \Omega$, the real parts of the eigenvalues of $M_l'(u)$ are non negative. This condition is also sufficient in the scalar case $K = 1$. The monotonicity of M can be interpreted as a subcharacteristic condition. Let us show it on an example.

Example

As already remarked in [5], flux vector splitting schemes for systems of conservation laws own a kinetic interpretation. They can be extended to more general situations. Suppose that for all $u \in \Omega$

$$A_j(u) = A_j^+(u) - A_j^-(u), \quad j = 1, 2$$

in such a way that the eigenvalues of $(A_j^+)'(u)$ and $(A_j^-)'(u)$ are non negative. We take $\lambda_1 > 0$, $\lambda_2 > 0$. We put

$$M_1 = \frac{A_1^+}{\lambda_1}, \quad M_2 = \frac{A_2^+}{\lambda_2}, \quad M_4 = \frac{A_1^-}{\lambda_1}, \quad M_5 = \frac{A_2^-}{\lambda_2},$$

and

$$M_3(u) = u - (M_1(u) + M_2(u) + M_4(u) + M_5(u)) - \frac{3B(u)}{\theta^2}.$$

The characteristic velocities are

$$\lambda^{(1)} = \lambda_1 \,{}^t(1,0,0,-1,0)\,, \quad \lambda^{(2)} = \lambda_2 \,{}^t(0,1,0,0,-1)\,.$$

The compatibility conditions (4) are satisfied. The maxwellian function is monotone if the real parts of the eigenvalues of the matrix

$$I - \frac{(A_1^+)'(u) + (A_1^-)'(u)}{\lambda_1} - \frac{(A_2^+)'(u) + (A_2^-)'(u)}{\lambda_2} - \frac{3B'(u)}{\theta^2}$$

are non negative. This is a generalized subcharacteristic condition.

3 The numerical schemes

3.1 The Principle of the Method

The general idea is to construct a discretization of the BGK system (2). This discretization depends on the parameter ε in such a way that when ε tends to zero, a consistent discretization of system (15) is obtained.

Taking $U^\varepsilon = (f_1^\varepsilon, \ldots, f_N^\varepsilon, u^\varepsilon, v^\varepsilon)$ as unknown, system (6) is equivalent to (2) and can be written in the more synthetic form:

$$\partial_t U^\varepsilon + C^{(1)} \partial_x U^\varepsilon + C^{(2)} \partial_y U^\varepsilon = Q_\varepsilon(U^\varepsilon, \partial_x U^\varepsilon, \partial_y U^\varepsilon). \tag{7}$$

The $C^{(d)}$ do not depend on ε:

$$C^{(d)} = \begin{pmatrix} Diag(\lambda_{ld})_{1\leq l \leq N} & 0 \\ D^{(d)} & \Sigma^{(d)} \end{pmatrix}, \quad d = 1,2. \tag{8}$$

The blocks $D^{(d)}$ and $\Sigma^{(d)}$ are detailed below.

Notations

We consider a computational domain V of $\mathbb{R}^2$ composed of polygonal cells C_α which are either the elements of a mesh (triangles or quadrangles) or constructed from these elements. The measure of C_α is denoted $|C_\alpha|$.

The (possibly variable) time step is denoted Δt and the discrete time levels are $t_0 = 0$ and $t_{n+1} = t_n + \Delta t$. The numerical solution of (15) at time t_n is a vector $u^n = (u_\alpha^n)_\alpha$. Each u_α^n is an approximation of the mean value of $u(., t_n)$ on the cell C_α. At time $t_0 = 0$, if u_0 is the given initial value, we put

$$u_\alpha^0 = \frac{1}{|C_\alpha|} \int_{C_\alpha} u_0(x,y) dx dy.$$

As we want to approximate (6), we need for initial values for U. We take the maxwellian states:

$$f_{l,\alpha}^0 = M_l(u_\alpha^0)\,, \quad f_{N+m,\alpha}^0 = \frac{B(u_\alpha^0)}{\theta^2}$$

for $l = 1, \ldots, N$ and $m = 1, 2, 3$. In view of (11), (4), (5), this gives

$$v_\alpha^0 = A(u_\alpha^0).$$

Suppose now that for some $n \geq 0$, an approximation U^n of the relaxed limit U of U^ε is known.

We use a fractional step method. First, we solve on $[t_n, t_{n+1}]$ the linear, independant of ε system associated to the left hand side:

$$\partial_t U + \sum_{d=1}^{2} C^{(d)} \partial_{x_d} U = 0. \tag{9}$$

We want to obtain an approximate value $U^{n+1/2}$ at time t_{n+1}. This is possible because we know that the linear system (9) is hyperbolic diagonalizable [1]. For all $\xi \in \mathbb{R}^2$ such that $\xi_1^2 + \xi_2^2 = 1$, let us denote

$$C_\mu(\xi) = \xi_1 C^{(1)} + \xi_2 C^{(2)}.$$

It is highly desirable to know the eigenvalues of $C_\mu(\xi)$ analytically and we show below that it is the case here. Consequently, we may easily design an upwind (Godunov) scheme or a higher order extension:

$$U_\alpha^{n+1/2} = U_\alpha^n - \frac{\Delta t}{|C_\alpha|} \sum_{e \subset C_\alpha, e = \Gamma_{\alpha\beta}} |e| \Phi(U_\alpha^n, U_\beta^n, n_e). \tag{10}$$

Here we denote $\Gamma_{\alpha\beta}$ the part of the boundary of C_α which is common to the cells C_α and C_β, and n_e is the unit normal pointing in the direction of C_β. The function Φ is the numerical flux approximating $\frac{1}{|e|} \int_{\Gamma_{\alpha\beta}} C_\mu(n_e) U ds$ and

$$\Phi = (\mathcal{F}_1, \ldots, \mathcal{F}_N, \mathcal{V}, \Phi_{N+2}, \Phi_{N+3})^T.$$

Then, with initial value $U^{n+1/2}$ at time t_n, we solve exactly the nonlinear system related to the right hand side:

$$\partial_t U^\varepsilon = Q_\varepsilon(U^\varepsilon, \partial_x U^\varepsilon, \partial_y U^\varepsilon). \tag{11}$$

We proved in [1] that the obtained solution $U^\varepsilon(t_{n+1})$ has the following limit $\overline{U^{n+1}}$ when ε goes to zero:

$$\overline{u^{n+1}} = u^{n+1/2}, \quad \overline{f_l^{n+1}} = M_l(\overline{u^{n+1}}), \quad 1 \leq l \leq N,$$
$$\overline{v^{n+1}} = A(\overline{u^{n+1}}) - \nabla B(\overline{u^{n+1}}).$$

Hence, a discretization $\nabla_h B(u)$ of $\nabla B(u)$ being given, we put:

$$u^{n+1} = u^{n+1/2}, \quad f_l^{n+1} = M_l(u^{n+1}), \quad 1 \leq l \leq N,$$
$$v^{n+1} = A(u^{n+1}) - \nabla_h B(u^{n+1}).$$

The obtained scheme is a consistent discretization of system (15) which takes the form:

$$\begin{cases} u_\alpha^{n+1} = u_\alpha^n - \dfrac{\Delta t}{|C_\alpha|} \sum_{e \subset C_\alpha, e = \Gamma_{\alpha\beta}} |e| \mathcal{V}(U_\alpha^n, U_\beta^n, n_e), \\ f_{\alpha,l}^{n+1} = M_l(u_\alpha^{n+1}), \qquad l = 1, \ldots, N, \\ v_\alpha^{n+1} = A(u_\alpha^{n+1}) - (\nabla_h B(u))_\alpha^{n+1}. \end{cases} \quad (12)$$

3.2 The effective schemes

In view of formulas (12), we need only for the $\mathcal{V}$ component of the numerical flux and for $\nabla_h B(u)$.

The latest can be obtained by several ways, depending on the type of finite volumes. For instance, suppose that we use a vertex centered method where the edges of the control volumes are the perpendicular bisectors of the edges of the triangles. The $(u_\alpha)_\alpha$ being given, one can construct the related P1 approximation of u on the primal mesh and then compute an approximation of $\nabla B(u)$ on the control volumes.

Let us now show that we can compute $\mathcal{V}$ explicitly by Godunov's scheme. To that aim, we fix

$$\begin{pmatrix} \sigma_{11} \\ \sigma_{21} \\ \sigma_{31} \end{pmatrix} = \begin{pmatrix} \frac{1}{\sqrt{2}} \\ -\frac{1}{\sqrt{2}} \\ 0 \end{pmatrix}, \quad \begin{pmatrix} \sigma_{12} \\ \sigma_{22} \\ \sigma_{32} \end{pmatrix} = \begin{pmatrix} -\frac{1}{\sqrt{6}} \\ -\frac{1}{\sqrt{6}} \\ \frac{2}{\sqrt{6}} \end{pmatrix}.$$

Then the blocks of the matrix $C_\mu(\xi)$ are given by:

$$\begin{cases} D_{\mu,1l}(\xi) = 0, \\ D_{\mu,2l}(\xi) = \xi_1 \left[\lambda_{l1}^2 - \dfrac{\mu^2}{3} + \dfrac{\mu \lambda_{l2}}{\sqrt{6}} \right] + \xi_2 \left[\lambda_{l1} \lambda_{l2} + \dfrac{\mu \lambda_{l1}}{\sqrt{6}} \right], \quad l = 1, \ldots, N, \\ D_{\mu,3l}(\xi) = \xi_1 \left[\lambda_{l1} \lambda_{l2} + \dfrac{\mu \lambda_{l1}}{\sqrt{6}} \right] + \xi_2 \left[\lambda_{l2}^2 - \dfrac{\mu^2}{3} - \dfrac{\mu \lambda_{l2}}{\sqrt{6}} \right], \end{cases}$$

and

$$\Sigma_\mu(\xi) = \begin{pmatrix} 0 & \xi_1 & \xi_2 \\ \dfrac{\xi_1 \mu^2}{3} & -\dfrac{\xi_2 \mu}{\sqrt{6}} & -\dfrac{\xi_1 \mu}{\sqrt{6}} \\ \dfrac{\xi_2 \mu^2}{3} & -\dfrac{\xi_1 \mu}{\sqrt{6}} & \dfrac{\xi_2 \mu}{\sqrt{6}} \end{pmatrix}$$

The eigenvalues of this last matrix are $a_1(\xi) = -\dfrac{\mu}{\sqrt{6}}(\xi_2 + \xi_1 \sqrt{3})$, $a_2(\xi) = \xi_2 \dfrac{2\mu}{\sqrt{6}}$ and $a_3(\xi) = \dfrac{\mu}{\sqrt{6}}(-\xi_2 + \xi_1 \sqrt{3})$. Moreover, it turns out that the associated right eigenvectors do not depend on ξ and can be written as:

$$\Pi(\mu) = \begin{pmatrix} 1 & 1 & 1 \\ \dfrac{-\mu}{\sqrt{2}} & 0 & \dfrac{\mu}{\sqrt{2}} \\ \dfrac{-\mu}{\sqrt{6}} & \dfrac{2\mu}{\sqrt{6}} & \dfrac{-\mu}{\sqrt{6}} \end{pmatrix}.$$

Hence, the matrix of right eigenvectors for $C_\mu(\xi)$ is

$$Q(\mu) = \begin{pmatrix} I & 0 \\ R & \Pi(\mu) \end{pmatrix}, \quad R = \begin{pmatrix} 1 & \cdots & 1 \\ \lambda_{11} & \cdots & \lambda_{N1} \\ \lambda_{12} & \cdots & \lambda_{N2} \end{pmatrix}.$$

This allows us to write down explicitly the numerical fluxes. We denote ξ the outward normal vector to C_α on $\Gamma_{\alpha\beta}$, and $\lambda_l(\xi) = \xi_1 \lambda_{l1} + \xi_2 \lambda_{l2}$. Taking into account the expression of U_α and U_β as functions of u_α and u_β:

$$\begin{cases} \mathcal{V}(U_\alpha, U_\beta, \xi) = \displaystyle\sum_{l=1}^{N} \left[\lambda_l^-(\xi) M_l(u_\beta) + \lambda_l^+(\xi) M_l(u_\alpha) \right] + \mathcal{R}(U_\alpha, U_\beta, \xi), \\ \mathcal{R}(U_\alpha, U_\beta, \xi) = \displaystyle\sum_{m=1}^{3} \left[a_m^-(\xi) z_{m,\beta} + a_m^+(\xi) z_{m,\alpha} \right] \end{cases} \quad (13)$$

with

$$z_1 = \tfrac{B(u)}{\theta^2} + \tfrac{1}{\mu\sqrt{2}}(\partial_x B(u))_h + \tfrac{1}{\mu\sqrt{6}}(\partial_y B(u))_h,$$
$$z_2 = \tfrac{B(u)}{\theta^2} - \tfrac{2}{\mu\sqrt{6}}(\partial_y B(u))_h,$$
$$z_3 = \tfrac{B(u)}{\theta^2} - \tfrac{1}{\mu\sqrt{2}}(\partial_x B(u))_h + \tfrac{1}{\mu\sqrt{6}}(\partial_y B(u))_h.$$

Due to the compatibility conditions (4), the terms of the sum over l in (13) are a consistent approximation of the hyperbolic flux associated to A. Let us point out the fact that (unusually) those terms depend generally also on B.

As an example, we now deal with a cartesian grid with elements C_{ij} and complete the expression of $\mathcal{R}$ in this case. We obtain (with obvious notations):

$$\begin{cases} \mathcal{R}_{i+1/2,j} = -\tfrac{1}{2}\left[(\partial_x B)_{i+1,j} + (\partial_x B)_{ij}\right] - \tfrac{\mu}{\theta^2\sqrt{2}}[B(u_{i+1,j}) - B(u_{ij})] \\ \qquad\qquad - \tfrac{1}{2\sqrt{3}}\left[(\partial_y B)_{i+1,j} - (\partial_y B)_{ij}\right], \\ \mathcal{R}_{i,j+1/2} = -\tfrac{1}{3}\left[(\partial_y B)_{i,j+1} + 2(\partial_y B)_{ij}\right] - \tfrac{2\mu}{\theta^2\sqrt{6}}[B(u_{i,j+1}) - B(u_{ij})]. \end{cases}$$

In these expressions the second and third terms are diffusion terms. As the only condition on μ is its positivity, we can make it tend to zero and then suppress the second term.

The first term of $\mathcal{R}_{i+1/2,j}$ provides a centered approximation of $\partial_{xx}^2 B$ while in the second expression, the first term provides a non-symmetric approximation of $\partial_{yy}^2 B$.

Our method provides spatial discretizations. One can then modify the related time discretization but this is out of the scope of this paper.

4 Numerical experiments

As the main difficulty here comes from the parabolic part of the problem, we test our scheme on the porous media equation, which owns analytical (Barenblatt) solutions. This is a nonlinear degenerate parabolic equation.

We consider a cartesian grid and solve the one-dimensional porous media equation in the direction X, making angle $\pi/3$ with the x−axis:

$$\partial_t u = \frac{1}{2}\partial^2_{XX} u^2 \ .$$

The computational domain is $[0,1] \times [0,1]$. The approximate and exact solutions in the direction X are depicted in In Fig. 1. In the left, we take $\Delta x = 1/51$ and $\Delta y = 1/61$, in the right, we take $\Delta x = 1/301$ and $\Delta y = 1/266$. The results are satisfactory and are confirmed by others tests.

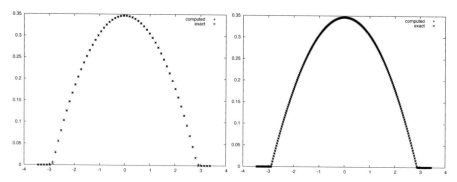

Fig. 1. Two-dimensional computations on porous media equation.

References

1. Aregba-Driollet, D., Natalini, R., Tang, S.: Explicit diffusive kinetic schemes for nonlinear degenerate parabolic systems. Math. Comp., **73**, 63–94 (2004)
2. Bouchut, F., Guarguaglini, F., Natalini, R.: Diffusive BGK Approximations for Nonlinear Multidimensional Parabolic Equations, Indiana Univ. Math. J., **49**, 723–749, (2000)
3. Cavalli, F., Naldi, G., Semplice, M.: Parallel algorithms for non-linear diffusion by using relaxation approximation. ENUMATH2005
4. Eymard, R., Gallouet, T., Herbin, R., Michel, A: Convergence of a finite volume scheme for nonlinear degenerate parabolic equations. Numer. Math., **92**, no. 1, 41–82 (2002)
5. Harten, A., Lax, P.D., van Leer, B.: On upstream differencing and Godunov-type schemes for hyperbolic conservation laws. SIAM Rev., **25**, no. 1, 35–61 (1983).
6. Jin, S., Xin, Z.: The relaxation schemes for systems of conservation laws in arbitrary space dimensions, Comm. Pure Appl. Math **48**, 235–277 (1995).
7. Lattanzio, C., Natalini, R.:Convergence of diffusive BGK approximations for parabolic systems, Proc. Roy. Soc. Edinburgh Sect. A, **132**, no. 2, 341–358 (2002)
8. Michel, A., Vovelle, J.: Entropy formulation for parabolic degenerate equations with general Dirichlet boundary conditions and application to the convergence of FV methods. SIAM J. Numer. Anal. **41**, no. 6, 2262–2293 (2003)

Anisotropic Doubly Nonlinear Degenerate Parabolic Equations

Mostafa Bendahmane and Kenneth H. Karlsen

Centre of Mathematics for Applications, University of Oslo P.O. Box 1053, Blindern, N–0316 Oslo, Norway
(mostafab@math.uio.no, kennethk@math.uio.no)

Summary. The purpose of this note is to review recent results by the authors on the well-posedness of entropy and renormalized entropy solutions for anisotropic doubly nonlinear degenerate parabolic equations.

1 Introduction

We are interested in the uniqueness of entropy and renormalized entropy solutions of anisotropic doubly nonlinear degenerate parabolic problems:

$$\begin{cases} \partial_t u + \sum_{i=1}^{d} \partial_{x_i} f_i(u) = \sum_{i=1}^{d} \partial_{x_i}\left(|\partial_{x_i} A_i(u)|^{p_i - 2} \partial_{x_i} A_i(u) \right) \text{ in } Q_T, \\ u|_{t=0} = u_0 \text{ in } \Omega \text{ and } u = 0 \text{ on } (0, T) \times \partial\Omega, \end{cases} \quad (1)$$

where $u(t, x) : Q_T \to \mathbf{R}$ is the unknown function, $Q_T = (0, T) \times \Omega$, $T > 0$ is a fixed time, $\Omega \subset \mathbf{R}^d$ is a bounded domain with smooth boundary $\partial\Omega$, and $p_i > 1$ for $i = 1, \ldots, d$. We always asume that

$$A_i \in \text{Lip}_{\text{loc}}(\mathbf{R}), \ A_i(\cdot) \text{ is nondecreasing, } A_i(0) = 0, \ i = 1, \ldots, d, \quad (2)$$

and

$$f(u) \in \text{Lip}_{\text{loc}}(\mathbf{R}; \mathbf{R}^d) \text{ and } f(0) = 0. \quad (3)$$

In [2] we proved the uniqueness of entropy solutions of the problem (1). In that paper we did not prove the existence of entropy solutions; This problem still remains open, essentially bacause Minty's argument does not apply to this highly anisotropic problem. In this note we review instead recent progress on the existence question for the following simplified anisotropic problem

$$\begin{cases} \partial_t u + \sum_{i=1}^{d} \partial_{x_i} f_i(u) = \sum_{i=1}^{d} \partial_{x_i}\left(|\partial_{x_i} A(u)|^{p_i - 2} \partial_{x_i} A(u) \right) \text{ in } Q_T, \\ u|_{t=0} = u_0 \text{ in } \Omega \text{ and } u = 0 \text{ on } (0, T) \times \partial\Omega, \end{cases} \quad (4)$$

where A and f satisfy the assumptions (2) and (3) respectively.

Global solutions of (1) and (4) are in general discontinuous and it is well-known that discontinuous weak solutions are not uniquely determined by their data. Consequently, it is more challenging to devise reasonable solution concepts and to prove uniqueness/stability results.

Let us state a closely related problem, namely the following one containing an "isotropic" second order operator:

$$\begin{cases} \partial_t u + \operatorname{div} f(u) = \operatorname{div}\left(|\nabla A(u)|^{p-2} \nabla A(u)\right) & \text{in } Q_T, \\ u|_{t=0} = u_0 \text{ in } \Omega \text{ and } u = 0 \text{ on } (0,T) \times \partial\Omega, \end{cases} \quad (5)$$

where $p > 1$ and $A(\cdot)$ is a scalar nondecreasing Lipschitz function with $A(0) = 0$. Note that when $p_i = p \neq 2$ and $A_i \equiv A$ for all i, the anisotropic problem (1) does not coincide with (5) (but it does when $p = 2$).

When the data $u_0 \in L^1 \cap L^\infty$, Igbida and Urbano [10] prove existence and uniqueness results for weak solutions of the isotropic problem (5), under the additional structure condition

$$f(u) = F(A(u)), \quad \text{for some Lipschitz function } F: \mathbf{R} \to \mathbf{R}^d,\ F(0) = 0. \quad (6)$$

Uniqueness of weak solutions is obtained by verifying that any weak solution is also an entropy solution and then using the doubling of the variables approach developed by Carrillo [6].

In this contribution we review recent results [1, 2, 3] by the authors on a solution theory that avoids any structure condition like (6) and is able to encompass the anisotropic problem (1). Carrillo's approach (as used in [10]) is a good one when the second order differential operator is isotropic. However, it is not applicable to an anisotropic problem like (1). Recently, in [2, 3] we have developed well-posedness theory based on a notion of entropy solutions for the bounded (L^∞) data case and a notion of renormalized entropy solutions for the unbounded (L^1) data case. A similar theory can be found in [1] for the Cauchy problem for the equation

$$\partial_t u + \operatorname{div} f(u) = \operatorname{div}\left(a(u) \nabla u\right), \quad a(u) = \sigma(u)\sigma(u)^\top,$$

where $\sigma(u) \in L^\infty_{\text{loc}}(\mathbf{R}; \mathbf{R}^{d \times K})$, $1 \leq K \leq d$. The paper [1], which uses Kružkov approach, is inspired by Chen and Perthame [8] and their study of the same equation using the kinetic approach. We recall that the notion of renormalized solutions was introduced by DiPerna and Lions in the context of Boltzmann equations [9]. This notion was then adapted to nonlinear elliptic and parabolic equations with L^1 (or measure) data by various authors. We refer to [5] for recent results in this context and relevant references. Bénilan, Carrillo, and Wittbold [4] introduced a notion of renormalized entropy solutions for scalar conservation laws in unbounded domains with L^1 data and proved the well-posedness of such solutions (see [7] for bounded domains).

2 Entropy solution

For $1 \leq i \leq d$, we set

$$\zeta_i(u) = \int_0^u (A'_i(\xi))^{\frac{p_i-1}{p_i}} d\xi, \qquad \zeta(u) = (\zeta_1(u), \ldots, \zeta_d(u)),$$

and for any $\psi \in L^\infty_{\text{loc}}(\mathbf{R})$

$$\zeta_i^\psi(u) = \int_0^u \psi(\xi)(A'_i(\xi))^{\frac{p_i-1}{p_i}} d\xi, \qquad \zeta^\psi(u) = \left(\zeta_1^\psi(u), \ldots, \zeta_d^\psi(u)\right).$$

Definition 1. *We call (η, q), with $\eta : \mathbf{R} \to \mathbf{R}$ and $q = (q_1, \ldots, q_d) : \mathbf{R} \to \mathbf{R}^d$, an entropy-entropy flux pair if*

$$\eta \in C^2(\mathbf{R}), \qquad \eta'' \geq 0, \qquad q' = \eta' f'.$$

If, in addition,

$$\eta(0) = 0, \qquad \eta'(0) = 0, \qquad q(0) = 0,$$

we call (η, q) a boundary entropy-entropy flux pair.

The following notion of an entropy solution is used in [2]:

Definition 2 (entropy solution). *A entropy solution of (1) is a measurable function $u : Q_T \to \mathbf{R}$ satisfying the following conditions:*
(D.1) $u \in L^\infty(Q_T)$ and $\partial_{x_i} \zeta_i(u) \in L^{p_i}(Q_T)$ for any $i = 1, \ldots, d$
(D.2) (interior entropy condition) For any entropy-entropy flux pair (η, q),

$$\partial_t \eta(u) + \sum_{i=1}^d \partial_{x_i} q_i(u) - \sum_{i=1}^d \partial_{x_i} \left(\eta'(u) |\partial_{x_i} A_i(u)|^{p_i-2} \partial_{x_i} A_i(u)\right) \qquad (7)$$

$$\leq -\sum_{i=1}^d \eta''(u) |\partial_{x_i} \zeta_i(u)|^{p_i} \quad \text{in } \mathcal{D}'([0,T) \times \Omega).$$

that is, for any $0 \leq \phi \in \mathcal{D}([0,T) \times \Omega)$,

$$\int_{Q_T} \left(\eta(u)\partial_t\phi + \sum_{i=1}^d q_i(u)\partial_{x_i}\phi \right.$$
$$\left. - \sum_{i=1}^d \eta'(u) |\partial_{x_i} A_i(u)|^{p_i-2} \partial_{x_i} A_i(u) \partial_{x_i}\phi\right) dx\, dt$$
$$+ \int_\Omega \eta(u_0)\phi(0,x)\, dx \geq \int_{Q_T} \sum_{i=1}^d \eta''(u) |\partial_{x_i} \zeta_i(u)|^{p_i} \phi\, dx\, dt.$$
$$(8)$$

(D.3) (boundary entropy condition) For any boundary entropy-entropy flux pair (η, q) and for any $0 \leq \phi \in \mathcal{D}([0,T) \times \overline{\Omega})$, (8) holds.
(D.4) For any $\psi \in L^\infty_{\text{loc}}(\mathbf{R})$,

$$\int_{Q_T} \left(\partial_{x_i} \zeta_i^\psi(u) \phi + \zeta_i^\psi(u) \partial_{x_i}\phi\right) dx\, dt = 0, \quad \forall \phi \in \mathcal{D}([0,T) \times \overline{\Omega}),$$

for $i = 1, \ldots, d$.

Remark 1. In the case $A_i = A$ for $i = 1, \ldots, d$, we can replace (**D**.4) by $A(u) \in L^p(0, T; W_0^{1,p}(\Omega))$, where $p = \min(p_1, \ldots, p_d)$.

Remark 2. We will make repeated use of the following chain rule property. Let u be an entropy solution to (1) and fix $\psi \in L_{\text{loc}}^\infty(\mathbf{R})$. We have for a.e. $t \in (0, T)$,

$$\partial_{x_i} \zeta_i^\psi(u(t,x)) = \psi(u(t,x))\partial_{x_i}\zeta_i(u(t,x)), \tag{9}$$

for a.e. $x \in \Omega$ and in $L^{p_i}(\Omega)$, $i = 1, \ldots, d$.

Remark 3. By the chain rule (9) we have for a.e. $t \in (0, T)$

$$\partial_{x_i} A_i(u(t,x)) = (A_i'(u(t,x)))^{\frac{1}{p_i'}} \partial_{x_i}\zeta_i(u(t,x)),$$

a.e. in Ω and in $L^{p_i'}(\Omega) \cap L^1(\Omega)$, $p_i' = p_i/(p_i - 1)$, so that by (**D**.1) there holds $\partial_{x_i} A_i(u) \in L^{p_i'}(Q_T)$, $i = 1, \ldots, d$. This also implies

$$|\partial_{x_i} A_i(u)|^{p_i - 2} \partial_{x_i} A_i(u) \in L^{p_i'}(Q_T), \qquad p_i' = \frac{p_i}{p_i - 1},$$

for $i = 1, \ldots, d$, and thus (8) makes sense.

In [2] we prove the following theorem:

Theorem 1 (uniqueness). *Suppose (2) and (3) hold. Let u and v be two entropy solutions of (1) with initial data $u|_{t=0} = u_0 \in L^\infty(\Omega)$ and $v|_{t=0} = v_0 \in L^\infty((\Omega))$, respectively. Then for a.e. $t \in (0, T)$*

$$\int_\Omega (u(t,x) - v(t,x))^+ \, dx \leq \int_\Omega (u_0 - v_0)^+ \, dx.$$

The following existence result is proved in [3]:

Theorem 2 (existence). *Suppose (2)–(3) hold. Let $u_0 \in L^\infty(\Omega)$. Then there exists at least one entropy solution u of the problem (4).*

3 Renormalized entropy solution

Let us recall the definition of the truncation function $T_l : \mathbf{R} \to \mathbf{R}$ at height $l > 0$:

$$T_l(u) = \begin{cases} -l, & u < -l, \\ u, & |u| \leq l, \\ l, & u > l. \end{cases} \tag{10}$$

The following notion of an L^1 solution is suggested in [3]:

Definition 3 (renormalized entropy solution). *A renormalized entropy solution of (1) is a measurable function $u : Q_T \to \mathbf{R}$ satisfying the following conditions:*

(D.1) $u \in L^\infty(0,T; L^1(\mathbf{R}^d))$, $\partial_{x_i}\zeta_i(T_l(u)) \in L^{p_i}(Q_T)$, $i=1,\ldots,d$, for any $l > 0$.

(D.2) For any $l > 0$ and any entropy-entropy flux triple (η, q), there exists a nonnegative bounded Radon measure μ_l^u on $[0,T] \times \overline{\Omega}$ such that

$$\partial_t \eta(T_l(u)) + \sum_{i=1}^d \partial_{x_i} q_i(T_l(u))$$
$$- \sum_{i=1}^d \partial_{x_i} \left(\eta'(T_l(u)) |\partial_{x_i} A_i(T_l(u))|^{p_i-2} \partial_{x_i} A_i(T_l(u)) \right)$$
$$\leq - \sum_{i=1}^d \eta''(T_l(u)) |\partial_{x_i} \zeta_i(T_l(u))|^{p_i} + \mu_l^u \quad \text{in } \mathcal{D}'([0,T) \times \Omega). \tag{11}$$

that is, for any $0 \leq \phi \in \mathcal{D}([0,T) \times \Omega)$,

$$\int_{Q_T} \left(\eta(T_l(u))\partial_t \phi + \sum_{i=1}^d q_i(T_l(u))\partial_{x_i}\phi \right.$$
$$\left. - \sum_{i=1}^d \eta'(T_l(u)) |\partial_{x_i} A_i(T_l(u))|^{p_i-2} \partial_{x_i} A_i(T_l(u)) \partial_{x_i}\phi \right) dx\,dt$$
$$+ \int_\Omega \eta(T_l(u_0))\phi(0,x)\,dx$$
$$\geq \int_{Q_T} \sum_{i=1}^d \eta''(T_l(u)) |\partial_{x_i}\zeta_i(T_l(u))|^{p_i} \phi\,dx\,dt - \int_{Q_T} \phi\,d\mu_l^u(t,x). \tag{12}$$

(D.3) For any boundary entropy-entropy flux pair (η, q) and for any $0 \leq \phi \in \mathcal{D}([0,T) \times \overline{\Omega})$, (12) holds.

(D.4) For any $\psi \in L^\infty_{\text{loc}}(\mathbf{R})$,

$$\int_{Q_T} \left(\partial_{x_i} \zeta_i^\psi(T_l(u)) \phi + \zeta_i^\psi(T_l(u)) \partial_{x_i} \phi \right) dx\,dt = 0, \quad \forall \phi \in \mathcal{D}([0,T) \times \overline{\Omega}),$$

for $i = 1,\ldots,d$.

(D.5) The total mass of the renormalization measure μ_l^u vanishes as $l \uparrow \infty$, that is, $\lim_{l \uparrow \infty} \mu_l^u([0,T] \times \overline{\Omega}) = 0$.

Remark 4. Since $T_l(u) \in L^\infty(Q_T)$, the integrals in (12) are well defined. Moreover, if a renormalized entropy solution u belongs to $L^\infty(Q_T)$, then it is also an entropy solution in the sense of Definition 2 (let $l \uparrow \infty$ in of Definition 3).

Remark 5. The measure μ_l is supported on the set $\{|u| = l\}$ and encode information about the behavior of the "p_i-energies" on the set where $|u|$ is large. Condition (**D**.5) says that the p_i-energies should be small for large values of $|u|$, that is, the total mass of the measure μ_l should vanish as $l \to \infty$.

The proof of the following results can be found in [3]:

Theorem 3 (uniqueness). *Suppose conditions (2) and (3) hold. Let u and v be two renormalized entropy solutions of (1) with initial data $u|_{t=0} = u_0 \in L^1(\Omega)$ and $v|_{t=0} = v_0 \in L^1(\Omega)$, respectively. Then for a.e. $t \in (0,T)$*

$$\int_\Omega (u(t,x) - v(t,x))^+ \, dx \leq \int_\Omega (u_0 - v_0)^+ \, dx.$$

Theorem 4 (existence). *Suppose (2)–(3) hold. Let $u_0 \in L^1(\Omega)$. Then there exists at least one renormalized entropy solution u of the problem (4).*

Acknowledgement

This research is supported by an Outstanding Young Investigators Award from Research Concil of Norway.

References

1. Bendahmane, M., Karlsen, K.H.: Renormalized entropy solutions for quasilinear anisotropic degenerate parabolic equations. SIAM J. Math. Anal., **2**, 405–422 (2004)
2. Bendahmane, M., Karlsen, K.H.: Uniqueness of entropy solutions for doubly nonlinear anisotropic degenerate parabolic equations. Nonlinear Partial Differential Equations and Related Analysis, Contemporary Mathematics, American Mathematical Society (AMS): Providence, USA (2005)
3. Bendahmane, M., Karlsen, K.H.: Existence and uniqueness of Renormalized entropy solutions for doubly nonlinear anisotropic degenerate parabolic equations. In preparation (2006)
4. Bénilan, P., Carrillo, J., Wittbold, P.: Renormalized entropy solutions of scalar conservation laws. Ann. Scuola Norm. Sup. Pisa Cl. Sci., **4**, 29(2):313–327 (2000)
5. Blanchard, D., Murat, F., Redwane, H.: Existence and uniqueness of a renormalized solution for a fairly general class of nonlinear parabolic problems. J. Differential Equations, **177**(2), 331–374 (2001)
6. Carrillo, J.: Entropy solutions for nonlinear degenerate problems. Arch. Rational Mech. Anal., **147**(4), 269–361 (1999)
7. Carrillo, J., Wittbold, P.: Renormalized entropy solutions of scalar conservation laws with boundary condition. J. Differential Equations, **185**, 137–160 (2002)
8. Chen, G.-Q, Perthame, B.: Well-posedness for non-isotropic degenerate parabolic-hyperbolic equations. Ann. Inst. H. Poincaré Anal. Non Linéaire, **20**(4), 645–668 (2003)
9. DiPerna, R.J., Lions, P.-L.: On the Cauchy problem for Boltzmann equations: global existence and weak stability. Ann. of Math. (2), **130**(2), 321–366 (1989)
10. Igbida, N., Urbano, J.M.: Uniqueness for nonlinear degenerate problems. NoDEA Nonlinear Differential Equations Appl., **10**(3), 287–307 (2003)
11. Kružkov, S.N.: First order quasi-linear equations in several independent variables. Math. USSR Sbornik, **10**(2), 217–243 (1970)

A Multiresolution Method for the Simulation of Sedimentation-Consolidation Processes

Raimund Bürger[1] and Alice Kozakevicius[2]

[1] Departamento de Ingeniería Matemática, Universidad de Concepción, Casilla 160-C, Concepción, Chile
rburger@ing-mat.udec.cl
[2] Departamento de Matemática, Universidade Federal de Santa Maria, Faixa de Camobi, km 9, Campus Universitário, Santa Maria, RS, CEP 97105-900, Brazil
alicek@smail.ufsm.br

Summary. A multiresolution method for a one-dimensional strongly degenerate parabolic equation modeling sedimentation-consolidation processes is introduced. The method is based on the switch between central interpolation or exact evaluation of the numerical flux combined with a thresholded wavelet transform applied to point values of the solution to control the switch. A numerical example is presented.

1 Introduction

The multiresolution method has been devised to reduce the computational cost of high resolution methods for conservation laws, whose solutions are usually smooth on the major part of the computational domain but strongly vary in small regions near discontinuities. The method adaptively concentrates computational effort on the latter regions. It goes back to Harten [8] for conservation laws and was used in [2, 12] for parabolic equations.

In this contribution, we construct adaptive multiresolution schemes and present numerical results for the strongly degenerate parabolic equation

$$u_t + f(u)_x = A(u)_{xx}, \quad (x,t) \in Q_T := (0,L) \times (0,T], \tag{1}$$

where $f, A : \mathbb{R} \to \mathbb{R}$ are piecewise smooth and Lipschitz continuous, and $A(\cdot)$ is nondecreasing. On intervals $[\alpha, \beta]$ with $A(u) = \text{const.}$ for all $u \in [\alpha, \beta]$, equation (1) degenerates into a conservation law. Equation (1) arises from a sedimentation-consolidation model for suspensions [1]; see [5] for other applications. Since $A = 0$ on u-intervals of positive length, (1) is called *strongly degenerate*. Its solutions are in general discontinuous.

The multiresolution method reduces the number of exact flux evaluations required by a high resolution scheme. To this purpose, point values or cell averages of the numerical solution are defined on a hierarchical sequence of nested diadic meshes, where the initially given mesh is the finest. The sequence of coefficients for all meshes forms the *multiresolution representation* of the solution. Since multiresolution coefficients are small on smooth regions, data can be compressed by *thresholding*, i.e. setting to zero those multiresolution coefficients which are in absolute value smaller than a prescribed tolerance.

The multiresolution representation can be used to locate discontinuities of the numerical solution, since multiresolution coefficients measure its local regularity. Harten converted this idea [8] into a sensor to decide at which fine-mesh positions fluxes should be exactly evaluated, or can otherwise be obtained more cheaply by interpolation from coarser scales. Our multiresolution scheme combines the switch between central interpolation and exact evaluation of both convective and diffusive numerical fluxes with a thresholded wavelet (multiresolution) transform applied to solution point values to control the switch. The first alternative is performed on smooth regions, while the second applies near strong variations. Instead of calculating a wavelet transform for cell averages as in [8], we use here the interpolatory framework (point values) to analyze the smoothness of the solution. This slight change improves the efficiency of the algorithm, since the multiresolution representation is cheaply obtained as in [9]. The efficiency of the multiresolution method is measured in terms of the data compression rate and CPU time.

Our scheme can be extended to multidimensional problems by a multidimensional wavelet transform [3] or by dimensional splitting (see e.g. [6]).

In this work, we consider the zero-flux initial-boundary value problem for a bounded domain $\Omega := [0, L]$ with the conditions

$$u(x,0) = u_0(x), \quad x \in \Omega; \quad f(u) - A(u)_x = 0, \quad x \in \{0, L\}, \; t \in (0, T]. \quad (2)$$

Solutions of strongly degenerate parabolic PDEs are in general discontinuous and must be defined as entropy solutions. Recent works on the analysis and numerics of these PDEs include [4, 10, 11], see [5] for further references.

2 The multiresolution scheme

Let $(G^0, G^1, \ldots, G^{L_c})$ denote a family of uniform nested grids on $I := [a, b]$, where $G^0 := (x_0^0, x_1^0, \ldots, x_{N_0}^0)$, $N_0 = 2^m$, $m \in \mathbb{N}$ is the finest resolution level, and $h_0 := (b-a)/N_0$. The values of a function u on G^0 are the input data. The remaining diadically coarsened grids are obtained as follows: given G^{k-1}, we obtain G^k by removing the even-indexed grid points. Therefore $G^{k-1} \setminus G^k = (x_{2j-1}^{k-1})_{j=1,\ldots,N_k}$, $G^{k-1} \cap G^k = G^k$, and $x_j^k = x_{2j}^{k-1}$ for $0 \leqslant j \leqslant N_k = 2^{m-k}$, $k = 1, \ldots, L_c$. The representation of u on any coarser grid $G^1, \ldots, G^{L_c}$ can be obtained directly from G^0: $u_j^k = u(x_j^k) = u(x_{2^k j}^0) = u_{2^k j}^0$ for $0 \leqslant j \leqslant N_k$. To recover the representation of u on G^{k-1} from its representation on G^k, we need an interpolation operator $\mathcal{I}(u^k, x)$ of u on G^k to obtain approximations for the missing points of G^{k-1}. The function value at x_{2j-1}^{k-1} is obtained from the $(r-1)$-th degree polynomial interpolating $(u_{j-s}^k, \ldots, u_{j+s-1}^k)$. Therefore

$$\widetilde{u}_{2j-1}^{k-1} = \mathcal{I}(u^k, x_{2j-1}^{k-1}) = \sum_{l=1}^{s} \beta_l (u_{j+l-1}^k + u_{j-l}^k), \quad r = 2s, \tag{3}$$

with $\beta_1 = 1/2$ for $r = 2$ and $\beta_1 = 9/16$, $\beta_2 = -1/16$ for $r = 4$. The interpolation errors, known as *details* or *wavelet coefficients*, are $d_j^k = u_{2j-1}^{k-1} - \widetilde{u}_{2j-1}^{k-1}$ for $1 \leq j \leq N_k$. Thus, from $u^k := (u_0^k, u_1^k, \ldots, u_{N_k}^k)$ and $d^k := (d_0^k, d_1^k, \ldots, d_{N_k}^k)$, we can exactly recover the representation of u on G^{k-1}. The pair of vectors (u^k, d^k) is the *multiresolution representation* of u^{k-1}. Applying successively this procedure for $1 \leq k \leq L_c$, we can recover the values of u on G^0 from its values on G^{L_c} and the sequence of all details from levels L_c to 1:

$$u^0 \leftrightarrow (d^1, u^1) \leftrightarrow (d^1, d^2, u^2) \leftrightarrow \cdots \leftrightarrow (d^1, d^2, \ldots, d^{L_c}, u^{L_c}) =: u_M, \tag{4}$$

where u_M is the *multiresolution representation* of $u^0 \equiv u$. The details d^k contain information on the smoothness of u, and will be used to flag the non-smooth parts of the solution in the adaptive numerical method.

Standard interpolation results imply that if u at a given point x has $p-1$ continuous derivatives and a discontinuity in its p-th derivative, then $d_j^k \sim (h_k)^p [u^{(p)}]$ for $0 \leq p \leq \bar{r}$ and $d_j^k \sim (h_k)^{\bar{r}} u^{(\bar{r})}$ for $p > \bar{r}$, for x_j^k near x, where $\bar{r} := r - 1$ is the order of accuracy of the approximation and $[\cdot]$ denotes the jump. Therefore $|d_{2j}^{k-1}| \approx 2^{-\bar{p}} |d_j^k|$, if the k-th level is fine enough, where $\bar{p} := \min\{p, \bar{r}\}$. Thus, away from discontinuities of u, the wavelet coefficients d_j^k diminish as the levels of resolution become finer.

We see that near a discontinuity of the function, the wavelet coefficients remain of the same size for all levels of refinement. Thus, data compression and reduction of computational effort can be attained by discarding wavelet coefficients that are smaller than a prescribed tolerance. This operation is known as *thresholding* or *truncation*. To define it, let us denote by $\text{tr}_{\varepsilon_k}$ the hard thresholding operator with ε_k as threshold parameter:

$$\hat{d}_j^k = \text{tr}_{\varepsilon_k}(d_j^k) = \begin{cases} 0 & \text{if } |d_j^k| \leq \varepsilon_k, \\ d_j^k & \text{if } |d_j^k| > \varepsilon_k, \end{cases} \quad 1 \leq j \leq N_k, \quad 1 \leq k \leq L_c. \tag{5}$$

Consequently, $\hat{u}_M := (\hat{d}^1, \ldots, \hat{d}^{L_c}, u^{L_c})$ is the thresholded multiresolution representation. Let $\widetilde{u}^0$ be the data recovered from $\hat{u}_M$. Harten proved in [8] that $\|u^0 - \widetilde{u}^0\| \leq c_1(\varepsilon_1 + \cdots + \varepsilon_{L_c}) \leq \varepsilon$, where the constant c_1 is independent of L_c. Hence, given a tolerance ε, we can compress data by truncating u_M. Clearly, the actual compression rate depends on the chosen *strategy* $(\varepsilon_1, \ldots, \varepsilon_{L_c})$.

In contrast to the evaluation on a *sparse point representation* [9], we evaluate the differential operator on the uniform fine grid but adapt the manner of flux computation to the significant coefficients of $\widetilde{u}^0$, as is done in [8, 2, 6]. This strategy does not provide memory savings, but we have a better compression rate, and consequently, a smaller number of exact flux evaluations.

Finally, the index set of significant coefficients in each time step, $\mathcal{D}^n$, needs to capture the finite speed of propagation of information and the formation of shock waves. For this reason, Harten [8] proposed an algorithm to extend $\mathcal{D}^n$ after thresholding, including so-called *safety points* near positions associated with significant details, which yields an extended index set $\widetilde{\mathcal{D}}^n$. We here utilize a version of Harten's algorithm [8, Alg. (6.1)], see [5, Alg. 2.1] for details.

For the time discretization of $u_t = \mathcal{L}(u) \equiv -f(u)_x + A(u)_{xx}$ we use an explicit 3-step third-order Runge-Kutta TVD (RKTVD) scheme [7]. A general n_{RK}-step explicit RKTVD scheme has the form

$$u_j^{(0)} = u_j^n, \quad u_j^{(i)} = \sum_{k=0}^{i-1}\left(\alpha_{ik}u_j^{(k)} + \Delta t \beta_{ik}\mathcal{L}_j(u^{(k)})\right), \quad u_j^{n+1} = u_j^{(n_{\mathrm{RK}})}, \quad (6)$$

$i = 1, \ldots, n_{\mathrm{RK}}$, where $\mathcal{L}_j(u)$ contains the flux and diffusion terms. We distinguish between the interior operators $\mathcal{L}_1, \ldots, \mathcal{L}_{N_0-1}$ and the boundary operators $\mathcal{L}_0$ and $\mathcal{L}_{N_0}$, which include the boundary conditions.

Point values of the initial solution of (1) are given on a uniform fine grid G^0, $u_i = u(x_i)$, and the index set $\widetilde{\mathcal{D}}^n$ is considered already built. Then a conservative semi-discrete scheme is given by

$$\dot{u}_0(t) = \bar{\mathcal{L}}_0(u(t)) := -(\bar{F}_{1/2} - \bar{D}_{1/2})/\Delta x,$$
$$\dot{u}_j(t) = \bar{\mathcal{L}}_j(u(t)) := -(\bar{F}_{j+1/2} - \bar{F}_{j-1/2} - (\bar{D}_{j+1/2} - \bar{D}_{j-1/2}))/\Delta x, \quad (7)$$
$$\dot{u}_{N_0}(t) = \bar{\mathcal{L}}_{N_0}(u(t)) := (\bar{F}_{N_0-1/2} - \bar{D}_{N_0-1/2})/\Delta x,$$

where $k = 1, \ldots, N_0$, $u(t) := (u_0(t), \ldots, u_{N_0}(t))$, and the numerical fluxes $\bar{F}_{i+1/2}$ and $\bar{D}_{i+1/2}$ contain the advective and diffusive terms, respectively.

If $i \in \widetilde{\mathcal{D}}^n$, then we use a Lax-Friedrichs splitting [13] with a third-order ENO interpolation for $\bar{F}_{i+1/2}$ and add a fine-grid finite difference of the diffusion term. If $i \notin \widetilde{\mathcal{D}}^n$, the numerical flux is approximated by interpolation of fluxes previously evaluated on a coarser level. On our finite domain, the interpolator (3) is replaced by $\mathcal{I}^{\mathrm{L}}(\bar{F}^k, x_{2j+3/2}^{k-1}) = \widetilde{\mathcal{I}}_{k,j}^{\mathrm{L}}/16$, where

$$\widetilde{\mathcal{I}}_{k,j}^{\mathrm{L}} := \begin{cases} 5\bar{F}_{1/2}^k + 15\bar{F}_{3/2}^k - 5\bar{F}_{5/2}^k + \bar{F}_{7/2}^k, & j = 0, \\ -\bar{F}_{j-1/2}^k + 9\bar{F}_{j+1/2}^k + 9\bar{F}_{j+3/2}^k - \bar{F}_{j+5/2}^k, & j = 1, \ldots, N_k - 2, \\ \bar{F}_{N_k-7/2}^k - 5\bar{F}_{N_k-5/2}^k + 15\bar{F}_{N_k-3/2}^k + 5\bar{F}_{N_k-1/2}^k, & j = N_k - 1. \end{cases}$$

By construction, all positions from the coarsest level L_c are in $\widetilde{\mathcal{D}}^n$. Therefore all fluxes on level L_c are always exactly evaluated. The u-values required for the flux computation are taken from the finest level, $k = 0$.

The convective fluxes in (7) are given by $\bar{F}_{i+1/2}^k = (\hat{f}^+)_{i+1/2}^k + (\hat{f}^-)_{i+1/2}^k$, $i = 0, \ldots, N_0 - 1$, where $f^+(u_i) = (f(u_i) + \alpha u_i)/2$ and $f^-(u_i) = (f(u_i) - \alpha u_i)/2$ for $0 \leq i \leq N_0$, where $\alpha := \max_u |f'(u)|$ [13] and $(\hat{f}^+)_{i+1/2}^k$, $(\hat{f}^-)_{i+1/2}^k$ are approximations obtained by the ENO interpolator of each splitting component f^+ and f^-, evaluated at cell boundaries. The diffusive fluxes at level k are calculated by $D_{i+1/2}^k := (A(u_{2^ki+1}^0) - A(u_{2^ki}^0))/\Delta x$.

The stability condition is the same as that of the difference finite scheme on the finest grid. i.e. $\mathrm{CFL} := \max_u |f'(u)|(\Delta t/\Delta x) + 2\max_u |a(u)|(\Delta t/\Delta x^2) \leq 1$.

Since the ENO interpolator needs six points to search the least oscillatory four-point stencil for the flux calculation, we extrapolate the solution across the boundaries of I. We summarize the flux computation procedure as follows.

Algorithm 1

1. Compute $\bar{F}_{i+1/2}^{L_c}$ and $\bar{D}_{i+1/2}^{L_c}$, as stated above, for $i = 0, \ldots, N_{L_c} - 1$.
2. **do** $k = L_c, L_c - 1, \ldots, 1$ *(compute fluxes for all other levels)*
   do $j = 0, \ldots, N_k - 1$
   $\bar{F}_{2j+1/2}^{k-1} \leftarrow \bar{F}_{j+1/2}^{k}, \quad \bar{D}_{2j+1/2}^{k-1} \leftarrow \bar{D}_{j+1/2}^{k}$
   if $(j,k) \in \widetilde{\mathcal{D}}^n$ **then** *(flux/diffusion terms are computed explicitly)*
   $\bar{F}_{2j+3/2}^{k-1} \leftarrow (\hat{f}^+)_{2j+3/2}^{k-1} + (\hat{f}^-)_{2j+3/2}^{k-1}$
   $\bar{D}_{2j+3/2}^{k-1} \leftarrow \left(A\big(u_{2^{k-1}(2j+1)+1}^{0}\big) - A\big(u_{2^{k-1}(2j+1)}^{0}\big)\right)/\Delta x$
   else *(flux/diffusion terms are computed by interpolation)*
   $\bar{F}_{2j+3/2}^{k-1} \leftarrow \mathcal{I}^{\mathrm{L}}\big(\bar{F}^k, x_{2j+3/2}^{k-1}\big), \quad \bar{D}_{2j+3/2}^{k-1} \leftarrow \mathcal{I}^{\mathrm{L}}\big(\bar{D}^k, x_{2j+3/2}^{k-1}\big)$
   endif
   enddo
   enddo

The final multiresolution scheme for calculating the approximate solutions $u^{1,0}, \ldots, u^{\mathcal{N},0}$, where $\mathcal{N}$ is the number of time steps, is the following algorithm.

Algorithm 2

1. Create the initial set of significant positions $\widetilde{\mathcal{D}}^0$ [5, Algorithm 2.1]
2. **do** $n = 1, \ldots, \mathcal{N}$
   $u_j^{(0)} \leftarrow u_j^{n,0}, \quad j = 0, \ldots, N_0$
   do $i = 1, \ldots, n_{\mathrm{RK}}$
      do $k = 0, \ldots, i-1$
         if $\beta_{ik} \neq 0$ **then**
            using $u_0^{(k)}, \ldots, u_{N_0}^{(k)}$ as input data for Algorithm 1, calculate
            $\bar{\mathcal{L}}_0(u^{(k)}) \leftarrow -(\bar{F}_{1/2}^0 - \bar{D}_{j+1/2}^0)/\Delta x$,
            $\bar{\mathcal{L}}_j(u^{(k)}) \leftarrow -(\bar{F}_{j+1/2}^0 - \bar{F}_{j-1/2}^0 - (\bar{D}_{j+1/2}^0 - \bar{D}_{j-1/2}^0))/\Delta x$,
            $j = 1, \ldots, N_0 - 1$,
            $\bar{\mathcal{L}}_{N_0}(u^{(k)}) \leftarrow (\bar{F}_{N_0-1/2}^0 - \bar{D}_{N_0-1/2}^0)/\Delta x$
         endif
      enddo
      calculate $u_0^{(i)}, \ldots, u_{N_0}^{(i)}$ by (6), with $\mathcal{L}_j$ replaced by $\bar{\mathcal{L}}_j$
   enddo
   $u_j^{n+1,0} \leftarrow u_j^{(n_{\mathrm{RK}})}, \quad j = 0, \ldots, N_0$, compute u_{M}^{n+1},
   determine $\widetilde{\mathcal{D}}^{n+1}$ using [5, Alg. 2.1]; apply data compression to u_{M}^{n+1}
   enddo

3 Sedimentation-consolidation processes

We limit ourselves here to batch settling of a suspension of initial concentration $u_0 = u_0(x)$ in a column of height L, where $u_0(x) \in [0, u_{\max}]$ and $u_{\max}$ is a maximum solids volume fraction. The relevant initial and boundary conditions are (2). The unknown is the solids concentration u as a function of time t and depth x. The suspension is characterized by the hindered settling function $f(u)$ and the integrated diffusion coefficient $A(u)$, which models the sediment compressibility. The function $f(u)$ is assumed to satisfy $f(u) > 0$ for $u \in (0, u_{\max})$ and $f(u) = 0$ for $u \le 0$ and $u \ge u_{\max}$. A typical example is

$$f(u) = v_\infty u(1-u)^C \text{ for } u \in (0, u_{\max}), \ C > 0, \quad f(u) = 0 \text{ otherwise,} \qquad (8)$$

where $v_\infty > 0$ is the settling velocity of a single particle in unbounded fluid. Moreover, we have $A(u) = \int_0^u a(s)ds$, where $a(u) := f(u)\sigma_e'(u)/(\Delta_\varrho g u)$. Here, $\Delta_\varrho > 0$ is the solid-fluid density difference, g is the acceleration of gravity, and $\sigma_e'(u)$ is the derivative of the effective solid stress function $\sigma_e(u)$. We assume that the solid particles touch each other at a critical concentration $0 \le u_c \le u_{\max}$, and that $\sigma_e(u), \sigma_e'(u) = 0$ for $u \le u_c$ and $\sigma_e(u), \sigma_e'(u) > 0$ for $u > u_c$. This implies that $a(u) = 0$ for $u \le u_c$, such that for this application, (1) is indeed strongly degenerate parabolic. A typical function is

$$\sigma_e(u) = 0 \text{ for } u \le u_c, \quad \sigma_e(u) = \sigma_0[(u/u_c)^\beta - 1] \text{ for } u > u_c, \qquad (9)$$

with $\sigma_0 > 0$ and $\beta > 1$. In our numerical example, the suspension is characterized by the parameters $v_\infty = 2.7 \times 10^{-4}$ m/s, $C = 21.5$, $u_{\max} = 0.5$, $\sigma_0 = 1.2$ Pa, $u_c = 0.07$, $\beta = 5$, $\Delta_\varrho = 1660$ kg/m^3 and $g = 9.81$ m/s^2.

4 Numerical results

We consider a suspension of concentration $u_0 \equiv 0.06$ in a column of depth $L = 0.16$ m. Figure 1 shows the numerical solution. The finest and coarsest levels are $N_0 := 2^{11}$ and $N_{L_c} := 2^3$, respectively. The thresholding strategy is $\varepsilon_1 = 1.9 \times 10^{-7}$ and $\varepsilon_k = 2.99\varepsilon_{k-1}$ for $k \ge 2$. We used the parameters CFL $= 0.075$, $\Delta t = 0.0491898$ h, $\Delta x = L/N_0$ and a final time $t = 4000$ s. The CPU time for this simulation was 503 min (user time) against 1852 min when all fluxes are calculated on G^0 without multiresolution. The example involves the formation of a stationary type-change interface (the sediment level). Figure 1 also displays the grid positions of the significant wavelet coefficients of the solution. The marked positions are the current elements of $\widetilde{\mathcal{D}}^n$, at which fluxes are evaluated explicitly. At unmarked positions, these terms have been obtained by a simple cubic interpolation. Figure 1 illustrates how the scheme concentrates significant multiresolution coefficients near the downwards propagating shock (Figures 1 (a) and (b)), near the parabolic-hyperbolic type change interface, and near $x = 0$ and $x = L$. Figure 2 shows the number of

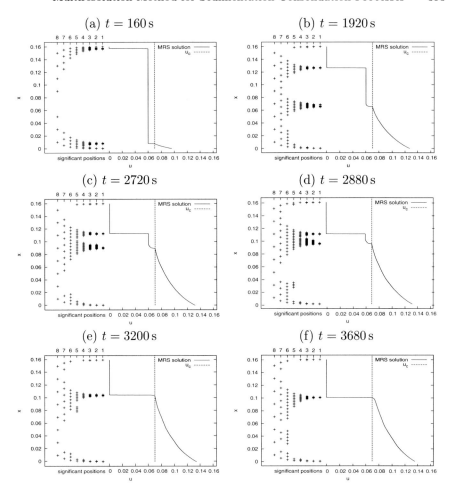

Fig. 1. Numerical solution to the batch settling problem, including significant positions of the wavelet coefficients of the solution per transformation level

significant wavelet coefficients of the solution in each time step of the simulation and the corresponding compression rate $N_0/\#\widetilde{\mathcal{D}}^n$. This simulation starts from a very high compression rate, since the initial solution is constant all over the domain, having a discontinuity near the boundary. As time evolves, the solution varies rapidly, and through the multiresolution analysis, this causes a variation of the density of significant positions.

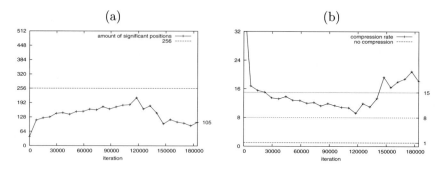

Fig. 2. Number of significant wavelet coefficients and compression rate per iteration

Acknowledgements

RB acknowledges support by Fondecyt project 1050728 and Fondap in Applied Mathematics. AK has been supported by FAPERGS, Brazil, by the ARD project 0306981, and Fondap in Applied Mathematics.

References

1. Berres, S., Bürger, R., Karlsen, K.H., Tory, E.M.: Strongly degenerate parabolic-hyperbolic systems modeling polydisperse sedimentation with compression. SIAM J. Appl. Math. **64**, 41–80 (2003)
2. Bihari, B.L., Harten, A.: Application of generalized wavelets: a multiresolution scheme. J. Comp. Appl. Math. **61**, 275–321 (1995)
3. Bihari, B.L., Harten A.: Multiresolution schemes for the numerical solution of 2-D conservation laws I, SIAM J. Sci. Comput. **18**, 315–354 (1997)
4. Bürger, R., Coronel, A., Sepúlveda, M.: A semi-implicit monotone difference scheme for an initial- boundary value problem of a strongly degenerate parabolic equation modelling sedimentation-consolidation processes. Math. Comp., to appear.
5. Bürger, R., Kozakevicius, A., Sepúlveda, M.: Multiresolution schemes for strongly degenerate parabolic equations, Preprint 2005-14, Departamento de Ingeniería Matemática, Universidad de Concepción, Concepción, Chile.
6. Chiavassa, G., Donat, R.: Point value multiscale algorithms for 2D compressive flows. SIAM J. Sci. Comput. **23**, 805–823 (2001)
7. Gottlieb, S., Shu, C.-W.: Total variation diminishing Runge-Kutta schemes. Math. Comp. **67**, 73–85 (1998)
8. Harten, A.: Multiresolution algorithms for the numerical solution of hyperbolic conservation laws. Comm. Pure Appl. Math. **48**, 1305–1342 (1995)
9. Holmström, M.: Solving hyperbolic PDEs using interpolating wavelets. SIAM J. Sci. Comp. **21**, 405–420 (1999)
10. Mascia, C., Porretta, A., Terracina, A.: Nonhomogeneous Dirichlet problems for degenerate parabolic-hyperbolic equations. Arch. Rat. Mech. Anal. **163**, 87–124 (2002)

11. Michel, A., Vovelle, J.: Entropy formulation for parabolic degenerate equations with general Dirichlet boundary conditions and application to the convergence of FV methods. SIAM J. Numer. Anal. **41**, 2262–2293 (2003)
12. Roussel, O., Schneider, K., Tsigulin, A., Bockhorn, H.: A conservative fully adaptive multiresolution algorithm for parabolic PDEs. J. Comp. Phys. **188**, 493–523 (2003)
13. Shu, C.: Essentially non-oscillatory and weighted essentially non-oscillatory schemes for hyperbolic conservation laws. In: Cockburn, B., Johnson, C., Shu, C.-W. and Tadmor, E., Advanced Numerical Approximation of Nonlinear Hyperbolic Equations (Quarteroni, A. (ed)), Lecture Notes in Mathematics vol. 1697, Springer-Verlag, Berlin, 325–432 (1998)

Diffusive Relaxation Limit for Hyperbolic Systems

Corrado Lattanzio

Sezione di Matematica per l'Ingegneria
Dipartimento di Matematica Pura ed Applicata
Università di L'Aquila
Piazzale E. Pontieri, 2, Monteluco di Roio
I–67040 L'Aquila, Italy
corrado@univaq.it

Summary. The aim of this paper is to collect some results concerning relaxation limits of hyperbolic systems of balance laws toward parabolic equilibrium systems. More precisely, we will discuss BGK approximations for strongly parabolic systems in the case of weak solutions, by means of compensated compactness techniques. Moreover, we will study the case of a semilinear relaxation approximation to a 2×2 hyperbolic-parabolic equilibrium system, with applications to viscoelasticity, in the case of classical solutions in one and several space variables. The latter case will be used as a case study to apply the modulated energy estimates.

1 Introduction

In this paper we collect some of the personal contribution to the theory of diffusive relaxation limits [8, 4, 5].

We shall focus our attention firstly in the following BGK system

$$\begin{cases} f_t^\varepsilon + \left(a(\xi) + \dfrac{b(\xi)}{\sqrt{\varepsilon}}\right) f_x^\varepsilon = \dfrac{1}{\varepsilon}(M_{f^\varepsilon} - f^\varepsilon), & (x, t; \xi) \in \mathbb{R} \times \mathbb{R}^+ \times \Xi \\ f^\varepsilon(x, 0; \xi) = f_0(x; \xi), & (x; \xi) \in \mathbb{R} \times \Xi. \end{cases} \quad (1)$$

In (1), $f^\varepsilon \in \mathbb{R}^N$, $a, b \in L^\infty(\Xi)$ and M_{f^ε} is a function defined as follows

$$M_{f^\varepsilon} = M(u^\varepsilon; \xi, \varepsilon),$$

where $M : \mathcal{U} \subset \mathbb{R}^N \times \Xi \times \mathbb{R}^+ \to \mathbb{R}^N$ is a C^2 function with respect to u, measurable and bounded with respect to ξ and

$$u^\varepsilon(x, t) = \int_\Xi f^\varepsilon(x, t; \xi) d\xi. \quad (2)$$

The formal limit of (1) is given by the following system

$$\begin{cases} u_t + A(u)_x = B(u)_{xx}, & (x,t) \in \mathbb{R} \times \mathbb{R}^+ \\ u(x,0) = u_0(x), & x \in \mathbb{R}, \end{cases} \quad (3)$$

with $u \in \mathbb{R}^N$ and $A : \mathbb{R}^N \to \mathbb{R}^N$ is a regular function, provided the function M is a *local Maxwellian* (see the corresponding compatibility conditions in Section 2). We shall assume the equilibrium system (3) is strongly parabolic, that is $B(u)$ verifies

$$\nabla B(u) + (\nabla B(u))^* > 0 \quad (4)$$

for any $u \in \mathcal{U} \subset \mathbb{R}^N$ and $u_0 \in L^\infty(\mathbb{R})$. In [8], under additional conditions on the Maxwellian, we proved the L^∞ stability of (1) and the strong convergence in L^p_{loc}, $p < +\infty$, of u^ε given by (2) toward the distributional solution of (3), by means of compensated compactness techniques.

The paper [4] is devoted to the study of the following semilinear hyperbolic system with relaxation term

$$\begin{cases} u_t - v_x = 0 \\ v_t - z_x = 0 \\ \varepsilon^2 z_t - \mu v_x = -z + \sigma(u). \end{cases} \quad (5)$$

Clearly, the latter system reduces for $\varepsilon = 0$ to

$$\begin{cases} u_t - v_x = 0 \\ v_t - \sigma(u)_x = \mu v_{xx}, \end{cases} \quad (6)$$

namely, an *incompletely parabolic system*, since the (constant) diffusion matrix $\begin{pmatrix} 0 & 0 \\ 0 & \mu \end{pmatrix}$ is positive semidefinite. Using standard energy estimates, in [4] we prove the strong convergence in L^p_{loc}, $p \geq 2$ of smooth solutions of (5) toward smooth solutions of its limit (6), provided the relaxation term is globally Lipschitz, namely

$$\sup_{u \in \mathbb{R}} |\sigma'(u)| < +\infty. \quad (7)$$

We emphasize that, due to the lack of parabolicity of the limit, the compensated compactness techniques do not have a straightforward generalization to this case. The only result of singular convergence toward a degenerate parabolic equilibrium has been proved in [1], but it applies only for BGK systems approximating a scalar equation and our result is the first in the case of an incompletely parabolic equilibrium. These results are summarized in Section 3.

Finally, in Section 4 we present some preliminary results [5] concerning the three–dimensional version of (5), namely

$$\partial_t F_{i\alpha} - \partial_\alpha v_i = 0$$
$$\partial_t v_i - \partial_\alpha S_{i\alpha} = 0 \qquad (8)$$
$$\varepsilon \partial_t S_{i\alpha} - \mu \partial_\alpha v_i = -S_{i\alpha} + T_{i\alpha}(F),$$

where v_i, $i = 1, \ldots 3$ is the velocity of the motion, $F_{i\alpha}$ and $S_{i\alpha}$, $i, \alpha = 1, \ldots 3$, the deformation gradient and the stress tensor. In particular, we prove the L^2 stability and convergence of its solutions toward the solutions of its equilibrium

$$\partial_t F_{i\alpha} - \partial_\alpha v_i = 0$$
$$\partial_t v_i - \partial_\alpha T_{i\alpha}(F) = \mu \partial_\alpha \partial_\alpha v_i \qquad (9)$$

in the smooth regime and for *polyconvex* stored energy, that is for

$$T_{i\alpha}(F) = \frac{\partial}{\partial F_{i\alpha}} W(F)$$

and the equilibrium internal energy $W(F)$ a convex function of the minors of the matrix F, namely

$$W(F) = g(\Phi(F)), \quad \Phi(F) := (F, \operatorname{cof} F, \det F),$$

with $g = g(F, Z, w) = g(A)$ a convex function of $A \in \mathbb{R}^{19}$. To apply the *modulated energy* technique, we rewrite system (8) as an approximation via the wave operator

$$\partial_t F_{i\alpha} - \partial_\alpha v_i = 0$$
$$\partial_t v_i - \partial_\alpha T_{i\alpha}(F) = \mu \partial_\alpha \partial_\alpha v_i - \varepsilon \partial_t^2 v_i \qquad (10)$$

and we correct the energy of (9) by higher order contributions of acoustic waves, in such a way the resulting energy dissipates along the relaxation process. This method has been successfully applied in [9] for the hyperbolic–hyperbolic stress relaxation limit of (8) or (10), namely for $\mu = \bar{\mu}\varepsilon$.

There is a wide literature of papers concerning the study of nonhomogeneous hyperbolic system with an underlying parabolic behavior. Among all, we recall the papers of Kurtz [7] and McKean [11], where for the first time this feature for hyperbolic systems has been put into evidence. Then, we recall the papers of Marcati with various collaborators (see [12, 6] and the references therein), where the above scaling has been used for different systems and the convergence has been obtained for weak solutions with the aid of the compensated compactness. In the framework of Boltzmann kinetic models with a finite number of velocities, we recall the paper of Lions and Toscani [10], where the same parabolic behavior has been pointed out, proving in particular the convergence toward the porous media equation. Finally, we recall the paper of Brenier, Natalini and Puel [2], where modulated energy techniques have been used in the diffusive relaxation for the incompressible Navier–Stokes equation.

2 BGK approximation of strongly parabolic systems

In this section we review the results of [8] concerning BGK approximation of the form (1) for equilibrium systems (3) which are strongly parabolic in the sense of (1) for any $u \in \mathcal{U} \subset \mathbb{R}^N$ and with L^∞ initial datum. In (1), u^ε is given by (2) and the Maxwellian $M_{f^\varepsilon} = M(u^\varepsilon; \xi, \varepsilon)$ verifies the following compatibility conditions:

$$\int_\Xi M(w; \xi, \varepsilon) d\xi = w, \qquad \int_\Xi \left(a(\xi) + \frac{b(\xi)}{\sqrt{\varepsilon}} \right) M(w; \xi, \varepsilon) d\xi = A(w),$$

$$\int_\Xi b(\xi)^2 M(w; \xi, \varepsilon) d\xi = B(w),$$

for any $w \in \mathcal{U}$ and $\varepsilon \in (0, 1]$ and $M(w; \xi, \varepsilon) \to M(w; \xi, 0)$ as $\varepsilon \downarrow 0$, uniformly in $w \in \mathcal{U}$, a.e. in ξ. We shall prove the convergence of the approximating sequence u^ε by means of compensated compactness [3]. Thus, we shall first obtain the stability of this approximating sequence in L^∞ and the control of its deviation from the equilibrium manifold, namely the difference $M_{f^\varepsilon} - f^\varepsilon$. The former property can be proved in the context of Maxwellian functions independent from ε and under appropriate assumptions (see [8] for details). Since here we focus our attention only in the rigorous proof of the relaxation limit, we put ourselves in the following framework [8]:

(**H$_1$**) the function M are Maxwellians such that, for any $\varepsilon \in (0, 1]$, a.e. in ξ, $M(\cdot; \xi, \varepsilon) : \mathcal{U} \to \mathcal{U}_{\xi,\varepsilon}$ are global diffeomorphisms and $\mathcal{U}, \mathcal{U}_{\xi,\varepsilon}$ are compact, convex sets. Moreover, $f^\varepsilon(x, t; \xi)$ is a global-in-time solution of (1), uniformly bounded, such that $f^\varepsilon : \mathbb{R} \times \mathbb{R}^+ \to \mathcal{F}^\varepsilon$;

(**H$_2$**) there exist a function $F : \mathcal{U} \to \mathbb{R}^N$ and a positive constant α, independent from ε, such that the matrix $(\nabla M)^* \nabla F$ is symmetric and it verifies $(\nabla M)^* \nabla F \geq \alpha > 0$, for any $u \in \mathcal{U}$, for any $\varepsilon \in (0, 1]$, a.e. in ξ.

Condition (**H$_1$**) implies $u^\varepsilon \in \mathcal{U}$, namely, the desired L^∞ stability, and condition (**H$_2$**), together with an appropriate finite–energy condition for $f_0(x; \xi)$, gives the L^2 control of the relaxation limit, that is $\|M_{f^\varepsilon} - f^\varepsilon\|_{L^2(\mathbb{R} \times [0,T] \times \Xi)} = O(\sqrt{\varepsilon})$, for any fixed $T > 0$. We the aid of this estimate and thanks to compensated compactness, we obtain the strong convergence of the relaxation limit as follows [8].

Theorem 1. *Let hypotheses (**H$_1$**) and (**H$_2$**) hold. Then, if the initial data $f_0(x; \xi)$ verifies an appropriate integrability condition, $u^\varepsilon \to u$ strongly in $L^p_{loc}(\mathbb{R} \times \mathbb{R}^+)$ for any $p < +\infty$, where $u(x, t)$ verifies (3) in the sense of distributions.*

The crucial part in the proof of the previous theorem is the reformulation of the BGK model as the *non–closed* system of the first two moments of f^ε

$$\begin{cases} u^\varepsilon_t + v^\varepsilon_x = 0 \\ v^\varepsilon_t + z^\varepsilon_x + \dfrac{1}{\sqrt{\varepsilon}} w^\varepsilon_x + \dfrac{1}{\varepsilon} \widetilde{B}^\varepsilon_x = \dfrac{1}{\varepsilon} (A(u^\varepsilon) - v^\varepsilon), \end{cases}$$

where

$$z^\varepsilon = \int_\Xi a^2(\xi) f^\varepsilon(\xi) d\xi, \quad w^\varepsilon = \int_\Xi a(\xi) b(\xi) f^\varepsilon(\xi) d\xi, \quad \widetilde{B}^\varepsilon = \int_\Xi b^2(\xi) f^\varepsilon(\xi) d\xi.$$

Then, the above L^2 estimate implies that the deviation of that system from (3) is compact in a negative Sobolev space, which allows us to apply the compensated compactness techniques.

3 One–dimensional semilinear model in viscoelasticity

This section is devoted to the study of the one–dimensional model (5) and its relaxation limit toward the incompletely parabolic equilibrium system (6) [4]. It is worth observing that this relaxation limit can be viewed as the passage from a model of viscoelasticity with memory to a model of viscoelasticity of the rate type. Indeed, the stress z in (5) is given by $z = \frac{\mu}{\varepsilon^2} u - \int_{-\infty}^t \frac{1}{\varepsilon^2} e^{-\frac{t-\tau}{\varepsilon^2}} \left(\frac{\mu}{\varepsilon^2} u - \sigma(u) \right)(\tau) d\tau$, while in the limit (6), it becomes $z = \sigma(u) + \mu v_x$. We shall prove the H^1 stability of solutions to the semilinear system (5) by means of standard energy estimates, which in turns implies the global existence of these solutions and their strong convergence in L^p_{loc}, for any $p \geq 2$ [4].

Theorem 2. *Let $(u_\varepsilon, v_\varepsilon, z_\varepsilon)(x,t)$ be the solution to (5) with initial condition $(u, v, z)(\cdot, 0) \in H^1(\mathbb{R})$. Suppose that the function σ satisfies condition (7). Then, the following inequality holds for any $t > 0$*

$$\varepsilon^2 \|z_\varepsilon(t)\|^2_{H^1(\mathbb{R})} + \|u_\varepsilon(t)\|^2_{H^1(\mathbb{R})} + \|v_\varepsilon(t)\|^2_{H^1(\mathbb{R})} + \int_0^t \|z_\varepsilon(s)\|^2_{H^1(\mathbb{R})} ds$$
$$\leq \left[\varepsilon^2 \|z(0)\|^2_{H^1(\mathbb{R})} + \|u(0)\|^2_{H^1(\mathbb{R})} + \|v(0)\|^2_{H^1(\mathbb{R})} \right] e^{Ct}, \tag{11}$$

where C is a positive constant depending only on $\sup_{u \in \mathbb{R}} |\sigma'(u)|$. In particular, $u_\varepsilon \to u$, $v_\varepsilon \to v$ strongly in $L^p_{loc}([0,T] \times \mathbb{R})$, for any $2 \leq p < +\infty$, and (u, v) is the solution of (6) with $(u, v)(\cdot, 0)$ as initial condition.

Proof (Sketch). The system (5) admits a symmetrizer, which is positive definite for small values of ε, namely

$$\mathcal{B}^\varepsilon = \begin{pmatrix} \frac{\mu}{\varepsilon^2} & 0 & -1 \\ 0 & \frac{\mu}{\varepsilon^2} - 1 & 0 \\ -1 & 0 & 1 \end{pmatrix}.$$

Hence, denoting with W^ε the (column) vector of the solutions $(u_\varepsilon, v_\varepsilon, z_\varepsilon)$, we define the energy

$$\mathcal{E}^\varepsilon(W^\varepsilon) = (\mathcal{B}^\varepsilon W^\varepsilon, W^\varepsilon)_{L^2(\mathbb{R})} = \int_{-\infty}^{+\infty} \left[\frac{\mu}{\varepsilon^2} u_\varepsilon^2 - 2 u_\varepsilon z_\varepsilon + \left(\frac{\mu}{\varepsilon^2} - 1 \right) v_\varepsilon^2 + z_\varepsilon^2 \right] dx$$

and we obtain (11) from the energy estimates for $\mathcal{E}^\varepsilon(W^\varepsilon)$ and $\mathcal{E}^\varepsilon(W^\varepsilon_x)$.

The last part of the theorem comes from standard compactness arguments and form the uniqueness of the solutions of (6). ∎

Finally, let $u^\varepsilon, v^\varepsilon, z^\varepsilon$ be solutions of (5) with initial data $u_0^\varepsilon, v_0^\varepsilon, z_0^\varepsilon \in H^1(\mathbb{R})$ and (u,v) be solutions of (6) with initial data $u_0, v_0 \in H^2(\mathbb{R})$. Then the differences $\bar{u} = u^\varepsilon - u$, $\bar{v} = v^\varepsilon - v$, $\bar{z} = z^\varepsilon - z$, $z = \sigma(u) + \mu v_x$, satisfies

$$\begin{cases} \bar{u}_t - \bar{v}_x = 0 \\ \bar{v}_t - \bar{z}_x = 0 \\ \bar{z}_t - \frac{\mu}{\varepsilon^2}\bar{v}_x = -z_t - \frac{1}{\varepsilon^2}\left[\bar{z} - (\sigma(\bar{u}+u) - \sigma(u))\right]. \end{cases}$$

The above system has the same principal part of (5) and therefore we can repeat the arguments of Theorem 2 to control directly the differences $\bar{u}, \bar{v}$ and $\bar{z}$ in L^2 and justify the relaxation limit, for *well–prepared* initial data, that is for $\|u_0^\varepsilon - u_0\|_{L^2(\mathbb{R})} + \|v_0^\varepsilon - v_0\|_{L^2(\mathbb{R})} + \varepsilon\|z_0^\varepsilon - z_0\|_{L^2(\mathbb{R})} \to 0$ as $\varepsilon \downarrow 0$.

4 Multidimensional viscoelasticity and modulated energy

In this section we examine the three–dimensional model (10) by means of modulated energy techniques. Indeed, we shall prove that, for $\varepsilon \ll 1$, the following high–order energy gives a natural tool to control the relaxation limit

$$\mathcal{E}_m = \frac{1}{2}\left(|v|^2 + |F|^2\right) + \varepsilon v_i \partial_t v_i + \frac{1}{2}\varepsilon^2 \lambda |\partial_t v|^2 + \frac{1}{2}\varepsilon \lambda \mu |\nabla_\alpha v|^2 + \varepsilon \lambda \partial_\alpha v_i T_{i\alpha}(F).$$

Theorem 3. *Any smooth solution of (8) verifies*

$$\partial_t \mathcal{E}_m - \partial_\alpha[v_i T_{i\alpha}(F) + \mu v_i \partial_\alpha v_i + \varepsilon \lambda \mu \partial_t v_i \partial_\alpha v_i + \varepsilon \lambda \partial_t v_i T_{i\alpha}(F)]$$
$$+ \left(\mu|\nabla_\alpha v|^2 - \varepsilon \lambda \partial_\alpha v_i \frac{\partial T_{i\alpha}(F)}{\partial F_{j\beta}}\partial_\beta v_j\right) + \varepsilon(\lambda - 1)|\partial_t v|^2 = \partial_\alpha v_i(F_{i\alpha} - T_{i\alpha}(F)),$$
(12)

where λ is an arbitrary constant.

Moreover, if $\nabla_F T(F) = \nabla_F^2 W(F) \leq \Gamma I$ for any F, then, for $\varepsilon < \varepsilon_0(\mu, \Gamma)$,

$$\psi(t) := \int_{\mathbb{R}^3} \left(|v(x,t)|^2 + |F(x,t)|^2 + \varepsilon^2|\partial_t v(x,t)|^2 + \varepsilon|\nabla_\alpha v(x,t)|^2\right) dx \leq O(T),$$
(13)

for any $t \in [0,T]$.

Proof (Sketch). We multiply $(10)_1$ by $F_{i\alpha}$ and $(10)_2$ by v_i to obtain

$$\partial_t\left[\frac{1}{2}\left(|F|^2 + |v|^2\right) + \varepsilon v_i \partial_t v_i\right] - \partial_\alpha[v_i T_{i\alpha}(F) + v_i \partial_\alpha v_i] + \mu|\nabla_\alpha v|^2 - \varepsilon|\partial_t v|^2$$
$$= \partial_\alpha v_i(F_{i\alpha} - T_{i\alpha}(F)).$$
(14)

The above relation does not give a coercive energy which dissipates along the relaxation and we need to correct it with an higher order acoustic energy. To this end, we multiply (10)$_2$ by $\varepsilon\lambda\partial_t v_i$ and we have

$$\partial_t \frac{1}{2}\left[\varepsilon^2\lambda|\partial_t v|^2 + \varepsilon\lambda\mu|\nabla_\alpha v|^2\right] - \partial_\alpha\left[\varepsilon\lambda\mu\partial_t v_i\partial_\alpha v_i\right] + \varepsilon\lambda|\partial_t v|^2 - \varepsilon\lambda\partial_t v_i\partial_\alpha T_{i\alpha}(F) = 0. \tag{15}$$

We interchange the t and x derivatives in the last term of (15) as follows

$$-\varepsilon\partial_t v_i\partial_\alpha T_{i\alpha}(F) = -\varepsilon\partial_\alpha v_i\partial_t T_{i\alpha}(F) + \varepsilon\partial_t[\partial_\alpha v_i T_{i\alpha}(F)] - \varepsilon\partial_\alpha[\partial_t v_i T_{i\alpha}(F)]$$
$$= -\varepsilon\partial_\alpha v_i \frac{\partial T_{i\alpha}(F)}{\partial F_{j\beta}}\partial_\beta v_j + \varepsilon\partial_t[\partial_\alpha v_i T_{i\alpha}(F)] - \varepsilon\partial_\alpha[\partial_t v_i T_{i\alpha}(F)]$$

and we sum the resulting identity to (6) to obtain (12).

Moreover, if $\varepsilon < \min\{\frac{\mu}{\Gamma}, \frac{\mu}{\Gamma^2}\}$ and $\lambda > 1$ is properly chosen, we have

$$\left(\mu|\nabla_\alpha v|^2 - \varepsilon\lambda\partial_\alpha v_i\frac{\partial T_{i\alpha}(F)}{\partial F_{j\beta}}\partial_\beta v_j\right) > O(1)|\nabla_\alpha v|^2, \quad \int_{\mathbb{R}^3}\mathcal{E}_m \geq O(1)\psi(t).$$

Thus, the Gronwall Lemma implies (13), that is, the L^2 stability of the relaxation process for smooth solutions of (8). ∎

Finally, we can repeat the above arguments to control the *relative modulate energy*

$$\mathcal{E}_{rm} = \frac{1}{2}\left(|v-\hat{v}|^2 + |F-\hat{F}|^2\right) + \varepsilon(v_i - \hat{v}_i)\partial_t(v_i - \hat{v}_i) + \frac{1}{2}\varepsilon^2\lambda|\partial_t(v-\hat{v})|^2$$
$$+ \frac{1}{2}\varepsilon\lambda\mu|\nabla_\alpha(v-\hat{v})|^2 + \varepsilon\lambda\partial_\alpha(v_i-\hat{v}_i)(T_{i\alpha}(F) - T_{i\alpha}(\hat{F}))$$

between smooth solutions (F,v) of (10) and smooth solutions $(\hat{F},\hat{v})$ of its equilibrium (9). The resulting estimate reads as follows

$$\partial_t\mathcal{E}_{rm} - \partial_\alpha[(v_i - \hat{v}_i)(T_{i\alpha}(F) - T_{i\alpha}(\hat{F})) + \mu(v_i - \hat{v}_i)\partial_\alpha(v_i - \hat{v}_i)$$
$$+ \varepsilon\lambda\mu\partial_t(v_i - \hat{v}_i)\partial_\alpha(v_i - \hat{v}_i) + \varepsilon\lambda\partial_t(v_i - \hat{v}_i)(T_{i\alpha}(F) - T_{i\alpha}(\hat{F}))]$$
$$+ \left(\mu|\nabla_\alpha(v-\hat{v})|^2 - \varepsilon\lambda\partial_\alpha(v_i - \hat{v}_i)\frac{\partial T_{i\alpha}(F)}{\partial F_{j\beta}}\partial_\beta(v_j - \hat{v}_j)\right)$$
$$+ \varepsilon(\lambda - 1)|\partial_t(v-\hat{v})|^2$$
$$= \partial_\alpha(v_i - \hat{v}_i)(F_{i\alpha} - \hat{F}_{i\alpha} - (T_{i\alpha}(F)) - T_{i\alpha}(\hat{F})))$$
$$- \varepsilon\partial_t^2\hat{v}_i(v_i - \hat{v}_i) - \varepsilon^2\lambda\partial_t^2\hat{v}_i\partial_t(v_i - \hat{v}_i)$$
$$+ \varepsilon\lambda\partial_\alpha(v_i - \hat{v}_i)\left(\frac{\partial T_{i\alpha}(F)}{\partial F_{j\beta}} - \frac{\partial T_{i\alpha}(\hat{F})}{\partial F_{j\beta}}\right)\partial_t\hat{F}_{j\beta},$$

which has the same structure of (12).

Thus, proceeding as before, we obtain the rigorous justification in L^2 of the relaxation limit in the regime of smooth solutions and for *well-prepared* initial data, that is for $\psi_d(0) \to 0$ as $\varepsilon \downarrow 0$, where

$$\psi_d(t) := \int_{\mathbb{R}^3} \left(|v - \widehat{v}|^2 + |F - \widehat{F}|^2 + \varepsilon^2 |\partial_t (v - \widehat{v})|^2 + \varepsilon |\nabla_\alpha (v - \widehat{v})|^2 \right) dx.$$

Remark 1. It is worth to observe that the lack of convexity of the stored energy $W(F)$ does not play any role in the present relaxation limit. This is due to the fact that, thanks to the (partial) diffusion present in the limit, it is possible to construct the modulated energy $\mathcal{E}_m$ (and the relative modulated energy $\mathcal{E}_{rm}$) multiplying $(10)_1$ by $F_{i\alpha}$ instead of $T_{i\alpha}(F)$. In this way we obtain a coercive energy disregarding the nature of $W(F)$ and the extra error we produce is indeed controlled in terms of $\|F\|_{L^2}$ and $\|\nabla_\alpha v\|_{L^2}$.

References

1. Bouchut, F., Guarguaglini, F., Natalini, R.: Discrete kinetic approximation to multidimensional parabolic equations. Indiana Univ. Math. J., **49**, 723–749 (2000)
2. Brenier, Y., Natalini, R., Puel, M.: On a relaxation approximation of the incompressible Navier–Stokes equations. Proc. Amer. Math. Soc., **132**, 1021–1028 (2004)
3. Dacorogna, B.: Weak continuity and weak lower semicontinuity of nonlinear functionals. Lecture Notes in Mathematics, **922**. Springer-Verlag, Berlin-New York (1982)
4. Di Francesco, M., Lattanzio, C.: Diffusive relaxation 3×3 model for a system of viscoelasticity. Asymptot. Anal., **40**, 235–253 (2004)
5. Donatelli, D., Lattanzio, C.: On the diffusive stress relaxation for multidimensional viscoelasticity. Preprint 2006 (in preparation)
6. Donatelli, D., Marcati, P.: Convergence of singular limits for multi-d semilinear hyperbolic systems to parabolic systems. Trans. Amer. Math. Soc., **356**, 2093–2121 (2004)
7. Kurtz, T.: Convergence of sequences of semigroups of nonlinear operators with an application to gas kinetics. Trans. Amer. Math. Soc., **186**, 259–272 (1973)
8. Lattanzio, C., Natalini, R.: Convergence of Diffusive BGK Approximations for Nonlinear Strongly Parabolic Systems. Proc. Roy. Soc. Edinburgh Sect. A, **132**, 341–358 (2002)
9. Lattanzio, C., Tzavaras, A.E.: Structural properties of stress relaxation and convergence from viscoelasticity to polyconvex elastodynamics. Arch. Ration. Mech. Anal., to appear
10. Lions, P.L., Toscani, G.: Diffusive limit for finite velocity Boltzmann kinetic models. Rev. Mat. Iberoamericana, **13**, 473–513 (1997)
11. McKean, H.P.: The central limit theorem for Carleman's equation. Israel J. Math., **21**, 54–92 (1975)
12. Marcati, P., Rubino, B.: Hyperbolic to Parabolic Relaxation Theory for Quasilinear First Order Systems. J. Differential Equations, **162**, 359–399 (2000)

Parallel Algorithms for Nonlinear Diffusion by Using Relaxation Approximation

Fausto Cavalli[1], Giovanni Naldi[1] and Matteo Semplice[1]

Department of Mathematics, University of Milano, via Saldini 50 Milano (Italy)
{cavalli,naldi,semplice}@mat.unimi.it

Summary. It has been shown that the equation of diffusion, linear and nonlinear, can be obtained in a suitable scaling limit by a two-velocity model of the Boltzmann equation [7] . Several numerical approximations were introduced in order to discretize the corresponding multiscale hyperbolic systems [8, 1, 4]. In the present work we consider relaxed approximations for multiscale kinetic systems with asymptotic state represented by nonlinear diffusion equations. The schemes are based on a relaxation approximation that permits to reduce the second order diffusion equations to first order semi-linear hyperbolic systems with stiff terms. The numerical passage from the relaxation system to the nonlinear diffusion equation is realized by using semi-implicit time discretization combined with ENO schemes and central differences in space. Finally, parallel algorithms are developed and their performance evaluated. Application to porous media equations in one and two space dimensions are presented.

1 Relaxation approximation of nonlinear diffusion

The main aim of this work is to approximate solutions of a nonlinear, degenerate parabolic equation

$$\frac{\partial u}{\partial t} = \Delta(g(u)) \qquad (1)$$

for $x \in \Omega \subset \mathbb{R}^d$, $d \geq 1$, $t \geq 0$, with suitable boundary conditions and initial condition $u(x,0) = u_0(x)$, where g is a non-decreasing Lipschitz continuous function on $\mathbb{R}$, the degenerate case corresponding to $g(0) = 0$. This framework is so general that it includes the porous medium equation and the Stefan problem as well as a wide class of mildly nonlinear parabolic equations. Using the same idea which is at the basis of the well-known relaxation schemes for hyperbolic conservation laws [5], it is possible to develop stable numerical schemes for diffusion and for transport reaction-diffusion equations. In the case of the nonlinear diffusion operator, by introducing an additional variable $\mathbf{v}(x,t) \in \mathbb{R}^d$ and the positive parameter ϵ, we have the following relaxation system

$$\begin{cases} \frac{\partial u}{\partial t} + \operatorname{div}(\mathbf{v}) = 0 \\ \frac{\partial \mathbf{v}}{\partial t} + \frac{1}{\epsilon}\nabla g(u) = -\frac{1}{\epsilon}\mathbf{v} \end{cases} \quad (2)$$

Formally, in the small relaxation limit, $\epsilon \to 0^+$, the system (2) approximates to leading order equation (1). In order to have a non singular transport operator, by using a suitable parameter Φ we can rewrite system (2) as

$$\begin{cases} \frac{\partial u}{\partial t} + \operatorname{div}(\mathbf{v}) = 0 \\ \frac{\partial \mathbf{v}}{\partial t} + \Phi^2 \nabla g(u) = -\frac{1}{\epsilon}\mathbf{v} + \left(\Phi^2 - \frac{1}{\epsilon}\right)\nabla g(u) \end{cases} \quad (3)$$

Then, using an auxiliary variable $w(x,t) \in \mathbb{R}$ we get

$$\begin{cases} \frac{\partial u}{\partial t} + \operatorname{div}(\mathbf{v}) = 0 \\ \frac{\partial \mathbf{v}}{\partial t} + \Phi^2 \nabla w = -\frac{1}{\epsilon}\mathbf{v} + \left(\Phi^2 - \frac{1}{\epsilon}\right)\nabla w \\ \frac{\partial w}{\partial t} + \operatorname{div}(\mathbf{v}) = -\frac{1}{\epsilon}(w - g(u)) \end{cases} \quad (4)$$

In the previous systems the parameter ϵ has physical dimensions of a time and represents the relaxation time, i.e. the characteristic time to reach the equilibrium point in the evolution of the variable $\mathbf{v}$ governed by the stiff second equation of (4). For consistency, w has the same dimensions as u, while each component of $\mathbf{v}$ has the dimension of u times a velocity; finally Φ is a velocity. Equations (4) form a semilinear hyperbolic system with characteristic velocities $0, \pm\Phi$. The parameter Φ allows the use of this system with non-stiff velocities (in fact, when $\Phi = 0$ these velocities are instead $0, \pm\frac{1}{\epsilon}$).

One of the main advantages of this approach resides in the semilinearity of the system, that is all the nonlinearities are in the (stiff) source terms, while the differential operator is linear. Moreover we point out that degenerate parabolic equations often model physical situations with free boundaries or discontinuities: we expect that schemes for hyperbolic systems will be able to reproduce faithfully these details of the solution. Finally, the relaxation approximation does not exploit the form of the nonlinear function g and hence it gives rise to a numerical scheme that, to a large extent, is independent of it, resulting in a very versatile tool.

In the following section we will describe the scheme that we used to integrate the relaxed version of (4), i.e. when $\varepsilon = 0$, and we will discuss its properties. Numerical results for 1D and 2D cases will be presented in section 3 together with the performance of the parallel implementation.

2 The numerical scheme

For simplicity, we will consider a regular rectangular grid on $\mathbb{R}^d$. For $d = 1$ this consists of a set of equally spaced grid points $x_{i+1/2}$, $i = ..., -1, 0, 1, ...,$ with uniform mesh width $\Delta x = x_{i+1/2} - x_{i-1/2}$. When $d > 1$, we consider the obvious generalization and i will represent a multi-index of d integers. The discrete time levels t^n for $n = 0, 1, 2, \ldots$ ($t^0 = 0$) are also spaced uniformly with time step Δt. As usual we denote by U_i^n the approximate value of U at the centre of the cell $[x_{i-1/2}, x_{i+1/2}]$ at time t^n. The spatial discretization of the relaxation system and the corresponding numerical fluxes is realized here by using ENO techniques [10]. This allows us to get highly accurate spatial reconstruction of the solution. The resulting semi-discrete approximation of (4) is of the form

$$\frac{\partial z_i(t)}{\partial t} + F(z_{i-r}(t), \ldots, z_{i+s}(t)) = G(z_{i-r}(t), \ldots, z_{i+s}(t)) \qquad (5)$$

where z denotes the collection $(u, v_1, \ldots, v_d, w)$ of all the variables appearing in (4) and the stencil used for approximating z_i is $r+s+1$ points wide. In (5), F is a discretization of the linear differential operator appearing on the left hand side of (4), while G is a discretization of the non-linear and stiff source terms of (4).

In order to avoid severe restrictions on the time step, we need to couple this high order in space scheme with a time integrator of equal accuracy. Moreover, due to the structure of (5) sketched above, we wish to treat implicitly the time integration of G, which is stiff, and explicitly the one of F, which is linear. This is achieved with IMEX schemes tailored to relaxation systems [2, 9].

Numerical tests [5, 8] suggest that the difference between the results obtained with $\varepsilon \ll 1$ and those with $\varepsilon = 0$ are negligible. However the relaxed scheme (with $\varepsilon = 0$) gives immediately the projection of the solution onto the equilibrium state in the relaxation step; hence it is simpler to implement. For this reason we consider here only the relaxed scheme.

2.1 Implicit relaxed step

The structure of the system to be solved implicitly is of particular importance. By using simply the backward Euler formula, the values $(u_i^{(1)}, v_i^{(1)}, w_i^{(1)})$ of the solutions of the system $z_t = G(z)$ with initial data (u_i^n, v_i^n, w_i^n) may be computed by solving

$$\begin{cases} \frac{u^{(1)} - u^n}{\Delta t} = 0 \\ \frac{v^{(1)} - v^n}{\Delta t} = -\frac{v^{(1)}}{\epsilon} + \left(\Phi^2 - \frac{1}{\epsilon}\right) \hat{\nabla} w^{(1)} \\ \frac{w^{(1)} - w^n}{\Delta t} = -\frac{w^{(1)} - g(u^{(1)})}{\epsilon} \end{cases}$$

where we have suppressed the spatial index i for clarity and $\hat{\nabla}$ is a suitable discretization of the spatial gradient. Formally, in the limit $\epsilon \to 0^+$, this reduces to

$$u^{(1)} = u^n, \quad v^{(1)} = -\hat{\nabla} w^{(1)}, \quad w^{(1)} = g(u^{(1)}). \qquad (6)$$

We note here that the first equation is immediately solved and the remaining two decouple. Hence even in the implicit step, we do not need an implicit solver. Up to second order precision is space, the usual 3-point central difference approximation of the derivative is suitable. For degree 3 and 4 one needs however the 5-point approximation or otherwise the quality of the solution and rate of convergence is degraded. Formula (6) represent the relaxation step of our relaxed schemes.

2.2 Explicit step

For each explicit step of the IMEX scheme one has to advance from time t^n to time t^{n+1} the system $z_t + F(z) = 0$, i.e.

$$\frac{\partial}{\partial t}\begin{pmatrix} u \\ v \\ w \end{pmatrix} + \frac{\partial}{\partial x}\begin{pmatrix} 0 & 1 & 0 \\ 0 & 0 & \Phi^2 \\ 0 & 1 & 0 \end{pmatrix}\begin{pmatrix} u \\ v \\ w \end{pmatrix} = 0 \qquad (7)$$

with initial data set to the values $(u_i^{(1)}, v_i^{(1)}, w_i^{(1)})$ obtained from the relaxation step, as described previously. The characteristic variables U, V, W move at speed $\Phi, -\Phi$ and 0 respectively. By changing variables to diagonalize the system we need to reconstruct via ENO only the two fields $U(x)$ and $V(x)$ and calculate the corresponding numerical fluxes.

In two space dimensions, the above system generalizes as

$$\frac{\partial}{\partial t}\begin{pmatrix} u \\ v_{(1)} \\ v_{(2)} \\ w \end{pmatrix} + \frac{\partial}{\partial x}\begin{pmatrix} 0 & 1 & 0 & 0 \\ 0 & 0 & 0 & \Phi^2 \\ 0 & 0 & 0 & 0 \\ 0 & 1 & 0 & 0 \end{pmatrix}\begin{pmatrix} u \\ v_{(1)} \\ v_{(2)} \\ w \end{pmatrix} + \frac{\partial}{\partial y}\begin{pmatrix} 0 & 1 & 0 & 0 \\ 0 & 0 & 0 & 0 \\ 0 & 0 & 0 & \Phi^2 \\ 0 & 1 & 0 & 0 \end{pmatrix}\begin{pmatrix} u \\ v_{(1)} \\ v_{(2)} \\ w \end{pmatrix} = 0 \qquad (8)$$

where $\mathbf{v} = (v_{(1)}, v_{(2)})$ and similarly in higher dimensions. We note that only one of the fields $v_{(i)}$ appear in the differential operator along the ith direction. One may then calculate the fluxes separately for each spatial direction by using the aforementioned ENO reconstructions on the fields $v_{(i)}$ for $i = 1, \ldots, d$. In order to analyse some stability properties of the scheme we consider the linear case $g(u) = u$. Moreover we adopt the simplest IMEX scheme represented by the combination of a backward Euler timestep for $z_t = G(z)$ followed by a forward Euler timestep for $z_t + F(z) = 0$. For example when F is approximated with upwind fluxes and linear reconstructions and the spatial derivative in G with the central differences formula, this results in a scheme for the variable u for which a simple Von Neumann analysis reveals the necessity of a CFL condition of the form $\Delta t \approx C\Delta x^2$ with the numerical estimate $C \leq 0.875$ for the constant C. In Figure 1 we show an example of stability regions obtained with the Von Neumann analysis cited above.

3 Numerical results

As a numerical test we consider the porous media equation, which corresponds to the choice $g(u) = u^2$ in (1) and we use relaxed schemes, i.e. $\epsilon = 0$. We adopt

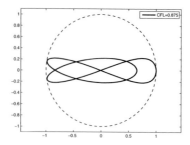

Fig. 1. An example of the stability regions for linear diffusion, $g(u) = u$.

periodic boundary conditions but more general boundary conditions may be easily implemented. In order to perform some test for the accuracy of the proposed schemes we consider numerical solutions in comparison with the exact 2D Barenblatt solution. The contour plot of the numerical solution at time $T = 2$ with 200×200 grid points is given in In Figure 3 we show the

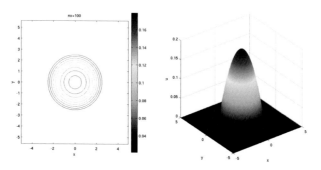

Fig. 2. Contour plot of the numerical solution for porous media equation.

behaviour of the free boundary, points for the passage from positive to zero value, of the numerical solution and of the Barenblatt solution. In Figure 3 the continuous line represents the exact solution, while the stars are the numerical front. The error between the true front and the approximate front appears of order Δx (the dashed line is at distance Δx from the exact front).

3.1 Parallel code

The parallel implementation has been realized through a decomposition of the computational domain by a balanced subdivision of the nodes among the processors set at the beginning of the program. Each processor solves its local problem, using MPI communications to get the boundary data needed. Since both the ENO subroutines and the program to solve (1) with the relaxed scheme do not involve nonlinear solvers of any kind (see Section 3), we expect

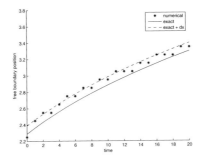

Fig. 3. Approximation of the free boundary with 200 grid points along a cross section.

a linear scaling of the solution time with the number of processors. For portability and easy extension, both the ENO library and the program are written exploiting the PETSC libraries In Figure 4 we report the scaling plot for our parallel code. The algorithm shows a good, almost linear, scaling behaviour,

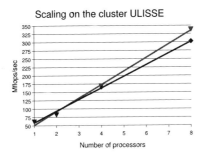

Fig. 4. Scaling of the parallel code on a Linux cluster with 72 Intel Xeon processors. When the subproblems assigned to each processor become too small, the time spent exchanging MPI messages among the processors become predominant and the overhead of MPI communications shows up as reduced speedup on the smaller grid.

until the subproblems assigned to each processor become too small and MPI communications slow the code down.

3.2 Comparison with another method

We compared our numerical results with those obtained with a linear method proposed and studied, among others, by Berger, Brezis, Rogers [3] and by Magenes, Nochetto and Verdi [7]. Their method is based on the non-linear Chernoff formula and it does not give an explicit formula for the solution, which is instead found by solving a linear problem. It is thus more costly then

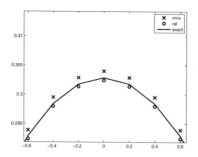 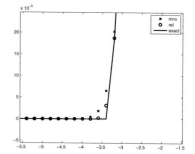

Fig. 5. Comparison of two numerical solutions of the porous media equation obtained with the relaxation and the BBRMNV method, together with the exact Barenblatt solution. 100 grid points were used. We show particulars of the areas around the maximum point (left hand figure) and around the moving front (right hand figure).

N	BBRMNV	rel
100	4.21e-3	2.75e-3
200	7.73e-4	2.58e-4
400	1.98e-4	6.51e-5
800	5.05e-5	1.83e-5

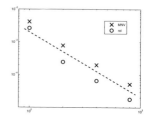

Fig. 6. Comparison of the 1-norm of the error of the relaxation and the BBRMNV method. The exact Barenblatt solution at time $t = 2$ was used as reference to compute the errors. The dotted line is a reference decay of second order.

our method where only matrix-vector products are needed. In the following we refer to this method as the BBRMNV method. In Figure 5 we present a comparison of two numerical solutions of the porous medium equation ($g(u) = u^2$) and the exact self-similar solution due to Barenblatt. The final time of all the simulations is $T = 2$. One may see that the higher numerical diffusion of the BBRMNV method shows up both as a lesser accuracy in the neighbourhood of the maximum at $x = 0$ and as a lower precision in the neighbourhood of the front. The solution represented by dots in Figure 5 was obtained with the relaxation method described in this paper, using spatial ENO reconstructions of degree 2 and an IMEX timestepping procedure with the same accuracy. Both methods under comparison are of second order, as shown by the table and the plot of the 1-norm of the error against the number of grid points that is shown in Figure 6. In any case we point out that the BBRMNV method, however, has been studied more extensively than our technique based on relaxation and adaptive versions should now be implementable. Finally, both methods are easily generalizable to different functions $g(u)$ and also to equations of the form $u_t - \Delta(g(u)) = f$.

4 Concluding remarks

In this work we briefly described a class of relaxed schemes for nonlinear and degenerate diffusion problems in any space dimension and in a rectangular domain. Using suitable relaxation approximation we are able to formulate a general diffusion equation into the form of a semilinear system. Then we coupled ENO schemes and IMEX approach for spatial and, respectively, time discretization. We can develop high order methods and we proposed a "blackbox" scheme: it does not exploit the form of nonlinear term $g(u)$ for the solution. Moreover, in this scheme we avoid the use of nonlinear solvers and, in the relaxed IMEX version, we don't need solvers at all. The numerical methods can be easily extended from 1D to higher dimensions and parallelized. A comparison with another method is also shown and differences in computational cost and accuracy are outlined. In forthcoming works we will perform theoretical study of the stability of relaxed schemes in the nonlinear case and we will explore the possibility to introduce non structured (rectangular) grids near fronts. Finally, the parallel implementation of WENO reconstructions are under study with interesting preliminary numerical results.

References

1. Aregba-Driollet D., Natalini R., Tang S.Q.: Diffusive kinetic explicit schemes for nonlinear degenerate parabolic systems. Quaderno IAC N. 26/2000, Roma (2000)
2. Asher U., Ruuth S., Spiteri R.J.: Implicit-explicit Runge-Kutta methods for time dependent Partial Differential Equations. Appl. Numer. Math. **25**, 151–167 (1997)
3. Berger A.E., Brezis H., Rogers J.C.W: A numerical method for solving the problem $u_t - \Delta f(u) = 0$. RAIRO numerical analysis 13, 297–312 (1979)
4. Jin S., Pareschi L., Toscani G.: Diffusive relaxation schemes for multiscale discrete velocity kinetic equations. SIAM J. Numer. Anal., **35**, 2405–2439 (1998)
5. Jin S., Xin, Z.: The relaxation schemes for systems of conservation laws in arbitrary space dimension. Comm. Pure and Appl. Math., **48**, 235–276 (1995)
6. Lions P.L., Toscani G.: Diffusive limit for two-velocity Boltzmann kinetic models. Rev. Mat. Iberoamericana, **13**, 473–513 (1997)
7. Magenes E., Nochetto R.H., Verdi C.: Energy error estimates for a linear scheme to approximate nonlinear parabolic problems. RAIRO Modl. Math. Anal. Numr. 21, 655–678 (1987)
8. Naldi G., Pareschi L.: Numerical schemes for hyperbolic systems of conservation laws with stiff diffusive relaxation. SIAM J. Numer. Anal., **37**, 1246–1270 (2000)
9. Pareschi L., Russo G.: Implicit-explicit Runge-Kutta schemes and applications to hyperbolic systems with relaxation. J. Sci. Comp. (to appear)
10. Shu C.W.: ENO end WENO schemes for hyperbolic conservation laws. In Quarteroni A. (ed) Advanced Numerical Approximation of Nonlinear Hyperbolic Equations. LN in Mathematics, **1697**, Springer, Berlin Heidelberg New York (1998)

On a Degenerated Parabolic-Hyperbolic Problem Arising From Stratigraphy

Guy Vallet

Lab. Applied Math., UMR CNRS 5142, IPRA BP1155, 64013 Pau Cedex (France)
guy.vallet@univ-pau.fr

Summary. In this communication, presented in the minisymposium on Degenerated Parabolic Equations, we are interested in the mathematical analysis of a stratigraphic model concerning geologic basin formation. Firstly, we present the physical model and the mathematical formulation, which lead to an original degenerated parabolic - hyperbolic conservation law. Then, the definition of a solution and some mathematical tools in order to resolve the problem are given. At last, we present numerical illustrations in the $1-D$ case and we give some open problems.

1 Introduction and presentation of the model

In this paper, we are interested in the mathematical study of a stratigraphic model. It concerns geologic basin formation by the way of erosion and sedimentation and leads to mathematical questions within the framework of degenerated parabolic - hyperbolic free-boundary problems.

By taking into account large scale in time and space and by knowing *a priori*, the tectonics, the eustatism and the sediments flux at the basin boundary, the model has to state about the transport of sediments.

Let us consider in the sequel a sedimentary basin with base Ω considered as a smooth, bounded domain in $\mathbb{R}^d$ ($d = 1, 2$); for any positive T, note $Q =]0, T[\times \Omega$ and denote by u the topography of the basin.

Then, the model proposed initially by R. Eymard et al. [4] and D. Granjeon et al. [6] is based on two considerations:

i) In the meaning of Darcy, the sediments flux $\vec{q}$ is assumed to be proportional to ∇u,

and

ii) the erosion speed, $\partial_t u$ in its nonpositive part, is underestimated by $-E$, where E is a given nonnegative bounded measurable function in Q (a weathering limited process depending on the climate and the age of the sediments): i.e. $\partial_t u + E \geq 0$ a.e. in Q.

In order to join together the constraint and a conservative formulation, D. Granjeon et al. propose in [6] to correct the diffusive flux $-\nabla u$ by introducing a dimensionless multiplier λ. One gets a new definition of the flux, given by $-\lambda\nabla u$, where λ is an unknown function with values *a priori* in $[0,1]$.

In order to simplify this academic study, one considers in the sequel homogeneous Dirichlet conditions on the boundary Γ. Therefore, the mathematical modelling has to express respectively:

the mass balance of the sediment: $\partial_t u - div(\lambda \nabla u) = 0$ in Q, (1)

the boundary condition: $u = 0$ on $]0,T[\times\Gamma$, (2)

the moving obstacle condition: $\partial_t u \geq -E$ in Q, (3)

the initial condition: $u(0,.) = u_0$ in Ω, (4)

where u_0 is assumed to be in $H_0^1(\Omega) \cap L^\infty(\Omega)$.

In order to give a mathematical modelling of λ, R. Eymard et al. propose in [4] to consider the following global constraint:

$$\partial_t u + E \geq 0, \quad 1 - \lambda \geq 0 \quad \text{and} \quad (\partial_t u + E)(1 - \lambda) = 0 \quad \text{in } Q. \quad (5)$$

It means that if the erosion rate constraint is inactive, the flux is equal to the diffusive one. Obviously, the boundary of the set $\{(t,x) \in Q, \lambda = 1\}$ is a free one and such a constraint (5) is non standard.

Then S. N. Antontsev et al. in [1, 2], G. Gagneux et al. in [5] and G. Vallet [7] propose the following original conservative formulation that contains implicitly the constraint (5): If H denotes the maximal monotone graph of the Heaviside function, then (u,λ) is formally a solution to:

$$0 = \partial_t u - div(\lambda \nabla u) \quad \text{where} \quad \lambda \in H(\partial_t u + E) \quad \text{in } Q. \quad (6)$$

In other words, since H is a graph, one considers the differential inclusion:

$$0 \in \partial_t u - div\{H(\partial_t u + E)\nabla u\} \quad \text{in } Q.$$

Let us give a remark on the equation: $0 = \partial_t u - div\{a(\partial_t u)\nabla u\}$, where, for example, a is assumed to be a continuous function such that $a(x) = 0$ if $x \leq 0$ and $0 < a(x) \leq 1$ for any positive real x with $a(x) = 1$ for any $x \geq 1$ (imagine a continuous approximation of the heaviside function H).

2 A locally hyperbolic behaviour

Note that informally, $0 = \partial_t u - \{a'(\partial_t u)\nabla u\}\nabla \partial_t u - a(\partial_t u)\Delta u$. Thus, the discriminant Δ satisfies $-\frac{|a'(\partial_t u)\nabla u|^2}{4} \leq 0$ and the equation is of degenerated hyperbolic type on free boundaries since Δ may vanish if $a'(\partial_t u) = 0$ or $\nabla u = 0$ in a non-negligible $\mathcal{L}^{d+1}$ subset in Q.

Let us illustrate this remark by considering travelling-wave solutions of equation

$$\partial_t u = \partial_x\{a(\partial_t u)\partial_x u\},\ x \in \Omega =\ ]-1,1[. \tag{7}$$

For any real ξ, let us note $f(\xi) = \xi a(\xi)$, and consider $\beta = f^{-1}$ in $]0,+\infty[$, $\beta(s) = 0$ if $s \leq 0$ and $B(y) = \int_0^{\frac{y}{\lambda^2}} \frac{\lambda^2 dx}{\beta(x)}$.

Then, if one assumes that $\int_0^1 \frac{dx}{\beta(x)} < +\infty$, $y = B^{-1}$ is a global classical solution to the ordinary differential equation

$$y' = \beta\left(\frac{y}{\lambda^2}\right) \quad \text{in } R \quad \rightrightarrows \approx \quad y(0) = 0 \tag{8}$$

such that $y(\xi) > 0$ for any positive ξ.

Remark 1. Note that the hypothesis $\int_0^1 \frac{dx}{\beta(x)} < +\infty$ is really observed in practice when $a(x) = \min(kx^\alpha, 1)$ in R^+ for any positive α and k.

Then, for any positive λ, $u(t,x) = y(t+\lambda x)$ is a travelling-wave solution to (7) for the initial datum given by $u_0(x) = y(\lambda x)$ and the boundary conditions

$$u(t,-1) = y(t-\lambda) \quad \text{and} \quad u(t,1) = y(t+\lambda).$$

At last, note that $u_0(x) = 0$ in $]-1,0]$ and that u possesses the property of finite speed of propagation (from nonzero disturbances) in the following sense:

$$u(x,t) = 0,\ \frac{t}{\lambda} \leq -x < 1.$$

3 Definition of a solution and existence results for a discretized problem

Definition 1. *For any u_0 in $H_0^1(\Omega) \cap L^\infty(\Omega)$, a solution to the Cauchy-Dirichlet problem (2-4-6) is a pair (u,λ) in $[H^1(Q) \cap L^\infty(0,T;H_0^1(\Omega))] \times L^\infty(Q)$ such that:*

$$\lambda \in H(\partial_t u + E),\quad u(t=0) = u_0 \text{ in } \Omega,\quad \partial_t u + E \geq 0 \text{ in } Q, \tag{9}$$

$$\forall v \in H_0^1(\Omega),\ \int_\Omega \{\partial_t uv + \lambda \nabla u \nabla v\}\, dx = 0 \quad a.e.\ t \in]0,T[. \tag{10}$$

Remark 2. Assume that $E = 0$, $u_0 \leq 0$ with $\Delta u_0 \geq 0$ in $D'(\Omega)$. Then:
on the one hand, the pair $(u_0, 0)$ is a solution to (9)–(10),
on the other hand, a second solution is given by $(\theta, 1)$, where θ is the solution in $H^1(Q) \cap L^\infty(0,T;H_0^1(\Omega))$ of the heat equation with initial datum u_0.
Thus, the solution to the problem is not unique and in the meaning of the entropic solution, a physically relevant solution would be the one given by λ as close as possible to 1; i.e. the diffusive flux has to be the less possible corrected. Then, in the above definition of a solution, one is looking for a maximal λ in the sense: if (w,μ) is another solution then $\mu \leq \lambda$; mathematical result difficult to obtain.

3.1 The sedimentation case and some surprising behaviours

Let us consider in this subsection the sedimentation process, i.e. $E = 0$, and give some qualitative properties of the solutions. In particular, for understanding the total degeneracy of an equation that seems to be parabolic.

Proposition 1. *Assume that (u, λ) is any solution to (9)–(10). Then, $\lambda \nabla u^+ = 0$ a.e. in Q and for any t, $u^+(t, .) = u_0^+$ a.e. in Ω.*

Proof. It is proved by using the test-function $v = u^+$ and since $\partial_t u \geq 0$. ∎

Note that, if $u_0 \geq 0$ in Ω, for any t, $u^+(t, .) = u_0$ a.e. in Ω. Thus, for any solution (u, λ) to (9)–(10), one gets, $u(t, .) = u_0$ a.e. in Ω. Moreover, if $\nabla u_0 \neq 0$ in Ω, the problem has to degenerate by taking null values for λ.

Corollary 1. *(Barrier effect and dead-zone) Assume that there exists a compact set K and an open set ω with $K \subset \omega \subset \Omega$ and $\omega \backslash K \subset \{u_0 \geq 0\}$; then, for any t, $u(t, .) = u_0$ in ω.*

Proof. By considering any v in $H_0^1(\Omega)$ such that $1_K \leq v \leq 1_\omega$, one gets that $\int_K \partial_t u \, dx \leq 0$ and $\partial_t u = 0$ a.e. in ω. ∎

This expresses that any zone surrounded by a zone where $u_0 \geqslant 0$ is stationary in time. One may find in the last section some numerical illustrations of this total degeneracy.

3.2 An implicit time discretization method

A standard way to prove the existence of a solution is to consider an implicit time - discretization scheme. One proposes in this section the analysis of such a method, coupled with a technique of artificial viscosity and of regularisation of H. One invites the reader interested by the details of the demonstrations to consult G. Gagneux et al. [5].

In the sequel one assumes that $E \in L^2(0, T, H^1(\Omega))$. Let us consider two positive real parameters ε and $h = \frac{T}{N}$, where N is an integer whose vocation is to tend to infinity. Denotes by $E_k = \frac{1}{2h} \int_{](k-1)h, (k+1)h[} E(s, .) \, ds$ and by H_ε any lipschitzian-continuous function satisfying

$$\forall x \in R, \quad \max[\varepsilon, \min(\frac{(1-\varepsilon)x}{\varepsilon}+1, 1)] \leq H_\varepsilon(x) \leq \max[\varepsilon, \min(\frac{(1-\varepsilon)x}{\varepsilon}+\varepsilon, 1)].$$

At last, set $A_\varepsilon(x) = \int_0^x H_\varepsilon(\sigma) \, d\sigma$, $x \in R$.

Let us first focus our attention on the construction of iteration (u_1, λ_1).

Proposition 2. *For any u_0 in $H_0^1(\Omega) \cap L^\infty(\Omega)$ and any nonnegative E in $H^1(\Omega)$, there exists a unique u_ε in $H_0^1(\Omega)$ such that*

$$\forall v \in H_0^1(\Omega), \quad \int_\Omega \left\{ \frac{u_\varepsilon - u_0}{h} v + H_\varepsilon\left(\frac{u_\varepsilon - u_0}{h} + E \right) \nabla u_\varepsilon . \nabla v \right\} dx = 0. \quad (11)$$

Moreover, $\inf_\Omega \operatorname{ess} u_0 \leq u_\varepsilon \leq \sup_\Omega \operatorname{ess} u_0$.

Proof. The existence of a solution is obtained by using Schauder-Tykonov fixed point theorem in the framework of separable Hilbertian spaces and the uniqueness is proved by using a classical L^1 T - contraction method. ∎

By passing to limits with respect to ε, one gets that

Proposition 3. *For any u_0 in $H_0^1(\Omega) \cap L^\infty(\Omega)$ and any nonnegative E in $H^1(\Omega)$, there exists (u, λ) in $H_0^1(\Omega) \times L^\infty(\Omega)$ such that $\lambda \in H(\frac{u - u_0}{h} + E)$ and*

$$\forall v \in H_0^1(\Omega), \quad \int_\Omega \left\{ \frac{u - u_0}{h} v + \lambda \nabla u . \nabla v \right\} dx = 0. \quad (12)$$

Moreover, $\inf_\Omega \operatorname{ess} u_0 \leq u \leq \sup_\Omega \operatorname{ess} u_0$ *and* $u \geq u_0 - hE$ *a.e. in* Ω.

In order to prove this result, let us first give some *a priori* estimates.

Proof. Denote by $w_\varepsilon = \frac{u_\varepsilon - u_0}{h} + E$.
On the one hand, by using $v = w_\varepsilon - E$ as a test function, one gets that (w_ε) and $(A_\varepsilon(w_\varepsilon))$ are bounded generalised sequences respectively in $L^2(\Omega)$ and $H^1(\Omega)$.
Then, on the other hand, since $v = -(w_\varepsilon + \varepsilon)^- \in H_0^1(\Omega)$, one proves that $(w_\varepsilon)^-$ converges towards 0 in $L^2(\Omega)$ when ε goes to 0^+.
Then, sub-sequences can be extracted and passing to limits leads to the conclusion (see. G. Gagneux *et al.* [5]). ∎

Given that u_1 has got the same properties as u_0, by induction the following result holds:

Proposition 4. *There exists a sequence $(u^k, \lambda_k)_k$ in $H_0^1(\Omega) \times L^\infty(\Omega)$ such that $\lambda_k \in H(\frac{u^k - u^{k-1}}{h} + E_k)$, $u^0 = u_0$, $u^k \geq u^{k-1} - hE_k$ a.e. in Ω,*

$$\forall v \in H_0^1(\Omega), \quad \int_\Omega \left\{ \frac{u^k - u^{k-1}}{h} v + \lambda_k \nabla u^k . \nabla v \right\} dx = 0. \quad (13)$$

Moreover, $\inf_\Omega \operatorname{ess} u_0 \leq \inf_\Omega \operatorname{ess} u^{k-1} \leq u^k \leq \sup_\Omega \operatorname{ess} u^{k-1} \leq \sup_\Omega \operatorname{ess} u_0$.

3.3 About the existence of a solution

Let us give information concerning the passing to limits with respect to the time-step parameter h.

Thanks to the test-function $v = \dfrac{u^k - u^{k-1}}{h}$ and the discrete Gronwall lemma, one gets that,

Lemma 1. *Independently of h, for any integer n, one computes that*

$$\frac{2}{h}\sum_{k=1}^{n} ||u^k - u^{k-1}||^2_{L^2(\Omega)} + ||u^n||^2_{H^1_0(\Omega)} + \sum_{k=1}^{n} ||u^k - u^{k-1}||^2_{H^1_0(\Omega)} \leq C.$$

Therefore, if one denotes by

i) $\hat{u}_h(t,x) = \sum_{k=0}^{N} [\dfrac{u^k - u^{k-1}}{h}(t-kh) + u^{k-1}] 1_{[kh,(k+1)h]}$ where $u^{-1} = u^0$,

ii) $\lambda_h(t,x) = \sum_{k=0}^{N} \lambda^k 1_{[kh,(k+1)h[}$ and $E_h = \sum_{k=0}^{N} E_k I_{[kh,(k+1)h[}$,

the following result holds:

Proposition 5. *The sequence $(\hat{u}_h)$ is bounded in $H^1(Q) \cap L^\infty(0,T; H^1_0(\Omega))$. Thus, it is relatively compact in $C([0,T], L^2(\Omega))$.*
The sequence (λ_h) is bounded in $L^\infty(Q)$. Moreover,

$$\lambda_h \in H(\partial_t \hat{u}_h + E_h), \quad \partial_t \hat{u}_h + E_h \geq 0 \quad a.e. \text{ in } Q,$$

and for any v in $L^2(0,T, H^1_0(\Omega))$, one has the approximating equation of continuity:

$$\int_Q \{\partial_t \hat{u}_h v + \lambda^h \nabla \hat{u}_h . \nabla v\} \, dxdt = o(h). \tag{14}$$

On the one hand, each accumulation point provides a "mild solution" in the sense of Ph. Bénilan *et al.* [3]; on the other hand, the double weak convergence does not allow us to pass to limits in the diffusion term $\int_Q \lambda^h \nabla \hat{u}_h . \nabla v \, dxdt$. Therefore,

Proposition 6. *If one conjectures that λ_h may converge a.e. in Q to λ for a sub-sequence, then (u,λ) is a solution of the problem in the sense of the definition 1.*

Proof. This result comes from the Lebesgue's theorem, and an argument of positivity. ∎

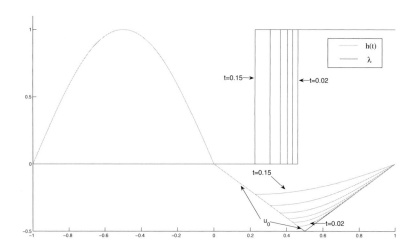

Fig. 1. Numerical simulation of case 1

4 Some numerical illustrations

Assume in this section that $u_0 \geq 0$ in $]-1,0]$, $u_0 \leq 0$ and convex in $[0,1[$ and let us consider some numerical illustrations of u_k, at different time steps k, obtained by a fixed point technique on the second order operator (see G. Gagneux et al. [5]).

i) The first simulation is obtained with $E = 0$.

In particular, one is able to see explicitly the changing type of the equation since one observes: a total degeneracy of the problem in $]-1,0[$ (a deadzone); a parabolic behaviour with infinite speed of propagation in $]0.5, 1[$; and a hyperbolic behaviour with finite speed of propagation and a front given by a free boundary in $]0, 0.5[$.

ii) The second illustration is obtained with a positive E.

5 Conclusion and open problems

In this paper, a new conservation law coming from geological problematic has been presented. Its general study remains still open. The solution of the problem presented in the definition 1 is not unique in general, but, besides the research of a maximal solution mentioned in Remark 2, an important point lies in the obtaining of a variational solution (*i.e.* a solution to (9-10)).

In this paper, an heuristic simplification of the real problem has been presented. One has now to consider the case of a realistic geological problem with relevant boundary conditions.

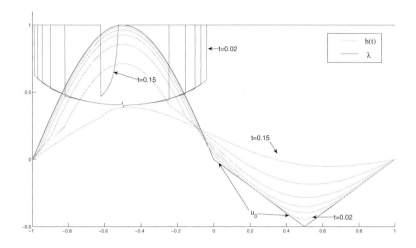

Fig. 2. Numerical simulation of case 2

At last, an other problem concerns the numerical simulation and analysis of this geological phenomenon in situations of practical importance. What kind of method one has to use when, on the one hand, the diffusive coefficient is a nonlinear function of the time-derivative of the unknown; on the other hand, one has a nonlinear type changing degenerated equation involving a maximal monotone graph?

Even if some improvement has been obtained in the $1-D$ case, a general procedure must be devoted to the construction of accurate schemes for approximating such non standard free-boundary problems.

References

1. Antontsev, S.N., Gagneux, G., Vallet, G.: Analyse mathématique d'un modèle d'asservissement stratigraphique. Approche gravitationnelle d'un processus de sédimentation sous une contrainte d'érosion maximale. Publication interne du Laboratoire de Mathématiques Appliquées UMR-CNRS 5142, n°2001/23, Pau (2001)
2. Antontsev, S.N., Gagneux, G., Vallet, G.: On some stratigraphic control problems, *Prikladnaya Mekhanika Tekhnicheskaja Fisika (Novosibirsk)*, **44(6)**, 85-94 (2003) (in russian) and *Journal of Applied Mechanics and Technical Physics (New York)*, **44(6)**, 821-828 (2003)
3. Bénilan, Ph., Crandall, M.G., Pazy, A.: Bonnes solutions d'un problème d'évolution semi-linéaire. C. R. Acad. Sci. Paris, Sér. I, **306**, 527-530 (1988)
4. Eymard, R., Gallouët, T., Granjeon, D., Masson, R., Tran, Q. H.: Multi-lithology stratigraphic model under maximum erosion rate constraint. Internat. J. Numer. Methods Eng., **60(2)**, 527-548 (2004)

5. Gagneux, G., Luce R., Vallet, G.: A non standard free-boundary problem arising from stratigraphy. Publication interne du Laboratoire de Mathématiques Appliquées UMR-CNRS 5142, n°04/33, Pau (2004)
6. Granjeon, D., Huy Tran, Q., Masson, R., Glowinski, R.: Modèle stratigraphique multilithologique sous contrainte de taux d'érosion maximum. Publication interne de l'Institut Français du Pétrole (2000)
7. Vallet, G.: Sur une loi de conservation issue de la géologie. C. R. Acad. Sci. Paris, Sér. I, **337**, 559-564 (2003)

Discontinuous Galerkin

Schwarz Domain Decomposition Preconditioners for Interior Penalty Approximations of Elliptic Problems

Paola F. Antonietti[1], Blanca Ayuso[2] and Luca Heltai[3]

[1] Dipartimento di Matematica, Università degli Studi di Pavia, via Ferrata 1, 27100 Pavia, Italy
 paola.antonietti@unipv.it
[2] Istituto di Matematica Applicata e Tecnologie Informatiche – CNR, Via Ferrata 1, 27100 Pavia, Italy
 blanca@imati.cnr.it
[3] Dipartimento di Matematica, Università degli Studi di Pavia, via Ferrata 1, 27100 Pavia, Italy
 luca.heltai@unipv.it

Summary. We present a two-level non-overlapping additive Schwarz method for Discontinuous Galerkin approximations of elliptic problems. In particular, a two level-method for both symmetric and non-symmetric schemes will be considered and some interesting features, which have no analog in the conforming case, will be discussed. Numerical experiments on non-matching grids will be presented.

1 Introduction

In the past twenty years extensive research has been done on developing domain decomposition (DD) methods for solving efficiently the large algebraic linear systems arising from various discretization of partial differential equations. Although the theory of DD techniques for finite elements (FE) methods (conforming, non conforming and mixed) is by now well understood (see, e.g., [9]), only a few results can be found in the literature for discontinuous Galerkin (DG) approximations (see [5, 7, 1]). Based on discontinuous FE spaces, DG methods have deserved a substantial attention due to their flexibility in handling meshes with hanging nodes and their high degree of locality.

In this paper we consider, for the case of the family of DG *Interior Penalty* (IP) approximations (including both the symmetric [2] and the non-symmetric schemes [8, 4]), the additive Schwarz method proposed in [1]. In particular, a non-symmetric additive Schwarz preconditioner for a diffusion problem, that was proposed in [1] for the very first time, will be considered.

An outline of the paper is as follows. In Sect. 2 we recall the DG approximations of a diffusion problem. In Sects. 3 and 4, we provide the construction and the analysis of the additive Schwarz method. Finally, in Sect. 5 we present some numerical results on meshes with hanging nodes.

2 Discontinuous Galerkin methods for elliptic problems

For the sake of simplicity we restrict ourselves to the model problem,

$$-\Delta u = f \quad \text{in } \Omega, \qquad u = 0 \quad \text{on } \partial\Omega, \tag{1}$$

where $\Omega \subset \mathbb{R}^d$, $d = 2, 3$ is assumed to be (a smooth domain or) a convex polygon or polyhedron and f a given function in $L^2(\Omega)$.

Let $\{\mathcal{T}_h, h > 0\}$ be a family of shape-regular and locally quasi-uniform partitions of the domain Ω made of d-simplices or parallelograms (if $d = 2$) or parallelepipeds (if $d = 3$), with possible hanging nodes. Denoting by h_T the diameter of the element $T \in \mathcal{T}_h$, we define the mesh size $h := \max_{T \in \mathcal{T}_h} \{h_T\}$.

An interior face (if $d = 2$, "face" means "edge") of $\mathcal{T}_h$ is the (non-empty) interior of $\partial T^+ \cap \partial T^-$, T^+ and T^- being two adjacent elements of $\mathcal{T}_h$, not necessarily matching. Similarly, a boundary face of $\mathcal{T}_h$ is the (non-empty) interior of $\partial T \cap \partial \Omega$, where T is a boundary element of $\mathcal{T}_h$. We denote by $\mathcal{E}^I$ and $\mathcal{E}^B$ the sets of all interior and boundary faces of $\mathcal{T}_h$, respectively, and set $\mathcal{E} = \mathcal{E}^I \cup \mathcal{E}^B$. We define the local mesh size $\mathbf{h}(\mathbf{x}) := \min\{h_{T^+}, h_{T^-}\}$, if $\mathbf{x} \in \partial T^+ \cap \partial T^-$, and $\mathbf{h}(\mathbf{x}) := h_T$ if $\mathbf{x} \in \partial T$ is on the boundary. Let $e \in \mathcal{E}^I$ be an interior face shared by two elements T^+ and T^- with outward normal unit vectors $\mathbf{n}^{\pm}$. We denote by $v^{\pm}$ and $\boldsymbol{\tau}^{\pm}$ the traces of piecewise smooth scalar-valued and vector-valued functions v and $\boldsymbol{\tau}$, respectively, taken from the interior of $\partial T^{\pm}$, and we define the following trace operators:

$$\{\!\{v\}\!\} := (v^+ + v^-)/2, \qquad \{\!\{\boldsymbol{\tau}\}\!\} := (\boldsymbol{\tau}^+ + \boldsymbol{\tau}^-)/2,$$
$$[\![v]\!] := v^+ \mathbf{n}^+ + v^- \mathbf{n}^-, \qquad [\![\boldsymbol{\tau}]\!] := \boldsymbol{\tau}^+ \cdot \mathbf{n}^+ + \boldsymbol{\tau}^- \cdot \mathbf{n}^-.$$

On $e \in \mathcal{E}^B$, we set $\{\!\{v\}\!\} := v$, $\{\!\{\boldsymbol{\tau}\}\!\} := \boldsymbol{\tau}$, $[\![v]\!] := v\mathbf{n}$ and $[\![\boldsymbol{\tau}]\!] := \boldsymbol{\tau}\cdot\mathbf{n}$. Finally, we define the discontinuous FE space $V_h := \{v \in L^2(\Omega) : v|_T \in \mathcal{M}^{\ell_h}(T), \forall T \in \mathcal{T}_h\}$, where $\mathcal{M}^{\ell_h}(T)$ is the space $\mathcal{P}^{\ell_h}(T)$ of polynomials of degree at most $\ell_h \geq 1$ on T, for T a d-simplex, and the space $\mathcal{Q}^{\ell_h}(T)$ of polynomials of degree at most ℓ_h in each variable on T, if T is a parallelogram or a parallelepiped.

The family of *Interior Penalty* (IP) approximations for problem (1) reads: Find $u \in V_h$ such that $\mathcal{A}_h(u, v) = \int_\Omega fv$, $\forall v \in V_h$, where

$$\mathcal{A}_h(u, v) := \sum_{T \in \mathcal{T}_h} \int_T \nabla u \cdot \nabla v - \sum_{e \in \mathcal{E}} \int_e \{\!\{\nabla u\}\!\} \cdot [\![v]\!]$$

$$-(1-\gamma)\sum_{e\in\mathscr{E}}\int_e [\![u]\!]\cdot\{\!\{\nabla v\}\!\} + \sum_{e\in\mathscr{E}}\int_e \alpha_e h^{-1} [\![u]\!]\cdot [\![v]\!] \ . \quad (2)$$

where for $\gamma = 0, 1$ or 2 we obtain, respectively, the symmetric interior penalty (SIP) method [2], the incomplete interior penalty (IIP) method [4] and the non-symmetric interior penalty (NIP) method [8]. The penalty parameter $\alpha_e > 0$ is independent of h and, for the first two methods, it should be taken large enough to guarantee the coerciveness of $\mathcal{A}_h$ (see [3] for further details).

We shall denote by $a_h(\cdot, \cdot)$ the symmetric part of the bilinear form[4] $\mathcal{A}_h(\cdot, \cdot)$, i.e., $a_h(u,v) = (\mathcal{A}_h(u,v) + \mathcal{A}_h(v,u))/2$.

3 Non-overlapping Schwarz methods

In this section, we present our two-level algorithm for the family of the IP methods. Let $\mathcal{T}_{N_S}$ be the family of partitions of Ω into N_s non-overlapping subdomains $\Omega = \bigcup_{i=1}^{N_s} \Omega_i$ and let $\{\mathcal{T}_H, H > 0\}$ and $\{\mathcal{T}_h, h > 0\}$ be the families of coarse and fine partitions, respectively, with mesh sizes H and h. All partitions are assumed to be shape-regular and locally quasi-uniform and we further assume that they are related by $\mathcal{T}_{N_s} \subseteq \mathcal{T}_H \subseteq \mathcal{T}_h$.

For each subdomain Ω_i of $\mathcal{T}_{N_s}$ we denote by $\mathscr{E}_i$ the set of all faces of $\mathscr{E}$ belonging to $\overline{\Omega}_i$, and by $\Gamma := \bigcup_{i=1}^{N_s} \Gamma_i$ where $\Gamma_i := \{e \in \mathscr{E}_i : e \in \partial\Omega_i \setminus \partial\Omega\}$. For $i = 1, \dots, N_s$, we define the *local spaces* $V_h^i := \{v \in V_h : v \equiv 0 \text{ in } \Omega \setminus \overline{\Omega}_i\}$, and the *prolongation* operators[5]: $R_i^T : V_h^i \longrightarrow V_h$, $\mathcal{D}_i^T : \nabla V_h^i \longrightarrow \nabla V_h$. Both operators set to zero the degrees of freedom outside $\overline{\Omega}_i$ while on $e \in \Gamma_i$ their action is defined in the following way:

$$R_i^T v_i := \begin{cases} (R_i^T v_i)^+ = v_i \ , \\ (R_i^T v_i)^- = 0 \ , \end{cases} \qquad \mathcal{D}_i^T \tau_i := \begin{cases} (\mathcal{D}_i^T \tau_i)^+ = \tau_i \ , \\ (\mathcal{D}_i^T \tau_i)^- = \tau_i \ , \end{cases} \quad (3)$$

where we denoted by $(\cdot)^\pm$ the traces from the interior of the elements $T^\pm$ sharing the face[6] e. From (3) it follows that $[\![R_i^T v_i]\!] = v_i \mathbf{n}_i$ and $\{\!\{\mathcal{D}_i^T \tau_i\}\!\} = \tau_i$ on $e \in \Gamma_i$. Notice also that $V_h = \oplus_{i=1}^{N_s} R_i^T V_h^i$. The restriction operators R_i, $\mathcal{D}_i$ are defined as the transpose of R_i^T and $\mathcal{D}_i^T$ with respect to the Euclidean scalar product. For each $i = 1, \dots, N_s$, we define the *local-solvers* by considering the IP approximation to the problem: $-\Delta u_i = R_i f$ on Ω_i, $u_i = 0$ on $\partial\Omega_i$. Thus, in view of (2), the bilinear forms of the local solvers are defined by:

$$\mathcal{A}_i(u_i, v_i) := \sum_{T \in \Omega_i} \int_T \nabla u_i \cdot \nabla v_i - \sum_{e \in \mathscr{E}_i} \int_e \{\!\{\nabla u_i\}\!\} \cdot [\![v_i]\!]$$

---

[4] Obviously, for the SIP method, one has $a_h(u,v) = \mathcal{A}_h(u,v)$.
[5] With a small abuse of notation, we shall also denote by ∇v_h the elementwise gradient of $v_h \in V_h$.
[6] Taking into account (3), it follows that $\mathcal{D}_i^T \nabla v_i \neq \nabla R_i^T v_i, \forall v_i \in V_h^i, i = 1, \dots, N_s$.

$$- (1-\gamma) \sum_{e \in \mathcal{E}_i} \int_e [\![u_i]\!] \cdot \{\!\{\nabla v_i\}\!\} + \sum_{e \in \mathcal{E}_i} \int_e \alpha_e \mathrm{h}^{-1} [\![u_i]\!] \cdot [\![v_i]\!] \ , \quad (4)$$

for $u_i, v_i \in V_h^i$. Moreover, taking into account the definition of R_i^T and $\mathcal{D}_i^T$ acting on scalar- and vector-valued functions, respectively, it follows (see [1])

$$\mathcal{A}_i(u_i, v_i) = \mathcal{A}_h(R_i^T u_i, R_i^T v_i) \ , \quad \forall u_i, v_i \in V_h^i, \quad \forall i = 1, \ldots, N_s \ . \quad (5)$$

The last step is the construction of the *coarse solver*. The coarse space is defined as $V_H = V_h^0 := \{v_H \in L^2(\Omega) : v_H|_T \in \mathcal{M}^{\ell_H}(T), \forall T \in \mathcal{T}_H\}$, with $0 \leq \ell_H \leq \ell_h$. The prolongation operators $R_0^T : V_h^0 \longrightarrow V_h$, $\mathcal{D}_i^T : \nabla V_h^0 \longrightarrow \nabla V_h$ are defined as before[7]. We define the coarse solver $\mathcal{A}_0 : V_h^0 \times V_h^0 \longrightarrow \mathbb{R}$ as the restriction of $\mathcal{A}_h$ to $V_h^0 \times V_h^0$, i.e., $\mathcal{A}_0(u_0, v_0) = \mathcal{A}_h(R_0^T u_0, R_0^T v_0), \forall u_0, v_0 \in V_h^0$.

Algebraic Formulation and Projection Operators

We denote by $\mathbb{A}_h$, $\mathbb{A}_i$ and $\mathbb{A}_0$ the stiffness matrices associated with the global, local and coarse bilinear forms $\mathcal{A}_h$, $\mathcal{A}_i$ and $\mathcal{A}_0$, respectively. The additive Schwarz preconditioner is defined as $\mathbb{B} := \sum_{i=0}^{N_s} R_i^T \mathbb{A}_i^{-1} R_i$. For $i = 0, \ldots, N_s$, we define the $\mathcal{A}_h$-projection like operators: $P_i : V_h \longrightarrow R_i^T V_h^i \subset V_h$, by $\mathcal{A}_h(P_i u, R_i^T v_i) = \mathcal{A}_h(u, R_i^T v_i), \forall v_i \in V_h^i$. Notice that, for all $i = 0, \ldots, N_s$, $P_i = [R_i^T \mathbb{A}_i^{-1} R_i] \mathbb{A}_h$ are well-defined since both the local $\mathcal{A}_i$ and coarse $\mathcal{A}_0$ bilinear forms are coercive. Moreover, the preconditioned matrix $\mathbb{B}\mathbb{A}_h$ is equal to the Schwarz operator:

$$P_{\mathrm{ad}} := \sum_{i=0}^{N_s} P_i = \sum_{i=0}^{N_s} [R_i^T \mathbb{A}_i^{-1} R_i] \mathbb{A}_h \ . \quad (6)$$

The additive Schwarz method consists in replacing the original discrete problem $\mathcal{A}_h u = f$ by the preconditioned system $P_{\mathrm{ad}} u = g$ with $g = \sum_{i=0}^{N_s} g_i$, $g_i = P_i g$ being the solution of $\mathcal{A}_i(g_i, v) = (f, v), \forall v \in V_h^i, i = 0, \ldots, N_s$. This last system is solved by means of a suitable iterative method.

On the one hand, for the SIP method, P_{ad} is symmetric and we use the Conjugate Gradient (CG) method, for which the following upper bound on the error reduction property at the k-th iteration, is known (see [6]):

$$\|\mathbf{e}_k\|_2 \leq 2 \|\mathbf{e}_0\|_2 \left(\frac{\sqrt{\kappa(P_{\mathrm{ad}})} - 1}{\sqrt{\kappa(P_{\mathrm{ad}})} + 1} \right)^k , \quad (7)$$

where $\mathbf{e}_k = \mathbf{u}_k - \mathbf{u}$ is the error, $\mathbf{u}$ being the exact solution and $\|\cdot\|_2$ the standard Euclidean norm. By $\kappa(P_{\mathrm{ad}})$ we denote the condition number of P_{ad}. On the other hand, the lack of symmetry of the NIP and IIP methods, implies that their Schwarz operators P_{ad}, as defined in (6), fail to be self-adjoint

---

[7] Notice that, in this case, they coincide with the natural injection operators.

w.r.t. $\mathcal{A}_h$. Hence, to solve the resulting preconditioned system, we use the Generalized Minimal Residual (GMRES) method, for which one has

$$\|\mathbf{r}_k\|_2 \leq \left(1 - \frac{c_p^2}{C_p^2}\right)^{k/2} \|\mathbf{r}_0\|_2 \; .$$

where $\mathbf{r}_k := g - P_{\mathrm{ad}} u_k$ is the residual at the k-th iterate and c_p and C_p are:

$$c_p(P_{\mathrm{ad}}) := \inf_{u \neq 0} \frac{a_h(u, P_{\mathrm{ad}} u)}{a_h(u,u)} \;, \quad C_p(P_{\mathrm{ad}}) := \sup_{u \neq 0} \frac{a_h(P_{\mathrm{ad}} u, P_{\mathrm{ad}} u)}{a_h(u,u)} \;, \quad (8)$$

We stress that $c_p(P_{\mathrm{ad}}) > 0$ is a necessary and sufficient condition for convergence of GMRES in a finite number of iterations. See [6] for further details.

4 Convergence analysis

The following two results, which have been proved in [1], provide the convergence for the considered two-level Schwarz method. The former one (Theorem 1) concern the SIP method and provides an estimate on the condition number of the preconditioned system. Consequently, in view of (7), the convergence rates of the corresponding preconditioned iterative solver are fully determined. The latter one (Theorem 2) provides lower and upper bounds for the quantities c_p and C_p defined in (8) and applies to the Schwarz preconditioner for the non-symmetric NIP and IIP methods. In both results, N_c denotes the maximum number of adjacent subdomains, and C a positive constant depending only on the shape regularity of $\mathcal{T}_h$ and the polynomial degrees ℓ_h and ℓ_H.

Theorem 1. *Set $\gamma = 0$ in (2) and let $\mathcal{A}_h$ be the bilinear form of the SIP method. Let P_{ad} be its additive Schwarz operator as defined in (6). Then,*

$$\kappa(P_{\mathrm{ad}}) \leq C[2 + N_c] H h^{-1} \; . \qquad (9)$$

For the proof, we refer to [1] where the result is shown for all the symmetric DG methods for elliptic problems present in the literature.

Theorem 2. *Let P_{ad} be the additive Schwarz method for the non-symmetric NIP ($\gamma = 2$) and IIP ($\gamma = 1$) methods. Then, there exist $C_0^2 = O(H/h)$ and $C_{min} > 0$ such that if $\alpha_* = \min_{e \in \mathscr{E}} \alpha_e \geq C_2 H h^{-1}$ with $C_2 > C_{min}$, then:*

$$C C_0^{-2} a_h(u,u) \leq a_h(u, P_{\mathrm{ad}} u) \;, \qquad a_h(P_{\mathrm{ad}} u, P_{\mathrm{ad}} u) \leq C(N_c + 1)^2 a_h(u,u) \;.$$

In [1], the proof is accomplished under a technical assumption on the penalty parameter α_*. Nevertheless, as it will be shown in the numerical experiments such a restriction is not required in practice (see [1] for further details).

5 Numerical results

We present some numerical experiments to illustrate the performance of the proposed non-overlapping Schwarz method on non-matching meshes (extensive numerical experiments on matching grids are contained in [1]). We take $\Omega = (0, \pi)^2$ and we choose f such that the exact solution of the model problem (1) is given by $u(x,y) = \sin(x)\sin(y)$. The subdomain partitions consist of N_s squares, $N_s = 4, 16$ (see Fig. 1 for $N_s = 4$). The initial coarse and fine non-matching Cartesian grids are depicted in Fig. 1, where we have denoted by H_0 and h_0 the corresponding mesh sizes, respectively. We consider n successive global uniform refinements of these initial grids so that the resulting mesh sizes are $H_n = H_0/2^n$ and $h_n = h_0/2^n$, respectively, with $n = 1, 2, 3$. For all the tests (except the last one), we set the penalty parameter $\alpha_e = \alpha = 10$ $\forall e \in \mathcal{E}$. The iterative solvers used are the CG method for the symmetric scheme SIP, and the GMRES method for the non-symmetric NIP and IIP methods. The tolerance is set to 10^{-6} and we allow for a maximum of 200 iterations (for the non-preconditioned ones we admit at most 1000 iterations). All computations have been performed in MATLAB.

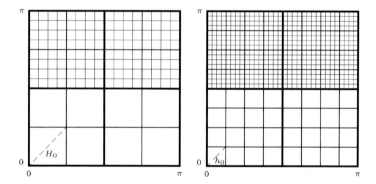

Fig. 1. Subdomain partition ($N_s = 4$) with the initial coarse (left) and fine (right) meshes.

We first address the scalability of the proposed Schwarz method, that is the independence of the convergence rate of the number of subdomains. In Tables 1(a) and 1(c) we report the condition number estimates for the SIP method on the two different subdomain partitions ($N_s = 4, 16$) by using piecewise bilinear polynomials both for the fine and coarse mesh spaces ($\ell_h = \ell_H = 1$). The corresponding iteration counts are given in Tables 1(b) and 1(d), respectively. The dashes indicate that $\mathcal{T}_H \not\subset \mathcal{T}_h$ and therefore it is meaningless to build the preconditioner. As predicted by Theorem 1, our preconditioner seems to be substantially insensitive on the number of the subdomains, and the convergence rates are clearly achieved. Notice that, if we refine both the

Table 1. SIP method: $\ell_h = 1$, $\ell_H = 1$, $\alpha = 10$.

(a) Condition Number: $N_s = 4$.

	h_0	$h_0/2$	$h_0/4$	$h_0/8$
H_0	31.4	65.9	137.4	278.8
$H_0/2$	6.3	32.5	67.4	139.1
$H_0/4$	-	6.4	32.0	67.1
$H_0/8$	-	-	6.5	32.0
$\kappa(\mathbb{A}_h)$	4.3e3	1.7e4	7.0e4	2.8e5

(b) Iteration Counts: $N_s = 4$.

	h_0	$h_0/2$	$h_0/4$	$h_0/8$
H_0	28	44	76	112
$H_0/2$	15	33	52	82
$H_0/4$	-	16	36	54
$H_0/8$	-	-	16	34
#iter($\mathbb{A}_h$)	128	226	442	877

(c) Condition Number: $N_s = 16$.

	h_0	$h_0/2$	$h_0/4$	$h_0/8$
H_0	29.3	65.5	139.6	285.2
$H_0/2$	6.1	31.5	65.2	135.0
$H_0/4$	-	6.4	31.3	66.0
$H_0/8$	-	-	6.4	31.8
$\kappa(\mathbb{A}_h)$	4.3e3	1.7e4	7.0e4	2.8e5

(d) Iteration Counts: $N_s = 16$.

	h_0	$h_0/2$	$h_0/4$	$h_0/8$
H_0	29	48	75	117
$H_0/2$	15	33	52	81
$H_0/4$	-	16	35	54
$H_0/8$	-	-	16	35
#iter($\mathbb{A}_h$)	128	226	442	877

Table 2. SIP method: $\ell_h = 2$, $\alpha = 10$.

(a) Condition Number: $\ell_H = 2$.

	h_0	$h_0/2$	$h_0/4$	$h_0/8$
H_0	78.0	172.2	350.9	705.1
$H_0/2$	6.4	81.1	175.5	356.6
$H_0/4$	-	6.4	82.8	177.6
$H_0/8$	-	-	6.5	83.6
$\kappa(\mathbb{A}_h)$	3.4e4	1.4e5	5.5e5	2.2e6

(b) Condition Number: $\ell_H = 1$.

	h_0	$h_0/2$	$h_0/4$	$h_0/8$
H_0	135.2	284.7	574.0	1152.2
$H_0/2$	61.9	130.7	272.3	548.1
$H_0/4$	-	62.0	133.5	278.1
$H_0/8$	-	-	63.1	133.9
$\kappa(\mathbb{A}_h)$	3.4e4	1.4e5	5.5e5	2.2e6

fine and the coarse meshes keeping the ratio H/h constant, we observe that both the condition numbers and the iteration counts remain substantially unchanged. In Tables 2(a) and 2(b) we have reported the iteration counts for the SIP method with $\ell_h = \ell_H = 2$ (piecewise biquadratic polynomials for both the fine and the coarse mesh spaces), and with $\ell_h = 2$ and $\ell_H = 1$, respectively. Notice that, in both cases, the convergence rates predicted by Theorem 1 are clearly achieved, although, it can be observed that, by choosing $\ell_h = \ell_H = 2$ our preconditioner performs better that with $\ell_h = 2$ and $\ell_H = 1$. Now, we address the scalability of the preconditioner for the non-symmetric IIP method. The theoretical estimates given in Theorem 2 can be clearly observed from Tables 3(a) and 3(b), where the iteration counts on two subdomain partitions $N_s = 4$ and $N_s = 16$, respectively, are reported. The crosses (also in

Table 3. IIP method: $\ell_h = 1$, $\ell_H = 1$, $\alpha = 10$.

(a) Iteration Counts: $N_s = 4$.

	h_0	$h_0/2$	$h_0/4$	$h_0/8$
H_0	26	43	69	108
$H_0/2$	14	33	52	80
$H_0/4$	-	16	35	52
$H_0/8$	-	-	15	34
#iter($\mathbb{A}_h$)	117	207	389	x

(b) Iteration Counts: $N_s = 16$.

	h_0	$h_0/2$	$h_0/4$	$h_0/8$
H_0	30	45	70	107
$H_0/2$	15	33	52	78
$H_0/4$	-	16	35	52
$H_0/8$	-	-	16	35
#iter($\mathbb{A}_h$)	117	207	389	x

Table 4. NIP method: $\ell_h = 1$, $\ell_H = 1$, $N_s = 16$.

(a) $\alpha = 10$.

	h_0	$h_0/2$	$h_0/4$	$h_0/8$
H_0	30	46	70	106*
$H_0/2$	15	33	52	78
$H_0/4$	-	15	34	51
$H_0/8$	-	-	15	34
#iter($\mathbb{A}_h$)	122	210	388	x

(b) $\alpha = 2$.

	h_0	$h_0/2$	$h_0/4$	$h_0/8$
H_0	18	25	35	49
$H_0/2$	13	20	26	35
$H_0/4$	-	15	20	26
$H_0/8$	-	-	15	20
#iter($\mathbb{A}_h$)	65	107	198	x

Table 4, below) indicate that we were not able to solve the non-preconditioned system due to the excessive GMRES memory storage requirements. In Table 4 the iteration counts for the NIP method with $\ell_h = \ell_H = 1$ and $\alpha = 10$ (left) and $\alpha = 2$ (right) are given; the choice $\alpha = 2$ (and also the starred value for $\alpha = 10$ with $H = H_0$ and $h = h_0/8$) actually violates the assumption in Theorem 2, but it can be clearly seen that, also in this case, the optimal convergence rates are achieved.

References

1. Antonietti, P.F., Ayuso, B.: Schwarz domain decomposition preconditioners for discontinuous Galerkin approximations of elliptic problems: non-overlapping case. Tech. Rep. IMATI-CNR, PV-20 (2005). Submitted
2. Arnold, D.N.: An interior penalty finite element method with discontinuous elements. SINUM, **19**, 742–760 (1982)
3. Arnold, D.N., Brezzi, F., Cockburn, B., Marini, L.D.: Unified analysis of discontinuous Galerkin methods for elliptic problems. Comput. Methods Appl. Mech. Engrg., **193**, 2565–2580 (2004)
4. Dawson, C., Sun, S., Wheeler, M.F.: Compatible algorithms for coupled flow and transport. Comput. Methods Appl. Mech. Engrg. , **193**, 2565–2580 (2004)

5. Feng, X., Karakashian, K.: Two-level additive Schwarz methods for a discontinuous Galerkin approximation of second order elliptic problems. SIAM J. Numer. Anal., **39**, 1343–1365 (2001)
6. Golub, G., Van Loan, Ch.: Matrix computations. Third edition. Johns Hopkins Studies in the Mathematical Sciences, Baltimore, MD (1996)
7. Lasser, C., Toselli, A.: An overlapping domain decomposition preconditioner for a class of discontinuous Galerkin approximations of advection-diffusion problems. Math. Comp., **72**, 1215–1238 (2003)
8. Rivière, B., Wheeler, M.F., Girault, V.: Improved energy estimates for interior penalty, constrained and discontinuous Galerkin methods for elliptic problems, Part. Computational Geosciences, **3**, 337–360 (1999)
9. Toselli, A., Widlund, O.: Domain decomposition methods—algorithms and theory. Springer Series in Computational Mathematics, **34** (2005)

Higher Order Semi-Implicit Discontinuous Galerkin Finite Element Schemes for Nonlinear Convection-Diffusion Problems*

Vít Dolejší

Charles University Prague, Faculty of Mathematics and Physics, Sokolovská 83, 186 75 Prague, Czech Republic
dolejsi@karlin.mff.cuni.cz

Summary. We deal with the numerical solution of a scalar nonstationary nonlinear convection-diffusion equation. We present a scheme which uses a discontinuous Galerkin finite element method for a space semi-discretization and the resulting system of ordinary differential equations is discretized by backward difference formulae. The linear terms are treated implicitly whereas the nonlinear ones by a higher order explicit extrapolation which preserves the accuracy of the schemes and leads to a system of linear algebraic equations at each time step. Thenumerical examples presented verify expected orders of convergence.

1 Introduction

Our aim is to developed a sufficiently efficient, robust and accurate numerical scheme for simulation of unsteady viscous compressible flow which is described by the system of Navier–Stokes equations. During the last years, the so-called discontinuous Galerkin method (DGM) became very popular for the solution of the Navier–Stokes equations, see e.g., [2, 3, 10]. DGM is based on a piecewise polynomial but discontinuous approximation where the interelement continuity is replaced by additional stabilization terms.

For time-dependent problems, it is possible to use a discontinuous approximation also for the time discretization (see [11]) but the most standard approach is the method of lines. In this case, Runge-Kutta methods are very popular due to their simplicity and a high order of accuracy, see [2, 4, 5], but their drawback is a strong restriction to the choice of the time step. To avoid this disadvantage it is convenient to use an implicit time discretization. Fully implicit schemes lead to a need to solve a nonlinear system of algebraic

---

* This work is a part of the research project MSM 0021620839 financed by the Ministry of Education of the Czech Republic and it was partly supported by the Grant No. 201/05/0005 of the Czech Grant Agency.

equations in each time step, which is rather expensive. Therefore, we proposed in [6] a semi-implicit method for a scalar convection-diffusion equation where the backward and forward Euler methods were applied to the linear and nonlinear terms, respectively. This scheme was analysed in [7] and a priori error estimates of order $O(h^p + \tau)$ in the L^2-norm and the H^1-seminorm were derived. Here h and τ denote the space and time steps, respectively, and p is the degree of polynomial approximation in space.

In this paper we introduce a generalization of the semi-implicit scheme from [7] with a nth-order ($n \geq 1$) time discretization. The formal order of accuracy of this scheme is $O(h^p + \tau^n)$. In Section 2 we state the definition of the method and in Section 3 we investigate the numerical orders of convergence with respect to τ and h. In Section 4 we give a short conclusion. The numerical analysis of these schemes and an extension to the system of the Navier–Stokes equations will be the subject of forthcoming papers.

2 Scalar equation

2.1 Continuous problem

Let us consider the following nonstationary nonlinear convection-diffusion problem: Find $u : Q_T = \Omega \times (0,T) \to \mathbb{R}$ such that

$$\text{a)} \quad \frac{\partial u}{\partial t} + \sum_{s=1}^{d} \frac{\partial f_s(u)}{\partial x_s} = \varepsilon \Delta u + g \quad \text{in } Q_T, \tag{1}$$

$$\text{b)} \quad u|_{\partial \Omega \times (0,T)} = u_D,$$

$$\text{c)} \quad u(x,0) = u^0(x), \quad x \in \Omega.$$

We assume that $\Omega \subset \mathbb{R}^d$, $d = 2, 3$, is a bounded polygonal (if $d = 2$) or polyhedral (if $d = 3$) domain with Lipschitz-continuous boundary $\partial \Omega$ and $T > 0$. The diffusion coefficient $\varepsilon > 0$ is a given constant, $g : Q_T \to \mathbb{R}$, $u_D : \Gamma_D \times (0,T) \to \mathbb{R}$, and $u^0 : \Omega \to \mathbb{R}$ are given functions, $f_s \in C^1(\mathbb{R})$, $s = 1, \ldots, d$, are prescribed inviscid fluxes.

2.2 Space discretization

Let $\mathcal{T}_h$ ($h > 0$) denote a triangulation of the closure $\overline{\Omega}$ of the domain Ω into a finite number of closed triangles (if $d = 2$) or tetrahedra (if $d = 3$) K with mutually disjoint interiors. In [8] we analysed the use of more general even nonconvex elements.

We set $h = \max_{K \in \mathcal{T}_h} \text{diam}(K)$. All elements of $\mathcal{T}_h$ will be numbered so that $\mathcal{T}_h = \{K_i\}_{i \in I}$, where I is a suitable index set. If two elements K_i, $K_j \in \mathcal{T}_h$ contain a nonempty open part of their faces, we call them *neighbours*. In this case we put $\Gamma_{ij} = \Gamma_{ji} = \partial K_i \cap \partial K_j$. For $i \in I$ we set

$s(i) = \{j \in I; K_j \text{ is a neighbour of } K_i\}$. The boundary $\partial\Omega$ is formed by a finite number of faces of elements K_i adjacent to $\partial\Omega$. We denote all these boundary faces by S_j, where $j \in I_b$ is a suitable index set and put $\gamma(i) = \{j \in I_b; S_j \text{ is a face of } K_i\}$, $\Gamma_{ij} = S_j$ for $K_i \in \mathcal{T}_h$ such that $S_j \subset \partial K_i$, $j \in I_b$. For K_i not containing any boundary face S_j we put $\gamma(i) = \emptyset$. Moreover we put $S(i) = s(i) \cup \gamma(i)$ and $\boldsymbol{n}_{ij} = ((n_{ij})_1, \ldots, (n_{ij})_d)$ is the unit outer normal to ∂K_i on the face Γ_{ij}.

Over the triangulation $\mathcal{T}_h$ we define the *broken Sobolev space*

$$H^k(\Omega, \mathcal{T}_h) = \{v; v|_K \in H^k(K) \ \forall K \in \mathcal{T}_h\}, \tag{2}$$

where $H^k(K) = W^{k,2}(K)$ denotes the (classical) Sobolev space on element K. For $v \in H^1(\Omega, \mathcal{T}_h)$ we set

$$v|_{\Gamma_{ij}} = \text{trace of } v|_{K_i} \text{ on } \Gamma_{ij}, \quad v|_{\Gamma_{ji}} = \text{trace of } v|_{K_j} \text{ on } \Gamma_{ji}, \tag{3}$$

$$\langle v \rangle_{\Gamma_{ij}} = \frac{1}{2}\left(v|_{\Gamma_{ij}} + v|_{\Gamma_{ji}}\right) \quad \text{and} \quad [v]_{\Gamma_{ij}} = v|_{\Gamma_{ij}} - v|_{\Gamma_{ji}},$$

denoting the *traces, average* and *jump of the traces* of v on $\Gamma_{ij} = \Gamma_{ji}$, respectively.

2.3 Space semidiscretization

We use the so-called nonsymmetric interior penalty Galerkin method (NIPG) which does not give an optimal a priori order of convergence in the L^2-norm but its advantage is a coercivity property for any positive penalty coefficient σ, see [1, 4]. This is important for the case of the Navier–Stokes equations when numerical analysis is impossible and the choice of σ is rather heuristic. A detailed definition of NIPG can be found, e.g., in [6, 8] so we present here only the definition of an approximate solution. For $u, v \in H^2(\Omega, \mathcal{T}_h)$, $u \in L^\infty(\Omega)$ we define the forms

$$a_h(u, \varphi) = \varepsilon \sum_{i \in I} \Bigg\{ \int_{K_i} \nabla u \cdot \nabla \varphi \, \mathrm{d}x \tag{4}$$
$$- \sum_{\substack{j \in s(i) \\ j < i}} \int_{\Gamma_{ij}} \left(\langle \nabla u \rangle \cdot \boldsymbol{n}_{ij}[\varphi] - \langle \nabla \varphi \rangle \cdot \boldsymbol{n}_{ij}[u]\right) \mathrm{d}S$$
$$- \sum_{j \in \gamma(i)} \int_{\Gamma_{ij}} \left(\nabla u \cdot \boldsymbol{n}_{ij}\, \varphi \, \mathrm{d}S - \nabla \varphi \cdot \boldsymbol{n}_{ij}\, u\right) \mathrm{d}S \Bigg\},$$

$$b_h(u, \varphi) = \sum_{i \in I} \Bigg\{ \sum_{j \in S(i)} \int_{\Gamma_{ij}} H\left(u|_{\Gamma_{ij}}, u|_{\Gamma_{ji}}, \boldsymbol{n}_{ij}\right) \varphi|_{\Gamma_{ij}} \, \mathrm{d}S$$
$$- \int_{K_i} \sum_{s=1}^{d} f_s(u) \frac{\partial \varphi}{\partial x_s} \, \mathrm{d}x \Bigg\},$$

$$J_h^\sigma(u,\varphi) = \sum_{i \in I}\left\{\sum_{\substack{j \in s(i) \\ j < i}} \int_{\Gamma_{ij}} \sigma[u]\,[\varphi]\,\mathrm{d}S + \sum_{j \in \gamma(i)} \int_{\Gamma_{ij}} \sigma\,u\,\varphi\,\mathrm{d}S\right\},$$

$$\ell_h(\varphi)(t) = \int_\Omega g(t)\,\varphi\,\mathrm{d}x + \varepsilon \sum_{i \in I}\sum_{j \in \gamma}\int_{\Gamma_{ij}} (\nabla\varphi \cdot \boldsymbol{n}_{ij}\,u_D + \sigma\,u_D\,\varphi)\,\mathrm{d}S,$$

where σ is defined by $\sigma|_{\Gamma_{ij}} = 1/\mathrm{diam}(\Gamma_{ij})$, $j \in S(i)$, $i \in I$ and $(\cdot,\cdot)$ denotes a L^2-scalar product.

The convective terms are approximated with the aid of a *numerical flux* $H = H(u, v, \boldsymbol{n})$ known from the theory of finite volume methods, see e.g., [9].

The approximate solution of problem (1), a)–c) is sought in the space of discontinuous piecewise polynomial functions S_h defined by

$$S_h = S^{p,-1}(\Omega, \mathcal{T}_h) = \{v; v|_K \in P^p(K)\ \forall K \in \mathcal{T}_h\},$$

where p is a positive integer and $P^p(K)$ denotes the space of all polynomials on K of degree at most p. Obviously, $S_h \subset H^2(\Omega, \mathcal{T})$.

Now we can introduce the *semidiscrete problem*.

Definition 1. *Function u_h is a semidiscrete solution of the problem (1), if*

a) $u_h \in C^1([0,T]; S_h)$, \hfill (5)

b) $\left(\dfrac{\partial u_h(t)}{\partial t}, \varphi_h\right) + b_h(u_h(t), \varphi_h) + a_h(u_h(t), \varphi_h) + \varepsilon J_h^\sigma(u_h(t), \varphi_h)$
$= \ell_h(\varphi_h)(t) \qquad \forall\varphi_h \in S_h,\ \forall t \in (0, T),$

c) $u_h(0) = u_h^0,$

where $u_h^0 \in S_h$ denotes an S_h-approximation of the initial condition u^0.

The above discrete problem has been obtained by means of the *method of lines*, i.e. the *spatial semidiscretization*. In the next section we discus the full space-time discretization.

2.4 Space-time discretization

First-order scheme

In [7] we analysed the following full space-time variant of (5), a)–c). Let $0 = t_0 < t_1 < \cdots < t_r = T$ be a partition of the time interval $(0, T)$ and $\tau_k = t_{k+1} - t_k$, $k = 0, \ldots, r-1$.

Definition 2. *We define the approximate solution of problem (1) as functions u_h^k, $t_k \in [0, T]$, satisfying the conditions*

a) $u_h^{k+1} \in S_h$, (6)

b) $\left(\dfrac{u_h^{k+1} - u_h^k}{\tau_k}, v_h\right) + a_h(u_h^{k+1}, v_h) + b_h(u_h^k, v_h)$

$+ \varepsilon J_h^\sigma(u_h^{k+1}, v_h) = \ell_h(v_h)(t_{k+1}) \quad \forall v_h \in S_h, \ \forall t_{k+1} \in (0, T]$,

c) u_h^0 is S_h approximation of u^0.

The function u_h^k is called the approximate solution at time t_k.

This means that the linear and nonlinear terms are discretized implicitly and explicitly with respect to the time, respectively. Therefore, for each time step we solve a system of linear algebraic equations. Numerical experiments show that the scheme (6) is practically unconditionally stable with respect to the choice of τ_k.

We derived in [7] the following a priori error estimates in the L^2-norm and the H^1-seminorm (for a constant time step $\tau \equiv \tau_k$, $k = 0, \ldots, r-1$):

$$\|u - \Pi^1 u_h\|_{L^\infty((0,T);L^2(\Omega))} = O(h^p + \tau),$$ (7)
$$\|u - \Pi^1 u_h\|_{L^2((0,T);H^1(\Omega))} = O(h^p + \tau),$$

where u is the exact solution of (1) and $\Pi^1 u_h : (0, T) \to S_h$ is piecewise linear function such that $\Pi^1 u_h(t_k) = u_h^k$, $k = 0, \ldots, r$.

Higher order schemes

Our aim is now to increase the degree of approximation with respect to time in (5), b). We define a sum of all linear forms by

$$A_h(u_h^k, v_h) \equiv a_h(u_h^k, v_h) + \varepsilon J_h^\sigma(u_h^k, v_h) - \ell_h(v_h)(t_k), \quad u_h^k, v_h \in S_h.$$ (8)

The time derivative term in (6), b) is approximated by a high degree multi-step approximation and for the first argument of the nonlinear form $b_h(\cdot, \cdot)$ in (6), b) we use an explicit higher order extrapolation. Therefore, we define a n-step scheme ($n \in \mathbb{N}$, $n \geq 1$) by Definition 2, where relation (6), b) is replaced by

$$\dfrac{1}{\tau_k}\left(\sum_{l=0}^n \alpha_l u_h^{k+1-l}, v_h\right) + A_h(u_h^{k+1}, v_h) + b_h\left(\sum_{l=1}^n \beta_l u_h^{k+1-l}, v_h\right) = 0,$$ (9)

where the coefficients α_l, $l = 0, \ldots, n$ and β_l, $l = 1, \ldots, n$ depend on the time steps τ_{k-l}, $l = 0, \ldots, n-1$. Since the choice of α_l and β_l for nonconstant time step is rather complicated and the final relations are long, we present only the values of α_l and β_l for constant time steps ($\tau \equiv \tau_k$, $k = 0, \ldots, r-1$) and $n = 1, 2, 3$, see Table 2.

Based on results from [7] obtained for $n = 1$, we expect that the formal orders of convergence of scheme (9) are

Table 1. Values of the coefficients α_l and β_l for constant time step

n	α_l, $l = 0, \ldots, n$				β_l, $l = 1, \ldots, n$		
1	1,	-1			1		
2	$\frac{3}{2}$,	-2,	$\frac{1}{2}$		2,	-1	
3	$\frac{11}{6}$,	-3,	$\frac{3}{2}$,	$-\frac{1}{3}$	3,	-3,	1

$$\|u - u_h\|_{L^\infty((0,T);L^2(\Omega))} = O(h^p + \tau^n), \tag{10}$$
$$\|u - u_h\|_{L^2((0,T);H^1(\Omega))} = O(h^p + \tau^n).$$

A rigorous numerical analysis of these schemes will be the subject of future papers. Here we present a numerical verification of the estimates (10).

3 Numerical results

3.1 Convergence with respect to τ

We solve the problem (1), a) – c) with $\Omega = (0,1)^2$, $f_s(u) = u^2/2$, $s = 1, 2$, $T = 1$, $\varepsilon = 0.01$ and the functions u_D, u_0 and g are chosen in such a way that the exact solution has the form

$$u(x_1, x_2, t) = 16 \frac{e^{10t} - 1}{e^{10} - 1} x_1(1 - x_1)x_2(1 - x_2). \tag{11}$$

The computations were carried out on a triangular mesh having 4219 elements with piecewise cubic approximation in space and for 6 different time steps: 1/20, 1/40, 1/80, 1/160, 1/320, 1/640. Fig. 1 shows the computational errors at $t = T$ and the corresponding orders of convergence with respect to τ in L^2-norm and H^1-seminorm for schemes (9) with $n = 1$, $n = 2$ and $n = 3$. The expected order of convergence $O(\tau^n)$ is observed in each case.

3.2 Convergence with respect to h

We solve the problem (1), a) – c) as in Section 3.1 but with $\varepsilon = 0.1$ and the functions u_D, u_0 and g are chosen in such a way that the exact solution has the form

$$u(x_1, x_2, t) = (1 - e^{-10t}) \left[x_1 x_2^2 - x_2^2 \exp\left(\frac{2(x_1 - 1)}{\varepsilon}\right) \right.$$
$$\left. - x_1 \exp\left(\frac{3(x_2 - 1)}{\varepsilon}\right) + \exp\left(\frac{2x_1 + 3x_2 - 5}{\varepsilon}\right) \right].$$

The computations were carried out by a third order scheme with respect to time on 7 different triangular meshes having 148, 289, 591, 1056, 2360, 4219

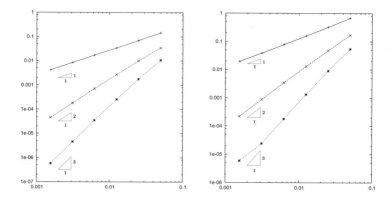

Fig. 1. Computational errors and orders of convergence in the L^2–norm (left) and the H^1–seminorm (right) for schemes (9) with $n = 1$ (full line), $n = 2$ (dashed line) and $n = 3$ (dotted line).

and 9872 elements. Fig. 2 shows the computational errors at $t = T$ and the corresponding orders of convergence with respect to h in the L^2–norm and the H^1–seminorm for schemes (9) with piecewise linear P_1, quadratic P_2 and cubic P_3 approximations. We observe the order of convergence $O(h^{p+1})$ for $p = 1, 3$ and $O(h^p)$ for $p = 2$ in L^2–norm and $O(h^p)$ in H^1–seminorm. These results are in agreement with those of other authors, see [1].

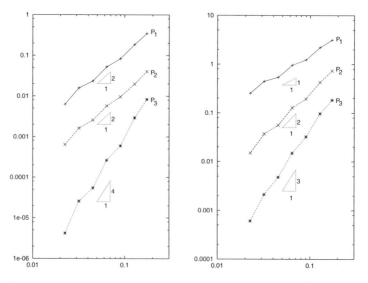

Fig. 2. Computational errors and orders of convergence in the L^2–norm (left) and the H^1–seminorm (right) for schemes (9) with P_1 (full line), P_2 (dashed line) and P_3 (dotted line) approximations.

4 Conclusion

We presented a higher order method with respect to space and time for a scalar convection-diffusion equation. The scheme is stable without an essential restriction for a time step and at each time level we solve only linear system of equations. Numerical experiments verify the expected orders of convergence.

References

1. Arnold, D.N., Brezzi, F., Cockburn, B., Marini, L.D.: Unified analysis of discontinuous Galerkin methods for elliptic problems. SIAM J. Numer. Anal. **39**(5), 1749–1779 (2002)
2. Bassi, F., Rebay, S.: A high-order accurate discontinuous finite element method for the numerical solution of the compressible Navier–Stokes equations. J. Comput. Phys. **131**, 267–279 (1997)
3. Baumann, C.E., Oden, J.T.: A discontinuous hp finite element method for the Euler and Navier-Stokes equations. Int. J. Numer. Methods Fluids **31**(1), 79–95 (1999)
4. Cockburn, B.: Discontinuous Galerkin methods for convection dominated problems. In: T.J. Barth, H. Deconinck (eds.) High–Order Methods for Computational Physics, Lecture Notes in Computational Science and Engineering 9, pp. 69–224. Springer, Berlin (1999)
5. Dolejší, V.: On the discontinuous Galerkin method for the numerical solution of the Navier–Stokes equations. Int. J. Numer. Methods Fluids **45**, 1083–1106 (2004)
6. Dolejší, V., Feistauer, M.: Error estimates of the discontinuous Galerkin method for nonlinear nonstationary convection-diffusion problems. Numer. Funct. Anal. Optim. **26**(25-26), 2709–2733 (2005)
7. Dolejší, V., Feistauer, M., Hozman, J.: Analysis of semi-implicit DGFEM for nonlinear convection-diffusion problems. Comput. Methods Appl. Mech. Engrg. (submitted). Preprint No.MATH-knm-2005/4, Charles University Prague, School of Mathematics, 2005
8. Dolejší, V., Feistauer, M., Sobotíková, V.: A discontinuous Galerkin method for nonlinear convection–diffusion problems. Comput. Methods Appl. Mech. Engrg. **194**, 2709–2733 (2005)
9. Feistauer, M., Felcman, J., Straškraba, I.: Mathematical and Computational Methods for Compressible Flow. Oxford University Press, Oxford (2003)
10. Lomtev, I., Quillen, C.B., Karniadakis, G.E.: Spectral/hp methods for viscous compressible flows on unstructured 2d meshes. J. Comput. Phys **144**(2), 325–357 (1998)
11. van der Vegt, J.J.W., van der Ven, H.: Space-time discontinuous Galerkin finite element method with dynamic grid motion for inviscid compressible flows. I: General formulation. J. Comput. Phys. **182**(2), 546–585 (2002)

On Some Aspects of the Discontinuous Galerkin Method*

Miloslav Feistauer

Charles University Prague, Faculty of Mathematics and Physics, Sokolovská 83, 186 75 Praha 8, Czech Republic
feist@karlin.mff.cuni.cz

Summary. The paper is concerned with some aspects of the discontinuous Galerkin finite element method (DGFEM) for the numerical solution of convection-diffusion problems and compressible flow. In particular, theoretical analysis of the space-time discontinuous Galerkin discretization is briefly discussed. The robustness of the DGFEM is demonstrated by its application to the simulation of compressible low Mach number flows.

1 Continuous problem

The DGFEM uses piecewise polynomial approximations of the sought solution on a FE mesh without any requirement on the continuity between neighbouring elements and can be considered as a generalization of finite volume and finite element methods. It allows to construct higher order schemes for the solution of conservation laws and singularly perturbed problems in a natural way. For a survey of DG methods, see e.g. [1].

Here we shall apply the DGFEM to the numerical solution of the following initial-boundary value convection-diffusion-reaction problem. Let $\Omega \subset \mathbb{R}^d$ ($d = 2$ or 3) be a bounded polyhedral domain and $T > 0$. We want to find $u : Q_T = \Omega \times (0, T) \to \mathbb{R}$ such that

$$\frac{\partial u}{\partial t} + \boldsymbol{v} \cdot \nabla u - \varepsilon \Delta u + cu = g \quad \text{in } Q_T, \tag{1}$$

$$u = u_D \quad \text{on } \partial\Omega^- \times (0, T), \tag{2}$$

$$\varepsilon \frac{\partial u}{\partial \boldsymbol{n}} = u_N \quad \text{on } \partial\Omega^+ \times (0, T), \tag{3}$$

$$u(x, 0) = u^0(x), \quad x \in \Omega. \tag{4}$$

We assume that $\partial\Omega = \partial\Omega^- \cup \partial\Omega^+$, where the sets $\partial\Omega^+$ and $\partial\Omega^-$ are defined

---

* This work is a part of the research project MSM 0021620839 financed by the Ministry of Education of the Czech Republic and partly supported by the grant No. 201/04/1503 of the Czech Grant Agency.

in such a way that $\boldsymbol{v}(x,t) \cdot \boldsymbol{n}(x) < 0$ on $\partial \Omega^-$ and $\boldsymbol{v}(x,t) \cdot \boldsymbol{n}(x) \geq 0$ on $\partial \Omega^+$, for all $t \in [0,T]$. Here $\boldsymbol{n}(x)$ is the unit outer normal to the boundary $\partial \Omega$ of Ω. In the case $\varepsilon = 0$ we put $u_N = 0$ and ignore the Neumann condition (3).

Assumptions on data (A)
We assume that the data satisfy the following conditions:
a) $g \in C([0,T]; L^2(\Omega))$,
b) $u_0 \in L^2(\Omega)$,
c) u_D is the trace of some $u^* \in C([0,T]; H^1(\Omega)) \cap L^\infty(Q_T)$ on $\partial \Omega^- \times (0,T)$,
d) $\boldsymbol{v} \in C([0,T]; W^{1,\infty}(\Omega))$,
e) $c \in C([0,T]; L^\infty(\Omega))$,
f) $c - \frac{1}{2}\operatorname{div}\boldsymbol{v} \geq \gamma_0 > 0$ in Q_T with a constant γ_0,
g) $u_N \in C([0,T]; L^2(\partial \Omega^+))$,
h) $\varepsilon \geq 0$.

2 Discretization of the problem

Let $\mathcal{T}_h = \bigcup_{i \in i_h} K_i$ (where $i_h \subset \{0, 1, 2, ...\}$ is a suitable index set) be a standard triangulation of the closure of the domain Ω into a finite number of closed triangles ($d = 2$) or tetrahedra ($d = 3$). If $K_i \cap K_j = \Gamma_{ij} = \Gamma_{ji}$ is a common face of K_i and K_j, we call these elements neighbours. We denote all boundary faces on $\partial \Omega$ by S_j, where $j \in I_{bh} \subset \{-1, -2, ...\}$. For $i \in i_h$ we set

$$s(i) = \{j \in i_h; K_j \text{ is a neighbour of } K_i\}, \tag{5}$$
$$\Gamma_{ij} = S_j \text{ for } K_i \in \mathcal{T}_h \text{ such that } S_j \subset \partial K_i \cap \partial \Omega, j \in I_{bh}. \tag{6}$$

For $K \in \mathcal{T}_h$, we denote by h_K and ρ_K the diameter of K and the diameter of the largest ball inscribed in K, respectively. We set $h = \max_{K \in \mathcal{T}_h} h_K$.

We introduce the so-called broken Sobolev space

$$H^k(\Omega, \mathcal{T}_h) = \{\varphi; \varphi|_K \in H^k(K) \; \forall K \in \mathcal{T}_h\} \tag{7}$$

and define the seminorm

$$|\varphi|_{H^k(\Omega, \mathcal{T}_h)} = \left(\sum_{K \in \mathcal{T}_h} |\varphi|^2_{H^k(K)} \right)^{1/2}. \tag{8}$$

For $\varphi \in H^1(\Omega, \mathcal{T}_h)$ we introduce the following notation:
$\varphi|_{\Gamma_{ij}} = $ the trace of $\varphi|_{K_i}$ on Γ_{ij}, $\varphi|_{\Gamma_{ji}} = $ the trace of $\varphi|_{K_j}$ on $\Gamma_{ji} = \Gamma_{ij}$,
$$\langle \varphi \rangle_{\Gamma_{ij}} = \frac{1}{2}\left(\varphi|_{\Gamma_{ij}} + \varphi|_{\Gamma_{ji}}\right), \quad [\varphi]_{\Gamma_{ij}} = \varphi|_{\Gamma_{ij}} - \varphi|_{\Gamma_{ji}}, \tag{9}$$
$\boldsymbol{n}_{ij} = $ the unit outer normal to ∂K_i on the face Γ_{ij}.

Further, for $i \in i_h$ we set $\partial K_i^-(t) = \{x \in \partial K_i; \boldsymbol{v}(x,t) \cdot \boldsymbol{n}(x) < 0\}, \partial K_i^+(t) = \{x \in \partial K_i; \boldsymbol{v}(x,t) \cdot \boldsymbol{n}(x) \geq 0\}$. (Here, $\boldsymbol{n}$ denotes the unit outer normal to ∂K_i.)

2.1 Space semidiscretization

In [6] we analyzed the spatially semidiscrete problem of finding a function $u_h \in C^1([0,T]; S_h^p)$ such that

$$\left(\frac{\partial u_h}{\partial t}, \varphi_h\right) + A_h(u_h(t), \varphi_h) = l_h(\varphi_h)(t) \quad \forall \varphi_h \in S_h^p \ \forall t \in (0,T), \quad (10)$$

$$(u_h(0), \varphi_h) = (u^0, \varphi_h) \quad \forall \varphi_h \in S_h^p, \quad (11)$$

where

$$A_h(u, \varphi) = a_h(u, \varphi) + b_h(u, \varphi) + c_h(u, \varphi) + \varepsilon J_h(u, \varphi), \quad (12)$$
$$S_h^p = \{\varphi \in L^2(\Omega); \varphi|_K \in \mathcal{P}^p(K) \ \forall K \in \mathcal{T}_h\}, \quad (13)$$

$p \geq 1$ is an integer and $\mathcal{P}^p(K)$ is the space of polynomials of degree at most p on K. The bilinear forms $(\cdot, \cdot), a_h, b_h, c_h, J_h$ are defined as follows:

$$(u, \varphi) = \int_\Omega u\varphi \, dx, \quad (14)$$

$$a_h(u, \varphi) = \varepsilon \sum_{i \in i_h} \int_{K_i} \nabla u \cdot \nabla \varphi \, dx$$

$$- \varepsilon \sum_{i \in i_h} \sum_{j \in s(i), j<i} \int_{\Gamma_{ij}} (\langle \nabla u \rangle \cdot \boldsymbol{n}_{ij} [\varphi] - \langle \nabla \varphi \rangle \cdot \boldsymbol{n}_{ij} [u]) \, dS$$

$$- \varepsilon \sum_{i \in i_h} \int_{\partial K_i^- \cap \partial \Omega} ((\nabla u \cdot \boldsymbol{n})\varphi - (\nabla \varphi \cdot \boldsymbol{n})u) \, dS, \quad (15)$$

$$b_h(u, \varphi) = \sum_{i \in i_h} \int_{K_i} (\boldsymbol{v} \cdot \nabla u)\varphi \, dx - \sum_{i \in i_h} \int_{\partial K_i^- \cap \partial \Omega} (\boldsymbol{v} \cdot \boldsymbol{n}) u\varphi \, dS$$

$$- \sum_{i \in i_h} \int_{\partial K_i^- \setminus \partial \Omega} (\boldsymbol{v} \cdot \boldsymbol{n})[u]\varphi \, dS, \quad (16)$$

$$c_h(u, \varphi) = \int_\Omega cu\varphi \, dx \quad (17)$$

$$J_h(u, \varphi) = \sum_{i \in i_h} \sum_{j \in s(i)} \mathrm{diam}(\Gamma_{ij})^{-1} \int_{\Gamma_{ij}} [u][\varphi] \, dS$$

$$+ \sum_{i \in i_h} \sum_{j: \Gamma_{ij} \subset \partial \Omega^-} \mathrm{diam}(\Gamma_{ij})^{-1} \int_{\Gamma_{ij}} u\varphi \, dS, \quad (18)$$

$$l_h(\varphi)(t) = \int_\Omega g(t)\varphi \, dx + \sum_{i \in i_h} \int_{\partial K_i^+ \cap \partial \Omega} u_N(t)\varphi \, dS$$

$$+ \varepsilon \sum_{i \in i_h} \int_{\partial K_i^- \cap \partial \Omega} \sigma u_D(t)\varphi \, dS + \varepsilon \sum_{i \in i_h} \int_{\partial K_i^- \cap \partial \Omega} u_D(t)(\nabla \varphi \cdot \boldsymbol{n}) \, dS$$

$$- \sum_{i \in i_h} \int_{\partial K_i^- \cap \partial \Omega} (\boldsymbol{v} \cdot \boldsymbol{n}) u_D(t)\varphi \, dS. \quad (19)$$

This means that the nonsymmetric interior and boundary penalty formulation of the diffusion terms is used.

In [6] we proved that the error $e_h = u_h - u$ satisfies the estimate

$$\max_{t \in (0,T)} \|e_h(t)\|_{L^2(\Omega)} + \sqrt{\varepsilon} \sqrt{\int_0^T |e_h(\vartheta)|^2_{H^1(\Omega,\mathcal{T}_h)} \, d\vartheta + \int_0^T J_h(e_h(\vartheta), e_h(\vartheta)) \, d\vartheta}$$
$$\leq Ch^p(\sqrt{\varepsilon} + \sqrt{h}) \tag{20}$$

with a constant C independent of $h > 0$ and $\varepsilon \geq 0$.

2.2 Discontinuous Galerkin discretization in space and time

In practical computations it is necessary to carry out time discretization as well. In computational fluid dynamics explicit Runge-Kutta schemes are popular, but they are conditionally stable and the length of the time step is strongly limited by the CFL condition.

In order to construct a stable, high-order accurate time discretization, it is possible to apply the discontinuous Galerkin method in space as well as in time. For this purpose, we consider a partition $0 = t_0 < t_1 < \ldots < t_M = T$ of the time interval $[0,T]$ and define $I_m = (t_{m-1}, t_m), \tau_m = t_m - t_{m-1}, m = 1, \ldots, M$. For a function φ defined in $[0,T]$, discontinuous in general at $t_m, m = 1, \ldots, M-1$, we introduce the notation $\varphi_m^\pm = \varphi(t_m\pm) = \lim_{t \to t_m\pm} \varphi(t)$ and $\{\varphi\}_m = \varphi_m^+ - \varphi_m^-$. For each time interval $I_m, m = 1, \ldots, M$, we shall consider, in general, a different triangulation $\mathcal{T}_{h,m} = \{K_i\}_{i \in i_{h,m}}$ of the domain Ω. Therefore, for different intervals I_m we have different $S^p_{h,m}, a_{h,m}, b_{h,m}, J_{h,m}, l_{h,m}, A_{h,m}$, etc. The definition of $S^p_{h,m}$ now becomes

$$S^p_{h,m} = \{\varphi \in L^2(\Omega); \varphi|_K \in \mathcal{P}^p(K) \ \forall K \in \mathcal{T}_{h,m}\} \tag{21}$$

and in the definitions of $A_{h,m}$ and $l_{h,m}$ "$i \in i_h$" is changed into "$i \in i_{h,m}$". We set $h_m = \max_{K \in \mathcal{T}_{h,m}} h_K$, $h = \max_{m=1,\ldots,M} h_m$ and $\tau = \max_{m=1,\ldots,M} \tau_m$. Let $p, q \geq 1$ be integers. We define the space

$$S^{p,q}_{h,\tau} = \left\{ \varphi \in L^2(Q_T); \varphi|_{I_m} = \sum_{i=0}^q t^i \varphi_i \text{ with } \varphi_i \in S^p_{h,m}, m = 1, \ldots, M \right\} \tag{22}$$

and the forms

$$B(u,v) \tag{23}$$
$$= \sum_{m=1}^M \int_{I_m} \left(\left(\frac{\partial u}{\partial t}, v \right) + A_{h,m}(u,v) \right) dt + \sum_{m=2}^M (\{u\}_{m-1}, v_{m-1}^+) + (u_0^+, v_0^+),$$

$$L(v) = \sum_{m=1}^M \int_{I_m} l_{h,m}(v) \, dt + (u_0, v_0^+).$$

Then the *space-time DG approximate solution* is defined as a function $U \in S_{h,\tau}^{p,q}$ satisfying

$$B(U, \varphi) = L(\varphi) \quad \forall \varphi \in S_{h,\tau}^{p,q}. \tag{24}$$

3 Error estimates

In order to estimate the error $e = U - u$, we consider a system of triangulations $\mathcal{T}_{h,m}$, $m = 1, ..., M$, $h \in (0, h_0)$, which is *shape regular*: there exists a constant C_T independent of K, m and h such that

$$\frac{h_K}{\rho_K} \leq C_T, \quad K \in \mathcal{T}_{h,m}, \, m = 1, ..., M, \, h \in (0, h_0). \tag{25}$$

The derivation of the error estimates is rather technical. We can mention here only some of the most important steps. (Details can be found in [5].) As important tools we use the multiplicative trace inequality (see [3]), the inverse inequality and the $S_{h,\tau}^{p,q}$-interpolation. It is defined similarly as in [8]:

$$\pi u \in S_{h,\tau}^{p,q}, \tag{26}$$

$$\int_{I_m} (\pi u - u, \varphi^*)\, dt = 0 \quad \forall \varphi^* \in S_{h,\tau}^{p,q-1},$$

$$\pi u(t_m^-) = \Pi_m u(t_m^-),$$

for $m = 1, ..., M$, where Π_m is L^2-projection on $S_{h,m}^p$ in space. Let us set $\eta = u - \pi u$.

The basis for the error analysis is the following abstract error estimate.

Theorem 1. *Let us denote*

$$\|\varphi\|_{\boldsymbol{v},\Gamma} = \left(\int_\Gamma |\boldsymbol{v}\cdot\boldsymbol{n}|\,\varphi^2\, dt\right)^{1/2} \quad \text{for } \Gamma \subset \partial K_i, \tag{27}$$

$$\sigma_m = \Big(\varepsilon|\eta|_{H^1(\Omega,\mathcal{T}_{h,m})}^2 + \gamma_0\|\eta\|_{L^2(\Omega)}^2 + \varepsilon J_{h,m}(\eta,\eta) \tag{28}$$

$$+ \frac{1}{2}\sum_{i \in i_{h,m}} (\|\eta\|_{\boldsymbol{v},\partial K_i \cap \partial \Omega}^2 + \|[\eta]\|_{\boldsymbol{v},\partial K_i^-\setminus \partial \Omega}^2)\Big)^{1/2} + \sqrt{\varepsilon}h|\eta|_{H^2(\Omega,\mathcal{T}_{h,m})}$$

$$+ \left(\sum_{i \in i_{h,m}} \|\eta^-\|_{\boldsymbol{v},\partial K_i^-\setminus \partial \Omega}^2\right)^{1/2} + \left(\sum_{i \in i_{h,m}} h_{K_i}^{-2}\|\eta\|_{L^2(K_i)}^2\right)^{1/2}$$

and

$$\|v\|_{E,m}^2 = \varepsilon|v|_{H^1(\Omega,\mathcal{T}_{h,m})}^2 + \gamma_0\|v\|_{L^2(\Omega)}^2 + \varepsilon J_{h,m}(v,v) \tag{29}$$

$$+\frac{1}{2}\sum_{i\in i_{h,m}}(\|v\|_{\boldsymbol{v},\partial K_i\cap\partial\Omega}^2+\|[v]\|_{\boldsymbol{v},\partial K_i^-\setminus\partial\Omega}^2).$$

Then
$$\sum_{m=1}^M\int_{I_m}\|e\|_{E,m}^2\,dt\le C\sum_{m=1}^M\int_{I_m}\sigma_m^2(\eta)\,dt+C\sum_{m=1}^{M-1}\|\eta_m^-\|_{L^2(\Omega)}^2. \tag{30}$$

The estimation of the right-hand side in (30) is carried out under the assumption that the exact solution satisfies the regularity condition

$$u\in\mathcal{H}=H^{q+1}(0,T;H^1(\Omega))\cap C([0,T];H^{p+1}(\Omega)) \tag{31}$$

and that there exist constants $C_S,\hat{C}_S$ such that

$$\frac{1}{\hat{C}_S}h_K\le\tau_m\le C_S h_K,\quad K\in\mathcal{T}_{h,m},\quad m=1,...,M,\quad h\in(0,h_0). \tag{32}$$

Then
$$\int_{I_m}|\eta|_{H^1(\Omega,\mathcal{T}_{h,m})}^2\,dt\le Ch^{2p}|u|_{L^2(I_m;H^{p+1}(\Omega))}^2+C\tau_m^{2q+2}|u|_{H^{q+1}(I_m;H^1(\Omega))}^2,$$

$$\int_{I_m}\|\eta\|_{L^2(K)}^2\,dt\le Ch_K^{2p+2}|u|_{L^2(I_m;H^{p+1}(K))}^2+C\tau_m^{2q+2}|u|_{H^{q+1}(I_m;L^2(K))}^2,$$
$$K\in\mathcal{T}_{h,m},$$

$$\int_{I_m}J_{h,m}(\eta,\eta)\,dt\le Ch^{2p}|u|_{L^2(I_m;H^{p+1}(\Omega))}^2+C\tau_m^{2q}(|u|_{H^{q+1}(I_m;L^2(\Omega))}^2$$
$$+\tau_m^2|u|_{H^{q+1}(I_m;H^1(\Omega))}^2),$$

$$\int_{I_m}\sum_{i\in i_{h,m}}(\|\eta\|_{\boldsymbol{v},\partial K_i\cap\partial\Omega}^2+\|[\eta]\|_{\boldsymbol{v},\partial K_i^-\setminus\partial\Omega}^2)\,dt\le Ch^{2p+1}|u|_{L^2(I_m;H^{p+1}(\Omega))}^2$$
$$+C\tau_m^{2q+1}\{|u|_{H^{q+1}(I_m;L^2(\Omega))}^2+h\tau_m|u|_{H^{q+1}(I_m;H^1(\Omega))}^2\},$$

$$\int_{I_m}|\eta|_{H^2(\Omega,\mathcal{T}_{h,m})}^2\,dt\le Ch^{2(p-1)}|u|_{L^2(I_m;H^{p+1}(\Omega))}^2+C\tau_m^{2q}|u|_{H^{q+1}(I_m;H^1(\Omega))}^2,$$

$$\sum_{m=1}^{M-1}\|\eta_m^-\|_{L^2(\Omega)}^2\le Ch^{2p+1}\|u\|_{C([0,T];H^{p+1}(\Omega))}^2. \tag{33}$$

The abstract error estimate and the above relations imply the main result.

Theorem 2. *Let the assumptions (A) on the data, and (13), (31) and (32) be satisfied. Then the error $e=U-u$ satisfies the estimate*

$$\sum_{m=1}^M\int_{I_m}\|e\|_{E,m}^2\,dt$$
$$\le Ch^{2p}|u|_{C([0,T];H^{p+1}(\Omega))}^2+C\tau^{2q}|u|_{H^{q+1}(0,T;H^1(\Omega))}^2. \tag{34}$$

The estimate holds true even if $\varepsilon=0$ (hyperbolic case).

4 Application of the DGFEM to compressible flow with a wide range of mach numbers

Standard finite volume methods have difficulties with the solution of flows with very low Mach numbers. Therefore, various modifications of the Euler (Navier–Stokes) equations have been introduced in order to enable the finite volume solution of compressible flow at the incompressible limit. See, e. g. [7]. In [2] a robust, efficient DG technique for the solution of compressible flow is presented. This method has been extended so that it allows the solution of high-speed flow as well as low Mach number flow at the incompressible limit, using conservative variables without any modification of the governing equations. The main ingredients of this technique are the semi-implicit version of the DGFEM, GMRES method with diagonal preconditioning for the solution of large linear algebraic systems, the use of the homogeneity of inviscid fluxes and the use of the Vijayasundaram numerical flux, characteristic treatment of the boundary conditions in inviscid terms, hp approach to the limiting of order of accuracy in order to avoid the Gibbs phenomenon, proposed in [4] and the use of isoparametric elements at curved boundaries.

As an example we consider a stationary inviscid flow past a circular cylinder with far field velocity parallel to the axis x_1 and Mach number $M_\infty = 10^{-4}$. We present here the comparison of the DG approximate compressible solution with exact incompressible flow. The maximal density variation and the maximum of the density gradient of the approximate solution are $\rho_{\max} - \rho_{\min} = 2.3 \cdot 10^{-8}$ and $\max_{K \in \mathcal{T}_h} |\nabla \rho_h|_K| < 1.99 \cdot 10^{-6}$, respectively, which indicates that the compressible approximate solution behaves practically as an incompressible one.

5 Conclusion

We tried to show that the DGFEM is a robust, accurate method for the numerical solution of convection-diffusion problems and compressible flow.

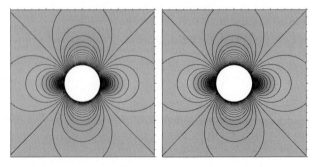

Fig. 1. Velocity isolines: compressible flow (left) incompressible flow (right)

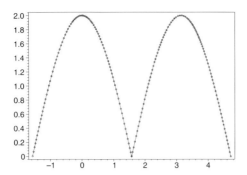

Fig. 2. Velocity distribution along the cylinder (full line – compressible flow, dotted line – icompressible flow)

There are still some open questions as, for example
- investigation of the optimality of the obtained error estimates,
- development of the space-time hp adaptivity,
- analysis of the effect of the numerical integration,
- extension of the semi-implicit schemes to nonstationary compressible viscous flow.

References

1. Cockburn, B., Karniadakis, G.E., Shu, C.W. (eds): Discontinuous Galerkin Methods, Lecture Notes in Computational Science and Engineering 11, Springer, Berlin (2000)
2. Dolejší, V., Feistauer, M.: A semi-implicit discontinuous Galerkin finite element method for the numerical solution of inviscid compressible flow. J. Comput. Phys., **198**, 727–746 (2004)
3. Dolejší, V., Feistauer, M., Schwab, C.: A finite volume discontinuous Galerkin scheme for nonlinear convection-diffusion problems. Calcolo, **39**, 1–40 (2002)
4. Dolejší, V., Feistauer, M., Schwab, C.: On some aspects of the discontinuous Galerkin finite element method for conservation laws. Math. Comput. Simul., **61**, 333–346 (2003)
5. Feistauer, M., Hájek, J., Švadlenka, K.: Space-time discontinuous Galerkin method for solving nonstationary convection-diffusion-reaction problems. SIAM J. Numer. Anal. (submitted). Preprint No. MATH-knm-2005/2, School of Mathematics, Charles University Prague, Faculty of Mathematics and Physics (2005)
6. Feistauer, M., Švadlenka, K.: Discontinuous Galerkin method of lines for solving nonstationary singularly perturbed linear problems, J. Numer. Math., **2**, 97–117 (2004)
7. Klein, R.: Semi-implicit extension of a Godunov-type scheme based on low Mach number asymptotics 1: one-dimensional flow. J. Comput. Phys., **121**, 213–237 (1995)
8. Schötzau, D.: hp-DGFEM for Parabolic Evolution Problems. Applications to Diffusion and Viscous Incompressible Fluid Flow, PhD Dissertation ETH No. 13041, Zürich (1999)

Mixed Discontinuous Galerkin Methods with Minimal Stabilization

Daniele Marazzina

Dipartimento di Matematica, Universita degli Studi di Pavia, Via Ferrata 1, 27100 Pavia, Italy
`daniele.marazzina@unipv.it`

Summary. We will address the problem of finding the minimal necessary stabilization for a class of Discontinuous Galerkin (DG) methods in mixed form. In particular, we will present a new stabilized formulation of the Bassi-Rebay method (see [2] for the original unstable method) and a new formulation of the Local Discontinuous Galerkin (LDG) method (see [5] for the original LDG method).

It will be shown that, in order to reach stability, it is enough to add jump terms only over a part of the boundary of the domain, instead of over all the skeleton of the mesh, as it is usually done (see [1], for instance).

1 DG methods

We consider the model problem:

$$-\Delta u = f \text{ in } \Omega, \quad u = g \text{ on } \partial\Omega,$$

where Ω is assumed to be a convex polygonal domain, $\Omega \subseteq \mathbb{R}^d$, $d = 1, 2$; $f \in L^2(\Omega)$ and $g \in H^{1/2}(\partial\Omega)$ are given.

To obtain the associated weak formulation we rewrite the above problem as:

$$\boldsymbol{\sigma} = -\nabla u \text{ in } \Omega, \quad \text{div } \boldsymbol{\sigma} = f \text{ in } \Omega, \quad u = g \text{ on } \partial\Omega. \quad (1)$$

If $I \in \mathcal{T}_h$, then from (1) we get

$$\int_I \boldsymbol{\sigma} \cdot \boldsymbol{\tau} \, dx = \int_I \nabla u \cdot \boldsymbol{\tau} \, dx,$$
$$-\int_I \text{div}(\boldsymbol{\sigma}) \, v \, dx = \int_I f \, v \, dx,$$

where $\boldsymbol{\tau}$ and v are smooth test functions.

Thus, integrating by parts, we obtain

$$\int_I \boldsymbol{\sigma} \cdot \boldsymbol{\tau} \, dx + \int_I u \, \mathrm{div}(\boldsymbol{\tau}) \, dx - \int_{\partial I} u \, \boldsymbol{\tau} \cdot \mathbf{n} \, ds = 0, \quad (2)$$

$$\int_I \boldsymbol{\sigma} \cdot \nabla v \, dx - \int_{\partial I} \boldsymbol{\sigma} \cdot v \, \mathbf{n} \, ds = \int_I f \, v \, dx.$$

See [1] for further details.

1.1 Discrete formulation

In order to introduce the formulation of the Bassi-Rebay and the LDG methods, we need to define the *numerical fluxes*, which are discrete approximations of $\boldsymbol{\sigma}$ and u on the skeleton of $\mathcal{T}_h$.

Let $\mathcal{E} := \bigcup_{I \in \mathcal{T}_h} \partial I$ be the skeleton of $\mathcal{T}_h$ (i.e., the union of all the subdivision points or edges, if $d = 1, 2$, respectively, of our mesh), $\mathcal{E}^0 = \mathcal{E} \setminus \partial \Omega$ and $\mathcal{E}^\partial = \mathcal{E} \setminus \mathcal{E}^0$; if $q \in T(\mathcal{E}) := \prod_{I \in \mathcal{T}_h} L^2(\partial I)$, we define the average $\{\{q\}\}$ and the jump $[[q]]$ operators on $\mathcal{E}^0$ as follows: if K_1 and K_2 are elements of $\mathcal{T}_h$ sharing a point (edge) $e \in \mathcal{E}^0$, $\mathbf{n}_i$ is the unit normal vector pointing exterior to K_i and $\mathbf{q}_i = \mathbf{q}|_{\partial K_i}$, $i = 1, 2$; if $\mathbf{q}$ is a vector-valued function we set

$$\{\{\mathbf{q}\}\} = \frac{\mathbf{q}_1 + \mathbf{q}_2}{2}, \quad [[\mathbf{q}]] = \mathbf{q}_1 \cdot \mathbf{n}_1 + \mathbf{q}_2 \cdot \mathbf{n}_2 \quad \text{on } e;$$

if q is a scalar valued function we set

$$\{\{q\}\} = \frac{q_1 + q_2}{2}, \quad [[q]] = q_1 \, \mathbf{n}_1 + q_2 \, \mathbf{n}_2 \quad \text{on } e,$$

(see [1] or [4] for further details).

So we are ready to define the numerical fluxes on $\mathcal{E}^0$:

$$\begin{pmatrix} \widehat{\boldsymbol{\sigma}} \\ \widehat{u} \end{pmatrix}(x) = \begin{pmatrix} \{\{\boldsymbol{\sigma}\}\} \\ \{\{u\}\} \end{pmatrix}(x) - \begin{pmatrix} C_{11} & \mathbf{C}_{12} \\ -\mathbf{C}_{12} & 0 \end{pmatrix} \begin{pmatrix} [[u]] \\ [[\boldsymbol{\sigma}]] \end{pmatrix}(x),$$

where C_{11} and $\mathbf{C}_{12}$ are coefficients that will be suitably chosen.

The Dirichlet boundary conditions are imposed through particular choices of the numerical fluxes on $\mathcal{E}^\partial$:

$$\widehat{\boldsymbol{\sigma}} = \boldsymbol{\sigma}^+ - C_{11} \left(u^+ \, \mathbf{n}^+ + g \, \mathbf{n}^- \right), \quad \widehat{u} = g,$$

where the superscripts $+$ and $-$ stand for the interior and exterior of the domain, respectively.

In order to deal with the discrete formulation of our model problem, we define the following spaces

$$V_h = \{ v_h \in L^2(\Omega) \text{ s. t. } v_h|_T \in P^k(T) \, \forall \, T \in \mathcal{T}_h \}, \quad \Sigma_h = [V_h]^d.$$

We also define the averages and jumps on $\mathcal{E}^\partial$ in a suitable way: if w_h is an approximation in V_h of the solution u of (1), we define

$$\{\{w_h\}\} = \frac{w_h + g}{2}, \quad [[w_h]] = (w_h - g)\,\mathbf{n} \quad \text{on } e \in \mathcal{E}^\partial,$$

where $\mathbf{n}$ is the unit normal vector external to Ω; if v_h is a test function we set

$$\{\{v_h\}\} = \frac{v_h}{2}, \quad [[v_h]] = v_h\,\mathbf{n} \quad \text{on } e \in \mathcal{E}^\partial.$$

Starting from (2) and replacing the traces on the skeleton of $\mathcal{T}_h$ by the numerical fluxes, we obtain the so-called *flux formulation*:

Find $(\boldsymbol{\sigma}_h, u_h) \in \Sigma_h \times V_h$ s.t. $\forall\,(\boldsymbol{\tau}_h, v_h) \in \Sigma_h \times V_h$ (3)

$$\int_\Omega \boldsymbol{\sigma}_h \cdot \boldsymbol{\tau}_h\, dx - \sum_{I \in \mathcal{T}_h} \int_I u_h\, \mathrm{div}(\boldsymbol{\tau}_h)\, dx - \int_{\mathcal{E}^0} \widehat{u}_h\, [[\boldsymbol{\tau}_h]]\, ds$$
$$- \int_{\partial\Omega} \widehat{u}_h\, \boldsymbol{\tau}_h \cdot \mathbf{n}\, ds = 0,$$

$$\int_\Omega \boldsymbol{\sigma}_h \cdot \nabla_h(v_h)\, dx - \int_{\mathcal{E}^0} \widehat{\boldsymbol{\sigma}}_h\, [[v_h]]\, ds - \int_{\partial\Omega} \widehat{\boldsymbol{\sigma}}_h \cdot v_h \mathbf{n}\, ds = \int_\Omega f v_h\, dx.$$

As in [1] and [3], we define the following *lifting operators*:
$R : \left[L^1(\mathcal{E})\right]^d \to \Sigma_h$ is defined by

$$\int_\Omega R([[u_h]]) \cdot \boldsymbol{\tau}_h\, dx = -\sum_{e \in \mathcal{E}} \int_e [[u_h]] \cdot \{\{\boldsymbol{\tau}_h\}\}\, ds \quad \forall\, \boldsymbol{\tau}_h \in \Sigma_h,$$

$L_\beta : \left[L^1(\mathcal{E})\right]^d \to \Sigma_h$ is defined by

$$\int_\Omega L_\beta([[u_h]]) \cdot \boldsymbol{\tau}_h\, dx = -\sum_{e \in \mathcal{E}^0} \int_e \beta \cdot [[u_h]]\, [[\boldsymbol{\tau}_h]]\, ds \quad \forall\, \boldsymbol{\tau}_h \in \Sigma_h,$$

with $\beta \in \mathbb{R}^d$. So using these lifting operators, it is easy to see that (3) is equivalent to the so-called *primal formulation*:

Find $u_h \in V_h$ s.t. $\forall\, v_h \in V_h$ (4)

$$\int_\Omega (\nabla_h u_h + R([[u_h]]) + L_\beta([[u_h]])) \cdot (\nabla_h v_h + R([[v_h]]) + L_\beta([[v_h]]))\, dx$$
$$+ \int_\mathcal{E} C_{11}\, [[u_h]] \cdot [[v_h]]\, ds = \int_\Omega f v_h\, dx,$$

with $\beta = -\mathbf{C_{12}}$ from now on.

The Bassi-Rebay method

If we choose $C_{11} = 0$ and $\mathbf{C_{12}} = \mathbf{0}$, (3) and (4) define the original Bassi-Rebay method (see [2]). Such a method is unstable.

The local discontinuous Galerkin method

If we choose $C_{11} = C$ or $C_{11} = \frac{C}{h}$, with $C > 0$, and $\mathbf{C_{12}} \in \mathbb{R}^d$, then (3) and (4) are the flux and the primal formulation of the Local Discontinuous Galerkin method, respectively (see [1, 4, 5] and [6]).

2 Minimal stabilization

We say that a bilinear form $B(\,\cdot\,,\,\cdot\,)$ is stable respect to a norm $||\cdot||$ in a space V if
$$\exists\, C > 0 \text{ s.t. } B(v, v) \geq C||v||^2 \quad \forall\, v \in V.$$

If we define $||u_h||_B = \sqrt{B(u_h, u_h)}$, where

$$B(u_h, v_h) := \int_\Omega (\nabla_h u_h + R([[u_h]]) + L_\beta([[u_h]])) \cdot (\nabla_h v_h + R([[v_h]]) + L_\beta([[v_h]])) dx$$
$$+ \int_\mathcal{E} C_{11}\, [[u_h]] \cdot [[v_h]] ds \quad \forall\, u_h,\, v_h \in V_h,$$

it is clear that if $||\cdot||_B$ is a norm, then $B(\,\cdot\,,\,\cdot\,)$ is stable with respect to this norm in the space V_h. The following result holds true.

Proposition 1. *If $C_{11} = 0$, then there is u_h in V_h, $u_h \neq 0$, such that $||u_h||_B = 0$.*

Proof. As in [1], we define the kernel of the bilinear form $B(\cdot,\cdot)$ as

$$Ker(B) = \{v_h \in V_h \text{ s.t. } B(v_h, z_h) = 0 \,\forall\, z_h \in V_h\}$$
$$= \{v_h \in V_h \text{ s.t. } \nabla_h v_h + R([[v_h]]) + L_\beta([[v_h]]) = 0\}.$$

Therefore $v_h \in Ker(B(.,.))$ if, and only if,

- $\int_I v_h q\, dx = 0 \quad \forall q \in P^{k-1}(I) \,\forall I \in \mathcal{T}_h,$
- $\{\{v_h\}\} - \boldsymbol{\beta} \cdot [[v_h]]|_e = 0 \quad \forall e \in \mathcal{E}^0.$

These conditions do not imply $v_h = 0$, as it is easy to prove considering the degrees of freedom of an element of the space V_h and imposing the two conditions above (see Fig. 1). ∎

Theorem 1. *(Stability: the one-dimensional case). Let $d = 1$ and $\Omega = [a, b]$ be the domain of our model problem; if $C_{11} > 0$ in Γ, $C_{11} = 0$ in $\mathcal{E} \setminus \Gamma$, then the method is stable for the following choices of Γ*

$$\Gamma = \{b\} \text{ if } C_{12} \geq 0 \quad \text{or} \quad \Gamma = \{a\} \text{ if } C_{12} \leq 0.$$

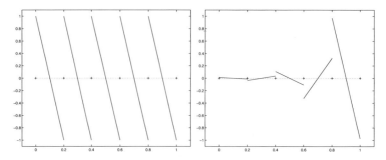

Fig. 1. Function $v_h \in Ker(B)$ with $\Omega = [0,1]$ and (a) $C_{11} = C_{12} = 0$, (b) $C_{11} = 0$, $C_{12} = 1$

Proof. If $\|u_h\|_B = 0$, then $u_h = 0$ on Γ because of the hypothesis on C_{11}.
Thus Proposition 1 and easy calculations show that $u_h \equiv 0$, and this completes the proof. ∎

In order to study the two-dimensional case, we consider a domain $\Omega \subseteq \mathbb{R}^2$ and a triangulation $\mathcal{T}_h$ and define the set G in the following way

$$G = \bigcup_{T \in \mathcal{T}_h} \left\{ e \in \partial T : \mathbf{C_{12}} \cdot \mathbf{n} < 0 \right\}$$

where $\mathbf{n}$ is the unit normal vector pointing exterior to the element T.

We are now ready to show the following results.

Theorem 2. *(Stability of the Bassi-Rebay method: the-two-dimensional case).* Let $\Omega \subseteq \mathbb{R}^2$ be the domain of our model problem, if $\Gamma = \{e\}, e \in \mathcal{E}^\partial$, and we choose $C_{11} > 0$ in Γ, $C_{11} = 0$ in $\mathcal{E} \setminus \Gamma$, $\mathbf{C_{12}} = \mathbf{0}$ in $\mathcal{E}$, then the method is stable.

Theorem 3. *(Stability of the LDG method: the two-dimensional case).* Let $\Omega \subseteq \mathbb{R}^2$ be the domain of our model problem and $int(\Omega) := \Omega \setminus \partial \Omega$ *(i.e. the interior of the domain), if we define*

$$\Gamma = \bigcup_{T \in \mathcal{T}_h} \{ e \in \partial T \cap \partial \Omega : \partial T \cap int(\Omega) \subseteq G \}$$

and we choose $C_{11} > 0$ in Γ, $C_{11} = 0$ in $\mathcal{E} \setminus \Gamma$, $\mathbf{C_{12}} \neq \mathbf{0}$ in $\mathcal{E}$, then the method is stable.

Proof. The proof is analogous to the one of Theorem 1 extended to the two-dimensional case. ∎

Remark 1. If we consider an element $T \in \mathcal{T}_h$ and an edge $e \in \partial T \cap \mathcal{E}^0$, we can rewrite the condition $\{\{v_h\}\} - \boldsymbol{\beta} \cdot [[v_h]]|_e = 0$, introduced in the proof of Prop. 1, as $v_h^I|_e = \alpha\, v_h^E|_e$, where I and E mean interior and exterior to the element T, respectively, and α depends on the coefficient C_{12}. We speak about *outflow stabilization* if $|\alpha| > 1$ (i.e., if $|v_h^I|_e| < \epsilon$, then $|v_h^E|_e| < \epsilon$) and we choose Γ considering all the element $T \in \mathcal{T}_h$ such that for all the edges $e \in \partial T \cap \mathcal{E}^0$ the condition of outflow stabilization holds. If $C_{12} = 0$ this condition never holds, so we are free to choose $\Gamma = \{e\}$, for any $e \in \mathcal{E}^\partial$.

Remark 2. In Theorems 1, 2 and 3 we have found conditions on the parameters C_{11} and $\mathbf{C_{12}}$ which make the method stable. Unlike the usual stabilizations of the Bassi-Rebay method and the standard LDG method, we consider the stability term only over a part of the boundary of the domain, instead of over the skeleton of the entire mesh, that is why we call it Minimal Stabilization. Thus we speak of Bassi-Rebay method with Minimal Stabilization if $\mathbf{C_{12} = 0}$, and of a Local Discontinuous Galerkin method with Minimal Stabilization otherwise.

3 Numerical results

We define the following norm in $V(h) = H^1(\mathcal{T}_h) + V_h$:

$$|||v|||^2 = \sum_{K \in \mathcal{T}_h} ||\nabla v||^2_{0,K} + \sum_{e \in \mathcal{E}} \frac{1}{h} \int_e [[v]]^2 (s)\, ds,$$

where $H^1(\mathcal{T}_h) = \{v \in L^2(\Omega) : v|_T \in H^1(T)\ \forall\, T \in \mathcal{T}_h\}$.

We now show numerical results about the error in the above discrete norm and in the L^2-norm.

3.1 One-dimensional case

We consider the problem

$$\begin{cases} -u''(x) = \sin(x) & \text{in } [0, \pi], \\ u(0) = u(\pi) = 0. \end{cases}$$

First of all we study the Bassi-Rebay method with Minimal Stabilization; we suppose

$$\Gamma = \{0\};\quad C_{11} = 1 \text{ on } \Gamma;\quad C_{12} = 0.$$

From the numerical results shown in Table 1 and in others not reported here, it is clear that the orders of convergence are the following:

$$\begin{array}{ll}
||u - u_h||_0 \le C\, h^k, & |||u - u_h||| \le C\, h^{k-1} \quad \text{if } k \text{ is odd,} \\
||u - u_h||_0 \le C\, h^{k+1}, & |||u - u_h||| \le C\, h^k \quad \text{if } k \text{ is even.}
\end{array}$$

Table 1. BR method with Minimal Stabilization: error taking $k = 1, 2$.

h	$k=1, \|\cdot\|_0$	$k=1, \|\|\cdot\|\|$	$k=2, \|\cdot\|_0$	$k=2, \|\|\cdot\|\|$
0.196349541	0.027218431	0.469291917	0.000037743	0.000871445
0.098174770	0.013655816	0.476612166	0.000004415	0.000220331
0.049087385	0.006833809	0.479697877	0.000000540	0.000055372
0.024543693	0.003417640	0.481097048	0.000000067	0.000013878
0.012271846	0.001708912	0.481760837	0.000000008	0.000003474
0.006135923	0.000854467	0.482083807	0.000000001	0.000000869
0.003067962	0.000427235	0.482243064	0.000000000	0.000000217

This difference between odd and even polynomial degrees was also noticed in [2] for the original method.

Now we consider the Local Discontinuous Galerkin with Minimal Stabilization: we take

$$\Gamma = \{\pi\}; \quad C_{11} = 1 \text{ on } \Gamma; \quad C_{12} = \frac{1}{2}.$$

In the numerical results of Table 2 and in others not reported here, we obtain the same orders of convergence as the original LDG, i.e. $\|u - u_h\|_0 \leq Ch^{k+1}$ and $\|\|u - u_h\|\| \leq C\,h^k$, $\forall k \geq 1$.

Table 2. LDG method with Minimal Stabilization: error taking $k = 1, 2$.

h	$k=1, \|\cdot\|_0$	$k=1, \|\|\cdot\|\|$	$k=2, \|\cdot\|_0$	$k=2, \|\|\cdot\|\|$
0.196349541	0.003099825	0.042309854	0.000045792	0.000780704
0.098174770	0.000752208	0.020831067	0.000005726	0.000198182
0.049087385	0.000185748	0.010334288	0.000000718	0.000049938
0.024543693	0.000046182	0.005146926	0.000000090	0.000012534
0.012271846	0.000011515	0.002568429	0.000000011	0.000003140
0.006135923	0.000002875	0.001282959	0.000000001	0.000000786
0.003067962	0.000000718	0.000641166	0.000000000	0.000000197

3.2 Two-dimensional case

We consider the problem

$$\begin{cases} -\Delta u(x) = \frac{\Pi^2}{2} \sin(\frac{\pi}{2}(x+1))\sin(\frac{\pi}{2}(y+1)) & \text{in } \Omega := [-1,1] \times [-1,1], \\ u = 0 & \text{on } \partial\Omega. \end{cases}$$

First of all we consider the BR method with Minimal Stabilization: given an unstructured mesh and Γ (depending on the mesh) as for example shown in

Fig. 2(a), choosing $C_{11} = 1$ on Γ, 0 otherwise, we obtain the errors shown in Table 3(a) for $k = 1$.

Next we present numerical experiments for the LDG method with Minimal Stabilization: consider a structured mesh, if we choose $\mathbf{C_{12}} = (1, 1/2)$ and $C_{11} = 1$ on Γ, Γ chosen according to Theorem 3 (see Fig. 2(b) for an example), we obtain the results shown in Table 3(b) for $k = 1$.

Table 3. Error (a) for the BR method with Minimal Stabilization, (b) for the LDG with Minimal Stabilization.

h	$\|\cdot\|_0$	$\|\|\cdot\|\|$	h	$\|\cdot\|_0$	$\|\|\cdot\|\|$
0.3536	0.294685	2.010469	0.5	3.946351	21.470925
0.1768	0.068335	0.940656	0.25	0.146931	1.634253
0.0884	0.017151	0.468730	0.125	0.052945	1.190417
0.0442	0.004372	0.237118	0.0625	0.010127	0.454469
0.0221	0.001095	0.118248	0.0313	0.002685	0.243431

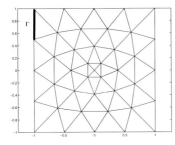
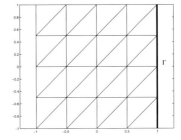

Fig. 2. Example of (a) an unstructured mesh and (b) a structured mesh used in the numerical experiments.

From this numerical results and from others not reported here, we can assume that for both methods the following inequality holds:

$$\|u - u_h\|_0 \leq Ch^{k+1} \quad , \quad \|\|u - u_h\|\| \leq C h^k \qquad \forall \, k \geq 1.$$

References

1. Arnold, D.N., Brezzi, F., Cockburn, B., Marini, L.D.: Unified analysis of discontinuous Galerkin methods for elliptic problems. SIAM J. Num. Anal, **39** (2002)
2. Bassi, F., Rebay, S.: A high-order accurate discontinuous Galerkin finite element method for the numerical solution of the compressible Navier-Stokes equations. Journal of Computational Physics, **131** (1997)

3. Brezzi, F., Hughes, T.J.R., Marini, L.D., Masud, A.: Mixed discontinuous Galerkin methods for Darcy flow. In: ICES Report 04-17, University of Texas at Austin (2004)
4. Castillo, P., Cockburn, B., Perugia, I., Schotzau, D.: An a priori error analysis of the LDG method for elliptic problems. SIAM J. Numer.Anal., **38**, No.5, 1676–1706 (2000)
5. Cockburn, B., Shu, C.-W.: The local discontinuous Galerkin method for time-dependent convection-diffusion systems. SIAM Journal on Numerical Analysis, **35** (1998)
6. Perugia, I., Schötzau, D.: An hp-analysis of the Local Discontinuous Galerkin method for diffusion problems. J. Sci. Comp., **17**, 561–571 (2002)

Discontinuous Galerkin Finite Element Method for a Fourth-Order Nonlinear Elliptic Equation Related to the Two-Dimensional Navier–Stokes Equations

Igor Mozolevski[1]*, Endre Süli[2] and Paulo Rafael Bösing[3]†

[1] Mathematics Department, Federal University of Santa Catarina, Campus Universitário, 88040-900, Trindade, Florianópolis, SC, Brazil
igor@mtm.ufsc.br
[2] University of Oxford, Computing Laboratory, Wolfson Building, Parks Road, Oxford OX1 3QD, UK
Endre.Suli@comlab.ox.ac.uk
[3] Department of Applied Mathematics, IME, University of São Paulo, Rua do Matão, 1010, 05508-090, São Paulo, SP, Brazil
paulo@ime.usp.br

Summary. We develop an hp-version discontinuous Galerkin method for a nonlinear biharmonic equation corresponding to the two-dimensional incompressible Navier–Stokes equations in the stream-function formulation. We linearize the equation and then we solve the resulting linear problem using a combination of the nonsymmetric discontinuous Galerkin finite element method for the biharmonic part of the equation, and a discontinuous Galerkin finite element method with a jump-penalty term for the hyperbolic part of the equation. Numerical experiments are presented to demonstrate the accuracy of the method for a wide range of Reynolds numbers.

1 Introduction

One of the most important challenges that must be addressed in the design of high-order finite element approximations for practical problems is the construction of efficient hp-adaptive algorithms, capable of delivering accurate numerical approximations in a reliable and robust manner. The latter objective has led in recent years to the intensive study of discontinuous Galerkin finite element methods (DGFEM) for the Navier–Stokes equations, with the aim to develop high-order numerical algorithms for industrially relevant CFD problems (see, for example, [5] for aerodynamic simulations).

* Partially supported by CNPq-Brazil.
† Grant from CNPq-Brazil.

The papers of Cockburn and co-workers (see [4] for a review) have introduced, analyzed and numerically tested local discontinuous Galerkin methods for linear incompressible fluid flow. A family of DGFEMs for Stokes and Navier–Stokes problems was formulated and analyzed recently in [3]. The discontinuous Galerkin method is a stabilized mixed finite element method, which is locally conservative, offers high-order accuracy and is very robust for a wide range of Reynolds numbers. The fact that the finite element space consists of discontinuous piecewise polynomial functions makes the method ideally suitable for the design of hp-adaptive finite element algorithms on irregular meshes which admit any number of hanging nodes.

A critical consideration in the construction of mixed finite element approximations of the incompressible Navier–Stokes equations in the primitive-variable (i.e. velocity–pressure) formulation is that the finite element spaces for the velocity and the pressure need to be compatible in the sense that a Babuška–Brezzi type inf-sup condition is satisfied, — preferably, independent of the discretization parameters. An alternative, at least in two space dimensions, is to use the stream-function formulation of the incompressible Navier–Stokes equations. This ensures that the incompressibility constraint is automatically satisfied, though the system of Navier–Stokes equations is transformed into a scalar nonlinear fourth-order partial differential equation (cf. [2]). A number of authors have used this approach to solve some practical problems and to demonstrate the effectiveness of the finite element method in a range of geometries.

Having said this, the application of *conforming* finite element methods to fourth-order partial differential equations suffers from the disadvantage that only elements achieving global C^1 continuity may be employed. This difficulty is easily avoided by using nonconforming globally C^0 finite element methods, or — even more extremely — discontinuous Galerkin finite element methods (DGFEMs). Indeed, the use of completely discontinuous finite element approximations leads to easy implementation of locally high-order finite elements, without the need for enforcing global regularity requirements; see [9].

In this paper we present the construction, validation, and application of an hp-version discontinuous Galerkin finite element method for the numerical solution of the Navier–Stokes equations governing two-dimensional stationary incompressible flows.

We linearize the equation using a Picard-type fixed-point iteration and we then solve the resulting linear problem by combining a discontinuous Galerkin finite element method for the biharmonic equation proposed in [9] with a discontinuous Galerkin finite element approximation of the advective terms developed in [1].

The paper consists of four sections. Using a stream-function formulation, in Section 2 we reduce the system of Navier–Stokes equations to a single fourth-order nonlinear partial differential equation and consider a linearization of this equation. Then, in Section 3, we introduce the discontinuous finite element space, define the discontinuous Galerkin finite element method for a fourth-order linear advective equation and present the main result — an hp-version error bound for the method. In Section 4, we confirm numerically the order of convergence of the method for a wide range of Reynolds numbers, on the so-called Kovasznay solution, [8], which, for a given Reynolds number, is a two-dimensional analytical solution of the incompressible Navier–Stokes equations. Finally, we consider the application of our method to the two-dimensional lid-driven cavity problem.

2 Mathematical formulation

Let $\Omega \subset \mathbf{R}^2$ be a bounded convex polygonal domain with boundary $\partial\Omega$. We consider in Ω a steady, two-dimensional incompressible fluid flow which is governed by the Navier–Stokes equation:

$$-\mu \triangle \mathbf{v} + \mathbf{v} \cdot \nabla \mathbf{v} + \nabla p = \mathbf{F} \quad \text{in } \Omega \tag{1}$$

and the continuity equation

$$\nabla \cdot \mathbf{v} = 0 \quad \text{in } \Omega, \tag{2}$$

subject to the boundary condition

$$\mathbf{v} = \mathbf{g} \quad \text{on } \partial\Omega. \tag{3}$$

In these equations $\mathbf{v}$ is the velocity field, p is the pressure, μ is the kinematic viscosity of the fluid, $\mu = 1/\text{Re}$, Re is Reynolds number, $\mathbf{F}$ is a prescribed external body force. We shall suppose that $f = \frac{\partial \mathbf{F}_2}{\partial x_1} - \frac{\partial \mathbf{F}_1}{\partial x_2} \in \mathrm{L}^2(\Omega)$ and that the Dirichlet data $\mathbf{g}$ are sufficiently smooth and satisfy the compatibility condition $\int_{\partial\Omega} \mathbf{g} \cdot \mathbf{n} \, ds = 0$, where $\mathbf{n}$ is the outward normal unit vector to $\partial\Omega$.

Let ψ be a stream-function related to the velocity field $\mathbf{v}$ as follows: $v_1 = \frac{\partial \psi}{\partial x_2}$, $v_2 = -\frac{\partial \psi}{\partial x_1}$; then, the Navier–Stokes equations can be reduced to

$$\frac{1}{\text{Re}} \triangle^2 \psi + \left(\frac{\partial}{\partial x_2} \triangle \psi\right) \frac{\partial \psi}{\partial x_1} - \left(\frac{\partial}{\partial x_1} \triangle \psi\right) \frac{\partial \psi}{\partial x_2} = f \quad \text{in } \Omega. \tag{4}$$

Note that in this formulation the incompressibility constraint is automatically satisfied and the pressure is excluded. However, this approach is valid in 2D only. Due to the nonlinearities, this equation is solved iteratively by a Picard-type fixed point iteration or by the Newton–Raphson method, using a linearization of the equation (4). Therefore, in this work we focus on the following boundary-value problem for the linear fourth-order elliptic equation with advective term:

$$\frac{1}{\text{Re}} \triangle^2 \psi + \mathbf{b} \cdot \nabla \psi = f \quad \text{in } \Omega, \tag{5}$$
$$\psi = g_0 \quad \text{on } \partial\Omega,$$
$$\mathbf{n} \cdot \nabla \psi = g_1 \quad \text{on } \partial\Omega,$$

where $f \in \mathrm{L}^2(\Omega)$, $\mathbf{b} = (b_1, b_2) \in \mathrm{C}^1(\overline{\Omega}) \times \mathrm{C}^1(\overline{\Omega})$.

3 DGFEM for a 4th-order advective PDE

Let us consider a shape-regular family of triangulations $\{\mathcal{K}_h\}$ of Ω of granularity h, $h = \max h_K$, $h_K = \text{diam}(K)$, $K \in \mathcal{K}_h$, such that each $K \in \mathcal{K}_h$ is an affine image of the master element $\widehat{K} = (0,1) \times (0,1) \subset \mathbf{R}^2$: $K = F_K(\widehat{K})$, $K \in \mathcal{K}_h$. Let e denote the interior of any edge in the triangulation and let $\mathcal{E}$ be the set of (open) edges e of all elements in the mesh. Let $\mathcal{E}_{\text{int}} = \{e \in \mathcal{E} : e \subset \Omega\}$ be the set of all interior edges, and $\mathcal{E}_\partial = \{e \in \mathcal{E} : e \subset \partial\Omega\}$ the set of all boundary edges. In what follows we will use the standard Discontinuous Galerkin nomenclature. For example, we define the (mesh-dependent) broken Sobolev space equipped with the corresponding broken Sobolev norm:

$$\mathrm{H}^{\mathbf{s}}(\Omega, \mathcal{K}_h) = \left\{ \psi \in \mathrm{L}^2(\Omega) : \psi|_K \in \mathrm{H}^{s_K}(K) \quad \forall K \in \mathcal{K}_h \right\},$$

where $s_K \geq 0$ is the local Sobolev index; we introduce the finite element space $S^{\mathbf{p}}(\Omega, \mathcal{K}_h, \mathbf{F}) = \left\{ \psi \in \mathrm{L}^2(\Omega) : \psi|_K \circ F_K \in \mathcal{Q}_{p_K}(\widehat{K}) \, \forall K \in \mathcal{K}_h \right\}$, where $\mathcal{Q}_p(\widehat{K}) = \text{span}\{x_1^{\alpha_1} x_2^{\alpha_2} : 0 \leq \alpha_1, \alpha_2 \leq p\}$, and p_K is the local polynomial approximation degree in K for each $K \in \mathcal{K}_h$.

Let us introduce the hp-version of the interior penalty discontinuous Galerkin finite element method for the boundary value problem (5). Following the ideas presented in [9], we shall consider the nonsymmetric formulation corresponding to the biharmonic operator and we shall use the stabilised discontinuous Galerkin method introduced in [1] to approximate the advective terms of the equation. Thus, we consider the bilinear form

$$B_{\text{DG}}(\psi, \phi) = \frac{1}{\text{Re}} \Bigg[\int_\Omega \Delta\psi \Delta\phi \, dx + \int_\mathcal{E} \Big(\{\nabla(\Delta\psi)\} \cdot [\![\phi]\!] - [\![\psi]\!] \cdot \{\nabla(\Delta\phi)\} \Big) ds$$
$$- \int_\mathcal{E} \Big(\{\Delta\psi\}[\![\nabla\phi]\!] - [\![\nabla\psi]\!]\{\Delta\phi\} \Big) ds + \int_\mathcal{E} \Big(\alpha[\![\psi]\!] \cdot [\![\phi]\!] + \beta[\![\nabla\psi]\!][\![\nabla\phi]\!] \Big) ds \Bigg]$$
$$+ \int_\Omega \mathbf{b} \cdot \nabla\psi\phi \, dx - \int_{\mathcal{E}_{\text{int}}} [\![\psi]\!] \cdot \{\mathbf{b}\phi\} ds + \int_{\mathcal{E}_{\text{int}} \cup \mathcal{E}_{\partial_-}} c_\mathcal{E} [\![\psi]\!] \cdot [\![\phi]\!] ds$$

on $[S^{\mathbf{p}}(\Omega, \mathcal{K}_h, \mathbf{F})]^2$. Here we have used the following notation

$$\int_\mathcal{A} \psi \, ds = \sum_{e \in \mathcal{A}} \int_e \psi \, ds$$

for the integral over any subset $\mathcal{A}$ of the skeleton $\mathcal{E}$; $[\![\cdot]\!]$ and $\{\cdot\}$, respectively, denote the jump and the mean-value of a vector- or scalar-function across an interior or boundary edge; on a boundary edge, the function to which $[\![\cdot]\!]$ and $\{\cdot\}$ are applied is defined to be zero outside the set Ω. The functions α and β are defined, edge-wise, by the formulae $\alpha|_e = \alpha_e$, $\beta|_e = \beta_e$ for all $e \in \mathcal{E}$, $\alpha_e = \sigma_\alpha \frac{\{p^6\}_e}{h_e^3}$, $\beta_e = \sigma_\beta \frac{\{p^2\}_e}{h_e}$, $c_\mathcal{E} \geq \theta |\mathbf{b} \cdot \mathbf{n}|$ on $e \in \mathcal{E}_{\text{int}}$, $c_\mathcal{E} = |\mathbf{b} \cdot \mathbf{n}|$ on $e \in \mathcal{E}_\partial$, where

σ_α, σ_β and θ are positive constants independent of e; $\theta = 1/2$ corresponds to upwinding. Let us also define the linear functional $l(\cdot)$ on $S^\mathbf{p}(\Omega, \mathcal{K}_h, \mathbf{F})$:

$$l(\phi) = \frac{1}{\mathrm{Re}}\left[\int_\Omega f\phi\,\mathrm{d}x - \int_{\mathcal{E}_\partial}\Big(g_0(\mathbf{n}\cdot\nabla(\Delta\phi)) - g_1\Delta\phi\Big)\mathrm{d}s \right.$$
$$\left. + \int_{\mathcal{E}_\partial}\Big(\alpha g_0\phi + \beta g_1(\nu\cdot\nabla\phi)\Big)\mathrm{d}s\right] + \int_{\mathcal{E}_\partial} c_\mathcal{E} g_0\phi\,\mathrm{d}s.$$

We introduce the following interior penalty Discontinuous Galerkin method.

IPDGM: Find $\psi_{\mathrm{DG}} \in S^\mathbf{p}(\Omega, \mathcal{K}_h, \mathbf{F})$ such that

$$B_{\mathrm{DG}}(\psi_{\mathrm{DG}}, \phi) = l(\phi) \qquad \forall \phi \in S^\mathbf{p}(\Omega, \mathcal{K}_h, \mathbf{F}). \tag{6}$$

In order to ensure the consistency of the method (and thus the Galerkin orthogonality property), we suppose that the solution ψ to the boundary value problem (5) is sufficiently smooth: namely $\psi \in \mathrm{H}^\mathbf{s}(\Omega, \mathcal{K}_h)$ with $s_K \geq 4$ for all $K \in \mathcal{K}_h$ and $[\![\nabla(\Delta\psi)]\!] = [\![\nabla\psi]\!] = [\![\Delta\psi]\!] = [\![\psi]\!] = 0$ on all edges e in $\mathcal{E}_{int}$.

Let us consider the norm $|\!|\!|\cdot|\!|\!|_{\mathrm{DG}}$ associated with the bilinear form $B_{\mathrm{DG}}(\cdot,\cdot)$:

$$|\!|\!|\phi|\!|\!|^2_{\mathrm{DG}} = \frac{1}{\mathrm{Re}}\left[\|\Delta\phi\|^2_{0,\Omega} + \left\|\sqrt{\alpha}[\![\phi]\!]\right\|^2_{0,\mathcal{E}} + \left\|\sqrt{\beta}[\![\nabla\phi]\!]\right\|^2_{0,\mathcal{E}}\right] + \|\phi\|^2_{0,\Omega} + \left\|\sqrt{c_\mathcal{E}}[\![\phi]\!]\right\|^2_{0,\mathcal{E}}$$

where $\phi \in S^\mathbf{p}(\Omega, \mathcal{K}_h, \mathbf{F})$. To proceed, we adopt the following hypotheses.

Hypothesis H1. There exists a positive constant γ_0 such that

$$-\nabla\cdot b \geq \gamma_0 \quad \text{a.e. in} \quad \Omega.$$

Hypothesis H2. $b\cdot\nabla\phi \in S^\mathbf{p}(\Omega, \mathcal{K}_h, \mathbf{F})$ for all $\phi \in S^\mathbf{p}(\Omega, \mathcal{K}_h, \mathbf{F})$.

Now, combining the hp-version error analyses of the nonsymmetric interior-penalty DGFEM for the biharmonic equation from [9] and of a stabilized version of the DGFEM for first-order hyperbolic equations from [1] and [7], we arrive at the following *a priori* error bound for the IPDGM (3).

Theorem 1. *Let $\mathbf{p} = (p_K, K \in \mathcal{K})$, $p_K \geq 2$, be an arbitrary polynomial degree vector of bounded local variation. Let us suppose that the exact solution ψ to the problem belongs to $\mathrm{H}^\mathbf{s}(\Omega, \mathcal{K}) \cap \mathrm{H}^4(\Omega)$ and let ψ_{DG} be the solution to the discrete problem **IPDGM**. Moreover, let us assume that $\sigma_\alpha > 0$, $\sigma_\beta > 0$ and that Hypotheses 1 and 2 are valid. Then, the following error bound holds:*

$$|\!|\!|\psi - \psi_{\mathrm{DG}}|\!|\!|^2_{\mathrm{DG}} \leq C\sum_{K\in\mathcal{K}_h}\left[\frac{1}{\mathrm{Re}}\frac{h_K^{2t_K-4}}{p_K^{2s_K-7}} + \frac{h_K^{2t_K-1}}{p_K^{2s_K-1}}\theta_0 + \|\nabla\cdot\mathbf{b}\|_{\infty,K}\frac{h_K^{2t_K}}{p_K^{2s_K}}\right]\|\psi\|^2_{\mathrm{H}^{s_K}(K)},$$

where $\theta_0 = \max(\frac{1}{\theta}, 1)$, $2 \leq t_K \leq \min(p_K+1, s_K)$, and C depend only on the space-dimension, the shape-regularity constant, and on $s = \max\limits_{K\in\mathcal{K}_h} s_K$, $s_K \geq 4$.

4 Numerical results

We begin by presenting a numerical experiment to confirm the *a priori* error estimates derived above. For this purpose we choose the two-dimensional exact solution to the incompressible Navier–Stokes equations derived by Kovasznay [8]. Hence, in $\Omega = (-0.5, 1.5) \times (0.0, 2.0)$ we solve the fourth-order linear advective equation with the coefficients

$$b_1 = \exp(\lambda x_1)\cos(2\pi x_2)(4\pi^2 - \lambda^2), \quad b_2 = \frac{\lambda}{2\pi}\exp(\lambda x_1)\sin(2\pi x_2)(\lambda^2 - 4\pi^2)$$

with right-hand side f and Dirichlet boundary data g_0, g_1 which correspond to Kovasznay's exact solution: $\Psi(x_1, x_2) = x_2 - \frac{1}{2\pi}\exp(\lambda x_1)\sin(2\pi x_2)$, where $\lambda = \mathrm{Re}/2 - \sqrt{\mathrm{Re}^2/4 + 4\pi^2}$.

For the purposes of calculating the order of h-convergence, quadrilateral meshes generated by consecutive refinements of the original computational region were used. In each refinement, each grid cell is divided into four similar cells by connecting the midpoints of opposite edges. In Table 1, orders of convergence of the method with respect to the DG norm $\|\cdot\|_{\mathrm{DG}}$ and the H^1 seminorm $|\cdot|_{H^1}$ are presented, for a wide range of Reynolds numbers. These orders were calculated on the refinement-levels $L = 3, 4, 5$, for polynomial degrees $p = 3, 4, 5$. As one can see from the table, the method seems robust and the numerical results confirm the theoretical orders of convergence for all values of the Reynolds number considered. We note that the order of convergence

Table 1. Observed errors in the H^1 seminorm and the DG norm.

p	Level	Re = 10 $\|\cdot\|_{H^1}$	$\|\cdot\|_{\mathrm{DG}}$	Re = 10^2 $\|\cdot\|_{H^1}$	$\|\cdot\|_{\mathrm{DG}}$	Re = 10^3 $\|\cdot\|_{H^1}$	$\|\cdot\|_{\mathrm{DG}}$	Re = 10^4 $\|\cdot\|_{H^1}$	$\|\cdot\|_{\mathrm{DG}}$
3	3	3.041	2.344	3.189	2.348	2.878	2.488	2.820	2.577
	4	2.820	2.126	2.886	2.039	2.889	2.051	2.850	2.098
	5	2.923	2.196	2.897	2.064	2.963	2.056	2.967	2.057
4	3	1.606	0.965	1.665	1.044	1.686	1.077	1.751	1.075
	4	3.912	3.405	3.954	3.147	3.948	3.132	3.855	3.154
	5	3.948	3.213	3.986	3.060	3.983	3.051	3.973	3.053
5	3	5.264	4.585	5.344	4.500	5.348	4.491	5.216	4.556
	4	4.911	4.234	4.945	4.100	4.958	4.086	4.946	4.090
	5	4.953	4.256	4.923	4.094	4.983	4.081	4.984	4.081

of the method for the velocity field components is p (cf. $|\cdot|_{H^1}$ columns in Table 1), which is by one unit less than the corresponding order of the local discontinuous Galerkin method for the (linear) Oseen equation obtained in [4]. On the other hand, the local discontinuous Galerkin mixed finite element method from [4] involves many more unknowns than our method here. In our second

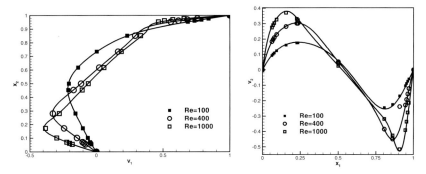

Fig. 1. Profiles of the velocity component v_1 along the vertical mid-line (left) and the velocity component v_2 along the horizontal mid-line (right), in comparison with the results from [6].

example, we demonstrate the potentials of the DGFEM described above when applied to the nonlinear equation (4). We consider the problem of simulating a two-dimensional lid-driven cavity flow, a model problem which is frequently used in the CFD literature for validation purposes. The flow-domain of interest is the unit square $\Omega = (0,1) \times (0,1)$ with the upper horizontal lid moving with uniform velocity $\mathbf{v} = (1,0)^\top$, which corresponds to the Dirichlet boundary condition $\psi = 0, \mathbf{n} \cdot \nabla \psi = 1$; the homogeneous Dirichlet boundary condition $\psi = 0, \mathbf{n} \cdot \nabla \psi = 0$ is applied on all the other (static) walls. We compute numerically the flow using a nonuniform rectangular mesh, refined at each corner, composed of 2340 elements and with Q_3 discontinuous polynomial approximation. We have used the Newton–Raphson method for solving the global nonlinear system $A(b)b = f$. We chose zero as the initial guess for b for Re = 100 and Re = 400, and for Re = 1000 the initial guess was taken from the previous result corresponding to Re = 700. In all cases considered no more than ten iterations were needed to obtain an approximate solution with relative error $e_i = \|b_i - b_{i-1}\|/\|b_{i-1}\| \leq 10^{-7}$, where b_i is the numerical solution from iteration i. In each Newton–Raphson step the linear system was solved using an LU factorization. Fig. 1 shows the v_1 velocity profile along the vertical mid-line and the v_2 velocity profile along the horizontal mid-line of the square, calculated for Re = 100, 400 and 1000, in comparison with the results from [6], in which the data were obtained on a 129×129 uniform grid using a second-order accurate finite-difference scheme for both the streamfunction and vorticity equations. For all Reynolds numbers considered, the velocity profiles computed by our method are in excellent agreement with the tabulated data from that work. In Fig. 2 we compare velocity profiles, computed by our method for p-refined and h-refined meshes with the same number of degrees of freedom, DOF= 14400. We see that the results corresponding to p-refinement perfectly coincide with the data from Ghia et al. [6], while the results corresponding to h-refinement are less accurate for larger Reynolds

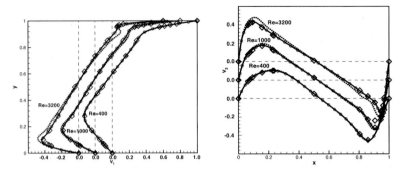

Fig. 2. Profiles of the velocity components v_1 (left) and v_2 along the vertical and horizontal mid-lines respectively for different mesh enrichments: $30 \times 30 \times \mathcal{Q}_3$ - dotted line, $20 \times 20 \times \mathcal{Q}_5$ - circle, $15 \times 15 \times \mathcal{Q}_7$ - solid line. The results from [6] - diamond.

numbers. So high-order DGFEMs presented here seem more appropriate for high Reynolds numbers simulations.

To conclude, we introduced a new hp-version interior-penalty DGFEM for the two-dimensional incompressible Navier–Stokes equations in streamfunction formulation and demonstrated the high accuracy of the method when solving a classical benchmark problem.

References

1. Brezzi, F., Marini, L., Süli, E.: Discontinuos Galerkin methods for first-order hyperbolic problems. Math. Models Methods Appl. Sci. **14**, 1893–1903 (2004)
2. Ciarlet, P.G.: Basic Error Estimates for Elliptic Problems. In: Handbook of Numerical Analysis, Vol.2. North-Holland, Amsterdam (1991)
3. Girault, V., Rivière, B., Wheeler, M.-F.: A discontinuous Galerkin method with nonoverlapping domain decomposition for the Stokes and Navier–Stokes problems. Math. Comp. **74**, 53–84 (2005)
4. Cockburn, B., Kanschat, G., Schötzau, D.: The local discontinuous Galerkin method for linear incompressible fluid flow: A review. **34**(4-5), 491–506, (2005)
5. Fidkowski, K. J Darmofal, D. L.: Development of a higher-order solver for aerodynamic applications. 42nd AIAA Aerospace Sciences Meeting and Exhibit, AIAA Paper 2004-0436 (2004)
6. Ghia, U., Ghia, K.N., Shin, C.T.: High-Resolutions for incompressible flow using the Navier–Stokes equation and a multigrid method. J. Comput. Phys., **48**, 387–411 (1982)
7. Houston, P., Schwab, C., Süli, E.: Stabilized hp-finite element methods for hyperbolic problem. SIAM J. Numer. Anal., **37**, 1618–1643 (2000)
8. Kovasznay, L.: Laminar flow behind a two-dimensional grid. Proc. Camb. Philos. Soc., **44**, 58–62, (1948)
9. Mozolevski, I. Süli, E.: A priori error analysis for the hp-version of the discontinuous Galerkin finite element method for the biharmonic equation. Comput. Meth. Appl. Math. **3**(4), 596–607, (2003)

Domain Decomposition

Fourier Method with Nitsche-Mortaring for the Poisson Equation in 3D

Bernd Heinrich and Beate Jung

Technische Universität Chemnitz, Fakultät für Mathematik, D-09107 Chemnitz
b.heinrich@mathematik.tu-chemnitz.de, b.jung@mathematik.tu-chemnitz.de

The paper deals with a combination of the Nitsche-mortaring with the Fourier-finite-element method. The approach is applied to the Dirichlet problem of the Poisson equation in three-dimensional axisymmetric domains with non-axisymmetric data. The approximating Fourier method yields a splitting of the 3D-problem into 2D-problems on the meridian plane treated by the Nitsche-finite-element method (as a mortar method). Some important properties of the approximation scheme as well as error estimates in some H^1-like norm as well as in the L_2-norm are derived.

1 Introduction

For the efficient numerical treatment of boundary value problems (BVP) in 3D, domain decomposition methods as well as dimension decomposition methods are widely used in science and engineering. Both methods are convenient for parallelization of the numerical solution of partial differential equations.

In this paper, we shall present a combination of the so-called Fourier-finite-element method with the Nitsche-finite-element method as a mortar method. The approach is applied to the Dirichlet problem of the Poisson equation,

$$-\Delta_3 \hat{u} := -\sum_{i=1}^{3} \frac{\partial^2 \hat{u}}{\partial x_i^2} = \hat{f} \quad \text{in } \widehat{\Omega}, \quad \hat{u} = 0 \quad \text{on } \partial\widehat{\Omega}, \quad \widehat{\Omega} \in \mathbb{R}^3, \tag{1}$$

where the domain $\widehat{\Omega}$ is bounded and axisymmetric with respect to the x_3-axis. The data and the solution $\hat{u}$ of the BVP in 3D are non-axisymmetric. If we denote the part of the x_3-axis contained in $\widehat{\Omega}$ by Γ_0, then the set $\widehat{\Omega} \setminus \Gamma_0$ is generated by rotation of a plane polygonal meridian domain Ω_a about the x_3-axis.

The two methods to be combined can be characterized as follows. The Fourier-finite-element method (FFEM, see e.g. [4, 5, 7, 8, 11]) is based on the well-known approximating Fourier method and on the finite-element method.

That is, trigonometric polynomials of degree $\leq N$ are used in one space direction, here with respect to the rotational angle φ. This yields an approximate splitting of the 3D-problem into $2N+1$ problems on the 2D domain Ω_a for the parameter $k = 0, \pm 1, ..., \pm N$, with solutions u_k being the Fourier coefficients of u.

Furthermore, we employ the Nitsche-finite-element discretization as a mortar method for solving numerically the 2D-problems on the meridian domain Ω_a (cf. [2, 6, 9, 12], also [1, 3, 13] for general aspects). Along the interface Γ of the domain decomposition of Ω_a, non-matching meshes (cf. Fig. 1(b)) as well as discontinuities of the approximated solutions are admitted. But compared with the papers cited previously, the differential operator depends now on the parameter k and has a more general form.

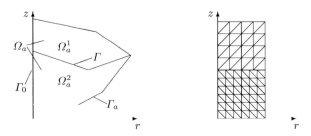

Fig. 1. (a) Domain Ω_a with subdomains; (b) Non-matching triangulation

The aim of this paper is to present the combined method, which seems to be new. This method has the advantage that the dimension of the problem is reduced and that we have a natural parallelization of the solution process. Moreover, the handling of non-matching meshes of triangles on the meridian domain Ω_a is easier than of non-matching meshes and elements in 3D. In the following, it is analyzed how the approximation schemes in 2D generate the mortar approximation in 3D. Important properties of the approximation schemes as well as results for convergence $u_{hN} \to u$ of the Nitsche-Fourier-finite-element approximation u_{hN} with respect to $N \to \infty$ and $h \to 0$ (N: length of the Fourier sum, h: mesh size on Ω_a) are presented, where N and h can be chosen independently from each other. In some H^1-like norm and for regular solutions u, the convergence rate is proved to be of the type $\mathcal{O}(h + N^{-1})$, in the L_2-norm like $\mathcal{O}(h^2 + N^{-2})$.

Since the domain $\widehat{\Omega}$ is axisymmmetric, we employ cylindrical coordinates r, φ, z ($x_1 = r\cos\varphi$, $x_2 = r\sin\varphi$, $x_3 = z$), with $r > 0$ and $\varphi \in (-\pi, \pi]$. Here, r is the distance of a point to the z-axis, φ the rotational angle. For each function $\hat{v}(x)$ with $x \in \widehat{\Omega} \setminus \Gamma_0$, some function v is defined on $\Omega := \Omega_a \times (-\pi, \pi]$ by

$$v(r, \varphi, z) := \hat{v}(r\cos\varphi, r\sin\varphi, z). \tag{2}$$

The boundary part Γ_a is defined by $\Gamma_a := \partial\Omega_a \setminus \overline{\Gamma}_0$, where $\partial\Omega_a \in C^{0,1}$ is the boundary of Ω_a, see Fig. 1(a). For getting regular solutions $\hat{u} \in H^2(\widehat{\Omega})$ of the

BVP (1) for $\hat{f} \in L_2(\hat{\Omega})$ ($H^s(\hat{\Omega})$: the usual Sobolev-Slobodetskiĭ space with $s \geq 0$, s real, $H^0 = L_2$), it is sufficient to assume that the interior angles θ at the corners of $\partial\Omega_a$ satisfy $\theta < \pi$, at the x_3-axis even $\theta < 0.72616\pi$ (cf. [4]).

We denote by $X^l_{1/2}(\Omega)$ the Sobolev-type spaces of functions periodic with respect to $\varphi \in (-\pi, \pi]$ and with the weights $r^{\frac{1}{2}}$ received by the mapping (2): $H^l(\hat{\Omega}) \to X^l_{1/2}(\Omega)$ ($l = 0, 1, 2$), where $X^l_{1/2}(\Omega)$ represents the space $H^l(\hat{\Omega})$ in terms of cylindrical coordinates, for details we refer to [11, 8]. According to (2), the variational formulation of the BVP (1) in cylindrical coordinates is given as follows. Find $u \in V_0(\Omega) := \{u \in X^1_{1/2}(\Omega) : u|_{\Gamma_a \times (-\pi,\pi]} = 0\}$:

$$b(u,v) = f(v) \quad \forall v \in V_0(\Omega), \quad \text{with} \tag{3}$$

$$b(u,v) := \int_\Omega \left\{ \frac{\partial u}{\partial r}\overline{\frac{\partial v}{\partial r}} + \frac{1}{r^2}\frac{\partial u}{\partial \varphi}\overline{\frac{\partial v}{\partial \varphi}} + \frac{\partial u}{\partial z}\overline{\frac{\partial v}{\partial z}} \right\} r\,dr\,d\varphi\,dz, \quad f(v) := \int_\Omega f\,\bar{v}\,r\,dr\,d\varphi\,dz.$$

2 Fourier decomposition and mortaring in 2D

For $u(r, \varphi, z)$, $u \in X^1_{1/2}(\Omega)$, and for $f(r, \varphi, z)$, $f \in X^0_{1/2}(\Omega)$, resp., we employ partial Fourier analysis with respect to the rotational angle φ taking the system of trigonometric functions $\{e^{ik\varphi}\}_{k \in \mathbb{Z}}$ ($i^2 = -1$; $\mathbb{Z} = \{0, \pm 1, \pm 2, \ldots\}$):

$$u(r, \varphi, z) = \sum_{k \in \mathbb{Z}} u_k(r, z)\,e^{ik\varphi}, \quad u_k(r, z) := \frac{1}{2\pi}\int_{-\pi}^{\pi} u(r, \varphi, z)\,e^{-ik\varphi}\,d\varphi \quad \text{for } k \in \mathbb{Z}. \tag{4}$$

Using the functionals

$$b_k(u_k, v_k) = \int_{\Omega_a} \left\{ \frac{\partial u_k}{\partial r}\overline{\frac{\partial v_k}{\partial r}} + \frac{\partial u_k}{\partial z}\overline{\frac{\partial v_k}{\partial z}} + \frac{k^2}{r^2} u_k \bar{v}_k \right\} r\,dr\,dz, \quad f_k(v_k) = \int_{\Omega_a} f_k \bar{v}_k\,r\,dr\,dz$$

for $k \in \mathbb{Z}$, the BVP (3) can be decomposed into a family of decoupled BVPs in 2D written in the variational form as follows (see e.g. [4, 7, 8, 11]):

$$\begin{aligned}
k = 0: &\text{ find } u_0 \in V^a_0: \quad b_0(u_0, w) = f_0(w) \ \forall w \in V^a_0, \\
k \in \mathbb{Z}_0 := \mathbb{Z}\setminus\{0\}: &\text{ find } u_k \in W^a_0: \quad b_k(u_k, w) = f_k(w) \ \forall w \in W^a_0,
\end{aligned} \tag{5}$$

with $V^a_0 := \{v \in H^1_{1/2}(\Omega_a) : v|_{\Gamma_a} = 0\}$, $W^a_0 := \{v \in V^a_0 : v \in L_{2,-1/2}(\Omega_a)\}$. Here, $H^l_\alpha(\Omega_a)$ (resp. $L_{2,\alpha}(\Omega_a)$) denote the spaces of functions with power weights r^α (α real): $H^l_\alpha(\Omega_a) := \{w = w(r,z) : r^\alpha D^\beta w \in L_2(\Omega_a), 0 \leq |\beta| \leq l\}$ for $l \in \{0, 1, 2\}$. It is important to note that the solutions u_k ($k \in \mathbb{Z}$) of (5) are the Fourier coefficients of u from (3), i.e., solving the 2D problems (5) we get the solution u of (3), or $\hat{u}$ of (1).

For simplicity, for the Nitsche-finite-element discretization we shall employ a decomposition of the domain Ω_a into two polygonal subdomains Ω^1_a, Ω^2_a with

$\overline{\Omega}_a = \overline{\Omega}_a^1 \cup \overline{\Omega}_a^2$, $\Omega_a^1 \cap \Omega_a^2 = \emptyset$, $\Gamma = \overline{\Omega}_a^1 \cap \overline{\Omega}_a^2$, see Fig. 1(a) above. In view of the subdivision of Ω_a we introduce the restrictions $v^i := v|_{\Omega_a^i}$ of some function v on Ω_a^i as well as the vectorized form $v = (v^1, v^2)$, i.e. $v^i(x) = v(x)$ holds for $x \in \Omega_a^i$ ($i = 1, 2$). It should be noted that for simplicity we use here the same symbol v for denoting the function on Ω_a as well as the vector (v^1, v^2). Using this notation we obtain that for each $k \in \mathbb{Z}$ and sufficiently regular u_k the solution of the BVPs (5) is equivalent to the solution of the following problems: Find (u_k^1, u_k^2) such that

$$-\left\{\frac{\partial^2 u_k^i}{\partial r^2} + \frac{\partial^2 u_k^i}{\partial z^2} + \frac{1}{r}\frac{\partial u_k^i}{\partial r}\right\} + \frac{k^2}{r^2} u_k^i = f_k \text{ in } \Omega_a^i, \quad i = 1, 2,$$

$$\frac{\partial u_k^1}{\partial n_1} + \frac{\partial u_k^2}{\partial n_2} = 0 \text{ on } \Gamma, \quad u_k^1 = u_k^2 \text{ on } \Gamma \text{ for } k \in \mathbb{Z}, \tag{6}$$

are satisfied, where n_i ($i = 1, 2$) denotes the outward normal to $\partial \Omega_a^i \cap \Gamma$. Boundary conditions are given by $u_k^i = 0$ on $\partial \Omega_a^i \cap \Gamma_a$, $u_k^i = 0$ on $\partial \Omega_a^i \cap \Gamma_0$ (only for $k \in \mathbb{Z}_0$), and the boundary condition for u_0^i on $\partial \Omega_a^i \cap \Gamma_0$ is given only in the variational context of (5).

Now, the solutions $u_k = (u_k^1, u_k^2)$ ($|k| \leq N$) of the 2D-BVPs (6) shall be approximated by the Nitsche-finite-element method, cf. also [2, 6, 9, 12]. First we cover Ω_a^i ($i = 1, 2$) as usual by a conforming triangulation $\mathcal{T}_h^i$ ($i = 1, 2$) consisting of shape regular triangles T, which are non-matching at the interface Γ. Let h_T denote the diameter of T, $h = \max\{h_T, T \in \mathcal{T}_h^1 \cup \mathcal{T}_h^2\}$ the mesh parameter. Introduce 'broken' finite element spaces $V_{ah} := V_{ah}^1 \times V_{ah}^2$ and $W_{ah} := W_{ah}^1 \times W_{ah}^2$, with $V_{ah}^i := \{v_h^i \in C(\overline{\Omega}_a^i) : v_h^i \in \mathbb{P}_1(T) \, \forall T \in \mathcal{T}_h^i, v_h^i|_{\partial \Omega_a^i \cap \Gamma_a} = 0\}$, $W_{ah}^i := \{v_h^i \in V_{ah}^i \text{ and } v_h^i|_{\partial \Omega_a^i \cap \Gamma_0} = 0\}$ for $i = 1, 2$, i.e., we employ linear finite element functions which are in general not continuous across Γ. Further we introduce some triangulation $\mathcal{E}_h$ of the interface Γ of domain decomposition by intervals E ($E = \overline{E}$), i.e., $\Gamma = \cup_{E \in \mathcal{E}_h} E$, where h_E denotes the diameter of E. A natural choice for the triangulation $\mathcal{E}_h$ is $\mathcal{E}_h := \mathcal{E}_h^1$ or $\mathcal{E}_h := \mathcal{E}_h^2$, where $\mathcal{E}_h^i$ ($i = 1, 2$) denotes the trace of the triangulation $\mathcal{T}_h^i$ on Γ, cf. Fig. 2. The triangulations $\mathcal{T}_h^1$, $\mathcal{T}_h^2$ and $\mathcal{E}_h$ should be consistent on Γ in a local sense, cf. [10]. We now define the Nitsche-finite-element approximation of the solutions of the family of BVPs (6) following the ideas for BVPs in 2D as given e.g. in [2, 6, 9, 12]. They are to be adapted to the new

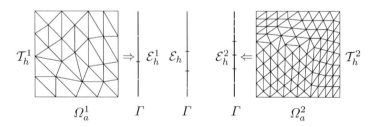

Fig. 2. Triangulation of the mortar interface

situation: here we have spaces with power weights r^α and, moreover, owing to the parametrization of the derivative $\frac{\partial}{\partial\varphi}$ a new term containing the parameter $k \in \mathbb{Z}$ occurs now in the sesquilinear form.

For $k = 0$ and $u_h, v_h \in V_{ah}$ as well as for $k \in \mathbb{Z}_0$ and $u_h, v_h \in W_{ah}$, we introduce sesquilinear forms $\mathcal{B}_{h,k}(\cdot,\cdot)$ and linear forms $\mathcal{F}_{h,k}(\cdot)$ depending on $k \in \mathbb{Z}$ and on real parameters $\alpha_1, \alpha_2 \geq 0$, $\alpha_1 + \alpha_2 = 1$:

$$\mathcal{B}_{h,k}(u_h, v_h) :=$$
$$\sum_{i=1}^{2}\{(\nabla u_h^i, \nabla v_h^i)_{1/2,\Omega_a^i} + k^2(u_h^i, v_h^i)_{-1/2,\Omega_a^i}\} - \langle \alpha_1 \frac{\partial u_h^1}{\partial n_1} - \alpha_2 \frac{\partial u_h^2}{\partial n_2}, v_h^1 - v_h^2 \rangle_{1/2,\Gamma}$$
$$- \langle \alpha_1 \frac{\partial v_h^1}{\partial n_1} - \alpha_2 \frac{\partial v_h^2}{\partial n_2}, u_h^1 - u_h^2 \rangle_{1/2,\Gamma} + \gamma \sum_{E \in \mathcal{E}_h} h_E^{-1}(u_h^1 - u_h^2, v_h^1 - v_h^2)_{1/2,E} \quad (7)$$

$$\mathcal{F}_{h,k}(v_h) := \sum_{i=1}^{2}(f_k^i, v_h^i)_{1/2,\Omega_a^i}.$$

Here, $\langle\cdot,\cdot\rangle_{1/2,\Gamma}$ denotes a convenient duality pairing (cf.[10, p.5]), $(\cdot,\cdot)_{1/2,E}$ is the weighted $L_{2,1/2}(E)$-scalar product, and for $v_h = (v_h^1, v_h^2) \in V_{ah}$, the pairing $\langle\cdot,\cdot\rangle_{1/2,\Gamma}$ can be represented by the $L_{2,1/2}(\Gamma)$-scalar product. Moreover, γ is a sufficiently large positive constant to be restricted subsequently.

The Nitsche-finite-element approximations $u_{0h} = (u_{0h}^1, u_{0h}^2) \in V_{ah}$ and $u_{kh} = (u_{kh}^1, u_{kh}^2) \in W_{ah}$, $k \in \mathbb{Z}_0$, of the Fourier coefficients $u_k = (u_k^1, u_k^2)$ being the solution of (6) are defined to be the solutions of the equations

$$\mathcal{B}_{h,k}(u_{kh}, v_h) = \mathcal{F}_{h,k}(v_h) \;\forall v_h \in W_{ah}, \; k \in \mathbb{Z}_0 \; (\forall v_h \in V_{ah}, \; k = 0, \text{ resp.}). \quad (8)$$

First we observe the consistency of the solutions u_k ($k \in \mathbb{Z}$) from (5) with the variational equations (8) in the sense of $\mathcal{B}_{h,k}(u_k, v_h) = \mathcal{F}_{h,k}(v_h) \;\forall v_h \in W_{ah}$, $k \in \mathbb{Z}_0$ ($\forall v_h \in V_{ah}, k = 0$, resp.), cf. [10]. Secondly, it can be shown that

$$\sum_{E \in \mathcal{E}_h} h_E \left\| \alpha_1 \frac{\partial v_h^1}{\partial n_1} - \alpha_2 \frac{\partial v_h^2}{\partial n_2} \right\|^2_{L_{2,1/2}(E)} \leq C_I \sum_{i=1}^{2} \alpha_i \|\nabla v_h^i\|^2_{L_{2,1/2}(\Omega_a)} \text{ for } v_h \in V_{ah}$$

holds. In the following we use the norms $\|\cdot\|_{1,h,k}$ ($k \in \mathbb{Z}$), which depend on h and $k \in \mathbb{Z}$, and compared with [2, 6, 9, 12], weighted norms and an additional term $k^2\|\cdot\|_{L_{2,-1/2}}$ occur:

$$\|v_h\|^2_{1,h,k} := \sum_{i=1}^{2}\{\|\nabla v_h^i\|^2_{L_{2,1/2}(\Omega_a^i)} + k^2\|v_h^i\|^2_{L_{2,-1/2}(\Omega_a^i)}\} + \sum_{E \in \mathcal{E}_h} h_E^{-1}\|v_h^1 - v_h^2\|^2_{L_{2,1/2}(E)}.$$

If the constant γ in (7) is chosen independently of h and k and satisfies $\gamma > C_I$, then the inequality $\mathcal{B}_{h,k}(v_h, v_h) \geq \mu_1 \|v_h\|^2_{1,h,k} \;\forall v_h \in W_{ah}, \; k \in \mathbb{Z}_0$ ($v_h \in V_{ah}, k = 0$, resp.) holds with a positive constant μ_1, cf. [10].

3 Fourier-nitsche-finite-element approximation in 3D

In order to define the Fourier-Nitsche-finite-element approximation for the 3D-BVP (3), also for (1), we introduce the family of spaces V_{hN} and for

$u, v \in X^1_{1/2}(\Omega^1) \times X^1_{1/2}(\Omega^2)$ the forms $\mathcal{B}^N_h$, $\mathcal{F}^N_h$ as follows:

$$V_{hN} := \{v(r,\varphi,z) = \sum_{|k| \leq N} v_{kh}(r,z)\, e^{ik\varphi} : v_{0h} \in V_{ah},\ v_{kh} \in W_{ah},\ 1 \leq |k| \leq N\},$$

$$\mathcal{B}^N_h(u,v) := 2\pi \sum_{|k| \leq N} \mathcal{B}_{h,k}(u_k, v_k), \quad \mathcal{F}^N_h(v) := 2\pi \sum_{|k| \leq N} \mathcal{F}_{h,k}(v_k),$$

with the domain decomposition in 3D: $\Omega^j := \Omega^j_a \times (-\pi, \pi]$, $j = 1, 2$. Then for treating the BVP (1), i.e. (3), in 3D, the combined Fourier-Nitsche-finite-element method is defined by the Galerkin approach

find $u_{hN} \in V_{hN}$ such that $\mathcal{B}^N_h(u_{hN}, v_{hN}) = \mathcal{F}^N_h(v_{hN}) \quad \forall v_{hN} \in V_{hN}.$ (9)

The solution u_{hN} of (9) is given and can be calculated numerically by $u_{hN} = (u^1_{hN}, u^2_{hN})$ with $u^j_{hN} = \sum_{|k| \leq N} u^j_{kh}(r,z)\, e^{ik\varphi}$ for $j = 1, 2$, where $u_{kh} = (u^1_{kh}, u^2_{kh})$ are the solutions of the 2D-problems (8).

The error $u - u_{hN}$ (u from (3)) is measured in the L_2-norm $\|\cdot\|_{L_2(\widehat{\Omega})} = \|\cdot\|_{X^0_{1/2}(\Omega)}$ as well as in the H^1-like norm $\|\cdot\|_{1,h,\Omega}$ which is defined by

$$\|v\|^2_{1,h,\Omega} := \sum_{j=1}^{2} |v^j|^2_{X^1_{1/2}(\Omega^j)} + \sum_{E \in \mathcal{E}_h} h_E^{-1} \|v^1 - v^2\|^2_{X^0_{1/2}(E \times (-\pi, \pi])}, \quad v^j \in X^1_{1/2}(\Omega^j),$$

with $\|v^1 - v^2\|^2_{X^0_{1/2}(E \times (-\pi, \pi])} := 2\pi \sum_{k \in \mathbb{Z}} \|v^1_k - v^2_k\|^2_{L_{2,1/2}(E)}$ (cf. [10]).

Theorem 1. *Let u be the solution of the BVP (3), with $u \in X^2_{1/2}(\Omega)$, and u_{hN} its approximation given by (9). Then the error $e_{hN} := u - u_{hN}$ satisfies*

$$\|e_{hN}\|_{1,h,\Omega} \leq C(h + N^{-1})\|f\|_{X^0_{1/2}(\Omega)}, \quad \|e_{hN}\|_{X^0_{1/2}(\Omega)} \leq C(h^2 + N^{-2})\|f\|_{X^0_{1/2}(\Omega)}.$$

In order to show that the combined methods works and for observing the convergence rates of the discretization we consider the following examples. The meridian domain Ω_a generating $\widehat{\Omega}$ is a pentagon with the vertices $(0,0)$, $(1,0)$, $(2,1)$, $(1,2)$, and $(0,2)$, cf. Fig. 3(a), (b). The right-hand side f is chosen so that the solution of the BVP (3) is

$$u(r,\varphi,z) = -r^2\, (r - z - 1)\, (r + z - 3)\, (z^2 - 2z)\, \Phi(\varphi)$$

with $\Phi(\varphi) := -[|\varphi|(\pi + \varphi)]^{1.51}$ for $\varphi \in (-\pi, 0]$ and $\Phi(\varphi) := [\varphi(\pi - \varphi)]^{1.51}$ for $\varphi \in (0, \pi]$. For the first example, the subdomains of Ω_a are given by $\Omega^1_a = \{(r,z) \in \Omega_a : z > 1\}$ and $\Omega^2_a = \{(r,z) \in \Omega_a : z < 1\}$, cf. Fig. 3(a). In this case, we have $\partial \Omega^i_a \cap \Gamma_a \neq \emptyset$ for $i = 1, 2$ and $\Gamma \cap \Gamma_0 \neq \emptyset$. For the second example we employ the subdomains $\Omega^1_a = \Omega_a \setminus \overline{\Omega^2_a}$, $\Omega^2_a = (0.5, 1) \times (0.5, 1.5)$, cf. Fig. 3(b). Here, the mortar interface does not touch the boundary of Ω_a, i.e. $\Gamma \cap \Gamma_0 = \emptyset$ and $\Gamma \cap \Gamma_a = \emptyset$ hold. For the experiments, the initial meshes shown

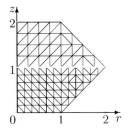 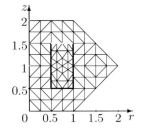

Fig. 3. (a) Triangulation (first example)   (b) Triangulation (second example)

in Fig. 3(a), (b) are refined globally by dividing each triangle into four equal triangles such that the mesh parameters form a sequence $\{h_1, \ldots, h_5\}$ given by $h_{i+1} = 0.5\, h_i$. The mortar parameters are chosen as follows: $\mathcal{E}_h := \mathcal{E}_h^1$, $\alpha_1 = 1$ ($\alpha_2 = 0$), and $\gamma = 4$. For the discretization with respect to N, we employ five levels N_i, where $N_1 = 8$ and $N_{i+1} = 2\, N_i$ for $i = 1, \ldots, 4$. According to Theorem 1, the expected convergence rate in the $X_{1/2}^0(\Omega)$-norm is of the type $\mathcal{O}(h^{\beta_0} + N^{-\delta_0})$, with $\beta_0 = 2$ and $\delta_0 = 2$; in the $\|\cdot\|_{1,h,\Omega}$-norm with $\beta_1 = 1$ and $\delta_1 = 1$. In the experiments we observed for both examples the following values on the highest levels of refinement: $\beta_{obs,0} \approx 2.0$, $\delta_{obs,0} \approx 2.02$, $\beta_{obs,1} \approx 1.0$, and $\delta_{obs,1} \approx 1.1$, i.e., the values are very close to the theoretically expected values.

The numerical example also illustrates that for problems in 3D the Fourier-finite-element method combined with Nitsche mortaring is a suitable mortar approach for the numerical treatment of non-matching meshes and discontinuous (near some interface) finite element approximations. In comparison with 3D-mortaring, an advantage of the described method is the easier implementation because the mortar interface is only one-dimensional. Moreover, it should be mentioned that the Fourier-finite-element method combined with Nitsche mortaring (in connection with local mesh refinement in 2D) is also convenient for solving BVPs with non-regular solutions (i.e. $\hat{u} \in H^{1+\delta}(\widehat{\Omega})$, $\frac{1}{2} < \delta < 1$). This will be discussed in a forthcoming paper.

References

1. Arnold, D.N., Brezzi, F., Cockburn, B., Marini, D.: Unified analysis of discontinuous Galerkin methods for elliptic problems. SIAM J. Numer. Anal., **39**, 1749-1779 (2002)
2. Becker, R., Hansbo, P., Stenberg, R.: A finite element method for domain decomposition with non-matching grids. M2AN Math. Model. Numer. Anal., **37(2)**, 209-225 (2003)
3. Ben Belgacem, F.: The Mortar finite element method with Lagrange multipliers. Numer. Math., **84**, 173-197 (1999)

4. Bernardi, C., Dauge, M., Maday, Y.: Spectral methods for axisymmetric domains. Series in Applied Mathematics, Gauthiers-Villars, Paris and North Holland, Amsterdam (1999)
5. Canuto, C., Hussaini, M. Y., Quarteroni, A., Zang, T. A.: Spectral Methods in Fluid Dynamics. Springer-Verlag, New York (1988)
6. Fritz, A., Hüeber, S., Wohlmuth, B.: A comparison of Mortar and Nitsche Techniques for linear elasticity. Calcolo, **41**, 115-137 (2004)
7. Heinrich, B.: The Fourier-finite-element method for elliptic problems in axisymmetric domains. In: Hackbusch, W. et al.(eds.) Numerical treatment of coupled systems. Proceedings of the 11th GAMM-Seminar, Kiel, Germany, January 20-22, 1995. Vieweg, Wiesbaden (1995)
8. Heinrich, B.: The Fourier-finite-element-method for Poisson's equation in axisymmetric domains with edges. SIAM J. Numer. Anal., **33**, 1885-1911 (1996)
9. Heinrich, B., Pietsch, K.: Nitsche Type Mortaring for some Elliptic Problem with Corner Singularities. Computing, **68**, 217-238 (2002)
10. Heinrich, B., Jung, B.: The Fourier finite-element method with Nitsche mortaring. Preprint SFB393/04-11, Technische Universität Chemnitz (2004)
11. Mercier, B., Raugel, G.: Résolution d'un problème aux limites dans un ouvert axisymétrique par éléments finis en r, z et séries de Fourier en θ. R.A.I.R.O. Analyse numérique, **16(4)**, 405-461 (1982)
12. Stenberg, R.: Mortaring by a method of J. A. Nitsche. In: Computational mechanics. New trends and applications. Centro Internac. Métodos Numér. Ing., Barcelona (1998)
13. Wohlmuth, B. I.: Discretization Methods and Iterative Solvers Based on Domain Decomposition. Lecture Notes in Computational Science and Engineering 17, Springer-Verlag, Berlin Heidelberg (2001)

Substructuring Preconditioners for the Bidomain Extracellular Potential Problem

Micol Pennacchio[1] and Valeria Simoncini[2,1]

[1] IMATI - CNR, via Ferrata, 1, 27100 Pavia, Italy
   micol@imati.cnr.it
[2] Dipartimento di Matematica, Università di Bologna, Piazza di Porta S. Donato, 5, 40127 Bologna, and CIRSA Ravenna, Italy
   valeria@dm.unibo.it

Summary. We study the efficient solution of the linear system arising from the discretization by the mortar method of mathematical models in electrocardiology. We focus on the bidomain extracellular potential problem and on the class of substructuring preconditioners. We verify that the condition number of the preconditioned matrix only grows polylogarithmically with the number of degrees of freedom as predicted by the theory and validated by numerical tests. Moreover, we discuss the role of the conductivity tensors in building the preconditioner.

1 Introduction

A macroscopic model accounting for the excitation process in the myocardium is the "bidomain" model that yields the following Reaction-Diffusion (R-D) system of equations for the intra-, extracellular and transmembrane potential u_i, u and $v = u_i - u$: find $(v(\boldsymbol{x}, t), u(\boldsymbol{x}, t))$, $\boldsymbol{x} \in \Omega$, $t \in [0, T]$ such that

$$\begin{array}{ll} c_m \partial_t v - \mathrm{div}\ M_i \nabla v + I(v) = \mathrm{div}\ M_i \nabla u + I_{app} & \text{in } \Omega \times]0, T[\\ -\mathrm{div}\ M \nabla u = \mathrm{div}\ M_i \nabla v & \text{in } \Omega \times]0, T[\end{array} \qquad (1)$$

with $M_i, M_e, M = M_i + M_e$ conductivity tensors modeling the cardiac fibers, I_{app} an applied current used to initiate the process, c_m the surface capacitance of the membrane. The function $I(v)$ is the transmembrane ionic current which is assumed for simplicity to depend only on v and to be a cubic polynomial (see [7]). The general R-D system can be more complex, including additional ordinary differential equations that govern the evolution of v.

These models are computationally challenging because of the different space and time scales involved; realistic three-dimensional simulations with uniform grids yield discrete problems with more than $O(10^7)$ unknowns at every time step.

To improve computational efficiency, we consider a non–conforming non–overlapping domain decomposition, within the mortar finite element method. This allows us to concentrate the computational work only in regions of high electrical activity; in addition, the matching of different discretizations on adjacent subdomains is weakly enforced. In [8, 9] we compared this technique to the classical conforming FEM verifying its better performance.

In this paper, we focus on the problem of the efficient solution of the linear system arising from this discretization and here for simplicity, we concentrate on the problem with the elliptic equation of (1): *for each time instant t find* $u(\boldsymbol{x},t)$, *solution of*:

$$\begin{cases} -\operatorname{div} M\nabla u = \operatorname{div} M_i \nabla v & \text{in } \Omega \\ \mathbf{n}^T M \nabla u = -\mathbf{n}^T M_i \nabla v & \text{on } \Gamma. \end{cases} \quad (2)$$

Such problem is of interest in its own right, as it represents a separate model for the bidomain extracellular potential [8].

We consider substructuring preconditioners and we report our numerical experience on solving problem (2). Our experiments confirm the theory depicting polylogarithmic bound for the condition number of the preconditioned matrix. Moreover, attention is devoted to tuning the preconditioner so as to take into account the conductivity tensor M in (2).

2 Mortar method

The computational domain Ω is decomposed as the union of L subdomains $\Omega_1, \ldots, \Omega_L$ (see, e.g., [10]). We set $\Gamma_{\ell_n} = \partial\Omega_n \cap \partial\Omega_\ell$, $\mathcal{S} = \cup \Gamma_{\ell_n}$ and we denote by $\gamma_\ell^{(i)}$, $i = 1, \ldots, 4$ the i-th side of the ℓ-th domain so that $\partial\Omega_\ell = \bigcup_{i=1}^4 \gamma_\ell^{(i)}$. Here we consider only a geometrically conforming decomposition, i.e. each edge $\gamma_l^{(i)}$ coincides with Γ_{ln} for some n.

The Mortar Method is applied by choosing a splitting of the skeleton $\mathcal{S}$ as the disjoint union of a certain number of subdomain sides $\gamma_l^{(i)}$, called *mortar* or *slave* sides: we fix an index set $I \subset \{1, \ldots, L\} \times \{1, \ldots, 4\}$ such that $\mathcal{S} = \bigcup_{(l,i)\in I} \gamma_l^{(i)}$. The index–set corresponding to *trace* or *master* sides will be denoted by I^*: $I^* \subset \{1, \ldots, L\} \times \{1, \ldots, 4\}$, $I^* \cap I = \emptyset$ and $\mathcal{S} = \bigcup_{(l,i)\in I^*} \gamma_l^{(i)}$.

Let the spaces X and T be $X = \prod_\ell H^1(\Omega_\ell)$, $T = \prod_\ell H^{1/2}(\partial\Omega_\ell)$ with the broken norms: $\|u\|_X^2 = \sum_\ell \|u\|_{1,\Omega_\ell}^2$ and $\|\eta\|_T^2 = \sum_\ell \|\eta\|_{1/2,\partial\Omega_\ell}^2$. For each ℓ, let also V_h^ℓ be a family of finite dimensional subspaces of $H^1(\Omega_\ell) \cap C^0(\mathrm{ar}\Omega_\ell)$, depending on a parameter $h = h_\ell > 0$, $X_h = \prod_{\ell=1}^L V_h^\ell \subset X$, $T_h^\ell = V_h^\ell|_{\partial\Omega_\ell}$ and $T_h = \prod_{\ell=1}^L T_h^\ell \subset T$. Then we define two composite bilinear forms $a_X, a_X^i : X \times X \longrightarrow \mathbb{R}$ as:

$$a_X(u,\phi) = \sum_\ell \int_{\Omega_\ell} \nabla\phi_l^T M \nabla u_\ell \, dx, \quad a_X^i(u,\phi) = \sum_\ell \int_{\Omega_\ell} \nabla\phi_l^T M_i \nabla u_\ell \, dx. \quad (3)$$

Since these bilinear forms are not coercive on X, we consider proper subspaces of X consisting of functions satisfying a suitable *weak continuity* constraint, leading to the following *constrained* approximation and trace spaces

$$\mathcal{X}_h = \{v_h \in X_h, \int_S [v_h]\lambda\, ds = 0,\ \forall \lambda \in M_h\} \tag{4}$$

$$\mathcal{T}_h = \{\eta \in T_h, \int_S [\eta]\lambda\, ds = 0,\ \forall \lambda \in M_h\}, \tag{5}$$

with M_h a suitably chosen finite dimensional multiplier space. We can write the discrete problem, whose solution existence was proved in [8, Theorem 3.1]:

Problem 1. Find $u_h \in \mathcal{X}_h$ such that for all $\phi_h \in \mathcal{X}_h$

$$a_X(u_h, \phi_h) = -a_X^i(v_h, \phi_h). \tag{6}$$

We remark that Problem 1 admits a solution unique up to an additive constant related to the reference potential chosen. In this paper we consider as reference potential the one given by the potential at a reference point $\boldsymbol{x}_0 \in \Omega$.

3 Substructuring preconditioners

A key aspect of substructuring preconditioners is that they distinguish among three types of degrees of freedom: *interior* (corresponding to basis functions vanishing on the skeleton and supported on one sub-domain), *edge* and *vertex* degrees of freedom [4, 1]. Thus, each function $u \in \mathcal{X}_h$ can be written as the sum of three suitably defined components: $u = u^0 + u^E + u^V$. More specifically, let $w = (w_\ell)_{\ell=1,\cdots,L} \in X_h$ be any discrete function, then

$$w = w^0 + R_h(w), \qquad w^0 \in X_h^0, \tag{7}$$

with $w^0 \in X_h^0$ *interior* function and $R_h(w)$ a discrete lifting, i.e. $R_h(w) = (R_h^\ell(w_\ell))_{\ell=1,\ldots,K}$, where $R_h^\ell(w_\ell)$ is the unique element in V_h^ℓ satisfying $R_h^\ell(w_\ell) = w_\ell$ on Γ_ℓ and

$$\int_{\Omega_\ell} \sum_{i,j} M \frac{\partial}{\partial x_i} \frac{\partial}{\partial x_j} R_h^\ell(w_\ell) v_h^\ell\, dx = 0, \quad \forall v_h \in V_h^\ell. \tag{8}$$

Consequently, the spaces X_h and $\mathcal{X}_h$ can be split as:

$$X_h = X_h^0 \oplus R_h(T_h) \qquad \mathcal{X}_h = \mathcal{X}_h^0 \oplus R_h(\mathcal{T}_h)$$

and it can be verified that

$$a_X(w,v) = a_X(w^0, v^0) + a_X(R_h(w), R_h(v)) = a_X(w^0, v^0) + s(\eta(w), \eta(v)), \tag{9}$$

where the *discrete Steklov-Poincaré* operator $s : T_h \times T_h \to \mathbb{R}$ is defined by

$$s(\xi,\eta) := \sum_{\ell} \int_{\Omega_\ell} (M(\boldsymbol{x})\nabla R_h^\ell(\xi)) \cdot \nabla R_h^\ell(\eta). \tag{10}$$

Furthermore the space of constrained skeleton functions $\mathcal{T}_h$ can be split as the sum of *vertex* and *edge* functions. More specifically, denoting by $\mathfrak{L} \subset \prod_{\ell=1}^{L} H^{1/2}(\partial\Omega_\ell)$ the space $\mathfrak{L} = \{(\eta_\ell)_{\ell=1,\cdots,L},\ \eta_\ell \text{ is linear on each edge of } \Omega_\ell\}$, then we can define the space of constrained *vertex* functions as

$$\mathcal{T}_h^V = \mathcal{P}_h \mathfrak{L} \tag{11}$$

with $\mathcal{P}_h$ the correction operator imposing the constraint. We make the (not restrictive) assumption $\mathfrak{L} \subset \mathcal{T}_h$, which yields $\mathcal{T}_h^V \subset \mathcal{T}_h$, and we introduce the space of constrained *edge* functions $\mathcal{T}_h^E \subset \mathcal{T}_h$ defined by

$$\mathcal{T}_h^E = \{\eta = (\eta_\ell)_{\ell=1,\cdots,L} \in \mathcal{T}_h,\ \eta_\ell(A) = 0,\ \forall \text{ vertex } A \text{ of } \Omega_\ell\}. \tag{12}$$

We can easily verify that $\mathcal{T}_h = \mathcal{T}_h^V \oplus \mathcal{T}_h^E$.

Then we will consider a block Jacobi type preconditioner $\widetilde{s} : \mathcal{T}_h \times \mathcal{T}_h \longrightarrow \mathbb{R}$ defined as

$$\widetilde{s}(\eta,\xi) = b^V(\eta^V,\xi^V) + b^E(\eta^E,\xi^E) \tag{13}$$

with blocks related to the following edge and vertex global bilinear forms

$$\begin{array}{ll} b^E : \mathcal{T}_h^E \times \mathcal{T}_h^E \longrightarrow \mathbb{R} & \text{such that} \quad b^E(\eta^E,\eta^E) \simeq s(\eta^E,\eta^E) \\ b^V : \mathcal{T}_h^V \times \mathcal{T}_h^V \longrightarrow \mathbb{R} & \text{such that} \quad b^V(\eta^V,\eta^V) \simeq s(\eta^V,\eta^V). \end{array} \tag{14}$$

3.1 Matrix form

In this section we derive the matrix form of the discrete Steklov-Poincaré operator s in (10). Equation (6) yields the following linear system of equations:

$$A\mathbf{u} = \mathbf{b} \quad \text{with} \quad \mathbf{b} = -A_i \mathbf{v}, \tag{15}$$

where A, A_i are the stiffness matrices associated to the discretization of a_X, a_X^i defined in (3). It can be shown that the matrix A is positive semidefinite and the system is consistent.

We reorder the vector of unknowns as: $\mathbf{u} = (\mathbf{u}_0, \mathbf{u}_E, \mathbf{u}_V, \mathbf{u}_S)^T$, with $\mathbf{u}_0, \mathbf{u}_E, \mathbf{u}_V, \mathbf{u}_S$ interior, edge, vertex and slave nodes, respectively. From the mortar condition, it follows that the interior nodes of the multiplier sides are not associated with genuine degrees of freedom in the FEM space. Indeed, the value of the coefficients $\mathbf{u}_S$ corresponding to basis functions "living" on slave sides is uniquely determined by the remaining coefficients through the jump (mortar) condition and can be eliminated from the global vector $\mathbf{u}$, i.e.

$$C_S \mathbf{u}_S = -C_E \mathbf{u}_E - C_V \mathbf{u}_V \qquad \mathbf{u}_S =: Q_E \mathbf{u}_E + Q_V \mathbf{u}_V \tag{16}$$

where $Q_E = -C_S^{-1} C_E$, $Q_V = -C_S^{-1} C_V$. The entries of C_S, C_E, C_V are given by $c_{ij} = \int_{\gamma_m} [\phi_j] \lambda_i\, ds$, $\lambda_i \in M_h$ with ϕ_j corresponding to the different nodal basis functions on the slave and master side and associated with the vertices. Since biorthogonal basis functions are employed, the square matrix C_S is diagonal and easily invertible (cf. [11]). The reduction in (16) may be written in matrix form as

$$\mathbf{u} = Q \begin{pmatrix} \mathbf{u}_O \\ \mathbf{u}_E \\ \mathbf{u}_V \end{pmatrix} \quad \text{with} \quad Q = \begin{pmatrix} I_0 & 0 & 0 \\ 0 & I_E & 0 \\ 0 & 0 & I_V \\ 0 & Q_E & Q_V \end{pmatrix} \quad (17)$$

where Q is a global "switching" matrix. The resulting reduced system is thus given by

$$\widetilde{\mathcal{A}} \mathbf{u}_M = \widetilde{\mathbf{b}} \quad (18)$$

with $\widetilde{\mathcal{A}} = Q^T \mathcal{A} Q$ and $\widetilde{\mathbf{b}} = Q^T \mathbf{b}$. We note that the (1,1) block in $\widetilde{\mathcal{A}}$ is cheaply invertible. Therefore, the Schur complement of the system relative to the (1,1) block can readily be obtained, yielding the further reduced system

$$S \begin{pmatrix} \mathbf{u}_E \\ \mathbf{u}_V \end{pmatrix} = \begin{pmatrix} \widehat{\mathbf{b}}_E \\ \widehat{\mathbf{b}}_V \end{pmatrix}.$$

The Schur complement S represents the matrix form of the Steklov-Poincaré operator $s(\cdot, \cdot)$. To obtain the matrix form of $\widetilde{s}(\cdot, \cdot)$ we consider the space $\mathfrak{L}$ of linear functions, used in the splitting of the trace space (11). Then, we introduce an interpolation map denoted by R_H^T (say piecewise interpolation) from the nodal value on V (vertices) onto all nodes of $\mathcal{S}$. The matrix R_H can be viewed as the weighted restriction map from $\mathcal{S}$ onto V. By defining the square matrix $J = \left(\begin{pmatrix} I_E \\ O \end{pmatrix} R_H \right)$, with I_E the $n_E \times n_E$ identity matrix, we can derive the new Schur complement matrix $\widetilde{S}$, after the "vertex" correction,

$$\widetilde{S} = J^T S J = \begin{pmatrix} \widetilde{S}_E & \widetilde{S}_{EV} \\ \widetilde{S}_{EV}^T & \widetilde{S}_V \end{pmatrix}. \quad (19)$$

3.2 The preconditioner

We describe a generalization of a known optimal preconditioner for $\widetilde{S}$, and some more computationally effective variants. The matrix discretization of the form $\widetilde{s}$ yields the following (block-Jacobi type) diagonal preconditioner

$$P = \begin{pmatrix} P_E & 0 \\ 0 & P_V \end{pmatrix},$$

where P_E, P_V are the matrix counterparts of the bilinear forms b^E and b^V in (14), respectively.

It can be verified that the preconditioned matrix $P^{-1}S$ satisfies the theory developed in [1, 3] so that

$$\operatorname{cond}(P^{-1}\widetilde{S}) \lesssim \left(1 + \log\left(\frac{H}{h}\right)\right)^2. \tag{20}$$

with H size of the subdomains and h finest meshsize of the finite element spaces used. Moreover, if an auxiliary coarse mesh is chosen for the vertex block with mesh size $\delta > h$ as studied in [3], then a similar estimate can be obtained but with a factor $\left(1 + \log\left(\frac{H}{h}\right)\right)^3$.

The next three variants make the preconditioner above computationally more appealing with no essential loss of optimality. This goal is achieved by replacing either or both the edge and vertex blocks P_E and P_V with more convenient approximations. In their construction, we were inspired by a similar approach first proposed in [4, 1, 3] for elliptic problems. For later considerations, we recall here an important bound for the condition number of the preconditioned matrix, expressed in terms of the preconditioning quality of the two diagonal blocks. More precisely, let $P = \operatorname{diag}(P_1, P_2)$ be a Jacobi-type preconditioner, and let $\mu_M = \max\{\lambda_{\max}(P_1^{-1}\widetilde{S}_E), \lambda_{\max}(P_2^{-1}\widetilde{S}_V)\}$, $\mu_m = \min\{\lambda_{\min}(P_1^{-1}\widetilde{S}_E), \lambda_{\min}(P_2^{-1}\widetilde{S}_V)\}$. Then

$$\operatorname{cond}(P^{-1}\widetilde{S}) \leq \frac{1+\gamma}{1-\gamma}\frac{\mu_M}{\mu_m} \qquad \gamma \leq 1, \tag{21}$$

where $1 + \gamma$ is the largest eigenvalue of the preconditioned matrix obtained by using the block diagonal of $\widetilde{S}$ as preconditioner P (see, e.g., [6]).

Following [1, 4] a simple approach consists in dropping all couplings between different edges and between edges and vertex points: P_E is replaced by its block diagonal part with one block for each mortar. This simplification provides our first variant,

$$P_1 = \begin{pmatrix} P_E^{diag} & 0 \\ 0 & P_V \end{pmatrix}.$$

Assembling the edge and vertex block preconditioner with such a choice could be quite expensive. A more efficient preconditioner may be obtained by approximating the edge block P_E of P as

$$P_E^{(R)} = \alpha R$$

where R is the square root of the stiffness matrix associated on each edge to the discretization of the operator $-d^2/dx^2$ with homogeneous Dirichlet conditions at the extrema [4, 5, 10]. The choice of the parameter α is discussed below. Thus our second variant is

$$P_2 = \begin{pmatrix} P_E^{(R)} & 0 \\ 0 & P_V \end{pmatrix}.$$

It can be easily verified that (cf., e.g., [4, 5])

$$c_1 \mathbf{v}^T \widetilde{S}_E \mathbf{v} \leq \mathbf{v}^T R \mathbf{v} \leq c_2 \left(1 + \log\left(\frac{H}{h}\right)\right)^2 \mathbf{v}^T \widetilde{S}_E \mathbf{v} \qquad (22)$$

where c_1, c_2 are independent of H, h but may depend on the coefficients of $M(\mathbf{x})$. Since $P_V = \widetilde{S}_V$, in (21) we obtain $\mu_M = \max\{\lambda_{\max}((\alpha R)^{-1}\widetilde{S}_E), 1\} \leq \max\{\alpha^{-1}c_2 \left(1 + \log\left(\frac{H}{h}\right)\right)^2, 1\}$. Analogously, $\mu_m \geq \min\{\alpha^{-1}c_1, 1\}$. Therefore, the determination of μ_M, μ_m is influenced by the magnitude of c_1, c_2 and of α. The anisotropic conductivity tensor $M = M_i + M_e$ is given as $M_s = M_s(\mathbf{x}) = \sigma_t^s I + (\sigma_l^s - \sigma_t^s)\mathbf{a}\mathbf{a}^T$, $s = i, e$, where $\mathbf{a} = \mathbf{a}(\mathbf{x})$ is the unit vector tangent to the cardiac fiber at a point $\mathbf{x} \in \Omega$, I is the identity matrix and σ_l^s, σ_t^s for $s = i, e$ are the conductivity coefficients along and across fiber, in the (i) and (e) media, assumed constant with $\sigma_l^s > \sigma_t^s > 0$. As already mentioned, the magnitude of c_1, c_2 depends on the conductivity coefficients. Therefore, to minimize the bound on the condition number in (21), it is standard practice to select α of the same order of magnitude as the conductivity coefficients [5]. In our case, by choosing α as $\|M\| \leq 2\sigma_t + \sigma_\ell =: \alpha$ we optimize the upper bound μ_M with respect to the conductivity coefficients. Since $\alpha \ll 1$, this value of α usually also leads to the lower estimate $\mu_m = 1$. Numerical experiments validated this choice.

Our third variant copes with the already mentioned fact that building P_V becomes expensive when grid refinements are required. Various choices have been discussed in the literature [10]; for instance, in [3] the vertex preconditioner was chosen as the vertex block of the Schur complement matrix on a fixed auxiliary coarse mesh, independent of the space discretization. We thus approximate P_V with the matrix P_{Vc} obtained with a fixed *coarse* mesh, yielding

$$P_3 = \begin{pmatrix} P_E^{(R)} & 0 \\ 0 & P_{Vc} \end{pmatrix}.$$

This variant leads to very moderate (close to unit) values of $\lambda_{\min}(P_{Vc}^{-1}\widetilde{S}_V)$ and $\lambda_{\max}(P_{Vc}^{-1}\widetilde{S}_V)$ to achieve an estimate in (21). Therefore, we maintained the selection of α as discussed above.

In Table 1 we report numerical experiments with the preconditioners P_1, P_2 and P_3 for the Schur complement system associated with the matrix in (19). The results are in close agreement with the theory: the condition number of the preconditioned matrix grows at most polylogarithmically with the number of degrees of freedom per subdomain, as indicated by (20). The columns with $\alpha = 1$ refer to such parameter selection in $P_E^{(R)}$. This corresponds to discarding information on the conductivity tensor M in (2). The worse convergence validates our choice and shows the importance of an appropriate choice of the parameter. Note that the selected value of α was more effective than other choices of similar magnitude. Indeed choosing $\alpha = \sigma_l$ and $\alpha = \sigma_t$ for $N^2 = 256$, $n = 5, 10, 20, 40$ (fourth row in the table) and P_3, we obtained $29, 31, 33, 34$ and $42, 45, 48, 49$, respectively. In summary, our experiments demonstrate that the proposed variants allow us to limit the computational cost (the cost of forming $P_E^{(R)}$ and P_{V_c} is much lower than that for their original counterparts), with basically no loss in convergence rate, for the appropriate scaling factor.

Table 1. Number of conjugate gradient iterations needed to reduce the residual of a factor 10^{-5} with the preconditioners P_1, P_2, P_3 and P_2, P_3 with $\alpha = 1$. $K = N^2$: # of subdomains. n^2 : # of elements per subdomain. Symbol '*': the preconditioner P_1 could not be built due to memory constraints.

	P_1				P_2				$P_2\ \alpha = 1$			P_3				$P_3\ \alpha = 1$			
$N^2\backslash n$	5	10	20	40	5	10	20	40	5	10	20	40	5	10	20	40	5	10	20	40
16	26	26	29	31	26	26	28	31	24	28	33	37	26	27	27	30	24	50	64	83
64	25	26	27	29	25	27	27	29	24	29	35	39	25	27	28	32	24	51	71	89
144	25	27	27	30	25	28	28	29	27	33	38	42	25	27	28	32	27	54	75	94
256	25	28	29	30	25	28	28	29	29	36	41	46	25	27	28	30	29	55	75	95
400	25	27	29	*	25	28	28	29	31	38	44	50	25	27	28	30	31	56	75	95
576	24	27	28	*	25	27	27	29	34	41	47	53	25	27	27	29	34	57	77	97
784	25	26	27	*	25	25	27	28	36	44	50	56	25	25	27	29	36	55	78	98

References

1. Achdou, Y., Maday, Y., Widlund, O.: Substructuring preconditioners for the mortar method in dimension two. SIAM J. Numer. Anal., **36**, 551–580 (1999)
2. Bernardi, C., Maday, Y., Patera, A.T: Domain decomposition by the mortar element method. In: Kaper, H. et al. (eds) Asymptotic and Numerical Method for Partial Differential Equations with Critical Parameters. Dordrecht: Reidel, (1993)
3. Bertoluzza, S., Pennacchio, M.: Preconditioning the mortar method by substructuring: the high order case. Appl. Num. Anal. Comp. Math., **1**, 434-454 (2004)
4. Bramble, J. H., Pasciak, J.E., Schatz, A. H.: The construction of preconditioners for elliptic problems by substructuring, I. Math. Comp., **47**, 103–134 (1986)
5. Chan, T. F., and Mathew, T. P., Domain Decomposition Algorithms, In: Acta Numerica 1994, Cambridge University Press, 61–143 (1994)
6. Y. Notay, Algebraic multigrid and algebraic multilevel methods: a theoretical comparison. Numer. Lin. Alg. Appl., **12**, 419–451 (2005)
7. Pennacchio, M. and Simoncini, V.: Efficient algebraic solution of reaction–diffusion systems for the cardiac excitation process. J. Comput. Appl. Math., **145** (1): 49–70 (2002)
8. Pennacchio, M. : The mortar finite element method for the cardiac "bidomain" model of extracellular potential, J. Sci. Comput., **20**, n.2, pag. 191-210 (2004)
9. Pennacchio, M.: The Mortar Finite Element Method for Cardiac Reaction-Diffusion Models, Computers in Cardiology 2004; IEEE Proc., **31**: 509-512 (2004)

10. Toselli, A., Widlund, O. Domain decomposition methods—algorithms and theory, Springer Series in Computational Mathematics, volume **34** Springer-Verlag, Berlin (2005)
11. Wohlmuth, B.: Discretization Methods and Iterative Solvers Based on Domain Decomposition, Lecture Notes in Computational Science and Engineering, volume **17**, Springer (2001)

Edge Oriented Interior Penalty Procedures in Computational Fluid Dynamics

A Face Penalty Method for the Three Fields Stokes Equation Arising from Oldroyd-B Viscoelastic Flows

Andrea Bonito[1,2] and Erik Burman[1]

[1] Department of Mathematics, Ecole Polytechnique Fédérale de Lausanne, Switzerland
{andrea.bonito,erik.burman}@epfl.ch
[2] Supported by the Swiss National Science Foundation

Summary. We apply the continuous interior penalty method to the three fields Stokes problem. We prove an inf-sup condition for the proposed method leading to optimal a priori error estimates for smooth exact solutions. Moreover we propose an iterative algorithm for the separate solution of the velocities and the pressures on the one hand and the extra-stress on the other. The stability of the iterative algorithm is established.

1 Introduction

Numerical modeling of viscoelastic flows is of great importance for complex engineering applications involving foodstuff, blood, paints or adhesives. When considering viscoelastic flows, the velocity, pressure and stress must satisfy the mass and momentum equation, supplemented with a constitutive equation involving the velocity and stress. The simplest model is the so-called Oldroyd-B constitutive relation which can be derived from the kinetic theory of polymer dilute solutions, see for instance [1, 12]. The unknowns of the Oldroyd-B model are the velocity u, the pressure p, the extra-stress σ (the non Newtonian part of the stress due to polymer chains for instance) which must satisfy :

$$\rho \frac{\partial u}{\partial t} + \rho(u \cdot \nabla)u - 2\eta_s \nabla \cdot \epsilon(u) + \nabla p - \nabla \cdot \sigma = f, \qquad \nabla \cdot u = 0,$$

$$\sigma + \lambda\left(\frac{\partial \sigma}{\partial t} + (u \cdot \nabla)\sigma - (\nabla u)\sigma - \sigma(\nabla u)^T\right) - 2\eta_p \epsilon(u) = 0.$$

Here ρ is the density, f a force term, η_s and η_p are the solvent and polymer viscosities, λ the relaxation time, $\epsilon(u) = \frac{1}{2}(\nabla u + \nabla u^T)$ the strain rate tensor, $(\nabla u)\sigma$ denotes the matrix-matrix product between ∇u and σ.

When solving viscoelastic flows with finite element methods, the following points should be addressed:

i) the presence of the quadratic term $(\nabla u)\sigma + \sigma(\nabla u)^T$ which prevents a priori estimates to be obtained and therefore existence to be proved for any data;
ii) the presence of a convective term $(u \cdot \nabla)\sigma$ which requires the use of numerical schemes suited to transport dominated problems;
iii) the finite element spaces used to approximate the velocity, the pressure and the extra-stress fields can not be chosen arbitrarily, an inf-sup condition has to be satisfied [10, 11, 14, 15];
iv) the case $\eta_s = 0$ which requires either a compatibility condition between the finite element spaces for u and σ or the use of adequate stabilization procedures.

In this paper, we will focus on points $iii)$, $iv)$ and propose an alternative to the EVSS method [2, 9, 13, 16]. We will consider the stationary linear problem, say $\rho = 0$, $\lambda = 0$ and $\eta_p = 0$. This is

$$\begin{aligned} -\nabla \cdot \sigma + \nabla p = f & \quad \text{in } \Omega, & \nabla \cdot u = 0 & \quad \text{in } \Omega, \\ \sigma - 2\eta_p \epsilon(u) = 0 & \quad \text{in } \Omega, & u = 0 & \quad \text{on } \partial\Omega. \end{aligned} \quad (1)$$

There are a vast number of finite element spaces satisfying the inf-sup condition for the pressure velocity coupling. For the extra-stress however the situation is much less clear even though the relation is trivial in the continuous case. A number of different stabilized methods have therefore been proposed in order to get a stable approximation using equal order approximation for the velocities and the extra-stress, see [7, 10, 11, 14, 15]. In this paper we propose to extend the recently introduced continuous interior penalty method (CIP), or Edge stabilization method, of [3, 5, 8] to the case of the three fields Stokes equation. The case of the Stokes-Darcy problem was treated in [6] and the generalized Oseen's problem was considered in [4]. Advantages of the present method is the unified way of stabilizing different phenomena: for each case the jump in the gradient, or the jump of the non-symmetric operator in question, over element faces is penalized in the L^2 sense. This yields a method with optimal convergence properties for all polynomial degrees that is completely flexible with respect to time-stepping schemes and which does not give rise to any artificial boundary conditions. The price to pay are some added couplings in the stiffness matrix since the penalty operator couples all the degrees of freedom in adjacent elements. However note that in the case of the three fields Stokes equation the stabilization only acts on the pressure and the velocities, hence keeping down the additional memory cost to a moderate factor of 1.5 in two and three space dimensions compared to (the unstable) standard Galerkin formulation. For more complex cases as those encountered in viscoelastic flows where also convection of the extra-stress has to be stabilized, on the other hand one must expect to pay a factor two in the case of two space dimensions and a factor three in the case of three space dimensions due to the fact that the stabilization has to act also on the extra-stress.

2 A finite element formulation

Let Ω be a bounded, polygonal (respectively polyhedral) and connected open set of $\mathbb{R}^d$, $d \geq 2$. We will use the notation $(.,.)$ for the $L^2(\Omega)$ scalar product for scalars, vectors, tensors and $\langle u, v \rangle_x = \int_x u \cdot v \, ds$. Let $\mathcal{T}_h$ be a conforming triangulation of Ω, $\mathcal{E}$ be the set of interior faces in $\mathcal{T}_h$ and $[x]_e$ be the jump of the quantity x on the face e. We shall henceforth assume the local quasiuniformity of the mesh, we assume there exists a constant $C_q > 0$ such that for all $\mathcal{T}_h$ and all vertices $S_i \in \mathcal{T}_h$, we have

$$\max_{e \in \Omega_i} h_e \leq C_q \min_{e \in \Omega_i} h_e. \tag{2}$$

Here Ω_i denotes the macro-element formed by elements $K \in \mathcal{T}_h$ sharing vertex S_i. Let $\mathbf{W_h} = \{w_h : w_h|_K \in \mathbb{P}_k(K)\}$ and $\mathbf{V_h} = \mathbf{W_h} \cap H^1(\Omega)$ We introduce the interior penalty operators

$$j_p(p_h, q_h) = \gamma_p \sum_{e \in \mathcal{E}} \left\langle \frac{h^3}{\eta_p} [\nabla p_h], [\nabla q_h] \right\rangle_e \tag{3}$$

and

$$j_u(u_h, v_h) = \gamma_u \sum_{e \in \mathcal{E}} \langle 2\eta_p h_e [\nabla u_h], [\nabla v_h] \rangle_e + \gamma_b \left\langle \frac{\eta_p}{h} u_h, v_h \right\rangle_{\partial \Omega}, \tag{4}$$

where γ_p, γ_u and γ_b are positive constants to be determined. Moreover let us introduce the bilinear forms

$$a(\sigma_h, v_h) = (\sigma_h, \epsilon(v_h)) - \langle \sigma_h \cdot n, v_h \rangle_{\partial \Omega} \tag{5}$$

$$b(p_h, v_h) = -(p_h, \nabla \cdot v_h) + \langle p_h, v_h \cdot n \rangle_{\partial \Omega}. \tag{6}$$

The method we propose then takes the form, find $(u_h, \sigma_h, p_h) \in \mathbf{V_h^d} \times \mathbf{V_h^{d \times d}} \times \mathbf{V_h}$ such that

$$a(\sigma_h, v_h) + b(p_h, v_h) - b(q_h, u_h) - a(\tau_h, u_h) + \left(\frac{1}{2\eta_p} \sigma_h, \tau_h \right) + j_p(p_h, q_h)$$
$$+ j_u(u_h, v_h) = (f, v_h), \quad \text{for all } (v_h, \tau_h, q_h) \in \mathbf{V_h^d} \times \mathbf{V_h^{d \times d}} \times \mathbf{V_h}. \tag{7}$$

For ease of notation we will also consider the following compact form, introducing the variables $U_h = (u_h, \sigma_h, p_h)$ and $V_h = (v_h, q_h, \tau_h)$ and the finite element space $\mathbf{X_h} = \mathbf{V_h^d} \times \mathbf{V_h^{d \times d}} \times \mathbf{V_h}$

$$A(U_h, V_h) = a(\sigma_h, v_h) + b(p_h, v_h) - b(q_h, u_h) - a(\tau_h, u_h) + \left(\frac{1}{2\eta_p} \sigma_h, \tau_h \right)$$

and

$$J(U_h, V_h) = j_p(p_h, q_h) + j_u(u_h, v_h), \quad F(V_h) = (f, v_h)$$

yielding the compact formulation find $U_h \in \mathbf{X_h}$ such that

$$A(U_h, V_h) + J(U_h, V_h) = F(V_h) \quad \text{for all } V_h \in \mathbf{X_h}.$$

Clearly, since $j_u(U, V_h) = 0$, this formulation is strongly consistent for $(u, \sigma, p) \in H^2(\Omega)^d \times H^1(\Omega)^{d \times d} \times H^2(\Omega)$.

3 The inf-sup condition

For the numerical scheme (7) to be well posed it is essential that there holds an inf-sup condition uniformly in the mesh size h.

Consider the triple norm given by

$$|||U|||^2 = |||(u,\sigma,p)|||^2 = \frac{1}{2\eta_p}\|\sigma\|_{0,\Omega}^2 + 2\eta_p\|\epsilon(u)\|_{0,\Omega}^2 + \frac{1}{2\eta_p}\|p\|_{0,\Omega}^2$$

and the following corresponding discrete triple norm

$$|||U_h|||_h^2 = \frac{1}{2\eta_p}\|\sigma_h\|_{0,\Omega}^2 + 2\eta_p\|\epsilon(u_h)\|_{0,\Omega}^2 + \frac{1}{2\eta_p}\|p_h\|_{0,\Omega}^2 + j_u(u_h,u_h) + j_p(p_h,p_h).$$

Then the following inf-sup condition is satisfied for the discrete form.

Theorem 1. *Assume that the mesh satisfies the local quasiuniformity condition (2). Then for the formulation (7) there holds for all $U_h \in \mathbf{V}_h^d \times \mathbf{V}_h^{d \times d} \times V_h$*

$$|||U_h|||_h \leq \sup_{V_h \neq 0} \frac{A(U_h, V_h) + J(U_h, V_h)}{|||V_h|||_h}.$$

4 A priori error estimates

A priori error estimates follow from the previously proved inf-sup condition together with the proper continuities of the bilinear forms and the approximation properties of the finite element space.

Theorem 2. *Assume that the mesh satisfies the local quasiuniformity condition (2) and that all the components of $U := (u, p, \sigma)$ are in $H^{k+1}(\Omega)$ then there holds*

$$|||U - U_h|||_h \leq Ch^k \left(\eta_p^{1/2} \|u\|_{k+1,\Omega} + \frac{1}{\eta_p^{1/2}} h \|\sigma\|_{k+1,\Omega} + \frac{1}{\eta_p^{1/2}} h \|p\|_{k+1,\Omega} \right)$$

where $k \geq 1$ is the polynomial order of the finite element spaces.

Theorem 3. *Assume that the mesh satisfies the local quasiuniformity condition (2) and that $U := (u, p, \sigma) \in H^2(\Omega)^d \times H^1(\Omega) \times H^1(\Omega)^{d \times d}$ then there holds*

$$|||U - U_h||| \leq Ch \left(\eta_p^{1/2} \|u\|_{2,\Omega} + \frac{1}{\eta_p^{1/2}} \|\sigma\|_{1,\Omega} + \frac{1}{\eta_p^{1/2}} \|p\|_{1,\Omega} \right).$$

5 A stable iterative algorithm

A similar iterative method as in [9, 10] and [2] is presented. The aim of such an algorithm is to de-couple the velocity-pressure computation from the extra stress computation for solving (7).

Each subiteration of the iterative algorithm consists of two steps. Firstly, using the Navier-Stokes equation, the new approximation (u_h^n, p_h^n) is determined using the value of the extra stress at previous step σ_h^{n-1}. Then the new approximation σ_h^n is computed by using the constitutive relation using the value u_h^n. More precisely, assuming that $(u_h^{n-1}, \sigma_h^{n-1}, p_h^{n-1})$ is the known approximation of (u_h, σ_h, p_h) after $n-1$ steps. The first step consists on finding (u_h^n, p_h^n) such that

$$(A+J)((u_h^n, \sigma_h^{n-1}, p_h^n), (v_h, 0, q_h)) + K(u_h^n, u_h^{n-1}, v_h) = (f_h, v_h) \quad (8)$$
$$\forall (v_h, p_h) \in \mathbf{V_h^d} \times \mathbf{V_h},$$

and in the second step we find σ_h^n such that

$$A((u_h^n, \sigma_h^n, p_h^n), (0, \tau_h, 0)) = 0 \quad \forall \tau_h \in \mathbf{V_h^{d \times d}}. \quad (9)$$

Hereabove $K : H^1(\Omega)^d \times H^1(\Omega)^d \times H^1(\Omega)^d \to \mathbb{R}$ is defined for all $u_1, u_2, v \in H^1(\Omega)^d$ by $K(u_1, u_2, v) := 2\eta_p(\epsilon(u_1 - u_2), \epsilon(v))$. The term $K(u_h^n, u_h^{n-1}, v)$, which vanishes at continuous level, has been added to (7) in (8) in order to obtain a stable iterative algorithm.

Lemma 1 (Stability). *Assume that the mesh satisfies the local quasiuniformity condition (2). Let $(u_h^n, \sigma_h^n, p_h^n)$ be the solution of (8),(9) with $f = 0$. There exists γ_u^* and $\gamma_b^* > 0$ independent of h such that for all $\gamma_p > 0$, $\gamma_u \geq \gamma_u^*$ and $\gamma_b \geq \gamma_b^*$, there exists a constant $C > 0$ such that*

$$\eta_p \left\| \epsilon(u_h^n) \right\|^2 + \frac{3}{16\eta_p} \left\| \sigma_h^n \right\|^2 + C \left\| p_h^n \right\|^2 \leq \eta_p \left\| \epsilon(u_h^{n-1}) \right\|^2 + \frac{3}{16\eta_p} \left\| \sigma_h^{n-1} \right\|^2.$$

6 Preliminary numerical results

For all the numerical experiments we choose $\eta_p = 1\ [Pa.s]$ and consider $\mathbb{P}_1$ approximations for the velocity, the pressure and the stress.

6.1 Poiseuille flow

Consider a rectangular pipe of dimensions $[0, L_1] \times [0, L_2]$ in the $x-y$ directions, where $L_1 = 0.15\ [m]$ and $L_2 = 0.03\ [m]$. The boundary conditions are the following. On the top and bottom sides ($y = 0$ and $y = L_2$), no-slip boundary conditions apply. On the inlet ($x = 0$) the velocity and the extra-stress are given by

$$\mathbf{u}(0,y) = \begin{pmatrix} u_x(y) \\ 0 \end{pmatrix}, \qquad \sigma(0,y) = \begin{pmatrix} 0 & \sigma_{xy}(y) \\ \sigma_{xy}(y) & 0 \end{pmatrix}, \qquad (10)$$

with $u_x(y) = (L_2 + y)(L_2 - y)$ and $\sigma_{xy}(y) = -2\eta_p y$. On the outlet ($x = L_1$) the velocity and the pressure are given by

$$\mathbf{u}(L_1, y) = \begin{pmatrix} u_x(y) \\ 0 \end{pmatrix}, \qquad p(L_1, y) \equiv 0.$$

The velocity and extra-stress must satisfy (10) in the whole pipe. Three unstructured meshes are used to check convergence (coarse: 50×10, intermediate: 100 × 20, fine: 200 × 40). In Fig. 1, the error in the L^2 norm of the velocity u, the pressure p and extra-stress components σ_{xx}, σ_{xy} is plotted versus the mesh size. Clearly order one convergence rate is observed for the pressure (in fact superconvergence is observed for the pressure) and the extra-stress whilst the convergence rate of the velocity is order two, this being consistent with theoretical predictions.

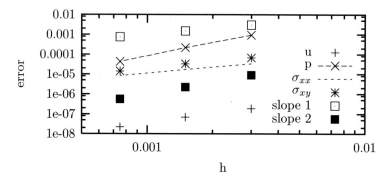

Fig. 1. Poiseuille flow: convergence orders.

6.2 The 4:1 planar contraction

Numerical results of computation in the 4:1 abrupt contraction flow case are presented and comparison with the EVSS (see for instance [2, 10]) method is performed. This test case underlines the importance of the stabilization of the constitutive equation. The symmetry of the geometry is used to reduce the computational domain by half, as shown in Fig. 2 (left). Zero Dirichlet boundary conditions are imposed on the walls, the Poiseuille velocity profile $u_x(y) = 64(L_0 - y)(L_0 + y)$ is imposed at the inlet with $L_0 = 0.025[m]$, natural boundary conditions on the symmetry axis and at the outlet of the domain.

The results applying only GLS stabilization for the pressure are shown in Fig. 2 (right). Similar results obtained using the EVSS method (see [2] for a detailed description) and the CIP formulation are presented in Fig. 3.

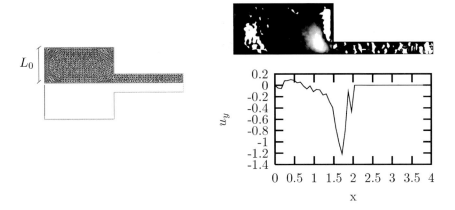

Fig. 2. (left) Computational domain for the 4:1 contraction, (right-top) 20 isovalues of the GLS method only for the pressure from -0.9 (black) to 0.06 (white) and (right-bottom) profile of $u_y(x, 0.025)$.

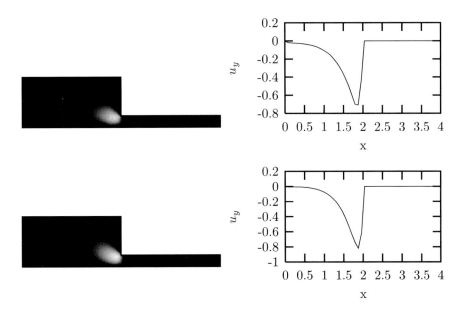

Fig. 3. Left column: 20 isovalues from -0.9 (black) to 0.06 (white), right column: profile of $u_y(x, 0.025)$ (top: EVSS, bottom: CIP).

References

1. Bird, R., Curtiss, C., Armstrong, R., Hassager, O.: Dynamics of polymeric liquids, **1** & **2**. John Wiley & Sons, New-York (1987)

2. Bonvin, J., Picasso, M., Stenberg, R.: GLS and EVSS methods for a three-field Stokes problem arising from viscoelastic flows. Comput. Methods Appl. Mech. Engrg., **190**(29-30), 3893–3914 (2001)
3. Burman, E.: A unified analysis for conforming and non-conforming stabilized finite element methods using interior penalty. SIAM, J. Numer. Anal., **45**, 2012–2033 (2005)
4. Burman, E., Fernández, M.A., Hansbo, P.: Edge Stabilization: an Interior Penalty Method for the Incompressible Navier-Stokes Equation. Technical Report 23.2004, Ecole Polytechnique Federale de Lausanne (2004)
5. Burman, E., Hansbo, P.: Edge stabilization for Galerkin approximations of convection-diffusion problems. Comput. Methods Appl. Mech. Engrg., **193**, 1437–2453 (2004)
6. Burman, E., Hansbo, P.: Edge stabilization for the generalized Stokes problem: a continuous interior penalty method. Comput. Methods Appl. Mech. Engrg., **195**, 2392–2410 (2006)
7. Codina, R.: Finite element approximation of the three field formulation of the elasticity problem using stabilization. Computational Mechanics. Tsinghua University Press & Springer-Verlag, 276–281 (2004)
8. Douglas, J., Dupont, T.: Interior penalty procedures for elliptic and parabolic Galerkin methods. Computing methods in applied sciences (Second Internat. Sympos., Versailles, 1975). Lecture Notes in Phys., **58**. Springer, Berlin, 207–216 (1976)
9. Farhloul, M., Fortin, M.: A new mixed finite element for the Stokes and elasticity problems. SIAM J. Numer. Anal., **30**(4), 971–990 (1993)
10. Fortin, M., Guénette, R., Pierre, R.: Numerical analysis of the modified EVSS method. Comput. Methods Appl. Mech. Engrg., **143**(1-2), 79–95 (1997)
11. Franca, L.P., Stenberg, R.: Error analysis of Galerkin least squares methods for the elasticity equations. SIAM J. Numer. Anal., **28**(6), 1680–1697 (1991)
12. Öttinger, H. C.: Stochastic processes in polymeric fluids. Springer-Verlag, Berlin (1996)
13. Rajagopalan, D., Armstrong, R.C, Brown, R.A.: Finite-element methods for calculation of steady, viscoelastic flow using constitutive-equations with newtonian viscosity. J. Non Newtonian Fluid Mech., **36**, 159–1992 (1990)
14. Ruas, V.: Finite element methods for the three-field Stokes system in $\mathbf{R}^3$: Galerkin methods. RAIRO Modél. Math. Anal. Numér., **30**(4), 489–525 (1996)
15. Sandri, D.: Analyse d'une formulation à trois champs du problème de Stokes. RAIRO Modél. Math. Anal. Numér., **27**(7), 817–841 (1993)
16. Yurun, F.: A comparative study of discontinuous galerkin and continuous supg finite element methods for computation of viscoelastic flows. Comput. Methods Appl. Mech. Engrg., **141**, 47–65 (1997)

Anisotropic H^1-Stable Projections on Quadrilateral Meshes

Malte Braack

Institute of Applied Mathematics, Heidelberg University, INF 294, 69120 Heidelberg
malte.braack@iwr.uni-heidelberg.de

Summary. In this work we analyze a projector of non-smooth functions on anisotropic quadrilateral meshes. In particular, a stability result and an upper bound for the approximation error is derived. It turns out that the partial derivatives become well combined with the corresponding mesh size parameters.

1 Introduction

Projections of H^1 functions onto finite element spaces are of fundamental interest in numerical analysis. In particular, they are necessary to prove stability and a priori estimates of stabilized finite element schemes. Examples of such projections on isotropic elements are the Clement interpolation [4], and the variant of Scott and Zhang [6] in order to maintain Dirichlet conditions. For an overview of anisotropic interpolation operators we refer to the book [1] where several H^1-stable projections on tensor grids are addressed.

In this work we use a projection operator $B_h : H^1(\Omega) \to Q_h$ developed in [2] which is suitable for anisotropic quadrilateral meshes obtained by bilinear transformations. Originally it was designed and analyzed for meshes aligned with the coordinate axes. In this work, this interpolation operator is considered for a much general class of quadrilateral meshes, which introduces further couplings between the partial derivatives. In particular, we allow for bilinear transformations, where the nonlinear contribution and the shearing should be limited by the mesh sizes into the particular directions of anisotropy. The results of this work are used in [3].

We consider anisotropic meshes without any restriction of grid alignment with the coordinate axis. The transformation T_K from the reference cells $\widehat{K}$ to the physical cell K is allowed to be bilinear. Such a transformation can be expressed as a composition of translation, rotation, shearing and stretching, augmented with the pure bilinear term,

$$T_K \begin{pmatrix} \widehat{x} \\ \widehat{y} \end{pmatrix} = \begin{pmatrix} \widehat{x_0} \\ \widehat{y_0} \end{pmatrix} + \begin{pmatrix} \cos\theta & -\sin\theta \\ \sin\theta & \cos\theta \end{pmatrix} \left[\begin{pmatrix} 1 & s \\ 0 & 1 \end{pmatrix} \begin{pmatrix} h_x & 0 \\ 0 & h_y \end{pmatrix} \begin{pmatrix} \widehat{x} \\ \widehat{y} \end{pmatrix} + \begin{pmatrix} \alpha \widehat{xy} \\ \beta \widehat{xy} \end{pmatrix} \right].$$

There are eight free parameters, $x_0, y_0, \theta, s, h_x, h_y, \alpha, \beta$, depending usually on the specific cell, i.e., $s = s_K$, $h_x = h_{K,x}$, $h_y = h_{K,y}$, $\alpha = \alpha_K$ and $\beta = \beta_K$ but the subscript K will be suppressed in order to simplify notations.

Throughout this work we use the notation $a \lesssim b$ for $a \leq Cb$ with a constant C independent of the specific cell K (and hence independent of the eight local parameters). The expression $a \sim b$ is used if there holds $a \lesssim b$, and $b \lesssim a$ as well. Without loss of generality we may assume $h_y \leq h_x$.

We formulate three assumptions with are supposed to be fulfilled throughout the entire work without stating this explicitly:

(A1) The shearing parameter s is bounded by a constant $s_0 \geq 0$ (ind. of K):
$$|s| \leq s_0.$$

(A2) Interior angle conditions for neighbour cells $K, L \in \mathcal{T}_h$:
$$h_{K,x} \sim h_{L,x} \quad \text{and} \quad h_{K,y} \sim h_{L,y}.$$

(A3) A restriction to the parameters α, β:
$$|\alpha| \leq \frac{h_x}{4} \quad \text{and} \quad |\beta| \leq \frac{1}{4} \min\left\{h_y, \frac{h_x}{s_0}\right\}.$$

Note that assumption (A1) allows for moderate stretching, because the stretching s is coupled to the extend of anisotropy. The conditions of (A3) bound the influence of the pure bilinear part of the transformation. As a consequence of the assumptions (A1) and (A3) we know the determinant of the transformation:

Lemma 1. *By T_K^η we denote the Jacobian of the transformation T_K at $\eta \in \hat{K}$. Its determinant can be estimated by:* $\det T_K^\eta \sim h_x h_y$.

Proof. The determinant of the rotational part is equal to 1. Hence, we can assume $\theta = 0$. In the case of vanishing bilinear contribution, $\alpha = \beta = 0$ or $\eta = (0,0)^T$, the Jacobian becomes:
$$T_K^\eta = \begin{pmatrix} 1 & s \\ 0 & 1 \end{pmatrix} \begin{pmatrix} h_x & 0 \\ 0 & h_y \end{pmatrix} = \begin{pmatrix} h_x & sh_y \\ 0 & h_y \end{pmatrix}.$$

In this case the assertion is obvious. For not vanishing α or β the impact of the nonlinearity is maximal at $\eta^* = (1,1)^T$. Hence it is sufficient to determine the determinant of the matrix:
$$T_K^{\eta^*} = \begin{pmatrix} h_x + \alpha & sh_y + \alpha \\ \beta & h_y + \beta \end{pmatrix}. \tag{1}$$

Its determinant
$$\det T_K^{\eta^*} = (h_x + \alpha)(h_y + \beta) - (sh_y + \alpha)\beta$$

can be estimated from above and from below by $h_x h_y$ times appropriate constants due to (A3), i.e.,

$$\frac{3}{8} h_x h_y \leq \det T_K^{\eta^*} \leq \frac{15}{8} h_x h_y.$$

∎

As a consequence of (A2) it holds for neighbour cells K and L:

$$\det T_K \sim \det T_L. \tag{2}$$

2 Anisotropic H^1-stable projectors

In order to maintain the presentation in this section easier we will neglect rotation and translation fo a moment. Hence, T_K^η for $\eta \in \widehat{K}$ can always be expressed by (1), where $|\alpha|$ and $|\beta|$ are bounded as in (A3). By $P(K)$ we denote the patch of cells having one node in common with K.

Lemma 2. *Assuming (A1) and (A2) there exists for each cell $K \in \mathcal{T}_h$ and $u \in H^1(P(K))$ a constant $c \in \mathbb{R}$ so that:*

$$\|u - c\|_{P(K)} \lesssim (h_x + s_0 h_y)\|\partial_x u\|_{P(K)} + h_y \|\partial_y u\|_{P(K)}. \tag{3}$$

Proof. For the particular case that the grid is aligned with the coordinate axes (i.e. $s_0 = \alpha = \beta = 0$) this result is proven in [1]. Here, we need the generalization to parallelogram meshes. Theorem 4.2 in [5] ensures the existence of a constant c so that on the reference patch holds:

$$\|\hat{u} - c\|^2_{P(\widehat{K})} = \|\partial_{\hat{x}} \hat{u}\|^2_{P(\widehat{K})} + \|\partial_{\hat{y}} \hat{u}\|^2_{P(\widehat{K})}.$$

Due to the consequence (2) of assumption (A2) we obtain now:

$$\|u - c\|^2_{P(K)} = \sum_{L \in P(K)} \det T_L \|\hat{u} - c\|^2_{\widehat{L}}$$
$$\lesssim \det T_K (\|\partial_{\hat{x}} \hat{u}\|^2_{P(\widehat{K})} + \|\partial_{\hat{y}} \hat{u}\|^2_{P(\widehat{K})}).$$

The gradients with respect to the reference element can be expressed by:

$$\partial_{\hat{x}} \hat{u}(\hat{x}) = (h_x + \alpha)\partial_x u(x) + \beta \partial_y u(x) \tag{4}$$
$$\partial_{\hat{y}} \hat{u}(\hat{x}) = (sh_y + \alpha)\partial_x u(x) + (h_y + \beta)\partial_y u(x). \tag{5}$$

The L^2-norms of the partial derivatives become:

$$\|\partial_{\hat{x}} \hat{u}\|^2_{P(\widehat{K})} = \det T_K^{-1} \left((h_x + \alpha)^2 \|\partial_x u\|^2_{P(K)} + \beta^2 \|\partial_y u\|^2_{P(K)}\right) \tag{6}$$
$$\|\partial_{\hat{y}} \hat{u}\|^2_{P(\widehat{K})} \lesssim \det T_K^{-1}((h_y s + \alpha)^2 \|\partial_x u\|^2_{P(K)} + (h_y + \beta)^2 \|\partial_y u\|^2_{P(K)}). \tag{7}$$

Due to condition (A3) it holds:
$$\|u - c\|_{P(K)}^2 \lesssim (h_x^2 + s^2 h_y^2)\|\partial_x u\|_{P(K)}^2 + h_y^2 \|\partial_y u\|_{P(K)}^2.$$

Using assumption (A1) and taking the square root leads to (3). ∎

Lemma 3. *The trace theorem on the cell K with an edge Γ with the transformation of type (1) reads for $u \in H^1(K)$:*
$$\|u\|_\Gamma \lesssim h_y^{-1/2}\left(\|u\|_K + (h_x + s_0 h_y)\|\partial_x u\|_K + h_y \|\partial_y u\|_K\right).$$

Proof. Since $|\Gamma|$ is of size $O(h_x)$, transformation on the reference patch $P(\hat{K})$ and the trace theorem on $P(\hat{K})$ gives:
$$\|u\|_\Gamma^2 \lesssim h_x \|\hat{u}\|_{\hat{\Gamma}}^2 \lesssim h_x(\|\hat{u}\|_{\hat{K}}^2 + \|\partial_{\hat{x}} \hat{u}\|_{\hat{K}}^2 + \|\partial_{\hat{y}} \hat{u}\|_{\hat{K}}^2)$$
$$\lesssim h_x \det T_K^{-1}\left[\|u\|_K^2 + h_x^2 \|\partial_x u\|_K^2 + s^2 h_y^2 \|\partial_x u\|_K^2 + h_y^2 \|\partial_y u\|_K^2\right]$$
$$= h_y^{-1}\left[\|u\|_K^2 + (h_x^2 + s^2 h_y^2)\|\partial_x u\|_K^2 + h_y^2 \|\partial_y u\|_K^2\right].$$
∎

2.1 Definition and existence

We introduce a certain class of projections $\mathcal{I}_h : H^1(\Omega) \to V_h$ suitable for anisotropic estimates. For this, we use the notation $\hat{\mathcal{I}}_h$ for the transformation onto the reference cell K, defined by $\hat{\mathcal{I}}_h \hat{u}(\hat{x}) := \mathcal{I}_h u(x)$.

Definition 1. *We call a projector $\mathcal{I}_h : H^1(\Omega) \to V_h$ "anisotropic H^1-stable", if it holds (the set of those projectors will be denoted by $\mathcal{A}_h$):*

$$\|\mathcal{I}_h u\|_K \lesssim \|u\|_{P(K)} + (h_x + s_0 h_y)\|\partial_x u\|_{P(K)} + h_y \|\partial_y u\|_{P(K)}, \tag{8}$$

$$\|\partial_{\hat{x}} \hat{\mathcal{I}}_h \hat{u}\|_{\hat{K}} \lesssim \|\partial_{\hat{x}} \hat{u}\|_{P(\hat{K})} + h_x^{-1} h_y \|\partial_{\hat{y}} \hat{u}\|_{P(\hat{K})}, \tag{9}$$

$$\|\partial_{\hat{y}} \hat{\mathcal{I}}_h \hat{u}\|_{\hat{K}} \lesssim \|\partial_{\hat{y}} \hat{u}\|_{P(\hat{K})}. \tag{10}$$

Proposition 1. *It holds $\mathcal{A}_h \neq \emptyset$, i.e., there is at least one anisotropic H^1-stable projector.*

Proof. The operator $B_h : H^1(\Omega) \to V_h$ proposed by Becker [2] fulfills the estimates (9) and (10) on the reference cells. In order to show (8) we need a closer look onto its construction. B_h can be considered as a generalization of the interpolation operator of Scott & Zhang [6] to anisotropic meshes. In order to obtain nodal values averaging is performed along the long edges of the elements. To be more specific, we associate to each node N_i, $1 \leq i \leq N$, the longest edge Γ_i: $\Gamma_i := \arg\max_{\Gamma \in \mathcal{E}_i} |\Gamma|$. If the maximum is not unique, we may chose any of the largest edges. Furthermore, we associate to i a linear function $\psi_i : \Gamma_i \to \mathbb{R}$, characterized by the product with the nodal basis functions ϕ_j:

$$\int_{\Gamma_i} \psi_i \phi_j = \delta_{ij} \quad \forall 1 \leq j \leq N.$$

If N_k is the node which is connected by Γ_i with node N_i, this condition is equivalent to $\int_{\Gamma_i} \psi_i \varphi_k = 0$ and $\int_{\Gamma_i} \psi_i \varphi_i = 1$, because the remaining nodal functions ϕ_j, $j \neq i, k$ vanish on Γ_i. Due to this two conditions, the linear function ψ_i is uniquely defined. Hence, it is easy to check $\|\psi_i\|_\infty \sim h_i^{-1}$ and therefore $\|\psi_i\|_{\Gamma_i} \lesssim h_i^{-1/2}$. Now, B_h is given by

$$B_h u(x) := \sum_{i=1}^{N} \phi_i(x) \int_{\Gamma_i} u\psi_i \, ds. \tag{11}$$

Since B_h is the identity on V_h it is a projection. In particular $B_h c = c$ for constants c. Using the definition (11) of B_h and the Cauchy-Schwarz inequality, we obtain: $\|B_h u\|_K \leq \sum_{i=1}^{N} \|\phi_i\|_{P(K)} \|\psi_i\|_{\Gamma_i} \|u\|_{\Gamma_i}$. Now we use $\|\psi_i\|_{\Gamma_i} \lesssim h_i^{-1/2} = h_x^{-1/2}$, and $\|\phi_i\|_{P(K)} \lesssim |P(K)|^{1/2} \sim (h_x h_y)^{1/2}$, in order to obtain:

$$\|B_h u\|_K \lesssim h_y^{1/2} \max_{i \in \mathcal{N}(P(K))} \|u\|_{\Gamma_i}.$$

The trace theorem (Lemma 3) finally gives the desired result. ∎

2.2 Stability properties

The next Proposition gives an upper bound for the partial derivatives of "anisotropic H^1-stable projectors":

Proposition 2. *For $\mathcal{I}_h \in \mathcal{A}_h$ it holds:*

$$\|\partial_x \mathcal{I}_h u\|_K \lesssim (1 + s_0 h_x^{-1} h_y) \|\partial_x u\|_{P(K)} + h_x^{-1} h_y \|\partial_y u\|_{P(K)}, \tag{12}$$
$$\|\partial_y \mathcal{I}_h u\|_K \lesssim (1 + s_0)(s_0 \|\partial_x u\|_{P(K)} + \|\partial_y u\|_{P(K)}). \tag{13}$$

Proof. We use the inverse of T_K. In the case $\alpha = \beta = 0$ it becomes:

$$T_K^{-1} = \begin{pmatrix} \partial \hat{x}/\partial x & \partial \hat{x}/\partial y \\ \partial \hat{y}/\partial x & \partial \hat{y}/\partial y \end{pmatrix} = \begin{pmatrix} h_x^{-1} & -s h_x^{-1} \\ 0 & h_y^{-1} \end{pmatrix}.$$

In the general bilinear case under assumption (A3) it is easy to verify:

$$\partial \hat{x}/\partial x \sim h_x^{-1}, \quad \partial \hat{y}/\partial y \sim h_y^{-1},$$
$$|\partial \hat{y}/\partial x| \leq h_x^{-1}, \quad |\partial \hat{x}/\partial y| \leq |s| h_x^{-1} + \alpha h_y^{-1} h_x^{-1} \lesssim s_0 h_x^{-1} + h_y^{-1}.$$

Transformation to $\hat{K}$ back and forth, and use of (6) and (7) gives:

$$\|\partial_x \mathcal{I}_h u\|_K = \det T_K^{1/2} \|\partial_x \hat{\mathcal{I}}_h \hat{u}\|_{\hat{K}}$$
$$= \det T_K^{1/2} (\|(\partial_x \hat{x}) \partial_{\hat{x}} \hat{\mathcal{I}}_h \hat{u} + (\partial_x \hat{y}) \partial_{\hat{y}} \hat{\mathcal{I}}_h \hat{u}\|_{\hat{K}})$$
$$= \det T_K^{1/2} h_x^{-1} (\|\partial_{\hat{x}} \hat{\mathcal{I}}_h \hat{u}\|_{\hat{K}} + \|\partial_{\hat{y}} \hat{\mathcal{I}}_h \hat{u}\|_{\hat{K}}).$$

Now, we use the properties (9)–(10) and (1)–(2):

$$\|\partial_x \mathcal{I}_h u\|_K \lesssim \det T_K^{1/2} h_x^{-1} (\|\partial_{\hat{x}} \hat{u}\|_{P(\hat{K})} + (1 + h_x^{-1} h_y) \|\partial_{\hat{y}} \hat{u}\|_{P(\hat{K})})$$
$$\lesssim (1 + s h_x^{-1} h_y) \|\partial_x u\|_{P(K)} + h_x^{-1} h_y \|\partial_y u\|_{P(K)}.$$

With assumption (A1) we get (3). Analogously, due to property (9) and (10):

$$\|\partial_y \mathcal{I}_h u\|_K \lesssim \det T_K^{1/2} ((|s| h_x^{-1} + \alpha h_x h_y^{-1}) \|\partial_{\hat{x}} \hat{u}\|_{\hat{K}} + h_y^{-1} \|\partial_{\hat{y}} \hat{\mathcal{I}}_h \hat{u}\|_{\hat{K}})$$
$$\lesssim \det T_K^{1/2} \left(s_0 h_x^{-1} \|\partial_{\hat{x}} \hat{u}\|_{P(\hat{K})} + (h_y^{-1} + s_0 h_x^{-1}) \|\partial_{\hat{y}} \hat{u}\|_{P(\hat{K})} \right)$$
$$\lesssim s_0 (1 + s_0) \|\partial_x u\|_{P(K)} + (s_0 + 1 + s_0 h_x^{-1} h_y) \|\partial_y u\|_{P(K)}.$$

We obtain (4) due to $h_y h_x^{-1} \leq 1$. ∎

2.3 Local approximation properties

Proposition 3. *For $\mathcal{I}_h \in \mathcal{A}_h$ it holds:*

$$\|u - \mathcal{I}_h u\|_K \lesssim (1 + s_0) h_x \|\partial_x u\|_{P(K)} + h_y \|\partial_y u\|_{P(K)}. \tag{14}$$

Proof. Due to the stability property (8) it holds for an arbitrary $c \in \mathbb{R}$:

$$\|u - \mathcal{I}_h u\|_K \leq \|u - c\|_K + \|c - \mathcal{I}_h u\|_K$$
$$= \|u - c\|_K + \|\mathcal{I}_h (c - u)\|_K$$
$$\lesssim \|u - c\|_{P(K)} + (1 + s_0) h_x \|\partial_x (u - c)\|_{P(K)} + h_y \|\partial_y (u - c)\|_{P(K)}.$$

The assertion follows due to $\partial_x c = \partial_y c = 0$ and Lemma 2. ∎

Lemma 4. *For $\mathcal{I}_h \in \mathcal{A}_h$ and $\hat{u} \in H^2(P(\hat{K}))$ it holds:*

$$\|\partial_{\hat{x}} (\hat{u} - \hat{\mathcal{I}}_h \hat{u})\|_{\hat{K}} \lesssim \|\partial_{\hat{x}} \hat{\nabla} \hat{u}\|_{P(\hat{K})} + h_x^{-1} h_y \|\partial_{\hat{y}} \hat{\nabla} \hat{u}\|_{P(\hat{K})}, \tag{15}$$
$$\|\partial_{\hat{y}} (\hat{u} - \hat{\mathcal{I}}_h \hat{u})\|_{\hat{K}} \lesssim \|\partial_{\hat{y}} \hat{\nabla} \hat{u}\|_{P(\hat{K})}. \tag{16}$$

Proof. On the reference element, it holds for the nodal interpolant $\hat{I} u$:

$$\|\partial_x (\hat{u} - \hat{I} \hat{u})\|_{\hat{K}} \lesssim \|\partial_{\hat{x}}^2 \hat{u}\|_{\hat{K}} + \|\partial_{\hat{y}} \partial_{\hat{x}} \hat{u}\|_{\hat{K}} \leq \|\partial_{\hat{x}} \hat{\nabla} \hat{u}\|_{\hat{K}},$$

$$\|\partial_y(\hat{u} - \hat{I}\hat{u})\|_{\hat{K}} \lesssim \|\partial_{\hat{y}}^2 \hat{u}\|_{\hat{K}} + \|\partial_{\hat{x}}\partial_{\hat{y}}\hat{u}\|_{\hat{K}} \leq \|\partial_{\hat{y}}\hat{\nabla}\hat{u}\|_{\hat{K}}.$$

$$\|\partial_{\hat{x}}(\hat{u} - \hat{\mathcal{I}}_h\hat{u})\|_{\hat{K}} \leq \|\partial_{\hat{x}}(\hat{u} - \hat{I}\hat{u})\|_{\hat{K}} + \|\partial_{\hat{x}}(\hat{I}\hat{u} - \hat{\mathcal{I}}_h\hat{u})\|_{\hat{K}}$$
$$= \|\partial_{\hat{x}}(\hat{u} - \hat{I}\hat{u})\|_{\hat{K}} + \|\partial_{\hat{x}}\hat{\mathcal{I}}_h(\hat{I}\hat{u} - \hat{u})\|_{\hat{K}}.$$

The first term on the right hand side can obviously bounded by the right hand side of (15). Hence, it remains to show that $\|\partial_{\hat{x}}\hat{\mathcal{I}}_h(\hat{I}\hat{u} - \hat{u})\|_{\hat{K}}$ can be bounded properly. But this is an immediate consequence of (9). The arguments for showing (16) are analogous. ∎

Lemma 5. *It holds for $\mathcal{I}_h \in \mathcal{A}_h$ and $u \in H^2(P(K))$:*

$$\|\partial_x(u - \mathcal{I}_h u)\|_K \lesssim (1 + s_0)^2 h_x \|\partial_x^2 u\|_{P(K)} + (1 + s_0) h_y \|\partial_{xy} u\|_{P(K)} \quad (17)$$
$$+ h_x^{-1} h_y^2 \|\partial_y^2 u\|_{P(K)},$$

$$\|\partial_y(u - \mathcal{I}_h u)\|_K \lesssim (1 + s_0)^2 h_x \|\partial_x^2 u\|_{P(K)} + (1 + s_0)^2 h_x \|\partial_{xy} u\|_{P(K)} \quad (18)$$
$$+ (1 + s_0) h_y \|\partial_y^2 u\|_{P(K)}.$$

Proof. We begin with the x–derivative of the interpolation error:

$$\|\partial_x(u - \mathcal{I}_h u)\|_K^2 = \det T_K \int_{\hat{K}} (\partial_x(u - \mathcal{I}_h u)(T_K \hat{x}))^2 d\hat{x}$$
$$= h_x^{-2} \det T_K \int_{\hat{K}} \left[(\partial_{\hat{x}}(\hat{u} - \hat{\mathcal{I}}_h \hat{u})(\hat{x}))^2 + (\partial_{\hat{y}}(\hat{u} - \hat{\mathcal{I}}_h \hat{u})(\hat{x}))^2\right] d\hat{x}$$
$$\lesssim h_x^{-2} \det T_K \left(\|\partial_{\hat{x}}(\hat{u} - \hat{\mathcal{I}}_h \hat{u})\|_{\hat{K}}^2 + \|\partial_{\hat{y}}(\hat{u} - \hat{\mathcal{I}}_h \hat{u})\|_{\hat{K}}^2\right).$$

For the y–derivative we use $|\partial_y \hat{x}| \lesssim s_0 h_x^{-1} + h_y^{-1}$ and (16):

$$\|\partial_y(u - \mathcal{I}_h u)\|_K^2$$
$$\lesssim \det T_K \int_{\hat{K}} (h_y^{-1} \partial_{\hat{y}}(\hat{u} - \hat{\mathcal{I}}_h \hat{u})(\hat{x}) + (s_0 h_h^{-1} + h_y^{-1})(\partial_{\hat{x}}(\hat{u} - \hat{\mathcal{I}}_h \hat{u})(\hat{x}))^2 d\hat{x}$$
$$\lesssim \det T_K (h_y^{-2} \|\partial_{\hat{y}}(\hat{u} - \hat{\mathcal{I}}_h \hat{u})\|_{\hat{K}}^2 + (s_0 h_x^{-1} + h_y^{-1})^2 \|\partial_{\hat{x}}(\hat{u} - \hat{\mathcal{I}}_h \hat{u})\|_{\hat{K}}^2).$$

With help of Lemma 4 and (A1) we obtain:

$$\|\partial_x(u - \mathcal{I}_h u)\|_K \lesssim \det T_K^{\frac{1}{2}} h_x^{-1} \left(\|\partial_{\hat{x}} \hat{\nabla} \hat{u}\|_{P(\hat{K})} + \|\partial_{\hat{y}} \hat{\nabla} \hat{u}\|_{P(\hat{K})}\right),$$

$$\|\partial_y(u - \mathcal{I}_h u)\|_K \lesssim \det T_K^{\frac{1}{2}} \left((s_0 h_x^{-1} + h_y^{-1}) \|\partial_{\hat{x}} \hat{\nabla} \hat{u}\|_{P(\hat{K})}\right.$$
$$\left. + (h_y^{-1} + s_0 h_x^{-2} h_y) \|\partial_{\hat{y}} \hat{\nabla} \hat{u}\|_{P(\hat{K})}\right).$$

In order to bound the right hand sides we need the second derivatives of $\hat{u}$:

$$|\partial_{\hat{x}}^2 \hat{u}| \lesssim h_x^2 |\partial_x^2 u| + h_x h_y |\partial_{xy} u| + h_y^2 |\partial_y^2 u|,$$

$$|\partial_{\hat{x}\hat{y}}\hat{u}| = |\partial_{\hat{y}\hat{x}}\hat{u}| \lesssim (s_0 h_y + \alpha) h_x |\partial_x^2 u| + (s_0 h_y + \alpha) h_y |\partial_{xy} u| + h_y^2 |\partial_y^2 u|,$$
$$|\partial_{\hat{y}}^2 \hat{u}| \lesssim (s_0 h_y + \alpha)^2 |\partial_x^2 u| + (s_0 h_y + \alpha) h_y |\partial_{xy} u| + h_y^2 |\partial_y^2 u|.$$

There is a constant s_1 with $1 \leq s_1 \leq 1 + s_0$ so that the L^2 norms of the second derivatives of $\hat{u}$ can now be bounded by

$$\|\partial_{\hat{x}} \hat{\nabla} \hat{u}\|_{\hat{K}} \lesssim \det T_K^{-1/2} \left(s_1 h_x^2 \|\partial_x^2 u\|_{\hat{K}} + s_1 h_x h_y \|\partial_{xy} u\|_K + h_y^2 \|\partial_y^2 u\|_K \right),$$
$$\|\partial_{\hat{y}} \hat{\nabla} \hat{u}\|_K \lesssim \det T_K^{-1/2} \left(s_1^2 h_x^2 \|\partial_x^2 u\|_K + s_1 h_x h_y \|\partial_{xy} u\|_K + h_y^2 \|\partial_y^2 u\|_K \right).$$

Hence, for the x-derivative of the approximation error:

$$\|\partial_x (u - \mathcal{I}_h u)\|_K \lesssim s_1^2 h_x \|\partial_x^2 u\|_{P(K)} + s_1 h_y \|\partial_{xy} u\|_{P(K)} + h_x^{-1} h_y^2 \|\partial_y^2 u\|_{P(K)}.$$

This verifies (17). For the y-derivative of the approximation error we obtain:

$$\|\partial_y (u - \mathcal{I}_h u)\|_K \lesssim \left((s_0 h_x^{-1} + h_y^{-1}) s_1 h_x^2 + (h_y^{-1} + s_0 h_x^{-2} h_y) s_1^2 h_x^2 \right) \|\partial_x^2 u\|_{P(K)}$$
$$+ \left((s_0 h_x^{-1} + h_y^{-1} + s_0 h_x^{-2} h_y) s_1 h_x h_y \right) \|\partial_{xy} u\|_{P(K)}$$
$$+ (s_0 h_y + (h_y^{-1} + s_0 h_x^{-1}) h_y^2) \|\partial_y^2 u\|_{P(K)}$$

The assertion (18) follows immediately. ∎

Corollary 1. *It holds for* $\mathcal{I}_h \in \mathcal{A}_h$ *and* $u \in H^2(P(K))$:

$$\|\nabla(u - \mathcal{I}_h u)\|_K \lesssim (1 + s_0)^2 h_x \|\partial_x^2 u\|_{P(K)} + (1 + s_0)^2 h_x \|\partial_{xy} \nabla u\|_{P(K)}$$
$$+ (1 + s_0) h_y \|\partial_y^2 u\|_{P(K)}$$

Proof. Combining (17) and (18) gives the result. ∎

3 General result

Finally, we interpret the results for transformation including the rotation, i.e., $\theta \neq 0$. In this case, we have to replace the partial derivatives along the coordinate axes, ∂_x, ∂_y, by appropriate directional derivatives: By $\eta_1 = Te_1 \in \mathbb{R}^2$ we denote the unit vector aligned with the longest side of K, and $\eta_2 \in \mathbb{R}^2$ is orthogonal to it. Furthermore, h_i is the length of K in direction of η_i, $i = 1, 2$.

With these notations, we assemble the results of Lemma 1, 3, Proposition 1 and Corollary 1 as follows:

Corollary 2. *There is a linear operator* $B_h : H^1(\Omega) \to V_h$ *with the following features for all* $K \in \mathcal{T}_h$ *(with* $s_1 := 1 + s_0$*):*

(i) Stability:

$$\|\partial_x B_h u\|_K \lesssim s_1 \|\partial_{\eta_1} u\|_{P(K)} + h_1^{-1} h_2 \|\partial_{\eta_2} u\|_{P(K)},$$
$$\|\partial_{\eta_2} B_h u\|_K \lesssim s_1 \left(s_0 \|\partial_{\eta_1} u\|_{P(K)} + \|\partial_{\eta_2} u\|_{P(K)} \right).$$

(ii) Approximation:

$$\|u - B_h u\|_K \lesssim s_1 h_1 \|\partial_{\eta_1} u\|_{P(K)} + h_2 \|\partial_{\eta_2} u\|_{P(K)}.$$

(iii) Approximation for $u \in H^2(P(K))$:

$$\|\nabla(u - B_h u)\|_K \lesssim s_1^2 h_1 \|\partial_{\eta_1}^2 u\|_{P(K)} + s_1^2 h_1 \|\partial_{\eta_1 \eta_2} u\|_{P(K)} + s_1 h_2 \|\partial_{\eta_2}^2 u\|_{P(K)}.$$

References

1. Apel,T: Anisotropic finite elements: Local estimates and applications. Advances in Numerical Mathematics. Teubner, Stuttgart (1999)
2. Becker, R.: An adaptive finite element method for the incompressible Navier-Stokes equation on time-dependent domains. PhD Dissertation, SFB-359 Preprint 95-44, Universität Heidelberg (1995)
3. Braack, M., Richter,T.: Local projection stabilization for the stokes system on anisotropic quadrilateral meshes. In Numerical Mathematics and Advanced Applications, ENUMATH 2005. Springer (2006)
4. Clément, P.: Approximation by finite element functions using local regularization. R.A.I.R.O. Anal. Numer., **9**, 77–84 (1975)
5. Dupont, T., Scott, R.: Polynomial approximation of functions in Sobolev spaces. Math. Comp., **34**(150), 441–463 (1980)
6. L. Scott and S. Zhang. Finite element interpolation of nonsmooth functions satisfying boundary conditions. Math. Comp., **54**(190), 483–493 (1990)

Continuous Interior Penalty hp-Finite Element Methods for Transport Operators

Erik Burman[1] and Alexandre Ern[2]

[1] Institut d'Analyse et Calcul Scientifique (CMCS/IACS), Ecole Polytechnique Fédérale de Lausanne, 1015 Lausanne, Switzerland
[2] CERMICS, Ecole nationale des ponts et chaussées, Champs sur Marne, 77455 Marne la Vallée Cedex 2, France

Summary. A continuous interior penalty hp-finite element method that penalizes the jump of the gradient of the discrete solution across mesh interfaces is introduced and analyzed. Error estimates are presented for first-order transport equations. The analysis relies on three technical results that are of independent interest: an hp-inverse trace inequality, a local discontinuous to continuous hp-interpolation result, and hp-error estimates for continuous L^2-orthogonal projections.

1 Introduction

Interior penalty procedures for finite element methods utilizing continuous functions have been introduced in the pioneering works of Babuška and Zlámal [1] for the biharmonic operator and of Douglas and Dupont [9] for second-order elliptic and parabolic problems. The common feature of these methods consists of penalizing the jump of the gradient of the discrete solution at mesh interfaces, but the motivations behind [1] and [9] are different. The goal pursued in [1] was to weakly enforce C^1-continuity. Because of a non-consistency in the formulation, a superpenalty procedure had to be applied, leading to suboptimal convergence rates. The subsequent work of Baker [2], valid for general $2m$th order coercive operators, designed a consistent interior penalty method, utilizing *discontinuous* functions, that was shown to be optimally convergent.

The motivation behind the work of Douglas and Dupont was different, namely to keep *continuous* finite element methods because they were standard for elliptic problems and, at the same time, to cope with the difficulties encountered by such methods in problems where the first-order (advection) terms dominate the second-order (diffusion) terms. However, one of the main issues at stake, namely the robustness of the error estimates with respect to the cell Péclet numbers, was not addressed in [9]. This issue has been addressed only recently for linear finite elements, namely in the work of Burman and Hansbo in 2004 [7].

The goal of this paper is to present, for the first time, an hp-convergence analysis for a high-order CIP finite element method applied to first-order transport equations. No proofs will be given here. For brevity we focus on tensor product finite elements. We refer to the report [3] for full detail including the extension to simplicial finite elements and to advection dominated second order transport equations.

2 Continuous interior penalty finite element methods

Let Ω be an open bounded and connected set in $\mathbb{R}^d$ with Lipschitz boundary $\partial\Omega$ and outer normal n, let $\beta \in [W^{1,\infty}(\Omega)]^d$ be a vector field, and let $\sigma \in L^\infty(\Omega)$. Let $f \in L^2(\Omega)$, let $\partial\Omega^\pm = \{x \in \partial\Omega; \pm\beta(x) \cdot n(x) > 0\}$, assume that $\partial\Omega^-$ and $\partial\Omega^+$ are well separated, and consider the problem

$$\begin{cases} \sigma u + \beta \cdot \nabla u = f, \\ u|_{\partial\Omega^-} = 0. \end{cases} \quad (1)$$

Define $W = \{w \in L^2(\Omega); \beta \cdot \nabla w \in L^2(\Omega)\}$ and observe that functions in W have traces in $L^2(\partial\Omega; \beta \cdot n)$. Consider the operator $A : W \ni w \mapsto \sigma w + \beta \cdot \nabla w \in L^2(\Omega)$. Henceforth, it is assumed that there is $\sigma_0 > 0$ such that

$$\sigma - \tfrac{1}{2}\nabla \cdot \beta \geq \sigma_0, \quad \text{a.e. in } \Omega. \quad (2)$$

Then, letting $V = \{w \in W; w|_{\partial\Omega^-} = 0\}$, $A : V \to L^2(\Omega)$ is an isomorphism, i.e., (1) is well-posed; see, e.g., [5].

Let $\mathcal{K}$ be a subdivision of Ω into non-overlapping rectangular cells $\{\kappa\}$. For $\kappa \in \mathcal{K}$, h_κ denotes its diameter. Set $h = \max_{\kappa \in \mathcal{K}} h_\kappa$. Assume that (i) $\mathcal{K}$ covers $\overline{\Omega}$ exactly, (ii) $\mathcal{K}$ does not contain any hanging nodes, and (iii) $\mathcal{K}$ is quasi-uniform in the sense that there exists a constant $\rho > 0$, independent of h, such that $\rho h \leq \min_{\kappa \in \mathcal{K}} h_\kappa$. Each $\kappa \in \mathcal{K}$ is an affine image of the unit hypercube $\widehat{\kappa} = [-1, 1]^d$, i.e., $\kappa = F_\kappa(\widehat{\kappa})$. Let $\mathcal{F}$ denote the set of interior faces ($(d-1)$-manifolds) of the mesh, i.e., the set of faces that are not included in the boundary $\partial\Omega$. For $F \in \mathcal{F}$, h_F denotes its diameter.

Let $p \geq 1$ and let $\mathbb{Q}_{p,d}(\widehat{\kappa})$ be the space of polynomials of degree at most p in each variable. Introduce the continuous and discontinuous finite element spaces

$$V_h^p = \{\, v_h \in C^0(\overline{\Omega}); \forall \kappa \in \mathcal{K}, v_h|_\kappa \circ F_\kappa \in \mathbb{Q}_{p,d}(\widehat{\kappa})\,\}, \quad (3)$$

$$W_h^p = \{\, w_h \in L^2(\Omega); \forall \kappa \in \mathcal{K}, w_h|_\kappa \circ F_\kappa \in \mathbb{Q}_{p,d}(\widehat{\kappa})\,\}. \quad (4)$$

For a subset $R \subset \Omega$, $(\cdot,\cdot)_R$ denotes the $L^2(R)$–scalar product, $\|\cdot\|_R = (\cdot,\cdot)_R^{\frac{1}{2}}$ the corresponding norm, and $\|\cdot\|_{s,R}$ the $H^s(R)$–norm. For $s \geq 1$, let $H^s(\mathcal{K})$ be the space of piecewise H^s functions. For $v \in H^2(\mathcal{K})$ and an interior face $F = \kappa_1 \cap \kappa_2$, where κ_1 and κ_2 are two distinct elements of $\mathcal{K}$

with respective outer normals n_1 and n_2, introduce the (scalar-valued) jump $[\nabla v \cdot n]_F = \nabla v|_{\kappa_1} \cdot n_1 + \nabla v|_{\kappa_2} \cdot n_2$ (the subscript F is dropped when there is no ambiguity). Similarly, for $v \in H^1(\mathcal{K})$, define the (scalar-valued) jump $[v]_F = v|_{\kappa_1} - v|_{\kappa_2}$ (the arbitrariness in the sign of $[v]_F$ can be avoided by considering the vector-valued jump $[v]_F = v|_{\kappa_1} n_1 + v|_{\kappa_2} n_2$; nothing that is stated hereafter depends on this arbitrariness).

On $H^1(\Omega) \times H^1(\Omega)$ define the standard Galerkin bilinear form

$$a(v,w) = ((\sigma - \nabla \cdot \beta)v, w)_\Omega - (v, \beta \cdot \nabla w)_\Omega + (\beta \cdot nv, w)_{\partial \Omega^+}, \tag{5}$$

and on $H^q(\mathcal{K}) \times H^q(\mathcal{K})$, $q > \frac{3}{2}$, define the CIP bilinear form

$$j(v,w) = \sum_{F \in \mathcal{F}} \frac{h_F^2}{p^\alpha} |\beta \cdot n|_F ([\nabla v \cdot n], [\nabla w \cdot n])_F, \tag{6}$$

where $|\beta \cdot n|_F$ denotes the L^∞-norm of the normal component of β on F. Since $W^{1,\infty}(\Omega) \subset C^0(\overline{\Omega})$, the field β is continuous by assumption and, therefore, the quantity $\beta \cdot n$ is single-valued on all interior faces $F \in \mathcal{F}$. The exponent α will be determined by the convergence analysis in mS4; see (18).

The finite element approximation to (1) consists of seeking $u_h \in V_h^p$ such that

$$a(u_h, v_h) + j(u_h, v_h) = (f, v_h)_\Omega, \quad \forall v_h \in V_h^p. \tag{7}$$

For $v \in H^q(\mathcal{K})$, $q > \frac{3}{2}$, consider the norm

$$\|v\|_{a,j}^2 = \|\sigma_0^{\frac{1}{2}} v\|_\Omega^2 + \frac{1}{2}\||\beta \cdot n|^{\frac{1}{2}} v\|_{\partial\Omega}^2 + j(v,v). \tag{8}$$

The well-posedness of the approximate problem (7) results from the following

Lemma 1. *For all $v \in H^q(\mathcal{K})$, $q > \frac{3}{2}$, $a(v,v) + j(v,v) \geq \|v\|_{a,j}^2$.*

3 Technical results

Henceforth, we use the notation $A \lesssim B$ to represent the inequality $A \leq cB$ where the constant c is independent of p and h, but can depend on the space dimension d and the quasi-uniformity parameter ρ. All the results stated below are simplified using the practical assumption that $d \leq 3$.

3.1 *hp*-trace inequalities

Let $\{g_j\}_{0 \leq j \leq p}$ be the Gauß–Lobatto nodes in the unit interval $[-1,1]$. Set $I_{p,d} = \{0,\ldots,p\}^d$ and $I_{p,d}^0 = \{1,\ldots,p-1\}^d$. For a multi-index $(i) = (i_1,\ldots,i_d) \in I_{p,d}$, the tensor-product Gauß–Lobatto node $a_{\hat{\kappa},(i)}$ in the unit hypercube $\hat{\kappa}$ is the point with coordinates equal to $(g_{i_1},\ldots,g_{i_d})$.

Let $\kappa \in \mathcal{K}$. Introduce the tensor-product Gauß–Lobatto nodes in K such that $a_{\kappa,(i)} = F_\kappa(a_{\widehat{\kappa},(i)})$ for all $(i) \in I_{p,d}$ and define the space

$$\mathbb{Q}^0_{p,d}(\kappa) = \{\, v \in \mathbb{Q}_{p,d}(\kappa);\ \forall (i) \in I^0_{p,d},\ v(a_{\kappa,(i)}) = 0 \,\}. \tag{9}$$

In other words, $\mathbb{Q}^0_{p,d}(\kappa)$ is the subspace of $\mathbb{Q}_{p,d}(\kappa)$ spanned by the polynomials that vanish at all the interior tensor-product Gauß–Lobatto nodes in κ.

Lemma 2. *The following trace and inverse trace inequalities hold:*

$$\forall v \in \mathbb{Q}_{p,d}(\kappa), \quad \|v\|_{\partial\kappa} \lesssim \left(\frac{p^2}{h_\kappa}\right)^{\frac{1}{2}} \|v\|_\kappa, \tag{10}$$

$$\forall v \in \mathbb{Q}^0_{p,d}(\kappa), \quad \|v\|_\kappa \lesssim \left(\frac{h_\kappa}{p^2}\right)^{\frac{1}{2}} \|v\|_{\partial\kappa}. \tag{11}$$

An important observation is that the inverse trace inequality (11) is optimal (asymptotically in p) with respect to the trace inequality (10).

3.2 Continuous hp-interpolation

The goal is to construct an operator $\mathcal{I}_{\mathrm{Os}} : W^p_h \to V^p_h$ endowed with a local interpolation property.

Let $\kappa \in \mathcal{K}$. For a node ν in κ, set $\mathcal{K}_\nu = \{\kappa' \in \mathcal{K};\ \nu \in \kappa'\}$; then, for $w_h \in W^p_h$, define $\mathcal{I}_{\mathrm{Os}} w_h$ locally in κ by the value it takes at all the tensor-product Gauß–Lobatto nodes by setting

$$\mathcal{I}_{\mathrm{Os}} w_h(\nu) = \frac{1}{\operatorname{card}(\mathcal{K}_\nu)} \sum_{\kappa \in \mathcal{K}_\nu} w_h|_\kappa(\nu). \tag{12}$$

Clearly, $\mathcal{I}_{\mathrm{Os}} w_h \in V^p_h$.

Lemma 3. *The following estimate holds for all $\kappa \in \mathcal{K}$,*

$$\forall w_h \in W^p_h, \quad \|w_h - \mathcal{I}_{\mathrm{Os}} w_h\|_\kappa \lesssim \left(\frac{h_\kappa}{p^2}\right)^{\frac{1}{2}} \sum_{F \in \mathcal{F}(\kappa)} \|[w_h]\|_F, \tag{13}$$

where $\mathcal{F}(\kappa) = \{F \in \mathcal{F};\ F \cap \kappa \neq \emptyset\}$.

3.3 hp-error estimate for continuous $L^2(\Omega)$-orthogonal projection

Let $\Pi_h : L^2(\Omega) \to V^p_h$ be the $L^2(\Omega)$-orthogonal projector onto V^p_h. The purpose of this section is to investigate the approximation properties of Π_h in the L^2- and the H^1-norm.

First, we recall the following local hp-approximation property [5, 9]. Let $\Pi^*_h : L^2(\Omega) \to W^p_h$ be the $L^2(\Omega)$-orthogonal projector onto W^p_h. Then, for all $\kappa \in \mathcal{K}$ and all $w \in H^q(\mathcal{K})$, $q \geq 1$,

$$\|w - \Pi_h^* w\|_\kappa \lesssim \left(\frac{h}{p}\right)^s \|w\|_{s,\kappa}, \tag{14}$$

$$\|\nabla(w - \Pi_h^* w)\|_\kappa \lesssim p^{\frac{1}{2}} \left(\frac{h}{p}\right)^{s-1} \|w\|_{s,\kappa}, \tag{15}$$

with $s = \min(p+1, q)$. The global counterpart of (14)–(15) for the continuous $L^2(\Omega)$-orthogonal projector Π_h is the following.

Lemma 4. *For all $w \in H^q(\Omega)$, $q \geq 1$, the following holds with $s = \min(p+1, q)$,*

$$\|w - \Pi_h w\|_\Omega \lesssim \left(\frac{h}{p}\right)^s \|w\|_{s,\Omega}, \tag{16}$$

$$\|\nabla(w - \Pi_h w)\|_\Omega \lesssim p^{\frac{1}{2}} \left(\frac{h}{p}\right)^{s-1} \|w\|_{s,\Omega}. \tag{17}$$

4 Convergence analysis

Let u solve (1) and let u_h solve (7). Henceforth, it is assumed that the exact solution u is smooth enough, i.e., $u \in H^q(\Omega)$, $q > \frac{3}{2}$. Bounds on the approximation error $u - u_h$ are obtained in the spirit of the Second Strang Lemma by establishing consistency and boundedness properties for the discrete setting. Recall that the discrete setting satisfies the stability property stated in Lemma 1.

Theorem 1. *Let $u \in H^q(\Omega)$, $q > \frac{3}{2}$, solve (1) and let u_h solve (7). Take*

$$\alpha = \frac{7}{2}. \tag{18}$$

Then,

$$\|u - u_h\|_{a,j} \lesssim (p^{\frac{1}{4}} + p^{\frac{3}{2}} h^{\frac{1}{2}}) \left(\frac{h}{p}\right)^{s-\frac{1}{2}} \|u\|_{s,\Omega}, \tag{19}$$

with $s = \min(p+1, q)$. Hence, if $h \leq p^{-\frac{5}{2}}$,

$$\|u - u_h\|_{a,j} \lesssim p^{\frac{1}{4}} \left(\frac{h}{p}\right)^{s-\frac{1}{2}} \|u\|_{s,\Omega}. \tag{20}$$

5 Numerical results

In this section we present some numerical experiments using the method proposed above in one space dimension. We use the Lagrange polynomials based on the Gauß–Lobatto interpolation nodes and integration at the same nodes. The linear systems are solved using the direct solver of Matlab.

5.1 Convergence results, smooth solutions

We consider problem (1) in one space dimension with $\Omega = (0,1)$, $\sigma = 0$, and $\beta = 1$. Boundary data and right-hand side f are chosen to yield the exact solutions $u = \sin(2\pi x)$ (test case I) and $u = \arctan(\frac{x-0.5}{\epsilon})$ with $\epsilon = 0.01$ (test case II).

The first row in Figure 1 displays a sequence of standard Galerkin approximations ($\gamma = 0$) for test case II using different polynomial orders, but the same number of degrees of freedom. Upstream node to node oscillations are present in the three cases. The same solutions with stabilization are displayed in the second row, showing that oscillations are essentially eliminated. For both standard and stabilized Galerkin approximations, the $\mathbb{P}_2$ approximation suffers from poor resolution of the source term (because of insufficient quadrature accuracy) leading to an incorrect size of the jump. Figure 2 displays convergence curves for the two test cases. The left plot presents h-convergence curves for $\mathbb{P}_2$, $\mathbb{P}_3$, and $\mathbb{P}_5$ polynomials for test case I (the error is of machine precision already on the coarsest mesh for higher order polynomials). Optimal convergence orders are observed. The middle plot presents the h-convergence curves for $\mathbb{P}_2$, $\mathbb{P}_5$, and $\mathbb{P}_8$ polynomials for test case II. Optimal convergence orders are observed up to the finest meshes where accuracy is affected by the precision of the quadrature. The right plot presents p-convergence curves for both test cases on a fixed mesh with 64 elements. Both test cases yield exponential convergence under p refinement in agreement with the estimate (20). The slope for test case I is larger because $\|u\|_{p+1,\Omega}$ does not depend on p.

5.2 Conditioning and scatter plots of system matrix eigenvalues

In this section we study the condition number of the system matrix and the distribution of its eigenvalues in the complex plane for different values of the

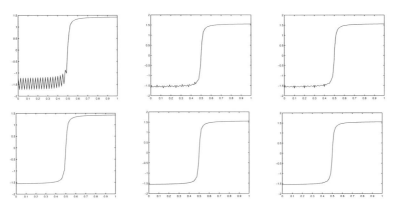

Fig. 1. Above: standard Galerkin approximation with 81 degrees of freedom. From left to right: 40 $\mathbb{P}_2$-elements, 16 $\mathbb{P}_5$-elements, and 10 $\mathbb{P}_8$-elements. Below: solutions computed with stabilization ($\gamma = 0.2$).

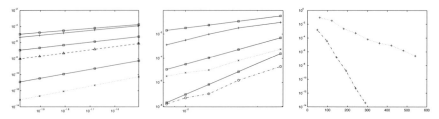

Fig. 2. Convergence curves. Left: case I, h-convergence for $\mathbb{P}_2('+')$, $\mathbb{P}_3('\triangle')$, and $\mathbb{P}_5('x')$. Center: case II, h-convergence for $\mathbb{P}_2('+')$, $\mathbb{P}_5('x')$, and $\mathbb{P}_8('o')$. Right: p-convergence for case I (dashed) and case II (dotted). In the left and middle plots, full lines with square markers correspond to theoretical slopes.

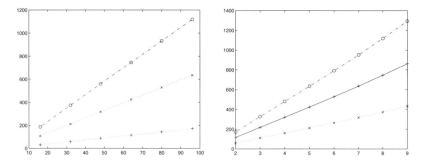

Fig. 3. Left: condition number against h^{-1}: $\mathbb{P}_2('+')$, $\mathbb{P}_5('x')$, and $\mathbb{P}_8('o')$. Right: condition number against p: $h = 32^{-1}('x')$, $h = 64^{-1}('+')$, and $h = 96^{-1}('o')$.

stabilization parameter γ. Figure 3 presents the condition number against h^{-1} (left) and p (right). The condition number scales as h^{-1} keeping p fixed and as p (and not as p^2 as could be heuristically expected from inverse inequalities) keeping h fixed. The scaling in h agrees with the theoretical results of [8].

Finally Figure 4 presents scatter plots of the eigenvalues for the three hp-discretizations in the second row of Figure 1 and with three different values of the stabilization parameter ($\gamma = 0$, $\gamma = 0.1$, and $\gamma = 1.0$). For the three hp-discretizations, the eigenvalues produced by the standard Galerkin method are the closest to the imaginary axis. The effect of stabilization is to shift their real part away from zero. Comparing the case $\gamma = 0.1$ and $\gamma = 1.0$ shows that increased stabilization leads to a separation of the eigenvalues into two subsets, one shifted closer to the standard Galerkin spectrum (presumably corresponding to the C^1 modes) and the other shifted to higher values of the real part. In the case of $\mathbb{P}_2$ approximation, the range on the imaginary axis decreases as the stabilization increases. This effect is not present for higher order polynomials and can be attributed to the fact that only a few C^1 modes are present with $\mathbb{P}_2$ polynomials.

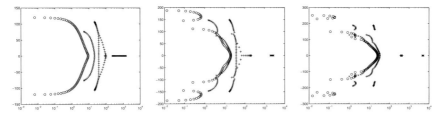

Fig. 4. Scatter plots of the eigenvalues of the system matrix corresponding to a discretization with 81 degrees of freedom: $\gamma = 0('o')$, $\gamma = 0.1('+')$, and $\gamma = 1.0('x')$. From left to right: $\mathbb{P}_2$, $\mathbb{P}_5$, and $\mathbb{P}_8$ polynomials.

References

1. Babuška, I, Zlámal, M.: Nonconforming elements in the finite element method with penalty. SIAM J. Numer. Anal., **10**, 863–875 (1973)
2. Baker, G.A.: Finite element methods for elliptic equations using nonconforming elements. Math. Comp., **31**(137), 45–59 (1977)
3. Burman, E., Ern, A.: Continuous interior penalty hp-finite element methods for advection and advection–diffusion equations. Math. Comp., submitted. Technical Report 02.2005, Ecole Polytechnique Federale de Lausanne (2005)
4. Burman, E., Hansbo, P.: Edge stabilization for galerkin approximations of convection–diffusion–reaction problems. Comput. Methods Appl. Mech. Engrg., **193**, 1437–1453 (2004)
5. Canuto, C., Quarteroni, A.: Approximation results for orthogonal polynomials in Sobolev spaces. Math. Comp., **38**, 67–86 (1982)
6. Douglas Jr., J., Dupont, T.: Interior Penalty Procedures for Elliptic and Parabolic Galerkin Methods, In R. Glowinski and J.-L. Lions, editors, Computing Methods in Applied Sciences, **58** of Lecture Notes in Physics, 207–216. Springer-Verlag, Berlin (1976)
7. Ern, A., Guermond,J.-L.: Theory and Practice of Finite Elements, **159** Applied Mathematical Sciences. Springer-Verlag, New York, NY (2004)
8. Ern, A., Guermond,J.-L.: Evaluation of the condition number in linear systems arising in finite element approximations. Math. Model. Numer. Anal. (M2AN) (2005) In press
9. Houston, P., Schwab, C., Süli, E.: Discontinuous hp-finite element methods for advection–diffusion–reaction problems. SIAM J. Numer. Anal., **39**(6), 2133–2163 (2002)

A Nonconforming Finite Element Method with Face Penalty for Advection–Diffusion Equations

L. El Alaoui,[1,2] A. Ern[2] and E. Burman[3]

[1] Department of Mathematics, Imperial College, London, SW7 2AZ, UK
l.elalaoui@imperial.ac.uk
[2] CERMICS, ENPC, Champs sur Marne, 77455 Marne la Vallée Cedex 2, France
ern@cermics.enpc.fr
[3] CMCS/IACS, EPFL, Lausanne, Switzerland
erik.burman@epfl.ch

Summary. We present a nonconforming finite element method with face penalty to approximate advection–diffusion–reaction equations. The a priori error analysis leads to (quasi-)optimal error estimates in the mesh-size keeping the Péclet number fixed. The a posteriori error analysis yields residual-type error indicators that are semi-robust in the sense that the lower and upper bounds of the error differ by a factor bounded by the square root of the Péclet number. Finally, to illustrate the theory, numerical results including adaptively generated meshes are presented.

1 Introduction

Let Ω be a polygonal domain of $\mathbb{R}^d$ with Lipschitz boundary $\partial\Omega$ and outward normal n. Let $\epsilon > 0$, $\beta \in [\mathcal{C}^{0,\frac{1}{2}}(\Omega)]^d$, and $\nu \in L^\infty(\Omega)$ be, respectively, the diffusion coefficient, the velocity field, and the reaction coefficient. Set $\partial\Omega_{\text{in}} = \{x \in \partial\Omega : \beta \cdot n < 0\}$ and $\partial\Omega_{\text{out}} = \{x \in \partial\Omega : \beta \cdot n \geq 0\}$. Let $f \in L^2(\Omega)$ and $g \in L^2(\partial\Omega_{\text{in}})$. Consider the following advection–diffusion–reaction problem with mixed Robin–Neumann boundary conditions:

$$\begin{cases} -\epsilon \Delta u + \beta \cdot \nabla u + \nu u = f & \text{in } \Omega, \\ -\epsilon \nabla u \cdot n + \beta \cdot n\, u = g & \text{on } \partial\Omega_{\text{in}}, \\ \nabla u \cdot n = 0 & \text{on } \partial\Omega_{\text{out}}. \end{cases} \quad (1)$$

Without loss of generality, we assume that (1) is non-dimensionalized so that $\|\beta\|_{[L^\infty(\Omega)]^d}$ and the length scale of Ω are of order unity; hence, the parameter ϵ is the reciprocal of the Péclet number.

Under the assumption that there is $\sigma_0 > 0$ such that $\sigma = \nu - \frac{1}{2}\nabla\cdot\beta \geq \sigma_0$ in Ω and that $\nabla\cdot\beta \in L^\infty(\Omega)$, it is straightforward to verify using the Lax–Milgram Lemma that the following weak formulation of (1) is well posed:

$$\begin{cases} \text{Seek } u \in H^1(\Omega) \text{ such that} \\ a(u,v) = \int_\Omega fv - \int_{\partial\Omega_{\text{in}}} gv \quad \forall v \in H^1(\Omega), \end{cases} \qquad (2)$$

where

$$a(u,v) = \int_\Omega \epsilon \nabla u \cdot \nabla v + \int_\Omega (\nu - \nabla \cdot \beta) uv - \int_\Omega u(\beta \cdot \nabla v) + \int_{\partial\Omega_{\text{out}}} (\beta \cdot n) uv. \qquad (3)$$

Advection–diffusion–reaction equations are encountered in many applications, including pollutant transport and the Oseen equations. It is well-known that the standard Galerkin approximation to these equations leads to non-physical oscillations in the advective–dominated regime. To stabilize this phenomenon, several well–established techniques have been proposed and analyzed in a conforming setting (e.g., streamline–diffusion [2], subgrid viscosity [8], and residual free bubbles [3]), in a nonconforming setting (e.g., streamline–diffusion [10]), and in a discontinuous setting [9].

The purpose of this paper is twofold. First, to design and analyze a nonconforming finite element method to approximate advection–diffusion–reaction equations. Drawing from ideas in [3, 7], the method is stabilized by penalizing the jumps across interfaces of the gradient of the discrete solution. The advantage of using the face penalty technique rather than streamline–diffusion is that the former involves a single user-dependent parameter which is independent of the diffusion coefficient. Moreover, the face penalty technique is readily extendable to time-dependent problems. The second goal of this paper is to derive an a posteriori error estimator which is semi–robust, namely the ratio between the upper and lower bounds for the error is bounded by the square root of the Péclet number. Semi-robust error estimators for advection–diffusion equations are derived in [12] in a conforming setting, but to our knowledge, no such results are available in a nonconforming setting. The derivation of robust error estimators (in which the ratio in question is independent of the Péclet number) such as those obtained in [11, 13] in a conforming setting, goes beyond the present scope.

This paper is organized as follows. Section 2 presents the nonconforming finite element method with face penalty and the a priori error analysis. Section 3 is devoted to the a posteriori error analysis, which is of residual type. Only the main results are stated; we refer to [6] for details and proofs. Finally, Section 4 contains numerical results.

2 The nonconforming finite element scheme with face penalty

In this section we design a nonconforming finite element approximation to (1) with face penalty stabilization. The convergence analysis leads to an a priori error estimate which is (quasi-)optimal in the mesh-size keeping the Péclet number fixed (the estimate is sub–optimal of order $\frac{1}{2}$ in the L^2–norm and optimal in the broken graph norm for quasi–uniform meshes).

2.1 The discrete setting

Let $(\mathcal{T}_h)_h$ be a shape–regular family of simplicial affine meshes of Ω. For an element $T \in \mathcal{T}_h$, let h_T denote its diameter and set $h = \max_{T \in \mathcal{T}_h} h_T$. Let $\mathcal{F}_h$, $\mathcal{F}_h^i$, and $\mathcal{F}_h^\partial$ denote respectively the set of faces, internal, and external faces in $\mathcal{T}_h$. Let $\mathcal{F}_h^{\text{in}}$ and $\mathcal{F}_h^{\text{out}}$ be the set of faces belonging respectively to $\partial \Omega_{\text{in}}$ and to $\partial \Omega_{\text{out}}$ such that $\mathcal{F}_h^\partial = \mathcal{F}_h^{\text{in}} \cup \mathcal{F}_h^{\text{out}}$. For a face $F \in \mathcal{F}_h$, let h_F denote its diameter and $\mathcal{T}_F$ the set of elements in $\mathcal{T}_h$ containing F. For an element $T \in \mathcal{T}_h$, let $\mathcal{F}_T$ denote the set of faces belonging to T.

For an integer $k \geq 1$, let $H^k(\mathcal{T}_h) = \{v \in L^2(\Omega); \forall T \in \mathcal{T}_h, v|_T \in H^k(T)\}$. We introduce the discrete gradient operator $\nabla_h : H^1(\mathcal{T}_h) \to [L^2(\Omega)]^d$ such that for all $v \in H^1(\mathcal{T}_h)$ and for all $T \in \mathcal{T}_h$, $(\nabla_h v)|_T = \nabla(v|_T)$. Let $F \in \mathcal{F}_h^i$; then, there are $T_1(F)$ and $T_2(F) \in \mathcal{T}_h$ such that $F = T_1(F) \cap T_2(F)$. Conventionally, choose n_F to be the unit normal vector to F pointing from $T_1(F)$ towards $T_2(F)$. For $v \in H^1(\mathcal{T}_h)$, define its jump across F as

$$[\![v]\!]_F = v|_{T_1(F)} - v|_{T_2(F)} \quad \text{a.e. on } F. \tag{4}$$

A similar notation is used for the jumps of vector-valued functions, the jump being taken componentwise. Nothing that is said hereafter depends on the arbitrariness in the sign of the jump.

For a subset $R \subset \Omega$, $(\cdot, \cdot)_{0,R}$ denotes the $L^2(\Omega)$–scalar product, $\|\cdot\|_{0,R}$ the associated norm, and $\|\cdot\|_{k,R}$ the $H^k(R)$–norm for $k \geq 1$. Consider the Crouzeix-Raviart finite element space $P^1_{\text{nc}}(\mathcal{T}_h)$ defined as [5, 5]

$$P^1_{\text{nc}}(\mathcal{T}_h) = \{v_h \in L^2(\Omega); \forall T \in \mathcal{T}_h, v_h|_T \in P^1(T); \forall F \in \mathcal{F}_h^i, \int_F [\![v_h]\!]_F = 0\},$$

where $P^1(T)$ denotes the vector space of polynomials on T with degree less than or equal to 1.

2.2 The discrete problem

Set $V = H^2(\mathcal{T}_h) \cap H^1(\Omega)$ and $V(h) = V + P^1_{\text{nc}}(\mathcal{T}_h)$ and equip $V(h)$ with the norm

$$\|v\|_{\epsilon\beta\sigma,\Omega} = \|\epsilon^{\frac{1}{2}} \nabla_h v\|_{0,\Omega} + \|\sigma^{\frac{1}{2}} v\|_{0,\Omega} + \||\beta \cdot n|^{\frac{1}{2}} v\|_{0,\partial\Omega}. \tag{5}$$

Introduce the bilinear form a_h defined on $V(h) \times V(h)$ by

$$a_h(v, w) = \int_\Omega \epsilon \nabla_h v \cdot \nabla_h w + \int_\Omega (\nu - \nabla \cdot \beta) v w - \int_\Omega v(\beta \cdot \nabla_h w)$$
$$+ \sum_{F \in \mathcal{F}_h^i} \int_F \beta \cdot n_F [\![vw]\!]_F + \int_{\partial\Omega_{\text{out}}} (\beta \cdot n) v w. \tag{6}$$

To control the jump of the discrete solution accross mesh interfaces, let us introduce on $V(h) \times V(h)$ the bilinear form

$$j_h(v,w) = \sum_{F \in \mathcal{F}_h^i} \int_F (\beta \cdot n_F)[\![v]\!]_F w^{\downarrow}, \tag{7}$$

where $w^{\downarrow}$ is the so-called downwind value of w defined as $w^{\downarrow} = w|_{T_2(F)}$ if $\beta \cdot n_F \geq 0$ and $w^{\downarrow} = w|_{T_1(F)}$ otherwise. To control the advective term, let us introduce on $V(h) \times V(h)$ the bilinear form

$$s_h(v,w) = \sum_{F \in \mathcal{F}_h^i} \int_F \gamma \frac{h_F^2}{\beta_{F,\infty}} [\![\beta \cdot \nabla_h v]\!]_F [\![\beta \cdot \nabla_h w]\!]_F, \tag{8}$$

where $\gamma > 0$ is independent of ϵ and $\beta_{F,\infty} = \|\beta\|_{[L^\infty(F)]^d}$ (the contribution of a face $F \in \mathcal{F}_h^i$ is conventionally set to zero if $\beta_{F,\infty} = 0$).

The discrete problem we consider is the following:

$$\begin{cases} \text{Seek } u_h \in P_{\mathrm{nc}}^1(\mathcal{T}_h) \text{ such that for all } v_h \in P_{\mathrm{nc}}^1(\mathcal{T}_h), \\ a_h(u_h, v_h) + j_h(u_h, v_h) + s_h(u_h, v_h) = (f, v_h)_{0,\Omega} - (g, v_h)_{0,\partial\Omega_{\mathrm{in}}}. \end{cases} \tag{9}$$

It is readily infered that the bilinear form $(a_h + j_h + s_h)$ is $\|\cdot\|_{\epsilon\beta\sigma,\Omega}$–coercive; hence, (9) is well–posed owing to the Lax–Milgram Lemma.

Henceforth, c denotes a generic positive constant, independent of h and ϵ, whose value can change at each occurrence. Since the advection–diffusion problem has been non-dimensionalized so that the field β is of order unity, the dependency on β in the error estimates can be hidden in the constants. The same is done for the function ν since we are not interested in the asymptotics of strong reaction regimes. Finally, without loss of generality, we assume that $h \leq 1$ and $\epsilon \leq 1$.

2.3 A priori error estimate

The error analysis is performed in the spirit of the Second Strang Lemma. Define on $V(h)$ the following norm:

$$\|w\|_{A,\Omega} = \|w\|_{\epsilon\beta\sigma,\Omega} + \left(\sum_{F \in \mathcal{F}_h^i} \||\beta \cdot n_F|^{\frac{1}{2}}[\![w]\!]_F\|_{0,F}^2 \right)^{\frac{1}{2}} + s_h(w,w)^{\frac{1}{2}}. \tag{10}$$

Theorem 1 (Convergence). *Let u be the unique solution to (2) and let u_h be the unique solution to (9). Assume that $u \in H^2(\Omega)$. Then, there exists a constant c such that*

$$\|u - u_h\|_{A,\Omega} \leq c h (\epsilon^{\frac{1}{2}} + h^{\frac{1}{2}}) \|u\|_{2,\Omega}. \tag{11}$$

Remark 1. The a priori error estimate (11) shows that when keeping the Péclet number ϵ fixed, the convergence order in the mesh-size for the error $\|u - u_h\|_{A,\Omega}$ is 1 in the diffusion-dominated regime and $\frac{3}{2}$ in the advection-dominated regime. This estimate is similar to the usual estimates derived for stabilized schemes in conforming settings; see, e.g., [3, 2, 7, 8].

3 A posteriori error estimates

In this section we present a semi-robust a posteriori error estimator of residual type for the discrete problem (9).

Let f_h, β_h, and ν_h denote the L^2–orthogonal projection of f, β, and ν onto the space of piecewise constant functions on $\mathcal{T}_h$ respectively, and let g_h be the L^2–orthogonal projection of g onto the space of piecewise constant functions on $\mathcal{F}_h^\partial$. Furthermore, define

$$\alpha_S = \min(\epsilon^{-\frac{1}{2}} h_S, 1), \qquad (12)$$

where S belongs to $\mathcal{T}_h$ or to $\mathcal{F}_h$. For $T \in \mathcal{T}_h$, let Δ_T denote the union of elements of $\mathcal{T}_h$ sharing at least a vertex with T, and set $\mathcal{F}_T^{(1)} = \mathcal{F}_T \cap \{\mathcal{F}_h^i \cup \mathcal{F}_h^{\text{out}}\}$ and $\mathcal{F}_T^{(2)} = \mathcal{F}_T \cap \mathcal{F}_h^{\text{in}}$.

Theorem 2 (Upper bound). *Let u be the unique solution to (2) and let u_h be the unique solution to (9). Then, there is $c > 0$ such that*

$$c\|u - u_h\|_{\epsilon\beta\sigma,\Omega} \leq \left(\sum_{T \in \mathcal{T}_h} [\eta_T(u_h)^2 + \delta_T(u_h)^2] + \sum_{F \in \mathcal{F}_h^i} \eta_F(u_h)^2 \right)^{\frac{1}{2}}, \qquad (13)$$

with local data error indicators

$$\delta_T(u_h) = \alpha_T (\|f - f_h\|_{0,T} + \|(\beta - \beta_h) \cdot \nabla u_h\|_{0,T} + \|(\nu - \nu_h) u_h\|_{0,T})$$
$$+ \sum_{F \in \mathcal{F}_T^{(2)}} \epsilon^{-\frac{1}{4}} \alpha_F^{\frac{1}{2}} \|g - g_h + (\beta - \beta_h) \cdot n u_h\|_{0,F}, \qquad (14)$$

and local residual error indicators

$$\eta_T(u_h) = \alpha_T \|f_h - \beta_h \cdot \nabla u_h - \nu_h u_h\|_{0,T} + \sum_{F \in \mathcal{F}_T^{(1)}} \epsilon^{-\frac{1}{4}} \alpha_F^{\frac{1}{2}} \|\epsilon [\![\nabla_h u_h]\!]_F\|_{0,F}$$
$$+ \sum_{F \in \mathcal{F}_T^{(2)}} \epsilon^{-\frac{1}{4}} \alpha_F^{\frac{1}{2}} \|g_h + \epsilon \nabla u_h \cdot n - \beta_h \cdot n_F u_h\|_{0,F}, \qquad (15)$$

$$\eta_F(u_h) = h_F^{\frac{1}{2}} \max(\alpha_F, \epsilon^{\frac{1}{2}}) \|[\![\nabla_h u_h]\!]_F\|_{0,F}. \qquad (16)$$

Theorem 3 (Lower bound). *In the above framework, the following holds:*

$$\forall T \in \mathcal{T}_h, \quad \eta_T(u_h) \leq c \sum_{T' \in \Delta_T} \left((1 + \epsilon^{-\frac{1}{2}} \alpha_{T'}) \|u - u_h\|_{\epsilon\beta\sigma,T'} + \delta_{T'}(u_h) \right), \qquad (17)$$

$$\forall F \in \mathcal{F}_h^i, \quad \eta_F(u_h) \leq c \epsilon^{-\frac{1}{2}} \alpha_F (\|u - u_h\|_{\epsilon\beta\sigma,\mathcal{T}_F} + \inf_{z_h \in [P_c^1(\mathcal{T}_h)]^d} \|\epsilon^{\frac{1}{2}} (\nabla u - z_h)\|_{0,\mathcal{T}_F}). \qquad (18)$$

Remark 2. Keeping ϵ fixed, the quantities $\delta_T(u_h)$ and $\inf_{z_h \in [P_c^1(\mathcal{T}_h)]^d} \|\epsilon^{\frac{1}{2}}(\nabla u - z_h)\|_{0,\mathcal{T}_F}$ should converge at least with the same order as, respectively, the quantities $\|u - u_h\|_{\epsilon\beta\sigma,T}$ and $\|u - u_h\|_{\epsilon\beta\sigma,\mathcal{T}_F}$.

4 Numerical results

In this section two test cases are presented. In both cases, $\Omega = (0,1) \times (0,1)$ and a shape–regular family of unstructured triangulations is considered with mesh-size $h_i = h_0 \times 2^{-i}$ with $h_0 = 0.1$ and $i \in \{0, \cdots, 4\}$. The diffusion coefficient ϵ takes values in $\{10^{-2}, 10^{-4}, 10^{-6}\}$, the reaction coefficient ν is set to 1, and the parameter γ in (8) is set to 0.005.

4.1 Test case 1

The goal of this test case is to illustrate the convergence of the scheme. Let $\beta = (1,0)^T$ and choose the data f and g such that the exact solution of (1) is

$$u(x,y) = \frac{1}{2}\left(1 - \tanh(\tfrac{0.5-x}{a_w})\right), \tag{19}$$

with internal layer width $a_w = 0.01$.

Table 1 presents the convergence results for the error $\|u - u_h\|_{A,\Omega}$; N_{fa} denotes the number of degrees of freedom (i.e., the number of mesh faces) and ω denotes the convergence order with respect to the mesh–size. In the advection–dominated regime ($\epsilon = 10^{-4}$ and $\epsilon = 10^{-6}$), the error decreases as $h^{\frac{3}{2}}$. In the intermediate regime ($\epsilon = 10^{-2}$), the convergence order changes from $\frac{3}{2}$ to 1 as the mesh is refined. These results are in agreement with the estimate derived in Theorem 1.

Table 1. Numerical errors and convergence orders for the different values of ϵ

Mesh		$\epsilon = 10^{-2}$		$\epsilon = 10^{-4}$		$\epsilon = 10^{-6}$	
i	N_{fa}	$\|u - u_h\|_{A,\Omega}$	ω	$\|u - u_h\|_{A,\Omega}$	ω	$\|u - u_h\|_{A,\Omega}$	ω
0	374	$4.17\,10^{-1}$	-	$4.07\,10^{-1}$	-	$4.03\,10^{-1}$	-
1	1441	$1.45\,10^{-1}$	1.57	$1.35\,10^{-1}$	1.65	$1.33\,10^{-1}$	1.64
2	5621	$5.41\,10^{-2}$	1.45	$4.64\,10^{-2}$	1.57	$4.56\,10^{-2}$	1.57
3	22330	$2.12\,10^{-2}$	1.36	$1.64\,10^{-2}$	1.51	$1.60\,10^{-2}$	1.52
4	88961	$8.62\,10^{-3}$	1.30	$5.85\,10^{-3}$	1.49	$5.69\,10^{-3}$	1.49

4.2 Test case 2

The goal of this test case for which the mesh is illustrate how the a posteriori error estimates can be used to generate adaptively refined meshes at internal layers. Let Γ_1 denote the lower horizontal edge of Ω and let Γ_2 denote its left vertical edge. Set $\beta = (2,1)^T$, $f = 0$, and g such that

$$g(x,y) = \begin{cases} \varphi(x) & \text{on } \Gamma_1 \\ \varphi(-y) & \text{on } \Gamma_2 \end{cases} \quad \text{where } \varphi(s) = \tfrac{1}{2}\left(\tanh(\tfrac{s}{h_0}) + 1\right), \tag{20}$$

with $h_0 = 0.1$. Figure 1 presents the contour lines of the computed solution for the different values of ϵ.

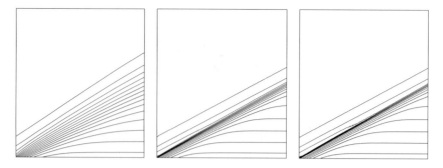

Fig. 1. Contour lines of the solution for test case 2. Left: $\epsilon = 10^{-2}$; center: $\epsilon = 10^{-4}$; right: $\epsilon = 10^{-6}$

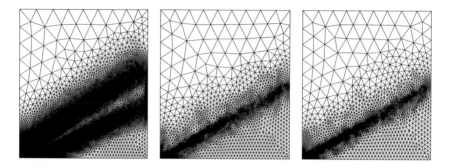

Fig. 2. Adaptive meshes after five iterations. Left: $\epsilon = 10^{-2}$ and $N_{\text{fa}} = 18157$; center: $\epsilon = 10^{-4}$ and $N_{\text{fa}} = 7145$; right: $\epsilon = 10^{-6}$ and $N_{\text{fa}} = 6934$

To refine the mesh adaptively using the local error indicator $\eta_T(u_h)$ defined in (15), we consider an adaptive algorithm where the error indicators are equi-distributed; see [6] for more details. Figure 2 presents the adaptively refined meshes after five iterations of the adaptive algorithm. For the three values of the diffusion coefficient, the mesh is refined at the origin. In the diffusion–dominated regime, the mesh is refined around the inner layer and at the outflow layer. In the advection–dominated regime, the meshes are refined along the inner layer. The refined zone becomes smaller as the diffusion coefficient ϵ takes smaller values, indicating that the local error indicator $\eta_T(u_h)$ alone can detect the inner layer.

Acknowledgment

This work was partly supported by the GdR MoMaS (CNRS–2439, ANDRA, BRGM, CEA, EdF).

References

1. Brezzi, F., Russo, A.: Choosing bubbles for advection–diffusion problems. Math. Models Meth. Appl. Sci. **4**, 571–587 (1994)
2. Brooks, A., Hughes, T.: Streamline upwind/Petrov–Galerkin formulations for convective dominated flows with particular emphasis on the incompressible Navier-Stokes equations. Comput. Methods Appl. Mech. Engrg. **32**, 199–259 (1982)
3. Burman, E.: A unified analysis for conforming and non-conforming stabilized finite element methods using interior penalty. to appear in SIAM, J. Numer. Anal. (2005)
4. Burman, E., Hansbo, P.: Edge stabilization for Galerkin approximations of convection–diffusion–reaction problems. Comput. Methods Appl. Mech. Engrg. **193**, 1437–1453 (2004)
5. Crouzeix, M., Raviart, P.-A.: Conforming and nonconforming mixed finite element methods for solving the stationary Stokes equations I. RAIRO Modél Math. Anal. Numér. **3**, 33–75 (1973)
6. El Alaoui, L., Ern, A., Burman E.: A priori and a posteriori error analysis of nonconforming finite elements with face penalty for advection–diffusion–reaction equations. Submitted (2005) [CERMICS Technical Report 2005–289]
7. Ern, A., Guermond, J.-L.: Theory and Practice of Finite Elements, Springer, New York, 2004
8. Guermond, J.-L.: Subgrid stabilization of Galerkin approximations of linear monotone operators. IMA, Journal of Numerical Analysis. **21**, 165–197 (2001)
9. Johnson, C. and Pitkäranta, J.: An analysis of the discontinuous Galerkin method for a scalar hyperbolic equation. Math. Comput. **46**, 1–26 (1986)
10. Matthies, G. and Tobiska, L.: The streamline–diffusion method for conforming and nonconforming finite elements of lowest order applied to convection–diffusion problems. Computing. **66**, 343–364 (2001)
11. Sangalli, G.: On robust a posteriori estimators for the advection-diffusion-reaction problem. Technical Report 04–55, ICES, (2004)
12. Verfürth, R.: A posteriori error estimators for convection–diffusion equations. Numer. Math. **80**, 641–663 (1998)
13. Verfürth, R.: Robust a posteriori error estimates for stationary convection–diffusion equations. SIAM, J. Numer. Anal., (2004) (submitted)

Efficient Multigrid and Data Structures for Edge-Oriented FEM Stabilization

Abderrahim Ouazzi and Stefan Turek

Institute of Applied Mathematics, University of Dortmund, 44227 Dortmund, Germany
ouazzi@math.uni-dortmund.de, ture@featflow.de

Summary. We study edge-oriented FEM stabilizations w.r.t. linear multigrid solvers and data structures with the goal to examine the efficiency of such stabilizations due to the extending matrix stencil which is not supported by standard FEM data structures. A new edge-oriented data structure has been developed to support the additional coupling. So, the local element-wise and edge-wise matrices are easily deduced from the global ones. Accordingly, efficient Vanka smoothers are introduced, namely a full cell-oriented and an edge-oriented Vanka smoother so that it becomes possible to privilege edge-oriented stabilization for CFD simulations.

1 Introduction

1.1 Problem formulation

As a model problem we consider incompressible flow problrms:

$$\frac{\partial \mathbf{u}}{\partial t} + \mathbf{u} \cdot \nabla \mathbf{u} - \nu \triangle \mathbf{u} + \nabla p = \mathbf{f}, \quad \text{div } \mathbf{u} = 0 \tag{1}$$

where p is the pressure and $\mathbf{u}$ being the velocity. Let us consider the non-stationary (or stationary, without the term $\frac{\partial \mathbf{u}}{\partial t}$) Navier-Stokes problem 1 in a bounded domain $\Omega \subset \mathbf{R}^2$, first discretized in time by a standard numerical solution method for ODEs. The θ-scheme, as for instance backward Euler or Crank-Nicholson or the Fractional-step-θ-scheme, yields a sequence of boundary value problems of the following form [2]:
Given $\mathbf{u}^n$, compute $\mathbf{u} = \mathbf{u}^{n+1}$ and $p = p^{n+1}$ by solving

$$[\alpha\mathbf{l} + \theta(\mathbf{u} \cdot \nabla - \nu\triangle]\mathbf{u} + \nabla p = [\alpha I - \theta_1(\mathbf{u}^n \cdot \nabla - \nu\triangle]\mathbf{u}^n + \theta_2\mathbf{f}^{n+1} + \theta_3\mathbf{f}^n \tag{2}$$

subject to the incompressibility constraint $\nabla \cdot \mathbf{u} = 0$.

Here, $(\cdot)^n$ indicates the value of the generic quantity $(\cdot)$ at time step t_n for time-dependent problems or the n-th iteration for the steady-state formulation. The time-dependent problem is defined for $\alpha = 1/\Delta t$, while the steady-state formulation is recovered for $\alpha = 0$, $\theta = \theta_1 = \theta_3 = 1$, and $\theta_2 = 0$.

For the spatial discretization let V_h and Q_h be approximative spaces of $H_0^1(\Omega)$, and $L_2(\Omega)$ respectively, then the resulting discrete problems reads: *Compute* $\mathbf{u}$ *and* p *by solving*:

$$\mathsf{A}\mathbf{u} + \mathsf{B}p = \mathbf{g} \quad , \quad \mathsf{B}^T\mathbf{u} = 0 \quad \text{where} \tag{3}$$

$$\mathbf{g} = [\alpha\mathsf{M} - \theta_1\mathsf{L} - \theta_1\mathsf{N}(\mathbf{u}^n)]\,\mathbf{u}^n + \theta_2\mathbf{f}^{n+1} + \theta_3\mathbf{f}^n \tag{4}$$

Here, M is the (consistent or lumped) mass matrix, B is the discrete gradient operator and $-\mathsf{B}^T$ is the associated divergence operator. Furthermore,

$$\mathsf{A}\mathbf{u} = [\alpha\mathsf{M} + \theta\mathsf{L} + \theta\mathsf{N}(\mathbf{u})]\,\mathbf{u}, \tag{5}$$

where L is the viscous term and $\mathsf{N}(\mathbf{u})$ is the nonlinear transport operator. Furthermore, the discretized equations (2) as well as the linear subproblems can be solved within the outer iteration loop by a fixpoint defect correction or Newton method. In this paper, we employ the stable $\widetilde{Q}_1/Q_0$ finite element pair. In the two-dimensional case, the nodal values are the mean or midpoint values of the velocity vector over the element edges, and the mean values of the pressure over the elements (see [2]). There are two well-known situations for nonconforming finite element methods when severe numerical problems may arise: Firstly, the lack of coercivity for nonconforming low order approximations for symmetric deformation tensor formulations, mainly visible for small Re numbers. Secondly, convection dominated problems, for instance for medium and high Re numbers or for the treatment of pure transport problems. Then, the standard Galerkin formulation fails and may lead to numerical oscillations or convergence problems of the iterative solvers, too (see[1, 4]).

Among the stabilization methods existing in the literature for these types of problems, we use the proposed one in [4] which is based on the penalization of the gradient jumps over element boundaries. It takes the following form (with $h_E = |E|$)

$$\langle \mathsf{J}\mathbf{u}, \mathbf{v}\rangle = \sum_{\text{edge } E} \max(\gamma^*\nu h_E, \gamma h_E^2) \int_E [\nabla\mathbf{u}] : [\nabla\mathbf{v}]\,d\sigma, \tag{6}$$

and will be added to the original bilinear form in order to cure numerical instabilities when computing incompressible flow problems using low order nonconforming finite elements. Moreover, only one generic stabilization term takes care of all instabilities (see [4]).

2 Sparsity of the matrix

Sparse matrices are an integral part of the FEM analysis for incompressible flow problems which may lead to huge and ill-conditioned systems so that very fast solvers of Krylov-space or particularly of multigrid type are required.

In addition the introduced edge-oriented stabilization techniques destroy the typical local sparsity properties since this approach involves more than the adjacent elements: The corresponding rows and columns for the new stiffness matrices J may contain 23 nonzero matrix elements, in contrast to the usual 7 for the non-stabilized case in 2D (see Fig. 1), and 61 nonzero matrix elements in contrast to the usual 11 for the non-stabilized case in 3D.

2.1 Storage in the same FEM data structure

To overcome the problem of storing the new matrix J – coming from $\langle J\mathbf{u}, \mathbf{v}\rangle$ – with regard to the standard FEM data structures, the matrix J is written as a sum of two matrices J^* and J_{rest}, $J = J^* + J_{rest}$, where J^* has the same sparsity structure as the usual corresponding finite element matrix; then, $J_{rest} = J - J^*$. Hence, J^* can be handled with the same linear algebra techniques which are typically used for the treatment of the standard nonconforming finite element approach; J_{rest} is the complementary part and will be used as a correction for the calculation of the residuals inside of the linear solvers only. Then, given any approximation $\mathbf{v}$, and by A denoting the standard stiffness matrix from (5) without the new stabilization matrices, we can write the complete residual as:

$$\mathbf{f} - (A + J)\mathbf{v} = \mathbf{f} - (A + J^*)\mathbf{v} - J_{rest}\mathbf{v} \qquad (7)$$

Consequently, only the partial matrix $A + J^*$ has to be stored in the complete stiffness matrix so that the first part of the residual can be obtained via standard matrix-vector multiplication while the second part is assembled via elementwise operations. Moreover, the construction of preconditioners for the corresponding linear systems may only include parts of the (sub)matrix $A+J^*$, too, which will be explained in the following (see also [3] for more details).

2.2 Storage in a special edge-oriented data structure

A data structure for the storage of the stiffness matrix for edge-oriented stabilization is not common in FEM community. Fortunately, it is not difficult to develop one from the available FEM storage techniques. In fact, each edge E_i has two surrounding elements with n_i edges $(E_{i,j})_{j=1}^{n_i}$, then by the intermediate of the edges $(E_{i,j})_{j=1}^{n_i}$ the other contributed elements and edges $(E_{i,j_k})_{k=1}^{m_j}$ required for edge-oriented stabilization techniques are obtained (see Fig. 1). This is exactly the graph of the extended matrix: In fact, let the index i be assimilated to any matrix row and the index j be the corresponding nonzero columns in the standard FEM data structure, the extension will consist of the corresponding nonzero columns j_k to the rows j.

Edge-oriented storage algorithm

Based on the standard Compressed Sparse Row CSR-FEM storage technique, let N_A be the number of entries in the matrix A, N_{Eq} be the number of

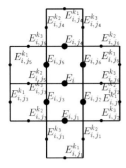

Storage Technique Level	two-dimensional mesh		three-dimensional mesh		
	FEM	EO		FEM	EO
	Elements	Matrix-entries	Elements	Matrix entries	
1	4	60 128	27	918	3474
2	16	232 628	216	7236	32436
3	64	912 2732	1728	57456	277632
4	246	3616 11356	13824	457920	2292768
5	1024	14400 46268	110592	3656448	18627264
6	4096	57472 186748	884736	29223936	150155136

Fig. 1. An illustration for edge-oriented storage technique and the total number of nonzero matrix entries for the $\widetilde{Q}_1$ element on a unit square.

equations, $P_c(N_A)$ a vector with dimension N_A to be the column pointer and $P_r(N_{Eq}+1)$ a vector with dimension $N_{Eq}+1$ to be the pointer row. Then, the edge-oriented storage technique is deduced from the standard FEM storage technique as following

$$\widetilde{N}_A = 1 \quad , \quad l_1 = 1. \tag{8}$$

For each $i = 1, .., N_{eq}$ the corresponding nonzero columns are given by the following nested loops

$$\widetilde{P}_r(i) = l_i. \tag{9}$$

1. In standard FEM storage we get

$$i_j = P_c(l), \quad P_r(i) \leq l \leq P_r(i+1) - 1. \tag{10}$$

2. For each i_j the extension consists of the nonzero corresponding column in the standard FEM storage which is given by

$$k_{i_j} = P_c(l), \quad P_r(i_j) \leq l \leq P_r(i_j+1) - 1$$
$$\widetilde{N}_A = \widetilde{N}_A + 1; \quad l_i = l_i + 1; \quad \widetilde{P}_c(l_i) = k_{i_j}. \tag{11}$$

Here, $\widetilde{N}_A$ denotes the number of entries in the matrix A, $\widetilde{P}_r$ is the row pointer and $\widetilde{P}_c$ is the column pointer in the edge-oriented storage. In practice we consider the so-called **edge-oriented patches** Ω_i which consist of the neighboring elements sharing the same edge

$$\Omega_i = \cup \{T, T \in \mathcal{T}_h \wedge \cap_{T \in \mathcal{T}_h} = E_i\}. \tag{12}$$

All our elementary operations will be based on Ω_i. Looking more carefully at the resulting matrix stencils for the terms $\int_E [\nabla \phi_i][\nabla \phi_j] d\sigma$, the matrix structure can be seen in Fig. 2. While the matrix stencils are always increased, leading to couplings between FEM basis functions which do not have common local support, it is also visible that reduced integration, for instance via midpoint rule, may lead to a different amount of additional memory requirements.

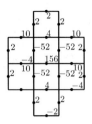
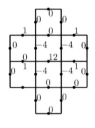

Level	Quadrature technique	
	exact Gauss	1x1 Gauss
1	128	76
2	628	328
3	2732	1360
4	11356	5536
5	46268	22336
6	186748	89728

Fig. 2. Stencil for $\int_E [\nabla \phi_i][\nabla \phi_j] d\sigma$ with exact (left), and with 1x1 Gauss quadrature (middle); total number of nonzero matrix entries (right) for the $\tilde{Q}_1$ element with midpoints as degree of freedom on the unit square.

We can see this reduction for the $\tilde{Q}_1$ element with midpoint values on edges as degree of freedom, which shows that the connections for the edge-oriented finite element methods can be chosen optimally which will lead to moderately increased matrix stencils. A more detailed analysis will be performed in a forthcoming paper.

3 Local pressure Schur complement approach

Local Pressure Schur complement schemes (see [2]) as generalization of so-called Vanka smoothers are simple iterative methods for coupled systems

$$\begin{pmatrix} A+J & B \\ B^T & 0 \end{pmatrix} \begin{bmatrix} \mathbf{u} \\ p \end{bmatrix} = \begin{bmatrix} \mathbf{Res_u} \\ \mathbf{Res}_p \end{bmatrix}, \quad (13)$$

of saddle point type which are acting directly on element level and which are embedded into an outer block Jacobi/Gauss-Seidel iteration. The local character of this procedure together with a global defect-correction mechanism is crucial for our approach. If $\mathbf{Res_u}$ and $\mathbf{Res}_p$ denote the residuals for the (complete) discrete momentum and continuity equations which include the complete stabilization term due to J as described in (6), then, two types of Vanka smoothers can be applied with respect to the decomposition of the domain Ω to patches $\{\Omega_i, i = 1, ..., I\}$ which is not required to be disjoined.

3.1 Cell-oriented Vanka smoother

In this case the *patches* Ω_i may consist of only one element and the index I is the total number of elements, which means that the global stiffness matrix is restricted to the single cells/quadrilaterals of the mesh. It is straightforward to deduce the element stiffness matrix from the global stiffness matrix as follows

$$K = \sum_{T \in \mathcal{T}_h} K_T, \quad (14)$$

where K and K_T denote the global and element stiffness matrices respectively:

$$\begin{aligned}
{[K_T]_{ij}} &= [\mathsf{A}_{|T}]_{ij} + [\mathsf{J}_{|T}]_{ij} &&\text{for } 1 \leq i,j \leq 4 \\
[K_T]_{i5} &= [\mathsf{B}_{|T}]_i &&\text{for } 1 \leq i \leq 4 \\
[K_T]_{5i} &= \left[\mathsf{B}_{|T}^T\right]_i &&\text{for } 1 \leq i \leq 4 \\
[K_T]_{55} &= 0.
\end{aligned} \quad (15)$$

With the standard FEM data structure (without the extension of the matrix) the contribution of the jump term will be restricted to J^*. Then, there holds

$$[K_T^*]_{ij} = [\mathsf{A}_{|T}]_{ij} + \left[\mathsf{J}_{|T}^*\right]_{ij} \quad \text{for } 1 \leq i,j \leq 4. \quad (16)$$

Then one smoothing step can be described as follows

$$\begin{bmatrix} \mathbf{u}^{n+1} \\ p^{n+1} \end{bmatrix} = \begin{bmatrix} \mathbf{u}^n \\ p^n \end{bmatrix} + \omega^n \sum_{T \in \mathcal{T}_h} \left[\widetilde{K_T^*}\right]^{-1} \begin{bmatrix} \mathbf{Res}_\mathbf{u}(\mathbf{u}^n, p^n) \\ \mathbf{Res}_p(\mathbf{u}^n, p^n) \end{bmatrix}_{|T} \quad (17)$$

where the matrix $\widetilde{K_T^*}$ is easily invertible and remains close to K_T^*. Related to the choice of the matrix $\widetilde{K_T^*}$ two types of Vanka smoothers are described, namely diagonal Vanka smoother and full Vanka smoother.

(a) *Diagonal Vanka smoother:* The diagonal Vanka smoother updates the velocity and the pressure values connected to the element T by

$$\begin{bmatrix} \mathbf{u}^{n+1} \\ p^{n+1} \end{bmatrix} = \begin{bmatrix} \mathbf{u}^n \\ p^n \end{bmatrix} + \omega^n \sum_{T \in \mathcal{T}_h} [\text{diag}(K_T^*)]^{-1} \begin{bmatrix} \mathbf{Res}_\mathbf{u}(\mathbf{u}^n, p^n) \\ \mathbf{Res}_p(\mathbf{u}^n, p^n) \end{bmatrix}_{|T}. \quad (18)$$

(b) *Full Vanka smoother:* The full Vanka smoother updates the velocity and the pressure values connected to the element T by

$$\begin{bmatrix} \mathbf{u}^{n+1} \\ p^{n+1} \end{bmatrix} = \begin{bmatrix} \mathbf{u}^n \\ p^n \end{bmatrix} + \omega^n \sum_{T \in \mathcal{T}_h} [K_T^*]^{-1} \begin{bmatrix} \mathbf{Res}_\mathbf{u}(\mathbf{u}^n, p^n) \\ \mathbf{Res}_p(\mathbf{u}^n, p^n) \end{bmatrix}_{|T}. \quad (19)$$

As can be seen, for the preconditioning step only parts of the matrix (here: $\mathsf{A} + \mathsf{J}^*$) are involved while the residual contains all parts of the matrix. Consequently, when this approach converges, the result is the solution of the stabilized version while the preconditioning steps only determine the speed of the overall iteration procedure.

3.2 Edge-oriented Vanka smoother

To incorporate the full jump J into the preconditioning step we use the edge-oriented patches Ω_i. This will keep the size of the local problem small and the full matrix J will be used for the preconditioning steps. The extension of the matrix to support the jump term leads to a 5×5 FEM matrix block of the type (15). To keep the size of the local problem small, the element matrix is disassembled to its edge contributions

$$K_T = \sum_{i=1}^{m} K_T^{E_i}, \qquad (20)$$

where $K_T^{E_i}$ is the contribution of the edge E_i to K_T and m is the number of the edges on the cell T. From the definition of edge-oriented patches (12), the edge stiffness matrix may contain the contributions of all sharing elements

$$K^{E_i} = \sum_{T \in \Omega_i} K_T^{E_i} = K_{\Omega_i}^{E_i}. \qquad (21)$$

Then, one basic iteration can be described as follows

$$\begin{bmatrix} \mathbf{u}^{n+1} \\ p^{n+1} \end{bmatrix} = \begin{bmatrix} \mathbf{u}^n \\ p^n \end{bmatrix} + \omega^n \sum_{i \in I} \left[K_{\Omega_i}^{E_i} \right]^{-1} \begin{bmatrix} \mathbf{Res}_\mathbf{u}(\mathbf{u}^n, p^n) \\ \mathbf{Res}_p(\mathbf{u}^n, p^n) \end{bmatrix}_{|\Omega_i}, \qquad (22)$$

where I is the total number of internal edges. This blocking strategy is different from that used in [2] to generate isotropic subdomains for stabilizing strong mesh anisotropy. Indeed, for the edge-oriented patches the number of block matrices is only depending on the number of edges and not on the number of patches itself. The global defect restricted to a single patch Ω_i is given by

$$\begin{bmatrix} \mathbf{Res}_\mathbf{u}(\mathbf{u}^n, p^n) \\ \mathbf{Res}_p(\mathbf{u}^n, p^n) \end{bmatrix}_{|\Omega_i} = \left(\begin{bmatrix} \mathsf{L} + \tilde{\mathsf{N}} + \mathsf{J} & \mathsf{B} \\ \mathsf{B}^T & 0 \end{bmatrix} \begin{bmatrix} \mathbf{u}^n \\ p^n \end{bmatrix} - \begin{bmatrix} \mathbf{f} \\ 0 \end{bmatrix} \right)_{|\Omega_i}. \qquad (23)$$

In practice the following auxiliary problem

$$\left[K_{\Omega_i}^{E_i} \right] \begin{bmatrix} \mathbf{v}_i^{n+1} \\ q_i^{n+1} \end{bmatrix} = \begin{bmatrix} \mathbf{Res}_\mathbf{u}(\mathbf{u}^n, p^n) \\ \mathbf{Res}_p(\mathbf{u}^n, p^n) \end{bmatrix}_{|\Omega_i} \qquad (24)$$

is solved, and then the new iterates $\mathbf{u}^{n+1}$ and p^{n+1} are computed

$$\begin{bmatrix} \mathbf{u}^{n+1} \\ p^{n+1} \end{bmatrix} = \begin{bmatrix} \mathbf{u}^n \\ p^n \end{bmatrix} + \omega^n \sum_{i \in I} \begin{bmatrix} \mathbf{v}_i^{n+1} \\ q_i^{n+1} \end{bmatrix}. \qquad (25)$$

The resulting local MPSC method corresponds to a simple block-Jacobi iteration for the mixed problem (13) and to a block-Gauss-Seidel method by using the updated solution for the computation of the local defect (23).

4 Numerical example

The realistic evaluation of the efficiency of the edge-oriented FEM storage technique versus the standard one is difficult to handle because of the interplay of different components. Here, we restrain the numerical examples to the

DFG benchmark of flow around cylinder (see [4]). Our numerical test (Stokes problem) is concerned with the symmetric deformation tensor formulation to show the advantage of using the edge-oriented stabilization with special storage technique. We also present the gradient formulation for comparison since it does not require any stabilization. In Table 1, we list the total number of multigrid sweeps (MG) and the total CPU time for both storage techniques with and without using the jump terms.

Table 1. Vanka smoother coupled with standard and edge-oriented FEM for the symmetric deformation tensor and gradient formulations

	edge-oriented storage technique				standard FEM storage technique			
	without jump stab.		with jump stab.		without jump stab.		with jump stab.	
Level	MG	Time	MG	Time	MG	Time	MG	Time
			the gradient formulation					
4	12	37	12	44	12	32	12	180
5	12	153	11	166	12	128	12	780
6	12	634	11	676	12	531	11	2594
			the deformation tensor formulation					
4	191	542	8	28	191	442	8	115
5	535	6209	9	133	535	5426	9	524
6	1225	63614	8	525	1225	49502	8	1905

The results in Table 1 for several mesh levels show that the edge-oriented storage technique moderately increases the CPU cost. Moreover, the need for edge-oriented stabilization for the deformation tensor is cleary visible.

Summarising, we have developed new techniques to make edge-oriented FEM stabilizations more advantageous for CFD simulations. However, more research is required concerning the corresponding time-accurate methods and approximate preconditioners for global Pressure Schur Complement schemes.

References

1. Burman, E., Hansbo, P.: A stabilized non-conforming finite element method for incompressible flow, J. Comput. Methods. Appl. Mech. Engrg. (2004) accepted
2. Turek, S.: Efficient solvers for incompressible flow problems: An algorithmic and computational approach. Springer, Berlin-Heidelberg (1999)
3. Turek, S., Ouazzi, A., Schmachtel, R.: Multigrid method for stabilized nonconforming finite elements for incompressible flow involving the deformation tensor formulation, JNM, **10**, 235–248 (2002)
4. Turek, S., Ouazzi, A.: Unified edge–oriented stabilization of nonconforming finite element methods for incompressible flow problems: Numerical investigations, JNM, (2005) (accepted)

Electromagnetism

Adaptive Methods for Dynamical Micromagnetics

Ľubomír Baňas*

Ghent university, Ghent, Belgium
l.banas@imperial.ac.uk

Summary. We propose a space-time adaptive algorithm for two iterative numerical methods for the solution of nonlinear time depended Landau-Lifshitz-Gilbert equation of micromagnetism. The first method is derived from implicit backward Euler time discretisation, the second method is based on midpoint rule. The space discretisation is done by linear finite elements. The resulting nonlinear systems are solved by an iterative fixed-point technique. The performance of the proposed adaptive strategy is demonstrated by numerical experiments.

1 Introduction

The Landau-Lifshitz-Gilbert (LLG) equation plays an important role in applications which require simulation of nonlinear magnetic behaviour on microscale such as, e.g., magnetic recording. The time dependent LLG equation takes the form [13]

$$\partial_t \boldsymbol{m} = \boldsymbol{h}_T \times \boldsymbol{m} + \alpha \boldsymbol{m} \times (\boldsymbol{h}_T \times \boldsymbol{m}) \quad \text{in} \quad \Omega \times (0,T) \tag{1}$$

where $\boldsymbol{m} \in \mathbb{R}^3$ is the magnetisation vector, Ω is a bounded domain with sufficiently smooth boundary; α is so called damping constant. The total field $\boldsymbol{h}_T$ from (1) can consist of several contributions, here we take

$$\boldsymbol{h}_T = \Delta \boldsymbol{m} + \boldsymbol{h},$$

where $\Delta \boldsymbol{m}$ is exchange field. The magnetic field $\boldsymbol{h}$ can be obtained from the Maxwell's equations, for simplicity we treat it as a known vector field throughout the rest of the paper.

---

* Currently with: Imperial College, London, UK

We consider a homogeneous Neumann boundary condition at the boundary Γ i.e.
$$\frac{\partial \boldsymbol{m}}{\partial \nu} = 0 \quad \text{on} \quad \partial \Omega.$$
and initial condition $\boldsymbol{m}(0) = \boldsymbol{m}_0$ in Ω.

A scalar multiplication of (1) by $\boldsymbol{m}$ gives
$$\partial_t \boldsymbol{m} \cdot \boldsymbol{m} = \frac{1}{2} \partial_t |\boldsymbol{m}|^2 = 0. \tag{2}$$

This implies conservation of magnitude of magnetisation $|\boldsymbol{m}(t)| = |\boldsymbol{m}_0| = 1$, which is an important conservation property of the LLG equation.

By combining (2) with the standard vector cross-product formula
$$\boldsymbol{a} \times (\boldsymbol{b} \times \boldsymbol{c}) = (\boldsymbol{a} \cdot \boldsymbol{c})\boldsymbol{b} - (\boldsymbol{a} \cdot \boldsymbol{b})\boldsymbol{c},$$
we obtain the following identity
$$\boldsymbol{m} \times (\boldsymbol{m} \times \Delta \boldsymbol{m}) = -\Delta \boldsymbol{m} - |\nabla \boldsymbol{m}|^2 \boldsymbol{m}.$$
From this, we see, that for sufficiently smooth solutions, (1) is equivalent to
$$\begin{aligned}\partial_t \boldsymbol{m} - \alpha \Delta \boldsymbol{m} &= \alpha |\nabla \boldsymbol{m}|^2 \boldsymbol{m} + \Delta \boldsymbol{m} \times \boldsymbol{m} \\ &+ \alpha(\boldsymbol{h} - (\boldsymbol{h} \cdot \boldsymbol{m})\boldsymbol{m}) + \boldsymbol{h} \times \boldsymbol{m}.\end{aligned} \tag{3}$$

This implies a close relation of LLG equation to the harmonic maps equation.

Another equivalent formulation of (1), the so-called Gilbert form of the LLG equation ([10]) is given by
$$\boldsymbol{m}_t - \alpha \boldsymbol{m} \times \boldsymbol{m}_t = (1 + \alpha^2) \boldsymbol{m} \times \boldsymbol{h}_T. \tag{4}$$

The numerical solution of (1) will be based on formulations (3) and (4).

2 Numerical methods

We define the following spaces of vector functions: $\mathbf{L}_2(\Omega) = (L_2(\Omega))^3$, $\mathbf{H}^1(\Omega) = (H^1(\Omega))^3$, where $L_2(\Omega)$ and $H^1(\Omega)$ are the usual function spaces. We denote the $\mathbf{L}_2(\Omega)$-inner product by $(\boldsymbol{a}, \boldsymbol{b}) = \int_\Omega (\boldsymbol{a} \cdot \boldsymbol{b})$. The discrete inner product is defined as $(\boldsymbol{a}, \boldsymbol{b})_h = \int_\Omega \mathcal{I}^h(\boldsymbol{a} \cdot \boldsymbol{b})$ where $\mathcal{I}^h$ is the usual interpolation operator. The notation $\|\cdot\|$ stands for the $\mathbf{L}_2$ norm and $\|\cdot\|_1$ is $\mathbf{H}^1$ norm.

We divide the time interval $(0, T)$ into subintervals (t_i, t_{i+1}), $i = 0, \ldots, n$ with variable time step size $\tau_{i+1} = t_{i+1} - t_i$. We denote by $\mathcal{T}^i$ a quasi-uniform partition of Ω into simplices (see [7]) on time level i. The triangulation $\mathcal{T}^i$ is obtained from $\mathcal{T}^{i-1}$ by refinement or coarsening. Given a triangle $K \in \mathcal{T}^i$, h_K stands for its diameter. We also denote by $\mathcal{E}^i$ the set of all edges e from $\mathcal{T}^i$, h_e denotes the size of $e \subset E^i$ and by a_j, $j = 1, \ldots, N_i$ the set of all vertices from $\mathcal{T}_i$. The space $\mathbf{V}_i^h \subset \mathbf{H}^1$ is the space of finite element functions that are piecewise linear on $\mathcal{T}^i$.

2.1 Backward Euler projection scheme

The implicit backward Euler time discretisation method for the LLG equation is derived from the formulation (3). It is a known fact, that the backward Euler discretisation violates (2), therefore a projection step is needed to enforce the constraint explicitely in the numerical approximation. The continuous variational formulation of (3) reads as follows

$$(m_t, \psi) + \alpha(\nabla m, \nabla \psi) = \alpha(|\nabla m|^2 m, \psi) - (m \times \nabla m, \nabla \psi) \quad \forall \psi \in \mathbf{H}^1(\Omega).$$

Then the implicit backward Euler projection scheme based on the above variational formulation consists of two steps

- solve

$$\left(\frac{m_{i+1}^h - m_i^{h,*}}{\tau_{i+1}}, v\right)_h + \alpha(\nabla m_{i+1}^h, \nabla v) = \alpha(|\nabla m_{i+1}^h|^2 m_{i+1}^h, v)_h$$
$$-(m_{i+1}^h \times \nabla m_{i+1}^h, \nabla v)$$
$$+\alpha(h_{i+1}^h - (h_{i+1}^h \cdot m_{i+1}^h) m_{i+1}^h)$$
$$+h_{i+1}^h \times m_{i+1}^h$$
$$\forall v \in \mathbf{V}_{i+1}^h(\Omega). \quad (5)$$

  project the solution

-

$$m_{i+1}^{h,*}(a_j) = \frac{m_{i+1}^h(a_j)}{|m_{i+1}^h(a_j)|} \quad j = 1, \ldots, k.$$

The discrete system (5) is nonlinear. We solve the system by a fixed point technique. Starting with $k = 0$, $m_{i,0}^h = m_i^h$ we compute

$$\left(\frac{m_{i+1,k+1}^h - m_i^{h,*}}{\tau_{i+1}}, v\right)_h + \alpha(\nabla m_{i+1,k+1}^h, \nabla v) = \alpha(|\nabla m_{i+1,k}^h|^2 m_{i+1,k+1}^h, v)_h$$
$$-(m_{i+1,k}^h \times \nabla m_{i+1,k+1}^h, \nabla v)$$
$$+\alpha(h_{i+1}^h - (h_{i+1}^h \cdot m_{i+1,k}^h)$$
$$m_{i+1,k+1}^h)$$
$$+h_{i+1}^h \times m_{i+1,k+1}^h$$
$$\forall v \in \mathbf{V}_{i+1}^h(\Omega). \quad (6)$$

until the difference

$$\|m_{i+1,k+1}^h - m_{i+1,k}^h\|_h < TOL$$

where TOL is a sufficiently small prescribed tolerance.

2.2 Midpoint rule

We introduce some additional notation. The midpoint values of the numerical solution are denoted by $\boldsymbol{m}^h_{i+1/2} = \frac{1}{2}(\boldsymbol{m}^h_{i+1} + \boldsymbol{m}^h_i)$, the discrete Laplacian $\Delta_h : \mathbf{H}^1(\Omega) \to \mathbf{V}^h$ is represented by the formula

$$(\Delta_h \boldsymbol{u}, \boldsymbol{v})_h = (\nabla \boldsymbol{u}, \nabla \boldsymbol{v}) \quad \forall \boldsymbol{v} \in \mathbf{V}^h_{i+1}(\Omega).$$

The midpoint rule in the context of micromagnetism was studied in a number of works, e.g., [14, 18, 3]. We will use the formulation from [3] which reads as

$$\left(\frac{\boldsymbol{m}^h_{i+1} - \boldsymbol{m}^h_i}{\tau_{i+1}}, \boldsymbol{v}\right)_h + \alpha \left(\boldsymbol{m}^h_i \times \frac{\boldsymbol{m}^h_{i+1} - \boldsymbol{m}^h_i}{\tau_{i+1}}, \boldsymbol{v}\right)_h$$
$$= (1+\alpha^2)(\boldsymbol{m}^h_{i+1/2} \times \Delta_h \boldsymbol{m}^h_{i+1/2}, \boldsymbol{v})_h \quad \forall \boldsymbol{v} \in \mathbf{V}^h_{i+1}(\Omega). \tag{7}$$

By taking $\boldsymbol{v} = (\boldsymbol{m}^h_{i+1} + \boldsymbol{m}^h_i)\varphi^j$ ($\varphi^j \in \mathbf{V}^h_i$ is a base function which satisfies $\varphi^j(a_j) = 1$) in (7) we immediately see that $\boldsymbol{m}^h_{i+1}(a_j)| = 1$.

Similarly as in the previous case we solve the nonlinear system (7) by a fixed-point technique (cf. [3]). We compute

$$\left(\frac{\boldsymbol{m}^h_{i+1,k+1} - \boldsymbol{m}^h_i}{\tau_{i+1}}, \boldsymbol{v}\right)_h + \alpha \left(\boldsymbol{m}^h_i \times \boldsymbol{m}^h_{i+1,k+1}, \boldsymbol{v}\right)_h$$
$$- \frac{(1+\alpha^2)}{4}(\boldsymbol{m}^h_{i+1,k+1} \times \Delta_h \boldsymbol{m}^h_{i+1,k}, \boldsymbol{v})_h - \frac{(1+\alpha^2)}{4}(\boldsymbol{m}^h_{i+1,k+1} \times \Delta_h \boldsymbol{m}^h_i, \boldsymbol{v})_h$$
$$- \frac{(1+\alpha^2)}{4}(\boldsymbol{m}^h_i \times \Delta_h \boldsymbol{m}^h_{i+1,k+1}, \boldsymbol{v})_h$$
$$= \frac{(1+\alpha^2)}{4}(\boldsymbol{m}^h_i \times \Delta_h \boldsymbol{m}^h_i, \boldsymbol{v})_h, \tag{8}$$

until

$$\|\boldsymbol{m}^h_{i+1,k+1} - \boldsymbol{m}^h_{i+1,k}\|_h < TOL,$$

where TOL is a prescribed tolerance.

3 Adaptive algorithm

Our adaptive algorithm makes use of the local error indicators μ^τ_{i+1} and $\mu^h_{K,i+1}$ for the time step control and mesh refinement, respectively (see e.g., [8, 16, 6, 9, 12, 15, 21] for related works). For adaptive techniques in micromagnetism see, e.g., [1, 17, 11, 20].

The local error indicators can be obtained from the a posteriori error estimates (cf. [5]) and take the following form

$$\mu^\tau_{i+1} = \|\boldsymbol{m}^h_{i+1} - \boldsymbol{m}^h_i\|^2_1 + \int_{t_i}^{t_{i+1}} \|(h - h_{i+1})\|^2, .$$

$$\mu_{K,i+1}^h = \sum_{e \subset K} h_e \| [\nabla m_{i+1}^h \cdot \nu_e] \|_{L_2(e)}^2 + \| h_K | \nabla m_{i+1}^h |^2 m_{i+1}^h \|_{L_2(K)}^2$$
$$+ \left\| h_K \frac{m_{i+1}^h - m_i^h}{\tau_{i+1}} \right\|_{L_2(K)}^2 + \| h_K (h_{i+1} - h_{i+1}^h) \|_{L_2(K)}^2$$

For a given tolerance TOL start with $\mathcal{T}_0$, τ_0, m_0^h.

1. until $t_{i+1} < T$ set $\tau_{i+1} = \tau_i$, $\mathcal{T}_{i+1} = \mathcal{T}_i$;

2. set $t_{i+1} = t_i + \tau_{i+1}$ and compute the discrete solution by (6) or (8), if $\mu_{i+1}^\tau \leq \varepsilon_\tau^r TOL$ proceed with the space refinement step 3, else decrease τ_{i+1} step and repeat step 1;

3. for all $K \in \mathcal{T}_{i+1}$, if $\mu_{K,i+1}^h > \varepsilon_h^r TOL/N_{i+1}$ mark K for refinement, if $\mu_{K,i+1}^h < \varepsilon_h^c TOL/N_{i+1}$ mark K for coarsening;

4. refine/coarsen mesh and compute new solution, if, $\mu_{i+1}^\tau \leq \varepsilon_\tau^c TOL$ increase τ_{i+1} and go to step 2 (this can be repeated several times, otherwise we proceed to the next time step with, i.e. we go to step 1).

The constants ε_τ^r, ε_h^c are chosen (e.g. 0.5, 0.5), N_{i+1} is the number of elements from $\mathcal{T}_{i+1}$.

4 Numerical experiment

In this numerical example we will apply our adaptive strategy to a problem from [2, 3]. There this problem has been studied on uniform meshes. The problem is computed in domain $\Omega = (0,1) \times (0,1)$ with $h \equiv 0$ and initial data ($x = (x_1 - 0.5, x_2 - 0.5)$)

$$m_0(x) = \begin{cases} (2xA, A^2 - |x|^2)/(A^2 + |x|^2) & x \leq 0.5 \\ (0, 0, -1) & x \geq 0.5. \end{cases}$$

where $A = (1 - 2|x|)^4/16$. The initial data is chosen in such a way that after a finite time a singularity (i.e. $\nabla m \notin L_\infty(\Omega)$) starts to form in the middle of the domain. We studied the problem on time interval $t \in (0, 0.31)$.

The initial mesh and mesh at final time are depicted in Figures 1 (49563 unknowns) and Figure 2 (34395 unknowns for midpoint method and 34491 unknowns for backward-Euler method). For this particular choice of parameters in adaptive algorithm, the meshes at the final time were graphically indistinguishable for both methods. The three components of the magnetisation near the time $t = 0.31$ are depicted in Figures 3-5. Again, the results were graphically identical for both methods. It is clear from the results that the adaptive algorithm correctly detect the position of the singularity and increases the efficiency of the computation.

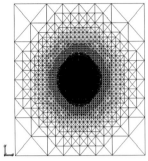 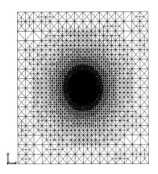

Fig. 1. Initial Mesh **Fig. 2.** Mesh at final time

Fig. 3. x-component of m **Fig. 4.** y-component of m **Fig. 5.** z-component of m

The time step size for both methods varied from $\mathcal{O}(1^{-5})$ to $\mathcal{O}(1^{-7})$. In [3] the authors need $\tau = \mathcal{O}(h^2)$ for the convergence of (8) on uniform meshes. With our adaptive strategy we attained numerical convergence of the fixed-point iterations (8) while using larger time steps for midpoint method. The time step sizes for the midpoint method were comparable to those used with the backward Euler method, which is robust with respect to mesh refinement (cf. [19, 4]). The evolutions of number of unknowns (i.e. vertices of the mesh) during the computation can be found in figures (6) and (7).

Although, the used adaptive algorithm was originally developed for backward-Euler method (see [5]), the presented numerical results indicate, that it can be successfully used with midpoint method. Moreover, the behaviour of both adaptive methods (e.g. mesh evolution and topology, time stepping) was very similar in our experiments.

Acknowledgements

The author would like to acknowledge the support of the IUAP project of the Ghent University and the EPSRC grant of the Imperial College.

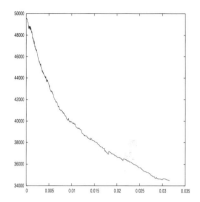

Fig. 6. Degrees of freedom for backward-Euler method

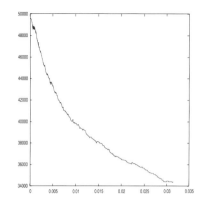

Fig. 7. Degrees of freedom for midpoint method

References

1. Bagnérés-Viallix, A., Baras, P., Albertini, J.B.: 2D and 3D calculations of micromagnetic wall structures using finite elements. IEEE Trans. Magn., **27**, 3819–3822 (1991)
2. Bartels, S., Ko, J., Prohl., A.: Numerical approximation of the Landau-Lifshitz-Gilbert equation and finite time blow-up of weak solutions. preprint, http://www.fim.math.ethz.ch/preprints, (2005)
3. Bartels, S., Prohl, A.: Convergence of an implicit finite element method for the Landau-Lifshitz-Gilbert eqauation. preprint, http://www.fim.math.ethz.ch/preprints, (2005)
4. Baňas, Ľ.: Numerical methods for the Landau-Lifshitz-Gilbert equation. In: Li, Z., Vulkov, L., Wasniewski, J. (eds) Numerical Analysis and Its Applications: Third International Conference, NAA 2004, Rousse, Bulgaria, June 29-July 3, 2004, Revised Selected Papers. Lecture Notes in Computer Science, **3401**, Springer (2005)
5. Baňas, Ľ.: On dynamical micromagnetism with magnetostriction. PhD thesis, Ghent University, Ghent (2005)
6. Chen, Z., Dai., S.: Adaptive galerkin methods with error control for a dynamical Ginzburg-Landau model in superconductivity. SIAM J. Numer. Anal., **38**, 1961–1985 (2001)
7. Ciarlet., P.G.: The finite element method for elliptic problems. North-Holland, Amsterdam (1978)
8. Eriksson, K., Johnson., C.: Adaptive finite element methods for parabolic problems I: A linear model problem. SIAM J. Numer. Anal., **28**, 43–77 (1991)
9. Eriksson, K., Johnson, C.: Adaptive finite element methods for parabolic problems IV: Nonlinear problems. SIAM J. Numer. Anal., **32**, 1729–1749 (1995)
10. Gilbert., T.L.: A Lagrangian formulation of gyromagnetic equation of the magnetic field. Phys. Rev., **100:1243**, (1955)
11. Hertel, R., Kronmüller, H.: Adaptive finite element mesh refinement techniques in three-dimensional micromagnetic modeling. IEEE Trans. Magn., **34**, 3922–3930 (1998)

12. Ivarsson, J.: A posteriori error analysis in $L_2(L_2)$ and $L_2(H^1)$ of the discontinuous Galerkin method for the time-dependent Ginzburg-Landau equations. preprint, (2001)
13. Landau, L.D., Lifshitz, E.M.: Electrodynamics of continuous media. Translated from the Russian by J.B. Sykes and J.S. Bell. Pergamon Press, Oxford-London-New York-Paris (1960)
14. Monk, P.B., Vacus, O.: Accurate discretization of a nonlinear micromagnetic problem. Comput. Methods Appl. Mech. Eng., **190**, 5243–5269 (2001)
15. Nochetto, R.H., Schmidt, A., Verdi, C.: A posteriori error estimation and adaptivity for degenerate parabolic problems. SIAM J. Numer. Anal., **69**, 1–24 (2000)
16. Picasso., M.: Adaptive finite elements for a linear parabolic problem. Comput. Meth. Appl. Mech. Eng., **167**, 223–237 (1998)
17. Scholz, W., Schrefl, T., Fidler, J.: Mesh refinement in Fe-micromagnetics for multi-domain $Nd_2Fe_{14}B$ particles. J. Magn. Magn. Mater., **196-197**, 933–934 (1999)
18. Serpico, C., Mayergoyz, I.D., Bertotti, G.: Numerical technique for integration of the Landau-Lifshitz equation. J. Appl. Phys., **89**, 6991–6993 (2001)
19. Suess, D., Tsiantos, V., Schrefl, T., Fidler, J., Scholz, W., Forster, R., Dittrich, H., Miles J.J.: Time resolved micromagnetics using a preconditioned time integration method. J. Magn. Magn. Mater., **248**, 298–311 (2002)
20. Tako, K.M., Schrefl, T., Wongsam, M.A., Chantrell, R.W.: Finite element micromagnetic simulations with adaptive mesh refinement. J. Appl. Phys., **81**, 4082–4084 (1997)
21. Verfürth, R.: A posteriori error estimates for nonlinear problems. $L^r(0,t;L^\rho(\Omega))$-error estimates for finite element discretizations of parabolic equations. Math. Comp., **67**, 1335–1360 (1998)

Stability for Walls in Ferromagnetic Nanowire

G. Carbou[1] and S. Labbé[2]

[1] MAB, Université Bordeaux 1, 351 cours de la Libération, 33405 Talence cedex (France)
carbou@math.u-bordeaux1.fr
[2] Laboratoire de Mathématiques, Bât. 425, Université Paris Sud, 91405 Orsay Cedex (France)
stephane.labbe@math.u-psud.fr

Summary. We study the stability of travelling wall profiles for a one dimensional model of ferromagnetic nanowire submitted to an exterior magnetic field. We prove that these profiles are asymptotically stable modulo a translation-rotation for small applied magnetic fields.

1 Model for ferromagnetic nanowires

Ferromagnetic materials are characterized by a spontaneous magnetization described by the magnetic moment u which is a unitary vector field linking the magnetic induction B with the magnetic field H by the relation $B = H + u$. The variations of u are described by the Landau-Lifschitz Equation

$$\frac{\partial u}{\partial t} = -u \wedge H_e - u \wedge (u \wedge H_e) \tag{1}$$

where the effective field is given by $H_e = \Delta u + h_d(u) + H_a$, and the demagnetizing field $h_d(u)$ is deduced from u solving the magnetostatic equations:

$$\text{div } B = \text{div}(H + u) = 0 \text{ and rot } H = 0$$

where H_a is an appplied magnetic field.

For more details on the ferromagnetism model, see [2, 9, 14] and [18]. For existence results about the Landau Lifschitz equations see [3, 4, 10, 8]. For numerical studies see [8, 12] and [13]. For asymptotic studies see [1, 5, 7, 15] and [16].

In this paper we consider an asymptotic one dimensional model of ferromagnetic nanowire submitted to an applied field along the axis of the wire. We denote by (e_1, e_2, e_3) the canonical basis of $\mathbb{R}^3$. The ferromagnetic nanowire is assimilated to the axis $\mathbb{R}e_1$. The demagnetizing energy is approximated by the formula $h_d(u) = -u_2 e_2 - u_3 e_3$ where $u = (u_1, u_2, u_3)$ (this approximation of the demagnetizing energy for a ferromagnetic wire is obtained using a BKW method by D. Sanchez, taking the limit when the diameter of the wire tends to zero in [16]). We assume in addition that an exterior magnetic field δe_1 is applied along the wire axis.

To sum up we study the following system

$$\begin{cases} \dfrac{\partial u}{\partial t} = -u \wedge h_\delta(u) - u \wedge (u \wedge h_\delta(u)) \\ \text{with } h_\delta(u) = \dfrac{\partial^2 u}{\partial x^2} - u_2 e_2 - u_3 e_3 + \delta e_1 \end{cases} \quad (2)$$

For $\delta = 0$, that is without applied field, we observe in physical experiments the formation of a wall breaking down the domain in two parts: one in which the magnetization is almost equal to e_1 and another in which the magnetization is almost equal to $-e_1$. Such a distribution is described in our one dimensional model by the following profile M_0:

$$M_0 = \begin{pmatrix} \operatorname{th} x \\ 0 \\ \dfrac{1}{\operatorname{ch} x} \end{pmatrix}. \quad (3)$$

This profile is a steady state solution of Equation (2) with $\delta = 0$. We prove in [6] the stability of the profile M_0 for Equation (2) without applied field (when $\delta = 0$).

When we apply a magnetic field in the direction $+e_1$ (that is with $\delta > 0$) since the Landau-Lifschitz Equation tends to align the magnetic moment with the effective field, we observe a translation of the wall in the direction $-e_1$. Furthermore, we observe a rotation of the magnetic moment around the wire axis. This phenomenon is described by the solution of (2)

$$U_\delta(t, x) = R_{\delta t}(M_0(x + \delta t)) \quad (4)$$

where R_θ is the rotation by an angle θ around the axis $\mathbb{R}e_1$:

$$R_\theta = \begin{pmatrix} 1 & 0 & 0 \\ 0 & \cos\theta & -\sin\theta \\ 0 & \sin\theta & \cos\theta \end{pmatrix}$$

We study in this paper the stability of U_δ, we prove that for a small δ, U_δ is stable for the H^2 norm and asympotically stable for the H^1 norm, modulo a translation in the variable x and a rotation around $\mathbb{R}e_1$. This result is claimed in the following theorem:

Theorem 1. *There exists $\delta_0 > 0$ such that for all δ with $|\delta| < \delta_0$ then for $\varepsilon > 0$ there exists $\eta > 0$ such that if $\|u(t = 0, x) - U_\delta(t = 0, x)\|_{H^2} < \eta$ then the solution u of Equation (2) with initial data $u(t = 0, x)$ satisfies:*

$$\forall\, t > 0,\, \|u(t,x) - U_\delta(t,x)\|_{H^2} < \varepsilon.$$

In addition there exists σ_∞ and θ_∞ such that

$$\|u(t,x) - R_{\theta_\infty}(U_\delta(t, x + \sigma_\infty))\|_{H^1} \longrightarrow 0 \text{ when } t \longrightarrow +\infty.$$

This result is a generalization of the stability result concerning the static walls when $\delta = 0$ in [6]. It looks like the theorems of stability concerning the travelling waves solutions for semilinear equations like the Ginzburg Landau Equation (see Kapitula [11]). Here we have three new difficulties. The first one is that the magnetic moment takes its values in the sphere and not in a linear space. In order to work with maps with values in a linear space we will use a mobile frame adapted to the Landau-Lifschitz equation and we will describe in Section 2 the magnetic moment in this mobile frame. The second difficulty is that we have here a two dimensional invariance family for Equation (2) whereas the Ginzburg-Landau Equation is only invariant by translation. This is the reason why we must use in the perturbations description the translations and the rotations (see Section 3). The last difficulty is that the Landau-Lifschitz Equation is quasilinear, and then we have to couple variational estimates and semi-group estimates to control the perturbations of our profiles. Section 4 is devoted to these estimates.

2 Landau-Lifschitz Equation in the mobile frame

2.1 First reduction of the problem

For u a solution of the Landau-Lifschitz Equation (2) we define v by $v(t,x) = R_{-\delta t}(u(t, x - \delta t))$ (that is $u(t,x) = R_{\delta t}(v(t, x + \delta t))$). A straightforward calculation gives that u satisfies (2) if and only if v satisfies

$$\begin{cases} \dfrac{\partial v}{\partial t} = -v \wedge h(v) - v \wedge (v \wedge h(v)) - \delta\left(\dfrac{\partial v}{\partial x} + v_1 v - e_1\right) \\ h(v) = \dfrac{\partial^2 v}{\partial x^2} - v_2 e_2 - v_3 e_3 \end{cases} \quad (5)$$

In addition U_δ is stable for (2) if and only if M_0 is stable for (5), that is we are led to study the stability of a static profile, which is more convenient.

2.2 Mobile frame

Let us introduce the mobile frame $(M_0(x), M_1(x), M_2)$, where

$$M_1(x) = \begin{pmatrix} \dfrac{1}{\operatorname{ch} x} \\ 0 \\ -\operatorname{th} x \end{pmatrix} \text{ and } M_2 = \begin{pmatrix} 0 \\ 1 \\ 0 \end{pmatrix}$$

Let $v : \mathbb{R}_t^+ \times \mathbb{R}_x \longrightarrow S^2 \subset \mathbb{R}^3$ be a small perturbation of M_0. We can decompose v in the mobile frame writing

$$v(t,x) = r_1(t,x)M_1(x) + r_2(t,x)M_2 + \sqrt{1 - r_1^2 - r_2^2}M_0(x).$$

Now we can obtain a new version of the Landau-Lifschitz Equation: v satisfies (5) if and only if $r = (r_1, r_2)$ satisfies

$$\frac{\partial r}{\partial t} = (\mathcal{L} + \delta l)r + G(r)\left(\frac{\partial^2 r}{\partial x^2}\right) + H(x, r, \frac{\partial r}{\partial x}) \tag{6}$$

where

- the linear operator $\mathcal{L}$ is given by $\mathcal{L} = JL$ with $J = \begin{pmatrix} -1 & -1 \\ 1 & -1 \end{pmatrix}$ and

$$L = -\frac{\partial^2}{\partial x^2} + 2\operatorname{th}^2 x - 1,$$

- the linear perturbation due to the presence of the applied magnetic field δe_1 is given by δl with $l = \frac{\partial}{\partial x} + \operatorname{th} x$,
- the higher degree non linear part is $G(r)(\frac{\partial^2 r}{\partial x^2})$, where $G(r)$ is a matrix depending on r with $G(0) = 0$,
- the last non linear term $H(x, r, \frac{\partial r}{\partial x})$ is at least quadratic in the variable $(r, \frac{\partial r}{\partial x})$.

In addition the stability of the profile M_0 for Equation (5) is equivalent to the stability of the zero solution for Equation (6).

3 A new system of coordinates

We remark that L is a self adjoint operator on $L^2(\mathbb{R})$, with domain $H^2(\mathbb{R})$. Furthermore, L is positive since we can write $L = l^* \circ l$ with $l = \frac{\partial}{\partial x} + \operatorname{th} x$, and Ker L is the one dimensional space generated by $\frac{1}{\operatorname{ch} x}$.

The matrix J being invertible, Ker $\mathcal{L}$ is the two dimensional space generated by v_1 and v_2 with

$$v_1(x) = \begin{pmatrix} 0 \\ \frac{1}{\operatorname{ch} x} \end{pmatrix}, \quad v_2(x) = \begin{pmatrix} \frac{1}{\operatorname{ch} x} \\ 0 \end{pmatrix}.$$

We introduce $mE = (\operatorname{Ker} \mathcal{L})^{\perp}$. We denote by Q the orthogonal projection onto mE for the $L^2(\mathbb{R})$ scalar product.

The Landau-Lifschitz equation (5) is invariant by translation in the variable x and by rotation around the axis e_1. Therefore for $\Lambda = (\theta, \sigma)$ fixed in $\mathbb{R}^2$, M_Λ defined by $M_\Lambda(x) = R_\theta(M_0(x - \sigma))$ is a solution of Equation (5). We introduce $R_\Lambda(x)$ the coordinates of $M_\Lambda(x)$ in the mobile frame $(M_1(x), M_2(x))$:

$$R_\Lambda(x) = \begin{pmatrix} M_\Lambda(x) \cdot M_1(x) \\ M_\Lambda(x) \cdot M_2 \end{pmatrix}$$

The map Ψ given by

$$\begin{aligned} \Psi : \mathbb{R}^2 \times mE &\longrightarrow H^2(\mathbb{R}) \\ (\Lambda, W) &\longmapsto r(x) = R_\Lambda(x) + W(x) \end{aligned}$$

is a diffeomorphism in a neighborhood of zero. Thus we can write the solution r of Equation (6) in the form :

$$r(t, x) = R_{\Lambda(t)}(x) + W(t, x)$$

where for all t, $W(t) \in mE$ and where $\Lambda : \mathbb{R}_t^+ \mapsto \mathbb{R}^2$.

We will re-write Equation (5) in the coordinates (Λ, W). Taking the scalar product of (5) with v_1 and v_2 we obtain the equation satisfied by Λ, and using Q the orthogonal projection onto mE, we deduce the equation satisfied by W. After this calculation we obtain that r is a solution of Equation (5) if and only if (Λ, W) satisfies the following system

$$\begin{cases} \dfrac{\partial W}{\partial t} = (\mathcal{L} + \delta l + \mathcal{K}_\Lambda)W + \mathcal{R}_1(x, \Lambda, W)(\dfrac{\partial^2 W}{\partial x^2}) + \mathcal{R}_2(x, \Lambda, W, \dfrac{\partial W}{\partial x}) \\ \dfrac{d\Lambda}{dt} = \mathcal{M}(W, \dfrac{\partial W}{\partial x}, \Lambda) \end{cases} \quad (7)$$

where

- $\mathcal{K}_\Lambda : H^2(\mathbb{R}) \longrightarrow mE$ is a linear map satisfying

$$\exists K_1, \ \forall \Lambda \in \mathbb{R}^2, \ \forall W \in mE, \ \|\mathcal{K}_\Lambda W\|_{L^2(\mathbb{R})} \leq K_1 |\Lambda| \|W\|_{H^2(\mathbb{R})} \quad (8)$$

- the non linear terms take their values in mE and satisfy that there exists a constant K_2 such that for $|\Lambda| \leq 1$ and for all $W \in mE$

$$\|\mathcal{R}_1(., \Lambda, W)(\dfrac{\partial^2 W}{\partial x^2})\|_{L^2(\mathbb{R})} \leq K_2 \|W\|_{H^1(\mathbb{R})} \|W\|_{H^2(\mathbb{R})}$$

$$\|\mathcal{R}_2(., \Lambda, W, \dfrac{\partial W}{\partial x})\|_{H^1(\mathbb{R})} \leq K_2 \|W\|^2_{H^1(\mathbb{R})} \quad (9)$$

- $\mathcal{M} : H^1(\mathbb{R}) \times L^2(\mathbb{R}) \times \mathbb{R}^2 \longrightarrow \mathbb{R}^2$ satisfies

$$\exists K_3, \ \forall \Lambda \text{ such that } |\Lambda| \leq 1, \ \forall W \in mE, \ |\mathcal{M}(W, \dfrac{\partial W}{\partial x}, \Lambda)| \leq K_3 \|W\|_{H^1(\mathbb{R})} \quad (10)$$

Theorem 1 is equivalent to the following Proposition:

Proposition 1. *There exists $\delta_0 > 0$ such that for δ with $|\delta| < \delta_0$, we have the following stability result for Equation (7): for $\varepsilon > 0$ there exists $\eta > 0$ such that if $|\Lambda_0| < \eta$ and if $\|W_0\|_{H^2} < \eta$ then the solution (Λ, W) of (7) with initial value (Λ_0, W_0) satisfies*

1. *for all $t > 0$, $\|W(t)\|_{H^2} \leq \varepsilon$ and $|\Lambda| \leq \varepsilon$,*
2. *$\|W(t)\|_{H^1}$ tends to zero when t tends to $+\infty$,*
3. *there exists $\Lambda_\infty \in \mathbb{R}^2$ such that $\Lambda(t)$ tends to Λ_∞ when t tends to $+\infty$.*

The last section is devoted to the proof of Proposition 1.

4 Estimates for the perturbations

4.1 Linear semi group estimates

On mE we have $Re\,(sp\,\mathcal{L}) \subset\,]-\infty, -1]$. In particular this fact implies that the H^2 norm is equivalent on mE to the norm $\|\mathcal{L}u\|_{L^2}$. Furthermore it implies good decreasing properties for the semigroup generated by $\mathcal{L}$. We first prove that this decreasing property is preserved for the linear part of the Equation on W in (7) for a small applied field, and if we assume that Λ remains small.

The operator l is an order one operator dominated on mE by $\mathcal{L}$, thus there exists $\delta_0 > 0$ such that if $|\delta| < \delta_0$, $Re\,(sp\,\mathcal{L} + \delta l) \subset\,]-\infty, -1/2[$.

Let us fix δ such that $|\delta| < \delta_0$. With Estimate (10), if Λ remains small, $\mathcal{K}_\Lambda$ is a small perturbation of $\mathcal{L} + \delta l$. This implies that for Λ small, the semigroup generated by $\mathcal{L} + \delta l + Q\mathcal{K}_\Lambda$ has the same good decreasing properties as $\mathcal{L}$, that is there exists $\nu_0 > 0$ such that if $|\Lambda(t)|$ remains less than ν_0 for all t, then there exists K_4 and $\beta > 0$ such that

$$\|S_\Lambda(t)W_0\|_{H^1} \leq K_4 e^{-\beta t}\|W_0\|_{H^1}$$
$$\leq K_4 \frac{e^{-\beta t}}{\sqrt{t}}\|W_0\|_{L^2}. \quad (11)$$

We can then use the Duhamel formula to solve the equation on W in (7):

$$W(t) = S_\Lambda(t)W_0 + \int_0^t S_\Lambda(t-s)\mathcal{R}_1(s)ds + \int_0^t S_\Lambda(t-s)\mathcal{R}_2(s)ds$$

and then using the estimates (9) and (11) we obtain that if $|\Lambda(t)|$ remains less than ν_0 then there exists K_5 such that

$$\|W(t)\|_{H^1} \leq K_5 e^{-\beta t}\|W_0\|_{H^1} + \int_0^t K_5 \frac{e^{-\beta(t-s)}}{\sqrt{t-s}}\|W(s)\|_{H^1}\|W(s)\|_{H^2}$$
$$+ \int_0^t K_5 e^{-\beta(t-s)}\|W(s)\|_{H^1}^2 \quad (12)$$

4.2 Variational estimates

We see that Estimate (12) is not sufficient to conclude since we have the H^2 norm of W in the right hand side of this estimate. In order to dominate this H^2 norm, we multiply the equation on W in (7) by $J^2 \mathcal{L}^2 W$ and we obtain that there exists a constant K_6:

$$\frac{d}{dt}\|LW\|_{L^2}^2 + \|L^{\frac{3}{2}}W\|_{L^2}^2 \left(1 - K_6\|LW\|_{L^2}\right) \leq 0$$

From this estimate we deduce that if $\|LW\|_{L^2} < \frac{1}{K_6}$, then $1 - K_6\|LW\|_{L^2}$ is positive, thus $\frac{d}{dt}\|LW\|_{L^2}^2$ is negative and $\|LW\|_{L^2}$ remains less than $\frac{1}{K_6}$. So if $\|LW_0\|_{L^2} < \frac{1}{K_6}$, then for all t $\|LW(t)\|_{L^2} \leq \|LW_0\|_{L^2}$. This property gives a bound for the H^2 norm of W since the H^2 norm is equivalent on mE to $\|LW\|_{L^2}$, and reducing the H^2 norm of W_0, we obtain the first part of the conclusion 1 in Proposition 1.

4.3 Conclusion

Let us assume that $\|LW_0\|_{L^2(\mathbb{R})} \leq \frac{1}{K_6}$. Then for all t, $\|W(t)\|_{H^2(\mathbb{R})} \leq C_1\|LW(t)\|_{L^2} \leq \|LW_0\|_{L^2} \leq C_2\|W_0\|_{H^2(\mathbb{R})}$, where C_1 and C_2 are constants.

Multiplying (12) by $(1+t)^2$, defining $G(t) = \max_{[0,T]}(1+s)^2\|W(s)\|_{H^1}$, we obtain that there exists a constant K_7 such that if $|\Lambda(t)|$ remains less than ν_0 we have:

$$G(t) \leq K_7 G(0) + K_7 G(t)\|W_0\|_{H^2} + K_7(G(t))^2$$

If we suppose in addition that $\|W_0\|_{H^2} \leq \frac{1}{2K_7}$ we obtain that

$$0 \leq K_7 G(0) - \frac{1}{2}G(t) + K_7(G(t))^2 := P(G(t)) \tag{13}$$

The polynomial map $P(\xi) = K_7\xi^2 - \frac{1}{2}\xi + K_7 G(0)$ has for $G(0)$ small enough two positive roots. We denote by $\xi(G(0))$ the smallest one. For $G(0)$ small enough we have $G(0) \leq \xi(G(0)) \leq 2K_7 G(0)$ (we can a priori assume that $K_7 \geq 1$ for example). Estimate (13) implies that for all t, $G(t) \leq \xi(G(0))$ that is

$$\forall\, t > 0, \|W(t)\|_{H^1(\mathbb{R})} \leq \frac{\xi(G(0))}{1+t^2} \leq \frac{2K_7 G(0)}{1+t^2}. \tag{14}$$

This implies that $\|W(t)\|_{H^1(\mathbb{R})}$ tends to zero when t tends to $+\infty$. It remains to prove that Λ remains less that ν_0 and admits a limit when t tends to $+\infty$.

Plugging Estimate (14) in the equation on Λ in (7) and using (10), we obtain that $\frac{d\Lambda}{dt}$ is integrable on $\mathbb{R}^+$, that is Λ admits a limit when t tends to $+\infty$. Furthermore, by integration we have

$$\forall\, t, \ |\Lambda(t)| \leq |\Lambda(0)| + \int_0^t K_3 \frac{2K_7 G(0)}{1+s^2}\,ds \leq |\Lambda(0)| + \pi K_3 K_7 G(0)$$

Reducing $|\Lambda_0|$ and $G(0) = \|W_0\|_{H^1(\mathbb{R})}$ we obtain that for all t, $|\Lambda(t)|$ remains less than ν_0, which justifies all our estimates a posteriori.

References

1. Alouges, F., Rivière, T., Serfaty, S.: Néel and cross-tie wall energies for planar micromagnetic configurations. Control, Optimisation and Calculus of Variations, **8**, 31–68 (2002)
2. Brown, F.: Micromagnetics. Wiley, New York (1963)
3. Carbou, G., Fabrie, P.: Time average in micromagnetism. J. Differential Equations, **147** (2), 383–409 (1998)
4. Carbou, G., Fabrie, P.: Regular solutions for Landau-Lifschitz equation in a bounded domain. Differential Integral Equations, **14** (2), 213–229 (2001)
5. Carbou, G., Fabrie, P., Guès, O.: On the ferromagnetism equations in the non static case. Comm. Pure Appli. Anal., **3**, 367–393 (2004)
6. Carbou, G., Labbé, S.: Stability for Static Walls in Ferromagnetic Nanowires. to appear in Discrete and Continuous Dynamical Systems
7. DeSimone, A., Kohn, R. V., Müller, S., Otto, F.: Magnetic microstructures— a paradigm of multiscale problems. In ICIAM 99 (Edinburgh), Oxford Univ. Press, Oxford, 175–190 (2000)
8. Haddar, H., Joly, P.: Stability of thin layer approximation of electromagnetic waves scattering by linear and nonlinear coatings. J. Comput. Appl. Math., **143**, 201–236 (2002)
9. Halpern, L., Labbé, S.: Modélisation et simulation du comportement des matériaux ferromagétiques. Matapli, **66**, 70–86 (2001)
10. Joly, J.-L., Métivier, G., Rauch, J.: Global solutions to Maxwell equations in a ferromagnetic medium. Ann. Henri Poincaré, **1**, 307-340 (2000)
11. Kapitula, T.: Multidimensional stability of planar travelling waves. Trans. Amer. Math. Soc., **349** (1), 257–269 (1997)
12. Labbé, S.: Simulation numérique du comportement hyperfréquence des matériaux ferromagnétiques. Thèse de l'Université Paris 13 (1998)
13. Labbé, S., Bertin, P.-I.: Microwave polarisability of ferrite particles with non-uniform magnetization. Journal of Magnetism and Magnetic Materials, **206**, 93–105 (1999)
14. Landau, L., Lifschitz,E.: Electrodynamique des milieux continues. cours de physique théorique, tome VIII (ed. Mir) Moscou (1969)
15. Rivière, T., Serfaty, S.: Compactness, kinetic formulation, and entropies for a problem related to micromagnetics. Comm. Partial Differential Equations, **28** (1-2), 249–269 (2003)
16. Sanchez, D.: Behaviour of the Landau-Lifschitz equation in a ferromagnetic wire. preprint MAB (2005)
17. Visintin, A.: On Landau Lifschitz equation for ferromagnetism. Japan Journal of Applied Mathematics, **1**, 69-84 (1985)
18. Wynled, H.: Ferromagnetism. Encyclopedia of Physics, Vol. XVIII / 2, Springer Verlag, Berlin (1966)

Continuous Galerkin Methods for Solving Maxwell Equations in 3D Geometries

Patrick Ciarlet, Jr[1] and Erell Jamelot[2]

[1] POEMS, UMR CNRS-ENSTA-INRIA 2706, ENSTA, 32 bd Victor, 75739 Paris Cedex 15, France
patrick.ciarlet@ensta.fr
[2] Same address
erell.jamelot@ensta.fr

Summary. Maxwell equations are easily resolved when the computational domain is convex or with a smooth boundary, but if on the contrary it includes geometrical singularities, the electromagnetic field is locally unbounded and globally hard to compute. The challenge is to find out numerical methods which can capture the EM field accurately. Numerically speaking, it is advised, while solving the coupled Maxwell-Vlasov system, to compute a continuous approximation of the field. However, if the domain contains geometrical singularities, continuous finite elements span a strict subset of all possible fields, which is made of the H^1-regular fields. In order to recover the total field, one can use additional ansatz functions or introduce a weight. The first method, known as the singular complement method [4, 3, 14, 2, 9, 15, 16] works well in $2D$ and $2D\frac{1}{2}$ geometries and the second method, known as the weighted regularization method [13] works in $2D$ and $3D$. In this contribution, we examine some recent developments of the latter method to solve instationary Maxwell equations and we provide numerical results.

1 Introduction and notations

Let $\Omega \subset \mathbb{R}^3$ be a bounded polyhedron with a Lipschitz boundary $\partial\Omega$. In order to simplify the presentation, we suppose that Ω is simply connected and $\partial\Omega$ is connected. Let **n** be the unit outward normal to $\partial\Omega$. The boundary $\partial\Omega$ may contain reentrant corners and/or edges, which are called geometrical singularities later on. Let c, ε_0 and μ_0 be respectively the light velocity, the dielectric permittivity and the magnetic permeability ($c \approx 3.10^8$ m.s^{-1}, $\varepsilon_0\mu_0 c^2 = 1$). Maxwell equations in vacuum read:

$$\partial_t \mathcal{E} - c^2 \operatorname{\mathbf{curl}} \mathcal{B} = -\mathcal{J}/\varepsilon_0, \tag{1}$$

$$\partial_t \mathcal{B} + \operatorname{\mathbf{curl}} \mathcal{E} = 0, \tag{2}$$

$$\operatorname{div} \mathcal{E} = \rho/\varepsilon_0, \tag{3}$$

$$\operatorname{div} \mathcal{B} = 0. \tag{4}$$

Above, $\mathcal{E}$ and $\mathcal{B}$ are the electric field and magnetic induction respectively, ρ and $\mathcal{J}$ are the charge and current densities which satisfy the charge conservation equation:
$$\operatorname{div}\mathcal{J} + \partial_t \rho = 0. \tag{5}$$
These quantities depend on the space variable $\mathbf{x}$ and on the time variable t.

The boundary is made up of two parts: $\partial\Omega = \overline{\Gamma}_C \cup \overline{\Gamma}_A$, where Γ_C is a perfectly conducting boundary, and Γ_A an artificial boundary. Note that we do not require that $\partial\Gamma_A \cap \partial\Gamma_C = \emptyset$. On Γ_C, we have:
$$\mathcal{E} \times \mathbf{n} = 0 \text{ on } \Gamma_C, \ \mathcal{B} \cdot \mathbf{n} = 0 \text{ on } \Gamma_C. \tag{6}$$
Since the choice of the location of Γ_A is free, it is located so that it does not cut nor contains any geometrical singularity [8]. Therefore the tangential trace of $\mathcal{E}$ and the normal trace of $\mathcal{B}$ are regular, and in addition the tangential trace $\mathcal{E} \times \mathbf{n}$ and the normal trace $\mathcal{B} \cdot \mathbf{n}$ vanish near the geometrical singularities. We further split the artificial boundary Γ_A into Γ_A^i and Γ_A^a. On Γ_A^i, we model incoming plane waves, whereas we impose on Γ_A^a an absorbing boundary condition. Both can be modelled [1] as a Silver-Müller boundary condition on Γ_A:
$$(c\mathcal{B}\mathcal{E} \times \mathbf{n}) \times \mathbf{n} = c\mathbf{b} \times \mathbf{n} \text{ on } \Gamma_A, \text{ where } \mathbf{b} \text{ is given.} \tag{7}$$
In order to solve equations (1-4), with boundary conditions (6) and (7), one needs to define initial conditions (for instance at time $t = 0$):
$$\mathcal{E}(\cdot, 0) = \mathcal{E}_0, \ \mathcal{B}(\cdot, 0) = \mathcal{B}_0, \tag{8}$$
where the couple $(\mathcal{E}_0, \mathcal{B}_0)$ depends only on the variable $\mathbf{x}$.

If we derive (1) in time and inject **curl** of (2) in it, we get a vector wavelike equation for $\mathcal{E}$. We consider then the following equivalent problem (PE): *Find $\mathcal{E}$ such that*
$$\partial_t^2 \mathcal{E} + c^2 \mathbf{curlcurl}\mathcal{E} = -\partial_t \mathcal{J}/\varepsilon_0, \text{ in } \Omega, \ t \in]0, T[, \tag{9}$$
$$\operatorname{div}\mathcal{E} = \rho/\varepsilon_0, \text{ in } \Omega, \ t \in]0, T[, \tag{10}$$
$$\mathcal{E} \times \mathbf{n}_{|\Gamma_C} = 0, \text{ and } (c\mathcal{B} + \mathcal{E} \times \mathbf{n}) \times \mathbf{n}_{|\Gamma_A} = c\mathbf{b} \times \mathbf{n}_{|\Gamma_A}, \ t \in]0, T[, \tag{11}$$
$$\mathcal{E}(\cdot, 0) = \mathcal{E}_0, \text{ in } \Omega, \tag{12}$$
$$\partial_t \mathcal{E}(\cdot, 0) = \mathcal{E}_1 := c^2(\mathbf{curl}\mathcal{B}_0 - \mu_0 \mathcal{J}(\cdot, 0)), \text{ in } \Omega. \tag{13}$$
The same procedure can be carried out on the magnetic field.

In addition to the usual Lebesgue and Sobolev spaces, the building of the ad hoc variational formulations requires to introduce some non-standard functional spaces [13, 8]. We suppose that Ω has N_{re} reentrant edges of dihedral angles $(\Theta_e = \pi/\alpha_e)_{e=1,\ldots,N_{re}}$, with $1/2 < \alpha_e < 1$. Let r_e denote the orthogonal distance to the reentrant edge e, and $r = \min_{e=1,\ldots,N_{re}} r_e$.

Let $L^2(D)$ be the usual Lebesgue space of square integrable functions over D, $D \in \{\Omega, \partial\Omega\}$, and $L^2_\gamma(\Omega)$ be the following weighted space, with $\|.\|_{0,\gamma}$ norm:

$$L^2_\gamma(\Omega) = \{v \in \mathcal{D}'(\Omega) \mid \int_\Omega w(r) \, v^2 \, d\Omega < \infty\}, \, \|v\|^2_{0,\gamma} = \int_\Omega w(r) \, v^2 \, d\Omega.$$

Above, the weight w is a function of the distance to the reentrant edges, namely $w(r) = \min(r^{2\gamma}, 1)$, with for instance $\gamma = 0.99$ (one may choose $\gamma \in [0, 1]$). Notice that this definition is slightly different than the general one given in [13]. $H^1(\Omega)$ will denote the space of $L^2(\Omega)$ functions with gradients in $L^2(\Omega)^3$. We now define variational spaces for vector fields, together with the associated norms:

$$\mathcal{H}(\mathbf{curl}, \Omega) := \{\mathcal{F} \in L^2(\Omega)^3 \mid \mathbf{curl}\mathcal{F} \in L^2(\Omega)^3\}, \, \|\mathcal{F}\|^2_{0,\mathbf{curl}} = \|\mathcal{F}\|^2_0 + \|\mathbf{curl}\mathcal{F}\|^2_0,$$
$$\mathcal{H}(\mathrm{div}_{(\gamma)}, \Omega) := \{\mathcal{F} \in L^2(\Omega)^3 \mid \mathrm{div}\mathcal{F} \in L^2_{(\gamma)}(\Omega)\}, \, \|\mathcal{F}\|^2_{0,\mathrm{div}_{(\gamma)}} = \|\mathcal{F}\|^2_0 + \|\mathrm{div}\mathcal{F}\|^2_{0,(\gamma)}.$$

The index $_{(\gamma)}$ means that one can choose to use weights or not.
Under suitable data assumptions, $\mathcal{E} \in \mathcal{X}^A_{\mathcal{E}(,\gamma)}$, with:

$$\mathcal{H}_A(\mathbf{curl}, \Omega) := \{\mathcal{F} \in \mathcal{H}(\mathbf{curl}, \Omega) \mid \mathcal{F} \times \mathbf{n}_{|\partial\Omega} \in \mathcal{L}^2_t(\partial\Omega), \, \mathcal{F} \times \mathbf{n}_{|\Gamma_C} = 0\},$$
$$\mathcal{X}^A_{\mathcal{E}(,\gamma)} := \mathcal{H}_A(\mathbf{curl}, \Omega) \cap \mathcal{H}(\mathrm{div}_{(\gamma)}, \Omega),$$

where $\mathcal{L}^2_t(\partial\Omega) := \{\mathbf{u} \in L^2(\partial\Omega)^3 \mid \mathbf{u} \cdot \mathbf{n} = 0 \text{ a. e.}\}$. When $\Gamma_C = \partial\Omega$, we write simply $\mathcal{X}^0_{\mathcal{E}(,\gamma)}$.

According to Costabel [12], and to Costabel-Dauge [13], the graph norm and the semi-norm: $\|\mathcal{F}\|^2_{\mathcal{X}^0_{\mathcal{E}(,\gamma)}} = \|\mathbf{curl}\mathcal{F}\|^2_0 + \|\mathrm{div}\mathcal{F}\|^2_{0,(\gamma)}$ are equivalent on $\mathcal{X}^0_{\mathcal{E}(,\gamma)}$. Note that, when there is a weight, this is true only if $\gamma < 1$. Moreover, $\exists \gamma_{min} \in]0, 1[$ such that for all $\gamma \in]\gamma_{min}, 1[$, $\mathcal{X}^0_{\mathcal{E},\gamma} \cap H^1(\Omega)^3$ is dense in $\mathcal{X}^0_{\mathcal{E},\gamma}$.

2 Variational formulations and discretization

Starting from the second order system of eqs. (9-13), we obtain a series of variational formulations, retracing the steps below:
- Multiply eq. (9) by $\mathcal{F} \in \mathcal{H}_A(\mathbf{curl}, \Omega)$, and integrate by parts over Ω. We get the variational formulation (VF): Find $\mathcal{E}(t) \in \mathcal{H}_A(\mathbf{curl}, \Omega)$ such that $\forall \mathcal{F} \in \mathcal{H}_A(\mathbf{curl}, \Omega), \forall t$,

$$(\mathcal{E}'', \mathcal{F})_0 + c^2(\mathbf{curl}\mathcal{E}, \mathbf{curl}\mathcal{F})_0 + c\int_{\Gamma_A} (\mathcal{E}' \times \mathbf{n}).(\mathcal{F} \times \mathbf{n}) \, d\Gamma$$
$$= -(\mathcal{J}'/\varepsilon_0, \mathcal{F})_0 + \int_{\Gamma_A} (c\mathbf{b}' \times \mathbf{n}).\mathcal{F} \, d\Gamma, \qquad (14)$$

- Add $c^2(\mathrm{div}\mathcal{E}, \mathrm{div}\mathcal{F})_{0,(\gamma)}$ on the LHS and $c^2(\rho/\varepsilon_0, \mathrm{div}\mathcal{F})_{0(,\gamma)}$ on the RHS to get the augmented VF (AVF): Find $\mathcal{E}(t) \in \mathcal{X}^A_{\mathcal{E}(,\gamma)}$ such that $\forall \mathcal{F} \in \mathcal{X}^A_{\mathcal{E}(,\gamma)}, \forall t$,

$$(\mathcal{E}'', \mathcal{F})_0 + c^2(\mathcal{E}, \mathcal{F})_{\mathcal{X}^0_{\mathcal{E}(,\gamma)}} + c\int_{\Gamma_A} (\mathcal{E}' \times \mathbf{n}).(\mathcal{F} \times \mathbf{n})\,\mathrm{d}\Gamma$$
$$= -(\mathcal{J}'/\varepsilon_0, \mathcal{F})_0 + c^2(\rho/\varepsilon_0, \mathrm{div}\mathcal{F})_{0(,\gamma)} + \int_{\Gamma_A} (c\mathbf{b}' \times \mathbf{n}).\mathcal{F}\,\mathrm{d}\Gamma, \quad (15)$$

- Add $(p, \mathrm{div}\mathcal{F})_{0(,\gamma)}$ on the LHS and consider a constraint on the divergence of $\mathcal{E}$ (cf. (17)). If $p \in L^2_{(\gamma)}(\Omega)$ is the Lagrange multiplier, we reach the mixed AVF (MAVF): Find $(\mathcal{E}(t), p(t)) \in X^A_{\mathcal{E}(,\gamma)} \times L^2_{(\gamma)}(\Omega)$ such that $\forall \mathcal{F} \in X^A_{\mathcal{E}(,\gamma)}$, $\forall t$,

$$(\mathcal{E}'', \mathcal{F})_0 + c^2(\mathcal{E}, \mathcal{F})_{\mathcal{X}^0_{\mathcal{E}(,\gamma)}} + (p, \mathrm{div}\mathcal{F})_{0(,\gamma)} + c\int_{\Gamma_A} (\mathcal{E}' \times \mathbf{n}).(\mathcal{F} \times \mathbf{n})\,\mathrm{d}\Gamma$$
$$= -(\mathcal{J}'/\varepsilon_0, \mathcal{F})_0 + c^2(\rho/\varepsilon_0, \mathrm{div}\mathcal{F})_{0(,\gamma)} + \int_{\Gamma_A} (c\mathbf{b}' \times \mathbf{n}).\mathcal{F}\,\mathrm{d}\Gamma, \quad (16)$$

and $\forall q \in L^2_{(\gamma)}(\Omega)$, $\forall t$,

$$(\mathrm{div}\mathcal{E}, q)_{0(,\gamma)} = (\rho/\varepsilon_0, q)_{0(,\gamma)}. \quad (17)$$

The constraint (17) is added to reinforce Gauss' law (3) and also to avoid numerical instabilities when the discrete charge conservation equation is not satisfied while solving the Maxwell-Vlasov system [5].

Theorem 1. *Suppose that $\partial_t \mathcal{J} \in L^2(0, T; L^2(\Omega)^3)$, $\partial_t \rho \in L^2(0, T; L^2_{(\gamma)}(\Omega))$, ρ and $\mathcal{J}$ satisfying (5). Suppose that $(\mathcal{E}_0, \mathcal{E}_1) \in X^A_{\mathcal{E}(,\gamma)} \times L^2(\Omega)^3$. Then, equations (16-17) are equivalent to problem (PE) and have a unique solution $(\mathcal{E}, p)$ such that $(\mathcal{E}, \partial_t \mathcal{E}) \in C^0(0, T; X^A_{\mathcal{E}(,\gamma)}) \times C^0(0, T; L^2(\Omega)^3)$ and $p = 0$.*

The proof can be found in [16]. Idem for the magnetic field.

To build a discretized (M)AVF, we use a leap-frog scheme in time, and either the continuous P_k Lagrange FE (no Lagrange mutliplier) or the P_{k+1}-P_k continuous Taylor-Hood FE in space. We choose an explicit scheme. Let Δt be the time step and $t_n = n\Delta t$, $n \in \mathbb{N}$. $u''(., t_{n+1})$ is approximated by: $u''(., t_{n+1}) \approx [u(., t_{n+1}) - 2u(., t_n) + u(., t_{n-1})]/\Delta t^2$. Recall that for an explicit scheme, one must satisfy a CFL-like condition. For the P_1 FE, we must have: $\Delta t \leq 0.5\,c\,\min_l h_l$, where h_l is the diameter of the lth tetrahedron. Let N_k (resp. N_{k+1}) be the number of P_k (resp. P_{k+1}) degrees of freedom.

Let $\mathrm{E}^n \in (\mathbb{R}^3)^{\mathrm{N}_{k+1}}$ be the discretized electric field and $\mathrm{p}^n \in \mathbb{R}^{\mathrm{N}_k}$ be the discretized Lagrange multiplier at time t_n. Let $\mathbb{M}_\Omega \in (\mathbb{R}^{3\times 3})^{\mathrm{N}_{k+1} \times \mathrm{N}_{k+1}}$ be the mass matrix, and $\mathbb{M}^{\|}_{\Gamma_A} \in (\mathbb{R}^{3\times 3})^{\mathrm{N}_{k+1} \times \mathrm{N}_{k+1}}$ be the boundary mass matrix on Γ_A. Let $\mathbb{C} \in (\mathbb{R}^{1\times 3})^{\mathrm{N}_k \times \mathrm{N}_{k+1}}$ be the constraint matrix. At a given time t_{n+1}, $n \in \mathbb{N}$, we have to solve:

$$(\mathbb{M}_\Omega + c\Delta t \mathbb{M}^{\|}_{\Gamma_A})\mathrm{E}^{n+1} + \mathbb{C}^T \mathrm{p}^{n+1} = \mathrm{RHS}^{n+1},$$
$$\mathbb{C}\mathrm{E}^{n+1} = \mathrm{G}^{n+1}. \quad (18)$$

Let $\mathbb{M} = \mathbb{M}_\Omega + c\Delta t \mathbb{M}^\|_{\Gamma_A}$. The algorithm is the following:
Solve first $\mathbb{M}\mathrm{E}_0^{n+1} = \mathrm{RHS}^{n+1}$, then $\mathbb{CM}^{-1}\mathbb{C}^T \mathrm{p}^{n+1} = \mathbb{C}\mathrm{E}_0^{n+1} - \mathrm{G}^{n+1}$, and finally $\mathbb{M}\mathrm{E}^{n+1} = \mathbb{M}\mathrm{E}_0^{n+1} - \mathbb{C}^T \mathrm{p}^{n+1}$. The Lagrange multiplier p^{n+1} may be computed with the Uzawa algorithm. Note that when there is no coupling with Vlasov equation, p^{n+1} remains small at all times, so that there is actually no need to compute it at all times. To speed up the resolution, one can lump $\mathbb{M}$ with $\widetilde{P}_1$ or $\widetilde{P}_2$ FE [11]. Both $\widetilde{P}_k$ FE preserve accuracy, at the cost of increasing the total number of degrees of freedom for the $\widetilde{P}_2$ FE.

3 Numerical results and conclusion

The numerical results are given for the following model problem (fig. 1): Ω has a single reentrant edge of dihedral angle $2\pi/3$, so that $\alpha = 2/3$. A current bar crosses the domain, with $\mathcal{J} = 10^{-5} \omega \sin(\pi z/L) \cos(\omega t)\mathbf{z}$ and $\rho = 10^{-5}(\pi/L)\cos(\pi z/L)\sin(\omega t)$, for $\omega = 2.5$ GHz. There is no incoming wave. The spatial wavelength associated to ω is of order 0.75 m, and the time period is of order 2.5 ns. It is clear that the dimensions of our domain are not realistic, however we made this choice in order to visualize oscillations. We report the results of computations made with the $\widetilde{P}_1$ FE (discretization of the AVF), and with 685 000 tetrahedra. We encoded the problem in Fortran 77.

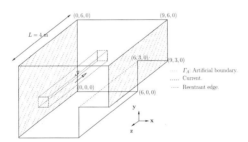

Fig. 1. The model problem.

On figures 2 and 3 the space evolution of the x and y-components of the electric field are represented in the plane $z = 2.5$ m, at times $T_1 = 1$ ns, $T_2 = 8$ ns, $T_3 = 15$ ns, $T_4 = 20$ ns. We can see that an electric wave is created by the current, that it propagates into the cavity with wavelength ≈ 0.75 m, and is reflected by the conductor as expected. At T_3, we observe a growing peak of intensity close to the reentrant corner.

On figure 4, we represented the space evolution of the z-component in the plane $z = 2.5$ m, at times T_i, $i = 1, 4$. Again, we observe the propagation of the wave with wavelength ≈ 0.75 m, and the reflections. Note that this component has a regular behaviour, which is due to the fact that the only

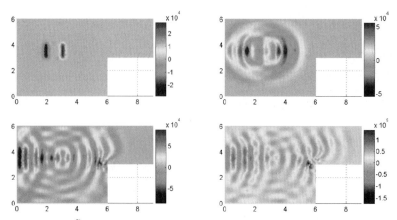

Fig. 2. $E_x^{\tilde{P}_1}$ component at times T_i, $i = 1$ to 4, in plane $z = 2, 5$ m.

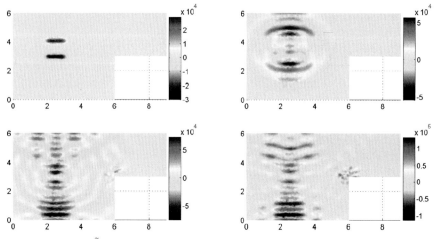

Fig. 3. $E_y^{\tilde{P}_1}$ component at times T_i, $i = 1$ to 4, in plane $z = 2, 5$ m.

geometrical singularity is along the z-axis [7, 9]. Moreover, it takes smaller (absolute) values than the x and y-components.

On figures 2, 3 and 4, one can see spurious reflections on Γ_A, due to the fact that the Silver-Müller boundary condition is simply of first order: only plane waves with normal incidence are absorbed, which is not our case. In addition, the spurious reflections appear more important for E_x than for E_y, since its values are more intense horizontally.

On figure 5, we present the time evolution of the x-component of the electric field at points $M_1 = (1,1,2)$, $M_2 = (5.5, 2.5, 2)$, $M_3 = (1,1,2)$, $M_4 = (8, 5.5, 2)$. It remains equal to zero until the electric wave reaches the point under consideration. Then the field oscillates with a period ≈ 2.5 ns, as expected.

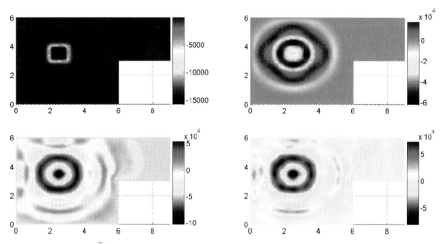

Fig. 4. $E_z^{\widetilde{P}_1}$ component at times T_i, $i = 1$ to 4, in plane $z = 2, 5$ m.

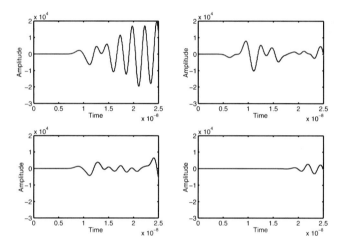

Fig. 5. $E_x^{\widetilde{P}_1}$ component at points M_i, $i = 1$ to 4.

To the authors' knowledge, this is the first time a $3D$ singular electric field is computed with continuous Lagrange FE. According to M. Dauge (private communication), the WRM can also be used to compute the magnetic field, with similar assumptions on γ. In order to avoid spurious reflections, we suggest to use perfectly matched layers [6]. For the resolution of $2D$ Maxwell equations with continuous Galerkin finite elements, we refer the reader to [10].

References

1. Assous, F., Degond, P., Heintzé, E., Raviart, P.-A., Segré, J.: On a finite element method for solving the three-dimensional Maxwell equations. J. Comput. Phys., **109**, 222–237 (1993)
2. Assous, F., Ciarlet, Jr, P., Labrunie, S., Segré, J.: Numerical solution to the time-dependent Maxwell equations in axisymmetric singular domains: The Singular Complement Method. J. Comput. Phys., **191**, 147–176 (2003)
3. Assous, F., Ciarlet, Jr, P., Segré, J.: Numerical solution to the time-dependent Maxwell equations in two-dimensional singular domains: the singular complement method. J. Comput. Phys., **161**, 218–249 (2000)
4. Assous, F., Ciarlet, Jr, P., Sonnendrücker, E.: Resolution of the Maxwell equations in a domain with reentrant corners. Modél. Math. Anal. Numér., **32**, 359–389 (1998)
5. Barthelmé, R.: Le problème de conservation de la charge dans le couplage des équations de Maxwell et de Vlasov. PhD thesis, Université Strasbourg I, France (2005)
6. Bérenger, J.-P.: A perfectly matched layer for the absorption of electromagnetic waves. J. Comput. Phys., **114**, 185–200 (1994)
7. Buffa, A., Costabel M., Dauge M.: Anisotropic regularity results for Laplace and Maxwell operators in a polyhedron. C. R. Acad. Sci. Paris, Ser I, **336**, 565–570 (2003)
8. Ciarlet, Jr, P.: Augmented formulations for solving Maxwell equations. Comp. Meth. Appl. Mech. and Eng., **194**, 559–586 (2005)
9. Ciarlet, Jr, P., Garcia, E., Zou, J.: Solving Maxwell equations in $3D$ prismatic domains. C. R. Acad. Sci. Paris, Ser. I, **339**, 721–726 (2004)
10. Ciarlet, Jr, P., Jamelot, E.: A comparison of nodal finite element methods for solving Maxwell equations. *In preparation.*
11. Cohen, G.: Higher-Order Numerical Methods for Transient Wave Equations. Scientific Computation Series. Springer-Verlag, Berlin (2002)
12. Costabel, M.: A coercive bilinear form for Maxwell's equations. J. Math. An. Appl., **157**, 527–541 (1991)
13. Costabel, M., Dauge, M.: Weighted regularization of Maxwell equations in polyhedral domains. Numer. Math., **93**, 239–277 (2002)
14. Garcia, E.: Solution to the instationary Maxwell equations with charges in nonconvex domains (in French). PhD thesis, Université Paris VI, France (2002)
15. Jamelot, E.: Éléments finis nodaux pour les équations de Maxwell. C. R. Acad. Sci. Paris, Sér. I, **339**, 809–814 (2004)
16. Jamelot, E.: Solution to Maxwell equations with continuous Galerkin finite elements (in French). PhD thesis, École Polytechnique, Palaiseau, France (2005)

Exponential Integrators

On the Use of the Gautschi-Type Exponential Integrator for Wave Equations

Volker Grimm

Heinrich–Heine–Universität, Mathematisches Institut, Lehrstuhl für Angewandte Mathematik, Universitätsstraße 1, 40225 Düsseldorf, Germany.
grimm@am.uni-duesseldorf.de

Summary. Wave equations are especially challenging for numerical integrators since the solution is often not smooth and there is no smoothing in time. The largest usable step size of standard integrators, as for example the often used Störmer-Verlet-Leap-Frog-scheme, depends on the space discretisation. The better the approximation in space, the smaller the required step size of the integrator. The presented exponential integrator allows for error bounds independent of the space discretisation but only dependent on constants arising from the original problem. This favourable property is demonstrated with the Sine–Gordon equation.

1 Introduction

Semilinear wave equations appear in many physical relevant applications. They can often be formulated as abstract evolutionary equations on a Hilbert space H:

$$u'' + Au = g(u), \qquad u(0) = u_0, u'(0) = u'_0,$$

with an unbounded linear operator A. The primes indicate time derivatives. Spacial discretisation usually leads to an ordinary differential equation on $I\!R^n$:

$$y'' + A_n y = g_n(y), \qquad y(0) = y_0, y'(0) = y'_0,$$

where $A_n = \Omega_n^2$ is a symmetric and positive semi-definite real matrix of large norm. If n refers to the number of freedoms in the space discretisation, the norm of A_n is tending to infinity for finer and finer space discretisations, reflecting the unbounded operator properly, while g_n and its derivatives remain bounded. The large norm of A_n restricts the step size of standard integrators and introduces an oscillatory solution. Choosing a "stiff" integrator for the semi-discretisation does not help for improving the accuracy, since the solution y is often not smooth. Due to these specific difficulties connected with the oscillatory solution, these differential equations are called oscillatory or highly-oscillatory. Oscillatory differential equations are a topic of current interest, see, e.g., [1, 2, 3, 4, 5, 6, 7, 8, 9, 10, 11, 12, 13, 14, 15, 16]).

To show that an integrator does not suffer from a step-size restriction introduced by the norm of A_n, one needs to prove error bounds independent of this norm or, in more physical terms, of the frequencies, which are the eigenvalues of Ω_n. Error bounds of this type have been proved for the *mollified impulse method*, proposed in [5], and the *Gautschi-type exponential integrator*, proposed an analysed in [11] and shown to be completely independent of the norm of A_n in [7]. These bounds are derived via a finite-energy condition. In this paper, it is shown that these methods provide error bounds in time which are independent of the refinement of the space discretisation. Rather than by writing down all technical details, the result is demonstrated for the Sine–Gordon equation and the Gautschi-type exponential integrator. But the results apply to general semilinear wave equations and more general exponential integrators.

This paper is organised as follows. In Section 2, the Sine–Gordon equation is given with emphasis on its natural properties that are important for understanding the performance of the Gautschi-type exponential integrator. Section 3 discusses how the properties of the abstract equation affect the ordinary differential equations arising from a semi-discretisation in space. The Gautschi-type integrator is introduced in Section 4 together with the main theorem on its error bounds for wave equations. Finally, in Section 5, the findings are numerically illustrated.

2 Sine–Gordon equation

The Sine–Gordon equation can be written as

$$u_{tt} = -Au - g(u), \qquad u(0) = u_0 \in V = H_0^1(0,1),\ u'(0) = u_0' \in H = L^2(0,1), \tag{1}$$

with $A = -u_{xx}$, $g(u) = \sin(u)$ and $V = \mathcal{D}(A^{\frac{1}{2}})$. The nonlinearity g is often smooth in semilinear wave equations and we assume the bounds $\|g\|_H \leq M_1$, $\|g_u\|_H \leq M_2$ and $\|g_{uu}\|_H \leq M_3$ in the Hilbert-space norm or the operator norms, respectively. For the Sine–Gordon equation, we have $M_1 = M_2 = M_3 = 1$. (The notation of the norms is quite compressed. Note that $\|\cdot\|_H$ designates different norms depending on the argument and that the norms are to be considered from V to H.)

The energy of the wave is given by

$$H(u, u') = \frac{1}{2}\|u'\|_H^2 + \frac{1}{2}\|A^{\frac{1}{2}}u\|_H^2 + G(u),$$

where

$$G(u) = \int_0^1 (f(tu), u)_H\, dt = \int_0^1 \int_{\Omega=[0,1]} \sin(tu(x))u(x)\, dx\, dt$$

Therefore and due to the bounds

$$-\|u\|_H^2 - C \le G(u) \le D + \|A^{\frac{1}{2}}u\|_H^2, \tag{2}$$

where C and D are moderate constants, g is called a gradient operator. For the one-dimensional Sine–Gordon equation, $C = 1/4$ and $D = 1/\pi^2$. The bounds (2) immediately imply that the *finite energy*

$$H_e(u, u') = \frac{1}{2}\|u'\|_H^2 + \frac{1}{2}\|A^{\frac{1}{2}}u\|_H^2 \le \frac{1}{2}K^2$$

is moderately bounded by a constant K whenever $H(u, u')$ is. These properties are intrinsic to wave equations and only these properties will be used in the following discussion.

3 Discretisation

Two different discretisations, namely pseudo-spectral and finite-difference, are considered for the (abstract) Sine–Gordon equation (1). The two of them lead to an ordinary differential equation

$$y'' + A_n y = g_n(y), \qquad y(0) = y_0, y'(0) = y'_0, \tag{3}$$

in $(\mathbb{R}^n, \|\cdot\|_\Delta)$, where $\|\cdot\|_\Delta$ is just an appropriately scaled Euclidean norm. The important point is that equation (3) inherits the properties of the abstract wave equation. The norm of A_n tends to infinity if n tends to infinity, reflecting the unbounded operator A. The remaining properties are summarised in the following proposition.

Proposition 1. *The initial values of (3) satisfy the finite-energy condition*

$$\frac{1}{2}\|y'(0)\|_\Delta^2 + \frac{1}{2}\|A_n^{\frac{1}{2}}y(0)\|_\Delta^2 \le \frac{1}{2}K^2,$$

with the same constant K as in the abstract formulation and the bounds $\|g_n\|_\Delta \le \|g\|_H \le M_1$, $\|g_{n,y}\|_\Delta \le \|g_u\|_H \le M_2$ and $\|g_{n,yy}\|_\Delta \le \|g_{uu}\|_H \le M_3$ hold.

Details on the discretisations and the proof of Proposition 1 are given in the following two subsections.

3.1 Pseudo-spectral discretisation

The eigenfunctions and eigenvalues of the operator A,

$$e_k = \sqrt{2}\sin\pi k x, \qquad \lambda_k = \pi^2 k^2,$$

form an orthonormal basis of H. Choosing $V_h = \{e_1, \cdots, e_n\}$ gives an n-dimensional subspace of H. With the projection P_n on this subspace, one is left with a system of ordinary differential equations in the space $\mathbb{R}^n$ with the norm $\|\cdot\|_\Delta = \|\cdot\|_2$, where $\|\cdot\|_2$ is the Euclidean norm. The system of ordinary differential equation reads

$$y'' = -A_n y + g_n(y), \qquad y(t_0) = y_0, \quad y'(t_0) = y'_0,$$

with $A_n = \mathrm{diag}(\lambda_1, \ldots, \lambda_n)$,

$$y_0 = \begin{bmatrix} (u_0, e_1) \\ \vdots \\ (u_0, e_n) \end{bmatrix}, \quad y'_0 = \begin{bmatrix} (u'_0, e_1) \\ \vdots \\ (u'_0, e_n) \end{bmatrix} \quad \text{and} \quad g_n(y) = F_n \sin(F_n^{-1} y),$$

where $F_n = \frac{\sqrt{2}}{n+1}\left(\sin\frac{kj\pi}{n+1}\right)_{k,j=1}^{n}$ is the matrix belonging to the Discrete Sine Transform (DST) and the evaluation of sin at a vector is to be understood pointwise.

The statements in the proposition above are easily verified. Namely,

$$\frac{1}{2}\|y'_0\|_\Delta^2 + \frac{1}{2}\|A_n^{\frac{1}{2}} y_0\|_\Delta^2 = \frac{1}{2}\|P_n u'_0\|_H^2 + \frac{1}{2}\|P_n A^{\frac{1}{2}} u_0\|_H^2 \le \frac{1}{2}\|u'_0\|_H^2 + \frac{1}{2}\|A^{\frac{1}{2}} u_0\|_H^2 \le \frac{1}{2}K^2$$

By differentiation, $\|g_n\|_\Delta \le \|g\|_H$, $\|g_{n,y}\|_\Delta \le \|g_u\|_H$ and $\|g_{n,yy}\|_\Delta \le \|g_{uu}\|_H$ follow.

3.2 Finite-difference approximation

For a finite-difference discretisation in space, a regular grid $x_i = ih$ for $i = 0, \ldots, n$ with $h = \frac{1}{n+1}$ is chosen. By setting $y_i(t) := u(x_i, t)$ for $i = 1, \ldots, n$ and approximating the second derivative by a symmetric finite-difference of order 2, one arrives at

$$y'' = -A_n y + g_n(y), \quad y(0) = y_0 = (u_0(x_1, 0), \ldots, u_0(x_n, 0))^T, \quad y'(0) = R_n u'(0),$$

where $A_n = \frac{-1}{h^2}\mathrm{tridiag}(1, -2, 1) \in \mathbb{R}^{n,n}$, $y \in \mathbb{R}^n$ and $R_n : L^2(0,1) \to \mathbb{R}^n$, with

$$R_n u = \left(\frac{1}{h}\int_{x_1-\frac{h}{2}}^{x_1+\frac{h}{2}} u(x)\,dx, \ldots, \frac{1}{h}\int_{x_n-\frac{h}{2}}^{x_n+\frac{h}{2}} u(x)\,dx\right)^T.$$

The restriction R_n is necessary since $u'(0)$ is not necessarily continuous. The properties of this discretisation, stated in Proposition 1, are readily justified. A_n is a symmetric positive definite Matrix and $\Omega_n := A_n^{\frac{1}{2}}$ is defined. With $\|y\|_\Delta^2 = h\|y\|_2^2$, often called discrete Sobolev norm, we have

$$\frac{1}{2}\|y'(0)\|_\Delta^2 + \frac{1}{2}\|\Omega_n y(0)\|_\Delta^2 \le \frac{1}{2}\|u'(0)\|_H^2 + \frac{1}{2}\|A^{\frac{1}{2}} u\|_H^2 \le \frac{1}{2}K^2,$$

since

$$\|y'(0)\|_{\Delta}^2 = h \sum_{i=1}^{n} \left(\frac{1}{h} \int_{x_i - \frac{h}{2}}^{x_i + \frac{h}{2}} u'(x,0) \, dx \right)^2 \leq \frac{1}{h} \sum_{i=1}^{n} \left(\int_{x_i - \frac{h}{2}}^{x_i + \frac{h}{2}} |u'(x,0)| \, dx \right)^2$$

$$\stackrel{\mathrm{CSU}}{\leq} \frac{1}{h} \sum_{i=1}^{n} h \int_{x_i - \frac{h}{2}}^{x_i + \frac{h}{2}} |u'(x,0)|^2 \, dx = \int_{x_1 - \frac{h}{2}}^{x_n + \frac{h}{2}} |u'(x,0)|^2 \, dx \leq \|u'(0)\|_H^2$$

and (with $y_{0,0} = y_{0,n+1} := 0$)

$$\|\Omega_n y(0)\|_{\Delta}^2 = h(A_n y_0, y_0) = -h \sum_{i=1}^{n} \frac{y_{0,i+1} - 2y_{0,i} + y_{0,i-1}}{h^2} \cdot y_{0,i}$$

$$= h \sum_{i=1}^{n} \frac{y_{0,i+1} - y_{0,i}}{h} \cdot \frac{y_{0,i+1} - y_{0,i}}{h} = \frac{1}{h} \sum_{i=0}^{n} \left(\int_{x_i}^{x_{i+1}} u_x(x,0) \, dx \right)^2$$

$$\leq \sum_{i=0}^{n} \int_{x_i}^{x_{i+1}} |u_x(x,0)|^2 \, dx = \|u_x(0)\|_H^2 = \|A^{\frac{1}{2}} u(0)\|_H^2.$$

The statements $\|g_n\|_{\Delta} \leq \|g\|_H$, $\|g_{n,y}\| \leq \|g_u\|_H$ and $\|g_{n,yy}\|_{\Delta} \leq \|g_{uu}\|_H$ follow by differentiation.

4 The Gautschi-type exponential integrator

In [11], Hochbruck and Lubich consider the Gautschi-type method for the solution of systems of oscillatory second-order differential equations like (3). The Gautschi-type method, which is based on the requirement that it solves exactly linear problems with constant inhomogeneity g, is given by

$$y_{m+1} - 2\cos(\tau\Omega) y_m + y_{m-1} = \tau^2 \operatorname{sinc}^2\left(\frac{\tau\Omega}{2}\right) g(\phi(\tau\Omega) y_m),$$

with the *filter function* ϕ whose purpose is to filter out resonant frequencies at integer multiples of π.

Combining the error bound for the Gautschi-type method, which is proved in [11], with the result in [7] and slight modifications of the proof give the following theorem.

Theorem 1. *If the solution of the abstract wave equation (1) satisfies the finite-energy condition at the starting values*

$$\frac{1}{2}\|u'(0)\|_H^2 + \frac{1}{2}\|A^{\frac{1}{2}} u(0)\|_H^2 \leq \frac{1}{2} K^2,$$

then the error of the Gautschi-type method, by application on the semi-discretised systems, for $0 \leq t_m = mh \leq T$ is bounded by

$$\|y(t_m) - y_m\|_{\Delta} \leq \tau^2 C,$$

where C only depends on T, K, $\|g\|_H, \|g_u\|_H, \|g_{uu}\|_H$ and ϕ.

This is an extraordinary error bound. The error of the discretisation in time is independent of the chosen discretisation in space. The bound only depends on constants that stem from the original formulation of the wave equation.

5 Numerical example

To illustrate the bound numerically, the Sine–Gordon equation is integrated with initial values

$$u(0) = 0, \quad \text{and} \quad u'(0) = 1_{[\frac{1}{4}, \frac{3}{4}]}(x),$$

where $1_I(x)$ denotes the indicator function for the interval I. Figure 1 shows the global error after an integration time of 1 versus the chosen step-size for the Gautschi-type method and the Verlet-scheme, which is one of the most used second-order standard integrators. The semi-discretisation is finite-differences with 34 points. The Verlet-scheme fails to integrate the differential equation for larger step-sizes due to its dependence of the space discretisation. The result was the same if a pseudo-spectral discretisation would have been chosen. The Verlet-scheme needs smaller time-steps for finer grids. For 128 grid points, the Verlet-scheme does not give a reasonable solution for the whole interval [0.01, 1] in Figure 1, whereas the graph of the error for the Gautschi-type method hardly changes. This impressively demonstrates the advantage of the presented error bounds.

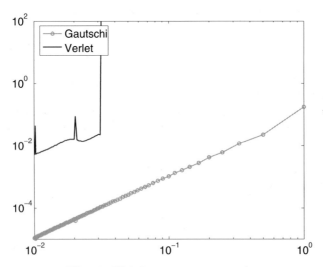

Fig. 1. Global error versus step-size

References

1. Ascher, U. M., Reich, S.: On some difficulties in integrating highly oscillatory Hamiltonian systems. In Proc. Computational Molecular Dynamics, Springer Lecture Notes, 281–296, (1999)
2. Cohen, D.: Conservation properties of numerical integrators for highly oscillatory hamiltonian systems. IMA J. Numer. Anal., **26**, 34–59 (2005)
3. Cohen, D., Hairer, E., Lubich, Ch.: Modulated fourier expansions of highly oscillatory differential equations. Foundations of Comput. Maths., **3**, 327–450 (2003)
4. Cohen, D., Hairer, E., Lubich, Ch.: Numerical energy conservation for multi-frequency oscillatory differential equations. BIT, **45**, 287–305 (2005)
5. García-Archilla, B., Sanz-Serna, J., Skeel, R.: Long-time-step methods for oscillatory differential equations. SIAM J. Sci. Comput., **30**, 930–963 (1998)
6. Grimm, V.: Exponentielle Integratoren als Lange-Zeitschritt-Verfahren für oszillatorische Differentialgleichungen zweiter Ordnung. PhD Thesis, Heinrich-Heine-Universität, Düsseldorf (2002)
7. Grimm, V.: A note on the Gautschi-type method for oscillatory second-order differential equations. Numer. Math., **102**, 61–66 (2005)
8. Grimm, V.: On error bounds for the Gautschi-type exponential integrator applied to oscillatory second-order differential equations. Numer. Math., **100**, 71–89 (2005)
9. Hairer, E., Lubich, Ch.: Long-time energy conservation of numerical methods for oscillatory differential equations. SIAM J. Numer. Anal., **38**, 414–441 (2000)
10. Hairer, E., Lubich, Ch., Wanner, G.: Geometric Numerical Integration. Springer-Verlag (2002)
11. Hochbruck, M., Lubich, Ch. A Gautschi-type method for oscillatory second-order differential equations. Numer. Math., **83**, 403–426 (1999)
12. Iserles, A.: On the global error of discretization methods for highly-oscillatory ordinary differential equations. BIT, **42**, 561–599 (2002)
13. Iserles, A.: Think globally, act locally: solving highly-oscillatory ordinary differential equations. Appld. Num. Anal., **43**, 145–160 (2002)
14. Iserles, A.: On the method of Neumann series for highly oscillatory equations. BIT, **44**, 473–488 (2004)
15. Lorenz, K., Jahnke, T., Lubich, Ch.: Adiabatic integrators for highly oscillatory second order linear differential equations with time-varying eigendecomposition. BIT, **45**, 91–115 (2005)
16. Petzold, L., Jay, L. and Yen, J.: Numerical Solution of Highly Oscillatory Ordinary Differential Equations. Acta Numerica, **6**, 437–484 (1997)

Positivity of Exponential Multistep Methods

Alexander Ostermann and Mechthild Thalhammer

Institut für Mathematik, Leopold-Franzens-Universität Innsbruck
Mailing address: Technikerstraße 13, A-6020 Innsbruck, Austria
{alexander.ostermann, mechthild.thalhammer}@uibk.ac.at

Summary. In this paper, we consider exponential integrators that are based on linear multistep methods and study their positivity properties for abstract evolution equations. We prove that the order of a positive exponential multistep method is two at most and further show that there exist second-order methods preserving positivity.

1 Introduction

Integration schemes that involve the evaluation of the exponential were first proposed in the 1960s for the numerical approximation of stiff ordinary differential equations. Nowadays, due to advances in the computation of the product of a matrix exponential with a vector, such methods are considered as practicable also for high-dimensional systems of differential equations. The renewed interest in exponential integrators is further enhanced by recent investigations which showed that they have excellent stability and convergence properties. In particular, they perform well for differential equations that result from a spatial discretisation of nonlinear parabolic and hyperbolic initial-boundary value problems, see [4, 9] and references therein.

However, aside from a favourable convergence behaviour, the usability of a numerical method for practical applications is substantially affected by its qualitative behaviour, and, in many cases, it is inevitable to ensure that certain geometric properties of the underlying problem are well preserved by the discretisation. In particular, it is desirable that the positivity of the true solution is retained by the numerical approximation. More precisely, if the solution of a linear abstract evolution equation

$$u'(t) = A\,u(t) + f(t), \quad 0 < t \leq T, \quad u(0) \text{ given}, \tag{1}$$

remains positive, the numerical solution should retain this property. Unfortunately, as proven by Bolley and Crouzeix [3], the order of positive rational one-step and linear multistep methods, respectively, is restricted by one.

The objective of the present paper is to investigate exponential multistep methods where the coefficients are combinations of the exponential and closely related functions. The general form of the considered schemes is introduced below in Section 3. Examples include Adams-type methods that were studied recently in [4, 9] for parabolic problems, see also the earlier works [8, 12].

The main result, which we deduce in Section 4, states that positive exponential multistep methods are of order two at most. Further, we show that there exist second-order methods which preserve positivity. Thus, the order barrier of [3] is raised by one. For exponential Runge–Kutta methods, a similar result has been obtained recently in [10].

Our analysis of exponential multistep methods for abstract evolution equations is based on an operator calculus which allows to define the Laplace-Stieltjes transform involving the generator of a positive C_0-semigroup. We refer to the subsequent Section 2, where the basic hypotheses on the differential equation and some fundamental tools of the employed analytical framework are recapitulated.

2 Analytical framework

In this section, we state the basic assumptions on the abstract initial value problem (1).

Throughout, we let $(V, \|\cdot\|)$ denote the underlying Banach space. Further, we suppose $A : D \subset V \to V$ to be a densely defined and closed linear operator on V that generates a *strongly continuous* semigroup $\left(e^{tA}\right)_{t\geq 0}$ of type (M, ω), that is, there exist constants $M \geq 1$ and $\omega \in \mathbb{R}$ such that the bound

$$\|e^{tA}\| \leq M e^{t\omega}, \qquad t \geq 0, \tag{2}$$

is valid. For a detailed treatment of C_0-semigroups, we refer to the monographs [6, 11].

The notion of positivity requires the Banach space V to be endowed with an additional order structure. In the present paper, to keep the analytical framework simple, we restrict ourselves to the consideration of the Lebesgue spaces and subspaces thereof, respectively, as it is then straightforward to define the positivity of an element pointwise.[1] In general, an appropriate setting is provided by the theory of Banach lattices treated in Yosida [13, Chap. XII]. Our results remain valid within this framework.

We recall that a bounded linear operator $B : V \to V$ is said to be *positive* if for any element $v \in V$ satisfying $v \geq 0$ it follows $Bv \geq 0$.

---

[1] A function $v : \Omega \subset \mathbb{R}^d \to \mathbb{R}$ in $L^p(\Omega)$, $1 \leq p \leq \infty$, is said to be *positive* if it is pointwise positive, i.e., $v(x) \geq 0$ for almost all $x \in \Omega$. In that case, we write $v \geq 0$ for short. We employ here the standard terminology, although the term *non-negative* would be more appropriate.

Example 1. We consider the differential operator ∂_{xx} subject to a mixed boundary condition on the Banach space of continuous functions, that is, for some $c_1, c_2 \in \mathbb{R}$ we set $A : D \to V : v \mapsto \partial_{xx} v$ where $V = \mathrm{C}([0,1])$ and $D = \{v \in \mathrm{C}^2([0,1]) : v'(0) + c_1 v(0) = 0 = v'(1) + c_2 v(1)\}$. It is shown in Arendt et al. [1, p. 134] that the associated semigroup $\left(\mathrm{e}^{tA}\right)_{t \geq 0}$ is positive.

Henceforth, we assume that the linear operator $A : D \to V$ is the generator of a positive semigroup $\left(\mathrm{e}^{tA}\right)_{t \geq 0}$ of type (M, ω), see (2). Then, from the formulation of the linear evolution equation (1) as a Volterra integral equation

$$u(t) = \mathrm{e}^{tA} u(0) + \int_0^t \mathrm{e}^{(t-\tau)A} f(\tau) \, \mathrm{d}\tau, \qquad 0 \leq t \leq T, \tag{3}$$

it is seen that the solution u remains positive, provided that the initial value $u(0)$ and the function f are positive.

Let $a \in \mathrm{BV}$ denote a function of bounded variation that is normalised at its discontinuities and satisfies $a(0) = 0$. The associated *Laplace-Stieltjes transform* is defined through

$$G(z) = \int_0^\infty \mathrm{e}^{tz} \, \mathrm{d}a(t), \tag{4}$$

see Hille and Phillips [6, Sect. 6.2]. We recall that a real-valued function G is said to be *absolutely monotonic* on an interval $I \subset \mathbb{R}$ if

$$G^{(j)}(x) \geq 0, \qquad x \in I, \quad j \geq 0.$$

The following result by Bernstein [2], which characterises absolutely monotonic functions of the form (4), is the basis of our analysis in Section 4.

Theorem 1 (Bernstein). *A function G is absolutely monotonic on the half line $(-\infty, \omega]$ iff it is the Laplace-Stieltjes transform of a non-decreasing function $a \in \mathrm{BV}$ such that*

$$\int_0^\infty \mathrm{e}^{t\omega} |\mathrm{d}a(t)| < \infty.$$

A well-known operational calculus described in Hille and Phillips [6, Chap. XV] allows to extend (4) to unbounded linear operators. More precisely, for A being the generator of a strongly continuous semigroup $\left(\mathrm{e}^{tA}\right)_{t \geq 0}$ on V, it holds

$$G(hA) v = \int_0^\infty \mathrm{e}^{thA} v \, \mathrm{d}a(t), \qquad h \geq 0, \quad v \in V, \tag{5}$$

where the integral is defined in the sense of Bochner. It is thus straightforward to deduce the following corollary from Theorem 1, see also Kovács [7].

Corollary 1. *Suppose that the linear operator A generates a positive and strongly continuous semigroup of type (M, ω). Assume further that the function G is absolutely monotonic on $(-\infty, h\omega]$ for some $h \geq 0$. Then, the linear operator $G(hA)$ defined by (5) is positive.*

Remark 1. We note that the converse of the above corollary is true as well. Namely, if $G(hA)$ is positive for any generator A of a positive and strongly continuous semigroup, then the function G is absolutely monotonic. The proof of this statement is in the lines of Bolley and Crouzeix [3, Proof of Lemma 1].

The construction of exponential integrators often relies on the variation-of-constants formula (3) and a replacement of the integrand f by an interpolation polynomial. As a consequence, the linear operators $\varphi_j(hA)$ defined through

$$\varphi_j(z) = \int_0^1 e^{tz} \frac{(1-t)^{j-1}}{(j-1)!} \, dt, \qquad j \geq 1, \qquad z \in \mathbb{C}, \tag{6}$$

naturally arise in the numerical schemes. By the above Theorem 1, these functions are absolutely monotonic, and thus the positivity of the associated operators $\varphi_j(hA)$ follows from Corollary 1.

3 Exponential multistep methods

In this section, we introduce the considered exponential multistep methods for the time integration of the linear evolution equation (1) and state the order conditions. The positivity properties of the numerical schemes are then studied in Section 4.

We let $t_j = jh$ denote the grid points associated with a constant stepsize $h > 0$. Besides, we suppose that the starting values $u_0, u_1, \ldots, u_{k-1} \in V$ are approximations the exact solution values of (1). Then, for integers $j \geq k$, the numerical solution values $u_j \approx u(t_j)$ are given by the k-step recursion

$$\sum_{\ell=0}^k \alpha_\ell(hA)\, u_{n+\ell} = h \sum_{\ell=0}^k \beta_\ell(hA)\, f(t_{n+\ell}), \qquad n \geq 0. \tag{7a}$$

Throughout, we choose $\alpha_k = 1$. Furthermore, we assume that the coefficient functions α_ℓ and β_ℓ are given as Laplace-Stieltjes transforms of certain functions a_ℓ and b_ℓ. Thus, it holds

$$\alpha_\ell(z) = \int_0^\infty e^{tz} \, da_\ell(t), \qquad \beta_\ell(z) = \int_0^\infty e^{tz} \, db_\ell(t), \qquad z \in (-\infty, \omega]. \tag{7b}$$

For simplicity, we require b_ℓ to be piecewise differentiable such that the left-sided limit of $b'_\ell(t)$ exist at $t = j$ for all integers $j \geq 0$. In particular, these assumptions are satisfied if the coefficients functions are (linear) combinations of the exponential and the related φ-functions (6). We therefore refer to (7) as an *exponential linear k-step method*. Due to (7b), the operators $\alpha_\ell(hA)$ and $\beta_\ell(hA)$ are bounded on V.

Examples that have recently been studied in literature for the time integration of semilinear evolution equations are exponential Adams-type methods. For the choice $\alpha_1 = \ldots = \alpha_{k-1} = 0$ and $\beta_k = 0$, the resulting methods are discussed in Calvo and Palencia [4]. On the other hand, the case

$\alpha_0 = \ldots = \alpha_{k-2} = 0$ and $\beta_k = 0$ generalising the classical Adams–Bashforth methods is covered by the analysis given in [9].

In the following, we derive the order conditions for the exponential k-step method. We note that the arguments given below extend to semilinear problems $u'(t) = A\,u(t) + F(t, u(t))$ by setting $f(t) = F(t, u(t))$. As usual, the numerical method (7) is said to be *consistent* of order p, if the local error

$$d(t,h) = \sum_{\ell=0}^{k} \alpha_\ell(hA)\, u(t+\ell h) - h \sum_{i=0}^{k} \beta_\ell(hA)\, f(t+\ell h) \tag{8}$$

is of the form $d(t, h) = \mathcal{O}(h^{p+1})$ for $h \to 0$, provided that the function f is sufficiently smooth, see Hairer, Nørsett, and Wanner [5, Chap. III.2].

In order to determine the leading h-term in $d(t, h)$, we make use of the variation-of-constants formula

$$u(t + \ell h) = e^{\ell h A}\, u(t) + \int_0^{\ell h} e^{(\ell h - \tau)A}\, f(t + \tau)\, d\tau,$$

see also (3). We expand all occurrences of f in Taylor series at t and apply the definition of the φ-functions (6). A comparison in powers of h finally yields the following result.

Lemma 1. *The order conditions for exponential multistep methods (7) are*

$$\sum_{\ell=0}^{k} \alpha_\ell(hA)\, e^{\ell h A} = 0, \tag{9a}$$

$$\sum_{\ell=1}^{k} \alpha_\ell(hA)\, \ell^q\, \varphi_q(\ell h A) = \sum_{\ell=0}^{k} \beta_\ell(hA)\, \frac{\ell^{q-1}}{(q-1)!}, \qquad 1 \leq q \leq p, \tag{9b}$$

where by definition $\ell^0 = 1$ *for* $\ell = 0$.

The first condition corresponds to the requirement that the exponential multistep method (7) is exact for the homogeneous equation $u'(t) = A\,u(t)$. By setting $A = 0$ in (9), the usual order conditions

$$\sum_{\ell=0}^{k} \alpha_\ell(0) = 0, \qquad \sum_{\ell=1}^{k} \alpha_\ell(0)\, \ell^q = q \sum_{\ell=0}^{k} \beta_\ell(0)\, \ell^{q-1}, \qquad 1 \leq q \leq p$$

for a linear multistep method with coefficients $\alpha_\ell(0)$ and $\beta_\ell(0)$ follow, see also [5, Chap. III.2].

4 Positivity and order barrier

In this section, we derive an order barrier for positive exponential multistep methods. According to Bolley and Crouzeix [3], the numerical method (7) is said to be *positive*, if the numerical solution values u_n remain positive for all $n \geq k$, provided that the semigroup $(e^{tA})_{t \geq 0}$, the function f, and further the starting values $u_0, u_1, \ldots, u_{k-1}$ are positive. We note that the requirement of positivity implies that the coefficients operators $\alpha_\ell(hA)$ satisfy

$$-\alpha_\ell(hA) \geq 0, \qquad 0 \leq \ell \leq k-1. \tag{10}$$

We next give the main result of the paper.

Theorem 2. *The order of a positive exponential k-step method is two at most.*

Proof. Our main tools for the proof of Theorem 2 are the representation (7b) of the coefficient functions as Laplace-Stieltjes transforms and further the characterisation of positivity given in Section 2. For the following, we set $a_\ell(t) = 0 = b_\ell(t)$ for $t \leq 0$. We note that due to Corollary 1, it is justified to work with the complex variable z instead of the linear operator hA. For the characteristic function of the interval $[r, s)$, we henceforth employ the abbreviation

$$Y_{[r,s)}(t) = \begin{cases} 1 & \text{if } r \leq t < s, \\ 0 & \text{else.} \end{cases}$$

(i) We first show that the validity of the first order condition (9a) together with the requirement (10) imply that the coefficient functions α_ℓ are of the form

$$\alpha_\ell(z) = -\mu_{k-\ell}\, e^{(k-\ell)z}, \qquad \mu_{k-\ell} \geq 0, \qquad 0 \leq \ell \leq k-1, \tag{11}$$

or, equivalently, that the associated functions a_ℓ are given by

$$a_\ell(t) = -\mu_{k-\ell}\, Y_{[k-\ell,\infty)}(t), \qquad \mu_{k-\ell} \geq 0, \qquad 0 \leq \ell \leq k-1. \tag{12}$$

Inserting (7b) into (9a) and applying $\alpha_k(z) = 1$, we get

$$e^{kz} = -\sum_{\ell=0}^{k-1} \alpha_\ell(z)\, e^{\ell z} = -\sum_{\ell=0}^{k-1} \int_0^\infty e^{tz}\, Y_{[\ell,\infty)}(t)\, \mathrm{d}a_\ell(t-\ell)$$

and furthermore conclude

$$Y_{[k,\infty)}(t) = -\sum_{\ell=0}^{k-1} a_\ell(t-\ell)\, Y_{[\ell,\infty)}(t). \tag{13}$$

From (10) and Remark 1 we deduce that the function $-\alpha_\ell$ is absolutely monotonic and thus Theorem 1 shows that $-a_\ell$ is non-decreasing. Due to the fact that $a_\ell(0) = 0$, we finally obtain (12). For the following considerations, accordingly to our choice $\alpha_k(z) = 1$, it is useful to define $\mu_0 = -1$. As a consequence, inserting (11) into (9a) we have

$$\sum_{\ell=1}^{k} \mu_\ell = -\mu_0 = 1. \tag{14}$$

(ii) We next reformulate the order conditions in terms of the functions a_ℓ and b_ℓ given by (7b). Inserting (11) into (9b), we have

$$-\sum_{\ell=1}^{k} \mu_{k-\ell}\, \ell^q\, e^{(k-\ell)z}\, \varphi_q(\ell z) = \sum_{\ell=0}^{k} \beta_\ell(z) \frac{\ell^{q-1}}{(q-1)!}, \qquad 1 \leq q \leq p.$$

For the following considerations, it is convenient to employ the abbreviation

$$\chi_{q;k-\ell,k}(t) = \frac{\ell^q - (k-t)^q}{q}\, Y_{[k-\ell,k)}(t) + \frac{\ell^q}{q}\, Y_{[k,\infty)}(t), \qquad (15)$$

Obviously, $\chi_{q;k-\ell,k}$ is a continuous function such that the support of its derivative is contained in the interval $[k-\ell, k)$. Therefore, making use of the fact that

$$e^{(k-\ell)z}\, \varphi_q(\ell z) = \frac{1}{\ell^q} \int_0^\infty e^{tz} \frac{(k-t)^{q-1}}{(q-1)!}\, Y_{[k-\ell,k)}(t)\, \mathrm{d}t,$$

see (6) for the definition of φ_q, we obtain

$$-\sum_{\ell=1}^{k} \mu_{k-\ell}\, \chi_{q;k-\ell,k}(t) = \sum_{\ell=0}^{k} \ell^{q-1} b_\ell(t), \qquad 1 \leq q \leq p. \qquad (16)$$

(iii) Exploiting the relations given above, we now show that the assumption $p \geq 3$ and the requirement of positivity, that is, the assumptions $\mu_\ell \geq 0$ for $1 \leq \ell \leq k$ and $b_\ell(t)$ a non-decreasing function for any $t \in \mathbb{R}$ and $0 \leq \ell \leq k$, lead to a contradiction. We recall that by definition $\mu_0 = -1$. Regarding the order conditions (16), restricting t to the first interval $[0, 1)$, we obtain the following relations for the derivatives[2]

$$\sum_{\ell=0}^{k} b'_\ell(t) = 1,$$
$$\sum_{\ell=0}^{k} \ell\, b'_\ell(t) = k - t, \qquad \sum_{\ell=0}^{k} \ell^2\, b'_\ell(t) = (k-t)^2. \qquad (17)$$

Taking a suitable linear combination of (17), it follows

$$\sum_{\ell=0}^{k} (t - k + \ell)^2\, b'_\ell(t) = 0.$$

Using that the functions b'_ℓ are non-negative, we conclude that they vanish on $[0, 1)$. This contradicts the first relation in (17). ∎

---

[2] As the function b_ℓ is non-decreasing, its derivative exists almost everywhere and is non-negative. Assertions involving b'_ℓ are thus valid for almost all t.

Remark 2. The order two barrier of Theorem 2 is sharp in the sense that there exist positive second-order schemes. A simple example is given by the exponential trapezoidal rule where $k=1$, $\alpha_0(z) = -e^z$, $\alpha_1 = 1$, $\beta_0 = \varphi_1 - \varphi_2$, and $\beta_1 = \varphi_2$.

For *analytic* semigroups it is well-known that the order conditions (9b) can be weakened, see e.g. [9]. Following the lines of [10] it can be shown that an order two barrier holds in this case, too. For instance, the exponential midpoint rule with $k=2$, $\alpha_0(z) = -e^{2z}$, $\alpha_1 = 0$, $\alpha_2 = 1$, $\beta_1(z) = 2\varphi_1(2z)$, and $\beta_0 = \beta_2 = 0$ has weak order two and preserves positivity.

References

1. Arendt, W., Grabosch, A., Greiner, G., Groh, U., Lotz, H.P., Moustakas, U., Nagel, R., Neubrander, F., Schlotterbeck, U.: One-parameter Semigroups of Positive Operators. Springer, Berlin (1980)
2. Bernstein, S., Sur les fonctions absolument monotones. Acta Mathematica **51**, 1–66 (1928)
3. Bolley, C., Crouzeix, M.: Conservation de la positivité lors de la discrétisation des problèmes d'évolution paraboliques. R.A.I.R.O. Anal. Numér. **12**, 237–245 (1978)
4. Calvo, M.P., Palencia, C.: A class of explicit multistep exponential integrators for semilinear problems. Numer. Math. **102**, 367–381 (2006)
5. Hairer, E., Nørsett, S.P., Wanner, G.: Solving Ordinary Differential Equations I. Nonstiff Problems. Springer, Berlin (1993)
6. Hille, E., Phillips, R.S.: Functional Analysis and Semi-Groups. American Mathematical Society, Providence (1957)
7. Kovács, M.: On positivity, shape, and norm-bound preservation of time-stepping methods for semigroups. J. Math. Anal. Appl. **304**, 115–136 (2005)
8. Nørsett, S.P.: An A-stable modification of the Adams–Bashforth methods. In: Conference on the Numerical Solution of Differential Equations, J. Morris, ed., Lecture Notes in Mathematics **109**, 214–219, Springer, Berlin (1969)
9. Ostermann, A., Thalhammer, M., Wright, W.: A class of explicit exponential general linear methods. To appear in BIT (2006)
10. Ostermann, A., Van Daele, M.: Positivity of exponential Runge–Kutta methods. Preprint, University of Innsbruck (2006)
11. Pazy, A.: Semigroups of Linear Operators and Applications to Partial Differential Equations. Springer, New York (1983)
12. Verwer, J.G.: On generalized linear multistep methods with zero-parasitic roots and an adaptive principal root. Numer. Math. **27**, 143–155 (1977)
13. Yosida, K.: Functional Analysis. Springer, Berlin (1965)

Fluid-Structure Interaction

Stability Results and Algorithmic Strategies for the Finite Element Approach to the Immersed Boundary Method

Daniele Boffi[1], Lucia Gastaldi[2] and Luca Heltai[3]

[1] Dipartimento di Matematica "F. Casorati", Via Ferrata 1, I-27100 Pavia, Italy
    daniele.boffi@unipv.it
[2] Dipartimento di Matematica, Via Valotti 9, I-25133 Brescia, Italy
    gastaldi@ing.unibs.it
[3] Dipartimento di Matematica "F. Casorati", Via Ferrata 1, I-27100 Pavia, Italy
    luca.heltai@unipv.it

Summary. The immersed boundary method is both a mathematical formulation and a numerical method for the study of fluid structure interactions. Many numerical schemes have been introduced to reduce the difficulties related to the non-linear coupling between the structure and the fluid evolution; however numerical instabilities arise when explicit or semi-implicit methods are considered. In this work we present a stability analysis based on energy estimates for the variational formulation of the immersed boundary method.

A two dimensional incompressible fluid and a boundary in the form of a simple closed curve are considered. We use a linearization of the Navier-Stokes equations and a linear elasticity model to prove the unconditional stability of the fully implicit discretization, achieved with the use of a backward Euler method for both the fluid and the structure evolution (BE/BE), and we present a computable CFL condition for the semi-implicit method where the fluid terms are treated implicitly while the structure is treated explicitly (FE/BE).

1 Introduction

The idea behind the immersed boundary (IB) method lies on the observation that the Navier-Stokes equations for incompressible fluids express nothing more than Newton's law $\mathbf{F} = m\mathbf{a}$ in an Eulerian and "fluid-specialized" framework. The IB method consists in adding to the Navier-Stokes equations some additional "internal" forces concentrated on the particles of the "fluid-solid" material to compensate the fluid behavior with the missing elastic part, in order to simulate efficiently the interaction between a fluid and an elastic material.

In this paper we will present a numerical analysis of the stability of the IB method applied to a one-dimensional volume-less and mass-less membrane, immersed in a two-dimensional fluid domain, modeled by the dynamic Stokes equations.

Numerical instabilities arise when computations are carried on using semi-implicit or explicit time-stepping techniques which require a careful choice of the discretization parameters. We will address the stability aspect of IB computations taking advantage of the natural energy estimates that arise from the use of a variational approach to the IB method, as introduced in [2, 3].

In Sect. 2 we briefly present the finite element IB method, as it was introduced in [2, 3] and the elasticity model that will be used throughout the paper. Section 3 describes the time stepping schemes that will be analyzed in Sect. 4, while Sect. 5 is dedicated to numerical validation and conclusions.

2 The finite element immersed boundary method

Let Ω be a two dimensional domain containing both the fluid and the elastic membrane. To be more precise, for all $t \in [0,T]$, let Γ_t be a simple closed elastic curve, the configuration of which is given in a parametric form, $\mathbf{X}(s,t)$, $0 \le s \le L$, $\mathbf{X}(0,t) = \mathbf{X}(L,t)$, where the parameter s marks a material point and L is related to the unstressed length of the boundary.

$\mathbf{X}(s,t)$ represents the position in Ω of the material point which was labeled by s at the initial time. We are interested in expressing formally the force exerted by the structure on the fluid in terms of the elastic force density $\mathbf{f}(s,t)$ generated by the deformation of the immersed material itself. In the IB method this is achieved by mean of the defining properties of the Dirac delta distribution δ:

$$\mathbf{F}(\mathbf{x},t) = \int_D \mathbf{f}(s,t)\delta(\mathbf{x} - \mathbf{X}(s,t))mDs, \quad \text{in } \Omega\times]0,T[. \qquad (1)$$

Here the Dirac delta is used as a way to pass from the Lagrangian to the Eulerian formulation by introducing an "implicit" change of variables.

The force generated by the element of boundary mDs on the fluid is $\mathbf{f}(s,t)mDs$. We will concentrate on the case of linearized hyper-elastic incompressible materials, characterized by the existence of a positive potential energy density Ψ associated with the deformation of the elastic material and which is independent on translations and (linearized) rotations of the material itself.

The relation between the potential energy density Ψ and the force $\mathbf{f}$ is given through the use of the deformation tensor $\mathbb{F}$ and the first Piola-Kirchoff stress tensor $\mathbb{P}$ as follows:

$$\mathbb{F}_{ij} := \frac{\partial \mathbf{X}_j(s,t)}{\partial s_i}, \quad \mathbb{P}_{ij}(s,t) = \frac{\partial \Psi}{\partial \mathbb{F}_{ij}}(s,t), \quad \mathbf{f}_j(s,t) = \frac{\partial \mathbb{P}_{ij}}{\partial s_i}(s,t),$$

where, in the last equation, summation is implied over repeated indices.

In the two-dimensional case $i = 1$ and $j = 1,2$, therefore all tensors become vectors. We will use a linear "fiber-like" formulation where the Piola-Kirchoff stress tensor is defined by a scalar tension $T = \kappa|\mathbb{F}|$ and the unit vector $\tau = \mathbb{F}/|\mathbb{F}|$ tangent to the immersed curve. The following equations hold

$$\mathbb{P} = T\tau = \kappa\mathbb{F}, \qquad \mathbf{f} = \kappa\frac{\partial^2 \mathbf{X}}{\partial s^2}, \qquad \Psi(\mathbb{F}) = \frac{\kappa}{2}|\mathbb{F}|^2, \qquad (2)$$

where κ is the elasticity constant of the material along the immersed boundary.

We observe that, in equation(1), the boundary force $\mathbf{f}$ is multiplied by a two-dimensional Dirac distribution, over a domain of dimension one, so that the resulting force density $\mathbf{F}$ is a one-dimensional Dirac distribution along Γ_t and the following Lemma holds true.

Lemma 1. *Assume that, for all $t \in [0,T]$, the immersed boundary Γ_t is Lipschitz continuous and that $\mathbf{f} \in L^2([0,L] \times]0,T[)$. Then for all $t \in]0,T[$, the force density $\mathbf{F}(t)$, defined formally in(1), is a distribution belonging to $H^{-1}(\Omega)^2$ defined as follows: for all $\mathbf{v} \in H_0^1(\Omega)^2$*

$$_{H^{-1}}\langle \mathbf{F}(t), \mathbf{v} \rangle_{H_0^1} = \int_0^L \mathbf{f}(s,t) \cdot \mathbf{v}(\mathbf{X}(s,t))\,ds \quad \forall t \in]0,T[. \qquad (3)$$

Let $\mathcal{T}_h$ be a subdivision of Ω into triangles or rectangles. We denote by h_x the biggest diameter of the elements of $\mathcal{T}_h$. We then consider two finite dimensional subspaces $\mathbf{V}_h \subseteq H_0^1(\Omega)^2$ and $Q_h \subseteq L_0^2(\Omega)$. It is well known that the pair of spaces $\mathbf{V}_h$ and Q_h need to satisfy the inf-sup condition in order to have existence, uniqueness and stability of the discrete solution of the Navier-Stokes problem (see [5]).

Next, let s_i, $i = 0, \ldots, m$ with $s_0 = 0$ and $s_m = L$, be $m+1$ distinct points of the interval $[0,L]$. We set $h_s = \max_{0 \leq i \leq m} |s_i - s_{i-1}|$. Let $\mathbf{S}_h$ be the finite element space of piecewise linear vectors defined on $[0,L]$ as follows

$$\mathbf{S}_h = \{\mathbf{Y} \in C^0([0,L];\Omega) : \mathbf{Y}|_{[s_{i-1},s_i]} \in \mathcal{P}^1([s_{i-1},s_i])^2, \ i = 1,\ldots,m, \\ \mathbf{Y}(s_0) = \mathbf{Y}(s_m)\} \qquad (4)$$

where $\mathcal{P}^1(I)$ stands for the space of affine polynomials on the interval I. For an element $\mathbf{Y} \in \mathbf{S}_h$ we shall use also the following notation $\mathbf{Y}_i = \mathbf{Y}(s_i)$ for $i = 0, \ldots, m$.

Taking into account Lemma 1, it is possible to show that

$$\langle \mathbf{F}_h(t), \mathbf{v} \rangle = \sum_{i=0}^{m-1} \kappa \left(\frac{\partial \mathbf{X}_{h\,i+1}}{\partial s}(t) - \frac{\partial \mathbf{X}_{h\,i}}{\partial s}(t) \right) \mathbf{v}(\mathbf{X}_{h\,i}(t)). \qquad (5)$$

Notice that the right hand side of (5) is meaningful, since $\mathbf{v}$ is continuous as it is required for the elements in $\mathbf{V}_h$.

The finite element discretization of the IB method reads:

Problem 1. *Given $\mathbf{u}_{0h} \in \mathbf{V}_h$ and $\mathbf{X}_{h0} \in \mathbf{S}_h$, for all $t \in]0,T[$, find $(\mathbf{u}_h(t), p_h(t)) \in \mathbf{V}_h \times Q_h$ and $\mathbf{X}_h(t) \in \mathbf{S}_h$, such that*

$$\rho\frac{d}{dt}(\mathbf{u}_h(t), \mathbf{v}) + \mu(\boldsymbol{\nabla}\mathbf{u}_h(t), \boldsymbol{\nabla}\mathbf{v}) - (\boldsymbol{\nabla}\cdot\mathbf{v}, p_h(t)) = \langle \mathbf{F}_h(t), \mathbf{v} \rangle \ \forall \mathbf{v} \in \mathbf{V}_h \quad (6)$$

$$(\nabla \cdot \mathbf{u}_h(t), q) = 0 \qquad \forall q \in Q_h \qquad (7)$$

$$<\mathbf{F}_h(t), \mathbf{v}> = \sum_{i=0}^{m-1} \kappa \left(\frac{\partial \mathbf{X}_{h\,i+1}}{\partial s}(t) - \frac{\partial \mathbf{X}_{h\,i}}{\partial s}(t) \right) \mathbf{v}(\mathbf{X}_{h\,i}(t)) \qquad \forall \mathbf{v} \in \mathbf{V}_h \qquad (8)$$

$$\frac{\partial \mathbf{X}_{h\,i}}{\partial t}(t) = \mathbf{u}_h(\mathbf{X}_{h\,i}(t), t) \qquad \forall i = 0, 1, \ldots, m \qquad (9)$$

$$\mathbf{u}_h(\mathbf{x}, 0) = \mathbf{u}_{0h}(\mathbf{x}) \qquad \forall \mathbf{x} \in \Omega \qquad (10)$$

$$\mathbf{X}_{h\,i}(0) = \mathbf{X}_0(s_i) \qquad \forall i = 1, \ldots, m. \qquad (11)$$

3 Time discretization by finite differences

In [9] it was shown how a fully implicit discretization in time for both the elasticity and the fluid equations appears to be unconditionally stable. We will not report numerical experiments on this approach (referred in the sequel as the Backward Euler/Backward Euler, or BE/BE scheme), we will however show that this approach is unconditionally stable.

A natural alternative to the fully implicit method is the use of a semi-implicit modification. We will refer to this time stepping technique, which couples the pressure and diffusion implicitly in a Stokes solve while treating the elastic terms explicitly, as the Forward Euler/Backward Euler (in short FE/BE) scheme, following the notations of [7]. We will show that this method is not unconditionally stable and we will give an appropriate CFL condition needed for it to remain stable.

Let Δt denote the time step and let us indicate by the superscript n an unknown function at time $t_n = n\Delta t$, so that the number of time steps needed to reach the final time T is N.

The two schemes can be formally described in a unified way:

Problem 2. Given $\mathbf{u}_{0h} \in \mathbf{V}_h$ and $\mathbf{X}_{0h} \in \mathbf{S}_h$, set $\mathbf{u}_h^0 = \mathbf{u}_{0h}$ and $\mathbf{X}_h^0 = \mathbf{X}_{0h}$, then for $n = 0, 1, \ldots, N-1$

Step 1. compute the source term

$$<\mathbf{F}_h^{n+1}, \mathbf{v}> = \sum_{i=0}^{m-1} \kappa \left(\frac{\partial \mathbf{Y}_{i+1}}{\partial s} - \frac{\partial \mathbf{Y}_i}{\partial s} \right) \mathbf{v}(\mathbf{Y}_i) \qquad \forall \mathbf{v} \in \mathbf{V}_h;$$

Step 2. find $(\mathbf{u}_h^{n+1}, p_h^{n+1}) \in \mathbf{V}_h \times Q_h$, such that

$$\rho \left(\frac{\mathbf{u}_h^{n+1} - \mathbf{u}_h^n}{\Delta t}, \mathbf{v} \right) + \mu(\nabla \mathbf{u}_h^{n+1}, \nabla \mathbf{v}) - (\nabla \cdot \mathbf{v}, p_h^{n+1})$$
$$= <\mathbf{F}_h^{n+1}, \mathbf{v}> \qquad \forall \mathbf{v} \in \mathbf{V}_h$$

$$(\nabla \cdot \mathbf{u}_h^{n+1}, q) = 0 \qquad \forall q \in Q_h$$

Step 3. find $\mathbf{X}_h^{n+1} \in \mathbf{S}_h$, such that

$$\frac{\mathbf{X}_{hi}^{n+1} - \mathbf{X}_{hi}^n}{\Delta t} = \mathbf{u}_h^{n+1}(\mathbf{Y}_i^n) \quad \forall i = 1, \ldots, m,$$

where $\mathbf{Y}$ is $\mathbf{X}_h^n$ in the FE/BE scheme and $\mathbf{X}_h^{n+1}$ in the BE/BE one.

The FE/BE scheme introduced in Problem 2 is computable, while the BE/BE scheme (here introduced only formally) requires the implementation of some sort of iterative scheme. We refer to [9] for the derivation of one such a scheme and we will only give some theoretical results about its unconditional stability.

4 Stability analysis by energy estimates

We prove here the unconditional stability of the fully implicit method and present the CFL conditions that need to be satisfied to preserve the stability of the semi-implicit numerical scheme.

In the following we will make extensive use of the total potential energy of the elastic material, which in our case is defined as

$$E[\mathbf{X}(t)] := \int_D \Psi(\mathbb{F}(s,t)) m Ds = \frac{\kappa}{2} \left\| \frac{\partial \mathbf{X}(t)}{\partial s} \right\|_{0,D}^2. \tag{12}$$

4.1 Stability of the continuous problem

The following stability estimate holds true for the solution of both the continuous and the space discretized problem:

Lemma 2. *For $t \in \,]0, T[$, let $\mathbf{u}_h(t) \in \mathbf{V}_h$, $p_h(t) \in Q_h$ and $\mathbf{X}_h(t) \in \mathbf{S}_h$ be a solution of Problem 1, then it holds:*

$$\frac{\rho}{2} \frac{d}{dt} \|\mathbf{u}_h(t)\|_{0,\Omega}^2 + \mu \|\boldsymbol{\nabla} \mathbf{u}_h(t)\|_{0,\Omega}^2 + \frac{\kappa}{2} \frac{d}{dt} \left\| \frac{\partial \mathbf{X}_h(t)}{\partial s} \right\|_{0,D}^2 = 0. \tag{13}$$

Using the same principles it is possible to provide some stability results also for the fully discretized case both in the BE/BE case,

Theorem 1. *Let $\mathbf{u}_h, \mathbf{X}_h$ be a solution of the BE/BE scheme in Problem 2. The following discrete energy inequality holds:*

$$\frac{\rho}{2\Delta t} \left(\|\mathbf{u}_h^{n+1}\|_{0,\Omega}^2 - \|\mathbf{u}_h^n\|_{0,\Omega}^2 \right) + \mu \|\boldsymbol{\nabla} \mathbf{u}_h^{n+1}\|_{0,\Omega}^2$$
$$+ \frac{\kappa}{2\Delta t} \left(\left\| \frac{\partial \mathbf{X}_h^{n+1}}{\partial s} \right\|_{0,D}^2 - \left\| \frac{\partial \mathbf{X}_h^n}{\partial s} \right\|_{0,D}^2 \right) \leq 0, \tag{14}$$

as well as in the FE/BE case:

Theorem 2. Let $\mathbf{u}_h^n, \mathbf{X}_h^n$ be a solution at time $t = n\Delta t$ of the FE/BE scheme of Problem 2 and let L^n be defined as

$$L^n := \max_{i=1,\ldots,m-1} |\mathbf{F}_i^n|, \tag{15}$$

then the following discrete energy inequality holds:

$$\frac{\rho}{2\Delta t}\left(\|\mathbf{u}_h^{n+1}\|_{0,\Omega}^2 - \|\mathbf{u}_h^n\|_{0,\Omega}^2\right) + \mu\|\nabla\mathbf{u}_h^{n+1}\|_{0,\Omega}^2$$
$$+\frac{\kappa}{2\Delta t}\left(\left\|\frac{\partial\mathbf{X}_h^{n+1}}{\partial s}\right\|_{0,D}^2 - \left\|\frac{\partial\mathbf{X}_h^n}{\partial s}\right\|_{0,D}^2\right) \leq \frac{C\kappa}{2}\frac{\Delta t}{h_x}L^n\|\nabla\mathbf{u}_h^{n+1}\|_{0,\Omega}^2. \tag{16}$$

Theorem 2 gives a quantitative estimate of the artificial energy introduced into the system by the FE/BE numerical discretization.

For the problem to remain stable, i.e. with bounded energy, it is evident that some care has to be taken on the choice of the time step size, the fluid mesh size and the immersed boundary mesh size. The following lemma summarizes the CFL condition needed in order to maintain the property of decreasing total energy:

Lemma 3. If there exists a positive K_0 such that for each $n = 0, \ldots, N-1$

$$\mu - \frac{C\kappa}{2}\frac{\Delta t}{h_x}L^n \geq K_0 > 0, \tag{17}$$

then the following discrete energy inequality holds:

$$\frac{\rho}{2}\|\mathbf{u}^n\|_{0,\Omega}^2 + \Delta t\sum_{k=1}^n K_0\|\nabla\mathbf{u}^k\|_{0,\Omega}^2 + \frac{\kappa}{2}\left\|\frac{\partial\mathbf{X}^n}{\partial s}\right\|_{0,D}^2$$
$$\leq \frac{\rho}{2}\|\mathbf{u}_0\|_{0,\Omega}^2 + \frac{\kappa}{2}\left\|\frac{\partial\mathbf{X}^0}{\partial s}\right\|_{0,D}^2. \tag{18}$$

Detailed proofs of Theorems 1, 2 and Lemma 3 can be found in [4].

5 Numerical results

To verify numerically the results stated in Theorem 2 and Lemma 3 we set up the extremely simple test problem of a balloon at rest inflated and immersed in the same fluid, which translate in our numerical framework in a circle with radius $R \leq .5$ immersed in the middle of the square domain $[0,1]^2$ (we used $R = .4$), with null initial velocity $\mathbf{u}$ and initial parametric representation given by

$$\mathbf{X}(s,t) = \begin{pmatrix} R\cos(s/R) + .5 \\ R\sin(s/R) + .5 \end{pmatrix} \quad s \in [0, 2\pi R]. \tag{19}$$

We are interested in showing the dependency of the stability on the CFL parameter given by

$$\eta^n = \frac{\kappa \Delta t}{h_x} L^n. \tag{20}$$

In Figure 1 we plotted the evolution of the normalized total energy of the system and of the η^n parameter during time for different values of κ and Δt. All the computations we performed show how the η^n parameter (20) is able to capture the instabilities as soon as they arise.

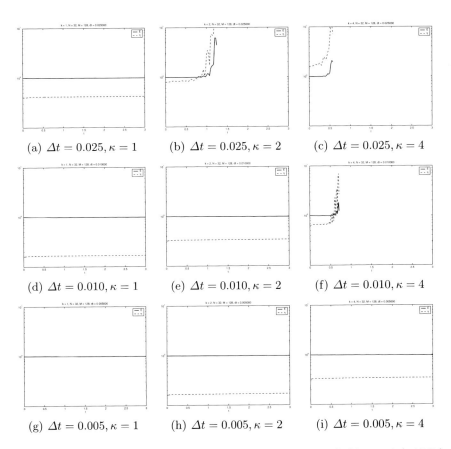

Fig. 1. Time versus normalized total energy and η, for $\kappa = 1$ (left), $\kappa = 2$ (middle) and $\kappa = 4$ (right) with $h_x = 1/32$ and $h_s = 1/128$.

It is evident that when the η^n parameter gets too close to a threshold, which here seems to be near .8, then the energy (which here is supposed to remain constant) explodes. The simulation in these cases stops without reaching the final time $t = 3$, because the immersed boundary starts oscillating too heavily and it ends outside the computational domain.

The program used to compute these examples has been written in $C++$ with the support of the *deal.II* libraries (see [1] for a technical reference).

6 Conclusions

We recalled the formulation of the finite element immersed boundary method as found in [2, 3]. Choosing the correct strategy for the approximation of the delta distribution has been one of the major challenges for the developers of the original IB method (see, for example, [6]). In this paper we presented a numerical stability analysis of the finite element IB method, which in particular does not depend on a regularization of the Dirac delta distribution. The unconditional stability for the fully implicit time stepping technique (referred to as the BE/BE scheme) and a CFL condition for the semi-implicit time stepping technique (FE/BE) were presented. Previous work in this direction was carried on in [8, 7], by analyzing the vibrational modes of immersed fibers and their influence on the time-stepping technique. Our approach follows a somewhat different path, by asking the numerical method to satisfy physical conditions like the conservation of the total energy of the system. The numerical experiments we performed show good agreement between the theoretical results and the instability that sometimes arise during the IB problem computations.

References

1. Bangerth, W., Hartmann, R., Kanschat, G.: deal.II Differential Equations Analysis Library, Technical Reference. http://www.dealii.org
2. Boffi, D., Gastaldi, L.: A finite element approach for the immersed boundary method. Comput. & Structures **81**(8-11), 491–501 (2003). In honor of Klaus-Jürgen Bathe
3. Boffi, D., Gastaldi, L., Heltai, L.: A finite element approach to the immersed boundary method. In: S. Saxe-Coburg Publications Stirling (ed.) Progress in Engineering Computational Technology, B.H.V. Topping and C.A. Mota Soares Eds., pp. 271–298 (2004)
4. Boffi, D., Gastaldi, L., Heltai, L.: Numerical stability of the finite element immersed boundary method. Mathematical Models and Methods in Applied Sciences (2005). Submitted
5. Brezzi, F., Fortin, M.: Mixed and hybrid finite element methods, *Springer Series in Computational Mathematics*, vol. 15. Springer-Verlag, New York (1991)
6. Peskin, C.S.: The immersed boundary method. In: Acta Numerica, 2002. Cambridge University Press (2002)
7. Stockie, J.M., Wetton, B.R.: Analysis of stiffness in the immersed boundary method and implications for time-stepping schemes. J. Comput. Phys. **154**(1), 41–64 (1999)
8. Stockie, J.M., Wetton, B.T.R.: Stability analysis for the immersed fiber problem. SIAM J. Appl. Math. **55**(6), 1577–1591 (1995)
9. Tu, C., Peskin, C.S.: Stability and instability in the computation of flows with moving immersed boundaries: a comparison of three methods. SIAM J. Sci. Stat. Comput. **13**(6), 1361–1376 (1992)

Heat Transfer and Combustion

A Comparison of Enthalpy and Temperature Methods for Melting Problems on Composite Domains

J.H. Brusche[1], A. Segal[1], C. Vuik[1] and H.P. Urbach[2]

[1] Department of Applied Mathematical analysis, Delft University of Technology, Mekelweg 4, 2628 CD Delft, The Netherlands,
    [j.h.brusche,a.segal,c.vuik]@ewi.tudelft.nl
[2] Philips Research Laboratories, Professor Holstlaan 4, 5656 AA Eindhoven, The Netherlands
    h.p.urbach@philips.com

Summary. In optical rewritable recording media, such as the Blu-ray Disc, amorphous marks are formed on a crystalline background of a phase-change layer, by means of short, high power laser pulses. It is of great importance to understand the mark formation process, in order to improve this data storage concept. The recording layer is part of a grooved multi-layered geometry, consisting of a variety of materials of which the material properties are assumed to be constant per layer, but may differ by various orders of magnitude in different layers. The melting stage of the mark formation process requires the inclusion of latent heat. In this study a comparison is made of numerical techniques for resolving the associated Stefan problem. The considered methods have been adapted to be applicable to multi-layers.

1 Introduction

In optical rewritable recording, a disk consists of various layers. The actual recording of data, stored as an array of amorphous regions in a crystalline background, takes place in a specific layer of the recording stack, containing a so-called phase-change material. The amorphous regions, called marks, are created as a result of very short high intensity pulses of a laser beam that is focused on this active layer. The light energy of the laser is transformed into heat, which locally causes the phase-change material to melt. As soon as the laser is switched off, the molten material solidifies. At the same time, recrystallization occurs in those regions where the temperature is below the melting temperature, but still above the recrystallization temperature. Since the cooling down is very rapid (quenching), almost no recrystallization occurs within the molten region, and thus a solid amorphous region is formed. The same laser beam, but at a lower power level, can be used in a similar way to fully recrystallize the amorphous regions. The recorded data is then erased.

Although much is understood about the concept of optical rewritable recording, many open questions remain. In order to gain better inside in for instance the influence of polarization and wavelength of the incident light or the geometry and composition of the stack on the shape and position of a mark, robust (numerical) modeling is essential. As a result, the occurrence of unwanted effects, such as cross-track cross-talk, could be minimized.

In this study we will focus on the melting phase of the mark formation process. In contrast to earlier work [1], this requires the contribution of latent heat to be taken into account in the computation of the temperature distribution in the optical recording disk. The mathematical model associated with the melting problem will be presented first. Two numerical techniques to resolve the mathematical problem will then be introduced. Emphasize is put on how these methods can be applied to multi-layered domains. Finally, a comparison with respect to accuracy, convergence behavior and computational demand is presented. For convenience, we will restrict ourselves to 1D and 2D test problems only.

2 Problem description

The melting of the phase-change material is a complex process. Material specific properties, such as the latent heat, greatly influence the melting behavior. Therefore, some assumptions are made with respect to several physical aspects of the melting process. First of all, it is assumed that the melting of the phase-change material occurs at a melting *point* T_m, rather than along a melting trajectory. In this way, the shape and size of mark are simply determined by the (sharp) free interface between the solid and liquid state of the (crystalline) phase-change material. Furthermore, the density ρ, latent heat L, heat capacity c, and conductivity κ are taken to be constant per phase. When needed, a subscript s or l is used to distinct between the solid and liquid state, respectively.

Because the position of the free interface evolves in time, depending on the heat distribution, the melting process is modeled as a free boundary problem. For an arbitrary bounded domain $\Omega \subset \mathbb{R}^n$ with fixed outer boundary $\delta\Omega$

Table 1. Performance with respect to Stefan number St, $\Delta t = 2000$, $n = 800$, $T_{\text{end}} = 20$ days

$\epsilon = 10^{-3}$			$St \approx 5 \times 10^{-4}$	$St \approx 5 \times 10^{-2}$	$St \approx 5 \times 10^{0}$	$St \approx 5 \times 10^{2}$
total #Newton iterations	Fachinotti	2322	2829	4957	6603
	Nedjar	78947	43045	14396	6271
total time (s)	Fachinotti	22.6	29.1	45.1	58.1
	Nedjar	355	195	132	32.3

and moving boundary $\Gamma(t)$, leading to two sub-domains Ω_s and Ω_l such that $\bar{\Omega} = \bar{\Omega}_s \cup \bar{\Omega}_l$ and $\Omega_s \cap \Omega_l = \emptyset$, the two-phase Stefan problem is given by:

$$\begin{cases} \rho c_{s,l} \dfrac{\partial T(\boldsymbol{x},t)}{\partial t} = \kappa_{s,l} \Delta T(\boldsymbol{x},t) + Q(\boldsymbol{x},t) & \forall \boldsymbol{x} \in \Omega_{s,l}, t > 0, \quad (1a) \\ \rho L v_n = \left[\kappa_{s,l} \dfrac{\partial T(\boldsymbol{x},t)}{\partial n} \right], \quad T(\boldsymbol{x},t) = T_m & \text{for } \boldsymbol{x} = \Gamma(t), t \geq 0, \quad (1b) \\ T(\boldsymbol{x},0) = \bar{T}_1(\boldsymbol{x}) & \forall \boldsymbol{x} \in \Omega_{s,l}, \quad (1c) \end{cases}$$

together with appropriate boundary conditions on the fixed outer boundary $\delta\Omega$. By $[\phi]$ we denote the jump in ϕ defined as:

$$[\phi] = \lim_{\substack{\boldsymbol{x} \to \Gamma(t) \\ \boldsymbol{x} \in \Omega_s(t)}} \phi(\boldsymbol{x},t) - \lim_{\substack{\boldsymbol{x} \to \Gamma(t) \\ \boldsymbol{x} \in \Omega_l(t)}} \phi(\boldsymbol{x},t). \quad (2)$$

At $t = 0$ the whole domain is taken to be solid. The spatial and time dependence of the temperature is omitted from this point onward.

Two fixed grid approaches to solve the Stefan problem given above are considered: the *enthalpy* formulation and the *temperature* formulation. The enthalpy $H(T)$ can be defined as:

$$H(T) = \begin{cases} \rho c_s (T - T_m), & T \leq T_m, \\ \rho c_l (T - T_m) + \rho L, & T > T_m. \end{cases} \quad (3)$$

In the enthalpy formulation the enthalpy H is treated as a second dependent variable besides the temperature T. Using relation (3), the heat conduction equation (1a) and the Stefan condition (1b) are replaced by the well-known enthalpy equation:

$$\dfrac{\partial H(T)}{\partial t} - \kappa_{s,l} \Delta T = Q. \quad (4)$$

In the temperature formulation, instead of separating the domain in a liquid and solid part, as via definition (3), the enthalpy is written according to its formal definition: as the sum of sensible and latent heat:

$$H(T) = H^{\text{sensible}} + H^{\text{latent}} = \rho c_{s,l}(T - T_m) + \rho L f_l(T), \quad (5)$$

where $f_l(T)$ denotes the liquid volume fraction, which in case of isothermal phase-change is equal to the Heavyside step function $\mathcal{H}(T - T_m)$. Definition (5) leads to the classical Fourier heat conduction equation, but with an additional term for the latent heat contribution:

$$\rho c_{s,l} \dfrac{\partial T}{\partial t} + \rho L \dfrac{\partial f_l}{\partial t} - \kappa_{s,l} \Delta T = Q. \quad (6)$$

3 Numerical methods

The resolution of the Stefan problem (1a-1c) by means of a numerical method is not trivial. On the one hand, the method should be applicable to multi-layered domains, with possibly large jumps in physical parameters. In addition, there are variations in the geometry (i.e., the grooved tracks), in three dimensional space. These demands also make that a finite element discretization is preferred. On the other hand, the method should allow for the breaking and merging of interfaces, as a result of for instance the inhomogeneity of the internal heating by the laser or the applied multi-pulse strategy. Furthermore, a future generalization of the method (e.g., non-isothermal melting; temperature dependent material properties) should be possible.

Two numerical approaches to resolve the Stefan problem (1a-1c) are described next. The first method is based on the work by Nedjar [4], in which a relaxed linearization of the temperature is used to solve the enthalpy equation (4). The governing equation for the second method is the temperature formulation (6). Key to this approach is the discontinuous integration across elements that are intersected by the free boundary, as proposed by Fachinotti et al. [3].

3.1 Relaxed linearization

Using standard Galerkin procedures and Euler backward discretization in time, a discretization of the enthalpy equation (4) can be written as follows:

$$M\frac{\mathbf{H}^{n+1} - \mathbf{H}^n}{\Delta t} + S\mathbf{T}^{n+1} = \mathbf{Q}^{n+1}, \tag{7}$$

where M and S denote the mass matrix and the stiffness matrix, respectively, and Δt is the time step.

In [4] it is described how to solve the above system of equations using a pseudo-Newton iterative procedure in terms of a temperature increment $\Delta \mathbf{T}^i$. In order to explain how this technique can be adapted to solve for the temperature distribution in a multi-layered domain, we will briefly repeat the three steps that form the key idea behind the proposed integration algorithm by Nedjar.

First, introduce the reciprocal function $\tau : \mathbb{R} \longrightarrow \mathbb{R}$ of (3), given by $\mathbf{T} = \tau(\mathbf{H})$ and define

$$\mathbf{H}^{i+1} = \mathbf{H}^i + \Delta \mathbf{H}^i, \tag{8}$$

$$\mathbf{T}^{i+1} = \mathbf{T}^i + \Delta \mathbf{T}^i. \tag{9}$$

Next, consider a linearization of the function $\tau(\mathbf{H})$:

$$\mathbf{T}^{i+1} = \mathbf{T}^i + \Delta \mathbf{T}^i = \tau(\mathbf{H}^i) + \tau'(\mathbf{H}^i)\Delta \mathbf{H}^i, \tag{10}$$

or, rewritten in terms of the enthalpy update $\Delta \mathbf{H}^i$:

$$\Delta \mathbf{H}^i = \frac{1}{\tau'(\mathbf{H}^i)} \left[\Delta \mathbf{T}^i + \left(\mathbf{T}^i - \tau(\mathbf{H}^i) \right) \right]. \tag{11}$$

Here, τ' denotes the derivative of τ with respect to its argument. Unfortunately, this derivative can be zero. This is resolved by approximating the fraction in equation (11) by a constant μ defined as:

$$\mu = \frac{1}{\max(\tau'(\mathbf{H}^i))}, \tag{12}$$

such that the relaxed enthalpy update becomes:

$$\Delta \mathbf{H}^i = \mu \left[\Delta \mathbf{T}^i + \left(\mathbf{T}^i - \tau(\mathbf{H}^i) \right) \right]. \tag{13}$$

If we now define

$$\widetilde{\mathbf{Q}} = \mathbf{Q}^{n+1} + \widetilde{M} \mathbf{H}^n, \qquad \widetilde{M} = \frac{1}{\Delta t} M, \tag{14}$$

then substitution of (8)-(10) and (13) into the discretized system (7) gives:

$$\widetilde{M} \left\{ \mathbf{H}^i + \mu \left[\Delta \mathbf{T}^i + \left(\mathbf{T}^i - \tau(\mathbf{H}^i) \right) \right] \right\} + S(\mathbf{T}^i + \Delta \mathbf{T}^i) = \widetilde{\mathbf{Q}}. \tag{15}$$

A rearrangement of terms finally leads to

$$\left(\mu \widetilde{M} + S \right) \Delta \mathbf{T}^i = \widetilde{\mathbf{Q}} - \left(\mu \widetilde{M} + S \right) \mathbf{T}^i - \widetilde{M} \left(\mathbf{H}^i - \mu \tau(\mathbf{H}^i) \right). \tag{16}$$

By rewriting the discretized heat conduction equation in terms of the temperature increment $\Delta \mathbf{T}^i$ for those layers of a recording stack that do not contain a phase change material, it is possible to build a system of equations for the multi-layer as a whole. This means, that any existing finite element code for heat conduction problems in composite domains, can easily be extended to include melting.

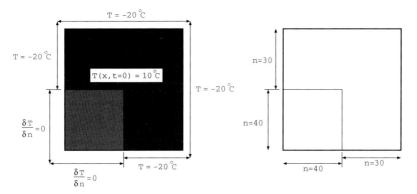

Fig. 1. Test case 2: initial and boundary conditions (left). Number of grid points used (right).

3.2 Discontinuous integration

A distinct feature of a temperature based model such as (6), is the use of discontinuous spatial integration. The key idea behind discontinuous integration, as for instance described by Fachinotti et al. [3] is that for elements intersected by the free interface, the integrals arising from a finite element discretization of the governing equation (6) are computed over the liquid and solid part separately. Because no regularization of the integrand is required, an accurate evaluation of the discrete balance equation is assured.

In the temperature based formulation (6), the term representing the latent heat contribution can be interpreted as an additional source for the classical heat conduction equation. Therefore, the discontinuous integration method can be easily included in any existing finite element code for multi-layered domains, to incorporate melting.

Remark that, in particular is case of isothermal phase-change, some care has to be taken when dealing with the Heavyside step function $\mathcal{H}(T - T_m)$. We refer to [3] for an elegant solution to this problem.

4 Numerical experiments

The application of the above methods to melting problems in a three dimensional grooved recording stack gives rise to large nonlinear systems. These systems are preferable solved using a (pseudo-) Newton-Raphson iterative procedure. For a given 3D melting problem, the choice of which numerical method is most suited, will rely heavily on computational efficiency and accuracy.

The two methods presented in the previous section have been applied to a 1D and a 2D test problem. A comparison is made of the performance of both methods with respect to number of iterations and computational time

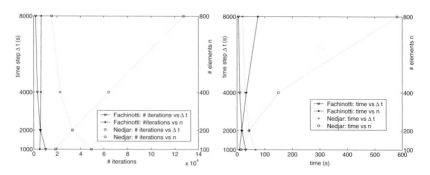

Fig. 2. Test case 1: Performance with respect to total time and total number of iterations. For varying Δt, a fixed number of elements $n = 200$ is used. For varying n, $\Delta t = 2000s$.

for varying time step size Δt and number of elements n. Since the numerical method should be applicable to a wide range of materials, performance is also compared for varying Stefan numbers $St = c_r(T_{\text{sat}} - T_m)/L$. Here, c_r is the relative conductivity and T_{sat} the saturation temperature.

4.1 Results

First, we compare the performance of the two methods for the melting of a single material in 1D. For a description of the problem, and the material parameters used, we refer to the test case 'unequal parameters', as described in [2]. In Figure 2 the required number of iterations and computational time are plotted for varying time step size Δt and number of elements n. Convergence of the (pseudo-) Newton-Raphson is said to be reached when the Euclidean norm of the residual vector becomes less than $\epsilon = 10^{-6}$. The method by Fachinotti clearly shows to be the computationally least demanding method, for the given level of accuracy. The method by Nedjar appears to be faster only for very small time steps on a coarse grid.

An interesting observation can be made from Table 1. For problems, where the melting front moves very rapidly, i.e. the Stefan number $St \approx O(10^2)$, relaxed linearization seems to be the method of choice. However, for the phase-change materials that are used in optical rewritable recording, a Stefan number of $O(10^{-3})$ is commonly found.

As a second test case, we consider the melting of a square region filled by a phase-change material, which on two sides is embedded by a non-melting material, see Figure 1. The material properties of the phase-change material are taken equal to those used in the first test case. For the embedding material, $\rho = 1$ kg/m^3, $C = 1 \times 10^6$ J/kg°C, $\kappa = 0.5$ W/m°C, $L = 2 \times 10^4$ J/kg and $T_m = 20$ °C. Table 2 shows that the computational demand of the Fachinotti method does not increase as rapidly for smaller values of the tolerance ϵ then that of the Nedjar method. This is caused by the damping factor μ in the pseudo Newton iteration process.

Figure 3 illustrates the difference in the captured interface position ($\epsilon = 10^{-3}$). Although the free interface is very smooth when using the temperature approach, Nedjar clearly shows wiggles.

Table 2. Computational load for 2D multi-layer test problem for varying tolerance ϵ: $\Delta t = 2000$, $T_{\text{end}} = 20$ days.

		$\epsilon = 10^{-2}$	$\epsilon = 10^{-4}$	$\epsilon = 10^{-6}$
total #Newton iterations	Fachinotti	1734	2599	3376
	Nedjar	10107	46020	88964
total time (s)	Fachinotti	116	166	208
	Nedjar	266	1093	2030

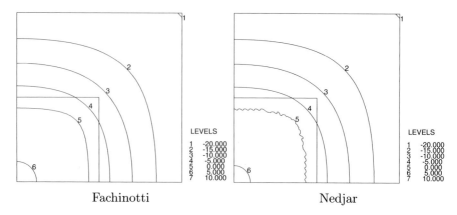

Fig. 3. Results for test case 2. Fachinotti: no wiggles; Nedjar: wiggles, $\Delta t = 2000$, $T_{end} = 20$ days.

5 Conclusions

Based on the results, the temperature based method seems to be the method of choice for small and medium range Stefan numbers. Not only can it easily be integrated into existing finite element codes for conduction problems in composite domains, it also outperforms the enthalpy based method with respect to accuracy, stability and computational load. For melting problems with Stefan number of $O(10^2)$ or larger, the enthalpy method might be the better choice.

References

1. Brusche, J.H., Segal, A., Urbach, H.P.: Finite-element model for phase-change recording. J. Opt. Soc. Am. A, **22**, 773–786 (2005)
2. Chun, C.K., Park, S.O.: A fixed-grid finite-difference method for phase-change problems. Numer. Heat Transfer, Part B, **38**, 59–73 (2000)
3. Fachinotti, V.D., Cardona, A., Huespe, A.E.: A fast convergent and accurate temperature model for phase-change heat conduction. Int. J. Numer. Meth. Engng., **44**, 1863–1884 (1999)
4. Nedjar, B.: An enthalpy-based finite element method for nonlinear heat problems involving phase change. Comput. Struct., **80**, 9–21 (2002)

Qualitative Properties of a Numerical Scheme for the Heat Equation

Liviu I. Ignat

Departamento de Matemáticas, Facultad de Ciencias,
Universidad Autónoma de Madrid, 28049 Madrid, Spain
liviu.ignat@uam.es

Summary. In this paper we consider a classical finite difference approximation of the heat equation. We study the long time behaviour of the solutions of the considered scheme and various questions related to the fundamental solutions. Finally we obtain the first term in the asymptotic expansion of the solutions.

1 Introduction

The main goal of this paper is the study of the long time behaviour of classical finite difference approximations of the heat equation.

Let us consider the linear heat equation on the whole space

$$\begin{cases} u_t - \Delta u = 0 \text{ in } \mathbf{R}^d \times (0, \infty), \\ u(0, x) = \varphi(x) \text{ in } \mathbf{R}^d. \end{cases}$$

By means of Fourier's transform, solutions can be represented as the convolutions between the fundamental solutions and the initial data:

$$u(t) = G(t, \cdot) * \varphi,$$

where

$$G(t,x) = \frac{1}{(4\pi t)^{-d/2}} mE^{-\frac{|x|^2}{4t}} = \frac{1}{(2\pi)^d}\int_{\mathbf{R}^d} mE^{ix\cdot\xi} mE^{-|\xi|^2 t} mD\xi.$$

The smoothing effect of the fundamental solutions $G(t, x)$ yields to the following behaviour of the solution (cf. [3], Ch. 3, p. 44):

$$\|u(t)\|_{L^p(\mathbf{R}^d)} \leq C(p,q)\, t^{-d/2\,(1/q-1/p)} \|\varphi\|_{L^q(\mathbf{R}^d)}, \quad t > 0, \ p \geq q. \tag{1}$$

A finer analysis is given in [4], where the authors consider initial data which decay polynomially at infinity. Duoandikoetxea & Zuazua [4] study how the mass of the solution is distributed as $t \to \infty$. They prove the existence of a positive constant $c = c(p, q, d)$ such that for any q and p satisfying $1 \leq q < d/(d-1)$, $d \geq 2$ ($1 \leq q < \infty$ for $d = 1$), $q \leq p < \infty$,

$$\left\| u(t, \cdot) - \left(\int_{\mathbf{R}^d} \varphi(x) mDx \right) G(t, \cdot) \right\|_{L^p(\mathbf{R}^d)} \leq c t^{-\frac{1}{2} - \frac{d}{2}(\frac{1}{q} - \frac{1}{p})} \| |x| \varphi \|_{L^q(\mathbf{R}^d)} \quad (2)$$

holds for all $t > 0$ and $\varphi \in L^1(\mathbf{R}^d)$ with $|x|\varphi(x) \in L^q(\mathbf{R}^d)$.

Let us consider the classical finite-difference scheme:

$$\begin{cases} \dfrac{du^h}{dt} = \Delta_h u^h, \ t > 0, \\ u^h(0) = \varphi^h. \end{cases} \quad (3)$$

Here u^h stands for the infinite unknown vector $\{u_{\mathbf{j}}^h\}_{\mathbf{j} \in \mathbf{Z}^d}$, $u_{\mathbf{j}}^h(t)$ being the approximation of the solution u at the node $x_{\mathbf{j}} = \mathbf{j}h$, and Δ_h is the classical second order finite difference approximation of Δ:

$$(\Delta_h u^h)_{\mathbf{j}} = \frac{1}{h^2} \sum_{k=1}^{d} (u_{\mathbf{j}+e_k}^h + u_{\mathbf{j}-e_k}^h - 2u_{\mathbf{j}}^h).$$

This scheme is widely used and satisfies the classical properties of consistency and stability which imply L^2-convergence (cf. [7], Ch. 13, p. 292).

It is interesting to know whenever the properties of the continuous problem are preserved by the numerical scheme. In the following we are concerned with the spatial shape of the discrete solution for large times. To do that we introduce the spaces $l^p(h\mathbf{Z}^d)$:

$$l^p(h\mathbf{Z}^d) = \left\{ \{u_{\mathbf{j}}\}_{\mathbf{j} \in \mathbf{Z}^d} : \|u\|_{l^p(h\mathbf{Z}^d)}^p = h^d \sum_{\mathbf{j} \in \mathbf{Z}^d} |u_{\mathbf{j}}|^p < \infty \right\}$$

and study the behaviour of $l^p(h\mathbf{Z}^d)$-norms of the solutions as $t \to \infty$.

The main tool in our analysis is the semi-discrete Fourier transform (SDFT):

$$\widehat{u}(\xi) = h^d \sum_{\mathbf{j} \in \mathbf{Z}^d} mE^{-\mathbf{ij} \cdot \xi h} u_{\mathbf{j}}, \ \xi \in \left[-\frac{\pi}{h}, \frac{\pi}{h} \right]^d$$

and its inverse

$$u_{\mathbf{j}} = \frac{1}{(2\pi)^d} \int_{[-\pi/h, \pi/h]^d} \widehat{u}(\xi) mE^{\mathbf{ij} \cdot \xi h} mD\xi, \ \mathbf{j} \in \mathbf{Z}^d.$$

We refer to [5] and [10] for a survey on this subject. By means of SDFT we compute the solutions of equation (3) in a similar way as in the continuous case, writing them as a convolution of a fundamental solution $K_t^{d,h}$ and the initial datum. This allows us to obtain decay rates of the solution in different $l^q - l^p$ norms analogous to (1). All the estimates are uniform with respect to the step size, h. This proves a kind of $l^q - l^p$ stability of our scheme:

Theorem 1. Let $1 \leq q \leq p \leq \infty$. Then there exists a positive constant $c(p,q,d)$ such that

$$\|u^h(t)\|_{l^p(h\mathbf{Z}^d)} \leq c(p,q,d)\, t^{-d/2\,(1/q-1/p)} \|\varphi^h\|_{l^q(h\mathbf{Z}^d)}$$

for all $t > 0$, uniformly in $h > 0$.

A similar approach in the case of the transport equation has been studied by Brenner and Thomée [2] and Trefethen [11]. They introduce a finite difference approximation and give conditions which guarantee the l^p-stability of the scheme.

Next we prove that the fundamental solutions $K_t^{d,h}$ of equation (3) are related to the modified Bessel function $I_\nu(x)$:

$$(K_t^{d,h})_{\mathbf{j}} = \left(\frac{\exp(-\frac{2t}{h^2})}{\pi h}\right)^d \prod_{k=1}^{d} I_{j_k}\left(\frac{2t}{h^2}\right), \quad \mathbf{j} = (j_1, j_2, \ldots, j_d) \in \mathbf{Z}^d. \quad (4)$$

This property proves the positivity and various properties regarding the monotonicity of the discrete kernel $K_t^{d,h}$.

Finally, we consider the weighted space $l^1(h\mathbf{Z}^d, |x|)$ and obtain the first term in the asymptotic expansion of the discrete solution. The weighted spaces $l^p(h\mathbf{Z}^d, |x|)$, $1 \leq p < \infty$ are defined as follows:

$$l^p(h\mathbf{Z}^d, |x|) = \left\{\{u_{\mathbf{j}}\}_{\mathbf{j} \in \mathbf{Z}^d} : \|u\|^p_{l^p(h\mathbf{Z}^d, |x|)} = h^d \sum_{\mathbf{j} \in \mathbf{Z}^d} |u_{\mathbf{j}}|^p |\mathbf{j}h|^p < \infty\right\}.$$

The following theorem gives us the first term of the asymptotic expansion of the solution u^h:

Theorem 2. Let $p \geq 1$. Then there exists a positive constant $c(p,d)$ such that

$$\left\|u^h(t) - \left(h \sum_{\mathbf{j} \in \mathbf{Z}^d} \varphi_{\mathbf{j}}^h\right) K_t^{d,h}\right\|_{l^p(h\mathbf{Z}^d)} \leq c(p,d)\, t^{-1/2-d/2\,(1-1/p)} \|\varphi^h\|_{l^1(h\mathbf{Z}^d, |x|)}$$

for all $\varphi^h \in l^1(h\mathbf{Z}^d, |x|)$ and $t > 0$, uniformly in $h > 0$.

This shows that for t large enough the solution behaves as the fundamental solution. In contrast with (2) our result is valid only for the initial data in the weighted space $l^1(h\mathbf{Z}, |x|)$. The extension of this result to general initial data, i.e. in $l^q(h\mathbf{Z}, |x|)$, $1 < q < p$, remains an open problem. In [6] we consider the first $k \geq 1$ terms of the asymptotic expansion of the discrete solution and obtain a similar result.

2 Proof of the results

By means of SDFT we obtain that $\hat{u}^h$ satisfies the following ODE:

$$\frac{d\widehat{u}^h}{dt}(t,\xi) = -\frac{4}{h^2}\sum_{k=1}^{d}\sin^2\left(\frac{\xi_k h}{2}\right)\widehat{u}^h(t,\xi), \quad t>0, \ \xi \in \left[-\frac{\pi}{h},\frac{\pi}{h}\right]^d.$$

In the Fourier space, the solution $\widehat{u}^h$ reads

$$\widehat{u}^h(t,\xi) = mE^{-tp_h(\xi)}\widehat{\varphi}^h(\xi), \quad \xi \in \left[-\frac{\pi}{h},\frac{\pi}{h}\right]^d,$$

where the function $p_h : [-\pi/h, \pi/h]^d \to \mathbf{R}$ is given by

$$p_h(\xi) = \frac{4}{h^2}\sum_{k=1}^{d}\sin^2\left(\frac{\xi_k h}{2}\right). \tag{5}$$

The solution of equation (3) is given by a discrete convolution between the fundamental solution $K_t^{d,h}$ and the initial datum:

$$u^h(t) = K_t^{d,h} * \varphi^h.$$

The inverse SDFT of the function $mE^{-tp_h(\xi)}$ gives us the following representation of the fundamental solution $K_t^{d,h}$:

$$(K_t^{d,h})_{\mathbf{j}} = \frac{1}{(2\pi)^d}\int_{[-\pi/h,\pi/h]^d} mE^{-tp_h(\xi)}mE^{i\mathbf{j}\cdot\xi h}mD\xi, \quad \mathbf{j}\in \mathbf{Z}^d.$$

We point out that for any $\mathbf{j} = (j_1, j_2, \ldots, j_d) \in \mathbf{Z}^d$ the kernel $K_t^{d,h}$ can be written as the product of one-dimensional kernels $K_t^{1,h}$:

$$(K_t^{d,h})_{\mathbf{j}} = \prod_{k=1}^{d}(K_t^{1,h})_{j_k}. \tag{6}$$

A simple change of variables in the explicit formula of $K_t^{1,h}$ relates it with the modified Bessel functions:

$$(K_t^h)_j = \frac{\exp(-\frac{2t}{h^2})}{\pi h}I_j\left(\frac{2t}{h^2}\right), \quad j\in \mathbf{Z}.$$

Separation of variables formula (6) proves (4). We recall that the modified Bessel's function $I_\nu(x)$ is positive for any positive x. Also for a fixed x, the map $\nu \to I_\nu(x)$ is even and decreasing on $[0,\infty)$ (cf. [8], Ch. II, p. 60). These properties prove that the kernel $K_t^{d,h}$ has the following properties:

Theorem 3. *Let $t>0$ and $h>0$. Then*
i) For any $\mathbf{j} = (j_1, j_2, \ldots, j_d) \in \mathbf{Z}^d$

$$(K_t^{d,h})_{\mathbf{j}} = \left(\frac{\exp(-\frac{2t}{h^2})}{\pi h}\right)^d \prod_{k=1}^{d} I_{j_k}\left(\frac{2t}{h^2}\right).$$

ii) For any $\mathbf{j} \in \mathbf{Z}^d$, the kernel $(K_t^{d,h})_{\mathbf{j}}$ is positive.

iii) The map $j \in \mathbf{Z} \mapsto (K_t^{1,h})_j$ is increasing for $j \leq 0$ and decreasing for $j \geq 0$.

iv) For any $\mathbf{a} = (a_1, a_2, \ldots, a_d) \in \mathbf{Z}^d$ and $\mathbf{b} = (b_1, b_2, \ldots, b_d) \in \mathbf{Z}^d$ satisfying

$$|a_1| \leq |b_1|, |a_2| \leq |b_2|, \ldots, |a_d| \leq |b_d|,$$

the following holds

$$(K_t^{d,h})_{\mathbf{b}} \leq (K_t^{d,h})_{\mathbf{a}}.$$

The long time behaviour of the kernel $K_t^{d,h}$ is similar to the one of its continuous counterpart.

Theorem 4. *Let $p \in [1, \infty]$. Then there exists a positive constant $c(p,d)$ such that*

$$\|K_t^{d,h}\|_{l^p(h\mathbf{Z}^d)} \leq c(p,d)\, t^{-d/2\,(1-1/p)} \quad (7)$$

holds for all positive times t, uniformly on $h > 0$.

Once Theorem 4 is proved, Young's inequality provides the decay rates of the solutions of equation (3) as stated in Theorem 1.

Proof (of Theorem 4). A scaling argument shows that $(K_t^{d,h})_{\mathbf{j}} = (K_{t/h^2}^{d,1})_{\mathbf{j}}$, reducing the proof to the case $h = 1$.

In the sequel we consider the band limited interpolator of the sequence $K_t^{d,1}$ (cf. [12], Ch. I, p. 13):

$$K_*^d(t,x) = \frac{1}{(2\pi)^d} \int_{[-\pi,\pi]^d} mE^{ix\cdot\xi} mE^{-tp_1(\xi)} mD\xi. \quad (8)$$

In [9] the authors prove the existence of a positive constant A such that for any function f with its Fourier transform supported in the cube $[-\pi, \pi]^d$ the following holds:

$$\sum_{\mathbf{j} \in \mathbf{Z}^d} |f(\mathbf{j})|^p \leq A^d \int_{\mathbf{R}^d} |f(x)|^p mDx, \; p \geq 1. \quad (9)$$

This reduces (7) to similar estimates on the $L^p(\mathbf{R}^d)$-norm of K_*^d. The interpolator K_*^d satisfies

$$\|D^\alpha K_*^d(t,\cdot)\|_{L^p(\mathbf{R})} \leq c(\alpha,p,d)\, t^{-|\alpha|/2 - d/2\,(1-1/p)} \quad (10)$$

for any multiindex $\alpha = (\alpha_1, \ldots, \alpha_d)$ and $1 \leq p \leq \infty$. Using (5) and (8), we reduce (10) to the one dimensional case. We consider the cases $p = 1$ and $p = \infty$. The general case, $1 < p < \infty$, follows by the Hölder inequality. The case $p = \infty$ easily follows by the rough estimate:

$$\|D^\alpha K_*^1(t,\cdot)\|_{L^\infty(\mathbf{R}^d)} \leq \frac{1}{2\pi} \int_{-\pi}^{\pi} |\xi|^\alpha \exp\left(-4t \sin^2 \frac{\xi}{2}\right) mD\xi \leq c(\alpha)\, t^{-(\alpha+1)/2}.$$

Finally, we apply Carlson-Beurling's inequality (cf. [1] and [2]):
$$\|\hat{a}\|_{L^1(\mathbf{R})} \le (2\|a\|_{L^2(\mathbf{R})}\|a'\|_{L^2(\mathbf{R})})^{1/2}$$
to the function $a(\xi) = |\xi|^\alpha \exp(-4t\sin^2 \xi/2)$. We obtain the existence of a positive constant C such that for all $t > 0$,
$$\|K_*^1(t)\|_{L^1(\mathbf{R})} \le C.$$
This proves Theorem 4. ∎

Now we sketch the proof of Theorem 2.

Proof (of Theorem 2). First, a scaling argument reduces the proof to the case $h = 1$. We consider the cases $p = 1$ and $p = \infty$, the other cases follow by interpolation. The solution $u^1(t)$ of equation (3) is given by:
$$u_\mathbf{j}^1(t) = (K_t^{d,1} * \varphi^1)_\mathbf{j} = \sum_{\mathbf{n} \in \mathbf{Z}^d} (K_t^{d,1})_{\mathbf{j}-\mathbf{n}} \varphi_\mathbf{n}^1.$$

Let us introduce the sequence $\{a_\mathbf{j}(t)\}_{\mathbf{j} \in \mathbf{Z}^d}$ as follows
$$a_\mathbf{j}(t) = \left(u^1(t) - K_t^{d,1} \sum_{\mathbf{n} \in \mathbf{Z}^d} \varphi_\mathbf{n}^1\right)_\mathbf{j} = u_\mathbf{j}^1(t) - (K_t^{d,1})_\mathbf{j} \sum_{\mathbf{n} \in \mathbf{Z}^d} \varphi_\mathbf{n}^1$$
$$= \sum_{\mathbf{n} \in \mathbf{Z}^d} (K_t^{d,1})_{\mathbf{j}-\mathbf{n}} \varphi_\mathbf{n}^1 - (K_t^{d,1})_\mathbf{j} \sum_{\mathbf{n} \in \mathbf{Z}^d} \varphi_\mathbf{n}^1$$
$$= \sum_{\mathbf{n} \in \mathbf{Z}^d} \varphi_\mathbf{n}^1 \left((K_t^{d,1})_{\mathbf{j}-\mathbf{n}} - (K_t^{d,1})_\mathbf{j}\right).$$

In the sequel we denote by c a constant that may change from one line to another. It remains to prove that
$$\sup_{\mathbf{j} \in \mathbf{Z}^d} |a_\mathbf{j}(t)| \le c t^{-(d+1)/2} \|\varphi^1\|_{l^1(\mathbf{Z}^d, |x|)} \tag{11}$$
and
$$\sum_{\mathbf{j} \in \mathbf{Z}^d} |a_\mathbf{j}(t)| \le c t^{-1/2} \|\varphi^1\|_{l^1(\mathbf{Z}^d, |x|)}.$$

The Taylor formula applied to the function K_*^d gives us
$$K_*^d(t, \mathbf{j} - \mathbf{n}) - K_*^d(t, \mathbf{j}) = \int_0^1 \sum_{|\alpha|=1} D^\alpha K_*^d(t, \mathbf{j} - s\mathbf{n})(-\mathbf{n})^\alpha m D s.$$

As a consequence, for any $\mathbf{j} \in \mathbf{Z}^d$ the sequence $a_\mathbf{j}(t)$ satisfies
$$|a_\mathbf{j}(t)| \le \sum_{\mathbf{n} \in \mathbf{Z}^d} |\varphi_\mathbf{n}^1| \sum_{|\alpha|=1} \int_0^1 |\mathbf{n}^\alpha| |D^\alpha K_*^d(t, \mathbf{j} - s\mathbf{n})| m D s$$

$$\leq c \sum_{\mathbf{n}\in\mathbf{Z}^d} |\varphi_\mathbf{n}^1||\mathbf{n}| \sum_{|\alpha|=1} \int_0^1 |D^\alpha K_*^d(t,\mathbf{j}-s\mathbf{n})| mDs$$

$$= c \sum_{\mathbf{n}\in\mathbf{Z}^d} |\varphi_\mathbf{n}^1||\mathbf{n}| \sum_{|\alpha|=1} b_{\mathbf{j},\mathbf{n}}^\alpha(t). \tag{12}$$

To prove inequality (11), which corresponds to $p=\infty$, it is sufficient to show that

$$b_{\mathbf{j},\mathbf{n}}^\alpha(t) \leq c t^{-(d+1)/2}$$

for all indices α with $|\alpha|=1$. Inequality (10) shows that

$$b_{\mathbf{j},\mathbf{n}}^\alpha(t) \leq \|D^\alpha K_*^d(t)\|_{L^\infty(\mathbf{R}^d)} \leq c t^{-|\alpha|/2-d/2} = c t^{-(d+1)/2}.$$

Now let us consider the case $p=1$. We sum on $\mathbf{j}\in\mathbf{Z}^d$ in inequality (12) and obtain:

$$\sum_{\mathbf{j}\in\mathbf{Z}^d} |a_\mathbf{j}(t)| \leq \sum_{\mathbf{j}\in\mathbf{Z}^d} \sum_{\mathbf{n}\in\mathbf{Z}^d} |\varphi_\mathbf{n}^1||\mathbf{n}| \sum_{|\alpha|=1} b_{\mathbf{j},\mathbf{n}}^\alpha(t)$$

$$= \sum_{\mathbf{n}\in\mathbf{Z}^d} |\varphi_\mathbf{n}^1||\mathbf{n}| \sum_{|\alpha|=1} \sum_{\mathbf{j}\in\mathbf{Z}^d} b_{\mathbf{j},\mathbf{n}}^\alpha(t).$$

It remains to prove that

$$\sum_{\mathbf{j}\in\mathbf{Z}^d} b_{\mathbf{j},\mathbf{n}}^\alpha(t) \leq c t^{-1/2} \tag{13}$$

for all $\mathbf{n}\in\mathbf{Z}^d$ and for any multiindex α with $|\alpha|=1$. Using the separation of variables, we get for all $\mathbf{j}=(j_1,\ldots,j_d)\in\mathbf{Z}^d$ and $\mathbf{n}=(n_1,\ldots,n_d)\in\mathbf{Z}^d$,

$$b_{\mathbf{j},\mathbf{n}}^\alpha(t) = \int_0^1 \prod_{k=1}^d |D^\alpha K_*^1(t,j_k-sn_k)| mDs$$

and hence,

$$\sum_{\mathbf{j}\in\mathbf{Z}^d} b_{\mathbf{j},\mathbf{n}}^\alpha(t) = \int_0^1 \prod_{k=1}^d \left(\sum_{j_k\in\mathbf{Z}} |D^{\alpha_k} K_*^1(t,j_k-sn_k)|\right) mDs$$

$$\leq \sup_{s\in\mathbf{R}} \prod_{k=1}^d \left(\sum_{j_k\in\mathbf{Z}} |D^{\alpha_k} K_*^1(t,j_k-s)|\right).$$

We prove that each term in the last product is dominated by $t^{-\alpha_k/2}$ and consequently the product will be bounded by $t^{-|\alpha|/2}$. Applying (9) to the function $K_*^1(t,\cdot-s)$, each of the above sums satisfies

$$\sum_{j_k\in\mathbf{Z}} |D^{\alpha_k} K_*^1(t,j_k-s)| \leq c \int_\mathbf{R} |D^{\alpha_k} K_*^1(t,x-s)| mDx$$

$$= c \int_{\mathbf{R}} |D^{\alpha_k} K_*^{1,1}(t,x)| m Dx \leq c t^{-|\alpha_k|/2}.$$

This proves inequality (13) and finishes the proof of Theorem 2. ∎

Acknowledgements

The author wishes to thank the guidance of his Ph.D advisor Enrique Zuazua.
    This work has been supported by the doctoral fellowship AP2003-2299 of MEC (Spain) and the grants MTM2005-00714 of the MEC (Spain), 80/2005 of CNCSIS (Romania), "Smart Systems" of the European Union.

References

1. Beurling, A.: Sur les intégrales de Fourier absolument convergentes et leur application à une transformation fonctionnelle. In 9. Congr. des Math. Scand., 345–366 (1939)
2. Brenner, P., Thomée, V.: Stability and convergence rates in L_p for certain difference schemes. Math. Scand., **27**, 5–23 (1970)
3. Cazenave, T., Haraux, A.: An introduction to semilinear evolution equations. Oxford Lecture Series in Mathematics and its Applications. 13. Oxford: Clarendon Press. xiv, (1998)
4. Duoandikoetxea, J., Zuazua, E.: Moments, masses de Dirac et décomposition de fonctions. (Moments, Dirac deltas and expansion of functions). C. R. Acad. Sci. Paris Sér. I Math., **315**, (6), 693–698 (1992)
5. Henrici, P.: Applied and computational complex analysis. Volume III: Discrete Fourier analysis, Cauchy integrals, construction of conformal maps, univalent functions. Wiley Classics Library. New York, Wiley, (1993)
6. Ignat, L.I.: Ph.D. thesis. Universidad Autónoma de Madrid. In preparation
7. Iserles, A.: A first course in the numerical analysis of differential equations. Cambridge Texts in Applied Mathematics. Cambridge Univ. Press, (1995)
8. Olver, F.W.J.: Introduction to asymptotics and special functions. Academic Press, (1974)
9. Plancherel, M., Pólya, G.: Fonctions entières et intégrales de Fourier multiples. II. Comment. Math. Helv., **10**, 110–163 (1937)
10. Trefethen, L.N.: Finite Difference and Spectral Methods for Ordinary and Partial Differential Equations. http://web.comlab.ox.ac.uk/oucl/work/nick.trefethen /pdetext.html.
11. Trefethen, L.N.: On l^p-instability and oscillation at discontinuities in finite difference schemes. Advances in Computer Methods for Partial Differential Equations, V, 329–331 (1984)
12. Vichnevetsky, R., Bowles, J.B.: Fourier analysis of numerical approximations of hyperbolic equations. SIAM Studies in Applied Mathematics, 5. SIAM, (1982)

Modeling Radiation and Moisture Content in Fire Spread

L. Ferragut[1], M.I. Asensio[1] and S. Monedero[1]

Departamento de Matemática Aplicada, Universidad de Salamanca,
pza. Merced s.n. 37008 Salamanca, Spain
ferragut@usal.es, mas@usal.es, smonedero@usal.es

Summary. A numerical method for a 2-dimensional surface fire model taking into account moisture content and radiation is developed. We consider the combustion of a porous solid, where a simplified energy conservation equation is applied. The effects of the moisture content and the endothermic pyrolysis of the vegetation are introduced in the model by means of a multivalued function representing the enthalpy. Its resolution is based on the Yosida approximation of a perturbation of this operator. The radiation term allows us to cope with wind and slope effects. In order to avoid heavy time consuming computations, this term is approximated using the characteristic method, combined with a discrete ordinate method. Finally, the approximate solution of the energy equation in the porous solid is obtained using a finite element method together with a semi-implicit Euler algorithm in time.

1 Introduction

Many existing physical models for fire spread in porous fuel bed use the principle of energy conservation applied to the preheated fuel. Generally, radiation is considered as the dominant mechanism of the fuel preheating [3]. Moreover, slope and wind effects as well as the initial vegetation moisture have to be taken into account in order to obtain reliable rates of fire spread. Physical models from fundamental conservation equations and complex physics have been developed [10]. These valuable approaches are computationally expensive and too slow to be used in real time mode, even with fast and parallel processing. Besides, several works have appeared recently where one or two dimensional physical models are considered in order to simulate fire spread in small computers, with moderate simulation times, see for example [4, 1, 7]. This paper is a contribution to generally applicable models of fire spread through fuel beds, by means of simple models, but taking into account nonlocal radiation and moisture content. The radiation model allows to cope with wind and slope effects. Particularly the influence of the moisture content and eventually heat absorption by pyrolysis, can be represented as two free boundaries, and is treated in this paper using a multivalued operator representing the enthalpy. The maximal monotone property of this operator allows the implementation of a numerical algorithm with well-known convergence properties. The main contribution of this paper is the use of a multivalued operator to model moisture content and the numerical solution of the radiation equation combining characteristics, finite elements and discrete ordinate method, avoiding the discretization of a convolution operator which is computationally costly. This simple model could represent fire spread in thin fuel layers, typical in laboratory experiments, where pyrolysis of the solid fuel gives rise to a gas fuel which burns above the layer. The radiation from the flame is the heat source, which first dries the fuel and later produces pyrolysis of the solid fuel in the surface neighborhood. The model considered is too simple to simulate propagation of fires in general situations, nevertheless the mathematical modeling and numerical methods used in this work to model moisture and non local radiation can be applied to more complex and realistic models. This model is a variant of the models in [3](chapter one), model I in [7] or model in [11] where we have introduced the influence of the moisture content and the heat absorption by pyrolysis by using the enthalpy multivalued operator, and a new method to compute the non local radiation term.

2 Physical model

Let $Q = [0, lx] \times [0, ly] \subset \text{Re}^2$ a rectangle and S be a surface defined by the mapping

$$S: Q \longmapsto \text{Re}^3$$
$$x, y \longmapsto (x, y, h(x, y))$$

representing the part of the terrain where the propagation of a fire can take place (Fig. 1). We will assume that vegetation can be represented by a given fuel load together with a moisture content defined over Q. Besides we will assume that the height of the flames in a particular fire are known and bounded by H. In order to take into account some three dimensional effects, and particularly the radiation from the flames above the surface S, we will consider the following three dimensional domain

$$D = \{(x, y, z) \in \text{Re}^3 : \quad (x, y) \in Q, \quad h(x, y) < z < h(x, y) + H\}.$$

In the following sections we develop a model for fire propagation considering the energy and mass conservation equations in the surface S, and the radiation equation in D.

3 Governing equations

3.1 Energy equations

As the front of pyrolysis and the front of drying are assumed to be sharp we have neglected heat conduction in the vegetation. Energy conservation is described by the equations:

$$\partial_t e + \alpha u = r \quad \text{in } S \quad t \in (0, t_{max}), \tag{1}$$
$$e \in G(u) \quad \text{in } S \quad t \in (0, t_{max}). \tag{2}$$

The initial condition is given by $u(\mathbf{x}, 0) = u_0(\mathbf{x})$, $\mathbf{x} \in S$. The unknowns e and u are the non-dimensional enthalpy and the non-dimensional temperature. The non-dimensional enthalpy e is an element of a multivalued maximal monotone operator G, given by

$$G(u) = \begin{cases} u & if \quad u < u_v \\ [u_v, u_v + \lambda_v] & if \quad u = u_v \\ u + \lambda_v & if \quad u_v < u < u_p \\ [u_p + \lambda_v, u_p + \lambda_v + \lambda_p] & if \quad u = u_p \\ u + \lambda_v + \lambda_p & if \quad u > u_p \end{cases}$$

where u_v and u_p, are the non-dimensional evaporation temperature of the water and the non-dimensional pyrolysis temperature of the solid fuel, respectively. The quantities λ_v and λ_p are the non-dimensional evaporation heat and pyrolysis heat respectively.

It should be noticed that in the burnt zone the multivalued operator does not exactly represent the physical phenomena as the water vapor is no longer in the porous medium. This drawback can be circumvented setting $\lambda_v = 0$ and $\lambda_p = 0$ in the burnt area. The term αu represents the energy lost by convection in the vertical direction.

3.2 Fuel equation

The mass fraction of solid fuel y, is given by

$$\partial_t y = -g(u)y \quad \text{in } S \quad t \in (0, t_{max}), \tag{3}$$
$$y(x, 0) = y_0(\mathbf{x}) \quad \mathbf{x} \in S. \tag{4}$$

Equation (3) represents the fuel mass variation due to pyrolysis and (4) is the corresponding initial condition. g is given by the Arrhenius law

$$g(u) = (u > u_p)(y > y_e)\beta exp(-\gamma/(1 + u)),$$

where y_e is the mass fraction lower bound of extinction and the logical expressions are equal to 1 if the expression is true and 0 if the expression is false. γ is related to the activation energy.

3.3 Radiation

The right hand side of equation (1) describes the thermal radiation reaching the surface S from the flame above the layer. The intensity is defined as the radiation energy passing through an area per unit time, per unit of projected area and per unit of solid angle. The projected area is formed by taking the area that the energy is passing through and projecting it normal to the direction of travel. The unit elemental solid angle is centered about the direction of travel and has its origin at the area element.

After adimensionalization, the radiation equations in the direction Ω can be written as

$$\Omega.\nabla i + ai = \delta(1+u_g)^4 \quad \text{in } D, \tag{5}$$
$$i = 0 \quad \text{on } \partial D \cap \{\mathbf{x}; \, \Omega.\mathbf{n} < 0\}, \tag{6}$$

where i, a and u_g are the non dimensional radiation intensity, absorbtion coefficient and flame temperature respectively. In a first approximation we have considered a gray body and neglected the scattering. The right hand side represents the total emissive power of a blackbody. The incident energy at a point $\mathbf{x}(x,y,h(x,y))$ of the surface S due to radiation from the flame above the surface per unit time and per unit area will be obtained summing up the contribution of all directions Ω, that is

$$r(\mathbf{x}) = \int_{\omega=0}^{2\pi} i(\mathbf{x},\Omega)\Omega.\mathbf{n}\, d\omega, \tag{7}$$

where we have only considered the hemisphere above the fuel layer.

4 Numerical method

4.1 Time integration

Let $\Delta t = t^{n+1} - t^n$ a time step and let y^n, e^n and u^n denote approximations at time step t^n, to the exact solution y, e and u respectively.

We consider a semi-implicit scheme. At each time step we solve,

$$\frac{e^{n+1}-e^n}{\Delta t} + \alpha u^{n+1} = r^n, \tag{8}$$

$$e^{n+1} \in G(u^{n+1}), \tag{9}$$

$$\frac{y^{n+1}-y^n}{\Delta t} = -y^{n+1}g(u^{n+1}). \tag{10}$$

The basic idea is to treat implicitly the positive terms. The non local radiation term r, depends strongly on the temperature u and on the fuel mass y, therefore, it will be evaluated explicitly at time t^n and its computation is explained in subsection 4.3. Once the radiation r^n is given, problem (8-10) is non linear due to the multivalued operator G. However, the solution of this problem can be reduced to explicit calculations as it is explained in the next subsection.

4.2 Solution at each time step

The multivalued operator in (9) is maximal monotone, then its resolvent $J_\lambda = (Id + \lambda G)^{-1}$ for any $\lambda > 0$ is a well defined univalued operator. Moreover the Yosida approximation of G, $G_\lambda = \frac{Id - J_\lambda}{\lambda}$ is a Lipschitz operator and the inclusion (9) is equivalent for all $\lambda > 0$ to the equation

$$e^{n+1} = G_\lambda(u^{n+1} + \lambda e^{n+1}) \quad \text{or} \quad u^{n+1} = J_\lambda(u^{n+1} + \lambda e^{n+1}), \tag{11}$$

on the other hand, rearranging (8) we have

$$u^{n+1} + \frac{1}{\alpha \Delta t} e^{n+1} = \frac{1}{\alpha \Delta t} e^n + \frac{1}{\alpha} r^n, \tag{12}$$

taking $\lambda = 1/(\alpha \Delta t)$ by substitution into (11) we obtain

$$u^{n+1} = J_{1/\alpha \Delta t}(\frac{1}{\alpha \Delta t} e^n + \frac{1}{\alpha} r^n). \tag{13}$$

For a given $b = \frac{1}{\alpha \Delta t} e^n + \frac{1}{\alpha} r^n$, the solution of $s = J_{1/\alpha \Delta t}(b)$, is equivalent to solving

$$(\alpha \Delta t\, Id + G)s \ni \bar{b} = \alpha \Delta t\, b. \tag{14}$$

Once u^{n+1} has been obtained, we calculate e^{n+1} and y^{n+1} explicitly

$$e^{n+1} = e^n - \alpha \Delta t u^{n+1} + \Delta t r^n, \tag{15}$$

$$y^{n+1} = \frac{y^n}{1 + \Delta t g(u^{n+1})}. \tag{16}$$

4.3 Numerical solution of the radiation equation

The radiation term r in the energy equation (1) is computed by numerical integration of (7). More precisely, at each point $(x, y, h(x, y))$ on the surface S we consider the tangent plane, its corresponding unit tangent vectors and unit normal.

$$\tau_x = \frac{(1, 0, \frac{\partial h}{\partial x})^t}{\sqrt{1 + (\frac{\partial h}{\partial x})^2}}, \quad \tau_y = \frac{(0, 1, \frac{\partial h}{\partial y})^t}{\sqrt{1 + (\frac{\partial h}{\partial y})^2}}, \quad \mathbf{n} = \frac{(-\frac{\partial h}{\partial x}, -\frac{\partial h}{\partial y}, 1)^t}{\sqrt{1 + (\frac{\partial h}{\partial x})^2 + (\frac{\partial h}{\partial y})^2}}.$$

In the corresponding axes the directions Ω can be expressed as

$$\Omega = (1 - \mu^2)^{1/2} \frac{\gamma \sqrt{1 - \varsigma^2} - \varsigma \sqrt{1 - \gamma^2}}{\sqrt{1 - \varsigma^2}} \tau_x + (1 - \mu^2)^{1/2} \frac{\sqrt{1 - \gamma^2}}{\sqrt{1 - \varsigma^2}} \tau_y + \mu \mathbf{n}, \tag{17}$$

where $\mu = \cos \theta$, $\gamma = \cos \phi$, and $\varsigma = \cos \alpha = \tau_x \cdot \tau_y$, α is the angle between the two tangent vectors and (θ, ϕ) with $0 \leq \theta \leq \pi/2$ being the angle of Ω with $\mathbf{n}$

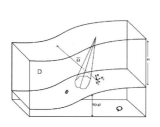
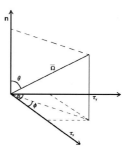

Fig. 1. Left:Fire Domain   Right: Tangent Space

(polar angle) and $0 \leq \phi \leq 2\pi$ the angle of the projection of Ω on the tangent plane (azimuthal angle), with τ_x (Fig. 1).

To compute the incident radiation in a certain direction we consider the cartesian coordinates $\Omega = (\Omega_1, \Omega_2, \Omega_3)$ at a point on the surface S $(\bar{x}, \bar{y}, \bar{z})$, with $\bar{z} = h(\bar{x}, \bar{y})$ and consider the characteristic line

$$\xi \longrightarrow (x(\xi) = \bar{x} + \xi\Omega_1,\ y(\xi) = \bar{y} + \xi\Omega_2,\ z(\xi) = \bar{z} + \xi\Omega_3)$$

On the characteristic, the equation (5) becomes

$$\frac{di}{d\xi} + ai = \delta(1 + u_g)^4 \quad \text{together with} \quad \lim_{\tau \to \infty} i(\xi) = 0. \tag{18}$$

Finally summing up for all the solid angles

$$r(\bar{\mathbf{x}}) = \int_{\theta=0}^{\theta=\pi/2} \int_{\phi=0}^{\phi=2\pi} i(\bar{\mathbf{x}}, \theta, \phi) \cos\theta \sin\theta \, d\theta d\phi =$$

$$\int_{\mu=0}^{\mu=1} \int_{\gamma=-1}^{\gamma=1} \frac{i_+(\bar{\mathbf{x}}, \mu, \gamma)\mu}{\sqrt{1-\gamma^2}} d\mu d\gamma + \int_{\mu=0}^{\mu=1} \int_{\gamma=-1}^{\gamma=1} \frac{i_-(\bar{\mathbf{x}}, \mu, \gamma)\mu}{\sqrt{1-\gamma^2}} d\mu d\gamma \tag{19}$$

where i_+ (resp. i_-) stands for the radiation intensity i corresponding to an angle ϕ such that $0 \leq \phi < \pi$ (resp. $\pi \leq \phi < 2\pi$). The integrals in (19) are computed using Gauss-Legendre quadrature with respect to μ and Gauss-Chebyshev quadrature with respect γ in order to cope with the singular weight $\frac{1}{\sqrt{1-\gamma^2}}$. That is

$$r(\bar{\mathbf{x}}) \approx \sum_{k,l} W_{kl}\, i_+(\bar{\mathbf{x}}, \mu_k, \gamma_l)\mu_k + \sum_{k,l} W_{kl}\, i_-(\bar{\mathbf{x}}, \mu_k, \gamma_l)\mu_k. \tag{20}$$

Problem (18) is solved by an Euler implicit method or Crank-Nicolson method with variable step size. To do so, we need to evaluate de gas temperature u_g in the domain D based on the temperature over the surface S given by (1),(2) and (3). We compute this extended field assuming a convective transport due to the wind, that is,

$$\widetilde{u}(x,y,z) = u(x - (z - h(x,y))\frac{v_x}{v_z}, y - (z - h(x,y))\frac{v_y}{v_z}, h(x,y)).$$

Where (v_x, v_y, v_z) stands for the wind velocity field which we suppose to be known. Otherwise a three-dimensional velocity field can be computed involving only two-dimensional computations using for example the model in [8].

5 Numerical results

5.1 Adjustment of the radiation model parameters

First of all, we compare the results of the radiative model with the experimental results in [5, 6]. In order to match the experimental setup, we consider a linear fire front and three different tilt angles for the flames: $\pi/2$, $\pi/4$ and $5\pi/12$ (with respect to the normal and in the direction perpendicular to the fire front). A semi-genetic algorithm is used to search for the flame temperature, absortion coefficient, flame height and fire front width that fit best the experimental results.

Both, the parameters found and the radiation profile generated by the model resemble quite well the physical parameters and the radiation profile of the experimental data, all with a low computational cost. The left side of Fig. 2 shows the radiation profile perpendicular to the fire front (the flame is tilted towards the right) together with the experimental results.

5.2 Propagation of fire on a striped surface

We consider a striped surface given by $h(x,y) = 0.2max(0, \cos(3\pi(x - 1)))$ resembling a sequence of small parallel hills. The numerical calculations correspond to a square fuel bed of $2 \times 2\ m^2$ composed with Pinus Pinaster with

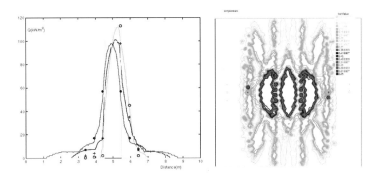

Fig. 2. LEFT: Radiation profile (continuous lines). Experimental measurements o,+,* RIGHT:Temperature contours over a striped surface

a fuel load of 1 kg/m^2 and no wind profile. Fire is ignited at the center of the domain (Fig. 2), it spreads out over the top of the surface and ignites the hills on its sides through the radiative term before spreading downhill.

Acknowledgement

Supported by the research projects: CGL2004-06171-c03-03/cl1 from the Education and science Ministry (Spain) and FEDER funds (European Union). SA078A05 from the Castilla y Leon government.

References

1. Balbi, J.H., Morandini F., Santoni, P.A., Simenoni, A.: Two dimensional fire spread model including long-range radiation and simplied flow. Int. Forest Fire Research and Wildland Fire Safety, Luso, Coimbra, Viegas(eds). Millpress: Rotterdam (2002)
2. Bermúdez, A., Moreno, C.: Duality methods for solving variational inequalities. Comp. and Math. Appl. **7**, 43-58 (1981)
3. Cox, G.: Combustion Fundamentals of Fire. Academic Press, London, (1995)
4. Dupuy, J.L., Larini, M.: Fire spread through a porous forest fuel bed: a radiation and convective model including fire-induced flow effects. Int. J. of Wildland Fire, **9** 155-172 (1999)
5. Ventura, J., Mendes-Lopes, J., Ripado, L.: Temperature-Time curves in fire propagating in beds of pine needles. III International Confer. on Forest Fires Research
6. Ventura, J., Mendes-Lopes, J., Amaral, J.: Rate of spread and flame characteristics in a bed of pine needles. III International Confer. on Forest Fires Research
7. Margerit, J., Séro Guillaume, O.: Modelling forest fires. Part II: Reduction to two-dimensional models and simulation of propagation. Int. J. Heat and Mass Transfer, **45** 1723-1737 (2002)
8. Asensio, M.I., Ferragut, L., Simon, J.:A convection model for fire spread simulation. Applied Mathematics letters, **18**, 673-677 (2005)
9. Pironneau, O.: On the Transport-Difussion Algorithm and its applications to the Navier-Stokes Equations. Numer. Math. **38**, 309-332 (1982)
10. Linn, R. R.: Transport model for Prediction of Wildland Behaviour. Los Alamos National Laboratory, Scientific Report, LA1334-T, 1997
11. Simeoni, A., Larini, M., Santoni P.A., Balbi J.H.: Coupling of a simplified flow with a phenomenological fire spread model. C.R. Mecanique **330**, 783-790 (2002)
12. Siegel, Howell: Thermal Radiation Heat Transfer. McGraw-Hill, New York (1971)
13. Viegas, DX, Pita, LP, Matos, L, Palheiro, P.: Slope and wind effects on fire spread. Int. Forest Fire Research and Wildland Fire Safety, Luso, Coimbra, Viegas (eds). Millpress: Rotterdam, (2002)
14. http://www.freefem.org [10 april 2003]

Fast Multipole Method for Solving the Radiosity Equation

J. Morice[1,2], K. Mer-Nkonga[2] and A. Bachelot[1]

[1] Université Bordeaux 1, MAB, Avenue des facultés, 33405 Talence, France
   Alain.Bachelot@math.u-bordeaux1.fr
[2] CEA-CESTA, BP. 2, 33114 Le Barp, France
   katherine.nkonga@cea.fr, jacques.morice@cea.fr

Summary. For the radiosity equation, we investigate iterative solutions and acceleration by the Fast Multipole Method (FMM). In this paper, a new FMM for general kernels is proposed to solve this equation, inspired by the method proposed by Gimbutas and Rokhlin [SIAM J. Sci. Comput. **24**, 796–817 (2002)]. Finaly, a new theoretical and numerical comparison of different FMM methods for an integral kernel is presented in the context of the radiosity equation for $1/r^4$.

1 Introduction

A new fast method for solving the radiosity equation is considered by using the Fast Multipole Method (FMM), in the context of heat transfert calculations. This equation, which is an integral equation, models radiative exchanges between gray diffuse surfaces without participating media [7]. The radiosity equation plays also an important role in obtaining realistic image in computer graphics [11]. After discretization of the whole surface by finite elements, the size of the system generated can be quite large, and consequently the cost of solving this system is important in time (with an iterative method: $O(N^2)$ where N is the number of elements) and memory. Two classes of fast methods to solve this problem have been developed in computer graphics. Firstly, classical hierarchical methods (HM) for sets of plane surfaces and their generalization to initial curved surfaces with clustering [11], and secondly methods based on an expansion of the integral kernel: panel clustering and FMM [1, 5]. The main drawback of HM and panel clustering is their limitation to sets of plane surfaces. Furthermore, the former class encounter problems of iterative robustness and prediction of accuracy which are unacceptable for application in radiative heat transfert calculations. To accelerate iterative solution of the radiosity equation, we propose to use FMM as in [5]. This method was introduced for N-body problems [8] and used in many other physical applications. Based on an expansion of the kernel of the integral equation, this method reduces the interaction generated by the kernel between elements of the mesh to interactions between multipole boxes and so accelerates matrix-vector products of iterative methods. By using the multi-level FMM (MLFMM), we can evaluate solution system with a cost of $O(N \ln(N))$. The radiosity kernel depends on the surface, due to the normal. So we investigate FMM expansion for $1/r^4$. A FMM method based on a Taylor expansion for smooth kernels have been proposed in [12]. A multipole expansion based on the expansion of $1/r^\gamma$ with the Gegenbauer polynomials was used to solve the Fokker-Planck-Landau's equation ($\gamma = 3$) [6] and the radiosity equation ($\gamma = 4$) [4] with Spherical Harmonics (SH). With the multipole expansion in [4] and a formula in [10], Karapurkar et al. introduced a FMM expansion for the radiosity kernel with SH [5]. The Rotational Coaxial Translation Decomposition (RTCD) of [3], primary used for Laplace's equation ($\gamma = 1$), uses properties of SH to accelerate transfers between boxes. The RTCD is, in this paper, extended to the radiosity kernel to improve the FMM proposed in [5]. Furthermore, we introduce a new fast method for general kernels inspired by [2]. The optimizations, we provide compare to [2] are based on: the use of a reduced SVD, and the use of symetries allowed by this reduced SVD which cannot be used with method of [2].

In Section 2, we present the radiosity equation and its numerical solution (iterative solution, fast methods). In Section 3, we present FMM expansions for $1/r^\gamma$ and in Section 4, the new fast method. Finaly, in Section 5, we compare numerically and theoricaly the FMMs investigated, with a new comparison process for surface interaction problems that takes into account the empty boxes in the octree, before concluding.

2 Radiosity equation and numerical solution

The radiosity equation is a mathematical model for the radiative exchanges between gray diffuse surfaces without participating media. In computer graphics, it modelizes the light transport in the same condition. This equation writes for a polyhedral surface S of $\mathcal{R}^3$:

$$\forall x \in S, \ B(x) = \epsilon(x)\sigma T^4(x) + \rho(x) \int_S V(x,y) \, G(x,y) B(y) d\sigma_y, \qquad (1)$$

with B the radiosity, ϵ the emissivity, ρ the reflectivity, T the temperature and σ the Stefan-Boltzmann constant. The visibility $V(x,y)$ equals 1 if the two points x and y see each other ($[x,y] \cap S = \{x,y\}$) and cancels otherwise. The radiosity kernel G is given by: $G(x,y) = \frac{(x-y).n_x (y-x).n_y}{|x-y|^4}$, where n_x is the inner unit normal to the surface S at point x. A condition to the existence of a unique solution to (1) is $|\rho|_{L^\infty(S)} < 1$, which is a physical condition [1].

Finite elements discretization: After discretization of the unknown B by finite elements, we obtain with the Galerkin method:

$$(I - M)B = E, \quad \text{with } M_{ij} = \rho_i \, F_{ij}. \qquad (2)$$

Table 1. Number of iterations and CPU time for a relative residual of 10^{-6}.

	Southwell	GS	Hybrid GS	GMRES
iter	148703	22	9	8
time (second)	9.48	8.16	3.86	3.72

F is the shape factor matrix defined by: $S_i F_{ij} = \int_{S_i} \int_{S_j} V(x,y) G(x,y) d\sigma_x d\sigma_y$.

Iterative solutions: We have compared four different iterative methods for system (2): Southwell, Gauss-Seidel (GS), Hybrid GS [9] and GMRES. The Southwell method is extensively used in computer graphics, but it has no matrix vector product that could be accelerated by FMM. For our comparisons, we consider a rectangular box in the center of a sphere (radius= 1) with the characteristics $l_x = 0.4, l_y = 0.4, l_z = 0.02$. There are 7000 elements, $\rho = 0.7$ for both surfaces with $T = 80K$ for the sphere and $T = 80.2K$ for the box. The results in Table 1 show that GMRES and Hybrid GS are the fastest iterative solutions. For different configurations (ρ, T, S), we obtain the same conclusion. So there are the best methods to be used with the FMM.

Fast methods: If a good accuracy is needed for the radiosity problem, system (2) can become quite large. For the unoccluded case ($V \equiv 1$), the cost of solving (2) by an iterative method is $O(N^2)$, where N is the number of elements. If $V \neq 1$, this cost is $O(N^3)$ due to the calculus of F. Two classes of fast methods to solve (2) except FMM have been developed for computer graphics applications: panel clustering [1] and hierarchical methods (HM) [11]. In the latter, a hierarchical representation of interactions between initial plane surfaces (or shape factor matrix) is constructed by adaptively subdividing planar surfaces into sub-surfaces according to a local error of interaction between two surfaces, to have a multiresolution solution of (2). The cost of this method is linear with respect to the refined surfaces, but quadratic in the initial plane surfaces. An improvement of HM, clustering has a quasi-linear cost in the initial plane surfaces by grouping elements into volume clusters. The drawback of all these methods are their limitation to sets of larger plane surface. Furthermore, HM and clustering encounter problems of iterative robustness and prediction of accuracy which are unacceptable in radiative heat transfert.

3 FMM and kernel expansion

We propose to use the FMM applied to the radiosity equation as in [5]. The FMM was introduced for N-body problems [8] and used in many other physical applications. The potentiel field is decomposed into far field and near field using an octree decomposition of the 3D domain. The near field for a body target corresponds to the box containing the target and the neighbours of this box. A classical calculus of interaction is used for this field. Based on an expansion of the potential of interaction (an integral kernel in our case), the FMM reduces the far interactions to interactions between multipole boxes. The radiosity kernel depend on the surface, due to the normal. So we investigate FMM expansion for $1/r^4$ as in [5]. We consider here a more general potential $1/r^\gamma$, $\gamma \in \mathbb{N}^*$. If the visibilty $V \equiv 1$, we can evaluate solution system with a quasi-linear cost with the MLFMM.

In the sequel, we present the different steps of MLFMM or FMM methods with the following abreviations: S: source, M: multipole, L: local, and T: target. A FMM expansion for a kernel K approximates interactions between two points x and y inside boxes B_{x_0} and B_{y_0} respectively of centers x_0 and y_0, by a separation of x and y: mainly $K(x,y) \simeq \sum_i L_i(x,x_0) T_i(x_0,y_0) M_i(y,y_0)$, where T_i is the transfer or M2L operator. To obtain a MLFMM expansion, we need an operator between father and child boxes of centers x_0^f and x_0 respectively for functions L_i (L2L operator) and M_i (M2M operator). For example, for function L_i, we have $L_i(x, x_0^f) \simeq \sum_{i'} L_{i,i'}(x_0^f, x_0) L_{i'}(x, x_0)$. All FMM expansions presented here are also MLFMM expansions (see [12, 5]).

A Taylor FMM expansion of order L' [12] for a smooth kernel is given by:

$$\forall x \in B_{x_0}, \forall y \in B_{y_0}, K(x,y) \simeq \sum_{\alpha+\beta \leq L'-1} \frac{D_x^\alpha D_y^\beta K(x_0,y_0)}{\alpha!\beta!}(x-x_0)^\alpha (y-y_0)^\beta.$$

Alternatively, it is possible to construct a multipole method with an expansion with the Gegenbauer polynomials used by Lemou to accelerate the solution of Fokker–Planck–Landau's equation ($\gamma = 3$) [6]. With [10], we obtain the same expansion with Spherical Harmonics (SH) used by Hausner [4] for light transport: $\frac{1}{|x-y|^\gamma} \simeq \sum_{l=0}^{L_1-1} \sum_{j=0}^{[l/2]} \sum_{m=-(l-2j)}^{(l-2j)} O_{l,j,m}^\gamma(x-y_0) I_{l,j,m}^\gamma(y-y_0)$, where $I_{l,j,m}^\gamma$ and $O_{l,j,m}^\gamma$ are functions of SH. To obtain a FMM, we need to introduce the second center x_0. In [10], there is a formula that allows this. So we obtain a SH based FMM expansion which is used for Laplace's equation in 3D ($\gamma = 1$) [3] and for the radiosity equation ($\gamma = 4$) by Karapurkar et al. [5]. For $\gamma = 1$, RTCD was introduced to speed up the original FMM [3] and is based on the properties of SH. We extended here the RTCD for $\gamma = 4$.

A MLFMM for general kernels was introduced in [2]. They approximate the kernel by an expansion based on a tensor product of Legendre polynomials, at each level of the octree, and use a SVD of these approximations to construct multipole and local expansions. We introduce in the next part a new fast method inspired by [2].

4 A new fast method

In [2], a MLFMM for non oscillatory kernel was developed in $\mathcal{R}^d$. We present here a new multilevel method in the non-adaptive case. The kernel K is, as in [2], approximated by an expansion of tensor product of Legendre polynomials

$\widetilde{K}_l(x,y)$ for each level l of the octree. A SVD of these approximations defined for x in box b and for y in the interaction list of b is used to perform M2L operations. For simplicity, we present these methods in the case $d = 3$ and for a symetrical kernel $K(x,y) = K(y,x)$.

In this part, we will assume that all charges (q) are located in $D = [0,1]^3$. We introduce for a box of the octree $Y^b = \prod_{h=1}^{3} [b_h, b'_h]$ the multi-index $b = (l, I)$ where l is the level and I is the index of the box. In the paper, we will denote also b for the box; X^b the union of box at level l which are not b and immediate neighbours of b at level l, and X_2^b the union of all boxes at level l whose father is a neighbour of b's father and which is not immediate neighbours of b (*interaction list*, see for more details [2](*List* 2)). We have $X^b = \bigcup_{c \in a(b)} X_2^c$, where $a(b)$ is b and all b's ancestors. We introduce also $P_m^{\alpha,\beta}$ the mth Legendre polynomial on the interval $[\alpha, \beta]$ (degree m).

At level l, we approximate K by $\widetilde{K}_l$ which is defined on $Y^b \times Y^c$ with a tensor product of Legendre polynomials of maximum degree $n-1$ by direction:

$$\widetilde{K}_l(x,y) = \sum_{j=0}^{n^3-1} \sum_{j'=0}^{n^3-1} \left(\int_{Y^b \times Y^c} K(x',y') P_j^b(x') P_{j'}^c(y')\, dx' dy' \right) P_j^b(x)\, P_{j'}^c(y),$$

where $P_j^b(x) = \prod_{h=1}^{3} P_{j_h}^{b_h, b'_h}(x_h)$ with $j = j_1 + (n-1)j_2 + (n-1)^2 j_3$ and $x = (x_1, x_2, x_3)$. It corresponds also to an interpolation by tensor product of Lagrange's polynomials defined by the n Gauss points per direction of the two boxes [13]. We introduce $\mathcal{P}_n(Y^b) = \text{span}\{(P_j^b)_{j=0, n^3-1}\}$. A SVD is then considered to compress the resulting representation of $\widetilde{K}_l$ by a truncation with p terms of the SVD (exact for $p = n^3$). Then for b in level l, a SVD of $\widetilde{K}_l$ on $Y^b \times X^b$ (for [2]) and $Y^b \times X_2^b$ (for the new method) is precalculated and stored: $\widetilde{K}_l(x,y) \simeq \sum_{k=1}^{p} u_k^b(x) s_k^b v_k^b(y)$ (for [2]) $\simeq \sum_{k=1}^{p} u_k^b(x) s_k^b v_k^b(y)$ (for the new method). The calculus of the SVD of $\widetilde{K}_l$ is developed in [13] and is equivalent to a calculus of interactions, for each box b, between n^3 and $n_b n^3$ particules, where $n_b = \text{card}(X^b) \geq 8^{l-1} - 27$ for the method in [2] and $n_b = \text{card}(X_2^b) = 189$ for the new method (refered also to reduced SVD method). If K is invariant by translation, it is possible to use only one SVD at each level l between one box and $n_l (> n_b)$ boxes for the two methods ($n_l = 316$ for the new method). Furthermore for the new method, if the kernel satisfies $K(\lambda x, \lambda y) = g(\lambda) K(x,y)$, we only need to perform a unique SVD between a single box and 316 boxes which is not valid for [2]. For $K(x,y) = 1/|x-y|^4$, due to six symetries of K_l (the same as K) which doesn't change the cube Y^b (x), it is possible to reduce again the SVD between one box to 16 boxes (refered to the 16 reduced SVD method). This allows to reduce the truncation p and the cost of the original reduced SVD. In the new method, we need two other expansions: a pseudo multipole expansion (PM) and a pseudo local expansion (PL).

Now, we describe how to obtain with the original reduced SVD method and [2] an approximation of the far field potential at the point x in b: $\int_{X^b} K(x,y)q(y)dy$.

New method: **S2PM**: we transform the sources for finest boxes ($l = l_{\max}$) in pseudo multipole expansion γ_j^b defined by $\gamma_j^b = \sum_{i=1}^{s_b} P_j^b(x_i)q_i \simeq \int_{Y^b} P_j^b(x)q(x)dx$, where s_b is the number of sources in the box b.

PM2PM: calculus of γ_j^b for all boxes.

PM2M: conversion of γ_j^b into multipole expansion for all boxes defined by $M_k^b = \sum_{j=0}^{n^3-1} a_j^{k,b} \gamma_j^b$, where $u_k^b(x) = \sum_{j=0}^{n^3-1} a_j^{k,b} P_j^b(x)$.

In [2]: **S2M**: calculus of the multipole expansion for finest boxes due to the sources in this box defined by $M_k^b = \sum_{i=1}^{s_b} u_k^b(x_i)q_i \simeq \int_{Y^b} u_k^b(x)q(x)dx$.

M2M: calculus of an approximation of the multipole expansion M_k^b for all boxes in the octree.

After, for the two methods: **M2L**: conversion of the multipole expansion into local expansion which is an approximation of $L_k^b \simeq \int_{X_2^b} v_k^b(y)q(y)dy$ (for [2]) and $L_k^b \simeq \int_{X_b^2} v_k^b(y)q(y)dy$ (for the new method), respectively. The operator M2L is based on a projection of v_k^b into Y^c on the pth first terms of $(u_k^c)_k$ (a basis of $P_n(Y^c)$) for the new method (in the same manner for [2]). So we obtain these approximations: $\forall x \in Y^b$, $\int_{X_2^b} k(x,y)q(y)dy \simeq \sum_{k=1}^{p} u_k^b(x) s_k^b L_k^b \simeq \sum_{k=1}^{p} u_k^b(x) s_k^b L_k^b$.

In [2]: the authors approximate $\int_{X^b} k(x,y)q(y)dy \simeq \sum_{k=1}^{p} u_k^b(x) s_k^b \widetilde{L}_k^b$, where $\widetilde{L}_k^b \simeq \int_{X^b} v_k^b(y)q(y)dy$ is the local expansion due to all charges in X^b.

L2L: calculus of $\widetilde{L}_k^b$ for all boxes.

L2T: calculus of the far field potential with the latest formula for finest boxes in the target particle.

New method: **L2PL**: conversion of L_k^b into Γ_j^b the pseudo local expansion due to charges in X_2^b (use the fact that $u_k^b \in P_n(Y^b)$).

PL2PL: calculus of the pseudo local expansion $\widetilde{\Gamma}_j^b$ due to charges in X^b: $\int_{X^b} k(x,y)q(y)dy \simeq \sum_{c \in a(b)} \sum_{j=0}^{n^3-1} \Gamma_j^c P_j^c(x) = \sum_{j=0}^{n^3-1} \widetilde{\Gamma}_j^b P_j^b(x)$.

PL2T: evaluation of the far field potential with this latest formula for finest boxes in the target particle.

Numerical complexity: With B_l the number of non empty boxes at level l and C_{nf} the average number of boxes in the nearfield for the finest boxes, the cost of the new method is: $C_{\mathrm{nf}} Ns + 2C'Nn^3 + \left(\sum_{l=4}^{l_{\max}} B_l\right) 3n^3 \frac{n+1}{2} + \left(\sum_{l=3}^{l_{\max}} B_l\right)(\alpha p^2 + 2pn^3)$, where $\sum_{l=3}^{l_{\max}} B_l \simeq \sum_{l=4}^{l_{\max}} B_l \simeq 8N/(7s)$, with s the average number of charges in the finest boxes, and α the average number of boxes in the interaction list. The different terms are respectively the cost of near interactions, S2PM+PL2T, PM2PM+PL2PL, PM2M+L2PL and M2L.

For memory cost, the new methods need more memory for expansion coefficients: pseudo multipole expansions: $8Nn^3/(7s)$, multipole expansions for [2]: $8Np/(7s)$.

For kernels invariant by translation, the method in [2] has an important cost for SVDs and operator L2L due to n_l, and the cost of $u_k^b(x_i)$ is $O(Npn^3)$. The new method solves these problems (the cost of $P_j^b(x_i)$ is only $O(Nn^3)$).

5 Numerical results

In the litterature, most of the theoretical comparisons between different FMMs make implicity, the assumption that there is no empty box in the octree. This comparison is not correct for surface interaction problems, because there are a lot of empty boxes in the octree (see for example [3]). Here, we present a theoretical comparison which takes this fact into account.

If we choose a finest level $l_{\max}$, the cost of near interactions is the same for all FMMs. In this part, we suppose that we have fixed $l_{\max}$ for all methods. We are interested in the cost of steps from M2M to L2L (from PL2PL to PM2PM for the new method). If $\sum_{l=3}^{l_{\max}} B_l \simeq \sum_{l=4}^{l_{\max}} B_l$ is valid, the cost of these steps depend linearly on α the average number of boxes in the interaction list. So we do our comparisons between methods on α, because it takes into account the empty boxes. We introduce for method 1 and method 2, the maximum numbers of boxes of the interaction list for which method 1 is faster than method 2, denoted by α_{1-2}. The cost of steps at finest level: S2M, L2T, S2PM and PL2T are for the moment not considered because these steps depend on the cost of evaluation of SH functions or Legendre polynomials.

We use the error $\varepsilon_{L^2}^2 = (\sum_{j=1}^{N}(\Phi_e(x_j) - \Phi_a(x_j))^2)/\sum_{j=1}^{N}\Phi_e(x_j)^2$ where $\Phi_e(x) = \sum_{j=1}^{N} k(x, y_j)q_j$ and Φ_a is an approximation with FMM. For our tests, we consider a uniform source intensity ($q_j \equiv q$) and interactions between a cube of size 1 of center (0,0,0) and the other cubes in the interaction list with N random particles in each cube. For $\gamma = 4$, to obtain the error $\epsilon_{L^2} \leq 10^{-3}$, we need to consider the parameters $n = 6, p = 50$ for the 16 reduced SVD, $L = 8$ for SH or RTCD and $L' = 11$ for Taylor. The truncation $p = 50$ allows to have the asymptotic error given by $p = n^3$ (see Fig. 1). However, Fig. 1 demonstrates that it is possible to take a truncation p much smaller. We see in the Table 2 that the new method is always better than the SH method, and better than the RTCD and Taylor for $\alpha > 27$. We note here that we have not considered the version of Taylor proposed by Tausch [12].

Table 2. Value of α_{1-2} where the method 2 is the 16 reduced SVD method for ($n=6, p=50$), for different truncation, case $\gamma=4$. (+) means that the method 1 is faster.

$n=6, p=50$	7	8	9	10	11	12
SH	4	1	0	0	0	0
RTCD	88	27	12	6	3	2
Taylor	+	+	+	55	24	12

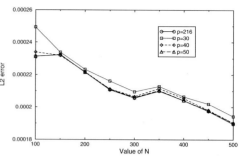

Fig. 1. L2 error between the box (0,0,0) and the boxes in the interaction list for $\gamma=4$ and $n=6$ for different values of N.

6 Conclusion

In this paper, we have investigated iterative methods and the FMM to accelerate the iterative solution of the radiosity equation: Taylor or SH expansions, the RTCD optimization and a new method. This new method, inspired by [2], is for a general kernel and is presented in this paper in the non adaptive case. A comparison of FMMs, with respect to the number of non-empty boxes in the interaction lists, is presented for $1/r^4$ and leads to the result that the new method is faster than the one of [5] for most of the steps of the algorithm, and also in some configurations of the octree than the RTCD and Taylor FMMs. In the future, we plan to take into account the other steps (those at the finest level) in the comparison, to determinate the optimum truncation parameter of the SVD and to compare these FMMs in a real case for the radiosity equation.

References

1. Atkinson, K., Chien, D.: A fast matrix-vector multiplication for solving the radiosity equation. Adv. Comput. Math., **12**, 51–74 (2000)
2. Gimbutas, Z., Rokhlin, V.: A generalized fast multipole method for nonoscillatory kernels. SIAM J. Sci. Comput., **24**, 796–817 (2002)
3. Gumerov, N.A., Duraiswami R: comparison of the efficiency of translation operators used in the fast multipole method for the 3D Laplace equation. UMIACS TR 2005-09 , University of Maryland, College Park (2005)
4. Hausner, A.: Multipole expansion of light. IEEE Transactions on vizualisation and computer graphics, **3**, 12–22 (1997)
5. Karapurkar, A., Goel, N., Chandran S.: The Fast Multipole Method for Global Illumination. ICVGIP2004, 119–125 (2004)
6. Lemou, M.: Multipole expansion of the Fokker-Planck-Landau operator. Numer. Math., **78**, 597–618 (1997)
7. Modest, M.F.: Radiative heat transfer . Academic Press second edition (2003)

8. Rokhlin, V.: Rapid solution of integral equations of classical potential theory. J. Comput. Phys., **60**, 187–207 (1985)
9. Rousselle, F., Lebond, M., Renaud C.: Radiosité progressive par groupes. In Journées AFIG'99, Reims, 323–331 (1999) (in French)
10. Sack, R.A.: Three dimensional addition theorem for arbitrary functions involving expansions in spherical harmonics. J. Math. Phys., **5**, 252–259 (1964)
11. Sillion, F.: Clustering and volume scattering for hiearchical radiosity calculations. In Fifth Eurographics Workshop on rendering, Darmstadt, 105–117 (2004)
12. Tausch, J.: The Variable Order Fast Multipole Method for Boundary Integral Equations of the Second Kind. Computing, **72**, 267–291 (2004)
13. Yarvin, N., Rokhlin, V.: Generalized Gaussian quadratures and singular value decomposition of integral operators. SIAM J. Sci. Comput., **20**, 699–718 (1998)

Numerical Modelling of Kinetic Equations

J. Banasiak, N. Parumasur[1] and J.M. Kozakiewicz[2]

[1] University of KwaZulu-Natal, Durban, 4041, South Africa
    banasiak@ukzn.ac.za, parumasurn1@ukzn.ac.za
[2] University of Zululand, Kwadlangezwa, 3886, South Africa
    jkozakie@pan.uzulu.ac.za

Summary. We consider using a modified asymptotic procedure for the numerical modelling of various kinetic equations.

1 Introduction

Many important models in kinetic theory, e.g., the telegraph equation and the linear Boltzmann equation, share a common mathematical form

$$\partial_t u + \mathcal{A}u + \mathcal{S}u + \frac{1}{\varepsilon}\mathcal{C}u = 0, \qquad (1)$$

where u is the particle distribution, ∂_t is the time derivative and the operators $\mathcal{A}$, $\mathcal{S}$, $\mathcal{C}$ describe attenuation, streaming and collisions of particles, respectively. Due to the appearance of $1/\varepsilon$ multiplying the collision operator it is evident that the collisions play a dominant role in the time evolution of the system. It is useful then to employ a strategy based on separating the solution into components for the kinetic and hydrodynamic parts. The original solution is then made up of a sum of these separate components. Of particular importance is the case when the hydrodynamic space is the null-space of the collision operator $\mathcal{C}$ and the projection of u onto this space is the hydrodynamic part of the solution that is expected to have a slow evolution. Hence, let $\mathcal{P}$ be the projection of the state space onto the hydrodynamic space of the collision operator $\mathcal{C}$, and let $\mathcal{Q} = \mathcal{I} - \mathcal{P}$ be the complementary projection. Accordingly, by $\mathcal{P}u = v$ we denote the hydrodynamic part of the solution u and by $\mathcal{Q}u = w$ the kinetic part. Applying these projections on both sides of (1) we get

$$\partial_t v = \mathcal{P}(\mathcal{A} + \mathcal{S})\mathcal{P}v + \mathcal{P}(\mathcal{A} + \mathcal{S})\mathcal{Q}w$$
$$\varepsilon \partial_t w = \varepsilon \mathcal{Q}(\mathcal{A} + \mathcal{S})\mathcal{Q}w + \varepsilon \mathcal{Q}(\mathcal{S} + \mathcal{A})\mathcal{P}v + \mathcal{QCQ}w, \qquad (2)$$

with the initial conditions

$$v(0) = \overset{o}{v}, \ w(0) = \overset{o}{w},$$

where $\overset{o}{v} = \mathcal{P}\overset{o}{u}, \ \overset{o}{w} = \mathcal{Q}\overset{o}{u}$. We have kept the superfluous symbols $\mathcal{P}v$ and $\mathcal{Q}w$ for the sake of notational symmetry. The projected operators $\mathcal{PSP}$, $\mathcal{PAQ}$ and $\mathcal{QAP}$ vanish for most types of linear equations so we obtain the following form of (2)

$$\partial_t v = \mathcal{PAP}v + \mathcal{PSQ}w,$$
$$\varepsilon \partial_t w = \varepsilon \mathcal{QSP}v + \varepsilon \mathcal{QSQ}w + \varepsilon \mathcal{QAQ}w + \mathcal{QCQ}w, \quad (3)$$
$$v(0) = \overset{o}{v}. \ w(0) = \overset{o}{w},$$

Following the standard asymptotic approach, we consider the solution of (3) as a sum of the bulk and the initial layer parts:

$$v(t) = \bar{v}(t) + \tilde{v}(\tau), \qquad w(t) = \bar{w}(t) + \tilde{w}(\tau), \quad (4)$$

where the variable τ in the initial layer part is given by $\tau = t/\varepsilon$. In the next section we describe a modified asymptotic method which is more suitable for the numerical treatment of (3) and is very suitable for dealing with models from chemistry [4] and nuclear reactor kinetics [2].

2 Asymptotic method

The following algorithm describes the modified asymptotic procedure proposed in [5]:

Algorithm 1

1. The bulk approximation $\bar{v}$ is not expanded into powers of ε.
2. The bulk approximation $\bar{w}$ is explicitly written in terms of $\bar{v}$ and expanded in powers of ε.
3. The time derivative $\partial_t \bar{v}$ and the initial value $\bar{v}(0)$ are expanded into powers of ε.

To illustrate the key aspects of the above algorithm we revisit the example considered in [6] and given by the initial value problem

$$\varepsilon \frac{d^2 x}{dt^2} + A \frac{dx}{dt} + f(x) = 0,$$
$$x(0) = \alpha, \ \frac{dx}{dt}(0) = \beta, \quad (5)$$

where $t \in [0, t_1]$, $t_1 > 0$, $x : [0, t_1] \to \mathbf{R}^n$, $n \geq 1$, $f : \mathbf{R}^n \to \mathbf{R}^n$, $\alpha, \beta \in \mathbf{R}^n$, ε is a small positive parameter and A is a matrix whose eigenvalues have all positive real parts. This can be converted to first order system in a standard way

$$\varepsilon \frac{dz}{dt} = -Az - f(x), \qquad \frac{dx}{dt} = z;$$
$$x(0) = \alpha, \quad z(0) = \beta. \tag{6}$$

We will apply the asymptotic expansion method to the system, truncating the expansions at first order terms. Consider the system (6). In the first order approximation we obtain

$$z(t) = \bar{z}^{(1)}(t) + \tilde{z}^{(1)}(\tau) + O(\varepsilon^2), \qquad x(t) = w(t) + \tilde{x}^{(1)}(\tau) + O(\varepsilon^2) \tag{7}$$

$$\bar{z}^{(1)}(t) = \bar{z}_0(t) + \varepsilon \bar{z}_1(t), \quad \tilde{z}^{(1)}(\tau) = \tilde{z}_0(\tau) + \epsilon \tilde{z}_1(\tau), \quad \tilde{x}^{(1)}(\tau) = \tilde{x}_0(\tau) + \varepsilon \tilde{x}_1(\tau)$$

where $\tau = t/\varepsilon$, is the scaled time variable applicable in the initial layer and w denotes the first order bulk approximation to the solution x of the original system (5). According to Algorithm 1 the bulk solution w is not expanded. The bulk solution for z depend on time through its functional dependence on w. Thus introducing the functions ϕ_0 and ϕ_1 we can write

$$\bar{z}_0(t) = \phi_0(w(t)), \qquad \bar{z}_1(t) = \phi_1(w(t)). \tag{8}$$

Substituting these into the first equation of (6), and retaining only the terms of the first and second order, we obtain

$$\varepsilon \frac{d\phi_0}{dw} \frac{dw}{dt}\bigg|_0 = -A\phi_0(w) - \varepsilon A \phi_1(w) - f(w), \tag{9}$$

where $\frac{dw}{dt}\big|_0$ is the zero order term of the expansion of $\frac{dw}{dt}$ (the derivative of w is expanded into powers of ε). Equating like powers in ε we obtain the following expressions for the unknown functions ϕ

$$\phi_0(w) = -A^{-1} f(w), \quad \phi_1(w) = -A^{-1} \frac{d\phi_0}{dw} \frac{dw}{dt}\bigg|_0 = -A^{-2} \frac{df}{dw}(w) A^{-1} f(w).$$

Substituting $z^{(1)} = \phi_0 + \varepsilon \phi_1$ for z and w for x in the first equation in (6) we obtain the first order approximation to x

$$\frac{dw}{dt} = -A^{-1}\left(I - \varepsilon A^{-1}\frac{df}{dw}(w)A^{-1}\right) f(w). \tag{10}$$

Now we derive the initial condition for (10) taking into account that the initial layer functions satisfy:

$$\frac{d\tilde{x}_0}{d\tau} = 0, \qquad \frac{d\tilde{z}_0}{d\tau} = -A\tilde{z}_0, \qquad \frac{d\tilde{x}_1}{d\tau} = \tilde{z}_0.$$

They have to decay exponentially with τ. Hence

$$\tilde{x}_0(\tau) \equiv 0, \qquad \tilde{z}_0(\tau) = e^{-A\tau} \tilde{z}_0(0) \tag{11}$$

$$\widetilde{x}_1(\tau) = -\int_\tau^\infty \widetilde{z}_0(s)\,ds = -\int_\tau^\infty e^{-As}\widetilde{z}_0(0)\,ds = -A^{-1}e^{-A\tau}\widetilde{z}_0(0). \tag{12}$$

These equations together with (7) and (6) yield

$$w(0)\big|_0 = \alpha, \qquad w(0)\big|_1 = A^{-1}(\beta + A^{-1}f(\alpha)).$$

The initial condition for (10) is then

$$w(0) = w(0)\big|_0 + \varepsilon w(0)\big|_1 = \alpha + \varepsilon A^{-1}(\beta + A^{-1}f(\alpha)). \tag{13}$$

With sufficiently smooth function f such that the solution of the original equation exists over a certain time interval $[0, t_1]$, we require that the matrix A has all the eigenvalues with positive real parts, so that the initial layer solutions are exponentially decaying. From (7) we have the first order asymptotic solution for x defined by (6)

$$x^{(1)}(t) = w(t) + \varepsilon \widetilde{x}_1(\tau), \tag{14}$$

where w is the solution of (10) with the initial condition (13) and $\widetilde{x}_1$ is given by (12). The function $x^{(1)}$ is uniformly convergent to x so that

$$\|x(t) - x^{(1)}(t)\| = C_1 \varepsilon^2, \quad t \in [0, t_1], \tag{15}$$

where C_1 is a constant independent of t and $\|\ \|$ is an arbitrary norm. If we are not interested in the behaviour of x inside the initial layer, then we can replace $x^{(1)}$ with w we have

$$\|x(t) - w(t)\| = D_1 \varepsilon^2, \quad t \in [t_0, t_1], \tag{16}$$

where D_1 is a constant and t_0 is an arbitrary number such that $0 < t_0 < t_1$.

Following a similar approach we can obtain the equation for the hydrodynamic solution of (3). We give a brief description here as complete details may be found in [1]. To this end, according to Algorithm 1 we make the following expansions

$$\bar{w} = \bar{w}_0 + \varepsilon \bar{w}_1, \quad \bar{v} = \bar{v}_0 + \varepsilon \bar{v}_1, \quad \widetilde{w} = \widetilde{w}_0 + \varepsilon \widetilde{w}_1. \tag{17}$$

Substituting the expansion for $\bar{w}$ into (3) and comparing terms of the same powers of ε yield the equation for the hydrodynamic variable

$$\partial_t \bar{v} = \mathcal{P}\mathcal{A}\mathcal{P}\bar{v} - \varepsilon \mathcal{P}\mathcal{S}\mathcal{Q}(\mathcal{Q}\mathcal{C}\mathcal{Q})^{-1}\mathcal{Q}\mathcal{S}\mathcal{P}\bar{v}. \tag{18}$$

For the initial layer a similar procedure yields

$$\widetilde{v}_0(\tau) \equiv 0, \quad \widetilde{v}_1(\tau) = \mathcal{P}\mathcal{S}\mathcal{Q}(\mathcal{Q}\mathcal{C}\mathcal{Q})^{-1} e^{\tau \mathcal{Q}\mathcal{C}\mathcal{Q}} \overset{o}{w}, \tag{19}$$

This leads to the initial condition for (5)

$$\bar{v}(0) = \overset{o}{v} - \varepsilon \mathcal{P}\mathcal{S}\mathcal{Q}(\mathcal{Q}\mathcal{C}\mathcal{Q})^{-1} \overset{o}{w}. \tag{20}$$

In the next section we consider two numerical examples from neutron transport theory.

3 Numerical examples

3.1 Linear Boltzmann equation of neutron transport theory

We consider a special case of the linear Boltzmann equation in a slab geometry [3],

$$\partial_t u(x,\mu,t) = -\mu \partial_x u(x,\mu,t) - \frac{1}{\varepsilon}u + \frac{1}{\varepsilon}\int_{-1}^{1}\left(\frac{1}{2}+\mu'\mu\right)u(x,\mu',t)d\mu', \quad (21)$$

where $u(x,\mu,t)$ is the distribution of neutrons, $x \in [0,1]$ is the spatial position of particles, $\mu = \cos\varphi$, φ is the angle between the velocity of a neutron and the positive x-axis, and ε a small positive parameter related to the mean free path. Here, the operators $\mathcal{A}$ and $\mathcal{S}$, are defined by

$$\mathcal{A}u = 0, \qquad \mathcal{S}u(x,\mu) = -\mu\partial_x u(x,\mu),$$

and $\mathcal{C}$ is given by

$$\mathcal{C}u(x,\mu) = -u(x,\mu) + \int_{-1}^{1}\left(\frac{1}{2}+\mu'\mu\right)u(x,\mu')\,d\mu'.$$

We assume that u satisfies periodic boundary conditions and use the following initial condition

$$u(x,\mu,0) = \frac{1}{\sqrt{2}}[x^3(1-x)^3+1] + \sqrt{\frac{3}{2}}\mu x^3(1-x)^3. \quad (22)$$

We made four kinds of comparisons and calculated appropriate errors. First we compared the exact solution v with the solution of the diffusion equation with uncorrected initial conditions $\hat{\rho}(t)$. The error is denoted by

$$E = |v(t) - \hat{\rho}(t)|.$$

Next we compared v with the solution of the diffusion equation with corrected initial condition $\bar{\rho}$ in this case the error is denoted by

$$E_{IC} = |v(t) - \bar{\rho}(t)|.$$

Then v was compared with $\hat{\rho}$ supplemented with the initial layer corrector

$$\tilde{v} = \varepsilon S \overset{o}{w} e^{-\frac{t}{\varepsilon}},$$

and the resulting error is

$$E_{IL} = |v(t) - (\hat{\rho}(t) + \tilde{v})|.$$

Finally v is compared with the solution to the diffusion equation with the corrected initial condition $\bar{\rho}(t)$ supplemented with the initial layer corrector. The error is

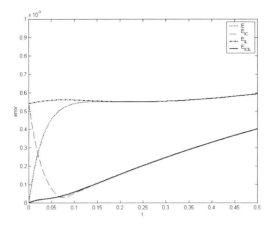

Fig. 1. Errors for Linear Boltzmann Equation wiht $\varepsilon = 0.01$

$$E_{ICIL} = |v(t) - (\bar{\rho}(t) + \widetilde{v})|.$$

The errors are presented in Figure 1. It is seen that the best results are obtained when (5) was applied together with both initial condition corrector (20) and the initial layer corrector (19). The following algorithm gives the details for the implementation of the above numerical procedure in a Matlab program.

Algorithm 2

1. Initialisation
   - Input of chosen parameters like ε, time, number of grid points for spatial variable and cosine of the velocity.
   - Calculation of initial values of the transport equation using (22).
2. Evaluation of the solution of the transport equation
   - Calling the ODE solver *ode15s* which uses *ode-tr* an ODE file for numerical integration of the transport equation.
3. Evaluation of the solution of the diffusion equation with uncorrected initial conditions.
   - Calculation of initial values of the diffusion equation.
   - calling the ODE solver *ode45* which uses *ode-dif* an ODE file for numerical integration of the diffusion equation.
4. Calculation of the corrected initial values of the diffusion equation.
5. Evaluation of the solution of the diffusion equation with corrected initial conditions.
   - Calculation of corrected initial values of the diffusion equation.
   - Calling the ODE solver *ode45* which uses *ode-dif* an ODE file for numerical integration of the diffusion equation.
6. Evaluation of the errors and initial layer corrector.

- Evaluation of the error of the uncorrected diffusion approximation.
- Evaluation of the error of the corrected diffusion approximation.
- Evaluation of the initial layer corrector.
- Evaluation of the error of the uncorrected diffusion approximation with initial layer corrector.
- Evaluation of the error of the corrected diffusion approximation with initial layer corrector.

7. Output printing in the form of tables and plots.

Remarks:

- to compute the spatial derivatives in transport equation the appropriate formula depends on the sign of μ. The procedure *ode-tr* uses both NN (for $\mu < 0$) and NP (for $\mu > 0$) matrices where NN define backward and NP forward difference scheme respectively. Number of μ-gridpoints is chosen to be even so that we do not have $\mu = 0$.

3.2 Linear Boltzmann equation of semiconductor theory

We consider the linear Boltzmann equation which describes the time evolution of the spatially dependent electron distribution function $u(x, \mu, t)$ under the influence of spatially uniform constant electric field [3]. Here we use a scaling corresponding to a weak external field. The equation is of the form

$$\partial_t u(x,\mu,t) = -\mu \partial_x u(x,\mu,t) - a\partial_\mu u(x,\mu,t) - \frac{1}{\varepsilon} u(x,\mu,t)$$
$$+ \frac{1}{\varepsilon} m(\mu) \int_{-\infty}^{+\infty} u(x,\mu',t) d\mu', \qquad (23)$$

where a is the acceleration due to the electric field, and $m(\mu) = \sqrt{\frac{\beta}{\pi}} \exp(-\beta\mu^2)$ is the normalized Maxwellian distribution, with $\beta = m/(2Tk)$, T the temperature of the background, m mass of the particles and k the Boltzmann constant. The initial condition is

$$u(x,\mu,0) = f_0(\mu)(1 + A\sin x),$$

with $0 \leq x \leq 2\pi$, $v \in \mathbf{R}$ and $A \leq 1$ and the function f_0 is such that $\int_{-\infty}^{+\infty} f_0(\mu) d\mu = 1$ and $\int_{-\infty}^{+\infty} \mu f_0(\mu) d\mu = s_0 \neq 0$ As in the previous case we assume periodic boundary conditions. The operators for attenuation, streaming are

$$\mathcal{A}u(x,\mu) = 0, \quad \mathcal{S}u(x,\mu) = -\mu\partial_x u(x,\mu) - a\partial_\mu u(x,\mu),$$

and the collision operator is

$$\mathcal{C}u(x,\mu) = -u(x,\mu) + m(\mu)\int_{-\infty}^{+\infty} u(x,\mu')d\mu'$$

Similarly, the projection and complementary operators are given by

$$\mathcal{P}u(x,\mu)=m(\mu)\int_{-\infty}^{+\infty}u(x,\mu)d\mu, \quad \mathcal{Q}u(x,\mu)=u(x,\mu)-m(\mu)\int_{-\infty}^{+\infty}u(x,\mu)d\mu.$$

The calculations are the same as the previous example and it seen that the best results are obtained when both correctors are applied.

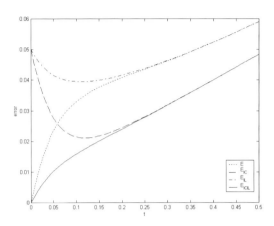

Fig. 2. Errors for Linear Boltzmann Equation of Semiconductor Theory with $\varepsilon = 0.05$

References

1. Banasiak, J., Kozakiewicz, J.M., Parumasur, N.: Diffusion Approximation of Linear Kinetic Equations with Non-equilibrium Dat - Computational Experiments. Transport Theory Statist. Phys., (accepted)
2. Beauwens, R., Mika J.: On the improved prompt jump approximation. 17th IMACS World Congress, Paris, France, July 11-15 2005 http://imacs2005.ec-lille.fr
3. Demeio, L., Frosali, G.: Diffusion Approximations of the Botzmann Equation: Comparison results for linear model problems. Atti Sem. Mat. Fis. Univ. Modena., **XLVI**, 653-675 (1998)
4. Galli, M., Groppi, M., Riganti, R., Spiga, G.: Singular perturbation techniques in the study of a diatomic gas with reactions of dissociation and recombination. Applied Mathematics and Computation, **146**, 509-531 (2003)
5. Mika, J.R., Palczewski, A.: Asymptotic Analysis of Singularly Perturbed Systems of Ordinary Differential Equations. Comput. Math. Appl., **21**, 13–32 (1991)
6. Mika, J.R., Kozakiewicz, J.M.: First Order Asymptotic Expansion Method for Singularly Perturbed Systems of Second-Order Ordinary Differential Equations. Comput. Math. Appl., **25**, 3–11 (1993)

Hyperbolic Equations

On a Subclass of Hölder Continuous Functions with Applications to Signal Processing

Sergio Amat[1], Sonia Busquier[2], Antonio Escudero[3] and J. Carlos Trillo[4]

[1] Departamento de Matemática Aplicada y Estadística. Universidad Politécnica de Cartagena. Paseo Alfonso XIII,52. 30203 Cartagena(Murcia). Spain
sergio.amat@upct.es
[2] sonia.busquier@upct.es
[3] antonio.escudero@upct.es
[4] jc.trillo@upct.es

Summary. In this paper a new family of Hölder continuous functions are presented. Using the properties of this family it is possible to generalized the classical Harten's subcell resolution theory and to apply it for the discretization of piecewise Hölder continuous functions. Some numerical experiments that confirm the theoretical results are presented.

1 Introduction

The most usual interpolatory techniques are based in polynomials. High order linear reconstructions associated to large support are affected by the presence of singularities in the considered signal. For instance, centered interpolation techniques produce Gibbs-like phenomenon in the presence of jump discontinuities.

To obtain good resolution near singularities we could consider nonlinear schemes. Essential Non-Oscillatory (ENO) methods, constructed by Harten, Osher, Engquist, and Chakravarthy ([8, 9, 10]), are a class of data-dependent interpolations. The most efficient implementation of ENO methods has been investigated by Shu and Osher ([12]-[13]), where they considered the point value framework. The goal of this interpolation is to enlarge the region of high accuracy by constructing piecewise polynomial interpolatory functions using only information from smoothness regions of the interpolated function. If the singularities are sufficiently well separate, this is possible for all the intervals except, of course, for the one containing the singularity. Some theoretical and numerical results in several dimensions are available in [1, 2].

A more precise strategy, presented by A.Harten in [7], improves on ENO schemes within such cells by Subcell Resolution (ENO-SR). Using ENO polynomial pieces at each side of the singularity we can recover, until some accuracy, the location of an isolate discontinuity in the derivative of a continuous function. This information is used to modify the polynomial piece corresponding to the cell with the singularity improving the accuracy. Using the cell-average or the hat-average frameworks it is possible to detect jump and δ singularities of the signal [5, 4].

On the other hand, the basic tool used in ENO-SR theory is Taylor's expansions. Then a great problem appears when it is not considered a piecewise differentiable signal, that is the case of Hölder continuous functions. In particular, M.-S.Lee shows [11] that the Hölder smoothness of a general two-dimensional image is only between 0.2 and 0.7.

The aim of this paper is to generalize the theory of the ENO-SR technique for non-piecewise differentiable functions. We introduce a family $\mathcal{X}^s$ of dense subspaces of C^s (Hölder continuous functions) where we can use a generalized version of Taylor's Theorem. In particular, we can extend the Harten's subcell resolution strategy.

The paper is organized as follows: in section 2 we present the family of subspaces $\mathcal{X}^s$ and their properties. In order to contemplate the case of piecewise Hölder continuous signals, we generalize the theory introduced by Harten only for piecewise smooth functions in section 3. Finally, these strategies are tested in section 4, allowing to compare the performances of linear and nonlinear approximations.

2 Hölder continuous spaces

Let be $s \in (0, 1)$, we consider

$$\mathcal{C}^s([a,b]) := \{f \in \mathcal{C}^0([a,b]) \ / \ \exists C > 0 \text{ s. t. } \forall x \in [a,b],$$
$$|f(x+h) - f(x)| \leq C|h|^s\}.$$

The functions $f \in \mathcal{C}^s([a, b])$ are called s-Hölder continuous.

For a fixed collection of increasing s-Hölder continuous functions $\{\gamma_s\}_{s \in (0,1)}$, such that $\gamma_s \in \mathcal{C}^s([a,b]) \setminus \mathcal{C}^{s+\epsilon}([a,b])$ for all $\epsilon > 0$, we introduce the space of s-differentiable continuous functions.

$$\mathcal{X}^s([a,b]) := \{f \in \mathcal{C}^0([a,b]) \ / \ \forall x \in [a,b],$$
$$\exists f^{(s}(x) := \lim_{h \to 0} \frac{f(x+h) - f(x)}{\gamma_s(x+h) - \gamma_s(x)}$$
$$\text{and } f^{(s} \in C^0([a,b])\}.$$

A first example of these $\{\gamma_s\}_{s \in (0,1)}$ functions is given by x^s, $s \in (0, 1)$.

Notice you, this generalized s-derivative in the definition of $\mathcal{X}^s$ is linear and zero for constants. For these space we generalize Taylor's Theorem.

Proposition 1. (Generalized Rolle's Theorem)
  Let be $f \in \mathcal{X}^s([a,b])$. If $f(a) = f(b) = 0$ then exists $\xi \in (a,b)$ such that $f^{(s}(\xi) = 0$.

Proof

If $f(x) = 0$ for all $x \in [a,b]$ then $f^{(s}(x) = 0$ for all $x \in (a,b)$.
Let ξ be such that $f(\xi) := \max_{x \in [a,b]} |f(x)|$. Then

$$\frac{f(\xi+h) - f(\xi)}{\gamma_s(\xi+h) - \gamma_s(\xi)} \cdot \frac{f(\xi-h) - f(\xi)}{\gamma_s(\xi-h) - \gamma_s(\xi)} \leq 0$$

and from definition $f^{(s}(\xi) = 0$ (γ_s is increasing). ∎

Definition 1. *We define $\mathcal{X}_n^s([a,b])$ as the subspace of n-times s-Hölder differentiable functions, that is, denoting by $f_0^{(s} := f$, $f_1^{(s} := f^{(s}$, ..., $f_n^{(s} := f^{(s^{(n)}(s}$, a function $f \in \mathcal{X}_n^s([a,b])$ if and only if exits $f_n^{(s} \in \mathcal{X}^s([a,b])$.*

Proposition 2. (Generalized Taylor's Theorem)
Let $f \in \mathcal{X}_n^s([a,b])$. Then for all $x, x_0 \in (a,b)$ exists $\xi \in (\min(x,x_0), \max(x,x_0))$ such that

$$f(x) = f_0^{(s}(x_0) + \frac{f_1^{(s}(x_0)}{1!}(\gamma_s(x) - \gamma_s(x_0))$$

$$+ \frac{f_2^{(s}(x_0)}{2!}(\gamma_s(x) - \gamma_s(x_0)) \cdot (\gamma_s(x) - \gamma_s(x_0))$$

$$+ \ldots + \frac{f_n^{(s}(x_0)}{n!}(\gamma_s(x) - \gamma_s(x_0)) \stackrel{(n)}{\ldots} (\gamma_s(x) - \gamma_s(x_0))$$

$$+ \frac{f_{n+1}^{(s}(\xi)}{(n+1)!}(\gamma_s(x) - \gamma_s(x_0)) \stackrel{(n+1)}{\ldots} (\gamma_s(x) - \gamma_s(x_0))$$

Proof

We denote by

$$P_n(x) = f_0^{(s}(x_0) + \frac{f_1^{(s}(x_0)}{1!}(\gamma_s(x) - \gamma_s(x_0))$$

$$+ \frac{f_2^{(s}(x_0)}{2!}(\gamma_s(x) - \gamma_s(x_0)) \cdot (\gamma_s(x) - \gamma_s(x_0))$$

$$+ \ldots + \frac{f_n^{(s}(x_0)}{n!}(\gamma_s(x) - \gamma_s(x_0)) \stackrel{(n)}{\ldots} (\gamma_s(x) - \gamma_s(x_0))$$

Let $\Psi(t) = f(t) - P_n(t) + \mathcal{K}(\gamma_s(t) - \gamma_s(x_0)) \stackrel{(n)}{\ldots} (\gamma_s(t) - \gamma_s(x_0))$ where $\mathcal{K}$ is such that $\Psi(x) = 0$. Applying proposition 1 to $\Psi(x)$ and its n-first s-derivatives the proposition holds. ∎

Proposition 3. *For all $s \in (0,1)$, $\mathcal{X}^s([a,b])$ is a dense subset of $\mathcal{C}^s([a,b])$.*

Proof

$$\overline{(\mathcal{X}^s([a,b]))}^{\mathcal{C}^s([a,b])} = \overline{(\mathcal{X}^s([a,b]))}^{\mathcal{C}^0([a,b])} \bigcap \mathcal{C}^s([a,b])$$

$$= \mathcal{C}^0([a,b]) \bigcap \mathcal{C}^s([a,b])$$

$$= \mathcal{C}^s([a,b])$$

∎

3 Generalized Harten's Subcell resolution technique

The error decay in approximation theory is related with the function smoothness. If the function has some isolate singularity, the error coefficients, corresponding to approximation which stencil cross the singularity, will have poor decay. To obtain good resolution everywhere we need to work with nonlinear reconstruction. A possible track for such improvements is the ENO-SR scheme.

We start with a finite number of values $\bar{f}_i$ which represent sampling of weighted-averages of a function $f(x)$ corresponding to a uniform grid x_i ($h = x_i - x_{i-1}$) of $[0,1]$.

$$\bar{f}_i = \frac{1}{h}\int f(x)\omega(\frac{x-x_i}{h})$$

Some of the functions $\omega(x)$ are the following:
a) Point value $\omega(x) = \delta(x)$.
b) Cell average

$$\omega(x) = \begin{cases} 1 & x \in [-1,0) \\ 0 & otherwise \end{cases}$$

The essential feature of the ENO interpolatory technique is a stencil selection procedure that attempts to choose the interpolated stencil S_j within a smoothness region of an interpolated function $f(x)$. For each interval $[x_{j-1}, x_j]$, we consider all possible stencils of $r \geq 2$ points that include x_{j-1} and x_j,

$$\{x_{j-r+1}, \ldots, x_j\}, \cdots, \{x_{j-1}, \ldots, x_{j+r-2}\}$$

and assign to it the stencil for which $f(x)$ is "smoothest" in some sense. For notational purposes, we assume that $i(j)$ is the first point in the final stencil. We consider the following stencil selection algorithm (see [3])

Algorithm. Non hierarchical choice of the stencil

Choose $i(j)$ such that
$|f[x_{i(j)}, \ldots, x_{i(j)+r-1}]| = \min\{|f[x_l, \ldots, x_{l+r-1}]|, \quad j-r+1 \leq l \leq j-1\}$

where $f[x_{l-1}, \ldots, x_{l+r-2}]$ denote the divided differences of f.

Definition 2. *We said that a function f has a s-singularity ($s \in (0,1)$) at x_d if $f \in \mathcal{X}^s([a,b])$, for all $(a,b) \subset [0,1] \setminus \{x_d\}$.*

Assuming that $f(x)$ has a s-singularity $x_d \in (x_{j-1}, x_j)$ and denoting by $q_{j-1}(x)$ and $q_{j+1}(x)$ the ENO interpolatory function in $[x_{j-2}, x_{j-1}]$ and $[x_j, x_{j+1}]$ respectively, the location of the s-singularity, x_d, can be recovered by the following function:

$$G_j(x) = q_{j+1}(x) - q_{j-1}(x)$$

Using Generalized Taylor's expansion in regions of s-smoothness

$$G_j(x_{j-1}) \times G_j(x_j) = [f^{(s)}]_{x_d}^2 (\gamma_s(x_j-h) - \gamma_s(x_j-ah))(\gamma_s(x_j) - \gamma(x_j-ah)) + \cdots$$

where $x_d = x_j - ah$, $0 < a < 1$ and $[f^{(s)}]_{x_d}$ denotes the jump of the s-derivative at x_d.

Therefore, if h is sufficiently small, there is a root of G_j in (x_{j-1}, x_j) be such that $G_j(\theta_j) = 0$.

Remark 1. If the function is a piecewise polynomial with a s-singularity in x_d then $x_d = \theta_j$.

The new piecewise polynomial interpolatory function has the following form

$$I^{SR}(x) = \begin{cases} q_l(x) & x \in [x_{l-1}, x_l] \\ q_{j-1}(x) & x \in [x_{l-1}, \theta_j] \\ q_{j+1}(x) & x \in [\theta_j, x_j] \end{cases}$$

thus, it uses extrapolation and interpolation techniques.
The accuracy of linear approximations is improved.

A key remark

Given the point-values discretization $f(x_k)$ of a s-Hölder continuous function, we can define, in each subinterval $[x_k, x_{k+1}]$, an increasing $\{\gamma_s\}$ function by

$$\gamma_s(x_k + \rho h) = \gamma_s(x_k) + C(x_k, \rho h)(f(x_k + \rho h) - f(x_k))$$

where $sgn(C(x,h)) = sign(f(x+h) - f(x_n))$ and $0 < \rho \leq 1$. In this situation $f \in \mathcal{X}^s([a,b])$.

4 Numerical experiments

In order to see the performances of the introduced reconstruction, we have tested it on some piecewise Hölder continuous signals. We present a comparison with linear reconstructions.

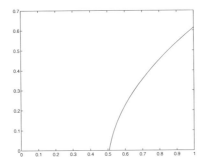

Fig. 1. Piecewise C^∞ signal.

Table 1. l_∞ and l_1 prediction errors for the signal of figure 1.

N_c	N_f	SR	Linear
17	33	5.39 $e-04$, 4.95 $e-05$	7.64 $e-02$, 4.82 $e-03$
33	65	1.22 $e-03$, 4.39 $e-05$	4.03 $e-02$, 1.30 $e-03$
65	129	3.00 $e-04$, 6.87 $e-06$	1.44 $e-02$, 2.32 $e-04$

We start with the piecewise differentiable signal of figure 1. This is a typical signal considered in the analysis of SR by Harten.

We consider point-values reconstructions of 4 points [6]. In table 1, we compute the prediction error. We denote by N_c and N_f the number of coefficient in the coarsest and finest scales respectively. We can see a proper adaptation to the singularity in the case of SR. Nevertheless, the linear scheme is drastically affected by the singularity.

Next, we modify the original signal in order to remove its differentiability. We analyze the signal of figure 2.

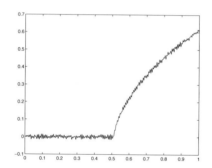

Fig. 2. Piecewise s-Hölder continuous signal.

In table 2, we compute the prediction error for different level steps. We use the same framework as before. The conclusions are similar, only good adaptation of the nonlinear scheme, see figure 3.

Table 2. l_∞ and l_1 prediction errors for the signal of figure 2.

N_c	N_f	SR	Linear
17	33	1.04 $e-02$, 6.33 $e-03$	8.47 $e-02$, 1.16 $e-02$
33	65	1.00 $e-02$, 6.15 $e-03$	4.68 $e-02$, 7.54 $e-03$
65	129	9.99 $e-03$, 6.40 $e-03$	1.81 $e-02$, 6.61 $e-03$

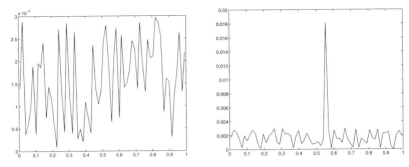

Fig. 3. Error for the signal of figure 2, $N_c = 65$, $N_f = 129$, left SR, right linear.

Using the cell-average framework we can consider also jump discontinuities [6]. We work with the jump of figure 4.

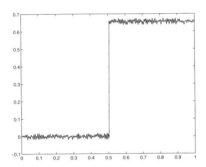

Fig. 4. Piecewise s-Hölder continuous signal.

In table 3 we observe that the SR identifies the singularity and improves the accuracy of the linear scheme. As before, only the Hölder perturbation produces the errors, see figure 5.

Table 3. l_∞ and l_1 prediction errors for the signal of figure 4.

N_c	N_f	SR	Linear
16	32	$3.64\,e-02$, $2.16\,e-02$	$4.54\,e-01$, $4.78\,e-02$
32	64	$3.14\,e-02$, $1.85\,e-02$	$4.60\,e-01$, $3.26\,e-02$
64	128	$3.94\,e-02$, $1.92\,e-02$	$4.59\,e-01$, $2.62\,e-02$

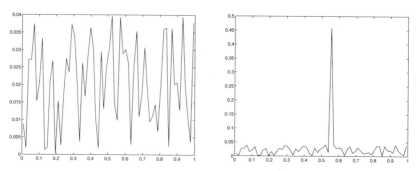

Fig. 5. Error for the signal of figure 4, $N_c = 65$, $N_f = 129$, left SR, right linear.

Acknowledgments

Research of the two first authors supported in part by the Spanish grants MTM2004-07114 and 00675/PI/04.

References

1. Amat, S., Aràndiga, F., Cohen, A., Donat, R.: Tensor product multiresolution analysis with error control for compact image representation. Signal Processing **82**(4), 587-608 (2002)
2. Amat, S., Aràndiga, F., Cohen, A., Donat, R., García, G., Von Oehsen, M.: Data compression with ENO schemes: A case study, Applied and Computational Harmonic Analysis **11** 273-288 (2001)
3. Aràndiga, F., Donat, R.: Nonlinear Multi-scale Decomposition: The Approach of A. Harten, Numerical Algorithms, **23**, 175-216 (2000)
4. Aràndiga, F., Donat, R., Harten, A.: Multiresolution Based on Weighted Averages of the Hat Function II: Nonlinear Reconstruction Operators, SIAM J. Sci. Comput. **20**(3), 1053-1099 (1999)
5. Harten, A.: Discrete multiresolution analysis and generalized wavelets, J. Appl. Numer. Math. **12**, 153-192 (1993)
6. Harten, A.: Multiresolution representation of data II, SIAM J. Numer. Anal. **33**(3), 1205-1256 (1996)
7. Harten, A.: ENO Schemes with Subcell Resolution, J. Comput. Phys. **83**, 148-184 (1989)
8. Harten, A., Osher, S. J.: Uniformly high order accurate nonoscillatory schemes I, SIAM J. Numer.Anal. **24** 279-309 (1987)
9. Harten, A., Osher, S. J.: B. Engquist and S. R. Chakravarthy, Some results on uniformly high-order accurate essentially non-oscillatory schemes, Appl. Numer. Math. **2**, 347-377 (1987)
10. Harten, A., Engquist, B., Osher, S. J., Chakravarthy, S. R.: Uniformly high order accurate essentially non-oscillatory schemes III, J. Comput. Phys. **71**, 231-303 (1987)
11. Lee, M. S.: Signal smoothness estimation in Hölder spaces, Intern. J. Computer Math. **73**, 321-332 (2000)

12. Shu, C. W., Osher, S. J.: Efficient implementation of essential non-oscillatory shock capturing schemes, J. Comput. Phys. **77**, 231-303 (1987)
13. Shu, C. W., Osher, S. J.: Efficient implementation of essential non-oscillatory shock capturing schemes II, J. Comput. Phys. **83**, 32-78 (1989)

Modelisation and Simulation of Static Grain Deep-Bed Drying

Aworou-Waste Aregba[1] and Denise Aregba-Driollet[2]

[1] TREFLE, ENSAM, Esplanade des Arts et Métiers, 33405 Talence cedex, France
aworou-waste.aregba@bordeaux.ensam.fr
[2] IMB, MAB, Université Bordeaux 1, 351 cours de la Libération, 33405 Talence cedex, France aregba@math.u-bordeaux1.fr

Summary. We study mathematical models for static grain deep-bed drying. These models take the general form of hyperbolic semilinear systems. The solutions vary strongly at the beginning of the drying process, so that the use of second order semi-implicit schemes is useful. Numerical experiments show a good qualitative behaviour of our approximations.

1 Introduction

Considerable amount of agricultural crops, and more particularly grains, are dried artificially using near ambient or high temperature air in various grain drying systems. The need for storage over a long period of time requires an accurate control of the properties of the product being dried. Excess moisture content can promote the growth of moulds and infestations by insects and hence spoil the stored products.

The speed and efficiency of drying depend on the drying air characteristics: relative humidity, temperature and velocity. With a high temperature and a small relative humidity for instance, one can increase the drying speed but might deteriorate the grain quality. It is important to control the grain temperature and moisture content during the drying process in order to determine safe drying conditions for a given product and a given type of drier.

In this paper, we study non-equilibrium models for static grain deep-bed drying: motionless piles of grain are exposed to drying air and there is no heat and mass equilibrium between the drying air and the grain through the bed. Those models have been developed mainly during the past fifty years and the more recent ones take into account complex constitutive laws and empirical correlations. In Sect. 2, we describe briefly the governing equations. In Sect. 4, we construct their numerical approximation. In Sect. 4 we present some numerical experiments and comparisons.

2 Models

Let us consider the drier as a vertical column in the x direction, $x > 0$. The drying air comes vertically from the bottom $x = 0$ with a constant temperature T_{ab} and a positive constant speed V_a. We denote respectively X_a and X_p the moisture contents of the air and of the product to dry, T_a and T_p the temperatures of the air and of the product. They are functions of (x,t) and have to take non negative values. All the functions here below are defined for non negative values of the unknowns.

The grain is supposed to be hygroscopic: when dried, its moisture content tends to a (nonzero) equilibrium state $X_{eq}(X_a)$, which is specific of each product. For cereals X_{eq} is a smooth strictly increasing function defined on a bounded interval $[0, X_{max}[$ and such that

$$X_{eq}(0) = 0, \quad \lim_{X \to X_{max}} X_{eq}(X) = +\infty.$$

The drying time is defined as the infimum of the times $t > 0$ such that

$$\|X_p(.,t) - X_{eq}(X_a(.,t))\|_\infty \leq \tau$$

where τ is a little tolerance value.

The drying kinetic is deduced from experiments on thin layers of the considered products. Here we have:

$$\partial_t X_p = -K(T_p)[X_p - X_{eq}(X_a)]$$

with $K(T_p) = d \exp(-c/T_p)$, d and c being positive constants.

Then, mass and energy balance give the following family of models:

$$\begin{cases} \partial_t X_a + \dfrac{V_a}{\epsilon} \partial_x X_a = \dfrac{\alpha(X_a, T_a)}{\epsilon} K(T_p)[X_p - X_{eq}(X_a)], \\ \partial_t T_a + \dfrac{V_a}{\epsilon} \partial_x T_a = -\dfrac{\beta(X_a, T_a, X_p, T_p)}{\epsilon}(T_a - T_p), \\ \partial_t X_p = -K(T_p)[X_p - X_{eq}(X_a)], \\ \partial_t T_p = \psi(X_a, T_a, X_p, T_p, \partial_x X_a). \end{cases} \quad (1)$$

The positive constant ϵ is the void fraction of the bed. We point out the fact that ϵ is fixed and is not small. The functions α, β and ψ are smooth given functions and α is positive. The system is supplemented with initial data $(X_{a0}, T_{a0}, X_{p0}, T_{p0})$ and boundary data

$$X_a(0,t) = X_{ab}, \quad T_a(0,t) = T_{ab}. \quad (2)$$

Spencer's model [7] is one of the oldest examples. Here, α is a positive constant, β depends only on (X_a, T_a) and is positive, and the function ψ does not depend on $\partial_x X_a$:

$$\psi(X_a, T_a, X_p, T_p) = \psi_1(X_a, T_a, X_p)(T_a - T_p) \quad (3)$$

$$-\psi_2(X_p,T_p)\left[X_p - X_{eq}(X_a)\right], \quad \psi_1 \geq 0.$$

Another model can be found in [4]: the functions α and β depend only on (X_a, T_a) and are positive, and

$$\psi(X_a, T_a, X_p, T_p, \partial_x X_a) = \psi_3(X_a, T_a, X_p, \partial_x X_a)(T_a - T_p) \tag{4}$$
$$+ \psi_4(X_a, T_a, X_p, T_p, \partial_x X_a).$$

See also [6] and references there in for others examples. The mathematical analysis of those systems is to be done. If ψ does not depend on $\partial_x X_a$, the system is semilinear hyperbolic. Otherwise, it is quasilinear and generally still hyperbolic. The general theory gives local existence of solutions. A first study of global existence and asymptotic behaviour is performed in [3].

The physical solutions vary quickly during a short transition period, (see Figs. 1, 2), and a low accuracy in this time interval leads to instabilities. Hence, even in this one-dimensional case, the computation is long and delicate. In most cases, authors prefer to discretize the following system, obtained by neglecting the accumulation terms in (2), see [4] for example:

$$\begin{cases} \partial_x X_a = \dfrac{\alpha(X_a, T_a)}{V_a} K(T_p)\left[X_p - X_{eq}(X_a)\right], \\ \partial_x T_a = -\dfrac{\beta(X_a, T_a, X_p, T_p)}{V_a}(T_a - T_p), \\ \partial_t X_p = -K(T_p)\left[X_p - X_{eq}(X_a)\right], \\ \partial_t T_p = \psi(X_a, T_a, X_p, T_p, \partial_x X_a). \end{cases} \tag{5}$$

In that case, $\partial_x X_a$ is known by the first equation, so that the source-term depends only on (X_a, T_a, X_p, T_p). All known models can then be written with a function ψ under the form

$$\psi = \psi_1(X_a, T_a, X_p, T_p)(T_a - T_p) - \psi_2(X_a, T_a, X_p, T_p)(X_p - X_{eq}(X_a)), \quad \psi_1 \geq 0. \tag{6}$$

The choice of ψ_1 and ψ_2 is not unique because a quadratic interaction $(T_a - T_p)(X_p - X_{eq})$ is possible.

The boundary conditions (2) remain the same while only the initial data (X_{p0}, T_{p0}) is requested. This Goursat problem admits local solutions. Here also, some global existence and asymptotic behaviour results can be obtained, see [3]. The problem can be discretized by ODE methods, making easy the use of time step control procedures.

3 The numerical schemes

In this section, we propose a discretization of systems (2) and (5) by second-order semi-implicit methods.

Notations

We consider a computational domain $V = [0, L]$ with an uniform mesh composed of cells $C_i = [x_{i-1/2}, x_{i+1/2}]$, $i = 1, \ldots, I$:

$$\Delta x = \frac{L}{I}, \quad x_{i+1/2} = i\Delta x, \quad i = 0, \ldots, I.$$

The (possibly variable) time step is denoted Δt and the discrete time levels are $t_0 = 0$ and $t_{n+1} = t_n + \Delta t$, $n \geq 0$.

3.1 Approximation of models (2)

Very few references exist on this problem. In [5] and [8], particular slightly different models are discretized. In [5] the method is not explicited. The implicit scheme used in [8] involves the numerical resolution of large linear systems. We propose another approach, which applies to all the considered systems.

Denoting $U = (X_a, T_a, X_p, T_p)$ and $\Lambda = diag(\frac{V_a}{\epsilon}, \frac{V_a}{\epsilon}, 0, 0)$, these models take the general form

$$\partial_t U + \Lambda \partial_x U = Q(U, \partial_x U). \tag{7}$$

We adopt a finite volume viewpoint: each U_i^n is an approximation of the mean value of $U(., t_n)$ on the cell C_i. If U_0 is the given initial value and U_b is the boundary value, we put for $i = 1, \ldots, I$ and $n \geq 0$:

$$U_i^0 = \frac{1}{\Delta x} \int_{C_i} U_0(x) dx, \quad U_0^n = \frac{1}{\Delta t} \int_{t_n}^{t_{n+1}} U_b(t) dt.$$

By a fractional step method, we split the system (7) into a set of linear transport equations:

$$\partial_t U + \Lambda \partial_x U = 0, \tag{8}$$

and a system of nonlinear equations:

$$\partial_t U = Q(U, \partial_x U). \tag{9}$$

Starting with U^n at time t_n, a second order MUSCL extension of upwind scheme is applied to system (8) over one time step. Thanks to linearity, this can be easily done and we do not detail the formulas. We denote T_Δ this scheme. The minmod limiter is used, so that positivity is preserved. We obtain

$$U^{n+1/2} = T_\Delta(U^n). \tag{10}$$

Then, $U^{n+1/2}$ is used as initial condition to solve (9). We observe that the source-term can be written as

$$Q(U, \partial_x U) = A(U, \partial_x U)U + G(U, \partial_x U)$$

where A is a smooth matrix valued function:

$$A(U, \partial_x U) = \begin{pmatrix} 0 & 0 & \frac{K\alpha}{\epsilon} & 0 \\ 0 & \frac{-\beta}{\epsilon} & 0 & \frac{\beta}{\epsilon} \\ 0 & 0 & -K & 0 \\ 0 & p & -q & -p \end{pmatrix}.$$

In the case where ψ is given by formula (3) (Spencer's model), $p = \psi_1$ and $q = \psi_2$. When ψ is given by formula (4), $p = \psi_3$ and $q = 0$. The source-term of all known models can be written in such a form. We can therefore apply a semi-implicit approximation:

$$U_i^{n+1} = U_i^{n+1/2} + \Delta t \left[A(U_i^{n+1/2}, V_i^{n+1/2}) U_i^{n+1} + G(U_i^{n+1/2}, V_i^{n+1/2}) \right] \quad (11)$$

where $V_i^{n+1/2}$ is an approximation of $\partial_x U$ on C_i computed by using a centered difference formula.

Denoting $A_i^{n+1/2} = A(U_i^{n+1/2}, V_i^{n+1/2})$ and $G_i^{n+1/2} = G(U_i^{n+1/2}, V_i^{n+1/2})$ the scheme results in an explicit formula:

$$U_i^{n+1} = (I - \Delta t A_i^{n+1/2})^{-1} \left[U_i^{n+1/2} + \Delta t G_i^{n+1/2} \right]. \quad (12)$$

As $V^{n+1/2}$ depends on $U^{n+1/2}$, this defines a function D_Δ:

$$U^{n+1} = D_\Delta(U^{n+1/2}). \quad (13)$$

Moreover, $(I - \Delta t A_i^{n+1/2})^{-1}$ is computed analytically. For example, with the choice (4), the final expression (where the $_i^{n+1/2}$ have been dropped) is:

$$\begin{cases} X_{a,i}^{n+1} = X_a + \dfrac{\Delta t \alpha K}{\epsilon(1+\Delta t K)}(X_p - X_{eq}(X_a)) \\ T_{a,i}^{n+1} = T_a + \dfrac{\Delta t \beta}{\epsilon(1+(\frac{\beta}{\epsilon}+\psi_3)\Delta t)}(T_p - T_a + \Delta t \psi_4) \\ X_{p,i}^{n+1} = X_p + \dfrac{\Delta t K}{1+\Delta t K}(-X_p + X_{eq}(X_a)) \\ T_{p,i}^{n+1} = T_p + \dfrac{\Delta t}{1+(\frac{\beta}{\epsilon}+\psi_3)\Delta t}(\psi_3(T_a - T_p) + \psi_4 + \Delta t \frac{\beta}{\epsilon}\psi_4). \end{cases} \quad (14)$$

As the first step of the algorithm is second order accurate in space, the obtained scheme is second order in space but first order in time. We reach second order in time by Heun's method:

$$W = D_\Delta(T_\Delta(U^n)), \quad Z = D_\Delta(T_\Delta(W)), \quad U^{n+1} = \frac{U^n + Z}{2}. \quad (15)$$

The function ψ_3 may be chosen positive, and β and K are positive, so that the scheme is well defined. Is is easy to see that the positivity of X_p is preserved. A further study of stability is difficult because the sign of ψ_4 is not known in general. We observed numerically that for a CFL condition between 0.2 and 0.5, the scheme is stable and provides physically relevant solutions. This is also true for others models.

3.2 Approximation of models (5)(6)

Those systems have been discretized by many authors, see [4, 7] and the extensive bibliography in [1]. Usually, the authors use explicit RK methods in each time and space direction. Here we derive the following semi-implicit scheme.

Denoting $U = (X_a, T_a, X_p, T_p)$, $U_a = (X_a, T_a)$, $U_p = (X_p, T_p)$ these models take the general form

$$\begin{cases} \partial_x U_a = A_a(U) U_a + G_a(U), \\ \partial_t U_p = A_p(U) U_p + G_p(U) \end{cases} \tag{16}$$

with

$$A_a(U) = \begin{pmatrix} 0 & 0 \\ 0 & \frac{-\beta}{V_a} \end{pmatrix}, \quad A_p(U) = \begin{pmatrix} -K & 0 \\ -\psi_2 & -\psi_1 \end{pmatrix}.$$

Here, we deal with numerical methods for ODE and each U_i^n is an approximation of $U(i\Delta x, t_n)$. If $U_{p,0}$ is the given initial value and U_{ab} is the boundary value, we put for $i = 0, \ldots, I$ and $n \geq 0$:

$$U_{p,i}^0 = U_{p,0}(i\Delta x), \quad U_{a,0}^n = U_{ab}(t_n).$$

We present a second order accurate method. A third order method has also been tested but it did not improve the results significantly.

We first define a spatial discretization R_x: let $t \geq 0$ be fixed and suppose that some approximation $U_{p,\Delta} = (U_{p,i})_i$ of U_p is known on $[0, L] \times \{t\}$. $R_x(U_{ab}, U_{p,\Delta})$ is an approximate solution $(U_{a,i})_{0 \leq i \leq I}$ of the problem

$$\partial_x U_a = Q_a(U_a, U_{p,\Delta}), \quad U_a(0, t) = U_{ab}.$$

To that aim, let P_x be the following semi-implicit approximation:

$$P_x(U_{a,i}, U_{p,i}) = (I - \Delta x A_a(U_{a,i}, U_{p,i}))^{-1}(U_{a,i} + \Delta x G_a(U_{a,i}, U_{p,i})) \tag{17}$$

It also reads as

$$P_x(U_{a,i}, U_{p,i}) = \begin{pmatrix} X_{a,i} + \Delta x \frac{\alpha_i K_i}{V_a}(X_{p,i} - X_{eq}(X_{a,i})) \\ T_{a,i} + \frac{\Delta x \beta_i}{(1 + \Delta x \frac{\beta_i}{V_a}) V_a}(T_{p,i} - T_{a,i}) \end{pmatrix}.$$

Then, denoting $U_{a,0} = U_{ab}$ and using Heun's scheme we put

$$U_{a,i+1} = \frac{1}{2}[U_{a,i} + P_x(P_x(U_{a,i}, U_{p,i}), U_{p,i+1})] = R_x(U_{ab}, U_{p,\Delta})_{i+1}.$$

Now we solve the problem in time. Applying again a semi-implicit procedure, we can define a first order step S_t:

$$\begin{cases} S_{t,1}(U_{a,\Delta}, U_{p,\Delta})_i = X_{p,i} - \dfrac{\Delta t}{1 + K_i \Delta t} K_i(X_{p,i} - X_{eq}(X_{a,i})) \\ S_{t,2}(U_{a,\Delta}, U_{p,\Delta})_i = T_{p,i} - \dfrac{\Delta t}{(1 + K_i \Delta t)(1 + \psi_{1,i} \Delta t)} \psi_{2,i}(X_{p,i} - X_{eq}(X_{a,i})) \\ \qquad\qquad\qquad\qquad + \dfrac{\Delta t}{(1 + \psi_{1,i} \Delta t)} \psi_{1,i}(T_{a,i} - T_{p,i}) \end{cases}$$

Now we proceed as follows: for $n = 0$ we compute $R_x(U_{ab}^0, U_{p,\Delta}^0) = U_{a,\Delta}^0$. Then for $n \geq 0$ we put

$$\begin{cases} U_p^{n+1/2} = S_t(U_a^n, U_p^n), \\ U_a^{n+1} = R_x(U_{ab}^n, U_p^{n+1/2}), \\ U_p^{n+3/2} = S_t(U_a^{n+1}, U_p^{n+1/2}), \\ U_p^{n+1} = \dfrac{1}{2}(U_p^n + U_p^{n+3/2}). \end{cases} \qquad (18)$$

The functions K and ψ_1 are positive. The scheme is well defined if β is positive, which is not the case for some models, see [6]. However, even in that case, we did not observe non physical solutions. As explained above, the physical solutions vary strongly in a first phase of the computation, enforcing the use of very small time steps, even with high order schemes. This rather short phase is followed by a longer one where the solutions are smooth. For these reasons, we use a time-step control procedure which reduces the computing times by 90%. There is no theoretical condition linking Δt and Δx. We fix initially Δt very small, for example $\Delta t = 0.05 \Delta x$, because the time step control procedure computes then automatically the optimal Δt, so that the computation is not affected by this choice.

4 Numerical experiments and comparisons

These numerical schemes have been tested in various situations. In [1], they are used to determine the conditions for which models (2) can be replaced by (5). It appeared that when the transfer time $L/V_a < 0.2$, the relative humidity $H_{ra,b}$ of the incoming air at $x = 0$ is more than 50% and when $T_{ab} \leq 30$ Celsius degrees, both models give similar results. Otherwise, models (5) underestimate the drying time and the computed solutions differ. As an example, we show the comparison of predictions of both models with the choice (4). Here $V_a = 2$ m/s, $T_{ab} = 45$ Celsius degrees, $H_{ra,b} = 30\%$, $L = 0.6$ m. X_{ab} is a known function of T_{ab} and $H_{ra,b}$, see [2] for instance. Initially, the temperature of the air and of the product is 25 Celsius degrees, and the moisture content of the product is $X_{p,0} = 0.6$. The void fraction is $\epsilon = 0.416$.

Figure 1 shows the time evolution of the air moisture content (left) and of the air temperature (right) at the position $x = 0.59$. Figure 2 shows the time evolution of the grain moisture content (left) and of the grain temperature at the same position. One can observe the strong variations at the beginning of the drying process. Here, we find that model (2) gives a drying time of

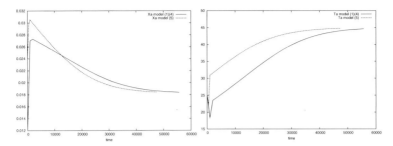

Fig. 1. Air moisture content (left) and temperature (right) for $x = 0.59$.

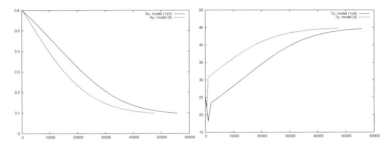

Fig. 2. Grain moisture content (left) and temperature (right) for $x = 0.59$.

about 57000 seconds, while model (5) computes a drying time around 48000 seconds. Moreover, the computed solutions differ.

References

1. Aregba, A.W., Nadeau, J.P.: Comparison of two non-equilibrium models for static grain deep-bed drying by numerical simulations. To appear in Journal of Food engineering (2006)
2. Aregba, A.W., Sebastian, P., Nadeau, J.P.: Stationary deep-bed drying: a comparative study between a logarithmic model and a non-equilibrium model. To appear in Journal of Food engineering (2005)
3. Aregba-Driollet, D.: Mathematical study of some models for deep-bed drying, in preparation.
4. Brooker, D.B., Bakker-Arkema, F.W., Hall, C.W.: Drying cereal grains. Westport, Connecticut, AVI (1974)
5. Chalabi, Z.S., Sun, Y., Pantelides, C.C.: Mathematical modelling and simulation of near ambient grain drying. Computers and Electronics in Agriculture **13**, 243–271 (1995)
6. Sharp, J.R.: A review of low temperature drying simulation models. J. Agric. Eng. Res. **27**, 169–190 (1982)
7. Spencer, H.B.: A mathematical simulation of grain drying. J. Agric. Eng. Res. **14** (3), 226–235 (1969)
8. Srivastava, V.K., John, J.: Deep bed drying modelling. Energy conversion and management **43**, 1689–1708 (2000)

Hybrid Godunov-Glimm Method for a Nonconservative Hyperbolic System with Kinetic Relations

Bruno Audebert[1] and Frédéric Coquel[2]

[1] ONERA, 29 Avenue de la Division Leclerc, 92322 Châtillon, France
bruno.audebert@ensta.org
[2] CNRS and Laboratoire Jacques-Louis Lions, UPMC, 75252 Paris cedex 05, France coquel@ann.jussieu.fr

Summary. We study the numerical approximation of a system from the physics of compressible turbulent flows, in the regime of large Reynolds numbers. The PDE model takes the form of a nonconservative hyperbolic system with singular viscous perturbations. Weak solutions of the limit system are regularization dependent and classical approximate Riemann solvers are known to grossly fail in the capture of shock solutions. Here, the notion of kinetic functions is used to derive a complete set of generalized jump conditions which keeps a precise memory of the underlying viscous mechanism. To enforce for validity these jump conditions, we propose a hybrid Godunov-Glimm method coupled with a local nonlinear correction procedure.

1 Introduction

We examine the numerical approximation of a system governing plane wave solutions of a second order closure model for compressible turbulence in two space dimensions [5]. This system has to be tackled in the regime of very large Reynolds numbers and can be given the following nonconservative form:

$$\partial_t \mathbf{u}^\epsilon + \mathcal{A}(\mathbf{u}^\epsilon)\partial_x \mathbf{u}^\epsilon = \epsilon \partial_x(\mathcal{D}(\mathbf{u}^\epsilon)\partial_x \mathbf{u}^\epsilon), \quad x \in I\!\!R, \ t > 0. \tag{1}$$

In the limit $\epsilon \to 0^+$, solutions involve in general shock waves which turn to be very sensitive with respect to the viscous tensor $\mathcal{D}$. Motivated by [2, 3, 4], we characterize all the entropy pairs of (1) so as to define precisely their associated dissipation rates (the so-called kinetic relations) when studying traveling waves of (1). This allows us to propose a complete set of generalized jump conditions which keeps full memory of the small scale sensitiveness. Properties of the available entropy pairs lead us to introduce an hybrid Godunov-Glimm method based on exact or approximate Riemann solutions. After [3, 4], this method then receives a local correction intending to keep all the discrete entropy rates in the exact balance prescribed by the kinetic relations.

2 The PDE model and main properties

The system under interest (see [1] for the physical background) reads :

$$\begin{cases} \partial_t \rho^\epsilon + \partial_x (\rho u)^\epsilon = 0, \\ \partial_t (\rho u)^\epsilon + \partial_x (\rho u^2 + p(\mathbf{u}) + R_{11})^\epsilon = \epsilon \partial_x (\mu \partial_x u^\epsilon), \\ \partial_t (\rho v)^\epsilon + \partial_x (\rho u v + R_{12})^\epsilon = \epsilon \partial_x (\nu \partial_x v^\epsilon), \\ \partial_t (\rho E)^\epsilon + \partial_x \{(\rho E + p(\mathbf{u}) + R_{11})u + R_{12}v\}^\epsilon = \epsilon \partial_x (\mu u^\epsilon \partial_x u^\epsilon + \nu v^\epsilon \partial_x v^\epsilon), \\ \partial_t R_{11}^\epsilon + \partial_x (R_{11} u)^\epsilon + 2 R_{11}^\epsilon \partial_x u^\epsilon = 0, \\ \partial_t R_{22}^\epsilon + \partial_x (R_{22} u)^\epsilon + 2 R_{12}^\epsilon \partial_x v^\epsilon = 0, \\ \partial_t R_{12}^\epsilon + \partial_x (R_{12} u)^\epsilon + R_{11}^\epsilon \partial_x v^\epsilon + R_{12}^\epsilon \partial_x u^\epsilon = 0, \end{cases} \quad (2)$$

where $\epsilon > 0$ denotes the inverse of a Reynolds number Re we address in the asymptotic regime $Re \to \infty$. Here, u and v denote the normal and tangential components of the Favre average $<U>$ of the instantaneous velocity U in the 2D compressible Navier-Stokes equations. Next, $R_{i,j}$, $1 \le i,j \le 2$, are the components of the symmetric Reynolds stress tensor $<U' \times U'>$ where U' stands for the departure of U from $<U>$. We assume a polytropic pressure law $p(\mathbf{u})$:

$$p(\mathbf{u}) = (\gamma - 1) \left\{ \rho E - \frac{(\rho u)^2 + (\rho v)^2}{2\rho} - \frac{R_{11} + R_{22}}{2} \right\}, \quad \gamma > 1, \quad (3)$$

while the viscosity coefficients μ and ν are two given positive constants satisfying $0 < \nu < \mu$. The natural phase space $\Omega_\mathbf{u} \subset \mathbb{R}^7$ is such that [1] :

$$\rho > 0, \ (u,v) \in \mathbb{R}^2, p(\mathbf{u}) > 0, \ R_{ii} \ge 0, \ i = 1, 2, \ R_{11} R_{22} - R_{12}^2 \ge 0. \quad (4)$$

The underlying first order system in (2) is hyperbolic for all $\mathbf{u} \in \Omega_\mathbf{u}$ with the following increasingly arranged eigenvalues [1] :

$$u - c(\mathbf{u}) < u - a(\mathbf{u}) \le u, \ u, \ u \le u + a(\mathbf{u}) < u + c(\mathbf{u}), \quad \mathbf{u} \in \Omega_\mathbf{u}. \quad (5)$$

where $a^2(\mathbf{u}) = \frac{R_{11}}{\rho} \ge 0$ and $c^2(\mathbf{u}) = \frac{\gamma p(\mathbf{u})}{\rho} + 3 a^2(\mathbf{u}) > 0$. The two extreme fields are genuinely nonlinear while all the other intermediate ones are linearly degenerate. It can be proved [1] that solely the discontinuities coming with the extreme fields give rise to ambiguities in the nonconservative products involved in the underlying hyperbolic system. Concerning these shock solutions, being given $\mathbf{u}_-$ in $\Omega_\mathbf{u}$ and a speed $\sigma \in \mathbb{R}$, it is well-known [2] that the right state $\mathbf{u}_+$ cannot be determined without an explicit reference to the precise shape of the viscous tensor $\mathcal{D}$ in (2). Here, the relevant definition of shock solutions follows from smooth solutions of (2) with $\epsilon = 1$, of the form :

$$\mathbf{u}(x,t) = \mathbf{w}(x - \sigma t) = \mathbf{w}(\xi), \quad \lim_{\xi \to \pm \infty} \mathbf{w}(\xi) = \mathbf{u}_\pm. \quad (6)$$

Then for fixed $\epsilon > 0$, the function $\mathbf{w}_\epsilon(\xi) = \mathbf{w}(\xi/\epsilon)$ leads to a traveling wave of (2) with again $\lim_{\xi \to \pm \infty} \mathbf{w}_\epsilon(\xi) = \mathbf{u}_\pm$. Since $||d_\xi \mathbf{w}_\epsilon||_{L^1(\mathbb{R})} = ||d_\xi \mathbf{w}||_{L^1(\mathbb{R})} < \infty$, the sequence $\{\mathbf{w}_\epsilon\}_{\epsilon > 0}$ is seen to converge strongly in $L^1_{loc}(\mathbb{R})$ as $\epsilon \to 0^+$ to :

$$\mathbf{u}(x,t) = \mathbf{u}_- + (\mathbf{u}_+(\sigma, \mathbf{u}_-; \mathcal{D}) - \mathbf{u}_-)H(x - \sigma t), \qquad (7)$$

referred to a shock solution [2] of the limit system in (2). The crucial issue stems from the very sensitiveness of $\mathbf{u}_+$ with respect to $\mathcal{D}$. Motivated by [2], we propose to encode this sensitiveness by investigating the existence of a change of unknown $\mathbf{u} \in \Omega_\mathbf{u} \to \mathbf{v}(\mathbf{u}) \in \Omega_\mathbf{v}$ so that the smooth solutions of (2) obey :

$$\partial_t \mathbf{v}(\mathbf{u}^\epsilon) + \partial_x \mathcal{F}(\mathbf{v}(\mathbf{u}^\epsilon)) = \epsilon \mathcal{R}_\mathcal{D}(\mathbf{v}(\mathbf{u}^\epsilon), \partial_x \mathbf{v}(\mathbf{u}^\epsilon), \partial_{xx} \mathbf{v}(\mathbf{u}^\epsilon)). \qquad (8)$$

Tackling the formal limit $\epsilon \to 0^+$ in (8) and assuming suitable estimates on the sequence $\mathbf{u}^\epsilon$ and its derivatives, the nonconservative term $\epsilon \mathcal{R}_\mathcal{D}(\mathbf{u}^\epsilon, \partial_x \mathbf{u}^\epsilon, \partial_{xx} \mathbf{u}^\epsilon)$ cannot be expected to converge to zero in the sense of measures as $\epsilon \to 0^+$ but instead to a bounded Borel measure $\Upsilon_\mathbf{u}$ concentrated on the shock discontinuities of the limit function $\mathbf{u}$. Thus and with (8), a shock solution (7) solves the following set of generalized Rankine-Hugoniot jump conditions :

$$-\sigma\left(\mathbf{v}(\mathbf{u})_+ - \mathbf{v}(\mathbf{u}_-)\right) + (\mathcal{F}(\mathbf{v}(\mathbf{u}_+)) - \mathcal{F}(\mathbf{v}(\mathbf{u}_-))) = \mathcal{K}_\mathcal{D}(\mathbf{u}_-, \sigma). \qquad (9)$$

where the so-called kinetic function $\mathcal{K}_\mathcal{D} : \Omega_\mathbf{u} \times \mathbb{R} \to \mathbb{R}^7$ denotes the mass of Borel measure $\Upsilon_\mathbf{u}$ given (with little abuse in the notations) from (6) by :

$$\mathcal{K}_\mathcal{D}(\mathbf{u}_-, \sigma) = <\Upsilon_\mathbf{w}, \mathbb{R}_\xi> = \int_{\xi \in \mathbb{R}} \mathcal{R}_\mathcal{D}(\mathbf{w}, \mathbf{w}', \mathbf{w}'') d\xi. \qquad (10)$$

Equivalent forms (8) clearly follow from the characterization of all the entropy pairs of (2). Besides useless nonlinear transforms, these are given by [1] :

$$\partial_t \{\rho E_t\}(\mathbf{u}^\epsilon) + \partial_x (\{\rho E_t\}(\mathbf{u}^\epsilon) u^\epsilon + R_{12}{}^\epsilon v^\epsilon) = \epsilon \partial_x (\nu v^\epsilon \partial_x v^\epsilon) - \epsilon \nu (\partial_x v^\epsilon)^2 \qquad (11)$$

with $\{\rho E_t\}(\mathbf{u}) = \rho \frac{v^2}{2} + \frac{R_{12}{}^2}{2R_{11}}$, the so-called tangential energy, then :

$$\partial_t \{\rho s\}(\mathbf{u}^\epsilon) + \partial_x \{\rho s\}(\mathbf{u}^\epsilon) u^\epsilon = \frac{\epsilon \mu}{T(\mathbf{u}^\epsilon)} (\partial_x u^\epsilon)^2 + \frac{\epsilon \nu}{T(\mathbf{u}^\epsilon)} (\partial_x v^\epsilon)^2, \qquad (12)$$

where $\{\rho s\}(\mathbf{u}) = \rho \log(\frac{p(\mathbf{u})}{\rho^\gamma})$ and at last :

$$\begin{array}{ll} \partial_t \{\rho \mathcal{W}\}(\mathbf{u}^\epsilon) + \partial_x \{\rho \mathcal{W}\}(\mathbf{u}^\epsilon) u^\epsilon = 0, & \{\rho \mathcal{W}\}(\mathbf{u}) = R_{22} - \frac{R_{12}{}^2}{R_{11}}, \\ \partial_t \{\rho \mathcal{I}\}(\mathbf{u}^\epsilon) + \partial_x \{\rho \mathcal{I}\}(\mathbf{u}^\epsilon) u^\epsilon = 0, & \mathcal{I}(\mathbf{u}) = R_{11} \tau^3. \end{array} \qquad (13)$$

Observe that these additionnal laws are well-defined when focusing on solutions with $R_{11}(x,t) > 0$ of sole real interest, i.e. for initial data with $R_{11}(x,0) > 0$, [1]. If in addition $R_{12}(x,0) > 0$ then $R_{12}(x,t) > 0$ but the sign of R_{12} may vary in the initial data (see (4)), thus neither ρE_t nor ρs can provide us with a suitable change of variable in full generality. Nevertheless, they are seen below to be valid if one pays attention to traveling wave solutions. Implementing the proposed framework first requires the study of traveling waves of (2), say for the first field from frame invariance reasons. Existence is proved [1] under the classical condition $u_- - c(\mathbf{u}_-) > \sigma$. The exit state $\mathbf{u}_+(\sigma, \mathbf{u}_-; \mathcal{D})$ heavily depends on the viscosity ratio ν/μ. Such a dependence can be tracked through the next generalized jump condition :

$$-\sigma\left[\{\rho E_t\}(\mathbf{u})\right] + [\{\rho E_t\}(\mathbf{u})u + R_{12}v] = \frac{<\Upsilon_{E_t}, I\!R_\xi>}{<\Upsilon_s, I\!R_\xi>}\left(-\sigma[\{\rho s\}(\mathbf{u})] + [\{\rho s\}(\mathbf{u})u]\right), \quad (14)$$

expressing that the jumps in ρE_t and ρs must evolve in the reported proportion. Modifying the ratio ν/μ directly affects the ratio $<\Upsilon_{E_t}, I\!R_\xi> / <\Upsilon_s, I\!R_\xi>$ (see (11)–(12)) and therefore $\mathbf{u}_+(\sigma, \mathbf{u}_-; \mathcal{D})$. This observation is of central importance hereafter. Next, for turbulent Mach numbers $\beta(\mathbf{u}_-) = a(\mathbf{u}_-)/c(\mathbf{u}_-)$ large enough (depending on γ in (3)), viscous profiles are seen to violate the geometric Lax conditions. They are indeed overcompressive :

$$u_+ - c(\mathbf{u}_+) < \sigma < u_- - c(\mathbf{u}_-), \quad \text{but with } u_+ - a(\mathbf{u}_+) < \sigma < u_- - a(\mathbf{u}_-). \quad (15)$$

In our non conservative setting, (15) implies the existence of an infinite number of traveling waves issuing from $\mathbf{u}_-$ and reaching as many distinct states $\mathbf{u}_+$. Choosing $\gamma = 1.4$, (15) arises for rather unexpectedly large values of $\beta(\mathbf{u}_-)$ [1] and we tacitly restrict ourselves to moderate values of $\beta(\mathbf{u}_-)$ to ensure uniqueness. Uniqueness implies that R_{12} keeps a constant sign in traveling waves in contrast with arbitrary solutions of (2). This important property clearly allows to express the generalized jump conditions in $\mathbf{v}(\mathbf{u}) = \{\rho, \rho u, \rho v, \rho E, \rho \mathcal{I}, \rho \mathcal{W}, \rho E_t\}$ while recovering the sign of $R_{12}{}^+$ from $R_{12}{}^-$. The kinetic function (10) reduces to one non trivial component Υ_{E_t}, it has been tabulated in [1] from a suitable numerical integration of (6). Its knowledge has allowed us to prove the monotony of the projections of the resulting shock curves in the plane $(u, p(\mathbf{u}) + R_{11})$. Consequently, the Riemann problem for the limit system in (2) is shown to admit an unique solution for initial data with moderate values of β. The general pattern of Riemann solutions is made of at most five simple waves separating six constant states :

$$\mathbf{u}_L \; 1-wave \; \mathbf{u}_1 \; 2-wave \; \mathbf{u}_2 \; 3-wave \; \mathbf{u}_3 \; 4-wave \; \mathbf{u}_4 \; 5-wave \; \mathbf{u}_R, \quad (16)$$

where $\mathbf{u}_L$ and $\mathbf{u}_R$ define the initial data. Here, the intermediate waves systematically coincide with contact discontinuities while across the extreme waves, R_{12} necessarily keeps a constant sign, namely :

$$R_{12}{}^{(1)} R_{12}{}^L \geq 0, \quad R_{12}{}^{(4)} R_{12}{}^R \geq 0, \quad (17)$$

with respectively strict inequality when $R_{12}{}^L \neq 0$ (resp. $R_{12}{}^R \neq 0$). In other words, the sign of R_{12} may only change within the fan of a given Riemann solution [1]. This property is of importance in the sequel.

3 Godunov-Glimm Hybrid method

We propose a Godunov-Glimm hybrid method to approximate the solutions of (2) in the limit $\epsilon \to 0^+$. In this brief paper, this procedure is exemplified on the exact Riemann solver for simplicity but the use of approximate solvers is under way. To motivate its introduction, we first emphasize the need for averaging ρE_t, whenever possible, in place of R_{12} within the frame of exact or approximate Godunov methods. In that aim, we approximate a strong shock solution (1) so that R_{12} keeps a constant sign and makes ρE_t to be an admissible change of variable. The results displayed in figure (2) highlight that averaging R_{12} (solid line) at each time step results in unacceptable errors in the capture of the exact shock solution (see [1] for the origin of the failure). By contrast, averaging directly ρE_t (dotted line) yields a much better agreement even if some error still persists. A correction is proposed in the last section but can only take place when suitably averaging ρE_t. Indeed for general initial

ρ_L	1	ρ_R	2.69197
u_L	890.20812	u_R	393.54227
v_L	0	v_R	−1.20350
p_L	10^5	p_R	4.65756 10^5
R_{11}^L	1443.29897	R_{11}^R	28155.86059
R_{22}^L	10^3	R_{22}^R	2695.73233
R_{12}^L	144.32990	R_{12}^R	1095.34000

Fig. 1. Initial data for a one shock wave

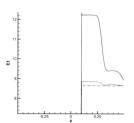

Fig. 2. Tangential energy

data, the sign of $R_{12}(x,t)$ may vary in the solution and prevents us from averaging ρE_t uniformly in each cell. But when approximating such a solution with a sequence of non interacting Riemann solutions, R_{12} keeps locally a constant sign through the extreme waves (responsible for shock solutions) in each successive fans. Roughly speaking, we take advantage of this property when averaging locally in each cell ρE_t wherever R_{12} keeps a constant sign. More precisely, we adopt a three-steps procedure based on a Lagrangian-Eulerian method to simplify the notations. A direct Eulerian method can be derived along the same lines. With classical notations, let be given a piecewise constant approximate solution $\mathbf{u}_{\Delta x}^n(x)$ at time t^n (the space step Δx being constant). This solution is evolved to the next date t^{n+1} according to :

1. *Evolution in time.* We solve the the Cauchy problem in Lagrangian coordinates with initial data $\mathbf{w}(y,0) = \mathbf{w}(\mathbf{u}_{\Delta x}^n(x))$ where $\mathbf{w} = (\tau, u, v, E, \mathcal{I}, R_{12}, \mathcal{W})$. Choosing a CFL less than $1/2$, the solution $\mathbf{w}(y,t)$ is made of a sequence of noninteracting Riemann solutions $w(., \mathbf{w}_i^n, \mathbf{w}_{i+1}^n)$ centered at each interface.
2. *Local averaging within each cell.* Each Lagrangian cell L_j^n is split into three domains depicted in figure (3). By construction $\mathbf{w}(y,t)$ is nothing but the constant state $(\mathbf{w}_i^{n+1-})_L = w(0+, \mathbf{w}_{i-1}^n, \mathbf{w}_i^n)$ in Ω_i^L : see indeed the wave

pattern (16) but expressed in Lagrangian coordinates. The same holds in Ω_i^R with $(\mathbf{w}_i^{n+1-})_R = w(0-, \mathbf{w}_i^n, \mathbf{w}_{i+1}^n)$. Then possible shock solutions only propagate within the domain Ω_i^C where R_{12} keeps a constant sign and this leads us to define an intermediate constant state when averaging the next PDEs over Ω_i^C :

$$\partial_t \mathbf{v} + \partial_y \mathcal{G}(\mathbf{v}) = \Upsilon_\mathbf{v}, \tag{18}$$

with $\mathbf{v} = (\rho, u, v, E, \mathcal{I}, E_t, \mathcal{W})$ (i.e. with E_t in place of R_{12}) to get :

$$\begin{cases} (\mathbf{v}_i^{n+1-})_C = \frac{1}{1-\lambda_l \Delta(\rho a)} \Big\{ \mathbf{v}_i^n - \lambda_l \Big((\rho a)_{i+\frac{1}{2}}^L \mathbf{v}_{i+\frac{1}{2}}(0^+) + (\rho a)_{i-\frac{1}{2}}^R \mathbf{v}_{i-\frac{1}{2}}(0^+) \\ + \mathcal{G}(\mathbf{v}_{i+\frac{1}{2}}(0^+)) - \mathcal{G}(\mathbf{v}_{i-\frac{1}{2}}(0^+)) \Big) + \lambda_l \sum_{\text{shock} \in L_i^n} < \Upsilon_\mathbf{v}, L_i^n > \Big\}, \end{cases} \tag{19}$$

with $\lambda_l = \frac{\Delta t}{\rho_j^n \Delta x}$ and $\Delta(\rho a) = (\rho a)_{i+\frac{1}{2}}^L - (\rho a)_{i-\frac{1}{2}}^R$. We then set $(\mathbf{w}_i^{n+1-})_C = \mathbf{w}((\mathbf{v}_i^{n+1-})_C)$ defining $(R_{12})_i^{n+1-}$ from $(E_t)_i^{n+1-}$ and the sign of $(R_{12})_i^n$.

3 **Eulerian step with sampling.** This step amounts to solve transport equations [1] with speed u. Let us introduce $x_{i+\frac{1}{2}}^* = x_{i+\frac{1}{2}} + \Delta t\, u_{i+\frac{1}{2}}^n$ so as to define the following piecewise constant function in each Eulerian cell at time t^{n+1-} (see figure (4) and [1] for the details):

$$\mathbf{w}_i^{n+1-}(x) = \begin{cases} (\mathbf{w}_{i-1}^{n+1-})_R, & \text{if } x_{i-\frac{1}{2}} < x \le x_{i-\frac{1}{2}}^*, \\ (\mathbf{w}_i^{n+1-})_L, & \text{if } x_{i-\frac{1}{2}}^* < x \le x_{i-\frac{1}{2}}^* + (a)_{i-\frac{1}{2}}^R \Delta t, \\ (\mathbf{w}_i^{n+1-})_C, & \text{if } x_{i-\frac{1}{2}}^* + (a)_{i-\frac{1}{2}}^R \Delta t < x \le x_{i+\frac{1}{2}}^* - (a)_{i+\frac{1}{2}}^L \Delta t, \\ (\mathbf{w}_i^{n+1-})_R, & \text{if } x_{i+\frac{1}{2}}^* - (a)_{i+\frac{1}{2}}^L \Delta t < x \le x_{i+\frac{1}{2}}^*, \\ (\mathbf{w}_{i+1}^{n+1-})_L, & \text{if } x_{i+\frac{1}{2}}^* < x \le x_{i+\frac{1}{2}}. \end{cases} \tag{20}$$

In order to define a unique constant state in each cell, we propose to sample the above function with a VanDerCorput sequence $\{\theta(n)\}_{n \ge 0}$, $\theta(n) \in [0,1]$, as in Glimm method [1] :

$$\mathbf{u}_i^{n+1} = \mathbf{u}(\mathbf{w}_i^{n+1-}(x_{i-1/2} + \theta(n)\Delta x)), \quad i \in \mathbb{Z}. \tag{21}$$

This completes the description of the hybrid method.

4 The nonlinear projection

To motivate the introduction of an additional correction step in the hybrid method, we first comment on the persistence of an error when averaging locally in each cell ρE_t (see figure (2)). Its roots are found in the property that exact shock solutions are regularization-dependent : the (global or local as well) averaging procedure in each cell induces artificial dissipation, distinct in nature from the exact one. This discrepancy tends to corrupt the discrete shock

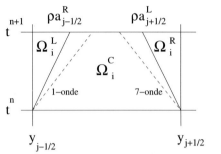
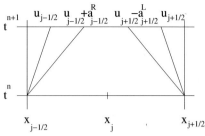

Fig. 3. The Lagrangian cell

Fig. 4. The Eulerian cell

profiles. Here we propose to enforce the artificial dissipation in the numerical method to mimic the exact dissipation mechanism. Kinetic functions play a central role in the correction procedure. It has been introduced by Berthon, Coquel [3] and then extended in Chalons, Coquel [4] for Navier Stokes equations with several independent entropies. In the present setting, it can be seen [1] that the Lagragian form of the generalized jump condition (14) is violated in the domain $(\Omega_i^n)_C$ of each Lagrangian cell, exactly where E_t is averaged at each time step. More precisely, E_t and $s(\mathbf{v})$ no longer evolve in the required proportion (14) because of spurious terms of size $\mathcal{O}(\|\mathbf{v}_{i+1}^n - \mathbf{v}_i^n\|^2)$. Clearly, errors are quite large in a discrete shock profile. Underlining again that (14) precisely reflects the sensitiveness of shock solutions to the exact dissipation mechanism, we are led to enforce for validity this generalized jump condition at the discrete level.

In that aim, let us again emphasize that incriminating averages only take place in each domain $(\Omega_i^n)_C$ in the Lagrangian step of the Hybrid method. Let us relabel $(\mathbf{v}_i^{n+1,=})_C$ the resulting prediction. The correction procedure then naturally takes place as a third step just before the sampling and last step. We propose to keep unchanged the local averages of the conservative variables:

$$
\begin{aligned}
(\tau_i^{n+1,-})_C &= (\tau_i^{n+1,=})_C, & (u_i^{n+1,-})_C &= (u_i^{n+1,=})_C, \\
(v_i^{n+1,-})_C &= (v_j^{n+1,=})_C, & (E_i^{n+1,-})_C &= (E_i^{n+1,=})_C, \\
(\mathcal{I}_i^{n+1,-})_C &= (\mathcal{I}_i^{n+1,=})_C, & (\mathcal{W}_i^{n+1,-})_C &= (\mathcal{W}_i^{n+1,=})_C,
\end{aligned}
\qquad (22)
$$

and we recalculate $((E_t)_i^{n+1,-})_C$ as the solution of the next nonlinear algebraic equation :

$$
((E_t)_i^{n+1,-})_C - (E_t)_i^{n+1;\star} = \frac{\sum_{\text{shock} \in L_i^n} <\Upsilon_{E_t}, L_i^n>}{\sum_{\text{shock} \in L_i^n} <\Upsilon_s, L_i^n>} \\
\times (\{s\} ((\mathbf{v}_i^{n+1,-})_C) - s_i^{n+1,\star})
\qquad (23)
$$

where

$$(E_t)_i^{n+1;\star} = C_l \left((E_t)_i^n - \lambda_l \left((\rho a)_{i+\frac{1}{2}}^L (E_t)_{i+\frac{1}{2}}(0^+) + (\rho a)_{i-\frac{1}{2}}^R (E_t)_{i-\frac{1}{2}}(0^+) + \Delta (R_{12}v)_{i+\frac{1}{2}}^n \right) \right), \qquad (24)$$

$$s_i^{n+1,\star} = C_l \left(\{s\}(\mathbf{v}_i^n) - \lambda_l \left((\rho a)_{i+\frac{1}{2}}^L s_{i-\frac{1}{2}}(0^+) + (\rho a)_{i-\frac{1}{2}}^R s_{i+\frac{1}{2}}(0^+) \right) \right), \qquad (25)$$

with $C_l = \frac{1}{1-\lambda_l \Delta(\rho a)_{i+\frac{1}{2}}^n}$ and $\lambda_l = \frac{\Delta t}{\rho_i^n \Delta x}$. Let underline from (22) that $\{s\}((\mathbf{v}_i^{n+1,-})_C))$ in (23) has to be understood as a nonlinear function of solely $((E_t)_i^{n+1,-})_C$. Solving (23) gives $(\mathbf{w}_i^{n+1-})_C = \mathbf{w}((\mathbf{v}_i^{n+1-})_C)$. We are then in a position to apply the last step in the hybrid method : namely, the Eulerian step with sampling. This concludes the algorithm. Figures (5) and (6) illustrate the benefit of the correction technique for a Riemann problem with initial data given below:

$\rho_L = 1, u_L = 100, v_L = 0.2, p_L = 10^5, R_{11}^L = 10^4, R_{22}^L = 7.10^3, R_{12}^L = 5.10^3,$
$\rho_R = 0.9, u_R = -100, v_R = 0, p_R = 10^5, R_{11}^R = 8.10^3, R_{22}^R = 8.10^3, R_{12}^R = 4.510^3.$

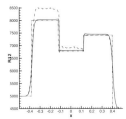

Fig. 5. Tangential component R_{12} **Fig. 6.** Transversal velocity

——— Correction   - - - - - Without correction   - - - - - Exact solution

References

1. Audebert, B.: Shock Wave-Turbulence Interactions in Reynolds Stress Models. PhD Thesis, ONERA and University Pierre et Marie Curie, France (2006)
2. C. Berthon, Coquel F., Lefloch, P.G.: Entropy dissipation measure and kinetic relation associated with nonconservative hyperbolic systems, to be submitted.
3. Berthon, C., Coquel, F.: Nonlinear projection methods for multi-entropies Navier-Stokes systems, Innovative methods for numerical solutions of partial differential equations, Hafez M.M. et al., World Scientific, 278–304 (2002)
4. Chalons, C., Coquel F.: The Riemann problem for the multi-pressure Euler system, Journal of Hyperbolic Differential Equations, **2**, 3, 745–782 (2005)
5. Brun, G., Hérard, J. M., Jeandel, D., Uhlmann, M.: Roe-type Riemann Solver for a Class of Realizable Second Order Closures, IJCFD, **13**, 223–249 (2000)

Cell-Average Multiwavelets Based on Hermite Interpolation*

F. Aràndiga, A. Baeza, and R. Donat

Departament de Matemàtica Aplicada, Universitat de València.
C/Doctor Moliner, 50, 46100, Burjassot (València), Spain.
{arandiga,antonio.baeza,donat}@uv.es

Summary. Harten's interpolatory multiresolution representation of data [2] has been extended in the case of point-value discretization to include Hermite interpolation by Warming and Beam in [3]. In this work we extend Harten's framework for multiresolution analysis to the vector case for cell-averaged data, focusing on Hermite interpolatory techniques. Some numerical experiments compare the algorithm with some well known scalar methods.

1 Introduction

In [4] Warming and Beam introduced a multiresolution analysis based on Hermite interpolation, within Harten's point-value multiresolution framework. Hermite interpolation uses both function and derivative point-values, hence multiple variables are required at each grid point, leading to the so-called vector multiresolution. Hermite-type reconstruction techniques allow for an increase in the order of accuracy of the reconstruction without increasing the support of the basis functions in the multiscale transformation, which is important in some applications.

In this paper we extend the construction in [4] to the cell-average framework. The organization of the paper is as follows: in Sect. 2 we briefly describe the general framework for multiresolution analysis of Harten and the particular case of discretization by cell-averages. In Sect. 3 we present our extension of the cell-average framework to the vector case. Finally some numerical experiments and comparisons are described in Sect. 4.

2 Harten's framework for multiresolution analysis

A multiresolution analysis, as described by Harten (see [3] and references therein) is defined by a nested sequence of linear vector spaces V^k, a sequence

---

* Research supported by EUCO Project HPRN-CT-2002-00286

of linear decimation operators $\{D_k^{k-1}\}_{k=n+1}^m$, with $D_k^{k-1} : V^k \longrightarrow V^{k-1}$, and a sequence of prediction operators $\{P_{k-1}^k\}_{k=n+1}^m$, $P_{k-1}^k : V^{k-1} \longrightarrow V^k$, satisfying the compatibility condition:

$$D_k^{k-1} P_{k-1}^k = I_{V^{k-1}}. \tag{1}$$

Given $v^k \in V^k$, the *prediction error* is defined as $e^k = Q_k v^k := (I_{V^k} - P_{k-1}^k D_k^{k-1}) v^k$. Each *error* vector, e^k belongs to $\mathcal{N}(D_k^{k-1})$ (the null space of D_k^{k-1}). Let $G_k : V^k \longrightarrow \mathcal{N}(D_k^{k-1})$ be the operator which assigns to each vector $e^k \in V^k$ the coefficients d^k of its representation in terms of a given basis, $\{\mu_j^k\}$, of $\mathcal{N}(D_k^{k-1}) \subset V^k$, and let E_k be the canonical injection $\mathcal{N}(D_k^{k-1}) \hookrightarrow V^k$. These operators verify

$$G_k E_k = I_{\mathcal{N}(D_k^{k-1})}, \quad E_k G_k = I_{V^k}.$$

The *non-redundant* information in the error vector is contained in the set of coefficients $\{d_j^k\}$, called the scale coefficients at level k. We note that v^k and $\{v^{k-1}, d^k\}$ have the same cardinality and contain the same information. Given $v^k \in V^k$ we evaluate

$$v^{k-1} = D_k^{k-1} v^k, \quad d^k = G_k(I_{V^k} - P_{k-1}^k D_k^{k-1}) v^k \tag{2}$$

and given v^{k-1} and d^k computed by (2) the vector v^k is recovered by the inverse formula $v^k = P_{k-1}^k v^{k-1} + E_k d^k$.

This gives the equivalence between v^k and $\{v^{k-1}, d^k\}$. By repeating step (2) for v^{k-1} one obtains its corresponding decomposition $\{v^{k-2}, d^{k-1}\}$, and iterating this process from $k = m$ to $n+1$ we find that a multiresolution setting $\{\{V^k\}_{k=n}^m, \{D_k^{k-1}\}_{k=n+1}^m\}$ and a sequence of corresponding prediction operators $\{P_{k-1}^k\}_{k=n+1}^m$ satisfying (1) define an invertible multiresolution transform.

The decimation and prediction operators can be built from a sequence of discretization operators $\mathcal{D}_k : \mathcal{F} \longrightarrow V^k$ and a sequence of reconstruction operators $\mathcal{R}_k : V^k \longrightarrow \mathcal{F}$. The reconstruction operators $\mathcal{R}_k$ have to be *compatible* with $\mathcal{D}_k$, i. e.

$$\mathcal{D}_k \mathcal{R}_k = I_{V^k}, \forall k \in \{n, \ldots, m\} \tag{3}$$

The decimation and prediction operators are thus built via the standard relations $D_k^{k-1} = \mathcal{D}_{k-1} \mathcal{R}_k$ and $P_{k-1}^k = \mathcal{D}_k \mathcal{R}_{k-1}$. The compatibility condition (1) is a consequence of (3).

In Harten's framework, the discretization process specifies the setting, then the choice of a reconstruction operator defines a multiresolution transformation whose properties are closely related to those of the reconstruction. From the point of view of data-compression applications, accuracy of the reconstruction is an important feature. Stability of the resulting transformation is also essential.

Consider the set of nested dyadic grids defined in $[0,1]$:

$$X^k = \{x_j^k\}_{j=0}^{J_k}, \quad J_k = 2^k, \quad x_j^k = jh_k, \quad h_k = \frac{1}{J_k}, \quad k = n, \ldots, m. \qquad (4)$$

In the case of discretization by cell-averages let $f \in \mathcal{F} = L^1([0,1])$ and consider the set of nested dyadic grids defined by (4). The cell-average discretization operator $\mathcal{D}_k : \mathcal{F} \longrightarrow V^k$ is defined in [3] as follows:

$$\bar{f}_j^k := (\mathcal{D}_k f)_j = \frac{1}{h_k} \int_{x_{j-1}^k}^{x_j^k} f(x)dx, \quad 1 \leq j \leq J_k.$$

The decimation is then computed by:

$$\bar{f}_j^{k-1} = \frac{1}{h_{k-1}} \int_{x_{j-1}^{k-1}}^{x_j^{k-1}} f(x)dx = \frac{1}{2h_k} \int_{x_{2j-2}^k}^{x_{2j}^k} f(x)dx = \frac{1}{2}(\bar{f}_{2j}^k + \bar{f}_{2j-1}^k).$$

Piecewise polynomial reconstructions can be built by considering $F(x) = \int_0^x f(y)dy \in \mathcal{C}([0,1])$, the *primitive function* of f, and its point-value discretization $\{F_j^k\}_{j=0}^{J_k}, F_j^k = F(x_j^k)$. If $\mathcal{I}_{k-1}(x, F^{k-1})$ is an interpolatory reconstruction of $\{F_j^{k-1}\}$ we can obtain a reconstruction operator in the cell-average setting by defining $(\mathcal{R}_{k-1}f^{k-1})(x) = \frac{d}{dx}\mathcal{I}_{k-1}(x, F^{k-1})$. The resulting prediction formulas [2] are:

$$\begin{aligned}(P_{k-1}^k \bar{f}^{k-1})_{2j-1} &= \frac{1}{h_k}\left(\mathcal{I}_{k-1}(x_{2j-1}^k, F^{k-1}) - F_{j-1}^{k-1}\right), \\ (P_{k-1}^k \bar{f}^{k-1})_{2j} &= 2\bar{f}_j^k - (P_{k-1}^k \bar{f}^{k-1})_{2j-1}.\end{aligned} \qquad (5)$$

When $\mathcal{I}_{k-1}(x, F^{k-1})$ is a piecewise polynomial interpolatory function, with polynomial pieces constructed by Lagrange interpolation, the prediction formulas (5) can be written without explicitly computing the values of F on X^{k-1}. See [2] for a more detailed description.

3 Vector multiresolution analysis for cell-averaged data

We extend here the ideas of [4] to the cell-average multiresolution framework. Recalling ideas from Sect. 2 A way to make such an extension is to apply the Hermite interpolatory technique to the primitive function F. To carry out our construction, let us start by considering a function $f \in \mathcal{C}([0,1])$. Then its primitive function, $F(x) = \int_0^x f(y)dy$, belongs to $\mathcal{C}^1([0,1])$.

The Hermite piecewise polynomial interpolatory reconstruction for F on the grid X^k, is computed as follows: A polynomial Q_j^k is computed in each cell $c_j^k = [x_{j-1}^k, x_j^k]$ by imposing the conditions:

$$Q_j^k(x_{j-1}^k) = F(x_{j-1}^k), \quad Q_j^{k-1}(x_j^k) = F(x_j^k),$$
$$\frac{d}{dx}Q_j^k(x_{j-1}^k) = F'(x_{j-1}^k), \quad \frac{d}{dx}Q_j^k(x_j^k) = F'(x_j^k),$$
(6)

and we define $Q^k(x) := Q_j^k(x), \quad \forall x \in c_j^k$.

In what follows we describe the multiresolution algorithms that result from using this piecewise polynomial function in the reconstruction process. The input data for our vector multiresolution analysis are the cell-averages and the point values of a function f on a given (fine) mesh. We then consider the vector discretization operator:

$$\mathbf{f}^k := \mathcal{D}_k f = \left[\{\bar{f}_j^k\}_{j=1}^{J_k}, \{f_j^k\}_{j=0}^{J_k} \right].$$
(7)

where $\bar{f}_j^k = \frac{1}{h_k} \int_{x_{j-1}^k}^{x_j^k} f(x)dx$ and $f_j^k = f(x_j^k)$.

A *compatible*, in the sense of (3), reconstruction operator for the discretization operator defined in (7) is given in terms of the piecewise polynomial function Q^k defined above, as follows

$$(\mathcal{R}_k \mathbf{f}^k)(x) = \frac{d}{dx}Q^k(x).$$
(8)

Notice that the function $Q^k(x)$ is constructed using the point-values of the primitive function (obtained from the cell-average data) and the point-values of the function itself. We recall that the values $F(x_j^k)$ and $F'(x_j^k)$ are related to the cell-averages and the point-values of the function $f(x)$, respectively, by $F(x_j^k) = h_k \sum_{l=0}^{j} \bar{f}_l^k$ and $F'(x_j^k) = f(x_j^k), \quad j = 0, \ldots, J_k$. Thus, the polynomial pieces Q^k can be constructed from the knowledge of both the cell-averages and the point-values of f on a given grid.

The compatibility condition $\mathcal{D}_k \mathcal{R}_k = I_{V^k}$ is a consequence of the following relations, whose proof is straightforward,

$$\bar{f}_j^k = \frac{1}{h_k} \int_{x_{j-1}^k}^{x_j^k} (\mathcal{R}_k \mathbf{f}^k)(x)dx,$$
$$(\mathcal{R}_k \mathbf{f}^k)(x_{j-1}^k) = f(x_j^k), \quad (\mathcal{R}_k \mathbf{f}^k)(x_j^k) = f(x_j^k).$$
(9)

In fact the piecewise polynomial reconstruction obtained by imposing the conditions (9) on the cell-averages and the point-values of f at each subinterval is the same reconstruction obtained by imposing the Hermite-type conditions (6) on the primitive function F at each subinterval and computing the reconstruction via (8), i.e., conditions (6) and (9) are equivalent.

In Harten's framework, the multiresolution transformations are perfectly defined once the discretization and prediction operators have been specified. The decimation operator corresponding to the discretization in (7) acts on sequences $\mathbf{f}^k$ as follows:

$$\mathbf{f}^{k-1} = (D_k^{k-1}\mathbf{f}^k) = \left[\left\{\frac{\bar{f}^k_{2j-1} + \bar{f}^k_{2j}}{2}\right\}_{j=1}^{J_{k-1}}, \{f^k_{2j}\}_{j=0}^{J_{k-1}}\right],$$

The corresponding (compatible) prediction operator is built, accordingly, as $P^k_{k-1} = \mathcal{D}_k \mathcal{R}_{k-1}$. The prediction operator has two components that we denote by $(P^k_{k-1}\mathbf{f}^k)^1$ and $(P^k_{k-1}\mathbf{f}^k)^2$. A simple calculation leads to:

$$(P^k_{k-1}\mathbf{f}^{k-1})^1_{2j} = 2\bar{f}^{k-1}_j - (P^k_{k-1}\mathbf{f}^{k-1})^1_{2j-1},$$
$$(P^k_{k-1}\mathbf{f}^{k-1})^2_{2j} = f^{k-1}_j.$$

Hence, to completely define the multiscale transformations, we only need to obtain an explicit formula for $(P^k_{k-1}\mathbf{f}^{k-1})^1_{2j-1}$ and $(P^k_{k-1}\mathbf{f}^{k-1})^2_{2j-1}$, where

$$(P^k_{k-1}\mathbf{f}^{k-1})^1_{2j-1} = \frac{1}{h_k}\int_{x^k_{2j-2}}^{x^k_{2j-1}}(\mathcal{R}_{k-1}\mathbf{f}^{k-1})(x)dx,$$
$$(P^k_{k-1}\mathbf{f}^{k-1})^2_{2j-1} = (\mathcal{R}_{k-1}\mathbf{f}^{k-1})(x^k_{2j-1}).$$

In practice one does not need to compute the reconstruction $\mathcal{R}_{k-1}\mathbf{f}^{k-1}(x)$ explicitly and, as in the scalar case, all computations can be carried out without computing the values of the primitive function. In fact, a straightforward calculation leads to

$$(P^k_{k-1}\mathbf{f}^{k-1})^1_{2j-1} = \bar{f}^{k-1}_j - \frac{1}{4}(f^{k-1}_j - f^{k-1}_{j-1}).$$
$$(P^k_{k-1}\mathbf{f}^k)^2_{2j-1} = \frac{3}{2}\bar{f}^{k-1}_j - \frac{1}{4}(f^{k-1}_j + f^{k-1}_{j-1}). \tag{10}$$

The direct and inverse multiresolution algorithms are finally as follows:

Direct

$$\begin{cases} \text{for } k = m, \ldots, n+1 \\ \bar{f}^{k-1}_j = \frac{\bar{f}^k_{2j} + \bar{f}^k_{2j-1}}{2} \\ f^{k-1}_j = f^k_{2j} \\ (d^k_{\bar{f}})_j = \bar{f}^k_{2j-1} - \bar{f}^{k-1}_j + \frac{1}{4}\left(f^k_{2j} - f^k_{2j-2}\right) \\ (d^k_f)_j = f^k_{2j-1} - \frac{3}{2}\bar{f}^{k-1}_j + \frac{1}{4}\left(f^k_{2j} + f^k_{2j-2}\right) \\ \text{end} \end{cases}$$

If $f(x)$ is at least three times differentiable, the reconstruction is third order accurate. Hence, good compression properties are expected for smooth functions. In the presence of a singularity, the fact that the reconstruction technique is compact, implies that the area in which the accuracy of the reconstruction is lost is reduced to the interval in which the singularity is located.

Inverse
$$\begin{cases} \textbf{for } k = n+1, \ldots, m \\ \quad \bar{f}_{2j-1}^{k} = (d_{\bar{f}}^{k})_j + \bar{f}_j^{k-1} - \frac{1}{4}\left(f_j^{k-1} - f_{j-1}^{k-1}\right) \\ \quad \bar{f}_{2j}^{k} = 2\bar{f}_j^{k-1} - \bar{f}_{2j-1}^{k} \\ \quad f_{2j-1}^{k} = (d_f^{k})_j + \frac{3}{2}\bar{f}_j^{k-1} - \frac{1}{4}\left(f_j^{k-1} + f_{j-1}^{k-1}\right) \\ \quad f_{2j}^{k} = f_j^{k-1} \\ \textbf{end} \end{cases}$$

4 Numerical experiments

Data compression is an important application of multiresolution transformations. Many scale coefficients are often very small. Setting these to zero leads to a compressed version of the signal that, after reconstruction, is expected to provide an accurate approximation to the original data. We will compare the vector multiresolution algorithm described in Sect. 3 with the same-order, linear, scalar algorithm based on centered Lagrange interpolation, and a non-linear scalar algorithm of the same order based on *essentially non-oscillatory* (ENO henceforth) interpolation (see [2] for details on these multiscale transformations). The compression procedure used here for the scalar algorithms is truncation or hard-thresholding. It consists of setting to zero those scale coefficients whose module is smaller than a prescribed tolerance [3]. In the vector case we have two sets of scale coefficients. The approach used here to compress a multiscaled signal consists of setting to zero a pair of scale coefficients (corresponding to the same spatial location) if their modules are both below the given tolerance.

Let T_m be the number of nonzero elements of the compressed data. The compression ratio is defined by:

$$C_r = \frac{J_m}{T_m} \geq 1. \tag{11}$$

In order to compare the performance of the vector algorithm versus the scalar algorithms, we start with a sequence of (cell-averaged) data and implement the vector algorithm computing approximations to the point-values using a nonlinear, ENO-based algorithm applied on the finest grid. The algorithm is developed in [1] for point-value discretized data, and is applied here to the primitive function. See [1] for further details. This ensures that the stored, compressed, data will have the same size in all cases.

Our first test function is Harten's function:

$$f(x) = \begin{cases} \frac{1}{2}\sin(3\pi x) & \text{if } x \leq \frac{1}{3}, \\ |\sin(4\pi x)| & \text{if } \frac{1}{3} < x \leq \frac{2}{3}, \\ -\frac{1}{2}\sin(3\pi x) & \text{if } x > \frac{2}{3}. \end{cases} \tag{12}$$

This function presents two different types of singularities (jump discontinuities and a corner), and is often used as test function for multiresolution algorithms.

The second test function is the *chirp* function:

$$f(x) = \sin(\alpha\pi(x-0.5)^3) \tag{13}$$

It is a smooth function with high variations at some regions, with which linear methods often perform very well and nonlinear methods typically exhibit a quite unstable behavior. In our tests we have used $\alpha = 64$. In Figs. 1 and 2 the dots represent the computed approximation and the solid line represents the original function.

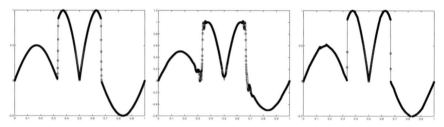

Fig. 1. Vector algorithm (left) linear scalar algorithm (center) and ENO scalar algorithm(right). $m = 10, n = 3, C_r = 32$ ($J_m = 1024, T_m = 32$) in all cases. Test function (12)

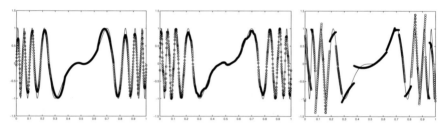

Fig. 2. Vector algorithm (left) linear scalar algorithm (center) and ENO scalar algorithm(right). $m = 10, n = 3, C_r = 32$ ($J_m = 1024, T_m = 32$) in all cases. Test function (13)

In Fig. 1 we examine the advantages of using the compact reconstruction of the vector algorithm versus the centered Lagrange interpolation of the linear algorithm and the ENO algorithm for non-smooth functions such as (12). Notice that the performance of the vector algorithm is clearly superior respect to the linear algorithm for high compression ratios: The blurry,

Gibbs-like effects observed in the signal reconstructed with the linear algorithm are absent in the reconstruction obtained with the vector algorithm.

The vector algorithm also outperforms the scalar ENO algorithm for large compression ratios. It can be observed that the smooth areas in the signal are better reconstructed when the vector algorithm is used.

In Fig. 2 we observe that the vector algorithm produces results similar to the linear method and does not present artifacts as the ENO method does.

Finally, we present in Fig. 3 a plot that display the 2-norm of the difference between the original signal and the reconstruction from its compressed representation against the compression ratio, for each one of the test functions and algorithms under consideration. We see that the evaluation of performance is similar for the vector and ENO methods with the first test function, and better than the linear method, except for small compression ratios, in which the three methods have similar behavior. With the second test function the vector algorithm is competitive against the linear algorithm and clearly outperforms the ENO method, which has stability problems for high compression ratios.

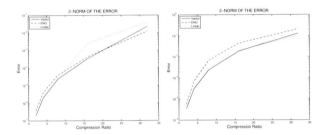

Fig. 3. 2-Norm of the error for a given compression ratio. Test functions (12) (left) and (13) (right). $m = 10, n = 3$.

References

1. Aràndiga, F., Baeza, A., Donat, R.: Discrete multiresolution analysis using Hermite interpolation: computing derivatives, Communications in Nonlinear Science and Numerical Simulation, **31**, 263–273 (2004)
2. Aràndiga, F., Donat, R.: Nonlinear multiscale decompositions: The approach of A. Harten, Num. Alg., **23**, 175–216 (2000)
3. Harten, A.: Multiresolution representation of data II: General framework, SIAM J. Numer. Anal., **33**, 1205–1256 (1996)
4. Warming, R., Beam, R.: Discrete multiresolution analysis using Hermite interpolation: biorthogonal multiwavelets, SIAM J. Sci. Comp., **22**, 1269–1317 (2000)

On a General Definition of the Godunov Method for Nonconservative Hyperbolic Systems. Application to Linear Balance Laws

M.J. Castro[1], J.M. Gallardo[1], M.L. Muñoz[2] and C. Parés[1]

[1] Dpt. Análisis Matemático, Facultad de Ciencias, Universidad de Málaga, Campus de Teatinos s/n, 29071 Málaga, Spain
castro, gallardo, pares@anamat.cie.uma.es
[2] Dpt. Matemática Aplicada, E.T.S.I. Telecomunicación, Universidad de Málaga, Campus de Teatinos s/n, 29071 Málaga, Spain
munoz@anamat.cie.uma.es

Summary. This work is concerned with the numerical approximation of Cauchy problems for one-dimensional nonconservative hyperbolic systems, for which it is assumed that each characteristic field is either genuinely nonlinear or linearly degenerate. The theory developed by Dal Maso, LeFloch and Murat [1] is used to define the concept of weak solutions of these systems, giving a sense to nonconservative products as Borel measures, based on the choice of a family of paths in the phases space. We establish some basic hypotheses concerning this family of paths which ensure the fulfilling of some good properties for weak solutions. A family of paths satisfying these hypotheses can be constructed at least for states that are close enough. In particular, we prove that the choice of such a family allows to write the Godunov method for a nonconservative system in a simple and general manner. The previous results are applied to a linear balance law, for which the Godunov method can be explicitly written and easily implemented.

1 Introduction

This work is concerned with the numerical approximation of Cauchy problems for one-dimensional nonconservative hyperbolic systems:

$$W_t + \mathcal{A}(W)W_x = 0, \quad x \in \mathbb{R}, \ t > 0, \tag{1}$$

where $W(x,t)$ belongs to Ω, an open convex subset of $\mathbb{R}^N$, and $W \in \Omega \mapsto \mathcal{A}(W) \in \mathcal{M}_N(\mathbb{R})$ is a smooth locally bounded map. We suppose that system (1) is strictly hyperbolic, that is, for each $W \in \Omega$ the matrix $\mathcal{A}(W)$ has N real distinct eigenvalues $\lambda_1(W) < \cdots < \lambda_N(W)$, with associated eigenvectors $R_1(W), \ldots, R_N(W)$. We also suppose that for each $i = 1, \ldots, N$, the characteristic field $R_i(W)$ is either genuinely nonlinear or linearly degenerate.

In general, the nonconservative product $\mathcal{A}(W)W_x$ does not make sense as a distribution. After the theory developed by Dal Maso, LeFloch and Murat ([2]) it is possible to give a definition of weak solutions associated to the choice of a family of paths in Ω. A *family of paths* in Ω is a locally Lipschitz map $\Phi\colon [0,1]\times \Omega \times \Omega \to \Omega$ which satisfies

$$\Phi(0; W_L, W_R) = W_L \text{ and } \Phi(1; W_L, W_R) = W_R, \text{ for any } W_L, W_R \in \Omega,$$

together with certain smoothness hypotheses. Once a family of paths Φ has been chosen, the nonconservative product $\mathcal{A}(W)W_x$ can be interpreted as a Borel measure for $W \in (L^\infty(\mathbb{R}\times\mathbb{R}^+)\cap BV(\mathbb{R}\times\mathbb{R}^+))^N$, denoted by $[\mathcal{A}(W)W_x]_\Phi$. A function $W \in (L^\infty(\mathbb{R}\times\mathbb{R}^+)\cap BV(\mathbb{R}\times\mathbb{R}^+))^N$ which is piecewise $\mathcal{C}^1$ is said to be a *weak solution* of (1) if it satisfies the equality

$$W_t + [\mathcal{A}(W)W_x]_\Phi = 0.$$

When no confusion arises, the dependency on Φ will be dropped.

Across a discontinuity, a weak solution must satisfy the generalized Rankine-Hugoniot condition:

$$\int_0^1 \left(\sigma\mathcal{I} - \mathcal{A}(\Phi(s;W^-,W^+))\right)\frac{\partial \Phi}{\partial s}(s;W^-,W^+)\,ds = 0, \tag{2}$$

where σ is the speed of propagation of the discontinuity, $\mathcal{I}$ is the identity matrix, and W^- and W^+ are the left and right limits of the solution at the discontinuity.

In the particular case of a system of conservation laws, that is, when $\mathcal{A}(W)$ is the Jacobian matrix of some flux function $F(W)$, the definition of the nonconservative product as a Borel measure does not depend on the choice of paths, and the generalized Rankine-Hugoniot condition reduces to the usual one.

As it occurs in the conservative case, not any discontinuity is admissible. Therefore, we must also assume a concept of entropic solution, as the one due to Lax or one related to an entropy pair.

Once a notion of entropy is chosen the theory of simple waves of hyperbolic systems of conservation laws and the results concerning the solutions of Riemann problems can be extended to systems of the form (1).

The choice of the family of paths is important because it determines the speed of propagation of shocks. The simplest choice is given by the family of segments, that corresponds to the definition of nonconservative products proposed by Volpert ([7]). In practical applications, it has to be based on the physical background of the problem. Even if the family of paths can be chosen arbitrarily, it is natural from the mathematical point of view to require this family to satisfy some hypotheses concerning the relation of the paths with the integral curves of the characteristic fields. The first goal of this paper is to establish three basic hypotheses of this nature. This is done in Section 2 where we show that, when these hypotheses are satisfied, there is a strong relation between the path linking two close states and the total Borel measure associated to the solution of the Riemann problem having these states as initial condition. A family of paths satisfying these hypotheses can always be constructed at least for states that are near enough in a sense to be determined.

In this work we are concerned with Godunov's methods for systems of the form (1). In Section 3 we show that, when the family of paths satisfies the above mentioned hypotheses, these methods can be written under a natural form that generalizes the classical expression of the Godunov method for systems of conservation laws.

In order to verify in practice the properties of Godunov's methods, in Section 5 a linear balance law is considered, i.e. a system

$$W_t + AW_x = CW\frac{d\sigma}{dx}, \tag{3}$$

where A is a diagonalizable and regular matrix. For this particular case, a Godunov method based on a family of paths satisfing the basic hypotheses (at least for close states) can be explicitly written and easily implemented.

The complete proofs of the results presented here and more details can be found in [6].

2 Choice of paths

The choice of the family of paths Φ is important in the definition of weak solutions of system (1) because it determines the speed of propagation of shocks. We will suppose that the family of paths satisfies the following hypotheses:

(H1) Given two states W_L and W_R belonging to the same integral curve γ of a linearly degenerate field, the path $\Phi(\cdot; W_L, W_R)$ is a parametrization of the arc of γ linking W_L and W_R.
(H2) Given two states W_L and W_R belonging to the same integral curve γ of a genuinely nonlinear field, R_i, and such that $\lambda_i(W_L) < \lambda_i(W_R)$, the path $\Phi(\cdot; W_L, W_R)$ is a parametrization of the arc of γ linking W_L and W_R.
(H3) Let us denote by $\mathcal{RP} \subset \Omega \times \Omega$ the set of pairs (W_L, W_R) such that the Riemann problem

$$\begin{cases} W_t + \mathcal{A}(W)W_x = 0, \\ W(x,0) = \begin{cases} W_L & \text{if } x < 0, \\ W_R & \text{if } x > 0, \end{cases} \end{cases} \tag{4}$$

has a unique self-similar weak solution composed by at most N simple waves (i.e. entropic shocks, contact discontinuities or rarefaction waves) connecting $J+1$ intermediate constant states

$$W_0 = W_L, W_1, \ldots, W_{J-1}, W_J = W_R,$$

with $J \leq N$. Then, given $(W_L, W_R) \in \mathcal{RP}$, the curve described by the path $\Phi(\cdot; W_L, W_R)$ in Ω is equal to the union of those corresponding to the paths $\Phi(\cdot; W_{j-1}, W_j)$, $j = 1, \ldots, J$.

These hypotheses allow to prove the three following properties:

Proposition 1. *Let us assume that the concept of weak solutions of (1) is defined on the basis of a family of paths satisfying hypotheses (H1)-(H3). Then:*

(i) Given two states W_L and W_R belonging to the same integral curve of a linearly degenerate field, the contact discontinuity given by

$$W(x,t) = \begin{cases} W_L & \text{if } x < \sigma t, \\ W_R & \text{if } x > \sigma t, \end{cases}$$

where σ is the (constant) value of the corresponding eigenvalue through the integral curve, is an entropic weak solution of (1).

(ii) Let (W_L, W_R) be a pair belonging to $\mathcal{RP}$ and let W be the solution of the Riemann problem (4). The following equality holds for every $t > 0$:

$$\langle \mathcal{A}(W(\cdot, t)) W_x(\cdot, t), 1 \rangle = \int_0^1 \mathcal{A}(\Phi(s; W_L, W_R)) \frac{\partial \Phi}{\partial s}(s; W_L, W_R) \, ds.$$

Consequently, the total mass of the Borel measure $\mathcal{A}(W(\cdot, t)) W_x(\cdot, t)$ does not depend on t.

(iii) Let (W_L, W_R) be a pair belonging to $\mathcal{RP}$ and W_j any of the intermediate states involved by the solution of the Riemann problem (4). Then:

$$\int_0^1 \mathcal{A}(\Phi(s; W_L, W_R)) \frac{\partial \Phi}{\partial s}(s; W_L, W_R) \, ds$$
$$= \int_0^1 \mathcal{A}(\Phi(s; W_L, W_j)) \frac{\partial \Phi}{\partial s}(s; W_L, W_j) \, ds$$
$$+ \int_0^1 \mathcal{A}(\Phi(s; W_j, W_R)) \frac{\partial \Phi}{\partial s}(s; W_j, W_R) \, ds.$$

Notice that the mass associated to a stationary shock wave or contact discontinuity (that is, $\sigma = 0$) is equal to zero (see (2)).

A general procedure to construct a family of paths satisfying hypotheses (H1)-(H3), *at least for the class $\mathcal{RP}$*, can be given, extending the theory of simples waves of hyperbolic systems of conservation laws and the results concerning the solutions of Riemann problems to hyperbolic nonconservative systems.

3 Godunov's method

We consider system (1) under the conditions stated in Section 1, with initial condition $W(x,0) = W_0(x)$, $x \in \mathbb{R}$. We assume that the family of paths Φ used in the definition of the nonconservative product $\mathcal{A}(W)W_x$ satisfies the hypotheses (H1)-(H3).

For the discretization of the system, computing cells $I_i = [x_{i-1/2}, x_{i+1/2}]$ are considered. For simplicity, we suppose that these cells have constant size Δx and that $x_{i+\frac{1}{2}} = i \Delta x$. Define $x_i = (i - 1/2)\Delta x$, the center of the cell I_i. Let Δt be the constant time step and define $t^n = n \Delta t$.

We denote by W_i^n the approximation of the cell averages of the exact solution provided by the numerical scheme:

$$W_i^n \cong \frac{1}{\Delta x} \int_{x_{i-1/2}}^{x_{i+1/2}} W(x,t^n)\, dx.$$

Suppose that the averages W_i^n at time $t = t^n$ are known. As it is usual in Godunov-type methods, we approximate the solution at time t^{n+1} by

$$W_i^{n+1} = \frac{1}{\Delta x}\left(\int_{x_{i-1/2}}^{x_i} W^{i-1/2}(x,t^{n+1})\, dx + \int_{x_i}^{x_{i+1/2}} W^{i+1/2}(x,t^{n+1})\, dx\right),$$

where $W^{i+1/2}$ is the solution of the Riemann problem linking the states W_i^n and W_{i+1}^n at the intercell $x = x_{i+1/2}$.

If a CFL-1/2 condition is assumed we obtain the following expression for the Godunov method:

$$W_i^{n+1} = W_i^n - \frac{\Delta t}{\Delta x}\left(\int_0^1 \mathcal{A}(\Phi(s, W_{i-1/2}^n, W_i^n))\frac{\partial \Phi}{\partial s}(s; W_{i-1/2}^n, W_i^n)\, ds \right.$$
$$\left. + \int_0^1 \mathcal{A}(\Phi(s; W_i^n, W_{i+1/2}^n))\frac{\partial \Phi}{\partial s}(s; W_i^n, W_{i+1/2}^n)\, ds\right), \quad (5)$$

where $W_{i+1/2}^n$ is the constant value of $W^{i+1/2}$ at the intercell $x = x_{i+1/2}$. Notice that $W^{i+1/2}$ could be discontinuous at $x = x_{i+1/2}$. In that case, the discontinuity at $x = x_{i+1/2}$ has to be stationary and therefore the mass associated to the corresponding jump is zero. Thus, in the scheme we can replace $W_{i+1/2}^n$ either by the limit of $W^{i+1/2}$ to the left, $W_{i+1/2}^{n,-}$, or to the right of $x_{i+1/2}$, $W_{i+1/2}^{n,+}$. As in the case of system of conservation laws, a CFL-1 condition is used in practice, as this condition ensures the linear stability of the method.

This method is well-balanced, i.e., it solves correctly steady state solutions (see [5] for more details).

4 Application to linear balance laws

We consider now system (3), where $W \in \Omega \subset \mathbb{R}^N$, $A, C \in \mathcal{M}_N(\mathbb{R})$ and $\sigma(x)$ is a known function. This system can be interpreted as a linear system of conservation laws with source term.

Assume that the system is strictly hyperbolic, that is, the matrix A has N real distinct eigenvalues $\lambda_1 < \cdots < \lambda_N$, and let $R_1, \ldots, R_N$ be the associate eigenvectors.

If we add to the system (3) the trivial equation

$$\frac{\partial \sigma}{\partial t} = 0$$

we obtain the following system in nonconservative form:

$$\widetilde{W}_t + \widetilde{\mathcal{A}}(\widetilde{W})\widetilde{W}_x = 0, \qquad (6)$$

where $\widetilde{W}$ is the augmented vector

$$\widetilde{W} = \begin{bmatrix} W \\ \sigma \end{bmatrix}$$

and the block structure of $\widetilde{\mathcal{A}}(\widetilde{W})$ is given by

$$\widetilde{\mathcal{A}}(\widetilde{W}) = \begin{bmatrix} A & -CW \\ 0 & 0 \end{bmatrix}.$$

Assume that the eigenvalues λ_i of the matrix A satisfy $\lambda_i \neq 0$, $i = 1, \ldots, N$. Then, for every $\widetilde{W}$, $\widetilde{\mathcal{A}}(\widetilde{W})$ has $N+1$ real distinct eigenvalues $\lambda_1, \ldots, \lambda_N, \lambda_* = 0$, with associated eigenvectors $\widetilde{R}_1(\widetilde{W}), \ldots, \widetilde{R}_N(\widetilde{W}), \widetilde{R}_*(\widetilde{W})$ given by

$$\widetilde{R}_i(\widetilde{W}) = \begin{bmatrix} R_i \\ 0 \end{bmatrix}, \quad i = 1, \ldots, N; \qquad \widetilde{R}_*(\widetilde{W}) = \begin{bmatrix} A^{-1}CW \\ 1 \end{bmatrix}.$$

All the fields are linearly degenerate and their integral curves can be easily obtained.

In order to make a choice of paths, first, hypothesis (H1) is used in order to define the path linking two states belonging to the same integral curve of the characteristic fields. Next, given two arbitrary states $\widetilde{W}_L$ and $\widetilde{W}_R$, where

$$\widetilde{W}_L = \begin{bmatrix} W_L \\ \sigma_L \end{bmatrix}, \quad \widetilde{W}_R = \begin{bmatrix} W_R \\ \sigma_R \end{bmatrix},$$

we consider the Riemann problem

$$\begin{cases} \widetilde{W}_t + \widetilde{\mathcal{A}}(\widetilde{W})\widetilde{W}_x = 0, \\ \widetilde{W}(x,0) = \begin{cases} \widetilde{W}_L & \text{if } x < 0, \\ \widetilde{W}_R & \text{if } x > 0. \end{cases} \end{cases} \quad (7)$$

When this problem has a unique solution consisting of at most $N+2$ intermediate states linked by contact discontinuities, hypothesis (H3) is used in order to define the path linking the states $\widetilde{W}_L$ and $\widetilde{W}_R$. Concerning the existence and uniqueness of problem (7), we can prove that, for $[\sigma] = \sigma_R - \sigma_L$ small enough, the Riemann problem (7) has a unique solution consisting of at most $N+2$ intermediate states linked by contact discontinuities.

If we consider system (3) with initial condition $W(x,0) = W_0(x)$, or equivalenty, system (6) with initial condition $\widetilde{W}(x,0) = \widetilde{W}_0(x)$, $x \in \mathbb{R}$, where

$$\widetilde{W}_0 = \begin{bmatrix} W_0 \\ \sigma \end{bmatrix},$$

it can be proved that the Godunov method can be written under the form

$$\widetilde{W}_i^{n+1} = \widetilde{W}_i^n - \frac{\Delta t}{\Delta x}\left(\widetilde{F}_{i+1/2}^- - \widetilde{F}_{i-1/2}^+\right), \quad (8)$$

where

$$\widetilde{F}_{i+1/2}^\pm = \begin{bmatrix} F_{i+1/2}^\pm \\ 0 \end{bmatrix},$$

with

$$F_{i+1/2}^- = A \sum_{j=1}^{I} s_j^{i+1/2} R_j, \quad F_{i+1/2}^+ = -A \sum_{j=I+1}^{N} s_j^{i+1/2} R_j,$$

being I the maximum value of i for which $\lambda_i < 0$, and $s_j^{i+1/2}$, $j = 1, \ldots, N$, the solutions of

$$W_{i+1}^n - e^{A^{-1}C(\sigma_{i+1} - \sigma_i)} W_i^n = e^{A^{-1}C(\sigma_{i+1} - \sigma_i)} \sum_{j=1}^{I} s_j^{i+1/2} R_j + \sum_{j=I+1}^{N} s_j^{i+1/2} R_j, \quad (9)$$

provided that there exists a unique solution of this linear system.

Some numerical experiments have been designed to compare Godunov's method with the Q-scheme of Roe upwinding the source term introduced in [1] for general systems of balance laws. In particular, we have considered the scalar equation

$$u_t + au_x = c\sigma'(x)u \quad (10)$$

with discontinuous data

$$\sigma(x) = \begin{cases} -1 & \text{if } x < 0, \\ 1 & \text{if } x > 0, \end{cases} \qquad (11)$$

and initial condition

$$u(x,0) = \begin{cases} 0.5 & \text{if } x < 0, \\ -0.5 & \text{if } x > 0, \end{cases} \qquad (12)$$

in the interval $-1 < x < 1$ with $\Delta x = 0.02$, CFL=1, $a = 2$ and $c = 1$. As it can be observed in Figure 1, the Godunov scheme captures the solution exactly, while the Q-scheme of Roe produces an incorrect jump. If we consider regular data σ, both methods approximate correctly the solution, although the Godunov scheme produces more precise results.

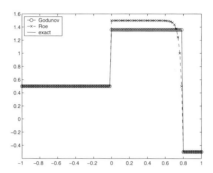

Fig. 1. Solution of problem (10) with initial condition (12) at time $t = 0.4$. Comparison between Godunov's method, Roe's method and the exact solution.

Acknowledgements

This research has been partially supported by the Spanish Government Research project BFM2003-07530-C02-02.

References

1. Bermúdez, A., Vázquez, M.E.: Upwind methods for hyperbolic conservation laws with source terms. Computers and Fluids, **23**(8), 1049–1071 (1994)
2. Dal Maso, G., LeFloch, P.G., Murat, F.: Definition and weak stability of nonconservative products. J. Math. Pures Appl. **74**, 483–548 (1995)
3. Godlewski, E., Raviart, P.A.: Numerical Approximation of Hyperbolic Systems of Conservation Laws. Springer (1996)

4. Gosse, L.: A well-balanced scheme using non-conservative products designed for hyperbolic system of conservation laws with source terms. Math. Models Methods Appl. Sci. **11**, 339–365 (2001)
5. Parés, C., Castro, M.: On the well-balance property of Roe's method for non-conservative hyperbolic systems. Applications to shallow water systems. ESAIM: M2AN **38**(5), 821–852 (2004)
6. Parés, C., Gallardo, J.M., Castro, M.J., Muñoz, M.L.: Godunov's method for non-conservative hyperbolic systems. Application to Linear Balance Laws. In preparation.
7. Volpert, A.I.: The space BV and quasilinear equations. Math. USSR Sbornik **73**(115), 225–267 (1967)

Sequential Flux-Corrected Remapping for ALE Methods

Pavel Váchal[1] and Richard Liska[1]

Czech Technical University in Prague, Břehová 7, 115 19 Prague 1, Czech Republic
vachal@galileo.fjfi.cvut.cz, liska@siduri.fjfi.cvut.cz

Summary. A new FCT-based algorithm is presented for conservative, local bounds preserving interpolations, necessary in the remapping step of Arbitrary Lagrangian-Eulerian (ALE) simulations. To avoid overrestriction of high-order fluxes, caused by separate processing of variables, the method incorporates particular conservation laws incrementally. Contrary to popular a posteriori correction methods, it utilizes physical information about the modeled process already during the remapping step. Moreover, extension to multiple dimensions is trivial.

1 Introduction

Many numerical methods for fluid dynamics are based on a combination of high-order schemes in smooth regions of the solution with low-order schemes near discontinuities. One of the simplest is the Flux-Corrected Transport (FCT), proposed by Boris and Book [1], as a way to combine the high-order and low-order fluxes for each cell interface separately, without creating new local extrema or amplifying the existing ones. Later, Zalesak [2] suggested a modification which improves the resolution of peaks and proposed a formalism for simple and straightforward extension to multiple dimensions. Recently, Schär and Smolarkiewicz [3] introduced an iterative version improving especially extrema in smooth solutions. A good review of evolution and state of the art of the FCT-based methods can be found in [4], together with references to several implementations for systems of conservation laws. Most of the current approaches either treat each state variable separately and adjust the limiters a posteriori, or do not consider all conservation laws.

Our motivation was to develop a FCT-based algorithm for conservative, local bounds preserving interpolations, necessary in the remapping step of ALE methods [5]. Because of our application, we prefer to call the family of methods Flux-Corrected Remapping (FCR). To avoid unnecessary restriction of high-order fluxes, our method processes the conservation laws incrementally (first mass, then momentum and finally total energy), always using results from the

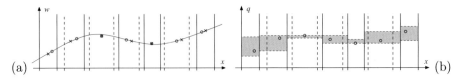

Fig. 1. (a): Discretization of initial condition (curve). Cell-centered values on the old mesh (dashed line, circles) and the new mesh (solid line, cross markers). (b): Local bounds for new values given by range of old values in the immediate neighborhood

previous step. Instead of a posteriori correction, we impose the constraints to keep certain variables in local bounds already during the remapping phase. For simplicity, the method will be presented in 1D, but it will be shown, that it is very easily extensible to multiple dimensions, since many of the FCR formulas are independent on dimension and mesh structure.

Let us denote the conservative variables, that is mass, momentum and total energy as $W = \{m, \mu, E\}$ and their distributions $w = \{\rho, \rho u, \rho \varepsilon + \rho u^2/2\}$ where ε is the specific internal energy. On the old mesh, the cell-related values are given either by discretization of the initial condition, or from the previous time step (see Fig. 1(a)). Then the mesh changes and our aim is to remap the values onto the new mesh so, that the accuracy is as high as possible, the variables W stay conservative, and selected variables stay in local bounds. To make the method conservative by default, we use the update step which can be written in the flux form

$$W^{\text{new}}_{i+\frac{1}{2}} = W^{\text{old}}_{i+\frac{1}{2}} + F^W_{i+1} - F^W_i. \tag{1}$$

As usual, the numerical intercell fluxes will be approximated by reconstruction of the discrete values by some piecewise polynomial function which is then integrated over the *swept regions* given by difference of the old and new mesh. Using polynomials of first or higher order without further correction may lead to violation of local bounds. As for the last requirement, we want to constrain density, velocity and specific internal energy $q = \{\rho, u, \varepsilon\}$, so that their cell-based values stay inside local bounds $[q^{\min}, q^{\max}]$ defined by minimal resp. maximal old value in the immediate neigborhood, that is in 1D

$$\min_{k \in \{i-\frac{1}{2}, i+\frac{1}{2}, i+\frac{3}{2}\}} q^{\text{old}}_k \stackrel{def.}{=} q^{\min}_{i+\frac{1}{2}} \leq q^{\text{new}}_{i+\frac{1}{2}} \leq q^{\max}_{i+\frac{1}{2}} \stackrel{def.}{=} \max_{k \in \{i-\frac{1}{2}, i+\frac{1}{2}, i+\frac{3}{2}\}} q^{\text{old}}_k. \tag{2}$$

The range of possible new values given by this definition is shown in Fig. 1(b).

One of the possibilities to treat overshoots and undershoots caused by high-order fluxes is the a posteriori Repair technique [6], which redistributes them to the immediately neighboring cells. If there is not enough space available, the stencil is increased and checked again, etc. For scalar quantities, this method is guaranteed to work and practical implementation has been presented for general unstructured meshes in 2D and 3D [7]. The repair techniques have been improved in [8, 9]. However, all versions of the repair method correct the results a posteriori, simply redistributing the quantities to nearest available cells. Our intention is to utilize data about physical behavior of the system.

2 Sequential FCR method

2.1 FCT-based approach for density

Like many methods, FCR combines the low-order and high-order schemes $W^{\text{L}}_{i+1/2} = W^{\text{old}}_{i+1/2} + F^{W,\text{L}}_{i+1} - F^{W,\text{L}}_{i}$, $W^{\text{H}}_{i+1/2} = W^{\text{old}}_{i+1/2} + F^{W,\text{H}}_{i+1} - F^{W,\text{H}}_{i}$ according to local smoothness. Suppose we have a low-order flux $F^{W,\text{L}}$ which preserves local bounds by default (for example the donor defined by piecewise constant reconstruction). We start with this flux and then try to approach to the higher-order flux $F^{W,\text{H}}$ as close as possible without violating the local bounds. This is done by adding to each intercell flux some portion of the difference of high-order and low-order fluxes dF^W_i, referred to as *antidiffusive flux*. The amount of used antidiffusive flux is triggered by *limiter* $0 \leq C^W_i \leq 1$. To summarize, the FCR flux has the form

$$F^{W,\text{FCR}}_i = F^{\text{L}}_i + C^W_i \left(F^{W,\text{H}}_i - F^{W,\text{L}}_i \right) = F^{\text{L}}_i + C^W_i dF^W_i \qquad (3)$$

Clearly, the updated value is always in bounds for $C = 0$, which corresponds to the low-order flux. But to achieve higher accuracy, one wants to move C as close to 1 as possible. Since the algorithm proceeds interface by interface, the method is local and very simple. Due to its construction, the FCR method follows the high-order result on smooth regions of the solution, but approaches to the (diffusive) low-order result near singularities to avoid violation of local bounds.

Let us first describe the method for density, which exactly follows the original FCT approach [1] as it was modified by Zalesak [2]. Multiplying the constraints (2) for density ρ by cell volumes, we have an equivalent requirement for mass, which can be expressed in the the flux form (1) as

$$m^{\min}_{i+\frac{1}{2}} \leq m^{\text{L}}_{i+\frac{1}{2}} + \left(C^m_{i+1} dF^m_{i+1} - C^m_i dF^m_i \right) \leq m^{\max}_{i+\frac{1}{2}}. \qquad (4)$$

Defining the maximum available space for mass change $Q^{m,+}_{i+1/2} = m^{\max}_{i+1/2} - m^{\text{L}}_{i+1/2}$, $Q^{m,-}_{i+1/2} = m^{\text{L}}_{i+1/2} - m^{\min}_{i+1/2}$ (note that both parameters are nonnegative), we can formulate the task in the following way: Find such limited antidiffusive fluxes dF, that

$$-Q^{m,-}_{i+\frac{1}{2}} \leq C^m_{i+1} dF^m_{i+1} - C^m_i dF^m_i \leq Q^{m,+}_{i+\frac{1}{2}} \qquad (5)$$

Now we define *signed fluxes*, where positive sign means mass incoming to cell $i + 1/2$ and negative sign means outgoing mass: $\Phi^m_{i,i-1/2} = dF^m_i$, $\Phi^m_{i,i+1/2} =$

$-dF_{i+1}^m$. If we suppose that $C_i^m = 1$ for all i, then the constraint (5) can be written as

$$-Q_{i+\frac{1}{2}}^{m,-} \leq \sum_k \Phi_{k,i+\frac{1}{2}}^m \leq Q_{i+\frac{1}{2}}^{m,+} \qquad (6)$$

where k goes over all faces of cell $i+1/2$, in our 1D case $k \in \{i, i+1/2\}$. Note, that this formula is the same in any dimension and with any mesh topology. With this trick, a substantial part of the following computation becomes independent on mesh geometry details. This is one of the best properties of FCR. In the next step, we collect all incoming and all outgoing fluxes for cell $i+1/2$:

$$P_{i+\frac{1}{2}}^{m,+} = \sum_k \max\left(\Phi_{k,i+\frac{1}{2}}^m, 0\right), \qquad P_{i+\frac{1}{2}}^{m,-} = -\sum_k \min\left(\Phi_{k,i+\frac{1}{2}}^m, 0\right) \qquad (7)$$

where the sum goes over all faces of the cell, that is in our 1D case $k \in \{i, i+1\}$. Now the constraint (6) becomes

$$-Q_{i+\frac{1}{2}}^{m,-} \leq P_{i+\frac{1}{2}}^{m,+} - P_{i+\frac{1}{2}}^{m,-} \leq Q_{i+\frac{1}{2}}^{m,+} \qquad (8)$$

Note that (8) is equivalent to (5) only if $C_i^m = 1$ for all i, that is if the pure high-order fluxes do not violate any bounds. If it is not the case, we have to restrict the flux totals in (8) by additional parameters $0 \leq R_{i+1/2}^{m,\pm} \leq 1$ to keep the values inside:

$$-Q_{i+\frac{1}{2}}^{m,-} \leq R_{i+\frac{1}{2}}^{m,+} P_{i+\frac{1}{2}}^{m,+} - R_{i+\frac{1}{2}}^{m,-} P_{i+\frac{1}{2}}^{m,-} \leq Q_{i+\frac{1}{2}}^{m,+} \qquad (9)$$

We suppose the *worst case scenario* and thus require that two inequalities

$$Q_{i+\frac{1}{2}}^{m,-} \geq R_{i+\frac{1}{2}}^{m,-} P_{i+\frac{1}{2}}^{m,-}, \qquad Q_{i+\frac{1}{2}}^{m,+} \geq R_{i+\frac{1}{2}}^{m,+} P_{i+\frac{1}{2}}^{m,+} \qquad (10)$$

hold, which imply satisfaction of (9). These inequalities (with equality for nonzero P's) hold if we define $R_{i+1/2}^{m,-}$ and $R_{i+1/2}^{m,+}$ by

$$R_{i+\frac{1}{2}}^{m,\pm} = Q_{i+\frac{1}{2}}^{m,\pm}/P_{i+\frac{1}{2}}^{m,\pm} \text{ if } P_{i+\frac{1}{2}}^{m,\pm} > 0, \quad R_{i+\frac{1}{2}}^{m,\pm} = 1 \text{ if } P_{i+\frac{1}{2}}^{m,\pm} = 0 \qquad (11)$$

For a given interface i, we have four constraints, two from each connected cell: One to avoid overshoot in density ($R^{m,+}$) and another to avoid undershoot ($R^{m,-}$). However, since it does not make sense to use all of them, the definition of limiter uses only the two *active* ones, according to direction of the antidiffusive flux:

$$\overline{C_i^m} = \min\left(R_{i-\frac{1}{2}}^{m,-}, R_{i+\frac{1}{2}}^{m,+}, 1\right) \text{ if } dF_i^m \leq 0, \quad \min\left(R_{i-\frac{1}{2}}^{m,+}, R_{i+\frac{1}{2}}^{m,-}, 1\right) \text{ else} \qquad (12)$$

2.2 Extension to systems of conservation laws

Now we want to constrain also velocity and internal energy by restricting intercell fluxes of momentum and total energy. From Sect. 2.1 we know, that density stays in bounds, if the mass flux limiters C_i^m satisfy $0 \le C_i^m \le \overline{C_i^m}$ with $\overline{C_i^m}$ given by (12). One of the simplest ways to keep density in bounds is to continue with restricted antidiffusive fluxes $\overline{C_i^m} dF_i^m$ instead of the unrestricted dF_i^m and compute factors $0 \le \overline{C_i^\mu} \le 1$, that velocity is in bounds for $F_i^{m,L} + \overline{C_i^\mu}\,\overline{C_i^m} dF_i^m$ and $F_i^{\mu,L} + \overline{C_i^\mu} dF_i^\mu$. Note, that since $\overline{C_i^\mu}\,\overline{C_i^m} \le \overline{C_i^m}$, density constraints are satisfied automatically and we do not need to care about them anymore. Similarly, we find such factors $0 \le \overline{C_i^E} \le 1$, that internal energy is in bounds for

$$F_i^{m,\text{FCR}} = F_i^{m,L} + \overline{C_i^E}\,\overline{C_i^\mu}\,\overline{C_i^m} dF_i^m, \qquad F_i^{\mu,\text{FCR}} = F_i^{\mu,L} + \overline{C_i^E}\,\overline{C_i^\mu} dF_i^\mu,$$
$$F_i^{E,\text{FCR}} = F_i^{E,L} + \overline{C_i^E} dF_i^E. \qquad (13)$$

This corresponds to the FCR notation (3) with limiters $C_i^E = \overline{C_i^E}$, $C_i^\mu = \overline{C_i^E}\,\overline{C_i^\mu}$ and $C_i^m = \overline{C_i^E}\,\overline{C_i^\mu}\,\overline{C_i^m}$.

2.3 Sequential FCR: velocity constraints

Let us now require preservation of local bounds for velocity. The constraints given by (2) and expressed in terms of conservative FCR values are $u_{i+1/2}^{\min} \le \mu_{i+1/2}^{\text{FCR}}/m_{i+1/2}^{\text{FCR}} \le u_{i+1/2}^{\max}$ and, using the fluxes (13), we have

$$u_{i+\frac{1}{2}}^{\min} \le \frac{\mu_{i+\frac{1}{2}}^L + \left(\overline{C_{i+1}^\mu} dF_{i+1}^\mu - \overline{C_i^\mu} dF_i^\mu\right)}{m_{i+\frac{1}{2}}^L + \left(\overline{C_{i+1}^\mu}\,\overline{C_{i+1}^m} dF_{i+1}^m - \overline{C_i^\mu}\,\overline{C_i^m} dF_i^m\right)} \le u_{i+\frac{1}{2}}^{\max} \qquad (14)$$

Let us recall that local bounds for density (4) are preserved automatically due to multiplication of $\overline{C_i^\mu}$ by $\overline{C_i^m}$ computed in (12), which i.a. implies positive denominator in (14). In this section, we are looking for $\overline{C_i^\mu}$. Using the same logic as in the density case above and using definitions

$$Q_{i+\frac{1}{2}}^{\mu,-} = \mu_{i+\frac{1}{2}}^L - u_{i+\frac{1}{2}}^{\min} m_{i+\frac{1}{2}}^L, \qquad Q_{i+\frac{1}{2}}^{\mu,+} = u_{i+\frac{1}{2}}^{\max} m_{i+\frac{1}{2}}^L - \mu_{i+\frac{1}{2}}^L, \qquad (15)$$
$$\widetilde{\Phi^u}_{i,i\pm\frac{1}{2}}^{\min} = \mp\left(dF_i^\mu - u_{i\pm\frac{1}{2}}^{\min} dF_i^m\right), \qquad \widetilde{\Phi^u}_{i,i\pm\frac{1}{2}}^{\max} = \mp\left(dF_i^\mu - u_{i\pm\frac{1}{2}}^{\max} dF_i^m\right), \qquad (16)$$
$$P_{i+\frac{1}{2}}^{\mu,-} = -\sum_k \min\left(\widetilde{\Phi^u}_{k,i+\frac{1}{2}}^{\min}, 0\right), \qquad P_{i+\frac{1}{2}}^{\mu,+} = \sum_k \max\left(\widetilde{\Phi^u}_{k,i+\frac{1}{2}}^{\max}, 0\right), \qquad (17)$$

with k going over all faces of the cell ($k \in \{i, i+1\}$ in 1D), then in the worst case scenario the sufficient condition for constraints (14) has the form (10) with all m in superscript replaced by μ and can be fulfilled by definition of $R^{\mu,\pm}$ analogous to (11). For each interface, this defines four velocity constraints, two for each connected cell. Out of these, we need to select the *active* ones, since we do not want to impose the constraints which cannot technically be violated. Unlike the simple density case, where only zero or two of the constraints were necessary (according to sign of dF_i^m, see (12)), now up to all four constraints can be needed and thus the following algorithm is to be carried out:

1. Start with $\overline{C_i^\mu} = 1$.
2. For the first connected cell ($k = i - 1/2$ in 1D)
   - If $\widetilde{\Phi^u}_{i,k}^{\min} < 0$, activate undershoot constr.: set $\overline{C_i^\mu} = \min\left(\overline{C_i^\mu}, R_k^{\mu,-}\right)$
   - If $\widetilde{\Phi^u}_{i,k}^{\max} > 0$, activate overshoot constr.: set $\overline{C_i^\mu} = \min\left(\overline{C_i^\mu}, R_k^{\mu,+}\right)$
3. Repeat Step 2 for the other connected cell ($k = i + 1/2$ in 1D)

2.4 Constraining internal energy in sequential FCR

Finally, let us treat local bounds for internal energy. The constraints given by (2) in cell $i + 1/2$ are

$$\varepsilon_{i+\frac{1}{2}}^{\min} \leq \left(E_{i+\frac{1}{2}}^{\text{FCR}} - \frac{1}{2}(\mu_{i+\frac{1}{2}}^{\text{FCR}})^2 / m_{i+\frac{1}{2}}^{\text{FCR}}\right) / m_{i+\frac{1}{2}}^{\text{FCR}} \leq \varepsilon_{i+\frac{1}{2}}^{\max} \qquad (18)$$

We now use the antidiffusive mass and momentum fluxes, which have been already restricted in Sects. 2.1 and 2.3. Thus we are looking for limiters $\overline{C_i^E}$ to get restricted fluxes in the form (13). Let us again define signed antidiffusive fluxes and their totals

$$\widetilde{\Phi^m}_{i,i\pm\frac{1}{2}} = \mp \overline{C_i^\mu}\,\overline{C_i^m}\,dF_i^m, \quad \widetilde{\Phi^\mu}_{i,\pm\frac{1}{2}} = \mp \overline{C_i^\mu}\,dF_i^\mu, \quad \widetilde{\Phi^E}_{i,\pm\frac{1}{2}} = \mp dF_i^E, \qquad (19)$$

$$P_{i+\frac{1}{2}}^{W,+} = \sum_k \max\left(\widetilde{\Phi^W}_{k,i+\frac{1}{2}}, 0\right), \quad P_{i+\frac{1}{2}}^{W,-} = -\sum_k \min\left(\widetilde{\Phi^W}_{k,i+\frac{1}{2}}, 0\right) \qquad (20)$$

where $W = (m, \mu, E)$ and in 1D case $k \in \{i, i+1\}$. Considering for example the left inequality of (18), we are looking for such $0 \leq R_{i+1/2}^E \leq 1$, that satisfies even the worst scenario given by

$$\left[E_{i+\frac{1}{2}}^L - R_{i+\frac{1}{2}}^E P_{i+\frac{1}{2}}^{E,-}\right] \geq \left[m_{i+\frac{1}{2}}^L + R_{i+\frac{1}{2}}^E P_{i+\frac{1}{2}}^{m,+}\right]\varepsilon_{i+\frac{1}{2}}^{\min} + \frac{1}{2}\frac{\left[\mu_{i+\frac{1}{2}}^L + R_{i+\frac{1}{2}}^E P_{i+\frac{1}{2}}^\mu\right]^2}{\left[m_{i+\frac{1}{2}}^L - R_{i+\frac{1}{2}}^E P_{i+\frac{1}{2}}^{m,-}\right]} \qquad (21)$$

where

$$P_{i+\frac{1}{2}}^\mu = \begin{cases} P_{i+\frac{1}{2}}^{\mu,+} & \text{if } \left(\mu_{i+\frac{1}{2}}^L + R_{i+\frac{1}{2}}^E P_{i+\frac{1}{2}}^{\mu,+}\right)^2 \geq \left(\mu_{i+\frac{1}{2}}^L - R_{i+\frac{1}{2}}^E P_{i+\frac{1}{2}}^{\mu,-}\right)^2 \\ -P_{i+\frac{1}{2}}^{\mu,-} & \text{if } \left(\mu_{i+\frac{1}{2}}^L + R_{i+\frac{1}{2}}^E P_{i+\frac{1}{2}}^{\mu,+}\right)^2 < \left(\mu_{i+\frac{1}{2}}^L - R_{i+\frac{1}{2}}^E P_{i+\frac{1}{2}}^{\mu,-}\right)^2 \end{cases} \qquad (22)$$

In the last term, we need to distinguish two cases, since momentum value in numerator can be positive as well as negative and thus its squared value can be maximized in two ways. Since all low-order values $\{m,\mu,E\}^{\text{L}}$, all flux totals $P^{\{m,\mu,E\},\pm}$ and internal energy bound $\varepsilon^{\min}$ are known, relation (21) is a pair of quadratic inequalities with fixed coefficient w.r.t. $R^E_{i+1/2}$, each of them corresponding to one line of (22). In practice, the only difference caused by this nonlinearity is, that in the space of $\overline{C^E_i}$, we deal with a union of intervals rather than with a single interval to satisfy each of the (up to four) active constraints.

3 Numerical examples

To show properties of the FCR method, let us now present results of three cyclic remapping tests from [10]. In each test, we start on an equidistant grid ($k = 0$) of n nodes with cell-centered values of the state variables. Then we generate a new mesh ($k = 1$) with node positions slightly changed according to certain periodic function of k and remap the state variables onto this new mesh. After $k_{\max}$ such remaps, the nodes return to the original locations and we evaluate the error of the result w.r.t. the original profile. Convergence properties have been tested using various resolutions given by n and $k_{\max}$. In all tests, F^{L} was computed from piecewise constant reconstruction and F^{H} from piecewise linear extrapolation of the cell centered finite differences.

In Test 1, (ρ, u, ε) are given by $(8, 1, 0.1)$ for $0 \le x \le 0.5$ and $(1, 0, 1.5)$ for $0.5 \le x \le 1$. Test 2 is the popular exponential shock. Results after 1280 remaps on 257 grid nodes are shown in Fig. 2(a) for Test 1 and (b) for Test 2. Obviously, the high-order scheme resolves the discontinuities much better than the low-order one, but produces overshoots and undershoots in variables q. The FCR results closely follow the high-order solution, but avoid the unwanted oscillations. The accuracy of repaired high-order scheme and FCR is approximately the same and about twice the precision of repaired MinMod. Finally, an additional check on the smooth region in the left part of Test 2 confirmed, that if the high-order flux produces no overshoots or undershoots, then the limiters $C^W_i = 1$ for all variables W and all interfaces i, and therefore the FCR results are identical to the high-order results.

Generally, we found that in such simple tests the accuracy of FCR and repaired high-order scheme does not differ substantially. However there are cases where repair can fail for internal energy, while FCR is guaranteed to work always (giving low-order scheme in worst case). Also, differences should appear for more physical problems, where repair distributes the overshoots or undershoots "blindly" to all directions.

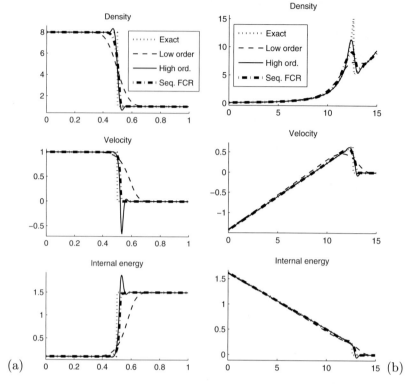

Fig. 2. Results for (a) Test 1 and (a) Test 2. $n = 257, k_{\max} = 1280$

Acknowledgments

The authors would like to thank M. Shashkov for fruitful collaboration and B. Wendroff and R. Loubere for helpful discussions and comments. This work has been partly supported by the Czech Grant Agency grant GAČR 202/03/H162 and Czech Ministry of Education grants FRVŠ 1987G1/2005, MSM 6840770022 and LC528.

References

1. Boris, J., Book, D.: Flux-Corrected Transport I: SHASTA, a Fluid Transport Algorithm That Works. J. Comp. Phys., **11**, 38–69 (1973)
2. Zalesak, S.T.: Fully Multidimensional Flux-Corrected Transport Algorithms for Fluids. J. Comp. Phys., **31**, 335–362 (1979)
3. Schär, C., Smolarkiewicz, P.K.: A Synchronous and Iterative Flux-Correction Formalism for Coupled Transport Equations. J. Comp. Phys., **128**, 101–120 (1996)

4. Kuzmin, D., Löhner, R., Turek, S. (eds.): Flux-Corrected Transport. Principles, Algorithms and Applications. Springer, Berlin Heidelberg (2005)
5. Margolin, L.G.: Introduction to "An Arbitrary Lagrangian-Eulerian Computing Method for All Flow Speeds". J. Comp. Phys., **135** (2), 198–202 (1997)
6. Kuchařík, M., Shashkov, M., Wendroff, B.: An efficient Linearity-and-bound-preserving Remapping Method. J. Comp. Phys., **188**, 462–471 (2003)
7. Garimella, R., Kuchařík, M., Shashkov, M.: An Efficient Linearity and Bound Preserving Conservative Interpolation (Remapping) on Polyhedral Meshes. Computers and Fluids. Accepted.
8. Shashkov, M., Wendroff, B.: The Repair Paradigm and Application to Conservation Laws. J. Comp. Phys., **198** (1), 265–277 (2004)
9. Loubere, R., Staley, M., Wendroff, B.: Technical Report LA-UR-05-6320, Los Alamos National Laboratory, Los Alamos, NM, USA (2005)
10. Margolin, L.G., Shashkov, M.: Remapping, Recovery and Repair on a Staggered Grid. Comp. Meth. Appl. Mech. Engrg., **193**, 4139–4155 (2004)

hp-Finite Elements

Orthogonal *hp*-FEM for Elliptic Problems Based on a Non-Affine Concept

Pavel Šolín,[1] Tomáš Vejchodský[2] and Martin Zítka[1]

[1] Department of Mathematical Sciences, University of Texas at El Paso, El Paso, Texas 79968-0514, USA
solin@utep.edu, mzitka@utep.edu
[2] Mathematical Institute, Academy of Sciences, Žitná 25, 11567 Praha 1, Czech Republic
vejchod@math.cas.cz

Summary. In this paper we propose and test a new non-affine concept of hierarchic higher-order finite elements (*hp*-FEM) suitable for symmetric linear elliptic problems. The energetic inner product induced by the elliptic operator is used to construct partially orthonormal shape functions which automatically eliminate all internal degrees of freedom from the stiffness matrix. The stiffness matrix becomes smaller and better-conditioned compared to standard types of higher-order shape functions. The orthonormalization algorithm is elementwise local and therefore easily parallelizable. The procedure is extendable to nonsymmetric elliptic problems. Numerical examples including performance comparisons to other popular sets of higher-order shape functions are presented.

1 Introduction and historical remarks

Hierarchic higher-order finite element methods (*hp*-FEM) are increasingly popular in computational engineering and science for their excellent approximation properties and the potential of reducing the size of finite element models significantly. The resulting discrete problems usually are much smaller compared to standard lowest-order FEM, but they may exhibit rather high condition numbers unless quality higher-order shape functions are used.

The concept of the *p*-FEM as well as the historically first hierarchic higher-order shape functions were introduced by A.G. Peano in the group of B. Szabó [3] in the mid-1970s. The strong dependence of the condition number of the stiffness and mass matrices on the choice of higher-order shape functions was discovered later, after more advanced *p*-FEM and *hp*-FEM computations were performed [2]. In 2001, an affine-equivalent family of well-conditioned hierarchic finite elements based on integrated Legendre polynomials was proposed by M. Ainsworth and J. Coyle [1].

In this paper we show that the affine concept of finite elements is an obstacle on the way to optimal higher-order shape functions. To obtain optimal shape functions, finite elements have to be constructed in the physical mesh.

2 Preliminaries

Let $\Omega \subset \mathbb{R}^2$ be an open bounded connected set with a piecewise-linear boundary. The weak formulation of a general symmetric linear elliptic problem in Ω reads: Find $u \in V$ such that

$$a(u,v) = l(v) \qquad \text{for all } v \in V, \tag{1}$$

where V is a suitable Sobolev space and $l \in V'$. The symmetric V-elliptic bilinear form $a : V \times V \to \mathbb{R}$ defines an energetic inner product on $V \times V$,

$$(u,v)_e = a(u,v) \qquad \text{for all } u, v \in V. \tag{2}$$

Let the domain Ω be covered with a triangular finite element mesh $\mathcal{T}_{h,p} = \{K_1, K_2, \ldots, K_M\}$, where every element K_i is equipped with a polynomial degree $p_i = p(K_i) \geq 1$. Let $V_{h,p} \subset V$ be the corresponding piecewise-polynomial finite element space with a basis $\mathcal{B} = \{v_1, v_2, \ldots, v_N\}$.

The discrete counterpart of (1) reads: Find $u_{h,p} \in V_{h,p}$ such that

$$a(u_{h,p}, v_{h,p}) = l(v_{h,p}) \qquad \text{for all } v_{h,p} \in V_{h,p}. \tag{3}$$

Using the basis $\mathcal{B}$, (3) translates into a system of N linear algebraic equations of the form $SY = F$ where S is the stiffness matrix, F is the load vector and Y is the vector of unknown expansion coefficients of the approximate solution $u_{h,p}$ to the basis $\mathcal{B}$. The stiffness matrix S has the form $S = \{s_{ij}\}_{i,j=1}^N$, $s_{ij} = a(v_j, v_i)$. If the basis $\mathcal{B}$ was orthonormal under the energetic inner product (2), then S would be the identity matrix. However, it is not possible to construct an orthonormal basis in the space $V_{h,p}$ that at the same time would have local and hierarchic structure required by the hp-FEM [4].

The hierarchic basis of the space $V_{h,p}$ consists of vertex, edge and bubble functions. Typically, these basis functions are constructed by means of hierarchic shape functions defined on a suitable triangular reference domain and suitable affine reference maps (an affine concept). There are three types of hierarchic shape functions for triangular elements, which are depicted in Fig. 1.

The corresponding hierarchic structure of the basis functions splits the finite element space $V_{h,p}$ into a direct sum $V_{h,p} = V_V \oplus V_E \oplus V_B$. At the same time, it induces a block structure of the stiffness matrix S shown in Fig. 2.

In Fig. 2, the VV-block contains the products of vertex basis functions with vertex test functions, the VE-block contains vertex-edge products, etc. For symmetric problems the VV-, EE-, and BB-blocks are symmetric, and it holds $VE = EV^T$, $VB = BV^T$, and $EB = BE^T$.

Fig. 1. Hierarchic structure of higher-order shape functions: Vertex, Edge and Bubble functions.

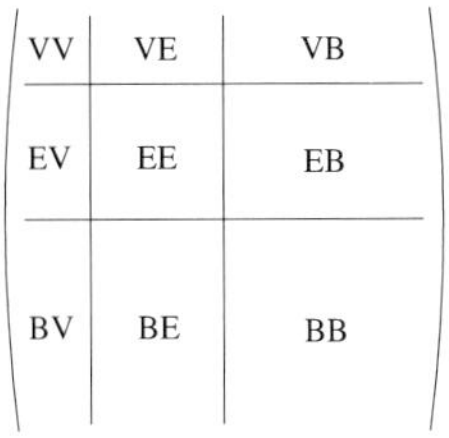

$$\begin{pmatrix} VV & VE & VB \\ EV & EE & EB \\ BV & BE & BB \end{pmatrix}$$

Fig. 2. Typical block structure of the stiffness matrices in the hp-FEM.

3 Numerical example

For illustration, let us consider an electrostatics problem defined in [5] (Paragraph B.2.6, Example 3). The problem is discretized via the hp-FEM. Fig. 3 shows the sparsity structure of the corresponding stiffness matrix with nz= 1010464 nonzero entries.

The left part of Fig. 3 makes a false impression that the BB-block is diagonal. In reality this block only is block-diagonal, as shown in the right part. The kth block on the diagonal of the BB-block has the size $(p_k - 1)(p_k - 2)/2$ and it contains all bubble-bubble products on the element K_k. In general, these blocks are dense for linear elliptic operators discretized via standard choices of bubble functions [1, 2, 3, 4]. It is worth mentioning that (in the

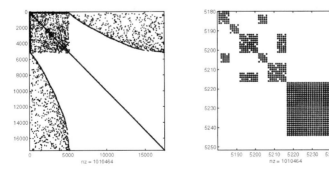

Fig. 3. Sparsity structure of the stiffness matrix. Global view and detail of the lower-right corner of the EE-block/upper-left corner of the BB-block.

case of uniform polynomial degree p in elements) when the polynomial degree p is increased, the size of the BB-block grows as $O(p^2)$, while the size of the EE-block only grows as $O(p)$.

4 Construction of basis functions

On a mesh element K_k we have $(p_k-1)(p_k-2)/2$ bubble functions which span the polynomial space $P_0^{p_k}(K_k) \subset H_0^1(K_k)$. In Paragraph 4.2 we construct a basis of this space which is orthonormal under the energetic inner product (2). With such basis on every K_k, the BB-block becomes the identity matrix. In Paragraph 4.3 we further construct vertex and edge functions which are normal to all bubble functions under the energetic inner product. With these vertex and edge functions, the VB- and EB-blocks become zero.

4.1 Limitations of the affine concept

Let us recall that within the affine concept of the FEM, every triangular mesh element K_k is mapped onto a suitable reference triangle $\hat{K}$ via an affine reference map $\boldsymbol{x}_{K_k} : \hat{K} \to K_k$. The reference domain $\hat{K}$ is equipped with a set of hierarchic shape functions. Basis functions on the element K_k are defined to be the images of these shape functions through the reference map $\boldsymbol{x}_{K_k}$. The affine concept facilitates the computer implementation, but at the same time it hinders the construction of optimal basis functions for the hp-FEM:

Let $\Omega \subset \mathbb{R}^2$ be a bounded domain. Consider a function $c \in L^\infty(\Omega)$ which is greater than some positive constant c_0 in Ω. The elliptic equation

$$-\Delta u + cu = f, \qquad (4)$$

equipped with homogeneous Dirichlet boundary conditions, yields the energetic inner product

$$(u,v)_e = \int_\Omega \nabla u \cdot \nabla v \, d\boldsymbol{x} + \int_\Omega cuv \, d\boldsymbol{x}, \qquad u,v \in H_0^1(\Omega). \qquad (5)$$

On every mesh element K_k, the inner product (5) can be restricted to the polynomial space $P_0^{p_k}(K_k)$ and transformed to the reference element $\hat{K}$, where it becomes an inner product in the space $P_0^{p_k}(\hat{K})$,

$$(\tilde{u},\tilde{v})_{e,K_k} = \int_{\hat{K}} \nabla \tilde{u} \cdot \left(\frac{D\boldsymbol{x}_{K_k}}{D\boldsymbol{\xi}}\right)^{-1} \left(\frac{D\boldsymbol{x}_{K_k}}{D\boldsymbol{\xi}}\right)^{-T} \nabla \tilde{v} \, d\boldsymbol{\xi} + \int_{K_k} c\tilde{u}\tilde{v} \, d\boldsymbol{\xi} \qquad (6)$$

(here $\tilde{u} = u \circ \boldsymbol{x}_{K,k}$ and $\tilde{v} = v \circ \boldsymbol{x}_{K_k}$). For every element $K_k \in \mathcal{T}_{h,p}$, any basis of $P_0^{p_k}(\hat{K})$ which is orthonormal under the inner product (6) yields through the affine map $\boldsymbol{x}_{K_k}$ a basis in the space $P_0^{p_k}(K)$ which is orthonormal under the inner product (5). Such basis eliminates all off-diagonal entries corresponding to internal degrees of freedom on the element K_k from the BB-block of the stiffness matrix S. Unfortunately, the inner product (6) differs from element to element, and thus it is not possible to define a single reference basis in the space $P_0^{p_{max}}(\hat{K})$ that would yield through the reference maps $\boldsymbol{x}_{K_k}$ orthonormal bases in all spaces $P_0^{p_k}(K_k)$, $k = 1, 2, \ldots, M$. To overcome this problem and make the BB-block an identity matrix, we have to construct an orthonormal basis in each polynomial space $P_0^{p_k}(K_k)$ separately.

4.2 Elimination of the BB-block

There are various ways to construct an orthonormal basis in the space $P_0^{p_k}(K_k)$, $K_k \in \mathcal{T}_{h,p}$, among which we prefer the modified Gram-Schmidt procedure [6]. We use the original Peano bubble functions [3] $\lambda_1^i \lambda_2 \lambda_3^j$, $1 \leq i, j$, $i + j \leq p_k - 1$, as the underlying basis for the orthonormalization. By φ_1^b, φ_2^b, $\ldots, \varphi_{m_k}^b$, where $m_k = (p_k - 1)(p_k - 2)/2$, we denote these functions in the lexicographic order. The functions $\varphi_i^v = \lambda_i$ are the barycentric coordinates corresponding to the triangle K_k (affine vertex functions on K_k).

In what follows, by $(u, v)_{e, K_k}$ we denote the restriction of the bilinear form $a(u, v)$ to the element K_k. The standard Gram-Schmidt procedure reads

$$\psi_r^b = \frac{\varphi_r^b - \sum_{s=1}^{r-1}(\varphi_r^b, \psi_s^b)_{e, K_k} \psi_s^b}{\|\varphi_r^b - \sum_{s=1}^{r-1}(\varphi_r^b, \psi_s^b)_{e, K_k} \psi_s^b\|_{e, K_k}}, \quad r = 1, 2, \ldots, m_k, \qquad (7)$$

where $\|v\|_{e, K_k}^2 = (v, v)_{e, K_k}$. This algorithm is known to be unstable for large m_k. Although we have not encountered any problems with the original procedure (7), to be on the safe side we use its more stable modification,

$$\psi_r^b = \frac{\varphi_r^b - \sum_{s=1}^{r-1}(\varphi_{r,s}^b, \psi_s^b)_{e, K_k} \psi_s^b}{\|\varphi_r^b - \sum_{s=1}^{r-1}(\varphi_{r,s}^b, \psi_s^b)_{e, K_k} \psi_s^b\|_{e, K_k}}, \quad r = 1, 2, \ldots, m_k. \qquad (8)$$

Here,

$$\varphi_{r,s}^b = \varphi_r^b - \sum_{l=1}^{s-1}(\varphi_{r,l}^b, \psi_l^b)_{e, K_k} \psi_l.$$

The new bubble functions satisfy

$$(\psi_i^b, \psi_j^b)_{e, K_k} = \delta_{ij} \quad 1 \leq i, j \leq m_k, \qquad (9)$$

where δ_{ij} is the Kronecker delta. The new block structure of the stiffness matrix is depicted in Fig. 4 (left).

4.3 Elimination of VB- and EB-blocks

Let the space $P_0^{p_k}(K_k)$ on each element K_k be equipped with the orthonormal basis $\psi_1^b, \psi_2^b, \ldots, \psi_{m_k}^b$ from the previous paragraph. By φ_i^v, $i = 1, 2, 3$, we denote the affine vertex functions on K_k. Let us define a new set of vertex functions,

$$\psi_i^v = \varphi_i^v - \sum_{j=1}^{m_k} c_{ij}\psi_j^b, \qquad 1 \leq i \leq 3. \tag{10}$$

The coefficients c_{ij} are defined by

$$c_{ij} = (\varphi_i^v, \psi_j^b)_{e,K_k}, \qquad 1 \leq i \leq 3,\ 1 \leq j \leq m_k. \tag{11}$$

Using (10), (11), and (9), it is easy to verify that $(\psi_i^v, \psi_j^b)_{e,K_k} = 0$ for all $1 \leq i \leq 3, 1 \leq j \leq m_k$. Since the linear combination of the bubble functions ψ_j^b in (10) has zero trace on the boundary ∂K_k, the traces of the original and new vertex functions on ∂K_k are identical. Therefore the replacement of the standard affine vertex functions with the new ones does not affect the global continuity of the approximation. The new block structure of the stiffness matrix is depicted in Fig. 4 (middle).

An analogous procedure is used to modify the standard edge functions to be normal to the new bubble functions (8): By φ_i^e, $i = 1, 2, \ldots, n_k$, denote the standard edge functions [4] on K_k. We define a new set of edge functions

$$\psi_i^e = \varphi_i^e - \sum_{j=1}^{m_k} d_{ij}\psi_j^b, \qquad 1 \leq i \leq n_k, \tag{12}$$

where the constants d_{ij} are given by

$$d_{ij} = (\varphi_i^e, \psi_j^b)_{e,K_k} \qquad \text{for all } 1 \leq i \leq n_k,\ 1 \leq j \leq m_k. \tag{13}$$

Using (12), (13), and (9), we obtain $(\psi_i^e, \psi_j^b)_{e,K_k} = 0$ for all $1 \leq i \leq n_k, 1 \leq j \leq m_k$. The new structure of the stiffness matrix is depicted in Fig. 4 (right).

The new edge functions (13) have values identical to the standard edge functions on element interfaces, and thus the global continuity of the approximation is not affected. The replacement of the standard basis functions with the new basis functions (10), (12), (9) only represents a change of basis in the

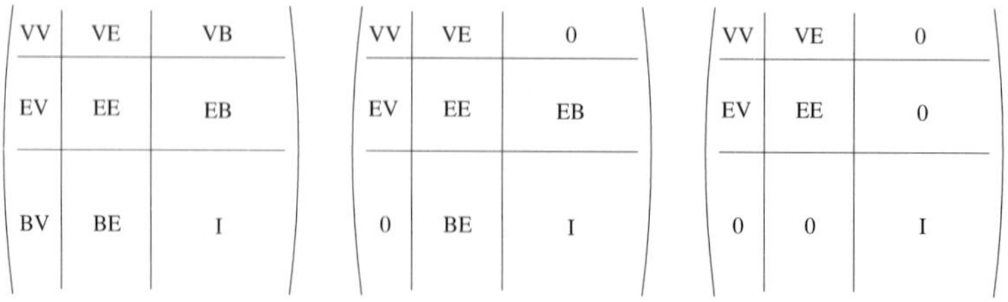

$$\begin{pmatrix} VV & VE & VB \\ EV & EE & EB \\ BV & BE & I \end{pmatrix} \begin{pmatrix} VV & VE & 0 \\ EV & EE & EB \\ 0 & BE & I \end{pmatrix} \begin{pmatrix} VV & VE & 0 \\ EV & EE & 0 \\ 0 & 0 & I \end{pmatrix}$$

Fig. 4. The bubble functions (8) make the BB-block an identity matrix (left). The vertex functions (10) eliminate the VB-block (middle). The edge functions (12) eliminate the EB-block (right).

finite element space $V_{h,p}$, and thus the solution $u_{h,p}$ of problem (3) remains unchanged.

The modified Gram–Schmidt procedure (8) as well as the adjustment of the vertex and edge functions (10) and (12) can be performed for every mesh element $K_k \in \mathcal{T}_{h,p}$ separately, in parallel.

Notice that to construct the new shape functions, it is not necessary to evaluate more energetic inner products of the underlying shape functions than it is needed for assembling of the standard stiffness matrix. Hence, the speed and memory requirements of the construction of the new shape functions are comparable to the standard approach (which yields larger matrices with higher condition numbers).

5 Numerical example continued

Let us return to the numerical example from Section 3, and discretize the problem on the same hp-mesh using the new partially orthonormal basis functions. The sparsity structure of the new stiffness matrix is shown in Fig. 5 (compare to Fig. 3). The original as well as the new system of linear algebraic equations are solved using a Conjugate Gradient solver. Fig. 6 shows the residuum in each case as the function of the number of iterations. We can see that the performance of the CG solver improves dramatically when the new set of partially orthonormal basis functions is used. Moreover, each step of the CG method is faster with the new shape functions, since the reduced stiffness matrix contains fewer nonzero entries (the reduced stiffness matrix contains 196452 nonzero entries while the original one had 1010464).

Let us remark that the procedure presented here also works for nonsymmetric linear elliptic problems where an energetic product is not available.

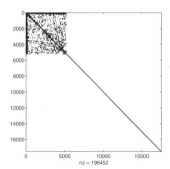 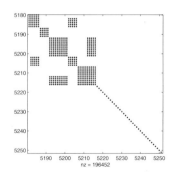

Fig. 5. Sparsity structure of the stiffness matrix. Global view and detail of the lower-right corner of the EE-block/upper-left corner of the BB-block.

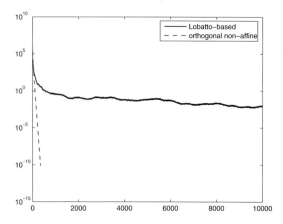

Fig. 6. Convergence history of a CG solver for the stiffness matrix from Fig. 3 (based on the Ainsworth–Coyle shape functions [1]) and for the reduced stiffness matrix from Fig. 5 (obtained with the new partially-orthonormal basis functions).

Acknowledgment

The first and third authors acknowledge the financial support of the Grant Agency of the Czech Republic under the Project No. 102/05/0629. The second author was partially supported by the Grant Agency of the Czech Republic under the Project No. 201/04/P021 and by the Academy of Sciences of the Czech Republic Institutional Research Plan No. AV0Z10190503. Computations were performed in part on a Cray XD1 supercomputer at the University of Texas at El Paso which was funded by the U.S. Department of Defense, Grant No. 05PR07548-00. This support is gratefully acknowledged.

References

1. Ainsworth, M., Coyle, J.: Hierarchic hp-Edge Element Families for Maxwell's Equations on Hybrid Quadrilateral/Triangular Meshes. Comput. Methods Appl. Mech. Engrg. **190**, 6709–6733 (2001)
2. Babuška, I., Szabó, B., Katz, I.N.: The p-Version of the Finite Element Method. SIAM J. Numer. Anal., **18**, 515–545 (1981)
3. Peano, A.G.: Hierarchies of Conforming Finite Elements for Plane Elasticity and Plate Bending. Computers Math. Appl., **2**, 211–224 (1976)
4. Šolín, P., Segeth, K., Doležel, I.: Higher-Order Finite Element Methods. Chapman & Hall/CRC Press, Boca Raton (2003)
5. Šolín, P.: Partial Differential Equations and the Finite Element Methods, J. Wiley & Sons (2005)
6. Zítka, M., Šolín, P., Vejchodský, T., Ávila, F.: Imposing Orthogonality to Hierarchic Higher-Order Finite Elements. Math. Comp. Sim., to appear (2005)

On Some Aspects of the hp-FEM for Time-Harmonic Maxwell's Equations

Tomáš Vejchodský,[1] Pavel Šolín[2] and Martin Zítka[2]

[1] Mathematical Institute, Academy of Sciences, Žitná 25, 11567 Praha 1, Czech Republic
vejchod@math.cas.cz

[2] Department of Mathematical Sciences, University of Texas at El Paso, El Paso, Texas 79968-0514, USA
solin@utep.edu, zitka@math.utep.edu

Summary. It is well known that the design of suitable higher-order shape functions is essential for the performance of the hp-FEM. In this paper we propose a new family of hierarchic higher-order edge elements for the time-harmonic Maxwell's equations which are capable of reducing the condition number of the stiffness matrices dramatically compared to the currently best known hierarchic edge elements. The excellent conditioning properties of the new elements are illustrated by numerical examples.

1 Introduction

Early applications of the finite element method (FEM) to the Maxwell's equations were based on continuous vector-valued approximations of the electric field in the space $[H^1(\Omega)]^d$. This approach, however, led to spurious oscillations and other unwanted phenomena (see e.g. [8, 10] or [16]). Later it was found that the electric field can exhibit stronger singularities than the space $[H^1(\Omega)]^d$ admits, and the space $\mathbf{H}(\mathrm{curl}, \Omega)$ came into the play. Finite element approximations conforming to $\mathbf{H}(\mathrm{curl}, \Omega)$ are discontinuous in general, but their tangential components are required to be continuous across element interfaces. Finite elements conforming to the space $\mathbf{H}(\mathrm{curl}, \Omega)$ are called edge elements.

The lowest order edge elements (with constant tangential components on element interfaces) were originally introduced in the late 1950s by Whitney [17] in the context of geometrical integration theory. The Whitney elements were later independently rediscovered and applied to the Maxwell's equations by numerous authors (see e.g. [1, 4] or [5]).

The Whitney elements can be extended to higher-order edge elements in both the nodal and hierarchic fashions [14]. Nowadays the most popular nodal edge elements are the Nédélec elements [11, 12]. Among the best known hierarchic edge elements are those proposed by Ainsworth and Coyle [2, 3].

The hierarchic shape functions on these elements were designed in a sophisticated way, exploiting the integrated Legendre polynomials (Lobatto shape functions). This technique can be viewed as a heuristic attempt to make the shape functions orthogonal in the curl-curl product. In the present paper we build on this approach by applying a systematic orthonormalization to the higher-order shape functions. This approach yields a new class of edge elements with significantly better conditioning properties.

The organization of the paper is as follows: Section 2 introduces the notation, formulates the time-harmonic Maxwell's equations, and mentions their hp-FEM discretization. In Section 3 we apply an orthonormalization process to construct a new set of hierarchic shape functions. Section 4 presents the results of numerical experiments which compare the conditioning properties of the new partially-orthonormal shape functions with the shape functions [2]. Finally, Section 5 contains a conclusion and mentions some additional exciting problems that remain to be addressed.

2 Formulation of the problem

Consider a bounded polygonal domain $\Omega \subset \mathbb{R}^2$. The time-harmonic Maxwell's equations usually are equipped with perfect conducting boundary conditions on a part Γ_P of the boundary $\partial \Omega$ and/or with impedance boundary conditions on a boundary part $\Gamma_\mathrm{I} \subset \partial \Omega$. The system is written as follows:

$$\mathbf{curl}\left(\mu_\mathrm{r}^{-1} \operatorname{curl} \mathbf{E}\right) - \kappa^2 \epsilon_\mathrm{r} \mathbf{E} = \mathbf{F} \quad \text{in } \Omega,$$
$$\mathbf{E} \cdot \tau = 0 \quad \text{on } \Gamma_\mathrm{P},$$
$$\mu_\mathrm{r}^{-1} \operatorname{curl} \mathbf{E} - i\kappa\lambda \mathbf{E} \cdot \tau = \mathbf{g} \cdot \tau \quad \text{on } \Gamma_\mathrm{I}.$$

Here, $\mathbf{curl}\, a = (\partial a / \partial x_2, -\partial a / \partial x_1)^\top$ and $\operatorname{curl} \mathbf{E} = \partial E_2 / \partial x_1 - \partial E_1 / \partial x_2$ are the standard vector and scalar curl operators, $\tau = (-\nu_2, \nu_1)^\top$ is the positively oriented unit tangent vector to $\partial \Omega$. The notation is summarized in Table 1:

Table 1. Quantities used in the time-harmonic Maxwell's equations

$\mathbf{E} = \mathbf{E}(x) \in \mathbb{C}^2$	phasor of the electric field strength (unknown)
$\mu_\mathrm{r} = \mu_\mathrm{r}(x) \in \mathbb{R}$	relative permeability
$\epsilon_\mathrm{r} = \epsilon_\mathrm{r}(x) \in \mathbb{C}^{2 \times 2}$	relative permittivity
$\kappa = \text{const.} \in \mathbb{R}$	wave number
$\lambda = \lambda(x) > 0$	impedance
$\mathbf{F} = \mathbf{F}(x) \in \mathbb{C}^2$	right-hand side of the equation
$\mathbf{g} = \mathbf{g}(x) \in \mathbb{C}^2$	right-hand side of the impedance boundary condition

The weak formulation of the problem is stated as follows: Find $\mathbf{E} \in V = \{\mathbf{E} \in \mathbf{H}(\operatorname{curl}, \Omega) : \mathbf{E} \cdot \tau = 0 \text{ on } \Gamma_\mathrm{P}\}$ such that

Aspects of the hp-FEM for Time-Harmonic Maxwell's Equations

$$a(\mathbf{E}, \mathbf{\Phi}) = \mathcal{F}(\mathbf{\Phi}) \quad \forall \mathbf{\Phi} \in V,$$

where the sesquilinear form $a(\cdot, \cdot)$ and the antilinear functional $\mathcal{F}$ are given by

$$a(\mathbf{E}, \mathbf{\Phi}) = \int_\Omega \mu_r^{-1} \operatorname{curl} \mathbf{E} \operatorname{curl} \overline{\mathbf{\Phi}} \, dx - \kappa^2 \int_\Omega (\epsilon_r \mathbf{E}) \cdot \overline{\mathbf{\Phi}} \, dx$$
$$- i\kappa \int_{\Gamma_I} \lambda (\mathbf{E} \cdot \tau)(\overline{\mathbf{\Phi}} \cdot \tau) \, ds,$$
$$\mathcal{F}(\mathbf{\Phi}) = \int_\Omega \mathbf{F} \cdot \overline{\mathbf{\Phi}} \, dx + \int_{\Gamma_I} (\mathbf{g} \cdot \tau)(\overline{\mathbf{\Phi}} \cdot \tau) \, ds.$$

The problem is discretized by the hp-FEM as follows: Define a triangulation $\mathcal{T}_{hp}$ of the domain Ω. Assign a polynomial degree $p_j \geq 0$ to every mesh element $K_j \in \mathcal{T}_{hp}$. Consider the piecewise polynomial space

$$V_{hp} = \{\mathbf{E}_{hp} \in V : \mathbf{E}_{hp}|_{K_j} \in [P^{p_j}(K_j)]^2\} \subset V,$$

where $P^{p_j}(K_j)$ stands for the space of polynomials of degree less than or equal to p_j on the triangle K_j. The finite element space V_{hp} is used to define the hp-FEM problem: Find $\mathbf{E}_{hp} \in V_{hp}$ such that

$$a(\mathbf{E}_{hp}, \mathbf{\Phi}_{hp}) = \mathcal{F}(\mathbf{\Phi}_{hp})$$

holds for all $\mathbf{\Phi}_{hp} \in V_{hp}$.

3 Shape functions

The basis of the space $V_{h,p}$ is constructed in a standard way [14], by means of suitable shape functions defined on a reference element and affine reference maps. The reference domain we use is depicted in Fig. 1.

On this triangle we define the lowest-order shape functions known as Whitney functions [17] and the first-order functions:

$$\hat{\psi}_0^{e_1} = \frac{1}{\|e_1\|} \left(\frac{\lambda_3 \mathbf{n}_2}{\mathbf{n}_2 \cdot \mathbf{t}_1} + \frac{\lambda_2 \mathbf{n}_3}{\mathbf{n}_3 \cdot \mathbf{t}_1} \right), \quad \hat{\psi}_1^{e_1} = \frac{1}{\|e_1\|} \left(\frac{\lambda_3 \mathbf{n}_2}{\mathbf{n}_2 \cdot \mathbf{t}_1} - \frac{\lambda_2 \mathbf{n}_3}{\mathbf{n}_3 \cdot \mathbf{t}_1} \right),$$
$$\hat{\psi}_0^{e_2} = \frac{1}{\|e_2\|} \left(\frac{\lambda_1 \mathbf{n}_3}{\mathbf{n}_3 \cdot \mathbf{t}_2} + \frac{\lambda_3 \mathbf{n}_1}{\mathbf{n}_1 \cdot \mathbf{t}_2} \right), \quad \hat{\psi}_1^{e_2} = \frac{1}{\|e_2\|} \left(\frac{\lambda_1 \mathbf{n}_3}{\mathbf{n}_3 \cdot \mathbf{t}_2} - \frac{\lambda_3 \mathbf{n}_1}{\mathbf{n}_1 \cdot \mathbf{t}_2} \right),$$
$$\hat{\psi}_0^{e_3} = \frac{1}{\|e_3\|} \left(\frac{\lambda_2 \mathbf{n}_1}{\mathbf{n}_1 \cdot \mathbf{t}_3} + \frac{\lambda_1 \mathbf{n}_2}{\mathbf{n}_2 \cdot \mathbf{t}_3} \right), \quad \hat{\psi}_1^{e_3} = \frac{1}{\|e_3\|} \left(\frac{\lambda_2 \mathbf{n}_1}{\mathbf{n}_1 \cdot \mathbf{t}_3} - \frac{\lambda_1 \mathbf{n}_2}{\mathbf{n}_2 \cdot \mathbf{t}_3} \right).$$

Here the symbols $\lambda_i, \mathbf{n}_i, \mathbf{t}_i$, where $i = 1, 2, 3$, denote the barycentric coordinates and unit outer normal and tangent vectors to the corresponding edges of the reference triangle. The symbol $\|e_i\|$ stands for the length of the edge.

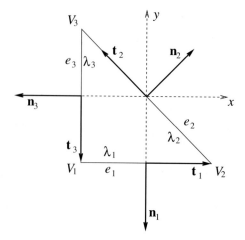

Fig. 1. The reference element $\hat{K}$ and the corresponding notation

Similarly, the higher-order edge functions are defined by

$$\hat{\psi}_k^{e_1} = \frac{2k-1}{k} L_{k-1}(\lambda_3 - \lambda_2)\hat{\psi}_1^{e_1} - \frac{k-1}{k} L_{k-2}(\lambda_3 - \lambda_2)\hat{\psi}_0^{e_1},$$

$$\hat{\psi}_k^{e_2} = \frac{2k-1}{k} L_{k-1}(\lambda_1 - \lambda_3)\hat{\psi}_1^{e_2} - \frac{k-1}{k} L_{k-2}(\lambda_1 - \lambda_3)\hat{\psi}_0^{e_2},$$

$$\hat{\psi}_k^{e_3} = \frac{2k-1}{k} L_{k-1}(\lambda_2 - \lambda_1)\hat{\psi}_1^{e_3} - \frac{k-1}{k} L_{k-2}(\lambda_2 - \lambda_1)\hat{\psi}_0^{e_3},$$

$k \geq 2$, where L_k denotes the Legendre polynomial of degree k. These shape functions are called edge functions since their tangent components are nonzero on one edge only. This construction is done in such a way that the tangent component of $\hat{\psi}_k^{e_i}$ is equal to the transformed and scaled Legendre polynomial of degree k on the edge e_i and it vanishes on the remaining edges.

To complete the basis for higher-order elements ($p \geq 2$), we also have to define the bubble functions (interior modes) whose tangent component vanishes on the whole boundary of the reference triangle. There are two groups of bubble functions which we call edge-based,

$$\hat{\psi}_k^{b,e_1} = \lambda_3 \lambda_2 L_{k-2}(\lambda_3 - \lambda_2)\mathbf{n}_1,$$
$$\hat{\psi}_k^{b,e_2} = \lambda_1 \lambda_3 L_{k-2}(\lambda_1 - \lambda_3)\mathbf{n}_2, \qquad (1)$$
$$\hat{\psi}_k^{b,e_3} = \lambda_2 \lambda_1 L_{k-2}(\lambda_2 - \lambda_1)\mathbf{n}_3, \quad k = 2, 3, \ldots,$$

and genuine,

$$\hat{\psi}^{b,1}_{n_1,n_2} = \lambda_1\lambda_2\lambda_3 L_{n_1-1}(\lambda_3 - \lambda_2) L_{n_2-1}(\lambda_2 - \lambda_1)\mathbf{e}_1, \tag{2}$$

$$\hat{\psi}^{b,2}_{n_1,n_2} = \lambda_1\lambda_2\lambda_3 L_{n_1-1}(\lambda_3 - \lambda_2) L_{n_2-1}(\lambda_2 - \lambda_1)\mathbf{e}_2, \quad 1 \le n_1, n_2.$$

Here, the symbols $\mathbf{e}_1$ and $\mathbf{e}_2$ stand for the canonical vectors in $\mathbb{R}^2$.

It is easily seen that the degree of the shape functions $\hat{\psi}^{e_i}_k$ and $\hat{\psi}^{b,e_i}_k$ is equal to k. The degree of the genuine bubbles $\hat{\psi}^{b,1}_{n_1,n_2}$ and $\hat{\psi}^{b,2}_{n_1,n_2}$ is n_1+n_2+1. The basis formed by the above-mentioned shape functions is hierarchic in the sense that it is enough to add shape functions of degree p to the basis of the space $[P^{p-1}(\hat{K})]^2$ in order to get a basis of $[P^p(\hat{K})]^2$. Even though these shape functions contain the Legendre polynomials in their formulae, their conditioning properties for the time-harmonic Maxwell's equations are rather poor, cf. Section 4. A new set of shape functions with better conditioning properties is constructed in the following:

Construction of orthonormal bubble functions

The number of bubble functions, $p^2 - 1$, grows quadratically with the polynomial degree p, while the number of edge functions, $3(p+1)$, only grows linearly. It follows from here that when finite elements of high polynomial degrees are used, the condition number of the stiffness matrix is mainly influenced by the bubble functions.

In order to obtain well-conditioned bubble functions, we propose to make them orthonormal under the inner product

$$(\Phi, \Psi) = \int_{\hat{K}} \operatorname{curl} \Phi \operatorname{curl} \Psi \, d\xi + \int_{\hat{K}} \Phi \cdot \Psi \, d\xi$$

on the reference domain $\hat{K}$. There are various ways to construct orthonormal basis in an inner product space, among which we prefer the modified Gram–Schmidt algorithm [6, 15].

Note that the shape functions resulting from the Gram–Schmidt algorithm depend on the underlying basis used for the orthonormalization as well as on the ordering of the basis functions. We have chosen bubble functions (1) and (2) in the natural ordering,

$p = 2:$ $\quad \hat{\psi}^{b,e_1}_2, \hat{\psi}^{b,e_2}_2, \hat{\psi}^{b,e_3}_2,$

$p = 3:$ $\quad \hat{\psi}^{b,e_1}_3, \hat{\psi}^{b,e_2}_3, \hat{\psi}^{b,e_3}_3, \hat{\psi}^{b,1}_{1,1}, \hat{\psi}^{b,2}_{1,1},$

$p = 4:$ $\quad \hat{\psi}^{b,e_1}_4, \hat{\psi}^{b,e_2}_4, \hat{\psi}^{b,e_3}_4, \hat{\psi}^{b,1}_{1,2}, \hat{\psi}^{b,1}_{2,1}, \hat{\psi}^{b,2}_{1,2}, \hat{\psi}^{b,2}_{2,1},$

$p = 5:$ $\quad \hat{\psi}^{b,e_1}_5, \hat{\psi}^{b,e_2}_5, \hat{\psi}^{b,e_3}_5, \hat{\psi}^{b,1}_{1,3}, \hat{\psi}^{b,1}_{2,2}, \hat{\psi}^{b,1}_{3,1}, \hat{\psi}^{b,2}_{1,3}, \hat{\psi}^{b,2}_{2,2}, \hat{\psi}^{b,2}_{3,1}, \ldots$

The actual orthonormalization procedure was implemented in Maple for polynomial degrees $p = 1, 2, \ldots, 10$. The complete list of the 99 resulting hierarchic shape functions can be download from the web page http://servac.math.utep.edu/fem_group/publications.

4 Numerical experiments

Let us solve the time-harmonic Maxwell's equations in the L-shape domain $\Omega = (-1,1)^2 \setminus (-1,0)^2$, shown in Fig. 2.

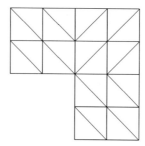

Fig. 2. The L-shape domain and its triangulation

For simplicity, the material parameters $\mu_r = 1$, $\epsilon_r = I$ (2×2 identity matrix), $\kappa = 1$, $\lambda = 1$ are used. Perfect conducting boundary conditions are considered on edges meeting at the reentrant corner and impedance boundary conditions are prescribed on the rest of the boundary in agreement with the exact solution, which is defined as $\mathbf{E} = \nabla u$,

$$u = r^{\frac{2}{3}} \sin\left((2\theta + \pi)/3\right).$$

Here, r and θ stand for the standard polar coordinates. The right-hand side $\mathbf{F}$ is chosen to agree with the exact solution, i.e., $\mathbf{F} = -\mathbf{E}$.

The problem was solved several times by our modular finite element system HERMES [14] on the mesh shown in Figure 2. First we ran the computation with lowest-order elements ($p = 0$ everywhere). Then we gradually increased the polynomial degree of all elements up to $p = 10$. In all cases we computed the condition number of the stiffness matrix using the Matlab function cond. The results are summarized in Table 2 and plotted in Fig. 3.

Table 2 shows that the rate of growth of the condition number of the stiffness matrix as a function of the polynomial degree p is dramatically slower for the new orthonormal bubble functions than for the bubble functions [2]. In the case $p = 10$, the condition numbers differ by *eight-orders of magnitude*.

5 Conclusion and outlook

We have presented a new set of higher-order shape functions for hierarchic edge elements for the time-harmonic Maxwell's equations. We used a numerical experiment to confirm that the new shape functions yield much better

Table 2. The condition number of the stiffness matrix for uniformly increasing elements' orders

		Condition number		
p	# DOF	old bubbles	new bubbles	improvement
0	80	$3.9 \cdot 10^2$	$3.9 \cdot 10^2$	$1.0 \cdot 10^0$
1	160	$1.0 \cdot 10^3$	$1.0 \cdot 10^3$	$1.0 \cdot 10^0$
2	384	$7.8 \cdot 10^3$	$3.3 \cdot 10^3$	$2.3 \cdot 10^0$
3	704	$3.3 \cdot 10^5$	$7.5 \cdot 10^3$	$4.4 \cdot 10^1$
4	1120	$4.9 \cdot 10^6$	$1.9 \cdot 10^4$	$2.6 \cdot 10^2$
5	1632	$9.1 \cdot 10^7$	$4.7 \cdot 10^4$	$1.9 \cdot 10^3$
6	2240	$1.3 \cdot 10^9$	$9.8 \cdot 10^4$	$1.3 \cdot 10^4$
7	2944	$3.2 \cdot 10^{10}$	$1.8 \cdot 10^5$	$1.8 \cdot 10^5$
8	3744	$7.0 \cdot 10^{11}$	$3.6 \cdot 10^5$	$1.9 \cdot 10^6$
9	4640	$1.4 \cdot 10^{13}$	$5.9 \cdot 10^5$	$2.4 \cdot 10^7$
10	5632	$3.3 \cdot 10^{14}$	$9.3 \cdot 10^5$	$3.5 \cdot 10^8$

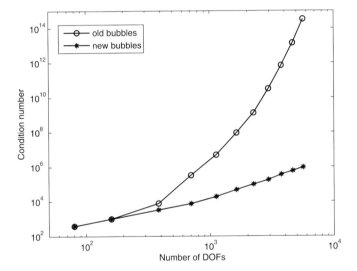

Fig. 3. The log-log plot of the condition number against the number of degrees of freedom (cf. Table 2)

conditioned stiffness matrices than other commonly used higher-order shape functions for hierarchic edge elements.

It should be noted that the result of the Gram–Schmidt algorithm depends both on the underlying basis used for the orthonormalization, as well as on the ordering of the basis functions. Thus there still remains freedom for further optimization of these shape functions. The choice of suitable optimization criteria is an exciting open question. Among other points that have not been fully resolved yet are the choice of an optimal inner product for the orthonormalization, and further possible adjustments of the edge functions. These topics are now in the middle of our research, and we hope to report new results soon.

Acknowledgment

This work was supported by the Grant Agency of the Czech Republic, grants No. 201/04/P021 and No. 102/05/0629, and by the Academy of Sciences of the Czech Republic, Institutional Research Plan No. AV0Z10190503. Computations were performed in part on a Cray XD1 supercomputer at the University of Texas at El Paso which was funded by the U.S. Department of Defense, Grant No. 05PR07548-00. This support is gratefully acknowledged.

References

1. Ahagon, A., Fujiwara, K., Nakata, T.: Comparison of various kinds of edge elements for electromagnetic field analysis. IEEE Trans. Magn., **32**, 898–901 (1996)
2. Ainsworth, M., Coyle, J.: Hierarchic hp-edge element families for Maxwell's equations on hybrid quadrilateral/triangular meshes. Comput. Methods Appl. Mech. Engrg., **190**, 6709–6733 (2001)
3. Ainsworth, M., Coyle, J.: Hierarchic finite element bases on unstructured tetrahedral meshes. Internat. J. Numer. Methods Engrg., **58**, 2103–2130 (2003)
4. Barton, M.L., Cendes, Z.J.: New vector finite elements for three-dimensional magnetic computation. J. Appl. Phys., **61**, 3919–3921 (1987)
5. Bespalov, A.M.: Finite element method for the eigenmode problem of a RF cavity. Sov. J. Numer. Anal. Math. Model., **3**, 163–178 (1988)
6. Björck, mrA.: Numerical methods for least squares problems. SIAM, Philadelphia (1996)
7. Davis, T.A.: Algorithm 832: UMFPACK V4.3 – an unsymmetric-pattern multifrontal method. ACM Trans. Math. Software, **30**, 196–199 (2004)
8. Hano, M.: Finite element analysis of dielectric-loaded waveguides. IEEE Trans. Microwave Theory Tech., **32**, 1275–1279 (1984)
9. Monk, P.: Finite element methods for Maxwell's equations. Oxford University Press, New York (2003)
10. Mur, G.: Edge elements, their advantages and their disadvantages. IEEE Trans. Magn., **30**, 3552–3557 (1994)
11. Nédélec, J.C.: A new family of mixed finite elements in $\mathbb{R}^3$. Numer. Math., **50**, 57–81 (1986)
12. Nédélec, J.C.: Mixed finite elements in $\mathbb{R}^3$. Numer. Math., **35**, 315–341 (1980)
13. Šolín, P., Segeth, K., Doležel, I.: Higher-order finite element methods. Chapman & Hall/CRC, Boca Raton (2004)
14. Šolín, P.: Partial differential equations and the finite element methods. J. Wiley & Sons (2005)

15. Stewart, G.W.: Matrix algorithms, Volume I: Basic decompositions. SIAM, Philadelphia (1998)
16. Sun, D., et al.: Spurious modes in finite element methods. IEEE Trans. Antennas and Propagation, **37**, 12–24 (1995)
17. Whitney, H.: Geometric Integration Theory. Princeton University Press, Princeton (1957)

Interface Problems

Numerical Simulation of Phase-Transition Front Propagation in Thermoelastic Solids

A. Berezovski[1] and G.A. Maugin[2]

[1] Centre for Nonlinear Studies, Institute of Cybernetics at Tallinn University of Technology, Akadeemia tee 21, Tallinn 12618, Estonia
Arkadi.Berezovski@cs.ioc.ee

[2] Laboratoire de Modélisation en Mécanique, Université Pierre et Marie Curie, 4 place Jussieu, case 162, 75252, Paris Cedex 05, France
gam@ccr.jussieu.fr

Summary. A thermodynamically consistent finite-volume numerical algorithm for martensitic phase-transition front propagation is described in the paper. The proposed numerical method generalizes the wave-propagation algorithm to the case of moving discontinuities in thermoelastic solids.

1 Introduction

It is well-known that initial-boundary-value problems, formulated according to the usual principles of continuum mechanics, can suffer from a lack of uniqueness of the solution when the body is composed of a multiphase material (e.g. [1]). The solution in this case involves a propagating phase boundary which separates the austenite from the martensite; the speed of this interface remains undetermined by the usual continuum theory. A nucleation criterion and a kinetic relation for the velocity of the phase boundary are needed as well as the construction of a proper numerical algorithm.

From a thermodynamic point of view, a phase transition is a non-equilibrium process; entropy is produced at the moving phase boundary. To perform simulations of practical examples, we need to move to a numerical approximation. In this case, we face a non-equilibrium behavior of finite-size discrete elements or computational cells. It is clear that the local equilibrium approximation is not sufficient to describe such a behavior. We have proposed to determine all the needed fluxes by means of non-equilibrium jump relations at the phase boundary [2]. These jump relations are connected with the contact quantities following from the thermodynamics of discrete systems [3].

In what follows we consider the simplest possible one-dimensional setting of the problem of impact-induced phase transformation front propagation in a shape-memory alloy (SMA) bar. Both martensitic and austenitic phases are considered as isotropic materials. The change in cross-sectional area of the bar is neglected. Since thermal expansion coefficient of SMA's is around $10^{-5} K^{-1}$, the thermal strain in the material is negligible under the variation up to $100\,K$. Therefore, the isothermal case is considered. The phase-transition front is viewed as an ideal mathematical discontinuity surface. However, the problem remains nonlinear even in this simplified description that requires a numerical solution.

Extensive study of 1-D dynamic phase-transition front propagation in materials with transformation softening behavior has been conducted [4]-[9]. In spite of using different constitutive models, all of them have demonstrated the ability to reproduce the observed behavior of shape memory alloys. However, the used constitutive models are not sufficient to describe the phase-transition front propagation. Therefore, we need to turn to the non-equilibrium description of the phase-transition front propagation [10].

The main focus of the paper is the construction of a numerical scheme for the propagation of phase-transition fronts.

2 Formulation of the problem

We consider the boundary value problem of the tensile impact loading of a 1-D, SMA bar that is initially in an austenitic phase and that has uniform cross-sectional area A_0 and temperature θ_0. The bar occupies the interval $0 < x < L$ in a reference configuration and the boundary $x = 0$ is subjected to the tensile shock loading

$$\sigma(0,t) = \hat{\sigma}(t) \quad \text{for} \quad t > 0. \tag{1}$$

The bar is assumed to be long compared to its diameter so it is under uniaxial stress state and the stress $\sigma(x,t)$ depends only on the axial position and time. Supposing the temperature is constant during the process, it is characterized by the displacement field $u(x,t)$, where x denotes the location of a particle in the reference configuration and t is time. Linearized strain is further assumed so the axial component of the strain $\varepsilon(x,t)$ and the particle velocity $v(x,t)$ are related to the displacement by

$$\varepsilon = \frac{\partial u}{\partial x}, \quad v = \frac{\partial u}{\partial t}. \tag{2}$$

The density of the material ρ is assumed constant. All field variables are averaged over the cross-section of the bar.

At each instant t during a process, the strain $\varepsilon(x,t)$ varies smoothly within the bar except at phase boundaries; across a phase boundary, it suffers jump discontinuity. The displacement field is assumed to remain continuous throughout the bar. Away from a phase boundary, balance of linear momentum and kinematic compatibility require that

$$\rho \frac{\partial v}{\partial t} = \frac{\partial \sigma}{\partial x}, \quad \frac{\partial \varepsilon}{\partial t} = \frac{\partial v}{\partial x}. \tag{3}$$

Suppose that at time t there is a moving discontinuity in strain or particle velocity at $x = \mathbf{S}(t)$. Then one also has the corresponding jump relations (cf. [10])

$$\rho V_{\mathbf{S}}[v] + [\sigma] = 0, \quad V_{\mathbf{S}}[\varepsilon] + [v] = 0, \quad V_{\mathbf{S}}\theta[S] = f_{\mathbf{S}} V_{\mathbf{S}}, \tag{4}$$

where $V_{\mathbf{S}}$ is the material velocity of the discontinuity, square brackets denote jumps, S is the entropy per unit volume, and the driving traction $f_{\mathbf{S}}(t)$ at the discontinuity is defined by (cf. [10])

$$f_{\mathbf{S}} = -[W] + <\sigma> [\varepsilon], \tag{5}$$

where W is the free energy per unit volume. The second law of thermodynamics requires that

$$f_{\mathbf{S}} V_{\mathbf{S}} \geq 0 \tag{6}$$

at strain discontinuities. If $f_{\mathbf{S}}$ is not zero, the sign of $V_{\mathbf{S}}$, and hence the direction of motion of discontinuity, is determined by the sign of $f_{\mathbf{S}}$.

Assuming that Hooke's law holds for each phase

$$\sigma = (\lambda_a + 2\mu_a)\varepsilon, \quad \sigma = (\lambda_m + 2\mu_m)(\varepsilon - \varepsilon_{tr}), \tag{7}$$

where subscripts "a" and "m" denote austenite and martensite, respectively, and ε_{tr} is the transformation stress, we can then rewrite the relevant bulk equations of inhomogeneous linear isotropic elasticity as follows:

$$\frac{\partial \varepsilon}{\partial t} = \frac{\partial v}{\partial x}, \quad \rho \frac{\partial v}{\partial t} = (\lambda(x) + 2\mu(x)) \frac{\partial \varepsilon}{\partial x}. \tag{8}$$

Here λ and μ are the Lame coefficients, values of which are constant but different depending on the martensitic or austenitic state.

It is easy to see that the cross-differentiation of equations (8) leads to the conventional wave equation, solution of which is well-known if corresponding fields are smooth. The difficulties relate to an unknown motion of the phase boundary and to the jump relations across it. That is why we need to develop a numerical scheme which is compatible with the non-equilibrium jump relations at the moving phase boundary.

3 Conservative wave propagation algorithm

The system of equations (8) can be expressed in the form of conservation law

$$\frac{\partial}{\partial t} q(x,t) + \frac{\partial}{\partial x} f(q(x,t)) = 0, \tag{9}$$

where

$$q(x,t) = \begin{pmatrix} \varepsilon \\ \rho v \end{pmatrix}, \quad \text{and} \quad f(x,t) = \begin{pmatrix} -v \\ -\rho c^2 \varepsilon \end{pmatrix}, \tag{10}$$

and $c = \sqrt{(\lambda + 2\mu)/\rho}$ is the sound velocity. In the linear homogeneous case, equation (9) can be rewritten in the form

$$\frac{\partial}{\partial t} q(x,t) + A \frac{\partial}{\partial x} q(x,t) = 0, \quad A = \begin{pmatrix} 0 & -1/\rho \\ -\rho c^2 & 0 \end{pmatrix}. \tag{11}$$

In finite volume numerical methods [11], the solution of the conservation law (9) is obtained in terms of averaged quantities at each time step

$$Q = \frac{1}{\Delta x} \int_{\Delta x} q(x,t) dx, \tag{12}$$

and numerical fluxes at the boundaries of each element

$$F^{\pm} \approx \frac{1}{\Delta t} \int_{t_k}^{t_{k+1}} f^{\pm}(q(x,t)) \, dt. \tag{13}$$

The corresponding finite-volume numerical scheme for a uniform grid (n) can be presented as follows (k denotes time steps)

$$Q_n^{k+1} - Q_n^k = -\frac{\Delta t}{\Delta x} \left((F^+)_n^k + (F^-)_n^k \right), \tag{14}$$

where superscripts "+" and "-" denote inflow and outflow parts in the flux decomposition. Numerical fluxes are determined by means of the solution of the Riemann problem at interfaces between cells [11]. In the considered case, the solution of the Riemann problem at the interface between cells $n-1$ and n consists of two waves, which we denote $\mathcal{L}_n^I$ and $\mathcal{L}_n^{II}$. The left-going wave $\mathcal{L}_n^I$ moves into cell $n-1$, the right-going wave $\mathcal{L}_n^{II}$ moves into cell n. In the linear case, these waves are proportional to eigenvectors r^I and r^{II} of the matrix A:

$$\mathcal{L}_n^I = \beta_n^I r_{n-1}^I, \quad \mathcal{L}_n^{II} = \beta_n^{II} r_n^{II}. \tag{15}$$

In the conservative wave-propagation algorithm [12], the solution of the generalized Riemann problem is obtained by using the decomposition of flux difference $f_n(Q_n) - f_{n-1}(Q_{n-1})$

$$\mathcal{L}_n^I + \mathcal{L}_n^{II} = f_n(Q_n) - f_{n-1}(Q_{n-1}), \tag{16}$$

and the corresponding numerical scheme has the form

$$Q_n^{k+1} - Q_n^k = -\frac{\Delta t}{\Delta x} \left(\mathcal{L}_n^{II} + \mathcal{L}_{n+1}^I \right). \tag{17}$$

Coefficients β^I and β^{II} are determined from the solution of the system of linear equations

$$\begin{pmatrix} 1 & 1 \\ \rho_{n-1}c_{n-1} & -\rho_n c_n \end{pmatrix} \begin{pmatrix} \beta_n^I \\ \beta_n^{II} \end{pmatrix} = \begin{pmatrix} -(v_n - v_{n-1}) \\ -(\rho c^2 \varepsilon_n - \rho c^2 \varepsilon_{n-1}) \end{pmatrix}. \qquad (18)$$

However, our main goal is the phase-transition front propagation, where it is difficult even to formulate a Riemann problem at the moving phase boundary. Fortunately, we have a tool for the determination of numerical fluxes at the phase boundary. This is nothing else but the *non-equilibrium jump relations* [2], which should be fulfilled for each pair of adjacent discrete elements.

4 Contact quantities and numerical fluxes

In the non-equilibrium case, we decompose the free energy density into two terms [13]

$$W = \bar{W} + W_{ex}. \qquad (19)$$

Then contact stress Σ and an excess entropy S_{ex} can be introduced [10]

$$\Sigma = \frac{\partial W_{ex}}{\partial \varepsilon}, \qquad S_{ex} = -\frac{\partial W_{ex}}{\partial \theta}, \qquad (20)$$

similarly to conventional definition of averaged (local equilibrium) stress and entropy

$$\bar{\sigma} = \frac{\partial \bar{W}}{\partial \varepsilon}, \qquad \bar{S} = -\frac{\partial \bar{W}}{\partial \theta}. \qquad (21)$$

Here overbars denote averaged quantities. In considered one-dimensional case, the non-equilibrium jump relations [2] take on the following form

$$[\bar{\sigma} + \Sigma] = 0, \quad \text{in the bulk} \qquad (22)$$

$$\left[\bar{\theta} \left(\frac{\partial S}{\partial \varepsilon} \right)_\sigma + \bar{\sigma} + \Sigma \right] = 0, \quad \text{at the phase boundary.} \qquad (23)$$

What we need now is to determine the values of contact quantities.

4.1 Contact quantities in the bulk

In the bulk we apply the non-equilibrium jump relation (22), which can be rewritten at the interface between elements (n) and $(n-1)$ as

$$(\Sigma^+)_{n-1} - (\Sigma^-)_n = (\bar{\sigma})_n - (\bar{\sigma})_{n-1}, \qquad (24)$$

This jump relation should be complemented by the kinematic condition between material and physical velocity [14] which can be rewritten in the one-dimensional case as follows

$$[\bar{v} + \mathcal{V}] = 0. \tag{25}$$

Assuming that the jump of contact velocity is determined by the second term of the last relation

$$[\mathcal{V}] = [V], \tag{26}$$

we obtain in the one-dimensional case

$$(\mathcal{V}^+)_{n-1} - (\mathcal{V}^-)_n = (\bar{v})_n - (\bar{v})_{n-1}. \tag{27}$$

Using relations between contact stresses and contact velocities

$$\Sigma_n^+ = \rho_n c_n \mathcal{V}_n^+, \quad \Sigma_n^- = -\rho_n c_n \mathcal{V}_n^-, \tag{28}$$

we obtain then a system of linear equations for contact velocities

$$\begin{pmatrix} \dfrac{1}{\rho_{n-1}c_{n-1}} & \dfrac{1}{-\rho_n c_n} \end{pmatrix} \begin{pmatrix} -\mathcal{V}_{n-1}^+ \\ \mathcal{V}_n^- \end{pmatrix} = \begin{pmatrix} -(\bar{v}_n - \bar{v}_{n-1}) \\ -(\rho c^2 \bar{\varepsilon}_n - \rho c^2 \bar{\varepsilon}_{n-1}) \end{pmatrix}. \tag{29}$$

Comparing the obtained equation with (18), we conclude that

$$\beta_n^I = -\mathcal{V}_{n-1}^+, \quad \beta_n^{II} = \mathcal{V}_n^-. \tag{30}$$

This means that the contact quantities correspond to numerical fluxes. Therefore, the conservative wave propagation numerical scheme (17) can be rewritten in terms of contact quantities

$$\bar{\varepsilon}_n^{k+1} - \bar{\varepsilon}_n^k = \frac{\Delta t}{\Delta x}\left(\mathcal{V}_n^+ - \mathcal{V}_n^-\right), \quad (\rho\bar{v})_n^{k+1} - (\rho\bar{v})_n^k = \frac{\Delta t}{\Delta x}\left(\Sigma_n^+ - \Sigma_n^-\right). \tag{31}$$

This means that the introduced non-equilibrium jump relations are consistent with conservation laws. From another point of view, this means that the wave-propagation algorithm is thermodynamically consistent.

4.2 Contact quantities at the phase boundary

At the phase boundary we keep the continuity of contact stresses at the phase boundary [10]

$$[\Sigma] = 0, \tag{32}$$

which yields

$$\Sigma_{p-1}^+ - \Sigma_p^- = 0. \tag{33}$$

To determine the contact stresses at the phase boundary completely, the relation (33) should be complemented by the coherency condition [15] which can be expressed in the small-strain approximation as follows

$$[\mathcal{V}] = 0. \tag{34}$$

We still keep the relations between contact stresses and contact velocities (28). This means that in terms of contact stresses equation (34) yields

$$\frac{(\Sigma^+)_{p-1}}{\rho_{p-1} c_{p-1}} + \frac{(\Sigma^-)_p}{\rho_p c_p} = 0. \tag{35}$$

It follows from the conditions (33) and (35) that the values of contact stresses and velocities vanish at the phase boundary

$$(\Sigma^+)_{p-1} = (\Sigma^-)_p = 0, \quad (\mathcal{V}^+)_{p-1} = (\mathcal{V}^-)_p = 0. \tag{36}$$

Now all the contact quantities at the phase boundary are determined, and we can update the state of the elements adjacent to the phase boundary by means of the numerical scheme (31).

4.3 Velocity of the phase boundary

After having the solution of a particular initial-boundary value problem, the material velocity at a moving discontinuity can be determined by means of the jump relation for linear momentum (4)

$$V_{\mathrm{S}}[\rho \bar{v}] + [\bar{\sigma}] = 0, \tag{37}$$

where $\bar{v}$ is the averaged velocity, ρ is the density. The application of the Maxwell-Hadamard lemma gives [16]

$$[\bar{v}] = -[\bar{\varepsilon}] V_{\mathrm{S}}, \tag{38}$$

and the jump relation for linear momentum (37) can be rewritten in the form that is more convenient for the calculation of the velocity at singularity

$$\rho V_{\mathrm{S}}^2 [\bar{\varepsilon}] = [\bar{\sigma}]. \tag{39}$$

The direction of the front propagation is determined by the positivity of the entropy production (6). We also apply the initiation criterion for the stress-induced martensitic phase transformation established in [10].

5 Conclusions

Success in numerical simulations of moving discontinuities in solids depends crucially on the jump relations at the discontinuities. These jump relations should be specified before the construction of a numerical scheme. Since conventional continuum theory does not provide the corresponding jump relations, a non-equilibrium description of the phase-transition front propagation is adopted in the paper. It appears that the non-equilibrium description can serve as a basis in the construction of a numerical scheme [17, 18], which is very close to the conservative wave-propagation algorithm [12] based on the solution of a generalized Riemann problem at interfaces between computational cells. Moreover, the non-equilibrium jump relations at the phase boundary can be successfully implemented in the developed numerical scheme. Examples of the phase-transition front propagation simulations in thermoelastic media by means of the formulated algorithm can be found in [10, 19].

Acknowledgment

Support of the Estonian Science Foundation under contract No.5756 is gratefully acknowledged.

References

1. Abeyaratne, R., Knowles, J.K.: A continuum model od a thermoelastic solid capable of undergoing phase transitions. J. Mech. Phys. Solids, **41**, 541–571 (1993)
2. Berezovski, A., Maugin, G.A.: On the thermodynamic conditions at moving phase-transition fronts in thermoelastic solids. J. Non-Equilib. Thermodyn., **29**, 37–51 (2004)
3. Muschik, W.: Fundamentals of non-equilibrium thermodynamics. In: Muschik, W. (ed.), Non-Equilibrium Thermodynamics with Application to Solids. Springer, Wien (1993) pp. 1–63.
4. Chen, Y.-C., Lagoudas, D.C.: Impact induced phase transformation in shape memory alloys. J. Mech. Phys. Solids, **48**, 275–300 (2000)
5. Bekker, A., Jimenez-Victory, J.C., Popov, P., Lagoudas, D.C.: Impact induced propagation of phase transformation in a shape memory alloy rod. Int. J. Plasticity **18**, 1447–1479 (2002)
6. Shaw, J.A.: A thermomechanical model for a 1-D shape memory alloy wire with propagating instabilities. Int. J. Solids Struct., **39**, 1275–1305 (2002)
7. Stoilov, V., Bhattacharyya, A.: A theoretical framework of one-dimensional sharp phase fronts in shape memory alloys. Acta Mater., **50**, 4939–4952 (2002)
8. Lagoudas, D.C., Ravi-Chandar, K., Sarh, K., Popov, P.: Dynamic loading of polycrystalline shape memory alloy rods. Mech. Materials, **35**, 689–716 (2003)
9. Dai, X., Tang, Z.P., Xu, S., Guo, Y., Wang, W.: Propagation of macroscopic phase boundaries under impact loading. Int. J. Impact Engineering, **30**, 385–401 (2004)
10. Berezovski, A., Maugin, G.A.: Stress-induced phase-transition front propagation in thermoelastic solids. Eur. J. Mech. - A/Solids, **24**, 1–21 (2005)
11. LeVeque, R.J.: Finite Volume Methods for Hyperbolic Problems. Cambridge University Press, Cambridge (2002)
12. Bale, D.S., LeVeque, R.J., Mitran, S., Rossmanith, J.A.: A wave propagation method for conservation laws and balance laws with spatially varying flux functions. SIAM J. Sci. Comp., **24**, 955–978 (2003)

13. Muschik, W., Berezovski, A.: Thermodynamic interaction between two discrete systems in non-equilibrium. J. Non-Equilib. Thermodyn., **29**, 237–255 (2004)
14. Maugin, G.A. : Material Inhomogeneities in Elasticity. Chapman and Hall, London (1993)
15. Maugin, G.A.: On shock waves and phase-transition fronts in continua. ARI, **50**, 141–150 (1998)
16. Maugin, G.A., Trimarco, C.: The dynamics of configurational forces at phase-transition fronts. Meccanica, **30**, 605–619 (1995)
17. Berezovski, A., Engelbrecht, J., Maugin, G.A.: Thermoelastic wave propagation in inhomogeneous media. Arch. Appl. Mech., **70**, 694–706 (2000)
18. Berezovski, A., Maugin, G.A.: Simulation of thermoelastic wave propagation by means of a composite wave-propagation algorithm. J. Comp. Physics, **168**, 249–264 (2001)
19. Berezovski, A., Engelbrecht, J., Maugin, G.A.: Numerical simulation of thermoelastic wave and phase-transition front propagation. In: Cohen, G.C., Heikkola, E., Joly, P., Neittaanmäki, P. (eds.) Mathematical and Numerical Aspects of Wave Propagation. Springer, Berlin 759–764 (2003)

The Level Set Method for Solid-Solid Phase Transformations

E. Javierre[1], C. Vuik[1], F. Vermolen[1], A. Segal[1] and S. van der Zwaag[2,3]

[1] Delft Institute of Applied Mathematics, Delft University of Technology, The Netherlands
  e.javierre,c.vuik,f.j.vermolen,a.segal@ewi.tudelft.nl
[2] Laboratory of Materials Science, Delft University of Technology, The Netherlands
  s.vanderzwaag@tnw.tudelft.nl
[3] Netherlands Institute for Metals Research (N.I.M.R.), Delft, The Netherlands

Summary. In this work we consider the homogenization process in Aluminum alloys, in which inhomogeneities dissolve. This process is governed by diffusion, and mass conservation leads to the Stefan condition on the moving interface. The Level Set Method is used to model this problem, due to its convenience to handle merging/breaking interfaces, compared with other available methods. In binary alloys, the interface concentration is the solid solubility predicted from thermodynamics. However, in multicomponent alloys, the interface concentrations must satisfy a hyperbolic coupling, and therefore, have to be found as part of the solution. In this work we present a computational method to solve three-dimensional dissolution of binary alloys, and we study its extension to multicomponent alloys. In this respect, we restrict ourselves to one-dimensional problems and we focus our attention in the solution of the nonlinear coupled system of diffusion equations.

1 Introduction

Heat treatment of metals is often used to optimize mechanical properties. During heat treatment, the metallurgical state of the alloy changes. This change can involve the phase present at a given location or the morphology of the various phases. Whereas equilibrium phases can be predicted quite accurately from thermodynamic models, there are no general models for microstructural changes nor for the kinetics of these changes. In the latter cases, both the initial morphology and the transformation mechanisms have to be prescribed explicitly. One of these processes, which is both of large industrial and scientific interest and amenable to modeling, is precipitate dissolution. Several physical models have been developed to describe the dissolution of precipitates, incorporating the effects of long-distance diffusion [1] and nonequilibrium conditions at the interface [2].

In binary alloys, the dissolution of second-phase particles is governed by Fickian diffusion in the diffusive phase

$$\frac{\partial c}{\partial t}(\mathbf{x},t) = D\Delta c(\mathbf{x},t), \quad \mathbf{x} \in \Omega_{dp}(t),\ t > 0, \qquad (1)$$

The concentration in the particle $\Omega_{part}(t)$ equals a given constant c^{part}. The concentration at the interface Γ, separating Ω_{part} and Ω_{dp}, is the solid solubility c^{sol} predicted from thermodynamics. In order to preserve mass, no flux of concentration is allowed through the boundary not being Γ, and the normal component v_n of the interface velocity is given by

$$\left(c^{part} - c^{sol}\right)v_n(\mathbf{x},t) = D\frac{\partial c}{\partial \mathbf{n}}(\mathbf{x},t), \quad \mathbf{x} \in \Gamma(t),\ t > 0, \qquad (2)$$

where $\mathbf{n}$ denotes the unit normal vector on the interface pointing outward with respect to Ω_{part}. Further, the initial position of the interface $\Gamma(0)$ and initial concentration profile are given. The above problem constitutes a so-called *(scalar)* Stefan problem. Stefan problems also arise in applications like dendritic solidification [3] and grain growth [4].

The two mainstreams concerning numerical solution methods for moving boundary problems are front-tracking and front-capturing methods. In the first case, the interface is identified with a set of points that should be updated each time step to define the new interface position. In the second case, the interface is identified by means of a mark function. A comparison of these methods is presented in [5].

Front capturing methods have shown to be the most adequate for moving boundary problems, especially when topological changes occur. The Level Set method [6, 7] captures the interface as the zero level set of a continuous function, the so-called level set function. This method has been used in our research to simulate the dissolution of a particle in a binary alloy in two- and three-space dimensions. The motion of the interface follows then from an advection equation for the level set function. The velocity field used for this advection should be a continuous extension of the front velocity [15]. Finite difference and finite element methods are used in the numerical solution. Both background meshes share the same mesh points. Finite difference schemes are used in the solution of hyperbolic equations arising from the level set formulation. Finite elements are used to solve the diffusion problem. The cut-cell method is used to adapt the triangulation to the interface location. See [8] for further details. A similar method [9] has recently been applied to dendritic growth. Figure 1 presents the dissolution of a plate-shaped particle. A number of cracks have been prescribed in the surface of the particle, which yields the breaking of the particle in successive subparticles.

Addition of secondary alloying elements can influence the dissolution kinetics strongly [10]. The addition of chemical species to the primary phase has been considered in [11] (for ternary alloys) and in [12] (for multi-component alloys). Geometrical assumptions are normally taken, reducing the problem

Fig. 1. Dissolution of a plate-shaped particle with cracks on the surface. Time evolution follows from left to right and up to down.

to a one-space dimension. Numerical methods for higher dimensions are still lacking. Our aim is to generalize the results obtained for dissolution of precipitates in binary alloys to multi-component alloys. This generalization of the problem requires the solution of a nonlinear coupling of the concentrations on the moving interface. In this work we consider the dissolution of stoichiometric multi-component particles in multicomponent alloys. In this study we limit ourselves to one spatial co-ordinate only. Several procedures to solve the nonlinear problem are compared. Our aim is to find an efficient and robust numerical method, able to handle the nonlinearity of the problem, and of which the extension to higher dimensional problems is computationally tractable.

2 The physical problem

The as-cast microstructure is simplified to a representative cell Ω containing a diffusive phase Ω_{dp} and a particle Ω_{part}. Let p be the number of chemical species in the alloy. The particle dissolves due to Fickian diffusion of the chemical species in the diffusive phase

$$\frac{\partial c_i}{\partial t}(\mathbf{x}, t) = D_i \Delta c_i(\mathbf{x}, t), \quad \mathbf{x} \in \Omega_{dp}(t), \ t > 0, \ i = 1, \ldots, p, \qquad (3)$$

where D_i denotes the diffusivity constant of the ith specie. Cross-diffusion effects [13] have not been considered in the present work. Furthermore, the particle remains stoichiometric during the dissolution process. Hence, the particle concentration of the ith alloying element equals c_i^{part}, and the concentrations on the moving interface Γ satisfy the hyperbolic relation

$$\prod_{i=1}^{p}\left(c_i^{sol}(\mathbf{x},t)\right)^{\tilde{n}_i} = \mathcal{K}, \quad \mathbf{x} \in \Gamma(t),\ t > 0, \tag{4}$$

where $\tilde{n}_i$ denotes the stoichiometric number of the ith alloying element and $\mathcal{K}$ is a given constant. Note that in this case the interface concentratios may depend on time and/or space. Mass conservation for all the chemical species implies that the velocity of Γ is given by

$$\left(c_i^{part} - c_i^{sol}(\mathbf{x},t)\right)v_n(\mathbf{x},t) = D_i \frac{\partial c_i}{\partial \mathbf{n}}(\mathbf{x},t), \quad \mathbf{x} \in \Gamma(t),\ t > 0,\ i = 1,\ldots,p. \tag{5}$$

The above equations constitute a so-called vector Stefan problem. Note that Eqs. (5) implicitly impose that

$$\frac{D_i}{c_i^{part} - c_i^{sol}(\mathbf{x},t)} \frac{\partial c_i}{\partial \mathbf{n}}(\mathbf{x},t) = \frac{D_{i-1}}{c_{i-1}^{part} - c_{i-1}^{sol}(\mathbf{x},t)} \frac{\partial c_{i-1}}{\partial \mathbf{n}}(\mathbf{x},t), \quad \mathbf{x} \in \Gamma(t),\ t > 0, \tag{6}$$

for $i = 2,\ldots,p$.

3 The computational method

The level set method is used to follow the moving interface Γ. The interface velocity Eq. (5) is continuously extended, i.e. advected, into the whole computational domain Ω. An iterative method is used to solve the nonlinear coupling Eqs. (4) and (6) of the interface concentrations.

3.1 Level set method

The level set function ϕ is used to capture the moving interface: $\phi(\mathbf{x},t) = 0 \iff \mathbf{x} \in \Gamma(t)$. This function is initialized as the signed distance function to the interface, being positive in $\Omega_{dp}(t)$. The advection of the interface is carried out by

$$\frac{\partial \phi}{\partial t}(\mathbf{x},t) + \mathbf{v}(\mathbf{x},t) \cdot \nabla \phi(\mathbf{x},t) = 0, \quad \mathbf{x} \in \Omega,\ t > 0, \tag{7}$$

where $\mathbf{v}$ denotes a continuous extension [8] of the front velocity v_n onto Ω. After solving Eq. (7), if necessary, the level set function is reinitialized to a signed distance function [14].

3.2 The nonlinear coupling of the interface concentrations

The nonlinear coupling in the boundary conditions Eqs. (4) and (6) is reformulated as the zero of a function $\mathbf{f} : \mathbb{R}_+^p \to \mathbb{R}^p$ given by

$$\begin{cases} f_1(\mathbf{c}^{sol}) = \prod_{i=1}^{p} (c_i^{sol})^{\tilde{n}_i} - \mathcal{K} \\ f_i(\mathbf{c}^{sol}) = \dfrac{D_i}{c_i^{part} - c_i^{sol}} \dfrac{\partial c_i}{\partial \mathbf{n}} - \dfrac{D_{i-1}}{c_{i-1}^{part} - c_{i-1}^{sol}} \dfrac{\partial c_{i-1}}{\partial \mathbf{n}}, \quad i = 2, \ldots, p \end{cases} \quad (8)$$

where the vectorial notation has been embraced for $\mathbf{c}^{sol}$. A numerical method is used to find the zero of the function $\mathbf{f}$. We compare the solution by using Newton's method, with central differences to approximate the Jacobian, Broyden's method and a fixed-point iteration. For the last case we define $\mathbf{c}^{sol}(\cdot, q+1) = \mathbf{g}(\mathbf{c}^{sol}(\cdot, q))$ where $\mathbf{g}$ is given by

$$\begin{cases} g_1(\mathbf{c}^{sol}) = \dfrac{\mathcal{K}}{(c_1^{sol})^{\tilde{n}_1 - 1} \prod_{i=2}^{p} (c_i^{sol})^{\tilde{n}_i}} \\ g_i(\mathbf{c}^{sol}) = c_i^{sol} + \delta \left(D_i (c_{i-1}^{part} - c_{i-1}^{sol}) \dfrac{\partial c_i}{\partial \mathbf{n}} - D_{i-1}(c_i^{part} - c_i^{sol}) \dfrac{\partial c_{i-1}}{\partial \mathbf{n}} \right), \end{cases}$$

for $i = 2, \ldots, p$, where a relaxation parameter δ is used.

4 Numerical results

In this section we compare the performance of the Picard, Broyden and Newton iterations proposed to solve the nonlinear coupling in the diffusion problem during the first time-step. We will investigate the influence of moderate and large ratios between the diffusivity constants. As a test-problem we take $p = 2$, $\mathbf{c}^{part} = (5,5)^t$, $\mathbf{c}_0 = (0,0)^t$, $\tilde{\mathbf{n}} = (1,1)^t$, $\mathcal{K} = 2$, $\Gamma(0) = 0.75$, $\Omega = [0,2]$ and $\Omega_{dp}(t) = (\Gamma(t), 2]$. The mesh width is $\Delta x = 0.02$. For this test-problem we consider the initial discontinuous concentration and we solve the (nonlinear) diffusion problem once, without moving the interface, with initial time step $\Delta t_0 = 0.7\Delta x^2$. The numerical iteration is stopped when $||\mathbf{f}(\mathbf{c}^{sol})||_1 < 10^{-6}$. The parameter δ in the Picard iteration is necessary to make $\mathbf{g}$ a contraction. It has been chosen linearly dependent of Δx, but also depends, at least, on the diffusivity constants.

Table 1 shows the results for various ratios D_1/D_2, where $D_2 = 1$ is fixed. The initial guess to c_2^{sol}, denoted by $c_{2,0}^{sol}$, is chosen $\sqrt{2}$ by default. Therefore, $c_{1,0}^{sol} = \mathcal{K}/c_{2,0}^{sol}$. However, when D_1 is large, neither Broyden nor Newton converged, and the initial guess had to be adapted accordingly. For the Picard iteration we use $\delta = 5 \times 10^{-2}\Delta x$ as a default value. Hence, the region where $\mathbf{g}$ is a contraction is reduced as D_1 increases. To compensate this, reducing δ seems a good strategy. The number of iterations for $D_1 = 10^8$ is dropped to 26 if $\delta = 10^{-2}\Delta x$ is used. The Picard method requires more iterations to converge, followed by Broyden and Newton methods respectively, independently of D_1. However it is the fastest, in terms of CPU-time, since each of its iterations requires only one function evaluation, in place of the

Table 1. Interface concentrations and number of iterations for various ratios of the diffusivity constants.

D_1	$c_{2,0}^{sol}$	c_2^{sol}	Picard	Broyden	Newton
2	$\sqrt{2}$	1.641680	21	5	3
4	$\sqrt{2}$	1.886870	23	6	4
8	$\sqrt{2}$	2.149723	25	7	4
16	$\sqrt{2}$	2.428018	26	7	5
10^2	4	3.200103	25	8	5
10^4	4	4.443626	27	8	8
10^8	4	4.513020	320	11	7

solution of a system of equations and/or computing an approximation to the Jacobian.

As a second test-problem we take five species, i.e., $p = 5$, $\mathbf{c}^{part} = (5, 5, 5, 5, 5)^t$, $\mathbf{c}^0 = (0, 0, 1, 2, 1.5)^t$, $\tilde{\mathbf{n}} = (1, 1, 1, 1, 1)^t$, $D = (10, 5, 1, 0.5, 0.25)^t$, $\mathcal{K} = 2$, $\Gamma(0) = 0.75$, $\Omega = [0, 2]$ and $\Omega_{dp}(t) = (\Gamma(t), 2]$. The mesh width is $\Delta x = 0.02$ and the time step is given by $\Delta t = \min(0.5\Delta x^2, \Delta t_{CFL})$, where $\Delta_{CFL} = 0.5 \frac{\Delta x}{\max |v|}$ denotes the stability condition. Figure 2 shows the interface position and interface concentrations as a function of time. Agreement between the similarity solution (see [12]) and the numerical solution is observed at the early stages of the dissolution. When the time evolves, the numerical solution diverges from the similarity solution due to the boundedness of the computational domain. Furthermore, convergence to the equilibrium interface position $\Gamma_{eq} = 0.592574$ is obtained. The interface concentrations also converge to the equilibrium values

$$\mathbf{c}_{eq}^{sol} = \begin{pmatrix} 0.559269 \\ 0.559269 \\ 1.447415 \\ 2.335562 \\ 1.891488 \end{pmatrix}$$

as time evolves. The equilibrium values of the interface position and interface concentrations, determined from a balance of mass argument, are the solution of the system

$$\begin{cases} \left(c_i^{part} - c_{i,eq}^{sol}\right)\Gamma_{eq} + Lc_{i,eq}^{sol} = \left(c_i^{part} - c_i^0\right)\Gamma_0 + Lc_i^0, \\ \prod_{i=1}^{p} \left(c_{i,eq}^{sol}\right)^{\tilde{n}_i} = \mathcal{K}, \end{cases}$$

where $L = 2$ denotes the length of the domain and Γ_0 the initial position of the interface.

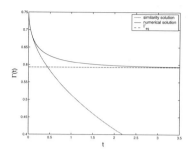 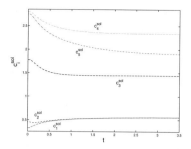

Fig. 2. Left: interface position vs time. Right: interface concentrations vs time.

5 Conclusions

In a previous study [8] a model, based on a Stefan problem, has been presented to predict dissolution kinetics in binary alloys. This model has been shown to be able to deal with general particle shapes and topological changes during the dissolution. The extension of this model to precipitate dissolution in multi-component alloys is aimed. This extension adds extra difficulties: the interface concentrations, which should be found as a part of the solution, should satisfy a nonlinear coupling. This nonlinearity is handled with an iterative method. Hence, an efficient, robust and of which extension to higher dimensional problems is affordable iterative method is sought here. We have considered here Newton, Broyden and Picard methods for a one-dimensional vector Stefan problem. This method will, then, be implemented into our three-dimensional model.

Newton provides fast convergence, but requires the numerical computation of the Jacobian, which is done here with central differences. This requires in total the solution of $2p^2$ diffusion problems for one evaluation of the Jacobian in one interface point, and the solution of a $p \times p$ system of equations per interface point and per iteration. Broyden method requires lower computational cost per iteration than Newton, since the approximation of the Jacobian by central differences is eliminated. The CPU-time per iteration though is of importance, since it has to solve a $p \times p$ system of equations per interface point and per iteration. Finally, Picard method is the cheapest per iteration, since it only requires one function evaluation per interface point and per iteration. This obviously pays the slow convergence back, especially for higher dimensional problems. The relaxation parameter δ dependents on the physical problem, and an arbitrary choice might delay the convergence. Furthermore, Picard's method seems to be the most robust with respect to the starting value of the interface concentrations.

References

1. Whelan, M.J.: On the kinetics of particle dissolution, Metals Sci. J., **3**, 95–97 (1969)
2. Aaron, H.B., Kotler, G.R.: Second phase dissolution, Metall. Trans., **2**, 393–407 (1971)
3. Gibou, F., Fedkiw, R., Caflisch, R., Osher, S.: A Level Set Approach for the Numerical Simulation of Dendritic Growth, J. Sci. Comp., **19**, 183–199 (2003)
4. Warren, J.A., Boettinger W.J.: Prediction of dendritic growth and microsegregation patterns in a binary alloy using the phase-field method, Acta Metall. Mater., **43**, 689–703 (1995)
5. Javierre, E., Vuik, C., Vermolen, F.J., Zwaag, S. van der: A comparison of numerical models for one-dimensional Stefan problems, Paper accepted in J. Comp. Appl. Math
6. Osher, S., Sethian, J.A.: Fronts propagating with curvature-dependent speed: Algorithms based on Hamilton-Jacobi formulations, J. Comput. Phys., **79**, 12–49 (1988)
7. Osher, S., Fedkiw, R.: Level Set Methods and Dynamic Implicit Surfaces. Springer-Verlag, New York (2003)
8. Javierre, E., Vuik, C., Vermolen, F.J., Segal, A.: A level set method for particle dissolution in a binary alloy, DIAM report 05-07, Delft University of Technology (2005)
9. Tan, L., Zabaras, N.: A level set simulation of dendritic solidification with combined features of front-tracking and fixed-domain methods, J. Comp. Phys., **211**, 36–63 (2006)
10. Reiso, O., Ryum, N., Strid, J.: Melting and dissolution of secondary phase particles in AlMgSi-alloys, Metall. Trans. A, **24A**, 2629–2641 (1993)
11. Hubert, R.: Modelisation numerique de la croissance et de la dissolution des precipites dans l'acierm, ATB Metallurgie, **34-35**, 5–14 (1995)
12. Vermolen, F.J., Vuik, C.: A mathematical model for the dissolution of particles in multi-component alloys, J. Comp. and Appl. Math., **126**, 233–254 (2000)
13. Vermolen, F.J., Vuik, C.: Solution of vector Stefan problems with cross-diffusion, J. Comp. and Appl. Math., **176**, 179–201 (2005)
14. Peng, D., Merriman, B., Osher, S., Zhao, H., Kang, M.: A PDE-Based Fast Local Level Set Method, J. Comput. Phys., **155**, 410–438 (1999)
15. Chen, S., Merriman, B., Osher, S., Smereka, P.: A Simple Level Set Method for Solving Stefan Problems, J. Comput. Phys., **135**, 8–29 (1997)

Level-Set Methods, Hamilton-Jacobi Equations and Applications

A Non-Monotone Fast Marching Scheme for a Hamilton-Jacobi Equation Modelling Dislocation Dynamics*

Elisabetta Carlini[1], Emiliano Cristiani[2] and Nicolas Forcadel[3]

[1] Dipartimento di Matematica, Università di Roma "La Sapienza", P.le Aldo Moro 2, Roma
`carlini@mat.uniroma1.it`
[2] Dipartimento Me.Mo.Mat., Università di Roma "La Sapienza", Via A. Scarpa 16, 00161 Roma
`cristiani@dmmm.uniroma1.it`
[3] CERMICS, ENPC, 6 et 8 avenue Blaise Pascal Cité Descartes – Champs sur Marne 77455 Marne la Vallée Cedex 2 (FRANCE)
`forcadel@cermics.enpc.fr`

Summary. In this paper we introduce an extension of the Fast Marching Method introduced by Sethian [6] for the eikonal equation modelling front evolutions in normal direction. The new scheme can deal with a *time-dependent* velocity without *any restriction on its sign*. This scheme is then used for solving dislocation dynamics problems in which the velocity of the front depends on the position of the front itself and its sign is not restricted to be positive or negative.

1 Introduction

In this paper, we propose a new Fast Marching Method for the following eikonal equation

$$\begin{cases} u_t(x,y,t) = c(x,y,t)|\nabla u(x,y,t)| & Q \subset \mathbb{R}^2 \times (0,T) \\ u(x,y,0) = u^0(x,y) & Q \subset \mathbb{R}^2. \end{cases} \quad (1)$$

This equation describes the propagation of a front $\Gamma_t = \partial \Omega_t$, where $\Omega_t = \{(x,y) \in Q \text{ s.t. } u(x,y,t) \geq 0\}$, with a normal speed $\mathbf{c} = c(x,y,t)\mathbf{n}$.

We consider here the case where the velocity $c(x,y,t)$ depends on time *without any restrictions on its sign*.

---

* The first two authors have been partially supported by the MIUR Project 2003 "Modellistica Numerica per il Calcolo Scientifico ed Applicazioni Avanzate". The last author was supported by the contract JC 1025 called "ACI jeunes chercheuses et jeunes chercheurs" of the French Ministry of Research (2003-2005).

The main objective is to extend the Fast Marching Method to the following non-local Hamilton-Jacobi equation

$$\begin{cases} u_t(x,y,t) = c^0(x,y) \star [u](x,y,t) |\nabla u(x,y,t)| & Q \subset \mathbb{R}^2 \times (0,T) \\ u(x,y,0) = u^0(x,y) & Q \subset \mathbb{R}^2. \end{cases} \quad (2)$$

The 0-level set of the solution of (2) represents a dislocation line in a 2D plane, here the kernel c^0 depends only on the space and $\star$ denotes the convolution in space (see [4] for a physical presentation of the model for dislocation dynamics).

To approach this problem, we first attack equation (1) generalizing the Fast Marching Method (FMM), introduced by Sethian [6], to fronts propagating with local speed $c(x,y,t)$ without any restriction on its sign . We will come back to equation (2) in the numerical section and in a future work.

It is well known that FMM is based on the following equation

$$c(x,y) |\nabla T(x,y)| = 1 \quad (3)$$

which is the stationary version of the equation (1) when $c = c(x,y) > 0$ or $c = c(x,y) < 0$ (see [6]). The front can be recovered as the level sets of the function $T(x,y)$.

In classical FMM the computation of the solution proceeds in an increasing order accepting at each iteration the smallest value of the nodes in the current *narrow band* (see [6] and [5]). The minimal value of the *narrow band* can be considered *exact* (within the discretization error) in the sense that it can not be improved in the following iterations. This result allows us to deal easily with a *time-dependent* speed function using the current minimal value of the *narrow band* as time t and then to evaluate the speed function $c(x,y,t)$ during the computation. Using this basic idea, Vladimirsky [8] extended FMM to a *signed* explicit time-depending function $c = c(x,y,t)$ and proved that in this case the evolution of front can be recovered as the level set of the time-independent function $T(x,y)$ which is the unique viscosity solution of the equation

$$c(x,y,T(x,y)) |\nabla T(x,y)| = 1. \quad (4)$$

In order to treat the *non-monotone* case in which speed is allowed to have different signs in different regions and/or to change sign in time, we introduce some important modifications to the classical scheme.

1) We perform a slight modification of the function c. If there are two or more regions with different sign for c at the same time, we force the speed to be exactly zero on the boundaries of these regions so that the evolution of the front in each region can be considered completely separate. The modified function will refer to as *numerical speed* and it will be indicated with $\hat{c}$.

2) Our new *narrow band* is the set of nodes which are going to be reached by the front *and* the nodes just reached by the front. This allows to deal with changes of sign of the velocity in time.

2 The FMM algorithm for unsigned velocity

In this section we give details for our FMM algorithm for unsigned velocity. We describe the evolution of the front Γ_t using an auxiliary function:

$$\theta(x,y,t) = \begin{cases} 1 & \text{if } u(x,y,t) \geq 0 \\ -1 & \text{otherwise.} \end{cases}$$

Notations and preliminary definitions

We consider a grid $Q_\Delta = \{(i,j) \in \mathbb{Z}^2 : (x_i, y_j) = (i\Delta, j\Delta) \in Q\}$ with space step Δ, we indicate with $0 < t_1 < ... < t_n < ... < t_N \leq T$ a non uniform grid on $[0,T]$, where t_n is the physical evolution time computed in each iteration of the FMM. We note that the partition of the time interval is not known *a priori*.
We introduce some definitions which will be useful in the sequel.

Definition 1. *We define* **neighborhood of the node** (i,j) *the set* $V(i,j) \equiv \{(l,m) \in Q_\Delta \text{ such that } |(l,m) - (i,j)| = 1\}.$

Definition 2. *Given the speed* $c_{i,j}^n \equiv c(x_i, y_j, t_n)$ *we define the* **numerical speed**
$$\hat{c}_{i,j}^n \equiv \begin{cases} 0 & if \text{ there exists } (l,m) \in V(i,j) \text{ such that} \\ & (c_{i,j}^n c_{l,m}^n < 0 \text{ and } |c_{i,j}^n| \leq |c_{l,m}^n|), \\ c_{i,j}^n & otherwise. \end{cases}$$

Definition 3. *Given* $\theta_{ij}^n = \theta(x_i, y_j, t_n)$ *we define the* **fronts** F_+^n *and* F_-^n *by*

$$F_\pm^n \equiv V(E) \backslash E, \quad \text{where } E = \{(i,j) \in Q_\Delta : \theta_{i,j}^n = \mp 1\} \quad \text{and} \quad F^n = F_+^n \cup F_-^n.$$

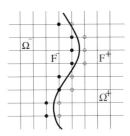

Fig. 1. The two Fronts

Remark 1. We should point out that the main difference with respect to the classical FMM algorithm is the presence of two *fronts*: F_+ and F_- (see Fig. 1).

If the speed is positive (negative) the front propagates using only the information coming from F_+ (F_-).

Description of the FMM algorithm for unsigned velocity

We need a discrete function T_I to indicate the approximate physical time for the front propagation on the nodes $I = (i, j)$ of the fronts.

Initialization

Step 1. $n = 1$
Step 2. *Initialization of the matrix* θ^0
$$\theta_I^0 = \begin{cases} 1 & \text{if } (x_i, y_j) \in \Omega_0 \\ -1 & \text{if } (x_i, y_j) \in Q \setminus \Omega_0 \end{cases}$$
Step 3. *Initialization of the time on the fronts*
$T_I^0 = 0$ for all $I \in F^0$

Main cycle

Step 4. *Computation of* $\widetilde{T}_I^{n-1}$.

We define $\hat{T}_{\pm,J}^{n-1} = \begin{cases} T_J^{n-1} & \text{if } J \in F_\pm^{n-1} \\ \infty & \text{else} \end{cases}$ Let $I \in F_\mp^{n-1}$, then

a) if $\pm \hat{c}_I^{n-1} \leq 0$, $\widetilde{T}_I^{n-1} = \infty$,
b) if $\pm \hat{c}_I^{n-1} > 0$, then we compute $\widetilde{T}_I^{n-1}$ as the greater solution of the following second order equation:

$$\sum_{k=1}^2 \left(\max_{\pm} \left(0, \widetilde{T}_I^{n-1} - \hat{T}_{+, I^{k,\pm}}^{n-1} \right) \right)^2 = \frac{(\Delta x)^2}{|\hat{c}_I^{n-1}|^2} \text{ if } I \in F_-^{n-1}$$

$$\sum_{k=1}^2 \left(\max_{\pm} \left(0, \widetilde{T}_I^{n-1} - \hat{T}_{-, I^{k,\pm}}^{n-1} \right) \right)^2 = \frac{(\Delta x)^2}{|\hat{c}_I^{n-1}|^2} \text{ if } I \in F_+^{n-1}$$

where
$$I^{k,\pm} = \begin{cases} (i \pm 1, j) & \text{if } k = 1 \\ (i, j \pm 1) & \text{if } k = 2 \end{cases}$$

Step 5. $\hat{t}_n = \min \{\widetilde{T}_I^{n-1}, I \in F^{n-1}\}$
Step 6. $\widetilde{t}_n = \begin{cases} \hat{t}_n & \text{if } \hat{t}_n < \infty \\ t_{n-1} + \delta & \text{if } \hat{t}_n = \infty \end{cases}$
where δ is a small constant, see following *Remark 1*
Step 7. $t_n = \max(t_{n-1}, \widetilde{t}_n)$
Step 8. if $t_n = t_{n-1} + \delta$ go to 4 with $n := n + 1$

Step 9. *Initialization of new accepted points*
$$NA_{\pm}^n = \{I \in F_{\pm}^{n-1}, \widetilde{T}_I^{n-1} = \widetilde{t}_n\}, NA^n = NA_+^n \cup NA_-^n$$
Step 10. *Reinitialization of θ^n*
$$\theta_I^n = \begin{cases} -1 & \text{if } I \in NA_+^n \\ 1 & \text{if } I \in NA_-^n \\ \theta_I^{n-1} & \text{else} \end{cases}$$
Step 11. *Reinitialization of T^n*
  a) If $I \in F^n \setminus V(NA^n)$ then $T_I^n = T_I^{n-1}$
  b) If $I \in NA^n$ then $T_I^n = t_n$
  c) If $I \in (F^{n-1} \cap V(NA^n)) \setminus (NA^n)$, then $T_I^n = T_I^{n-1}$
  d) If $I \in V(NA^n) \setminus F^{n-1}$ then $T_I^n = t_n$
Step 12. Go to 4 with $n := n+1$

Remark 1. The time computed in step 5 is the physical time, instead $\widetilde{t}$ in step 6 is an artificial time that allows to advance in time in any case. For example, if at the iteration n we have $\hat{c}_I^{n-1} = 0 \; \forall I \in F_{\pm}^{n-1}$, then there will not be new accepted point. As consequence the algorithm will be blocked. The term δ have to be small enough (like $\frac{\Delta}{|\hat{c}^{n-1}|}$). More details will be given in a paper in preparation focused on the convergence of the scheme.

Possible large time step could be computed when the speed is close to zero so that $\widetilde{t}_n$ could be very large. For this reason in step 8 we bound the size of the time step by δ.

Remark 2. In step 11 we change T_I^n only if a point of the neighborhood of I has been accepted.

Boundary conditions on ∂Q_Δ

The management of the boundary conditions is quite simple. As in the classical FMM (see for example [5]), we can assign to the nodes of the boundary a value, like $+\infty$, such that these nodes will not contribute at the computations. Then, at the end of the algorithm, these nodes will be cut off.

3 Numerical tests

3.1 Given velocity

We present some simulations which show the good behaviour of this new scheme. We propose a first test regarding the rotation of a line. We consider the square $[-1,1]^2$ and we approximate the evolution of a line crossing the $\{x = 0\}$ axis with the velocity $c(x,y,t) = -x$. We set $\theta^0 = -1$ above the line and $\theta^0 = 1$ below. Formally, one expects that a straight line remains a straight line for all $t > 0$ and that it rotates around the axis $\{x = 0\}$ (where the velocity is zero). Indeed, let us consider a generic straight line $r = \{(x,y) : y = ax+b\}$ and a point $(x_0, y_0) \in r$. We denote by (x_1, y_1) the image of (x_0, y_0) after the time Δt. A first order expansion gives

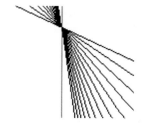

Fig. 2. Rotation of a line

$$(x_1, y_1) = (x_0, y_0) - \frac{\Delta t x_0}{\sqrt{1+a^2}}(-a, 1)$$

Then

$$y_1 = \left(\frac{a - \Delta t}{1 + a\Delta t}\right) x_1 + b,$$

since (x_1, y_1) satisfies the equation of a line we deduce that a straight line always remains a straight line. Fig. 2 shows that our algorithm computes what one expects.

Moreover, one can observe that the velocity of rotation of the line decreases when it approaches the axis $\{x = 0\}$. This is due to the fact that the velocity decreases near this axis.

We propose a second test regarding the evolution of a circle centered in the origin, with a speed $c(x, y, t) = 0.1t - x$. As shown in Fig. 1, the circle translates on the left and propagates in a self similar way. This test is run with $\Delta x = 2\pi/300$. The front is plotted every 0.5 physical time iterations with final time $T = 5$ and the solution is compared with that approximated by the classical finite difference scheme for equation (1).

Fig. 3. Evolution of a circle with FD

Fig. 4. Evolution of a circle with the FMM

Finally we propose a third test regarding the evolution of two circles. We set $\theta^0 = 1$ inside the circles and $\theta^0 = -1$ outside. We choose a velocity which changes sign in time, $c(x, y, t) = 1 - t$. This test is run with $\Delta x = 2\pi/300$. The front is plotted every 0.2 physical time with final time $T = 2.6$.

In Fig. 5 and 6 we show the result and we compare it with the approximation computed by the classical finite difference scheme for (1).

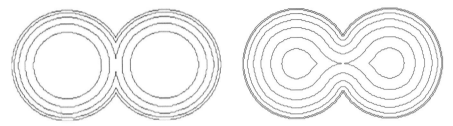

Fig. 5. Increasing (left) and decreasing (right) evolution of two circles with FMM

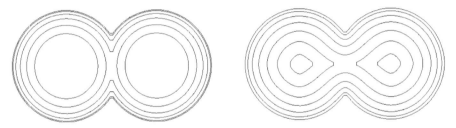

Fig. 6. Increasing (left) and decreasing (right) evolution of two circles by classical FD scheme

3.2 Dislocation dynamics

As we said in the introduction, our method can be extended to dislocations dynamic problems. In this case, we introduce a time step Δt and we consider a uniform grid over the time interval $[0, T]$, $\Delta t = \frac{T}{N}$ with $t_m = m\Delta t$, $m = 0, ..., N$.

We consider the function c as *time-independent* in each interval $\Delta t_m = [m\Delta t, (m + 1)\Delta t]$. Once the algorithm completed the computation for the front's evolution in every interval Δt_m, it updates the speed function. We need to fix the velocity on each time interval Δt_m, since it depends on the front itself. This avoid spurious oscillations which arise when the velocity is updated before all the nodes of the front are evolved for a physical time Δt. We propose one test regarding the relaxation of a dislocation line with sinusoidal shape. The Problem (2) is approximated in $[-1, 1]^2$ with

$$u(x,y,0) = \begin{cases} -1 & \text{if } y + 0.3\sin(x\pi) \leq 0 \\ 1 & \text{otherwise.} \end{cases}$$

We refer to [3] for the computation of the discrete convolution and for the description of the physical kernel. We remark that in this case the front has speed with different sign. In fact the upper part of the sinusoidal line moves to the left and the other part moves to the right. The three points where the line changes convexity do not move since they have speed equal to 0.

This test has been run with $\Delta x = 0.01$, $\Delta t = 0.1$ for a final time T=4. Fig. 7 represents the 0-level set of the discrete function θ plotted every 10 time iterations.

Fig. 7. Relaxation of a sinusoidal dislocation line

Acknowledgments

The authors would like to thank M. Falcone and R. Monneau for fruitful discussions during the preparation of this paper.

References

1. Alvarez, O., Cardaliaguet, P., Monneau R.: Existence and uniqueness for dislocation dynamics with nonnegative velocity. Interfaces and Free Boundaries, **7** (4), 415-434, (2005).
2. Alvarez, O., Carlini, E., Monneau, R., Rouy, E.: Convergence of a first order scheme for a non local eikonal equation. Accepted to IMACS Journal "Applied Numerical Mathematics".
3. Alvarez, O., Carlini, E., Monneau, R., Rouy, E.: A convergent scheme for a non local Hamilton-Jacobi equation modeling dislocation dynamic. Submitted to Numerische Mathematik.

4. Alvarez, O., Hoch, P., Le Bouar, Y., Monneau, R.: Dislocation dynamics driven by the self-force: short time existence and uniqueness of the solution. Accepted to Archive for Rational Mechanics and Analysis
5. Cristiani, E., Falcone, M.: Fast semi-Lagrangian schemes for the Eikonal equation and applications. Submitted.
6. Sethian, J. A.: A fast marching level set method for monotonically advancing fronts. Proc. Natl. Acad. Sci. USA, **93**, 1591–1595 (1996)
7. Sethian, J. A.: Level Set Methods, Evolving interfaces in Geometry, Fluid Mechanics, Computer Vision, and Material Science. Cambridge University Press (1996)
8. Vladimirsky, A.: Static PDEs for time-dependent control problems. Accepted by Interfaces and Free Boundaries; to appear.

A Time–Adaptive Semi–Lagrangian Approximation to Mean Curvature Motion

Elisabetta Carlini[1], Maurizio Falcone[1] and Roberto Ferretti[2]

[1] Dipartimento di Matematica, Università di Roma "La Sapienza", P.le Aldo Moro, 2, I - 00185 – Roma (Italy)
`carlini,falcone@mat.uniroma1.it`
[2] Dipartimento di Matematica, Università di Roma Tre, L.go S. Leonardo Murialdo, 1, I - 00146 – Roma (Italy)
`ferretti@mat.uniroma3.it`

Summary. We study the problem of time–step adaptation in semi–Lagrangian schemes for the approximation of the level–set equation of Mean Curvature Motion. We try to present general principles for time adaptivity strategies applied to geometric equations and to make a first attempt based on local truncation error. The efficiency of the proposed technique on classical benchmarks is discussed in the last section.

1 Introduction

Semi–Lagrangian (SL) schemes, whose use has been restricted for a long time to the Numerical Weather Prediction community (see the review paper [10]), have recently gained a certain popularity also in the field of first and second order Hamilton–Jacobi equations. This is due to the fact that they allow for larger time steps with respect to finite differences schemes. This feature is particularly interesting in geometric equations related to the "level set" formulation of moving interfaces (see e.g. [9] for a recent review of "level set" models, and [11, 4, 5] , [1] for the application of SL techniques). One of the most classical problems in this framework is the equation of Mean Curvature Motion (MCM), which reads

$$\begin{cases} v_t(x,t) = \operatorname{div}\left(\frac{Dv(x,t)}{|Dv(x,t)|}\right) |Dv(x,t)| & \text{in } \mathbb{R}^N \times (0,T) \\ v(x,0) = v_0(x). \end{cases} \qquad (1)$$

In (1), $v_0(x)$ can be any uniformly continuous function having the initial interface Γ_0 as its zero level set (or a prescribed level set) with nonzero gradient on the curve. The evolution Γ_t of the interface is then tracked by taking the same level set of the solution $v(x,t)$ of (1). In the sequel, we will only consider for simplicity the case $N = 2$.

A large time–step scheme for (1) has been proposed in [5] and extensively studied in [1]. In $\mathbb{R}^2$, it takes the form

$$v_j^{n+1} = \frac{1}{2}\left(I[V^n](x_j + \sigma_j^n\sqrt{\Delta t}) + I[V^n](x_j - \sigma_j^n\sqrt{\Delta t})\right) \qquad (2)$$

where $I[V](x)$ is a numerical reconstruction performed at the point x using the vector V of node values, and σ_j^n is defined by

$$\sigma_j^n = \frac{\sqrt{2}}{|D_j^n|}\begin{pmatrix} D_{2,j}^n \\ -D_{1,j}^n \end{pmatrix} \qquad (3)$$

with $D_{1,j}^n$, $D_{2,j}^n$ and D_j^n suitable numerical approximations of respectively $v_{x_1}(x_j, t_n)$, $v_{x_2}(x_j, t_n)$ and $Dv(x_j, t_n)$. Note that the numerical domain of dependence of v_j^{n+1} is given by the two regions around the points $x_j \pm \sigma_j^n\sqrt{\Delta t}$ which are about $2\sqrt{2\Delta t}$ apart.

The cost–effective use of large time–step schemes in PDEs typically results in a low resolution of smaller scales. In "level set" methods, this roughly means that it is difficult to track small structures of the interfaces, e.g. corners and cusps. Resolving such structures would require smaller time steps, as shown in Figure 1 where the evolution of a square is tracked up to $t = 0.1$ with respectively 10 time steps (left) and a single time step (right). On the other hand, once the solution is smoothed out (which usually happens with MCM equation (1)), efficiency would be increased by larger steps and this strongly motivates the introduction of an adaptive time–stepping strategy.

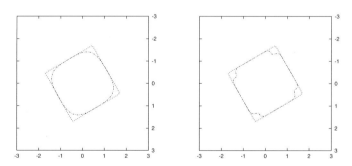

Fig. 1. Evolution of a square at $t = 0.1$, $\Delta t = 0.01$ and $\Delta t = 0.1$

It should be mentioned that space/time adaptive strategies have been extensively studied for linear PDEs mainly in the finite element community (see e.g. the survey paper [3] and the references therein). A lesser amount of literature exists for adaptive methods in nonlinear PDEs, and mainly for first order equations as in [2, 6] and [7]. Using (1) as a model problem, we will focus in this paper on evolutive nonlinear equation of *geometric* type, that is, of the form $u_t + H(x, u, Du, D^2u) = 0$ with H satisfying in particular the condition

$$H(x, u, \lambda p, \lambda X) = \lambda H(x, u, p, X), \quad \text{for } \lambda > 0. \tag{4}$$

where $p \in \mathbb{R}^N$ and $X \in S^N$, the space of symmetric $N \times N$ matrices. We recall that in a geometric equation the evolution of the interface Γ_t corresponding to a prescribed level set of the solution is independent of the choice of the initial condition v_0. Here, we will assume the space discretization to be fine enough for our needs, although our final goal is of course to study adaptivity in both time and space, as well as the application to more general situations. We will first try in Section 2 to single out a set of reasonable general axioms on the adaptation algorithm. In Section 3 we propose an adaptive strategy for the scheme proposed in [5], obtained by a suitable reworking of a classical technique for ODE schemes. Finally, in section 4 we will present some numerical tests.

2 General requirements on the adaptation strategy for geometric equations

We expect that some common requirements should be satisfied by any good adaptation strategy for the problem into consideration. Our plan is to define an adaptive time–stepping strategy which could correctly track the evolution of Γ_t. Let us then sketch some general assumptions for such a strategy.

Geometric behaviour: the time–step adaptation should depend on the behaviour of the interface, but not on how the interface is represented, i.e. not on the function v_0.

Global error control: the adaptation should take into account the error introduced both at the current and at future time steps. In other terms, a derefinement should not introduce at future time steps an error larger then the one introduced at the time step at which it is performed.

Range control: adaptation of the time–step should be bounded above and below by suitable relationships between Δt and Δx. In particular, for the problem at hand, a refinement of Δt should not go below the parabolic CFL condition $\Delta t = O(\Delta x^2)$ and a derefinement should in principle lead to the relationship maximizing the consistency rate.

Moreover, in order to increase efficiency and avoid a huge storage of data (the number of grid nodes can be quite large to get an accurate resolution of smaller scales), the indicator used to decide on the increase or decrease of the time step should be reasonably fast to compute and should only take into account informations from the last iteration. In the next section we will describe an adaptation strategy based on the local truncation error control as it is usually done for ODEs. We will show how such a technique can fit our requirements and give accurate results.

3 A strategy based on local truncation error

Following [5], and restricting to the non-degenerate case ($Dv \neq 0$), we write the local truncation error of the scheme (2), (3) as

$$L_{\Delta x, \Delta t}(x_j, t_n) = O\left(\Delta t^{1/2} + \frac{\Delta x^r}{\Delta t} + \frac{\Delta x^q}{\Delta t^{1/2}}\right). \tag{5}$$

In (5), the first term is related to time discretization, the second term to space discretization (with r denoting the rate of convergence of the space interpolation) and the third term to the approximation of the gradient (q being the rate of convergence of the finite-difference approximation of Dv). On a single time step, the perturbation introduced by the scheme on a smooth solution is therefore asymptotically given by

$$\epsilon_{\Delta x, \Delta t}(x_j, t_n) = \Delta t L_{\Delta x, \Delta t}(x_j, t_n) \sim C_1 \Delta t^{3/2} + C_2 \Delta x^r + C_3 \Delta t^{1/2} \Delta x^q. \tag{6}$$

On the other hand, following the same arguments used in [5], we can show that the same perturbation, when computed on a Lipschitz continuous solution, is in fact

$$\epsilon_{\Delta x, \Delta t}(x_j, t_n) \sim C_4 \Delta t^{1/2} + C_5 \Delta x. \tag{7}$$

Note that, in deriving (7), we have taken into account that the discrete gradient is simply bounded (so that, formally, $q = 0$) and that any polynomial reconstruction converges with order $r = 1$ on Lipschitz continuous functions.

In both (6) and (7), the symbols C_k should be intended as $C_k(x_j, t_n)$ (the dependence on the solution has been dropped for simplicity). In particular, by a careful examination of the consistency analysis in [5], it is possible to check that

$$C_1(x_j, t_n) \sim Dv(x_j, t_n) \cdot \frac{\partial}{\partial t}\sigma(x_j, t_n) = Dv \cdot \frac{\partial}{\partial t}\left[\frac{\sqrt{2}}{|Dv|}\begin{pmatrix}-v_{x_2}\\v_{x_1}\end{pmatrix}\right] \tag{8}$$

and at last, after some basic calculus,

$$C_1(x_j, t_n) \sim \frac{\sqrt{2}}{|Dv|}\eta \cdot \begin{pmatrix} v_{x_1}\eta \cdot \frac{\partial}{\partial t}(v_{x_2}, v_{x_1}) \\ v_{x_2}\eta \cdot \frac{\partial}{\partial t}(-v_{x_2}, v_{x_1}) \end{pmatrix} \tag{9}$$

where $\eta = Dv/|Dv|$ and all the functions have been computed at (x_j, t_n). On the other hand, one can also show that C_4 behaves like $const \cdot |Dv|$, and therefore both C_1 and C_4 are homogeneous of degree 1 in Dv.

According to the general framework outlined in the introduction, we will assume that the error term related to space and gradient discretization is negligible with respect to the term related to time discretization. As we said, the coupling of time and space adaptivity will be the object of a forthcoming study.

Following a classical technique for ODEs (see e.g. [5]), we compare numerical solutions obtained starting from $v(t_n)$ (in practice, from V^n) and using respectively two steps $\Delta t/2$ and a single step Δt, namely $v_{\Delta t/2} \approx v(t_{n+1}) + 2\epsilon_{\Delta t/2}$, and $v_{\Delta t} = v(t_{n+1}) + \epsilon_{\Delta t}$. Neglecting the space and gradient discretization terms and using (6) we get for the smooth case:

$$v_{\Delta t} - v_{\Delta t/2} \sim \epsilon_{\Delta t} - 2\epsilon_{\Delta t/2} \sim \left(1 - \frac{1}{\sqrt{2}}\right)\epsilon_{\Delta t} \qquad (10)$$

and hence

$$\epsilon_{\Delta t} \sim \frac{\epsilon_{\Delta t} - 2\epsilon_{\Delta t/2}}{1 - \frac{1}{\sqrt{2}}} \sim \frac{v_{\Delta t} - v_{\Delta t/2}}{1 - \frac{1}{\sqrt{2}}}, \qquad (11)$$

where the local errors ϵ are associated to the numerical solutions $v_{\Delta t}$ and $v_{\Delta t/2}$, and all the quantities refer to (x_j, t_n). In the Lipschitz case, following the same arguments and notations, and using (7) instead of (6), we obtain

$$\epsilon_{\Delta t} \sim \frac{v_{\Delta t/2} - v_{\Delta t}}{\sqrt{2} - 1}. \qquad (12)$$

For the problem under consideration, it is reasonable to assume that a time–step refinement should necessarily be performed in a nonsmooth situation (and therefore using (12) as a local truncation error estimate), and vice versa a derefinement would require a smooth solution (estimating in turn the local error by (11)).

There are still two questions to be addressed. First, in order to have a geometric behaviour of the algorithm, we should take into account that $\epsilon_{\Delta t}(x_j, t_n)$ is homogeneous of degree 1 in $|Dv|$, so that it should be divided by $|Dv(x_j, t_n)|$ itself (in practice, by the approximation $|D_j^n|$) to make it independent of the gradient of the solution (we recall that we assumed $Dv \neq 0$). Second, the refinement threshold should properly depend on Δx. The more natural choice is to set this threshold proportional to Δx so that, in the nonsmooth case, a local truncation error of order $O(\Delta t^{1/2})$ would be compared with a threshold $O(\Delta x)$ (this leads to the parabolic CFL condition $\Delta t = O(\Delta x^2)$). In order to preserve the relative scaling, the derefinement threshold will also be set proportional to Δx.

Summing up, the algorithm estimates (for every node in a suitable set S) $\epsilon_{\Delta t}$ via (11) or (12)). If

$$\left\| \frac{v_{\Delta t} - v_{\Delta t/2}}{|Dv|} \right\|_{L^\infty(S)} > (\sqrt{2} - 1)\tau_U \Delta x \qquad (13)$$

the time step is halved and a better approximation is computed. If in turn

$$\left\| \frac{v_{\Delta t} - v_{\Delta t/2}}{|Dv|} \right\|_{L^\infty(S)} < \left(1 - \frac{1}{\sqrt{2}}\right)\tau_L \Delta x, \qquad (14)$$

then the scheme advances to $v_{\Delta t/2}$ and the time step is doubled for the following iteration. Else, the scheme advances to $v_{\Delta t/2}$ and the time step is kept.

4 Numerical tests

We test our adaptive strategy on two standard problems for the MC flow in $\mathbb{R}^2$. Throughout this section, we will use a $\mathbb{P}_1$ (piecewise linear) space reconstruction and a centered–difference estimation of the gradient, so that in the smooth case we obtain $r = q = 2$. We will also use $\tau_L = \tau_U = 1$.

Our goal is to approximate only one level curve of interest, representing the front. When choosing to track the level curve $\Gamma_t^C \equiv \{(x,y) : u(x,y,t) = C\}$, the adaptation strategy will only consider the errors for nodes belonging to a band around such a level curve, given by

$$NB_\delta(\Gamma_t^C) \equiv \{(x,y) : C - \delta \leq u(x,y,t) \leq C + \delta\}.$$

(and the algorithm will use (13), (14) with $S = NB_\delta(\Gamma_t^C)$). This allows to reduce the computational complexity by collecting only the information around the level set of interest, although it is not clear whether a degradation of the solution might be caused at future times (see the requirements in Section 2).

Evolution of a circle. The first numerical test shows the shrinking of a circle. In this case the solution is smooth and we expect that the fixed time step scheme could also work efficiently. Moreover, it is easy to compute the exact solution so that we can compare both schemes in terms of the L^∞ error on the solution v. The following simulations has been run in $\Omega = [-4, 4]^2$ choosing

$$u(x,y,0) = \max((4^2 - x^2 - y^2)^4/4^8, 0).$$

We approximate the front $\Gamma_t^{0.2}$ and choose a band around the front with $\delta = 0.1$. The initial choice of the time step is $\Delta t = 0.1$.

Table 1 compares errors obtained at $T = 3.1$ in the band $NB_{0.1}(\Gamma_T^{0.2})$ for differents choices of the space step. The comparison has been performed by making the fixed step scheme run with the same overall number of steps as the adaptive scheme. Although in this test the fixed step scheme has remarkably good performances, we can see that the adaptive schemes has a comparable accuracy.

Different choices for the tolerance have been compared, yet obtaining similar behaviours, since in this case the local error is always under the more restrictive threshold. For any choice of the tolerance, the sequence of time steps is given in Table 2.

Table 1. Errors of the adaptive vs. fixed step scheme for the collapse of a circle

Δx	Adaptive Δt	Fixed Δt
$1.428 \cdot 10^{-1}$	$1.581 \cdot 10^{-3}$	$1.066 \cdot 10^{-3}$
$7.070 \cdot 10^{-2}$	$3.872 \cdot 10^{-4}$	$2.587 \cdot 10^{-4}$
$3.517 \cdot 10^{-2}$	$1.039 \cdot 10^{-4}$	$6.725 \cdot 10^{-5}$

Table 2. Sequence of time steps for the collapse of a circle

Iteration	Adaptive Δt
1	0.1
2	0.2
3	0.4
4	0.8
5	1.6

Evolution of a square. The second test shows the shrinking of a square, following the same test case in [5]. This is a situation in which a nonsmooth initial solution is smoothed out by the MC flow. Here, we really expect the adaptive scheme to be useful, although we cannot any longer compare the two schemes in terms of L^∞ errors. The simulations have been run in $\Omega = [-1.5, 1.5]^2$ choosing

$$u(x, y, 0) = 1.5 - |\xi(x, y) - \eta(x, y)| - |\xi(x, y) + \eta(x, y)|,$$

where $\xi(x, y) = 1/2(\sqrt{3}\, x + y)$ and $\eta(x, y) = 1/2(\sqrt{3}\, y - x)$ (this rotation avoids an alignment between the square and the grid). We approximate the front Γ_t^0 with $\Delta x = 0.03$, $\delta = 0.2$. Figure 2 shows 11 iterations of the numerical solutions obtained by the two methods up to time $T = 0.3$ (the corresponding fixed step is $\Delta t = 0.027$), with an apparent advantage of the adaptive scheme. Figure 3 shows, as functions of the iteration number, the values of the left–hand side of (13) and (14) (plotted by +) compared with upper and lower thresholds given by the respective right–hand sides (dotted lines), along with the resulting adapted time step (solid line).

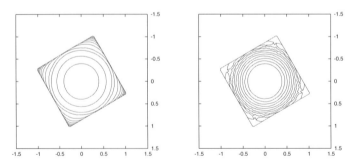

Fig. 2. Collapse of a square, adaptive vs. fixed time step

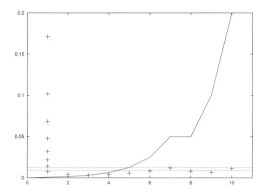

Fig. 3. Local error estimate and adaptive time step vs. iteration number

References

1. Carlini, E.: Semi-Lagrangian schemes for first and second order Hamilton-Jacobi equations, Ph.D Thesis, Università di Roma "La Sapienza", Roma (2004)
2. Johnson, C.: Adaptive finite element methods for conservation laws. In: Cockburn, B., Johnson, C. , Shu, C.-W., Tadmor, E., Advanced numerical approximation of nonlinear hyperbolic equations. Papers from the C.I.M.E. Summer School held in Cetraro, June 23–28, 1997. Edited by Alfio Quarteroni. Lecture Notes in Mathematics, 1697. Springer-Verlag, Berlin (1998)
3. ÊJohnson, C.: Adaptive computational methods for differential equations. ICIAM 99 (Edinburgh), 96–104, Oxford Univ. Press, Oxford (2000)
4. M. Falcone, R. Ferretti: Semi–Lagrangian schemes for Hamilton–Jacobi equations, discrete representation formulae and Godunov methods. J. Comp. Phys., **175**, 559–575 (2002)
5. Falcone, M., Ferretti, R.: Consistency of a large time–step scheme for mean curvature motion. In: Brezzi, F., Buffa, A., Corsaro, S., Murli, A., (eds) Numerical Mathematics and Advanced Applications – ENUMATH 2001. Springer-Verlag, Milano (2003)
6. Grüne, L.: An adaptive grid scheme for the discrete Hamilton–Jacobi–Bellman equation. Numer. Math., **75**, 319–337 (1997)
7. Grüne, L.: Adaptive grid generation for evolutive Hamilton-Jacobi-Bellman equations. In: Numerical methods for viscosity solutions and applications (Heraklion, 1999), 153–172, Ser. Adv. Math. Appl. Sci., 59, World Sci. Publishing, River Edge (2001)
8. Hairer, E., Nørsett, S. P., Wanner, G.: Solving ordinary differential equations. I. Nonstiff problems. Second edition. Springer Series in Computational Mathematics, 8. Springer-Verlag, Berlin, (1993)
9. ÊOsher, S., Fedkiw, R.: Level set methods and dynamic implicit surfaces. Applied Mathematical Sciences, 153. Springer-Verlag, New York (2003)
10. Staniforth, A.N., Côtè, J.: Semi–Lagrangian integration schemes for atmospheric models – a review. Mon. Wea. Rev., **119**, 2206–2223 (1991)
11. Strain, J.: Semi-Lagrangian methods for level set equations. J. Comput. Phys., **151**, 498–533 (1999)

Multiscale Methods

Heterogeneous Multiscale Methods with Quadrilateral Finite Elements

Assyr Abdulle

University of Basel, Department of Mathematics, Rheinsprung 21, CH-4051 Basel, Switzerland
`Assyr.Abdulle@unibas.ch`

Summary. The Heterogeneous Multiscale Method (HMM) applied to elliptic homogenization problems has been analyzed for simplicial elements in [E, Ming, Zhang, J. Amer. Math. Soc. 18, pp. 121-156, 2005] and [Abdulle, SIAM, Multiscale Model. Simul., Vol. 4, No 2, pp.195–220, 2005.]. In this paper we discuss and analyze the use of quadrilateral (hexahedral) finite elements for the HMM applied to elliptic homogenization problem. We give H^1 and L^2 a priori estimates and discuss a strategy to recover the microscopic information. Numerical examples confirm our error estimates.

1 Introduction

The numerical solution of multiscale elliptic problems is a basic problem for many applications. Solving these problems with a standard finite element method (FEM) is often difficult or even impossible, due to the computational work and the amount of memory needed to solve the small scale.

The heterogeneous multiscale method (HMM) introduced in [9] is a general framework for designing numerical methods for problems with multiple scales. For elliptic homogenization problems, this method couples a macro and a micro FEM in the following way: the macroscopic problem with unknown input data is computed by performing micro calculations on small sub-domains of a macroscopic mesh. This allows to assemble the macroscopic stiffness matrix without knowing or deriving beforehand the macroscopic equation [1, 2, 3, 5, 8, 9]. Several other approaches have been proposed for elliptic multiscale problems, as for example in [11] and [12]. We refer also to [9] for further references.

The finite element heterogeneous multiscale method (FE-HMM) for elliptic problems has been analyzed in [8, 3] for simplicial finite elements. In [3], the first fully discrete analysis of the method has been given, taking into account the error in the micro FE solver. For flexibility in applications, it is important to be able to use other elements, besides simplicial ones, as quadrilateral or hexahedral elements. In this paper we analyze the method for quadrilateral (hexahedral) finite elements. For simplicity, we present the method for piecewise bilinear polynomials in two dimensions. These results can be extended to higher dimensions with similar ideas. They can also be extended to higher orders FEM and convergence rates, provided higher order regularity of the true solutions is present.

The outline of the paper is as follows. In section 2 we recall the HMM for elliptic multiscale problems and extend the formulation for quadrilateral elements. In section 3 we derive the convergence results for the FE-HMM and give in section 4 numerical examples illustrating the sharpness of our estimates.

2 HMM with quadrilaterals finite elements

In a bounded domain $\Omega \subset \mathbb{R}^d$, for $f \in L^2(\Omega)$, we consider the elliptic problem

$$-\nabla \cdot (a^\varepsilon \nabla u^\varepsilon) = f \text{ in } \Omega, \quad u^\varepsilon = 0 \text{ on } \partial\Omega, \qquad (1)$$

where we assume that the tensor $a^\varepsilon(x) = a(x, \frac{x}{\varepsilon}) = a(x,y)$ is symmetric, coercive and periodic with respect to each component of $y = x/\varepsilon$ in the unit cube $Y = (0,1)^d$ (the superscript on the solution u emphasizes its dependence on ε). We further assume that $a_{ij}(x, \cdot) \in L^\infty(\mathbb{R}^d)$ and that $x \to a_{ij}(x, \cdot)$ is smooth from $\bar{\Omega} \to L^\infty(\mathbb{R}^d)$. It is known that u^ε converges (usually in a weak sense) to a "homogenized" function u^0, solution of the homogenized problem

$$-\nabla \cdot (a^0(x) \nabla u^0) = f(x) \in \Omega, \quad u^0 = 0 \quad \text{on } \partial\Omega, \qquad (2)$$

where the homogenized diffusion tensor a^0 is a smooth matrix with coefficients given by $a^0_{ij}(x) = \int_Y \left(a_{ij}(x,y) + \sum_{k=1}^n a_{ik}(x,y) \frac{\partial \chi^j}{\partial y_k}(x,y) \right) dy$ [6, 10]. Here, $\chi^j(x, \cdot)$ denote the solutions of the cell problems given in (15) below, but for the space

$$W^1_{per}(Y) = \{ v \in H^1_{per}(Y); \int_Y v dx = 0 \}, \qquad (3)$$

where $H^1_{per}(Y)$ is defined as the closure of $C^\infty_{per}(Y)$ (the subset of $C^\infty(\mathbb{R}^d)$ of periodic functions in the unit cube $Y = (0,1)^d$) for the H^1 norm. In the sequel, we assume that the solutions χ^j of the cell problems (15) satisfy $\chi^j(x_k, \cdot) \in W^{2,\infty}(Y)$, for a fixed first variable $x = x_k \in \Omega$, and

$$\| D_x^\alpha \left(\chi^j(x_k, x/\varepsilon) \right) \|_{L^\infty(K_\varepsilon)} \leq C \, \varepsilon^{-|\alpha|}, \ |\alpha| \leq 2, \ \alpha \in \mathbb{N}^d. \qquad (4)$$

As discussed earlier, we concentrate here for simplicity on piecewise bilinear continuous FEM in the micro and in the macro spaces. We therefore consider a macro finite element space defined by

$$S_H(\Omega) := S(\Omega, \mathcal{T}_H) := \{u^H \in H_0^1(\Omega);\ u^H|_K \in \mathcal{Q}^1(K),\ \forall K \in \mathcal{T}_H\}, \quad (5)$$

where $\mathcal{Q}^1(K)$ is the space of bilinear polynomials on the quadrilateral K, and $\mathcal{T}_H$ is a quasi-uniform partition of $\Omega \subset \mathbb{R}^d$ of shape regular quadrilaterals K. By "macro finite elements" we mean that H, the size of the macro elements K, can be much larger than the length scale ε. For a function $v^H \in S_H$, we will consider its linearization around a point $x_{l_i} \in K_i \in \mathcal{T}_H$

$$v_{l_i}^H(x)|_{K_i} = v^H(x_{l_i}) + (x - x_{l_i})\nabla v^H(x_{l_i}). \quad (6)$$

We also consider a micro finite element space $S_{h,per} \subset W_{per}^1(K_\varepsilon)$ defined by

$$S_{h,per}(K_\varepsilon) := S_{per}^1(K_\varepsilon, \mathcal{T}_h) := \{z^h \in W_{per}^1(K_\varepsilon);\ z^h|_T \in \mathcal{P}^1(T),\ T \in \mathcal{T}_h\}, \quad (7)$$

where $\mathcal{P}^1(T)$ is the space of linear polynomials on the triangle T.

Remark 1. We could have used also the space of bilinear polynomials $\mathcal{Q}^1(T)$ on quadrilaterals in (7). We consider here linear polynomials for the micro FE so that we can use the results derived in [3] for the micro FEM.

The FE-HMM for the elliptic homogenization problem, based on the macro space $S_H(\Omega)$ is defined by a *modified macro bilinear form* [9, 2],

$$B(u^H, v^H) = \sum_{K_i \in \mathcal{T}_H} \sum_{l_i=1}^n \frac{\omega_i}{|K_{\varepsilon, l_i}|} \int_{K_{\varepsilon, l_i}} \nabla u_{l_i}^h\, a(x_{l_i}, x/\varepsilon)(\nabla v_{l_i}^h)^T dx, \quad (8)$$

where $K_{\varepsilon, l_i} = x_{l_i} + \varepsilon[-1/2, 1/2]^d$ is a sampling sub-domain centered at the point $x_{l_i} \in K_i$, $|K_i|, |K_{\varepsilon, l_i}|$ denote the measure of K_i and K_{ε, l_i}, respectively, and where $u_{l_i}^h$ (resp. $v_{l_i}^h$) is the solution of the following *micro problem*: find $u_{l_i}^h$ such that $(u_{l_i}^h - u_{l_i}^H) \in S_{h,per}(K_{\varepsilon, l_i})$ and

$$\int_{K_{\varepsilon, l_i}} \nabla u_{l_i}^h\, a(x_{l_i}, x/\varepsilon)(\nabla z^h)^T dx = 0 \quad \forall z^h \in S_{h,per}(K_{\varepsilon, l_i}). \quad (9)$$

The macro FE-HMM solution is then defined by the following variational problem: find $u^H \in S_H(\Omega)$ such that

$$B(u^H, v^H) = \langle f, v^H \rangle, \quad \forall v^H \in S_H(\Omega). \quad (10)$$

Remark 2. The set $\{\omega_{l_i}, x_{l_i}\}_{l_i=1}^n$ is a quadrature formula on the element K_i. We will discuss in the Section 3 how to chose it.

Remark 3. We will see in the sequel that it is crucial to define the micro solution $u_{l_i}^h$ constrained (through periodic boundary conditions) by the *linearized* part of the corresponding macro solution u^H. Such an idea has been proposed in [8] to extend the HMM for higher order macroscopic solver on triangles using numerical quadrature schemes.

3 Error analysis

We first discuss the choice of the quadrature formula $\{\omega_{l_i}, x_{l_i}\}_{l_i=1}^n$. It is tempting to chose $\omega_{l_i} = K_i$, $x_{l_i} = x_k$, $n = 1$, where x_k is located at the barycenter of the element K_i (one point Gauss rule), since this formula is exact for bilinear polynomials. But with such a quadrature formula, the bilinear form (8) cannot be coercive (see Remark 5 below). Therefore, we chose $n = 4$ and the two points Gauss quadrature rule $\omega_{l_i} = |K_i|/4$, $x_{l_i} = F_i(1/2 \pm \sqrt{3}/6, 1/2 \pm \sqrt{3}/6)$, where $x = F_i(\xi)$ the affine mapping, which maps $[0,1]^2$ onto $K_i \in \mathcal{T}_H$. We next show that with this quadrature formula the problem (10) is well posed.

Proposition 1. *With the above quadrature formula, the problem (10) has a unique solution which satisfies*

$$\|u^H\|_{H^1(\Omega)} \leq C\|f\|_{L^2(\Omega)}. \tag{11}$$

Proof. We note that for $v^H \in S_H(\Omega)$, the gradient of the linearized part $\nabla v_{l_i}^H$ is constant over a macro element K_i. Hence, the following two relations between the micro solution of (9) and its corresponding linearized macro function $v_{l_i}^H$ hold.

$$\int_{K_{\varepsilon,l_i}} |\nabla v_{l_i}^h|^2 dx = \int_{K_{\varepsilon,l}} |\nabla v_{l_i}^h - \nabla v_{l_i}^H|^2 dx + \int_{K_{\varepsilon,l_i}} |\nabla v_{l_i}^H|^2 dx, \tag{12}$$

$$\int_{K_{\varepsilon,l_i}} \nabla v_{l_i}^h \, a(x_k, x/\varepsilon)(\nabla v_{l_i}^h - \nabla v_{l_i}^H)^T dx = 0. \tag{13}$$

Observe that equality (12) yields $\|\nabla v_{l_i}^h\|_{L^2(K_{\varepsilon,l_i})} \geq \|\nabla v_{l_i}^H\|_{L^2(K_{\varepsilon,l_i})}$. Next, since $\nabla v_{l_i}^H$ is constant, we have $\int_{K_{\varepsilon,l_i}} \|\nabla v_{l_i}^H\|_{L^2(K_{\varepsilon,l_i})}^2 dx = \nabla v_{l_i}^H(x_{l_i}) \cdot \nabla v_{l_i}^H(x_{l_i})$. Using these observations and the coercivity of a^ε we find

$$B(v^H, v^H) \geq C \sum_{K_i \in \mathcal{T}_H} \sum_{l_i=1}^{4} \frac{|K_i|}{4} \left(\nabla v_{l_i}^H(x_{l_i}) \cdot \nabla v_{l_i}^H(x_{l_i})\right) = C \sum_{K_i \in \mathcal{T}_H} \|\nabla v^H\|_{L^2(K_i)}^2,$$

where we used that the quadrature formula is exact for quadratic polynomials. Thus, the bilinear form B is coercive. Using (13) we see that the bilinear form B is bounded. Finally, the existence and uniqueness of a solution u^H of problem (10) as well as (11) follow from the Lax-Milgram theorem. ∎

Remark 4. It can be seen in (12) why we introduced the linearization $v_{l_i}^H$ of a function $v^H \in S_H(\Omega)$. For functions $v^H \in S_H(\Omega)$, (12) does not hold.

The following lemma can be shown similarly as in [3].

Lemma 1. *The function $u_{l_i}^h \in S_{h,per}(K_{\varepsilon,l})$, solution of problem (9), can be represented as*

$$u_{l_i}^h = u_{l_i}^H + \varepsilon \sum_{j=1}^{d} \chi^{j,h}(x_{l_i}, x/\varepsilon) \frac{\partial u_{l_i}^H(x_{l_i})}{\partial x_j}, \qquad (14)$$

where $\chi^{j,h}(x_{l_i}, y)$ are the (unique) solutions of the cell problems

$$\int_Y \nabla \chi^{j,h} a(x_{l_i}, y)(\nabla z^h)^T dy = -\int_Y e_j^T a(x_{l_i}, y)(\nabla z^h)^T dy\ , \forall z^h \in S_{h,per}(Y), \qquad (15)$$

where $Y = (0,1)^2$ and $\{e_j\}_{j=1}^{2}$ is the standard basis of $\mathbb{R}^2$.

Remark 5. It can be shown similarly as in [1, 2, 3, 4], that

$$\frac{1}{|K_{\varepsilon,l_i}|} \int_{K_{\varepsilon,l_i}} \nabla u_{l_i}^h\, a(x_{l_i}, x/\varepsilon)(\nabla v_{l_i}^h)^T dx = \frac{1}{|K_i|} \int_{K_i} \nabla u_{l_i}^H\, a(x_{l_i}, x/\varepsilon)(\nabla v_{l_i}^H)^T dx. \qquad (16)$$

Using the above formula, it can be seen that the bilinear form (10) defined upon the one point quadrature formula, located at the barycenter of each quadrilateral, would be indefinite (consider for example for $u^H(x) = (x_1 - 1/2) \cdot (x_2 - 1/2)$).

Following the lines of Lemma 3.3 and Proposition 3.5 of [3], we find

Theorem 1. *Let u^0 be the solution of the homogenized problem (2) and assume that u^0 is H^2-regular. Let u^H be the solution of problem (10) and suppose that (4) holds. Then*

$$\|u^0 - u^H\|_{L^2(\Omega)} \leq C(H^2 + (h/\varepsilon)^2)\|f\|_{L^2(\Omega)}, \qquad (17)$$
$$\|u^0 - u^H\|_{H^1(\Omega)} \leq C(H + (h/\varepsilon)^2)\|f\|_{L^2(\Omega)}, \qquad (18)$$

where H is the size of the mesh of the macro FE space $S_H(\Omega)$ and h is the size of the mesh of the micro FE space $S_{h,per}(K_\varepsilon)$.

Remark 6. If we denote by $M = \dim S_{h,per}(K_\varepsilon)$ and $N = \dim S_H(\Omega)$ the degrees of freedom of the micro and the macro FE spaces, respectively, the estimates (17) and (18) can be rewritten as $\|u^0 - u^H\|_{L^2(\Omega)} \leq (N^{-\frac{2}{d}} + M^{-\frac{2}{d}})$ and $\|u^0 - u^H\|_{H^1(\Omega)} \leq (N^{-\frac{1}{d}} + M^{-\frac{2}{d}})$, respectively, emphasizing that the quantity h/ε is independent of ε. The above estimates show that both, micro and macro meshes, have to be refined simultaneously.

Using Theorem 1 it can be shown that

$$\|u^\varepsilon - u^H\|_{L^2(\Omega)} \leq C(H^2 + (h/\varepsilon)^2 + \varepsilon)\|f\|_{L^2(\Omega)} \qquad (19)$$

(see [3] and also [8]). Convergence in the H^1 norm is however not possible in general (see [11, 4.4]). In [1, 3, 9], a procedure was given allowing to recover the small scale information of u^ε and H^1 estimates between the reconstructed solution and u^ε were obtained. Some care is required to extend this procedure for quadrilateral elements. In the sequel we explain how this can be done.

Let $\hat{K} = [0,1]^2$ be the reference quadrilateral and $\hat{x}_l = (1/2 \pm \sqrt{3}/6, 1/2 \pm \sqrt{3}/6)$, $l = 1,\ldots,4$ the two points Gauss quadrature formula. We divide $\hat{K}$ in four quadrilaterals joining the barycenter of $\hat{K}$ with the midpoint of each edges. We denote $\hat{K}_l$ the sub-quadrilateral with $x_l \in \hat{K}_l$. For each macro quadrilateral $K_i \in \mathcal{T}_H$ consider the quadrilateral K_{l_i} defined by $K_{l_i} = F_i(\hat{K}_l)$, where F_i is the affine mapping which maps $\hat{K}$ onto K_i. We define by $\Gamma_H = \{K_{l_i},\ i = 1,\ldots,4;\ K \in \mathcal{T}_H\}$ the new partition obtained by the subdivision of the partition $\mathcal{T}_H$ in sub-quadrilaterals as explained above. For each K_{l_i} we define

$$u_p^\varepsilon(x)|_{K_{l_i}} = u_{l_i}^H(x) + \left(u_{l_i}^h(x) - u_{l_i}^H(x)\right)|_{K_{l_i}}^P \quad \text{for } x \in K_{l_i} \in \Gamma_H, \quad (20)$$

where $|_{K_{l_i}}^P$ denotes the periodic extension of the fine scale solution $(u_{l_i}^h - u_{l_i}^H)$, available in K_{ε,l_i} on each element K_{l_i}. Since u_p^ε can be discontinuous across the macro elements K_{l_i}, we define a broken H^1 norm by

$$\|u\|_{\bar{H}^1(\Omega)} := \Big(\sum_{K_{l_i} \in \Gamma_H} \|\nabla u\|_{L^2(K_{l_i})}^2 \Big)^{1/2}. \quad (21)$$

The techniques used in [3, Thm. 3.11] can now be adapted to the above situation and we obtain

Theorem 2. *Let u_p^ε be defined by (20) and u^ε be the solution of (1). Suppose that (4) holds. Then*

$$\|u^\varepsilon - u_p^\varepsilon\|_{\bar{H}^1(\Omega)} \le C(\sqrt{\varepsilon} + H + h/\varepsilon)\|f\|_{L^2(\Omega)}, \quad (22)$$

where H is the size of the mesh of the macro FE space $S_H(\Omega)$ and h is the size of the mesh of the micro FE space $S_{h,per}(K_\varepsilon)$.

Note again (see Remark 6) that h/ε is independent of ε.

4 Numerical experiments

In this last section, we present numerical experiments to illustrate the sharpness of our error estimates. We consider the following model problem

$$-\nabla \cdot \left(a(\tfrac{x}{\varepsilon})\nabla u^\varepsilon\right) = f(x) \quad \text{in } \Omega = (0,1)^2 \quad (23)$$

$$u^\varepsilon|_{\Gamma_D} = 0 \quad \text{on } \Gamma_D := \{x_1 = 0\} \cup \{x_1 = 1\} \quad (24)$$

$$n \cdot \left(a(\tfrac{x}{\varepsilon})\nabla u^\varepsilon\right)|_{\Gamma_N} = 0 \quad \text{on } \Gamma_N := \partial\Omega \setminus \Gamma_N, \quad (25)$$

where $a(\tfrac{x}{\varepsilon}) = (\cos \tfrac{2\pi x_1}{\varepsilon} + 2)I$, I is the identity matrix and $f(x) \equiv 1$. The exact solution as well as the homogenized tensor can be derived analytically (see [3]). In the following, we compute the solution of the problem (23–25) with the FE-HMM for several macromeshes $H = (1/2)^d$, $d = 1, \ldots, 6$ and micro meshes $h_M = 1/5, 1/10, 1/20, 1/40$, where $h_M \simeq M^{-\frac{1}{2}}$ and M is the number of degrees of freedom of the micro FE space. The comparison of these numerical solutions with the homogenized solution in the L^2 and H^1 norm is given in Fig. 1 (left picture) and Fig. 1 (right picture), respectively. We give the result for $\varepsilon = 10^{-5}$. Note that the convergence rates are independent of ε (see Theorem 1). It can be seen, as predicted by Theorem 1, that the convergence rates are quadratic and linear for the L^2 and H^1 norm, respectively, provided that the micro mesh is also refined. The influences of the micro problems are in accordance with our fully discrete analysis. Micro and macro meshes have to be refined at the same speed for the L^2 norm while for the H^1 norm the micro mesh can be refined at a slower rate.

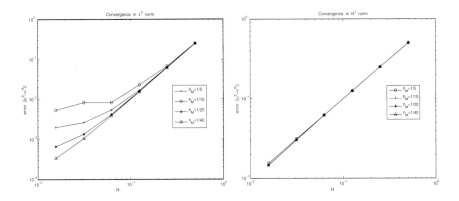

Fig. 1. Convergence rates of the error between the macro solution of the FE-HMM for the problem (23–25), with $\varepsilon = 10^{-5}$ and decreasing macro and micro meshes, and the homogenized solution of the problem (23–25).

Finally, in Fig. 2 (left picture) we compare, in the L^2 norm, the fine scale solution u^ε with the macro FE-HMM solution. We see that we have now a dependency towards ε as predicted by (19). To improve the results one has to decrease ε as shown in Fig. 2 (right picture), where $\varepsilon = 10^{-3}$.

Acknowledgment

This work is partially supported by the Swiss National Foundation under grant 200021-103863/1.

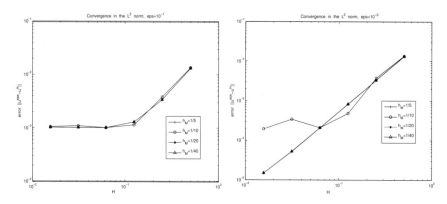

Fig. 2. Convergence rates of the error between the macro solution of the FE-HMM for the problem (23–25), with decreasing macro and micro meshes, and the fine scale solution of the problem (23–25), $\varepsilon = 10^{-1}$ (left picture) and $\varepsilon = 10^{-3}$ (right picture).

References

1. Abdulle, A., E, W.: Finite difference HMM for homogenization problems. J. Comput. Phys., **V191**, 18–39 (2003)
2. Abdulle, A., Schwab, C.: Heterogeneous multiscale FEM for diffusion problem on rough surfaces. SIAM, Multiscale Model. Simul., **3**, 195–220 (2005)
3. Abdulle, A.: On a priori error analysis of fully discrete heterogeneous multiscale FEM. SIAM MMS Multiscale Model. Simul., **4**, 195–220 (2005)
4. Abdulle, A.: Multiscale methods for advection-diffusion problems. Discrete Contin. Dyn. Syst. B, Suppl. Vol., 11-21 (2005)
5. Abdulle, A.: Analysis of a heterogeneous multiscale FEM for problems in elasticity. Math. Mod. Meth. Appl. Sci. (M3AS), **16**, 1-21 (2006)
6. Bensoussan, A., Lions, J.-L., Papanicolaou, G.: Asymptotic analysis for periodic structures. North Holland Amsterdam (1978)
7. Cioranescu, D., Donato, P.: An introduction to homogenization. Oxford University Press (1999)
8. E, W., Ming, P., Zhang, P.: Analysis of the heterogeneous multi-scale method for elliptic homogenization problems. J. Amer. Math. Soc., **18**, 121-156 (2005)
9. E, W., Engquist, B.: The Heterogeneous multiscale methods. Commun. Math. Sci., **1**, 87–132 (2003)
10. Jikov, V.V., Kozlov, S.M., Oleinik, O.A.: Homogenization of differential operators and integral functionals. Springer Berlin Heidelberg (1994)
11. Hou, T.-Y., Wu, X.-H., Cai, Z.: Convergence of a multi-scale finite element method for elliptic problems with rapidly oscillating coefficients. Math. of Comput., **68**, 913–943 (1999)

12. Matache, A.M., Schwab, C.: Two-scale FEM for homogenization problems. R.A.I.R.O. Anal. Numerique, **36**, 537–572 (2002)
13. Oden, J.T., Vemaganti, K.S.: Estimation of local modeling error and global-oriented adaptive modeling of heterogeneous materials: error estimates and adaptive algorithms. J. Comput. Phys., **164**, 22–47 (2000)

Stabilizing the $\mathbb{P}^1/\mathbb{P}^0$ Element for the Stokes Problem via Multiscale Enrichment

Rodolfo Araya[1]*, Gabriel R. Barrenechea[1][†] and Frédéric Valentin[2]

[1] Depto. de Ingeniería Matemática, Universidad de Concepción, Casilla 160-C, Concepción, Chile
   `raraya, gbarrene@ing-mat.udec.cl`
[2] Laboratório Nacional de Computação Científica, Av. Getúlio Vargas, 333, 25651-070 Petrópolis - RJ, Brasil
   `valentin@lncc.br`

Summary. This work concerns the derivation of new stabilized finite element methods for the Stokes problem. Starting from pairs of spaces which are not stable, they are made stable by enriching them with multiscale functions, i.e., functions which are local, but not bubble-like, arising from the solution of local problems at the element level. This general methodology is applied to stabilize the non-stable $\mathbb{P}^1/\mathbb{P}^0$ pair.

1 Introduction

In the mid nineties there was a growing interest in giving theoretical justifications for the different stabilized finite element methods for the Stokes problem. One answer came from the fact that the usual Galerkin method enriched with bubble functions led to the GLS method (see [2]). Also, this methodology of enriching the standard polynomial space with bubble functions led to a new class of methods, namely the Residual-Free Bubble method (cf. [3, 8]).

The imposition of a zero boundary condition on the element boundary for RFB presented some numerical problems. One possible solution for some of these problems is the multiscale finite element method (see [7, 6]). A particularity of such methods is that a Petrov-Galerkin strategy is proposed, in which the test function space is enriched with bubble functions in order to have a local problem containing the residual of the momentum equation on the right hand side. A special boundary condition (related to the one used in [9]) is imposed in order to solve these local problems analytically. The purpose of this work is to use the multiscale approach from [7, 6], combined with

---

* This author is partially supported by FONDECYT Project No. 1040595.
† This author is partially supported by CONICYT-Chile through FONDECYT Project No. 1030674 and FONDAP Program on Applied Mathematics.

the static condensation procedure, in order to propose new stabilized finite element methods for the Stokes problem. We proceed as in [7], defining an enrichment function for the trial space for the velocity that no longer vanishes on the element boundary (and hence it is not a bubble function). This enrichment function is statically condensed, and hence, in the particular case of the $\mathbb{P}^1/\mathbb{P}^0$ element, this procedure leads to an edge-based stabilized finite element method, containing non-standard jump terms on the interelement boundaries. Moreover, since we know exactly the trace of the enrichment function on the element boundary, the stabilization parameter associated with the jump terms is known exactly. Up to the authors' knowledge, this is the first time that the exact stabilization parameter for an edge-based stabilized method is known.

The plan of the paper is as follows: in Section 2 we present the general framework and derive a general form of the method. Afterward, in Section 3 this framework is applied to derive a stabilized finite element method for $\mathbb{P}^1/\mathbb{P}^0$ elements, where optimal order a-priori error estimates are derived for the natural norms of the unknowns, plus some extra control on the norm of the jumps presented in the formulation. Finally, numerical experiments confirming the theoretical results are presented in Section 4.

2 The model problem and the general framework

Let Ω be an open bounded domain in $\mathbb{R}^2$ with polygonal boundary, $\boldsymbol{f} \in L^2(\Omega)^2$ and let us consider the following Stokes problem:

$$-\nu \Delta \boldsymbol{u} + \nabla p = \boldsymbol{f}, \quad \nabla \cdot \boldsymbol{u} = 0 \quad \text{in } \Omega, \tag{1}$$
$$\boldsymbol{u} = \boldsymbol{0} \quad \text{on } \partial \Omega,$$

where $\nu \in \mathbb{R}^+$ is the fluid viscosity.

Let now $\{\mathcal{T}_h\}_{h>0}$ be a family of regular triangulations of $\overline{\Omega}$, build up using triangles K with boundary ∂K. Let also $\mathcal{E}_h$ be the set of internal edges of the triangulation, $h_K := diam(K)$ and $h := \max\{h_K : K \in \mathcal{T}_h\}$. Let V_h be the usual finite element space of continuous piecewise polynomials of degree k, $1 \leq k \leq 2$, with zero trace on $\partial \Omega$. Let also Q_h be a space of piecewise polynomials of degree l, $0 \leq l \leq 1$, which may be continuous or discontinuous in Ω and who belong to $L_0^2(\Omega)$. Let $H^m(\mathcal{T}_h)$ and $H_0^m(\mathcal{T}_h)$ ($m \geq 1$) be the spaces of functions whose restriction to $K \in \mathcal{T}_h$ belongs to $H^m(K)$ and $H_0^m(K)$, respectively. Furthermore, $(\cdot, \cdot)_D$ stands for the inner product in $L^2(D)$ (or in $L^2(D)^2$ or $L^2(D)^{2 \times 2}$, when necessary), and we denote by $\|\cdot\|_{s,D}$ ($|\cdot|_{s,D}$) the norm (seminorm) in $H^s(D)$ (or $H^s(D)^2$, if necessary). As usual, $H^0(D) = L^2(D)$, and $|\cdot|_{0,D} = \|\cdot\|_{0,D}$.

In order to propose a Petrov-Galerkin method for Stokes problem (1), let $E_h \subset H_0^1(\Omega)$ be a finite dimensional space, called multiscale space, such that $V_h \cap E_h = \{0\}$. Then, we propose the following Petrov-Galerkin scheme for (1): Find $\boldsymbol{u}_1 + \boldsymbol{u}_e \in [V_h \oplus E_h]^2$ and $p \in Q_h$ such that

$$\nu(\nabla(\boldsymbol{u}_1 + \boldsymbol{u}_e), \nabla \boldsymbol{v}_h)_\Omega - (p, \nabla \cdot \boldsymbol{v}_h)_\Omega + (q, \nabla \cdot (\boldsymbol{u}_1 + \boldsymbol{u}_e))_\Omega = (\boldsymbol{f}, \boldsymbol{v}_h)_\Omega,$$

for all $\boldsymbol{v}_h \in [V_h \oplus H_0^1(\mathcal{T}_h)]^2$ and all $q \in Q_h$. Now, this Petrov-Galerkin scheme is equivalent to the following system:

$$\nu(\nabla(\boldsymbol{u}_1 + \boldsymbol{u}_e), \nabla \boldsymbol{v}_1)_\Omega - (p, \nabla \cdot \boldsymbol{v}_1)_\Omega + (q, \nabla \cdot (\boldsymbol{u}_1 + \boldsymbol{u}_e))_\Omega = (\boldsymbol{f}, \boldsymbol{v}_1)_\Omega, \quad (2)$$

for all $(\boldsymbol{v}_1, q) \in V_h^2 \times Q_h$, and

$$\nu(\nabla(\boldsymbol{u}_1 + \boldsymbol{u}_e), \nabla \boldsymbol{v}_b)_K - (p, \nabla \cdot \boldsymbol{v}_b)_K = (\boldsymbol{f}, \boldsymbol{v}_b)_K, \quad (3)$$

for all $\boldsymbol{v}_b \in H_0^1(K)^2$ and all $K \in \mathcal{T}_h$. Equation (3) above is equivalent to the following strong problem

$$-\nu \Delta \boldsymbol{u}_e = \boldsymbol{f} + \nu \Delta \boldsymbol{u}_1 - \nabla p \quad \text{in } K. \quad (4)$$

Now, this differential problem above must be completed with boundary conditions. For reasons that will become clear in the sequel, we will impose the following boundary condition on $\boldsymbol{u}_e$:

$$\boldsymbol{u}_e = \boldsymbol{g}_e \quad \text{on each } Z \subset \partial K, \quad (5)$$

where $\boldsymbol{g}_e = \boldsymbol{0}$ if $Z \subset \partial \Omega$, and $\boldsymbol{g}_e$ is the solution of

$$-\nu \partial_{ss} \boldsymbol{g}_e = \frac{1}{h_Z} [\![\nu \partial_n \boldsymbol{u}_1 + p\mathbf{I} \cdot \boldsymbol{n}]\!] \quad \text{in } Z, \quad (6)$$

$$\boldsymbol{g}_e = \boldsymbol{0} \quad \text{at the nodes},$$

on the internal edges, where $h_Z = |Z|$, $\boldsymbol{n}$ is the normal outward vector on ∂K, ∂_s and ∂_n are the tangential and normal derivative operators, respectively, $[\![v]\!]$ stands for the jump of v across Z, and $\mathbf{I}$ is the $\mathbb{R}^{2\times 2}$ identity matrix.

Next, since the enriched part $\boldsymbol{u}_e$ is fully identified we can perform statical condensation to derive a stabilized finite element method for our problem (1). First, integrating by parts, we have, on each $K \in \mathcal{T}_h$,

$$\nu(\nabla \boldsymbol{u}_e, \nabla \boldsymbol{v}_1)_K = -\nu(\boldsymbol{u}_e, \Delta \boldsymbol{v}_1)_K + (\boldsymbol{u}_e, \nu \partial_n \boldsymbol{v}_1)_{\partial K},$$
$$(q, \nabla \cdot \boldsymbol{u}_e)_K = -(\boldsymbol{u}_e, \nabla q)_K + (\boldsymbol{u}_e, q\mathbf{I} \cdot \boldsymbol{n})_{\partial K}.$$

Using these identities we can rewrite (2) in the following way

$$\nu(\nabla \boldsymbol{u}_1, \nabla \boldsymbol{v}_1)_\Omega + \sum_{K \in \mathcal{T}_h} \left[-(\boldsymbol{u}_e, \nu \Delta \boldsymbol{v}_1)_K + (\boldsymbol{u}_e, \nu \partial_n \boldsymbol{v}_1)_{\partial K} \right] - (p, \nabla \cdot \boldsymbol{v}_1)_\Omega$$
$$+ (q, \nabla \cdot \boldsymbol{u}_1)_\Omega + \sum_{K \in \mathcal{T}_h} \left[-(\boldsymbol{u}_e, \nabla q)_K + (\boldsymbol{u}_e, q\mathbf{I} \cdot \boldsymbol{n})_{\partial K} \right] = (\boldsymbol{f}, \boldsymbol{v}_1)_\Omega, \quad (7)$$

which implies

$$\nu(\nabla \boldsymbol{u}_1, \nabla \boldsymbol{v}_1)_\Omega - (p, \nabla \cdot \boldsymbol{v}_1)_\Omega + (q, \nabla \cdot \boldsymbol{u}_1)_\Omega$$
$$+ \sum_{K \in \mathcal{T}_h} \Big[-(\boldsymbol{u}_e, \nu \Delta \boldsymbol{v}_1 + \nabla q)_K + (\boldsymbol{u}_e, \nu \partial_n \boldsymbol{v}_1 + q\mathbf{I}\cdot \boldsymbol{n})_{\partial K} \Big] = (\boldsymbol{f}, \boldsymbol{v}_1)_\Omega. \quad (8)$$

Finally, defining the linear bounded operator $\mathcal{B}_K : L^2(\partial K) \to L^2(\partial K)$ such that $\boldsymbol{u}_e\big|_{\partial K} = \frac{1}{\nu}\mathcal{B}_K(\llbracket \nu \partial_n \boldsymbol{u}_1 + p\mathbf{I}\cdot \boldsymbol{n}\rrbracket)$, then (8) becomes

$$\nu(\nabla \boldsymbol{u}_1, \nabla \boldsymbol{v}_1)_\Omega - (p, \nabla \cdot \boldsymbol{v}_1)_\Omega + (q, \nabla \cdot \boldsymbol{u}_1)_\Omega + \sum_{K \in \mathcal{T}_h} \Big[(\boldsymbol{u}_e, \nu \Delta \boldsymbol{v}_1 + \nabla q)_K$$
$$+ \frac{1}{\nu}\left(\mathcal{B}_K(\llbracket \nu \partial_n \boldsymbol{u}_1 + p\mathbf{I}\cdot \boldsymbol{n}\rrbracket), \nu \partial_n \boldsymbol{v}_1 + q\mathbf{I}\cdot \boldsymbol{n}\right)_{\partial K} \Big] = (\boldsymbol{f}, \boldsymbol{v}_1)_\Omega. \quad (9)$$

Using this form, in the next section we will present a stabilized finite element method for the simplest possible pair, i.e., $\mathbb{P}^1/\mathbb{P}^0$ elements.

3 Application to the $\mathbb{P}^1/\mathbb{P}^0$ pair

For this case, the finite element spaces are given by

$$\boldsymbol{V}_h := \{\boldsymbol{v} \in C^0(\overline{\Omega})^2 : \boldsymbol{v}|_K \in \mathbb{P}^1(K)^2, \forall K \in \mathcal{T}_h\} \cap H_0^1(\Omega)^2,$$
$$Q_h^0 := \{q \in L_0^2(\Omega) : q|_K \in \mathbb{P}^0(K), \forall K \in \mathcal{T}_h\},$$

for the velocity and pressure, respectively. Using these spaces, we propose the following stabilized method: Find $(\boldsymbol{u}_1, p_0) \in \boldsymbol{V}_h \times Q_h^0$ such that

$$\mathbf{B}_0((\boldsymbol{u}_1, p_0), (\boldsymbol{v}_1, q_0)) = \mathbf{F}_0(\boldsymbol{v}_1, q_0) \quad \forall (\boldsymbol{v}_1, q_0) \in \boldsymbol{V}_h \times Q_h^0, \quad (10)$$

where

$$\mathbf{B}_0((\boldsymbol{u}_1, p_0), (\boldsymbol{v}_1, q_0)) := \nu(\nabla \boldsymbol{u}_1, \nabla \boldsymbol{v}_1)_\Omega - (p_0, \nabla \cdot \boldsymbol{v}_1)_\Omega + (q_0, \nabla \cdot \boldsymbol{u}_1)_\Omega$$
$$+ \sum_{Z \in \mathcal{E}_h} \tau_Z (\llbracket \nu \partial_n \boldsymbol{u}_1 + p_0 \mathbf{I}\cdot \boldsymbol{n}\rrbracket, \llbracket \nu \partial_n \boldsymbol{v}_1 + q_0 \mathbf{I}\cdot \boldsymbol{n}\rrbracket)_Z, \quad (11)$$

$$\mathbf{F}_0(\boldsymbol{v}_1, q_0) := (\boldsymbol{f}, \boldsymbol{v}_1)_\Omega, \quad (12)$$

and τ_Z is given by

$$\tau_Z := \frac{h_Z}{12\nu}. \quad (13)$$

3.1 Derivation of the method

First we note that using spaces $\boldsymbol{V}_h$ and Q_h^0, equation (9) reduces to: Find $(\boldsymbol{u}_1, p_0) \in \boldsymbol{V}_h \times Q_h^0$ such that

$$\nu(\nabla u_1, \nabla v_1)_\Omega - (p_0, \nabla \cdot v_1)_\Omega + (q_0, \nabla \cdot u_1)_\Omega$$
$$+ \sum_{Z \in \mathcal{E}_h} \frac{1}{\nu} \left(\mathcal{B}_K \left([\![\nu \partial_n u_1 + p_0 \mathbf{I} \cdot n]\!] \right), [\![\nu \partial_n v_1 + q_0 \mathbf{I} \cdot n]\!] \right)_Z = (f, v_1)_\Omega, \quad (14)$$

for all $(v_1, q_0) \in V_h \times Q_h^0$.

We exploit next the fact that $[\![\nu \partial_n u_1 + p_0 \mathbf{I} \cdot n]\!] \big|_Z$ is a constant function. To this end, we define the (matrix) function $\boldsymbol{b}_K^u := (\mathcal{B}_K(e_1) | \mathcal{B}_K(e_2))$, where e_1, e_2 are the canonical vectors in $\mathbb{R}^2$, and we remark that, from its definition, $\boldsymbol{b}_K^u = b_K^u \mathbf{I}$, where b_K^u is the solution of

$$-\partial_{ss} b_K^u(s) = \frac{1}{h_Z} \quad \text{in } Z, \quad b_K^u = 0 \quad \text{at the nodes}, \quad (15)$$

in each $Z \subseteq \partial K \cap \Omega$. We further remark that the solution of (15) may be calculated explicitly and it is not difficult to realize that

$$\frac{(b_K^u, 1)_Z}{|Z|} = \frac{h_Z}{12}. \quad (16)$$

Finally, since $[\![\nu \partial_n u_1 + p_0 \mathbf{I} \cdot n]\!] \big|_Z$ is a constant function we obtain

$$(\mathcal{B}_K([\![\nu \partial_n u_1 + p_0 \mathbf{I} \cdot n]\!]), [\![\nu \partial_n v_1 + q_0 \mathbf{I} \cdot n]\!])_Z$$
$$= \frac{(b_K^u, 1)_Z}{|Z|} ([\![\nu \partial_n u_1 + p_0 \mathbf{I} \cdot n]\!], [\![\nu \partial_n v_1 + q_0 \mathbf{I} \cdot n]\!])_Z,$$

and hence replacing this into (14) and using (16) we obtain the method (10).

Remark 1. One of the drawbacks of RFB method for the Stokes problem is that, due to the zero boundary condition on the element boundary, there is not a bubble-based enrichment that makes stable the $\mathbb{P}^1/\mathbb{P}^0$ element (see [4] for a discussion), and hence, the use of a different boundary condition makes possible to stabilize the $\mathbb{P}^1/\mathbb{P}^0$ element.

3.2 Error analysis

From now on, C will denote a positive constant independent of h and ν, and that may change its value whenever it is written in two different places. Moreover, defining the mesh-dependent norm

$$\|(v, q)\|_h := \left[\nu |v|_{1,\Omega}^2 + \sum_{Z \in \mathcal{E}_h} \tau_Z \| [\![\nu \partial_n v + q \mathbf{I} \cdot n]\!] \|_{0,Z}^2 \right]^{\frac{1}{2}}, \quad (17)$$

we have the following continuity, well posedeness and consistency result.

Lemma 1. *Let be* $(\boldsymbol{v},q),(\boldsymbol{w},r) \in [H^2(\mathcal{T}_h) \cap H_0^1(\Omega)]^2 \times [H^1(\mathcal{T}_h) \cap L_0^2(\Omega)]$. *Then, the bilinear form* $\mathbf{B}_0$ *satisfies*

$$\mathbf{B}_0((\boldsymbol{v},q),(\boldsymbol{w},r)) \leq \|(\boldsymbol{v},q)\|_h \|(\boldsymbol{w},r)\|_h + (\nabla \cdot \boldsymbol{v}, r)_\Omega - (q, \nabla \cdot \boldsymbol{w})_\Omega, \quad (18)$$

$$\mathbf{B}_0((\boldsymbol{v},q),(\boldsymbol{v},q)) = \|(\boldsymbol{v},q)\|_h^2, \quad (19)$$

and hence the problem (10) *is well posed. Also, if* $(\boldsymbol{u},p) \in [H^2(\Omega) \cap H_0^1(\Omega)]^2 \times [H^1(\Omega) \cap L_0^2(\Omega)]$ *is the weak solution of* (1) *then*

$$\mathbf{B}_0((\boldsymbol{u}-\boldsymbol{u}_1, p-p_0),(\boldsymbol{v}_1, q_0)) = 0 \quad \text{for all } (\boldsymbol{v}_1, q_0) \in \boldsymbol{V}_h \times Q_h^0. \quad (20)$$

Proof. The result follows immediately from the definition of $\mathbf{B}_0$. The consistency of (10) follows by noting that $[\![\nu \partial_n \boldsymbol{u} + p \mathbf{I} \cdot \boldsymbol{n}]\!] = 0$ a.e. across all the internal edges. ∎

In order to perform the numerical analysis of this method, we will consider the Lagrange interpolation operator $I_h : C^0(\overline{\Omega}) \to V_h$ (with the obvious extension to vector-valued functions) to approximate the velocity, and we approximate the pressure considering $\Pi_h : L^2(\Omega) \to Q_h^0$, the $L^2(\Omega)$-projection onto Q_h^0.

Theorem 1. *Let* $(\boldsymbol{u},p) \in [H^2(\Omega) \cap H_0^1(\Omega)]^2 \times [H^1(\Omega) \cap L_0^2(\Omega)]$ *be the solution of* (1) *and* $(\boldsymbol{u}_1, p_0)$ *the solution of* (10). *Then, the following error estimates hold*

$$\|(\boldsymbol{u}-\boldsymbol{u}_1, p-p_0)\|_h \leq Ch\left(\sqrt{\nu} \, |\boldsymbol{u}|_{2,\Omega} + \frac{1}{\sqrt{\nu}} |p|_{1,\Omega}\right), \quad (21)$$

$$\|p-p_0\|_{0,\Omega} \leq Ch\left[\nu |\boldsymbol{u}|_{2,\Omega} + |p|_{1,\Omega}\right]. \quad (22)$$

Moreover, if Ω is a convex polygon, the following error estimate holds:

$$\|\boldsymbol{u}-\boldsymbol{u}_1\|_{0,\Omega} \leq Ch^2 \left(|\boldsymbol{u}|_{2,\Omega} + \frac{1}{\nu}|p|_{1,\Omega}\right).$$

Proof. Let $(\widetilde{\boldsymbol{u}}_h, \widetilde{p}_h) := (I_h(\boldsymbol{u}), \Pi_h(p)) \in \boldsymbol{V}_h \times Q_h^0$. From Lemma 1 we know that

$$\|(\boldsymbol{u}-\boldsymbol{u}_1, p-p_0)\|_h^2 = \mathbf{B}_0((\boldsymbol{u}-\boldsymbol{u}_1, p-p_0),(\boldsymbol{u}-\boldsymbol{u}_1, p-p_0))$$
$$= \mathbf{B}_0((\boldsymbol{u}-\boldsymbol{u}_1, p-p_0),(\boldsymbol{u}-\widetilde{\boldsymbol{u}}_h, p-\widetilde{p}_h))$$
$$\leq C \|(\boldsymbol{u}-\boldsymbol{u}_1, p-p_0)\|_h \, \|(\boldsymbol{u}-\widetilde{\boldsymbol{u}}_h, p-\widetilde{p}_h)\|_h$$
$$+ (\nabla \cdot (\boldsymbol{u}-\boldsymbol{u}_1), p-\widetilde{p}_h)_\Omega - (\nabla \cdot (\boldsymbol{u}-\widetilde{\boldsymbol{u}}_h), p-p_0)_\Omega.$$

Now, since $\boldsymbol{u}$ is a solenoidal field and $\nabla \cdot \boldsymbol{u}_1 \in Q_h^0$ we obtain

$$(\nabla \cdot (\boldsymbol{u}-\boldsymbol{u}_1), p-\widetilde{p}_h)_\Omega = -(\nabla \cdot \boldsymbol{u}_1, p-\widetilde{p}_h)_\Omega = 0. \quad (23)$$

On the other hand, integrating by parts element-wise and applying standard interpolation inequalities (cf. [5]) it is not difficult to realize that

$$(\nabla \cdot (\boldsymbol{u} - \widetilde{\boldsymbol{u}}_h), p - p_0)_\Omega \leq Ch^2((1+\gamma)\nu\,|\boldsymbol{u}|_{2,\Omega}^2 + \frac{1}{\nu}|p|_{1,\Omega}^2)$$
$$+ \frac{1}{\gamma} \sum_{Z \in \mathcal{E}_h} \frac{h_Z}{\nu} \|[\![(p - p_0)\mathbf{I} \cdot \boldsymbol{n}]\!]\|_{0,Z}^2\,,$$

where $\gamma > 0$. Now, using a local trace theorem and the fact that $\boldsymbol{V}_h$ is constituted by linear polynomials we arrive at

$$\sum_{Z \in \mathcal{E}_h} \frac{h_Z}{\nu} \|[\![(p - p_0)\mathbf{I} \cdot \boldsymbol{n}]\!]\|_{0,Z}^2 \leq \widetilde{C}\,\|(\boldsymbol{u} - \boldsymbol{u}_1, p - p_0)\|_h^2 + C\nu h^2\,|\boldsymbol{u}|_{2,\Omega}^2\,.$$

Hence, choosing $\gamma = 2\widetilde{C}$ and applying standard interpolation inequalities we obtain

$$\frac{1}{2} \|(\boldsymbol{u} - \boldsymbol{u}_1, p - p_0)\|_h^2 \leq Ch^2(\nu\,|\boldsymbol{u}|_{2,\Omega}^2 + \frac{1}{\nu}|p|_{1,\Omega}^2)\,, \tag{24}$$

and the result follows by extracting square root. The error estimate for the pressure follows by using the continuous inf-sup condition and the consistency of the method. The $\|\cdot\|_{0,\Omega}$ error estimate for the velocity follows by a duality argument (for details, see [1]). ∎

4 Numerical validations

In this section we perform a convergence analysis for method (10) using $\mathbb{P}^1/\mathbb{P}^0$ elements. We consider as domain the square $\Omega = (0,1) \times (0,1)$, $\nu = 1$, and $\boldsymbol{f}$ is set such as the exact solution of our Stokes problem is given by

$$u_1(x,y) = -256x^2(x-1)^2 y(y-1)(2y-1)\quad,\quad u_2(x,y) = -u_1(y,x)\,,$$
$$p(x,y) = 150(x - 0.5)(y - 0.5)\,.$$

We depict in Figures 1-2 the convergence history for method (10). The results reproduce our theoretical results showing an $O(h)$ order of convergence for $|\boldsymbol{u} - \boldsymbol{u}_1|_{1,\Omega}$, the jump terms, and $\|p - p_0\|_{0,\Omega}$, and an $O(h^2)$ convergence for $\|\boldsymbol{u} - \boldsymbol{u}_1\|_{0,\Omega}$.

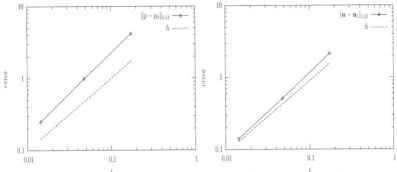

Fig. 1. Convergence history for $\|p - p_0\|_{0,\Omega}$ and $|\overset{h}{\boldsymbol{u}} - \boldsymbol{u}_1|_{1,\Omega}$.

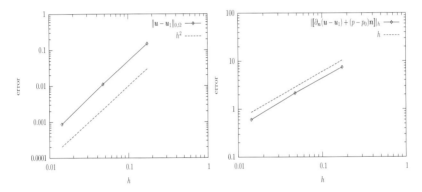

Fig. 2. Convergence history for $\|\boldsymbol{u} - \boldsymbol{u}_1\|_{0,\Omega}$ and $|[\![\nu\partial_n(\boldsymbol{u} - \boldsymbol{u}_1) + (p - p_0)\boldsymbol{n}]\!]|_h :=$
$\left[\sum_{Z \in \mathcal{E}_h} \tau_Z \|[\![\nu\partial_n(\boldsymbol{u} - \boldsymbol{u}_1) + (p - p_0)\mathbf{I}\cdot\boldsymbol{n}]\!]\|_{0,Z}^2\right]^{\frac{1}{2}}$.

References

1. Araya, R., Barrenechea, G.R., Valentin, F.: Stabilized finite element methods based on multiscale enrichment for the Stokes problem. SIAM J. Numer. Anal., **44**(1), 322–348 (2006)
2. Baiocchi, C., Brezzi, F., Franca, L.P.: Virtual bubbles and Galerkin-least-squares type methods (Ga. L. S.). Comput. Methods Appl. Mech. Engrg., **105**, 125–141 (1993)
3. Brezzi, F., Russo, A.: Choosing bubbles for advection-diffusion problems. Math. Models Methods Appl. Sci., **4**, 571–587 (1994)
4. Brezzi, F.: Recent results in the treatment of subgrid scales. In: ESAIM: Proceedings, Actes du 32ème congrès d'Analyse Numérique: CANUM 2000, (2000)

5. Ern, A., Guermond, J.-L.: Theory and Practice of Finite Elements, Springer-Verlag, New York (2004)
6. Franca, L., Madureira, A., Tobiska, L., Valentin, F.: Convergence analysis of a multiscale finite element method for singularly perturbed problems. SIAM Multiscale Model. Simul., to appear (2006)
7. Franca, L., Madureira, A., Tobiska, L., Valentin, F.: Towards multiscale functions: enriching finite element spaces with local but not bubble-like functions. Comput. Methods Appl. Mech. Engrg., **194**, 3006–3021 (2005)
8. Franca, L.P., Russo, A.: Approximation of the Stokes problem by residual-free macro bubbles. East-West J. Numer. Math., **4**, 265–278 (1996)
9. Hou, T., Wu, X.-H.: A multiscale finite element method for elliptic problems in composite materials and porous media. J. Compt. Phys., **134**, 169–189 (1997)

Adaptive Multiresolution Methods for the Simulation of Shocks/Shear Layer Interaction in Confined Flows

L. Bentaleb[1], O. Roussel[2] and C. Tenaud[1]

[1] LIMSI-UPR CNRS 3251, BP 133, Campus d'Orsay, F-91403 Orsay Cedex, France
bentaleb@limsi.fr, tenaud@limsi.fr
[2] ITCP, Universität Karlsruhe (TH), Kaiserstr. 12, 76128 Karlsruhe, Germany
roussel@ict-uni.karlsruhe.de

Summary. The main objective of this work is to study the ability of a multiresolution method based on wavelet approximation to predict unsteady shocked flows. A correct prediction of shock wave phenomena is often crucial in flow simulations for many industrial configurations such as in air intakes of supersonic vehicles or shock tube facilities where moving shock waves interact with shear layers. To capture these very fine and localized structures, many shock capturing schemes have been developed in the last decades that work with adequat robustness. However, shock wave/shear layer interactions generate unsteady vortical flows with separation that need adaptive multiresolution technics to achieve correct predictions.

1 Introduction

The unsteady vortical regime in the shock wave/shear layer interaction needs adaptive mesh refinement to correctly capture the smallest scales produced and consequently save as much as possible grid points in the other regions. To achieve this objective, we use Harten's multiscale decomposition [1]. Unlike Harten's strategy for which the solution is calculated at each time step on the finest grid, the solution is here predicted on a dynamic tree structure which evolves in time.

In this study, the multiscale analysis is performed on both compressible Euler and Navier-Stokes equations in conservative formulation. These equations are solved by means of finite volume discretization technics. We present two adaptions : a grid adaption and a scheme adaption. The main objective of this paper is to assess the ability of the biorthogonal wavelet analysis to the prediction of shocked flows.

The adaptive refinement technics consist in generating a hierarchy of nested grids. The solution is known from the coarsest to the finest grid. The use of biorthogonal wavelet basis [2] allows the mesh to be locally refined near discontinuities. The solution is decomposed into a sum of approximations on a coarse grid (j) and a fluctuation calculated on a finer one ($j+1$). This fluctuation stands for the difference between two successive levels of resolution [3, 1]. This value is called a *detail*, and is compared to a threshold value ε. All the details d_j at a level j smaller than a prescribed tolerance ε_j are removed from the memory. Hence, this refinement is linked to an *a-priori* error estimation. The threshold procedure allows us to reduce both CPU time and memory requirements [4]. In the next section, some numerical results for 1D inviscid and 2D viscous test cases are presented.

2 Numerical results

To validate this approach, a code named CARMEN was used [5, 6]. Several classical 1D inviscid and 2D viscous test cases have been checked such as 1D Lax shock tube problem or the viscous interaction between a vortex and a shock wave. The numerical fluxes are evaluated in both multiresolution and fine grid approach with several schemes such as ENO [8], MacCormack [9] and OS (One Step) schemes [7]. The last one was developped jointly at SINUMEF and LIMSI-CNRS. When we perform the calculation at level (or scale) j, the number of points is 2^j in each space direction. These results obtained by using these schemes were compared with those obtained by means of reference methods.

2.1 A 1D inviscid test case: The Lax shock tube

We consider a 1D tube equipped with a diaphragm at its center which separates two regions at different pressures. The diaphragm is initially broken and the propagation of a rarefaction wave on the left, a contact discontinuity and a shock wave on the right occur [10]. The 1D Euler equations are solved on the spatial domain $x \in [0, 2]$. The density distribution is studied at a dimensionless time $t = 0.32$ with 1024 grid points (scale $j = 10$). The CFL number is equal to 0.5. The solution is initially prescribed as:

$$\begin{cases} \rho = 0.445, \ u = 0.698, \ p = 3.528 \text{ if } x < 1, \\ \rho = 0.5, \quad \ u = 0, \quad \ \ p = 0.571 \text{ if } x \geq 1. \end{cases}$$

The performances of the multiscale approach (MR) with OSMP3 scheme and different threshold values ε are summarized in Table 1. The fine grid (FG) approach is taken as the reference in term of performance. The compression rate of both CPU time and memory usages are compared with the FG computation. These performances decrease when ε increases. Note that the value of $\varepsilon = 0$ corresponds to the complete tree structure without suppression of cells, i. e., the finest grid is recovered. Nevertheless, when choosing $\varepsilon = 0$, the

Table 1. Level 10, 1321 Iterations, $\Delta t = 2.42 \cdot 10^{-4}$, OSMP3 scheme, FG=Fine Grid. MR=Multiresolution

ε	CPU Time	CPU cost	Memory used
FG	180.53 s	100%	100%
MR 0	256.22 s	137.75%	200%
MR 10^{-5}	123.23 s	68.75%	99.45%
MR 10^{-4}	79.14 s	43.33%	66.74%
MR 10^{-3}	57.48 s	31.47%	49.86%
MR 10^{-2}	49.94 s	27.34%	44.08%
MR 10^{-1}	40.65 s	22.25%	36.04%

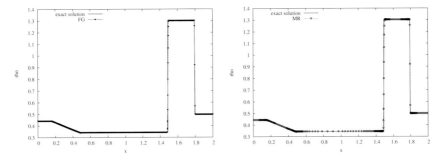

Fig. 1. Lax shock tube problem : density distribution, scale=10, t=0.32, CFL=0.5, scheme: OSMP3

multiresolution is more expensive than the FG approach. To illustrate these results, the solution using $\varepsilon = 10^{-2}$ and scale $j = 10$ is plotted and compared to the exact solution (Fig. 1). A very good agreement is recovered between the two approaches (Fig. 1). While the solution obtained with the MR method seems to fit well with the exact solution, an error analysis is nevertheless needed to study the robustness of the multiresolution analysis.

Several numerical schemes were used to perform the error analysis. The first one is a separated time and space integration which involves a 3rd order ENO scheme and a 3rd order TVD Runge-Kutta time integration. The second one is a coupled time and space integration based on a 3rd order Lax-Wendroff approach (OSMP3). Finally a combination between the OSMP3 scheme and a cheap MacCormack 2-4 scheme (2nd order in time and 4th order in space) is applied. The first one is applied on the finest scale, while the second one is applied on the coarsest scales, i. e., only on the smoothest regions of the solution. This combination represents a scheme adaptive procedure and is performed in order to gain in computational performance. We can see the evolution of the approximation error calculated in L_1- norm versus ε on the Fig. 2. Following the analysis conducted in [11], the error is decomposed into the discretization error τ_n^j and the perturbation error π_n^j estimated at level

j and at a prescribed integration time n. The former exhibits the difference between the solution given by the reference scheme and the exact solution. The latter is the difference between MR and FG approaches. We note that ENO3 scheme is less accurate than the OSMP3 scheme (see Fig. 2). The approximation error increases with ε. As expected, the combination of schemes (MacCormack 2-4 and OSMP3) is less accurate than the OSMP3 scheme. Therefore, a compromise must be reached to balance the loose of accuracy by a gain of performance which is done successfully here.

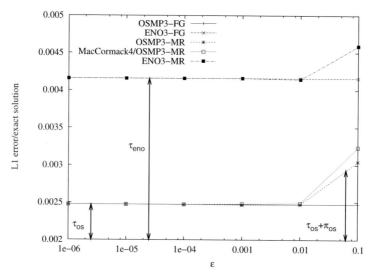

Fig. 2. Evolution of approximation error $\tau^{10} + \pi^{10}$ on density distribution at scale 10, Lax's test case

2.2 A 2D viscous test case: shock-vortex interaction

Among the 2D cases already studied, we present here the results of a viscous shock-vortex weak interaction ($Re = 2000$ and $Ma = 1.1588$) [7]. The spatial domain is $[0, 2] \times [0, 2]$. Initially, the vortex is centred at $x_0 = 0.5$, $y_0 = 0.5$ and a stationary plane weak shock wave is initiated at $x = 1$. The vortex is convected to interact with the stationary shock. Periodic boundary conditions are applied in y direction. We investigate the ability of numerical schemes to predict the transport of acoustic waves produced by this interaction.

The performances of the multiscale approach is summarized in Table 2. We remark that the MR algorithm is very efficient in this test case. A very good agreement between the two solutions is obtained (see Fig. 3 and Fig. 4). In Fig. 3, we can see the vortex/shock interaction at a dimensionless time

Table 2. Level 8, 378 Iterations, $\Delta t = 1.85 \cdot 10^{-3}$, OSMP3 scheme, FG=Fine Grid. MR=Multiresolution

ε	CPU Time	CPU cost	Memory used
FG	2 h 10 min 26 s	100%	100%
MR 0	2 h 38 min 51 s	127.38%	133.33%
MR 10^{-5}	1 h 31 min 12 s	73.13%	79.52%
MR 10^{-4}	1 h 20 min 50 s	64.82%	71.14%
MR 10^{-3}	53 min 55 s	43.15%	49.35%
MR 10^{-2}	25 min 46 s	20.25%	25.41%
MR 10^{-1}	10 min 34 s	8.53%	10.96%

$t = 0.7$ and at scale $j = 8$. The reference is calculated with an OSMP7 scheme on level $j = 9$ (512 × 512 gris points). The solution is shown for $\varepsilon = 10^{-3}$. The performance of the grid adaption procedure can be seen in Fig. 3 (bottom-right side). The grid points are concentrated around the shock and the vortex. We have extracted a slice of pressure along the line $y = 1$. The pressure distribution is compared with the one obtained by the FG computation, both with OSMP3 and MacCormack4+OSMP3 schemes. The OSMP3 scheme predicts with a great accuracy the transport of the acoustic wave (see Fig. 4), whereas the MacCormack4+OSMP3 scheme combination is slightly less accurate. The error analysis confirms the difference of accuracy between these schemes.

The error analysis is performed on the pressure distribution along the line $y = 1$. Fig. 5 presents the evolution of the approximation error in L_1-norm versus the rate of CPU compression. We study the advantage of the multiresolution approach. The black dots show the CPU cost of the fine grid at different scales (6, 7 and 8) versus the L_1-error. We have presented the discretization error τ^j for these three levels to measure the error obtained on the fine grid. For a finer grid, the error is lower, but the computational cost is higher. For the multiresolution approach, we have performed the calculation only on the level 8. In this figure, we remark that the error increases with ε. The combination of the schemes is not here as efficient as in the first test case. The error is slightly larger than the one of the OSMP3 scheme. If we consider the solution at $\varepsilon = 10^{-3}$, scale $j = 8$ (see Fig. 5), the CPU time is two times larger than the FG computation at scale $j = 7$ with a much lower error. Besides, this rate of CPU compression still profitable (CPU compression < 50%) with an equivalent order of accuracy in comparison with the FG computation at scale $j = 8$. We therefore consider that the multiresolution approach is a very powerful technique.

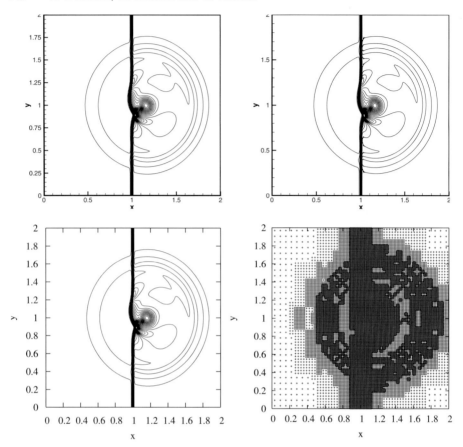

Fig. 3. Shock-vortex interaction, pressure contours at t=0.7 : (49 contours from 0.527 to 0.845), OSMP3 scheme. (Top left) Fine grid : 256 x 256 points. (Top right) Reference solution : 512 x 512 grid points, OSMP7 scheme. (Bottom left) Multiresolution at scale 8 and $\varepsilon = 10^{-3}$. (Bottom right) Adapted mesh.

3 Conclusion and perspectives

In the present paper, we have presented an adaptive technique based on Harten's multiresolution to simulate shock/shear layer interaction in confined flows. In the method presented here, we used a combination of a cheap scheme in the smooth regions (coarse levels) with an expensive, but efficient shock-capturing scheme near steep gradients (finest level). This strategy was chosen to obtain gains in CPU time. It turns out that it works very well in 1D configurations but the performance on 2D cases is more questionable depending

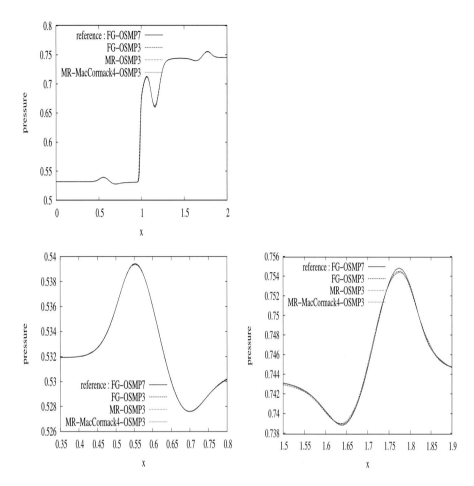

Fig. 4. Pressure along line y=1 at t=0.7 for several schemes. Level 8 for fine grid, multiresoltion with $\varepsilon = 10^{-3}$. CFL=0.5. (Up left) Global view. (Bottom left) Zoom on upstream zone. (Bottom right) Zoom on downstream zone.

on the test case considered. Nevertheless, an acceptable compromise between efficiency and accuracy can be reached using such kinds of methods.

As perspectives, we plan to extend this method to 3D configurations together with more physical boundary conditions, in order to simulate compressible flows in real-world configurations.

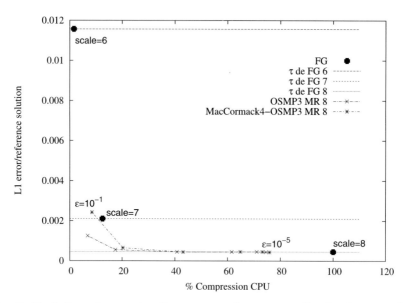

Fig. 5. Evolution of approximation error on the pressure distribution at scale 8: $(\tau^8 + \pi^8)$, shock/vortex

References

1. Harten, A.: Multiresolution algorithms for the numerical solution of hyperbolic conservation laws. Comm. Pure Appl. Math., **48**, 1305–1342 (1995)
2. Cohen A., Daubechies I., Feauveau J.C.: Biorthogonal bases of compactly supported wavelets. Comm. Pure and Appl. Math., **45**, 485–560 (1992)
3. Harten, A.: Adaptive multiresolution schemes for shock computations. J. Comput. Phys., **115**, 319–338 (1994)
4. Cohen, A., Kaber, S.M., Müller, S., Postel, M.: Fully adaptive multiresolution finite volume schemes for conseravtion laws. J. Comput. Phys., **161**, 493–523 (2000)
5. Roussel, O., Schneider, K., Tsigulin, A., Bockhorn, H.: A conservative fully adaptive multiresolution algorithm for parabolic PDEs. J. Comput. Phys., **188**, 264–286 (2003)
6. Roussel, O., Schneider, K.: An adaptive multiresolution method for combustion problems: application to flame ball - vortex interaction. Comp. Fluids, **34**(7), 817–831 (2005)
7. Daru, V., Tenaud, C.: High order one-step monotonicity-preserving schemes for unsteady compressible flow calculations. Math. Comp., **71**, 231–303 (1987)
8. Harten, A., Engquist, B., Osher, S., Chakravarthy, S.: Uniformaly high order essentially non-oscillatory schemes III. Math. Comp., **193**, 563–594 (2004)
9. Gottlieb, D., Turkel, E.: Dissipative two-four methods for time-dependent problems. J. Comput. Phys., **30**, 703-723 (1976)
10. Woodward, P., Colella, P.: The Numerical simulation of two-dimensional fluid flow with strong shocks. J. Comput. Phys., **54**, 115-173 (1984)

11. Müller, S.: Adaptive multiscale schemes for conservation laws, Lecture Notes in Computational Science and Engineering. Habilitation Thesis. Springer, RWTH Aachen (2003)
12. Cohen, A., Kaber, S. M., Müller, S., Postel, M.: Fully adaptive multiresolution finite volume schemes for conservation laws. Math. Comp., **72**, 183–225 (2003)

Local Projection Stabilization for the Stokes System on Anisotropic Quadrilateral Meshes

Malte Braack and Thomas Richter

Institute of Applied Mathematics, Heidelberg University, INF 294, 69120 Heidelberg
malte.braack@iwr.uni-heidelberg.de

Summary. The local projection stabilization for the Stokes system is formulated for anisotropic quadrilateral meshes. Stability is proven and an error analysis is given.

1 Introduction

The local projection stabilization (LPS) is suitable to stabilize the saddle point structure of the Stokes system when equal-order finite elements are used, as well as convective terms for Navier-Stokes. Hence, it has already been applied with large success to different fields of computational fluid dynamics, e.g., in 3D incompressible flows [7], compressible flows [12], reactive flows [8], parameter estimation [4, 5] and optimal control problems [11].

Although locally refined meshes have been used for this stabilization technique in all of these applications, the meshes have been isotropic so far. The solution of partial differential equations on anisotropic meshes are of substantial importance for efficient solutions of problems with interior layers or boundary layers, as for instance in fluid dynamics at higher Reynolds number. It is well known that stabilized finite element schemes, e.g. streamline upwind Petrov-Galerkin (SUPG), see [10], or pressure stabilized Petrov-Galerkin (PSPG), see [9], must be modified in the case of anisotropy. Becker has shown in [2] how the PSPG stabilization should be modified on anisotropic Cartesian grids.

In this work, we make the first step of formulating LPS on *anisotropic quadrilateral meshes* by considering the Stokes system in the domain $\Omega \subset \mathbb{R}^2$ for velocity v and pressure p:

$$-\Delta v + \nabla p = f, \quad \operatorname{div} v = 0,$$

together with appropriate boundary conditions for v on $\partial\Omega$. The right hand side f is supposed to be in the Hilbert space $L^2(\Omega)$. The corresponding

Galerkin formulation is known to be unstable for equal-order interpolation due to the violation of the discrete inf-sup condition [9]. By V_h and Q_h we denote the discrete test spaces for v and p, respectively, consisting of piecewise polynomials of degree $r = 1$ on quadrilaterals, (Q_1 elements).

After a short presentation of LPS on isotropic meshes, we will generalize this technique to the case of anisotropic meshes obtained by bilinear transformations from a reference quadrilateral. A stability proof an a priori estimate will be given.

2 Local projection stabilization on isotropic meshes

The mesh $\mathcal{T}_h$ is supposed to be constructed by patches of quadrilaterals. The coarser mesh $\mathcal{T}_{2h}$ is obtained by one global coarsening of $\mathcal{T}_h$. The correspondence between these two meshes is as follows: Each quadrilateral $P \in \mathcal{T}_{2h}$ is cut into four new quadrilaterals (dividing all lengths of edges of P by 2) in order to obtain the fine partition $\mathcal{T}_h$. The space Q_{2h}^{disc} consists of patch-wise polynomials of degree $r - 1$, but discontinuous across patches $P \in \mathcal{T}_{2h}$. The projection

$$\pi_h : L^2(\Omega) \to Q_{2h}^{disc}$$

is defined as the L^2-orthogonal projection:

$$(\pi_h q, \xi) = (q, \xi) \quad \forall \xi \in Q_{2h}^{disc}.$$

The idea of LPS, see [3], consists of adding the stabilization term involving the difference between the identity I and π_h to the Galerkin form:

$$s_h(p_h, \xi) := ((I - \pi_h)\nabla p_h, \alpha \nabla \xi).$$

The stabilization parameter is chosen as $\alpha \sim h^2$ on isotropic meshes. Hence, the discrete system becomes: Find $\{v_h, p_h\} \in V_h \times Q_h$ so that

$$a(v_h, p_h; \phi, \xi) + s_h(p_h, \xi) = (f, \phi) \quad \forall \{\phi, \xi\} \in V_h \times Q_h,$$

with the linear form:

$$a(v, p; \phi, \xi) = (\nabla v, \nabla \phi) - (p, \operatorname{div} \phi) + (\operatorname{div} v, \xi).$$

In [3] the following a priori estimate was shown for piecewise bilinear elements on quasi-uniform meshes:

$$\|\nabla(v - v_h)\| + \|p - p_h\| \leq Ch(\|\nabla p\| + \|\nabla^2 v\|), \tag{1}$$

with a constant $C \geq 0$ and the maximal mesh size h. The norms above denote the L^2−norms over Ω. This estimate is optimal on isotropic meshes. On anisotropic meshes, it is suboptimal: if the solution has much larger gradients in, e.g., y−direction the mesh size should be chosen as $h_y \ll h_x$. In (1), $h = h_x$, while the derivatives of p and v would be large in y direction. Much more suitable would be an estimate where the partial mesh sizes h_x and h_y are multiplied by the corresponding spatial derivatives.

3 Anisotropic affine linear meshes

3.1 Notations and assumptions

We consider anisotropic meshes without the restriction of grid alignment with the coordinate axis. The transformation T_K from the reference cells $\hat{K}$ to the physical cell K is allowed to be affine linear so that $\mathcal{T}_h$ consists of parallelograms K. Such a transformation can be expressed as a composition of translation, rotation, shearing and stretching augmented with the pure bilinear term,

$$T_K \begin{pmatrix} \hat{x} \\ \hat{y} \end{pmatrix} = \begin{pmatrix} x_0 \\ y_0 \end{pmatrix} + \begin{pmatrix} \cos\theta & -\sin\theta \\ \sin\theta & \cos\theta \end{pmatrix} \left[\begin{pmatrix} 1 & \sigma \\ 0 & 1 \end{pmatrix} \begin{pmatrix} h_x & 0 \\ 0 & h_y \end{pmatrix} \begin{pmatrix} \hat{x} \\ \hat{y} \end{pmatrix} + \begin{pmatrix} \alpha \widehat{xy} \\ \beta \widehat{xy} \end{pmatrix} \right].$$

The parameters usually depend on the specific cell, i.e., $\sigma = \sigma_K$, $h_x = h_{K,x}$, $h_y = h_{K,y}$, $\alpha = \alpha_K$ and $\beta = \beta_K$ but the subscript K will be suppressed in order to simplify notations. In [6] it was shown that the determinant of this transformation can be estimated by

$$\det T_K \sim h_x h_y.$$

Throughout this work we use the notation $a \lesssim b$ for $a \leq Cb$ with a constant C independent of h_x, h_y and σ and (for local estimates) independent of the specific cell K. The expression $a \sim b$ is used if there holds $a \lesssim b$, and $b \lesssim a$ as well. For ease of presentation, we just write h_x and h_y in local estimates on a cell K. Without loss of generality we may assume $h_y \leq h_x$.

We formulate the two assumptions with are supposed to be fulfilled throughout the entire work:

(A1) There is a $\sigma_0 \geq 0$ (independent of K) so that the shearing parameter σ is bounded by

$$|\sigma| \leq \sigma_0.$$

(A2) Interior angle conditions for neighbour cells $K, L \in \mathcal{T}_h$:

$$h_{K,x} \sim h_{L,x} \quad \text{and} \quad h_{K,y} \sim h_{L,y}.$$

(A3) A restriction to the parameters α, β:

$$|\alpha| \leq \frac{h_x}{4} \quad \text{and} \quad |\beta| \leq \frac{1}{4} \min\left\{h_y, \frac{h_x}{\sigma_0}\right\}.$$

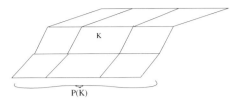

Fig. 1. Patch $P(K)$ of elements surrounding cell K

Note that assumption (A1) allows for moderate stretching, because the stretching s is coupled to the extend of anisotropy. As a consequence of (A2) it holds for neighbour cells K and L:

$$\det T_K \sim \det T_L. \tag{2}$$

3.2 Anisotropic H^1-stable projection

For an overview of anisotropic interpolation operators we refer to the book [1] where triangular as well as quadrilateral meshes aligned with the coordinate axis are considered. In particular, H^1-stable projections for tensor grids are addressed. In this work we use an "anisotropic H^1-stable" projection operator $B_h : H^1(\Omega) \to Q_h$ developed in [2] which is suitable for anisotropic meshes. Originally it was also designed for meshes aligned with the coordinate axes. In [6], this interpolation operator is analyzed also on meshes fulfilling assumptions (A1)-(A3), where further couplings between the partial derivatives have to be taken into account due to shearing.

For ease of presentation, we may neglect rotation in the transformation T_K, i.e. $\theta = 0$.

Proposition 1. *There is a linear operator $B_h : H^1(\Omega) \to V_h$ with the following features for all $K \in \mathcal{T}_h$:*
(i) Stability:

$$\|\partial_x B_h u\|_K \lesssim (1+\sigma_0)\|\partial_x u\|_{P(K)} + h_x^{-1} h_y \|\partial_y u\|_{P(K)}, \tag{3}$$

$$\|\partial_y B_h u\|_K \lesssim (1+\sigma_0)\left(\sigma_0 \|\partial_x u\|_{P(K)} + \|\partial_y u\|_{P(K)}\right). \tag{4}$$

(ii) Approximation:

$$\|u - B_h u\|_K \lesssim (1+\sigma_0) h_x \|\partial_x u\|_{P(K)} + h_y \|\partial_y u\|_{P(K)}. \tag{5}$$

(iii) Approximation for $u \in H^2(P(K))$:

$$\|\nabla(u - B_h u)\|_K \lesssim (1+\sigma_0)^2 h_x \|\partial_x^2 u\|_{P(K)} + (1+\sigma_0)^2 h_x \|\partial_{xy} u\|_{P(K)}$$
$$+ (1+\sigma_0) h_y \|\partial_y^2 u\|_{P(K)}.$$

Here, $P(K)$ denotes the patch of cells having one node in common with K, see Figure 1.

Proof. The construction of B_h is introduced in [2]. The estimates above are given in [6]. ∎

4 Local projection stabilization on anisotropic meshes

4.1 Definition of the local projection

By $\eta_1 = Te_i \in \mathbb{R}^2$ we denote the unit vector aligned with the longest side of K, and $\eta_2 \in \mathbb{R}^2$ is orthogonal to it. Furthermore, h_i is the length of K in direction of η_i, $i = 1, 2$.

The natural extension of LPS to anisotropic meshes is the use of a modified stabilization term proposed in [13]:

$$s_h(p_h)(\xi) := \sum_{i=1}^{2} ((I - \pi_h)(\partial_{\eta_i} p_h, h_i^2 \partial_{\eta_i} \xi). \tag{6}$$

This formulation recovers obviously the original formulation in the case of isotropic meshes, $h_1 \sim h_2$. For making implementation easier it is useful to express (6) in Cartesian coordinates:

$$s_h(p_h)(\xi) = (M(I - \pi_h)\nabla p_h, M\nabla \xi),$$

with the matrix M given by

$$M = \begin{pmatrix} h_x \cos\theta & h_x \sin\theta \\ -h_y \sin\theta & h_y \cos\theta \end{pmatrix}.$$

4.2 Stability

In some parts of the following analysis we will neglect rotation and translation as long as this is justified. Furthermore, it is justified to consider the Jacobian of T_K at the point $\eta = (1, 1)^T$ where the effect of the nonlinerity becomes maximal:

$$T_K^\eta = \begin{pmatrix} h_x + \alpha & \sigma h_y + \alpha \\ \beta & h_y + \beta \end{pmatrix}. \tag{7}$$

Although this is no restriction it simplifies the presentation substantially, because we can replace h_1 and h_2 by h_x and h_y, respectively, and the derivatives ∂_{η_1} and ∂_{η_2} by ∂_x and ∂_y, respectively. Hence, the stabilizing term (6) can be written as:

$$s_h(p, \xi) := ((I - \pi_h)(\partial_x p), h_x^2 \partial_x \xi) + ((I - \pi_h)(\partial_y p), h_y^2 \partial_y \xi). \tag{8}$$

The following Proposition states the stability in the norm

$$\|\{v, p\}\| := \left(\|\nabla v\|^2 + \|p\|^2 + s_h(p, p)\right)^{1/2}.$$

Proposition 2. *There is a h-independent constant $\gamma > 0$ so that for every $v_h \in V_h$ and $p_h \in Q_h$ there holds*

$$\sup_{\{\phi, \xi\} \in X_h} \frac{a(v_h, p_h; \phi, \xi) + s_h(p_h, \xi)}{\|\{\phi, \xi\}\|} \geq \gamma \|\{v_h, p_h\}\|.$$

Proof. Taking into account that $V_h - Q_{2h}$ is a stable pair for the Stokes system and following the results in [3] it is sufficient that the existence of an interpolation operator $i_h : Q_h \to Q_{2h}$ is ensured, so that

$$\|i_h p\| \lesssim \|p\| \qquad \forall p \in Q_h,$$
$$\|p - i_h p\|^2 \lesssim s_h(p, p) \qquad \forall p \in Q_h.$$

This conditions are fulfilled for the nodal interpolation onto Q_{2h}, due to the scaling argument on a patch $P \in \mathcal{T}_{2h}$. This can be verified easily for the case $\theta = 0$:

$$\|p - i_h p\|_P^2 = h_x h_y \|\widehat{p} - \widehat{i_h p}\|_{\widehat{K}}^2$$
$$\leq h_x h_y \|(I - \pi_h) \widehat{\nabla} \widehat{p}\|_{\widehat{K}}^2$$
$$= h_x h_y (\|(I - \pi_h) \partial_{\widehat{x}} \widehat{p}\|_{\widehat{K}}^2 + \|(I - \pi_h) \partial_{\widehat{y}} \widehat{p}\|_{\widehat{K}}^2)$$
$$\lesssim (h_x + \sigma h_y)^2 \|(I - \pi_h) \partial_x p\|_K^2 + h_y^2 \|(I - \pi_h) \partial_y p\|_K^2.$$

Due to Assumption (A1), $h_x + \sigma h_y \leq (1 + \sigma_0) h_x$, it follows

$$\|p - i_h p\|^2 \leq (1 + \sigma_0)^2 \, s_h(p, p).$$

For $\theta \neq 0$, the arguments are the same. ∎

4.3 A priori estimate

In the remainder of this work, expressions as for instance $h_x \|\partial_x u\|$, can always be replaced by $\sum_{K \in \mathcal{T}_{2h}} h_{K,x} \|\partial_x u\|_K$. At first we need to bound the stabilization term applied to the anisotropic H^1-stable projection B_h:

Lemma 1. *The stabilization term has the following interpolation property:*

$$s_h(B_h p, B_h p)^{1/2} \lesssim (1 + \sigma_0)^2 h_x \|\partial_x p\| + (1 + \sigma_0) h_y \|\partial_y p\|.$$

Proof.

$$s_h(B_h p, B_h p) = (M(I - \pi_h) \nabla B_h p, M \nabla B_h p)$$
$$= (M(I - \pi_h) \nabla B_h p, M(I - \pi_h) \nabla B_h p)$$
$$= \|M(I - \pi_h) \nabla B_h p\|^2$$

$$\leq h_x^2(\|\partial_x B_h p\|^2 + \|\pi_h \partial_x B_h p\|^2) + h_y^2(\|\partial_y B_h p\|^2 + \|\pi_h \partial_y B_h p\|^2).$$

Due to the L^2-stability of π_h and Proposition 1:

$$s_h(B_h p, B_h p)^{1/2} \lesssim h_x \|\partial_x B_h p\| + h_y \|\partial_y B_h p\|$$
$$\lesssim h_x(1+\sigma_0)\|\partial_x p\| + h_y^2 h_x^{-1}\|\partial_y p\| + (1+\sigma_0)h_x \sigma_0 \|\partial_x p\|$$
$$+ h_y(1+\sigma_0)\|\partial_y p\|)$$
$$\leq h_x(1+\sigma_0)^2 \|\partial_x p\| + h_y(h_y h_x^{-1} + 1 + \sigma_0)\|\partial_y p\|.$$

The assertion follows due to $h_y h_x^{-1} \leq 1$. ∎

Proposition 3. *Under the conditions (A1), (A2) and transformations of type (7) the following estimate holds for $v \in H^2(\Omega)$ and $p \in H^1(\Omega)$:*

$$\|\nabla(v - v_h)\| + \|p - p_h\| \lesssim \sigma_1^2 h_x \left\{\|\partial_x p\| + \|\partial_x^2 v\| + \|\partial_{xy} v\|\right\} +$$
$$\sigma_1 h_y \left\{\|\partial_y p\| + \|\partial_y^2 v\|\right\},$$

with $\sigma_1 := 1 + \sigma_0$.

Proof. As usual we split the error $\|\nabla(v - v_h)\| + \|p - p_h\|$ in the interpolation part $\|\nabla(v - B_h v)\| + \|p - B_h p\|$ and projection part $\|\nabla(v_h - B_h v)\| + \|p_h - B_h p\|$. Due to Proposition 1 the interpolation part can be bounded by the right hand side of the estimate in Proposition 3. What remains is to bound the projection error. Due to the stability of the bilinear form it holds:

$$\|\nabla(v_h - B_h v)\| + \|p_h - B_h p\|$$
$$\lesssim \sup_{\{\phi,\xi\} \in V_h \times Q_h} \frac{|a(v_h - B_h v, p_h - B_h p_h; \phi, \xi) + s_h(p_h - B_h p, \xi)|}{\|\{\phi, \xi\}\|}.$$

Using the perturbed Galerkin orthogonality for discrete ϕ, ξ,

$$a(v_h - v, p_h - p; \phi, \xi) = -s_h(p_h, \xi),$$

we obtain

$$a(v_h - B_h v, p_h - B_h p; \phi, \xi) + s_h(p_h - B_h p, \xi)$$
$$= a(v - B_h v, p - B_h p; \phi, \xi) - s_h(B_h p, \xi).$$

The last term is bounded by

$$|s_h(B_h p, \xi)| \leq s_h(B_h p, B_h p)^{1/2} \|\{0, \xi\}\| \leq s_h(B_h p, B_h p)^{1/2}.$$

Hence, Lemma 1 gives the desired bound for the stabilization part. Finally, the Galerkin part can be bounded as:

$$|a(v - B_h v, p - B_h p; \phi, \xi)| \lesssim (\|\nabla(v - B_h v)\| + \|p - B_h p\|) \|\{\phi, \xi\}\|,$$

consisting once more of the previously addressed interpolation part. ∎

This result separates the partial derivatives and partial mesh sizes much more properly than the isotropic version in the estimate (1). Let us shortly discuss this result in the situation of a boundary layer with the usual local property $|\partial_y^2 v| \gg |\partial_x^2 v| + |\partial_{xy} v|$ and $h_x \gg h_y$. Due to the multiplication of $|\partial_y^2 v|$ with the smaller mesh size h_y the estimate in Proposition 3 is properly tuned. The shearing parameter σ_1 enters moderately.

4.4 Summary and outlook

We extended the local projection stabilization (LPS) of the Stokes system to anisotropic quadrilateral meshes. In particular, we allow for high aspect ratios, shearing and bilinear effects. Stability and an a priori estimate is given. The result of this paper is still limited to the Stokes system. For the application to Navier-Stokes, also the convective term has to be stabilized. Although LPS is designed for doing so, the additional terms have to be estimated for the anisotropic version. This will be subject of forthcoming work.

References

1. Apel, T.: Anisotropic finite elements: Local estimates and applications. Advances in Numerical Mathematics. Teubner, Stuttgart (1999)
2. Becker, R.: An adaptive finite element method for the incompressible Navier-Stokes equation on time-dependent domains. PhD Dissertation, SFB-359 Preprint 95-44, Universität Heidelberg (1995)
3. Becker, R., Braack, M.: A finite element pressure gradient stabilization for the Stokes equations based on local projections. Calcolo, **38**(4), 173–199 (2001)
4. Becker, R., Braack, M., Vexler, B.: Numerical parameter estimaton for chemical models in multidimensional reactive flows. Combustion Theory and Modeling, **8**(4), 661–682 (2004)
5. Becker, R., Braack, M., Vexler, B.: Parameter identification for chemical models in combustion problems. Appl. Numer. Math., **54**(3-4), 519–536 (2005) accepted.
6. Braack, M.: Anisotropic H^1-stable projections on quadrilateral meshes. In Numerical Mathematics and Advanced Applications, ENUMATH 2005. Springer (2006)
7. Braack, M., Richter, T.: Solutions of 3D Navier-Stokes benchmark problems with adaptive finite elements. Computers and Fluids, **35**(4), 372–392 (2006)
8. Braack, M., Richter, T.: Stabilized finite elements for 3D reactive flow. Int. J. Numer. Methods Fluids, to appear 2006, online since 2005.
9. Hughes, T., Franca, L., Balestra, M.: A new finite element formulation for computational fluid dynamics: V. circumvent the Babuska-Brezzi condition: A stable Petrov-Galerkin formulation for the Stokes problem accommodating equal order interpolation. Comput. Methods Appl. Mech. Engrg., **59**, 89–99 (1986)
10. Johnson, C., Saranen, J.: Streamline diffusion methods for the incompressible euler and Navier-Stokes equations. Math. Comp., **47**, 1–18 (1986)
11. Kunisch, K., Vexler, B.: Optimal vortex reduction for instationary flows based on translation invariant cost functionals. submitted, 2005.

12. Paillere, H., Le Quere, P., Weisman, C., Vierendeels, J., Dick, E., Braack, M., Dabbene, F., Beccantini, A., Studer, E., Kloczko, T., Corre, C., D. M., Hosseinizadehand, S.: Modelling of natural convection flows with large temperature differences: A benchmark problem for low Mach number solvers. Part 2. Contributions to the June 2004 conference. Modél. Math. Anal. Numér., **39**(3), 617–621 (2005)
13. Richter, T.: Parallel multigrid for adaptive finite elements and its application to 3D flow problem. PhD Dissertation, Universität Heidelberg (2005)

An Interior Penalty Variational Multiscale Method for High Reynolds Number Flows

Erik Burman

Institut d'Analyse et Calcul Scientifique (CMCS/IACS), Ecole Polytechnique Fédérale de Lausanne, 1015 Lausanne, Switzerland

Summary. In this paper we present a framework using C^0 interior penalty methods for computations of the Navier-Stokes equations at high Reynolds number. The method is motivated by a formal scale separation argument and then justified by a priori error estimates. As a possible measure of solution quality we propose to monitor the ratio between the artificial dissipation induced by the numerical method and the computed physical dissipation. We prove that for our method the artificial dissipation serves as an a posteriori error estimator.

1 Introduction

The interior penalty method for continuous finite element spaces was originally introduced by Douglas and Dupont [9] for elliptic and parabolic problems. Recently the method, a.k.a. face (or edge in 2D) oriented stabilization has been extended first to advection dominated elliptic problems [7, 3] and then to the Navier-Stokes equations of incompressible flow [6, 5]. Here we will show how scale separation on the continuous level leads to a class of stabilized finite element methods using a least squares perturbation based on the projected residual. This class includes the type of stabilization advocated by Codina [8], Braack, Becker and Burman, [1] and Burman et al. [6] and is also related to the concept of minimal stabilisation procedures discussed in [2]. The variational multiscale method as a general methodology for multiscale problems was introduced in [6]. The basic idea is to perform scale separation in a variational framework and use some model to include the effect of the fine scales on the coarse scales.

It was then proposed for the computation of turbulent flow in [7]. The idea here was to let the turbulence model, act only on the fine scales of the solution so as to avoid the unphysical damping of large scales.

At high Reynolds number, to assure stability, some artificial dissipation must be introduced. Either in the form of stabilizing terms or in the form of a dissipative turbulence model. This is a consequence of the conservation properties of the Galerkin scheme and the fact that the nonlinear coupling pushes energy (enstrophy in 2D) to higher and higher frequencies. Here we will focus on the stabilization using the jump of the gradient of velocities and pressures over element edges. We show that this stabilization operator may be derived by a formal scale separation argument combined with an interpolation result between discrete spaces and that it is the dominating residual in the a posteriori error estimate for the flow equations at high Reynolds number.

2 The equations of incompressible flow

In this paper we will mainly be concerned with the time-dependent incompressible Navier-Stokes with homogeneous boundary conditions

$$\begin{cases} \partial_t u + u \cdot \nabla u - \nu \Delta u + \nabla p = f & \text{in} \quad \Omega \times (0,T), \\ \nabla \cdot u = 0 & \text{in} \quad \Omega \times (0,T), \\ u = 0 & \text{on} \quad \partial\Omega \times (0,T), \\ u(\cdot,0) = u_0 & \text{in} \quad \Omega. \end{cases} \quad (1)$$

These equations describe the motion of a viscous incompressible fluid confined in Ω. In (1), $\nu > 0$ corresponds to the kinematic fluid viscosity coefficient, $\mathbf{f}: \Omega \times (0,T) \longrightarrow \mathbb{R}^d$ represents a given source term and $u_0 : \Omega \longrightarrow \mathbb{R}^d$ stands for the initial velocity.

The scalar product in $L^2(\Omega)$ is denoted by $(\cdot,\cdot)$ and its norm by $\|\cdot\|_{0,\Omega}$. The scalar product on the boundary of Ω is denoted by $\langle\cdot,\cdot\rangle$ with associated norm $\|\cdot\|_{0,\partial\Omega}$. The closed subspaces $H_0^1(\Omega)$, consisting of functions in $H^1(\Omega)$ with zero trace on $\partial\Omega$, and $L_0^2(\Omega)$, consisting of function in $L^2(\Omega)$ with zero mean in Ω, will also be used.

Let the given functions $\mathbf{f}$ and u_0 have the following regularity properties $\mathbf{f} \in L^\infty(0,T;[L^2(\Omega)]^d)$, $u_0 \in [L^2(\Omega)]^d$. For sufficiently regular functions u and p, the equality (1) holds if

$$\begin{cases} (\partial_t u, v) + c(u;u,v) + a(u,v) + b(p,v) = (\mathbf{f},v), & \text{a.e. in} \quad (0,T) \\ b(q,u) = 0, & \text{a.e. in} \quad (0,T), \\ u(0) = u_0, & \text{a.e. in} \quad \Omega, \end{cases} \quad (2)$$

for all $(v,q) \in [H_0^1(\Omega)]^d \times L_0^2(\Omega)$, where

$$c(w;u,v) \stackrel{\text{def}}{=} (w \cdot \nabla u, v), \quad a(u,v) \stackrel{\text{def}}{=} (\nu \nabla u, \nabla v), \quad b(p,v) \stackrel{\text{def}}{=} -(p, \nabla \cdot v).$$

3 Separation of scales and stabilized finite element methods

We let X_h^D be the space of elementwise affine, discontinuous functions, on a locally quasi uniform, shape regular mesh and $X_h = X_h^D \cap C^0(\bar{\Omega})$ be its continuous subspace. We then introduce the orthogonal decomposition

$$L^2(\Omega) \equiv X_h \oplus \widetilde{X}.$$

For the scale separation argument we consider the incompressible Euler equations in $\Omega = (0,1) \times (0,1)$ with periodic boundary conditions. Find $U = (\boldsymbol{u}, p) \in W \stackrel{\text{def}}{=} [L^2(\Omega)]^3 \cap H_{\text{div}}(\Omega) \times L_0^2(\Omega)$ that satisfies

$$\partial_t \boldsymbol{u} + L(\boldsymbol{u})U = 0$$
$$\nabla \cdot \boldsymbol{u} = 0 \quad (3)$$
$$\boldsymbol{u}(x,0) = \boldsymbol{u}_0,$$

where $L(\boldsymbol{w})U = (\boldsymbol{w} \cdot \nabla)\boldsymbol{u} + \nabla p$. If $\boldsymbol{u}_0 \in [L^2(\Omega)]^2$ with $\nabla \cdot \boldsymbol{u}_0 = 0$ and $\nabla \times \boldsymbol{u}_0 \in L^\infty(\Omega)$ then the problem (3) admits a unique weak solution $\boldsymbol{u} \in [C([0,\infty], W^{1,\infty}(\Omega))]^2$ and $p \in C([0,\infty], W^{1,\infty}(\Omega))$ (see [13, Theorem 4.1, page 126]). For the solution U there holds

$$(\partial_t \boldsymbol{u} + (\boldsymbol{u} \cdot \nabla)\boldsymbol{u}, \boldsymbol{v}) - (p, \nabla \cdot \boldsymbol{v}) + (\nabla \cdot \boldsymbol{u}, q) = 0, \quad \text{for all } (\boldsymbol{v}, q) \in W. \quad (4)$$

We now separate the scales of the solution U into the L^2-projection onto $W_h \stackrel{\text{def}}{=} [X_h]^2 \times X_h$ and the L^2-orthogonal complement $\widetilde{W}$, $U = \pi_h U + (I - \pi_h)U = U_h + \widetilde{U}$. The idea is to derive an effective equation for $U_h = (\boldsymbol{u}_h, p_h)$ which represents the scales that can be resolved on the scales represented by the computational mesh and to specify the contributions from the fine scales $\widetilde{U} = (\widetilde{\boldsymbol{u}}, \widetilde{p})$ that have to be modelled. Rewriting the equation (4) as an equation for the coarse scales and an equation for the fine scales by separating the test function $V = (\boldsymbol{v}, q)$ into the L^2-projection onto the finite element space $V_h = (\boldsymbol{v}_h, q_h)$ and the orthogonal complement $\widetilde{V} = (\widetilde{\boldsymbol{v}}, \widetilde{q})$ we have

$$(\partial_t \boldsymbol{u}_h, \boldsymbol{v}_h) - (\boldsymbol{u}_h, (\boldsymbol{u}_h \cdot \nabla)\boldsymbol{v}_h) - (p_h, \nabla \cdot \boldsymbol{v}_h) + (\nabla \cdot \boldsymbol{u}_h, q_h) = (\widetilde{p}, \nabla \cdot \boldsymbol{v}_h)$$
$$+ (\widetilde{\boldsymbol{u}}, (\boldsymbol{u}_h \cdot \nabla)\boldsymbol{v}_h + \nabla q_h) + (\boldsymbol{u}, (\widetilde{\boldsymbol{u}} \cdot \nabla)\boldsymbol{v}_h),$$

$$(\partial_t \widetilde{\boldsymbol{u}} + (\boldsymbol{u} \cdot \nabla)\widetilde{\boldsymbol{u}} + (\widetilde{\boldsymbol{u}} \cdot \nabla)\boldsymbol{u}_h, \widetilde{\boldsymbol{v}}) - (\widetilde{p}, \nabla \cdot \widetilde{\boldsymbol{v}}) + (\nabla \cdot \widetilde{\boldsymbol{u}}, \widetilde{q}) = -(\nabla \cdot \boldsymbol{u}_h, \widetilde{q})$$
$$- ((\boldsymbol{u}_h \cdot \nabla)\boldsymbol{u}_h + \nabla p_h, \widetilde{\boldsymbol{v}}),$$

$$\boldsymbol{u}_h(0) = \pi_h \boldsymbol{u}(0), \quad \widetilde{\boldsymbol{u}}(0) = (I - \pi_h)\boldsymbol{u}(0).$$

In the following we will assume that the mesh is sufficiently fine so that $(I - \pi_h)\boldsymbol{u}(0)$ may be neglected. Note that the fine scales are convected with the exact solution with an additional reaction term depending on the gradient of the discrete solution. This equation is the forward perturbation equation. The adjoint equation of this problem will later be used for duality based a posteriori error estimation. Using now the orthogonality property of $\widetilde{W}$ we have the following equations

$$(\partial_t \boldsymbol{u}_h, \boldsymbol{v}_h) - (\boldsymbol{u}_h, (\boldsymbol{u}_h \cdot \nabla)\boldsymbol{v}_h) - (p_h, \nabla \cdot \boldsymbol{v}_h) + (\nabla \cdot \boldsymbol{u}_h, q_h)$$
$$= (\widetilde{p}, (I - \pi_h^*)\nabla \cdot \boldsymbol{v}_h) + (\widetilde{\boldsymbol{u}}, (I - \pi_h^*)L(\boldsymbol{u}_h)\boldsymbol{V}_h) + (\boldsymbol{u}, (\widetilde{\boldsymbol{u}} \cdot \nabla)\boldsymbol{v}_h),$$

$$(\partial_t \widetilde{\boldsymbol{u}} + (\boldsymbol{u} \cdot \nabla)\widetilde{\boldsymbol{u}} + (\widetilde{\boldsymbol{u}} \cdot \nabla)\boldsymbol{u}_h, \widetilde{\boldsymbol{v}}) - (\widetilde{p}, \nabla \cdot \widetilde{\boldsymbol{v}}) + (\nabla \cdot \widetilde{\boldsymbol{u}}, \widetilde{q})$$
$$= -((I - \pi_h^*)(L(\boldsymbol{u}_h)\boldsymbol{U}_h), \widetilde{\boldsymbol{v}}) - ((I - \pi_h^*)\nabla \cdot \boldsymbol{u}_h, \widetilde{q}).$$

Where π_h^* denotes a mapping from X_h^D to X_h to be defined later. It follows that the fine scales are driven by the projected residual of the coarse scales. We now "solve" for the fine scale solution $\widetilde{\boldsymbol{U}} = T^{-1}\{-(I - \pi_h^*)(L(\boldsymbol{u}_h)\boldsymbol{U}_h), -(I - \pi_h^*)\nabla \cdot \boldsymbol{u}_h\}$. This results in an effective equation for $\boldsymbol{U}_h$. Here we choose the crudest possible approximation of T^{-1}: a scaled, constant, diagonal matrix. The justification of this is (somewhat ad hoc) that we want the terms modelling the fine scales to lead to a scheme for the coarse scales for which a linear problem gives a linear numerical method and that is

- dissipative (the Bernoulli hypothesis),
- consistent (Galerkin orthogonality).

Hence we have, with $\pi^\perp \stackrel{\text{def}}{=} (I - \pi_h^*)$, $\widetilde{\boldsymbol{u}} = -\delta_u \pi^\perp (L(\boldsymbol{u}_h)\boldsymbol{U}_h)$ and $\widetilde{p} = -\delta_p \pi^\perp \nabla \cdot \boldsymbol{u}_h$. Inserting these expressions in the equation for the coarse scales gives the following equation

$$(\partial_t \boldsymbol{u}_h, \boldsymbol{v}_h) - (\boldsymbol{u}_h, (\boldsymbol{u}_h \cdot \nabla)\boldsymbol{v}_h) - (p_h, \nabla \cdot \boldsymbol{v}_h) + (\nabla \cdot \boldsymbol{u}_h, q_h)$$
$$= -(\delta_p \pi^\perp \nabla \cdot \boldsymbol{u}_h, \pi^\perp \nabla \cdot \boldsymbol{v}_h) - (\delta_u \pi^\perp L(\boldsymbol{u}_h)\boldsymbol{U}_h, \pi^\perp L(\boldsymbol{u}_h)\boldsymbol{V}_h) + (\boldsymbol{u}, (\widetilde{\boldsymbol{u}} \cdot \nabla)\boldsymbol{v}_h).$$

The first two fine to coarse scale interaction terms now take the form of a least squares type stabilized method based on the projected residual. The above system is not closed. The last term still contains fine to coarse scale interaction. Instead of using the approximation $\widetilde{\boldsymbol{u}} = -\delta_u \pi^\perp (L(\boldsymbol{u}_h)\boldsymbol{U}_h)$ in this expression as well, here we will simply neglect it.

If this simplification is granted we arrive at the following stabilized finite element method: Find $\boldsymbol{U}_h \in W_h$ such that

$$(\partial_t \boldsymbol{u}_h, \boldsymbol{v}_h) - (\boldsymbol{u}_h, (\boldsymbol{u}_h \cdot \nabla)\boldsymbol{v}_h) - (p_h, \nabla \cdot \boldsymbol{v}_h) + (\nabla \cdot \boldsymbol{u}_h, q_h)$$
$$+ (\delta_p \pi^\perp \nabla \cdot \boldsymbol{u}_h, \pi^\perp \nabla \cdot \boldsymbol{v}_h) + (\delta_u(\pi^\perp (L(\boldsymbol{u}_h)\boldsymbol{U}_h)), \pi^\perp L(\boldsymbol{u}_h)\boldsymbol{V}_h) = 0,$$

for all $\boldsymbol{V}_h \in W_h$ This method is strongly consistent since $\pi^\perp(\partial_t \boldsymbol{u}_h) = 0$. We have arrived at a class of stabilized methods that are based on the projected residual.

3.1 Face oriented stabilization

We now let π^* be the following interpolation operator

Definition 1. *For each node x_i, let n_i be the number of elements containing x_i as a node. We define a quasi-interpolant $\pi_h^* : X_h^D \to X_h$ by*

$$\pi_h^* v(x_i) \stackrel{\text{def}}{=} \frac{1}{n_i} \sum_{\{K : x_i \in K\}} v_{|K}(x_i), \quad \text{with } v \in X_h^D.$$

Recall the following discrete interpolation lemma of [6] in a form suitable for our needs.

Lemma 1. *Let $\mathcal{E}$ denote the set of all faces f sharing at least one vertex with K. For the interpolation operator $\pi_h^* : X_h^D \to X_h$ there exist a constant $c > 0$, depending only on the local mesh geometry, such that*

$$\|(I - \pi_h^*)L(\boldsymbol{u}_h)\boldsymbol{U}_h\|_K^2 \leq c \sum_{f \in \mathcal{E}} \int_f h_f |[\![L(\boldsymbol{u}_h)\boldsymbol{U}_h]\!]_f|^2 \, ds \qquad (5)$$

where $[\![x]\!]_f$ denotes the jump of quantity x over the interior face f. $[\![x]\!]_f = 0$ on boundary faces.

We conclude that the element stabilization of the projected residual may be replaced by a term stabilizing the jumps of the residual over element faces. Using finally the triangle inequality in (5) and denoting by n a unit normal vector of face f with arbitrary but fixed orientation

$$\sum_{f \in \mathcal{E}} \int_f h_f [\![L(\boldsymbol{u}_h)\boldsymbol{U}_h]\!]^2 ds \leq 2 \sum_{f \in \mathcal{E}} \int_f h_f (|\boldsymbol{u}_h \cdot \boldsymbol{n}|^2 |[\![\nabla \boldsymbol{u}_h]\!]|^2 + |[\![\nabla p_h]\!]|^2) ds$$

we arrive at the face oriented stabilization formulation. For the Navier-Stokes equations the space semi-discretized scheme we propose reads: For all $t \in (0, T)$, find $(\boldsymbol{u}_h(t), p_h(t)) \in W_h$ such that

$$(\partial_t \boldsymbol{u}_h, \boldsymbol{v}_h) + \mathbf{A}\big[\boldsymbol{u}_h; (\boldsymbol{u}_h, p_h), (\boldsymbol{v}_h, q_h)\big] + \mathbf{J}\big[\boldsymbol{u}_h; (\boldsymbol{u}_h, p_h), (\boldsymbol{v}_h, q_h)\big] = (\mathbf{f}(t), \boldsymbol{v}_h), \qquad (6)$$

for all $(\boldsymbol{v}_h, q_h) \in W_h$, equipped with the following initial condition

$$(\boldsymbol{u}_h(0), \boldsymbol{v}_h) = (\boldsymbol{u}_0, \boldsymbol{v}_h), \quad \forall \boldsymbol{v}_h \in [X_h]^d. \qquad (7)$$

In (6) we used the following notations

$$\mathbf{A}\big[\boldsymbol{w}_h; (\boldsymbol{u}_h, p_h), (\boldsymbol{v}_h, q_h)\big] \stackrel{\text{def}}{=} c_h(\boldsymbol{w}_h; \boldsymbol{u}_h, \boldsymbol{v}_h) + a_h(\boldsymbol{u}_h, \boldsymbol{v}_h) \\ + b_h(p_h, \boldsymbol{v}_h) - b_h(q_h, \boldsymbol{u}_h), \qquad (8)$$

$$c_h(\boldsymbol{w}_h; \boldsymbol{u}_h, \boldsymbol{v}_h) \stackrel{\text{def}}{=} c(\boldsymbol{w}_h; \boldsymbol{u}_h, \boldsymbol{v}_h) + \frac{1}{2}(\nabla \cdot \boldsymbol{w}_h \boldsymbol{u}_h, \boldsymbol{v}_h) \\ - \frac{1}{2}\langle \boldsymbol{w}_h \cdot \boldsymbol{n} \boldsymbol{u}_h, \boldsymbol{v}_h \rangle, \qquad (9)$$

$$a_h(\boldsymbol{u}_h, \boldsymbol{v}_h) \stackrel{\text{def}}{=} a(\boldsymbol{u}_h, \boldsymbol{v}_h) - \langle 2\nu \nabla \boldsymbol{u}_h \boldsymbol{n}, \boldsymbol{v}_h \rangle - \langle \boldsymbol{u}_h, 2\nu \nabla \boldsymbol{v}_h \boldsymbol{n} \rangle \\ + \left\langle \gamma \frac{\nu}{h} \boldsymbol{u}_h, \boldsymbol{v}_h \right\rangle + \langle \boldsymbol{u}_h \cdot \boldsymbol{n}, \boldsymbol{v}_h \cdot \boldsymbol{n} \rangle, \qquad (10)$$

$$b_h(p_h, \boldsymbol{v}_h) \stackrel{\text{def}}{=} b(p_h, \boldsymbol{v}_h) + \langle p_h, \boldsymbol{v}_h \cdot \boldsymbol{n} \rangle, \qquad (11)$$

$$\mathbf{J}\bigl[\boldsymbol{w}_h;(\boldsymbol{u}_h,p_h),(\boldsymbol{v}_h,q_h)\bigr] \stackrel{\text{def}}{=} \gamma j_{\boldsymbol{w}_h}(\boldsymbol{u}_h,\boldsymbol{v}_h) + \gamma_u j_n(\boldsymbol{u}_h,\boldsymbol{v}_h) + \gamma_p j(p_h,q_h),$$

with $j_{\boldsymbol{x}}(\boldsymbol{u}_h,\boldsymbol{v}_h) \stackrel{\text{def}}{=} \sum_{K \in \mathcal{T}_h} \int_{\partial K} h_K^2 |\boldsymbol{x} \cdot \boldsymbol{n}|^2 [\![\nabla \boldsymbol{u}_h]\!] : [\![\nabla \boldsymbol{v}_h]\!] \, \mathrm{d}s$, and
$j(p_h,q_h) \stackrel{\text{def}}{=} \sum_{K \in \mathcal{T}_h} \int_{\partial K} h_K^2 [\![\nabla p_h]\!] \cdot [\![\nabla q_h]\!] \, \mathrm{d}s$. Here γ, γ_u, γ_p are positive constants independent of h, but not of the problem data. An a priori error estimate for this formulation was proved in [5].

4 A posteriori error estimation

In this section we will propose an a posteriori error estimate for the case of piecewise affine approximation. First we introduce the dual problem. Let $\Psi \in [L^2(Q)]^d$, $\|\Psi\|_{0,Q}^2 = 1$, where $Q = \Omega \times (0,T)$. The dual problem is then given by

$$-\partial_t \varphi - (\boldsymbol{u} \cdot \nabla)\varphi + \frac{1}{2}(\nabla \boldsymbol{u}_h)^T \varphi - \frac{1}{2}(\nabla \varphi)^T \boldsymbol{u}_h - \nabla r - \nu \Delta \varphi = \Psi \text{ in } Q,$$
$$\nabla \cdot \varphi = 0 \text{ in } Q, \quad \varphi(\cdot,T) = 0 \text{ in } \Omega, \quad \varphi = 0 \text{ on } \partial \Omega. \tag{12}$$

Where $\boldsymbol{u}$ is solution of (1) and $\boldsymbol{u}_h \in [X_h]^d$ is the solution of (6). For each fixed h this problem admits a unique solution. We will also assume that the following regularity estimate holds

$$\int_0^T (\|\nabla r\|_{0,\Omega}^2 + \|\nu D^2 \varphi\|_{0,\Omega}^2) \, \mathrm{d}t \leq C(\boldsymbol{u},\boldsymbol{u}_h) \int_0^T \|\Psi\|_0^2 \, \mathrm{d}t, \tag{13}$$

for some constant $C(\boldsymbol{u},\boldsymbol{u}_h)$ independent of h. We may then prove the following a posteriori error estimate (see [4]).

Theorem 1. *Let $(\boldsymbol{u}_h,p_h) \in W_h$ be a piecewise affine finite element solution to the formulation (6) with data $\mathbf{f}, \boldsymbol{u}_0 \in [X_h]^d$ and assume that the dual solution satisfies (13), then the following holds:*

$$\int_0^T (\boldsymbol{u}-\boldsymbol{u}_h,\Psi) \, \mathrm{d}t \leq C \left(\int_0^T \sum_K \sum_{i=1}^3 \alpha_{i,K} \eta_{i,K}^2 \, \mathrm{d}t \right)^{\frac{1}{2}},$$

where $\alpha_{1,K} = \max(\frac{h_K^3}{\nu^2},h_K)$, $\alpha_{2,K} = \alpha_{3,K} = \frac{h_K^3}{\nu^2}$ and the error indicators are given by

$$\eta_{1,K} = \left(\|h_K \xi_K [\![\nabla \boldsymbol{u}_h]\!]\|_{\partial K \setminus \partial \Omega} + \|h_K [\![\nabla p_h]\!]\|_{\partial K \setminus \partial \Omega} \right),$$

$$\eta_{2,K} = \|\xi_K \boldsymbol{u}_h \cdot \boldsymbol{n}\|_{\partial K \cap \partial \Omega} \text{ and } \eta_{3,K} = \|\tfrac{\nu}{h_K}\boldsymbol{u}_h\|_{\partial K \cap \partial \Omega},$$

with $\xi_K = \max(|\boldsymbol{u}_h \cdot \boldsymbol{n}|,1)$. The constant C depends on the constant in the stability estimate (13) and interpolation constants.

5 A numerical result

We consider a Kelvin-Helmholz instability at Reynolds number 10000 in the unit square on a sequence of structured finite element meshes using BDF2 timestepping and a very small timestep, for details on the problem data see [4, 10]. This problem is often considered as a model problem for 2D turbulence. Four vortices form in the mixing layer and then merge in a two step transition process to one large vortex. We define the dissipation ratio as the relation between the artificial dissipation and the physical dissipation, corresponding to

$$D = \frac{\int_0^T \mathbf{J}\left[\boldsymbol{u}_h;(\boldsymbol{u}_h,p_h),(\boldsymbol{v}_h,q_h)\right]\mathrm{d}t}{\int_0^T \|\nu^{\frac{1}{2}}\nabla \boldsymbol{u}_h\|^2 \mathrm{d}t}.$$

Note the close relation between D and the error estimator η_1^2. In Table 1 we give the size of D and the average number of GMRES iterations per timestep for the various computations. In Figure 1 we compare vorticity plots for three different discretizations. Note that the transition times seem to start to be accurately captured when $D \approx 1$.

Table 1. Convergence of the dissipation ratio D for computations of the mixing layer using the $P1/P1$ interior penalty method and the $P2/P2$ interior penalty method. The average number of GMRES iterations per timestep is also presented

P1 el. per side	D	$O(h^\alpha)$	GMRES	P2 el. per side	D	$O(h^\alpha)$	GMRES
80	5.6	-	38	40	0.38	-	93
160	1.4	2	66	80	0.11	1.79	148
320	0.3	2.22	123	160	0.025	2.0	251

References

1. Braack, M., Burman, E.: Local projection stabilization for the Oseen problem and its interpretation as a variational multiscale method. SIAM, J. Numer. Anal, **43**(6), 2544–2566 (2006)
2. Brezzi, F., Fortin, M.: A minimal stabilisation procedure for mixed finite element methods. Numer. Math., **89**(3), 457–491 (2001)
3. Burman, E.: A unified analysis for conforming and nonconforming stabilized finite element methods using interior penalty. SIAM, J. Numer. Anal, **43**(5), 2012–2033 (2005)
4. Burman, E.: An interior penalty variational multiscale method for high Reynolds number flows. Technical report EPFL/IACS 18.2005 (2005)

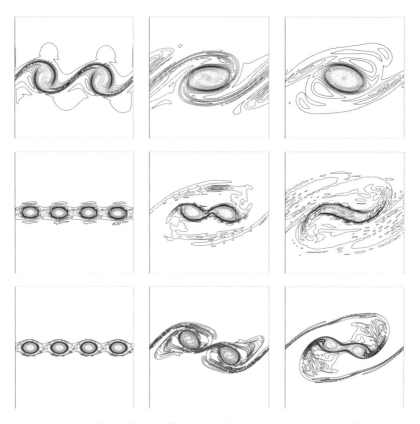

Fig. 1. Vorticity field at the nondimensional times $t = 50, 100, 120$, first line: $80 \times 80/P_1$ mesh, $40 \times 40/P_2$ mesh, , $160 \times 160/P_2$ mesh.

5. Burman, E., Fernàndez, M.: A finite element method with edge oriented stabilization for the time-dependent Navier-Stokes equations: space discretization and convergence. Technical Report 11.2005, Ecole Polytechnique Federale de Lausanne, (2005)
6. Burman, E., Fernández, M.A., Hansbo, P.: Continuous interior penalty method for the Oseen problem. SIAM, J. Numer. Anal, in press 2006
7. Burman, E., Hansbo, P.: Edge stabilization for Galerkin approximations of convection-diffusion problems. Comput. Methods Appl. Mech. Engrg., **193**, 1437–2453 (2004)
8. Codina, R.: Stabilized finite element approximation of transient incompressible flows using orthogonal subscales. Comput. Methods Appl. Mech. Engrg., **191**(39-40), 4295–4321 (2002)
9. Douglas Jr., J., Dupont, T.: Interior Penalty Procedures for Elliptic and Parabolic Galerkin Methods, In R. Glowinski and J.-L. Lions, editors, Computing Methods in Applied Sciences, **58** of Lecture Notes in Physics, pages 207–216. Springer-Verlag, Berlin (1976)

10. Gravemeier, V., Wall, W. A., Ramm, E.: Large eddy simulation of turbulent incompressible flows by a three-level finite element method. Internat. J. Numer. Methods Fluids, **48**(10), 1067–1099 (2005)
11. Hughes, T. J. R., Feijóo, G. R., Mazzei, L., Quincy, J.-B.J.-B.: The variational multiscale method—a paradigm for computational mechanics. Comput. Methods Appl. Mech. Engrg., **166**(1-2), 3–24 (1998)
12. Hughes, T. J. R., Mazzei, L, Jansen, K.E.: Large eddy simulation and the variational multiscale method. Comput. Vis. Sci., **3**, 47–59 (2000)
13. Lions, P. L.: Mathematical topics in fluid mechanics. Vol. 1. Incompressible models. Oxford University Press, New York (1996)

Variational Multiscale Large Eddy Simulation of Turbulent Flows Using a Two-Grid Finite Element or Finite Volume Method

Volker Gravemeier

Technical University of Munich, Chair for Computational Mechanics,
Boltzmannstr. 15, D-85748 Garching, Germany
vgravem@lnm.mw.tum.de

Summary. In this article, variational multiscale large eddy simulation based on multigrid scale-separating operators is presented. Two different scale-separating operators, which are basically applicable within both a finite element and a finite volume method, are proposed for separating large resolved scales and small resolved scales. One of these operators is a projector. Using the multigrid operators for scale separation, dynamic and non-dynamic subgrid-scale modeling approaches are applied to the challenging test case of turbulent flow in a diffuser. Variational multiscale large eddy simulation using a projective multigrid scale-separating operator provides remarkable results already in combination with a simple non-dynamic subgrid-scale modeling approach. Furthermore, this methodical combination turns out to be very efficient with regard to the important aspect of computational cost.

1 Introduction

The variational multiscale method represents a general approach for problems in computational mechanics which give rise to broad ranges of scales, see [6]. The basic concept differentiates a predefined number of scale groups. This theoretical framework was also applied to the problem of the incompressible Navier-Stokes equations in [7], in order to facilitate large eddy simulation (LES) of turbulent flows. Apart from the initial separation and potentially different treatment of the respective scale ranges, two important aspects characterize the variational multiscale large eddy simulation (VMLES). Firstly, a variational projection separates scale ranges within the VMLES rather than a spatial filter in the traditional LES. Secondly, the (direct) influence of the subgrid-scale model, which is introduced to represent the effect of the unresolved scales on the resolved scales, is confined to the small resolved scales. Thus, the larger scales are solved as a direct numerical simulation (DNS) (i.e., without any (direct) influence of the modeling term). Of course, the large resolved scales are still indirectly influenced by the subgrid-scale model due to the inherent coupling of all scales. The interested reader may consult, e.g., [7] or a recent review article [3] for a detailed description of the VMLES.

The particular implementation within the variational multiscale framework to be presented in this article is based on the scale-separating approach developed in [2]. A general class of scale-separating operators based on combined multigrid operators in a two-grid procedure was proposed in that study, in order to replace spatial filters, which are widely used in the traditional LES. One particular representative of this class is a projector. A projector of this type was also used in [8]. The scale-separating operators were implemented into the CDP-α code, the flagship LES code of the Center for Turbulence Research. Underlying this code is a second-order accurate energy-conserving finite volume method particularly suited for applications on unstructured grids, see, e.g., [5] for some basic features of CDP-α. This new multigrid-based approach for VMLES was initially tested for turbulent flow in a channel, a flow example exhibiting one direction of inhomogeneity, see [2] for results. Afterwards, it was applied to turbulent flow in a planar asymmetric diffuser in [4]. Turbulent flow in such a diffuser is a representative of the group of flow problems exhibiting more than one direction of inhomogeneity. Not only for this reason, it is a more challenging flow problem. Several features of this flow indicate its higher complexity (i.e., a large unsteady separation bubble due to an adverse pressure gradient, a sudden change of the streamwise pressure gradient from slightly favorable to strongly adverse at the diffuser throat, and a slowly growing internal layer emerging at the upper flat wall in the relaxation zone downstream of the sharp variation in the streamwise pressure gradient).

The remainder of the present article is organized as follows. In Sect. 2, the multiscale formulation using multigrid-based scale-separating operators is presented. Some numerical results from the application of this approach to turbulent flow in a diffuser are then provided in Sect. 3. Finally, conclusions are drawn in Sect. 4.

2 Multiscale formulation

A weighted residual formulation of the Navier-Stokes equations is given as follows: find $\{\mathbf{u}, p\} \in \mathcal{S}_{\mathbf{u}p}$, such that

$$B_{\text{NS}}(\mathbf{v}, q; \mathbf{u}, p) = (\mathbf{v}, \mathbf{f})_\Omega \qquad \forall \{\mathbf{v}, q\} \in \mathcal{V}_{\mathbf{u}p}, \qquad (1)$$

where $\mathbf{v}$ and q denote the weighting functions. $\mathcal{S}_{\mathbf{u}p}$ and $\mathcal{V}_{\mathbf{u}p}$ represent the combined formulation of the solution and weighting function spaces for velocity and pressure: $\mathcal{S}_{\mathbf{u}p} := \mathcal{S}_{\mathbf{u}} \times \mathcal{S}_p$ and $\mathcal{V}_{\mathbf{u}p} := \mathcal{V}_{\mathbf{u}} \times \mathcal{V}_p$. The form $B_{\text{NS}}(\mathbf{v}, q; \mathbf{u}, p)$ on the left hand side of (1) is defined as

$$B_{\text{NS}}(\mathbf{v}, q; \mathbf{u}, p) = \left(\mathbf{v}, \frac{\partial \mathbf{u}}{\partial t}\right)_\Omega + (\mathbf{v}, \nabla \cdot (\mathbf{u} \otimes \mathbf{u}))_\Omega + (\mathbf{v}, \nabla p)_\Omega$$
$$- (\mathbf{v}, 2\nu \nabla \cdot \varepsilon(\mathbf{u}))_\Omega + (q, \nabla \cdot \mathbf{u})_\Omega. \qquad (2)$$

The characteristic length scale h of the discretization chosen in large eddy simulations is usually considerably larger than the smallest length scale of the problem under investigation (i.e., by far not all scales of the problem can be resolved). Therefore, the subgrid viscosity approach, a usual way of taking into account the (dissipative) effect of unresolved scales in the traditional LES, is applied. According to this, a subgrid viscosity term is added to (1). In general form, this results in

$$B_{NS}\left(\mathbf{v}^h, q^h; \mathbf{u}^h, p^h\right) - \left(\mathbf{v}^h, \nabla \cdot \left(2\nu_T \varepsilon\left(\mathbf{u}^h\right)\right)\right)_\Omega = \left(\mathbf{v}^h, \mathbf{f}\right)_\Omega, \quad (3)$$

where ν_T denotes the subgrid viscosity. Note that the subgrid viscosity term is added to all resolved scales of the problem in (3). For the actual variational FE or FV formulation, the appropriate integration-by-parts procedures have to be applied to both the weighted residual form (2) and the subgrid viscosity term, see [3].

The resolved velocity vector $\mathbf{u}^h$ is separated into a large-scale part and a small-scale part subject to

$$\mathbf{u}^h = \left(\bar{\mathbf{u}} + \mathbf{u}'\right)^h. \quad (4)$$

With respect to this complete resolution level, a large-scale resolution level is identified a priori. This level is characterized by the characteristic discretization length $\bar{h}$, where $\bar{h} > h$, and, accordingly, yields a large-scale velocity $\bar{\mathbf{u}}^{\bar{h}}$. The small-scale velocity is consistently defined on the complete resolution level, characterized by the length h, as

$$\mathbf{u}'^h = \mathbf{u}^h - \bar{\mathbf{u}}^h, \quad (5)$$

where $\bar{\mathbf{u}}^h$ denotes the large-scale value transfered to this level. The scale separation used in the present study relies on multigrid operators. At the outset of the numerical simulation, two grids are created: a coarser grid, which is called the "parent" grid, and a finer grid, which is called the "child" grid. The child grid is obtained by an isotropic hierarchical subdivision of the parent grid. In the simulations of the present study, a subdivision by a factor of two in each spatial direction is exclusively applied. For more details concerning the implementation, it is refered to [2].

The general class of scale-separating operators based on multigrid operators is formulated as

$$\bar{\mathbf{u}}^h = S^m\left[\mathbf{u}^h\right] = P \circ R\left[\mathbf{u}^h\right] = P\left[\bar{\mathbf{u}}^{\bar{h}}\right], \quad (6)$$

where the multigrid scale-separating operator S^m consists of the sequential application of a restriction operator R and a prolongation operator P. Applying the restriction operator on $\mathbf{u}^h$ yields a large-scale velocity $\bar{\mathbf{u}}^{\bar{h}}$ defined at the degrees of freedom of the parent grid, which is then prolongated, in order to obtain a large-scale velocity $\overline{\mathbf{u}}^h$ defined at the degrees of freedom of the child grid. Various restriction as well as prolongation operators may be used in (6). Two special combinations of restriction and prolongation operators in the context of a colocated finite volume method were analyzed and compared to discrete smooth filters, which are widely used in traditional LES, in [2]. It was shown that these two multigrid scale-separating operators represent computationally efficient operators for separating the resolved scales of the problem in comparison to discrete smooth filters. Both multigrid scale-separating operators rely on the same restriction operator, but apply different prolongation operators afterwards. Corresponding operators may be defined for a finite element method in a straightforward manner, see [3].

The restriction operator is defined to be a volume-weighted average over all child control volumes within one parent control volume subject to

$$\overline{\mathbf{u}}_j^h = \frac{\sum_{i=1}^{n_{\text{cop}}} |\Omega_i| \mathbf{u}_i^h}{\sum_{i=1}^{n_{\text{cop}}} |\Omega_i|}, \qquad (7)$$

where $\overline{\mathbf{u}}_j^h$ denotes the large-scale velocity at the center of the parent control volume $\overline{\Omega}_j$ and n_{cop} the number of child control volumes in $\overline{\Omega}_j$. The first prolongation operator P^{p} yields a constant prolongation, which is given as

$$\overline{\mathbf{u}}_i^h = P^{\text{p}} \left[\overline{\mathbf{u}}_j^h \right]_i = \overline{\mathbf{u}}_j^h \qquad \forall \, \Omega_i \subset \overline{\Omega}_j \qquad (8)$$

and zero elsewhere. It was shown in [2] that the scale-separating operator defined as $S^{\text{pm}} := P^{\text{p}} \circ R$ has the property of a projector, which is indicated by the additional superscript "p". This projector is exactly the operator also used in [8], although it was not derived from the general formulation (6) and, thus, not split up into a restriction and prolongation operator in that study.

The second prolongation operator considered here yields a linear prolongation subject to

$$\overline{\mathbf{u}}_i^h = P^{\text{s}} \left[\overline{\mathbf{u}}_j^h \right]_i = \overline{\mathbf{u}}_j^h + \left(\nabla^{\overline{h}} \overline{\mathbf{u}}_j^h \right) \cdot (\mathbf{r}_i - \overline{\mathbf{r}}_j) \qquad \forall \, \Omega_i \subset \overline{\Omega}_j \qquad (9)$$

and zero elsewhere. The vectors $\mathbf{r}_i$ and $\overline{\mathbf{r}}_j$ denote geometrical vectors pointing to the centers of the child control volume Ω_i and the parent control volume $\overline{\Omega}_j$, respectively. The operator $\nabla^{\overline{h}}$ describes the discrete gradient operator on the parent grid. Due to this, values from neighbouring parent control volumes and, consequently, child control volumes contained in these neighbouring parent control volumes influence the final large-scale value in the child control volume Ω_i. The prolongation P^{s} does not provide us with a projective scale-separating operation. It rather produces a smoothing prolongation, which is, at least, smoother than the prolongation produced by P^{p}. Thus, it is indicated by the additional superscript "s", and the complete scale-separating operator is defined as $S^{sm} := P^s \circ R$.

A separation of the velocity weighting function analogous to the separation of the velocity solution function enables a decomposition of the variational FE or FV equation, respectively, into a large- and a small-scale equation. The coupled system of large- and small-scale equation, resulting from an initial three-scale separation, may be found, e.g., in [3]. These two equations may eventually be reunified to *one* final equation. In this final equation, the scale separation based on S^m remains perceptible only with respect to the subgrid viscosity term. Thus, the multiscale weighted residual formulation is given as

$$B_{NS}\left(\mathbf{v}^h, q^h; \mathbf{u}^h, p^h\right) - \left(\mathbf{v}'^h, \nabla \cdot \left(2\nu'_T \varepsilon\left(\mathbf{u}'^h\right)\right)\right)_\Omega = \left(\mathbf{v}^h, \mathbf{f}\right)_\Omega, \quad (10)$$

respectively, where $\mathbf{v}'^h$ denotes the small-scale part of the velocity weighting function and ν'_T the subgrid viscosity depending on the small resolved scales. As in (3), the appropriate integration-by-parts procedures have to be applied to both the weighted residual form (2) and the subgrid viscosity term for the actual variational FE or FV formulation, see [3].

3 Numerical results for turbulent flow in a diffuser

The diffuser geometry, which basically matches the experimental configuration in [1] ("Buice-experiment") and [9] ("Obi-experiment") as well as the numerical setup in [10] ("Wu-LES"), is shown in Fig. 1. In the inflow channel, the inflow velocity $\mathbf{u}^{in}(t)$ for the actual diffuser is generated. No-slip boundary conditions are assumed at the upper and lower walls Γ_w, a convective boundary condition is prescribed at the outflow boundary Γ_{out}, and periodic boundary conditions are assumed on the boundaries Γ_{per} in x_3-direction. The diffuser, including inlet and outlet channel, is discretized using 290, 64, and 80 control volumes in x_1-, x_2-, and x_3-direction, respectively. The control volumes are uniformly distributed in the spanwise direction. In the wall-normal direction, a cosine function for refinement towards the walls for the parent grid is used, with the isotropic hierarchical subdivision procedure subsequently applied. In the streamwise direction, the following control volume distribution is employed: in the inlet channel, h_1 decreases linearly from 0.15 to 0.05, in the asymmetric diffuser section, h_1 increases linearly from 0.05 to 0.475, in the first section of the outlet channel (up to $x_1 = 74.5$), h_1 increases linearly from 0.475 to 0.825, and in the remaining section of the outlet channel, the control volumes are uniformly distributed with $h_1 = 0.825$. Comparing the discretization of the diffuser to the finer discretization in the Wu-LES, which employed 590, 100, and 110 control volumes in x_1-, x_2-, and x_3-direction, it is stated that less than 23% the number of control volumes are used in the present case. More details concerning the numerical setup can be found in [4].

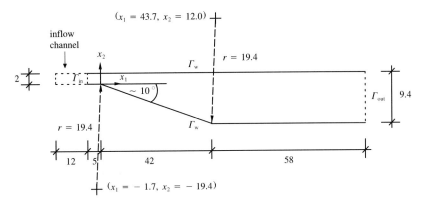

Fig. 1. Diffuser geometry in x_1-x_2-plane

All numerical simulations are conducted using the CDP-α code, see, e.g., [5] for details of the code.

Three different methods are investigated: the dynamic Smagorinsky (DS) model in a non-multiscale application, the constant-coefficient-based Smagorinsky model within the multiscale environment (CMS), and the dynamic Smagorinsky model within the multiscale environment (DMS). All of these methods are analyzed for the scale-separating operator S^{pm}. The abbreviation "DMS-PM", for instance, indicates the variational multiscale LES incorporating a dynamic Smagorinsky model, with the scale-separating operator S^{pm} applied. The scale-separating operator S^{sm} is only investigated for CMS, since this method revealed the most notable differences between the scale-separating operators for the test case in [2]. Results are also reported for simulations using no model at all (NM), which represents a coarse (i.e., not sufficiently resolved) DNS. The Wu-LES, which the results are compared to, applied the same dynamic Smagorinsky model in a traditional non-multiscale LES (i.e., DS using smooth filters for scale separation). Evaluating the necessary computational effort provides the following numbers. Setting the computational effort for NM to 1.0, the relative measures for CMS-PM, CMS-SM, DS-PM and DMS-PM are approximately 1.08, 1.34, 1.27, and 1.32, respectively. These numbers are even more impressively in favor of CMS-PM than the ones for the channel in [2]. Thus, it is confirmed that CMS in combination with PM is a very efficient method computationally, in the present case substantially more efficient than, for instance, DS. Using the scale-separating operator SM, the numbers increase drastically for CMS. Less effort is required for PM compared to SM for reasons explained in [2] and [4].

As one sample of the flow parameters investigated, Fig. 2 depicts the results for the skin friction coefficient along the upper wall of the diffuser. Results for further flow parameters can be found in [4]. It is stated that all methods tend to underpredict C_f compared to the results from the Wu-LES and the Buice-experiment. The worst results are produced by CMS-SM. The profile for NM

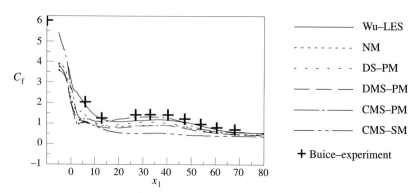

Fig. 2. Skin friction coefficient (factor 1000) along the upper wall of the diffuser

is closest to the ones from the Wu-LES and the Buice-experiment immediately behind the diffuser throat, but gets worse in its prediction further downstream. DS-PM yields a fairly good prediction throughout the diffuser, and DMS-PM produces worse results than DS-PM. Although the results for CMS-PM are worse than the ones for NM immediately behind the diffuser throat, the predicition is the best overall. It is the only method yielding results which almost match the experimental results in the section of the diffuser between $x_1 \approx 18$ and $x_1 \approx 46$. In this part of the diffuser, which is approximately the region where the flow is separated, CMS-PM appears to produce even better results than the substantially finer discretized Wu-LES. Furthermore, it seems to be the only one of the present methods which would have been able to predict the first point from the Buice-experiment at $x_1 \approx -10$, if the inlet channel had been elongated.

4 Conclusions

Variational multiscale large eddy simulation based on multigrid scale-separating operators has been investigated. Two different scale-separating operators, which are basically applicable within both a finite element and a finite volume method, have been used for separating large resolved scales and small resolved scales. One of these scale-separating operators is a projector. The scale-separating operators have been implemented in a second-order accurate energy-conserving finite volume method. Dynamic and non-dynamic subgrid-scale modeling approaches have been tested in combination with the multigrid scale-separating operators for the case of turbulent flow in a diffuser. Turbulent flow in a diffuser represents a challenging test case, in particular due to the appearance of flow separation, which is caused by an adverse pressure gradient, and subsequent reattachment. The results obtained by the various approaches have been compared to results from a recent non-multiscale LES with dynamic subgrid-scale modeling, performed on an approximately 5 times finer grid, and experimental results. In particular, the method using the simple constant-coefficient-based Smagorinsky model in combination with the projective operator has shown remarkable results. Furthermore, it turns out to be a very efficient methodical combination with regard to the important aspect of computational cost.

Acknowledgements

The work presented in this article was done by the author during his Postdoctoral research stay at the Center for Turbulence Research, Stanford University and NASA Ames Research Center. The support through a Feodor Lynen Fellowship, which was jointly funded by the Center for Turbulence Research and by the Alexander von Humboldt-Foundation, Germany, is gratefully acknowledged.

References

1. Buice, C.U., Eaton, J.K.: Experimental investigation of flow through an asymmetric plane diffuser. TSD-107, Department of Mechanical Engineering, Stanford University (1997)
2. Gravemeier, V.: Scale-separating operators for variational multiscale large eddy simulation of turbulent flows. J. Comput. Phys., in press (2005)
3. Gravemeier, V.: The variational multiscale method for laminar and turbulent flow. Arch. Comput. Meth. Engrg., in press (2005)
4. Gravemeier, V.: Variational multiscale large eddy simulation of turbulent flow in a diffuser. Preprint, submitted for publication in Comput. Mech. (2005)
5. Ham, F., Apte, S., Iaccarino, G., Wu, X., Herrmann, M., Constantinescu, G., Mahesh, K., Moin, P.: Unstructured LES of reacting multiphase flows in realistic gas turbine combustors. Annual Research Briefs - 2003, Center for Turbulence Research, Stanford University and NASA Ames Research Center, 139–160 (2003)
6. Hughes, T.J.R., Feijoo, G.R., Mazzei, L., Quincy, J.-B.: The variational multiscale method - a paradigm for computational mechanics. Comput. Methods Appl. Mech. Engrg. **166**, 3–24 (1998)
7. Hughes, T.J.R., Mazzei, L., Jansen, K.E.: Large eddy simulation and the variational multiscale method. Comput. Visual. Sci. **3**, 47–59 (2000)
8. Koobus, B., Farhat, C.: A variational multiscale method for the large eddy simulation of compressible turbulent flows on unstructured meshes - application to vortex shedding. Comput. Methods Appl. Mech. Engrg. **193**, 1367–1383 (2004)
9. Obi, S., Aoki, K., Masuda, S.: Experimental and computational study of turbulent separating flow in an asymmetric plane diffuser. Ninth Symp. on Turbulent Shear Flows, Kyoto, Japan, August 16-19 (1993)
10. Wu, X., Schlueter, J., Moin, P., Pitsch, H., Iaccarino, G., Ham, F.: Computational study on the internal layer in a diffuser. Annual Research Briefs - 2004, Center for Turbulence Research, Stanford University and NASA Ames Research Center, 169–182 (2004)

Issues for a Mathematical Definition of LES

Jean-Luc Guermond[1] and Serge Prudhomme[2]

[1] Texas A&M University, College Station TX 77843, USA, and LIMSI, CNRS UPR 3251, BP 133 Orsay Cedex, France
guermond@math.tamu.edu
[2] ICES, The University of Texas at Austin, TX 78712, USA
serge@ices.utexas.edu

Summary. The mathematical foundations of Large Eddy Simulation (LES) for three-dimensional turbulent incompressible viscous flows are discussed and the notion of suitable approximations is introduced.

1 Introduction

1.1 What is LES?

Since the early work of [7], Large Eddy Simulation (LES) has become over the years an increasingly popular method, as evidenced by the vast amount of publications on the subject in the literature, and is now considered a tool of choice for simulating three-dimensional incompressible viscous flows at large Reynolds numbers. Heuristically speaking, Large Eddy models are obtained by applying a low-pass filter to the Navier–Stokes equations. The filtered equations are then similar to the original equations but for the presence of the so-called subgrid scale stresses accounting for the influence of the small scales onto the large ones. Assuming that the behavior of the small scales is almost universal, the objective of LES is to model the subgrid scale stresses (the so-called closure problem) and to compute the dynamics of the large scales by using the filtered equations. Although this description of LES is widely accepted, it nevertheless falls short of an unambiguous mathematical theory. Our impression is that LES is at the present time a fuzzy concept. Some authors think of LES as the solution to the filtered equations whereas others think of it as finite-dimensional approximations thereof. Others expect LES to reproduce the statistics of the large scales instead of approximating individual solutions. It is also common practice to invoke the filtering of length scales without defining the filter being used or to outright ignore the concept of filter when modeling the subgrid scale tensor. Another common unjustified practice consists of assuming that the filtering length scale is equal to the mesh size of the approximation method that is used, regardless on the method in question.

In an attempt to address some of the above issues, we are currently developing a research program aiming at constructing a framework for a mathematical theory of LES. The present paper makes a first step is this direction by introducing the concept of suitable approximation (see mS 2.1). We show that the construction of suitable approximations shares many heuristic features with what is often referred to in the engineering literature as LES modeling. The proposal made in this paper is that the notion of suitable approximations be a concept that, together with other mathematical criteria yet to be clearly identified, should be seriously considered as part of any future mathematical definition of LES.

1.2 Suitable weak solutions

Let $\Omega \subset \mathbb{R}^3$ be an open smooth, bounded, connected domain occupied by a viscous fluid. Let $(0,T)$ be a time interval. It is generally accepted that the Navier–Stokes equations accurately model the behavior of turbulent incompressible flows of the fluid in Ω:

$$\begin{cases} \partial_t \mathbf{u} + \mathbf{u}\cdot\nabla\mathbf{u} + \nabla p - \nu\nabla^2\mathbf{u} = \mathbf{f} & \text{in } Q_T, \\ \nabla\cdot\mathbf{u} = 0 \quad \text{in } Q_T, \qquad \mathbf{u}|_\Gamma = 0 \text{ or } \mathbf{u} \text{ is periodic}, \qquad \mathbf{u}|_{t=0} = u_0, \end{cases} \quad (1)$$

where $\mathbf{u}$ and p are the velocity and the pressure respectively, $Q_T = \Omega \times (0,T)$, Γ is the boundary of Ω, $\mathbf{u}_0$ the solenoidal initial data, $\mathbf{f}$ a source term, ν the viscosity, and the density is chosen equal to unity. The problem is nondimensionalized, i.e., ν is the inverse of the Reynolds number.

To implicitly account for boundary conditions, we introduce

$$\mathbf{X} = \begin{cases} \mathbf{H}_0^1(\Omega) & \text{If Dirichlet conditions,} \\ \mathbf{H}_\#^1(\Omega) = \{\mathbf{v} \in \mathbf{H}^1(\Omega), \, \mathbf{v} \text{ periodic}\} & \text{If periodic conditions.} \end{cases} \quad (2)$$

$$\mathbf{V} = \{\mathbf{v} \in \mathbf{X}, \, \nabla\cdot\mathbf{v} = 0\}, \qquad \mathbf{H} = \overline{\mathbf{V}}^{\mathbf{L}^2} \quad (3)$$

Henceforth we focus our interest on suitable weak solutions to (1), [10].

Definition 1. *A weak solution to the Navier–Stokes equation $(\mathbf{u}, p)$ is suitable if $\mathbf{u} \in L^2(0,T;\mathbf{X}) \cap L^\infty(0,T;\mathbf{L}^2(\Omega))$, $p \in L^{\frac{5}{4}}(Q_T)$ and the local energy balance*

$$\partial_t(\tfrac{1}{2}\mathbf{u}^2) + \nabla\cdot((\tfrac{1}{2}\mathbf{u}^2 + p)\mathbf{u}) - \nu\nabla^2(\tfrac{1}{2}\mathbf{u}^2) + \nu(\nabla\mathbf{u})^2 - \mathbf{f}\cdot\mathbf{u} \leq 0 \quad (4)$$

is satisfied in the distributional sense in Q_T.

To the present time, the best partial regularity result available for (1) is the so-called Caffarelli-Kohn-Nirenberg Theorem [1] proving that the one-dimensional Hausdorff measure of the set of singularities of a suitable weak solution is zero. By analogy with nonlinear conservative laws, (4) can be viewed as an entropy-like condition which may (hopefully?) selects the physical solutions of (1). Whether suitable weak solutions are indeed classical is not known. Moreover, despite the fact that the result of the CKN Theorem also holds for weak solutions[6], it is not known whether weak solutions are in fact suitable.

2 Suitable approximations

2.1 Suitable approximations

A general definition for LES is out of the scope of the present paper, but we believe that a reasonable definition should at least be founded on the following criteria: (1) A LES approximation should be finite-dimensional, i.e., it should be computable; (2) A LES approximation should solve a problem which is consistent with the Navier–Stokes equations; (3) A sequence of LES approximations should select a physical solution of the Navier–Stokes equations under the appropriate limiting process, i.e., one which is suitable.

We collect the above three criteria by defining the notion of suitable approximation as follows:

Definition 2. *A sequence* $(\mathbf{u}_\gamma, p_\gamma)_{\gamma>0}$ *with* $\mathbf{u}_\gamma \in L^\infty(0,T;\mathbf{L}^2(\Omega)) \cap L^2(0,T;\mathbf{X})$ *and* $p_\gamma \in \mathcal{D}'((0,T), L^2(\Omega))$ *is said to be a suitable approximation to (1) if*

i. *There are two finite-dimensional vectors spaces* $\mathbf{X}_\gamma \subset \mathbf{X}$ *and* $M_\gamma \subset L^2(\Omega)$ *such that* $\mathbf{u}_\gamma \in \mathcal{C}^0([0,T];\mathbf{X}_\gamma)$ *and* $p_\gamma \in L^2((0,T); M_\gamma)$ *for all* $T > 0$.
ii. *The sequence converges (up to subsequences) to a weak solution of (1), say* $\mathbf{u}_\gamma \rightharpoonup \mathbf{u}$ *weakly in* $L^2(0,T;\mathbf{X})$ *and* $p_\gamma \to p$ *in* $\mathcal{D}'((0,T), L^2(\Omega))$.
iii. *The weak solution* $(\mathbf{u}, p)$ *is suitable.*

2.2 Practical construction of suitable approximations

In practice, the construction of a suitable approximations can be decomposed into the following three steps:

(1) Construction of what we hereafter call the pre–LES–model. This step consists of regularizing the Navier–Stokes equations by introducing a regularization parameter ε associated with some filtering of the Navier–Stokes equations. This parameter is a user-defined length scale of the smallest eddies that are allowed to be nonlinearly active in the flow. The purpose of the regularization technique is to yield a well-posed problem for all times. Moreover, the limit solution of the pre–LES–model must be a weak solution to the Navier–Stokes equations as $\varepsilon \to 0$ and should be suitable. The pre–LES–model can be thought of as a filtered version of the Navier–Stokes equations where the subgrid scale stresses have been modeled in such a way that the resulting PDE is well-posed and yields a unique weak solution that converges (up to subsequences) to a suitable weak solution to the Navier–Stokes equations.

(2) Discretization of the pre–LES–model. This step introduces the mesh-size parameter h associated with the size of the smallest scale that can be represented in the finite-dimensional spaces $\mathbf{X}_\gamma$, M_γ; roughly $\dim(\mathbf{X}_\gamma) = \mathcal{O}((L/h)^3)$ where $L = \mathrm{diam}(\Omega)$.

(3) Determination of a (possibly maximal) relationship between ε and h. The large eddy scale ε and the mesh size h must be selected in such a way that the sequence of discrete solutions is ensured to converge to a suitable solution of the Navier–Stokes equations when $\varepsilon \to 0$ and $h \to 0$. In the above definition the parameter γ is a yet to be specified combination of the two parameters h and ε that reminds us that the process $\lim_{\varepsilon \to 0, h \to 0}$ is a distinguished limit.

3 Review of existing pre–LES–models

We show in this section that some of the regularization techniques recognized in the literature as LES models are indeed pre–LES–models in the sense of our definition, i.e., they all select suitable solutions as $\varepsilon \to 0$.

3.1 Hyperviscosity

Lions [9] proposed the following hyperviscosity model:

$$\begin{cases} \partial_t \mathbf{u}_\varepsilon + \mathbf{u}_\varepsilon \cdot \nabla \mathbf{u}_\varepsilon + \nabla p_\varepsilon - \nu \nabla^2 \mathbf{u}_\varepsilon + \varepsilon^{2\alpha}(-\nabla^2)^\alpha \mathbf{u}_\varepsilon = \mathbf{f} & \text{in } Q_T, \\ \nabla \cdot \mathbf{u}_\varepsilon = 0 & \text{in } Q_T, \\ \mathbf{u}_\varepsilon|_\Gamma, \ldots, \partial_n^{\alpha-1} \mathbf{u}_\varepsilon|_\Gamma = 0, \quad \text{or } \mathbf{u}_\varepsilon \text{ is periodic} \quad \mathbf{u}|_{t=0} = \mathbf{u}_0, \end{cases} \quad (5)$$

where $\varepsilon > 0$ and α is an integer. Hyperviscosity models are frequently used in so-called LES simulations of oceanic and atmospheric flows or to control the Navier–Stokes equations. The appealing aspects of this regularization are that it yields a well-posed problem in the classical sense when $\alpha \geq \frac{5}{4}$ in three space dimensions and that limit solutions as $\varepsilon \to 0$ are suitable.

3.2 Leray mollification

A simple construction yielding suitable solutions has indeed been proposed by Leray [8] before this very notion was introduced in the literature.

Assume that Ω is the three-dimensional torus $(0, 2\pi)^3$ and let $(\phi_\varepsilon)_{\varepsilon > 0}$ be a sequence of non-negative mollifying functions. Leray suggested to regularize the Navier–Stokes equations as follows:

$$\begin{cases} \partial_t \mathbf{u}_\varepsilon + (\phi_\varepsilon * \mathbf{u}_\varepsilon) \cdot \nabla \mathbf{u}_\varepsilon + \nabla p_\varepsilon - \nu \nabla^2 \mathbf{u}_\varepsilon = \phi_\varepsilon * \mathbf{f}, \\ \nabla \cdot \mathbf{u}_\varepsilon = 0, \quad \mathbf{u}_\varepsilon \text{ is periodic}, \quad \mathbf{u}_\varepsilon|_{t=0} = \phi_\varepsilon * \mathbf{u}_0. \end{cases} \quad (6)$$

The mollification device has been introduced by Leray to prove the existence of weak solutions to (1). Quite amazingly not only the pair $(\mathbf{u}_\varepsilon, p_\varepsilon)$ converges to a weak solution to (1), but the weak solution in question is also suitable. Roughly speaking, the convolution process removes scales that are smaller than ε. Hence, by using $\phi_\varepsilon * \mathbf{u}_\varepsilon$ as the advection velocity, scales smaller than ε are not allowed to be nonlinearly active. This is a feature shared by most LES models.

3.3 Leray-α model

A variant of the Leray mollification consists of the so-called Leray–α model

$$\begin{cases} \partial_t \mathbf{u}_\varepsilon + \overline{\mathbf{u}_\varepsilon} \cdot \nabla \mathbf{u}_\varepsilon - \nu \nabla^2 \mathbf{u}_\varepsilon + \nabla \pi_\varepsilon = \mathbf{f}, & \mathbf{u}_\varepsilon|_\Gamma = 0 \text{ or } \mathbf{u}_\varepsilon \text{ is periodic}, \\ (I - \varepsilon^2 \nabla^2) \overline{\mathbf{u}_\varepsilon} = \mathbf{u}_\varepsilon, & \overline{\mathbf{u}_\varepsilon}|_\Gamma = 0 \text{ or } \overline{\mathbf{u}_\varepsilon} \text{ is periodic}, \\ \nabla \cdot \overline{\mathbf{u}_\varepsilon} = 0, & \mathbf{u}_\varepsilon|_{t=0} = \mathbf{u}_0, \end{cases} \quad (7)$$

as introduced in [2]. Once again, regularization yields existence and uniqueness in the large. Moreover, when periodic boundary conditions are enforced the pair $(\mathbf{u}_\varepsilon, p_\varepsilon)$ converges, up to subsequences, to a suitable solution.

3.4 Nonlinear Galerkin method (NLGM)

We focus in this section on the Nonlinear Galerkin Method as introduced in [3]. Let Ω be the torus $(0, 2\pi)^3$. Let $\mathbb{P}_N$ be the set of trigonometric polynomials of partial degree at most N: $\mathbb{P}_N = \left\{ p(\mathbf{x}) = \sum_{|\mathbf{k}|_\infty \leq N} c_\mathbf{k} e^{i\mathbf{k} \cdot \mathbf{x}}, c_\mathbf{k} = \overline{c}_{-\mathbf{k}} \right\}$, and denote by $\dot{\mathbb{P}}_N$ the subspace of $\mathbb{P}_N$ composed of the trigonometric polynomials of zero mean value. For any $\mathbf{k} \in \mathbb{Z}$, we denote by $|\mathbf{k}|$ the Euclidean norm of $\mathbf{k}$ and by $|\mathbf{k}|_\infty$ the maximum norm. We denote by $\overline{z}$ the conjugate of z. Let $\varepsilon > 0$ be a large eddy scale. Let us set $N = \frac{1}{\varepsilon}$ (or the integer the closest to $\frac{1}{\varepsilon}$). We now introduce the following finite-dimensional vector spaces:

$$\mathbf{X}_\varepsilon = \dot{\mathbb{P}}_N, \quad \text{and} \quad M_\varepsilon = \dot{\mathbb{P}}_N, \quad (8)$$

Let $P_\varepsilon : \mathbf{H}^1_\#(\Omega) \ni \sum_{\mathbf{k} \in \mathbb{Z}} \mathbf{v}_\mathbf{k} e^{i\mathbf{k} \cdot \mathbf{x}} \longmapsto \sum_{|\mathbf{k}|_\infty \leq N} \mathbf{v}_\mathbf{k} e^{i\mathbf{k} \cdot \mathbf{x}} \in \mathbb{P}_N$ be the usual truncation operator. All fields $\mathbf{v}$ can be decomposed as follows: $\mathbf{v} = P_\varepsilon \mathbf{v} + (1 - P_\varepsilon) \mathbf{v}$. The component $P_\varepsilon \mathbf{v}$ in $\mathbf{X}_\varepsilon$ is called the large scale component of $\mathbf{v}$ and the remainder $(1 - P_\varepsilon) \mathbf{v}$ is called the small scale component.

The nonlinear Galerkin method can be recast into the following form: Seek $\mathbf{u}_\varepsilon$ and p_ε in the Leray class such that

$$\begin{cases} \partial_t P_\varepsilon \mathbf{u}_\varepsilon - \nu \nabla^2 \mathbf{u}_\varepsilon + P_\varepsilon \mathbf{u}_\varepsilon \cdot \nabla \mathbf{u}_\varepsilon + \nabla p_\varepsilon, = \mathbf{f}, \\ \nabla \cdot \mathbf{u}_\varepsilon = 0, \quad P_\varepsilon \mathbf{u}_\varepsilon|_{t=0} = P_\varepsilon \mathbf{u}_0. \end{cases} \quad (9)$$

It is then possible to prove that (9) has a unique solution and that this solution converges, up to subsequences, to a suitable weak solution of (1).

4 Discretization

The purpose of this section is to introduce discrete versions of some of the pre–LES–models described above. In each case, we show that the requirement for the approximate solutions to be suitable approximations determines the relationship between the mesh size h and the large eddy scale ε, thus solving a question very often left open or simply heuristically answered in the LES literature.

4.1 The discrete hyperviscosity model

We turn our attention to the hyperviscosity model introduced in mS3.1 and we construct a Galerkin-Fourier approximation assuming that Ω is the torus $(0, 2\pi)^3$. Let $N \in \mathbb{N}\setminus\{0\}$ and introduce the meshsize and large eddy scale

$$h = N^{-1}, \qquad \varepsilon = h^\theta, \tag{10}$$

where $0 < \theta < 1$. We set $N_i = \frac{1}{\varepsilon} = N^\theta$. To approximate the velocity and the pressure fields we introduce the following finite-dimensional vector spaces:

$$\mathbf{X}_h = \dot{\mathbb{P}}_N, \quad \text{and} \quad M_h = \dot{\mathbb{P}}_N. \tag{11}$$

We introduce $Q(\mathbf{x}) = (2\pi)^{-3} \sum_{N_i \leq |\mathbf{k}|_\infty \leq N} |\mathbf{k}|^{2\alpha} e^{i\mathbf{k}\cdot\mathbf{x}}$ where $\alpha > \frac{5}{4}$. The spectral hyperviscosity model consists of the following: Seek $\mathbf{u}_h \in \mathcal{C}^0([0,T]; \mathbf{X}_h)$ and $p_h \in L^2([0,T]; M_h)$ such that $\forall \mathbf{v} \in \mathbf{X}_h$, $\forall q \in M_h$, and a.e. t in $(0, T)$,

$$\begin{cases} (\partial_t \mathbf{u}_h, v) + (\mathbf{u}_h \cdot \nabla \mathbf{u}_h, v) - (p_h, \nabla \cdot \mathbf{v}) + \nu(\nabla \mathbf{u}_h, \nabla \mathbf{v}) + \varepsilon_N^{2\alpha}(Q * \mathbf{u}_h, \mathbf{v}) = (\mathbf{f}, \mathbf{v}), \\ (\nabla \cdot \mathbf{u}_N, q) = 0, \forall t \in (0, T], \qquad (\mathbf{u}_N, \mathbf{v})|_{t=0} = (\mathbf{u}_0, \mathbf{v}). \end{cases} \tag{12}$$

The following result is proved in [5]:

Theorem 1. *Let* $\mathbf{f} \in L^2(0, T; \mathbf{L}^2(\Omega))$ *and* $\mathbf{u}_0 \in \mathbf{H}^\alpha(\Omega) \cap \mathbf{V}$. *Assume that* $0 < \theta < \frac{4\alpha - 5}{4\alpha}$ *if* $\alpha \leq \frac{3}{2}$, *or* $0 < \theta < \frac{2(\alpha - 1)}{2\alpha + 3}$ *otherwise, then the pair* $(\mathbf{u}_h, p_h)$ *is a suitable approximation to (1).*

4.2 The discrete Leray, Leray-α, and NLGM models

Let us keep the same notation as above; in particular, $h = N^{-1}$ and $\varepsilon = h^\theta$. Let us approximate $\phi_\varepsilon * \mathbf{u}_\varepsilon$ in (6) by the truncated Fourier series of $\mathbf{u}_\varepsilon$. Then, the discrete Leray model takes the following form: Seek $\mathbf{u}_h \in \mathcal{C}^0([0,T]; \mathbf{X}_h)$ and $p_h \in L^2([0,T]; M_h)$ such that $\forall \mathbf{v} \in \mathbf{X}_h$, $\forall q \in M_h$, and a.e. t in $(0, T)$,

$$\begin{cases} (\partial_t \mathbf{u}_h, \mathbf{v}) + (P_{\varepsilon_N} \mathbf{u}_h \cdot \nabla \mathbf{u}_h, \mathbf{v}) - (p_h, \nabla \cdot \mathbf{v}) + \nu(\nabla \mathbf{u}_h, \nabla \mathbf{v}) = (\mathbf{f}, \mathbf{v}), \\ (\nabla \cdot \mathbf{u}_h, q) = 0, \qquad (\mathbf{u}, \mathbf{v})|_{t=0} = (\mathbf{u}_0, \mathbf{v}). \end{cases} \tag{13}$$

Using again the Fourier setting, the discrete version of the Leray-α model (7) takes the following form: Seek $\mathbf{u}_h \in \mathcal{C}^0([0,T];\mathbf{X}_h)$ and $p_h \in L^2([0,T];M_h)$ such that for all $\mathbf{v} \in \mathbf{X}_h$, for all $q \in M_h$, and a.e. t in $(0,T)$,

$$\begin{cases} (\partial_t \mathbf{u}_N, \mathbf{v}) + (\bar{\mathbf{u}}_h \cdot \nabla \mathbf{u}_h, \mathbf{v}) - (p_h, \nabla \cdot \mathbf{v}) + \nu(\nabla \mathbf{u}_h, \nabla \mathbf{v}) = (\mathbf{f}, \mathbf{v}), \\ (\bar{\mathbf{u}}_h, \mathbf{v}) + \varepsilon^2(\nabla \bar{\mathbf{u}}_h, \nabla \mathbf{v}) = (\mathbf{u}_h, \mathbf{v}), \\ (\nabla \cdot \mathbf{u}_h, q) = 0, \qquad (\mathbf{u}_h, \mathbf{v})|_{t=0} = (\mathbf{u}_0, \mathbf{v}), \end{cases} \tag{14}$$

Still retaining the Fourier setting, the discrete version of NLGM (9) is as follows: Seek $\mathbf{u}_h \in \mathcal{C}^0([0,T];\mathbf{X}_h)$, and $p_h \in L^2(0,T;M_h)$ such that $\forall t \in (0,T]$, $\forall \mathbf{v} \in \mathbf{X}_h$, $\forall q \in M_h$, and a.e. t in $(0,T)$,

$$\begin{cases} (\partial_t P_\varepsilon \mathbf{u}_h, \mathbf{v}) + \nu(\nabla \mathbf{u}_h, \nabla \mathbf{v}) + (P_\varepsilon \mathbf{u}_h \cdot \nabla \mathbf{u}_h, \mathbf{v}) - (p_h, \nabla \cdot \mathbf{v}) = (\mathbf{f}, \mathbf{v}), \\ (\nabla \cdot \mathbf{u}_h, q) = 0, \qquad \mathbf{u}_h|_{t=0} = P_\varepsilon \mathbf{u}_0. \end{cases} \tag{15}$$

The following result holds for the three above approximation techniques:

Theorem 2. *Let $\mathbf{f} \in L^2(0,T;\mathbf{L}^2(\Omega))$ and $\mathbf{u}_0 \in \mathbf{H}$. If $0 < \theta < \frac{2}{3}$, the pair $(\mathbf{u}_h, p_h)$ is a suitable approximation to (1).*

4.3 The case of DNS

A natural question that comes to mind is whether a sequence of Direct Numerical Solutions (DNS) is a suitable approximation. To clarify this issue, let $\mathbf{X}_h \subset \mathbf{X}$ and $M_h \subset L^2(\Omega)$ be two finite-dimensional vector spaces and consider the following Galerkin approximation: Seek $\mathbf{u}_h \in \mathcal{C}^0([0,T];\mathbf{X}_h)$ and $p_h \in L^2([0,T];M_h)$ such that for all $\mathbf{v}_h \in \mathbf{X}_h$, all $q_h \in M_h$, and a.e. $t \in (0,T)$

$$\begin{cases} (\partial_t \mathbf{u}_h, \mathbf{v}) + b_h(\mathbf{u}_h, \mathbf{u}_h, \mathbf{v}) - (p_h, \nabla \cdot \mathbf{v}) + \nu(\nabla \mathbf{u}_h, \nabla \mathbf{v}) = (\mathbf{f}, \mathbf{v}), \\ (q, \nabla \cdot \mathbf{u}_h) = 0, \quad \text{and} \quad (\mathbf{u}_h|_{t=0}, \mathbf{v}) = (\mathbf{u}_0, \mathbf{v}), \end{cases} \tag{16}$$

where b_h accounts for the nonlinear term and can be written as follows:

$$b_h(\mathbf{u}, \mathbf{v}, \mathbf{w}) = \begin{cases} (\mathbf{u} \cdot \nabla \mathbf{v} + \frac{1}{2} \mathbf{v} \nabla \cdot \mathbf{u}, \mathbf{w}), & \text{or} \\ ((\nabla \times \mathbf{u}) \times \mathbf{v} + \frac{1}{2} \nabla(\mathcal{K}_h(\mathbf{u} \cdot \mathbf{v})), \mathbf{w}), \end{cases} \tag{17}$$

where $\mathcal{K}_h : L^2(\Omega) \longrightarrow M_h$ is a linear L^2-stable interpolation operator.

Owing to standard a priori estimates uniform in h, it is clear that the pair $(\mathbf{u}_h, p_h)$ complies with items (i) and (ii) of Definition 2. Although it is not known in general whether such a construction yields a suitable solution at the limit, it has been proved in [4] that it is indeed the case when low-order finite elements are used and periodic boundary conditions are enforced. More specifically, let $\pi_h : L^2(\Omega) \longrightarrow \mathbf{X}_h$ be the L^2-projection onto $\mathbf{X}_h$. We assume that there exists $c > 0$ independent of h such that

$$\forall q_h \in M_h, \quad \|\nabla q_h\|_{L^2} \le c \|\pi_h \nabla q_h\|_{L^2}. \tag{18}$$

This hypothesis is shown to hold in, at least, the following two situations (1) $\mathbf{X}_h$ is composed of $\mathbb{P}_1$–Bubble H^1-conforming finite elements and M_h is composed of $\mathbb{P}_1$ H^1-conforming finite elements; (2) $\mathbf{X}_h$ is composed of $\mathbb{P}_2$ H^1-conforming finite elements, M_h is composed of $\mathbb{P}_1$ H^1-conforming finite elements, and no tetrahedron has more than 3 edges on $\partial \Omega$.

Definition 3. *We say that $\mathbf{X}_h$ (resp. M_h) has the discrete commutator property if there exists $\mathcal{I}_h \in \mathcal{L}(\mathbf{H}^1_\#(\Omega); \mathbf{X}_h)$ (resp. $\mathcal{J}_h \in \mathcal{L}(L^2(\Omega); M_h)$) such that $\forall \phi$ in $W^{2,\infty}_\#(\Omega)$ (resp. $\forall \phi$ in $W^{1,\infty}_\#(\Omega)$) and $\forall v_h \in \mathbf{X}_h$ (resp. $\forall q_h \in M_h$)*

$$\|\phi v_h - \mathcal{I}_h(\phi v_h)\|_{H^l} \le c h^{1+m-l} \|v_h\|_{H^m} \|\phi\|_{W^{m+1,\infty}}, \quad 0 \le l \le m \le 1$$
$$\|\phi q_h - \mathcal{J}_h(\phi q_h)\|_{L^2} \le c h \|q_h\|_{L^2} \|\phi\|_{W^{1,\infty}}.$$

Standard H^1-conforming finite element spaces actually possess the discrete commutator property. This is not the case of Fourier-based approximation spaces since Fourier series do not have local interpolation properties.

The main result is the following (see [4] for details)

Theorem 3. *Under the above hypotheses, if $\mathbf{X}_h$ and M_h have the discrete commutator property, the pair $(\mathbf{u}_h, p_h)$ is a suitable approximation to (1).*

This result underlines that the nature of the approximation technique that is used plays a key role in the construction of suitable approximations. Low-order approximations seem to do the trick without requiring extra regularization provided the nonlinear term is written in skew-symmetric form, whereas spectral methods need smoothing or extra viscosities. This is related to the fact that spectral methods suffer from the Gibbs phenomenon. This result tends to confirm statements sometimes made in the literature that, when using low-order methods, it is preferable to let the "numerical diffusion do the job" than to perform any LES modeling. This result is also a cautionary notice to LES practitioners that heuristic arguments in the Fourier space may not be equivalent to arguments in the physical space. This point is important since a lot of heuristic LES argumentation is done in the Fourier space.

References

1. Caffarelli, L., Kohn, R., Nirenberg, L.: Partial regularity of suitable weak solutions of the Navier-Stokes equations. Comm. Pure Appl. Math., **35**(6), 771–831 (1982)
2. Cheskidov, A., Holm, D.D., Olson, E., Titi, E.S.: On a Leray-α model of turbulence. Royal Society London, Proceedings, Series A, Mathematical, Physical & Engineering Sciences, **461**, 629–649 (2005)
3. Foias, C., Manley, O., Temam, R.: Modelling of the interaction of small and large eddies in two-dimensional turbulent flows. RAIRO M^{2}AN, **22**, 93–114 (1988)

4. Guermond, J.-L.: Finite-element-based Faedo-Galerkin weak solutions to the Navier–Stokes equations in the three-dimensional torus are suitable. J. Math. Pures Appl., (2005) In press.
5. Guermond, J.-L., Prudhomme, S.: Mathematical analysis of a spectral hyperviscosity LES model for the simulation of turbulent flows. Math. Model. Numer. Anal. (M2AN), **37**(6), 893–908 (2003)
6. Cheng He.: On partial regularity for weak solutions to the Navier-Stokes equations. J. Funct. Anal., **211**(1), 153–162 (2004)
7. Leonard, A.: Energy cascade in Large-Eddy simulations of turbulent fluid flows. Adv. Geophys., **18**, 237–248 (1974)
8. Leray, J.: Essai sur le mouvement d'un fluide visqueux emplissant l'espace. Acta Math., **63**, 193–248 (1934)
9. Lions, J.-L.: Sur certaines équations paraboliques non linéaires. Bull. Soc. Math. France, **93**, 155–175 (1965)
10. Scheffer, V.: Hausdorff measure and the Navier-Stokes equations. Comm. Math. Phys., **55**(2), 97–112 (1977)

Stabilized FEM with Anisotropic Mesh Refinement for the Oseen Problem

Gert Lube[1], Tobias Knopp[2] and Ralf Gritzki[3]

[1] Math. Fakultät, Georg-August-Universität, D-37083 Göttingen, Germany
   `lube@math.uni-goettingen.de`
[2] DLR (German Aerospace Center), AS-NV, D-37073 Göttingen, Germany
   `Tobias.Knopp@dlr.de`
[3] Fakultät für Maschinenwesen, TGA, TU Dresden, D-01062 Dresden, Germany
   `gritzki@tga.tu-dresden.de`

Summary. Nonstationary incompressible flow problems can be split into auxiliary problems of Oseen type. We present the analysis of conforming stabilized Galerkin methods of SUPG/PSPG-type with equal-order interpolation of velocity/pressure and with emphasis on anisotropic mesh refinement in boundary layers. We prove a modified inf-sup condition with a constant independent of the viscosity and of critical parameters of the mesh. Numerical tests confirm the results.

1 Introduction

We consider the nonstationary, incompressible Navier-Stokes problem

$$\partial_t \mathbf{u} - \nu \Delta \mathbf{u} + (\mathbf{u} \cdot \nabla)\mathbf{u} + \nabla p = \mathbf{f} \tag{1}$$

$$\nabla \cdot \mathbf{u} = 0 \tag{2}$$

for velocity $\mathbf{u}$ and pressure p in a domain $\Omega \subset \mathbf{R}^d$, $d \leq 3$. In an outer loop, an A-stable low-order method (possibly with time step control) is applied. In an inner loop, we decouple and linearize the resulting system using a Newton-type iteration per time step. This leads to problems of Oseen type:

$$-\nu \Delta \mathbf{u} + (\mathbf{b} \cdot \nabla)\mathbf{u} + c\mathbf{u} + \nabla p = \mathbf{f} \quad \text{in } \Omega \tag{3}$$

$$\nabla \cdot \mathbf{u} = 0 \quad \text{in } \Omega. \tag{4}$$

We consider stabilized conforming finite element (FE) schemes with equal-order interpolation of velocity/pressure for problem (3)–(5) with emphasis on anisotropic mesh refinement in boundary layers. The classical streamline upwind and pressure stabilization (SUPG/PSPG) techniques for the incompressible Navier-Stokes problem for equal-order interpolation [4], together with additional stabilization of the divergence constraint (5), are well-understood on isotropic meshes [12].

Much less is known about the analysis in case of equal-order interpolation schemes with anisotropic mesh refinement for incompressible flow problems. The Stokes problem has been considered in [3] for the Q1/Q1-case and in [11] for the P1/P1-case. The extension to the Oseen problem seems to be new. Numerical experiments for the full Navier-Stokes problem, e.g. in [8, 6], show the applicability of anisotropic mesh refinement for low-order schemes.

The stabilized FEM for problem (3)–(5) is given in Sect. 2. In Sect. 3 we focus on hybrid meshes with anisotropic layer refinement of tensor product type and smooth transition to (unstructured) isotropic meshes away from the layer. Section 4 is devoted to error estimates and to the design of stabilization parameters. Numerical results are shown in Sect. 5. Full proofs are given in [2].

2 Stabilized FEM for linearized Navier-Stokes problem

We consider the Oseen model, for brevity with homogeneous Dirichlet data:

$$L_{os}(\mathbf{b};\mathbf{u},p) := -\nu\Delta\mathbf{u} + (\mathbf{b}\cdot\nabla)\mathbf{u} + c\mathbf{u} + \nabla p = \mathbf{f} \quad \text{in } \Omega, \tag{5}$$
$$\nabla\cdot\mathbf{u} = 0 \quad \text{in } \Omega, \tag{6}$$
$$\mathbf{u} = \mathbf{0} \quad \text{on } \partial\Omega \tag{7}$$

with $\mathbf{b} \in [H^1(\Omega)]^d$, $(\nabla\cdot\mathbf{b})(x) = 0$, $\mathbf{f} \in [L^2(\Omega)]^d$ and constants $\nu > 0$, $c \geq 0$. The variational formulation reads: find $U := \{\mathbf{u},p\} \in \mathbf{W} := \mathbf{V}\times\mathbf{Q} := [H_0^1(\Omega)]^d \times L_0^2(\Omega)$ with $L_0^2(\Omega) := \{q \in L^2(\Omega) \mid \int_\Omega q\,dx = 0\}$, s.t.

$$\mathcal{A}(\mathbf{b};U,V) = \mathcal{L}(V) \quad \forall V = \{\mathbf{v},q\} \in \mathbf{V}\times\mathbf{Q}, \tag{8}$$
$$\mathcal{A}(\mathbf{b};U,V) := (\nu\nabla\mathbf{u},\nabla\mathbf{v})_\Omega + ((\mathbf{b}\cdot\nabla)\mathbf{u} + c\mathbf{u},\mathbf{v})_\Omega$$
$$-(p,\nabla\cdot\mathbf{v})_\Omega + (q,\nabla\cdot\mathbf{u})_\Omega, \tag{9}$$
$$\mathcal{L}(V) := (\mathbf{f},\mathbf{v})_\Omega. \tag{10}$$

Let $\mathcal{T}_h$ be an admissible triangulation of the polyhedron Ω where each $T \in \mathcal{T}_h$ is a smooth bijective image $T = F_T(\hat{T})$ of a unit element $\hat{T}$ (unit simplex or hypercube in $\mathbf{R}^d$ or, for $d = 3$, the unit triangular prism). A mixture (with appropiate reference elements for each type) is admitted. Consider Lagrangian FE of order $r \in \mathbf{N}$, i.e., $\mathcal{P}_r(\hat{T})$ on $\hat{T}$ contains the polynomial set $\mathcal{P}_r$. We set

$$X_h^r = \{v \in C(\bar{\Omega}) \mid v|_T \circ F_T \in \mathcal{P}_r(\hat{T}) \;\forall T \in \mathcal{T}_h\} \tag{11}$$

and introduce conforming equal-order FE spaces for velocity and pressure

$$\mathbf{V}_h^r := \left[H_0^1(\Omega) \cap X_h^r\right]^d, \quad \mathbf{Q}_h^r := L_0^2(\Omega) \cap X_h^r, \quad r \in \mathbf{N}. \tag{12}$$

The Galerkin method reads: find $U = \{\mathbf{u},p\} \in \mathbf{W}_h^{r,r} := \mathbf{V}_h^r \times \mathbf{Q}_h^r$, s.t.

$$\mathcal{A}(\mathbf{b};U,V) = \mathcal{L}(V) \quad \forall V = \{\mathbf{v},q\} \in \mathbf{W}_h^{r,r}. \tag{13}$$

Well-known sources of instabilities of the Galerkin FEM (13) stem from dominating advection and from the violation of the discrete inf-sup or LBB-condition for $\mathbf{V}_h^r \times \mathbf{Q}_h^r$. Note that, in case of anisotropic elements, the discrete inf-sup constant is often not robust w.r.t. the maximal aspect ratio.

A standard approach to stabilize the Galerkin scheme is a combination of pressure stabilization (PSPG) with streamline-upwind stabilization (SUPG) together with a stabilization of the divergence constraint, the so-called grad-div stabilization. The method reads: find $U = \{\mathbf{u}, p\} \in \mathbf{W}_h^{r,r}$, s.t.

$$\mathcal{A}_s(\mathbf{b}; U, V) = \mathcal{L}_s(V) \quad \forall V = \{\mathbf{v}, q\} \in \mathbf{W}_h^{r,r}, \tag{14}$$

$$\mathcal{A}_s(\mathbf{b}; U, V) := \mathcal{A}(\mathbf{b}; U, V) + \sum_{T \in \mathcal{T}_h} \gamma_T \, (\nabla \cdot \mathbf{u}, \nabla \cdot \mathbf{v})_T$$

$$+ \sum_{T \in \mathcal{T}_h} (L_{os}(\mathbf{b}; \mathbf{u}, p), \, \delta_T ((\mathbf{b} \cdot \nabla)\mathbf{v} + \nabla q))_T \tag{15}$$

$$\mathcal{L}_s(V) := \mathcal{L}(V) + \sum_{T \in \mathcal{T}_h} (\mathbf{f}, \delta_T ((\mathbf{b} \cdot \nabla)\mathbf{v} + \nabla q))_T. \tag{16}$$

Remark 1. The stabilizing effect stems from control of the SUPG/ PSPG-term $\sum_T \delta_T \|(\mathbf{b} \cdot \nabla)\mathbf{u} + \nabla p\|^2_{[L^2(T)]^d}$ and of the term $\sum_T \gamma_T \|\nabla \cdot \mathbf{u}\|^2_{L^2(T)}$. Related variants are the GLS method [7] and the algebraic subgrid-scale method [5].

Consider a (possibly anisotropic) element $T \subset \mathbf{R}^d$, $d = 2, 3$, with sizes $h_{1,T} \geq \ldots \geq h_{d,T}$. A key point in the analysis is the local inverse inequality

$$\|\Delta \mathbf{w}\|_{[L^2(T)]^d} \leq \mu_{inv} h_{d,T}^{-1} \|\nabla \mathbf{w}\|_{[L^2(T)]^{d \times d}} \quad \forall \mathbf{w} \in \mathbf{V}_h^r. \tag{17}$$

Set

$$\||V\|| := \left(|[V]|^2 + \sigma \|q\|^2_{L^2(\Omega)} \right)^{1/2}, \tag{18}$$

$$|[V]|^2 := \|\sqrt{\nu} \nabla \mathbf{v}\|^2_{[L^2(\Omega)]^{d \times d}} + \|\sqrt{c}\mathbf{v}\|^2_{[L^2(\Omega)]^d}$$

$$+ \sum_{T \in \mathcal{T}_h} \left(\gamma_T \|\nabla \cdot \mathbf{v}\|^2_{L^2(T)} + \delta_T \|(\mathbf{b} \cdot \nabla)\mathbf{v} + \nabla q\|^2_{L^2(T)} \right) \tag{19}$$

where $\delta_T, \gamma_T, \sigma > 0$ are determined later on. For $\delta_T > 0$, $||\cdot||$ is a mesh-dependent norm on $\mathbf{W}_h^{r,r}$. The following result yields existence and uniqueness of the discrete solution without geometrical conditions on $\mathcal{T}_h$.

Lemma 1. *Assume the following conditions on the stabilization parameters*

$$0 < \delta_T \leq \frac{1}{2} \min \left(\frac{h_{d,T}^2}{\mu_{inv}^2 \nu}; \frac{1}{c} \right), \quad 0 \leq \gamma_T. \tag{20}$$

Then the bilinear form $\mathcal{A}_s(\mathbf{b}; \cdot, \cdot)$ *defined in (15) satisfies*

$$\mathcal{A}_s(\mathbf{b}; W_h, W_h) \geq \frac{1}{2} |[W_h]|^2, \quad \forall W_h \in \mathbf{W}_h^{r,r}. \tag{21}$$

3 Stability and convergence on hybrid meshes

We present a discrete inf-sup condition and a quasi-optimal error estimate w.r.t. $|||\cdot|||$. For the sake of clarity, we focus on hybrid meshes with anisotropic layer refinement of tensor product type (in the sense of [1, Chap. 3]) and smooth transition to a (unstructured) isotropic mesh away from layers. For simplicity, assume that the boundary layer is located at the hyperplane $x_d = 0$.

The advantage of such meshes is not only that the coordinate transformation is simplified in regions with anisotropic elements but also that certain edges/faces of the elements are orthogonal/parallel to coordinate axes.

Meshes of tensor product type in the boundary layer region consist of affine elements of tensor product type. That means the transformation of a reference element $\hat{T}$ to the element T shall have (block) diagonal form,

$$\mathbf{x} = \operatorname{diag}(A_T, \pm h_{d,T})\,\hat{\mathbf{x}} + \mathbf{a}_T \quad \text{for } d = 2,3,$$

where $\mathbf{a}_T \in \mathbf{R}^d$, $A_T = \pm h_T$ for $d = 2$ and $A_T \in \mathbf{R}^{2\times 2}$ with $|\det A_T| \sim h_{1,T}^2$, $\|A_T\| \sim h_{1,T}$, $\|A_T^{-1}\| \sim h_{1,T}^{-1}$ for $d = 3$. In this way, the element sizes $h_{1,T}, \ldots, h_{d,T}$ are implicitly defined; in particular $h_{1,T} \sim h_{2,T}$ for $d = 3$. Note further that under these assumptions the triangles/tetrahedra can be grouped into pairs/triples which form a rectangle/triangular prism of tensor product type.

Moreover, suppose that $h_{i,T} \sim h_{i,T'}$ for all T' with $\overline{T} \cap \overline{T'} \neq \emptyset$, $i = 1, \ldots, d$. This implies that the transision region between the structured and the unstructured mesh zones consists of isotropic elements only.

A critical point in the stability analysis is the following interpolation result for a modified Scott-Zhang quasi-interpolation operator $I_{h,r}^{qi} : H^1(\Omega) \to X_h^r$:

$$\|\nabla^m(v - I_{h,r}^{qi}v)\|_{L^2(T)} \leq C_{qi,m} h_{1,T}^{1-m} \|v\|_{H^1(\omega_T)}, \quad m = 0,1 \qquad (22)$$

where $\omega_T := \bigcup_{\overline{T'}\cap\overline{T}\neq\emptyset} T'$. (22) can be derived using ideas of [1, Chap. 3.4].

The error analysis requires a modified inf-sup condition w.r.t. $|||\cdot|||$, including control of the L^2-norm of the pressure. Note that, in contrast to the Galerkin method, the stability constant β below is independent of ν and of critical parameters of $\mathcal{T}_h$.

Lemma 2. *Assume the following conditions on the stabilization parameters*

$$0 < \mu_0 h_{1,T}^2 \leq \delta_T \leq \frac{1}{2}\min\left(\frac{h_{d,T}^2}{\mu_{inv}^2 \nu}; \frac{1}{c}\right), \quad 0 \leq \delta_T \|\mathbf{b}\|_{[L^\infty(T)]^d}^2 \leq \gamma_T. \qquad (23)$$

Then there exists a constant $\beta > 0$, independent of all relevant parameters s.t.

$$\inf_{W_h \in \mathbf{W}_h^{r,r}} \sup_{V_h \in \mathbf{W}_h^{r,r}} \frac{\mathcal{A}_s(\mathbf{b}; W_h, V_h)}{|||W_h|||\,|||V_h|||} \geq \beta \qquad (24)$$

with constant σ in (1) according to

$$\frac{1}{\sqrt{\sigma}} \sim \sqrt{\nu} + \sqrt{c}C_F + \frac{C_F \|\mathbf{b}\|_{[L^\infty(\Omega)]^d}}{\sqrt{\nu + cC_F^2}} + \max_T \frac{h_{1,T}\|\mathbf{b}\|_{[L^\infty(T)]^d}}{\sqrt{\nu}} + \sqrt{\gamma} + \frac{1}{\mu_0}. \quad (25)$$

Moreover, it denotes $\gamma = \max_T \gamma_T$ and C_F the Friedrichs constant.

Remark 2. The lower bound of δ_T in assumption (6) implicitly implies

$$\sqrt{\mu_0} \max_{T \in \mathcal{T}_h} \frac{h_{1,T}}{h_{d,T}} \leq \frac{1}{\mu_{inv}\sqrt{2\nu}}. \quad (26)$$

i.e., a restriction on the aspect ratio of T. A reasonable choice in boundary layers at a wall is $h_{d,T} \geq \sqrt{\nu}h_{1,T}$; thus leading to $\mu_0 = \mathcal{O}(1)$, see also Sect. 4.

We use from now on the notation $a \preceq b$, i.e., there exists a constant C, independent of all relevant parameters (ν, c, $\mathbf{h}_T$, aspect ratio, δ_T, γ_T), satisfying $a \leq C\, b$. The following continuity result shows the effect of stabilization:

Lemma 3. Let the assumptions (6) be valid. Then, for each $U = \{\mathbf{u}, p\} \in \mathbf{W}$ with $\Delta \mathbf{u}|_T \in [L^2(T)]^d$ $\forall T \in \mathcal{T}_h$ and $V_h = \{\mathbf{v}_h, q_h\} \in \mathbf{W}_h^{r,r}$ there holds

$$\mathcal{A}_s(\mathbf{b}; U, V_h) \preceq \mathcal{Q}_s(U)\, \|\|V_h\|\| \quad (27)$$

$$\mathcal{Q}_s(U) := \|[U]\| + \Big(\sum_{T \in \mathcal{T}_h} \frac{1}{\delta_T}\|\mathbf{u}\|^2_{[L^2(T)]^d}\Big)^{\frac{1}{2}} + \Big(\sum_{T \in \mathcal{T}_h} \frac{2}{\nu + \gamma_T}\|p\|^2_{L^2(T)}\Big)^{\frac{1}{2}}$$

$$+ \Big(\sum_{T \in \mathcal{T}_h} \delta_T \| -\nu \Delta \mathbf{u} + c\mathbf{u}\|^2_{[L^2(T)]^d}\Big)^{\frac{1}{2}}. \quad (28)$$

Using Lemmata 2 and 3, we obtain the following quasi-optimal error estimate.

Theorem 1. Let $U = \{\mathbf{u}, p\} \in \mathbf{W}$ and $U_h = \{\mathbf{u}_h, p_h\} \in \mathbf{W}_h^{r,r}$ be the solutions of (5)–(7) and of (14)–(16). Let $I_{h,r}U := \{I_{h,r}^u \mathbf{u}, I_{h,r}^p p\} \in \mathbf{W}_h^{r,r}$ be an appropriate interpolant for $\{\mathbf{u}, p\}$. Under assumption (6), we obtain

$$\|\|U - U_h\|\| \preceq \mathcal{Q}_s(U - I_{h,r}U). \quad (29)$$

4 Error estimates and design of stabilization parameters

Based on the quasi-optimal estimate in Theorem 1, we derive error estimates and design the parameters δ_T, γ_T with emphasis on the anisotropy of an element. Here, we assume that the solution of problem (5)–(7) is smooth enough such that the global Lagrangian interpolant can be used in Theorem 1.

Appropriate anisotropic interpolation estimates of the FE spaces X_h^r are required in order to compensate large derivatives in some direction x_d by the small element diameter $h_{d,T}$. We refer to [1] for a basic interpolation theory which relies on some geometrical conditions (maximal angle condition and the coordinate system condition) which are valid for the hybrid meshes introduced in Sec. 3. The anisotropic interpolation result for the Lagrangian interpolation operator $I_{h,r} : \mathbf{C}(\overline{T}) \to \mathcal{P}_r(T)$ reads as follows, see [1, Chap. 3].

Lemma 4. *Let $\mathcal{T}_h$ be a hybrid mesh as introduced in Section 3, and $T \in \mathcal{T}_h$. Assume that $v \in W^{\ell,p}(T)$, with $\ell \in \{1, \ldots, r+1\}$, $p \in [1, \infty]$, such that $p > 2/\ell$. Fix $m \in \{0, \ldots, \ell-1\}$. Then the following estimate holds*

$$\|v - I_{h,r}v\|_{W^{m,p}(T)} \leq C \sum_{|\alpha|=\ell-m} \mathbf{h}_T^\alpha \|D^\alpha v\|_{W^{m,p}(T)} \quad \mathbf{h}_T^\alpha := h_1^{\alpha_1} \ldots h_d^{\alpha_d}. \tag{30}$$

Corollary 1. *Let the assumptions of Theorem 1 be valid. Moreover, assume that the solution $U = \{\mathbf{u}, p\} \in \mathbf{W}$ is continuous and satisfies $\mathbf{u}|_T \in [H^k(T)]^d$, $p|_T \in H^k(T)$ with $k > 1$ for all $T \in \mathcal{T}_h$. Then, using the notation $l := \min(r, k-1)$ for the convergence order, we obtain*

$$\||U - U_h\||^2 \lesssim \sum_T \sum_{|\alpha|=l, |\beta|=1} \mathbf{h}_T^{2\alpha} \left(E_{T,\beta}^p \|D^{\alpha+\beta}p\|_{L^2(T)}^2 + E_{T,\beta}^u \|D^{\alpha+\beta}\mathbf{u}\|_{L^2(T)}^2 \right),$$
$$\tag{31}$$

$$E_{T,\beta}^p := \delta_T + \gamma_T^{-1}\mathbf{h}_T^{2\beta} \tag{32}$$

$$E_{T,\beta}^u := \nu + ch_{1,T}^2 + \gamma_T + \delta_T \|\mathbf{b}\|_{[L^\infty(T)]^d}^2 + \delta_T^{-1}\mathbf{h}_T^{2\beta}. \tag{33}$$

The mixed character of the problem requires a careful approach to fix the parameters δ_T, γ_T. . Using $\tilde{h}_T \in [h_{d,T}, h_{1,T}]$ and based on assumption (6), we propose to define the parameters according to

$$\delta_T \sim \min\left(\frac{h_{d,T}^2}{\mu_{inv}^2 \nu}; \frac{1}{c}; \frac{\tilde{h}_T}{\|\mathbf{b}\|_{(L^\infty(T))^d}} \right), \quad \gamma_T \sim \frac{\tilde{h}_T^2}{\delta_T}. \tag{34}$$

In the isotropic region Ω_{iso} away from the boundary layer, we propose to set $h_{1,T} \sim \tilde{h}_T$ which leads to the standard design and to the standard error contributions (see [7, 5]).

The parameter design in the boundary layer region Ω_{aniso} at $x_d = 0$ is more involved. From Prandtl's boundary layer theory for laminar flows, we know that p varies at most slowly with x_d, whereas $\mathbf{u}$ can have large gradients in x_d-direction. This motivates a mesh refinement in x_d-direction towards the wall by setting $h_{d,T} \sim g(x_d)h_{1,T}$ with a strongly increasing monitor function $g(\cdot)$ s.t. $g(x_d) \sim \sqrt{\nu}$ in the mesh layer nearest to the wall and $g(x_d) \sim 1$ in the transition region to the isotropic part of the hybrid mesh.

The velocity error part in the error contribution in Corollary 1 contains the critical term $\delta_T^{-1}h_T^{2\beta}$ which is at most of order $\mathcal{O}(1)$ in the mesh layer nearest to the wall at $x_d = 0$ since $h_{d,T} \sim \sqrt{\nu}h_{1,T}$. On the other hand, we observe that the stabilization parameters do not deteriorate there since $\nu^{-1}h_{d,T}^2 \sim h_{1,T}^2$.

It remains to discuss the choice of $\tilde{h}_T$. We obtain from (34) that an increasing $\tilde{h}_T$ implies an increasing γ_T, thus giving improved control of $\nabla \cdot \mathbf{u}$. On the other hand, the control parameter $\sqrt{\sigma}$ of $\|p - p_h\|_{L^2(\Omega)}$ behaves like $1/\sqrt{\sigma} \leq \max_T \sqrt{\gamma_T}$, i.e. the control of this norm gets worse with increasing

γ_T. Our favoured choice is $\tilde{h}_T = (\text{meas}(T))^{\frac{1}{d}}$, as a reasonable compromise to balance control of pressure and of divergence.

5 Application to channel flow

We present some numerical results for the Navier-Stokes problem using the research code *Parallel NS* with P1-approximations for velocity/pressure.

Consider the *laminar* stationary flow in the channel $\Omega = (0,1)^2$ with the data $\nu = 10^{-6}$, $\mathbf{b} = \mathbf{u}$, $c = 0$, $\mathbf{f} = \mathbf{0}$ and solution $p = \sqrt{\nu}(1-x)$, $\mathbf{u} = (1 - (e^{-y/\sqrt{\nu}} + e^{(y-1)/\sqrt{\nu}}), 0)^T$. The layer-adapted hybrid mesh is equidistant in x-direction and has a mesh grading in y-direction with $y_i = \frac{1}{2} + \frac{1}{2}\tanh(\frac{2i\gamma}{N_y-1})/\tanh(\gamma)$, $i = -\frac{1}{2}(N_y-1),\ldots,\frac{1}{2}(N_y-1)$. The parameter γ can be chosen such that condition (10) holds with $\mu_0 = \mathcal{O}(1)$.

In Fig. 1 (left), we show the pointwise error $(u_1 - u_{1,h})(\frac{1}{2}, y)$, $0 \le y \le 1$ for increasing values of N_y. In Fig. 1 (right), we present a zoom in a semilogarithmic scale for fixed $N_y = 129$ together with different values of γ (leading to different percentage of mesh points in the boundary layer regions $(0,1) \times (0, \delta_{99})$ and $(0,1) \times (1 - \delta_{99}, 1)$ where δ_{99} is given by $u_1(x, \delta_{99}) = (0.99, 0)$). On the grid with $N_y = 129$, the L^∞-error is reduced to $\le 0.2\%$ if 37.5 or 50 % of the grid points are located in the layer regions for resolving the gradient, whereas the solution on the corresponding uniform mesh has a L^∞-error of 10 %.

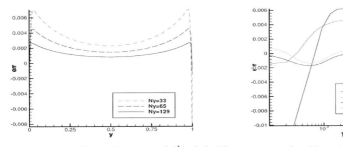

Fig. 1. Error $(u_1 - u_{1,h})(\frac{1}{2}, y)$ (left) and zoom for $N_y = 129$ (right)

Finally, we consider the *turbulent* 3d-channel flow in $\Omega = (0, H)^2 \times (0, L)$ with $H = 1$ [m] and $L = 5$ [m]. We apply the $k - \epsilon - \overline{v}^2 - \overline{f}$-model of Durbin in the "user-friendly" $\varphi - \overline{f}$-version [10] for the RANS version of problem (1)–(2) where the viscosity ν is replaced with $\nu_e = \nu + \nu_t$ based on the turbulent viscosity $\nu_t = c_\mu \, k\varphi \max\left(\frac{k}{\epsilon}, 6\sqrt{\frac{\nu}{\epsilon}}\right)$. The turbulent quantities $k, \epsilon, \varphi, \overline{f}$ are determined by a coupled nonlinear advection-diffusion-reaction system.

We compare the solution to DNS data of [9] for $Re_\tau = \frac{H u_\tau}{\nu} = 395$ based on the friction velocity $u_\tau = \sqrt{\tau_w} \equiv \sqrt{\nu \frac{\partial u_2}{\partial y}|_{\Gamma_w}} = 1.2087 \cdot 10^{-2}$. This corresponds

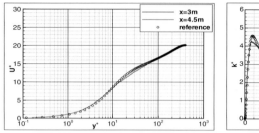 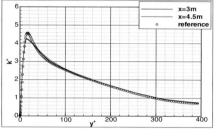

Fig. 2. Plot of u^+, k^+ vs. $y^+ := \frac{yu_\tau}{\nu}$ at $x \in \{3, 4.5\}$ [m]

to $Re_C = \frac{U_C H/2}{\nu} \approx 14.000$. Moreover, we have $\mathbf{f} = \frac{\tau_W}{H}\mathbf{e_x}$. Our calculations are performed on a FE-mesh with $33 \times 49 \times 65$ nodes. In y-direction, we use the above tanh-distribution with γ s.t. the first off-wall node is at $yu_\tau/\nu = 1$. The sets δ_T, γ_T are based on $\widetilde{h}_T = |\text{meas}(T)|^{\frac{1}{3}}$. In Fig. 2, we present the relevant quantities $u^+ = \frac{u_1}{u_\tau}$ and $k^+ = \frac{k}{u_\tau^2}$ in wall units at $x = 3[m]$ and $x = 4.5[m]$ in wall units. The results are in reasonable agreement with the DNS data and comparable to results in [10].

Acknowledgment

We thank Th. Apel for valuable discussions and M. Wannert for performing the numerical results for the laminar test problem.

References

1. Apel, Th.: Anisotropic finite elements: Local estimates and applications. Series "Advances in Numerical Mathematics", Teubner, Stuttgart (1999)
2. Apel, T., Knopp, T., Lube, G.: Stabilized finite element methods with anisotropic mesh refinement for the Oseen problem, submitted to APNUM.
3. Becker, R., Rannacher, R.: Finite element solution of the incompressible Navier-Stokes equations on anisotropically refined meshes, Notes Numer. Fluid Mech., **49**, 52–62 (1995)
4. Brooks, A.N., Hughes, T.J.R.: Streamline upwind Petrov-Galerkin formulation for convection dominated flows with particular emphasis on the incompressible Navier-Stokes equations, CMAME **32**, 199–259 (1982)
5. Codina, R.: Stabilization of incompressibility and convection through orthogonal subscales in finite element methods, CMAME **190** (13/14), 1579–1599 (2000)
6. Codina, R., Soto, O.: Approximation of the incompressible Navier-Stokes equations using orthogonal subscale stabilization and pressure segregation on anisotropic finite element meshes, CMAME **193**, 1403–1419 (2004)
7. Franca, L.P., Frey, S.L.: Stabilized finite element methods. II. The incompressible Navier-Stokes equations, CMAME **99**, 209–233 (1992)

8. Hughes, T.J.R., Jansen, K.: A stabilized finite element formulation for the Reynolds-averaged Navier-Stokes equations, Surv. Math. Ind. 4: 279–317 (1995)
9. Kim, J., Moin, P., Moser, R.: Turbulence statistics in fully developed channel flow at low Reynolds number, J. Fluid Mech. **177**, 133–166 (1987)
10. Laurence, L.R., Uribc, J.C., Utyuzhnikov, S.V.: A robust formulation of the $v2 - f$ model, J. Flow, Turbulence and Combustion **23**, 169–185 (2004)
11. Micheletti, S., Perotto, S., Picasso, M.: Stabilized finite elements on anisotropic meshes: A priori error estimates for the advection-diffusion and the Stokes problems. SINUM **41** 3, 1131–1162 (2003)
12. Tobiska, L., Verfürth, R.: Analysis of a streamline-diffusion finite element method for the Stokes and Navier-Stokes equations, SINUM **33**, 107–127 (1996)

Semi-Implicit Multiresolution for Multiphase Flows

N. Andrianov[1,3], F. Coquel[2], M. Postel[2] and Q. H. Tran[1]

[1] Département de Mathématiques Appliquées, Institut Français du Pétrole, 92852 Rueil-Malmaison Cedex, France
[2] Laboratoire Jacques-Louis Lions (UMR 7598), Université Pierre et Marie Curie, B.C. 187, 75252 Paris Cedex 05, France
[3] Currently at Schlumberger Research and Development, Taganskaya Str., 9, Moscow, 109147, Russia

Summary. In the context of multiphase flows we are faced with vector PDE solutions combining waves whose speeds are several orders of magnitude apart. The wave of interest is the transport one, and is relatively slow. The other fast acoustic waves are not interesting but impose a very restrictive CFL condition if a fully explicit in time scheme is considered. We therefore use a time semi-implicit conservative scheme where the fast waves are handled with a linearized implicit formulation and the slow wave remains explicitly solved. The CFL condition, governed by the explicit wave speed is then optimal. We combine this method with a multiscale analysis of the vector solution which enables to use a time varying adaptive grid based on the relevant smoothness properties of the discrete solution. In this short paper we compare different strategies to evaluate the fluxes at cells interfaces on a non uniform grid.

1 Introduction

In this paper we address the numerical approximation of systems of conservation laws in 1D, modeling physical problems where different waves arise with speeds separated by several orders of magnitude. Typically, a mixture of gas and oil moving along in a pipeline with a speed generated by a pumping system will generate two types of waves: very fast acoustic waves and slower transport waves. These slower waves are the only interesting ones from the oil production point of view since they model the front displacement of the gas mass fraction in the mixture. On the other hand the fast waves, of less practical interest, impose a severe stability restriction on the time step used in the numerical simulation, specially in the case of time explicit schemes.

The first answer to this numerical difficulty consists in using a conservative time semi-implicit scheme. Without going into details (see [4]), this amounts to writing the unknowns of the system in a basis where slow and fast waves action can be easily decoupled. The fast waves component of the solution will be evolved in time with an implicit scheme, therefore eliminating the most severe part of the CFL condition. The explicit scheme will treat only the slow waves component, therefore ensuring a much better resolution of contact discontinuities. Let us stress that these distinct time integration procedures are suitably combined so as to preserve the required conservation property in the method.

Next the discrete solution to be dealt with exhibits all the properties of a good candidate for an adaptive mesh refinement strategy. A realistic computation in a pipeline will consist in modeling the transport of a discontinuity in the gas mass fraction over ten kilometers, with an average speed of a few meters per second. Engineers will consider that a discretization made of several thousands grid points and a time simulation lasting over several thousands time steps are needed to compute accurately enough the transport of the mass fraction discontinuity to the other end of the pipeline. On the other hand, this very small cell size of a few meters is really necessary only in the vicinity of the discontinuity. Everywhere else the fluctuations in the solutions are due to acoustic waves which are not physically interesting. Hence a much coarser space grid suffices in the treatment of these waves. Note in addition that these acoustic waves, if sharp in the initial data, are smoothed out from the first time steps by the numerical dissipation due to the implicit time integration.

In answer to this observation, we have adapted the multiresolution techniques established for explicit schemes in [2] and based on ideas introduced in the context of systems of conservation laws by Harten [5] at the beginning of the nineties. The multiscale analysis of the solution is used to design an adaptive grid by selecting the correct level out of a hierarchy of nested grids according to the local smoothness of the solution. This non-uniform grid evolves with time, with a strategy based on the prediction of the displacement and formation of the singularities in the solution. The wavelet basis used to perform the multiscale analysis enables to reconstruct the solution at any time back to the finest level of discretization, within an error tolerance controlled by a threshold parameter.

The coupling of multiresolution with the semi-implicit scheme is detailed in [3]. In particular, to avoid excessive damping in the fast waves an additional CFL condition controlling the artificial dissipation is taken into account in the definition of the time step along with the CFL stability restriction in the slow waves. The robustness of the prediction of the adaptive grid from one time to the next has been studied in the above reference.

In this shorter paper we briefly recall the PDE model and the principles of the semi-implicit scheme in paragraph 2, then the multiresolution enhancement of the method in paragraph 3. We devote special attention to the comparison of available methods to compute the numerical fluxes on the adaptive grid.

2 Semi implicit scheme on uniform grid

The oil and gas mixture is supposed to move with the same velocity for both phases. The evolution in time of the density of the mixture ρ, velocity v and the gas mass fraction Y are related through the PDE system

$$\begin{cases} \partial_t(\rho) + \partial_x(\rho v) = 0, \\ \partial_t(\rho Y) + \partial_x(\rho Y v) = 0, \\ \partial_t(\rho v) + \partial_x(\rho v^2 + P) = 0. \end{cases}$$

The pressure law $P(\rho, \rho Y)$ can be in practice very costly to evaluate, which is one of the motivations to use an adaptive method. In the scope of this paper however, we will use for the numerical simulations the following thermodynamical closure law $P(\rho, \rho Y) = a_g^2 \rho_l \rho Y / (\rho_l - \rho(1 - Y))$, corresponding to a isothermal gas and incompressible liquid. The system is hyperbolic over a suitable phase space Ω with three distinct eigenvalues. The intermediate eigenvalue corresponds to the slow transport wave and is linearly degenerate, the two extreme ones are much larger and correspond to genuinely non linear waves of acoustic type.

We rewrite the PDE system under the generic form

$$\frac{\partial \mathbb{U}(x,t)}{\partial t} + \frac{\partial \mathcal{F}(\mathbb{U}(x,t))}{\partial x} = 0,$$

where $\mathbb{U}(x,t)$ denotes the conservative variables as a vector function from $\mathbb{R} \times \mathbb{R}+$ in Ω and the flux $\mathcal{F}(\mathbb{U})$ is a vector function from Ω in $\mathbb{R}^3$. We seek a finite volume discretization of this system on a grid $x_j = j \Delta x$ with $j \in \mathbb{Z}$. We denote by $\mathbb{U}_j$ the vector of the conservative unknowns on the cell $[x_j, x_{j+1}]$. The implicit scheme is defined by

$$\mathbb{U}_j^{n+1} = \mathbb{U}_j^n - \frac{\Delta t}{\Delta x}\left(\mathbb{F}_{j+1}^{n+1} - \mathbb{F}_j^{n+1}\right), \quad \text{with} \quad \mathbb{F}_j^n = \mathbb{F}(\mathbb{U}_{j-1}^n, \mathbb{U}_j^n).$$

The approximation of the flux $\mathbb{F}_j^n$ going through the interface at x_j between two cells is the standard Roe linearization. To bypass intractable nonlinearities the numerical flux at time $n+1$ is commonly approximated from the values at time n using a Taylor expansion in time to give birth to a linear problem in the unknown $\mathbb{U}_j^{n+1} - \mathbb{U}_j^n$. If the entering matrix is kept unchanged the method is implicit in time with respect to the three waves. We briefly report how to derive a mixed implicit explicit in time method. Cell by cell the eigenvalues of all 3×3 block matrices are actually related to the three eigenvalues of the PDE model. The fastest ones are kept unchanged but the intermediate eigenvalue is systematically set to zero; hence ensuring an explicit treatment of the corresponding wave. Since the slow wave is treated explicitly, one needs to impose a standard stability condition with a CFL number less than one. Furthermore, the damping undergone by the fast waves has to be somewhat controlled. To this effect, we impose an additional CFL criterion on the fast waves, but this time with a CFL number equal to 20.

3 Multiscale analysis of the explicit-implicit scheme

In [2], a multiresolution analysis was proposed for a system of hyperbolic conservation laws treated with an explicit scheme. We are now going to extend it to the case of the explicit-implicit scheme introduced in section 2.

3.1 Basics of multiresolution analysis

We consider a uniform mesh with step size Δx and starting from this finest discretization, we define a hierarchy of K nested grids by dyadic coarsening, with cell interfaces $x_j^k = 2^{K-k} j \Delta x$. Initially, the unknown function U is defined on the finest grid, numbered K, where it is represented by the sequence of its mean values $U^K = (u_j^K)_j$ on the cells $[x_j^K, x_{j+1}^K]$. The coarsening operator P_k^{k-1} consists in cell averaging from one grid to the coarser one, i.e.,

$$U^{k-1} = P_k^{k-1} U^k \quad \text{with} \quad u_{j,k-1} = \frac{1}{2}(u_{2j}^k + u_{2j+1}^k).$$

The inverse operator consists in recovering the mean values on grid level k, given the mean values on the coarser level $k-1$. This involves an approximation — or prediction — operator P_{k-1}^k. Among the infinite number of choices for the definition of P we use here the linear reconstruction based on a quadratic polynomial which obeys adequate locality and consistency rules therefore allowing for some analysis

$$\hat{U}^k = P_{k-1}^k U^{k-1} \quad \text{with} \quad \begin{cases} \hat{u}_{2j,k} = u_{j,k-1} - \frac{1}{8}(u_{j+1,k-1} - u_{j-1,k-1}), \\ \hat{u}_{2j+1,k} = u_{j,k-1} + \frac{1}{8}(u_{j+1,k-1} - u_{j-1,k-1}). \end{cases} \quad (1)$$

We define the prediction error E^k at a given level k as the difference between the solution on level k and its reconstruction $\hat{U}^k$ using the solution at level $k-1$. Thanks to the consistency property $e_{2j}^k = -e_{2j+1}^k$, we define the detail vector D^{k-1} with $d_j^{k-1} = u_{2j}^k - \hat{u}_{2j}^k$ and use it along with U^{k-1} to entirely recover U^k.

The two vectors U^k and (U^{k-1}, D^{k-1}) are of same length. Iterating this encoding operation from the finest level down to the coarsest provides the multiscale representation $M^K = (U^0, D^0, \ldots, D^{K-1})$. Using the local structure of the operators P_k^{k-1} and P_{k-1}^k, the multiscale transformation $\mathcal{M} : U^K \mapsto M^K$ and its inverse $\mathcal{M}^{-1}$ can be implemented with an optimal $\mathcal{O}(N_K)$ complexity, where N_K represents the dimension of the finest grid ∇^K. The interest of the multiscale representation lies in the fact that thanks to the consistency of the prediction operator, the local regularity of the function is reflected by the size of its details. We can use this property to compress the function in the multiscale domain by dropping all details below a given level-dependent threshold. To clarify this idea, we first define a threshold operator $\mathcal{T}_\Lambda$ acting on the multiscale representation M^K, depending on a subset $\Lambda \subset \nabla^K$ of indices $\lambda = (j, k)$, by

$$\mathcal{T}_\Lambda(d_\lambda) = \begin{cases} 0 & \text{if } \lambda \in \Lambda, \\ d_\lambda & \text{otherwise}. \end{cases}$$

Given level-dependent threshold values $\varepsilon = (\varepsilon_k)_k$, we introduce the subset $\Lambda_\varepsilon = \Lambda(\varepsilon_0, \varepsilon_1, \cdots, \varepsilon_K) := \{\lambda \text{ s.t. } |d_\lambda| \geq \varepsilon_{|\lambda|}\}$. This completes the definition of the threshold operator $\mathcal{T}_\varepsilon := \mathcal{T}_{\Lambda_\varepsilon}$ and gives rise to an approximating operator $\mathcal{A}_\varepsilon := \mathcal{M}^{-1} \mathcal{T}_\varepsilon \mathcal{M}$ acting on the physical domain representation. In practice, we take advantage of the fact that the remaining fine-scale details will be concentrated near singularities. This is not such a trivial result because the operator $\mathcal{A}_\varepsilon$ is nonlinear since Λ_ε depends on U^K through the threshold scheme. We refer to [1] for a thorough investigation of nonlinear approximation and the proof of the main result

$$\|U^K - \mathcal{A}_\varepsilon U^K\|_{L^1} < C\varepsilon_K$$

valid when $\varepsilon_k = 2^{k-K}\varepsilon_K$. This allows us to define an adaptive grid where the local size of the cell will be the grid step corresponding to the finest non negligible detail. The representation of $U_\varepsilon = \mathcal{A}_\varepsilon U^K$ on this adaptive grid is intermediate between the physical representation U_ε^K on the finest grid and the encoded multiscale representation M_ε^K. Note, in particular, that the representation by its mean value u_j^k on an intermediate level k does not mean that the function is locally constant on this cell of width $2^k h$, but simply that its mean values on the finest grid in this area can be recovered —within the ε accuracy— using the mean values on this intermediate level and the reconstruction operators P_{l-1}^l for $l = k+1, \ldots, K$. In the sequel, we will call *partial decoding* the algorithm that computes U_ε on this adaptive grid from M^K, and *partial encoding* the reverse transformation.

3.2 Prediction strategy for the tree

In the context of the semi implicit scheme presented in paragraph 2, the multiresolution representation of the solution must evolve with time. Its singularities will move with time. They can actually appear or disappear completely, which means that the tree Λ_ε depends on time and that its computation must be performed at each time step. Of course, we wish to compute this time-dependent tree without having to decode the solution back to the finest grid at each time step, since this would destroy all the benefits of the adaptive computation. This is possible thanks to the hyperbolic nature of the PDE's system, which ensures that the singularities of the solution move at finite speed. More specifically, if we denote by Λ_ε^n the graded tree obtained by applying $\mathcal{A}_\varepsilon$ to U_K^n, then Λ_ε^n can be inflated into $\widetilde{\Lambda}_\varepsilon^n$ containing $\Lambda_\varepsilon^{n+1}$ as well as Λ_ε^n, ensuring that both estimations

$$\|U_K^n - \mathcal{A}_{\widetilde{\Lambda}_\varepsilon^{n+1}} U_K^n\| \leq C\varepsilon_K \quad \text{and} \quad \|U_K^{n+1} - \mathcal{A}_{\widetilde{\Lambda}_\varepsilon^{n+1}} U_K^{n+1}\| \leq C\varepsilon_K$$

are satisfied. Setting $\widetilde{\Lambda}^{n+1}$ to S^K does the trick but it is not very interesting in practice. The inflated tree $\widetilde{\Lambda}^{n+1}$ should be as small as possible. The inflation strategy proposed by Harten [5] consists in adding immediate neighbors of cells where the detail is above the level-dependent threshold and the two subdivisions of cells where the detail is more than twice this threshold. This relies strongly on the CFL condition which is less than one in the case of an explicit scheme. In [3], we have extended this strategy to the time explicit-implicit scheme. This relies on an heuristic argument which can be justified in the case of linear systems of PDE's using Fourier analysis of the numerical solution.

Below is the actual adaptive algorithm we implemented.

1. Initialization: encoding of the initial solution and definition of Λ^0
2. Loop over time steps $n = 0, \ldots, N - 1$:
    - Prediction of $\widetilde{\Lambda}^{n+1}$ and partial decoding of $\mathbb{U}^n$
    - Evolution of $\mathbb{U}^n$ to $\mathbb{U}^{n+1}$ on the adaptive grid $\widetilde{\Lambda}^{n+1}$
    - Encoding of $\mathbb{U}^{n+1}$ and definition of Λ^{n+1}.
3. Decoding of $\mathbb{U}^N$ on the finest grid

Flux evaluation in the evolution step

An important point in this algorithm is the evaluation of the numerical fluxes between adjacent cells of the adaptive grid, which must be performed in order to update the solution $\mathbb{U}^n$ into $\mathbb{U}^{n+1}$. We compare here two methods in terms of complexity and accuracy.

Local reconstruction on the finest level. In the case where the underlying uniform scheme is of first order, the numerical flux is a function of the mean values on each side of the interface. If the adaptive grid data is used directly, the scheme is of first order with respect to the local grid size which can be quite large. Numerical experiments have shown that it is necessary to locally reconstruct the mean values on the finest grid on each side of the interface. As it is explained in [2], this can be done in $\mathcal{O}(N_\Lambda)$ operations, with N_Λ the number of cells in the adaptive grid, thanks to the linearity and uniformity of the reconstruction (1).

Direct evaluation of the flux on the adaptive grid. When the underlying uniform scheme is of higher order, the local reconstruction of the solution on the finest grid near the interfaces is less crucial since in the smooth regions the high order of numerical scheme will be able to compensate for the coarseness of the grid. The alternative consisting in applying the underlying high order scheme directly on the adaptive grid solution can be used. This can require some modification of the high order nonlinear reconstruction scheme, to take into account situations where interfaces can separate two cells of length in a ratio of two (see [6] for details). In our simulations the reference scheme on the uniform grid is of order two in space, achieved by a minmod limited linear reconstruction of the solutions on each side of the interface where the fluxes are computed. Note that this reconstruction is applied on the solution in primitive variable (ρ, Y and u) instead of the conservative form on which the multiresolution is performed.

This strategy has been implemented and tested on the cases already treated with the local reconstruction technique in [3]. We focus here on the most realistic case mimicking operating conditions at both ends of the pipeline. We prescribe gas and total mass flow rates at the inlet and monitor the pressure at the outlet as shown in figure 1. Besides the multiscale analysis of the solution at each time step which rules the local size of the space grid step, the variation of the time dependent boundary condition are also tested to decide whether fine cells should be added to the grid on the edges. The speed-up ratio between the uniform grid simulation and the fully adaptive simulation for a threshold $\varepsilon = 0.005$ is about 3.4.

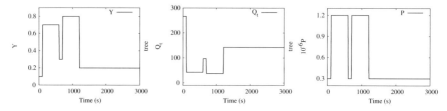

Fig. 1. Experimental gas fraction (left) and total flow (middle) at the inlet of pipeline and pressure at the outlet of pipeline (right).

We show in figure 2 the density at the final time $t = 3000s$ of the simulation along with the adaptive grid where active cells are indicated by $+$. The left hand side graph corresponds to the case when the solution is first locally reconstructed on the fine level. The right hand side graph corresponds to the case when the limited reconstruction is applied directly on the adaptive grid. It seems that the latter method produces a slightly more refined tree but the singularities of the solution are well localized in both cases. In figure 3 we see a

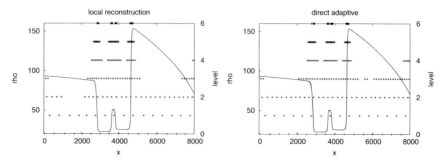

Fig. 2. Density and adaptive grid at final time.

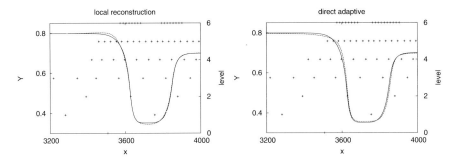

Fig. 3. Zoom on gas mass fraction at final time.

zoom of the gas mass fraction in a region of high variations. In these last graphs the differences are clearer: applying the scheme directly on the adaptive grid somewhat smoothes the solution around slope changes. On the other hand the alternative technique, involving several linear and non convex reconstructions before applying the limited second order reconstruction introduces a small overshoot which could be very damaging if the solution were bordering the physical constraints ($0 \leq Y \leq 1$, $\rho > 0$).

References

1. Cohen, A.: Numerical analysis of wavelet methods, volume 32 of Studies in Mathematics and its Applications. North-Holland Publishing Co., Amsterdam (2003)
2. Cohen, A., Kaber, S. M., Müller, S., Postel, M.: Fully adaptive multiresolution finite volume schemes for conservation laws. Math. Comp., **72**(241), 183–225 (2003)
3. Coquel, F., Postel, M., Poussineau, N., Tran, Q. H.: Multiresolution technique and explicit-implicit scheme for multicomponent flows. Publications du laboratoire Jacques-Louis Lions, R05026. Paris (2005)
4. Faille, I., Heintzé, E.: A rough finite volume scheme for modeling two phase flow in a pipeline. Computers and Fluids, **28**, 213–241 (1999)
5. Harten, A.: Multiresolution algorithms for the numerical solutions of hyperbolic conservation laws. Comm. on Pure and Appl. Math., **48**, 1305–1342 (1995)
6. Müller, A., Stiriba, Y.: Fully adaptive multiscale schemes for conservation laws employing locally varying time stepping. Technical report, IGPM, Report No. 238, RWTH Aachen (2004)

This work was supported by the Ministère de la Recherche under grant ERT-20052274: Simulation avancée du transport des hydrocarbures and by the Institut Français du Pétrole.

Numerical Simulation of Vortex-Dipole Wall Interactions Using an Adaptive Wavelet Discretization with Volume Penalisation

Kai Schneider[1] and Marie Farge[2]

[1] LMSNM–CNRS & CMI, Université de Provence, 39 rue Joliot–Curie,
 13453 Marseille cedex 13, France
 kschneid@cmi.univ-mrs.fr
[2] LMD–CNRS, Ecole Normale Supérieure, 24 rue Lhomond,
 75231 Paris cedex 05, France
 farge@lmd.ens.fr

Summary. We present an adaptive wavelet method for solving the incompressible Navier–Stokes equations in two space dimensions using the vorticity-stream function formulation. For time discretization a semi–implicit scheme of second order is used. The space discretization is based on a Petrov–Galerkin method, where orthogonal spline wavelets of 4th order are employed as trial functions and operator adapted wavelets as test functions. The no–slip boundary conditions are imposed using a volume penalisation method. As example we present adaptive simulations of vortex-dipole wall interactions.

1 Introduction

The mathematical properties of wavelets (see, *e.g.*, *Daub92*) motivate their use for the numerical solution of partial differential equations (PDEs). The localization of wavelets, both in scale and space, leads to effective sparse representations of functions and pseudo–differential operators (and their inverse) by performing nonlinear thresholding of the wavelet coefficients of the function and of the matrices representing the operators. Estimating the local regularity of the solution of the PDE auto–adaptive discretizations with local mesh refinements can be defined. The characterization of function spaces in terms of wavelet coefficients and the corresponding norm equivalences allow diagonal preconditioning of operators in wavelet space. Finally, the existence of the fast wavelet transform yields algorithms with optimal linear complexity.

The currently existing algorithms can be classified in different ways. We can distinguish between Galerkin, collocation schemes and algebraic wavelet methods. By the latter we mean algorithms which start from a classical discretization, e.g. by finite differences or finite volumes. Wavelets are then used to speed up the linear algebra and to define adaptive grids. On the other hand the former two schemes employ wavelets directly for the discretization of the solution and the operators. For an overview on wavelet methods we refer the reader to [5, 4].

Wavelet methods have been developed to solve Burger's equation, Stokes' equation, Kuramoto–Sivashinsky equation, the nonlinear Schrödinger equation, the Euler and Navier–Stokes equations.

In the follwing we present an adaptive wavelet algorithm of Galerkin type [10] to solve the two–dimensional Navier–Stokes equation in vorticity–stream function formulation. The boundary conditions are imposed using a volume penalisation technique. As application we present computations of a vortex dipole impinging on a no-slip wall in a square container at Reynolds number 1000, which is a challenging test case for numerical methods [12, 3]. Finally, we present some conclusions and perspectives for future work.

2 Adaptive wavelet discretization with volume penalisation

2.1 An adaptive wavelet scheme

The volume penalisation method has been proposed by Arquis and Caltagirone [1]. Its is based on the physical idea which consists in modelling solid walls or obstacles as porous media whose porosity η tends to zero. The geometry of the flow is described by a mask function χ. The incompressible Navier-Stokes equations are modified by adding a forcing term containing the mask function. Using vorticity ω and the stream function Ψ, which are both scalars in 2d, the equations are:

$$\partial_t \omega + \mathbf{v} \cdot \nabla \omega - \nu \nabla^2 \omega = \nabla \times \mathbf{F} \tag{1}$$
$$\nabla^2 \Psi = \omega \quad \text{and} \quad \mathbf{v} = \nabla^\perp \Psi \tag{2}$$

for $\mathbf{x} \in \Omega$, $t > 0$. The velocity is denoted by $\mathbf{v}$, $\nu > 0$ is the constant kinematic viscosity and $\nabla^\perp = (-\partial_y, \partial_x)$. The fluid region Ω_f is embedded in the enlarged domain Ω containing in addition a solid region Ω_s which is surrounding the fluid region. The penalisation term $\mathbf{F} = -\frac{1}{\eta} \chi_{\Omega_s} \mathbf{v}$ imposes no–slip boundary conditions on the walls, corresponding to the interface between the fluid and solid region, i.e. Ω_f and Ω_s, respectively. The mask function χ is defined as

$$\chi_{\Omega_s}(\mathbf{x}) = \begin{cases} 1 & \text{for} \quad \mathbf{x} \in \bar{\Omega}_s, \\ 0 & \text{elsewhere} \end{cases} \tag{3}$$

where Ω_s denotes the ensemble of solid obstacles. The above equations are completed with a suitable initial condition.

In [2] it has been shown rigorously that the penalised equations written in primitive variables converge towards the Navier–Stokes equations with no-slip boundary conditions, with order $\eta^{3/4}$ inside the obstacle and with order $\eta^{1/4}$ elsewhere, in the limit when η tends to zero. In numerical simulations an improved convergence of order η has been reported [2, 11].

Time discretization

Introducing a classical semi–implicit time discretization with a time step δt and setting $\omega^n(\mathbf{x}) \approx \omega(\mathbf{x}, n\delta t)$ we obtain

$$(1 - \nu \delta t \nabla^2)\omega^{n+1} = \omega^n + \delta t(\nabla \times F^n - \mathbf{v}^n \cdot \nabla \omega^n) \qquad (4)$$

$$\nabla^2 \Psi^{n+1} = \omega^{n+1} \quad \text{and} \quad \mathbf{v}^{n+1} = \nabla^\perp \Psi^{n+1} \qquad (5)$$

Hence in each time step two elliptic problems have to be solved and a differential operator has to be applied. Formally the above equations can be written in the abstract from $Lu = f$, where L is an elliptic operator with constant coefficients, corresponding to a Helmholtz type equation for ω with $L = (1 - \nu \delta t \nabla^2)$ and a Poisson equation for Ψ with $L = \nabla^2$.

In practice we use a time scheme composed of an Euler–Backwards scheme and an Adams–Bashforth scheme, both of second order [10].

Spatial discretization

For the spatial discretization we use the method of weighted residuals, *i.e.*, a Petrov–Galerkin scheme. The trial functions are orthogonal wavelets and the test functions are operator adapted wavelets. To solve the elliptic equations $Lu = f$ at time step t^{n+1} we develop u^{n+1} into an orthogonal wavelet series, *i.e.*, $u^{n+1} = \sum_\lambda \tilde{u}_\lambda^{n+1} \psi_\lambda$, where $\lambda = (j, i_x, i_y, d)$ denotes the multi–index containing scale, space and direction information. Requiring that the residuum vanishes with respect to all test functions $\theta_{\lambda'}$, we obtain a linear system for the unknown wavelet coefficients $\tilde{u}_\lambda^{n+1}$ of the solution u:

$$\sum_\lambda \tilde{u}_\lambda^{n+1} \langle L\psi_\lambda, \theta_{\lambda'} \rangle = \langle f, \theta_{\lambda'} \rangle. \qquad (6)$$

The test functions θ are defined such that the stiffness matrix turns out to be the identity. Therefore the solution of $Lu = f$ reduces to a change of the basis, *i.e.*, $u^{n+1} = \sum_\lambda \langle f, \theta_\lambda \rangle \psi_\lambda$.

The right-hand side f can then be developed into a biorthogonal operator adapted wavelet basis $f = \sum_\lambda \langle f, \theta_\lambda \rangle \mu_\lambda$, with $\theta_\lambda = L^{\star-1}\psi_\lambda$ and $\mu_\lambda = L\psi_\lambda$ (* denotes the adjoint operator). By construction θ and μ are biorthogonal, $\langle \theta_\lambda, \mu_{\lambda'} \rangle = \delta_{\lambda, \lambda'}$. It can be shown that both have similar localization properties in physical and Fourier space as has ψ and that they form a Riesz basis [10].

Adaptive discretization

To get an adaptive space discretization for the problem $Lu = f$ we consider only the significant wavelet coefficients of the solution. Hence we only retain coefficients $\tilde{u}_\lambda^n$ which have an absolute value larger than a given threshold ε, i.e., $|\tilde{u}_\lambda^n| > \varepsilon$. The corresponding coefficients are shown in Fig.1 (white area under the solid line curve).

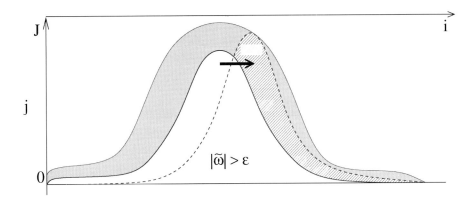

Fig. 1. Illustration of the dynamic adaption strategy in wavelet coefficient space.

Adaption strategy

To be able to integrate the equation in time we have to account for the evolution of the solution in wavelet coefficient space (indicated by the arrow in Fig. 1). Therefore we add at time step t^n the local neighbors to the retained coefficients, which constitute a security zone (grey domain in Fig.1). The equation is then solved in this enlarged coefficient set (white and grey region in Fig.1) to obtain $\tilde{u}_\lambda^{n+1}$. Subsequently we threshold the coefficients and retain only those with $|\tilde{u}_\lambda^{n+1}| > \varepsilon$ (coefficients under the dashed curve in Fig.1). This strategy is applied in each time step and allows hence to track automatically the evolution of the solution in scale and space.

Evaluation of the nonlinear term

For the evaluation of the nonlinear term $f(u^n)$, where the wavelet coefficients of u^n are given, there are two possibilities:

- evaluation in wavelet coefficient space.
  As illustration we consider a quadratic nonlinear term, i.e., $f(u) = u^2$. The wavelet coefficients of f can be calculated using the connection coefficients, i.e., one has to calculate the bilinear expression, $\sum_\lambda \sum_{\lambda'} \tilde{u}_\lambda T_{\lambda \lambda' \lambda''} \tilde{u}_{\lambda'}$

with the interaction tensor $T_{\lambda\lambda'\lambda''} = \langle \psi_\lambda \psi_{\lambda'}, \theta_{\lambda''} \rangle$. Although many coefficients of T are zero or very small, the size of T leads to a computation which is quite untractable in practice.
- evaluation in physical space.
  This approach is similar to the pseudo-spectral evaluation of nonlinear terms used in spectral methods, and therefore this method is called pseudo–wavelet technique. The advantage of this scheme is that more general nonlinear terms, e.g., $f(u) = (1-u)\, e^{-C/u}$, can be treated more easily. The method can be summarized as follows: starting from the significant wavelet coefficients of u, i.e., $|\tilde{u}_\lambda| > \varepsilon$, one reconstructs u on a locally refined grid, $u(x_\lambda)$. Then one can evaluate $f(u(x_\lambda))$ pointwise and the wavelet coefficients of f can be calculated using the adaptive decomposition to get $\tilde{f}_\lambda$.

Finally, we have to calculate those scalar products of the r.h.s f with the test functions θ, to advance the solution in time. We compute $\tilde{u}_\lambda = \langle f, \theta_\lambda \rangle$ belonging to the enlarged coefficient set (white and grey region in Fig. 1).

In summary the above algorithm is of $O(N)$ complexity, where N denotes the number of wavelet coefficients used in the computation.

3 Vortex-dipole wall interactions

To illustrate the above algorithm we present an adaptive wavelet computation of a vortex–dipole impinging on a no-slip wall at Reynolds number $Re = 1000$ in a square container with $Re = \frac{\bar{u}L}{\nu}$ and where $\bar{u}$ denotes the rms velocity of the flow, L the half-width of the container and ν the kinematic viscosity. The Navier-Stokes equations are solved in a periodic square domain of size 2.2 in which the square container $[-1,1]^2$ is imbedded. The no slip boundary conditions are imposed using a volume penalisation method. The porosity η is 10^{-3} and the maximal numerical resolution is 1024^2. The initial vorticity distribution of the two isolated monopoles is given by

$$\omega(r, t=0) = \omega_0 \left(1 - \left(\frac{r}{r_0}\right)^2\right) \exp\left(-\left(\frac{r}{r_0}\right)^2\right) \qquad (7)$$

where r is the distance from the center of the monopole. Following [3] we chose $r_0 = 0.1$ and $\omega_0 = \pm 320$. The initial position of the two isolated monopoles is $\{(x_1, y_1), (x_2, y_2)\} = \{(0, 0.1), (0, -0.1)\}$.

Figure 2 (left) shows snapshots of the vorticity field at times $t = 0.2, 0.4, 0.6$ and 0.8. We observe that the dipole is moving towards the wall and that strong vorticity gradients are created when it hits the wall. The computational grid is dynamically adapted during the flow evolution, since the nonlinear wavelet filter automatically refines the grid in regions where strong gradients develop.

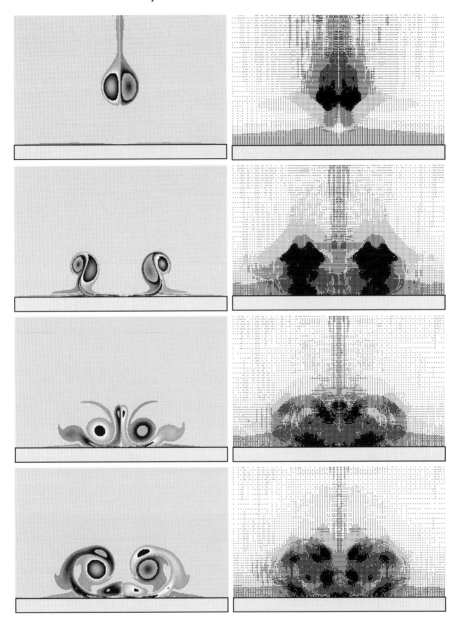

Fig. 2. Dipole–wall interaction at $Re = 1000$. Vorticity fields (left), corresponding centers of active wavelets (right), at $t = 0.2, 0.4, 0.6$ and 0.8 (from top to bottom).

Figure 2 (right) shows the centers of the retained wavelet coefficients at corresponding times. Note that during the computation only 5% out of 1024^2 wavelet coefficients are thus used. The time evolutions of total kinetic energy

$E(t) = \frac{1}{2}\int_{-1}^{1}\int_{-1}^{1}\mathbf{v}^2 dxdy$ and total enstrophy $Z(t) = \frac{1}{2}\int_{-1}^{1}\int_{-1}^{1}\omega^2 dxdy$ are plotted in Fig. 3 to illustrate the production of enstrophy and the dissipation of energy when the dipole is hitting the wall.

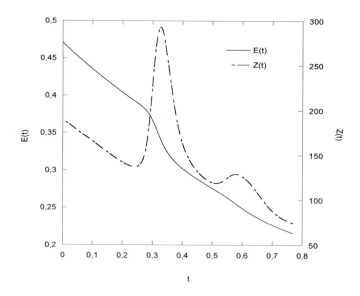

Fig. 3. Time evolution of energy (solid line) and enstrophy (dotted line).

4 Conclusions

In conclusion, we have checked the ability of the adaptive wavelet solver to track the evolution of the dipole and its nonlinear interaction with the no-slip wall. The utilisation of a volume penalisation method enables us to take into account complex geometries using a mask function, without modifying the numerical scheme and the underlying grid. The precision of the method is determined by the penalisation parameter η which can be chosen a priori. An explicit time discretization of the penalisation term implies, however, a time step smaller than η to guarantee stability of the numerical scheme. We have shown that this approach is suitable to model walls even in the case of strong interaction with vortices.

The adaptive wavelet method presented in this paper allows automatic grid generation and refinement near the wall and also in shear layers which develop during the flow evolution. Therewith, the number of required grid–points in the simulations is significantly reduced. We conjecture that the compression rate thus obtained increases with the Reynolds number. Current work is dealing with the development of adaptive local time stepping using a Runge–Kutta–Fehlberg method in order to control the error of the scheme in time [7].

In future work we will extend the penalisation scheme to compute three–dimensional flows and perform computations at high Reynolds numbers using the Coherent Vortex Simulation approach (CVS), proposed in [8, 9]. Applications to 3d turbulent mixing layers have been presented in [14].

Acknowledgements

We thankfully acknowledge financial support from the European Union project IHP on 'Breaking complexity'.

References

1. Arquis, E. Caltagirone, J.-P.: Sur les conditions hydrodynamiques au voisinage d'une interface milieu fluide – milieux poreux: application à la convection naturelle.
Comptes Rendus de l'Academie des Sciences, Paris, II **299**, 1–4 (1984)
2. Angot, P., Bruneau, C.-H., Fabrie, P.: A penalisation method to take into account obstacles in viscous flows. Num. Math., **81**, 497–520 (1999)
3. Clercx, H.J.H., van Heijst G.J.F.: On the dissipation of energy in 2D turbulence in a bounded domain, Adv. in Turbulence IX, CIMNE, Barcelona, 773–776 (2002)
4. Cohen, A.: Wavelet methods in numerical analysis. Handbook of Numerical Analysis. Eds. P.G. Ciarlet & J.L. Lions, Vol. 7, Elsevier, Amsterdam (2000)
5. Dahmen, W.: Wavelets and multiscale methods for operator equations. Acta Numerica, **6**, 55–228, Cambridge University Press (1997)
6. Daubechies, I.: Ten lectures on wavelets. SIAM (1992)
7. Domingues, M.O., Roussel, O., Schneider, K.: A time and space adaptive multiresolution scheme for parabolic PDEs. Preprint CMI, Marseille (2005), submitted.
8. Farge, M., Schneider, K., Kevlahan N.: Non–Gaussianity and Coherent Vortex Simulation for two–dimensional turbulence using an adaptive orthonormal wavelet basis, Phys. Fluids, **11**(8), 2187–2201 (1999)
9. Farge, M., Schneider, K.: Coherent Vortex Simulation (CVS), a semi–deterministic turbulence model using wavelets. Flow, Turbulence and Combustion **66**(4), 393–426 (2001)
10. Fröhlich, J., Schneider, K.: An adaptive wavelet–vaguelette algorithm for the solution of PDEs. J. Comput. Phys., **130**, 174–190 (1997)
11. Kevlahan, N., Ghidaglia, J.-M.: Computation of turbulent flow past an array of cylinders using a spectral method with Brinkman penalization. Eur. J. Mech./B, **20**, 333–350 (2001)
12. Orlandi, P.: Vortex dipole rebound from a wall. Phys. Fluids A **2**(8), 1429–1436 (1990)

13. Schneider, K., Farge, M.: Adaptive wavelet simulation of a flow around an impulsively started cylinder using penalisation. Appl. Comput. Harm. Anal., **12**, 374–380 (2002)
14. Schneider, K., Farge, M., Pellegrino, G., Rogers, M.: Coherent vortex simulation of three–dimensional turbulent mixing layers using orthogonal wavelets. J. Fluid Mech., **534**, 39–66 (2005)
15. Schneider, K.: Numerical simulation of the transient flow behaviour in chemical reactors using a penalisation method. Computers and Fluids, **34**, 1223–1238 (2005)

＃ Inviscid Flow on Moving Grids with Multiscale Space and Time Adaptivity

Philipp Lamby[1], Ralf Massjung[1], Siegfried Müller[1] and Youssef Stiriba[2]

[1] Institut für Geometrie und Praktische Mathematik, RWTH Aachen, D-52056 Aachen, Germany
   lamby,massjung,mueller@igpm.rwth-aachen.de
[2] Universitat "Rovira i Virgili" - ETSEQ, Departament Enginyeria Mecànica, Av. Paisos Catalans, 26, 43007 Tarragona, Spain
   youssef.stiriba@urv.cat

Summary. A fully adaptive multiscale finite volume scheme for solving the 2D compressible Euler equations on moving grids is presented. The scheme uses a multiscale analysis based on biorthogonal wavelets to adapt the grid in space. Refinement in time is performed using a locally varying time stepping strategy that has been recently developed. The CFL condition is satisfied locally and the number of grid adaptations is reduced. The performance of the scheme using global and local multilevel time stepping, respectively, is investigated by a flow past an oscillating boundary.

1 Introduction

The solutions of hyperbolic conservation laws typically exhibit locally steep gradients and large regions where they are smooth. To account for the highly nonuniform spatial behavior, we need numerical schemes that adequately resolve the different scales, i.e., use a high resolution only near sharp transition regions and singularities but a moderate resolution in regions with smooth, slowly varying behavior of the solution.

In [2, 7] multiresolution techniques have been used to construct locally refined meshes on which the discretization is performed. The basic idea is to represent the cell averages on a given highest level of resolution as cell averages

---

This work has been performed with funding by the Deutsche Forschungsgemeinschaft in the Collaborative Research Center SFB 401 "Flow Modulation and Fluid-Structure Interaction at Airplane Wings" of the RWTH Aachen, Germany and the "Ramón y Cajal" program of the Ministerio de Educacion y Ciencia, Spain.

on some coarse level where the fine scale information is encoded in arrays of *detail coefficients* of ascending resolution. If the detail information of a cell is small, the grid is locally coarsened. By now the fully adaptive multiresolution concept has been applied by several groups with great success to different real world applications, cf. [1] and references cited there.

So far a short-coming of this approach has been the lack of *temporal adaptivity*, i.e., all cell averages are evolved in time by the same time step size Δt satisfying the CFL condition for the cells on the *finest* mesh. Recently, a local time stepping strategy has been incorporated to the concept of fully adaptive multiresolution schemes, cf. [8, 5]. This has to be adjusted to the requirement that the resulting scheme provides an accuracy that is comparable to the accuracy of the reference mesh.

In the present work we apply this concept to 2D inviscid compressible fluid flows taking into account moving boundaries. This flow is governed by the arbitrary Lagrangian Eulerian (ALE) formulation of the Euler equations, cf. Sect. 2, that are discretized by a finite volume scheme, cf. Sect. 3. The efficiency of the reference scheme is improved by employing multiscale-based grid adaptation and local multilevel time stepping strategies, cf. Sect. 4. The adaptive scheme is applied to an oscillating boundary problem. Here we focus on the gain by the multilevel time stepping in comparison to the global time stepping, cf. Sect. 4.

2 The ALE formulation of the Euler equations

In the present study, inviscid fluid flow is described by the Euler equations for a compressible gas. In order to solve problems in time dependent domains, including moving boundaries, we consider the governing equations in its arbitrary Lagrangian Eulerian (ALE) formulation. Neglecting body forces and volume supply of energy, the conservation laws for any moving control volume $V \subset \Omega$ of the d-dimensional domain $\Omega \subset R^d$ with boundary ∂V and outward unit normal vector $\boldsymbol{n}$ on the surface element $dS \subset \partial V$ can be written in integral form as:

$$\frac{\partial}{\partial t} \int_{V(t)} \boldsymbol{u}\, dV + \oint_{\partial V(t)} \boldsymbol{f}(\boldsymbol{u}, \dot{\boldsymbol{x}}) \cdot \boldsymbol{n}\, dS = \boldsymbol{0}. \tag{1}$$

This system of conservation laws has to be supplemented by initial values and boundary conditions, respectively. Here $\boldsymbol{u} = (\rho, \rho\boldsymbol{v}, \rho E)^T$ denotes the vector of the unknown conserved quantities and $\boldsymbol{f}$ represents the convective flux:

$$\boldsymbol{f}(\boldsymbol{u}, \dot{\boldsymbol{x}}) = \begin{pmatrix} \rho(\boldsymbol{v} - \dot{\boldsymbol{x}}) \\ \rho(\boldsymbol{v} - \dot{\boldsymbol{x}}) \circ \boldsymbol{v} + p\,\boldsymbol{I} \\ \rho E(\boldsymbol{v} - \dot{\boldsymbol{x}}) + p\boldsymbol{v} \end{pmatrix} = \boldsymbol{f}(\boldsymbol{u}, \boldsymbol{0}) - \boldsymbol{u} \circ \dot{\boldsymbol{x}}, \tag{2}$$

where ρ denotes the density, p the static pressure, $\boldsymbol{v}$ the velocity vector of the fluid and E the total energy. Here $\circ$ is the dyadic product. The motion of the grid is considered by the convective fluxes, where $\dot{\boldsymbol{x}}$ expresses the grid velocity. The static pressure is related to the specific internal energy according to the equation of state for a perfect gas $p = \rho(\gamma - 1)(E - 1/2\,\boldsymbol{v}^2)$, where γ is the ratio of specific heats, which is taken as 1.4 for air.

3 Finite volume discretization

The balance equations (1) are solved approximately by a finite volume method. For this purpose the finite fluid domain $\Omega(t)$ is split into a finite set of moving subdomains, the cells $V_i(t)$, such that all $V_i(t)$ are disjoint at each instant of time and that their union gives $\Omega(t)$. Furthermore let $\mathcal{N}(i)$ be the set of cells that have a common edge with the cell i, and for $j \in \mathcal{N}(i)$ let $e_{ij}(t) := \partial V_i(t) \cap \partial V_j(t)$ be the interface between the cells i and j. The time interval is discretized by $t^{n+1} = t^n + \Delta t$ assuming a constant time step size. On this particular discretization the finite volume scheme can be written as

$$|V_i^{n+1}|\,\boldsymbol{v}_i^{n+1} = |V_i^n|\,\boldsymbol{v}_i^n - \Delta t \sum_{j \in \mathcal{N}(i)} |e_{ij}|\,\boldsymbol{F}(\boldsymbol{v}_i^n, \boldsymbol{v}_j^n, \dot{\boldsymbol{x}}_{ij}, \boldsymbol{n}_{ij}) \qquad (3)$$

using an explicit time discretization to compute the approximated cell averages $\boldsymbol{v}_i^{n+1}$ on the new time level. Here the numerical flux function $\boldsymbol{F}(\boldsymbol{u}, \boldsymbol{v}, \dot{\boldsymbol{x}}, \boldsymbol{n})$ is an approximation for the flux $\boldsymbol{f}(\boldsymbol{u}, \dot{\boldsymbol{x}}, \boldsymbol{n})$ in normal direction on the edge e_{ij}. It is assume to be *consistent*, i.e.,

$$\boldsymbol{F}(\boldsymbol{u}, \boldsymbol{u}, \dot{\boldsymbol{x}}, \boldsymbol{n}) = \boldsymbol{f}(\boldsymbol{u}, \dot{\boldsymbol{x}}, \boldsymbol{n}) := \boldsymbol{f}(\boldsymbol{u}, \dot{\boldsymbol{x}}) \cdot \boldsymbol{n}. \qquad (4)$$

For simplicity of presentation we neglect that due to higher order reconstruction the numerical flux usually depends on an enlarged stencil of cell averages.

3.1 Grid generation and grid movement

For the simulation of moving boundaries the grid generator has to cope with time dependent domain boundaries. To accomplish this task efficiently we employ for each time level t^n a *parametric mapping* $\mathbf{x} : [0,1]^2 \to \Omega$ from a logical space to the physical domain $\Omega(t^n)$. In this setting grid cells are the images of the corresponding cells in logical space, i.e., $V_i = \boldsymbol{x}(R_i)$ corresponding to the interval $R_i \subset [0,1]^2$. Then the discrete grid is determined simply by function evaluation.

For the representation of such a parameter mapping we use tensor product B-splines, i.e., $\mathbf{x}(u,v) = \sum_{i=0}^{N} \sum_{j=0}^{M} \boldsymbol{P}_{i,j}\, N_{i,p_u,U}(u)\, N_{j,p_v,V}(v)$. Here $N_{i,p,T}$ denotes the i-th normalized B-spline of order p with respect to the knot vector T. In our applications we usually choose cubic splines ($p = 4$), cf. [1].

The $\boldsymbol{p}_{ij}$ are the *control points* that are not to be confused with grid points. Typically, the number of control is much smaller than the number of grid points in the discrete grid. This makes grid deformation by parametric B-spline mappings highly efficient; only few control points have to be moved instead of all the grid points in the discrete grid. More elaborate details on grid generation via B-Splines can be found in [4].

From the grid functions we compute a space-time grid function that is realized by a two-level time discretization: before the timestep $t^n \to t^{n+1}$ is performed the grid generation module provides two grid representations $\boldsymbol{x}(\boldsymbol{\xi}, t^n)$ and $\boldsymbol{x}(\boldsymbol{\xi}, t^{n+1})$ at time levels t^n and t^{n+1}, respectively. Then for $t \in (t^n, t^{n+1})$ the grid function is determined by linear interpolation.

3.2 The geometric conservation law

The "geometric conservation laws" are discrete consistency conditions for the finite volume scheme. They stem from the requirement, that a reasonable numerical method should at least be able to maintain a constant flow field: if $\boldsymbol{u}(\boldsymbol{x}, t) = \boldsymbol{u}_\infty$ for all $(\boldsymbol{x}, t)$, then we require that the numerical solution fulfills $\boldsymbol{u}_i^n = \boldsymbol{u}_\infty$ for all index pairs (i, n), too. In the special case of a stationary grid we then get for each cell V_i the consistency condition for discretizations of the form (3)

$$\boldsymbol{0} = \sum_{j \in N(i)} |e_{ij}| \boldsymbol{n}_{ij}. \qquad (5)$$

What people usually understand to be "the" geometric conservation law stems from the requirement that the constant homogeneous flow should also be reproduced if the mesh is moving. If we assume equation (5) to be satisfied, we end up for each cell V_i with the condition

$$|V_i^{n+1}| - |V_i^n| = \Delta t \sum_{j \in N(i)} |e_{ij}| \kappa_{ij}. \qquad (6)$$

Here $\kappa_{ij} = \boldsymbol{n}_{ij} \cdot \dot{\boldsymbol{x}}_{ij}$ denotes the normal grid velocity on the face e_{ij}. The grid generator has to provide the quantities $|e_{ij}|, \boldsymbol{n}_{ij}, \kappa_{ij}$ and $|V_i|$ for the flow solver such that the consistency conditions (5) and (6) hold. On a curvilinear grid where these quantities are not uniquely defined this can be achieved by evaluating the integrals

$$\boldsymbol{N}_{ij} := \int_{e_{ij}(t)} \boldsymbol{n}(s, t) ds, \quad S_{ij} := \int_{t^n}^{t^{n+1}} \int_{e_{ij}(t)} \dot{\boldsymbol{x}}(s, t) \cdot \boldsymbol{n}_{ij}(s, t) \, ds dt.$$

exactly and then setting

$$|e_{ij}| := \|\boldsymbol{N}_{ij}\|_2, \quad \boldsymbol{n}_{ij} := \boldsymbol{N}_{ij}/|e_{ij}|, \quad \kappa_{ij} := S_{ij}/(\Delta t \, |e_{ij}|). \qquad (7)$$

3.3 The numerical flux

The fluxes in normal direction are approximated by an approximate Riemann solver. Since the cell edges are time-dependent we have to take into account the grid movement when solving the Riemann problem at the interfaces. For this purpose, we exploit the rotational and Galilean invariance of the underlying balance equations (1). Then we can rewrite the fluxes in normal direction as

$$f(u,\dot{x},n) = S\,f(S^{-1}u,0,n) \quad \text{with} \quad S = \begin{pmatrix} 1 & \mathbf{0}^T & 0 \\ \dot{x} & I & 0 \\ \frac{1}{2}\dot{x}^2 & \dot{x}^T & 1 \end{pmatrix},$$

cf. [6]. Carrying this identity over to the numerical flux we obtain

$$F(u_l, u_r, \dot{x}, n) = S\,F(S^{-1}u_l, S^{-1}u_r, 0, n). \tag{8}$$

Hence, we may derive a numerical flux over moving edges from standard numerical fluxes on stationary grids. Note that in the computations only the normal grid velocity κ is essentially needed. To perform the transformation (8) step by step it is sufficient to use $\kappa\,n$ instead of $\dot{x}$. This is admissible provided that the numerical flux is rotational invariant.

In the present work we use Roe's approximate Riemann solver. In order to avoid non-physical expansion shocks we use Harten's entropy fix. The spatial and temporal accuracy are improved by using a quasi one-dimensional second-order ENO reconstruction and Taylor expansion according to [3]. Here the reconstruction is applied to the characteristic variables.

4 Adaptive multiscale method

The efficiency of the reference finite volume scheme presented in Section 3 is significantly improved by employing recent multiscale-based grid adaptation techniques. Here we briefly summarize the basic conceptual ideas. For technical details we refer the reader to the book [7] and [1], respectively.

4.1 Multiscale-based spatial grid adaptation

Step 1: Multiscale analysis. The fundamental idea is to present the cell averages $\hat{u}_L$ representing the discretized flow field at fixed time level t^n on a given uniform highest level of resolution $l = L$ (*reference mesh*) associated with a given finite volume discretization (*reference scheme*) as cell averages on some coarsest level $l = 0$ where the fine scale information is encoded in arrays of *detail coefficients* d_l, $l = 0, \ldots, L-1$ of ascending resolution, see Figure 2.

The multiscale decomposition is performed on a hierarchy of *nested* grids $\mathcal{G}_l$ with increasing resolution $l = 0, \ldots, L$ determined by dyadic grid refinement of the logical space, see Figure 1. Note that this grid hierarchy can be efficiently realized by the parametric B-spline mappings in Section 3.1.

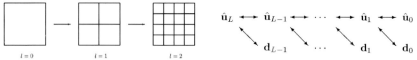

Fig. 1. Sequence of nested grids

Fig. 2. Multiscale transformation

Step 2: Thresholding. It can be shown that the detail coefficients become small with increasing refinement level when the underlying function is smooth. In order to compress the original data this motivates us to discard all detail coefficients $d_{l,k}$ whose absolute values fall below a level-dependent threshold value $\varepsilon_l = 2^{l-L}\varepsilon$. Let $\mathcal{D}_{L,\varepsilon}$ be the set of *significant details*. The ideal strategy would be to determine the threshold value ε such that the *discretization error* of the reference scheme, i.e., difference between exact solution and reference scheme, and the *perturbation error*, i.e., the difference between the reference scheme and the adaptive scheme, are balanced, see [2].

Step 3: Prediction and grading. Since the flow field evolves in time, grid adaptation is performed after each evolution step to provide the adaptive grid at the *new* time level. In order to guarantee the adaptive scheme to be *reliable* in the sense that no significant future feature of the solution is missed, we have to *predict* all significant details at the new time level $n+1$ by means of the details at the *old* time level n. Let $\widetilde{\mathcal{D}}_{L,\varepsilon}^{n+1} \supset \mathcal{D}_{L,\varepsilon}^{n} \cup \mathcal{D}_{L,\varepsilon}^{n+1}$ be the prediction set. The prediction strategy is detailed in [2]. In view of the grid adaptation step this set is additionally inflated such that it corresponds to a graded tree, i.e., the number of levels between two neighboring cells differs at most by 1.

Step 4: Grid adaptation. By means of the set $\widetilde{\mathcal{D}}_{L,\varepsilon}^{n+1}$ a locally refined grid is determined. For this purpose, we recursively check proceeding levelwise from coarse to fine whether there exists a significant detail to a cell. If there is one, then we refine the respective cell. We finally obtain the locally refined grid with hanging nodes represented by the index set $\mathcal{G}_{L,\varepsilon}$.

4.2 Multilevel time stepping

Since the reference scheme (3) is assumed to use an explicit time discretization, the time step size is bounded due to the CFL condition by the smallest cell in the grid. Hence Δt is determined by the highest refinement level L, i.e., $\Delta t = \tau_L$. However, for cells on the coarser scales $l = 0, \ldots, L-1$ we may use $\Delta t = \tau_l = 2^{L-l}\tau_L$ to satisfy locally the CFL condition. In [8] a multilevel time stepping strategy has been incorporated recently to the adaptive multiscale finite volume scheme as proposed in [7]. The basic idea is to save flux evaluations where the local CFL condition allows a large time step. The precise time evolution algorithm is schematically described by Fig. 3: In a global time stepping, i.e., using $\Delta t = \tau_L$ for all cells, each vertical line section appearing in Fig. 3 (left) represents a flux evaluation and each horizontal

 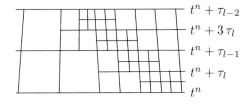

Fig. 3. Synchronized time evolution on space-time grid

line (dashed or drawn) represents a cell update of u due to the fluxes. In the multilevel time stepping a flux evaluation is only performed at vertical line sections that emanate from a point where at least one drawn horizontal line section emanates from. If a vertical line section emanates from a point, where two dashed horizontal sections emanate from, then we do not recompute the flux, but keep the flux value from the preceeding vertical line section. Hence fluxes are only computed for the vertical edges in Fig. 3 (right).

Note that on each intermediate time level (horizontal lines) u is updated for *all* cells and that grid adaptation is performed at each *even* intermediate time level, i.e., at $t^n + k\,\tau_L$ for k even. Hence it is possible to track, for instance, a shock movement on the intermediate time levels instead of a–priori refining the whole range of influence, see Fig. 3 (right).

Note further that τ_0 is the time scale at which the grid movement takes place. The grid boundary is only computed at the time levels $t^n, t^{n+1}, \ldots$ and on the intermediate time levels the grid movement is a linear interpolation between the grid positions at t^n and t^{n+1}. This means that the time step size $t^{n+1} - t^n$ is dictated either by the time step of the boundary movement or the time step size τ_0 according to the CFL condition on the coarsest spacial scale.

5 Numerical results

This example shows the inviscid flow over an oscillating plate with prescribed deformation in time. The flow domain extends from -5 to 5 in x-direction and from 0 to 5 in y-direction. At time $t = 0$ the lower boundary starts a periodic oscillation in the interval $[0,1]$ prescribed by a B-Spline representation $\boldsymbol{x}(\xi,t) = \sum_{i=0}^{12} \boldsymbol{p}_i(t) N_{i,4,T}(\xi)$. Here $T = (0,0,0,0,\frac{1}{10},\frac{2}{10},\ldots,\frac{9}{10},1,1,1,1)$ and the movement of the control points is given by $\boldsymbol{p}_0 = (0,0)^T$, $\boldsymbol{p}_{12} = (1,0)$, $\boldsymbol{p}_1 = (\frac{1}{30},0)^T$, $\boldsymbol{p}_{11} = (\frac{29}{30},0)^T$, and $\boldsymbol{p}_i(t) = (-\frac{1}{10}+\frac{i}{10},\frac{1}{5}\sin(t/t_{ref})\sin(\frac{\pi}{8}(i-2)))$ for $i = 2,\ldots,10$. Due to the simplicity of the geometry the grid deformation is performed using transfinite interpolation techniques. The flow enters the domain from the left hand side with free-stream conditions $\rho_\infty = 1.2929$ [kg/m³], $p_\infty = 101325$ [Pa], $\boldsymbol{v}_\infty = (165.619, 0)$ [m/s]. The reference time is determined by $t_{ref} = 1./\sqrt{p_\infty/\rho_\infty} = 279.947$ [m/s]. At the boundaries we impose slip conditions, i.e., $\dot{\boldsymbol{x}} \cdot \boldsymbol{n} = \boldsymbol{v} \cdot \boldsymbol{n}$, at the lower boundary

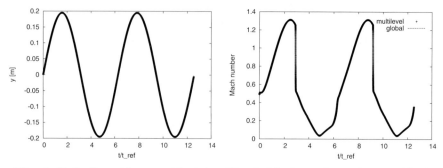

Fig. 4. Deflection at bump midpoint   **Fig. 5.** Mach number at bump midpoint

and characteristic boundary conditions elsewhere because of the subsonic free-stream conditions ($M_\infty = 0.5$). The grid is adapted after every timestep. The maximum refinement level is $L_{max} = 5$, the threshold $\varepsilon = 0.002$, the coarsest grid consist of 1375 cells. After two cycles of the boundary oscillation the number of grid cells varies around 40.000 grid points depending on the phase of the boundary movement.

The bump is moving periodically up and down which is reflected in Figure 4 where the deflection in the midpoint of the bump is shown. When the bump is moving upwards then a shock occurs at the leeward side because of the acceleration of the flow. The shock weakens and moves in upstream direction when the bump moves downward. This can be deduced from Figure 5 where the Mach number in the midpoint of the bump is plotted versus the dimensionless time t/t_{ref}. When the shock is passing a steep gradient can be seen.

The computation has been performed using the global and the multilevel time stepping strategy, respectively. Although we perform no grid deformation step for the intermediate time levels in the latter case the accuracy of the solution is not affected as can be concluded from Figure 5. On the other hand we gain a factor of 3.7 in comparison to a global time stepping strategy.

References

1. Bramkamp, F., Lamby, Ph. Müller, S.: An adaptive multiscale finite volume solver for unsteady and steady flow computations. Journal of Computational Physics, **197**, 460–490 (2004)
2. Cohen, A., Kaber, S.M., Müller, S., Postel, M.: Fully adaptive multiresolution finite volume schemes for conservation laws. Math. Comp, **72**, 183-225 (2003)
3. Harten, A., Engquist, B., Osher, S., Chakravarthy, SR.: Uniformly high order accurate essentially non–oscillatory schemes III. Journal of Computational Physics, **71**, 231–303 (1987)
4. Lamby, Ph.: Parametric Grid Generation using B-Spline Techniques. Phd Thesis, RWTH Aachen (2006)

5. Lamby, Ph., Müller, S., Stiriba, Y.: Solution of shallow water equations using fully adaptive multiscale schemes. International Journal for Numerical Methods in Fluids, **49**, 417–437 (2005)
6. Massjung, R.: Numerical Schemes and Well-Posedness in Nonlinear Aeroelasticity. Phd Thesis, RWTH Aachen (2002)
7. Müller, S.: Adaptive Multiscale Schemes for Conservation Laws. in: Lecture Notes on Computational Science and Engineering, vol. 27, Springer, Berlin Heidelberg New York (2002)
8. Müller, S., Stiriba, Y.: Fully adaptive multiscale schemes for conservation laws employing locally varying time stepping. IGPM-Report 238, RWTH Aachen (2004)

Multiphase Flow

A Relaxation Method for a Two Phase Flow with Surface Tension

C. Berthon[1,2], B. Braconnier[1,2,3], J. Claudel[3] and B. Nkonga[1,2]

[1] MAB, UMR 5466, Université Bordeaux 1, 351 cours de la libération, 33400 Talence, France.
[2] INRIA Futurs, projet ScAlApplix, Domaine de Voluceau-Rocquencourt, B.P. 105, 78153 Le Chesnay Cedex, France.
[3] CEA CESTA, route des Gargails, 33114 Le Barp.
braconnier.benjamin@math.u-bordeaux1.fr

Summary. The present work is devoted to the numerical approximation of a compressible two-phase flow. The phases are non-miscible and separated by an interface where capillary effects are considered. We use diffuse interface models having a single velocity and pressure. The surface tension forces are added with the CSF method [1]. We propose a Godunov type method based on a pressure relaxation procedure for the system including surface tension terms. Two numerical illustrations are performed showing the parasitic currents reduction and a liquid break-up.

1 Introduction

We are concerned with the numerical approximation of a two phase compressible flow. We assume the phases non-miscible and submitted to surface tension forces. For the sake of simplicity, the viscosity effects are neglected.

We consider an Eulerian formulation based on diffuse interface models. In this context, the interfaces are artificially smeared. The mathematical models [4, 6] consider the asymptotic when phases velocities and pressures are at equilibrium. Theses models write in the form:

$$\partial_t \mathbf{W} + \nabla \cdot (\mathbf{F}) + \underline{\mathbf{B}} \nabla \mathbf{W} = \mathbf{Q} \tag{1}$$

where:

$$\mathbf{W} = \begin{pmatrix} \alpha_1 \\ \alpha_1 \rho_1 \\ \alpha_2 \rho_2 \\ \rho \mathbf{u} \\ E \end{pmatrix}, \quad \mathbf{F} = \begin{pmatrix} 0 \\ \alpha_1 \rho_1 \mathbf{u} \\ \alpha_2 \rho_2 \mathbf{u} \\ \rho \mathbf{u} \otimes \mathbf{u} + p \\ (E+p)\mathbf{u} \end{pmatrix}, \quad \underline{\mathbf{B}} = \begin{pmatrix} \mathbf{u} & -\dfrac{\beta \mathbf{u}}{\rho} & -\dfrac{\beta \mathbf{u}}{\rho} & \dfrac{\beta}{\rho} & 0 \\ 0 & 0 & 0 & 0 & 0 \\ 0 & 0 & 0 & 0 & 0 \\ 0 & 0 & 0 & 0 & 0 \\ 0 & 0 & 0 & 0 & 0 \end{pmatrix} \tag{2}$$

and: $\mathbf{Q} = (0, 0, 0, \mathbf{q}, \mathbf{q} \cdot \mathbf{u})^T$. The variable $\alpha_1 \in (0,1)$ denotes the volume fraction for the first phase and $\alpha_2 = 1 - \alpha_1$ is recovered from the saturation constraint. For each phase k, we note $\rho_k > 0$ the partial density and $\varepsilon_k > 0$ the internal energy. Then, we define the mixture density, internal energy and total energy as follow:

$$\rho = \sum_{k=1}^{2} \alpha_k \rho_k, \quad \varepsilon = \sum_{k=1}^{2} \frac{\alpha_k \rho_k}{\rho} \varepsilon_k, \quad E = \rho \left(\varepsilon + \frac{1}{2} |\mathbf{u}|^2 \right) \tag{3}$$

where $\mathbf{u}$ is the fluid velocity and p is the pressure.

The parameter β is related to the compaction of the volume fraction. For the Kapila model [4], it is set to $\beta = \alpha_1 \alpha_2 \dfrac{\rho_1 c_1^2 - \rho_2 c_2^2}{\alpha_1 \rho_2 c_2^2 + \alpha_2 \rho_1 c_1^2}$. It governs the variation of the volume fraction across shock or rarefaction waves. There also exists a simplified model [6] obtained with $\beta = 0$. In this case, the volume fraction is constant except in the phases interface zone but the model does not respect exactly the physic of the fluid.

The source term $\mathbf{Q}$ is composed of two forces. We make the decomposition: $\mathbf{q} = \mathbf{f}_G + \mathbf{f}_{Sv}$. $\mathbf{f}_G = \rho \mathbf{g}$ stands for the gravitational force where $\mathbf{g}$ is the gravity. $\mathbf{f}_{Sv}$ is the surface tension force contribution. We focus on the method of Brackbill [1]. It requires a smooth color function noted Φ_1 to locate the interface. This function must be constant in the entire domain except in the neighborhood of the interface. In our case, the interfaces are artificially diffused and the variables α_1, $\alpha_1 \rho_1$ and $\alpha_2 \rho_2$ are smooth. Consequently, we propose the color function type: $\Phi_1 = \Phi_1(\alpha_1, \alpha_1 \rho_1, \alpha_2 \rho_2)$. Then the surface tension forces are recovered with the formulation:

$$\mathbf{f}_{Sv} = -\sigma \kappa \mathbf{n}_{Sv} \quad \text{with} \quad \mathbf{n}_{Sv} = \frac{\nabla(\phi_1)}{[\nabla(\phi_1)]} \quad \text{and} \quad \kappa = \nabla \cdot \left(\frac{\nabla(\phi_1)}{|\nabla(\phi_1)|} \right) \tag{4}$$

where σ is the surface tension coefficient, $\mathbf{n}_{Sv}$ the normal at the interface and κ the curvature. $[\nabla(\phi_1)]$ is the jump of the color function across the interface. For the closure of the system, we use a Stiffened gas equation of state:

$$p + \gamma P^\infty = (\gamma - 1) \rho \varepsilon \tag{5}$$

where the parameters γ and P^∞ are defined as follow:

$$\frac{1}{\gamma - 1} = \sum_k \frac{\alpha_k}{\gamma_k - 1} \quad \text{and} \quad \frac{\gamma P^\infty}{\gamma - 1} = \sum_k \frac{\alpha_k \gamma_k P_k^\infty}{\gamma_k - 1} \tag{6}$$

with γ_k and P_k^∞ are constant data of the phase k.

In order to perform the mathematical analysis of the non-conservative system (15), we use the projection of the system along the normal $\boldsymbol{\eta}$ and set $\mathbf{q} = 0$. It rewrites:

$$\partial_t \mathbf{W} + \underline{\mathbf{A}} \partial_\eta \mathbf{W} = 0 \quad \text{where} \quad \underline{\mathbf{A}} = \left(\frac{\partial \mathbf{F}}{\partial \mathbf{W}} + \underline{\mathbf{B}} \right) \cdot \boldsymbol{\eta} \tag{7}$$

The system is unconditionally hyperbolic with eigenvalues: $\mathbf{u} \cdot \boldsymbol{\eta} - c$, $\mathbf{u} \cdot \boldsymbol{\eta}$, $\mathbf{u} \cdot \boldsymbol{\eta} + c$ where the sound speed is given by the relation:

$$\begin{cases} \dfrac{1}{\rho c^2} = \dfrac{\alpha_1}{\rho_1 c_1^2} + \dfrac{\alpha_2}{\rho_2 c_2^2} & \text{for the Kapila model} \\[2ex] \rho c^2 = \dfrac{\alpha_1 \rho_1^2 c_1^2/(\gamma_1 - 1) + \alpha_2 \rho_2^2 c_2^2/(\gamma_2 - 1)}{\alpha_1 \rho_1/(\gamma_1 - 1) + \alpha_2 \rho_2/(\gamma_2 - 1)} & \text{for the simplified model} \end{cases} \tag{8}$$

2 Numerical approximation

2.1 Geometrical parameters

We consider a spatial domain $\Omega \in R^d$ and a mesh of elements (τ) (d-simplexes) defining the discrete space domain. Let us consider a decomposition C_i ($i = 1, .., Ns$) of the domain Ω into non-intersecting cells such that:

$$\bigcup_{i=1}^{Ns} C_i = \Omega \quad \text{and} \quad \dot{C}_i \bigcap \dot{C}_j = \emptyset \quad \text{if } i \neq j \tag{9}$$

We denote $\nu(i)$ the set of the neighboring vertices of the node i. Each cell C_i is decomposed in elementary area C_{ij} and its frontier ∂C_i is also decomposed in elementary frontiers ∂C_{ij} such that:

$$C_i = \bigcup_{j \in \nu(i)} C_{ij} \quad \partial C_i = \bigcup_{j \in \nu(i)} \partial C_{ij}^+ \tag{10}$$

We note $a_i = \int_{C_i} dx$ the area of the domain C_i and $a_{ij} = \int_{C_{ij}} dx$ the area of the domain C_{ij}. Then, we define:

$$\boldsymbol{\eta}_{ij} = \int_{\partial C_{ij}} \mathbf{n} dx \tag{11}$$

where $\mathbf{n}$ is the unit normal of ∂C_{ij} directed from C_i to C_j. We have the relation: $a_i = \sum_{j \in \nu(i)} a_{ij}$ and we use the convention $\boldsymbol{\eta}_{ij} = -\boldsymbol{\eta}_{ji}$.

2.2 Relaxation based Godunov type method

Godunov type schemes are based on 1D Riemann problems in the direction of cells interfaces normals $\boldsymbol{\eta}_{ij}$:

$$\partial_t \mathbf{V} + \underline{\mathbf{A}}\partial_\eta \mathbf{V} = 0 \quad \text{with} \quad \mathbf{V}(t^n, \eta) = \begin{cases} \mathbf{V}_i^n & \text{if } \eta < 0 \\ \mathbf{V}_j^n & \text{if } \eta > 0 \end{cases} \quad (12)$$

where η is a coordinate associated to η_{ij}. Then, the update state $\mathbf{V}_i^{n+1}$ defined on the cells C_i is determined as follow:

$$a_i \mathbf{V}_i^{n+1} = \sum_{j \in \nu(i)} a_{ij} \mathbf{V}_{ij}^{n+1} \quad \text{where} \quad \mathbf{V}_{ij}^{n+1} = \frac{1}{a_{ij}} \int_{C_{ij}} \mathcal{V}_{ij}(\eta) d\eta \quad (13)$$

where $\mathcal{V}_{ij}$ is the exact or approximate solutions of (12). The later relation is not usual for computation and we propose another relation closed to the usual fluxes balance formulation. In fact, we rewrite the updated solution with fluctuations over each cell interfaces:

$$\mathbf{V}_i^{n+1} = \mathbf{V}_i^n - \frac{\Delta t}{a_i} \sum_{j \in \nu(i)} \phi_{ij} \quad \text{where} \quad \phi_{ij} = \frac{1}{\Delta t} \int_{C_{ij}} (\mathbf{V}_i^n - \mathcal{V}_{ij}(\eta)) \, d\eta \quad (14)$$

In the general case of a non-conservative system, we have to determine two fluctuations at each cell interface as $\phi_{ij} \neq \phi_{ji}$. In the case of a conservative system, the fluctuations equal and the scheme becomes conservative.

3 Relaxation method for the simplified model with surface tension forces

After the works of Jin and Xin [3], Liu [5] and Suliciu [8], the relaxation method can be viewed as a well-established tool to approximate the solution of the compressible Euler equations of gas dynamics.

We propose a relaxation method for the simplified reduced system (15) with surface tension terms. In our physical context, we suppose the Mach number low such that $\nabla \cdot (\boldsymbol{u}) \approx 0$ and we focus on the simplified model (15) obtained with the approximation $\beta \nabla \cdot (\boldsymbol{u}) = 0$. Then, the volume fraction α_1 remains constant in the entire domain except at the interface so that we set $\Phi_1 = \alpha_1$ for the surface tension formulation (others choices are possible, Perigaud and Saurel use $\Phi_1 = \alpha_1 \rho_1/\rho$ in [7]). The curvature is locally frozen yielding a first order derivative formulation of the force.

Relaxation system

The major difficulty for the resolution of the Riemann problem associated to the system previously described, is related to the non-linearity of the conservative flux $\boldsymbol{F}$ in particular due to the pressure fields. In the context of relaxation methods, we introduce a new variable π supposed to tend toward the pressure p. We suppose this parameter free of thermodynamical property and governed by an equation closed to the pressure equation. Moreover, we want this equation to lead to a straightforward Riemann problem. We use:

$$\partial_t \pi + \boldsymbol{u} \cdot \nabla \pi + \frac{a^2}{\rho} \nabla \cdot (\boldsymbol{u}) = \frac{1}{\lambda}(p - \pi) \tag{15}$$

where the parameter a, governed by a transport equation, is detailed later on. Let note, on the second hand of the equation, the presence of a relaxation procedure making π tend toward p. The relaxation system writes:

$$\partial_t \mathbf{W}_R + \nabla \cdot (\boldsymbol{F}_R) + \underline{\boldsymbol{B}}_R \nabla \mathbf{W}_R = \frac{1}{\lambda} \mathbf{R} \tag{16}$$

where:

$$\mathbf{W}_R = \begin{pmatrix} \alpha_1 \\ \alpha_1 \rho_1 \\ \alpha_2 \rho_2 \\ \rho \boldsymbol{u} \\ E \\ a \\ \rho \pi \end{pmatrix}, \boldsymbol{F}_R = \begin{pmatrix} 0 \\ \alpha_1 \rho_1 \boldsymbol{u} \\ \alpha_2 \rho_2 \boldsymbol{u} \\ \rho \boldsymbol{u} \otimes \boldsymbol{u} + \pi \\ (E + \pi) \boldsymbol{u} \\ 0 \\ (\rho \pi + a^2) \boldsymbol{u} \end{pmatrix}, \underline{\boldsymbol{B}}_R = \begin{pmatrix} \boldsymbol{u} & 0 & 0 & 0 & 0 & 0 & 0 \\ 0 & 0 & 0 & 0 & 0 & 0 & 0 \\ 0 & 0 & 0 & 0 & 0 & 0 & 0 \\ \sigma \widetilde{\kappa} & 0 & 0 & 0 & 0 & 0 & 0 \\ \sigma \widetilde{\kappa} \boldsymbol{u} & 0 & 0 & 0 & 0 & 0 & 0 \\ 0 & 0 & 0 & 0 & \boldsymbol{u} & 0 \\ 0 & 0 & 0 & 0 & 0 & 0 & 0 \end{pmatrix} \tag{17}$$

and $\mathbf{R} = (0, 0, 0, 0, 0, 0, \rho(p - \pi))^T$. $\widetilde{\kappa}$ is a constant approximation of the curvature. Then we project the system (16) along the normal $\boldsymbol{\eta}$ and write it in the non-conservative form:

$$\partial_t \mathbf{W}_R + \underline{\boldsymbol{A}}_R \nabla \mathbf{W}_R = \frac{1}{\lambda} \mathbf{R} \quad \text{with} \quad \underline{\boldsymbol{A}}_R = \left(\frac{\partial \boldsymbol{F}_R}{\partial \mathbf{W}_R} + \underline{\boldsymbol{B}}_R \right) \cdot \boldsymbol{\eta} \tag{18}$$

With λ set to infinity, the system is unconditionally hyperbolic and linearly degenerated. We illustrate the Riemann invariants associated to each wave in the table (1). Let us point out that all the characteristic fields of the system

Table 1. Riemann invariant for the relaxation system

λ	I_λ^1	I_λ^2	I_λ^3	I_λ^4	I_λ^5	I_λ^6
$\boldsymbol{u} \cdot \boldsymbol{n} - \frac{a}{\rho}$	$\boldsymbol{u} \cdot \boldsymbol{n} - \frac{a}{\rho}$	$\pi + a \boldsymbol{u} \cdot \boldsymbol{n}$	$\pi^2 - 2a^2 \varepsilon$	α_1	a	$\boldsymbol{u} \cdot \boldsymbol{n}^\perp$
$\boldsymbol{u} \cdot \boldsymbol{n}$	$\boldsymbol{u} \cdot \boldsymbol{n}$	$\pi + \sigma \widetilde{\kappa} \alpha_1$				
$\boldsymbol{u} \cdot \boldsymbol{n} + \frac{a}{\rho}$	$\boldsymbol{u} \cdot \boldsymbol{n} + \frac{a}{\rho}$	$\pi - a \boldsymbol{u} \cdot \boldsymbol{n}$	$\pi^2 - 2a^2 \varepsilon$	α_1	a	$\boldsymbol{u} \cdot \boldsymbol{n}^\perp$

(16) are linearly degenerated and the resolution of the associated Riemann problem is straightforward.

Stability of the relaxation system

According to Liu [5] and Whitam [9] a compatibility condition must be satisfied by the relaxation system (16) to prevent numerical instabilities when λ

goes to zero. This condition leading to a stability requirement on the parameter a, is based on a first order asymptotic equilibrium system. In that way, we consider the Chapmann-Enskog expansion of a small departure π^λ from the equilibrium pressure p:

$$\pi = p + \lambda \pi^\lambda. \tag{19}$$

Substituting (19) into (16) and neglecting higher order terms, we end up with the following first order asymptotic equilibrium system:

$$\partial_t \mathbf{W} + \nabla \cdot (\mathbf{F}) + \underline{\mathbf{B}} \nabla \mathbf{W} = \lambda \mathbf{C} \nabla \cdot (\mathbf{D} \nabla \mathbf{W}) \tag{20}$$

where the matrices $\mathbf{C}$ and $\mathbf{D}$ are not detailed in the present paper. The stability condition comes from the requirement that the first-order correction operator in (20) must be dissipative relatively to the zero-order approximation. After some calculations, we find the stability condition:

$$a > \rho c \tag{21}$$

Numerical strategy

The numerical strategy is based on a splitting operator technique. Let us illustrate a time evolution from time t^n to $t^{n+1} = t^n + \Delta t$. First, we consider the system (16) with λ set to infinity. In this way, we use the formulation exposed in the section (2.2). The initial data of the Riemann problem are deduced from the equilibrium states: $\mathbf{W}_R = (\alpha_1, \alpha_1 \rho_1, \alpha_2 \rho_2, \rho \mathbf{u}, E, a, \rho p)^T$ at time t^n. We note $\mathbf{W}_R^{n+1,*}$ the solution obtained with the formula (14).

Secondly, we perform the relaxation procedure by resolving the ODE system:

$$\partial_t(\mathbf{W}_R) = \frac{1}{\lambda} \mathbf{R} \tag{22}$$

with initial data $\mathbf{W}_R^{n+1,*}$ and λ set to zero. The numerical approximation of gravity forces is standard and is not exposed in the paper.

3.1 Curvature evaluation

The evaluation of the curvature requires the knowledge of the gradient of the color function. We reconstruct this piecewise constant function on the entire domain and evaluate its gradient as follow:

$$\hat{\Phi}_1 = \sum_{i=1}^{Ns} (\Phi_1)_i \phi_i, \quad \nabla \hat{\Phi}_1 = \sum_{i=1}^{Ns} (\Phi_1)_i \nabla \phi_i \tag{23}$$

where ϕ_i are the basis function of the P^1 finite element formulation defined by $\phi_i(x_k) = 1$ if $i = k$ and 0 else. For the approximation of the curvature, we use the divergence formula for the function $\hat{s} = \nabla(\hat{\Phi}_1)/|\nabla(\hat{\Phi}_1)|$. It comes:

$$K_i = \frac{1}{a_i} \int_{C_i} \nabla \cdot (\hat{\boldsymbol{s}}) dx = \frac{1}{a_i} \int_{C_{ij}} \hat{\boldsymbol{s}} \cdot \boldsymbol{\eta}_{ij} d\eta \approx \frac{1}{a_i} \sum_{j \in \nu(i)} \boldsymbol{s}_{ij} \cdot \boldsymbol{\eta}_{ij} \qquad (24)$$

where $\boldsymbol{s}_{ij} = (1-\mu)\boldsymbol{s}_i + \mu \boldsymbol{s}_j$ ($\mu \in (0,1]$) is an approximate value of the function $\hat{\boldsymbol{s}}$ at the frontier ∂C_{ij}. In the present paper, we chose $\mu = 0.5$.

4 Numerical results

Parasitic currents are the major problem in the numerical approximation of the capillary effects by a volume force [2]. By way of illustration, we consider a square domain with a discus of light fluid surrounded of heavy fluid. The gravity effects are set to zero. Due to the surface tension force, the pressure is discontinuous at the fluid interface and its jump is governed by the Laplace law. We run this system at equilibrium and observe the parasitic currents generated. We plot in figure (1) the integral of the velocity L^2 norm over the domain. The curve denoted M1 is obtained when surface tension terms are resolved by a splitting technique and M2 when our method is used. Let emphasize that our method generates law parasitic currents.

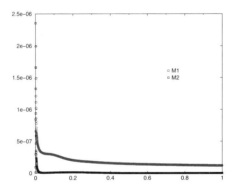

Fig. 1. $\int_\Omega |\boldsymbol{u}|_2(x,t) dx$ function of time $t \in [0,1]$. $M1$ splitting based resolution of surface tension, $M2$ relaxation method. The domain is $0.1m \times 0.1m$ and discretised with 40×40 points. At $t = 0$, we set: $\rho_1 = 100 kg.m^{-3}$, $\rho_2 = 1000 kg.m^{-3}$ and $u = 0 m.s^{-1}$ in the entire domain. The bubble center is $(0.5, 0.5)$ and radius is $0.03m$. Inside we set $\alpha_1 = 0.99$ $p = 10300 Pa$ and outside $\alpha_1 = 0.01$ $p = 10000 Pa$. $\sigma = 0.09 uSI$, $g = 0m.s^{-2}$.

The second illustration is a liquid break-up. We consider a rectangular domain filled with air and with a liquid layer at its base. At the initial time, a water bubble is at the top of the domain. The gravity forces are preponderant compared to the surface tension forces such that the bubble breaks and forms a liquid jet. Then, this liquid jet merges with the liquid layer. We plot in figure (2) the color function.

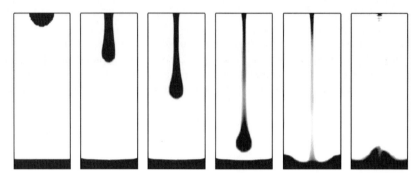

Fig. 2. Color function for a liquid break-up computation at times $t = 14, 126, 182, 239, 281$ and $478ms$. The domain is $0.15m \times 0.4m$ discretised with 45×120 points. We set $\rho_1 = 1 kg.m^{-3}$, $\rho_2 = 1000 kg.m^{-3}$, $u = 0 m.s^{-1}$ and $p = 100000 Pa$. At $t = 0$, liquid layer height $0.025m$, bubble shape: half discus with center $(0.075,4)$ and radius $0.034m$. In the air zone $\alpha_1 = 0.999$ and $\alpha_1 = 0.001$ in the water zone. $\sigma = 5.0 uSI$, $g = 20 m.s^{-2}$, contact angle $110°$.

5 Conclusion

A relaxation system for a two-phase flow with surface tension force has been devised and its numerical implementation has been performed. Two experiments have been run proving the relevance of the method. The future work is devoted to the extension of the method for high order MUSCL technique and viscous fluids.

References

1. Brackbill, J.U., Kothe, D.B., Zemach, C.: A Continuum Method for Modeling Surface Tension. Journal of Computational Physics, **100**, 335–354 (1992)
2. Jamet, D., Torres, D., Brackbill, J.U.: On the Theory and Computation of Surface Tension: The Elimination of Parasitic Currents through Energy Conservation in the Second-Gradient Method. Journal of Computational Physics, **182**, 262–276 (2002)

3. Jin, S., Xin, Z. P.: The relaxation schemes for systems of conservation laws in arbitrary space dimensions. Comm. Pure Appl. Math. **48**, 235–278 (1995)
4. Kapila, A.K., Menikoff, R., Stewart, D.S.: Two-Phase Modeling of Deflagration to Detonation Transition in Granular Materials: Reduced Equations. Physics of Fluids, **13**, 3002–3024 (2001)
5. Liu, T.P.: Hyperbolic conservation laws with relaxation. Comm. Math. Phys.,**108**, 153–175, (1987)
6. Massoni, J., Saurel, R., Nkonga, B., Abgrall, R.: Proposition de méthodes et modèles eulériens pour les problèmes à interfaces entre fluides compressibles en présence de transfert de chaleur. Journal of Heat and Mass Transfer, **45**, No. 6, 1287–1307 (2001)
7. Perigaud, G., Saurel, R.: A compressible flow model with capillary effects. Journal of Computational Physics, **209**, 139–178 (2005)
8. Suliciu, I.: Energy estimates in rate-type thermo-viscoplasticity. Int. J. Plast., **14**, No. 1-3, 227–224 (1998)
9. Whitham, J.: Linear and Nonlinear Waves. Wiley, New-York (1974)

Extension of Interface Coupling to General Lagrangian Systems

A. Ambroso[3], C. Chalons[2], F. Coquel[1], E. Godlewski[1], F. Lagoutière[2], P.-A. Raviart[1] and N. Seguin[1]

[1] Université Pierre et Marie Curie-Paris6, UMR 7598 LJLL, F-75005 France; CNRS, UMR 7598 LJLL, Paris, F-75005 France
  godlewski@ann.jussieu.fr
[2] Université Paris 7-Denis Diderot et UMR 7598 LJLL, F-75005 France
[3] CEA-Saclay, F-91191, Gif-sur-Yvette cedex, France

Summary. We study the coupling of two gas dynamics systems in Lagrangian coordinates at the interface $x = 0$. The coupling condition was formalized in [9, 10] by requiring that two boundary value problems should be well-posed, and it yields as far as possible the continuity of the solution at the interface. In this work we prove that we may choose the variables we transmit and extend the theory to Lagrangian systems of different sizes. The coupling condition is expressed in terms of Riemann problems. This is well suited for the numerical methods we are implementing and adapted to Lagrangian systems since the sign of the wave speeds is known, which enables us to solve the coupled Riemann problem.

1 Introduction

We are interested in the study of the coupling of two different hyperbolic systems at a fixed interface. In [9], a new coupling condition (CC in the sequel) is defined which results by expressing that two boundary value problems should be well-posed and the approach is justified in the scalar case. This CC resumes to impose as far as possible the continuity of the solution at the interface. The case of linear systems and ideas for the Euler system follow in [10]. Here, we show that we can choose the set of dependent variables which is transmitted and apply the result to systems in Lagrangian coordinates for which the solution of the coupled Riemann problem is given explicitly and illustrated numerically. We have expressed the boundary conditions in terms of Riemann problems. This approach is well suited for the two-flux method we are implementing and linked to the theoretical results concerning the convergence in the scalar case [9]. We first describe the theoretical settings and precise our notations. The case of the p−system is then detailed, with numerical illustrations and Lagrangian systems are considered in the next sections.

Let $\Omega \subset \mathrm{R}^p$ be the set of states and let $\mathbf{f}_\alpha, \alpha = L, R$, be two 'smooth' functions from Ω into R^p. Given a function $\mathbf{u}_0 : x \in \mathrm{R} \to \mathbf{u}_0(x)$, we want to find a function $\mathbf{u} : (x,t) \in \mathrm{R} \times \mathrm{R}_+ \to \mathbf{u}(x,t) \in \Omega$ solution of

$$\partial_t \mathbf{u} + \partial_x \mathbf{f}_L(\mathbf{u}) = \mathbf{0}, \quad x < 0, \ t > 0, \tag{1}$$

$$\partial_t \mathbf{u} + \partial_x \mathbf{f}_R(\mathbf{u}) = \mathbf{0}, \quad x > 0, \ t > 0, \tag{2}$$

satisfying the initial condition $\mathbf{u}(x,0) = \mathbf{u}_0(x), x \in \mathrm{R}$, and at the interface $x = 0$, a coupling condition CC which we now describe.

1.1 Coupling procedure

We have chosen this CC in order to obtain two well-posed initial boundary-value problems in $x > 0, t \geq 0$ and in $x < 0, t \geq 0$. This means that the trace $\mathbf{u}(0-,t)$ (resp. $\mathbf{u}(0+,t)$), $t \geq 0$, should be an admissible boundary condition at $x = 0$ for the system in $x > 0$ (resp. $x < 0$). We will assume that the systems are hyperbolic, i.e. for $\alpha = L, R$, the Jacobian matrix $\mathbf{f}'_\alpha(\mathbf{u})$ of $\mathbf{f}_\alpha(\mathbf{u})$ is diagonalizable with real eigenvalues $\lambda_{\alpha,k}(\mathbf{u})$ and corresponding eigenvectors $\mathbf{r}_{\alpha,k}(\mathbf{u}), 1 \leq k \leq p$. Then we introduce the solution of the Riemann problem for the system associated to the flux $\mathbf{f}_\alpha$, $\mathbf{u}(x,t) = \mathbf{W}_\alpha(x/t; \mathbf{u}_L, \mathbf{u}_R)$, i.e., the solution of $\partial_t \mathbf{u} + \partial_x \mathbf{f}_\alpha(\mathbf{u}) = \mathbf{0}$ with initial condition

$$\mathbf{u}(x,0) = \begin{cases} \mathbf{u}_L, & x < 0, \\ \mathbf{u}_R, & x > 0. \end{cases} \tag{3}$$

Following [7], we set for all $\mathbf{b} \in \Omega$, $\mathcal{O}_L(\mathbf{b}) = \{\mathbf{W}_L(0-; \mathbf{u}, \mathbf{b}); \mathbf{u} \in \Omega\}$, $\mathcal{O}_R(\mathbf{b}) = \{\mathbf{W}_R(0+; \mathbf{b}, \mathbf{u}); \mathbf{u} \in \Omega\}$ and we define admissible boundary conditions of the form $\mathbf{u}(0-,t) \in \mathcal{O}_L(\mathbf{b}(t))$, $t > 0$, for (1) (resp. $\mathbf{u}(0+,t) \in \mathcal{O}_R(\mathbf{b}(t))$, $t > 0$, for (2)). Hence natural coupling conditions for problem (1)–(2) consist in

$$\mathbf{u}(0-,t) \in \mathcal{O}_L(\mathbf{u}(0+,t)), \quad \mathbf{u}(0+,t) \in \mathcal{O}_R(\mathbf{u}(0-,t)). \tag{4}$$

The approach is thoroughly justified in the scalar case [9] and for linear systems [10]. In [9] it is shown that this is indeed a 'reasonable' way of coupling two conservation laws in the sense that, in meaningful situations, the coupled problem has a unique solution and the 'natural' numerical upwind scheme (the so called two-fluxes scheme) converges to this solution. Condition (4) resumes in a number of cases to the continuity of the solution at the interface

$$\mathbf{u}(0-,t) = \mathbf{u}(0+,t). \tag{5}$$

Thus we may interpret the coupling condition as a way of ensuring in a weak sense the continuity, we will say the *transmission* of the conservative variables.

1.2 Numerical coupling

We use a finite volume method for each system (1), (2). Let Δx, Δt, denote the uniform space and time steps, $\mu = \Delta t/\Delta x$, $t_n = n\,\Delta t$, $n \in \mathbb{N}$, $C_{j+1/2} = (x_j, x_{j+1})$, the cell with center $x_{j+1/2} = (j+1/2)\,\Delta x$, $j \in \mathbb{Z}$. The initial condition is discretized by $\mathbf{u}^0_{j+1/2} = \frac{1}{\Delta x}\int_{C_{j+1/2}} \mathbf{u}_0(x)dx$, $j \in \mathbb{Z}$. For the numerical coupling, we are given two numerical fluxes $\mathbf{g}_L$, $\mathbf{g}_R$ ($\mathbf{g}_\alpha$ is consistent with $\mathbf{f}_\alpha$) corresponding to 3-point monotone schemes (under some CFL condition), we set $\mathbf{g}^n_{\alpha,j} = \mathbf{g}_\alpha(\mathbf{u}^n_{j-1/2}, \mathbf{u}^n_{j+1/2})$ and define the scheme by

$$\mathbf{u}^{n+1}_{j-1/2} = \mathbf{u}^n_{j-1/2} - \mu\left(\mathbf{g}^n_{L,j} - \mathbf{g}^n_{L,j-1}\right), \quad j \le 0, n \ge 0,$$

$$\mathbf{u}^{n+1}_{j+1/2} = \mathbf{u}^n_{j+1/2} - \mu\left(\mathbf{g}^n_{R,j+1} - \mathbf{g}^n_{R,j}\right), \quad j \ge 0, n \ge 0,$$

(see also [1] in another context). So we have one fixed interface at $x = 0$ and two fluxes $\mathbf{g}^n_{\alpha,0}$. The choice $\mathbf{g}^n_{\alpha,0} = \mathbf{g}_\alpha(\mathbf{u}^n_{-1/2}, \mathbf{u}^n_{1/2})$, $\alpha = L, R$, corresponds to transmit the conservative variables. Namely, if $j \ge 0$, the scheme with flux $\mathbf{g}_R$ approximates the IBVP (2) with initial condition $u(x,0) = u_0(x)$, $x > 0$ and for boundary condition at $x = 0$, the scheme takes $\mathbf{u}^n_{-1/2}$. Since $\mathbf{g}^n_{L,0} \ne \mathbf{g}^n_{R,0}$, it is a nonconservative numerical approach, as for the continuous problem. For example, Godunov's scheme uses $\mathbf{g}^n_{R,0} = \mathbf{f}_R(\mathbf{W}_R(0+;\mathbf{u}^n_{-1/2},\mathbf{u}^n_{1/2}))$.

1.3 Choice of transmitted variables

When dealing with physical systems, we may prefer to transmit not the conservative variables but the *physical* variables, or even the flux. Now, assume that there exists a change of variables $\mathbf{v} \to \mathbf{u} = \varphi_\alpha(\mathbf{v})$, $\alpha = L, R$, from some set $\Omega_\mathbf{v} \subset \mathbb{R}^p$ onto Ω such that $\varphi'_\alpha(\mathbf{v})$ is an isomorphism of $\mathbb{R}^p$. If $\mathbf{c}$ is a given boundary *physical* data, we define $\mathcal{O}_\alpha(\varphi_\alpha(\mathbf{b}))$, $\alpha = L, R$, which are admissible boundary sets for the systems (1) and (2) respectively. Thus we now require

$$\mathbf{u}(0-,t) \in \mathcal{O}_L(\varphi_L(\mathbf{v}(0+,t))), \quad \mathbf{u}(0+,t) \in \mathcal{O}_R(\varphi_R(\mathbf{v}(0-,t))). \tag{6}$$

Since $\varphi_L(\mathbf{v}(0+,t)) \ne \varphi_R(\mathbf{v}(0+,t)) = \mathbf{u}(0+,t)$, the boundary sets in (4) and (6) are a priori distinct. Conditions (6) will ensure whenever possible the transmission of *physical* variables and their continuity instead of (5)

$$\mathbf{v}(0-,t) = \mathbf{v}(0+,t). \tag{7}$$

We are going to illustrate the two choices in the coupling procedure on the p–system and then for the full Euler system in Lagrangian coordinates. On the one hand, it is a simplified model of what we get when coupling more complex models associated to distinct systems whose closure laws are not always compatible, as will happen for instance in the context of thermal-hydraulics. On the other hand, the analysis will justify the use of Lagrange+projection schemes when coupling systems in Eulerian coordinates at a fixed interface.

Note however that for the Euler system in Lagrangian coordinates, the interface is characteristic and corresponds to a contact discontinuity. Hence, the coupling does not yield the continuity (5) or (7) for all the components. In our case, physical arguments, such as the continuity of some primitive quantities (for instance velocity and pressure) help defining the transmission. However, both theoretical considerations and numerical results obtained on some significant tests when coupling Euler systems (see [3, 4]) will prove that, if several CC based on continuity arguments are feasible, one cannot maintain all the conservation properties and we must choose which we want to be preserved.

2 Coupling two p−systems

We consider two systems (1) and (2) which are p-systems differing by the pressure law

$$\begin{cases} \mathbf{u} = (\tau, v)^T, \tau > 0 \\ \mathbf{f}_L(\mathbf{u}) = (-v, p)^T, \, p = p_L(\tau), \\ \mathbf{f}_R(\mathbf{u}) = (-v, p)^T, \, p = p_R(\tau). \end{cases} \quad (8)$$

We assume that $p'_\alpha < 0, p''_\alpha > 0$. The eigenvalues are $\pm\sqrt{-p'_\alpha}$.

We first transmit the conservative variables (τ, v). The study of the Riemann problem is needed in order to express the CC. We denote by $\mathcal{C}^i_\alpha(\mathbf{u}_-)$ the i−wave curve, i.e., the set of states that can be connected to a given state $\mathbf{u}_-$ by a $i-$ wave (either rarefaction or admissible shock) relative to the p−system with flux $\mathbf{f}_\alpha$. Expressing (4) gives that $\mathbf{u}(0-)$ is connected to $\mathbf{u}(0+)$ by a $2-L$ (positive) wave which means $\mathbf{u}(0+) \in \mathcal{C}^2_L(\mathbf{u}(0-))$ and similarly (for the right condition) by a $1-R$ (negative) wave. Thus $\mathbf{u}(0+) \in \mathcal{C}^2_L(\mathbf{u}(0-)) \cap \mathcal{C}^1_R(\mathbf{u}(0-))$ and $\mathbf{u}(0+) = \mathbf{u}(0-)$ because it is well known that the two wave curves intersect at only one point in the plane (τ, v) (see for instance [8]).

Now the IBVP's in both half planes are also well posed if one 'imposes' a given (v, p) on $x = 0$. Indeed, by assumption $p'_\alpha < 0$, we can define its inverse mapping $\tau_\alpha(p)$ for $\alpha = L, R$. Setting $\mathbf{v} = (v, p)^T$, we have an admissible change of variables: $\mathbf{u} = \varphi_\alpha(\mathbf{v})$ where

$$(v, p) \to \varphi_\alpha(v, p) \equiv (\tau, v) \quad (9)$$

is simply defined by $\tau = \tau_\alpha(p)$, for instance if $p_\alpha(\tau) = \tau^{-\gamma_\alpha}$, $\tau_\alpha(p) = p^{-1/\gamma_\alpha}$. We now transmit this set of variables (v, p). Expressing the coupling condition (6) yields that $\varphi_R(\mathbf{v}(0-, t))$ is connected to $\mathbf{u}(0+, t) = \varphi_R(\mathbf{v}(0+, t))$ by a $1 - R$ wave. We can parametrize the wave curves by p and project them onto the (v, p)−plane (see [8], Chapter I, section 7). If the $1 - R$ wave curve is $\mathcal{C}^1_R(\mathbf{u}(0-)) = \{(\tau, v); v = \Psi_{1,R}(\tau)\}$, then $\widetilde{\mathcal{C}}^1_R(\mathbf{v}(0-)) = \{(v, p); v = \Psi_{1,R}(\tau_R(p))\} = \{(v, p); \varphi_R(v, p) \in \mathcal{C}^1_R(\mathbf{u}(0-))\} = \varphi_R^{-1}(\mathcal{C}^1_R(\mathbf{u}(0-)))$ is its representation in the (v, p)−coordinates, we then have $\mathbf{v}(0+) \in \widetilde{\mathcal{C}}^1_R(\mathbf{v}(0-))$. Similarly, $\mathbf{u}(0-, t) \in \mathcal{O}_L(\varphi_L(\mathbf{v}(0+, t)))$ yields $\mathbf{v}(0+) \in \widetilde{\mathcal{C}}^2_L(\mathbf{v}(0-))$. We get

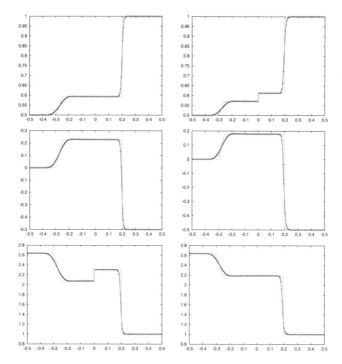

Fig. 1. Transmission of $\mathbf{u} = (\tau, v)$ left vs $\mathbf{v} = (v, p)$ right for the coupled p–system

$\mathbf{v}(0+) \in \tilde{\mathcal{C}}_R^1(\mathbf{v}(0-)) \cap \tilde{\mathcal{C}}_L^2(\mathbf{v}(0-))$ and it is easily proved that the two curves intersect at only one point in the plane (v, p) so that $\mathbf{v}(0+) = \mathbf{v}(0-)$. Hence we do have continuity of v, p, but not of τ since $\tau(0+) = p(0+)^{-1/\gamma_R} \neq p(0-)^{-1/\gamma_L} = \tau(0-)$. Let us illustrate the results on the solution of a Riemann problem (the exact solution is known).

We take a uniform grid, with 150 meshes and a first-order explicit Roe-type coupled scheme, the CFL is 0.5. The two pressure laws are $p_\alpha(\tau) = \tau^{-\gamma_\alpha}$ with $\gamma_L = 1.4, \gamma_R = 1.6$, and we represent in this order τ, v and p at a given time t (exact and approximate solution). We note in fig.1, left part, the continuity of τ, v the discontinuity of p at $x = 0$ while in the right part we note the discontinuity of τ and the continuity of v, p.

3 Coupling two Euler systems in Lagrangian coordinates

We consider the system of gas dynamics in Lagrangian coordinates

$$\partial_t \mathbf{u} + \partial_x \mathbf{f}(\mathbf{u}) = 0, \mathbf{u} = (\tau, v, e)^T, \mathbf{f}(\mathbf{u}) = (-v, p, pv)^T. \tag{10}$$

In (10), x stands for a mass variable, τ denotes the specific volume, v the velocity, $e = \varepsilon + \frac{1}{2}v^2$ the specific total energy, ε the specific internal energy, and we assume that the pressure p is a given function $p = p(\tau, \varepsilon)$. We study the coupling of two such systems at $x = 0$ thus at a contact discontinuity separating two fluids with different equations of state $p_\alpha(\tau, \varepsilon)$, $\alpha = L, R$. The corresponding flux functions are denoted by

$$\mathbf{f}_\alpha(\mathbf{u}) = (-v, p, pv)^T, \quad p = p_\alpha(\tau, \varepsilon), \alpha = L, R. \tag{11}$$

The eigenvalues of the Jacobian matrix of $\mathbf{f}(\mathbf{u})$ are $\lambda_1(\mathbf{u}) = -C < \lambda_2 = 0 < \lambda_3(\mathbf{u}) = C$, where $C = \sqrt{-p_\tau + pp_\varepsilon}$ denotes the Lagrangian sound speed. In this case, the interface $x = 0$ is characteristic ($\lambda = 0$ is an eigenvalue) hence, in general, the coupling does not yield the continuity (5) nor (7). However we have for each system one strictly positive and one strictly negative eigenvalue and we will see that it yields the continuity of a subset of two variables. When coupling the two systems (1) and (2) with $\mathbf{f}_\alpha$ given by (11), we may want to transmit also the velocity and the pressure. This corresponds to the CC (6) expressed in primitive variables

$$\mathbf{v} = (\tau, v, p)^T. \tag{12}$$

The change of variables $\mathbf{u} = (\tau, v, e)^T = \varphi_\alpha(\mathbf{v})$, is defined assuming that the functions $p = p_\alpha(\tau, \varepsilon)$ may be inverted in $\varepsilon = \varepsilon_\alpha(\tau, p)$, which is the case for instance for an ideal polytropic gas satisfying a γ-law

$$p_\alpha(\tau, \varepsilon) = (\gamma_\alpha - 1)\varepsilon/\tau. \tag{13}$$

More generally, we assume $\frac{\partial p}{\partial \varepsilon} > 0$.

3.1 Coupling with transmission of primitive variables

The Riemann problem for (10) is usually solved using primitive variable because the 'projection' of the wave curves on the (v, p)-plane are easily expressed. Let $\mathbf{u}_L$ and $\mathbf{u}_R$ be two given states. We denote by $\mathcal{S}^1_R(\mathbf{u}_L)$ the 1–wave curve consisting of states $\mathbf{u}$ which can be connected to $\mathbf{u}_L$ on the right by either a 1–shock or a 1–rarefaction wave corresponding to the equation of state $p = p_R(\tau, \varepsilon)$. Similarly, given a right state $\mathbf{u}_R$, we denote by $\mathcal{S}^3_L(\mathbf{u}_R)$, the (backward) 3–wave curve consisting of left states $\mathbf{u}$ which can be connected to $\mathbf{u}_R$ by a 3–shock or a 3–rarefaction wave corresponding to the equation of state $p = p_L(\tau, \varepsilon)$. We denote by $\mathbf{S}^1_R(\mathbf{v}_L)$ and $\mathbf{S}^3_L(\mathbf{v}_R)$ the 'projections' (in a sense to be precised below) onto the (v, p)-plane of the wave curves $\mathcal{S}^1_R(\mathbf{u}_L)$ and $\mathcal{S}^3_L(\mathbf{u}_R)$ respectively. In fact $\mathbf{S}^i(\mathbf{v}_L)$ is the projection of the i–wave curve $\varphi^{-1}(\mathcal{S}^i(\mathbf{u}_L))$ expressed in primitive variables $\mathbf{v} = (\tau, v, p)^T$ on the (v, p)-plane:

$$\varphi^{-1}(\mathcal{S}^i(\mathbf{u}_L)) = \{\mathbf{v} = (\tau, v, p)^T; \varphi(\mathbf{v}) \in \mathcal{S}^i(\mathbf{u}_L)\}$$

and

$$\mathbf{S}^i(\mathbf{v}_L) = \{(v, p); (\tau, v, p)^T \in \varphi^{-1}(\mathcal{S}^i(\mathbf{u}_L))\}.$$

Proposition 1. *In the case (12), the coupling conditions (6) are equivalent to*
$$v(0-,t) = v(0+,t),\ p(0-,t) = p(0+,t). \tag{14}$$

The proof consists as for the p-system in expressing the CC (6) in terms of solutions of Riemann problems and intersection of the projected wave curves. We assume that the curves $\mathbf{S}_R^1(\mathbf{v}_\ell)$ and $\mathbf{S}_L^3(\mathbf{v}_r)$ intersect at one point at most.

3.2 Transmission of conservative variables

In this case, the (v,p)-plane is not well suited, since p is no longer a transmitted variable. For two γ-laws we can think of the plane $(v, \pi = \varepsilon/\tau)$, since π is a variable independent of the pressure law. Indeed, following the above arguments while projecting on the (v, π)-plane, we can prove

Proposition 2. *Assuming (13), the coupling conditions (4) are equivalent to*
$$\begin{cases} v(0-,t) = v(0+,t), \\ \dfrac{\varepsilon}{\tau}(0-,t) = \dfrac{\varepsilon}{\tau}(0+,t). \end{cases} \tag{15}$$

We can easily extend the result to the case of pressure laws which can be written as a function of one dependent variable $\pi = \pi(\tau, \varepsilon)$ i.e. such that $p_\alpha(\tau, \varepsilon) = \bar{p}_\alpha(\pi(\tau, \varepsilon))$. The above argument will show that (v, π) is continuous at the interface $x = 0$. For general pressure laws, the velocity need not be continuous. This is in particular the case for two pressure laws of Grüneisen type
$$p_\alpha(\tau, \varepsilon) = (\gamma_\alpha - 1)\frac{\varepsilon}{\tau} + c_\alpha^2(\frac{1}{\tau} - \frac{1}{\tau_{ref,\alpha}}),\ \alpha = L, R, \tag{16}$$
such that $\frac{c_L^2}{\gamma_L - 1} \neq \frac{c_R^2}{\gamma_R - 1}$ (for details, we refer to [5]).

4 Coupling Lagrangian systems of different dimensions

We consider the p-system (8) in the left half-plane and the Euler system in Lagrangian coordinates (10) in the right half-plane (using in this section capital letters to distinguish the conservative variables)

$$\frac{\partial \mathbf{u}}{\partial t} + \frac{\partial}{\partial x}\mathbf{f}(\mathbf{u}) = \mathbf{0},\ x < 0,\ \mathbf{u} = (\tau, v)^T,\ \mathbf{f}_L(\mathbf{u}) = (-v, p)^T, p = p_L(\tau)$$

$$\frac{\partial \mathbf{U}}{\partial t} + \frac{\partial}{\partial x}\mathbf{F}_R(\mathbf{U}) = \mathbf{0},\ x > 0,\ \mathbf{U} = (\tau, v, e)^T,\ \mathbf{F}_R(\mathbf{U}) = (-v, p, pv)^T, p = p_R(\tau, \varepsilon).$$

The dimensions of the two systems are now different, but the physical context helps to give a meaning to the coupling since some state variables such as the specific volume τ, velocity v or pressure p are defined for each model. We write the CC using the variables (v,p) that are common to the two systems and which we have seen are good candidates for both. We reconstruct the missing variable for the smaller system in such a way that we may transmit the velocity and the pressure. Indeed, we can lift $\mathbf{v} = (v,p)^T$ by reconstructing τ when we transmit from the left to the right

$$\mathbf{v} = (v,p)^T \to \mathcal{L}(\mathbf{v}) = (\tau, v, p)^T, \tau = \tau_L(p), \tag{17}$$

where $p \to \tau_L(p)$ is the inverse of $p_L(\tau)$. And we easily project $\mathbf{V}$ when we transmit from the right to the left

$$\mathbf{V} = (\tau, v, p)^T \to \mathcal{P}(\mathbf{V}) = (v,p)^T. \tag{18}$$

Using the previously defined change of variables φ_α, the CC naturally writes

$$\mathbf{u}(0-,t) \in \mathcal{O}_L(\varphi_L(\mathcal{P}(\mathbf{V}(0+,t)))), \ \mathbf{U}(0+,t) \in \mathcal{O}_R(\varphi_R(\mathcal{L}(\mathbf{v}(0-,t)))). \tag{19}$$

We obtain the following result.

Proposition 3. *Assuming (17) with (18), the coupling conditions (19) are equivalent to*

$$v(0-,t) = v(0+,t), \ p(0-,t) = p(0+,t). \tag{20}$$

5 Conclusion

The extension of the previous approach to general Lagrangian systems requires some technical developments but is straightforward and presented in [5]. This work is part of an ingoing joint research program on multiphase flows between CEA and University Pierre et Marie Curie. Other topics encountered in the context of the coupling of two-phase flow models are developed in [2] [3] [4].

References

1. Abgrall R. , Karni, S.: Computations of compressible multifluids, Journal of Computational Physics, **169**, 594–623 (2001)
2. Ambroso, A., Chalons, C., Coquel, F., Godlewski, E., Lagoutière, F., Raviart, P.-A., Seguin, N.: Homogeneous models with phase transition: coupling by finite volume methods, FVCA IV proceedings, Hermes Science, 483–492 (2005)
3. Ambroso, A., Chalons, C., Coquel, F., Godlewski, E., Lagoutière, F., Raviart, P.-A.: Couplage de deux systèmes de la dynamique des gaz, 17ème congrès Français de mécanique (sept. 2005)
4. Ambroso, A., Chalons, C., Coquel, F., Godlewski, E., Lagoutière, F., Raviart, P.-A., Seguin, N., Hérard, J.-M.: Coupling of multiphase flow models, NURETH-11, International Topical Meeting on Nuclear Thermal-Hydraulics (2005)
5. Ambroso, A., Chalons, C., Coquel, F., Godlewski, E., Lagoutière, F., Raviart, P.-A., Seguin, N.: Coupling of general Lagrangian systems (in preparation)

6. Després, B.: Lagrangian systems of conservation laws. Invariance properties of Lagrangian systems of conservation laws, approximate Riemann solvers and the entropy condition. Numer. Math., **89**, Vol. 1, 99–134 (2001)
7. Dubois, F., Le Floch, P.: Boundary conditions for nonlinear hyperbolic systems of conservation laws, J. of Diff. Equations, **71**, 93–122 (1988)
8. Godlewski E., Raviart, P.-A.: Numerical approximation of hyperbolic systems of conservation laws. Appl. Math. Science 118, Springer, New York (1996)
9. Godlewski E., Raviart, P.-A.: The numerical interface coupling of nonlinear hyperbolic systems of conservation laws: I. The scalar case, Numer. Math. **97**, 81–130 (2004).
10. Godlewski, E., Le Thanh, K.-C., Raviart, P.-A.: The numerical interface coupling of nonlinear hyperbolic systems of conservation laws: II. The case of systems, ESAIM:M2AN, **39**, 649–692 (2005)

A Numerical Scheme for the Modeling of Condensation and Flash Vaporization in Compressible Multi-Phase Flows

Vincent Perrier[1,2], Rémi Abgrall[1,3] and Ludovic Hallo[2]

[1] MAB, 351 Cours de la Libération, 33405 Talence Cedex
[2] CELIA, 351 Cours de la Libération, 33405 Talence Cedex
[3] Institut Universitaire de France
   perrier@math.u-bordeaux1.fr

Summary. A thermodynamic model for phase transition is introduced. The equation of state (EOS) is not globally convex hence difficulties exist in solving the Riemann Problem (R.P.). This motivates another method for solving the R.P. where thermodynamically out-of-equilibrium states are taken into account in the framework of [2, 1, 9]. This method is tested numerically on several examples.

1 Introduction

This paper is a contribution to the modeling and the simulation of phase transition problems in compressible flows. This is a difficult problem that has already encountered lots of attention, see [9] and the references therein. We first describe a simplified thermodynamic model, then we recall the difficulties and propose a new kinetic closure relation. A numerical scheme is sketched and then two numerical examples are provided.

2 The thermodynamics of phase transition

Two phases of the same fluid are modelled by the two equations of state, $\varepsilon_l(P_l, T_l)$ and $\varepsilon_g(P_g, T_g)$, where ε is the internal energy. The subscripts l and g refer to quantities related with the liquid and gas phases. Assuming that the two phases are locally non–miscible, the optimization of the mixture entropy imposes pressure and temperature equilibrium and the equality of the chemical potential $\mu = \varepsilon + P\tau - Ts$ between the phases. From the equality $\mu_l(P, T) = \mu_g(P, T)$, P is a function of T in the mixture area: $P = P_{\mathrm{sat}}(T)$. The limits of stability between the mixture and the gas (resp. the liquid) are $\tau_g(P_{\mathrm{sat}}(T), T)$ (resp. $\tau_l(P_{\mathrm{sat}}(T), T)$). The mixture is stable only when $\{\tau_l \leq \tau \leq \tau_g\}$. This is the saturation area, see Figure 1. Above the saturation area, the fluid is a

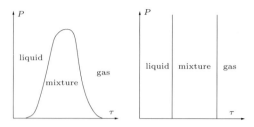

Fig. 1. The three areas in the general case (left) and in the case of two perfect gas EOS (right).

supercritical fluid: there is no more difference between vapor and liquid. In this paper, we assume that the fluid is never supercritical. Note that even if the convexity of equations of state is ensured in each area, the loss of derivative on the saturation boundaries leads to non global convexity.

In this paper, we chose to represent each phase by a perfect gas EOS, as in [6, 7]. The calculations are easy because the mixture EOS is explicit, contrarily to general case. The pressure is $P(\tau, \varepsilon) = \Gamma_2 \frac{\varepsilon}{\tau}$ if $\tau \leq \tau_2$, $P = \Gamma_2 \frac{\varepsilon}{\tau_2} = \Gamma_1 \frac{\varepsilon}{\tau_1}$ if $\tau_2 \leq \tau \leq \tau_1$ and $P = \Gamma_1 \frac{\varepsilon}{\tau}$ else.

3 The Riemann problem with phase transition at equilibrium

Our aim is to approximate the hyperbolic part of the Navier–Stokes equations, i.e. the Euler system

$$\begin{cases} \partial_t \rho + \partial_x (\rho u) = 0 \\ \partial_t (\rho u) + \partial_x (\rho u^2 + P) = 0 \\ \partial_t (\rho E) + \partial_x ((\rho E + P)u) = 0. \end{cases} \quad (1)$$

It is rewritten as $\partial_t \mathbf{U} + \partial_x \mathbf{F}(\mathbf{U}) = 0$. In (1), $E = \varepsilon + \frac{u^2}{2}$ is the total energy, The EOS links P, ρ, ε. The approximation via finite volume solvers necessitates the evaluation of the Riemann problem solution, i.e. (1) with the initial conditions

$$\mathbf{U}_0(x) = \begin{cases} \mathbf{U}_L & \text{if } x < 0 \\ \mathbf{U}_R & \text{if } x \geq 0 \end{cases}. \quad (2)$$

A classical way for solving problem (1)–(2) is to intersect the 1–wave and the 3–wave curves, projected in the (u, P) plane, see [5] for more details. If the solution of the Riemann problem is well understood for convex equations of state, several problems exist for non–convex EOS, as the one of section 2 (see [11]).

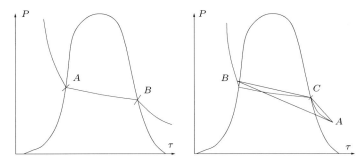

Fig. 2. **Left**: An isentropic curve in the (τ, P) plane crossing the whole saturation area. There is a loss of derivative in point A and B. **Right**: A Hugoniot curve in the (τ, P) plane. The shock from A to B can split into two shocks, and the most stable is the one with decomposition in B.

Regular simple waves are isentropic waves. When isentropes cross the saturation curve, a loss of derivative occur, and [11] shows that on the saturation curve $\gamma - \gamma_m > 0$, where γ (resp. γ_m) is the polytropic coefficient of the pure phase (resp. of the mixture), see Figure 2 for an illustration. At points A and B, the characteristic curves intersect when the isentropes enter the saturation curve, see Figure 3. This contradicts the regularity of the wave, which was assumed: there exist no regular wave connecting a point in the mixture area and a point in the gas area. Across a shock, the Rankine–Hugoniot relations write
$$M = \frac{u_2 - u_1}{\tau_2 - \tau_1}, \; M^2 = -\frac{p_2 - p_1}{\tau_2 - \tau_1} \text{ (Rayleigh line) } \varepsilon_2 - \varepsilon_1 + \frac{1}{2}(p_2 + p_1)(\tau_2 - \tau_1) = 0$$
(Crussard curve). If the EOS is convex, the shock splitting (i.e. replacing a shock by two successive shocks) is not stable because the second shock moves faster than the first one. If the EOS is not convex, a shock can split into two shocks, and the decomposition is stable and entropic: there may be several entropic solution for (1)–(2).

To summarize, there is a problem of non existence of regular waves, there may be non uniqueness of shock waves also. The uniqueness problem was

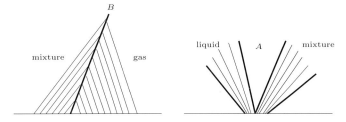

Fig. 3. Behavior of the characteristics in points A and B for regular waves.

solved in [10] : the correct solution is the one for which the entropy grows all along the viscous profile (decomposition in point C in the Figure 2).

In the following, we will consider that there is no more problem for condensation and we will focus on the problem of non existence for vaporization.

4 The Riemann problem with out of equilibrium EOS

When a liquid undergoes a strong rarefaction wave, it can be led into the saturation area without undergoing a phase transition [13]. The liquid is said to be *metastable*, or *overheated*. We suppose that a vaporization wave is a self–similar wave for which the Rankine–Hugoniot relations hold. The main difference with section 3 is that the equation of state of the upstream and downstream states are different. We can use the Chapman–Jouguet theory [5, 4]. A downstream state is defined as the intersection in the (τ, P) plane of the Crussard curve and the Rayleigh line. The main difference with classical shock relations is that the upstream state does not belong to the set of the downstream state. Thus, the Crussard curve is separated in three parts as in Figure 4 (left), depending on the slope of the Rayleigh line. Since a vaporization is a transformation in which the specific volume τ increases, we are interested only in its lower part. It is separated into two as on Figure 4(right). We are interested only in weak deflagrations in order to be consistent with the Lax characteristic condition : the wave is subsonic [5, 4]. Thus, the half–Riemann problem (i.e. all the waves on the same side of the contact discontinuity) is composed of a precursor sonic wave, shock or rarefaction wave, followed by a vaporization wave. If, given an initial state and a pressure $P^\star$ we want to compute the velocity $u^\star$, we see that we have seven unknowns (the velocity and two thermodynamic parameters for the state after the sonic

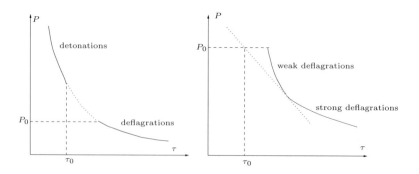

Fig. 4. The Crussard curve. On the left we can see the full curve, separated between detonations, deflagrations and a part not matching with the negative slope of the Rayleigh line. The right part is a zoom on the deflagration branch: it is separated into two parts (weak and strong deflagrations) by the Chapman Jouguet point

wave, the velocity and one thermodynamic parameter for the state $\star$, and the two wave velocities), but we only have six equations (three across each wave). Thus a kinetic closure is needed. In [9], it was proposed to choose the Chapman–Jouguet point. With the model of section 2, we find the following expressions for the CJ–point, depending whether it is in the saturation dome or not:

$$\tau_{CJ} = \tau_0 + \frac{2\tau_l}{\Gamma_l} \frac{\sqrt{\left(\frac{\tau_0}{\tau_l}\right)^2 - 1}}{\frac{\tau_0}{\tau_l} + 1 - \sqrt{\left(\frac{\tau_0}{\tau_l}\right)^2 - 1}} \quad \text{and} \quad \tau_{CJ} = \tau_0 \frac{\frac{\gamma_l+1}{\gamma_l-1}P_0 + P_{CJ}}{\frac{\gamma_g+1}{\gamma_g-1}P_{CJ} + P_0}$$

with

$$P_{CJ} = \frac{\Gamma_g P_0}{\Gamma_l} \left(1 - \sqrt{\left(1 - \frac{\Gamma_l}{\Gamma_g}\right)\left(1 + \frac{\Gamma_l}{\Gamma_g} + \frac{2\Gamma_l}{\Gamma_g(\gamma_g+1)}\right)}\right)$$

The downstream state function is drawn on the left of the Figure 5. We see that the closure of [9] leads to a solution of the Riemann problem that is not continuous with respect to the initial state. This is in contradiction with the very definition of hyperbolicity of the system ! This bad behavior was proved for the model of section 2, but it can also be proved for more realistic equation of states, like in [8, 9], provided the fluid is retrograde. This is not in contradiction with the experiments made in [13], because only partial evaporations were done. It is likely that the CJ closure is the right closure for partial evaporation waves, but not for total evaporation waves. In the following, we chose the closure of the right of Figure 5: the CJ closure for mixtures is continued by a line such that the set of downstream states is continuous. We note that this closure ensures that the downstream state is subsonic.

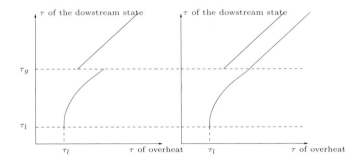

Fig. 5. Specific volume of the downstream state, function of the overheat. On the left figure, the closure is the one of [9], whereas on the right, the closure is the one chosen in that paper to ensure the continuity of the downstream state.

5 Numerical scheme

The derivation of the scheme uses the set averaging ideas of Drew and Passman [3] combined with the discretization principle introduced by Godunov. At each time step, the flow is described in each computational cell by the average $W_j = (\alpha_{1,j}, \alpha_{1,j}\rho_{1,j}, \alpha_{1,j}\rho_{1,j}u_{1,j}, \alpha_{1,j}\rho_{1,j}E_{1,j}, \alpha_{2,j}, \alpha_{2,j}\rho_{2,j}, \alpha_{2,j}\rho_{2,j}u_{2,j}, \alpha_{2,j}\rho_{2,j}E_{2,j})$. We consider a family of random subdivisions of the cell $C_j =]x_{j-1/2}, x_{j+1/2}[$. In each of the subcells of C_j, we randomly set the flow variables $W_{1,j} = (\rho_{1,j}, \rho_{1,j}u_{1,j}, \rho_{1,j}E_{1,j})$ or $W_{2,j} = (\rho_{2,j}, \rho_{2,j}u_{2,j}, \rho_{2,j}E_{2,j})$. The random process is done so that the average is W_j: the average length of the phase Σ_1 (resp. Σ_2) in C_j must be $\alpha_{1,j}(x_{j+1/2} - x_{j-1/2})$ (resp. $\alpha_{2,j}(x_{j+1/2} - x_{j-1/2})$).

The scheme is constructed in two steps. First for any realization, we evolve the conserved variables $W_{1,j}$ and $W_{2,j}$ following Godunov' principle. Then we make an ensemble average of the schemes. The results is precisely our discretization. The technicality is described in [2] to which we refer.

6 Numerical results

For both tests, the computation was made with $\gamma_g = 1.9$ and $\gamma_l = 1.2$. For these choice of adiabatic coefficients, the limit of stability of the liquid is $\rho = 2.5441\,\text{kg.m}^{-3}$ and the limit of stability of the gas is $\rho = 0.5654\,\text{kg.m}^{-3}$.

6.1 Liquefaction test

Initially, the left state is a gas at rest with a density $\rho = 0.5\,\text{kg.m}^{-3}$ and a pressure $P = 10^4\,\text{Pa}$. the right state is a liquid moving to the left with a velocity of $u = -60\,\text{m.s}^{-1}$. The liquid pressure is the gas one. The liquid density is $\rho = 3\,\text{kg.m}^{-3}$. The computed solution is compared with the analytical one in Figure 6.

The initial discontinuity induces two shocks, one of which is split into two parts because of the phase transition. The velocity, pressure and density computed perfectly agree with the analytical solution.

6.2 Vaporization test

Initially, the right state is a high pressure liquid ($P = 10^9\,\text{Pa}$) at rest, the density is $\rho = 3\,\text{kg.m}^{-3}$. The left state is a gas at atmospheric pressure ($P = 10^5\,\text{Pa}$). The density is $\rho = 0.5\,\text{kg.m}^{-3}$. The computed solution with 1000 points is compared with the analytical one in the Figure 7. The initial discontinuity induces a strong rarefaction wave that leads the liquid in a metastable state ($\rho \approx 1.156\,\text{kg.m}^{-3} < 2.5441\,\text{kg.m}^{-3}$). After the strong rarefaction wave, a vaporization wave occurs.

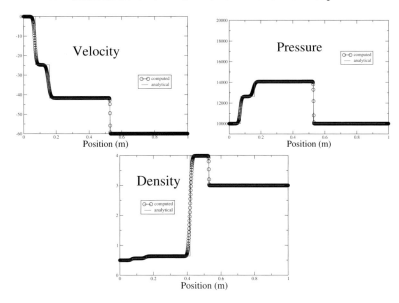

Fig. 6. Liquefaction test.

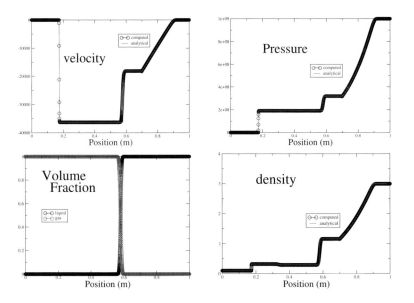

Fig. 7. Vaporization test.

7 Conclusions

We have sketched a numerical method able to simulate complex and strong phase transition phenomena. A second order version of the scheme has also been developed. We are currently extending this scheme to two dimensional situation. The kinetic closure is also discussed in length in [12].

Acknowledgments

This work was partially supported by the CEA through the contract number 123-C-BEFI.

References

1. Abgrall, R., Perrier, V.: Asymptotic expansion of a multiscale numerical scheme for compressible multiphase flow. Multiscale Model. Simul., **5**, 84–115 (2006).
2. Abgrall, R., Saurel, R.: Discrete equations for physical and numerical compressible multiphase mixtures. J. Comput. Phys., **186**, 361–396 (2003).
3. Drew, D.A., Passman, S.L.: Theory of Multicomponent fluids. Applied Mathematical Sciences **(135)**, Springer-Verlag, New York (1995).
4. Courant, R. and Friedrichs, K. O.: Supersonic Flow and Shock Waves. Interscience Publishers, Inc., New York, N. Y. (1948).
5. Godlewski, E., Raviart, P.-A.: Numerical approximation of hyperbolic systems of conservation laws. Applied Mathematical Sciences **(118)**, Springer-Verlag, New York (1996).
6. Helluy, P.,Barberon, T.: Finite volume simulation of cavitating flows. Computer & Fluids, **34 (7)**, 832–858 (2005).
7. Jaouen, S.: Étude mathématique et numérique de la stabilité pour des modèles hydrodynamiques avec transition de phase. Phd Thesis, Université Pierre et Marie Curie, Paris (2001).
8. Le Métayer, O., Massoni, J., Saurel, R.: Élaboration de lois d'état d'un liquide et de sa Vapeur pour les Modèles d'Écoulements Diphasiques. Int. J. Thermal. Sci., **43**, 265–276 (2003).
9. Le Métayer, O., Massoni, J.,Saurel, R.: Modelling evaporation fronts with reactive Riemann solvers. J. Comput. Phys., **205**, 567–610 (2005).
10. Liu, T.P.: The Riemann problem for general systems of conservation laws. J. Differential Equations, **18**, 218–234 (1975).
11. Menikoff, R. , Plohr, B.J.:The Riemann problem for fluid flow of real materials. Rev. Modern Phys., **61 (1)**, 75–130 (1989).
12. Perrier, V.: The Chapman–Jouguet closure for the Riemann Problem with vaporization. *In preparation.*
13. Simões-Moreira, J. R., Shepherd, J.E.: Evaporation waves in superheated dodecane. J. Fluid Mech., **382**, 63–86 (1999).

NAVIER-STOKES

An Adaptive Operator Splitting of Higher Order for the Navier-Stokes Equations

Jörg Frochte[1] and Wilhelm Heinrichs[2]

[1] Arbeitsgruppe Ingeneurmathematik, Universität Duisburg-Essen Campus Essen, Universitätsstr. 3, 45117 Essen
  joerg.frochte@uni-due.de
[2] Arbeitsgruppe Ingeneurmathematik, Universität Duisburg-Essen Campus Essen, Universitätsstr. 3, 45117 Essen
  wheinric@ing-math.uni-essen.de

Summary. This article presents an operator splitting for solving the time-dependent incompressible Navier-Stokes equations with Finite Elements. By using a postprocessing step the splitting method shows a reduction factor higher than second order. In this algorithm a gradient recovery technique is used to compute boundary conditions for the pressure and to achieve a higher convergence order for the gradient at different points of the algorithm.

1 Introduction

We consider the incompressible time dependent Navier-Stokes equations

$$\frac{\partial v}{\partial t} + (v \cdot \nabla)v - \nu \nabla^2 v + \nabla p = f \quad \text{in } \Omega, \ t \in [0, t_{\text{end}}] \tag{1}$$

$$\nabla \cdot v = 0 \text{ in } \Omega, \ t \in [0, t_{\text{end}}] \ ; \quad v = v_0 \text{ for } t = 0, \text{ in } \Omega, \tag{2}$$

$$v = h \text{ on } \partial\Omega, \ t \in [0, t_{\text{end}}]. \tag{3}$$

The solution of these equations on the time interval $[0, t_{\text{end}}]$ are the velocity v of a Newtonian fluid with the kinematic viscosity ν and the pressure p in a domain Ω. We assume that Ω is a bounded domain in $\mathbb{R}^2$ and that its boundary $\partial \Omega$ is polygonal. The boundary conditions are given by a time-dependent function h on $\partial \Omega$. To solve the Navier-Stokes equations we use a splitting technique with postprocessing. The algorithm without postprocessing, called base splitting algorithm, is related to the one published by Haschke and Heinrichs [2] for spectral methods. For linear Finite Elements in contradiction to the solution itself the convergence rate of the gradient is only of first order. To avoid this and to compute boundary conditions for the pressure which are always a challenge for splitting techniques a new gradient recovery technique is developed.

2 The Taylor based gradient recovery technique

Let $\mathcal{T}_h$ be a triangulation of Ω and $T \in \mathcal{T}_h$. Thus the linear Finite Element space is $V_h = \{u_h \in C(\bar{\Omega}) \,; u_{h|T} \in \mathcal{P}_1 \text{ for } T \in \mathcal{T}_h\}$. To motivate this gradient recovery technique we assume that $u \in C^2(\Omega)$ and $I_h u = u_h \in V_h$ with I_h as interpolation operator on V_h. To recover the gradient of u at a node a of $\mathcal{T}_h$ we use a second order Taylor approximation with the values of u_h at a and $n \geq 5$ nodes (x_j, y_j) in the neighbourhood of a:

$u_h(x_j, y_j) - u_h(x_a, y_a) = \mathbf{u_x(x_a, y_a)}(x_j - x_a) + \mathbf{u_y(x_a, y_a)}(y_j - y_a)$
$+ \frac{1}{2}(\mathbf{u_{xx}(x_a, y_a)}(x_j - x_a)^2 + \mathbf{u_{xy}(x_a, y_a)}(x_j - x_a)(y_j - y_a) + \mathbf{u_{yy}(x_a, y_a)}(y_j - y_a)^2)$

The bold marked terms are the unknowns that are to be computed by solving a $5 \times n$-least squares problem. Generally all neighbours of a and also their neighbours are chosen. Figure 2 shows an example for such a neighbourhood of a. The new Taylor-based recovery technique (TBR) uses the data from all displayed nodes while a technique like the Z^2 recovery [7] uses only the information from the nodes with filled circles. The greater database together with a proper weighting [1] improves the results, especially on

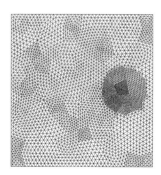

Fig. 1. Mesh with 3233 unknowns

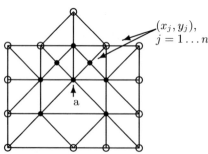

Fig. 2. Database of G_T/TBR and the Z^2 gradient recovery technique

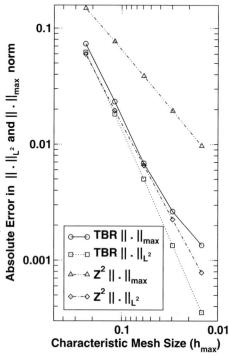

Fig. 3. Comparison of the Taylor-based-(TBR) and Z^2-gradient recovery technique

adaptive refined meshes and at the edges of Ω. Figure 3 shows the results of the two techniques recovering the partial derivation u_{hx} on a mesh like the one in figure 1. The data for the gradient recovery derives from a function $u_h \approx \sin(\pi(x-1)/2)\sin(\pi(y-1)/2)$ which is the solution of a Poisson equation, $-\nabla^2 u = f$. Figure 3 illustrates the fact that the TBR technique shows higher reduction rates in the L^2 norm. Very important for the computation of the needed boundary conditions for the pressure is the error in the nodal maximum norm because the maximum error often occurs at the edges of Ω. If this technique is used for all nodes of a triangulation we will use this according to the approximated nabla operator by $G_T u_h$.

3 The stabilized base splitting

For the approximation of $\frac{\partial}{\partial t}$ we use a BDF scheme of third order. The leading coefficient of the BDF scheme is denoted with β_0 and the time step size with Δt. Similar to the splitting for spectral methods [2] one time step of the splitting follows the scheme:

Time step in the base splitting

1. Compute a guess $(\bar{p}^{n+1})$ for the pressure
2. Based on the pressure compute an intermediate velocity $\tilde{v}^{n+1}$
3. Solve the Poisson equation (**) $-\nabla^2 p_{update} = -\frac{\beta_0}{\Delta t}\nabla\cdot\tilde{v}^{n+1}$; $p_{update} = 0$ on $\partial\Omega$ for the pressure and velocity update
4. Apply the update by $p^{n+1} = \bar{p}^{n+1} + p_{update}$; $v^{n+1} = \tilde{v}^{n+1} + \frac{\Delta t}{\beta_0}\nabla p_{update}$

In difference to [2] $\bar{p}^{n+1}$ is the solution of the following Poisson equation:

$$-\nabla^2 \bar{p} = -\nabla\cdot f + \nabla\cdot((v\cdot\nabla)v) \tag{4}$$

$$\underset{\nabla\cdot v=0}{\Leftrightarrow} -\nabla^2 \bar{p}^{n+1} = -(f^n_{1x}+f^n_{2y}) + v^n_{1x}v^n_{1x} + 2v^n_{2x}v^n_{1y} + v^n_{2y}v^n_{2y}. \tag{5}$$

All partial derivations on the right side were built with TBR. The Neumann boundary conditions are taken directly from the Navier-Stokes equations (1):

$$\nabla p = f - \underbrace{(\frac{\partial v}{\partial t} + (v\cdot\nabla)v - \nu\nabla^2 v)}_{(*)} \quad \text{on } \partial\Omega \tag{6}$$

The $(*)$ is zero for homogeneous zero Dirichlet boundary conditions. In the case that other boundary conditions are given $\frac{\partial v}{\partial t}$ is approximated with a BDF scheme of third order and the partial derivations are computed using G_T. The Laplace term is approximated by $G_T^2 v_1 = v_{1_{yy}} - v_{2_{yx}}$, $G_T^2 v_2 = v_{2_{xx}} - v_{1_{xy}}$. This formulation is more accurate than $v_{i_{xx}} + v_{i_{yy}}$ ($i=1,2$). The reconstruction of second order derivations at the edges still causes more problems than the recovery of the first order derivatives. But generally the quite small kinematic viscosity ν reduces the influence of this term heavily. With this procedure it is possible to add fitted boundary conditions for the pressure to the splitting and also to prevent an unstable behaviour of the algorithm for solutions of the type $v(t, x, y, z) = z(t)g(x, y, z)$. If $\bar{p}^{n+1}$ is simply set equal to p^n as in [2] the pressure update step would bump the same mesh based errors stepwise into the approximated pressure function. For small time step sizes the factor $\frac{\beta_0}{\Delta t}$ on the right side of the Poisson Equation amplifies this effect which is prevented with the above displayed procedure.

With the coefficients of the BDF scheme $\beta_j (j = 1..3)$ we set $\tilde{f} = f - G_T \bar{p}^{n+1} - \frac{1}{\Delta t} \sum_{j=1}^{3} \beta_j v^{m+1-j}$ and so the intermediate velocity can be computed explicitly

$$\left(-\nu \nabla^2 + \frac{\beta_0}{\Delta t} I\right) \tilde{v}_i^{m+1} = \tilde{f}^{n+1} - (v_e \cdot \nabla) v_e \qquad (7)$$

or implicitly

$$\left(-\nu \nabla^2 + \frac{\beta_0}{\Delta t} I\right) \tilde{v}_i^{m+1} + (v_e \cdot \nabla) \tilde{v}_i^{m+1} = \tilde{f}^{n+1} \qquad (8)$$

using a kind of Picard iteration ($v_e = \tilde{v}_i^{m+1}$) with the initial value $v_e = v^n$ and the stop criterion $\|\tilde{v}_i^{m+1} - \tilde{v}_{i-1}^{m+1}\| < \varepsilon_{Pic} = 10^{-3}$. The self-evident boundary conditions are taken from (3). The Finite Element spaces for the velocity and the pressure are chosen to fulfil the $\inf - \sup$ –condition, so we used triangle Taylor-Hood-Elements with linear and in the context of the postprocessing also with quadratic base functions.

4 The multi-grid postprocessing

The main reason for most splittings not to reach an order higher than two in time is that it seems not possible to compute a stable pressure approximation $\bar{p}$ of second order to compute $\tilde{v}$. With such an approximation the analysis done by Heinrichs in [3] would advise at least for the Stokes equations to get a scheme of third order. With a postprocessing step there is a stable way to compute an approximation of an order higher than one that can be used to compute $\tilde{v}$. To do this we use a set of nested Finite Elements spaces. Let $V_{h/2}$ be a Finite Element space that was built by a global regular refinement of the mesh of V_h. V_H is such a Finite Element space that V_h together with V_H satisfy the inf-sup-condition, e.g. quadratic base function of the same mesh or again a global refinement of V_h. Denote now $X_h = V_h \times V_h$ and $X_H = V_H \times V_H$. and set $V_{h,0}$ resp. $V_{h/2,0}$ as the subspace with the elements that satisfy $\int_\Omega u \, dx = 0$. First we compute $(v_{h/2}^{n+1}, p_h^{n+1})$ in $W_h = X_h \times V_{h,0}$ and use the results to perform a splitting step in $W_H = X_H \times V_{h/2,0}$. With this technique the number of Picard iterations in W_H can generally be reduced and the intermediate velocity can be computed with a pressure approximation of a higher order than in the base splitting. The following algorithm is an example for the use of linear base functions, so set $H = h/4$ and a full implicit treatment of the nonlinear term. Other variations based on this idea can be found in [1].

---

0. Compute an initial pressure p_h^0 for $t = 0$ with (5)

Time step with build-in postprocessing

1. Solve the PDE (8) for the intermediate velocity $\tilde{v}_{h/2}^{n+1}$ using p_h^n
2. Solve the Poisson equation (**) in V_h for the pressure and velocity update
3. Apply the update to the velocity $v_{h/2}^{n+1} = \tilde{v}_{h/2}^{n+1} + \frac{\Delta t}{\beta_0} \nabla p_{update}$
4. Solve the PDE (5) and use $\hat{v}_{h/2}^{n+1}$ on the right side to get $\bar{p}_{h/2}^{n+1}$
5. Solve the PDE (8) for the intermediate velocity $\tilde{v}_{h/4}^{n+1}$ with the initial value $v_e = P \hat{v}_{h/2}^{n+1}$ and $\bar{p}_{h/2}^{n+1}$ from step 4
6. Solve the Poisson equation (**) in $V_{h/2}$ for the pressure and velocity update: $-\nabla^2 p_{h/2_{update}} = -\frac{\beta_0}{\Delta t} \nabla \cdot \tilde{v}_{h/4}^{n+1}$; $p_{h/2_{update}} = 0$ on $\partial \Omega$
7. Apply the update to the velocity and the pressure
$p_{h/2}^{n+1} = \bar{p}_{h/2}^{n+1} + p_{h/2_{update}}$; $v_{h/4}^{n+1} = \tilde{v}_{h/4}^{n+1} + \frac{\Delta t}{\beta_0} \nabla p_{h/2_{update}}$
8. Compute the restrictions for the next splitting step:
$v_{h/2}^{n+1} = I_{h/2} v_{h/4}^{n+1}$, $p_h^{n+1} = I_h p_{h/2}^{n+1}$

The prolongation between the Finite Element spaces is done with the common prolongation and restriction from Multigridsolvers. Only in step 8 the interpolation operator is used. Because of the way the Finite Element spaces $V_{h,0} \subset V_h \subset V_{h/2} \subset V_{h/4}$ are nested in every part of the algorithm the inf-sup-condition is fulfilled. Another advantage of this procedure is that many tasks concerning adaptivity, especially adaptivity in time, can be answered in the coarser Finite Element spaces. This helps economising CPU costs. Adaptivity in space e. g. has been tested with the well-known Driven Cavity Problem, see [1] for further details.

5 Numerical results

The splitting with and without postprocessing was tested on various test-problems. Exemplarily the results of the test-problem **IV** from [1] are displayed. Here $\Omega = \{(x,y) \in \mathbb{R}^2 | 1 \leq r \leq 2\}$, $r = \sqrt{x^2 + y^2}$ is a spool. With a kinematic viscosity $\nu = 1/5000$ which is equivalent to a Reynolds' number of $Re \approx 1925$ the right side f and the boundary conditions are fitted so that the solution for the velocity is $v_1(x,y,t) = -y(0.25 - (r-1.5)^2) \sin(2\pi t)$, $v_2(x,y,t) = x(0.25 - (r-1.5)^2) \sin(2\pi t)$ and for the pressure $p(x,y,t) =$

Table 1. Comparison of the splitting with and without postprocessing

Δt	Degrees of freedom	with Postprocessing velocity (v_1) $\|u-u_h\|_{L^2}$	Quot.	pressure (p) $\|u-u_h\|_{L^2}$	Quot.	without Postprocessing velocity (v_1) $\|u-u_h\|_{L^2}$	Quot.	pressure (p) $\|u-u_h\|_{L^2}$	Quot.	Speed-up
1/8	29408	1.216e-01	-	1.907e-02	-	1.222e-01	-	5.087e-01	-	1.34
1/16	29408	1.768e-02	6.880	2.827e-03	6.746	4.035e-02	3.029	1.145e-01	4.443	1.12
1/32	116672	2.254e-03	7.843	4.260e-04	6.636	6.779e-03	5.952	2.735e-02	4.187	1.07
1/64	116672	3.026e-04	7.448	3.348e-04	1.273	2.247e-03	3.018	8.960e-03	3.052	1.35

$y\sin(x)\sin(2\pi t)$. At first glance the splitting with build-in postprocessing seems to be more expensive than the one without. But as table 1 shows the splitting technique with postprocessing is with the same number of unknowns in all numerical tests faster than the one without. This implicit postprocessing technique has been tested successfully for problems with a sufficiently smooth solution up to a Reynolds' number of 10,000. The unregulized driven cavity problem was solved with a Reynolds' number of 5,000. See [1] for further details.

'Flow around a cylinder'

A very popular benchmark problem the splitting was tested with is the 'Flow around a cylinder' defined by Schäfer and Turek in [6]. For the outflow Γ_3 we used like [5] the same time-dependent boundary conditions as for the inflow. To compute the drag (c_d) and the lift (c_l) coefficient we used an ansatz first published for the stationary Navier-Stokes equations in [4]. Applying it to the unstationary Navier-Stokes equations leads to the following equations:

$$c_d = -20 \int_\Omega \frac{\partial}{\partial t} v \cdot u_d + \nu \nabla v : \nabla u_d + (v \cdot \nabla) v \cdot u_d - p(\nabla \cdot u_d) \, d\Omega$$
$$c_l = -20 \int_\Omega \frac{\partial}{\partial t} v \cdot u_l + \nu \nabla v : \nabla u_l + (v \cdot \nabla) v \cdot u_l - p(\nabla \cdot u_l) \, d\Omega \, .$$

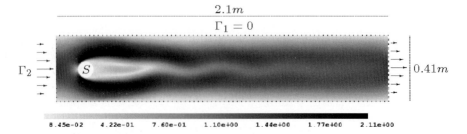

Fig. 4. 2D-3 4sec.

Type 2D-3 (unsteady)

There are different variations of the 'Flow around a cylinder'-Problem defined in [6]. For the 2D-3 the velocity is simulated over 8 seconds. The figures 5 and 6 show the results with 139344 unknowns for the velocity and 35048 for the pressure compared to the results computed by John in [5] with quadratic Taylor-Hood-Elements and 399616 unknowns in v and 50240 in p. John used a fractional-step-θ-scheme with a step size of 1/800. The intervals for the benchmark values defined in [6] are $c_{d,\max}^{\text{ref}} = [2.93, 2.97]$ and $c_{l,\max}^{\text{ref}} = [0.47, 0.49]$. Table 2 shows the good results which could be computed with a quite low number of unknowns.

Table 2. 'Flow around a cylinder' with postprocessing

Δt	$t(c_{d,\max})$	$c_{d.\max}$	$t(c_{l,\max})$	$c_{l,\max}$	$p_{\text{diff}}(8s)$
1/400	3.93	2.9509076	5.695	0.49461359	-0.11086049
1/1000	3.934	2.9478232	5.688	0.49117886	-0.11053843
1/1200	3.93	2.9465880	5.686667	0.49084030	-0.11048193
John:04	3.93625	2.9509216	5.6925	0.47811979	-0.11158097

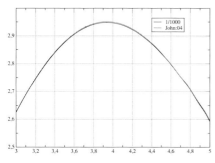

Fig. 5. 2D-3 : c_d

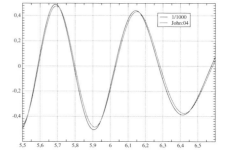

Fig. 6. 2D-3 : c_l

Type 2D-2 (periodic, unsteady)

The variation 2D-2 from [6] usually needs small time step sizes. An error indicator

$$e_t(t^m) \approx \frac{4\|v_{\frac{\Delta t_m}{2}}(t^m) - v_{\Delta t^m}(t^m)\|}{3\|v_{\frac{\Delta t_m}{2}}(t^m)\|} \qquad \Delta t_{m+1} = \sqrt{\frac{\varepsilon T tol}{e_t(t^m)}} \Delta t_m$$

Table 3. 2D-2: Results for an adaptive chosen time step size

ε_{Ttol}	$\bar{\triangle} t$	$c_{d.\,\max}$	$c_{l,\max}$	Strouhal
$1.0 \cdot 10^{-3}$	0.004246	3.2439	1.0104	0.29811
$7.5 \cdot 10^{-4}$	0.002479	3.2285	1.0031	0.30022

based on $v_{h/2}$ computed with $\triangle t$ and $\triangle t/2$ was used to choose a proper time step size in consideration of the stability of the BDF scheme. With this choice of time step sizes the splitting algorithm with the presented postprocessing and 555680 unknowns in v and 139344 in p computed the benchmark values displayed in table 3 with an average time step size $\bar{\triangle} t$. The tolerance intervals for c_d are [3.2200,3.2400], for c_l [0.9900,1.0100] and for the Strouhal number [0.2950, 0.3050].

6 Conclusions

The presented algorithm with build-in postprocessing shows an error reduction in the L^2 norm of an order $n > 2$ in time. It was successfully tested on analytic problems as well as on standard CFD problems. A very interesting aspect of the postprocessing with nested grids is that in all numerical experiments it caused no additional CPU costs. An extension of the techniques to three-dimensional problems as well as of the gradient recovery technique could be done straightforward and it is one of the future prospects. Beyond this further future prospects could be e.g. the integration of more levels together with the fourth order BDF scheme for the postprocessing and the use of Finite Elements of a higher order.

Acknowledgment

We would like to thank Volker John for providing his benchmark data for the 'Flow around a cylinder' 2D-3 case.

References

1. Frochte, J.: Ein Splitting-Algorithmus höherer Ordnung für die Navier-Stokes-Gleichung auf der Basis der Finite-Element-Methode [Diss./Phd-Thesis], Universität Duisburg-Essen (published in Dec. 2005)
2. Haschke, H., Heinrichs, W.: Splitting techniques with staggered grids for the Navier-Stokes equations in the 2d case, J. of Comp. Phys., **168**, 131–154 (2001)
3. Heinrichs, W.: Splitting techniques for the unsteady Stokes equations, SIAM J. Numer. Anal., **35**, 1646–1662 (1998)
4. John, V., Matthies, G.: Higher order Finite Element discretizations in a benchmark problem for the 3D Navier Stokes equations, Int. J. for Num. Methods in Fluid Mechanics, **40**, 775–798 (2002)

5. John, V.: Reference values for drag and lift of a two-dimensional time-dependent flow around a cylinder, Int. J. for Num. Methods in Fluids, **44**, 777–788 (2004)
6. Schäfer, M., Turek, S.: The benchmark problem 'flow around a cylinder'. In: Hirchel EH (ed.) Notes on Num. Fluid Mechanics. Vieweg Verlag Braunschweig, (1996)
7. Zienkiewicz, O. C., Zhu, J. Z.: The superconvergent patch recovery and a posteriori error estimates, Int. J. Num. Meth. Engrg., **33**, 1331–1382 (1992)

The POD Technique for Computing Bifurcation Diagrams: A Comparison among Different Models in Fluids

Pedro Galán del Sastre[1] and Rodolfo Bermejo[2]

[1] Departamento de Matemáticas. Facultad de Ciencias del Medio Ambiente. Universidad de Castilla la Mancha. Avda. Carlos III s/n, 45071 Toledo, Spain
pedro.galan@uclm.es

[2] Departamento de Matemática Aplicada a la Ingeniería Industrial. ETSII. Universidad Politécnica de Madrid. C/ José Gutiérrez Abascal 2, 28006 Madrid, Spain
rbermejo@etsii.upm.es

Summary. It is well know the importance of bifurcation diagrams in fluid models where any of the parameters has some kind of uncertainty (usually the Reynolds number in Navier-Stokes, or the Ekman parameter in geophysical models).

In this work we propose some modifications to the Proper Orthogonal Decomposition (POD) method (or Karhunen-Loeve expansions) in order to study this problem. Although some of this modifications have already been introduced in the literature, most of them are devoted to computing the first Hopf bifurcation. We show here how one can handle the bifurcation diagram also in periodic branches.

1 Introduction

This work is devoted to computing bifurcation diagrams in some fluid models by POD. Recently, there have been various authors [1, 2, 4, 6] who have made use of this technique for bifurcation studies. The POD is a decomposition technique to calculate a finite dimensional space from discrete sets of data obtained either via experimental measurements or numerical simulations. POD combined with a Galerkin projection provides a method to derive low-dimensional models from high-dimensional systems of ODEs.

The problem in using POD is to choose the good scheme to calculate the finite space. Although this is still unsolved, some improvements have been reported in [4, 7] (p-POD, SPOD) and [1, 2].

In this work, we improve the cut-off criterion used in the SPOD (Sequential POD) method, and analyze the results obtained by this technique in computing bifurcation diagrams for three fluid models.

2 The POD technique

The POD technique was first introduced in Statistics under the names of Karhunen-Loève decomposition or principal component analysis as a technique to analyse multidimennsional data. Starting with a number of vectors (data) that belong to a vector space, the technique provides an orthonormal basis for representing the data in a least square sense. One can generalize this idea in combination with a Galerkin projection procedure to provide a method for generating lower dimensional models of dynamical systems that have a very large or even infinite dimensional phase space. And it is in this context we use the POD method. To fix ideas, let us consider a Hilbert space $(H, (\cdot, \cdot)_H)$, a compact set $I \subset \mathbb{R}$ and the mapping $u : I \to H$; we are interested in finding an m-dimensional subspace $H_m \subset H$ such that for any m- dimensional subspace $Y_m \subset H$, $dist_{L^2(I,H)}(u, H_m) \leq dist_{L^2(I,H)}(u, Y_m)$. To do so, we shall calculate a basis of H_m looking for elements of $\phi \in H$ characterized by the property

$$\|(u(t), \phi)_H\|_{L^2(I,H)} = \max_{\varphi \in H, \|\varphi\|=1} \|(u(t), \varphi)_H\|_{L^2(I,H)}. \tag{1}$$

Hence, introducing the functional $J_\lambda : H \to \mathbb{R}$

$$\varphi \to \|(u(t), \varphi)_H\|_{L^2(I,H)} - \lambda \|\varphi\|_H, \lambda \in \mathbb{R},$$

we have that the solution of (1) is a critical point of J_λ. The following proposition ensures the existence of critical points.

Proposition 1. *Let $K = \{1, 2, \ldots, \dim(H)\}$ ($K = \mathbb{N}$ if $\dim(H) = \infty$). Then, there exist $\{(\lambda_j, \phi_j)\}_{j \in K} \subset \mathbb{R}^+ \times H$ such that ϕ_j is a critical point of J_{λ_j}. Futhermore, $\{\phi_j\}_{j \in K}$ is an orthonormal basis in H.*

Proof. If ϕ is a critical point of J_λ,

$$\lim_{h \to 0^+} \frac{J_\lambda(\phi + h\varphi) - J_\lambda(\varphi)}{h} = \\ = 2\left(((u(t), \phi)_H, (u(t), \varphi)_H)_{L^2(I)} - \lambda(\phi, \varphi)_H\right) = 0 \tag{2}$$

Then, one can define the operator $T : H \longrightarrow H$ by

$$(T\phi, \varphi)_H = ((u(t), \phi)_H, (u(t), \varphi)_H)_{L^2(I)} \quad \forall \varphi \in H. \tag{3}$$

One can prove [6] that T is a compact, positive and self-adjoint operator; so that, there exists a sequence of eigenvalues and associated eigenvectors of T, $(\lambda_j, \phi_j)_{j \in K} \in \mathbb{R}^+ \times H$, $\lambda_j \leq \lambda_{j+1}$, such that $\{\phi_j\}_{j \in K}$ is an orthonormal basis of H. The proof is finished because (2) implies that ϕ_j is a critical point of J_{λ_j}. ∎

The next proposition states that the orthonormal basis $\{\phi_j\}_{j \in K}$ is optimal in the sense defined above.

Proposition 2. *For any $m \in K$, let $H_m = \text{span}\{\phi_j\}_{j=1}^m$. Then,*

$$\|u(t) - P_{H_m}u(t)\|_{L^2(I;H)} \leq \|u(t) - P_Y u(t)\|_{L^2(I;H)} \qquad (4)$$

for all linear subspace Y such that $\dim Y = m$, where P_X denotes de projection operator onto X.

Proof. See [6]. ∎

Note that $m < \infty$ because one is always interested in finite dimensional subspaces.

An important issue is the evaluation of the error $\|u(t) - P_{H_m}u(t)\|_{L^2(I;H)}$ because this provides a computable criterion on the quality of the approximation of $P_{H_m}u(t)$ to $u(t)$. By virtue of Proposition 1 and (3) we have for all $t \in I$ and for all $j \in K$,

$$u(t) = \sum_{j \in K} (u(t), \phi_j) \phi_j \quad \text{and} \quad \|(u(t), \phi_j)_H\|_{L^2(I)}^2 = (T\phi_j, \phi_j)_H = \lambda_j;$$

hence, it readily follows that

$$\|u(t) - P_{H_m}u(t)\|_{L^2(I;H)} = \sum_{j > m} \lambda_j.$$

So that, the ratio

$$E(m) = \frac{\|u(t) - P_{H_m}u(t)\|_{L^2(I;H)}}{\|u(t)\|_{L^2(I;H)}} = \frac{\sum_{j>m} \lambda_j}{\sum_{j \in K} \lambda_j}$$

gives a computable measure of the goodness of the approximation. $E(m)$ is usually known as the energy of $u(t)$ in H_m.

The key idea of this work when using POD method is to compute a finite dimensional space that contains a specific (numerical) attractor. In this case, $u(t) \in V_h \subset L^2(\Omega)$ for all $t \in I$, where V_h is a finite element space, and $I = \{t_i\}_{i=1}^n$, t_i are time instants at which the numerical solution is computed ($u(t_i)$ are usually known as snapshots). The operator T defined in (3) is now the correlation matrix of these snapshots as shown in [8]; therefore, T is an $n \times n$ matrix and the calculation of its eigenvectors, $\{\phi_j\}_{j \in K}$, is independent of $\dim V_h$. Notice that n is usually much smaller than $\dim V_h$.

2.1 The POD method for computing bifurcation diagrams

We described a scheme to apply the POD technique to calculate bifurcation diagrams. Although this technique is able to compute quite accurately a basis for a single attractor, the problem arises when one wishes to compute a basis to describe different attractors (one can get some noise when using data that lie in different spaces). To overcome this problem, the key idea, formulated in [4, 7] under the name of SPOD (Sequential POD), is to compute a basis for each attractor and use the information they (the bases) share to put them together (some kind of orthogonalization). We follow this idea, improving upon the cut-off criterion, to obtain the dimension of a less noisy global basis.

Thus, suppose we are interested in computing a basis for several attractors, say $u^i \in L^2(I_i, H)$, for $i = 1, 2, \ldots, s$. The procedure can be formulated as follows:

1. Apply the POD technique to $\{u^1(t)\}_{t \in I_1}$ and obtain $\{\phi_j\}_{j=1}^{m_1}$.
2. For each $i = 1, 2, \ldots, s-1$, define

$$\hat{u}^{i+1}(t) = u^{i+1}(t) - \sum_{j=1}^{m_i} \left(u^{i+1}(t), \phi_j\right)_H \phi_j$$

and apply the POD method to $\{\hat{u}^{i+1}(t)\}_{t \in I_{i+1}}$ to get $\{\phi_j\}_{j=m_i}^{m_{i+1}}$.

To choose m_i, we specify a tolerance ε, usually $\varepsilon \leq 0.0001$, and define

$$E_i^*(m) = \frac{\|u^i(t) - P_{H_m} u^i(t)\|_{L^2(I;H)}^2}{\|u^i(t)\|_{L^2(I;H)}^2}, \quad (5)$$

we take m_i such that $E_i^*(m_i) < \varepsilon$.

3 Description of the examples

3.1 The obstacle problem

In this example we consider a 2D incompresible fluid in a periodic domain Ω with a square object in the middle. $\Omega = (-5/2, 5/2) \times (-1, 1) \setminus [-1/10, 1/10]^2$, with boundaries $\partial \Omega = \Gamma_0 \cup \Gamma_1 \cup \Gamma_2$, where $\Gamma_1 = \{-5/2\} \times (-1, 1)$, $\Gamma_2 = \{5/2\} \times (-1, 1)$ and $\Gamma_0 = \partial \Omega \setminus (\Gamma_1 \cup \Gamma_2)$. The equations of this model are

$$\begin{cases} \dfrac{\partial \mathbf{u}}{\partial t} + \mathbf{u} \cdot \nabla \mathbf{u} - \nu \Delta \mathbf{u} = -\nabla p + \mathbf{f}, & \text{in } \Omega \times (0, +\infty) \\ \operatorname{div} \mathbf{u} = 0, & \text{in } \Omega \times (0, +\infty) \\ \mathbf{u}_{|\Gamma_0} = 0, \mathbf{u}_{|\Gamma_1} = \mathbf{u}_{|\Gamma_2}, & \forall t \in (0, +\infty) \end{cases}$$

where $\mathbf{u}$ is the velocity, p is the pressure, $\nu = \dfrac{1}{Re}$, Re being the Reynolds number, and $\mathbf{f} = (2\nu, 0)$. For this model, ν will be the bifurcation parameter.

3.2 The vorticity-stream function model for a barotropic ocean with constant depth

We now consider a barotropic ocean model with constant depth. For this example we consider an idealized ocean enclosed in the domain $\Omega = [0, L] \times [0, 2L]$, $L = 1000 \, km$, with a constant depth of $800 \, m$ and forced by wind stress $\tau(x, y) = (-\tau_0 \cos(\pi y/L), 0)$. This is a classical problem in oceanography that has been long studied because although simple it retains much of the complexity of the ocean dynamics.

Using vorticity-stream function formulation and the β-plane approximation, the equations of this 2D model are:

$$\begin{cases} \dfrac{\partial \omega}{\partial t} + \dfrac{\partial \psi}{\partial x}\dfrac{\partial \omega}{\partial y} - \dfrac{\partial \psi}{\partial y}\dfrac{\partial \omega}{\partial x} + \beta\dfrac{\partial \psi}{\partial x} = A_H \Delta \omega - \gamma \omega + \dfrac{\operatorname{rot} \tau}{H}, & \text{in } \Omega \times (0, +\infty) \\ \Delta \psi = \omega, & \text{in } \Omega \times (0, +\infty) \\ \omega_{|\partial \Omega} = \psi_{|\partial \Omega} = 0, & \forall t \in (0, +\infty) \end{cases}$$

where ω is the vorticity of the fluid, ψ is the stream function, H denotes the depth of the ocean, β is the first order approximation to the Coriolis force, γ is the bottom friction coefficient and A_H denotes the horizontal viscosity playing the role of the bifurcation parameter.

3.3 A barotropic ocean model with realistic bottom topography applied to the North Atlantic Ocean

In this last example we consider again a barotropic ocean model, but now applied to the North Atlantic Ocean. This model includes realistic coastlines and bottom topography, and is forced by climatological wind stress. The model equations are:

$$\begin{cases} H\dfrac{\partial \mathbf{u}}{\partial t} + \mathbf{u} \cdot \nabla(H\mathbf{u}) - A_H \operatorname{div}(H\nabla \mathbf{u}) + f H \mathbf{u}^\perp = \\ \qquad\qquad = -\dfrac{1}{\rho_0} H \nabla p_s - \gamma H \mathbf{u} + \dfrac{\tau}{\rho_0}, & \text{in } \Omega \times (0, +\infty) \\ \operatorname{div}(H\mathbf{u}) = 0, & \text{in } \Omega \times (0, +\infty) \\ \mathbf{u}_{|\partial \Omega} = 0, & \forall t \in (0, +\infty) \end{cases}$$

where now $\mathbf{u}$ is the averaged horizontal fluid velocity over each water column, f is the Coriolis force ($f(y) = f_0 + \beta y$), ρ_0 is the (constant) density of the ocean, γ is the bottom friction coefficient, τ denotes the wind stress, H is a function describing the bottom topography and again A_H denotes the horizontal viscosity playing the role of bifurcation parameter.

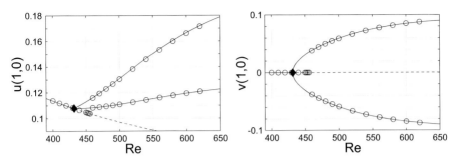

Fig. 1. Obstacle problem bifurcation diagram: DNS solution (*circle*) and POD reduced model (*line*) at one point of the domain (— stable and - - unstable branches); ◇ Hopf bifurcation point. Periodic attractor represented by branches of maximum and minimum values.

4 Computation of the bifurcation diagrams of the example models

The scheme to compute the bifurcation diagrams of these three models can be summarized as follows: (i) compute several attractors for different values of the bifurcation parameter, (ii) use the POD (SPOD) method with the cut-off criterion (5) to get a finite dimensional global basis for all the attractors to be studied, (iii) use the Galerkin projection to obtain an ODE system, and (iv) compute the bifurcation diagram of the ODE problem with, for instance, AUTO97.

Note that we need a large number of snapshots to generate a global basis for all the attractors we want to study to calculate the bifurcation diagram. For each model, the snapshots are produced by calculating the numerical solution with a finite element-semi-Lagrangian scheme, see [5, 3]. This numerical scheme is so efficient that allows to perform long term computations for different values of the bifurcation parameter.

We show in Table 1 the dimension m_i of the subbases at different values of the bifurcation parameter for the three models presented in the previous section. The union of these subbases form the global basis of the corresponding reduced model. In all the models, one can check that it is necesary to add just some few elements to the basis when data from one more attractor is added. This happens because the finite dimensional space spanned by the basis already calculated is very close to the new attractor. So, it is not necesary to add more data from new attractors to make use of Galerkin projection.

In the obstacle problem, as well as in the vorticity-stream function model, one can note that there is a good agreement between the solutions obtained by direct numerical simulations (DNS) and the bifurcation diagram computed by AUTO97, Figures 1 and 2. In particular, the stationary and periodic branches are very close to the DNS solutions, whereas the Hopf bifurcation is computed with an error lower than 2-3%.

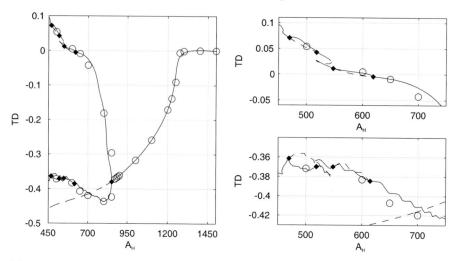

Fig. 2. Left panel: as Figure 1 (for model in Section 3.2). Right panels: detail of left panel.

Table 1. Description of the data u^i used to compute the global basis by POD and SPOD method for (a) obstacle problem, (b) vorticity-stream function model, and (c) barotropic model applied to North Atlantic Ocean. Note that *stat. att.* denotes a set made of stationary attractors.

(a)

i	Re	m_i
1	460	6
2	600	19
3	560	24
4	500	29
5	stat. att.	32

(b)

i	A_H	m_i
1	800	12
2	500	53
3	600	79
4	700	82
5	stat. att.	87

(c)

i	A_H	m_i
1	670	6
2	640	15
3	650	20
4	stat. att.	30

In Figure 3 we show the results obtained for the Atlantic Ocean model. As one can see, the stationary branch obtained by AUTO97 from the reduced model is also in good agreement with the DNS. Similarly, the Hopf bifurcation is also computed with an error of about 2-3%. However, one can note that the agreement between DNS and reduced model in the periodic branches is not as good as in the previous examples. We do not have a clear explanation for such a disagreement.

5 Conclusions

We have shown in this work how POD technique can be modified in a proper way to study bifurcation diagrams. Good agreements between DNS and the

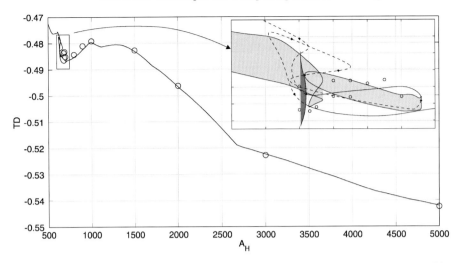

Fig. 3. As Figure 1 (for model in Section 3.3). Light gray means periodic unstable branches, and light gray periodic stable ones.

reduced model are obtained by POD, although some more effort must be made to study more complex models in periodic branches.

Acknowledgement

The authors acknowledge the financial support of Ministerio de Ciencia y Tecnología de España via the grant REN2002-03726.

References

1. Aubry, N., Lian, W.Y., Titi, E.S.: Preserving symmetries in the Proper Orthogonal Decomposition. SIAM J. Sci. Comput., **14**, 483–505 (1993)
2. Bangia, A.K., Batcho, P.F., Kevrekidis, I.G., Karniadakis, G.E.: Unsteady two-dimensional flows in complex geometries: comparative bifurcation studies with global eigenfunction expansions. SIAM J. Sci. Comput., **18**, 775–805 (1997)
3. Bermejo, R. Galán del Sastre, P.:Long term behavior of the wind driven circulation of a numerical North-Atlantic ocean circulation model. ECCOMAS 2004
4. Christensen, E.A., Brøns M., Sørensen J.N.: Evaluation of Proper Orthogonal Decomposition-based decomposition techniques applied to parameter dependent non-turbulent flows. SIAM J. Sci. Comput., **21**, 1419–1434 (2000)
5. Galán del Sastre, P.: Cálculo numérico del atractor en ecuaciones de Navier-Stokes aplicadas a la circulación del océno. PhD Thesis, Universidad Complutense de Madrid, Spain (2004)
6. Holmes, P., Lumley, J.L., Berkooz, G.: Turbulence, coherent structures, dynamical systems and symmetry. Cambridge Monographs on Mechanics. Cambridge: Cambridge University Press (1996)

7. Jørgensen, B.H.: Low-dimensional modeling and dynamics of the flow in a lid driven cavity with a rotating rod. PhD Thesis, Technical University of Denmark, Denmark (2000)
8. Sirovich, L.: Turbulence and the dynamics of coherent structures. Part I: Coherent structures. Quart. Appl. Math., **45**, 561–571 (1987)

Filtering of Singularities in a Marangoni Convection Problem

Henar Herrero[1], Ana M. Mancho[2] and Sergio Hoyas[3]

[1] Departamento de Matemáticas, Facultad de Ciencias Químicas,
Universidad de Castilla-La Mancha, 13071 Ciudad Real, Spain
Henar.Herrero@uclm.es
[2] Instituto de Matemáticas y Física Fundamental,
CSIC, Serrano 121, 28006 Madrid Spain
A.M.Mancho@mat.csic.es
[3] Universidad Politécnica de Madrid, 28080 Madrid, Spain
Sergio.Hoyas@uclm.es

Summary. The problem considered consists of a fluid within a cylindrical annulus heated laterally. As soon as a horizontal temperature gradient is applied a convective state appears. This state becomes unstable through stationary or oscillatory bifurcations as control parameters involved in the problem reach critical values. The problem is modelled with the incompressible Boussinesq Navier-Stokes equations and appropriate boundary conditions. In particular we consider lateral conducting walls and surface tension effects. This choice presents singularities at the point where free and solid surfaces meet, which consist on discontinuities on the temperature and its derivatives. These singularities are smoothed using a polynomial filtering. The main goal of this work is the study of the effect of this filtering in the stability problem. The filter improves the convergence of the numerical method. Convergence with the filtering scale depends on the Marangoni parameter.

1 Introduction

Instabilities and pattern formation in thermocapillary flows have been extensively studied in the last years. Classically heat is applied uniformly from below [1] where the conductive solution becomes unstable for vertical temperature gradients beyond a certain threshold. A more general set-up considers thermoconvective instabilities when a horizontal temperature gradient is imposed by heating the fluid through lateral conducting walls [4, 6]. This process displays many interesting instabilities. It has been treated from different points of view: experimental [4, 5] and theoretical both with semi-exact [11, 13] and numerical [12, 10]. This problem presents a viscous singularity at the points where free and solid surfaces meet. Finite difference methods solve the singularities by using local approximations of the derivatives and mesh refinements [2], and Chebyshev collocation methods [3] by means of a filter function [9]. In Ref. [9] the effect of a polynomial filter function in an evolution Chebyshev collocation scheme was studied. In this work the effect of this kind of filter is revised in a bifurcation problem. This point of view is important as one of the parameters responsible of the bifurcation appears at the singular boundary point.

The article is organized as follows. In section 2 the formulation of the problem is explained. In section 3 the numerical method, the filtering and the numerical solutions are detailed. Section 4 outlines the linear stability analysis and the influence of the filter on the eigenvalues. Finally, in section 5 conclusions are presented.

2 Formulation of the problem

The physical set-up consists of a horizontal fluid layer of depth d (z coordinate) in a container limited by two concentric cylinders of radii a and $a + \delta$ (r coordinate). The bottom plate is rigid and the top is open to the atmosphere. The inner cylinder has a temperature $T_{\max}$, the outer is at $T_{\min}$ and the environmental temperature is T_0. In the equations governing the system u_r, u_ϕ and u_z are the components of the velocity field u, Θ is the temperature, p is the pressure, $\mathbf{r}$ is the radio vector and t is the time. The system evolves according to momentum and mass balance equations and to the energy conservation principle, which in dimensionless form are (see Ref. [8]),

$$\nabla \cdot u = 0, \tag{1}$$

$$\partial_t \Theta + u \cdot \nabla \Theta = \nabla^2 \Theta, \tag{2}$$

$$\partial_t u + (u \cdot \nabla) u = Pr \left(-\nabla p + \nabla^2 u - \frac{R\rho}{\alpha \rho_0 \Delta T} e_z \right), \tag{3}$$

where the operators and fields are expressed in cylindrical coordinates and the Oberbeck-Boussinesq approximation has been used. Here e_z is the unit vector in the z direction, ρ is the density, α is the thermal expansion coefficient and ρ_0 is the mean density. These dimensionless numbers are introduced: the Prandtl number $Pr = \nu/\kappa$ and the Rayleigh number $R = g\alpha\Delta T d^3/\kappa\nu$, which represents the buoyant effect. In the definitions ν is the kinematic viscosity of the liquid, κ is the thermal diffusivity, g is the gravity constant and $\Delta T = T_{\max} - T_0$.

The boundary conditions (bc) are,

$$u_z = \frac{\partial u_r}{\partial z} + Ma\frac{\partial \Theta}{\partial r} = \frac{\partial u_\phi}{\partial z} + \frac{Ma}{r}\frac{\partial \Theta}{\partial \phi} = \frac{\partial \Theta}{\partial z} + B\Theta = 0 \text{ on } z = 1, \tag{4}$$

$$u_r = u_\phi = u_z = 0, \quad \Theta = \left(-\frac{r}{\delta^*} + \frac{a}{\delta}\right)\frac{\Delta T_h}{\Delta T} + 1 \text{ on } z = 0, \tag{5}$$

$$u_r = u_\phi = u_z = 0, \quad \Theta = 1 \quad \text{on} \quad r = a^*, \tag{6}$$
$$u_r = u_\phi = u_z = 0, \quad \Theta = 1 - \Delta T_h / \Delta T \quad \text{on} \quad r = a^* + \delta^*. \tag{7}$$

Here B is the Biot number which quantifies the heat exchange with the atmosphere, $a^* = a/d$, $\delta^* = \delta/d$, $\Delta T_h = T_{\max} - T_{\min}$ and $Ma = \gamma \Delta T d / (\kappa \nu \rho_0)$ is the Marangoni number which includes the surface tension coefficient γ. As explained later temperature conditions (5)–(7) together with the Marangoni condition (4) imply a singular boundary condition at the upper left and right corners.

3 Basic state

The horizontal temperature gradient at the bottom settles in a stationary convective motion with axial symmetry which is computed as it is indicated next.

3.1 Numerical method

We have solved numerically Eqs. (1)–(3) for any scalar field (X) together with the boundary conditions (5)–(7) translated into the domain $[-1,1]^2$ for the basic state which is stationary (i.e., $\partial_t X = 0$) and axisymmetric (i.e., $\partial_\phi X = 0$). We use a Chebyshev collocation method in the primitive variable formulation procedure as explained in Ref. [7]. The nonlinearity was treated with a Newton-like iterative method. In the first step the nonlinearity was neglected and a solution was found by solving the linear system: $u_r^0, u_z^0, p^0, \Theta^0$. This solution was corrected by perturbation fields: $u_r^1 = u_r^0 + \bar{u}_r$, $u_z^1 = u_z^0 + \bar{u}_z$, $p^1 = p^0 + \bar{p}$ and $\Theta^1 = \Theta^0 + \bar{\Theta}$. These expressions are introduced into Eqs. (1)–(7), which are linearized around the approach at step 0. The resulting linear system for the perturbations is solved and the first iteration solution, $u_r^1, u_z^1, p^1, \Theta^1$, is obtained. This process was undertaken in such a way that solutions at the $i+1$ step were obtained after solving Eqs. (1)–(7) linearized around the approach at step i.

Each step in the Newton-like method corresponds to solving a linear system of partial differential equations. The unknown fields at each step $\bar{u}_r(r,z)$, $\bar{u}_z(r,z)$, $\bar{p}(r,z)$ and $\bar{\Theta}(r,z)$ were expanded in a truncated series of orthonormal Chebyshev polynomials, i.e., $\bar{u}_r(r,z) = \sum_{n=0}^{N} \sum_{m=0}^{M} a_{nm} T_n(r) T_m(z)$. These expansions are substituted into the equations (1)–(3) and boundary conditions (5)–(7) posed in the domain $[-1,1]^2$. The $N+1$ Gauss-Lobatto points ($r_j = \cos(\pi(1-j/N))$, $j = 0, ..., N$) in the r axis and the $M+1$ Gauss-Lobatto points ($z_j = \cos(\pi(1-j/M))$, $j = 0, ..., M$) in the z axis were calculated. The previous equations were evaluated at these points according to the rules explained in Ref. [8] and, in this way $4(N+1)(M+1)$ equations are obtained with $4(N+1)(M+1)$ unknowns. The system does not have a maximun rank because pressure is only determined up to a constant value. Since this value does not influence the other physical magnitudes, the evaluation of the normal component of the momentum equations at $(x_{j=N} = 1, z_{j=4} = \cos(\pi(1 - 4/M)))$ was replaced by a value for the pressure at this point, for instance $p = 0$. The considered criterion of convergence is that the difference between two consecutive approximations in l^2 norm should be less than 10^{-9}.

3.2 Filtering

The singularities are present at the upper right and left corners of the domain $[-1, 1]^2$. They come from the boundary conditions,

$$\partial_z u_r + Ma\, \partial_r \Theta = 0, \quad \text{on } z = 1, \text{ and } \Theta = 1 \text{ on } r = -1, \tag{8}$$

$$\partial_z u_r + Ma\, \partial_r \Theta = 0, \quad \text{on } z = 1, \text{ and } \Theta = 1 - \Delta T_h/\Delta T \text{ on } r = 1. \tag{9}$$

On one hand Eqs. (6) and (7) imply that $\partial_z u_r = 0$ on lateral walls and this together with the Marangoni condition impose $\partial_r \Theta = 0$ at the upper right and left corners. On the other hand Eqs. (6) and (7) imply that those points have additional conditions for the temperature. The following regularization function is used

$$f_n(r) = (1 - r^{2n})^2, \quad -1 \le r \le 1, \tag{10}$$

in figure 1 several f_n functions for different n values are represented. This function is included in the Marangoni condition as

$$\partial_z u_r + Ma \partial_r \Theta f_n(r) = 0 \quad \text{on } z = 1, \tag{11}$$

in this way the discontinuity is avoided as Eq. (11) does not impose any requirement on $\partial_r \Theta$.

The filtering scale is defined as the distance from the solid boundaries r_n, at which $f_n(r) = 0.9$, and it behaves as $O(1/n)$ for large n (see figure 2).

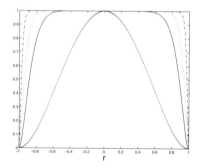

Fig. 1. Filter functions for several values of n.

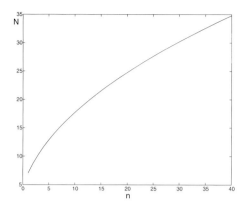

Fig. 2. Required order expansions vs. the n value.

3.3 Numerical solutions

Typically, in order to reach convergence at least three mesh points are necessary to fall into the sharp region of the filter (see Ref. [3]). This condition can be written as,

$$N > 3\pi/\text{acos}(r_n) = g(n), \qquad (12)$$

where r_n is such that $(1 - r_n^{2n})^2 = 0.9$. In figure 2 the function $g(n)$ displayed shows that higher order expansions are required for large n values.

Convergence of the numerical method strongly depends on the Marangoni number. If Ma number is small enough the same solution and bifurcation thresholds are reached with and without regularization. In figure 3 a basic solution for this case is shown. For large Marangoni numbers regularization is required in order to converge. In this case the regularization function must have large n. Figure 4 displays a solution with and without regularization which presents spurious oscillations. The filter prevents the presence of oscillations that disappear with the regularization.

4 Linear stability

Changes on the parameters appearing in Eqs. (1)–(7), may make the basic flow unstable leading to different bifurcations. The linear stability analysis supplies information on the critical values of the parameters at which these bifurcations occur and on the shape of growing modes. We study the stability by perturbing the basic solutions with fields depending on r, ϕ and z coordinates in a fully 3D analysis, following the numerical scheme of Refs. [8, 7]. Due to the periodical boundary conditions in the azimuthal coordinate and to the axial symmetry of the basic solution, the perturbations of any physical function X can be factorized and expanded by Fourier modes along the angular variable ϕ,

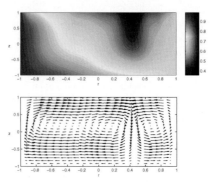

Fig. 3. Basic state for $B = 1$, $R = 2785$, $Ma = 16$ and $\Delta T_h/\Delta T = 0.5$.

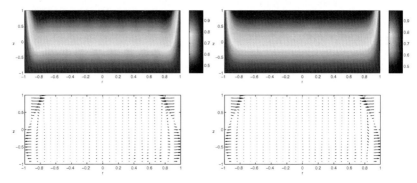

Fig. 4. Basic state for $B = 1.25$, $R = 1000$, $Ma = 92$ and $\Delta T_h/\Delta T = 0$ a) without regularization; b) with regularization.

$$X(r, \phi, z, t) = X(r, z)e^{im\phi + \lambda t}, \qquad (13)$$

where m is the wave number. The real part of the eigenvalue λ characterizes the instability, when it is negative the basic state is stable, but if it is positive the basic solution is unstable. In this case the imaginary part of λ may be zero and then the bifurcation is stationary, while if it is non zero the bifurcation is oscillatory.

4.1 Numerical eigenvalues

A test on the convergence of the numerical method is carried out by comparing the value of the maximum eigenvalue obtained for different order expansions in Chebyshev polynomials in the r (N) and z (M) coordinates. This study is shown in table I and it is obtained for a set of parameter values $B = 1$, $R = 348$, $Ma = 8$ and $\Delta T_h/\Delta T = 1$. Convergence depends on the filter scale, as while n increases larger order expansions are required (see figure 2), however as the order expansion increases the eigenvalue tends to a constant value.

Next we have calculated the maximum eigenvalues for solutions with two different Ma numbers. We use different regularizations increasing the degree of the polynomials and we compare the results with those obtained without any regularization. Figure 5a) shows that for Ma number small enough, by increasing n, the eigenvalue tends to the one without regularization. These results suggest that regularization is not necessary. Figure 5b) shows a similar result for larger Ma values where regularization is necessary. As n increases, the eigenvalue converge to a constant value that may be considered the maximum eigenvalue for the singular problem.

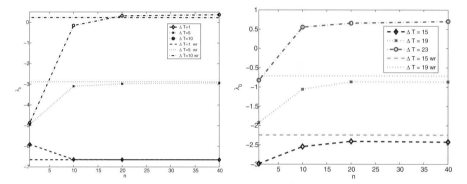

Fig. 5. Maximum eigenvalues as a function of the degree n of the polynomial filter; a) $B = 1$, $R = 348$, $Ma = 8$, $\Delta T_h = \Delta T$ and different values of ΔT. The horizontal lines correspond to the eigenvalue without regularization (wr); b) $B = 1.25$, $R = 1000$, $Ma = 92$, $\Delta T_h = 0$ and different values of ΔT. The horizontal lines correspond to the eigenvalue without regularization (wr).

Table 1. Convergence test for $B = 1$, $R = 348$, $Ma = 8$ and $\Delta T_h/\Delta T = 1$.

	17×13	21×17	25×21
$n = 20$	-4.3939	-4.4485	-4.4732
$n = 50$	-4.3282	-4.3816	-4.3959
$n = 100$	-4.3002	-4.3799	-4.4032

5 Conclusions

We have studied the effect of the use of a filtering function in a singular boundary condition that appears in a Marangoni convection problem. The filter improves the convergence of the method both for the basic state and its linear stability and allows the numerical resolution for larger Ma values. We find that for small and medium Marangoni numbers results without regularization are correct as these are recovered for sharp filetering functions. Convergence with the filtering scale depends on the parameters.

Acknowledgments

This work was partially supported by the Research Grants MCYT (Spanish Government) BFM2003-02832, MEC (Spanish Government) MTM2004-00797 and CCYT (JC Castilla-La Mancha) PAC-05-005 which include FEDER funds. AMM thanks MCYT for a Ramón y Cajal Research Fellowship.

References

1. Bénard, H.: Les tourbillons cellulaires dans une nappe liquide. Rev. Gén. Sci. Pures Appl., **11**, 1261-1268 (1900)
2. Canright, D.: Thermocapillary flows near a cold wall. Phys. Fluids, **6**, 1415-1424 (1994)
3. Canuto, C., Hussaini, M.Y., Quarteroni, A., Zang, T.A.: Spectral Methods in Fluid Dynamics. Springer-Verlag, Berlin, (1988)
4. Daviaud, F., Vince, J.M.: Traveling waves in afluid layer subjected to a horizontal temperature gradient. Phys. Rev. E, **48**, 4432-4436 (1993)
5. Garnier, N., Chiffaudel, A.: Two dimensional hydrothermal waves in an extended cylindrical vessel. Eur. Phys. J. B, **19**, 87-95 (2001)
6. Herrero, H., Mancho A.M.: Influence of aspect ratio in convection due to non-uniform heating. Phys. Rev. E, **57**, 7336-7339 (1998)
7. Herrero, H., Mancho, A.M.: On pressure boundary conditions for thermoconvective problems. Int. J. Numer. Meth. Fluids, **39**, 391-402 (2002)
8. Hoyas, S., Herrero, H., Mancho, A.M.: Thermal convection in a cylindrical annulus heated laterally. J. Phys. A: Math and Gen., **35**, 4067-4083 (2002)
9. Kasperski, G., Lebrosse G.: On the numerical treatment of viscous singularities in wall-confined thermocapillary convection, Phys. Fluids, **12**, 2695-2697 (2000)
10. Mancho, A.M., Herrero, H.: Instabilities in a lateraly heated liquid layer. Phys. Fluids, **12**, 1044-1051 (2000)
11. Mercier, J.F., Normand, C.: Buoyant-thermocapillary instabilities of differentially heated liquid layers. Phys. Fluids, **8**, 1433-1445 (1996)
12. Sim, B.C., Zebib, A., Schwabe, D.: Oscillatory thermocapillary convection in open cylindrical annuli. Part 2. Simulations. J. Fluid Mech., **491**, 259-274 (2003)
13. Smith, M.K., Davis, S.H.: Instabilities of dynamic thermocapillary layers. 1. Convective instabilities. J. Fluid Mech., **132**, 119-144 (1983)

On Application of Stabilized Higher Order Finite Element Method on Unsteady Incompressible Flow Problems

Petr Sváček[1] and Jaromír Horáček[2]

[1] Faculty of Mechanical Engineering, Czech Technical University in Prague, Department of Technical Mathematics, Karlovo na,. 13, 121 35 Praha 2
Petr.Svacek@fs.cvut.cz
[2] Institute of Thermomechanics, Czech Academy of Sciences, Dolejškova 5, 182 00 Praha 8
jaromirh@it.cas.cz

Summary. In this paper we address the problem of the numerical approximation of the incompressible flow around a vibrating airfoil. The robust higher order finite element method (FEM) for incompressible flow approximation is presented. The method is based on the combination of several techniques, e.g., the Arbitrary Lagrangian-Eulerian formulation of the Navier-Stokes equations, the stabilization of the finite element scheme and the linearization of the discrete nonlinear problem.

The main attention is paid to the proper stabilization of the higher order finite element method applied on incompressible flow problems. The stabilization procedure based on Galerkin Least-Squares (GLS) method is discussed. The numerical results are presented.

1 Mathematical model

Mathematical model for the relevant technical application consists of fluid and airfoil models. First, the fluid flow is described with the aid of incompressible Navier-Stokes system of equations written in Arbitrary Lagrangian-Eulerian (ALE) formulation, see, e.g., [5]. The ALE method combines the use of the classical Lagrangian and Eulerian reference frames (see, e.g. [3]).The fixed in space Eulerian reference frame is the typical framework used in the analysis of fluid mechanics problems. One of the disadvantages of the Eulerian system is that it does not track the path of any element, in particular the moving fluid-structure interface.

The Lagrangian reference frame is usually used in solid mechanics. It sets up the reference frame by fixing a grid to the material of interest. The material deformation causes also the grid deformation. On the other hand, the use of Lagrangian reference frame for fluid flow is not suitable as the fluid particles travel independent of each other, which causes excessive grid deformations. In what follows we start by introducing the ALE mapping $\mathcal{A}_t$. The ALE mapping is a generalization of the Lagrangian mapping, which follows motion of all particles of the original domain Ω_0, i.e. the Lagrangian mapping is the mapping $\mathcal{L}_t : \Omega_0 \to \Omega_t$, such that $\mathcal{L}_t(\xi) \in \Omega_t$ is the position of the fluid partical at time t originally located at the position $\xi \in \Omega_0$. The comparison of Lagrangian and ALE mappings is shown in Figure 1.

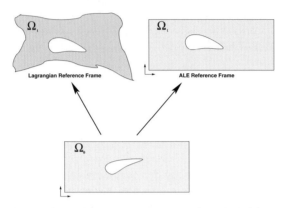

Fig. 1. Comparison of the Lagrangian mapping (on the left) and the Arbitrary Lagrangian Eulerian mapping (on the right).

The ALE mapping $\mathcal{A}_t$ maps the reference configuration Ω_0 onto the computational domain at time t Ω_t (i.e. the current configuration).

$$\mathcal{A}_t : \Omega_0 \mapsto \Omega_t,$$
$$Y \mapsto y(t, Y) = \mathcal{A}_t(Y).$$

By the differentiating of ALE mapping $\mathcal{A}_t$ with respect to time, the **domain velocity** $\mathbf{w}_g$ is computed in the reference coordinates $\widetilde{\mathbf{w}}_g(t, Y) = \frac{\partial y}{\partial t}(t, Y)$ and transformed to spatial coordinates y as $\mathbf{w}_g(t, y)$. The time derivative with respect to the original configuration is then called **ALE derivative**, it is denoted as $\frac{D^\mathcal{A} f}{Dt}$ and can be computed as

$$\frac{D^\mathcal{A} f}{Dt} = \frac{\partial f}{\partial t} + (\mathbf{w}_g \cdot \nabla) f. \qquad (1)$$

With the aid of the ALE derivative $\frac{D^\mathcal{A} f}{Dt}$ the Navier-Stokes system of equations can be rewritten as

$$\frac{D^{\mathcal{A}_t} \mathbf{u}}{Dt} - \nu \triangle \mathbf{u} + \Big((\mathbf{u} - \mathbf{w}_g) \cdot \nabla\Big)\mathbf{u} + \nabla p = 0, \quad \text{in } \Omega_t \times (0, T), \qquad (2)$$
$$\nabla \cdot \mathbf{u} = 0, \quad \text{in } \Omega_t \times (0, T),$$

where by Ω_t we denote the computational domain occupied by fluid at time t, $\mathbf{u}$ denotes the velocity vector, p - denotes the kinematic pressure (i.e. the dynamic pressure divided by the air density) and by $\mathbf{w}_g$ the domain velocity vector is denoted. On the boundary $\partial\Omega$ we prescribe suitable boundary conditions. First, the boundary $\partial\Omega$ is decomposed into three distinct parts, i.e. $\partial\Omega = \Gamma_{W_t} \cup \Gamma_D \cup \Gamma_O$. On Γ_D and Γ_{W_t} the Dirichlet boundary is prescribed, i.e.

$$\text{a)} \quad \mathbf{u} = \mathbf{u}_D \text{ on } \Gamma_D, \qquad \text{b)} \quad \mathbf{u} = \mathbf{w}_g \text{ on } \Gamma_{W_t}. \tag{3}$$

The latter part of boundary denoted by the symbol Γ_{W_t} is the only moving part of the boundary. The boundary Γ_O represents the outlet, where the following boundary condition is prescribed

$$\left[-(p - p_{ref})\mathbf{n} + \frac{1}{2}(\mathbf{u}\cdot\mathbf{n})^{-}\mathbf{u} + \nu\frac{\partial\mathbf{u}}{\partial\mathbf{n}}\right]\bigg|_{\Gamma_O} = 0, \tag{4}$$

where α^- means the negative part of α, i.e. $\alpha^- = \max(0, -\alpha)$. If Γ_O is the out-flowing part of the boundary, i.e. the negative part of the normal velocity is zero $((\mathbf{u}\cdot\mathbf{n})^- = 0)$, the condition (4) is equivalent to the well known *do-nothing* boundary condition. The weak formulation of the equation (2) then can be introduced: Find a velocity vector $\mathbf{u} \in \left(H^1(\Omega_t)\right)^2$ with Dirichlet boundary conditions (3) satisfied and a pressure $p \in L^2(\Omega_t)$, such that for all test functions $\mathbf{v} \in X \subset \left(H^1(\Omega_t)\right)^2$ (being zero on Dirichlet part of boundary) and for all pressure test functions $q \in Y = L^2(\Omega_t)$ the following equation is holds

$$\left(\frac{D^{\mathcal{A}}\mathbf{u}}{Dt}, \mathbf{v}\right) + \nu((\mathbf{u},\mathbf{v})) + c(\mathbf{u};\mathbf{u},\mathbf{v}) - \left((\mathbf{w}_g\cdot\nabla)\mathbf{u},\mathbf{v}\right) \tag{5}$$
$$-\left(p, \nabla\cdot\mathbf{v}\right) + \left(\nabla\cdot\mathbf{u}, q\right) + \int_{\Gamma_O} \frac{1}{2}(\mathbf{u}\cdot\mathbf{n})^+\mathbf{u}\cdot\mathbf{v}\, dS = 0$$

where

$$c(\mathbf{b};\mathbf{u},\mathbf{v}) = \int_{\Omega_t}\left(\frac{1}{2}(\mathbf{b}\cdot\nabla)\mathbf{u}\cdot\mathbf{v} - \frac{1}{2}(\mathbf{b}\cdot\nabla)\mathbf{v}\cdot\mathbf{u}\right)dx,$$

$$((\mathbf{u},\mathbf{v})) = \int_{\Omega_t}(\nabla\mathbf{u})\cdot(\nabla\mathbf{v})dx, \tag{6}$$

and by $(\cdot,\cdot)$ the scalar product on $L^2(\Omega_t)$ or $\left(L^2(\Omega_t)\right)^2$ is denoted.

The fluid flow description is then coupled with the nonlinear equations of motion for an flexibly supported body [6],

$$m\ddot{h} + S_\alpha\ddot{\alpha}\cos\alpha - S_\alpha\dot{\alpha}^2\sin\alpha + k_{hh}\,h = -L(t), \tag{7}$$
$$S_\alpha\ddot{h}\cos\alpha + I_\alpha\ddot{\alpha} + k_{\alpha\alpha}\,\alpha = M(t).$$

where h and α denotes the vertical and the rotational displacements, respectively, L and M denotes the aerodynamical lift force and aerodynamical torsional moment. Both mathematical models are coupled by the evaluation of the aerodynamical forces defined by

$$L = -\int_{\Gamma_{W_t}} \sum_{j=1}^{2} \sigma_{2j} n_j dS, \qquad M = -\int_{\Gamma_{W_t}} \sum_{i,j=1}^{2} \sigma_{ij} n_j r_i^{\text{ort}} dS, \qquad (8)$$

where $r_1^{\text{ort}} = -(x_2 - x_{EO2})$, $r_2^{\text{ort}} = x_1 - x_{EO1}$ and σ_{ij} denotes the fluid stress tensor , see, e.g., [3].

2 Time-spatial discretization

First, let start with the equidistant discretization of the time interval $[0, T]$ with the time step Δt, i.e. $t_k = k \cdot \Delta t$ for $k = 0, 1, 2, \ldots$. Let $\mathbf{u}_n, p_n$ denote the approximation of velocity vector $\mathbf{u}$ and pressure p evaluated at the time level t_n, i.e. $\mathbf{u}_n \approx \mathbf{u}(t_n)$ and $p_n \approx p(t_n)$. The ALE derivative of the velocity vector $\mathbf{u}$ then is approximated as

$$\frac{D^{\mathcal{A}_t} f}{Dt} \approx \frac{3\mathbf{u}_{n+1} - 4\hat{\mathbf{u}}_n + \hat{\mathbf{u}}_{n-1}}{2\Delta t}, \qquad (9)$$

where the velocity $\mathbf{u}_{n+1}$ denotes the approximate velocity at time t_{n+1} and the velocities $\hat{\mathbf{u}}_n, \hat{\mathbf{u}}_{n-1}$ are the velocities at previous time steps t_n and t_{n-1} transformed from domains $\Omega_{t_n}, \Omega_{t_{n-1}}$ on the current computational domain $\Omega_{t_{n+1}}$, i.e., $\hat{\mathbf{u}}_n \equiv \mathbf{u}_n \circ \mathcal{A}_{t_n} \circ \mathcal{A}_{t_{n+1}}^{-1}$, $\hat{\mathbf{u}}_{n-1} \equiv \mathbf{u}_{n-1} \circ \mathcal{A}_{t_{n-1}} \circ \mathcal{A}_{t_{n+1}}^{-1}$.

The approximate solution of the time discretized problem (5), (9) will be sought in the space of the triangular conforming piecewise polynomial elements. For the sake of clarity, we restrict ourselves on the time moment $t = t_{n+1}$ and we denote the computational domain $\Omega = \Omega_{t_{n+1}}$. Furthermore, we will use a triangulation τ_Δ of the domain Ω_t and on every element $K \in \tau_\Delta$ the local element spaces P_K and Q_K for velocity components and pressure are defined (in what follows the spaces are assumed to be polynomial spaces of degree lower or equal to $M > 0$). The space X_Δ of fluid velocity vectors is then introduced

$$X_\Delta = H_\Delta^2, \qquad H_\Delta = \{v \in C(\overline{\Omega}); v|_K \in P_K \subset P_k(K) \text{ for each } K \in \tau_\Delta\},$$

and the pressure space Y_Δ defined as

$$Y_\Delta = \{v \in C(\overline{\Omega}); v|_K \in Q_K \subset P_k(K) \text{ for each } K \in \tau_\Delta\}.$$

Moreover, we define the space of test functions being zero on the Dirichlet part of boundary

$$X_{\Delta,0} = \{\mathbf{v} \in X_\Delta : \mathbf{v}|_{\Gamma_D \cup \Gamma_{W_{t_{n+1}}}} = 0\}.$$

The standard Galerkin approximation of the weak formulation (5) may suffer from two sources of instabilities. One instability is caused by a possible incompatibility of pressure and velocity pairs. It can be overcome either by the use of the finite element velocity/pressure pair, that satisfy the Babuška-Breezi condition, or by the use of pressure stabilizing terms. Further, the dominating convection requires to introduce some stabilization of the finite element scheme, as, e.g. upwinding or streamline-diffusion method. In order to overcome both difficulties, the Galerkin Least Squares method can be applied, see, e.g. ([4]). First, we start with definition of standard Galerkin terms, SUPG/GLS stabilizing term and grad-div stabilizing terms, for details see [4, 6].

The Galerkin terms are defined as

$$\mathbf{a}(\mathbf{u}^*; U_\Delta, V_\Delta) = \frac{3}{2\Delta t}(\mathbf{u}, \mathbf{v})_\Omega + \nu(\nabla \mathbf{u}, \nabla \mathbf{v})_\Omega + c(\mathbf{u}^*; \mathbf{u}, \mathbf{v})$$
$$- ((\mathbf{w}_g^{n+1} \cdot \nabla)\mathbf{u}, \mathbf{v})_\Omega - (p, \nabla \cdot \mathbf{v})_\Omega + (\nabla \cdot \mathbf{u}, q)_\Omega, \qquad (10)$$

$$f(\mathbf{u}, \mathbf{v}) = \frac{1}{2\Delta t}(4\hat{\mathbf{u}}_n - \hat{\mathbf{u}}_{n-1}, \mathbf{v})_\Omega - \int_{\Gamma_O} p_{\text{ref}}(\mathbf{v} \cdot \mathbf{n})\, dS.$$

Next, we define the SUPG/GLS stabilizing terms

$$\mathcal{L}(\mathbf{u}^*; U_\Delta, V_\Delta) = \sum_{K \in T_\Delta} \delta_K \left(\frac{3}{2\Delta t}\mathbf{u} - \nu \triangle \mathbf{u} + ((\mathbf{u}^* - \mathbf{w}_g) \cdot \nabla)\mathbf{u} + \nabla p, \psi(\mathbf{u}^*, q)\right)_K,$$

$$\mathcal{F}(V_\Delta) = \sum_{K \in T_\Delta} \delta_K \left(\frac{1}{2\Delta t}(4\hat{\mathbf{u}}_n - \hat{\mathbf{u}}_{n-1}), \psi(\mathbf{u}^*, q)\right)_K, \qquad (11)$$

where $\psi(\mathbf{u}^*, q) \equiv ((\mathbf{u}^* - \mathbf{w}_g) \cdot \nabla)\mathbf{v} + \nabla q$. The grad-div stabilizing terms $\mathcal{P}(U_\Delta, V_\Delta)$ are defined as

$$\mathcal{P}(U_\Delta, V_\Delta) = \sum_{K \in T_\Delta} \tau_K (\nabla \cdot \mathbf{u}, \nabla \cdot \mathbf{v})_K, \qquad (12)$$

The stabilized discret problem: Find $U_\Delta = (\mathbf{u}, p) \in H_\Delta \times Y_\Delta$ such that $\mathbf{u}$ satisfies approximately the Dirichlet boundary conditions (3) and the equation

$$\mathbf{a}(\mathbf{u}; U_\Delta, V_\Delta) + \mathcal{L}(\mathbf{u}; U_\Delta, V_\Delta) + \mathcal{P}(U_\Delta, V_\Delta) = f(V_\Delta) + \mathcal{F}(V_\Delta), \qquad (13)$$

holds for all $V_\Delta = (\mathbf{v}, q) \in X_{\Delta,0} \times Y_\Delta$.

The choice of the parameters δ_K and τ_K depends on the chosen pair of local elements P_K/Q_K. Here, we distinguish between the Taylor-Hood family of finite element pairs and all the other finite element pairs.

In the case of the local element pair P_K, Q_K being of the Taylor-Hood family P^{m+1}/P^m the following choice of parameters is used

$$\tau_K = \tau_*, \qquad \delta_K = \delta^* h^2,$$

where $\tau^* > 0$ and $\delta^* > 0$ are fixed constants (e.g., we usually set $\tau^* = \delta^* = 1$). The local element size h depends on the local element, local stream velocity vector and the local element degree deg P_K of the velocity approximation.

In the case when the local element pair P_K/Q_K does not belong to the Taylor-Hood family $P^{m+1}(K)/P^m(K)$, the following choice of parameters is used

$$\tau_K = \nu \cdot \left(1 + Re^{loc} + \frac{h^2}{\nu \cdot \Delta t}\right), \qquad \delta_K = \frac{h^2}{\tau_K},$$

where the local Reynolds number is defined as $Re^{loc} = \frac{h\|\mathbf{u}\|_K}{2\nu}$.

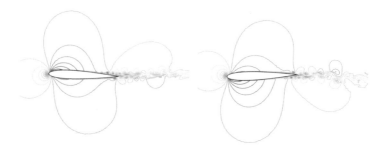

Fig. 2. The velocity distribution around the airfoil NACA 0012 for the angle of attack varying in time

3 Numerical solution and results

In order to find the solution of the nonlinear problem (13) coupled with (7), the strong coupling algorithm will be used on every time level t_{n+1}

- First, using the extrapolation of aerodynamical forces the system of ODE (7) is used, and the approximate computational domain $\Omega \approx \Omega_{t_{n+1}}$ is determined.
- Next, the problem (13) is solved on the domain $\Omega \approx \Omega_{t_{n+1}}$ using Oseen linearization.
- Using the obtained approximate velocity $\mathbf{u}_{n+1}$ and pressure p_{n+1} the aerodynamical forces are updated. We continue with the first step until convergence is obtained.

The system of ODEs (7) on time interval $[t_k, t_{k+1}]$ is solved by fourth order Runge-Kutta method, where the approximate values α_k and h_k are used instead of the exact ones $\alpha(t_k)$ and $h(t_k)$. The values of α_k and h_k determines the transformation of domain $\Omega_k \equiv \Omega_{t_k}$. In order to proceed from time level t_k to the time level t_{k+1} the approximate value of the aerodynamical lift force $\widetilde{L} \approx L(t_{k+1})$ and the approximate value of the aerodynamical torsional moment $\widetilde{M} \approx M(t_{k+1})$ are employed.

The solution of the nonlinear problem (13) is performed by Oseen linearizations, i.e. we start from approximation $U_\Delta^{(0)} = (\mathbf{u}^{(0)}, p^{(0)})$, and for $i = 0, \ldots, N_n - 1$ we solve the problem find $U_\Delta^{(i+1)} = (\mathbf{u}^{(i+1)}, p^{(i+1)})$

$$\mathbf{a}(\mathbf{u}^{(i)}, U_\Delta^{(i+1)}, V_\Delta) + \mathcal{L}(U_\Delta^{(i)}, U_\Delta^{(i+1)}, V_\Delta) + \mathcal{P}(U_\Delta^{(i+1)}, V_\Delta) = f(V_\Delta) + \mathcal{F}(V_\Delta),$$

then set the solution of the nonlinear problem $U_\Delta = U_\Delta^{(N_n)}$. In practical computation it is enough to compute 3-10 iterations.

Numerical results

The numerical simulation of flow over NACA 0012 airfoil, whose vibrations is either given analytically or obtain by the solution of the system of ODEs (7), are presented in Figures 3-5. In the first case, the numerical approximations of the airfoil surface values of the pressure coefficient

$$c_p = \frac{p - p_0}{\frac{1}{2}\rho U^2}$$

was compared with the experimental data from [1]. The time dependence of the rotational angle of the airfoil was prescribed as the periodical function with the frequency 30 Hz and the amplitude 3 degrees, the far field velocity is $U_\infty = 136 \text{ m s}^{-1}$ and the length of airfoil chord is $L = 0.1322$ m (see Figure 3).

Furthermore, the simulation of the coupled model (5) and (7) is presented in the case of the flexibly supported airfoil NACA 0012 in Figures 4-5. The solution was performed for far field velocities $U_\infty = 5 \text{m s}^{-1}$, $U_\infty = 26 \text{m s}^{-1}$ and the choice of parameter's values from report [2] was used.

This research was supported under grant No. 201/05/P142 of the Grant Agency of the Czech Republic and the project MPO ČR No. FT-TA/026 FOREMADE Theme 13.

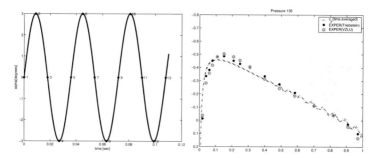

Fig. 3. Dependence of α on time t for vibrating airfoil with frequency 30 Hz and amplitude 3 degrees (on the left). On the right the comparison of the time averaged coefficient c_p along the profile NACA 0012 with the experimental data.

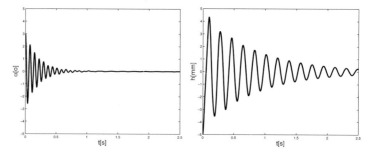

Fig. 4. The damped vibrations in h and α of flexibly supported airfoil NACA 0012 for far field velocity $U_\infty = 5 \text{m s}^{-1}$.

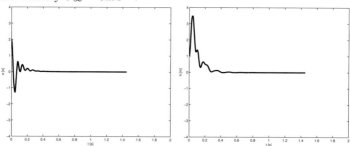

Fig. 5. The damped vibrations in h and α of flexibly supported airfoil NACA 0012 for far field velocity $U_\infty = 26 \text{m s}^{-1}$.

References

1. Benetka, J., Horáček, J.: Experimental pressure data on vibrating airfoils NACA 64A012M5 and NACA 0012. Technical report, Institute of Thermomechanics, Czech Academy of Sciences (2003)
2. Čečrdle, J., Maleček, J.: Verification FEM model of an aircraft construction with two and three degrees of freedom. Technical Report Research Report R-3418/02, Aeronautical Research and Test Institute, Prague, Letňany (2002)

3. Feistauer, M.: Mathematical Methods in Fluid Dynamics. Longman Scientific & Technical (1993)
4. Gelhard, T., Lube, G., Olshanskii, M. A.: Stabilized finite element schemes with LBB-stable elements for incompressible flows. Journal of Computational Mathematics, **177**, 243–267(2005)
5. Nomura, T., Hughes, T. J. R.: An arbitrary Lagrangian-Eulerian finite element method for interaction of fluid and a rigid body. Computer Methods in Applied Mechanics and Engineering, **95**, 115–138 (1992)
6. Sváček, P., Feistauer, M., Horáček, J.: Numerical simulation of flow induced airfoil vibrations with large amplitudes. Journal of Fluids and Structures (2004) (submitted).

Numerical Simulation of Coupled Fluid-Solid Systems by Fictitious Boundary and Grid Deformation Methods

Decheng Wan[1] and Stefan Turek[2]

[1] School of Naval Architecture, Ocean and Civil Engineering,
  Shanghai Jiao Tong University, Huashan Road 1954, 200030 Shanghai, China
  dcwan@sjtu.edu.cn
[2] Institute of Applied Mathematics LS III, University of Dortmund,
  Vogelpothsweg 87, 44227 Dortmund, Germany
  stefan.turek@math.uni-dortmund.de

Summary. Numerical simulations of coupled fluid-rigid solid problems by multigrid fictitious boundary and grid deformation methods are presented. The flow is computed by a special ALE formulation with a multigrid finite element solver. The solid body is allowed to move freely through the computational mesh which is adaptively aligned by a special mesh deformation method such that the accuracy for dealing with the interaction between the fluid and the solid body is highly improved. Numerical examples are provided to show the efficiency of the presented method.

1 Introduction

Efficient numerical solution of the coupled fluid-solid system is still a challenging task in many applications and a topic of current mathematical research. Different differential equations must be satisfied on each side of the interface between fluid and rigid body and the solutions are coupled through relationships or jump conditions that must hold at the interface. The movement of both the interface and the rigid body is unknown in advance and must be determined as part of the solution.

There are two types of methods to meet this challenge. The first approach is a generalized standard Galerkin finite element method [1, 2] in which both the fluid and solid equations of motion are incorporated into a single coupled variational equation. The computation is performed on an unstructured body-fitted grid, and an arbitrary Lagrangian-Eulerian (ALE) moving mesh technique is adopted to deal with the motion of the solid. In the case of 2D, the remeshing of body-fitted grid can be done by many available grid generation software tools, but in the more interesting case of a full 3D simulation, the problem of efficient, body-fitted grid generation is not solved in a satisfying manner yet. The second approach is based on the principle of embedded or fictitious domains, in which the fluid flow is computed as if the space occupied by the solid were filled with fluid, and the no-slip boundary condition on the solid boundaries is enforced as a constraint. The fictitious domain is discretized only once in the beginning. For example, the distributed Lagrange multiplier (DLM)/fictitious domain method developed by Glowinski, Joseph and coauthors [3], and our multigrid fictitious boundary method (FBM) [4, 5] belong to this class. An advantage of the fictitious domain method over the generalized standard Galerkin finite element method is that the fictitious domain method allows a fixed grid to be used, eliminating the need for remeshing. An underlying problem when adopting the fictitious domain method is that the boundary approximation is of low accuracy. One remedy could be to preserve the mesh topology, for instance as generalized tensorproduct or blockstructured meshes, while a local alignment with the physical boundary of the solid is achieved by a grid deformation process, such that the boundary approximation error can be significantly decreased.

Over the past decade, several grid adaptation techniques have been developed, namely the so-called h-, p- and r-methods. The first two do static remeshing, where the h-method does automatic refinement or coarsening of the spatial mesh based on a posteriori error estimates or error indicators and the p-method takes higher or lower order approximations locally as needed. In contrast, the r-method (also known as moving grid method) relocates grid points in a mesh having a fixed number of nodes in such a way that the nodes remain concentrated in regions of rapid variation of the solution or corresponding interfaces. The r-method is a dynamic method which means that it uses time stepping or pseudo-time stepping approaches to construct the desired transformation. Compared to the h- and p-methods in which often hanging nodes have to be dealt with, the r-method is much easier to incorporate into most CFD codes without the need for changing of system matrix structures and special interpolation procedures since in the r-method the data structures for mesh are fixed. The r-method has received considerably attention recently due to some new developments which clearly demonstrate its potential for problems such as those having moving interfaces [6, 7].

In this paper, we base on our multigrid fictitious boundary method (FBM) [4, 5] and the grid deformation method presented in [7] to solve numerically the coupled fluid-solid problems. As we have shown in [5], the use of the multigrid FBM does not require to change the mesh during the simulations when the solid bodies vary their positions. The advantage is that no expensive remeshing has to be performed. However, the accuracy for capturing the surfaces of solid bodies is only of first order which might lead to accuracy problems. For a better approximation of the solid surfaces, we adopt a deformed grid, created from an equidistant cartesian mesh in which the topology is preserved and only the grid spacing is changed such that the grid points are concentrated near the surfaces of the solid bodies. Here, only the solution of additional linear Poisson problems in every time step is required for generating the deformation grid, which means that the additional work is significantly less than the main fluid-solid part. The presented method is compared to the pure multigrid FBM using an equidistance mesh through a numerical simulation of a benchmark configuration of 2D flow around an airfoil in a channel. Its accuracy improvement is shown to be excellent.

2 Grid deformation method

In this section, we briefly describe the grid deformation method which will be adopted and coupled with our multigrid fictitious boundary method (FBM) (see the next section) to solve numerically the fluid-rigid solid problems. The details of the grid deformation method can be found in [7].

Grid deformation problems can be equated to construct a transformation ϕ, $x = \phi(\xi)$ from the computational space (with coordinate ξ) to the physical space (with coordinate x). There are two basic types of grid deformation methods, local based and velocity based, generally computing x by minimizing a variational form or computing the mesh velocity $v = x_t$ using a Lagrangian-like formulation. The grid deformation method we will employ belongs to the velocity-based method, which is based on Liao's work [6] and Moser's work [8]. This method has several advantages: only linear Poisson problems on fixed meshes are needed to be solved, the monitor function can be obtained directly from an error distribution, mesh tangling can be prevented, and the data structure is always the same as that for the starting mesh (see [7]).

Suppose $g(x)$ (area function) to be the area distribution on the undeformed mesh, while $f(x)$ (monitor function) in contrast describes the absolute mesh size distribution of the target grid, which is independent of the starting grid and chosen according to the need of physical problems. Then, the transformation ϕ can be computed via the following four steps:

1. Compute the scale factors c_f and c_g for the given monitor function $f(x) > 0$ and the area function g using

$$c_f \int_\Omega \frac{1}{f(x)} dx = c_g \int_\Omega \frac{1}{g(x)} dx = |\Omega|, \qquad (1)$$

where, $\Omega \subset \mathbb{R}^2$ is a computational domain, and $f(x) \approx$ local mesh area. Let $\tilde{f}$ and $\tilde{g}$ denote the reciprocals of the scaled functions f and g, i.e.,

$$\tilde{f} = \frac{c_f}{f}, \qquad \tilde{g} = \frac{c_g}{g}. \qquad (2)$$

2. Compute a grid-velocity vector field $v : \Omega \to \mathbb{R}^n$ by satisfying the following linear Poisson equation

$$-\operatorname{div}(v(x)) = \tilde{f}(x) - \tilde{g}(x), \quad x \in \Omega, \quad \text{and} \quad v(x) \cdot \mathfrak{n} = 0, \quad x \in \partial\Omega, \qquad (3)$$

where **n** being the outer normal vector of the domain boundary $\partial\Omega$, which may consist of several boundary components.

3. For each grid point x, solve the following ODE system

$$\frac{\partial \varphi(x,t)}{\partial t} = \eta(\varphi(x,t), t), \quad 0 \leq t \leq 1, \quad \varphi(x, 0) = x, \qquad (4)$$

with

$$\eta(y, s) := \frac{v(y)}{s\widetilde{f}(y) + (1-s)\widetilde{g}(y)}, \quad y \in \Omega, \ s \in [0,1]. \qquad (5)$$

4. Get the deformed grid points via

$$\phi(x) := \varphi(x, 1). \qquad (6)$$

3 Numerical solution of the fluid-solid system

In this section, we will describe how to solve numerically the coupled fluid-rigid solid problems by using our multigrid FBM and the above grid deformation method.

3.1 Governing equations in the frame of FBM

For a detailed description of the multigrid fictitious boundary method (FBM), the reader is referred to Refs. [4, 5]. The governing coupled fluid-solid system in the frame of the FBM can be given by

$$\begin{cases} \nabla \cdot \mathbf{u} = 0 & (a) \quad \text{for } \mathbf{X} \in \Omega_T, \\ \rho_f \left(\frac{\partial \mathbf{u}}{\partial t} + \mathbf{u} \cdot \nabla \mathbf{u} \right) - \nabla \cdot \sigma = 0 & (b) \quad \text{for } \mathbf{X} \in \Omega_f, \\ \mathbf{u}(\mathbf{X}) = \mathbf{U}_s + \omega_s \times (\mathbf{X} - \mathbf{X}_s) & (c) \quad \text{for } \mathbf{X} \in \bar{\Omega}_s, \end{cases} \qquad (7)$$

where σ is the total stress tensor in the fluid phase defined as

$$\sigma = -p\mathbf{I} + \mu_f \left[\nabla \mathbf{u} + (\nabla \mathbf{u})^T \right]. \qquad (8)$$

Here **I** is the identity tensor, fluid viscosity $\mu_f = \rho_f \cdot \nu$, ρ_f is the fluid density, p is the pressure and **u** is the fluid velocity. Ω_f is the domain occupied by the fluid, and Ω_s the domain occupied by the rigid bodies, $\Omega_T = \Omega_f \cup \Omega_s$ the entire computational domain.

The equations that govern the motion of the rigid body are the following Newton-Euler equations, i.e., the translational velocities $\mathbf{U}_s$ and angular velocities ω_s of the rigid body satisfy

$$M_s \frac{d\mathbf{U}_s}{dt} = (\Delta M_s)\mathbf{g} + \mathbf{F}_s, \quad \mathbf{I}_s \frac{d\omega_s}{dt} + \omega_s \times (\mathbf{I}_s \omega_s) = \mathbf{T}_s, \qquad (9)$$

where M_s is the mass of the rigid body; $\mathbf{I}_s$ is the moment of the inertia tensor; ΔM_s is the mass difference between the mass M_s and the mass of the fluid occupying the same volume; $\mathbf{g}$ is the gravity vector; $\mathbf{F}_s$ and T_s are the resultants of the hydrodynamic forces and the torque about the center of mass acting on the rigid body which are calculated by

$$\mathbf{F}_s = -\int_{\Omega_T} \sigma \cdot \nabla \alpha_s \, d\Omega, \qquad T_s = -\int_{\Omega_T} (\mathbf{X} - \mathbf{X}_s) \times (\sigma \cdot \nabla \alpha_s) \, d\Omega. \qquad (10)$$

Here the function α_s is defined by

$$\alpha_s(\mathbf{X}) = \begin{cases} 1 & \text{for } \mathbf{X} \in \bar{\Omega}_s, \\ 0 & \text{for } \mathbf{X} \in \Omega_T \setminus \Omega_s. \end{cases} \qquad (11)$$

The position $\mathbf{X}_s$ of the rigid body and its angle θ_s are obtained by integration of the kinematic equations

$$\frac{d\mathbf{X}_s}{dt} = \mathbf{U}_s, \qquad \frac{d\theta_s}{dt} = \omega_s. \qquad (12)$$

3.2 ALE formulation

When the grid deformation method described in section 2 is applied to the multigrid FBM, a mesh velocity $\mathbf{W}_m$ should be introduced in the convective term of Eq. (7b), i.e.,

$$\rho_f \left[\frac{\partial \mathbf{u}}{\partial t} + (\mathbf{u} - \mathbf{W}_m) \cdot \nabla \mathbf{u} \right] - \nabla \cdot \sigma = 0 \qquad \text{for } \mathbf{X} \in \Omega_f. \qquad (13)$$

In the literature this is referred to as an Arbitrary Lagrangian-Eulerian (ALE) formulation. Note that the mesh velocities $\mathbf{W}_m$ do not appear in the continuity equation, as a pressure-Poisson equation is solved to satisfy the continuity equation in an outer loop. Care has to be taken to satisfy the geometric conservation law (GCL), where the mesh velocity $\mathbf{W}_m$ must be equal to the movement of the mesh velocity $\Delta \mathbf{x}$ during the time step. Therefore, the mesh velocities $\mathbf{W}_m$ should be calculated according to the nodal movement from the previous time step by

$$\mathbf{W}_m = \frac{1}{\Delta t}(\mathbf{x}^{n+1} - \mathbf{x}^n), \qquad (14)$$

where Δt is the time step size $t^{n+1} - t^n$.

3.3 Numerical realization in FEATFLOW

A special solver has been developed to solve the coupled fluid-rigid solid problems in the ALE formulation by using the multigrid FBM and the grid deformation method. It is essentially based on the discrete projection type solver PP2D from the FEATFLOW package [9, 10].

In this solver, we first semi-discretize in time by a Fractional-step-θ-scheme (see [9]). Then, we obtain a sequence of generalized stationary Navier–Stokes equations with both prescribed boundary values and fictitious boundary conditions for the moving rigid body in the fluid in every time step, which is a nonlinear saddle point problem that has to be discretized in space. For the spatial discretization, we choose the nonconforming $\widetilde{Q}_1/Q_0$ element pair, in which the nodal values are the mean values of the velocity over the element edges and the mean values of the pressure over the elements. The non-linear one-step projection solution process is accelerated with a multigrid technique [9]. In each time step, a new deformation mesh is generated based on the starting mesh and two auxiliary routines are designed to update the system matrices and to calculate the mesh velocity according to the new position of the deformation mesh nodes, respectively. After the fluid part calculation, we can do the solid part calculation, including the calculation of the corresponding hydrodynamic forces and the torque acting on the rigid body as well as the updating of the new positions and velocities of the solid body.

4 Verification of the numerical techniques

The new solver is validated in this section by a benchmark configuration of 2D flow around an airfoil in a channel (see [3] for the details) to assess the accuracy and efficiency of the proposed combination based on the multigrid FBM and the grid deformation method compared to the results without using the grid deformation method.

We consider a NACA0012 airfoil that has a fixed center of mass and is induced to rotate freely around its center of mass due to hydrodynamical forces under the action of an incoming incompressible viscous flow in a channel. The channel is of width 20 and height 4. All values are in non-dimensional form. The density of the fluid is $\rho_f = 1$ and the density of the airfoil is $\rho_p = 1.1$. The viscosity of the fluid is $\nu_f = 10^{-2}$. The initial condition for the fluid velocity is $\mathbf{u} = 0$ and the boundary conditions are given as $\mathbf{u} = 0$ when $y = 0$ or 4 and $\mathbf{u} = 1$ when $x = 0$ or 20 for $t \geq 0$. Initial angular velocity and angle of incidence of the airfoil are zero. The airfoil length is 1.0089304 and the fixed center of mass of the airfoil is at $(0.420516, 2)$. Hence the Reynolds number is about 101 with respect to the length of the airfoil and the inflow speed.

The simulation is implemented on both fixed equidistance meshes and moving deformation meshes, each of them for two different levels, i.e., Level = 7 with 41409 nodes and 40960 elements, as well as Level = 8 with 164737 nodes and 163840 elements. The deformation mesh is generated in each time step in order to always keep grid alignment near the surface of the induced rotating airfoil. Fig. 1 (a) gives the starting equidistance mesh used to generate deformation meshes. In Fig. 1 (b), one deformation mesh at $t = 16.0$

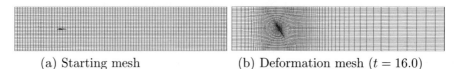

(a) Starting mesh (b) Deformation mesh ($t = 16.0$)

Fig. 1. Starting mesh and deformation meshes for a NACA0012 airfoil in a channel

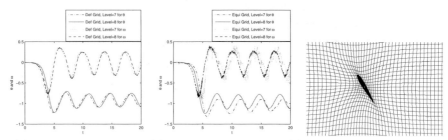

(a) Deformation meshes (b) Equidistance meshes (c) Zoom mesh at $t = 16.0$

Fig. 2. The time history of θ and ω of the rotating airfoil and zoom mesh

is presented, its local zoom is illustrated in Fig. 2 (c). Fig. 2 (a) and Fig. 2 (b) plot the time history of the angle of incidence θ and the angular velocity ω of the induced rotating airfoil calculated by using deformation meshes and equidistance meshes, each of them performed by two levels LEVEL = 7 and LEVEL = 8, respectively. The vector field and vorticity distribution at $t = 16.0$ are shown in Fig. 3. From these figures and pictures, we can see that the airfoil quickly reaches a periodic motion and intends to keep its broadside perpendicular to the in-flow direction which is a stable position for a noncircular rigid body settling in a channel at moderate Reynolds numbers. We observe that the results of the deformation meshes converge better to a mesh independent solution than those of the equidistance meshes, and are in excellent agreement with those obtained by Glowinski, Joseph and coauthors [3]. The results of the equidistance meshes exhibit too much numerical oscillation and lose stability since they cannot catch very well the velocity field close to the leading edge of the airfoil, which causes the numerical solution blew up near the leading edge of the airfoil. Obviously, good results and a significant accuracy improvement are achieved by using the grid deformation technique. It illustrates that the presented method can easily handle more complex shapes of rigid bodies and obtain more accurate and satisfying results than those without employing the grid deformation method.

5 Conclusions

We have presented the combination of the multigrid fictitious boundary method (FBM) and the grid deformation method for the simulations of 2D coupled fluid-rigid solid problems. Deformed grids, created from an arbitrary

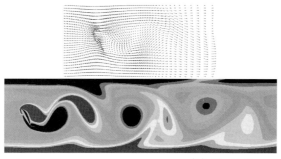

(a) Local vector field (b) Vorticity

Fig. 3. Induced rotation of a NACA0012 airfoil in a channel at $t = 16.0$

starting mesh, can preserve the topology of the mesh and the underlying data structure, while at the same time, the grid points can be concentrated and aligned near the surfaces of the solid bodies. Therefore, a better approximation of the solid surfaces is achieved. Incorporating the grid deformation method with the multigrid fictitious boundary method, an ALE formulation with a multigrid finite element solver is applied to solve the fluid flow, and the rigid body can move freely through the deformed meshes. Numerical examples have illustrated that the accuracy for dealing with the interaction between the fluid and the solid body can be significantly improved by the presented method. The presented method can be easily extended to the 3D case once 3D multigrid fictitious boundary method and 3D grid deformation method are available.

References

1. Hu, H.H., Joseph, D.D., Crochet, M.J.: Direct Simulation of Fluid Particle Motions. Theor. Comp. Fluid Dyn., **3**, 285–306 (1992)
2. Maury, B.: Direct Simulations of 2D Fluid-Particle Flows in Biperiodic Domains. J. Comput. Phy., **156**, 325–351 (1999)
3. Glowinski, R., Pan, T.W., Hesla, T.I., Joseph, D.D., Periaux, J.: A Fictitious Domain Approach to the Direct Numerical Simulation of Incompressible Viscous Flow Past Moving Rigid Bodies: Application to Particulate Flow. J. Comput. Phy., **169**, 363–426 (2001)
4. Turek, S., Wan, D.C., Rivkind, L.S.: The Fictitious Boundary Method for the Implicit Treatment of Dirichlet Boundary Conditions with Applications to Incompressible Flow Simulations. Challenges in Scientific Computing, Lecture Notes in Computational Science and Engineering, Vol. **35**, Springer, 37–68 (2003)
5. Wan, D.C., Turek, S.: Direct Numerical Simulation of Particulate Flow via Multigrid FEM Techniques and the Fictitious Boundary Method. Int. J. Numer. Method in Fluids, in press (2006)

6. Cai, X.X., Fleitas, D., Jiang, B., Liao, G.: Adaptive Grid Generation Based on Least–Squares Finite–Element Method. Computers and Mathematics with Applications, **48**(7-8), 1077–1086 (2004)
7. Grajewski, M., Köster, M., Kilian, S. and Turek, S.: Numerical Analysis and Practical Aspects of a Robust and Efficient Grid Deformation Method in the Finite Element Context. submitted to SISC, (2005)
8. Dacorogna, B., Moser, J.: On a Partial Differential Equation Involving the Jacobian Determinant. Annales de le Institut Henri Poincare, **7**, 1–26 (1990)
9. Turek, S.: Efficient Solvers for Incompressible Flow Problems. Springer Verlag, Berlin-Heidelberg-New York, (1999)
10. Turek, S., et al.: **FEATFLOW**–Finite element software for the incompressible Navier–Stokes equations: User Manual, Release 1.2, University of Dortmund, (1999)

Non-Linear PDE

An Iterative Method for Solving Non-Linear Hydromagnetic Equations

C. Boulbe,[1] T.Z. Boulmezaoud[2] and T. Amari[3]

[1] Laboratoire de mathématiques appliquées,
   Université de Pau et des Pays de l'Adour,
   IPRA - Av de l'Université, 64000 PAU, France
   `c.boulbe@etud.univ-pau.fr`
[2] Laboratoire de mathématiques appliquées,
   Université de Versailles Saint Quentin,
   45 av des Etats Unis, 75035 Versailles, France
   `boulmeza@math.uvsq.fr`
[3] Centre de Physique Théorique,
   Ecole Polytechnique, F91128 Palaiseau Cedex,France
   `amari@cpht.polytechnique.fr`

Summary. We propose an iterative finite element method for solving non-linear hydromagnetic and steady Euler's equations. Some three-dimensional computational tests are given to confirm the convergence and the high efficiency of the method.

1 Introduction. Statement of the problem

The understanding of plasma equilibria is one of the most important problems in magnetohydrodynamics and arises in several fields including solar physics and thermonuclear fusion. Such an equilibria is often governed by the well known steady hydromagnetic equations

$$\mathbf{curl\, B} \times \mathbf{B} + \nabla p = 0, \tag{1}$$

$$\mathrm{div}\, \mathbf{B} = 0, \tag{2}$$

which describe the balance of the Lorentz force by pressure. Here $\mathbf{B}$ and p are respectively the magnetic field and the pressure.

Notice that system (1)+(2) is quite similar to steady inviscid fluid equations

$$\mathbf{v}.\nabla \mathbf{v} + \nabla p = 0, \tag{3}$$

$$\mathrm{div}\, \mathbf{v} = 0. \tag{4}$$

This analogy is due to the vectorial identity

$$\mathbf{v}.\nabla\mathbf{v} - \nabla\frac{|\mathbf{v}|^2}{2} = \operatorname{\mathbf{curl}}\mathbf{v} \times \mathbf{v}.$$

System of equations (1)+(2) must be completed with some boundary conditions on $\mathbf{B}$ and p. Physical considerations suggest to prescribe the boundary normal field component:

$$\mathbf{B}.\mathbf{n} = g \text{ on } \partial\Omega \qquad (5)$$

where g satisfies the compatibility condition $\int_\Omega g = 0$ due to the equation $\operatorname{div}\mathbf{B} = 0$. Defining the inflow boundary as $\Gamma^- = \{\mathbf{x} \in \Omega, \; \mathbf{B}(\mathbf{x}).\mathbf{n}(\mathbf{x}) < 0\}$, one can also prescribe the normal component $\operatorname{\mathbf{curl}}\mathbf{B}.\mathbf{n}$ of the current density and the pressure p on Γ^-

$$\operatorname{\mathbf{curl}}\mathbf{B}.\mathbf{n} = h \text{ on } \Gamma^-, \qquad (6)$$

$$p = p_0 \text{ on } \Gamma^-. \qquad (7)$$

One can notice that if the pressure is neglected, equations (1)+(2) become

$$\operatorname{\mathbf{curl}}\mathbf{B} \times \mathbf{B} = \mathbf{0}, \qquad (8)$$
$$\operatorname{div}\mathbf{B} = 0. \qquad (9)$$

Equation (8) means that the magnetic field and its curl, which represents the current density, are everywhere aligned. The magnetic field is said *Beltrami* or *force-free* (FF). A usual way to tackle the problem (8) + (9) consists to rewrite equation (8) into the form

$$\operatorname{\mathbf{curl}}\mathbf{B} = \lambda(\mathbf{x})\mathbf{B}, \qquad (10)$$

where $\lambda(\mathbf{x})$ is a scalar function which can be a constant function or can depend on $\mathbf{x}$. In the former, the $\mathbf{B}$ field is said linear FF. In the latter, it is said non linear.

Some partial results concerning existence of 3D solutions of equations (1)+(2) in bounded domains are given in [1] and [11]. Linear force-free-fields were studied in [4]. For the existence of non-linear ones the reader can refer to [5, 3].

The numerical solving of equations (1)+(2) and equations (8)+(9) is of importance in magnetohydrodynamics studies and in solar physics. As it is known, the reconstruction of the coronal magnetic field has is of a great utility in observational and theoretical studies of the magnetic structures in the solar atmosphere. In this paper, we propose an iterative process for solving these equations (section 2). A finite element method is proposed for solving each one of the arising problems.

2 An iterative method for the magnetostatic system

Our objective here is to expose an iterative method for solving the non-linear equations (1)+(2) in a bounded and simply-connected domain. The starting idea of the method consists to split the current density $\boldsymbol{w} = \mathbf{curl\,B}$ into the sum

$$\boldsymbol{w} = \boldsymbol{w}_{||} + \boldsymbol{w}_{\perp}, \tag{11}$$

where the vector field $\boldsymbol{w}_{||} = \mu(\mathbf{x})\mathbf{B}$ is collinear to $\mathbf{B}$, while $\boldsymbol{w}_{\perp}$ is perpendicular to $\mathbf{B}$. The problem is decomposed formally into a curl-div system on $\mathbf{B}(\mathbf{x})$ and two first order hyperbolic equations on $\mu(\mathbf{x})$ and $p(\mathbf{x})$.

More precisely, writing $\boldsymbol{w}_{||}(\mathbf{x}) = \mu(\mathbf{x})\mathbf{B}(\mathbf{x})$ where μ is a scalar function and taking the divergence of (11), gives

$$\mathbf{B}.\nabla\mu = -\mathrm{div}\,\boldsymbol{w}_{\perp}. \tag{12}$$

Notice that the pressure satisfies a similar equation since

$$\mathbf{B}.\nabla p = 0. \tag{13}$$

Equation (1) becomes

$$\boldsymbol{w}_{\perp} \times \mathbf{B} = -\nabla p, \tag{14}$$

which means that $\boldsymbol{w}_{\perp}(\mathbf{x}) = \dfrac{1}{|\mathbf{B}(\mathbf{x})|^2}\nabla p(\mathbf{x}) \times \mathbf{B}(\mathbf{x})$ if $|\mathbf{B}(\mathbf{x})| \neq 0$.

In consideration of these remarks, we are going now to propose an iterative process to solve non-linear systems (1)+(2). In this process the transport equation (12) is perturbed by adding an artificial reaction term. Namely, we construct a sequence $(\mathbf{B}^{(n)}, p^{(n)})_{n\geq 0}$ as follows:

- The starting guess $\mathbf{B}_0 \in H^1(\Omega)$ is chosen as the irrotational field associated to g defined by

$$\mathbf{curl\,B}_0 = \mathbf{0}\text{ in }\Omega,\ \mathrm{div}\,\mathbf{B}_0 = 0\text{ in }\Omega\text{ and }\mathbf{B}_0.\mathbf{n} = g\text{ on }\partial\Omega. \tag{15}$$

This is a usual problem which can be reduced to a scalar Neumann problem since the domain is simply-connected.

- For all $n \geq 0$, $p^{(n)}$ is solution of the system

$$\begin{cases} \mathbf{B}^{(n)}.\nabla p^{(n)} + \eta p^{(n)} = \eta p^{(n-1)} \text{ in } \Omega, \\ p^{(n)} = p_0 \text{ on } \partial\Omega, \end{cases} \tag{16}$$

where η is a small parameter and $p^{(-1)} = 0$.

- For all $n \geq 0$, $\boldsymbol{w}_{\perp}^{(n)} = \dfrac{1}{|\mathbf{B}^{(n)}|^2}\nabla p^{(n)} \times \mathbf{B}^{(n)}$ and $\boldsymbol{w}_{||}^{(n)} = \mu^{(n)}\mathbf{B}^{(n)}$, where $\mu^{(n)}$ satisfies

$$\begin{cases} \mathbf{B}^{(n)}.\nabla\mu^{(n)} + \epsilon\mu^{(n)} = -\mathrm{div}\,\boldsymbol{w}_{\perp}^{(n)} + \epsilon\mu^{(n-1)} \text{ in } \Omega, \\ \mu^{(n)}(\mathbf{B}^{(n)}.\mathbf{n}) = h - \boldsymbol{w}_{\perp}^{(n)}.\mathbf{n} \text{ on } \Gamma^-. \end{cases} \tag{17}$$

Here $\mu^{(-1)} = 0$ and ϵ is a small parameter.

- For all $n \geq 0$, $\mathbf{B}^{(n+1)} = \mathbf{B}_0 + \mathbf{b}^{(n+1)}$, with $\mathbf{b}^{(n+1)}$ solution of

$$\begin{cases} \operatorname{curl} \mathbf{b}^{(n+1)} = \boldsymbol{\omega}^{(n)} + \nabla q^{(n)} \text{ in } \Omega, \\ \operatorname{div} \mathbf{b}^{(n+1)} = 0 \text{ in } \Omega, \\ \mathbf{b}^{(n+1)}.\mathbf{n} = 0 \text{ on } \partial\Omega, \end{cases}$$

where $\boldsymbol{\omega}^{(n)} = \boldsymbol{\omega}_{\|}{}^{(n)} + \boldsymbol{\omega}_{\perp}{}^{(n)}$ while $q^{(n)}$ is solution of the Laplace problem

$$-\Delta q^{(n)} = \operatorname{div} \boldsymbol{\omega}^{(n)} \text{ in } \Omega, \text{ and } q^{(n)} = 0 \text{ on } \partial\Omega. \qquad (18)$$

Notice that the appearance of the correction term $\nabla q^{(n)}$ is due to the fact that $\operatorname{div}(\boldsymbol{\omega}^{(n)})$ is not zero in general.

The convergence of this iterative process is not an easy matter. We conjecture that it converges if h is sufficiently small and $|\mathbf{B}_0(\mathbf{x})| \geq c > 0$ in Ω for some constant $c > 0$. Nevertheless, in the case of linear force-free fields (in that case the algorithm is simplified since at each iteration $p^{(n)} = 0$, $\boldsymbol{\omega}_{\perp}{}^{(n)} = \mathbf{0}$ and $\mu^{(n)}$ is a fixed real) Boulmezaoud and Amari [6] proved that this process is super-convergent. The proof of convergence in the general case is not given and remains an open question.

Notice that the same algorithm can be used for computing linear or non-linear force-free fields which are solutions of (9)+(10), provided that the computation of the pressure $p^{(n)}$ and the vector field $\boldsymbol{\omega}_{\perp}{}^{(n)}$ are dropped.

3 Finite element discretization

Here we give a short description of the finite elements methods we use for solving problems arising in the iterative process exposed above. Observe first that at each iteration of the algorithm one should solve two problems:
(a) A reaction-convection problem of the form: *find u solution of*

$$\begin{cases} \operatorname{div}(u\mathbf{B}) + \sigma u = f \text{ in } \Omega, \\ u = h \text{ on } \Gamma^-. \end{cases} \qquad (19)$$

(b) A vector potential problem: *find the pair $(\mathbf{b}, q)$ satisfying*

$$\begin{cases} \operatorname{curl} \mathbf{b} - \nabla q = \mathbf{j} \text{ in } \Omega, \\ \operatorname{div} \mathbf{b} = 0 \text{ in } \Omega, \\ \mathbf{b}.\mathbf{n} = 0 \text{ on } \partial\Omega, \\ q = 0 \text{ on } \partial\Omega. \end{cases} \qquad (20)$$

We begin with the approximation of (19).

It is well known that the direct application of a Galerkin finite elements method to the singularly perturbed problem (19) may lead to the appearance of spurious oscillations and instabilities. In the two last decades, several methods were proposed to remove this drawback (especially in the two dimensional case). Among these methods, one can recall the *streamline diffusion* method (see Brookes and Hughes [8], see also, e. g., Johnson et al. [10]), the *discontinuous* Galerkin method (see Lesaint [12]) and *bubble functions* methods (see, e. g., Brezzi et al. [7]). Here we shall use the method of streamline diffusion.

Thus, let us consider a family of regular triangulations $(\mathcal{T}_h)$ of Ω. The discrete problem we consider is

$$(\mathscr{P}_h) \begin{cases} \text{Find } u_h \in W_h \text{ such that} \\ a_h(u_h, w_h) = \ell_h(w_h), \forall w_h \in W_h, \end{cases}$$

where

$$a_h(u_h, w_h) = \int_\Omega (\mathbf{B}.\nabla u_h + \sigma u_h).(w_h + \delta_h \mathbf{B}.\nabla w_h) dx - \int_{\Gamma_-} u_h w_h (\mathbf{B}.\mathbf{n}) dx,$$

$$\ell_h(w_h) = \int_\Omega f(\mathbf{x})(\mathbf{B}.\nabla w_h + \delta_h w_h) - \int_{\Gamma_-} \alpha_0 w_h (\mathbf{B}.\mathbf{n}) dx.$$

Here W_h stands for the finite elements space

$$W_h = \{v_h \in H^1(\Omega); \, v_{|K} \in \mathbb{P}_k(K), \, \forall K \in \mathcal{T}_h\},$$

where for each $K \in \mathcal{T}_h$, $\mathbb{P}_k(K)$ denotes the space of polynomials of degree less or equal k.

One can prove that the problem $(\mathscr{P}_h)$ has a unique solution $u_h \in W_h$ when $\delta_h \sigma < 1$. Moreover, if $\delta_h = ch$ for some constant c and if $\mathbf{B} \in L^\infty(\Omega)^3 \cap H(\text{div}; \Omega)$ and $u \in H^{\ell+1}(\Omega)$ for some $\ell \geq 1$, then

$$(1 - \delta_h \sigma) |||u - u_h||| \leq Ch^{\ell+1/2} \|u\|_{H^{\ell+1}(\Omega)}, \tag{21}$$

where $|||w|||_\Omega^2 = \delta_h \|\mathbf{B}.\nabla w\|_{L^2(\Omega)}^2 + \sigma \|w\|_{L^2(\Omega)}^2 + \||\mathbf{B}.\mathbf{n}|^{1/2} w\|_{L^2(\partial \Omega)}^2$.

Now, we deal with the approximation of the curl-div system (20), which can be dispatched into two problems: a variational problem $(\mathscr{Q})$ in terms of $\mathbf{b}$ and the fictitious unknown $\theta = 0$, and Laplace equation (18) in terms of q. We only deal with the approximation of $\mathbf{b}$, since we shall see that the computation of the q is useless. Denote by $H(\mathbf{curl}; \Omega)$ the space

$$H(\mathbf{curl}; \Omega) = \{\mathbf{v} \in L^2(\Omega)^3; \, \mathbf{curl}\,\mathbf{v} \in L^2(\Omega)^3\}$$

equipped with its usual norm. The statement of problem (20) suggests the use of an $H(\mathbf{curl}; \Omega)$ approximation. Define M the space

$$M = \{v \in H^1(\Omega), \int_\Omega v \, dx = 0\}.$$

Let $X_h \subset H(\mathbf{curl}; \Omega)$, $M_h \subset M$ two finite-dimensional subspaces and set

$$V_h = \{\mathbf{v_h} \in X_h; \, (\mathbf{v_h}, \nabla \mu_h) = 0, \, \forall \mu_h \in M_h\}.$$

We make the following assumptions

($\mathcal{H}_1$) the inclusion $\{\nabla \mu_h, \mu_h \in M_h\} \subset X_h$ holds,
($\mathcal{H}_2$) there exists a constant C such that

$$\|\mathbf{v_h}\|_{0,\Omega} \leq C\|\mathbf{curl}\,\mathbf{v_h}\|_{0,\Omega}, \quad \forall \mathbf{v_h} \in V_h.$$

The discrete version of problem (20) writes

$$(\mathcal{Q}_h) \begin{cases} \text{Find } (\mathbf{b_h}, \theta_h) \in X_h \times M_h \text{ such as} \\ \forall \mathbf{v_h} \in X_h, \int_\Omega \mathbf{curl}\,\mathbf{b_h}.\mathbf{curl}\,\mathbf{v_h}\,dx + \int_\Omega \mathbf{v_h}.\nabla \theta_h\,dx = \int_\Omega \mathbf{j}.\mathbf{curl}\,\mathbf{v_h}\,dx, \\ \forall \mu_h \in M_h, \int_\Omega \mathbf{b_h}.\nabla \mu_h = 0. \end{cases}$$

According to Amrouche and al. [2], the problem ($\mathcal{Q}_h$) has one and only one solution $(\mathbf{b}_h, \theta_h)$ with $\theta_h = 0$, and

$$\|\mathbf{b} - \mathbf{b_h}\|_{H(\mathbf{curl};\Omega)} \leq C \inf_{\mathbf{v_h} \in X_h} \|\mathbf{b} - \mathbf{v_h}\|_{H(\mathbf{curl};\Omega)}. \tag{22}$$

A simple manner for constructing the spaces X_h and M_h is to use the $H(\mathbf{curl})$ conforming elements of Nédelec [13] (see Amrouche and al.). In that case, the following estimate holds

$$\|\mathbf{b} - \mathbf{b_h}\|_{H(\mathbf{curl};\Omega)} \leq Ch^\ell \{|\mathbf{b}|_{\ell,\Omega} + |\mathbf{b}|_{\ell+1,\Omega}\}, \tag{23}$$

which is valid if $\mathbf{b} \in H^{\ell+1}(\Omega)$.

An important feature of the discrete system ($\mathcal{Q}_h$) is that only the discrete vector field $\mathbf{b_h}$ is really unknown. Actually, we know that $\theta_h = 0$. This property can be exploited from a practical viewpoint to reduce the discrete system to a smaller one by eliminating θ_h. In term of matrices, the system writes

$$\begin{pmatrix} A^{curl} & B^T \\ B & 0 \end{pmatrix} \begin{pmatrix} X \\ Y \end{pmatrix} = \begin{pmatrix} C^{curl} \\ 0 \end{pmatrix} \tag{24}$$

where A^{curl} is a symmetric and positive square matrix (A^{curl} is not definite neither invertible). We can state the following

Lemma 1. *Let Λ be a square positive, definite and symmetric matrix having the same size as A. Then, the pair (X, Y) is solution of (24) if and only if $Y = 0$ and X is solution of*

$$(A^{curl} + B^T \Lambda B)X = C^{curl}. \tag{25}$$

Remark 1. In Lemma 1, the matrix A^{curl} and the RHS C^{curl} are not arbitrary. Indeed, if G denotes the matrix of the operator $\nabla : M_h \to X_h$, then necessarily $G^T A^{curl} = 0$ and $G^T C^{curl} = 0$. These identities are the discrete counterpart of the continuous relations $\text{div}\,(\mathbf{curl}.) = 0$ and $\text{div}\,\mathbf{j} = 0$.

A serious advantage of the new system (25) comparing with (24) is that number of unknowns is reduced.

4 Computational tests

In this last section, we expose some computational results we obtain with a 3D code. This code use the iterative method and the finite elements discretization exposed above to solve problem (1)+(2) and problem (10)+(9). We compare the exact solution and the numerical solution and we show the behavior of the errors in terms of h. Two exact solutions are used for the tests.

- **Test 1 (a non-linear force-free-field).**
  Let (r, θ, z) the cylindrical coordinates with respect to a point $(x_0, y_0, 0)$ ($x_0 = -3$ and $y_0 = -3$). The pair $(\mathbf{B}, p)$ is given $\mathbf{B} = \frac{1}{\sqrt{r}}(\mathbf{e}_\theta + \mathbf{e}_z)$, $p(\mathbf{x}) = 0$. This is a non-linear force-free field with $\lambda = \frac{1}{2r}$. Table 1 shows the behavior of the residue $\|\mathbf{B}^{(n+1)} - \mathbf{B}^{(n)}\|_{0,\Omega}$ and the product $\mathbf{curl}\,\mathbf{B}^{(n)} \times \mathbf{B}^{(n)}$ versus the iteration number. This example illustrates the superconvergence of the algorithm.

Table 1. Evolution of $\frac{\|\mathbf{B}^{(n+1)} - \mathbf{B}^{(n)}\|_{0,\Omega}}{\|\mathbf{B}^{(n)}\|_{0,\Omega}}$ and $\|\mathbf{curl}\,\mathbf{B}^{(n)} \times \mathbf{B}^{(n)}\|_\infty$.

n	$\frac{\|\mathbf{B}^{(n+1)} - \mathbf{B}^{(n)}\|_{0,\Omega}}{\|\mathbf{B}^{(n)}\|_{0,\Omega}}$	$\|\mathbf{curl}\,\mathbf{B}^{(n)} \times \mathbf{B}^{(n)}\|_\infty$
0	0.09912	6.740e-15
1	0.00566	0.06781
2	0.00036	0.01939
3	2.644e-05	0.01910

- **Test 2: (Bennet pinch)**. The pair $(\mathbf{B}, p)$ is given by

$$\mathbf{B} = \nabla A \times \mathbf{e}_y \text{ and } p = \frac{\lambda}{2} m E^{2A} \text{ with } A = -\ln(\frac{1 + \lambda k^2(x^2 + z^2)}{2k}).$$

In table 2, the relative L^2 errors on $\mathbf{B}$ and p after convergence of the algorithm are shown. These error decreases as $h^{1.8}$, which confirms the high accuracy of the method.

Table 2. Relative errors on $\mathbf{B}_h$ and p_h in norm L^2 (test 2).

h	$\frac{\|\mathbf{B}-\mathbf{B}_h\|_{0,\Omega}}{\|\mathbf{B}\|_{0,\Omega}}$	$\frac{\|p-p_h\|_{0,\Omega}}{\|p\|_{0,\Omega}}$
0.69282	0.03837	0.08648
0.23094	0.00492	0.01102
0.13856	0.00191	0.00396
0.09897	0.00108	0.00201

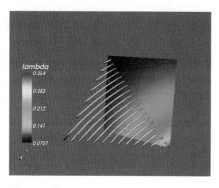 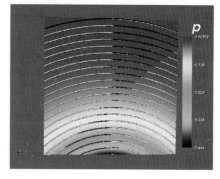

Fig. 1. Superposition of the the exact and the numerical solutions in the case of test 1 on the left and in a $(x-z)$ plane 2D cut for the test 2 on the right.

References

1. Alber, H.D.: Existence of three-dimensional, steady, inviscid, incompressible flows with non-vanishing vorticity. Math. Ann. **292**, 493–528 (1992)
2. Amrouche, C., Bernardi, C ., Dauge, M., Girault, V.: Vector potentials in three-dimensional non-smooth domains, Math. Methods Appl. Sci. **9**, 823–864 (1998)
3. Boulmezaoud, T.Z.: On the existence of non-linear Beltrami fields. Comptes Rendus de l'Académie des Sciences, **328**, 437–442 (1999)
4. Boulmezaoud, T.Z., Maday, Y., Amari, T.: On the linear Beltrami fields in bounded and unbounded three-dimensional domains: Mathematical Modelling and Numerical Analysis, **33**, 359–394 (1999)
5. Boulmezaoud, T.Z. , Amari, T.: On the existence of non-linear force-free fields in three-dimensional multiply-connected domains. à paraître dans Zeitschrift für Angewandte Mathematik und Physik (ZAMP)
6. Boulmezaoud, T. Z., Amari, T.: Approximation of linear force-free fields in bounded 3-D domains. Math. Comput. Modelling **31**, 109–129 (2000)
7. Brezzi, F., Hauke, G., Marini, L.D., Sangalli, G.: Link-Cutting Bubbles for the Stabilization of Convection-Diffusion-Reaction Problems, Math. Models Methods Appl. Sci., **13**, 445–461 (2003)
8. Brooks, A., Hughes,T. R. J.: Streamline upwind/Petrov-Galerkin formulations for convection dominated flows with particular emphasis on the incompressible Navier-Stokes equations . FENOMECH '81, Part I (Stuttgart, 1981). Comput. Methods Appl. Mech. Engrg. **32**, 199–259 (1982)

9. Girault, V., Raviart, P. A.: Finite element methods for Navier-Stokes equations. Springer-Verlag (1986)
10. Johnson, C., Nävert, U., Pitkäranta, J.: Finite element methods for linear hyperbolic problems, Comp. Meth. Appl. Mech. Engin. **45**, 285–312 (1984)
11. Laurence, P., Bruno, O.: Existence of 3D toroidal MHD equilibria with non-constant pressure, Communications on Pure and Applied Math, **49**, 717–764 (1996)
12. Lesaint, P.: Sur la résolution des systèmes hyperboliques du premier ordre par des méthodes d'éléments finis , Thèse de Doctorat, UPMC, Paris (1975)
13. Nédélec, J.C.: Mixed finite elements in R^3. Numer. Math. **35**, 315–341 (1980)

Mathematical and Numerical Analysis of a Class of Non-linear Elliptic Equations in the Two Dimensional Case

Nour Eddine Alaa[1], Abderrahim Cheggour[1] and Jean R. Roche[2]

[1] Département de Mathématiques et Informatique, Université des Sciences et Techniques Cadi Ayyad, B.P. 618, Guéliz, Marrakech, Maroc
alaa@fstg-marrakech.ac.ma
[2] I.E.C.N., Université Henri Poincaré, B.P. 239, 54506 Vandoeuvre lès Nancy, France
roche@iecn.u-nancy.fr

Summary. The aim of this paper is to show the existence and present a numerical analysis of weak solutions for a quasi-linear elliptic problem with Dirichlet boundary conditions in a domain Ω and data belonging to $L^1(\Omega)$. A numerical algorithm to compute a numerical approximation of the weak solution is described and analyzed. Numerical examples are presented and commented.

1 Introduction

The principal objective of this work is to give a result of existence and present a numerical analysis of weak solutions for the following quasi-linear elliptic problem:

$$\begin{cases} -\Delta u(x) + G(x, \nabla u(x)) = F(x, u(x)) + f(x) & \text{in } \Omega, \\ u(x) = 0 & \text{on } \partial\Omega \end{cases} \quad (1)$$

where G, F are Caratheodory non negative functions. The function $f \in L^1(\Omega)$ is given finite non negative. The domain $\Omega \subset \mathbb{R}^N$ is open and bounded. Such problems arise from biological, chemical and physical systems.

The two essential ingredients to the analysis of this problem are the convexity of $s \to G(x, s)$ and that $G(x, s)$ is sub-quadratic w.r.t. s namely:

$$G(x, s) \leq C(k(x) + \|s\|^2), \quad \text{where} \quad k(x) \in L^1(\Omega) \text{ and } C > 0. \quad (2)$$

Then the problem (1) has a solution in $W_0^{1,q}(\Omega)$ where $1 \leq q < N/(N-1)$, $N \geq 2$, provided that (1) has a super-solution in $W_0^{1,1}(\Omega)$.

In previous work [4] the authors show the existence of a weak solution in the one-dimensional case and with arbitrary growth of the non linearity and data measure.

We study a numerical method to compute the solution of the problem (1). In the first step we compute a super solution using a domain decomposition method. In the second step we compute a sequence of solutions of an intermediate problem obtained by using the Yosida approximation of G. This sequence converges to the weak solution of the problem (1).

2 Statement of the main result

Throughout this paper we suppose

$$f \in L^1(\Omega), \ f \geq 0. \tag{3}$$

The functions $G : \Omega \times \mathbb{R}^N \to [0, +\infty[$ and $F : \Omega \times \mathbb{R} \to [0, +\infty[$ are such that:

$$G, F \text{ are measurable, } r \to G(x, r) \text{ and } u \to F(x, u) \text{ are continuous.} \tag{4}$$

$$G \text{ is convex in } r \text{ and } F \text{ is nondecreasing in } u, \tag{5}$$

$$G(x, 0) = \min\{G(x, r), \ r \in \mathbb{R}^N\} = 0, \text{ and } F(x, 0) = 0, \tag{6}$$

$$G(x, r) \leq C(|r|^2 + K(x)), \tag{7}$$

$$F(x, u) \in L^1(\Omega) \text{ for every } u \in \mathbb{R} \tag{8}$$

with a constant $C > 0$ and $K \in L^1(\Omega)$.

We introduce now the notion of weak solutions of problem (1).

Definition 1. *A function u is said to be a weak solution of the problem (1), if*

$$\begin{cases} u \in W_0^{1,1}(\Omega), \ G(x, \nabla u) \text{ and } F(x, u) \in L^1(\Omega), \\ -\Delta u + G(x, \nabla u) = F(x, u) + f \text{ in } \mathcal{D}'(\Omega). \end{cases} \tag{9}$$

We will be interested in proving the existence of weak positive solutions of problem (1).

Theorem 1. *Under hypotheses (3)—(7), and assuming that there exists w such that*

$$\begin{cases} w \in W_0^{1,1}(\Omega), \ F(x, w) \in L^1(\Omega), \\ -\Delta w = F(x, w) + f \text{ in } \mathcal{D}'(\Omega), \end{cases} \tag{10}$$

the problem (1) has a positive weak solution.

3 Proof of theorem 1

3.1 Approximation scheme

We consider the sequence defined by $u_0 = w$ and for $n \geq 0$, u_{n+1} is the solution of the problem

$$\begin{cases} -\Delta u_{n+1} + G_{n+1}(x, \nabla u_{n+1}) = F(x, u_n) + f & \text{in } \mathcal{D}'(\Omega), \\ u_{n+1} \in W_0^{1,1}(\Omega), G_{n+1}(x, \nabla u_{n+1}) \in L^1(\Omega), \end{cases} \quad (11)$$

where $G_n(x, r)$ denotes the Yosida approximation of $G(x, r)$. The function $G_n(x, r)$ is convex in r, increases pointwise to $G(x, r)$ as n tends to ∞ and satisfies

$$G_n \leq G_{n+1} \leq G, \qquad \|G_{n,r}(x, r)\|_\infty \leq n, \quad (12)$$

where $G_{n,r}$ denotes a section of subdifferential of G_n with respect to r.

The classical works ([3, 6, 9]) combined with an induction argument can be applied to prove that (11) has a solution such that

$$0 \leq u_{n+1} \leq u_n \leq w. \quad (13)$$

3.2 Estimates and convergence

Let $\{u_n\}_n$ be a sequence defined as above. By integrating (11) in Ω and using (13) we obtain

$$\int_\Omega G_{n+1}(x, \nabla u_{n+1}) dx \leq \int_\Omega F(x, w) dx + \int_\Omega f(x) dx. \quad (14)$$

Therefore $\|\Delta u_{n+1}\|_{L^1(\Omega)}$ is bounded. Then there exists a subsequence still denoted by u_n for simplicity, such that u_n converges strongly to some u in $W_0^{1,q}(\Omega), 1 \leq q < N/(N-1)$, and $(u_n, \nabla u_n)$ converges to $(u, \nabla u)$ almost everywhere in Ω (see [3]).

Let us prove that u is in fact a solution of problem (1). According to the definition 2.1, we only have to show that

$$-\Delta u + G(x, \nabla u) = F(x, u) + f \quad \text{in } \mathcal{D}'(\Omega). \quad (15)$$

We know that $F(x, u_n) \longrightarrow F(x, u)$ strongly in $L^1(\Omega)$ and, for almost every x in Ω, there holds $G_{n+1}(x, \nabla u_{n+1}(x)) \longrightarrow G(x, \nabla u(x))$.

Then there exists a non-negative measure μ (see [7]) such that

$$-\Delta u_n + G_{n+1}(\nabla u_{n+1}) - F(u_n) - f \longrightarrow -\Delta u + G(\nabla u) - F(u) - f + \mu \text{ in } \mathcal{D}'(\Omega),$$

as n goes to ∞.

On the other hand

$$-\Delta u_{n+1} + G_{n+1}(x, \nabla u_{n+1}) = F(x, u_n) + f \longrightarrow F(x, u) + f \text{ in } L^1(\Omega). \quad (16)$$

Consequently
$$-\Delta u + G(x, \nabla u) \leq F(x, u) + f \text{ in } \mathcal{D}'(\Omega). \tag{17}$$

Therefore to conclude the proof of theorem 1, we must establish the opposite inequality. To this end we introduce the following test function

$$\psi \exp(-Cu_{n+1})H(\frac{u_{n+1}}{k}), \tag{18}$$

where $H \in C^1(\mathbb{R})$, $0 \leq H(s) \leq 1$, $H(s) = 0$ if $|s| \geq 1$ and $H(s) = 1$ if $|s| \leq \frac{1}{2}$, C is given by relation (7) and $\psi \leq 0$, $\psi \in H_0^1(\Omega) \cap L^\infty(\Omega)$. We multiply the equation satisfied by u_{n+1} in (11) by this test function and we integrate in Ω, to obtain

$$\int_\Omega (f_n + F(x, u_n))\psi \exp(-Cu_{n+1})H(\frac{u_{n+1}}{k}) \, dx = I_1 + I_2 + I_3,$$

where

$$\begin{aligned}
I_1 &= \int_\Omega \nabla u_{n+1} \nabla \psi \exp(-Cu_{n+1}) H(\frac{u_{n+1}}{k}) \, dx, \\
I_2 &= \frac{1}{k} \int_\Omega |\nabla u_{n+1}|^2 \psi \exp(-Cu_{n+1}) H'(\frac{u_{n+1}}{k}) \, dx, \\
I_3 &= \int_\Omega (G_{n+1}(x, \nabla u_{n+1}) - C|\nabla u_{n+1}|^2)\psi \exp(-Cu_{n+1}) H(\frac{u_{n+1}}{k}) \, dx.
\end{aligned} \tag{19}$$

By investigating separately each term, we get

$$\lim_{n \to \infty} I_1 = \int_\Omega \nabla u \nabla \psi \exp(-Cu) H(\frac{u}{k}) \, dx$$

and $\lim_{k \to \infty} I_2 = 0$ uniformly on n.

Now we investigate the remaining term I_3. Since G_{n+1} satisfies the inequality (7), $\psi \leq 0$, and by applying Fatou's lemma, we obtain

$$\lim_{n \to \infty} I_3 \geq \int_\Omega (G(x, \nabla u) - C|\nabla u|^2)\psi \exp(-Cu) H(\frac{u}{k}) dx. \tag{20}$$

Finally we have shown

$$\int_\Omega \nabla u \nabla \psi \exp(-Cu) H(\frac{u}{k}) dx + \int_\Omega \psi(G(x, \nabla u) - C|\nabla u|^2) \exp(-Cu) H(\frac{u}{k}) dx$$
$$+ \omega(\frac{1}{k}) \leq \int_\Omega (F(x, u) + f)\psi \exp(-Cu) H(\frac{u}{k}) dx,$$

where $w(\varepsilon)$ denotes a quantity that tends to 0 when ε tends to 0. Now we choose $\psi = -\varphi \exp(Cu) H(\frac{u}{k})$, where $\varphi \geq 0, \varphi \in \mathcal{D}(\Omega)$ and we replace ψ by this value in the previous inequality to get after appropriate calculations and using that the third term is equivalent to $w(\frac{1}{k})$

$$-\int_\Omega \nabla u \nabla \varphi H(\frac{u}{k})^2 dx - \int_\Omega \varphi G(x, \nabla u) H(\frac{u}{k})^2 dx + w(\frac{1}{k})$$
$$\leq -\int_\Omega (F(x,u) + f)\varphi H(\frac{u}{k})^2 dx.$$

We finally pass to the limit as k tends to infinity and we use the fact that $\lim_{k\to\infty} H(\frac{u}{k}) = 1$, to conclude for every $\varphi \geq 0, \varphi \in \mathcal{D}(\Omega)$ that

$$\int_\Omega \nabla u \nabla \varphi dx + \int_\Omega \varphi G(x, \nabla u) dx \geq \int_\Omega (F(x,u) + f)\varphi dx.$$

This finishes the proof of the theorem 1.

4 Numerical method

4.1 Introduction

In this section we present the numerical method to solve the equation (1) in $\mathbb{R}^2$. Formally the algorithm can be formulated in the following way:
1) Find $\overline{w} \in H_0^1(\Omega)$ such that:

$$-\Delta \overline{w}(x) \geq F(x, \overline{w}) + f \text{ in } \Omega. \tag{21}$$

2) Given $u_0 = \overline{w}$ we compute a sequence, $\{u_n\}_n$, solution in $H_0^1(\Omega)$ of the non linear equation:

$$-\Delta u_{n+1}(x) + G_{n+1}(x, \nabla u_{n+1}) = F(x, u_n) + f \text{ in } \Omega. \tag{22}$$

Both problems (21) and (22) are non-linear, and if (21) has a solution, in theorem 1 we have shown that (22) then also has a solution.

4.2 Numerical algorithm

This subsection summarizes the algorithm introduced in the previous subsection.

1) First step: given $\overline{w}^0 = 0$, iteratively for $k = 1$ until convergence we compute $\overline{w}^{k+1} = \overline{w}^k + \delta$ where at each iteration δ is the solution of the linear problem:

$$\begin{cases} -\Delta \delta(x) - \frac{\partial F(x,\overline{w}^k)}{\partial r}\delta(x) = \Delta \overline{w}^k(x) + F(x,\overline{w}^k) + f \text{ in } \Omega, \\ \delta(x) = 0 \text{ on } \partial\Omega. \end{cases} \quad (23)$$

To solve at each iteration the linear problem (23) we consider the domain decomposition method which will be introduced as follows:

a) We compute $c_\infty = \left\|\frac{\partial F(\overline{w}^k)}{\partial r}\right\|_\infty$. Determine an overlapping subdomain decomposition Ω_i, $i = 1, \ldots, m$ such that $\Omega = \bigcup_{i=1}^m \Omega_i$ and satisfies:

$$\max\{\text{mes}(\Omega_i), i = 1,\ldots, m\} < \min(\frac{c_0 \pi^2}{c_\infty}, \frac{\pi}{2\sqrt{c_\infty}}). \quad (24)$$

We denote by m the number of subdomains Ω_i and $\partial\Omega_i$ is the boundary of Ω_i.

b) Iteratively:
for $l = 1,\ldots$ until convergence and for $i = 1, \ldots, m$ we solve the following subdomain problems:

$$\begin{cases} -\Delta \delta_i^l(x) - \frac{\partial F(x,\overline{w}^k)}{\partial r}\delta_i^l(x) = \Delta \overline{w}^k(x) + F(x,\overline{w}^k) + f \text{ in } \Omega_i, \\ \delta_i^l(x) = \delta_j^{l-1}(x), \text{ on } \partial\Omega_i \cap \Omega_j, \\ \delta_i^l(x) = 0, \text{ on } \partial\Omega \cap \partial\Omega_i. \end{cases} \quad (25)$$

On each subdomain Ω_i we consider a finite element approximation method with N_i elements. At the end of the l-th loop we have computed an approximate discrete solution of the linear indefinite problem (23).

2) At this step for $u_0 = \overline{w}$, iteratively for $n = 1$, until convergence we solve the following non-linear problem

$$\begin{cases} -\Delta u_n(x) + G_n(x, \nabla u_n) = F(x, u_{n-1}) + f \text{ in } \Omega, \\ u_n(x) = 0 \text{ on } \partial\Omega. \end{cases} \quad (26)$$

At each n-th step the problem (26) is solved by using a Newton method. The discrete approximation of the solution of (1) is obtained at the end of the n-th loop.

4.3 Convergence of the domain decomposition method

To simplify, without lost of generality, we assume that we can consider a two-domain decomposition $\Omega = \Omega_1 \bigcup \Omega_2$ such that:

$$\max\{\text{mes}(\Omega_i), i = 1, 2\} < \min(\frac{c_0 \pi^2}{c_\infty}, \frac{\pi}{2\sqrt{c_\infty}}). \quad (27)$$

Now to prove the convergence of the Schwarz overlapping domain decomposition algorithm applied to problem (23), we consider two problems:

$$\begin{cases} -\Delta v_1(x) + c(x) v_1(x) = h(x) \text{ in } \Omega_1, \\ v_1(x) = 0 \text{ on } \partial\Omega \cap \partial\Omega_1; \ v_1(x) = v_2(x) \text{ on } \partial\Omega_1 \cap \Omega_2 \end{cases} \quad (28)$$

and

$$\begin{cases} -\Delta v_2(x) + c(x) v_2(x) = h(x) \text{ in } \Omega_2, \\ v_2(x) = v_1(x) \text{ on } \partial\Omega_2 \cap \Omega_1; \ v_2(x) = 0 \text{ on } \partial\Omega \cap \Omega_2. \end{cases} \quad (29)$$

Let v be

$$v = \begin{cases} v_1 \text{ in } \Omega_1, \\ v_2 \text{ in } \Omega_2, \end{cases} \quad (30)$$

$v_1 = v_2$ in $\Omega_1 \cap \Omega_2$.

With the restriction (27) we can suppose the existence of a solution of (28) in $W_0^{1,q}(\Omega_1)$ and a solution of (29) in $W_0^{1,q}(\Omega_2)$.

Then, if v^0 is an initialization function defined in Ω and vanishing in $\partial\Omega$, we define for $k \geq 0$ two sequences v_i^k, $i = 1, 2$ solving the following problems:

$$\begin{cases} -\Delta v_1^{k+1}(x) + c(x) v_1^{k+1}(x) = h(x) \text{ in } \Omega_1, \\ v_1^{k+1}(x) = 0 \text{ on } \partial\Omega \cap \partial\Omega_1; \ v_1^{k+1}(x) = v_2^k(x) \text{ on } \partial\Omega_1 \cap \Omega_2 \end{cases} \quad (31)$$

and

$$\begin{cases} -\Delta v_2^{k+1}(x) + c(x) v_2^{k+1}(x) = h(x) \text{ in } \Omega_2, \\ v_2^{k+1}(x) = v_1^k(x) \text{ on } \partial\Omega_2 \cap \Omega_1; \ v_2^{k+1}(x) = 0 \text{ on } \partial\Omega \cap \Omega_2. \end{cases} \quad (32)$$

Theorem 2. *Assume Ω_1 and Ω_2 with the restriction (27). Then the sequence v^k converges to v in $W_0^{1,q}(\Omega_1)$ and $W_0^{1,q}(\Omega_2)$.*

Proof. We give here an idea of the proof.

Let $d^k = v_1^k - v$ in Ω_1 and $e^k = v_2^k - v$ in Ω_2 then $d^k \in L^\infty(\Omega_1)$ and $e^k \in L^\infty(\Omega_2)$.

Thanks to the maximum principle we prove the following inequalities:

$$\|d^{k+2}\|_\infty \leq \gamma \|d^k\|_\infty \text{ and } \|e^{k+2}\|_\infty \leq \gamma \|e^k\|_\infty \quad (33)$$

where $\gamma < 1$.

But to be able to apply the maximum principle it will be necessary that the subdomains Ω_1 and Ω_2 verify the restriction (27). ∎

4.4 Numerical results

The algorithm introduced in the previous section has been implemented numerically for the model problem (1) where:
$G(x, r) = |r|^p = (r_1^2 + r_2^2)^{\frac{p}{2}}$ and $r = (r_1, r_2) \in \mathbb{R}^2$ for $1 < p < \infty$.
$F(x, s) = \eta s^q$ where $s \in \mathbb{R}^+$ and $1 < q < \infty$.
$f(x) = x_1^\alpha + x_2^\beta$ where $x = (x_1, x_2) \in \Omega$ and $-1 < \alpha, \beta < \infty$.

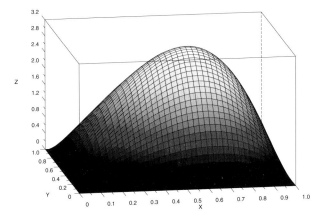

Fig. 1. $\eta = 45$, $p = q = 3$, $\alpha = \beta = 2$, $m = 36$

The number of subdomains is not fixed, it changes at each iteration according to the criterion (27). In figure 1 we can see the solution shape when the algorithm converges with $m = 36$ subdomains.

To study the convergence history of the numerical simulation plotted in figure 1 we consider two steps. In the first step, where we compute a super-solution, we observe the evolution of the number of subdomains: it goes from $m = 4$ subdomains to $m = 36$ subdomains in seven iterations according to criterion (27). Simulation stops after 34 iterations when the residual is less than 10^{-6}. In the second step, starting with the super-solution computed in the previous step we perform nine iterations of the Yosida approximation described in section 3 and the simulation stops when the correction computed is in uniform norm less than 10^{-6}.

References

1. Alaa, N., Maach, F., Mounir, I.: Existence for some quasilinear elliptic systems with critical growth nonlinearity and L^1 data. Journal of Applied Analysis, **11**, 81–94 (2005)
2. Alaa, N., Mounir, I.: Global existence for reaction-diffusion systems with mass control and critical growth with respect to the gradient. Journal of Mathematical Analysis and Applications, **253**, 532–557 (2001)
3. Alaa, N., Pierre, M.: Weak solution of some quasilinear elliptic equations with measures. SIAM J. Math.Anal, **24**, 23–35 (1993)
4. Alaa, N., Roche, J.R.: Theoretical and numerical analysis of a class of nonlinear elliptic equations. Mediterr. J. Math, **2**, 327–344 (2005)
5. Amann, H., Crandall, M.G.: On some existence theorems for semi linear equations. Indiana Univ. Math. J, **7**, 779–790 (1978)
6. Boccardo, L., Murat, F., Puel, J.P.: Existence de solutions non bornées pour certaines équations quasi-linéaires. Portugaliae Math., **41**, 507–534 (1982)

7. Brezis, H.: Analyse fonctionnelle appliquée, Masson, Paris (1980)
8. Brezis, H., W. Strauss, W.: Semilinear elliptic equation in L^1. J. Math. Soc. Japan, **25**, 565–590 (1973)
9. Lions, P.L.: Résolution de problèmes elliptiques quasilinéaires. Arch. Ration. Mech. Analysis, **74**, 335–353 (1980)
10. Porretta, A.: Existence for elliptic equations in L^1 having lower order terms with natural growth. Portugaliae Math., **57**, 179–190 (2000)

Numerical Linear Algebra and Approximation Methods

A s-step Variant of the Double Orthogonal Series Algorithm

J.A. Alvarez-Dios[1], J.C. Cabaleiro[2] and G. Casal[3]

[1] Departamento de Matemática Aplicada. Universidade de Santiago de Compostela
jaaldios@usc.es
[2] Departamento de Electrónica e Computación. Universidade de Santiago de Compostela
elcaba@usc.es
[3] Departamento de Matemática Aplicada. Universidade de Santiago de Compostela
gcasal@lugo.usc.es

Summary. We use the s-step technique proposed by Chronopoulos in [2, 3] for creating a s-step variant of the Double Orthogonal Series algorithm (s-DOS). The original Double Orthogonal Series algorithm, proposed by M. Amara and J. C. Nédélec [1], converges for any nonsingular coefficient matrix of the linear system in n iterations at most, where n is the order of the system. We prove the convergence of the new s-DOS method in the integer part of n/s iterations at most.

We go on to show numerical results for the comparison of the original method and s-DOS in parallel programming. Numerical tests for both methods have been carried out on the Beowulf Cluster in CESGA (*Centro de Supercomputación de Galicia*) formed by a total of 16 Pentium III processors running at 1GHz.

As a wrap up, we obtain that the s-step variant of the Double Orthogonal Series improves parallel programming efficiency and execution time with respect to the original method on large linear systems.

1 Introduction

In iterative method solvers for large linear systems most required computations are vector-vector and matrix-vector operations. Within the confines of the Basic Linear Algebra Subprogram Library (BLAS), they primarily translate as SDOTs (inner products) and SAXPYs (vector updates as a linear combination of two vectors) i.e. level 1 BLAS operations. On the other hand, BLAS 2 and BLAS 3 operations, based on blocks of submatrices, are much more efficient than BLAS 1 operations on parallel computers with optimized BLAS kernels. This is because the ratio between the number of operations performed and computer memory accesses increases as we raise the BLAS level. In order to improve the aforesaid ratio, an alternative approach using BLAS 3 operations in some iterative methods for linear systems called the *s*-step technique has been proposed by Chronopoulos [2, 3].

In [2] and [3] were given the *s*-step variants of some classical methods as the Conjugate Gradient algorithm, Orthomin(*m*) and other ones. None of them converge for any nonsingular matrix. Our objective is to obtain an *s*-step variant taking as original algorithm the Amara and Nedelec's Double Orthogonal Series which converges for every nonsingular matrix. This method has advantages over the normal equation approach, most notably that it does not degenerate due to the fact that in practice $A^t A$ may be rounded off to a non positive definite matrix if A is very ill-conditioned.

In this paper A is a square nonsingular matrix of order n, and b, x are column vectors of $\mathbb{R}^n$.

2 The algorithm of the Double Orthogonal Series

The Double Orthogonal Series method, presented by Amara and Nedelec in [1], is an iterative algorithm of resolution of the linear system $Ax = b$. It is based on the construction of two sequences of orthogonal vectors. The method converges for nonsingular matrix in n iterations maximum.

The algorithm of the Double Orthogonal Series is :

Algorithm. *(Double Orthogonal Series)*
Let $x_0 \in \mathbb{R}^n$ an initial vector:

1. **Start:**
   - $r_0 = b - Ax_0$
   - $p_0 = A^t r_0$
   - $q_0 = Ap_0$
   - $\alpha_0 = \dfrac{<r_0, r_0>}{<p_0, p_0>}$

2. **Iterations:** For $i = 0$ until convergence do
   - $x_{i+1} = x_i + \alpha_i p_i$
   - $r_{i+1} = b - Ax_{i+1}$
   - $\beta_i = -\dfrac{<q_i, q_i>}{<p_i, p_i>}$
   - $p_{i+1} = A^t q_i + \beta_i p_i$
   - $\gamma_i = -\dfrac{<p_{i+1}, p_{i+1}>}{<q_i, q_i>}$
   - $q_{i+1} = Ap_{i+1} + \gamma_i q_i$
   - $\alpha_{i+1} = \dfrac{<r_{i+1}, q_i>}{<p_{i+1}, p_{i+1}>}$

   EndFor

End

The following lemma is demonstrated in [1]:

Lemma 1. *Let be $k \in \mathbb{N}$, $1 \leq k \leq n$. If $r_i, p_i, q_i \neq 0$ then:*

$$< p_{k+1}, p_i > = 0; \qquad 1 \leq i \leq k \qquad (1)$$

$$< q_{k+1}, q_i > = 0; \qquad 1 \leq i \leq k \qquad (2)$$

3 s-Step methods

An attempt to gain efficiency in parallel programming with iterative methods is constituted by the s-step methods given by Chronopoulos and other authors basically in [2] and [3].

The idea is to construct in each iteration the Krylov subspace $K_s(A, r_i)$ generated by $\{r_i, Ar_i, \ldots, A^{s-1}r_i\}$, and to calculate the iterate which minimizes the residual norm or the error norm on the affine variety $x_i + K_s(A, r_i)$.

The s-step variants obtained by Chronopoulos et al. in their articles are the following ones:

- For a positive definite symmetric matrix, the s-step variant of the Conjugated Gradient algorithm, which converges within the integer part of n/s iterations at most.
- For an indefinite nonsymmetric matrix, the s-step variants of the Generalized Conjugate Residual (GCR), of the Minimal Residual (MR) and Orthomin(m) methods. They converge for any symmetric matrix, any positive definite nonsymmetric matrix, and a class of nonsymmetric indefinite matrices.

4 The s-step variant of the double orthogonal series

The s-step variant of the Double Orthogonal Series that we present in this work consists of taking as the original algorithm the DOS from Amara and Nedelec. Basing ourselves on the s-step methods of Chronopoulos, we obtain the s-step variant that we will denote by s-DOS.

By induction, we obtain that each iterate in DOS can be expressed in the following way:

$$x_{k+1} = x_0 + \sum_{i=0}^{k} \alpha_i p_i \qquad (3)$$

The following lemma verifies:

Lemma 2. *In the Double Orthogonal Series algorithm, it holds that:*

$$< \{p_0, ..., p_k\} > = < \{p_0, (A^t A)p_0, (A^t A)^2 p_0, ..., (A^t A)^k p_0\} > \qquad (4)$$

Proof. The inclusion "⊂" is proved by induction on k using the definitions of p_i and q_i. On the other hand $\{p_0, ..., p_k\}$ are linearly independent because of lemma 1, and this establishes the equality. ∎

Thus the iterate belongs to the affine variety:

$$x_{k+1} \in x_0 + <\{p_0, (A^t A)p_0, (A^t A)^2 p_0, ..., (A^t A)^k p_0\}> \quad (5)$$

Therefore the Krylov subspace that we will consider in our method is:

$$K_s(A^t A, A^t r_0) = <\{A^t r_0, (A^t A)A^t r_0, \ldots, (A^t A)^{s-1} A^t r_0\}> \quad (6)$$

where r_0 is the initial residual vector.

Let $s \in \mathbb{N}$, $M_{n \times s}$ be the set of matrices of order $n \times s$ and $A \in M_{n \times n}$. We denote by Δ_A^s the linear application:

$$\Delta_A^s : \mathbb{R}^n \longrightarrow M_{n \times s}$$

that assigns to each vector v of $\mathbb{R}^n$ the matrix of order $n \times s$ whose column vectors are $v, Av, A^2 v, \ldots, A^{s-1} v$, and then

$$\Delta_A^s(v) = (v, Av, A^2 v, \ldots, A^{s-1} v) \in M_{n \times s} \quad (7)$$

The s-step variant of the Double Orthogonal Series algorithm is:

Algorithm. (s-DOS)

Let $x_0 \in \mathbb{R}^n$ an initial vector and $s \in \mathbb{N}$:

1. **Start:**
   - $r_0 = b - Ax_0$
   - $P_0 = \Delta_{A^t A}(A^t r_0)$
   - $W_0 = (AP_0)^t AP_0$
   - $z_0 = (AP_0)^t r_0$
   - $y_0 = W_0^{-1} z_0$
   - $x_1 = x_0 + P_0 y_0$

2. **Iterations:** For $i = 1$ until convergence do
   - $r_i = r_{i-1} - AP_{i-1} y_{i-1}$
   - $R_i = \Delta_{A^t A}(A^t r_i)$
   - For $j = 0$ until $j = i - 1$:
     $B_j = -(P_j^t P_j)^{-1} P_j^t R_i$
   - EndFor
   - $P_i = R_i + \sum_{j=0}^{i-1} P_j B_j$
   - $W_i = (AP_i)^t AP_i$
   - $z_i = (AP_i)^t r_i$
   - $y_i = W_i^{-1} z_i$
   - $x_{i+1} = x_i + P_i y_i$

   EndFor

End

The minimal polynomial of a nonzero vector $v \in \mathbb{R}^n$ with respect to matrix A is the least degree monic polynomial $q(x)$ so that $q(A)v = 0$. To prove the convergence of the s-DOS we demonstrate the following lemma:

Lemma 3. *Let be $s, i \in \mathbb{N}$, $s \cdot i < n$, $< P_0, \ldots, P_{i-1} >$ the vector subspace generated by the column vectors of $P_0, \ldots, P_{i-1}$, and $p_0, \ldots, p_{s \cdot i}$ the direction vectors computed in the original DOS algorithm. Assume that the degree of the minimal polynomial of p_0 with respect to A is greater than $s \cdot i$ and W_j is nonsingular for all $j \leq i$. Then:*

a) $P_j^t P_k = 0$ for all $j \neq k$, $0 \leq j, k < i$.
b) $< P_0, \ldots, P_{i-1} > = < p_0, \ldots, p_{s \cdot i-1} >$.
c) $A^t r_i \notin < P_0, \ldots, P_{i-1} >$.

Proof. The definition of the matrices P_j and B_j implies a). Also, it is possible to check that $< P_0, \ldots, P_{i-1} >$ is contained in $< \{A^t r_0, \ldots, (A^t A)^{s \cdot i-1} A^t r_0\} >$ because every column vector of P_{i-1}, $0 \leq j < i$ can be written in terms of $(A^t A)^k A^t r_0$ with $k \leq s \cdot i - 1$. The nonsingularity of W_j implies the independence of the column vectors of P_j for all $j \leq i$. Then, by a) and the nonsingularity of W_j, the dimension of $< P_0, \ldots, P_{i-1} >$ is $s \cdot i$, and then both subspaces are equal. Using lemma 2 we obtain b).

To prove c) note that the first column vector of P_i is

$$p_i^0 = A^t r_i + \sum_{j=0}^{i-1} P_j b_{j1}$$

where b_{j1} are the first column elements of B_j. From a) and the nonsingularity of W_i we cannot have $A^t r_i \in < P_0, \ldots, P_{i-1} >$. ∎

Finally, we have the convergence of the s-DOS:

Theorem 1. *Under the assumptions of lemma 3, the s-DOS algorithm converges in at most $[n/s]$ iterations for every nonsingular matrix.*

Proof. The convergence in at most $[n/s]$ iterations follows from lemma 3, b) and c). ∎

The determination of an exponential upper bound for the error norm in each iteration is at present under study by the authors. Numerical tests seem to show that i iterations of the s-step method are roughly equivalent to $s \cdot i$ iterations of the sequential method.

5 Numerical results

For the numerical tests we programmed both algorithms, the Double Orthogonal Series (DOS) and its s-step variant, s-DOS. These tests were performed on a Beowulf Cluster at the Supercomputing Center of Galicia (CESGA), with 16 1GHz Pentium III processors, 512MB of memory and local 40GB ATA disk on each node.

We used the Portland Group Fortran 95 compiler and the Message Passing Interface (MPI).

All the matrices in the numerical tests are from the Harwell-Boeing Collection. More precisely, there are square real sparse matrices, nonsymmetric matrices, and matrices stored in three vectors according to *Compressed Sparse Column* format (CSC). The parallel programming strategy used is SPDM, the same program is to execute on all processors, although not necessarily the same instruction at each time.

In a first test we executed DOS and s-DOS for different values of s in sequential on several matrices (Table 1).

Table 1. Number of iterations

alg. \ matriz	gre185	gre343	gre512	gre1107
DOS	117	54	69	200
3-DOS	96	35	29	103
4-DOS	46	22	19	58
6-DOS	32	12	12	27
7-DOS	22	10	9	24

We verified that the number of iterations until convergence decreases simultaneously in all matrices with s-DOS(m) as the size of block s increases.

The selected matrices for this first test are not of large order and they condition numbers of a different order of magnitude (Table 2).

Table 2. Matrices

	n	nnz	condition number
gre185	185	1005	510000
gre343	343	1435	250
gre512	512	2192	380
gre1107	1107	5664	97000000

In table 3 we show the execution time on matrix fidapm11 of order 22294 executing programs DOS, 3-DOS, 5-DOS, 6-DOS and 7-DOS on several processors.

We obtain the best times for 6-DOS, which keep improving as we increase the number of processors.

Table 3. Execution time (seconds) for fiadpm11

N \ alg.	DOS	3-DOS	5-DOS	6-DOS	7-DOS
1	41.85	25.57	17.89	**16.67**	17.87
2	23.38	13.97	9.60	**8.98**	9.23
4	17.42	8.13	5.67	**5.23**	5.63
8	13.63	6.08	4.14	**3.73**	3.98
12	15.53	6.29	4.24	**3.69**	3,85
16	14.39	5.85	3.86	**3.47**	3.62

To be able to evaluate the improvement obtained with parallel programming on an algorithm two measures exist:

- The speedup is the quotient obtained between the execution time on a processor and the execution time on N processors:

$$S = \frac{T_1}{T_N}.$$

- The efficiency is the quotient between speedup and the number of processors:

$$E = \frac{S}{N}.$$

In table 4 are shown efficiencies for fidapm11.

Table 4. Efficiency for fiadpm11

alg. \ nproc	1	2	4	8	12	16
DOS	1	0.89	0.60	0.38	0.22	0.18
3-DOS	1	0.91	0.79	0.52	0.34	0.27
5-DOS	1	0.93	0.79	0.54	0.35	0.29
6-DOS	1	0.93	**0.80**	0.55	038	0.30
7-DOS	1	**0.97**	0.79	**0.56**	**0.39**	**0.31**

We see that the best efficiencies are obtained for 7-DOS, except in 4 processors that are obtained for 6-DOS. The efficiency decreases as the number of processors increases because of the time overhead resulting from the message passing.

In graphic 1 we see how the speedups vary with each algorithm as we increase the number of processors. It is possible to observe that the s-step methods have better speedups, which increase as s increases.

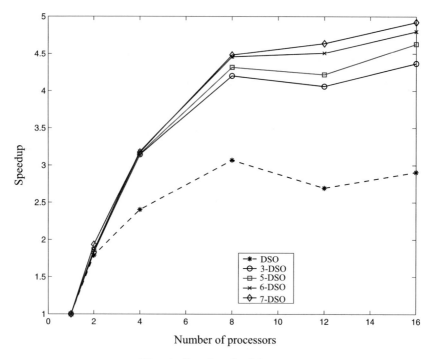

Fig. 1. Speedup for fidapm11

6 Conclusions

The s-step method allows us to obtain the exact solution of a general linear system in at most $[n/s]$ iterations. It improves the parallelism of the original DOS method and increases the BLAS 3 operation ratio, which results in fairly good numerical speedups.

Acknowledgment

The authors wish to thank the Galicia Supercomputing Center (CESGA) for their computational support.

References

1. Amara, M., Nédélec, J.C.: Résolution des systèmes matriciels indéfinis par une décomposition sur une double suite orthogonale. C. R. Acad. Sc. Paris, **295**, 309–312 (1982)
2. Chronopoulos, A.T.: s-Step iterative methods for (non)symmetric (in)definite linear systems. SIAM J. Numer. Anal., **28**, No.6, 1776–1789 (1991)
3. Chronopoulos, A.T., Swanson, C.D.: Parallel iterative s-step methods for unsymmetric linear systems. Parallel Computing, **22**, 623–641 (1996)

Linear Equations in Quaternions

Drahoslava Janovská[1] and Gerhard Opfer[2]

[1] Institute of Chemical Technology, Prague, Technická 5, 166 28 Prague 6, Czech Republic
`janovskd@vscht.cz`
[2] University of Hamburg, MIN Faculty, Department of Mathematics, Bundesstraσe 55, 20146 Hamburg, Germany
`opfer@math.uni-hamburg.de`

Summary. The aim is to solve a linear equation in quaternions namely, the equation $\sum_{j=1}^{j=\nu} a^{(j)} x b^{(j)} = e$, where $a^{(j)}$, $b^{(j)}$ and e are given quaternions, the quaternion x stands for the unknown solution. We give an algorithm based on a fixed point formulation.

1 Basic properties and definitions for quaternions

We start with some information on the algebra of quaternions. There are more details in our previous papers [3, 7]. General information are contained in a book [6], results concerning matrices with quaternion elements are surveyed in [9]. Applications to quantum mechanics are treated in [1, 2], and applications in chemistry are given in [8].

We denote by $\mathbb{H} = \mathbb{R}^4$ the skew field of quaternions. Let $a = (a_1, a_2, a_3, a_4)$, $b = (b_1, b_2, b_3, b_4) \in \mathbb{H}$. Then, addition is defined elementwise and multiplication is governed by the following rule:

$$ab := (a_1 b_1 - a_2 b_2 - a_3 b_3 - a_4 b_4, a_1 b_2 + a_2 b_1 + a_3 b_4 - a_4 b_3, \tag{1}$$
$$a_1 b_3 - a_2 b_4 + a_3 b_1 + a_4 b_2, a_1 b_4 + a_2 b_3 - a_3 b_2 + a_4 b_1).$$

The first component a_1 of $a = (a_1, a_2, a_3, a_4) \in \mathbb{H}$ is called the *real part* of a and denoted by $\operatorname{Re} a$. The second component a_2 is called the *imaginary part* of a and denoted by $\operatorname{Im} a$. A quaternion $a = (a_1, 0, 0, 0)$ will be identified with $a_1 \in \mathbb{R}$ and $a = (a_1, a_2, 0, 0)$ will be identified with $a_1 + \mathrm{i} a_2 \in \mathbb{C}$. The zero element $(0, 0, 0, 0) \in \mathbb{H}$ and the unit element $(1, 0, 0, 0) \in \mathbb{H}$ will be abbreviated by 0, 1, respectively. Let $a = (a_1, a_2, a_3, a_4) \in \mathbb{H}$. The *conjugate* of a, denoted by $\overline{a}$, will be defined by

$$\overline{a} := (a_1, -a_2, -a_3, -a_4).$$

The *absolute value* of a, denoted by $|a|$, will be defined by
$$|a| := \sqrt{a_1^2 + a_2^2 + a_3^2 + a_4^2}.$$
There are the following important rules:
$$\begin{aligned} \operatorname{Re}(ab) &= \operatorname{Re}(ba), \\ |ab| &= |ba| = |a||b|, \\ |a|^2 &= a\bar{a} = \bar{a}a, \\ \overline{ab} &= \bar{b}\bar{a}, \\ a^{-1} &= \frac{\bar{a}}{|a|^2}, \ a \neq 0, \\ (ab)^{-1} &= b^{-1}a^{-1}, \ a, b \neq 0. \end{aligned}$$
We denote by $\mathbb{H}^n$ the normed vector space of n-vectors formed by quaternions, where the norm of $\mathbf{x} := (x_1, x_2, \ldots, x_n) \in \mathbb{H}^n$ will be defined by
$$||\mathbf{x}|| := \sqrt{|x_1|^2 + |x_2|^2 + \cdots + |x_n|^2}.$$
Let $\mathbb{H}^{m \times n}$ be the set of all $(m \times n)$-matrices with elements from $\mathbb{H}$. We note here, that these matrices act as *linear mappings* $\ell : \mathbb{H}^n \to \mathbb{H}^m$ only in the following sense:
$$\begin{aligned} \ell(\mathbf{x} + \mathbf{y}) &= \ell(\mathbf{x}) + \ell(\mathbf{y}), \quad \mathbf{x}, \mathbf{y} \in \mathbb{H}^n, \\ \ell(\mathbf{x}\alpha) &= \ell(\mathbf{x})\alpha, \quad \mathbf{x} \in \mathbb{H}^n, \ \alpha \in \mathbb{H}. \end{aligned}$$
The converse is also true: A linear mapping ℓ defined by the above two properties is always represented by a matrix. This follows from standard arguments.

Let $\mathbf{A} \in \mathbb{H}^{m \times n}$. By $\mathbf{A}^T \in \mathbb{H}^{n \times m}$ we understand the *transposed matrix* of $\mathbf{A}$ where the rows and columns are exchanged. By $\overline{\mathbf{A}} \in \mathbb{H}^{m \times n}$ we understand the matrix which is formed by conjugation of all its elements. Finally,
$$\mathbf{A}^* := (\overline{\mathbf{A}})^T = \overline{\mathbf{A}^T}.$$
In case $\mathbf{A}^* = \mathbf{A}$, we call $\mathbf{A}$ *Hermitean*. The zero element of $\mathbb{H}^n$ and of $\mathbb{H}^{m \times n}$ will be denoted by $\mathbf{0}$. From the context it will become clear which zero element is meant. A matrix $\mathbf{A} \in \mathbb{H}^{n \times n}$ will be called *unitary* if $\mathbf{A}^*\mathbf{A} = \mathbf{A}\mathbf{A}^* = \mathbf{I}$, where $\mathbf{I}$ is the identity matrix. Unitary matrices $\mathbf{A}$ are characterized by $||\mathbf{A}\mathbf{x}|| = ||\mathbf{x}||$ for all $\mathbf{x} \in \mathbb{H}^n$.

Eigenvalue problems for $\mathbf{A} \in \mathbb{H}^{n \times n}$ have to be posed in the form
$$\mathbf{A}\mathbf{x} = \mathbf{x}\lambda \tag{2}$$
and similar matrices have the same set of eigenvalues. The set of eigenvalues is in general not finite. If λ is an eigenvalue, the whole *equivalence class*
$$[\lambda] := \{\sigma \in \mathbb{H} : \sigma = h\lambda h^{-1} \text{ for all } h \in \mathbb{H}\setminus\{0\}\}$$

consists of eigenvalues. The number of different equivalence classes is, however, at most n.

Lemma 1. *Two quaternions λ_1 and λ_2 are members of the same equivalence class if and only if $|\lambda_1| = |\lambda_2|$ and $\operatorname{Re}\lambda_1 = \operatorname{Re}\lambda_2$. As a consequence, two different complex numbers are equivalent if and only if they are conjugate two each other. Two real numbers are equivalent if and only if they coincide.*

Proof: See [3]. ∎

This lemma implies that in any equivalence class $[q]$ of quaternions there is exactly one complex quaternion $\widetilde{q}$ with $\operatorname{Re}\widetilde{q} \geq 0$. This will be called the *complex representative* of $[q]$. If $q = (q_1, q_2, q_3, q_4) \in [q]$, then $\widetilde{q} = (q_1, \sqrt{q_2^2 + q_3^2 + q_4^2}, 0, 0)$ is the complex representative of $[q]$.

We should note here, that Hermitean matrices have only real eigenvalues and that all eigenvalues λ of unitary matrices obey $|\lambda| = 1$.

2 Linear equations in quaternions

With the intention to obtain some insight for a forthcoming study of the multidimensional case, we study here first the simplest case with a linear equation in one variable.

At first, let us recall the following general matrix theorem.

Theorem 1. *Let $\mathbf{A}$ be a real, square matrix with the property*

$$\mathbf{A} + \mathbf{A}^{\mathrm{T}} = 2c\mathbf{I},$$

where $\mathbf{I}$ is the identity matrix of the same size as $\mathbf{A}$ and $c \in \mathbb{R}$. Then,

$$\operatorname{Re}\lambda(\mathbf{A}) = c$$

for all eigenvalues λ of $\mathbf{A}$.

Corollary 1. *Under the assumptions of the previous theorem let $c \neq 0$. Then, $\mathbf{A}$ is nonsingular.*

In the following equation, the two vectors $\mathbf{a} := (a^{(1)}, a^{(2)}, \ldots, a^{(\nu)})$, $\mathbf{b} := (b^{(1)}, b^{(2)}, \ldots, b^{(\nu)}) \in \mathbb{H}^\nu$, $\nu \in \mathbb{N}$, and the *right hand side* $e \in \mathbb{H}$ are given and $x \in \mathbb{H}$ stands for the unknown solution:

$$L^{(\nu)}(x) := \sum_{j=1}^{\nu} a^{(j)} x b^{(j)} = e, \quad e, x \in \mathbb{H}, \ a^{(j)}, b^{(j)} \in \mathbb{H}\backslash\{0\}, \ j = 1, \ldots, \nu. \quad (3)$$

The mapping $L^{(\nu)} : \mathbb{H} \to \mathbb{H}$ is additive, i. e. $L^{(\nu)}(x+y) = L^{(\nu)}(x) + L^{(\nu)}(y)$, but not homogeneous in general, i. e. $L^{(\nu)}(\alpha x) \neq \alpha L^{(\nu)}(x)$ and $L^{(\nu)}(x\alpha) \neq$

$L^{(\nu)}(x)\alpha$, $x \in \mathbb{H}$, $\alpha \in \mathbb{H}$. To call equation (3) a *linear equation* (for a fixed $\nu > 1$) is thus true only in a restricted sense. Without loss of generality, we assume that $a^{(1)} = b^{(\nu)} = 1$. So we have for $\nu = 1, 2, 3$ (simplifying the notation slightly)

$$L^{(1)}(x) := x; \quad L^{(2)}(x) := ax + xb; \quad L^{(3)}(x) := ax + cxd + xb. \quad (4)$$

Since the unknown x resides in the middle between c and d, we will call all terms of the type cxd *middle terms*. The problem $L^{(2)}(x) = e$ was treated, probably for the first time by Johnson in [5]. It is easy to find an explicit solution formula for $L^{(2)}(x) := ax + xb = e$. We assume that neither a nor b is real (including zero). We multiply $ax + xb = e$ from the left by $\bar{a}$ and from the right by $\bar{b}$ and divide by $|a|^2$ ($|b|^2$ would be possible, too). Then we add this equation to the original equation and obtain after some simple algebraic operations

$$\left(2\operatorname{Re} b + a + \frac{|b|^2}{|a|^2}\bar{a}\right)x = e + \frac{\bar{a}e\bar{b}}{|a|^2}.$$

Lemma 2. *Let $a = (a_1, a_2, a_3, a_4)$, $b = (b_1, b_2, b_3, b_4)$. Equation $L^{(2)}(x) = e$ (cf. (4)) has a unique solution for all choices of e if and only if $a_1 + b_1 \neq 0$ or $\sum_{j=2}^{4}(a_j^2 - b_j^2) \neq 0$.*

Proof: For simplicity in the sequel we put

$$s = a + b, \quad d = a - b. \quad (5)$$

Let the j-th component of s, d be denoted by s_j, d_j, respectively, $j = 1, 2, 3, 4$. If we use the multiplication rule (1), then $ax + xb = e$ is equivalent to the real 4×4 system

$$\mathbf{Ax} = \mathbf{e}, \quad \mathbf{A} := \begin{pmatrix} s_1 & -s_2 & -s_3 & -s_4 \\ s_2 & s_1 & -d_4 & d_3 \\ s_3 & d_4 & s_1 & -d_2 \\ s_4 & -d_3 & d_2 & s_1 \end{pmatrix}, \quad (6)$$

where $\mathbf{x}, \mathbf{e} \in \mathbb{R}^4$ have to be identified with $x, e \in \mathbb{H}$, respectively. We compute the determinant of $\mathbf{A}$ and find

$$\det(\mathbf{A}) = s_1^2(s_1^2 + s_2^2 + s_3^2 + s_4^2 + d_2^2 + d_3^2 + d_4^2) + (s_2 d_2 + s_3 d_3 + s_4 d_4)^2. \quad (7)$$

The determinant vanishes if and only if $s_1 := a_1 + b_1 = 0$ and $s_2 d_2 + s_3 d_3 + s_4 d_4 := a_2^2 + a_3^2 + a_4^2 - (b_2^2 + b_3^2 + b_4^2) = 0$. ∎

For solving the system (6), we compute the determinants of the j-th minors $\mathbf{A}_j := \mathbf{A}_{(1:j, 1:j)}$, $j = 1, 2, 3$: $\det(\mathbf{A}_1) = s_1$, $\det(\mathbf{A}_2) = s_1^2 + s_2^2$, $\det(\mathbf{A}_3) = s_1(s_1^2 + s_2^2 + s_3^2 + d_4^2)$. We see that $s_1 \neq 0$ implies that all four determinants (including that of $\mathbf{A}$) do not vanish, which implies that Gauss' elimination process can be carried out without pivoting. If however, $s_1 = 0$, the first and third minor have a vanishing determinant. In this case pivoting is necessary.

Corollary 2. *In the previous lemma let (i) $a = b$. The equation $L^{(2)}(x) = e$ (see (4)) has a unique solution if and only if $a_1 = \operatorname{Re} a \neq 0$. If $a_1 = 0$ but $a \neq 0$ the kernel of $\mathbf{A}$ is a two dimensional subspace of $\mathbb{R}^4$. (ii) Let $|a| = |b| \neq 0$. In this case $\mathbf{A}$, defined in (6) is singular if and only if $s_1 = a_1 + b_1 = 0$. If $s_1 = 0$ the kernel of $\mathbf{A}$ is a two dimensional subspace of $\mathbb{R}^4$, provided $a, b \notin \mathbb{R}$.*

Proof: We use formula (7). (i) In this case $d := a - b = 0$ implies $\det(\mathbf{A}) = 16a_1^2 |a|^2$, where $\mathbf{A}$ is defined in (6). If $a_1 = 0$, then

$$\mathbf{A} = 2 \begin{pmatrix} 0 & -a_2 & -a_3 & -a_4 \\ a_2 & 0 & 0 & 0 \\ a_3 & 0 & 0 & 0 \\ a_4 & 0 & 0 & 0 \end{pmatrix}. \tag{8}$$

(ii) In this case we have $\det(\mathbf{A}) = 4|a|^2 (a_1 + b_1)^2$ and $s_1 = 0$ implies

$$\mathbf{A} = \begin{pmatrix} 0 & -s_2 & -s_3 & -s_4 \\ s_2 & 0 & -d_4 & d_3 \\ s_3 & d_4 & 0 & -d_2 \\ s_4 & -d_3 & d_2 & 0 \end{pmatrix}. \tag{9}$$

Since $\mathbf{A} + \mathbf{A}^T = \mathbf{0}$, Theorem 1 implies that all four eigenvalues of $\mathbf{A}$ have vanishing real part. Since the eigenvalues appear pairwise conjugate, the rank of $\mathbf{A}$ is either 0, 2 or 4. If $\operatorname{rank}(\mathbf{A}) = 0$, then, $a_j = b_j = 0, j = 2, 3, 4$ and $a, b \in \mathbb{R}$. The case $\operatorname{rank}(\mathbf{A}) = 4$ was already excluded. A formula for the two dimensional kernel was given in [3]. ∎

The matrix $\mathbf{A}$ of (6) has the property that $\mathbf{A} + \mathbf{A}^T = 2(a_1 + b_1)\mathbf{I}$, where $\mathbf{I}$ is the (4×4) identity matrix. Theorem 1 implies that all eigenvalues of $\mathbf{A}$ have the same real part $a_1 + b_1$ and it implies (Corollary 1) that $a_1 + b_1 \neq 0$ is a sufficient condition for $\mathbf{A}$ being non singular.

We will develop a simple iterative algorithm for solving $L^{(2)}(x) := ax + xb = e$ in the original form under the assumption that both $a \neq 0, b \neq 0$. If $a = 0$ or $b = 0$, then finding the solution is trivial. We form two fixed point equations by multiplying $L^{(2)}(x) := ax + xb = e$ from the left by a^{-1} and another one by multiplying from the right by b^{-1}. This yields

$$T_1(x) := a^{-1}(e - xb) = x, \quad T_2(x) := (e - ax)b^{-1} = x. \tag{10}$$

Lemma 3. *Let $L^{(2)}(x) := ax + xb = e$ have a unique solution $\hat{x}$, regardless of the choice of e. Let $a \neq 0, b \neq 0$. If (i) $|a| > |b|$, let $q := |b|/|a| < 1$. The fixed point equation $T_1(x) = x$ is contractive and the sequence $\{x_j\}$ defined by $x_{j+1} := T_1(x_j), j = 0, 1, \ldots$ converges with geometric speed to the solution $\hat{x}$ regardless of the choice of the initial guess x_0. There is the error estimate*

$$|\hat{x} - x_j| \leq \min\left\{\frac{q^j}{1-q}|x_1 - x_0|, \frac{q}{1-q}|x_j - x_{j-1}|\right\}, j \geq 1.$$

If (ii) $|a| < |b|$, *let* $q := |a|/|b|$. *Then, the same is true for* T_2.

Proof: We treat the case (i). Then, $|T_1(x) - T_1(y)| = |a^{-1}(y-x)b| = |a^{-1}||b||x-y| = q|x-y|$. The remaining part follows from standard arguments. Case (ii) is analogue. ∎

Example 1. Take $a := (-2, -4, 7, -10)$, $b := (5, 9, 10, 6)$, $e := (-1, 0, -6, 3)$. Then, $q := |a|/|b| = 0.8357$, $\hat{x} = (-0.02825\,79443\,1218, 0.52768\,86450\,6780, -0.04595\,79753\,6487, 0.23548\,28692\,6819)$. Iteration with T_2 yields $|\hat{x} - x_{100}| \approx 9.3 \cdot 10^{-9}$ with error estimate $|\hat{x} - x_{100}| \leq 4.2 \cdot 10^{-8}$.

If $|a| = |b|$ and $a_1 + b_1 \neq 0$ the above iterations will in general not converge. If $|a|, |b|$ are different but close together, the convergence will be very slow.

We turn now to the case $L^{(3)}(x) = e$ where $L^{(3)}$ is defined in (4). Put $c = (c_1, c_2, c_3, c_4)$, $d = (d_1, d_2, d_3, d_4)$, $x = (x_1, x_2, x_3, x_4)$ and identify the column vector $\mathbf{x}$ with x. Then, the middle term cxd can be expressed as

$$cxd = \mathbf{M}\mathbf{x}, \quad \text{where}$$

$$\mathbf{M} := \begin{pmatrix} c_1d_1 - c_2d_2 - c_3d_3 - c_4d_4 & -c_1d_2 - c_2d_1 + c_3d_4 - c_4d_3 & -c_1d_3 - c_2d_4 - c_3d_1 + c_4d_2 & -c_1d_4 + c_2d_3 - c_3d_2 - c_4d_1 \\ c_1d_2 + c_2d_1 + c_3d_4 - c_4d_3 & c_1d_1 - c_2d_2 + c_3d_3 + c_4d_4 & c_1d_4 - c_2d_3 - c_3d_2 - c_4d_1 & -c_1d_3 - c_2d_4 + c_3d_1 - c_4d_2 \\ c_1d_3 - c_2d_4 + c_3d_1 + c_4d_2 & -c_1d_4 - c_2d_3 - c_3d_2 + c_4d_1 & c_1d_1 + c_2d_2 - c_3d_3 + c_4d_4 & c_1d_2 - c_2d_1 - c_3d_4 - c_4d_3 \\ c_1d_4 + c_2d_3 - c_3d_2 + c_4d_1 & c_1d_3 - c_2d_4 - c_3d_1 - c_4d_2 & -c_1d_2 + c_2d_1 - c_3d_4 - c_4d_3 & c_1d_1 + c_2d_2 + c_3d_3 - c_4d_4 \end{pmatrix} \in \mathbb{R}^{4 \times 4}.$$

Let us remark that $\det \mathbf{M} = |c|^2|d|^2 \neq 0$ if both $c \neq 0$ and $d \neq 0$. If $c = 0$ or $d = 0$, we come back to $L^{(2)}(x) = e$.

With $\mathbf{A}$ from (5), (6), the final (4×4) system has the form

$$(\mathbf{A} + \mathbf{M})\mathbf{x} = \mathbf{e}.$$

Under the assumption that all a, b, c and d are nonzero quaternions, the matrix $\mathbf{A} + \mathbf{M}$ is regular (we proved it by making use of Maple).

Similarly as in the previous case, we form this time three fixed point equations: we multiply $L^{(3)}(x) := ax + cxd + xb = e$ from the left by a^{-1} or multiply $L^3(x)$ from the right by b^{-1}. The last equation we obtain by multiplying the equation $cxd = e - ax - xb$ from the left by c^{-1} and from the right by d^{-1}. This yields

$$T_1(x) := a^{-1}(e - cxd - xb) = x, \quad T_2(x) := (e - cxd - ax)b^{-1} = x, \quad (11)$$

$$T_3(x) := c^{-1}(e - ax - xb)d^{-1} = x. \tag{12}$$

Lemma 4. *Let $L^{(3)}(x) := ax + cxd + xb = e$ have a unique solution $\hat{x}$, regardless of the choice of e. Let $a \neq 0$, $b \neq 0$, $c \neq 0$, $d \neq 0$. If (i) $|a| > |b|$ and $|c||d| < |a| - |b|$, let $q := \dfrac{|c||d| + |b|}{|a|} < 1$. The fixed point equation $T_1(x) = x$ is contractive and the sequence $\{x_j\}$ defined by $x_{j+1} := T_1(x_j)$, $j = 0, 1, \ldots$ converges with geometric speed to the solution $\hat{x}$ regardless of the choice of the initial guess x_0. If (ii) $|a| < |b|$ and $|c||d| < |b| - |a|$, let $q := \dfrac{|c||d| + |b|}{|a|} < 1$. Then, the same is true for T_2. If (iii) $|a| + |b| < |c||d|$, let $q := \dfrac{|a| + |b|}{|c||d|} < 1$. Then the fixed point equation $T_3(x) = x$ is contractive and the sequence $\{x_j\}$ defined by $x_{j+1} := T_3(x_j)$, $j = 0, 1, \ldots$ converges to the solution $\hat{x}$ regardless of the choice of the initial guess x_0.*
In all three cases, the error estimate is

$$|\hat{x} - x_j| \leq \min\left\{\frac{q^j}{1-q}|x_1 - x_0|, \frac{q}{1-q}|x_j - x_{j-1}|\right\}, j \geq 1.$$

Proof: We treat the case (iii). Then,

$$|T_3(x) - T_3(y)| = |c^{-1}(a(y-x) + (y-x)b)d^{-1}| = |c|^{-1}|a(y-x) + (y-x)b| \leq$$

$$\leq |c|^{-1}|d|^{-1}(|a(y-x)| + |(y-x)b|) = |c|^{-1}|d|^{-1}(|a||y-x| + |b||y-x|) =$$

$$= |c|^{-1}|d|^{-1}(|a| + |b|)|y-x| = q|x-y|.$$

The remaining part follows from standard arguments. Case (i) and (ii) is analogue to the proof of Lemma 3. ∎

Let us remark that the equation $ax + cxd + xb = e$ can be also transformed into the system of two equations for two unknown quaternions by introducing a new variable $u = cxd$:

$$\begin{aligned} ax + u + xb &= e \\ c^{-1}u - xd &= 0. \end{aligned} \tag{13}$$

For the general case of $L^{(\nu)}(x) = e$ introduced in (3) all middle terms define a matrix $\mathbf{M}_j$ of exactly the form of $\mathbf{M}$ so that the general case expressed as real equivalent is of the form

$$(\mathbf{A} + \sum_{j=2}^{\nu-1} \mathbf{M}_j)\mathbf{x} = \mathbf{e}. \tag{14}$$

The general case can also be transformed into a system of $\nu - 1$ equations in $\nu - 1$ unknowns by putting $u_j := a^{(j)}xb^{(j)}$, $j = 2, 3, \ldots, \nu - 1$. The system has then, the following form (assuming $a^{(1)} = b^{(\nu)} = 1$):

$$\begin{aligned} xb^{(1)} + u_2 + u_3 + \cdots + u_{\nu-1} + a^{(\nu)}x &= e, \\ (a^{(2)})^{-1}u_2 - xb^{(2)} &= 0, \\ (a^{(3)})^{-1}u_3 - xb^{(3)} &= 0, \\ &\vdots \\ (a^{(\nu-1)})^{-1}u_{\nu-1} - xb^{(\nu-1)} &= 0. \end{aligned} \tag{15}$$

We have multiplications from the left and from the right, but there are no middle terms with multiplication from the left and from the right, simultaneously.

Let us remark that the situation is more complicated in the general case of linear systems in quaternions. For example we have to define multiplication of the matrix by a quaternion from the left, to introduce left and right eigenvalues, etc. The aim of our future work will be to develop an algorithm (similar to the elimination procedure) for solving these systems.

Acknowledgment

The authors acknowledge with pleasure the support of the Grant Agency of the Czech Republic (grant No. 201/06/0356). The work is a part of the research project MSM 6046137306 financed by MSMT, Ministry of Education, Youth and Sports, Czech Republic.

References

1. Dongarra, J.J., Gabriel, J.R., Koelling, D.D., Wilkinson, J.H.: Solving the secular equation including spin orbit coupling for systems with inversion and time reversal symmetry. J. Comput. Phys., **54**, 278–288 (1984)
2. Dongarra, J.J., Gabriel, J.R., Koelling, D.D., Wilkinson, J.H.: The eigenvalue problem for hermitian matrices with time reversal symmetry. Linear Algebra Appl., **60**, 27–42 (1984)
3. Janovská, D., Opfer, G.: Givens' Transformation applied to Quaternion Valued Vectors. BIT Numerical Mathematics, **43**, 991–1002 (2003) [typographical error in title caused by publisher]
4. Janovská, D., Opfer, G.: Givens' reduction of a quaternion-valued matrix to upper Hessenberg form. In: Numerical Mathematics and Advanced Applications, ENUMATH 2003. Springer Verlag Berlin Heidelberg, 510–520 (2004)
5. Johnson, R.E: On the equation $\chi\alpha = \gamma\chi + \beta$ over an algebraic division ring. Bull. Amer. Math. Soc., **50**, 202–207 (1944)
6. Kuipers, J.B.: Quaternions and Rotation Sequences, a Primer with Applications to Orbits, Aerospace, and Virtual Reality. Princeton University Press, Princeton, NJ (1999)

7. Opfer, G.: The conjugate gradient algorithm applied to quaternion-valued matrices. ZAMM, **85**, 660–672 (2005)
8. Rösch, N.: Time-reversal symmetry, Kramers' degeneracy and the algebraic eigenvalue problem. Chemical Physics, **80**, 1–5 (1983)
9. Zhang, F.: Quaternions and matrices of quaternions. Linear Algebra Appl., **251**, 21–57 (1997)

Computing the Analytic Singular Value Decomposition via a Pathfollowing

Vladimír Janovský[1], Drahoslava Janovská[2] and Kunio Tanabe[3]

[1] Charles University, Faculty of Mathematics and Physics, Prague
   janovsky@karlin.mff.cuni.cz
[2] Institute of Chemical Technology, Prague
   janovskd@vscht.cz
[3] The Institute of Statistical Mathematics, Tokyo
   tanabe@ism.ac.jp

Summary. The aim is to compute ASVD for large sparse matrices. In particular we will consider branches of selected singular values and the corresponding left/right singular vectors. We apply a predictor-corrector algorithm with an adaptive stepsize control.

1 Introduction

A singular value decomposition (SVD) of a real matrix $A \in \mathbb{R}^{m \times n}$, $m \geq n$, is a factorization $A = U\Sigma V^T$, where $U \in \mathbb{R}^{m \times m}$ and $V \in \mathbb{R}^{n \times n}$ are orthogonal matrices and $\Sigma = \mathrm{diag}(s_1, \ldots, s_n) \in \mathbb{R}^{m \times n}$. The values s_i, $i = 1, \ldots, n$, are called singular values. They may be defined to be nonnegative and to be arranged in nonincreasing order, see [3].

Let A depend smoothly on a parameter $t \in \mathbb{R}$, $t \in [a, b]$. The aim is to construct a path of SVD's

$$A(t) = U(t)\Sigma(t)V(t)^T, \tag{1}$$

where $U(t)$, $\Sigma(t)$ and $V(t)$ depend smoothly on $t \in [a, b]$. If A is a real analytic matrix function on $[a, b]$, see [5], then there exists *Analytic Singular Value Decomposition* (ASVD), see [1]: There exists a factorization (1) that *interpolates* classical SVD defined at $t = a$ i.e.,

- the factors $U(t)$, $V(t)$ and $\Sigma(t)$ are real analytic on $[a, b]$,
- for each $t \in [a, b]$, both $U(t) \in \mathbb{R}^{m \times m}$ and $V(t) \in \mathbb{R}^{n \times n}$ are orthogonal matrices and $\Sigma(t) = \mathrm{diag}(s_1(t), \ldots, s_n(t)) \in \mathbb{R}^{m \times n}$ is a diagonal matrix,
- at $t = a$, the matrices $U(a)$, $\Sigma(a)$ and $V(a)$ are the factors of the classical SVD of the matrix $A(a)$.

Diagonal entries $s_i(t) \in \mathbb{R}$ of $\Sigma(t)$ are called *singular values*. Due to the requirement of smoothness, singular values may be negative and also their ordering may by arbitrary. Under certain assumptions, ASVD may be uniquely determined by the factors at $t = a$. For a theoretical background, see [4]. As far as the computation is concerned, an incremental technique is proposed in [1]: Given a point on the path, one computes a classical SVD for a neighboring parameter value. Next, one computes permutation matrices which link this result to the next point on the path. The procedure is approximative with a local error of order $O(h^2)$, where h is the step size.

An alternative technique for computing ASVD is presented in [7, 8]: A non-autonomous vector field $H : \mathbb{R} \times \mathbb{R}^N \to \mathbb{R}^N$ of a huge dimension $N = n + n^2 + m^2$ can be constructed in such a way that the solution of the initial value problem for the system $x' = H(t, x)$ is linked to the path of ASVD. Moreover, [7] contributes to the analysis of *non-generic points*, see [1], of the ASVD path. These points can be, in fact, interpreted as singularities of the vector field $\mathbb{R}^N$.

In [6], two methods for computing ASVD are presented and compared. The first one modifies the technique of [1]. The second method in [6] consists in solving ODE as in [7] but it uses an implicit integration technique. The comparison clearly prefers the former class of methods: The ODE integration, in spite of using an implicit scheme, lacks the precision.

Our aim is to present a new technique for computing ASVD. We formulate ASVD as a pathfollowing problem and apply a classical predictor–corrector algorithm. The main trick of our implementation consists in an efficient application of least–squares for solving of the corrector step (Section 3).

2 Formulation of the problem

As a preliminary, let us recall the notion of singular value of a matrix $A \in \mathbb{R}^{m \times n}$, $m \geq n$:

Definition 1. *We say that $s \in \mathbb{R}$ is a singular value of the matrix A if there exist $u \in \mathbb{R}^m$ and $v \in \mathbb{R}^n$ such that*

$$Av - su = 0, \quad A^T u - sv = 0, \quad \|u\| = \|v\| = 1. \qquad (2)$$

The vectors v and u are called the right and the left singular vectors of the matrix A.

Note that s is defined up to its sign: if the triplet (s, u, v) satisfies (2) then at least three more triplets

$$(s, -u, -v), \quad (-s, -u, v), \quad (-s, u, -v),$$

can be interpreted as singular values, left and right singular vectors of A.

Definition 2. *We say that $s \in \mathbb{R}$ is a simple singular value of a matrix A if there exist $u \in \mathbb{R}^m$ and $v \in \mathbb{R}^n$ such that*

$$(s, u, v), \quad (s, -u, -v), \quad (-s, -u, v), \quad (-s, u, -v)$$

are, for the given s, the only solutions to (2). *A singular value s which is not a simple singular value is called nonsimple singular value.*

Remark 1. $s = 0$ is a simple singular value of A if and only if $m = n$ and $\dim \operatorname{Ker} A = 1$.

Remark 2. Let s_i, s_j, $s_i \neq s_j$, be simple singular values of A. Then $s_i \neq -s_j$.

We will consider branches of *selected* singular values and corresponding left/right singular vectors $s_i(t)$, $U_i(t) \in \mathbb{R}^m$, $V_i(t) \in \mathbb{R}^n$:

$$A(t)V_i(t) = s_i(t)U_i(t), \quad A(t)^T U_i(t) = s_i(t)V_i(t),$$
$$U_i(t)^T U_i(t) = V_i(t)^T V_i(t) = 1$$

for $t \in [a, b]$. We will add the natural orthogonality conditions $U_i(t)^T U_j(t) = V_i(t)^T V_j(t) = 0$, $i \neq j$, $t \in [a, b]$. We are interested in p, $p \leq n$, selected singular values $S(t) = (s_1(t), \ldots, s_p(t)) \in \mathbb{R}^p$, and in the corresponding left/right singular vectors $U(t) = [U_1(t), \ldots, U_p(t)] \in \mathbb{R}^{m \times p}$, $V(t) = [V_1(t), \ldots, V_p(t)] \in \mathbb{R}^{n \times p}$ as $t \in [a, b]$.

In the operator setting, let

$$F : \mathbb{R} \times \mathbb{R}^p \times \mathbb{R}^{m \times p} \times \mathbb{R}^{n \times p} \to \mathbb{R}^{m \times p} \times \mathbb{R}^{n \times p} \times \mathbb{R}^{p \times p} \times \mathbb{R}^{p \times p} \quad (3)$$

be defined as

$$F(t, X) \equiv \left(A(t)V - U\Sigma, A^T(t)U - V\Sigma, U^T U - I, V^T V - I \right), \quad (4)$$

where $X \equiv (S, U, V) \in \mathbb{R}^p \times \mathbb{R}^{m \times p} \times \mathbb{R}^{n \times p}$ and $\Sigma = \operatorname{diag}(S)$. Under certain assumptions, the set of *overdetermined nonlinear equations*

$$F(t, X) = 0 \quad (5)$$

implicitly defines a curve in $\mathbb{R} \times \mathbb{R}^N$, where $\mathbb{R}^N$, $N = p(1 + m + n)$, and $\mathbb{R}^p \times \mathbb{R}^{m \times p} \times \mathbb{R}^{n \times p}$ are isomorphic. The image of F, namely $\mathbb{R}^{m \times p} \times \mathbb{R}^{n \times p} \times \mathbb{R}^{p \times p} \times \mathbb{R}^{p \times p}$, and $\mathbb{R}^M$, $M = p(m + n + 2p)$, are isomorphic.

The curve (5) can be parameterized by t i.e., $t \mapsto X(t) = (S(t), U(t), V(t))$ so that $F(t, X(t)) = 0$ as $t \in [a, b]$. Given a solution $X(t)$ at $t = a$, the curve is initialized. For this purpose, we may select p singular values and left/right singular vectors computed via the classical SVD of the matrix $A(a)$.

We have in mind mainly the application when $m \geq n$, n is large while p is comparatively small. We also want to exploit sparsity of $A(t)$ as $t \in [a, b]$.

We will apply *tangent continuation*, see [2], Algorithm 4.25, p.107. It is a predictor-corrector algorithm with an adaptive stepsize control. As far as the implementation is concerned, the corrector is crucial. We will discuss it in Sect. 3. This section represents the actual contribution of the paper. An example of the performance is given in Sect. 4. In Sect. 5 we will comment on the code and give some comparisons.

3 Solving defining equations

The role of our corrector is to find a root of $F(t, X) = 0$ for a fixed t. We intend to use Newton's method to find the root. In order to simplify notation, we think of F with a fixed t to be a mapping $F(t, \cdot) : \mathbb{R}^N \to \mathbb{R}^M$, where $N = p(1 + m + n)$ and $M = p(m + n + 2p)$. Let $G(t, X)$ be the partial differential $F_X(t, X)$. It is a linear operator $G(t, X) : \mathbb{R}^N \to \mathbb{R}^M$.

In order to find roots of $F(t, \cdot)$, we consider Gauss-Newton method for nonlinear least-squares problem namely, we define

$$X^\star = \arg \min_{X \in \mathbb{R}^N} \|F(t, X)\|_2^2 \,, \tag{6}$$

see [2] p.92, as a local minimizer on $\mathbb{R}^N$; $\| \; \|_2$ is the Euclidean norm on $\mathbb{R}^N$. The method approximates $X^\star$ by a sequence $\{X^{(j)}\}_{i=0}^\infty$ of $X^{(j)} \in \mathbb{R}^N$, which is defined by the recurrence

$$G(t, X^{(j)})^T G(t, X^{(j)}) \, \delta X = -G(t, X^{(j)})^T F(t, X^{(j)}) \,, \tag{7}$$
$$X^{(j+1)} = X^{(j)} + \delta X \,. \tag{8}$$

Solving the equation (7) for $\delta X \in \mathbb{R}^N$ represents a linear least-squares problem.

We say that the root $X = (S, U, V)$ of $F(t, \cdot)$ is *simple* provided that the differential $G(t, X)$ at X has full rank i.e., $\text{rank}(G(t, X)) = N$.

Theorem 1. *Let* $X^\star = (S, U, V)$, $S = (s_1, \ldots, s_p)$, *be a root of* $F(t, \cdot)$. *Then* $\text{rank}(G(t, X^\star)) = N$ *if and only if all singular values of* $A(t)$ *are simple (i.e., s_i is a simple singular value of $A(t)$ for each $i = 1, \ldots, p$.)*

Corollary 1. *If* $X^\star \in \mathbb{R}^N$ *is a simple root of* $F(t, \cdot)$ *then the iterations* (7)& (8) *are locally convergent. The rate of convergence is quadratic.*

Algorithms for solving linear least-squares problems i.e., the problem (7), are well known: let us quote Normal equation approaches in [3], namely the algorithms `CGNR` p.545 and `CGNE` p.546.

The matrix $A(t)$ was assumed to be sparse. Are the matrices $G(t, X) : \mathbb{R}^N \to \mathbb{R}^M$ and $G^T(t, X) : \mathbb{R}^M \to \mathbb{R}^N$ sparse? Actually, these matrices are not available in *cartesian coordinates* on $\mathbb{R}^N$ and $\mathbb{R}^M$. They are defined by actions as linear operators

$$G(t, \cdot) : \mathbb{R}^p \times \mathbb{R}^{m \times p} \times \mathbb{R}^{n \times p} \to \mathbb{R}^{m \times p} \times \mathbb{R}^{n \times p} \times \mathbb{R}^{p \times p} \times \mathbb{R}^{p \times p} \tag{9}$$

and

$$G^*(t,\cdot) : \mathbb{R}^{m\times p} \times \mathbb{R}^{n\times p} \times \mathbb{R}^{p\times p} \times \mathbb{R}^{p\times p} \to \mathbb{R}^p \times \mathbb{R}^{m\times p} \times \mathbb{R}^{n\times p}; \qquad (10)$$

$G^*(t,\cdot)$ is the relevant dual to $G(t,\cdot)$.

In particular, the action of $G(t,\cdot)$ is defined as

$$\delta S \in \mathbb{R}^p, \ \delta U \in \mathbb{R}^{m\times p}, \ \delta V \in \mathbb{R}^{n\times p} \longmapsto G = (G_1, G_2, G_3, G_4),$$

where

$$G_1 \equiv A\,\delta V - \delta U\,\Sigma - U\,\delta\Sigma,$$
$$G_2 \equiv A^T\,\delta U - \delta V\,\Sigma - V\,\delta\Sigma,$$
$$G_3 \equiv \delta U^T\,U + U^T\,\delta U,$$
$$G_4 \equiv \delta V^T\,V + V^T\,\delta V,$$

$\Sigma = \mathrm{diag}(S)$, $\delta\Sigma = \mathrm{diag}(\delta S)$ and $A = A(t)$.

The action of the dual is defined as

$$R \in \mathbb{R}^{m\times p}, \ Y \in \mathbb{R}^{n\times p}, \ W \in \mathbb{R}^{p\times p}, \ Z \in \mathbb{R}^{p\times p} \longmapsto G^* = (G_1^*, G_2^*, G_3^*),$$

where

$$G_1^* \equiv -\Big(\sum_{k=1}^m u_{k1} r_{k1} + \sum_{k=1}^n v_{k1} y_{k1}, \ldots, \sum_{k=1}^m u_{kp} r_{kp} + \sum_{k=1}^n v_{kp} y_{kp}\Big)^T,$$
$$G_2^* \equiv -R\Sigma^T + AY + U(W^T + W),$$
$$G_3^* \equiv -Y\Sigma^T + A^T R + V(Z^T + Z);$$

u_{kj}, r_{kj}, v_{kj} and y_{kj} are the relevant elements of matrices U, R, V and Y, and $\Sigma = \mathrm{diag}(S)$, $A = A(t)$.

Note that the actions of both $G(t,\cdot)$ and $G^*(t,\cdot)$ are composed from the actions of $A(t)$ and $A^T(t)$ on vectors from $\mathbb{R}^n$ and $\mathbb{R}^m$. Therefore, we may use the assumption that $A(t)$ and $A^T(t)$ are sparse when *evaluating* these actions. Observe that in the algorithms like CGNR or CGNE one needs to define just the **action** of $G(t,X)$ or $G^T(t,X)$ on a righthand side. In our code, we have used MATLAB-function LSQR, see MATLAB Function Reference, which is a modification of CGNE. One of the options is that you may define $G(t,X)$ and $G^T(t,X)$ by actions.

4 Experiments

We considered the homotopy

$$A(t) = t\,A2 + (1-t)\,A1, \quad t \in [0,1],$$

where the matrices $A1 \equiv$ well1033.mtx, $A2 \equiv$ illc1033.mtx were taken over from http://math.nist.gov/MatrixMarket/. Note that $A1, A2 \in \mathbb{R}^{1033 \times 320}$ are sparse, $A1$ and $A2$ are well and ill-conditioned, respectively.

The aim was to continue

- 10 largest singular values, left/right singular vectors of $A(t)$,
- 7 smallest singular values, left/right singular vectors of $A(t)$.

The continuation was initialized at $t = 0$: The initial decomposition of $A1$ was computed via SVDS, see MATLAB Function Reference.

Results of the continuation procedure are shown on Fig. 1 and Fig. 2. The relevant zooms, see Fig. 3, illustrate the stepsize control of the algorithm. Moreover, the zoom on the right shows quite satisfactory resolution of the cluster of small singular values (the scale 10^{-3}). Note that at the last continuation step the corrector failed to converge: The action of predictor is shown instead.

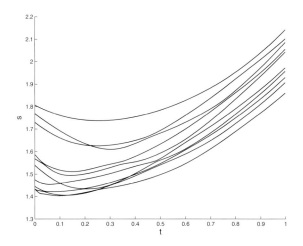

Fig. 1. Ten largest singular values s versus parameter t

The most time consuming part of the algorithm is the inner loop which consists of solving the linear least-squares problem (7)&(8). For example, consider continuation of ten largest singular values. If we introduce the natural cartesian coordinates (which we actually do not have to), the relevant matrix $G(t, \cdot) \in \mathbb{R}^{M \times N}$ has the size $M = 13730$, $N = 13530$. For sparsity pattern of this matrix family, see Fig. 4 as an example. If we solve (7)&(8), we need typically 250 iterations to reach precision 10^{-8}.

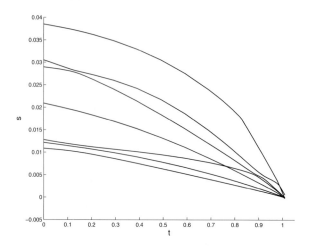

Fig. 2. Seven smallest singular values s versus parameter t

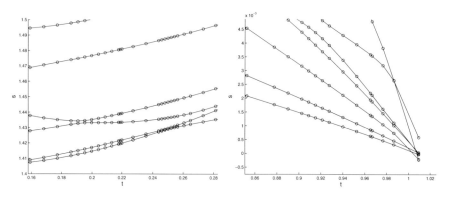

Fig. 3. Zooms: 10 largest s vs t, 7 smallest s vs t

5 Conclusions

We assume that the continuation is initialized at simple singular values i.e., $S(a) = (s_1(a), \ldots, s_p(a))$, $s_i(a)$ are simple for $i = 1, \ldots, p$. The branch $X(t)$ may break at that parameter value $t > a$ when a nonsimple singular value $s_i(t)$ turns up. In practice, the continuation may get stuck. Using an extrapolation strategy, the code incorporates an "early warning" of such a value of t. We try to go on starting from $t + \delta t$, where δt is a small positive increment. The initial value of $X(t + \delta t)$ is again obtained from an extrapolation and a simple asymptotic analysis. The mentioned extrapolation strategy is based on the assumption of a generic scenario that the branches $t \longmapsto s_i(t)$ of singular values, $i = 1, \ldots, n$, may intersect at isolated points only namely, at the points where $s_i(t) = s_j(t)$ or $s_i(t) = -s_j(t)$ for $i \neq j$, see [1], p 8. See also [7].

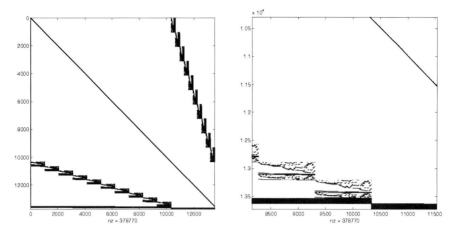

Fig. 4. Sparsity pattern of $G(1/2,\cdot)$ in natural cartesian coordinates with a zoom

The presented continuation algorithm could be linked to [7]. Actually, our predictor can be interpreted as Euler method applied on a vector field. This vector field differs from the one of [7], see Sect. 1. We believe that our construction is more robust since it is based on least-squares. Nevertheless, the main advantage is the corrector: You may obtain $X(t)$ with a prescribed precision. Finally, we can exploit the sparse structure of large matrices. For the techniques from [1] it is questionable. In [1], a stepsize control is also a problem.

Acknowledgments

The authors acknowledge with pleasure the support of the Grant Agency of the Czech Republic (grant No. 201/06/0356). The work is a part of the research projects MSM 0021620839 and MSM 6046137306 financed by MSMT, Ministry of Education, Youth and Sports, Czech Republic.

References

1. Bunse-Gerstner, A., Byers, R., Mehrmann, V., Nichols, N.K.: Numerical Computation of an Analytic Singular Value Decomposition of a Matrix Valued Function. Numer. Math., **60**, 1–39 (1991)
2. Deuflhart, P., Hohmann, A.: Numerical Analysis in Modern Scientific Computing. An Introduction. Springer Verlag, New York (2003)
3. Golub, G.H., van Loan, C.F.: Matrix Computations, 3rd ed. The Johns Hopkins University Press, Baltimore (1996)
4. Kato, T.: Perturbation Theory for Linear Operators, 2nd ed. Springer Verlag, New York (1976)

5. Krantz, S., Parks, H.: A Primer of Real Analytic Functions. Birkhauser, New York (2002)
6. Mehrmann, V., Rath, W.: Numerical Methods for the Computation of Analytic Singular Value Decompositions. Electronic Transactions on Numerical Analysis, **1**, 72–88 (1993)
7. Wright, K.: Differential equations for the analytic singular value decomposion of a matrix. Numer. Math. **63**, 283–295 (1992)
8. Wright, K.: Numerical solution of differential equations for the analytic singular value decomposion. In: Bainov, D. and Covachev, V. (eds) Proceedings of the 1st International Colloquium on Numerical Analysis, Plovdiv, Bulgaria, 1992. VSP, Utrecht, 131–140 (1993)

A Jacobi-Davidson Method for Computing Partial Generalized Real Schur Forms

Tycho van Noorden[1] and Joost Rommes[2]

[1] Department of Mathematics and Computer Science, Eindhoven University of Technology, P.O. Box 513, 5600 MB, Eindhoven, The Netherlands
t.l.v.noorden@tue.nl
[2] Department of Mathematics, Utrecht University, Budapestlaan 6, 3584 CD, Utrecht, The Netherlands
rommes@math.uu.nl

Summary. In this paper, a new variant of the Jacobi-Davidson method is presented that is specifically designed for *real unsymmetric* matrix pencils. Whenever a pencil has a complex conjugated pair of eigenvalues, the method computes the two dimensional real invariant subspace spanned by the two corresponding complex conjugated eigenvectors. This is beneficial for memory costs and in many cases it also accelerates the convergence of the JD method. In numerical experiments, the RJDQZ variant is compared with the original JDQZ method.

1 Introduction

Real unsymmetric matrices or real unsymmetric matrix pencils may have complex eigenvalues and corresponding eigenvectors. Therefore, the (partial generalized) Schur form may consist of complex matrices. In some situations [1], it is more desirable to compute a real (partial generalized) Schur form. For a matrix, this decomposition consists of an orthogonal real matrix and block upper triangular matrix, which has scalars or two by two blocks on the diagonal. The eigenvalues of such a two by two block correspond to two complex conjugated eigenvalues of the matrix (pencil) itself. Advantages of the real Schur form are that it requires less storage and that complex conjugated pairs of eigenvalues always appear together.

In this paper, a variant of the JDQZ method [5] is considered for the computation of a partial generalized real Schur form of a large matrix pencil. The original JDQZ method does not use the fact that the pencil is real: (1) it does not exploit the fact that eigenvalues are real or appear in complex conjugated pairs and (2) it needs complex arithmetic, even when only real eigenvalues appear. This is in contrast with other iterative eigenvalue solvers such as the Arnoldi method.

Algorithm 4.1 proposed in [2] solves the problem (1). This algorithm consists of an outer iteration in which the partial Schur form is expanded by a scalar block whenever the inner iteration, which consists of the Jacobi-Davidson method, returns a real eigenvalue, and with a two by two block if the inner iteration returns an eigenvalue with non-zero imaginary part. The algorithm does not solve problem (2): the inner iteration still needs complex arithmetic, even when only real eigenvalues are computed.

The variant of the Jacobi-Davidson method that is implemented in this paper does take into account in the inner iteration that either a real eigenvalue and eigenvector are computed, or a two dimensional invariant subspace corresponding to a pair of complex conjugated eigenvalues.

Other papers that mention real variants of the JD method are [3] and [5].

2 The RJDQZ method for real matrix pencils

A *generalized partial real Schur form* [5, 6] of the large real unsymmetric matrix pencil (A, B) is a decomposition of the form

$$AQ_k = Z_k S_k, \quad BQ_k = Z_k T_k, \qquad (1)$$

where Q_k and Z_k are *real* matrices with orthonormal columns, and S_k and T_k are *real* block upper triangular matrices with scalar or two by two diagonal blocks. The eigenvalues of the two by two diagonal blocks correspond to complex conjugated pairs of eigenvalues of the pencil (A, B).

Suppose that a partial generalized real Schur form (1) is computed already. It has been shown [5] that the following Schur vector q_{k+1} and the corresponding eigenvalue μ_{k+1} satisfy $Q_k^* q_{k+1} = 0$ and

$$(I - Z_k Z_k^*)(A - \mu_{k+1} B)(I - Q_k Q_k^*) q_{k+1} = 0. \qquad (2)$$

Note that this is again a generalized eigenvalue problem. Thus in order to build a Schur form, one has to solve a number of eigenvalue problems of the form (2).

The JDQZ method [5] uses the Jacobi-Davidson method to solve these eigenvalue problems: the JD method iteratively computes approximations to eigenvalues, and their corresponding eigenvectors, that are close to some specified target τ, of a generalized unsymmetric eigenvalue problem

$$Aq = \mu Bq, \qquad (3)$$

where A and B are in general unsymmetric $n \times n$ matrices. In each iteration, a search subspace colspan(V) and a test subspace colspan(W) are constructed. V and W are complex $n \times j$ matrices with $j \ll n$ and have orthonormal columns such that $V^*V = W^*W = I$. In the first part of a JD iteration, an approximation to an eigenvector of the generalized eigenvalues problem (3) is obtained from the projected eigenvalue problem

$$W^*AVu = \mu W^*BVu. \tag{4}$$

In the original JDQZ method, the search V and the test space W do not have to be real, and, therefore, the Petrov values (i.e. the eigenvalues of the projected eigenvalue problem) do not have to appear in complex conjugated pairs. This causes difficulties for the identification of complex pairs of eigenvalues of the original eigenvalue problem, see e.g. Chapter 8 in [8], and it also introduces additional rounding errors when a computed approximate eigenvalue with small imaginary part is replaced by its real part.

In the RJDQZ method this problem is avoided by keeping the search and test space real. Then the projected eigenvalue problem (4) is also real and eigenvalues are either real or form complex conjugated pairs. Since the eigenvalue problem (4) is small, all eigenvalues can be computed accurately and efficiently using a direct method like the QZ method. From these eigenvalues one eigenvalue (or complex conjugated pair) is selected with a given selection criterion (closest to a target value, or largest real part, etc.). Denote the selected Petrov value by $\widetilde{\mu}$ and the corresponding eigenvector by $\widetilde{u}$. An approximation $(\widetilde{\mu}, \widetilde{q})$ to an eigenpair of the full sized eigenvalue problem (3) can be constructed by computing $\widetilde{q} = Vu$. The residual vector r of the approximate eigenpair $(\widetilde{\mu}, \widetilde{q})$ is defined by

$$r := A\widetilde{q} - \widetilde{\mu}B\widetilde{q}.$$

The second part in a JD iteration is the expansion of the search and test space. The search space V is expanded by an approximate solution t of the linear equation

$$(I - \widetilde{z}\widetilde{z}^*)(A - \widetilde{\mu}B)(I - \widetilde{q}\widetilde{q}^*)t = -r. \tag{5}$$

This equation is called the Jacobi-Davidson correction equation. Here $\widetilde{z}$ is the vector $\widetilde{z} = (\kappa_0 A + \kappa_1 B)\widetilde{q}/\|(\kappa_0 A + \kappa_1 B)\widetilde{q}\|$ with e.g., $\kappa_0 = (1 + |\tau|^2)^{-1/2}$ and $\kappa_1 = -\tau(1 + |\tau|^2)^{-1/2}$ (harmonic Petrov value approach [5]). The test space W is expanded with the vector $w = (\kappa_0 A + \kappa_1 B)t$. This procedure is repeated until $\|r\|$ is small enough.

If the selected Petrov value $\widetilde{\mu}$ is real then the matrix and the right hand side in the correction equation (5) are both real. In this case the correction equation can be (approximately) solved using real arithmetic. This also holds in the preconditioned case, as long as the preconditioner is real. This means that if the selected Petrov value is real, then the search and test space are expanded with a real vector.

If the selected Petrov value $\widetilde{\mu}$ has non-zero imaginary part, then the matrix and the right hand side in the correction equation (5) also have non-zero imaginary parts. An (approximate) solution will in general consist of a complex vector v. In order to keep the search space real, it is expanded with the two dimensional real space $U = \text{span}\{\text{Re}(v), \text{Im}(v)\}$, which contains the vector v. It is easily seen that the space U is also spanned by v and its complex conjugate $\bar{v}$ and it is instructive to think of the space U as an approximation to a two dimensional generalized invariant subspace that can be spanned by two real vectors.

If a complex Petrov pair is selected, there are three ways to formulate the correction equation for the real variant of Jacobi-Davidson QZ:

(1) Complex correction equation: this is the formulation as in (5).

(2) Real variant of complex correction equation: let $\widetilde{\mu} = \nu + i\omega$, $t = u + iv$ and $r = x + iy$. Then the real equivalent of the correction equation (5) is

$$P_z \begin{bmatrix} A - \nu B & \omega B \\ -\omega B & A - \nu B \end{bmatrix} P_q \begin{bmatrix} u \\ v \end{bmatrix} = -\begin{bmatrix} x \\ y \end{bmatrix}, \qquad (6)$$

where (note that the tildes are dropped from z and q for convenience)

$$P_z = \begin{bmatrix} I - (\text{Re}(z)\text{Re}(z)^T + \text{Im}(z)\text{Im}(z)^T) & \text{Im}(z)\text{Re}(z)^T - \text{Re}(z)\text{Im}(z)^T \\ -(\text{Im}(z)\text{Re}(z)^T - \text{Re}(z)\text{Im}(z)^T) & I - (\text{Re}(z)\text{Re}(z)^T + \text{Im}(z)\text{Im}(z)^T) \end{bmatrix}.$$

P_q is defined similarly. By using $\|z\| = \|q\| = 1$ and some basic linear algebra, it can be shown that P_q and P_z are indeed projectors.

(3) Real generalized Sylvester equation: one can think of the vectors $\text{Re}(q)$ and $\text{Im}(q)$ as a basis of an approximate two dimensional invariant subspace. The corresponding residual for this invariant subspace $\begin{bmatrix} \text{Re}(q) & \text{Im}(q) \end{bmatrix}$ is

$$\begin{bmatrix} x & y \end{bmatrix} = A \begin{bmatrix} \text{Re}(q) & \text{Im}(q) \end{bmatrix} - B \begin{bmatrix} \text{Re}(q) & \text{Im}(q) \end{bmatrix} \begin{bmatrix} \nu & \omega \\ -\omega & \nu \end{bmatrix}.$$

The correction equation for $\begin{bmatrix} u & v \end{bmatrix}$ becomes a real generalized Sylvester equation

$$A \begin{bmatrix} u & v \end{bmatrix} - B \begin{bmatrix} u & v \end{bmatrix} \begin{bmatrix} \nu & \omega \\ -\omega & \nu \end{bmatrix} = -\begin{bmatrix} x & y \end{bmatrix}.$$

The equivalent block formulation of this Sylvester equation is

$$P_z \begin{bmatrix} A - \nu B & \omega B \\ -\omega B & A - \nu B \end{bmatrix} P_q \begin{bmatrix} u \\ v \end{bmatrix} = -\begin{bmatrix} x \\ y \end{bmatrix}, \qquad (7)$$

which is the same as the one obtained in (6).

The real and imaginary part of the exact solution of (5) and the exact solution of (6) span the same two dimensional real subspace. If the correction equation is solved exactly, it is more efficient to solve the complex correction equation: the solve of complex linear system of order n costs half the solve of a real linear system of order $2n$. In practice however, the correction equation is only solved approximately using an iterative linear solver like GMRES [7]. The rate of convergence of linear solvers depends, among others, on the condition number of the operator and the distribution of the eigenvalues.

Proposition 1. *Let $A, B \in \mathbb{R}^{n \times n}$, $\theta = \nu + i\omega \in \mathbb{C}$ and $(A - \theta B)v_j = \mu_j v_j$, $j = 1, \ldots, n$ with $v_j \in \mathbb{C}^n$ and $\mu_j \in \mathbb{C}$. Then the eigenpairs of*

$$C = \begin{bmatrix} A - \nu B & \omega B \\ -\omega B & A - \nu B \end{bmatrix} \in \mathbb{R}^{2n \times 2n} \tag{8}$$

are $(\mu_j, [v_j^T, (-iv_j)^T]^T)$, $(\bar{\mu}_j, [v_j^T, (-iv_j)^T]^*)$ for $j = 1, \ldots, n$. Furthermore $\mathrm{Cond}(C) = \mathrm{Cond}(A - \theta B)$.

From Proposition 1 it follows that no big differences in convergence are to be expected if the approximate solution is computed with a linear solver. This is also confirmed by numerical experiments [11].

If a preconditioner $K \approx A - \tau B$ is available for a target $\tau \in \mathbb{R}$, it can be used for the block systems as well:

$$\widetilde{K} = \begin{bmatrix} K & 0 \\ 0 & K \end{bmatrix}.$$

Using Proposition 1, the condition numbers of $K^{-1}(A - \theta B)$ and $\widetilde{K}^{-1}C$ are the same. So the use of a preconditioner also is not expected to cause big differences in speed convergence between the three approaches.

The three approaches may nevertheless lead to different approximate solutions because the inner product of two complex n-vectors is different from the inner product of the equivalent real $2n$-vectors. However, it will most likely require more steps of the linear solver to reduce the residual norm to a certain tolerance for a real problem of size $2n$ than for a complex problem of size n. This is confirmed by numerical experiments [11].

One can show that operator applications cost the same for all three approaches. The approach in (7) is the most elegant approach because no complex arithmetic is involved at all for the RJDQZ algorithm. If, however, an iterative method is used to solve the correction equation approximately, it is expected that within a fixed number of iterations, the approximate solution of the complex correction equation will be most accurate. Therefore, in practice the most efficient approach will be to solve the complex correction equation.

2.1 RJDQZ versus JDQZ

In this section the main differences in the costs between the JDQZ and the RJDQZ method are mentioned.

Memory costs: The orthonormal bases for the search and test space in the JDQZ method are expanded with one complex vector in each iteration. For the RJDQZ, the bases of the search and test space are expanded with one real vector if the selected Petrov value is real or with two real vectors if the selected Petrov value appears in a complex conjugated pair. This means that, although the dimension of the search and test space for the RJDQZ method can grow twice as fast as for the JDQZ method, the storage requirements are the same at most, and probably less.

Computational costs: (1) The correction equation: When in the RJDQZ method a real Petrov value is selected, the correction equation can be solved in real arithmetic. This approximately halves the number of (real) matrix-vector products that is needed for the approximate solution of the correction equation. When a Petrov value is selected that appears in a complex conjugated pair, then the JDQZ and the RJDQZ method need the same work for the approximate solution of the correction equation.

(2) The projected eigenproblem: In the RJDQZ method the real Schur forms of real projected eigenproblems are computed, but these may be twice as large as the complex projected eigenproblems that appear in the JDQZ method. Assume that computing a Schur form costs $\mathcal{O}(n^3)$ operations [6] and that an operation in complex arithmetic costs in average four operations in real arithmetic. Then it is easily deduced that computing the Schur form of a real eigenvalue problem costs about twice as much as computing the Schur form of a complex eigenvalue problem that is twice as small.

(3) Orthogonalization: One has to compare these two cases:

1. Orthogonalize a complex vector against k other complex vectors. This requires $4k$ real inner products for the projection plus 2 real inner products for the scaling.
2. Orthogonalize two real vectors against $2k$ other real vectors. This requires $2k$ inner products for projecting the first vector plus one inner product for scaling. For the next vector, $2k+1$ inner products are needed for the projection plus one for the scaling. This adds up to $4k+3$ inner products.

If the initial approximation is a real vector, then the cost of the extra inner product is eliminated, and the orthogonalization process in the RJDQZ method will cost at most as much as in the JDQZ method.

3 Numerical comparison

The RJDQZ method is compared with two variants of the JDQZ method: the original JDQZ method as described in [5] and a variant that is here denote by JDQZd, which is the method described for the standard eigenvalue problem in [2] translated to the generalized case. The experiments were performed on a Sunblade 100 workstation using Matlab 6. If not mentioned otherwise, the target value in each experiment equals 0, and the tolerance is set to 10^{-9}.

For all the numerical results presented in this section, the correction equation is solved approximately using at most 10 iterations of Bi-CGSTAB [10] or GMRES [7]. No restart strategy is used in the JD part of the algorithm.

The first test problem is the generalized eigenvalue problem BFW782 from the NEP collection at http://math.nist.gov/MatrixMarket. The eigenvalues and corresponding eigenvectors of interest are the ones with positive real

Table 1. Results for QZ methods

		Bi-CGSTAB			GMRES	
BFW782	JDQZ	JDQZd	RJDQZ	JDQZ	JDQZd	RJDQZ
iterations	28		28	33		33
dim. search sp.	18		18	23		23
mat.vec.	558		279	456		228
QG6468	JDQZ	JDQZd	RJDQZ	JDQZ	JDQZd	RJDQZ
iterations	87	64	52	106	72	58
dim. search sp.	69	51	76	88	59	87
mat.vec.	3176	2378	1744	2072	1412	1009
ODEP400	JDQZ	JDQZd	RJDQZ	JDQZ	JDQZd	RJDQZ
iterations	44	37	27	42	31	26
dim. search sp.	22	22	24	20	16	29
mat.vec.	1202	1180	570	462	386	243

parts. The matrix A is non-symmetric and B is symmetric positive definite. Six eigenvalues with largest real part are computed. The preconditioner that is used for the correction equations is the ILU factorization of A with drop tolerance 1e-3. In Table 1 the number of iterations, the number of matrix-vector products, and the dimension of the final search space is given.

The computed eigenvalues are all real. Observe that the JDQZ and the RJDQZ method both need exactly the same number of iterations and build a search space of the same dimension. From the intermediate iterations (not shown), it can be concluded that the two methods perform exactly the same steps, the only difference being that the RJDQZ method performs the steps in real arithmetic and the JDQZ method performs the steps using complex arithmetic. This explains why the JDQZ method needs twice as many matrix-vector products as the RJDQZ method.

The second test problem is the generalized eigenvalue problem QG6468, which arises in the stability analysis of steady states in the finite difference approximation of the QG model described in [4]. The eigenvalues and corresponding eigenvectors of interest are the ones with largest real parts. The matrix A is non-symmetric and B is symmetric positive semi definite. Six eigenvalues with largest real part are computed. The preconditioner that is used for the correction equations is the ILU factorization of A with drop tolerance 1e-7.

For this matrix pencil, two of the computed eigenvalues are real, and the other computed eigenvalues form four complex conjugated pairs. One sees that the RJDQZ method needs fewer iterations, but builds a larger search space than the JDQZ method. The storage requirements for the RJDQZ method is, however, still less than for the JDQZ method.

The third test problem is the generalized eigenvalue problem ODEP400 (also from the NEP collection at http://math.nist.gov/MatrixMarket/). Eigenvalues and corresponding eigenvectors with largest real part are computed. The matrix A is non-symmetric and B is symmetric positive semi definite. Six pairs of complex conjugate eigenvalues that are computed with largest real part. The preconditioner that is used for the correction equations is the LU factorization of A.

The computed eigenvalues are all complex, but still the RJDQZ method needs fewer iterations and fewer matrix-vector products than the JDQZ method. There can be two different reasons: the intermediate selected Petrov values can be real so that real arithmetic can be used in intermediate RJDQZ iterations, and secondly, in case the selected Petrov value is not real, then the dimension of the search space grows faster in the RJDQZ method, which may result in a faster convergence and thus fewer matrix-vector products.

4 Conclusions

An adapted version of the Jacobi-Davidson method is presented that is intended for real unsymmetric pencils. In all presented numerical experiments, this version needed fewer iterations, fewer matrix-vector products and less storage than the original JDQZ method. The difference is most pronounced when only real eigenvalues are computed. The better performance of the RJDQZ method can be attributed to two reasons: the method uses real arithmetic where possible, which results in fewer (real) matrix-vector products, and the dimension of the search space may grow twice as fast (while not using more storage) which accelerates the convergence, resulting in fewer iterations.

References

1. Bindel, D., Demmel, J.W., Friedman, M.J.: Continuation of Invariant Subspaces for Large and Sparse Bifurcations Problems. In: online proceedings of SIAM Conference on Applied Linear Algebra, Williamsburg (2003)
2. Brandts, J.: Matlab code for sorted real Schur forms. Numer. Linear Algebra Appl., **9**, 249-261 (2002)
3. Brandts, J.: The Riccati method for eigenvalues and invariant subspaces of matrices with inexpensive action. Linear Algebra Appl., **358**, 333–363 (2003)
4. Dijkstra, H.A., Katsman, C.A.: Temporal variability of the wind-driven quasi-geostrophic double gyre ocean circulation: basic bifurcation diagrams. Geophys. Astrophys. Fluid Dynamics, **85**, 195–232 (1997)
5. Fokkema, D.R., Sleijpen, G.L.G., Van der Vorst, H.A.: Jacobi-Davidson style QR and QZ algorithms for the partial reduction of matrix pencils. SIAM J. Sci. Comput., **20**, 94–125 (1998)
6. Golub, G.H., Van Loan, C.F.: Matrix Computations. The John Hopkins University Press, Baltimore, 2nd Ed. (1989)
7. Saad, Y., Schultz, M.H.: GMRES: a generalized minimal residual algorithm for solving nonsymmetric linear systems. SIAM J. Sci. Statist. Comput., **7**, 856–869 (1986)

8. Saad, Y.: Numerical methods for large eigenvalue problems. Manchester University Press, Manchester (1992)
9. Sleijpen, G.L.G., Van der Vorst, H.A.: A Jacobi-Davidson iteration method for linear eigenvalue problems. SIAM J. Matrix Anal. Appl., **17**, 401–425 (1996)
10. Van der Vorst, H.A.: Bi–CGSTAB: a fast and smoothly converging variant of Bi–CG for the solution of nonsymmetric linear systems. SIAM J. Sci. Statist. Comput., **13**, 631–644 (1992)
11. Van Noorden, T.L.: Computing a partial generalized real Schur form using the Jacobi-Davidson method. CASA-report 05-41, Department of Mathematics and Computer Science, Eindhoven University of Technology (2005)

NUMERICAL METHODS IN FINANCE

Pricing Multi-Asset Options with Sparse Grids and Fourth Order Finite Differences

C.C.W. Leentvaar and C.W. Oosterlee

Delft University of Technology, Mekelweg 4, 2628 CD, Delft, The Netherlands
{c.c.w.leentvaar,c.w.oosterlee}@ewi.tudelft.nl

Summary. We evaluate the sparse grid solution technique [9, 4] with 4th order discretization for pricing multi-asset options. Convergence in the sense of point-wise interpolation to a special point is considered. We also present a novel variant based on backward differentiation formula coefficients. In combination with the high order discretization we can solve five-dimensional option pricing problems satisfactorily on coarse grids.

1 Introduction

In this paper we evaluate sparse grid techniques [9, 4] for high-dimensional partial differential equations (PDEs) in the area of option pricing. We aim at minimizing the total number of unknowns in high dimensions. Therefore we employ 4th order finite differences and backward differentiation time integration [5]. Fourth order grid convergence with the sparse grid methods is theoretically not guaranteed. The solution's accuracy on reasonably sized grids may however already be satisfactory. Next to known sparse grid combination methods we present a novel variant based on backward differentiation formula coefficients. We use a 2D Poisson test problem to evaluate the combination of the 4th order stencil, sparse grid combination and interpolation technique. The methods are then used to evaluate the price of an option based on a geometric average. The outline of this paper is as follows. We describe in section 2 the model for an option on the geometric average of assets. In section 3, we set up the numerical techniques. The sparse grid method is discussed in section 4. Numerical results for PDE problems with up to five spatial dimensions, plus time, are presented in section 5.

2 Model

We solve multi-asset option pricing problems, defined by the multi-dimensional Black-Scholes equation [6]:

$$\frac{\partial u}{\partial t} + \frac{1}{2}\sum_{i=1}^{d}\sum_{j=1}^{d}\rho_{ij}\sigma_i\sigma_j S_i S_j \frac{\partial^2 u}{\partial S_i \partial S_j} + \sum_{i=1}^{d}(r-\delta_i)S_i \frac{\partial u}{\partial S_i} - ru = 0. \quad (1)$$

Solution u is the option price based on the underlying assets S_i with $i = 1,\ldots,d$, σ_i is the volatility of asset i, ρ_{ij} is the correlation coefficient between the assets with $\rho_{ii} = 1$, r is the risk-free interest rate, and δ_i is a continuous dividend yield. Equation (1) comes with a final condition, also called pay-off that determines the type of the option considered. In this paper we consider an option with a payoff based on the geometric average of the assets:

$$u(S_1, S_2, \ldots, S_d, T) = \max\{\prod_{i=1}^{d} S_i^{\frac{1}{d}} - K, 0\}. \quad (2)$$

This option contract is well-suited for evaluating high-dimensional techniques, as it is possible to rewrite the problem to a 1D equation, for which we know the exact solution. The new coordinate $x = \sqrt[d]{S_1 S_2 \cdots S_d}$, leads to the 1D Black-Scholes equation for a European Call option [1]:

$$\begin{cases} \frac{\partial u}{\partial t} + \frac{1}{2}\hat{\sigma}^2 x^2 \frac{\partial^2 u}{\partial x^2} + (r-\hat{\delta}) x \frac{\partial u}{\partial x} - ru = 0 \\ u(x,T) = \max\{x-K, 0\} \end{cases} \quad (3)$$

with

$$\hat{\sigma}^2 = \frac{1}{d^2}\sum_{i=1}^{d}\sum_{j=1}^{d}\rho_{ij}\sigma_i\sigma_j, \quad \hat{\delta} = \frac{1}{d}\sum_{i=1}^{d}\delta_i + \frac{1}{2}\left(\frac{1}{d}\sum_{i=1}^{d}\sigma_i^2 - \hat{\sigma}^2\right). \quad (4)$$

With the exact solution of (3) we are able to compare our numerical methods for any dimension d.

3 Discretization

The development of the numerical solution of (1) is based on 4th order long stencil finite differences. With Kronecker products we then generalize the discretization to the multi-dimensional situation. Therefore, the derivation of the stencils can be done in a 1D setting. The derivatives with respect to asset i are discretized on a grid with meshwidth h_i:

$$\frac{du}{dS_i} = \frac{-u_{k+2} + 8u_{k+1} - 8u_{k-1} + u_{k-2}}{12h_i} + \mathbf{O}(h_i^4), \quad (5)$$

$$\frac{d^2 u}{dS_i^2} = \frac{-u_{k+2} + 16u_{k+1} - 30u_k + 16u_{k-1} - u_{k-2}}{12h_i^2} + \mathbf{O}(h_i^4). \quad (6)$$

At the boundary, we prescribe the (known) exact solutions for the option from section 2. We can choose an extrapolation for virtual points u_{-1} and u_{N+1}, or backward differences. Both possibilities are evaluated in this paper. The extrapolation in point u_{-1} reads $u_{-1} = 4u_0 - 6u_1 + 4u_2 - u_3 + \mathbf{O}(h_i^4)$. If this is applied to the differences (5) and (6), we obtain:

$$\frac{du}{dS_i}\Big|_{n=1} = \frac{-2u_3 + 12u_2 - 6u_1 - 4u_0}{12h_i} + \mathbf{O}(h_i^3),$$

$$\frac{d^2u}{dS_i^2}\Big|_{n=1} = \frac{u_2 - 2u_1 + u_0}{h_i^2} + \mathbf{O}(h_i^2),$$

and analogously for the other boundary point. The other possibility is the use of backward differences at the boundary, for example, for the left point:

$$\frac{du}{dS_i}\Big|_{n=1} = \frac{-3u_0 - 10u_1 + 18u_2 - 6u_3 + u_4}{12h_i} + \mathbf{O}(h_i^4),$$

$$\frac{d^2u}{dS_i^2}\Big|_{n=1} = \frac{10u_0 - 15u_1 - 4u_2 + 14u_3 - 6u_4 + u_5}{12h_i^2} + \mathbf{O}(h_i^4).$$

Notice that with these choices the backward differences lead to higher accuracy, but also to awkward stencils with many positive off-diagonal elements. To obtain a semi-discrete equation, we use the Kronecker products [8]. If we want to express a stencil D of a derivative with respect to coordinate j in a d-dimensional way, we obtain:

$$D_j^d = \bigotimes_{i=j+1}^{d} \mathbf{I}_i \otimes D_j^1 \otimes \bigotimes_{i=1}^{j-1} \mathbf{I}_i, \qquad \bigotimes_{i=1}^{d} A_i = A_1 \otimes A_2 \otimes \ldots \otimes A_d, \qquad (7)$$

where $\mathbf{I}_i$ is the identity matrix of size $N_i \times N_i$ and D_j^1 the 1D-stencil as derived in (5) and (6). The cross-derivative can be written as:

$$\frac{\partial^2}{\partial S_2 \partial S_1} = \frac{\partial}{\partial S_2}\left(\frac{\partial}{\partial S_1}\right).$$

It follows that the cross-derivative stencil R with respect to i and $j > i$ reads:

$$R_{i,j}^d = \bigotimes_{k=j+1}^{d} \mathbf{I}_k \otimes \left(\frac{\partial}{\partial S}\right)_j^1 \otimes \bigotimes_{k=i+1}^{j-1} \mathbf{I}_k \otimes \left(\frac{\partial}{\partial S}\right)_i^1 \otimes \bigotimes_{k=1}^{i-1} \mathbf{I}_k. \qquad (8)$$

where $\left(\frac{\partial}{\partial S}\right)_j^1$ is the 1D discretized derivative (5) for coordinate S_j.

Finally, the semi-discrete PDE is written in a matrix vector notation:

$$\begin{cases} \dfrac{du}{dt} = Au + b(t) + \mathbf{O}(\sum_{i=1}^{d} h_i^p) \\ u(x,0) = u_0. \end{cases} \qquad (9)$$

We use the 4th order backward differentiation formula, BDF4, [5] for (9):

$$\frac{25}{12}u^{m+1} - 4u^m + 3u^{m-1} - \frac{4}{3}u^{m-2} + \frac{1}{4}u^{m-3} = \Delta t \cdot (Au^{m+1} + b^{m+1}), \quad (10)$$

with Δt the time-step. As initialization steps we use the sequence of BDF1, BDF2 and finally BDF3. The BDF4 scheme has a stability restriction in contrast to BDF1 and BDF2 [5], but, so far, we do not encounter stability problems.

4 Sparse grid combination technique

Multi-asset problems can lead to very large systems that are not easily solvable on nowadays machines. A 5D full grid problem with only 32 points in each direction leads to over 32 million unknowns. This is called the *curse of dimensionality*. To avoid this exponential growth of the number of unknowns, we use the sparse grid technique developed by Zenger [9] and Griebel et al. [4]. In combination with second order finite differences, this technique gives highly acceptable grid convergence for option pricing problems [7]. The basic technique to solve the PDE with sparse grids consists of two parts: the solution of many problems on small sized grids and the interpolation to a desired point or (sub-)grid. In [4] a combination technique has been developed based on a multi-index I. A multi-index I is a collection of d numbers, with d the problem dimension. Each number $i_k > 0$ is proportional to the size of the grid in that direction (k). The meshwidth in that direction is $h_k = 2^{-i_k}$ and the number of points is $N_k = h_k^{-1}$. The sum of the multi-index $|I|$ is used to determine which grids are needed in the sparse grid combination technique. If $i_k < 3$ the number of points available is to low to use the 4th order long stencil. Therefore, we multiply the number of grid-points by a factor 4 in this case. With the aid of the multi-index and its sum, we are able to define and compare three different combination techniques: the basic sparse grid method [4], a novel BDF-type sparse grid method which is a technique that employs one extra layer, and Reisinger's [7] 4th order sparse grid method which consists of a combination of multi-variate extrapolation and the basic sparse grid method.

4.1 Basic sparse grid combination technique

The basic combination technique [2, 4] reads:

$$u_n^c = \sum_{k=0}^{d-1}(-1)^{k+1}\binom{d}{k}\sum_{|I|=n+k} u_I. \qquad (11)$$

where u_n^c is the combined solution which can be compared to a full grid solution with meshwidth $h_n = 2^{-n}$ in each direction. u_I denotes a solution on a grid with its meshwidths based on the multi-index I. Note that the second sum is a summation over all solutions u_I for which $|I| = n+k$. For a Poisson equation, $\Delta u = f$, with exact solution u_{exact}, the upper bound error in the 2D case reads [4]

$$\|u_n^c - u_{exact}\| \leq K h_n^2 \left(1 + \frac{5}{4}\log_2(h_n^{-1})\right). \qquad (12)$$

The bound generalized to higher dimensions reads, for general d:

$$\|u_n^c - u_{exact}\| \approx \mathbf{O}(h_n^2(\log_2(h_m^{-1}))^{d-1}). \qquad (13)$$

4.2 BDF sparse grid technique

We now propose a BDF-type combination equation, that, in 2D reads:

$$u_n^c = \frac{3}{2}\sum_{|I|=n+2} u_I - 2\sum_{|I|=n+1} u_I + \frac{1}{2}\sum_{|I|=n} u_I, \qquad (14)$$

with the coefficients of a BDF2 scheme [5]. With vectors $\mathbf{b} = \frac{1}{2}[3, -4, 1]$ and:

$$\mathbf{w} = \left[\binom{d-2}{0}, \binom{d-2}{1}, \ldots, \binom{d-2}{d-2}\right], \qquad (15)$$

the d-dimensional combination technique in this setting is written as (11):

$$u_n^c = \sum_{k=0}^{d}(-1)^{k+1} a_k \sum_{|I|=n+k} u_I$$

with a_k the k-th element of $\mathbf{a} = \mathbf{b} \star \mathbf{w}$, where $\star$ is the convolution operator. Note that this method uses one additional finer layer compared to the basic sparse grid method (11). With the splitting of the error as given in (18), the error for the 2D Poisson case can be bounded by:

$$\|u_n^c - u_{exact}\| \leq \frac{35}{32} h_n^2 \log_2(h_n^{-1})$$

and similarly convergence ratio (13) for general d can be derived.

4.3 Reisinger's sparse grid method

The third technique evaluated is based on the basic combination technique and a multivariate extrapolation [7]. The combination equation then reads:

$$u_n^c = \sum_{\ell=n}^{n+2d-1} \sum_{|I|=\ell} u_I \left(\sum_{j=\max\{0,\ell-n-d\}}^{j=\min\{\ell-n,d-1\}} a_j \alpha_{\ell-n-d} \binom{N(I)}{\ell-n-j} \right), \qquad (16)$$

$$a_j = (-1)^{d-1-j}, \quad \alpha_j = (-4)^j (-3)^{-d}.$$

$N(I)$ is the number of nonzero elements in multi-index I. (The elements of the multi-index can be zero.) This method has an absolute error that is of almost 4th order accuracy for the d-dimensional Poisson equation [7]:

$$|u_n^c - u_{exact}| \leq \frac{10}{3} \frac{K}{(d-1)!} \left(\frac{85}{24}\right)^{d-1} \left(\log_2(h_n^{-1}) + 2d - 1\right)^{d-1} h_n^4. \qquad (17)$$

Truncation error estimates for 2D sparse grid methods are typically based on the result that a second order error of a solution of a Poisson problem can be split into [3]:

$$u_{i,j} - u_{exact} = C_1(x, h_i) h_i^2 + C_2(y, h_j) h_j^2 + + C(x, y, h_i, h_j) h_i^2 h_j^2 \qquad (18)$$

where $u_{i,j}$ is the numerical solution for meshwidths (h_i, h_j), u_{exact} the exact solution and C_1, C_2 and C are bounded constants. An error splitting as in (18) does not guarantee 4th order accuracy when employing 4th order finite differences within the sparse grid technique [7]. As mentioned, in our case, the number of grid-points to get a small-sized error is of higher importance than asymptotic convergence for $h \to 0$. Therefore, we evaluate the combination techniques with 4th order finite difference stencils.

A final remark on the three methods is on the number of underlying grids, and thus on the method's complexity. For a d-dimensional problem the basic sparse grid method uses $d-1$ "layers" of grids, the BDF sparse grid method employs d-, and Reisinger's method $2d-1$-layers. Reisinger's method uses the finest grids. However, the method is based on second order stencils which means that matrices are sparser than for 4th order.

5 Numerical results

Before we use the sparse grid techniques for the option pricing problem, we evaluate them first for a 2D Poisson equation with exact solution:

$$u = \exp \prod_{i=1}^{d} x_i, \quad (d = 2). \qquad (19)$$

Table 1. 2D Poisson test problem. Second order discretization: error in point $(\frac{1}{2},\frac{1}{2})$ with $h_n = 2^{-n}$ in the full grid case

	Full		Basic		BDF		Reisinger	
n	error	Conv	error	Conv	error	Conv	error	Conv
1	2.1×10^{-4}		2.1×10^{-4}		2.8×10^{-4}		3.3×10^{-5}	
2	1.1×10^{-4}	1.9	2.6×10^{-4}	0.8	7.2×10^{-5}	3.9	3.5×10^{-6}	9.4
3	3.2×10^{-5}	3.4	1.3×10^{-4}	1.9	1.2×10^{-5}	5.9	3.3×10^{-7}	10.6
4	8.4×10^{-6}	3.8	5.3×10^{-5}	2.5	9.3×10^{-7}	13.1	2.9×10^{-8}	11.2
5	2.1×10^{-6}	4.0	1.8×10^{-5}	2.9	3.7×10^{-7}	2.5	2.5×10^{-9}	11.7
6	5.3×10^{-7}	4.0	5.8×10^{-6}	3.1	2.5×10^{-7}	1.5	2.0×10^{-10}	12.3
7	1.3×10^{-7}	4.0	1.8×10^{-6}	3.3	1.8×10^{-7}	2.4	1.6×10^{-11}	12.4

Table 2. 2D Poisson test problem. Fourth order schemes: error in point $(\frac{1}{2},\frac{1}{2})$. B denotes backward differences at the boundary. $h_n = 2^{-n}$ in the full grid case

	Full		Basic		BDF		BDF B	
n	error	Conv	error	Conv	error	Conv	error	Conv
3	4.7×10^{-6}		4.7×10^{-6}		1.8×10^{-6}		7.4×10^{-9}	
4	3.0×10^{-7}	15.7	3.5×10^{-7}	13.5	1.4×10^{-7}	13.5	5.8×10^{-10}	12.7
5	1.9×10^{-8}	15.9	2.6×10^{-8}	13.4	1.0×10^{-8}	13.4	5.7×10^{-11}	10.1
6	1.2×10^{-9}	15.9	1.9×10^{-9}	13.6	7.5×10^{-10}	13.6	3.3×10^{-12}	17.5
7	7.4×10^{-11}	16.1	1.4×10^{-10}	13.9	5.6×10^{-11}	13.3	1.8×10^{-12}	-

In table 1, we compare the second order finite differences sparse grid techniques with the full-grid accuracy. Comparison takes place by checking the error in the solution at grid point $(1/2, 1/2)$. We see that, indeed, Reisinger's method performs very well. The improvement in error by an extra layer (14) compared to the basic combination technique is also visible in the column "BDF". We should, however, for fairness in costs compare a row in Basic with a previous row in BDF. In table 2, we present the results for the 4th order discretizations, and observe highly accurate solutions with a small number of points. Especially when backward differences are used at boundaries (indicated by 'B'), the coarse grid sparse solution corresponds very well with the exact solution. In the tables, we can observe the influence of the $\log_2(h_n^{-1})$-term on the asymptotic grid convergence, which will be more significant in higher dimensions. Disadvantage of backward differences is that the grid convergence is not smooth, probably due to several positive off-diagonal stencil elements. We now perform computations with the multi-D Black-Scholes equation and the geometric average as the pay-off. Model parameters used are $\sigma_i = 0.2$, $\rho_{ij} = 0.25$, $\delta_i = 0$, $r = 0.06$, $T = 0.5$ and $K = 40$. We calculate the numerical solution in the center of the S-domain, i.e. at $S_i = K$ $\forall i$. In table 3 up to 5D computations are presented. Values are compared to the 1D 'exact' solution, also presented in table 3. We confirm that the 4th order schemes

Table 3. Option problem based on a 32^d and a 64^d grid, i.e., n=5, 6, B denotes backward differencing at the boundary (NA means not available).

		2D		3D		4D		5D	
	Exact:	2.318		2.108		1.994		1.923	
Accuracy	Method	n=5	n=6	n=5	n=6	n=5	n=6	n=5	n=6
2	Full grid	2.302	2.315	2.100	2.106	1.991	NA	NA	NA
2	Basic	3.133	2.350	2.601	2,406	2,546	2,232	2,500	2,064
2	BDF	1.958	2.518	2.307	1,974	2,078	1,610	1,852	1,888
2	Reisinger	2.661	2.122	2.000	1,959	1,889	2,242	2,763	1.938
4	Full grid	2.311	2.317	2.106	2.107	1.994	NA	NA	NA
4	Basic	2.413	2.281	2.118	2.125	1,961	2,001	1,867	1,951
4	BDF	2.214	2.323	2.128	2.075	2,034	1,954	1,992	1.875
4	Basic B	2.413	2.281	2.118	2.125	1,961	2,001	1,867	1,951
4	BDF B	2.214	2.323	2.128	2.076	2.034	1,954	1,992	1.875

are performing very satisfactory as compared to the second order based finite difference schemes.

6 Conclusions

The grid convergence results for the test problem and the multi-asset option problem in the present paper give some insight in the accuracy that can be achieved by sparse grid methods for up to five spatial dimensions. We observe a satisfactorily convergence on relatively coarse grids, especially when using the 4th order stencils and backward differences at the boundaries. The BDF type sparse grid methods introduced here appear a valuable addition to the sparse grids family. The Basic sparse grid combination technique with 4th order discretization, however, seems to perform best in terms of cost. A more detailed convergence analysis for the sparse grid combination techniques, inclusion of stretched grids and more realistic multi-asset option contracts are part of a forthcoming report.

References

1. Berridge, S.J., Schumacher, J.M.: An irregular grid method for high-dimensional free-boundary problems in finance. Fut. Gener. Comp. Syst. **20**, 353-362 (2004)
2. Bungartz, H.J., Griebel, M. Sparse Grids, Acta Numerica, 1-123, Cambridge University Press (2004)
3. Bungartz, H.J., Griebel, M., Röschke, D., Zenger, C.: Pointwise convergence of the combination technique for Laplace's equation. East-West J. Num. Math., **2**, 21-45 (1994)

4. Griebel,M., Schneider, S., Zenger, C.: A combination technique for the solution of sparse grid problems. In: de Groen, P., Beauwens, R. (eds) Iterative Methods in Linear Algebra. IMACS, Elsevier, North-Holland (1992)
5. Hairer, E., Wanner, K.: Solving ordinary differential equations. Vol. 2. Stiff and differential-algebraic problems. Springer Verlag, Heidelberg (1996)
6. Kwok, Y.K.: Mathematical models of financial derivatives. Springer Verlag, Singapore (1998)
7. Reisinger, C.: Numerische Methoden für hochdimensionale parabolische Gleichungen am Beispiel von Optionspreisaufgaben. PhD Thesis, Ruprecht-Karls-Universität Heidelberg, Germany (2004)
8. Steeb, W.: Kronecker products and matrix calculus with applications. Wissenschaftsverlag Mannheim, Germany (1991)
9. Zenger, C.: Sparse grids, In: Hackbusch, W. (ed) Parallel Differential Equations, Notes on Num. Fluid Mech., **31**, Vieweg, Braunschweig, Germany (1991)

ODE and Fractional Step Methods

A Third Order Linearly Implicit Fractional Step Method for Semilinear Parabolic Problems

Blanca Bujanda[1] and Juan Carlos Jorge[2]

[1] Dpto. Matemática e Informática, Universidad Pública de Navarra
   blanca.bujanda@unavarra.es
[2] Dpto. Matemática e Informática, Universidad Pública de Navarra
   jcjorge@unavarra.es

Summary. In this paper a new efficient linearly implicit time integrator for semilinear multidimensional parabolic problems is proposed. This method preserves the advantages, in terms of computational cost reduction, of the classical fractional step methods for linear parabolic problems. We show some numerical tests for illustrating that this method combined with standard space discretization techniques, provides efficient numerical algorithms capable of computing stable numerical solutions without restrictions between the time step and the mesh size.

1 Introduction

In this paper we deal with the development of an efficient numerical method to integrate time dependent semilinear parabolic problems of the form:

$$\text{find } u(t) : [t_0, T] \to H \text{ solution of}$$
$$\begin{cases} \dfrac{du(t)}{dt} = L(t)u(t) + f(t) + g(t, u), \\ u(t_0) = u_0, \end{cases} \quad (1)$$

where H is a Hilbert space of functions defined on a certain domain $\Omega \subseteq \mathbb{R}^n$, $L(t) : \mathcal{D} \subseteq H \to H$ is a linear second order elliptic differential operator which contains the spatial derivatives of the solution, $f(t)$ is the source term and $g(t, u)$ is a non-linear function.

The obtaining of the solution of this problem can be realized by using a double process of discretization of the spatial and temporal variables. Thus, if we apply firstly a classical spatial discretization process, like finite differences or finite elements among others, we obtain a one parameter family of Initial Value Problems (IVP) of the form:

find $U_h(t) : [t_0, T] \to V_h$ solution of
$$\begin{cases} U_h'(t) = L_h(t)U_h(t) + f_h(t) + g_h(t, U_h(t)), \\ U_h(t_0) = u_{h0}; \end{cases} \quad (2)$$

typically, $h \in (0, h_0]$ denotes the size of the mesh used to discretize in space and it is destined to tend to zero; also, for every h we consider a finite dimensional space V_h which will be an space of discrete functions on a mesh in Finite Differences and it will be a subspace of H of piecewise polynomial functions in a classical Finite Element discretization.

In this work, we propose the use of a time integrator which is designed by combining, in an additive way, a Fractional Step Runge Kutta (FSRK) method for dealing with the linear non-homogeneous term, $L_h(t)U_h(t) + f_h(t)$, and a suitable explicit RK method for including the contribution of the non-linear term, $g_h(t, U_h(t))$. We decompose the spatial discretization of the elliptic operator and the source term in the form $L_h(t) = \sum_{i=1}^{n} L_{ih}(t)$ and $f_h(t) = \sum_{i=1}^{n} f_{ih}(t)$.

Then the numerical algorithm which we propose follows the scheme

$$\begin{cases} U_h^{m+1} = U_h^m + \tau \sum_{i=1}^{s} b_i^{k_i} \left(L_{k_i h}(t_{m,i}) U_h^{m,i} + f_{k_i h}(t_{m,i}) \right) \\ \qquad + \tau \sum_{i=1}^{s} b_i^{n+1} g_h(t_{m,i}, U_h^{m,i}), \text{ where} \\ U_h^{m,i} = U_h^m + \tau \sum_{j=1}^{i} a_{ij}^{k_j} \left(L_{k_j h}(t_{m,j}) U_h^{m,j} + f_{k_j h}(t_{m,j}) \right) \\ \qquad + \tau \sum_{j=1}^{i-1} a_{ij}^{n+1} g_h(t_{m,j}, U_h^{m,j}). \end{cases} \quad (3)$$

Here U_h^m denotes the approximation of $u(t_m)$ being $t_m = t_0 + m\tau$ (τ is the time step), $U_h^0 = U_h(t_0)$, $k_i, k_j \in \{1, \cdots, n\}$, n is the number of levels of the FSRK method, $t_{m,i} = t_0 + (m + c_i)\tau$ and the intermediate approximations $U_h^{m,i}$ for $i = 1, \cdots, s$ are the internal stages of the method. If we fill the last formulation with some null coefficients in this form

$$\mathcal{A}^k = (a_{ij}^k) \text{ where } a_{ij}^k = \begin{cases} a_{ij}^{n+1} & \text{if } k = n+1 \text{ and } i > j, \\ a_{ij}^{k_j} & \text{if } k = k_j \text{ and } i \geq j, \\ 0 & \text{otherwise,} \end{cases}$$

$$b^k = (b_j^k) \text{ where } b_j^k = \begin{cases} b_j^{n+1} & \text{if } k = n+1, \\ b_j^{k_j} & \text{if } k = k_j, \\ 0 & \text{otherwise,} \end{cases}$$

the coefficients of this method can be organized in a table, extending the notations introduced by Butcher for the standard RK methods, as follows

$$\begin{array}{c|c|c|c|c|c} \mathcal{C}\,e & \mathcal{A}^1 & \mathcal{A}^2 & \ldots & \mathcal{A}^n & \mathcal{A}^{n+1} \\ \hline & (b^1)^T & (b^2)^T & \ldots & (b^n)^T & (b^{n+1})^T \end{array}$$

where $\mathcal{C} = diag(c_1, \cdots, c_s)$ and $e = (1, \cdots, 1)$.

To refer abbreviately to these methods we will use the notation $(\mathcal{C}, \mathcal{A}^i, b^i)_{i=1}^{n+1}$ and we denote with $(\mathcal{C}, \mathcal{A}^i, b^i)$ each standard RK method which takes part in (3). Note that $\mathcal{A}^{n+1}$ is a strictly lower triangular matrix that we use for computing the contribution of the non-linear part, $g_h(t, U_h)$, of the derivative function.

In [1, 2] it is shown that the use of these new methods for discretizing in time permits us to avoid two of the main difficulties of the classical implicit schemes if a suitable partition of $L(t)$ is combined with appropriate spatial discretizations. On one hand, the convergence of the numerical scheme can be obtained without imposing too severe stability requirements; only a property of linear absolute stability is required for a stable numerical integration of (1) (see [2]). On the other hand, we achieve an important reduction in the computational cost with respect to classical implicit methods, because the use of iterative methods to resolve the internal stage equations $U_h^{m,i}$ is not necessary; in fact, its resolution involves only linear systems with very simple matrices.

2 A new third order linearly implicit FSRK method

In this section we show the construction process of a third order method of the class introduced in the previous section. This process is complicated due to the high number of order conditions that we must consider joint to the additional restrictions which must be imposed to preserve the stability. To design this method we have used the following order conditions (see [1]):

(r_1) $\mathcal{A}^1 e = \mathcal{C}\,e$ (r_2) $\mathcal{A}^2 e = \mathcal{C}\,e$ (r_3) $\mathcal{A}^3 e = \mathcal{C}\,e$
(α_1) $(b^1)^T e = 1$ (α_2) $(b^1)^T \mathcal{C}\,e = \frac{1}{2}$ (α_3) $(b^1)^T \mathcal{A}^1 \mathcal{C}\,e = \frac{1}{6}$
(α_4) $(b^1)^T \mathcal{C}\mathcal{C}\,e = \frac{1}{3}$ (β_1) $(b^2)^T e = 1$ (β_2) $(b^2)^T \mathcal{C}\,e = \frac{1}{2}$
(β_3) $(b^2)^T \mathcal{A}^2 \mathcal{C}\,e = \frac{1}{6}$ (β_4) $(b^2)^T \mathcal{C}\mathcal{C}\,e = \frac{1}{3}$ (γ_1) $(b^3)^T e = 1$
(γ_2) $(b^3)^T \mathcal{C}\,e = \frac{1}{2}$ (γ_3) $(b^3)^T \mathcal{A}^3 \mathcal{C}\,e = \frac{1}{6}$ (γ_4) $(b^3)^T \mathcal{C}\mathcal{C}\,e = \frac{1}{3}$
$(\times_{12})$ $(b^1)^T \mathcal{A}^2 \mathcal{C}\,e = \frac{1}{6}$ $(\times_{21})$ $(b^2)^T \mathcal{A}^1 \mathcal{C}\,e = \frac{1}{6}$ $(\times_{13})$ $(b^1)^T \mathcal{A}^3 \mathcal{C}\,e = \frac{1}{6}$
$(\times_{23})$ $(b^2)^T \mathcal{A}^3 \mathcal{C}\,e = \frac{1}{6}$ $(\times_{31})$ $(b^3)^T \mathcal{A}^1 \mathcal{C}\,e = \frac{1}{6}$ $(\times_{32})$ $(b^3)^T \mathcal{A}^2 \mathcal{C}\,e = \frac{1}{6}$

We have proven that a method of this class verifying such order conditions must have at least seven stages. In order to satisfy the order conditions (r_1), (r_2) and (r_3) we impose that $(\mathcal{C}, \mathcal{A}^1, b^1)$ has a first explicit stage. In this way the computational cost of the method will be of the same order that a six stage method. For stability reasons it is convenient to impose $(0, \cdots, 0, 1)^T \mathcal{A}^i = (b^i)^T$ for $i = 1, 2$ (see [2]) and to reduce a bit the computational cost of the final method we have also imposed that $(0,\cdots,0,1)^T \mathcal{A}^3 = (b^3)^T$. These restrictions will make that the calculus of the last stage provides directly $U_h^{m+1}(= U_h^{m,s})$. To simplify the study of the stability of the method $(\mathcal{C}, \mathcal{A}^i, b^i)_{i=1}^2$ we have chosen $a_{ii}^1 = a_{jj}^2 = a$, $i = 3, 5, 7$, and $j = 2, 4, 6$. By taking these premises the method has this structure:

0	0						0						0							
c_2	a_{21}^1	0					0	a					a_{21}^3	0						
c_3	a_{31}^1	0	a				0	a_{32}^2	0				a_{31}^3	a_{32}^3	0					
c_4	a_{41}^1	0	a_{43}^1	0			0	a_{42}^2	0	a			a_{41}^3	a_{42}^3	a_{43}^3	0				
c_5	a_{51}^1	0	a_{53}^1	0	a		0	a_{52}^2	0	a_{54}^2	0		a_{51}^3	a_{52}^3	a_{53}^3	a_{54}^3	0			
c_6	a_{61}^1	0	a_{63}^1	0	a_{65}^1	0	0	a_{62}^2	0	a_{64}^2	0	a	a_{61}^3	a_{62}^3	a_{63}^3	a_{64}^3	a_{65}^3	0		
c_7	b_1^1	0	b_3^1	0	b_5^1	0 a	0	b_2^2	0	b_4^2	0	b_6^2 0	b_1^3	b_2^3	b_3^3	b_4^3	b_5^3	b_6^3	0	
	b_1^1	0	b_3^1	0	b_5^1	0 a	0	b_2^2	0	b_4^2	0	b_6^2 0	b_1^3	b_2^3	b_3^3	b_4^3	b_5^3	b_6^3	0	

By solving the order conditions: (r_1), (r_2), $(\alpha_1), \cdots, (\alpha_4)$, $(\beta_1), \cdots, (\beta_4)$, $(\times_{12})$ and $(\times_{21})$ we obtain a family of FSRK methods $(\mathcal{C}, \mathcal{A}^i, b^i)_{i=1}^2$ of third order which depends of parameters a, c_3, c_4, c_5, c_6, a_{63}^1 and a_{65}^1.

In order to integrate efficiently Stiff IVP's of type (2) we must apply A-stable FSRK methods. To introduce this property for an FSRK method in the simplest way, we apply such method for the numerical integration of the test scalar ODE $y'(t) = (\lambda_1 + \lambda_2) y(t)$ with $Re(\lambda_i) \leq 0$, $i = 1, 2$, obtaining the recurrence $y_{m+1} = \left(1 + \sum_{i=1}^{2} \tau \lambda_i (b^i)^T \left(I - \sum_{j=1}^{2} \tau \lambda_j \mathcal{A}^j\right)^{-1} e\right) y_m$, substituting the values $\tau \lambda_i$ by z_i for $i = 1, 2$ the following rational complex function appears

$$R(z_1, z_2) = 1 + \sum_{i=1}^{2} z_i (b^i)^T \left(I - \sum_{j=1}^{2} z_j \mathcal{A}^j\right)^{-1} e,$$

which it is named the amplification function associated to the FSRK method. We say that an FSRK method is A-stable iff $|R(z_1, z_2)| \leq 1$, $\forall z_i \in \mathbb{C}$ with $Re(z_i) \leq 0$, $i = 1, 2$. Due to this method has seven stages, being the first one explicit, and it attains order 3 its amplification function admits this decomposition

$$R(z_1, z_2) = R_1(z_1) R_2(z_2) + Rest$$

where $R_i(z_i) = \frac{1 + (1-3a) z_i + \frac{1}{2}(1 - 6a + 6a^2) z_i^2 + \frac{1}{6}(1 - 9a + 18a^2 - 6a^3) z_i^3}{(1 - a z_i)^3}$ are the amplification functions of the RK methods $(\mathcal{C}, \mathcal{A}^i, b^i)$ for $i = 1, 2$ and

$$Rest = \frac{F_1 z_1 z_2^3 + F_2 z_1^2 z_2^2 + F_3 z_1^3 z_2 + G_2 z_1^2 z_2^3 + G_3 z_1^3 z_2^2 + H_3 z_1^3 z_2^3}{(1 - a z_1)^3 (1 - a z_2)^3}$$

with:
$$F_j = \sum_{\substack{\bar{i} \in \{1,2\}^4 \\ n_1(\bar{i}) = j}} (b^{i_1})^T \mathcal{A}^{i_2} \mathcal{A}^{i_3} \mathcal{A}^{i_4} e - \frac{1}{j!(4-j)!}, j = 1, 2, 3;$$

$$G_j = -3\,a\,F_{j-1} - 3\,a\,F_j + \sum_{\substack{\bar{i}\in\{1,2\}^5 \\ n_1(\bar{i})=j}} (b^{i_1})^T \mathcal{A}^{i_2} \mathcal{A}^{i_3} \mathcal{A}^{i_4} \mathcal{A}^{i_5} e - \frac{1}{12}, j = 1,2;$$

$$H_3 = 3\,a^2\,F_1 + 9\,a^2\,F_2 + 3\,a^2\,F_3 - 3\,a\,G_2 - 3\,a\,G_3 + \\ + \sum_{\substack{\bar{i}\in\{1,2\}^6 \\ n_1(\bar{i})=3}} (b^{i_1})^T \mathcal{A}^{i_2} \mathcal{A}^{i_3} \mathcal{A}^{i_4} \mathcal{A}^{i_5} \mathcal{A}^{i_6} e - \tfrac{1}{36};$$

being $n_1(\bar{i})$ the number of times that the index 1 appears in vector $\bar{i}$. The L-stability of each RK method $(\mathcal{C}, \mathcal{A}^i, b^i)$ for $i = 1,2$ is obtained by taking $a = 0.435866521508459$ (see [4]). To obtain that the contribution of the *Rest* to the amplification function be as small as possible we use the free parameters a_{63}^1 and a_{65}^1 to annihilate G_2, G_3 and H_3. By substituting the values obtained for these parameters in the terms F_i for $i = 1,2,3$ we obtain $|F_1| = |F_3| = 0.0179331$ and $F_2 = 0.0737263$. In this way it is possible to check that the family of third order FSRK methods developed here, which depends on c_3, c_4, c_5 and c_6, is A-stable.

Next, we consider the order conditions where the explicit part $(\mathcal{C}, \mathcal{A}^3, b^3)$ of the method is involved. Those ones are (r_3), (γ_1), $\cdots$, (γ_4), $(\times_{13})$, $(\times_{23})$, $(\times_{31})$ and $(\times_{32})$. These equations have been solved taking as free parameters c_3, c_4, c_5, c_6, a_{52}^3, a_{54}^3, a_{62}^3, a_{63}^3, a_{64}^3, a_{65}^3 and b_6^3 and fixing a_{21}^3, a_{31}^3, a_{41}^3, a_{51}^3, a_{61}^3, a_{32}^3, a_{42}^3, a_{53}^3, b_1^3, b_2^3, b_3^3, b_4^3 and b_5^3. With the remaining free parameters we have looked for obtaining not too large coefficients and minimizing somehow the main term of the local truncature error. In this way the chosen method was:

$$\left(\frac{\mathcal{A}^1}{(b^1)^T}\right) =$$

$$\begin{pmatrix}
0 \\
0.435866521508459 & 0 \\
-0.290577681005640 & 0 & 0.435866521508459 \\
1.096451335195801 & 0 & -0.878518074441571 & 0 \\
-0.617418271582877 & 0 & 0.617418271582877 & 0 & 0.435866521508459 \\
1.300883538855455 & 0 & -1.394886763694443 & 0 & 0.137589876989833 & 0 \\
\hline
-1.643545951341538 & 0 & 3.090808874540337 & 0 & -0.883129444707258 & 0 & 0.435866521508459 \\
-1.643545951341538 & 0 & 3.090808874540337 & 0 & -0.883129444707258 & 0 & 0.435866521508459
\end{pmatrix}$$

$$\left(\frac{\mathcal{A}^2}{(b^2)^T}\right) = \begin{pmatrix} 0 & & & & & & & \\ 0 & 0.435866521508459 & & & & & & \\ 0 & 0.145288840502819 & 0 & & & & & \\ 0 & -0.217933260754229 & 0 & 0.435866521508459 & & & & \\ 0 & 0.847478702563710 & 0 & -0.411612181055251 & 0 & & & \\ 0 & -0.637403153856101 & 0 & 0.245123284498488 & 0 & 0.435866521508459 & & \\ 0 & 2.480644479489623 & 0 & -2.963599628502850 & 0 & 1.482955149013227 & 0 & \\ 0 & 2.480644479489623 & 0 & -2.963599628502850 & 0 & 1.482955149013227 & 0 & \end{pmatrix}$$

$$\left(\frac{\mathcal{A}^3}{(b^3)^T}\right) =$$

$$\begin{pmatrix} 0 & & & & & & & \\ 0.435866521508459 & 0 & & & & & & \\ 0.098722266932037 & 0.046566573570781 & 0 & & & & & \\ 0.346958806704806 & -0.129025545950576 & 0 & 0 & & & & \\ -0.452615864750442 & 0 & 0.888482386258 9017 & 0 & 0 & & & \\ 0.043586652150845 & 0 & 0 & 0 & 0 & 0 & & \\ 0.963682249493564 & 0.549076317251797 & -2.771795439918187 & -0.325917606041921 & 1.584954479 2147467 & 1 & 0 & \\ 0.963682249493564 & 0.549076317251797 & -2.771795439918187 & -0.325917606041921 & 1.584954479 2147467 & 1 & 0 & \end{pmatrix}$$

$$\mathcal{C}\,e = \big(0, 0.435866521508459, 0.145288840502819, 0.217933260754229, 0.435866521508459,$$
$$0.043586652150845, 1\big)^T.$$

3 Numerical tests

In this section we show two numerical experiences where we have integrated some convection-diffusion-reaction problems with non-linear reaction term. In both cases we have carried out the time discretization by using the method described in previous section and we have discretized in space on a rectangular mesh by using a central difference scheme for the first example and an upwind scheme for the second one.

In [2] it is proven that the combination of methods of type finite differences of order q in the spatial discretization stage and pth–order time integrators of the type described in the first section of this paper provides a totally discrete scheme whose global error is $[u(t_m)]_h - U_h^m = \mathcal{O}(h^q + \tau^p)$; $[v]_h$ denotes the restriction of the function v to the mesh node.

The numerical maximum global errors have been estimated as follows

$$E_{N,\tau} = \max_{x_{1i}, x_{2i}, t_m} |U^{N,\tau}(x_{1i}, x_{2i}, t_m) - U^*|$$

where $U^{N,\tau}(x_{1i}, x_{2i}, t_m)$ are the numerical solutions obtained in the mesh point (x_{1i}, x_{2i}) in the time point $t_m = m\tau$, on a rectangular mesh with $(N+1) \times (N+1)$ nodes and with time step τ and U^* is some reference solution. Concretely, we use a numerical solution calculated in the same mesh points and time steps, but by using a spatial mesh with $(2N+1) \times (2N+1)$

nodes, halving the mesh size, and with time step $\frac{\tau}{2}$. We compute the numerical orders of convergence as $\log_2 \frac{E_{N,\tau}}{E_{2N,\tau/2}}$.

3.1 First example

We integrate the following non-linear convection-diffusion problem with a non-linear reaction term $r(u) = k_1 u + k_2 u + \frac{u^3}{(1+u^2)^2}$

$$\begin{cases} \frac{\partial u}{\partial t} = \Delta u - v_1 u_{x_1} - v_2 u_{x_2} - r(u) + f, & \forall (x_1, x_2, t) \in \Omega \times [0,5], \\ u(x_1, 0, t) = u(x_1, 1, t) = 0, & \forall x_1 \in [0,1] \text{ and } \forall t \in [0,5], \\ u(0, x_2, t) = u(1, x_2, t) = 0, & \forall x_2 \in [0,1] \text{ and } \forall t \in [0,5], \\ u(x_1, x_2, 0) = x_1^3(1-x_1)^3 x_2^3(1-x_2)^3, & \forall (x_1, x_2) \in \Omega, \end{cases}$$

with $\Omega = (0,1) \times (0,1)$, $v_1 = 1 + x_1 x_2 e^{-t}$, $v_2 = 1 + x_1 t$, $k_1 = k_2 = \frac{1+(x_1+x_2)^2 e^{-t}}{2}$ and $f = 10^3 e^{-t} x_1 (1-x_1) x_2 (1-x_2)$. We have carried out the space discretization on a uniform mesh.

To apply the integration method described in the last section we have taken $L_{ih}(t_{m,j}) U_h^{m,j} = \delta_{\overline{x_i} x_i} U_h^{m,j} - [v_i]_h \delta_{\hat{x}_i} U_h^{m,j} - [k_i]_h U_h^{m,j} - \frac{3(U_h^m)^2 - (U_h^m)^4}{2(1+(U_h^m)^2)^3} U_h^{m,j}$ for $i=1,2$ and $g_h(t_{m,j}, U_h^{m,j}) = \frac{3(U_h^m)^2 - (U_h^m)^4}{(1+(U_h^m)^2)^3} U_h^{m,j} - \frac{(U_h^{m,j})^3}{(1+(U_h^{m,j})^2)^2}$ for $j=1,\cdots,7$ where $\delta_{\overline{x_i} x_i}$ and $\delta_{\hat{x}_i}$ denote the classical central differences. Note that, by using this decomposition in each stage of (3) we must solve only one linear system whose matrix is tridiagonal; thus the computational complexity of computing the internal stages of this method is of the same order as an explicit method.

In table 1 we show the numerical errors and in table 2 their corresponding numerical orders of convergence. In order to obtain that the contribution to the error of the spatial and temporal part be of the same size we have taken the relation $\sqrt[3]{N^2 \tau} \equiv C = 0.1$ between the time step τ and the mesh size $\frac{1}{N}$.

Table 1. Numerical Errors ($E_{N,\tau}$)

N=8	N=16	N=32	N=64	N=128	N=256	N=512
2.2183E-2	5.6902E-3	1.4231E-3	3.5569E-4	8.8921E-5	2.2289E-5	5.6054E-6

Table 2. Numerical orders of convergence

N=8	N=16	N=32	N=64	N=128	N=256
1.9629	1.9995	2.0003	2.0000	1.9962	1.9915

3.2 Second example

Now, we show the numerical results obtained when we integrate the following problem:
$$\begin{cases} \frac{\partial u}{\partial t} = \varepsilon \Delta u - v_1 u_{x_1} - v_2 u_{x_2} - r(u) + f, \; \forall (x_1, x_2, t) \in \Omega \times [0,5], \\ u(x_1, 0, t) = u(x_1, 1, t) = 0, \; \forall x_1 \in [0,1] \text{ and } \forall t \in [0,5], \\ u(0, x_2, t) = u(1, x_2, t) = 0, \; \forall x_2 \in [0,1] \text{ and } \forall t \in [0,5], \\ u(x_1, x_2, 0) = h(x_1) h(x_2), \; \forall x_1, x_2 \in \overline{\Omega}, \end{cases}$$

with $\Omega = (0,1) \times (0,1)$, $v_1 = 1 + x_1 x_2 e^{-t}$, $v_2 = 1 + x_1 + x_2 t$, $r(u) = k_1 u + k_2 u + u^4 e^{u^4}$, $k_1 = k_2 = 1 + 2 x_1 + e^{-t}$, $h(z) = e^{\frac{-z}{\varepsilon}} + e^{\frac{-(1-z)}{\varepsilon}} - 1 - e^{\frac{-1}{\varepsilon}}$ and the source term $e^{-2t} x_1 (1 - x_1) x_2 (1 - x_2)$. In this case, we are chosen the same type of splitting $L_{ih}(t)$ as in the previous example excepting that the central differences $\delta_{\hat{x}_i}$ are substituted by first order backward differences.

In this problem a small parameter ε appears multiplying to the second order derivatives; this generates that the solution of this type of problems presents a multiscale character with some quick variation zones called boundary layers. Due to the existence of these layers we must realize a suitable space discretization; thus we use a mesh which concentrates points in the boundary layers. In [3] it is proven that it is not possible to obtain the ε-uniform convergence for some convection-diffusion problems of this type by using uniform meshes. We carry out the spatial discretization by using again the simple upwind scheme on a mesh which is constructed as tensorial product of onedimensional Shishkin meshes; in these onedimensional meshes the nodes are distributed as $x_{ik} = k \frac{4(1-\sigma)}{3N}$ if $k = 0, \cdots, \frac{3N}{4}$ and $x_{ik} = (1-\sigma) + (k - \frac{3N}{4}) \frac{4\sigma}{N}$ if $k = \frac{N}{4} + 1, \cdots, \frac{3N}{4}$, with $\sigma = \min\{\frac{1}{4}, \varepsilon \log N\}$ for $i = 1, 2$. By using these meshes we obtain a numerical order of convergence which tend to 1 as long as N increases according to the expected ε-uniformly convergent behaviour $(\tau^3 + N^{-1} \log N)$. In order to obtain that the contribution to the error of the spatial and temporal part are of the same order we have taken the relation $\sqrt[3]{N} \tau \equiv C = 0.1$. We have evaluated these errors from $t = 0.05$ up to $T = 5$. In the beginning of the time integration process an order reduction occurs because of data in $t = 0$ do not verify sufficient compatibility conditions to attain order $p = 3$.

Table 3. Numerical orders of convergence

ε	N=8	N=16	N=32	N=64	N=128	N=256	N=512
1	9.2059E-5	5.6752E-5	3.0109E-5	1.5864E-5	8.0341E-6	4.0479E-6	2.0316E-6
10^{-2}	6.4953E-2	4.6823E-2	3.6885E-2	2.7546E-2	1.6197E-2	1.0136E-2	6.5119E-3
10^{-4}	7.1269E-2	4.9975E-2	3.5713E-2	2.5482E-2	1.5825E-2	9.6886E-3	6.0157E-3
10^{-6}	7.1352E-2	5.0000E-2	3.5651E-2	2.5361E-2	1.5825E-2	9.6815E-3	6.0053E-3
10^{-8}	7.1353E-2	5.0000E-2	3.5650E-2	2.5361E-2	1.5822E-2	9.6815E-3	6.0052E-3
10^{-10}	7.1353E-2	5.0000E-2	3.5650E-2	2.5361E-2	1.5822E-2	9.6815E-3	6.0052E-3
$E_{N,\tau}^{max}$	7.1353E-2	5.0000E-2	3.6885E-2	2.7546E-2	1.6197E-2	1.0136E-2	6.5119E-3

Table 4. Numerical Errors $E_{N,T}$

ε	N=8	N=16	N=32	N=64	N=128	N=256
1	0.6979	0.9145	0.9245	0.9815	0.9889	0.9946
10^{-2}	0.4722	0.3442	0.4212	0.7661	0.6762	0.6384
10^{-4}	0.5121	0.4847	0.4869	0.6873	0.7079	0.6875
10^{-6}	0.5130	0.4880	0.4913	0.6808	0.7086	0.6890
10^{-8}	0.5130	0.4880	0.4913	0.6807	0.7086	0.6890
10^{-10}	0.5130	0.4880	0.4913	0.6807	0.7086	0.6890
Minimum	0.4722	0.3442	0.4212	0.6873	0.6762	0.6384

References

1. Bujanda, B., Jorge, J.C.: Order conditions for linearly implicit Fractional Step Runge-Kutta methods, submitted.
2. Bujanda, B., Jorge, J.C.: Efficient linearly implicit methods for nonlinear multi-dimensional parabolic problems. J. Comp. Appl. Math. **164**/165, 159–174 (2004)
3. Farrell, P.A., Miller, J.J.H., O'Riordan, E., Shishkin, G.I.: On the non-existence of ε-uniform finite difference methods on uniform meshes for semilinear two-point boundary value problems. Math. of Comp. **67**, 603–617 (1998)
4. Hairer, E., Wanner, G.: Solving ordinary differential equations II, Springer-Verlag (1987)

Numerical Solution of Optimal Control Problems with Sparse SQP-Methods

Georg Wimmer[1], Thorsten Steinmetz[2] and Markus Clemens[3]

Chair for Theory of Electrical Engineering and Computational Electromagnetics
Helmut-Schmidt-University, University of the Federal Armed Forces Hamburg, Germany
g.wimmer@hsu-hh.de[1], t.steinmetz@hsu-hh.de[2], m.clemens@hsu-hh.de[3]

Summary. Many physical processes can be modelled mathematically by ordinary differential equations. If such a process is governed by control variables an optimal control problem can be formulated. The basic problem is to choose the control variables such that some objective function is optimized while satisfying the differential equations. Approximating the control variables by linear functions and the state variables by low order Runge-Kutta schemes results in a nonlinear sparse constrained optimization problem. The inner iteration of a SQP-algorithm consists in solving an equality constrained quadratic optimization problem with a positive definite system matrix and a sparse constraint matrix. This optimization problem can be solved effectively by a projected cg-method when using a sparse LU decomposition of the constraint matrix. Since the system matrix is approximated by the ℓ-BFGS update scheme the matrix is not stored explicitly. Only the vectors which are used for the computation of the system matrix are stored. The cg-method simply needs a matrix vector product with the system matrix. Hence, the explicit computation of this matrix is not necessary. Instead, the matrix vector product is performed by a Neville like scheme. Numerical results are given for a problem in aerospace engineering.

1 Introduction

The basic problem in optimal control is to choose control variables such that some objective function is optimized while satisfying differential equations and boundary conditions. Optimal control problems can be solved either by the variational (indirect method) or by the nonlinear programming approach (direct method) [1], [5]. Indirect methods are characterized by using the calculus of variations. The optimal control is obtained by the maximum principle as a function of state and adjoint variables. Together with boundary conditions we get a nonlinear multi-point boundary value problem [6]. This approach has proven to be quite stable but there are at least two major drawbacks. It is necessary to derive an analytical expression for the control variable which can become quite time consuming for complicated problems. Furthermore the region of convergence for a root finding algorithm (Newton's method) may be extremely small and therefore good start values are needed. The direct method approximates the control in a finite dimensional space and results in a large finite dimensional nonlinear programming problem. Using a SQP-algorithm [8] the search direction is determined as the solution of an equality constrained quadratic optimization problem EQP. A Q-transformed SQP-method for example needs the factorization of the constraints and the system matrix is updated by a rank one correction [9]. Here, the solution of EQP is obtained by a projected cg-algorithm using a ℓ-BFGS update scheme [14]. Within this proceeding the whole Hessian matrix need not to be stored but only some vectors from which the Hessian can be calculated.

2 Optimal control problem

An optimal control problem (OCP) consists in finding a piecewise continuous control function $u : [t_0, t_f] \longrightarrow \mathbb{R}^k$ and the continous piecewise differentiable state function $y : [t_0, t_f] \longrightarrow \mathbb{R}^n$ which minimizes the functional $\phi(y(t_f))$ subject to differential equations and boundary conditions.

$$
\begin{aligned}
&\min_u \phi(y(t_f)) \\
\text{(OCP)} \quad &\text{subject to} \\
&\dot{y}(t) = f(t, y(t), u(t)), \quad t \in [t_0, t_f] \\
&r(y(t_0), y(t_f)) = 0
\end{aligned}
\tag{1}
$$

$\phi : \mathbb{R}^n \longrightarrow \mathbb{R}$, $f : [t_0, t_f] \times \mathbb{R}^n \times \mathbb{R}^k \longrightarrow \mathbb{R}^n$, $r : \mathbb{R}^n \times \mathbb{R}^n \longrightarrow \mathbb{R}^{n_r}$.

Approximation of the control function

The control function on the interval $[t_0, t_f]$ is approximated by a B-Spline $\mathcal{B}^p$ of degree p. $N_{i,p}$ $(i = 0, ..., l)$ denote the basis functions of the B-Spline [15].

$$\mathcal{B}^p : [0, 1] \to \mathbb{R}^{k+1}, \tau \mapsto \begin{pmatrix} t(\tau) \\ u(\tau) \end{pmatrix} = \begin{pmatrix} t(\tau) \\ u_1(\tau) \\ \vdots \\ u_k(\tau) \end{pmatrix}, \tag{2}$$

$$t(\tau) = t_0 + (t_f - t_0) \cdot \tau, \tag{3}$$

$$u_j(\tau) = \sum_{i=0}^{l} N_{i,p}(\tau) \cdot c_j^i, \quad j = 1, ..., k, \quad l \geq p. \tag{4}$$

Approximation of the state function

The discretized state variables $\bar{y}_i$ $(i = p, ..., l)$ at the knots $t(\tau_p), ..., t(\tau_{l+1})$ are obtained by integrating the differential equation in (1) with a one step method (OSM):

$$\bar{y}_p := y(t_0), \tag{5}$$
$$\bar{y}_{i+1} := \bar{y}_i + h_i \Phi(t(\tau_i), \bar{y}_i; h_i; f), \qquad i = p, \ldots, l. \tag{6}$$

The increment function is denoted by Φ and $h_i := t(\tau_{i+1}) - t(\tau_i)$ are the stepsizes of the OSM. In the example of Sect. 6 a linear continuous function as control ($p = 1$) and a Runge-Kutta method of order 4 are used as OSM. In order to obtain the gradients of the objective function and the Jacobian matrix of the constraints the sensitivity matrices with respect to the initial value and the parameters have to be calculated. This is achieved by the internal numerical differentiation described by Bock [4].

3 SQP-methods

After collecting the unknown intitial value $y(t_0)$ and the control parameters c_j^i into the vector $x \in \mathbb{R}^{n_x}$ a nonlinear programming problem (NLP) is obtained from (1) and (2)–(6):

$$\text{(NLP)} \quad \begin{array}{c} \min_x \ f(x) \\ \text{subject to} \\ c_j(x) = 0, \quad j = 1, \ldots, n_r \end{array} \tag{7}$$

with $f : \mathbb{R}^{n_x} \longrightarrow \mathbb{R}$, $c : \mathbb{R}^{n_x} \longrightarrow \mathbb{R}^{n_r}$ ($c = (c_1, \ldots, c_{n_r})^T$). At a constrained minimizer x^* the objective gradient ∇f can be written as a linear combination of the constraint gradients $\nabla c_1, \ldots, \nabla c_{n_r}$. This coefficients are called Lagrange multipliers λ^*. Given a starting vector x_0 a SQP-algorithm creates iterates (x_k, λ_k) which converge to (x^*, λ^*). The next iterate (x_{k+1}, λ_{k+1})

$$\begin{pmatrix} x_{k+1} \\ \lambda_{k+1} \end{pmatrix} := \begin{pmatrix} x_k \\ \lambda_k \end{pmatrix} + \alpha_k \begin{pmatrix} d_k \\ \mu_k - \lambda_k \end{pmatrix} \tag{8}$$

is obtained from an equality constrained quadratic optimization problem (EQP) where d_k and μ_k denote the solution and the corresponding Lagrange multiplier. The step restriction parameter is the result of a linesearch strategy to ensure a decrease of a suitable merit function [10].

$$\min_{d\in\mathbb{R}^n} \left[d^T \nabla_x f(x_k) + \frac{1}{2} d^T H_k d \right]$$

(EQP) subject to

$$(\nabla_x c_j(x_k))^T d + c_j(x_k) = 0, \quad j = 1, \ldots, n_r$$

(9)

Starting with the identity matrix $H_0 = I$ the positive definiteness of all H_k can be maintained by the BFGS-update

$$\bar{H} = U_{\mathrm{BFGS}}(s, y, H) := H + \frac{yy^T}{y^T s} - \frac{1}{s^T H s} H s s^T H, \quad s := x_{k+1} - x_k. \quad (10)$$

The matrix H_k is intented to be a positive definite approximation of the Hessian of the augmented Lagrange function

$$L(x, \lambda; \omega) := l(x, \lambda) + \frac{1}{2} \sum_{i=1}^{n_r} \omega_i (c_i(x))^2$$

$$= f(x) - \sum_{i=1}^{n_r} \lambda_i c_i(x) + \frac{1}{2} \sum_{i=1}^{n_r} \omega_i (c_i(x))^2. \quad (11)$$

The coefficients ω_i can be chosen to yield $(y^L)^T s > 0$, where

$$y := L_x(x_{k+1}, \mu_{k+1}) - L_x(x_k, \mu_{k+1}). \quad (12)$$

The BFGS-update is in this form a rank two correction. When using a Q-transformed Hessian method the transformed Cholesky factors are updated by a rank one correction [9].

4 Equality constrained quadratic subproblem

In order to solve EQP iteratively (9) is rewritten as follows:

$$\min_x \left[b^T x + \frac{1}{2} x^T H x \right]$$

(EQP) subject to

$$Wx = w$$

(13)

Extending the conjugate gradient method to (13) the algorithm depicted in Table 1 is obtained. The orthogonal projection in the kernel of W is denoted by P_Z.

Table 1. Projected cg-algorithm

1: Take start vector x_0 with $Wx_0 = w$
2: $k = 0$
3: $r_0 = -(b + Hx_0)$
4: $g_0 = P_Z r_0$
5: **while** $g_k \neq 0$ **do**
6: $\quad k = k + 1$
7: $\quad$ **if** $k = 1$ **then**
8: $\quad\quad p_1 = g_0$
9: $\quad$ **else**
10: $\quad\quad \beta_k = r_{k-1}^T g_{k-1} / r_{k-2}^T g_{k-2}$
11: $\quad\quad p_k = g_{k-1} + \beta_k p_{k-1}$
12: $\quad$ **end if**
13: $\quad \alpha_k = r_{k-1}^T g_{k-1} / p_k^T H p_k$
14: $\quad x_k = x_{k-1} + \alpha_k p_k$
15: $\quad r_k = r_{k-1} - \alpha_k H p_k$
16: $\quad g_k = P_Z r_k$
17: **end while**
---

5 Approximation of the Hessian

Given an initial approximation H_0, step vectors $s_0, \ldots, s_{p-1}$ and the gradients $y_0, \ldots, y_{p-1}$ of the augmented Lagrangian function [10] the Hessian matrix H_p is defined as

$$H_1 = H_0 + U(s_0, y_0, H_0), \qquad (14)$$
$$H_2 = H_1 + U(s_1, y_1, H_1), \qquad (15)$$
$$\vdots$$
$$H_p = H_{p-1} + U(s_{p-1}, y_{p-1}, H_{p-1}), \qquad (16)$$

$$U(s, y, H) := \frac{yy^T}{y^T s} - \frac{(Hs)(Hs)^T}{s^T H s}. \qquad (17)$$

In the algorithm of Table 1 the Hessian matrix does not have to be computed explicitly if its product with a vector is available. This matrix vector multiplication $H_p z$ can be calculated according to the following tableau:

$$
\begin{array}{ccccccc}
& y_{p-1} & & y_{p-2} & \cdots & y_1 & y_0 \\
\hline
H_p z \leftarrow H_{p-1}s_{p-1} & \stackrel{H_{p-2}s_{p-2}}{\longleftarrow} & H_{p-2}s_{p-1} & \cdots & H_1 s_{p-1} & \stackrel{H_0 s_0}{\longleftarrow} & H_0 s_{p-1} \\
\nwarrow & & & & & & \\
H_{p-1}z & \longleftarrow & H_{p-2}s_{p-2} & \cdots & H_1 s_{p-2} & \stackrel{H_0 s_0}{\longleftarrow} & H_0 s_{p-2} \\
& \nwarrow & & & & & \\
& H_{p-2}z & \cdots & H_1 s_{p-3} & \stackrel{H_0 s_0}{\longleftarrow} & H_0 s_{p-3} \\
& & \ddots & \vdots & \vdots & \vdots \\
& & & H_1 s_2 & \stackrel{H_0 s_0}{\longleftarrow} & H_0 s_2 \\
& & & H_1 s_1 & \stackrel{H_0 s_0}{\longleftarrow} & H_0 s_1 \\
& & & H_1 z & \longleftarrow & H_0 s_0 \\
& & & & \nwarrow & \\
& & & & & H_0 z
\end{array}
$$

If the Jacobian matrix of the contraints is sparse it is advantageous to use sparse matrix methods for the linear algebra [11]. The Schur complement method of Gill et. al [8] makes use of the fact that similar quadratic subproblems have to be solved. Using the ℓ-BFGS approximation of the Hessian matrix of the augmented Lagrangian function this matrix has to be computed explicitly. This can be avoided by applying the ℓ-BFGS update in connection with the cg-method. The projection $g = P_Z r$ can be obtained as the solution of the augmented system [12]

$$\begin{pmatrix} I_n & W^T \\ W & 0 \end{pmatrix} \begin{pmatrix} g \\ v \end{pmatrix} = \begin{pmatrix} r \\ 0 \end{pmatrix}. \tag{18}$$

Since it is important that the projection is performed accurately the projection scheme is applied several times in order to guarentee that the result lies with sufficient accuracy in the kernel of W. Therefore the sparse LU decomposition of (18) can be reused.

6 Example

The planar orbit transfer of a satellite from a near earth orbit to a geostationary orbit in minimum time [13], [5] is modelled by the two dimensional point mass equations of motion in an inverse square gravity field. The states of the system are radial position r, radial velocity w, circumferential velocity v and polar angle ϕ. The control is represented by the thrust direction angle ψ. The optimal control problem is stated in equation (19).

$$\min_{\psi} \left[t_f(1) \right]$$

$$\dot{r} = t_f \omega, \quad \dot{\phi} = t_f (v/r), \quad \dot{t}_f = 0$$

$$\dot{\omega} = t_f \left[\frac{v^2}{r} - \frac{r_\mu}{r^2} + \beta \sin\psi \right], \quad \dot{v} = t_f \left[-\frac{\omega v}{r} + \beta \cos\psi \right]$$

$$r(0) = r_0$$
$$\omega(0) = \omega_0 \qquad\qquad r(1) = r_f$$
$$v(0) = v_0 = \sqrt{\frac{r_\mu}{r_0}} \qquad \omega(1) = \omega_f$$
$$\phi(0) = \phi_0 \qquad\qquad v(1) = v_f = \sqrt{\frac{r_\mu}{r_f}}$$
$$t_f(0) = 0$$

(19)

The constants are given in table 2. For better numerical performance scaled values are used.

Table 2. Constants

constant	unscaled	scaled
r_0	$6.0 \cdot R_e$	6.00
r_f	$6.6 \cdot R_e$	6.60
ω_0	0 [rad]	0.00
ω_f	0 [rad]	0.00
ϕ_0	0 [rad]	0.00
β	1/637.82 [m/s]	0.01
R_e	6378.2 [km]	1.00
r_μ	$39.9 \cdot 10^4$ [km^3/s^2]	62.50

As starting values for the control $u \equiv 0$ and for the states the boundary condition at t_0 are chosen. Table 3 shows the performance of the proposed algorithm. The SQP-algorithm converges in 30 iterations due to the limitation

Table 3. Numerical results

number of SQP-iterations	30
total number of cg-iterations for (EQP)	185
total number of projections P_Z	219
average number of cg-iterations per SQP-step	6.2
average number projections P_Z per SQP-step	7.3

to only four BFGS vectors. However this drawback is compensated by the cheap matrix vector products in the cg-algorithm.

References

1. Barclay, A., Gill, P.E., Rosen, J.B.: SQP Methods and their Application to Numerical Optimal Control. In: Schmidt, W., Heier, K., Bittner, L., Bulirsch, R. (ed) Variational Calculus, Optimal Control and Applications, ISNM 124. Birkhäuser, Basel, 207–222 (1998)
2. Betts, J.T., Huffman, W.P.: Mesh Refinement in Direct Transcription Methods for Optimal Control. Optim. Control Appl. Meth., **22**, 1–21 (1998)
3. Bock, H.G., Plitt, K.J.: A Multiple Shooting Algorithm for Direct Solution of Optimal Control Problems. Proceedings of the 9th IFAC World Congress, Pergamon Press, Hungary, Budapest (1984)
4. Bock, H.G.: Randwertproblemmethoden zur Parameteridentifizierung in Systemen nichtlinearer Differentialgleichungen. Bonner Mathematische Schriften (1987)
5. Büskens, C., Maurer, H.: SQP-Methods for Solving Optimal Control Problems with Control and State Constraints: Adjoint Variables, Sensitivity Analysis and Real-Time Control. J. Comp. Appl. Math., **120**, 85–108 (2000)
6. Callies, R.: Entwurfsoptimierung und optimale Steuerung: Differentialalgebraische Systeme, Mehrgitter-Mehrzielansätze und numerische Realisierung. Habilitationsschrift, TU München (2000)
7. De Boor, C.: A Practical Guide to Splines. Springer-Verlag, New York (1978)
8. Gill, P.E., Murray W., Saunders, M.A.: SNOPT: An SQP Algorithm for Large-Scale Constrained Optimization. Siam J. Optim., **12**, 979–1006 (2002)
9. Gill, P.E., Golub, G.H., Murray, W., Saunders, M.A.: Methods for Modifying Matrix Factorizations. Math. Comp., **28**, 505–535 (1974)
10. Gill, P.E., Murray, W., Saunders, M.A., Wright, M.H.: Some Theoretical Properties of an Augmented Lagrangian Merit Function. In: Pardalos, P.M. (ed) Advances in Optimization and Parallel Computing, North-Holland, Amsterdam, 101–128 (1992)
11. Gill, P.E., Murray, W., Saunders, M.A., Wright, M.H.: Sparse Matrix Methods in Optimization. Siam J. Sci. Stat. Comput., **5**, 562–589 (1984)
12. Gould, N.I.M., Hribar, M.E., Nocedal, J.: On the Solution of Equality Constrained Quadratic Programming Problems Arising in Optimization. Siam J. Sci. Comput. **23**, 1376-1395 (2001)
13. Kluever, C.A.: Optimal Feedback Guidance for Low-Thrust Orbit Insertion. Opt. Control Appl. Methods, **16**, 155–173 (1995)
14. Liu, D.C., Nocedal, J.: On the Limited Memory BFGS Method for Large Scale Optimization. Math. Prog., **45**, 503–528 (1989)
15. Piegl, L., Tiller, W.: The Nurbs Book. Springer, Berlin (1997)

Optimization

Semi-Deterministic Recursive Optimization Methods for Multichannel Optical Filters

Benjamin Ivorra[1], Bijan Mohammadi[1], Laurent Dumas[2] and Olivier Durand[3]

[1] I3M, University of Montpellier 2, 34095 Montpellier, France
    ivorra@math.univ-montp2.fr
[2] Jacques-Louis Lions Laboratory, University of Paris 6, 75252 Paris, France
[3] Alcatel Research and Innovation, 91460 Marcoussis, France

Summary. In this paper, we reformulate global optimization problems in terms of boundary value problems. This allows us to introduce a new class of optimization algorithms. Indeed, many optimization methods, including non-deterministic ones, can be seen as discretizations of initial value problems for differential equations or systems of differential equations. Two algorithms included in this new class are applied and compared with a genetic algorithm for the design of multichannel optical filters.

1 Introduction

Global minimization (or maximization) problems are of great practical importance in many applications. For this reason, Genetic Algorithms (**GA**) have received a tremendous interest in recent years [1, 2, 5]. However, the main difficulties with these algorithms remain their computational time and their slow convergence.

Many minimization algorithms can be viewed as discrete forms of Cauchy problems for an ordinary differential equation (**ODE**) or a system of ODEs in the space of control parameters. We will see that if one introduces an extra information on the infimum, these algorithms can be formulated as Boundary Value Problems (**BVP**) for the same equations [3, 4]. A motivating idea is therefore to apply algorithms solving BVPs to global optimization. It is in particular shown that GAs be interpreted as a discrete form of BVPs for a set of coupled ODEs. Therefore, the BVP analysis has also been applied to them to improve their performances leading to the construction of a new algorithm called **HGSA**. All the algorithms issued from our BVP analysis, presented in section 2, are compared in section 3 to a classical GA for the design of a multichannel optical filter.

2 Global optimization methods

In this section, we consider a function $J : \Omega_{ad} \to \mathbb{R}$ to be minimized, where the optimization parameter x belongs to a compact admissible set $\Omega_{ad} \subset R^N$.

A unified dynamical system formulation is given for some stochastic and deterministic optimization algorithms. In particular, even if GAs are issued from evolutionary considerations, it is possible to associate to them a set of stochastic coupled ODEs (see subsection 2.2). A new class of global minimization methods is thus constructed, based on the solution of associated BVPs.

2.1 Semi-deterministic recursive optimization methods

We make here the following assumptions on the functional: $J \in C^2(\Omega_{ad}, \mathbb{R})$ and is coercive [3]. In this case, many deterministic minimization algorithms which perform the minimization of J can be seen as discretizations of the following dynamical system [3, 4]:

$$\begin{cases} M(\zeta)\dfrac{dx(\zeta)}{d\zeta} = -d(x(\zeta)) \\ x(0) = x_0 \end{cases} \quad (1)$$

where ζ is a fictitious parameter, M is a local metric transformation, d a direction in Ω_{ad} and $x_0 \in \Omega_{ad}$ is the initial condition.

For example if $d = \nabla J$ is the gradient of the functional J and $M = Id$, we recover the classical steepest descent method, while with $d = \nabla J$ and $M = \nabla^2 J$ the Hessian of J, we recover the Newton method.

A global optimization of J with system (1), called here *core optimization method*, is possible if the following boundary value problem has a solution:

$$\begin{cases} M(\zeta)\dfrac{dx(\zeta)}{d\zeta} = -d(x(\zeta)) \\ x(0) = x_0 \\ J(x(Z_{x_0})) = J_m \quad \text{with a finite } Z_{x_0} \in \mathbb{R} \end{cases} \quad (2)$$

where J_m denotes the minimum of J in Ω_{ad}. In practice, when J_m is unknown, we set J_m to a lower value (for example $J_m = 0$ for an inverse problem) and look for the best solution for a given complexity and computational effort.

The BVP (2) is over-determined as it includes two conditions and only one derivative. The over determination can be removed for instance by considering $x_0 = v$ in (1) as a new variable to be found by the minimization of the new functional:

$$h(v) = J(x_v(Z_v)) - J_m$$

where $x_v(Z_v)$ is the solution of (1) found at $\zeta = Z_v$ starting from v.

A new algorithm $A_1(v_1, v_2)$ with parameters v_1 and v_2 is then defined as:

1- $(v_1, v_2) \in \Omega_{ad} \times \Omega_{ad}$ **given**, $v_1 \neq v_2$

2- Find $v \in argmin_{w \in \mathcal{O}(v_2)} h(w)$ where $\mathcal{O}(v_2) = \mathbb{R}\overrightarrow{v_1 v_2} \cap \Omega_{ad}$
3- return the best v found during step 2

The line search minimization in A_1 might fail. For instance, a secant method degenerates on plateau and critical points. To avoid this problem, we add an external level to the algorithm A_1, keeping v_1 unchanged, and looking for v_2 by minimizing a new functional $w \mapsto h(A_1(v_1, w))$. This leads to the following two-level algorithm $A_2(v_1, v_2')$:

1- $(v_1, v_2') \in \Omega_{ad} \times \Omega_{ad}$ given, $v_1 \neq v_2'$
2- Find $v' \in argmin_{w \in \mathcal{O}(v_2')} h(A_1(v_1, w))$ where $\mathcal{O}(v_2') = \mathbb{R}\overrightarrow{v_1 v_2'} \cap \Omega_{ad}$
3- return the best v' found during step 2

The choice of the initial conditions in this algorithm is its only non-deterministic feature. The algorithm A_2 is thus called Semi-Determinist Algorithm (**SDA**). A mathematical background for this approach as well as a validation on academic test cases or on problems including solutions of nonlinear partial differential equations are available [3, 5, 7, 10].

Remarks:

- The construction can be pursued recursively considering

$$h^i(v_2^i) = \min_{v_2^i \in \Omega_{ad}} h^{i-1}(A_{i-1}(v_1, v_2^i))$$

with $h^1(v) = h(v)$ and where i denotes the external level, justifying the name of recursive optimization methods.

- In practice, this algorithm succeeds if the trajectory passes close enough to the infimum (i.e. in $B_\varepsilon(x_m)$ where ε defines the chosen accuracy in the capture of the infimum). Hence, in the algorithm above, $x_w(Z_w)$ is replaced by the best solution found over $[0, Z_w]$.

2.2 Genetic algorithms

Genetic algorithms approximate the global minimum (or maximum) of any functional $J : \Omega_{ad} \to \mathbb{R}$, also called fitness function, through a stochastic process based on an analogy with the Darwinian evolution of species [1]: a first family, called 'population', $X^0 = \{x_l^0 \in \Omega_{ad}, l = 1, ..., N_p\}$ of N_p possible solutions of the optimization problem, called 'individuals', is randomly generated in the search space Ω_{ad}. Starting from this population, we build recursively N_{gen} new populations $X^i = \{x_l^i \in \Omega_{ad}, l = 1, ..., N_p\}$ with $i = 1, .., N_{gen}$ through three stochastic steps, called selection, crossover and mutation. With these three basic evolution processes, it is generally observed that the best obtained individual is getting closer after each generation to the optimal solution of the problem [1]. An example of such stochastic processes is given below in order to show the analogy with the resolution of a discrete dynamical system.

We first rewrite X^i using the following (N_p, N)-real valued matrix form:

$$X^i = \begin{bmatrix} x_1^i \\ \vdots \\ x_{N_p}^i \end{bmatrix} \qquad (3)$$

Selection: each individual x_l^i is ranked with respect to its fitness value $J(x_l^i)$ and N_p elements are then selected among the population to become 'parents'.

Introducing $\mathcal{S}^n(J(X^n))$ a binary (N_p, N_p)-matrix with, for each line i, a value 1 on the jth row when the jth individual has been selected and 0 elsewhere, we define

$$X^{n+1/3} = \mathcal{S}^n(J(X^n))X^n \qquad (4)$$

Crossover: this process leads to a data exchange between two 'parents' and the apparition of two new individuals called 'childrens'.

Introduce $\mathcal{C}^n$ a real-valued (N_p, N_p)-matrix where for each couple of consecutive lines $(2i-1, 2i)$ ($1 \leq i \leq \frac{N_p}{2}$), the coefficients of the lth and kth rows are given by a 2×2 matrix of the form

$$\begin{bmatrix} \lambda_1 & 1-\lambda_1 \\ \lambda_2 & 1-\lambda_2 \end{bmatrix}$$

In this expression, $\lambda_1 = \lambda_2 = 1$ if no crossover is applied on the selected parents l and k or are randomly chosen in $[0, 1]$ in the other case (with a probability p_c). This step can be summarized as:

$$X^{n+2/3} = \mathcal{C}^n X^{n+1/3} \qquad (5)$$

Mutation: this process leads to new parameters values for some individuals of the population. More precisely, each children is modified (or mutated) with a fixed probability p_m.

Introduce for instance a random perturbation matrix $\mathcal{E}^n$ with a i-th line equal to 0 if no mutation is applied to the ith children and a random value $\epsilon_i \in \mathbb{R}^N$ in the other case. This step can then take the following form:

$$X^{n+1} = X^{n+2/3} + \mathcal{E}^n \qquad (6)$$

or more generally

$$X^{n+1} = f(X^{n+2/3}) \qquad (7)$$

for a certain stochastic operator f in the space of (N_p, N)-real valued matrices.

Therefore, GAs can be seen as discrete dynamical systems, writing for instance in the presented case:

$$X^{n+1} = \mathcal{C}^n \mathcal{S}^n(J(X^n))X^n + \mathcal{E}^n \tag{8}$$

which is a particular discretization of a set of nonlinear first order ODEs of the type:

$$\dot{X}(t) = \Lambda_1(t, J(X(t)), p_c, p_m)X(t) + \Lambda_2(t, p_c, p_m) \tag{9}$$

where $\{p_c, p_m\}$ are fixed parameters and the construction of Λ_1 and Λ_2 has been described above. Finally, GAs can been interpreted as solving the following BVP:

$$\begin{cases} \dot{X}(t) = \Lambda_1(t, J(X(t)), p_c, p_m)X(t) + \Lambda_2(t, p_c, p_m) \\ X(0) = X^0 \\ \widehat{J}(X(T)) = J_m \end{cases} \tag{10}$$

where $\widehat{J}(X) = min\{J(x_i)/1 \leq i \leq N_p\}$ for any $X = {}^t(x_1, ..., x_{N_p})$

Engineers like GAs because these algorithms do not require sensitivity computation, perform global and multi-objective optimization and are easy to parallelize. However, their drawbacks remain their weak mathematical background, their computational complexity and their slow convergence. As a fine convergence is difficult to achieve with GA based algorithms, it is recommended when it is possible, to complete the GA iterations by a descent method. This is especially useful when the functional is flat around the infimum (see [2] for more complex coupling of GAs with descent methods).

2.3 Hybrid genetic/semi-deterministic algorithm

It is interesting to notice that once GA is seen as a dynamical system (9) for the population, it can be used as a core optimization method in the way presented in subsection 2.1. The aim here is to find a compromise between the robustness of GAs and the low-complexity features of SDAs.

In order to reduce the GA population size while keeping the efficiency of the method, we couple it with a SDA. The SDA provides information on the choice of the initial population X^0 whereas the GA performs global optimization starting from this population. We call this approach **HGSA** (Hybrid Genetic/Semi-deterministic Algorithm).

3 Application to multichannel optical filters design

Many important developments in optical fiber devices for telecommunications have been done in the recent years. Among them are Fiber Bragg Gratings

(**FBG**) which are an attractive alternative in applications such as multichannel multiplexing. FBGs are optical fibers with a modulated refractive index which reflects a part of the wavelength band, called reflected spectrum, and let pass the complementary band called transmitted spectrum [8].

The inverse problem considered here is the design of a given optical filter based on a FBG. More precisely, the objective is to construct a multichannel filter with a reflected spectrum that consists of $N_{peaks} = 16$ totally reflective identical channels spaced of $\Delta\lambda = 0.8$nm. The optimization space consists of all possible FBG refractive index modulation profiles for a given length, namely $L = 103.9$mm.

These refractive index modulation profiles are generated by spline interpolation through a number of $N = 9$ points equally distributed along the first half of the FBG and completed by parity with a maximum refractive index amplitude of $\bar{n}_{max} = 5 \times 10^{-4}$. Thus the search space is defined by $\Omega_{ad} = [-5 \times 10^{-4}, 5 \times 10^{-4}]^9$.

The functional to be minimized in Ω_{ad} is defined by:

$$J(x) = \sum_{i=1}^{N_c} (r(x, \lambda_i) - r_{target}(x, \lambda_i))^2 \qquad (11)$$

where:

- $r(x, .)$ is the reflected spectrum of the FBG with a refractive index modulation profile associated to $x \in \Omega_{ad}$. It is a function defined from the transmission band $[1.530, 1.545]$ (in microns) to $[0, 1]$ which is determined by solving a certain direct problem [9].
- $r_{target}(x, .)$ denotes the nearest perfect reflected spectrum to $r(x, .)$:

$$r_{target}(x, \lambda) = \begin{cases} 1 \text{ if } \lambda \in \{\lambda_x, \lambda_x + \Delta\lambda, \ldots, \lambda_x + (N_{peaks} - 1)\Delta\lambda\} \\ 0 \text{ elsewhere} \end{cases}$$

for a certain λ_x in the transmission band.

Both functions $r(x)$ and $r_{target}(x)$ are evaluated on $N_c = 1200$ wavelengths equally distributed on the transmission band.

Results

The minimization of the cost function (11), has been tested with various algorithms presented in section 2, namely GA, HGSA and SDA algorithms.

The SDA method is applied with a core algorithm consisting of 10 iterations of a steepest descent method and a line search made of 5 iterations of a secant method for each level algorithm A_1 and A_2. The latter is initialized with random initial conditions v_1 and v'_2. As the minimal value of J is unknown, we set $J_m = 0$. Furthermore to reduce SDA computational time, gradient evaluations (representing 90% of this time) are done on a coarse mesh with $N_c = 300$ reducing the evaluation of a factor 4 (4s on a 3Ghz/512Mo Ram-Desktop computer) for a gradient variation of approximately 10%. Such method is called incomplete gradient approach [6].

HGSA and GA are applied with the following values: the population size is set to $N_p = 180$ for GA (respectively 10 for HGSA) and the generation number is set to $N_{gen} = 30$ for GA (resp. 10 for HGSA). The selection is a roulette wheel type [1, 2] proportional to the rank of the individual in the population. The crossover is barycentric in each coordinate with a probability $p_c = 0.45$. The mutation process is non-uniform with a probability $p_m = 0.15$ for GA (resp. 0.35 for HGSA). A one-elitism principle, that consists in keeping the current best individual in the next generation, has also been imposed. Finally, a steepest descent method is performed at the end of both algorithms.

The different optimization results are summarized in Table 1 whereas the corresponding reflected spectra obtained with the optimized profiles are presented on Figure 1. Although only SDA optimization produces 16 totally reflective peaks, GA and HGSA associated spectra are still industrially applicable due to the fact that in practice we only need 95%-reflective peaks [8]. The SDA method also gives the best result in terms of cost function minima and computational time. Note also that the HGSA technique over-perform a classical GA, providing an interesting alternative to SDA in cases where gradients cannot be evaluated.

Table 1. Optimization results

	SDA	HGSA	GA
minimal value of the cost function	3.0	4.3	5.8
Functional Evaluation Number	3000 (90% on coarse mesh)	2600	5500
Computational time	4h	11h	24h

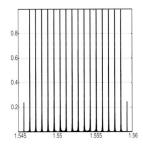
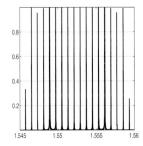
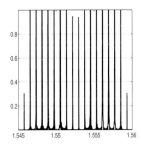

Fig. 1. Reflected spectra of optimized filters(reflexivity vs. wavelength (μm)) obtained with (**Left**) SDA, (**Center**) HGSA and (**Right**) GA.

4 Conclusions

A new class of semi-deterministic methods has been introduced. This approach allows us to improve deterministic and non-deterministic optimization algorithms. Both of them have been detailed and applied to the design of a multichannel optical filter for which the results obtained over-perform those obtained with a classical genetic algorithm.

It represents a new validation of theses methods on industrial problems involving multiple local minima after some previous others: temperature and pollution control in a bunsen flame [10], shape optimization of fast-microfluidic-mixer devices [5], shape optimization of under aerodynamic and acoustic constraints for internal and external flows [6].

References

1. Goldberg, D.: Genetic algorithms in search, optimization and machine learning. Addison Wesley (1989)
2. Dumas, L., Herbert, V., Muyl, F.: Hybrid method for aerodynamic shape optimization in automotive industry. Computers and Fluids. **33**(5), 849-858 (2004)
3. Mohammadi, B., Saiac, J-H.: Pratique de la simulation numérique. Dunod (2002)
4. Attouch, H., Cominetti, R.: A dynamical approach to convex minimization coupling approximation with the steepest descent method. Journal of Differential Equations. **128**(2), 519-540 (1996)
5. Ivorra, B., Mohammadi, B., Santiago, J-G., Hertzog, D-E.: Semi-deterministic and genetic algorithms for global optimization of microfluidic protein folding devices. International Journal of Numerical Methods in Engineering. **66**(2), 319-333 (2006)
6. Mohammadi, B., Pironneau, O.: Applied Shape Optimization for Fluids. Oxford University Press (2001)
7. Ivorra, B., Mohammadi, B., Redont, P., Dumas, L.: Semi-deterministic vs genetic algorithms for global optimization of multichannel Optical Filters. International Journal of Computational Science and Engineering. To be published. (2006)
8. Chow, J., Town, G., Eggleton, B., Ibsen, M., Sugden, K., Bennion, I.: Multiwavelength Generation in an Erbium-Doped Fiber Laser Using In-Fiber Com Filters. IEEE Photonic Technology Letter. **8**(1), 60-62 (1996)
9. Erdogan, T.: Fiber Grating Spectra. Journal of Lightwave Technology. **15**(8), 1277-1294 (1997)
10. Debiane, L., Ivorra, B., Mohammadi, B., Nicoud, F., Ern, A., Poinsot, T., Pitsch, H.: Temperature and pollution control in flames. Proceeding of the CTR Summer Program 2004. 367-375 (2004)

A Multigrid Method for Coupled Optimal Topology and Shape Design in Nonlinear Magnetostatics*

Dalibor Lukáš

Department of Applied Mathematics, VŠB–Technical University of Ostrava
dalibor.lukas@vsb.cz

Summary. Topology optimization searches for an optimal distribution of material and void without any restrictions on the structure of the design geometry. Shape optimization tunes the shape of the geometry, while the topology is fixed. In the paper we propose a sequential coupling so that a coarsely optimized topology is the initial guess for the following shape optimization. We aim at making this algorithm fast by using the adjoint sensitivity analysis to the Newton-method for the governing nonlinear state equation and a multigrid approach for the shape optimization. A finite element discretization method is employed. Numerical results are given for a 2–dimensional optimal design of a direct electric current electromagnet.

1 Introduction

In the process of development of industrial components one looks for the parameters to be optimal subject to a proper criterion. The geometry is usually crucial as far as the design of electromagnetic components is concerned. We can employ topology optimization, cf. [1], to find an optimal distribution of the material without any preliminary knowledge. Shape optimization, cf. [4, 5], is used to tune shapes of a known initial design. While in the structural mechanics topology optimization results in rather complicated structures the shapes of which are not needed to be then optimized, in magnetostatics we end up with simple topologies which, however, serve as very good initial guesses for the further shape optimization. The idea here is to couple them sequentially.

In [2] a connection between topological and shape gradient is shown and applied in structural mechanics. They proceed shape and topology optimization simultaneously so that at one optimization step both the shape and topology gradient are calculated. Then shapes are displaced and the elements with

---

* This research has been supported by the Czech Ministry of Education under the grant AVČR 1ET400300415 and by the Czech Grant Agency under the grant GAČR 201/05/P008.

great values of the topology gradient are removed, while introducing the natural boundary condition along the new parts, e.g. a hole. Here we are rather motivated by the approach in [6, 8]. In the latter they apply a similar algorithm as we do to structural mechanics, however, using re-meshing in a CAD software environment, which was computationally very expensive. Our aim here is to make the algorithm fast. Therefore, we additionally employ semianalytical sensitivity analysis and a multilevel method.

2 Topology optimization for magnetostatics

Let $\Omega \subset \mathbf{R}^2$ be a convex computational domain that is divided into a Lipschitz subdomain $\Omega_d \subset \Omega$, where the optimal distribution of the ferromagnetics and the air is to be find, and into the air Lipschitz subdomain $\Omega_0 := \Omega \setminus \overline{\Omega_d}$. Let further $Q := \{\rho \in L^2(\Omega_d) : 0 \leq \rho \leq 1, \int_{\Omega_d} \widetilde{\rho}(\rho) \, d\mathbf{x} \leq V_{\max}\}$ be a set of admissible material distributions, where $V_{\max} > 0$ is a maximal possible area occupied by the ferromagnetics and where $\widetilde{\rho} \in C^2((0,1))$ penalizes the values of $\rho \in [0, 1/2)$ to vanish and the values of $\rho \in (1/2, 1]$ to approach 1 as follows:

$$\widetilde{\rho}(\rho) \equiv \widetilde{\rho}_{p_\rho}(\rho) := \frac{1}{2}\left(1 + \frac{1}{\arctan(p_\rho)}\arctan(p_\rho(2\rho - 1))\right).$$

Finally, let $J : H^1(\Omega) \mapsto \mathbf{R}$ be a cost functional. We consider the following topology optimization problem:

$$\text{Find } \rho^* \in Q : J(u(\rho^*)) \leq J(u(\rho)) \quad \forall \rho \in Q \tag{1}$$

with respect to the 2-dimensional nonlinear magnetostatic state problem: Find $u(\rho) \in H^1_0(\Omega)$ so that

$$\forall v \in H^1_0(\Omega) : \int_\Omega \nu_0 \, \mathbf{grad}(u(\rho)) \cdot \mathbf{grad}(v) \, d\mathbf{x} + \int_{\Omega_d} \widetilde{\rho}(\rho) \, \nu(\|\mathbf{grad}(u(\rho))\|) \, \mathbf{grad}(u(\rho)) \cdot \mathbf{grad}(v) \, d\mathbf{x} = \int_\Omega Jv \, d\mathbf{x}, \tag{2}$$

where $\nu \in C^2((0,\infty))$ denotes a nonlinear material reluctivity of the ferromagnetics, $\nu_0 := 4\pi 10^{-7}$ [H/m] is the vacuum reluctivity constant and $J \in L^2(\Omega)$ is a current density. Note that in general, one has to pose an additional regularization of the topology ρ to avoid the so-called checkerboard effect. However, we are merely interested in a coarsely discretized problem, for which this ill-posedeness is neglectable.

2.1 Nonlinear state sensitivity analysis

When solving the problem (1), we use a nested approach, i.e. for a given design we eliminate the nonlinear state equation (2). The latter is discretized by the

finite element method using the linear Lagrange nodal elements on triangles, which reads as follows:
$$\mathbf{A}(\mathbf{u}(\boldsymbol{\rho}), \boldsymbol{\rho}) \cdot \mathbf{u}(\boldsymbol{\rho}) = \mathbf{f}, \tag{3}$$
where $\mathbf{A} \in \mathbb{R}^{n \times n}$ is the assembled reluctivity matrix, $\mathbf{f} \in \mathbb{R}^n$ is the right-hand side vector, $\mathbf{u} \in \mathbb{R}^n$ is the solution vector and $\boldsymbol{\rho} \in \mathbb{R}^m$ is the element-wise constant material function.

The problem (3) is solved by the Newton method. Moreover, the optimization algorithm under consideration requires the gradients of the cost functional with respect to $\boldsymbol{\rho}$. To this goal, we derived the corresponding adjoint Newton method by differentiating the original Newton method in the backward sense. The algorithms are depicted below. Note that for both of them the amount of computational work is the same.

Newton method
  Given $\boldsymbol{\rho}$
  $i := 0$
  Solve $\mathbf{A}(0, \boldsymbol{\rho}) \cdot \mathbf{u}^0 = \mathbf{f}$
  $\mathbf{f}^0 := \mathbf{f} - \mathbf{A}(\mathbf{u}^0, \boldsymbol{\rho}) \cdot \mathbf{u}^0$
  while $\|\mathbf{f}^i\|/\|\mathbf{f}\| > \text{prec}$ **do**
    $i := i + 1$
    Solve $\mathbf{A}'_\mathbf{u}(\mathbf{u}^{i-1}, \boldsymbol{\rho}) \cdot \mathbf{w}^i = \mathbf{f}^{i-1}$
    Find $\tau^i : \|\mathbf{f}^i(\tau^i)\| < \|\mathbf{f}^{i-1}\|$
    $\mathbf{u}^i := \mathbf{u}^{i-1} + \tau^i \mathbf{w}^i$
    $\mathbf{f}^i := \mathbf{f} - \mathbf{A}(\mathbf{u}^i, \boldsymbol{\rho}) \cdot \mathbf{u}^i$
    Store $\mathbf{w}^i$ and τ^i
  end while
  Store $\mathbf{u}^i$ and $k := i$
  Calculate objective $J(\mathbf{u}^i, \boldsymbol{\rho})$

Adjoint Newton method
  Given $\boldsymbol{\rho}, k, \mathbf{u}^k, \{\mathbf{w}^i\}_{i=1}^k$ and $\{\tau^i\}_{i=1}^k$
  $\boldsymbol{\lambda} := J'_\mathbf{u}(\mathbf{u}^k, \boldsymbol{\rho})$
  $\omega := 0$
  for $i := k, \ldots, 1$ **do**
    $\mathbf{u}^{i-1} := \mathbf{u}^i - \tau^i \mathbf{w}^i$
    Solve $\mathbf{A}'_\mathbf{u}(\mathbf{u}^{i-1}, \boldsymbol{\rho})^T \cdot \boldsymbol{\eta} = \boldsymbol{\lambda}$
    $\omega := \omega + \tau^i \mathbf{G}_{\boldsymbol{\rho}}(\mathbf{u}^{i-1}, \mathbf{w}^i, \boldsymbol{\rho})^T \cdot \boldsymbol{\eta}$
    $\boldsymbol{\lambda} := \boldsymbol{\lambda} + \tau^i \mathbf{G}_\mathbf{u}(\mathbf{u}^{i-1}, \mathbf{w}^i, \boldsymbol{\rho})^T \cdot \boldsymbol{\eta}$
  end for
  Solve $\mathbf{A}(0, \boldsymbol{\rho})^T \cdot \boldsymbol{\eta} = \boldsymbol{\lambda}$
  $\frac{dJ(\mathbf{u}^k(\boldsymbol{\rho}), \boldsymbol{\rho})}{d\boldsymbol{\rho}} := \omega + \mathbf{H}_{\boldsymbol{\rho}}(\mathbf{u}^0, \boldsymbol{\rho})^T \cdot \boldsymbol{\eta} + J'_{\boldsymbol{\rho}}(\mathbf{u}^k, \boldsymbol{\rho})$

The sensitivity information of the system matrix is involved in the following matrices:
$$\mathbf{G}_{\boldsymbol{\rho}}(\mathbf{u}, \mathbf{w}, \boldsymbol{\rho}) := -\left[\frac{\partial \mathbf{A}'_\mathbf{u}(\mathbf{u}, \boldsymbol{\rho})}{\partial \rho_1} \cdot \mathbf{w}, \ldots, \frac{\partial \mathbf{A}'_\mathbf{u}(\mathbf{u}, \boldsymbol{\rho})}{\partial \rho_m} \cdot \mathbf{w}\right]$$
$$-\left[\frac{\partial \mathbf{A}(\mathbf{u}, \boldsymbol{\rho})}{\partial \rho_1} \cdot \mathbf{u}, \ldots, \frac{\partial \mathbf{A}(\mathbf{u}, \boldsymbol{\rho})}{\partial \rho_m} \cdot \mathbf{u}\right],$$
$$\mathbf{G}_\mathbf{u}(\mathbf{u}, \mathbf{w}, \boldsymbol{\rho}) := -\left[\frac{\partial \mathbf{A}'_\mathbf{u}(\mathbf{u}, \boldsymbol{\rho})}{\partial u_1} \cdot \mathbf{w}, \ldots, \frac{\partial \mathbf{A}'_\mathbf{u}(\mathbf{u}, \boldsymbol{\rho})}{\partial u_n} \cdot \mathbf{w}\right] - \mathbf{A}'_\mathbf{u}(\mathbf{u}, \boldsymbol{\rho}),$$
$$\mathbf{H}_{\boldsymbol{\rho}}(\mathbf{u}, \boldsymbol{\rho}) := -\left[\frac{\partial \mathbf{A}(0, \boldsymbol{\rho})}{\partial \rho_1} \cdot \mathbf{u}, \ldots, \frac{\partial \mathbf{A}(0, \boldsymbol{\rho})}{\partial \rho_m} \cdot \mathbf{u}\right],$$

where $\mathbf{A}'_\mathbf{u}(\mathbf{u}, \boldsymbol{\rho})$ is the linearization of the nonlinear system matrix. We only need to implement their applications, which can be efficiently performed element-wise.

3 Sequential coupling of topology and shape optimization

We will use the optimal topology design as the initial guess for the shape optimization. The first step towards a fully automatic procedure is a shape identification, which we are doing by hand for the moment. For this purpose, one can use a binary image components recongnition based on boundary tracing, which is a well-known algorithm in the image processing, cf. [3]. The second step we are treating here is a piecewise smooth approximation of the boundaries by Bézier curves. Let $\rho^{\text{opt}} \in \mathcal{Q}$ be an optimized discretized material distribution. Recall that it is not a strictly 0-1 function. Let $\mathbf{p}_1 \in \mathbb{R}^{m_1}, \ldots, \mathbf{p}_s \in \mathbb{R}^{m_s}$ denote vectors of Bézier parameters of the shapes $\alpha_1(\mathbf{p}_1), \ldots, \alpha_s(\mathbf{p}_s)$ which form the interface between the air and ferromagnetic subdomains $\Omega_0(\alpha_1, \ldots, \alpha_s)$ and $\Omega_1(\alpha_1, \ldots, \alpha_s)$, respectively, i.e. $\Omega_1 \subset \Omega_d$, $\overline{\Omega} = \overline{\Omega_0} \cup \overline{\Omega_1}$ and $\Omega_0 \cap \Omega_1 = \emptyset$. Let further $\underline{\mathbf{p}_i}$ and $\overline{\mathbf{p}_i}$ denote the lower and upper bounds, respectively, and let $\mathcal{P} := \{(\mathbf{p}_1, \ldots, \mathbf{p}_s) \mid \underline{\mathbf{p}_i} \leq \mathbf{p}_i \leq \overline{\mathbf{p}_i} \text{ for } i = 1, \ldots, s\}$ be the set of admissible Bézier parameters. We solve the following least square fitting problem:

$$\min_{(\mathbf{p}_1, \ldots, \mathbf{p}_s) \in \mathcal{P}} \int_{\Omega_d} \left(\rho^{\text{opt}} - \chi(\Omega_1(\alpha_1(\mathbf{p}_1), \ldots, \alpha_s(\mathbf{p}_s))) \right)^2 d\mathbf{x}, \quad (4)$$

where $\chi(\Omega_1)$ is the characteristic function of Ω_1.

When solving (4) numerically, one encounters the problem of intersection of the Bézier shapes with the mesh on which ρ^{opt} is elementwise constant. In order to avoid it we use the property that the Bézier control polygon converges linearly to the curve, see Fig. 1, under the following refinement procedure:

$$[\mathbf{p}_i^{\text{new}}]_1 := [\mathbf{p}_i^{\text{old}}]_1$$
$$[\mathbf{p}_i^{\text{new}}]_j := \frac{j-1}{m_i+1} [\mathbf{p}_i^{\text{old}}]_{j-1} + \frac{m_i-j}{m_i+1} [\mathbf{p}_i^{\text{old}}]_j, \quad j = 2, \ldots, m_i,$$
$$[\mathbf{p}_i^{\text{new}}]_{m_i+1} := [\mathbf{p}_i^{\text{old}}]_{m_i}.$$

This procedure adds one control node so that the resulting Bézier curve remains unchanged. After a sufficient number of such refinements, the integration in (4) is replaced by a sum over the elements and we deal with intersecting the mesh with a polygon. Note that our least square functional is not twice differentiable whenever a shape touches the grid. This is still acceptable for the quasi-Newton optimization method that we apply.

4 Multilevel shape optimization

With the previous notation, the shape optimization problem under consideration is as follows:

$$\text{Find } (\mathbf{p}_1^*, \ldots, \mathbf{p}_s^*) \in \mathcal{P} : \quad \forall (\mathbf{p}_1, \ldots, \mathbf{p}_s) \in \mathcal{P} :$$
$$J(u(\mathbf{p}_1^*, \ldots, \mathbf{p}_s^*)) \leq J(u(\mathbf{p}_1, \ldots, \mathbf{p}_s)) \quad (5)$$

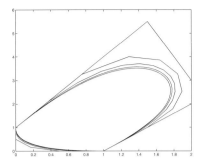

Fig. 1. Approximation of Bézier shapes by the refined control polygon

subject to the 2-dimensional nonlinear magnetostatics: Find $u(\mathbf{p}_1,\ldots,\mathbf{p}_s) \in H_0^1(\Omega)$:

$$\forall v \in H_0^1(\Omega): \int_{\Omega_0(\alpha_1(\mathbf{p}_1),\ldots,\alpha_s(\mathbf{p}_s))} \nu_0 \,\mathbf{grad}(u(\mathbf{p}_1,\ldots,\mathbf{p}_s)) \cdot \mathbf{grad}(v)\,d\mathbf{x}$$
$$+ \int_{\Omega_1(\alpha_1(\mathbf{p}_1),\ldots,\alpha_s(\mathbf{p}_s))} \nu(\|\mathbf{grad}(u(\mathbf{p}_1,\ldots,\mathbf{p}_s))\|)\,\mathbf{grad}(u(\mathbf{p}_1,\ldots,\mathbf{p}_s)) \cdot \mathbf{grad}(v)\,d\mathbf{x}$$
$$= \int_\Omega Jv\,d\mathbf{x}. \tag{6}$$

Concerning the finite element discretization, throughout the optimization we use the following moving grid approach: The control design nodes interpolate the Bézier shape and the remaining grid nodes displacements are given by solving an auxiliary discretized linear elasticity problem with the zero load and the nonzero Dirichlet boundary condition along the design shape that corresponds to the shape displacement. Then, we develop a fairly similar adjoint algorithm for the shape sensitivity analysis as in case of topology optimization.

Perhaps, the main reason for solving the coarse topology optimization as a preprocessing is that we get rid of a large number of design variables in case of fine discretized topology optimization. Once we have a good initial shape design, we will proceed the shape optimization in a multilevel way in order to speed up the algorithm as much as possible. We propose to couple the outer quasi-Newton method with the nested Newton method for eliminitaion of the nonlinear state problem, see the algorithm below. At each iteration of the nested Newton method we employ the conjugate gradient method preconditioned by a geometric multigrid (PCG) so that only one preconditioner per level is used for both the system matrix $\mathbf{A}^{(l)}$ as well as for the linearization $\mathbf{A}^{(l)}{'}_\mathbf{u}$, where $\mathbf{A}^{(l)} := \mathbf{A}^{(l)}(\mathbf{p}_1,\ldots,\mathbf{p}_s)$ denotes the reluctivity matrix assembled at the l-th level.

Multilevel shape optimization algorithm

Given $\mathbf{p}_1^{(1),\text{init}}, \ldots, \mathbf{p}_s^{(1),\text{init}}$

Discretize at the first level $\longrightarrow h^{(1)}, \mathcal{A}^{(1)}(\mathbf{p}_1^{(1),\text{init}}, \ldots, \mathbf{p}_s^{(1),\text{init}})$

Solve by a quasi-Newton method coupled with the nested Newton method, while using a nested direct solver: $\mathbf{p}_1^{(1),\text{init}}, \ldots, \mathbf{p}_s^{(1),\text{init}} \longrightarrow \mathbf{p}_1^{(1),\text{opt}}, \ldots, \mathbf{p}_s^{(1),\text{opt}}$

Store the first level preconditioner $\mathbf{C}^{(1)} := \left[\mathcal{A}^{(1)}(\mathbf{p}_1^{(1),\text{opt}}, \ldots, \mathbf{p}_s^{(1),\text{opt}})\right]^{-1}$

for $l = 2, \ldots$ do

    Refine: $h^{(l-1)} \longrightarrow h^{(l)}$

    Prolong: $\mathbf{p}_1^{(l-1),\text{opt}}, \ldots, \mathbf{p}_s^{(l-1),\text{opt}} \longrightarrow \mathbf{p}_1^{(l),\text{init}}, \ldots, \mathbf{p}_s^{(l),\text{init}}$

    Solve by a quasi-Newton method coupled with the nested Newton method, while using the nested conjugate gradients method preconditioned with $\mathbf{C}^{(l-1)}$: $\mathbf{p}_1^{(l),\text{init}}, \ldots, \mathbf{p}_s^{(l),\text{init}} \longrightarrow \mathbf{p}_1^{(l),\text{opt}}, \ldots, \mathbf{p}_s^{(l),\text{opt}}$

    Store the l-th level multigrid preconditioner $\mathbf{C}^{(l)} \approx \left[\mathcal{A}^{(l)}(\mathbf{p}_1^{(l)}, \ldots, \mathbf{p}_s^{(l)})\right]^{-1}$

end for

5 Numerical results

We consider a problem depicted in Fig. 2 **(a)**, which is a simplification of the direct electric current (DC) electromagnet depicted in Fig. 3 **(b)**. Some results on the usage and mathematical modeling of such electromagnets can be found in [7, 5], respectively. Our aim is to find a distribution of the ferromagnetic material so that the generated magnetic field is strong and homogeneous enough. Unfortunately, these assumptions are contradictory and we have to balance them. The cost functional reads as follows:

$$J(u) := \int_{\Omega_m} \|\mathbf{curl}(u) - B_m^{\text{avg}}(u)(0,1)\|^2 \, d\mathbf{x} + p_B \left(\min\{0, B_m^{\text{avg}}(u) - B^{\min}\}\right)^2,$$

where $\Omega_m \subset \Omega$ is the subdomain where the magnetic field should be homogeneous, $\mathbf{curl}(u) := (\partial u/\partial x_2, -\partial u/\partial x_1)$, $B_m^{\text{avg}}(u)$ is the mean value of the magnetic field component $-\partial u/\partial x_1$ over Ω_m, $B^{\min} := 0.12\,[\text{T}]$ is the required minimal field and $p_B := 10^6$ is the penalty of the minimal field constraint. There are 600 turns pumped by the current of $5\,[\text{A}]$, which is averaged into a current density J being constant in the coil subdomain and vanishing elsewhere. The nonlinear material reluctivity function is $\nu(\eta) = (\nu_0 - \nu_1)\left(\frac{\eta^4}{\eta^4 + \nu_0^{-1}} - 1\right)$, where $\nu_1 := \nu_0/5100$ is the linearized reluctivity of the used ferromagnetics.

The coarsely optimized topology of the quarter of the geometry is depicted in Fig. 2 **(b)**. We chose $p_\rho := 100$ and the very initial guess was $\rho := 0.5$ in Ω_d. In the coarse topology optimization problem there were 861 design, 1105 state variables and the optimization was done in 7 steepest descent iterations, which took 2.5 seconds. Further, we approximated the boundary of the black

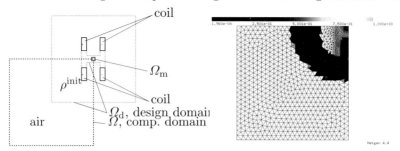

Fig. 2. Topology optimization: **(a)** initial design; **(b)** coarsely optimized design ρ^{opt}

domain by three Bézier curves described by 19 parameters in total. Solving the corresponding least square problem was finished in 8 quasi-Newton iterations, which took 26 seconds when using a forward numerical differentiation scheme. At the end, we proceeded with the multilevel shape optimization starting from the optimized curves of the previous fitting problem. The performance of this last step can be seen from Table 1. Note that from the sixth column of the table, we can see that the linear system $\mathbf{A}$ was solved almost in the optimal way (6 PCG iterations at worst), however, solution to the linearized system $\mathbf{A}'_{\mathbf{u}}$ deteoriates. This is due to the fact that we only used the preconditioner for the linear part, which did not bring any extra cost within one PCG iteration.

Table 1. Multilevel shape optimization

level	design variables	outer iters.	state variables	max. inner iters.	PCG steps lin./nonlin.	time
1	19	10	1098	3	direct	32s
2	40	15	4240	3	3/14–25	2min 52s
3	82	9	16659	4	4–5/9–48	9min 3s
4	166	10	66037	4	4–6/13–88	49min 29s
5	334	13	262953	5	3–6/20–80	6h 36min

The final result is depicted in Fig. 3 **(a)** and it is very similar to the existing geometry of the so-called O-Ring electromagnet, see Fig. 3 **(b)**.

6 Conclusion

This paper presented a method which sequentially combines topology and shape optimization. First, we solved a coarsely discretized topology optimization problem. Then, we approximated some chosen interfaces by Bézier shapes.

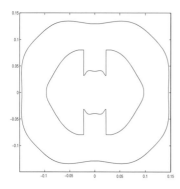 

Fig. 3. Multilevel shape optimization: **(a)** optimized geometry; **(b)** the O-Ring electromagnet

Finally, we proceeded with shape optimization in a multilevel way. We applied the method to a 2-dimensional optimal shape design of a DC electromagnet, for which we achieved fine optimized geometries in terms of minutes. It remains to analyze and improve the multigrid convergence, particularly, in case of the nonlinear state operator.

References

1. Bendsøe, M.P.: Optimization of Structural Topology, Shape and Material. Springer, Berlin, Heidelberg (1995)
2. Céa, J., Garreau, S., Guillaume, P., Masmoudi, M.: The shape and topological optimizations connection. Comput. Methods Appl. Mech. Eng. **188**, 713–726 (2000)
3. Gonzalez R.C., Woods R.E.: Digital Image Processing, Addison Wesley (1992)
4. Haslinger J., Neittaanmäki P.: Finite Element Approximation for Optimal Shape, Material and Topology Design. Wiley, Chinchester (1997)
5. Lukáš, D.: On solution to an optimal shape design problem in 3-dimensional magnetostatics. Appl. Math. **49**:5, 24 pp. (2004)
6. Olhoff, N., Bendsøe, M.P., Rasmussen, J.: On CAD-integrated structural topology and design optimization. Comp. Meth. Appl. Mech. Eng. **89**, 259–279 (1991)
7. Postava, K., Hrabovský, D., Pištora, J., Fert, A.R., Višňovský, Š., Yamaguchi, T.: Anisotropy of quadratic magneto-optic effects in reflection. J. Appl. Phys. **91**, 7293–7295 (2002)
8. Tang, P.-S., Chang, K.-H.: Integration of topology and shape optimization for design of structural components. Struct. Multidisc. Optim. **22**, 65–82 (2001)

Nonsmooth Optimization of Eigenvalues in Topology Optimization

K. Moritzen

Institute of Applied Mathematics, University of Dortmund
kay.moritzen@mathematik.uni-dortmund.de

Summary. During the last decade, topology optimization has become an important branch in engineering sciences, e.g., to save material or to optimize the heat distribution inside a structure. The modelling of a certain class of such problems (e.g. vibration analysis) leads to the optimization of suitable functions defined on the set of all eigenvalues of the corresponding differential operator. The resulting optimization problems are typically nonsmooth and require adequate nonsmooth optimization techniques. In this article an approach for the treatment of a typical class of eigenvalue optimization problems based on a nonsmooth bundle method is considered, and a mathematical framework for its analysis is developed.

Particular emphasis is laid on a suitable representation of subgradients for the occurring objective functions. This representation allows both, an efficient implementation of the method, and facilities of mathematical analysis. The resulting algorithmic scheme is compared to smooth optimization techniques.

1 Introduction

Topology optimization considers lay-out problems of mechanical structures. Thereby, it offers a combination of several features from structural optimization. Vibration problems were one of the early targets that engineers were concerned with in those days (s. [1], p. 310). Mathematical arguments within structural and topology optimization partially shifted to the foreground over the past decades. And still today the nonsmooth behavior of eigenfunctions keeps being an issue. In this report we focus on a special class of optimization problems where the objective functions inherit a special kind of eigenfunctions. This analysis uses terms from nonsmooth analysis (e.g. [9]).

This investigation ends with a numerical example considering the optimization of material structures due to robust vibration behavior (s. [1], mS2.1). We propose the use of a *bundle method* for nonsmooth nonlinear and nonconvex optimization problems (s. [10]).

2 Topology optimization and eigenproblems

Let Ω be some open connected subset of $\mathbb{R}^d$, $d \in \{1, 2, 3\}$, representing a *design domain*. The goal of topology optimization is to find optimal layouts $\Omega^* \subset \Omega$, where Ω^* represents a body of given elastic material. Here, optimality refers to some given objective function measuring the optimality of the design Ω^* subject to given constraints.

In this paper we consider objective functions characterizing properties related to eigenproblems occurring in mechanical analysis. The vibration behavior of a mechanical structure is one example of eigenvalue based problems in mechanics.

We start with the discretization of the wave equation. Its continuous formulation can be found in elementary textbooks of mechanics. The *discretized dynamical system equation* (e.g. [3]) reads as

$$(K - \lambda M)u = 0, \qquad (\lambda, u) \in \mathbb{R}_{\geq 0} \times \mathbb{R}^n_{\neq 0}, \qquad (1)$$

where $K, M \in \mathcal{S}_{>0}(n)$ are the positive definite *global stiffness* and the *global mass matrix* of same given structure. With $\mathcal{S}(n)$ we denote the space of all real symmetric $n \times n$-matrices, and $\mathcal{S}_{>0}(n)$ denotes the set of symmetric and positive definite $n \times n$-matrices. Each pair (λ, u) solving (1) we call an *eigenpair* with *eigenvalue* λ and *eigenvector* u.

The *system matrices* result from the discretization of the underlying variational problem defined on the domain Ω. These matrices are defined by the sums $K := \sum_{i=1}^m K_i$ and $M := \sum_{i=1}^m M_i$, while $(K_i)_i, (M_i)_i \subset \mathcal{S}_{\geq 0}(n)$, $i = 1, \ldots, m$, are the families of *local stiffness* and *mass matrices*. The number m denotes the number of cells, i.e., finite elements, discretizing Ω in the usual way.

For the process of topology optimization we now relax the system matrices by means of a so called *pseudo-density* or *design variable* $x \in \mathbb{I}^m := [0, 1]^m \subset \mathbb{R}^m$ introducing the following matrix functions

$$K, M : \mathbb{I}^m \longrightarrow \mathcal{S}_{\geq 0}(n), \qquad K(x) := \sum_{i=1}^m x_i^3 K_i, \qquad M(x) := \sum_{i=1}^m x_i M_i.$$

This relaxation scheme is called *SIMP* (structured isotropic material penalized, see [1]). Thus, any choice of a density vector $x \in \mathbb{I}^m$ one-to-one corresponds to a structure with stiffness matrix $K(x)$ and mass matrix $M(x)$. Optimizing over any (suitable) pseudo-densities to figure out some *best* structure depends on the objective of interest we have to determine. If an objective function is concerned involving eigenvalues, equation (1) extends to the *relaxed eigenproblem*

$$(K(x) - \lambda M(x))u = 0. \qquad (2)$$

based on the pseudo-density. Those eigenpairs $(\lambda, u) \in \mathbb{R} \times \mathbb{R}^n_{\neq 0}$ solving equation (2) have to be considered in dependency of the design x. We can make use of the abbreviation $(\lambda, u) := (\lambda(x), u(x))$.

Any optimization formulation considered here is of the following type

$$\min_{x \in \mathbb{R}^m} F(x)$$

$$\text{s.t.}: \text{(I)} \quad \sum_{j=1}^m x_j \text{Vol}_j = \text{Vol}_0, \tag{3}$$

$$\text{(II)} \quad (K(x) - \lambda M(x))u = 0,$$

$$\text{(III)} \quad \underline{x}_j \leq x_j \leq \overline{x}_j, \quad \underline{x}_j, \overline{x}_j \in \mathbb{I}, \quad j = 1, \ldots, m.$$

Here equation (I) defines the volume restriction with $\text{Vol}_0 \in \mathbb{R}_{>0}$ as the amount of total material volume (of the structure) to be distributed and $(\text{Vol}_j)_{j=1,\ldots,m} \subset \mathbb{R}_{>0}$ denotes the volumes of the material cells, i.e., finite elements. Equation (II) is the generalized eigenproblem from (2), and (III) is called *box constraint*.

The objective function is a given map $F : \mathbb{R}^m \longrightarrow \mathbb{R}$ where we assume that F can be written as the composition $F(\cdot) := (f \circ \text{eig} \circ \mathcal{AB})(\cdot)$. Here $\mathcal{AB}$ denotes the *matrix map* $\mathcal{AB} : \mathbb{R}^m \longrightarrow \mathcal{S}(n)^2$, $\mathcal{AB}(x) := (K(x), M(x))$, f is the function $f : \mathbb{R}^n \longrightarrow \mathbb{R}$, and eig denotes the *eigenvalue operator* eig $: \mathcal{S}(n)^2 \longrightarrow \mathbb{R}^n$. The eigenvalue operator maps any eigenvalues of eigenpairs solving equation (1) to a vector $\theta \in \mathbb{R}^n$ based on the system matrices $\mathcal{AB}(x)$ depending on the pseudo-density $x \in \mathbb{I}^m$. The components of θ are sorted in ascending order. Whenever the mass matrix $M(x)$ becomes singular then only $\hat{n} < n$ eigenpairs exist solving equation (2) (s. [8]). In this case the number of nonexisting eigenvalues $(n - \hat{n})$ are defined as the largest components of θ: $\theta_i := \infty$, $i = \hat{n}+1, \ldots, n$. Here, we neglect the case that not exactly n eigenpairs arise from eigenproblem (1), i.e. (2).

The goal of f is to collect certain eigenvalues and express a measure of these eigenvalues in terms of a real number, e.g., as a sum. The particular choice of f needs not be specified here so far. It can be arbitrary, and the choice of this function is the crucial point in modelling eigenvalue based optimization problems. From mechanical engineering we know some examples of optimization tasks.

- **Maximizing the fundamental eigenvalue.** A mechanical structure gains stability if its fundamental eigenvalue is large. If we search for a structure with this property, we define f as a simple projection on the smallest eigenvalue:

$$f(x) := -x_1 \quad \Longrightarrow \quad F(x) = -\text{eig}_1(\mathcal{AB}(x)). \tag{4}$$

- **Maximizing a spectral gap.** Here, the task is to design a structure preventing its eigenfrequencies getting close to some given scalar $t \in \mathbb{R}_{>0}$ with some given number $\bar{i} \in \mathbb{N}$, $\bar{i} \ll n$. The formulation of this hard problem, e.g., may be realized through minimizing the sum $f(x) := \sum_{i=1}^{\bar{i}} x_i(t - x_i)$,

3 Nonsmooth analysis

For optimization purposes it is quite natural to make use of gradient informations giving an optimization scheme steepest descent information to iteratively improve the design. We know that our objective function F becomes nonsmooth in the pseudo-density x, since the nonsmoothness is inherited from eigenvalues having multiplicities larger than one.

To formulate the desired nonsmooth optimization problem, it is necessary to assume proper continuity properties on F and its composition. To make our argumentation rigorous we first make the assumption on F to be Lipschitz-continuous in the point of interest $x \in \mathbb{R}^n$. It suffices that the component f in $\text{eig}(\mathcal{AB}(x))$ of F has to be assumed Lipschitz continuous as well. In [7] it is shown that the eig-operator is Lipschitz with Lipschitz-constant 1 for standard eigenproblems. It is easy to show that this is also true in the generalized case as long as M is regular. Especially, it suffices that $\mathcal{AB}$ in x is Fréchet-differentiable, which is obviously sharper than Lipschitz continuity.

Two main tools are essential for the discussion of nonsmoothness of functions. First we recall the usual *(classical) directional derivative* of f in x with direction $h \neq 0$ defined by

$$\mathrm{d}f(x;h) := \lim_{t \searrow 0} \frac{1}{t}\left(f(x+th) - f(x)\right).$$

The *(classical) subdifferential* generalizes the notions of "gradients" by means of the directional derivative and is defined by the set

$$\partial f(x) := \left\{ v \in \mathbb{R}^n \mid \langle v, h \rangle \leq \mathrm{d}f(x;h) \quad \forall h \in \mathbb{R}^n_{\neq 0} \right\},$$

where $\langle \cdot, \cdot \rangle$ denotes the euclidian scalar product. Each element of $\partial f(x)$ is called a *subgradient*. In case f is continuously differentiable in x its subdifferential will contain the usual gradient as a single point: $\partial f(x) = \{\nabla f(x)\}$.

Now, we give an answer to the question of how to describe the subdifferential of F in x. With [2] the inclusion

$$\partial F(x) \subset \partial f(\text{eig}(\mathcal{AB}(x))) \cdot \partial \text{eig}(\mathcal{AB}(x)) \cdot \nabla \mathcal{AB}(x) \tag{5}$$

holds by applying a chain rule (see below) for Lipschitz-continuous functions. We concentrate on the development of a suitable chain rule providing an expression that describes the subdifferential $\partial F(x)$ exactly. The unpleasant inclusion above is insufficient from a practical point of view. It can easily be seen, that linear combinations of the righthand side of (5) exist, that do not lie in the origin subdifferential $\partial F(x)$ as wished. In practise this can lead to wrong compuations of subgradinet of $\partial F(x)$.

As a first step we focus on the key problem of expressing the subdifferential for the eigenvalue operator $\partial \text{eig}(\cdot)$. We end up with the construction of the subdifferential for the composition F by introducing a special chain rule applied to F. The description of a subdifferential of eigenvalue functions $g : \mathcal{S}(n) \to \mathbb{R}$ has already been analyzed in [6, 7] for the case of standard eigenproblems. Nevertheless, generalized eigenproblems in addition with system matrices being maps in the pseudo-density require extended analyzing techniques.

Let $A, B \in \mathcal{S}(n)$ be matrices defining an eigenproblem (1), let $l := \#\mathrm{eig}(A,B)$ be the number of pairwise different eigenvalues, and $\mu \in \mathbb{R}^l$ the vector of eigenvalues to eigenproblem (2). The eigenvalues in μ are assumed to be in ascending order ($\mu_k < \mu_{k+1}$, $k = 1, \ldots, l-1$). The vector $\widetilde{n} \in \mathbb{N}^l$ denotes the multiplicities of the eigenvalues μ, i.e. the multiplicity of μ_k is $\widetilde{n}_k$ for all $k = 1, \ldots, l$. The eigenvectors of each eigenvalue μ_k are given by the column vectors of the matrix $\widetilde{V}_k \in \mathcal{O}(n, \widetilde{n}_k | B)$, where $\mathcal{O}(n, \widetilde{n}_k | B)$ denotes the set of all $(n \times \widetilde{n}_k)$-matrices with column vectors orthogonal w.r.t. the scalar product given by B as $\langle u, v \rangle_B := u^\top B v$. We set $V := \left(\widetilde{V}_1, \ldots, \widetilde{V}_l \right) \in \mathcal{O}(n|B)$ and $\mathcal{O}(n|B) := \mathcal{O}(n, n|B)$.

The *blockwise subdifferential* of the eigenvalue operator is defined by the set

$$\partial \widetilde{\mathrm{eig}}_k(A,B) := \left\{ \widetilde{Q} \in \mathcal{S}(n)^{2 \times \widetilde{n}_k} \ \middle| \ \begin{array}{l} (h_1^k, \ldots, h_{\widetilde{n}_k}^k) \in \mathcal{O}(\widetilde{n}_k), \\ \widetilde{Q}_i := \widetilde{V}_k h_i^k (h_i^k)^\top \widetilde{V}_k^\top, \quad i = 1, \ldots, \widetilde{n}_k, \\ \widetilde{Q} := \left(\begin{pmatrix} \widetilde{Q}_1 \\ -\mu_k \widetilde{Q}_1 \end{pmatrix}, \ldots, \begin{pmatrix} \widetilde{Q}_{\widetilde{n}_k} \\ -\mu_k \widetilde{Q}_{\widetilde{n}_k} \end{pmatrix} \right) \end{array} \right\}$$

for $k = 1, \ldots, l$. If the multiplicity $\widetilde{n}_k$ of the k-th eigenvalue is trivial, i.e. $\widetilde{n}_k = 1$, the k-th blockwise subdifferential $\partial \widetilde{\mathrm{eig}}_k(A,B)$ will contain only one point. In this case eig_k is Fréchet-differentiable in $\mathcal{AB}(x)$.

Theorem 1 (subdifferential of the eigenvalue operator). *We take the notation from above. The subdifferential for the eigenvalue operator in a point $(A,B) \in \mathcal{S}(n)^2$ is*

$$\partial \mathrm{eig}(A,B) \subseteq \partial \widetilde{\mathrm{eig}}(A,B) := \left\{ Q \in \mathcal{S}(n)^{2 \times n} \ \middle| \ \begin{array}{l} \widetilde{Q}_k \in \partial \widetilde{\mathrm{eig}}_k(A,B), \ k = 1, \ldots, l, \\ Q := \left(\widetilde{Q}_1, \ldots, \widetilde{Q}_l \right)^\top \end{array} \right\}$$

Proof: The proof can be found in [8], mS2. ∎

Based upon the last theorem we finally get the exact subdifferential expression for F in x

$$\partial F(x) = \partial f(\mathrm{eig}(\mathcal{AB}(x))) \bullet \partial \widetilde{\mathrm{eig}}(\mathcal{AB}(x)) \blacksquare \nabla \mathcal{AB}(x)$$

where the multiplications denoted by the symbols "$\blacksquare$" and "$\bullet$" are defined as follows. Let $Q \in \partial \mathrm{eig}(A,B)$, $\tau \in \mathbb{R}^{1 \times n}$ and $R \in \mathcal{S}(n)^2$. we define

$$\begin{aligned} \text{``}Q\blacksquare\text{''}: \quad & Q \blacksquare R := (\langle Q_1, R \rangle, \ldots, \langle Q_n, R \rangle)^\top \quad && \in \mathbb{R}^n, \\ \text{``}\bullet Q\text{''}: \quad & \tau \bullet Q := \sum_{i=1}^n \tau_i Q_i \quad && \in \mathcal{S}(n)^2, \end{aligned} \qquad (6)$$

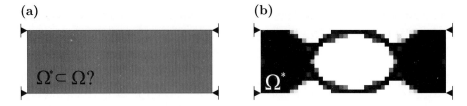

Fig. 1. Boundary condition of a plate clamped in any corners with aspect ration three to one (a). The maximization of the fundamental eigenvalue after hundred function evaluations with the bundle method leads to an almost clear black-and-white structure (b). The values of the three lowest eigenvalues of the final structure are listed in Table 1.

with $\langle Q_j, R \rangle = \langle (Q_j)_1, R_1 \rangle + \langle (Q_j)_2, R_2 \rangle$, $j = 1, \ldots, m$.

4 Numerical results

For numerical purposes it is necessary to cleverly implement the subdifferential of the eigenvalue operator. Therefore, we recommend to implement the

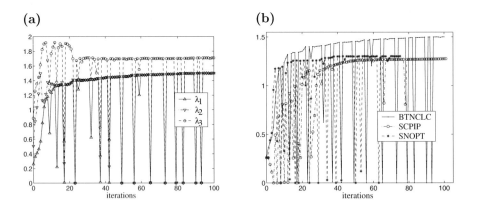

Fig. 2. The plot of the lowest three eigenvalues, optimized with the bundle method, shows that after about 15 iterations the fundamental eigenvalue gains a multiplicity of two (a). Utilizing smooth optimization programs leads to bad runtime behavior being a well known fact from theory. The plots in (b) show the fundamental eigenvalue optimized with the bundle method (BTNCLC) and two representatives of smooth optimizers (SCPIP, SNOPT). It is seen that the bundle method still improves the objective function while the smooth methods "get stuck" after few iterations on a lower level.

subdifferential of the composition $\Lambda : \mathbb{R}^m \longrightarrow \mathbb{R}^n$, $\Lambda(x) := (\text{eig} \circ \mathcal{AB})(x)$. The subdifferential $\partial \Lambda(x)$ can be expressed in a very compact way (s. [8]).

It should be emphasized that during optimization in each iteration step an eigenproblem has to be solved. In this problem possibly both matrices $K(x)$, $M(x)$ are positive semidefinite, i.e. are not invertible. Solving the eigenproblem constitutes a major part of the overall optimization process. Consequently, improvements on the eigensolver lead directly to a speed up of the whole optimization process.

We illustrate our theoretical findings presenting an eigenvalue problem taken from topology optimization. We focus on the goal to find a structure with maximal fundamental eigenvalue, i.e., for $f(x) = -x_1$ (if (4)). The subdifferential of F in $x \in \mathbb{I}^m$ simplifies as follows, while our chain rule introduced above has been applied $\nabla f(y) = -\mathbf{e}_1 := -(1, 0, \ldots, 0)^\top \in \mathbb{R}^n$, $\forall y \in \mathbb{R}^n$ in which follows $\partial F(x) = -\mathbf{e}_1^\top \widetilde{\partial \text{eig}}_1(\mathcal{AB}(x))\nabla \mathcal{AB}(x)$

The underlying optimizing problem is nonsmooth, nonlinear and nonconvex. Suitable nonsmooth optimization techniques are *bundle methods*. Bundle methods minimize a given objective function by locally approximating the function from below. The model function of these optimization techniques is convex and piecewise linear. The approximation of the model is purchased by spare (sub-)gradient informations that are computed under the restriction of linear constraints. In short words, each subgradient defines a hyperplane, and the intersection of all hyperplanes (the *bundle*) shape a polyhedra that locally approximates the origin from below. Clearly, if the optimization problem is of large scale the approximation will need plenty of subgradients. Further details on bundle methods gives e.g. [10].

In our numerical study we compare the runtime behavior of the bundle method BTNCLC from [10] with two smooth optimization programs: SCPIP from [11] is a MMA-type (method of moving asymptotes) optimization program, and SNOPT is a SQP method (sequential quadratic programming) by [4].

The Figures 1 and 2 (p. 1028) show a scenario that depicts the maximization of the fundamental eigenvalue. It has been made use of the optimization setup in (3). In Table 1 the fraction $\lambda_1^{(\infty)}/\lambda_1^{(0)}$ indicates the improvement of the fundamental eigenvalue comparing the values from initial and last iter-

Table 1. The values shown above reflect the runtime process of the optimization problem given by the scenario in figure 1. For maximizing the fundamental eigenvalue we compare the bundle method (BTNCLC) with two smooth optimization programs (SCPIP, SNOPT). One iteration step took some minutes on an 2.4GHz Intel-Computer.

				initializing			last iteration				optimization method
n_x	n_y	n	#It.	$\Lambda_1^{(0)}$	$\lambda_2^{(0)}$	$\lambda_3^{(0)}$	$\lambda_1^{(\infty)}$	$\lambda_2^{(\infty)}$	$\lambda_3^{(\infty)}$	$\frac{\lambda_1^{(\infty)}}{\lambda_1^{(0)}}$	
45	15	675	101	0.2610	0.5068	0.8871	1.2762	1.2777	1.2842	4.8899	SCPIP
45	15	675	75	0.2610	0.5068	0.8871	1.3021	1.3046	1.9165	4.9894	SNOPT
45	15	675	100	0.2610	0.5068	0.8871	1.4996	1.5003	1.7083	5.7459	BTNCLC

ation. The smooth methods (SCPIP, SNOPT) improve to about $\approx 490\%$ and $\approx 500\%$, whereby the bundle method (BTNCLC with upper bound of 20 Subgradients) improves with $\approx 575\%$ to almost 20%. The case study is taken from [8], mS6, where further details and an analysis of the underlying functions can be found.

5 Conclusion

We outlined a theoretical framework for the mathematical analysis of eigenvalue optimization problems and gave numerical evidence to it. The discussion of eigenvalue based optimization problems depends on suitable analytical tools which are at hand from nonsmooth analysis. Nevertheless, there exist plenty of examples in the literature, mainly in engineering sciences, that do without. This obviously leads to misunderstandings as well as to mistakes in many papers. Anyhow, if one is interested in mathematically proved optimal points the application of smooth methods on nonsmooth objectives can not be recommended.

References

1. Bendsøe, M.P., Sigmund, O.: Topology Optimization. Springer, Berlin Heidelberg New York (2003)
2. Clarke, F.H.: Optimization and Nonsmooth Analysis. Wiley, New York (1983)
3. Evans, L.C.: Partial Differential Equations. American Mathematical Society, Rhode Island, USA (1998)
4. Gill, A., Murray, W., Saunders, M.A.: Users's Guide For SnOpt 5.3 — A Fortran Package for Large-Scale Nonlinear Programming. Numerical Analysis Report **97**. University of California, USA (1997)
5. Krog, L.A., Olhoff, N.: Topology optimization of plate and shell structures with multiple eigenfrequencies. In: WCSMO-1 First World Congress of Structural and Multidisciplinary Optimization. 675–682 (1995)
6. Lewis, A.: Derivatives of spectral functions. Mathematics of Operations Research. **6**, 576–588 (1996)
7. Lewis, A.: Nonsmooth analysis of eigenvalues. Mathematical Programming. **84**, 1–24 (1999)
8. Moritzen, K.: Nichtglatte Analysis und Numerik von Eigenwerten zur Designoptimierung mechanischer Strukturen. Ph.D. Thesis in German, University of Dortmund, Germany (2005)
9. Rockafellar, R.T., Wets, R. J.-B.: Variational Analysis. Springer, Berlin Heidelberg New York (1998)
10. Schramm, H. , Zowe, J.: A Version of the Bundle Idea for Minimizing a Nonsmooth Function: Conceptual Idea, Convergence Analysis, Numerical Results. SIAM Journal on Optimization, **2(1)**, 121–152 (1992)
11. Zillober, C.: Software manual for SCPIP 2.2. Technical Report TR01-2, University of Bayreuth, Germany (2001)

Derivative Free Optimization of Stirrer Configurations

M. Schäfer[1], B. Karasözen[2], Ö. Uğur[3] and K. Yapıcı[4]

[1] Department of Numerical Methods in Mechanical Engineering, Darmstadt University of Technology, Petersenstr. 30, D-64287 Darmstadt, Germany
  schaefer@fnb.tu-darmstadt.de
[2] Department of Mathematics and Institute of Applied Mathematics, Middle East Technical University, 06531 Ankara, Turkey
  bulent@metu.edu.tr
[3] Institute of Applied Mathematics, Middle East Technical University, 06531 Ankara, Turkey
  ougur@metu.edu.tr
[4] Department of Chemical Engineering, Middle East Technical University, 06531 Ankara, Turkey

Summary. In the present work a numerical approach for the optimization of stirrer configurations is presented. The methodology is based on a parametrized grid generator, a flow solver, and a mathematical optimization tool, which are integrated into an automated procedure. The grid generator allows the parametrized generation of block-structured grids for the stirrer geometries. The flow solver is based on the discretization of the incompressible Navier-Stokes equations by means of a fully conservative finite-volume method for block-structured, boundary-fitted grids. As optimization tool the two approaches DFO and CONDOR are considered, which are implementations of trust region based derivative-free methods using multivariate polynomial interpolation. Both are designed to minimize smooth functions whose evaluations are considered expensive and whose derivatives are not available or not desirable to approximate. An exemplary application for a standard stirrer configuration illustrates the functionality and the properties of the proposed methods also involving a comparison of the two optimization algorithms.

1 Introduction

The mixing of different substances with stirrers is a process that is frequently used in many industries such as chemical, pharmaceutical, biotechnological, and food processing.

Important economic aspects for the stirring process are the minimization of the amount of energy needed for the creation of certain mixing conditions, the material costs for the stirrer, as well as the lifetime and the breakdown security of the system. These issues strongly depend on the various geometrical parameters of the stirrer and the vessel as well as on the rotation rate and the fluid properties. Numerical simulation techniques provide a great flexibility concerning the variations of these parameters. To employ such techniques for optimization purposes, an integrated approach combining geometry variation, flow simulation, and mathematical optimization is desirable. In the present work a corresponding methodology is presented, which is based on developments in [8] and [11].

2 Flow solver and numerical optimization tool

The major components of the proposed numerical approach for optimizing the stirrer configuration are: (i) a specially designed parameterized grid generator for stirrer geometries, (ii) an efficient parallel multigrid flow solver, and (iii) a derivative-free optimization method. We briefly outline the basic underlying concepts in the following.

2.1 Grid generator and flow solver

The grid generation tool involves an algebraic method based on transfinite interpolation for the generation of multi-block boundary-fitted grids, which facilitates the accurate representation of the complex geometries associated with stirrer configurations. In order to allow an easy design variation, the grid generation is parametrized with respect to the characteristic geometric quantities for the different stirrer types. Thus, the input parameters are radii of hub, shaft, disk, vessel or numbers and dimensions of blades, baffles and so on. After specifying the number of control volumes for different stirrer sections the grid is created automatically by respecting basic criteria with respect to grid quality, i.e. skewness, aspect ratio, and expansion factor (see e.g. [7]), as far as possible. Following this concept the geometrical input parameters can be used directly in an easy way as design parameters for the optimization purpose.

The flow solver (FASTEST, see [6]) is based on a fully conservative finite-volume method for non-orthogonal, boundary-fitted, block-structured grids, allowing a flexible discretization of even very complex stirrer geometries. The nonlinear algebraic equations are solved at each time step implicitly by a multigrid method with a pressure-correction smoother. For the parallelization a block-structured grid partitioning method with automatic load balancing and strongly implicit block coupling is used (see [5]). The grid movement of the stirrer against the vessel is handled by a clicking mesh approach. The solver was already applied for a variety of different problems in stirrer technology and has proven that it can compute complex problems on parallel computers with high numerical and parallel efficiency [12, 13].

2.2 Derivative free optimization

Concerning numerical optimization of fluid flows, by far the most work can be found in the field of aerodynamics. An overview of the subject is given by Mohammadi and Pironneau [9]. Gradient-free methods may offer advantages over gradient-based ones. The optimization tools we employ here are the DFO package developed by Conn and co-workers [2, 3, 4] and the CONDOR package developed by Berghen [1] which is based on the UOBYQA of Powell [10]. These methods are based on a derivative-free trust region method approximating the objective function by a second-order polynomial, which is then minimized by a sequential quadratic programming (SQP) method. The main steps of the employed algorithm can be summarized as follows:

Step 1. Choose a base for the multivariate interpolation; Lagrange or Newton polynomials,
Step 2. Build an interpolation set and construct a quadratic model based on multivariate interpolation,
Step 3. Minimize the model over the trust region centered at the base,
Step 4. Evaluate the true objective value at the solution of the trust region subproblem,
Step 5. Based on the quality of the achieved reduction and the predicted one, either
- accept the new point as the new base,
- or improve the model and update the trust region.

Particularly, Step 2 consists of interpolating the objective function f with a quadratic (or at least a linear in DFO) polynomial over the well-poised interpolation set. The algorithms differ in constructing the polynomial interpolation: DFO uses Newton polynomials while CONDOR uses Lagrange polynomials as the basis for the space of quadratic polynomials. If $\{\phi_i(\cdot)\}_{i=1}^p$ is a basis in the space of quadratic polynomials then the interpolation condition

$$\sum_{i=1}^p \alpha_i \phi_i(y_j) = f(y_j), \qquad j = 1, \ldots, p$$

emphasizes that the coefficient matrix $\Phi(\mathcal{Y}) = [\phi_i(y_j)]$ is closely affected by the chosen basis polynomials as well as the interpolation set $\mathcal{Y}$. For instance, the choice of Newton polynomials (in DFO) causes the matrix to be lower triangular with a special block diagonal structure, while the use of Lagrange polynomials (in CONDOR) results in an identity matrix. Conversely, the reduced forms of $\Phi(\mathcal{Y})$, which can be obtained by a procedure similar to that of the Gram-Schmidt orthogonalization process, provide the bases of fundamental polynomials.

Another essential difference between the DFO and CONDOR algorithms is that the former uses the smallest Frobenius norm of the Hessian matrix to minimize the local model, which may cause numerical instability, whereas

CONDOR uses the Euclidean norm which is more robust. Moreover, when the computation of the trust-region radius ρ is complete the checking of the validity (within ϵ) of the model around the point x_k in CONDOR is based on whether any of the following conditions

$$\|x_j - x_k\| \leq 2\rho \tag{1}$$

$$\frac{1}{6}M\|x_j - x_k\|^3 \max_d\{|P_j(x_j + d)| : \|d\| \leq \rho\} < \epsilon \tag{2}$$

holds for the Lagrange interpolating basis $\{P_j\}$. Here M is a nonnegative constant such that $|\psi'''(\alpha)| \leq M$, where $\psi(\alpha) = f(y + \alpha \bar{d})$, ($\alpha \in \mathbb{R}$). A complete reference of the variables can be found in [1]. Condition (1) prevents the algorithm from sampling the model at $N = (n+1)(n+2)/2$ new points. However, the checking of the validity of the model in DFO is mainly based on condition (2), which is not very often satisfied by the trust-region radius. Hence, in many cases more function evaluations are needed in order to rebuild the interpolation polynomial.

2.3 Automated procedure

The components described above are combined within an integrated optimization tool by means of a control script. After initializations, the procedure involves the following major steps [11]:

Optimizer: At each iteration the optimizer computes a new set of design variables unless it converges.

Grid Variation: When new design variables are available, the grid generation tool creates the new geometry and the corresponding numerical grid.

Flow Simulation: Flow solver computes the flow field and the corresponding objective function value for the new geometry.

Test for Convergence: If flow solver converges, then optimizer produces new geometrical design variables unless the objective value is found to be minimum.

For further details of the procedure and the modularity of the approach we refer to [11].

3 Results

In the following we consider a Rushton turbine as a representative test case for a practical stirrer configuration to illustrate the functionality of the proposed approach. The geometric parameters, which are considered to define the standard configuration, are given in Table 1. The working Newtonian fluid is a glucose solution with density $\rho = 1330 \, \text{kg/m}^3$ and viscosity $\mu = 0.105 \, \text{Pas}$.

Table 1. Geometrical parameters of standard stirrer configuration.

Parameter	Value
Tank diameter	$T = 0.15\,\mathrm{m}$
Impeller diameter	$D = T/3 = 0.05\,\mathrm{m}$
Bottom clearance	$C = H/2 = 0.075\,\mathrm{m}$
Height of the liquid	$H = T = 0.15\,\mathrm{m}$
Length of the baffles	$W = 3D/10 = 0.015\,\mathrm{m}$
Length of the blade	$\ell = D/4 = 0.0125\,\mathrm{m}$
Height of the blade	$w = D/5 = 0.01\,\mathrm{m}$
Disc thickness	$x = D/5 = 0.00175\,\mathrm{m}$
Diameter of the disk	$d = 3D/4 = 0.0375\,\mathrm{m}$

The numerical grid employed involves 22 blocks with a total number of 238 996 control volumes. 17 blocks are defined as rotating with the stirrer while the remaining 5 blocks are defined as stationary with the vessel. In Fig. 1 a sketch of the considered configuration and the corresponding surface grid of the stirrer is indicated. Concerning the computational requirements we remark that one time step approximately needs 20 seconds of computing time on an eight processor Redstone cluster machine. This results in about 8 hours computing time to reach a steady state flow in the sense of a frozen rotor computation, a criterion that we adopt for all cases.

As a characteristic reference quantity the (dimensionless) Newton number, Ne, is considered, which relates the resistance force to the inertia force:

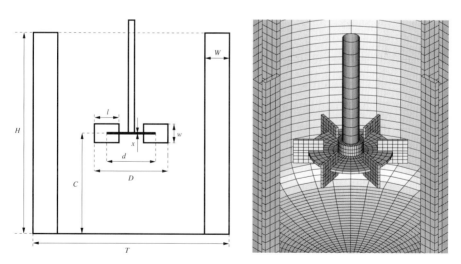

Fig. 1. Sketch of geometrical parameters of stirrer configurration (left) and its corresponding surface grid (right).

$$Ne = \frac{P}{\rho N^3 D^5}, \qquad P = -\int_S (pu_j + \tau_{ij} u_i)\, n_j\, \mathrm{d}S,$$

where N is the rotational speed of the impeller and P is the power computed from the flow quantities over the surface S of the impeller (pressure p, velocity u_i, stress τ_{ij}).

As an example we consider the minimization of the Newton number for a Reynolds number of $Re = 1000$. The design variables are the disk thickness x, the bottom clearance C, and the baffle length W, for which the inequality constraints $0.001 \leq x \leq 0.005$, $0.02 \leq C \leq 0.075$, and $0.005 \leq W \leq 0.03$ are prescribed, respectively. All other parameters are kept fixed according to the standard configuration.

Figure 2 shows the Newton number versus the number of cycles for the two optimization algorithms. Figure 3 depicts the corresponding changes of one of the design variables (bottom clearance) during the function evaluations. We remark that the two optimization tools differ in building the starting model: DFO starts to build the model objective function approximation (unique or not) from the very beginning. That is, its starting objective polynomial approximation is not fully quadratic. On the other hand, CONDOR starts with a fully quadratic model for the objective function. Despite its long initialization it turns out that CONDOR needs less function evaluations than DFO to reach an optimum point after completing the initial quadratic approximation. DFO, as the time passes, oscillates around the minimum although it approaches the

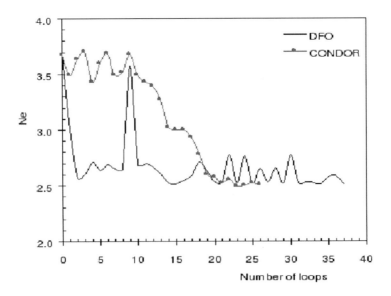

Fig. 2. Newton number versus number of the loops.

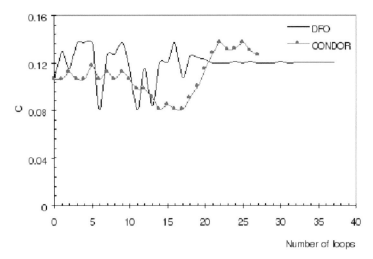

Fig. 3. Bottom clearance versus number of function evaluations.

minimum very sharply at the beginning. CONDOR waits for a full quadratic polynomial to build its model, and then gradually approaches the minimum, using all the power of a complete interpolating bases.

Both algorithms reach the same optimized Newton number which is significantly lower than the one for the standard tank configuration stated in Table 1.

4 Conclusion

In this study we have presented a numerical approach for optimizing practical stirrer configurations. The automated integrated procedure consists of a combination of a parametrized grid generator, a parallel flow solver, and a derivative-free optimization procedure. For the latter two different methods (i.e. DFO and CONDOR) have been investigated with repect to their characteristic convergence properties.

The numerical experiments have shown the principle applicability of the considered approach. For the considered Rushton turbine it has been possible to achieve a significant reduction of the Newton number with relatively low computational effort.

Of course, in a mixing process the power consumption is important but not the sole quantity in obtaining an optimum stirrer configuration. In particular, a satisfactory mixing should be achieved. Due to the generality and modularity of the considered approach, other objectives and/or other design variables can be handled straightforwardly in a similar way.

It should be remarked, since no rigorous convergence properties for globally optimal solutions are available, the optimal solutions obtained with both optimization tools must be considered as local ones. This aspect can be investigated by variations of the starting value and/or the trust region radius.

Acknowledgment

The financial support of the work by the Volkswagen-Stiftung is gratefully acknowledged.

References

1. Berghen, F.V.: CONDOR: a constrained, non-linear, derivative-free parallel optimizer for continuous, high computing load, noisy objective functions. PhD thesis, Université Libre de Bruxelles, Belgium (2004)
2. Conn, A.R., Scheinberg, K., Toint P.: On the convergence of derivative-free methods for unconstrained optimization. In: Iserles, A., Buhmann, M. (ed) Approximation Theory and Optimization: Tribute to M.J.D. Powell. Cambridge University Press, Cambridge, UK (1997)
3. Conn, A.R., Scheinberg, K., Toint P.: Recent progress in unconstrained nonlinear optimization without derivatives. Mathematical Programming, **79**, 397–414 (1997)
4. Conn, A.R., Toint, P.: An algorithm using quadratic interpolation for unconstrained derivative free optimization. In: Pillo G.D., Giannessi F. (ed) Nonlinear Optimization and Applications. Plenum Publishing, New York (1996)
5. Durst, F., Schäfer, M.: A Parallel Blockstructured Multigrid Method for the Prediction of Incompressible Flows. Int. J. for Num. Meth. in Fluids, **22**, 549–565 (1996).
6. FASTEST – User Manual, Department of Numerical Methods in Mechanical Engineering, Technische Universität Darmstadt (2004).
7. Ferziger J., Perić M.: Computational Methods for Fluid Dynamics. Springer, Berlin (1996)
8. Lehnhäuser, T.: Eine effiziente numerische Methode zur Gestaltsoptimierung von Strömungsgebieten. PhD thesis, Technische Universität Darmstadt, Germany (2003)
9. Mohammadi, B., Pironneau, O.: Applied Shape Optimization for Fluids. Oxford University Press, (2001)
10. Powell, M.J.D.: A direct search optimization method that models the objective and constraint functions by linear interpolation. In: Gomez S., Hennart J. (ed) Advances in Optimization and Numerical Analysis. Kluwer Academic, (1994)
11. Schäfer, M., Karasözen, B., Uludağ Y., Yapıcı K., Uğur Ö.: Numerical method for optimizing stirrer configurations. to be published in Computers & Chemical Engineering
12. Sieber, R., Schäfer, M., Lauschke G., Schierholz F.: Strömungssimulation in Wendel- und Dispersionsrührwerken. Chemie Ingenieur Technik, **71**, 1159–1163 (1999)

13. Sieber, R., Schäfer, M., Wechsler K., Durst F.: Numerical prediction of time-dependent flow in a hyperbolic stirrer. In: Friedrich R., Bontoux P. (ed) Computation and Visualization of Three-Dimensional Vortical and Turbulent Flows. volume 64 of *Notes on Numerical Fluid Mechanics*, Vieweg, Braunschweig (1998)

Mathematical Modelling and Numerical Optimization in the Process of River Pollution Control

L.J. Alvarez-Vázquez[1], A. Martínez[1], M.E. Vázquez-Méndez[2] and M. Vilar[2]

[1] Departamento de Matemática Aplicada II. ETSI Telecomunicación. Universidad de Vigo. 36200 Vigo. Spain
 lino@dma.uvigo.es, aurea@dma.uvigo.es
[2] Departamento de Matemática Aplicada. Escola Politécnica Superior. Universidad de Santiago de Compostela. 27002 Lugo. Spain
 ernesto@lugo.usc.es miguel@lugo.usc.es

Summary. Common methods of controlling river pollution include establishing water pollution monitoring stations located along the length of the river. The point where each station is located (*sampling point*) is of crucial importance and, obviously, depends on the reasons for the sample. Collecting data about pollution at selected points along the river is not the only objective; must also be extrapolated to know the characteristics of the pollution in the entire river. In this work we will deal with the optimal location of sampling points. A mathematical formulation for this problem as well as an efficient algorithm to solve it will be given. Finally, in last sections, we will present numerical results obtained by using this algorithm when applied to a realistic situation in the last sections of a river.

1 Introduction

For ages, people have used rivers as refuse sites to dispose of the waste which they have generated. As a consequence of the growing industrialization and population explosions in urban areas, wastewater discharges in our rivers have also increased, thus leading to serious water pollution problems. In the last century, developed countries became aware these problems and established strict legislative requirements concerning to the wastewater disposal in rivers. At present, all wastewater discharged into a river must first be treated in a purifying plant, in order to reduce its level pollutants. Depending on its source, wastewater in rivers is classified under two main types: domestic wastewater (that coming from a purifying plant which collect the water from a sewer system) and industrial wastewater (that coming from an industrial plant).

Wastewater purification at each plant must be strong enough so that the river basin is capable of assimilating all the wastewater disposed there. In order to be sure that the river is assimilating the discharges, we have to choose some concrete indicators of pollution levels and design an adequate sampling technique which gives us information about the values of these indicators along the river. For instance, if we want to control pollution in terms of pathogenic microorganisms coming from domestic wastewater, one of the most important indicators is the concentration ($units/m^3$) of faecal coliform bacteria because its concentration in wastewater discharges is much greater than any other microorganism concentrations. A common technique to control the concentration of coliform bacteria in rivers is to divide it into several sections, according to the morphology of the river basin and the number, type and location of the discharges, and to take samples of water at one point of each section. The point where the station sampling is located (sampling point) is of crucial importance if we want to obtain representative information about the pollution in the whole section. Thus, the optimal sampling point is that one at which the concentration of coliform bacteria over time is as similar as possible to the mean concentration in the whole section (an overview of the entire sampling process can be seen, for example, in [5]).

Several related problems have been analysed with the use of different mathematical techniques. An interesting example is the sentinel method (which is a convenient implementation of the least squares method, initially proposed by J.-L. Lions [7] and further developed by Kernévez [6] for environmental pollution problems). It can be used, for instance, for estimating the contributions of polluting sources in an aquifer, for identifying the time history of pollution releases, for estimating dispersion coefficients, or for identifying the sources of pollution. However, the problem studied here is quite different. The main goal of this work is to use mathematical modelling and numerical optimization to obtain the optimal sampling point in each section of a river. In the next section we will formulate the problem from a mathematical point of view, showing that, under suitable hypotheses, it can be formulated as an optimization problem where the cost function is obtained by solving two one-dimensional hyperbolic boundary value problems. In section 3 we will present a numerical algorithm to solve the complete model and to obtain the optimal sampling points, and, finally, in section 4 we will show the numerical results obtained in a realistic situation.

2 Mathematical formulation

We take a river L meters in length, and we consider E tributaries flowing into the river and V domestic wastewater discharges coming from purifying plants. We suppose the river is divided into N sections, consecutively numbered from the source, and we denote by Δ_i the length of the i-th zone ($i = 1, 2, \ldots, N$).

Since we are going to consider only one-dimensional changes along the direction of flow in the river, if we want to control pollution for T seconds, for each $(x, t) \in [0, L] \times [0, T]$, we denote by $\rho(x, t)$ the average coliform con-

centration in the transversal section, x meters from the source and t seconds from the moment the control is initiated. If we define $a_0 = 0$, $a_{i+1} = a_i + \Delta_i$,
$$c_i(t) = \frac{\int_{a_{i-1}}^{a_i} \rho(x,t)\,dx}{\Delta_i}, \text{ for } i = 1, 2, \ldots, N,$$ the problem consists of finding the sampling points $p_i \in [a_{i-1}, a_i]$, for $i = 1, 2, \ldots, N$, such that minimize the cost function
$$J(p) = \sum_{i=1}^{N} \int_0^T (\rho(p_i, t) - c_i(t))^2\,dt, \quad \text{with } p = (p_1, p_2, \ldots, p_N)$$

If we neglect the effect of molecular diffusion, the coliform concentration is given by solving the following hyperbolic boundary value problem:
$$\left.\begin{aligned}
\frac{\partial \rho}{\partial t} + u\frac{\partial \rho}{\partial x} + k\rho \\
= \frac{1}{A}\left(\sum_{j=1}^{V} m_j \delta(x - v_j) + \sum_{j=1}^{E} n_j \delta(x - e_j)\right) \text{ in } (0,L) \times (0,T), \\
\rho(0,t) = \rho_0(t) \qquad\qquad\qquad\qquad\qquad \text{in } [0,T], \\
\rho(x,0) = \rho^0(x) \qquad\qquad\qquad\qquad\qquad \text{in } [0,L],
\end{aligned}\right\} \quad (1)$$

where $\delta(x - b)$ denotes de Dirac measure at point b, $v_j \in (0, L)$ is the point where the j-th wastewater discharge is located and $m_j(t)$ is its mass coliform flow rate, $e_j \in (0, L)$ is the point where the mouth of the j-th tributary is located and $n_j(t)$ is its mass coliform flow rate, k is the loss rate for total coliform bacteria (experimental known) and, finally, $A(x,t)$ and $u(x,t)$ denote, respectively, the area of the section occupied by water (*wet section*) and the average velocity in that section. Functions $A(x,t)$ and $u(x,t)$ can be calculated by solving the following hyperbolic system, which is obtained by integrating the incompressible Euler equations on the wet section for each point (x,t)

$$\left.\begin{aligned}
\frac{\partial A}{\partial t} + \frac{\partial (Au)}{\partial x} = \sum_{j=1}^{E} q_j \delta(x - e_j) \qquad\qquad \text{in } (0,L) \times (0,T), \\
\frac{\partial (Au)}{\partial t} + \frac{\partial (Au^2)}{\partial x} + gA\frac{\partial \eta}{\partial x} \\
= \sum_{j=1}^{E} q_j U_j \cos(\alpha_j) \delta(x - e_j) + S_f \text{ in } (0,L) \times (0,T), \\
A(L,t) = A_L(t) > 0,\ u(0,t) = u_0(t) \qquad \text{in } [0,T], \\
A(x,0) = A^0(x),\ u(x,0) = u^0(x) \qquad \text{in } [0,L],
\end{aligned}\right\} \quad (2)$$

where $q_j(t)$ is the flow rate corresponding to the j-th tributary, $U_j(t)$ its velocity and α_j the angle between the j-th tributary and the main river; g is

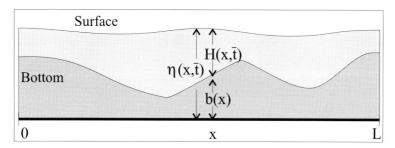

Fig. 1. Longitudinal cut of a river at time $\bar{t}$.

gravity, S_f denotes the bottom friction stress and $\eta(x,t) = H(x,t) + b(x)$ is the height of water with respect to a fixed reference level (see Figure 1).

Remark 1. At a first glance at system (2), three unknowns can be detected: $A(x,t)$, $u(x,t)$ and $\eta(x,t)$. However, it is obvious that, if the geometry of the river is known, $A(x,t)$ can be calculated from $\eta(x,t)$. In effect, for each $x \in [0,L]$, the geometry of the river gives us a strictly increasing and positive function $S(.,x)$ verifying

$$A(x,t) = S(H(x,t),x) \quad \text{in } [0,L] \times [0,T]. \tag{3}$$

In this way, if we write $\dfrac{\partial \eta}{\partial x}$ in terms of $A(x,t)$, the unknown $\eta(x,t)$ can be suppressed in (2). Explicitly, we have

$$\frac{\partial \eta}{\partial x}(x,t) = \frac{\partial}{\partial x} S^{-1}(A(x,t),x) + b'(x), \tag{4}$$

where, for each $x \in [0,L]$, $S^{-1}(.,x)$ denotes the inverse of the function $S(.,x)$.

3 Numerical solution

The first step in the numerical resolution of this problem is to obtain, by solving the system (2), the velocity field and the area of the wet section in each point and at each time. With this data we proceed to solve the problem (1), which give us the function $\rho(x,t)$ in $[0,L] \times [0,T]$ and, consequently, $c_i(t)$ in $[0,T]$ for $i = 1, 2, \ldots, N$. Then, the problem is reduced to solve N one-dimensional minimization problems.

In order to solve the systems (2) and (1) we consider the following discretization: For $N_T \in \mathbb{N}$ given, we define $\Delta t = T/N_T$ and take $t_n = n\Delta t$, for $n = 0, 1, \ldots, N_T$. Moreover, we choose τ_M a partition of $[0,L]$ in M subintervals $I_k = [x_{k-1}, x_k]$, $k = 1, 2, \ldots, M$, in such a way that, for each $i = 1, 2, \ldots, N$ the point a_i is equal to any node of the partition.

3.1 System (2)

We begin by making an implicit time discretization of the first equation of system (2). It leads us to write, for $n = 0, 1, \ldots, N_T - 1$

$$A(x, t_{n+1}) \approx A(x, t_n)$$

$$+ \Delta t \left(\sum_{j=1}^{E} q_j(t_{n+1}) \delta(x - e_j) - \frac{\partial (Au)}{\partial x}(x, t_{n+1}) \right) \quad \text{in } [0, L]. \tag{5}$$

For the time discretization of the second equation of system (2) we are going to use the method of characteristics (see [1]). It leads us to the following discretization: for each $n = 0, 1, \ldots, N_T - 1$

$$\frac{Q(x, t_{n+1})}{\Delta t} + g A(x, t_{n+1}) \left(\frac{\partial}{\partial x} S^{-1}(A(x, t_{n+1}), x) + b'(x) \right) \tag{6}$$

$$= \frac{Q(X^n(x), t_n) V^n(x)}{\Delta t} + \sum_{j=1}^{E} q_j(t_{n+1}) U_j(t_{n+1}) \cos(\alpha_j) \delta(x - e_j)$$

$$+ S_f(x, t_n) \quad \text{in } [0, L],$$

where $X^n(x) = X(x, t_{n+1}; t_n)$ and $V^n(x) = V(x, t_{n+1}; t_n)$ are given, respectively, by the solution of the following initial value problems:

- $\dfrac{dX}{d\tau}(x, t; \tau) = u(X(x, t; \tau), \tau), \qquad X(x, t; t) = x.$
- $\dfrac{dV}{d\tau}(x, t; \tau) = \dfrac{\partial u}{\partial x}(X(x, t; \tau), \tau) V(x, t; \tau), \qquad V(x, t; t) = 1.$

In previous discretization, if we use the approximation of $A(x, t_{n+1})$ given by (5), the unique unknown of equation (6) is $Q(x, t_{n+1})$. So, the problem given by (6) and the boundary conditions $u(0, t_{n+1}) = u_0(t_{n+1})$, $A(L, t_{n+1}) = A_L(t_{n+1})$ can be solved, for example, by using Lagrange P_1 finite elements (see [2] for further details).

3.2 System (1)

The system (1) can be solved by using an implicit upwind finite difference scheme, which does not need to solve any linear equations system. In order to do it, because of the Dirac measures characterizing the sources, we consider the following approximations: for each k, we define $\delta_{hk} : [0, L] \longrightarrow [0, \infty)$ by

$$\delta_{hk}(b) = \begin{cases} \dfrac{b - x_{k-1}}{(x_k - x_{k-1})^2}, & \text{if } b \in [x_{k-1}, x_k] \\ \dfrac{x_{k+1} - b}{(x_{k+1} - x_k)^2}, & \text{if } b \in [x_k, x_{k+1}] \\ 0, & \text{otherwise} \end{cases}$$

Taking $\{\rho_0^i = \rho_0(t_i),\ i = 0, 1, \ldots, N_T\}$ and $\{\rho_j^0 = \rho^0(x_j),\ j = 0, 1, \ldots, M\}$ as data, for each $n = 0, 1, \ldots, N_T - 1$ and $k = 1, \ldots, M$ we compute ρ_k^{n+1} from the following expression:

$$\frac{\rho_k^{n+1} - \rho_k^n}{\Delta t} + u_h^{n+1}(x_k) \frac{\rho_k^{n+1} - \rho_{k-1}^{n+1}}{x_k - x_{k-1}} + k(t_{n+1}, x_k)\rho_k^{n+1}$$

$$= \frac{1}{A_h^{n+1}(x_k)} \left(\sum_{j=1}^V m_j(t_{n+1})\delta_{hk}(v_j) + \sum_{j=1}^E n_j(t_{n+1})\delta_{hk}(e_j) \right)$$

Now, for each $n = 0, 1, \ldots, N_T$, we approach $\rho(x, t_n)$ by the unique function $\rho_h^n(x) \in \{y \in C^0([0, L]) \text{ such that } y_{|I_k} \in P_1,\ k = 1, 2, \ldots, M\}$ verifying $\rho_h^n(x_k) = \rho_k^n$, for each $k = 0, 1, \ldots, M$.

3.3 Minimization problem

With the above discretization, the problem of determining the optimal location of the sampling points is changed into the following N one-dimensional problems: for each $i = 1, 2, \ldots N$, we have to obtain the point $p_i \in [a_{i-1}, a_i]$, minimizing the functional

$$J_{ih}(x) = \frac{1}{2}\left((\rho_h^0(x) - c_{ih}^0)^2 + (\rho_h^{N_T}(x) - c_{ih}^{N_T})^2 \right) + \sum_{n=1}^{N_T - 1}(\rho_h^n(x) - c_{ih}^n)^2,$$

where

$$c_{ih}^n = \frac{\int_{a_{i-1}}^{a_i} \rho_h^n(x)\,dx}{\Delta_i}$$

Obviously, these N problems have solution, but no necessarily unique. The solution can be obtained by any simple one-dimensional optimization method. In this work we have used the golden-section direct search method. For minimizing a strictly unimodal function θ over the interval $[e_1, f_1]$ the algorithm can be easy summarized in the following way (see [3] for further details):

- **Initialization Step.-**
Choose an allowable final length of uncertainty $\epsilon > 0$. Let $[e_1, f_1]$ be the initial interval of uncertainty, and let $\lambda_1 = e_1 + (1 - \alpha)(f_1 - e_1)$ and

$\nu_1 = e_1 + \alpha(f_1 - e_1)$, where $\alpha = (1+\sqrt{5})/2$. Evaluate $\theta(\lambda_1)$ and $\theta(\nu_1)$, let $k = 1$, and go to the main step.

- **Main Step.-**
  1. If $f_k - e_k < \epsilon$, stop; the optimal solution lies in the interval $[e_k, f_k]$. Otherwise, if $\theta(\lambda_k) > \theta(\nu_k)$, go to 2, and if $\theta(\lambda_k) \leq \theta(\nu_k)$, go to 3.
  2. Let $e_{k+1} = \lambda_k$ and $f_{k+1} = f_k$. Furthermore, let $\lambda_{k+1} = \nu_k$, and let $\nu_{k+1} = e_{k+1} + \alpha(f_{k+1} - e_{k+1})$. Evaluate $\theta(\nu_{k+1})$, and go to 4.
  3. Let $e_{k+1} = e_k$ and $f_{k+1} = \nu_k$. Furthermore, let $\nu_{k+1} = \lambda_k$, and let $\lambda_{k+1} = e_{k+1} + (1-\alpha)(f_{k+1} - e_{k+1})$. Evaluate $\theta(\lambda_{k+1})$, and go to 4.
  4. Replace k by $k+1$, and go to 1.

When the solution is separated in an interval where J_{ih} is a strictly unimodal function, this procedure converges at a linear rate (the convergence rate is $2/(1+\sqrt{5})$).

4 Numerical results

In this section we present the numerical results obtained by using the previous proposed method to determine the optimal sampling points in a segment of a river which is divided into three sections. We suppose that the segment is $1000\,m$ in length and we consider 3 tributaries and 2 domestic wastewater discharges (the diagram and data can be seen in Figure 1). Moreover, we consider a parabolic river bed with a non-constant bottom in such a way that

$$S(H, x) = (4\sqrt{H^3})/3,$$

$$b(x) = \begin{cases} \dfrac{1}{200}(500 - x), & \text{if } 0 \leq x \leq 500 \\ 0, & \text{if } 500 \leq x \leq 1000 \end{cases}$$

Both initial and boundary conditions were taken as constant, particularly, $A_L(t) = (4\sqrt{125})/3\,m^2$, $u_0(t) = 1\,ms^{-1}$, $\rho_0(t) = 100\,um^{-3}$, $A^0(x) = (4\sqrt{125})/3\,m^2$, $u^0(x) = 1\,ms^{-1}$ and $\rho^0(x) = 100\,um^{-3}$. The time interval to control the pollution was 1 hour. Moreover, the loss rate for total coliform bacteria was considered constant ($k(x,t) = 10^{-4}s^{-1}$) and the bottom friction stress was neglected ($S_f = 0$).

The achieved sampling points were $p_1 = 38.6$, $p_2 = 599.0$ and $p_3 = 823.4$ first, second and third sections, respectively. Finally, in order to observe the goodness of the result, Figures 2-5 compare the concentration of coliform at each optimal sampling point with the average concentration in the corresponding section.

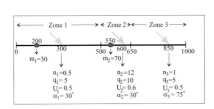

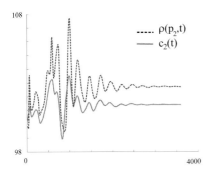

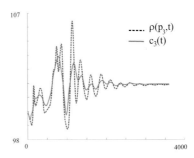

Fig. 2. Diagram of the river and data for the simulation.

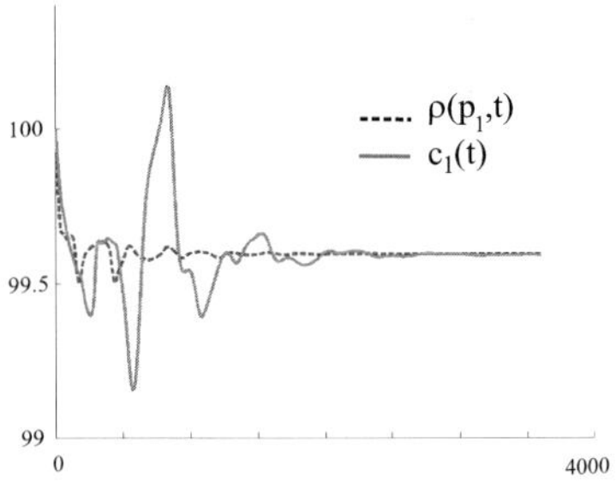

Fig. 3. Coliform concentration at p_1 and averaged concentration in zone 1.

Fig. 4. Coliform concentration at p_2 and averaged concentration in zone 2.

Fig. 5. Coliform concentration at p_3 and averaged concentration in zone 3.

Acknowledgements

The research contained in this work was supported by Project BFM2003-00373 of Ministerio de Ciencia y Tecnología (Spain).

References

1. Bercovier, M., Pironneau O., Sastri, V.: Finite elements and characteristics for some parabolic-hyperbolic problems. Appl. Math. Modelling, **7**, 89–96 (1983)
2. Bermúdez, A., Muñoz-Sola, R., Rodríguez, C., Vilar, M.A.: Theoretical and numerical study of an implicit discretization of an one dimensional inviscid model for river flows. Math. Mod. Meth. Appl. Sci. **16, n. 3**, 375–395 (2006)
3. Bazaraa, M. S., Shetty, C. M.: Nonlinear Programming. Theory and Algorithms. John Wiley & Sons, New York (1979)
4. Chapra, S. C.: Surface water quality modelling. McGraw-Hill, New York (1997)
5. Harsham, K. D.: Water sampling for pollution regulation. Gordon and Breach Publishers, Luxembourg (1995)

6. Kernévez, J.-P.: The sentinel method and its application to environmental pollution problems. CRC Press, Boca Raton, (1997)
7. Lions, J.-L.: Sentinelles pour les systèmes distribués à données incomplètes. Masson, Paris (1992)
8. Thomann, R. V., Mueller, J. A.: Principles of surface water quality modelling and control. Harper & Row, New York (1987)

PLATES

A Family of C^0 Finite Elements for Kirchhoff Plates with Free Boundary Conditions

L. Beirão da Veiga[1], J. Niiranen[2] and R. Stenberg[2]

[1] Dipartimento di Matematica "F. Enriques", via Saldini 50, 20133 Milano, Italy
beirao@mat.unimi.it
[2] Institute of Mathematics, Helsinki University of Technology, P.O.Box 1100, 02015 TKK, Finland
jarkko.niiranen@tkk.fi, rolf.stenberg@tkk.fi

Summary. A finite element method for the Kirchhoff plate bending problem is presented. This method has the twofold advantage of allowing low order polynomials and of holding convergence properties which does not deteriorate in the presence of free boundary conditions. Optimal a-priori and a-posteriori error estimates are shown without proof. Finally, some numerical tests are presented.

1 Introduction

In this contribution we first summarize the theoretical results already introduced in [4, 6], then we present some numerical tests. A family of finite elements for the bending problem of Kirchhoff plates is presented. This family is a modification of the stabilized method for Reissner-Mindlin plates of [6]. The introduced method has the advantage of requiring only a C^0 global regularity condition on the deflections, therefore allowing also a low order polynomial degree.

When the Kirchhoff plate bending problem is interpreted as a "zero thickness" limit of the Reissner-Mindlin problem, a difficulty arises in the presence of free boundary conditions, introducing a strong boundary inconsistency in the discrete method. In the finite element method here presented, this inconsistency is avoided by adding certain natural terms to the discrete bilinear form of the problem, leading to an optimally convergent method.

The paper is organized as follows. In Sections 2 and 3, we briefly introduce the problem and the proposed finite element method. In Sections 4 and 5, both a-priori and an a-posteriori error estimates are shown. Finally, in Section 6, we present a numerical example which shows on one hand the optimal rate of convergence of the proposed finite elements, and on the other hand the slow $O(h^{1/2})$ convergence of the "Reissner-Mindlin limit" method.

2 Kirchhoff plate bending problem

We consider the bending problem of an isotropic linearly elastic plate and assume that the midsurface of the undeformed plate is described by a given convex polygonal domain $\Omega \subset \mathbb{R}^2$. The plate is considered to be clamped on the part Γ_c of its boundary $\partial\Omega$, simply supported on the part $\Gamma_s \subset \partial\Omega$ and free on $\Gamma_f \subset \partial\Omega$. A transverse load $F = Gt^3 f$ is applied, where t is the thickness of the plate and G the shear modulus for the material.

Then, following the Kirchhoff plate bending model and assuming that the load is sufficiently regular, the deflection w of the plate can be found as the solution of the following well known biharmonic problem:

$$
\begin{aligned}
&D\Delta^2 w = Gf && \text{in } \Omega \\
&w = 0, \quad \tfrac{\partial w}{\partial n} = 0 && \text{on } \Gamma_c \\
&w = 0, \quad \boldsymbol{n}^T \boldsymbol{M} \boldsymbol{n} = 0 && \text{on } \Gamma_s \\
&\boldsymbol{n}^T \boldsymbol{M} \boldsymbol{n} = 0, \quad \tfrac{\partial}{\partial s} \boldsymbol{s}^T \boldsymbol{M} \boldsymbol{n} + (\mathbf{div}\,\boldsymbol{M}) \cdot \boldsymbol{n} = 0 && \text{on } \Gamma_f,
\end{aligned}
\qquad (1)
$$

where $\boldsymbol{n}$ and $\boldsymbol{s}$ are respectively the unit outward normal and the unit counterclockwise tangent to the boundary, while the scaled bending modulus and the bending moment are

$$
D = \frac{E}{12(1-\nu^2)}, \qquad \boldsymbol{M} = \frac{G}{6}\left(\varepsilon(\nabla w) + \frac{\nu}{1-\nu} \operatorname{div} \nabla w \, \boldsymbol{I}\right), \qquad (2)
$$

with E, ν the Young modulus and the Poisson ratio for the material, and ε the symmetric gradient operator. Note that it holds $G = \frac{E}{2(1+\nu)}$.

Due to the presence of the fourth order elliptic operator Δ^2, the natural space for the variational formulation of the problem (1) is the Sobolev space $H^2(\Omega)$. As a consequence, conforming finite element methods based on such a formulation need the C^1 regularity conditions. In order to keep minimal flexibility of the discrete space used, the C^1 regularity condition in turn requires a high order polynomial space, which may be preferable to avoid.

In the case of clamped and simply supported boundary conditions, the Kirchhoff problem can be treated as a penalty formulation for the Reissner–Mindlin plate bending problem with the thickness t interpreted as the thickness parameter; as a consequence, a non conforming Kirchhoff element (i.e. which uses globally C^0 deflections) can be obtained starting from any locking free Reissner–Mindlin element. On the other hand, in the presence of free boundary conditions, this is false because the two formulations (Reissner-Mindlin limit and Kirchhoff) are *not* equivalent (see for example [5, 3]).

In the sequel we will present a family of low order finite elements for Kirchhoff plates which avoids this difficulty; in particular, its rate of convergence to the Kirchhoff problem solution does not deteriorate in the presence of free boundary conditions.

3 Finite element formulation

In this section we introduce the numerical method, which is an extension of the method presented in [6]. For simplicity, we assume that the one-dimensional measure

$$\mathrm{meas}(\varGamma_\mathrm{c} \cup \varGamma_\mathrm{s}) > 0. \tag{3}$$

Let a regular family of triangular meshes on Ω be given. Given an integer $k \geq 1$, we then define the discrete spaces

$$W_h = \{v \in W \mid v_{|K} \in P_{k+1}(K) \ \forall K \in \mathcal{C}_h\}, \tag{4}$$
$$\boldsymbol{V}_h = \{\boldsymbol{\eta} \in \boldsymbol{V} \mid \boldsymbol{\eta}_{|K} \in [P_k(K)]^2 \ \forall K \in \mathcal{C}_h\}, \tag{5}$$

where $\mathcal{C}_h$ represents the set of all the triangles K of the mesh and $P_k(K)$ is the space of polynomials of degree k on K. In the sequel, we will indicate with h_K the diameter of each element K, while h will indicate the maximum size of all the elements in the mesh. Moreover, we will indicate with e a general edge of the triangulation and with h_e the length of e. Let two positive stability constants γ and α be assigned (for a discussion on the requirements of γ and α see [4]). Finally, we introduce the bilinear form

$$a(\boldsymbol{\phi},\boldsymbol{\eta}) = \frac{1}{6}\big((\boldsymbol{\varepsilon}(\boldsymbol{\phi}),\boldsymbol{\varepsilon}(\boldsymbol{\eta})) + \frac{\nu}{1-\nu}(\mathrm{div}\,\boldsymbol{\phi},\mathrm{div}\,\boldsymbol{\eta})\big) \quad \forall \boldsymbol{\phi} \in \boldsymbol{V},\ \boldsymbol{\eta} \in \boldsymbol{V}. \tag{6}$$

Then, the discrete problem reads:

Find $(w_h, \boldsymbol{\beta}_h) \in W_h \times \boldsymbol{V}_h$, such that

$$\mathcal{A}_h(w_h, \boldsymbol{\beta}_h; v, \boldsymbol{\eta}) = (f, v) \quad \forall (v, \boldsymbol{\eta}) \in W_h \times \boldsymbol{V}_h, \tag{7}$$

where the form $\mathcal{A}_h$ is

$$\mathcal{A}_h(z, \boldsymbol{\phi}; v, \boldsymbol{\eta}) = \mathcal{B}_h(z, \boldsymbol{\phi}; v, \boldsymbol{\eta}) + \mathcal{D}_h(z, \boldsymbol{\phi}; v, \boldsymbol{\eta}), \tag{8}$$

with

$$\mathcal{B}_h(z, \boldsymbol{\phi}; v, \boldsymbol{\eta}) = a(\boldsymbol{\phi}, \boldsymbol{\eta}) - \sum_{K \in \mathcal{C}_h} \alpha h_K^2 (\boldsymbol{L}\boldsymbol{\phi}, \boldsymbol{L}\boldsymbol{\eta})_K$$
$$+ \sum_{K \in \mathcal{C}_h} \frac{1}{\alpha h_K^2}(\nabla z - \boldsymbol{\phi} - \alpha h_K^2 \boldsymbol{L}\boldsymbol{\phi}, \nabla v - \boldsymbol{\eta} - \alpha h_K^2 \boldsymbol{L}\boldsymbol{\eta})_K \tag{9}$$

and

$$\mathcal{D}_h(z,\phi;v,\eta) = \sum_{e\in\Gamma_{\mathrm{f},h}} \Big((M_{ns}(\phi),[\nabla v - \eta]\cdot s)_e$$
$$+([\nabla z - \phi]\cdot s, M_{ns}(\eta))_e + \frac{\gamma}{h_e}([\nabla z - \phi]\cdot s, [\nabla v - \eta]\cdot s)_e\Big) \quad (10)$$

for all $(z,\phi) \in W_h \times \boldsymbol{V}_h$, $(v,\eta) \in W_h \times \boldsymbol{V}_h$, where $\Gamma_{\mathrm{f},h}$ represents the set of all the boundary edges in Γ_{f} and $M_{ns} = \boldsymbol{s}^T \boldsymbol{M}\boldsymbol{n}$. The bilinear form $\mathcal{B}_h$ constitutes essentially the original method of [6] with the thickness t set equal to zero, while the added form $\mathcal{D}_h$ is introduced to avoid the convergence deterioration in the presence of free boundaries.

Remark 1. If the original method of [6] without the additional form $\mathcal{D}_h$ is employed, in the presence of a free boundary an inconsistency term arises. In other words, if $(w,\boldsymbol{\beta})$ is the solution of problem (1), then

$$\mathcal{B}_h(w,\boldsymbol{\beta};v,\eta) = (f,v) + \sum_{e\in\Gamma_{\mathrm{f},h}} (M_{ns}(\boldsymbol{\beta}),[\nabla v - \eta]\cdot s)_e \quad (11)$$

for all $(v,\eta) \in W_h \times \boldsymbol{V}_h$. The second addendum in the right hand side is an inconsistency term of order $O(h^{1/2})$, which severely hinders the convergence of the method.

4 A-priori error estimates

For $(v,\eta) \in W_h \times \boldsymbol{V}_h$, we introduce the following mesh dependent norms:

$$|(v,\eta)|_h^2 = \sum_{K\in\mathcal{C}_h} h_K^{-2}\|\nabla v - \eta\|_{0,K}^2, \quad (12)$$

$$\|v\|_{2,h}^2 = \|v\|_1^2 + \sum_{K\in\mathcal{C}_h} |v|_{2,K}^2 + \sum_{e\in\mathcal{T}_h} h_K^{-1}\|[\![\frac{\partial v}{\partial n}]\!]\|_{0,e}^2, \quad (13)$$

$$|||(v,\eta)|||_h = \|\eta\|_1 + \|v\|_{2,h} + |(v,\eta)|_h, \quad (14)$$

and for $\boldsymbol{r} \in L^2(\Omega)$

$$\|\boldsymbol{r}\|_{-1,h} = \Big(\sum_{K\in\mathcal{C}_h} h_K^2 \|\boldsymbol{r}\|_{0,K}^2\Big)^{1/2}. \quad (15)$$

Given the space

$$\boldsymbol{V}_* = \{\eta \in [H^1(\Omega)]^2 \mid \eta = \boldsymbol{0} \text{ on } \Gamma_{\mathrm{c}},\ \eta\cdot s = 0 \text{ on } \Gamma_{\mathrm{f}} \cup \Gamma_{\mathrm{s}}\} \quad (16)$$

we also introduce the norm

$$\|\boldsymbol{r}\|_{-1,*} = \sup_{\eta\in\boldsymbol{V}_*} \frac{\langle \boldsymbol{r},\eta\rangle}{\|\eta\|_1}. \quad (17)$$

We then have the following a priori error estimate.

Theorem 1. *Let $(w, \boldsymbol{\beta})$ be the solution of the problem (1) and $(w_h, \boldsymbol{\beta}_h)$ the solution of the problem (7). Then it holds*

$$|||(w - w_h, \boldsymbol{\beta} - \boldsymbol{\beta}_h)|||_h + \|\boldsymbol{q} - \boldsymbol{q}_h\|_{-1,*} \leq Ch^s \|w\|_{s+2} \tag{18}$$

for all $1 \leq s \leq k$, where the exact and discrete "shear stresses" are

$$\boldsymbol{q}_{|K} = \frac{1}{\alpha h_K^2}(\nabla w - \boldsymbol{\beta} - \alpha h_K^2 \boldsymbol{L}\boldsymbol{\beta})_{|K} \quad \forall K \in \mathcal{C}_h \tag{19}$$

$$\boldsymbol{q}_{h|K} = \frac{1}{\alpha h_K^2}(\nabla w_h - \boldsymbol{\beta}_h - \alpha h_K^2 \boldsymbol{L}\boldsymbol{\beta}_h)_{|K} \quad \forall K \in \mathcal{C}_h. \tag{20}$$

5 A-posteriori error estimates

We now present the reliability and the efficiency results for an a-posteriori error estimator for our method. To this end, we introduce

$$\tilde{\eta}_K^2 := h_K^4 \|f_h + \operatorname{div} \boldsymbol{q}_h\|_{0,K}^2 + h_K^{-2} \|\nabla w_h - \boldsymbol{\beta}_h\|_{0,K}^2, \tag{21}$$

$$\eta_e^2 := h_e^3 \|[\![\boldsymbol{q}_h \cdot \boldsymbol{n}]\!]\|_{0,e}^2 + h_e \|[\![\boldsymbol{M}(\boldsymbol{\beta}_h)\boldsymbol{n}]\!]\|_{0,e}^2, \tag{22}$$

$$\eta_{s,e}^2 := h_e \|M_{nn}(\boldsymbol{\beta}_h)\|_{0,e}^2, \tag{23}$$

$$\eta_{f,e}^2 := h_e \|M_{nn}(\boldsymbol{\beta}_h)\|_{0,e}^2 + h_e^3 \|\frac{\partial}{\partial s} M_{ns}(\boldsymbol{\beta}_h) - \boldsymbol{q}_h \cdot \boldsymbol{n}\|_{0,e}^2, \tag{24}$$

where f_h is some approximation of the load f and $[\![\cdot]\!]$ represents the jump operator (which is assumed to be equal to the function value on boundary edges).

Given any element $K \in \mathcal{C}_h$, the local error indicator is

$$\eta_K := \left(\tilde{\eta}_K^2 + \frac{1}{2} \sum_{e \in \Gamma_{i,h} \cap \partial K} \eta_e^2 + \sum_{e \in \Gamma_{s,h} \cap \partial K} \eta_{s,e}^2 + \sum_{e \in \Gamma_{f,h} \cap \partial K} \eta_{f,e}^2 \right)^{1/2}, \tag{25}$$

where $\Gamma_{i,h}$ represents the set of all the internal edges, while $\Gamma_{c,h}, \Gamma_{s,h}$ and $\Gamma_{f,h}$ represent the sets of all the boundary edges in Γ_c, Γ_s and Γ_f, respectively. Finally, the global error indicator is defined as

$$\eta := \left(\sum_{K \in \mathcal{C}_h} \eta_K^2 \right)^{1/2}. \tag{26}$$

We assume the following saturation assumption:

Assumption. Given a mesh $\mathcal{C}_h$, let $\mathcal{C}_{h/2}$ be the mesh obtained by splitting each triangle $K \in \mathcal{C}_h$ into four triangles connecting the edge midpoints. Let $(w_{h/2}, \boldsymbol{\beta}_{h/2}, \boldsymbol{q}_{h/2})$ be the discrete solution corresponding to the mesh $\mathcal{C}_{h/2}$. We assume that there exists a constant ρ, $0 < \rho < 1$, such that

$$|||(w - w_{h/2}, \boldsymbol{\beta} - \boldsymbol{\beta}_{h/2})|||_{h/2} + \|\boldsymbol{q} - \boldsymbol{q}_{h/2}\|_{-1,*}$$
$$\leq \rho(|||(w - w_h, \boldsymbol{\beta} - \boldsymbol{\beta}_h)|||_h + \|\boldsymbol{q} - \boldsymbol{q}_h\|_{-1,*}) \qquad (27)$$

where by $||| \cdot |||_{h/2}$ we indicate the $||| \cdot |||_h$ norm with respect to the new mesh $\mathcal{C}_{h/2}$.

We then have the following results:

Theorem 2. *It holds*

$$|||(w - w_h, \boldsymbol{\beta} - \boldsymbol{\beta}_h)|||_h + \|\boldsymbol{q} - \boldsymbol{q}_h\|_{-1,*} \leq C \Big(\sum_{K \in \mathcal{C}_h} \eta_K^2 + h_K^4 \|f - f_h\|_{0,K}^2 \Big)^{1/2}.$$

Theorem 3. *It holds*

$$\eta_K \leq |||(w - w_h, \boldsymbol{\beta} - \boldsymbol{\beta}_h)|||_{h,\omega_K} + \|\boldsymbol{q} - \boldsymbol{q}_h\|_{-1,*,\omega_K} + h_K^2 \|f - f_h\|_{0,\omega_K},$$

where $||| \cdot |||_{h,\omega_K}$, $\| \cdot \|_{-1,*,\omega_K}$ and $\| \cdot \|_{0,\omega_K}$ represent respectively the norms $||| \cdot |||_h$, $\| \cdot \|_{-1,*}$ and $\| \cdot \|_0$ restricted to the domain ω_K, the set of all the triangles sharing an edge with K.

Theorems 2 and 3 prove, respectively, the reliability and efficiency of the error estimator.

6 Computational results

We consider the following Kirchhoff bending problem of a semi-infinite plate (see [2]). The plate midsurface and boundary are described by

$$\Omega = \{(x,y) \in \mathbb{R}^2 \mid y > 0\}, \quad \Gamma = \{(x,y) \in \mathbb{R}^2 \mid y = 0\}.$$

The plate is assumed to be free on Γ and subjected to the loading $g(x,y) = \cos x/G$, while the material constants are $E = 1$, $\nu = 0.3$. The exact x-periodic solution of this problem is given in [2].

We mesh the domain

$$\overline{D} = [0, \pi/2] \times [0, 3\pi/4]$$

setting the symmetry conditions on the vertical boundaries

$$\{x = 0, 0 \leq y \leq 3\pi/4\}, \quad \{x = \pi/2, 0 \leq y \leq 3\pi/4\},$$

while on the upper horizontal boundary

$$\{y = 3\pi/4, 0 \leq x \leq \pi/2\}$$

we use the non-homogeneus Dirichlet conditions adopting the exact solution as a reference. We show some sample meshes in Figure 1.

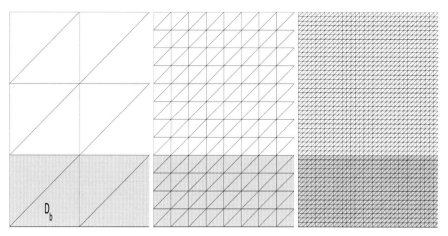

Fig. 1. Samples of the adopted meshes

Let D_b represent the boundary domain $[0, \pi/2] \times [0, \pi/4]$. In Figure 2 we show the convergence of the deflection w in the $|\!|\!| \cdot |\!|\!|_h$ norm restricted to D_b, for the polynomial degrees $k = 1, 2$. The dashed line shows the convergence graph for the original method (i.e. without the $\mathcal{D}_h$ correction) while the solid line refers to the improved method (7). In Figure 3 we show the convergence of the moment component M_{ns} in the $L^2(D_b)$ norm, for the polynomial degrees $k = 1, 2$, for both the original (dashed line) and improved (solid line) methods. As predicted by the theory, the original method shows a convergence rate of $O(h^{1/2})$ while the modified method follows an $O(h^k)$ error behavior.

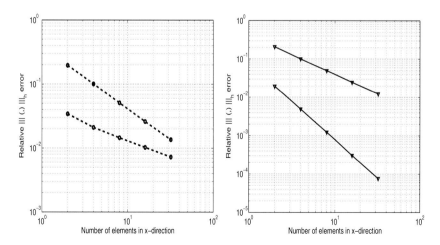

Fig. 2. Convergence of w in the $|\!|\!| \cdot |\!|\!|_h$ norm on D_b

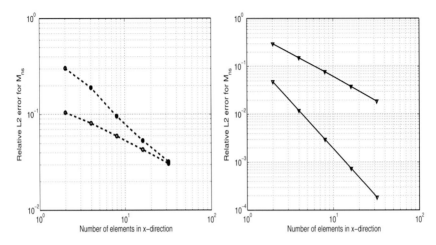

Fig. 3. Convergence of M_{ns} in the $L^2(D_b)$ norm

References

1. Arnold, D. N., Falk, R. S.: Asymptotic analysis of the boundary layer for the Reissner-Mindlin plate model. SIAM. J. Math. Anal., **27**, 486–514 (1996)
2. Arnold, D. N., Falk, R. S.: Edge effects in the Reissner-Mindlin plate theory. In: Noor, A. K., Belytschko, T., Simo, J. C. (ed) Analytic and Computational Models of Shells. ASME, New York, 71–90 (1989)
3. Beirão da Veiga, L.: Finite elements for a modified Reissner–Mindlin free plate model. Siam. J. Numer. Anal. **42** 1572–1591 (2004)
4. Beirão da Veiga, L., Niiranen, J., Stenberg, R.: A family of C^0 finite elements for Kirchhoff plates: theoretical analysis. Helsinki University of Technology, Insitute of Mathematics Research Report A483, **http://math.tkk.fi/reports**, Espoo (2005)
5. Destuynder, P., Salaun, M.: Mathematical Analysis of Thin Plate Models, Springer-Verlag, Berlin Heidelberg (1996)
6. Stenberg, R.: A new finite element formulation for the plate bending problem. In: Ciarlet, P.G., Trabucho, L. and Viaño, M. (ed) Asymptotic Methods for Elastic Structures, Proceedings of the International Conference, Lisbon, October 4-8, 1993. Walter de Gruyter & Co., Berlin New-York, 209–221 (1995)

A Postprocessing Method for the MITC Plate Elements

Mikko Lyly[1], Jarkko Niiranen[2] and Rolf Stenberg[2]

[1] CSC – Scientific Computing Ltd., P.O. Box 405, 02101 Espoo, Finland
   mikko.lyly@csc.fi
[2] Institute of Mathematics, Helsinki University of Technology, P.O. Box 1100, 02015 TKK, Finland
   jarkko.niiranen@tkk.fi, rolf.stenberg@tkk.fi

Summary. We summarize the main results obtained in [6]. For the MITC plate elements [2, 4] it is shown that the deflection has a superconvergence property. This is used in a local postprocessing method for obtaining an improved approximation for the deflection. The theoretical results are checked by various numerical computations.

1 Introduction

We consider the well-known MITC plate bending elements [2]. First, we review the improved error analysis performed in [5]. We then show that the difference between the approximate deflection and an interpolant to the exact solution is superconverging in the H^1-norm [6]. In the postprocessing method the superconvergence property is utilized in order to improve the accuracy of the approximation for the deflection. The new approximation is a piecewise polynomial of one degree higher than the original one. The postprocessing is done element by element which implies low computational costs.

In the next two sections we recall the Reissner–Mindlin plate model and the MITC finite elements. In Sects. 4 and 5 we formulate and discuss the superconvergence result, the postprocessing method and the improved error estimate. The benchmark computations in Sect. 6 illustrate and verify the results.

2 The Reissner–Mindlin plate model

We consider a linearly elastic and isotropic plate with the shear modulus G and the Poisson ratio ν. The midsurface of the undeformed plate is $\Omega \subset \mathbb{R}^2$ and the plate thickness t is constant. The boundary of the plate is divided into hard clamped, hard simply supported and free parts: $\partial\Omega = \Gamma_C \cup \Gamma_{SS} \cup \Gamma_F$. The soft clamped and soft simply supported cases would be possible as well. The spaces of kinematically admissible deflections and rotations are now

$$W = \{v \in H^1(\Omega) \mid v_{|\Gamma_C} = 0, \ v_{|\Gamma_{SS}} = 0\}, \tag{1}$$

$$\boldsymbol{V} = \{\boldsymbol{\eta} \in [H^1(\Omega)]^2 \mid \boldsymbol{\eta}_{|\Gamma_C} = \boldsymbol{0}, \ (\boldsymbol{\eta} \cdot \boldsymbol{\tau})_{|\Gamma_{SS}} = 0\}, \tag{2}$$

where $\boldsymbol{\tau}$ is the unit tangent to the boundary. For the bilinear form we define the bending energy and the linear strain tensor as

$$a(\boldsymbol{\phi}, \boldsymbol{\eta}) = \frac{1}{6}\{(\varepsilon(\boldsymbol{\phi}), \varepsilon(\boldsymbol{\eta})) + \frac{\nu}{1-\nu}(\operatorname{div} \boldsymbol{\phi}, \operatorname{div} \boldsymbol{\eta})\}, \tag{3}$$

$$\varepsilon(\boldsymbol{\eta}) = \frac{1}{2}(\nabla \boldsymbol{\eta} + (\nabla \boldsymbol{\eta})^T). \tag{4}$$

With these assumptions and notation the variational formulation for the Reissner–Mindlin plate model, under the transverse loading $g \in H^{-1}(\Omega)$, can be written in the following form [4, 6]: Find the deflection $w \in W$ and the rotation $\boldsymbol{\beta} \in \boldsymbol{V}$ such that

$$a(\boldsymbol{\beta}, \boldsymbol{\eta}) + \frac{1}{t^2}(\nabla w - \boldsymbol{\beta}, \nabla v - \boldsymbol{\eta}) = (g, v) \ \forall (v, \boldsymbol{\eta}) \in W \times \boldsymbol{V}. \tag{5}$$

For the analysis the problem is written in mixed form in which the shear force $\boldsymbol{q} = t^{-2}(\nabla w - \boldsymbol{\beta}) \in \boldsymbol{Q} = [L^2(\Omega)]^2$ is taken as an independent unknown [4, 6].

3 MITC finite element methods

For simplicity, we consider the triangular family but we emphasize that all the results are valid for quadrilateral families as well. By $\mathcal{C}_h$ we denote the triangulation of $\overline{\Omega}$. As usual, we denote $h = \max_{K \in \mathcal{C}_h} h_K$, where h_K is the diameter of the element K. The space of polynomials of degree k on K is denoted by $P_k(K)$. By C we denote positive constants independent of the thickness t and the mesh size h.

In the MITC methods [2, 4] the finite element subspaces $W_h \subset W$ and $\boldsymbol{V}_h \subset \boldsymbol{V}$ are defined for the polynomial degree $k \geq 2$ as

$$W_h = \{v \in W \mid v_{|K} \in P_k(K) \ \forall K \in \mathcal{C}_h\}, \tag{6}$$

$$\boldsymbol{V}_h = \{\boldsymbol{\eta} \in \boldsymbol{V} \mid \boldsymbol{\eta}_{|K} \in [P_k(K)]^2 \oplus [B_{k+1}(K)]^2 \ \forall K \in \mathcal{C}_h\}, \tag{7}$$

with the "bubble space"

$$B_{k+1}(K) = \{b = b_3 p \mid p \in \widetilde{P}_{k-2}(K), \ b_3 \in P_3(K), \ b_{3|E} = 0 \ \forall E \subset \partial K\}, \tag{8}$$

where $\widetilde{P}_{k-2}(K)$ is the space of homogeneous polynomials of degree $k-2$ on the element K. The rotated Raviart-Thomas space of order $k-1$ is defined as

$$Q_h = \{\, r \in H(\text{rot}; \Omega) \mid r_{|K} \in [P_{k-1}(K)]^2 \oplus (y, -x)\widetilde{P}_{k-1}(K) \,\forall K \in \mathcal{C}_h \,\}. \quad (9)$$

The reduction operator $R_h : H(\text{rot}; \Omega) \to Q_h$ is defined locally, with $R_K = R_{h|K}$, through the conditions

$$\langle (R_K \eta - \eta) \cdot \tau_E, p \rangle_E = 0 \quad \forall p \in P_{k-1}(E) \;\; \forall E \subset \partial K, \quad (10)$$

$$(R_K \eta - \eta, p)_K = 0 \quad \forall p \in [P_{k-2}(K)]^2, \quad (11)$$

where E denotes an edge to K and τ_E is the unit tangent to E. Notation $(\cdot, \cdot)_K$ and $\langle \cdot, \cdot \rangle_E$ are used for the L^2-inner products. Now the MITC method is defined as follows:

Method. *Find the approximations $w_h \in W_h$, $\beta_h \in V_h$, for the deflection and the rotation, such that*

$$a(\beta_h, \eta) + \frac{1}{t^2}(R_h(\nabla w_h - \beta_h), R_h(\nabla v - \eta)) = (g, v) \;\; \forall (v, \eta) \in W_h \times V_h. \quad (12)$$

The discrete shear force is $q_h = t^{-2} R_h(\nabla w_h - \beta_h) \in Q_h$.

For the error analysis we refer to [4, 7, 5]. In [5] we have analyzed the case of a convex domain with clamped boundaries for which a refined global estimate holds. For the interior and the boundary meshes, respectively, we use the notation $\Omega_i^h = \cup_{K \subset \Omega_i} K$, $\Omega_b^h = \Omega \setminus \Omega_i^h$. Then we denote the mesh size in the interior and near the boundary by $h_i = \max_{K \in \Omega_i^h} h_K$, $h_b = \max_{K \in \Omega_b^h} h_K$, respectively, and we can state the error estimate as follows [5]:

Theorem 1. *Let Ω be a convex polygon and suppose that the plate is clamped. For $g \in H^{k-2}(\Omega)$, $tg \in H^{k-1}(\Omega)$, with $k \geq 2$, it then holds*

$$\begin{aligned}
\|w - w_h\|_1 &+ \|\beta - \beta_h\|_1 + t\|q - q_h\|_0 \\
&\leq C\{h_i^k(\|g\|_{k-2} + t\|g\|_{k-1}) + h_b(\|g\|_{-1} + t\|g\|_0)\}, \\
\|w - w_h\|_0 &+ \|\beta - \beta_h\|_0 \\
&\leq Ch\{h_i^k(\|g\|_{k-2} + t\|g\|_{k-1}) + h_b(\|g\|_{-1} + t\|g\|_0)\}.
\end{aligned} \quad (13)$$

4 Superconvergence

For the superconvergence result we need the classical quasi-optimal interpolation operator [6]: With a vertex a and an edge E of the triangle K, the operator $I_h : H^s(\Omega) \to W_h$, $s > 1$, is defined through the conditions

$$(v - I_K v)(a) = 0 \;\; \forall a \in K, \quad (14)$$

$$\langle v - I_K v, p \rangle_E = 0 \;\; \forall p \in P_{k-2}(E) \;\; \forall E \subset K, \quad (15)$$

$$(v - I_K v, p)_K = 0 \quad \forall p \in P_{k-3}(K), \tag{16}$$

with $I_K = I_{h|K} \ \forall K \in \mathcal{C}_h$.

The key property in proving the superconvergence is the close connection between the interpolation and reduction operators [6]:

Lemma 1. *It holds*

$$\boldsymbol{R}_h \nabla v = \nabla I_h v \quad \forall v \in H^s(\Omega), s \geq 2. \tag{17}$$

This implies the superconvergence result [6]:

Theorem 2. *There is a positive constant C such that*

$$\|\nabla(I_h w - w_h)\|_{0,K} \leq Ch_K \|\boldsymbol{\beta} - \boldsymbol{\beta}_h\|_{1,K} + \|\boldsymbol{\beta} - \boldsymbol{\beta}_h\|_{0,K} + t^2 \|\boldsymbol{q} - \boldsymbol{q}_h\|_{0,K}$$
$$+ t^2 \|\boldsymbol{q} - \boldsymbol{R}_h \boldsymbol{q}\|_{0,K}. \tag{18}$$

This estimate gives a local improvement of order $h_K + t$. Furthermore, the corresponding global estimate implies for a convex domain with clamped boundaries the estimate [6]

$$\|w_h - I_h w\|_1 = \mathcal{O}\big((h+t)(h_i^k + h_b)\big). \tag{19}$$

Now we see that the convergence rate for $\|w_h - I_h w\|_1$ is by the factor $h + t$ better than the rates for both $\|w - w_h\|_1$ and $\|w - I_h w\|_1$.

Since $I_h w$ interpolates w at the vertices (cf. (14)) this also gives an indication that the vertex values of w_h are converging with an improved speed. The numerical results in Sect. 6 verify this.

5 Postprocessing method

In the postprocessing we construct an improved approximation for the deflection in the space

$$W_h^* = \{v \in W \mid v_{|K} \in P_{k+1}(K) \ \forall K \in \mathcal{C}_h\}. \tag{20}$$

For the postprocessing we first introduce the interpolation operator $I_h^* : H^s(\Omega) \to W_h^*$, $s > 1$, which follows the definitions (14)–(16) with $k + 1$ in place of k. Thus, the interpolation operators I_h^* and I_h are hierarchical, and the local spaces for the additional degrees of freedom are defined as

$$\widehat{W}(K) = \{v \in P_{k+1}(K) \mid I_K v = 0, \ (v,p)_K = 0 \ \forall p \in \widetilde{P}_{k-2}(K)\}, \tag{21}$$

$$\overline{W}(K) = \{v \in P_{k+1}(K) \mid I_K v = 0, \ \langle v,p \rangle_E = 0 \ \forall p \in \widetilde{P}_{k-1}(E) \ \forall E \subset K\}. \tag{22}$$

Now the method itself can be defined [6]:

Postprocessing scheme. *For all the triangles $K \in \mathcal{C}_h$ find the local postprocessed finite element deflection $w^*_{h|K} \in P_{k+1}(K) = P_k(K) \oplus \widehat{W}(K) \oplus \overline{W}(K)$ such that*

$$I_h w^*_{h|K} = w_{h|K}, \tag{23}$$

$$\langle \nabla w^*_h \cdot \boldsymbol{\tau}_E, \nabla \hat{v} \cdot \boldsymbol{\tau}_E \rangle_E = \langle (\boldsymbol{\beta}_h + t^2 \boldsymbol{q}_h) \cdot \boldsymbol{\tau}_E, \nabla \hat{v} \cdot \boldsymbol{\tau}_E \rangle_E \quad \forall E \subset \partial K, \ \forall \hat{v} \in \widehat{W}(K), \tag{24}$$

$$(\nabla w^*_h, \nabla \bar{v})_K = (\boldsymbol{\beta}_h + t^2 \boldsymbol{q}_h, \nabla \bar{v})_K \quad \forall \bar{v} \in \overline{W}(K). \tag{25}$$

We note that the postprocessed deflection is conforming since $(\boldsymbol{\beta}_h + t^2 \boldsymbol{q}_h) \cdot \boldsymbol{\tau}$ is continuous along inter element boundaries.

Next, we define the space $\boldsymbol{Q}^*_h$ by the definition (9) and the operator $\boldsymbol{R}^*_h$ by the definitions (10) and (11), both with $k+1$ in place of k. Then we have the following error estimate for the postprocessing method [6]:

Theorem 3. *There is a positive constant C such that*

$$\begin{aligned}\|\nabla(w - w^*_h)\|_{0,K} \leq C\{&h_K \|\boldsymbol{\beta} - \boldsymbol{\beta}_h\|_{1,K} + \|\boldsymbol{\beta} - \boldsymbol{\beta}_h\|_{0,K} + t^2 \|\boldsymbol{q} - \boldsymbol{q}_h\|_{0,K} \\ &+ \|\nabla(w - I^*_h w)\|_{0,K} + \|\boldsymbol{\beta} - \boldsymbol{R}^*_h \boldsymbol{\beta}\|_{0,K} \\ &+ t^2 \|\boldsymbol{q} - \boldsymbol{R}^*_h \boldsymbol{q}\|_{0,K} + t^2 \|\boldsymbol{q} - \boldsymbol{R}_h \boldsymbol{q}\|_{0,K}\}.\end{aligned} \tag{26}$$

This result is also local and it is made up of two parts: The first part is related to the error of the original method and the second part consists of interpolation estimates – both parts giving an improvement by the factor $h_K + t$ compared to the original approximation. The corresponding global estimate implies for the case of a convex domain with clamped boundaries the estimate [6]

$$\|w - w^*_h\|_1 = \mathcal{O}\big((h+t)(h_i^k + h_b)\big). \tag{27}$$

The numerical computations in the next section confirm that the improvement in the convergence rate is partially t-dependent, as seen in (27). However, let us also remark that, in practice, we are interested in the case $t < h$ since for a finer mesh the error in the model, with respect to the three dimensional structure, is greater than the discretization error.

6 Benchmark computations

We have performed the numerical computations for a test problem for which an analytical solution has been obtained in [1]. The domain is the semi-infinite region $\Omega = \{(x,y) \in \mathbb{R}^2 \mid y > 0\}$ and the loading is $g = \frac{1}{G} \cos x$. The Poisson ratio is $\nu = 0.3$, the shear modulus $G = 1/(2(1+\nu))$, the shear corrector factor $\kappa = 1$ and the thickness $t = 0.01$ (one specific case with $t = 1.0$). We have considered two different boundary conditions on $\Gamma = \{(x,y) \in \mathbb{R}^2 \mid y = 0\}$, hard clamped and free.

We have discretized the domain $D = (0, \pi/2) \times (0, 3\pi/2)$, with both uniform and general, non-uniform, meshes. For the uniform meshes the accuracy has been studied separately in the interior subdomain $D_i = \{(x,y) \in D \mid y \in (\pi, 5\pi/4)\}$ and in the boundary subdomain $D_b = \{(x,y) \in D \mid y \in (0, \pi/4)\}$, see Fig. 1. The number of elements in the x-direction is $N = 2, 4, 6$ or 8, and the mesh size is $h \approx 1/N$. The degree of the elements is $k = 2$ (quadratic) and $k = 3$ (cubic).

The relative errors, measured in the H^1-norm and in the L^2-norm, for both the original finite element deflection (dashed lines) and the postprocessed finite element deflection (solid lines) are on the logarithmic scale with respect to N. To study the convergence rate, a function of the form $CN^{-r} \approx Ch^r$ is fitted to the results by using least squares. The fitted lines with the slopes r and r^*, for the original and for the postprocessed deflections, are also shown in Fig. 2–5.

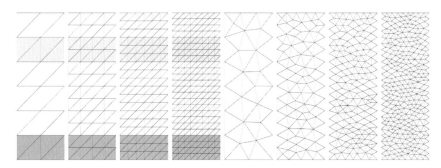

Fig. 1. The uniform and non-uniform meshes with $N = 2, 4, 6, 8$; Interior region D_i (light grey); Boundary region D_b (dark grey).

6.1 Uniform meshes

In the interior of the plate the numerical results are clearly in accordance with the theory: The convergence rate of the original finite element deflection in the H^1-norm is $r \approx k$, and the convergence rate of the postprocessed finite element deflection is $r^* \approx k+1 \approx r+1$, as seen in Fig. 2 (left). The behavior in the L^2-norm looks very similar, see Fig. 2 (right), although to rigorously prove the improvement in that case seems to be difficult – due to the t-dependency of the solution and the boundary layers.

Near the boundary of the plate the numerical results reflect the strength of the edge effect: For the clamped edge the convergence – especially in the H^1-norm – is almost as good as in the interior region. For the free edge the rate of convergence rapidly slows down, for both the original and the postprocessed deflection, see Fig. 3. The convergence rate seems to be of order $\mathcal{O}(h^{1/2})$, as proved in [8, 3] for the original approximation in the H^1-norm. But still, a

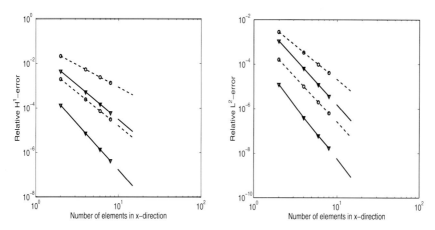

Fig. 2. Clamped edge; Interior region; Uniform mesh; Convergence of the relative H^1- and L^2-errors with $k = 2, 3$ (dashed line for the original, solid line for the postprocessed deflection).

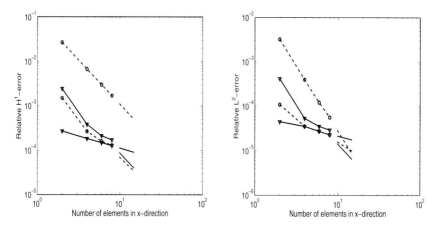

Fig. 3. Free edge; Boundary region; Uniform mesh; Convergence of the relative H^1- and L^2-errors with $k = 2, 3$ (dashed line for the original, solid line for the postprocessed deflection).

significant accuracy improvement is obtained, especially for coarse meshes and lower order elements. Furthermore, the superaccuracy of the vertex values is obvious, as seen in Fig. 4 (left).

To illustrate the partial t-dependence of the postprocessing estimate, we have used also the thickness value $t = 1.0$ for the case of clamped edge: In the interior of the plate the convergence rate of the postprocessed finite element deflection in the H^1-norm was $r^* \approx k + 1 \approx r + 1$ in Fig. 2 (left) for $t = 0.01$, but for $t = 1.0$ it is only $r^* \approx k \approx r$ in Fig. 4 (right) – there is only a shift, not a change in the slope, when comparing the dashed and solid lines.

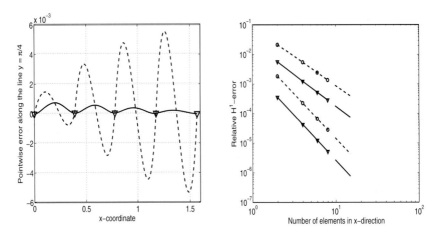

Fig. 4. *Left:* Free edge; Boundary region; Uniform mesh; Pointwise error along the line $x = \pi/4$ for $N = 4$ and $k = 2$ (dashed line for the original, solid line for the postprocessed deflection, triangles for the vertex values).
Right: $t = 1.0$; Clamped edge; Interior region; Uniform mesh; Convergence of the relative H^1-error with $k = 2, 3$ (dashed line for the original, solid line for the postprocessed deflection).

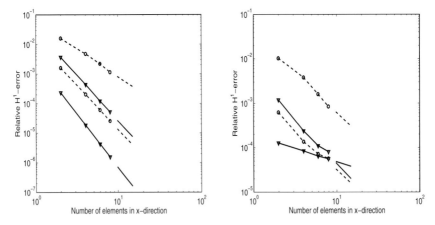

Fig. 5. Clamped (left) and free (right) edge; Whole domain; Non-uniform mesh; Convergence of the relative H^1-error with $k = 2, 3$ (dashed line for the original, solid line for the postprocessed deflection).

6.2 Non-uniform meshes

For the non-uniform meshes in Fig. 1 the relative errors have been computed in the whole discretized domain. As expected, the numerical results in Fig. 5 are essentially similar to those for the uniform meshes in the boundary region, compare Fig. 3 (left) to Fig. 5 (right). Thus, there is no essential dependence on the mesh distortion of reasonable order.

References

1. Arnold, D. N., Falk, R. S.: Edge effects in the Reissner–Mindlin plate theory. In: Noor, A. K., Belytschko, T., Simo, J. C. (ed) Analytic and Computational Models of Shells. ASME, New York, 71–90 (1989)
2. Bathe, K.-J., Brezzi, F., Fortin, M.: Mixed-interpolated elements for Reissner–Mindlin plates. Int. J. Num. Meths. Eng., **28**, 1787–1801 (1989)
3. Beirão da Veiga, L.: Finite element methods for a modified Reissner–Mindlin free plate model. SIAM J. Num. Anal., **42**, 1572–1591 (2004)
4. Brezzi, F., Fortin, M., Stenberg, R.: Error analysis of mixed-interpolated elements for Reissner–Mindlin plates. Math. Mod. Meth. Appl. Sci., **1**, 125–151 (1991)
5. Lyly, M., Niiranen, J., Stenberg, R.: A refined error analysis of MITC plate elements. Research Reports, A482, Helsinki University of Technology, Institute of Mathematics, http://math.tkk.fi/reports, Espoo (2005)
6. Lyly, M., Niiranen, J., Stenberg, R.: Superconvergence and postprocessing of MITC plate elements. Research Reports, A474, Helsinki University of Technology, Institute of Mathematics, http://math.tkk.fi/reports, Espoo (2005)
7. Peisker, P., Braess, D.: Uniform convergence of mixed interpolated elements for Reissner–Mindlin plates. M^2AN, **26**, 557–574 (1992)
8. Pitkäranta, J., Suri, M.: Design principles and error analysis for reduced-shear plate-bending finite elements. Numer. Math., **75**, 223–266 (1996)

A Uniformly Stable Finite Difference Space Semi-Discretization for the Internal Stabilization of the Plate Equation in a Square

Karim Ramdani[1], Takéo Takahashi[2] and Marius Tucsnak[3]

[1] IECN (Université Henri Poincaré) and INRIA
ramdani@loria.fr
[2] IECN (Université Henri Poincaré) and INRIA
takahash@iecn.u-nancy.fr
[3] IECN (Université Henri Poincaré) and INRIA
tucsnak@iecn.u-nancy.fr

Summary. We propose a finite difference space semi-discretization of the stabilized Bernoulli-Euler plate equation in a square. The scheme studied yields a uniform exponential decay rate with respect to the mesh size.

1 Statement of the main result

Consider a square plate $\Omega = (0, \pi) \times (0, \pi)$ subject to a feedback force distributed on a rectangular subdomain $\mathcal{O} = [a, b] \times [c, d]$ of Ω. If $\chi_\mathcal{O}$ denotes the characteristic function of $\mathcal{O}$, the stabilization problem considered reads:

$$\begin{cases} \ddot{w}(t) + \Delta^2 w(t) + \chi_\mathcal{O}\, \dot{w}(t) = 0, & x \in \Omega,\ t > 0, \\ w(t) = \Delta w(t) = 0, & x \in \partial\Omega,\ t > 0, \\ w(x, 0) = w_0(x), \quad \dot{w}(x, 0) = w_1(x), & x \in \Omega. \end{cases} \quad (1)$$

It is well known (cf. [3]) that the energy $E(t) = ||\dot{w}(t)||^2_{L^2(\Omega)} + ||\Delta w(t)||^2_{L^2(\Omega)}$ of system (1) decreases exponentially. The aim of this paper is to propose a space semi-discretization of this internal stabilization problem that ensures an exponential decay of the discretized energy $E_h(t)$ which is *uniform* with respect to the mesh size. This is not a trivial issue because of the possible appearance during the approximation process of high frequency spurious modes that cannot be damped by the feedback term. The appearance of such spurious modes in the approximation by finite differences or finite elements of control problems has been emphasized in several works (see, for instance [1, 2, 6] and the review paper [7]). Various solutions to overcome this difficulty have been proposed in the literature. The one followed in this paper is the one based on the introduction of an artificial numerical viscosity term.

Let us now precise the numerical scheme proposed. Given $N_1 \in \mathbb{N}$, denote by $h = \pi/(N_1+1)$, and assume that there exist integers $a(h), b(h), c(h), d(h)$ in $\{1, \ldots, N_1\}$ such that

$$a = a(h)h, \quad b = b(h)h, \quad c = c(h)h, \quad d = d(h)h. \quad (2)$$

Let $w_{j,k}$ denote for all $j, k \in \{0, N_1+1\}$ the approximation of the solution w of system (1) at the point $x_{j,k} = (jh, kh)$. We use the standard finite difference approximation of the laplacian, by setting for all $j, k \in \{1, \ldots, N_1\}$:

$$\Delta w(jh, kh) \approx \frac{1}{h^2}\left(w_{j+1,k} + w_{j-1,k} + w_{j,k+1} + w_{j,k-1} - 4w_{j,k}\right).$$

Set $V_h = \mathbb{R}^{(N_1)^2}$ and let $w_h \in V_h$ be the vector whose components are the $w_{j,k}$ for $1 \leq j, k \leq N_1$. In order to satisfy the boundary conditions in (1), we impose that

$$\forall k \in \{0, \ldots, N_1+1\}: \begin{cases} w_{0,k} = w_{k,0} = w_{N_1+1,k} = w_{k,N_1+1} = 0 \\ w_{-1,k} = -w_{1,k}, \ w_{N_1+2,k} = -w_{N_1,k}, \\ w_{k,-1} = -w_{k,1}, \ w_{k,N_1+2} = -w_{k,N_1}. \end{cases} \quad (3)$$

The matrix A_{0h} representing the discretization of the bilaplacian with hinged boundary conditions is defined via its square root $A_{0h}^{\frac{1}{2}}$ given by

$$\left(A_{0h}^{\frac{1}{2}} w_h\right)_{j,k} = -\frac{1}{h^2}\left(w_{j+1,k} + w_{j-1,k} + w_{j,k+1} + w_{j,k-1} - 4w_{j,k}\right),$$

for all $1 \leq j, k \leq N_1$. The finite-difference space semi-discretization for system (1) studied in this paper reads then

$$\begin{cases} \ddot{w}_{j,k} + (A_{0h} w_h)_{j,k} + (\chi_{\mathcal{O}} \dot{w}_h)_{j,k} + h^2 (A_{0h} \dot{w}_h)_{j,k} = 0, & 1 \leq j, k \leq N_1, \\ w_{j,k}(0) = w_{0h}, \quad \dot{w}_{j,k}(0) = w_{1h}, & 1 \leq j, k \leq N_1, \end{cases} \quad (4)$$

In the above equations, w_{0h} and w_{1h} are suitable approximations of the initial data w_0 and w_1 on the finite-difference grid and $\chi_{\mathcal{O}} \dot{w}_h$ denotes the vector of V_h whose components are the $\dot{w}_{j,k}$ if j and k are such that $x_{j,k} \in \mathcal{O}$, and 0 otherwise. The numerical viscosity term $h^2 A_{0h} \dot{w}_h$ in (4) is introduced in order to damp the high frequency modes. Our main result is the following.

Theorem 1. *The family of systems defined by (3)-(4) is uniformly exponentially stable, i.e. there exist constants C, α, $h^* > 0$ (independent of h, w_{0h} and w_{1h}) such that:*

$$\|\dot{w}_h(t)\|^2 + \left\|A_{0h}^{\frac{1}{2}} w_h(t)\right\|^2 \leq C e^{-\alpha t}\left(\|w_{1h}\|^2 + \left\|A_{0h}^{\frac{1}{2}} w_{0h}\right\|^2\right),$$

for all $h \in (0, h^*)$ and all $t > 0$.

In the above theorem and in the remaining part of this paper, we denote by $\|\cdot\|$ the Euclidean norm in $\mathbb{R}^m$ for various values of m. The proof of theorem 1 is based on the following frequency domain characterization for the uniform exponential stability of a sequence of semigroups (see [4, p.162]).

Theorem 2. *Let $(\mathbb{T}_h)_{h>0}$ be a family of semigroups of contractions on the Hilbert space V_h and A_h be the corresponding infinitesimal generators. The family $(\mathbb{T}_h)_{h>0}$ is uniformly exponentially stable if and only if the two following conditions are satisfied:*
i) For all $h > 0$, $i\mathbb{R} \subset \rho(A_h)$, where $\rho(A_h)$ denotes the resolvent set of A_h,
ii) $\sup\limits_{h>0, \omega \in \mathbb{R}} \|(i\omega - A_h)^{-1}\| < +\infty.$

2 Proof of Theorem 1

2.1 Abstract second and first order formulations

Let $U_h = \mathbb{R}^{(b(h)-a(h)+1) \times (d(h)-c(h)+1)}$ be the discretized input space, where the integers $a(h), b(h), c(h)$ and $d(h)$ are defined by (2). If $B_0 \in \mathcal{L}(L^2(\mathcal{O}), L^2(\Omega))$ denotes the restriction operator defined by $B_0 u = \chi_\mathcal{O} u$ for all $u \in L^2(\mathcal{O})$, we introduce its finite-difference approximation $B_{0h} \in \mathcal{L}(U_h, V_h)$ by setting for all $u_h \in U_h$: $(B_{0h} u_h)_{j,k} = u_{j,k}$ if j and k are such that $x_{j,k} \in \mathcal{O}$, and 0 otherwise. The adjoint $B_{0h}^* \in \mathcal{L}(V_h, U_h)$ of B_{0h} is then defined for all $w_h \in V_h$ by $(B_{0h}^* w_h)_{j,k} = w_{j,k}$ for all j, k such that $x_{j,k} \in \mathcal{O}$.

The finite-difference semi-discretization (3)-(4) admits the following abstract second order formulation:

$$\begin{cases} \ddot{w}_{j,k} + (A_{0h} w_h)_{j,k} + (B_{0h} B_{0h}^* \dot{w}_h)_{j,k} + h^2 (A_{0h} \dot{w}_h)_{j,k} = 0, & 1 \le j, k \le N_1, \\ w_{j,k} = \left(A_{0h}^{\frac{1}{2}} w_h \right)_{j,k} = 0, & j, k = 0, N_1 + 1, \\ w_{j,k}(0) = w_{0h}, \quad \dot{w}_{j,k}(0) = w_{1h}, & 0 \le j, k \le N_1 + 1. \end{cases} \quad (5)$$

It can be easily checked that the sequence $(\|B_{0h}\|_{\mathcal{L}(U_h, V_h)})$ is bounded and that the eigenvalues of $A_{0h}^{\frac{1}{2}}$ are

$$\lambda_{p,q,h} = \frac{4}{h^2} \left[\sin^2 \left(\frac{ph}{2} \right) + \sin^2 \left(\frac{qh}{2} \right) \right], \text{ for } 1 \le p, q \le N_1. \quad (6)$$

A corresponding sequence of normalized eigenvectors is given by the vectors $\varphi_{p,q,h} = \left(\varphi_{p,q,h}^{j,k} \right)_{1 \le j,k \le N_1}$, with components $\varphi_{p,q,h}^{j,k} = \frac{2h}{\pi} \sin(jph) \sin(kqh)$.

In order to apply theorem 2, we write system (5) as a first order system. Let us then introduce the space $X_h = V_h \times V_h$, which will be endowed with the norm $\|(\varphi_h, \psi_h)\|_{X_h}^2 = \|\varphi_h\|^2 + \left\| A_{0h}^{\frac{1}{2}} \psi_h \right\|^2$. Setting $z_h = \begin{bmatrix} w_h \\ \dot{w}_h \end{bmatrix}$, equations (5) can be easily written in the equivalent form

where $z_{0h} = \begin{bmatrix} w_{0h} \\ w_{1h} \end{bmatrix}$ and $A_h \in \mathcal{L}(X_h)$ is defined by

$$A_h = \begin{bmatrix} 0 & I \\ -A_{0h} & -h^2 A_{0h} - B_{0h} B_{0h}^* \end{bmatrix}. \tag{7}$$

It will be useful to introduce the operator $A_{1h} = \begin{bmatrix} 0 & I \\ -A_{0h} & 0 \end{bmatrix} \in \mathcal{L}(X_h)$ such that

$$A_h = A_{1h} - \begin{bmatrix} 0 & 0 \\ 0 & h^2 A_{0h} + B_{0h} B_{0h}^* \end{bmatrix}. \tag{8}$$

We will also need in the sequel the spectral basis of the operator A_{1h}. Moreover, it will be more convenient to number the eigenelements of A_{1h} using only one index m instead of the couple (p, q). To achieve this, let us first rearrange the sequence of eigenvalues $\lambda_{p,q} = p^2 + q^2$, $p, q \in \mathbb{N}^*$, of the continuous problem in nondecreasing order to obtain a new sequence $(\Lambda_m)_{m \in \mathbb{N}^*}$. Then, if

$$\Lambda_m = \lambda_{p,q} = p^2 + q^2, \qquad \forall\, m \in \mathbb{N}^*,\ \forall p, q \in \mathbb{N}^*, \tag{9}$$

then we set for all $1 \leq m \leq (N_1)^2$, and for all $1 \leq p, q \leq N_1$:

$$\Lambda_{m,h} = \lambda_{p,q,h}, \qquad \varphi_{m,h} = \varphi_{p,q,h}. \tag{10}$$

Let then $N_2(h) = (N_1)^2 = \left(\dfrac{\pi}{h} - 1\right)^2$ be the number of nodes of the finite-difference grid. If we extend the definition of $\Lambda_{m,h}$ and $\varphi_{m,h}$ to the values $m \in \{-1, \ldots, -N_2(h)\}$ by setting

$$\Lambda_{m,h} = -\Lambda_{-m,h}, \qquad \varphi_{m,h} = \varphi_{-m,h}, \tag{11}$$

then it can be easily checked that the eigenvalues of A_{1h} are $i\Lambda_{m,h}$, where $1 \leq |m| \leq N_2(h)$, and that an orthonormal basis of X_h formed by eigenvectors of A_{1h} is given by

$$\Phi_{m,h} = \frac{1}{\sqrt{2}} \begin{bmatrix} -\dfrac{i}{\Lambda_{m,h}} \varphi_{m,h} \\ \varphi_{m,h} \end{bmatrix}, \qquad 1 \leq |m| \leq N_2(h), \tag{12}$$

We are now in position to apply theorem 2.

2.2 Checking the assumptions of theorem 2

To prove condition $i)$ in theorem 2, we use a contradiction argument. Suppose that there exist $\begin{bmatrix} \varphi_h \\ \psi_h \end{bmatrix} \in X_h$ and $\omega \in \mathbb{R}$ such that: $A_h \begin{bmatrix} \varphi_h \\ \psi_h \end{bmatrix} = i\omega \begin{bmatrix} \varphi_h \\ \psi_h \end{bmatrix}$. Then, by using the definition (7) of A_h, we easily obtain that $\psi_h = i\omega\varphi_h$ and that

$$\left[\omega^2 - A_{0h} - i\omega(h^2 A_{0h} + B_{0h}B_{0h}^*)\right]\varphi_h = 0.$$

By taking the imaginary part of the inner product of this last relation with φ_h, we get that $\varphi_h = 0$, and thus $\psi_h = 0$. Therefore, for all $\omega \in \mathbb{R}$, $i\omega$ cannot be an eigenvalue of A_h. Thus, condition i) in theorem 2 holds true.

Now, we check condition ii) of theorem 2. Once again, we use a contradiction argument. Let us thus assume the existence for all $n \in \mathbb{N}$ of $h_n \in (0, h^*)$, $\omega_n \in \mathbb{R}$, $z_n = \begin{bmatrix} \phi_n \\ \psi_n \end{bmatrix} \in X_{h_n}$ such that

$$\|z_n\|^2 = \left\|A_{0h_n}^{\frac{1}{2}} \phi_n\right\|^2 + \|\psi_n\|^2 = 1 \qquad \forall\, n \in \mathbb{N} \tag{13}$$

$$\|i\omega_n z_n - A_{h_n} z_n\| \to 0. \tag{14}$$

To obtain a contradiction, the idea is to decompose z_n into a low frequency part and a high frequency part. Then, thanks to the numerical viscosity introduced in the scheme, we prove that the high frequency part tends to 0. Finally, we conclude by using a result on the uniform observability of low frequency packets of eigenvectors.

More precisely, for $0 < \varepsilon < 1$ and $h \in (0, h^*)$, we define the integer

$$M(h) = \max\left\{m \in \{1, \ldots, N_2(h)\} \mid h^2 (\Lambda_m)^2 \leq \varepsilon\right\}, \tag{15}$$

where the sequence $(\Lambda_m)_{m \in \mathbb{N}^*}$ defined in (9) constitutes the sequence of eigenvalues of the continuous problem. The eigenvalues $\Lambda_{m,h}$ for $1 \leq |m| \leq M(h)$ correspond to "low frequencies" and will be damped to zero by the feedback control term $B_{0h}B_{0h}^* \dot{w}_h$. The eigenvalues $\Lambda_{m,h}$ for $|m| > M(h)$ correspond to "high frequencies" and will be damped by the numerical viscosity term. To get the desired contradiction, we follow several steps.

Step 1

Let us prove the two relations

$$h_n^2 \left\|A_{0h_n}^{\frac{1}{2}} \psi_n\right\|^2 + \|B_{0h_n}^* \psi_n\|^2 \to 0, \tag{16}$$

$$\lim_{n \to \infty} \left\|A_{0h_n}^{\frac{1}{2}} \phi_n\right\|^2 = \lim_{n \to \infty} \|\psi_n\|^2 = \frac{1}{2}. \tag{17}$$

Relation (16) follows directly from (14) by taking the inner product in X_{h_n} of $i\omega_n z_n - A_{h_n} z_n$ by z_n and by considering only the real part. By using (14), (16), (8) and the fact that the operators B_{0h_n} are uniformly bounded we obtain that

$$\left\|i\omega_n z_n - A_{1h_n} z_n + \begin{bmatrix} 0 \\ h_n^2 A_{0h_n} \psi_n \end{bmatrix}\right\| \to 0. \tag{18}$$

It can be easily that the sequence (ω_n) is bounded away from zero for n large enough(use a contradiction argument). Therefore, taking the inner product in X_{h_n} of (18) by $\dfrac{1}{\omega_n} \begin{bmatrix} \phi_n \\ -\psi_n \end{bmatrix}$ and by considering the imaginary part, we obtain that $\lim_{n\to\infty} \left\| A_{0h_n}^{\frac{1}{2}} \phi_n \right\|^2 - \|\psi_n\|^2 = 0$. This last relation and (13) yield (17). Step 1 is thus complete.

In order to state the second step, let us introduce the modal decomposition of z_n on the spectral basis of $(\Phi_{m,h_n})_{1\leq|m|\leq N_2(h_n)}$ of A_{1h_n}. For all $n \in \mathbb{N}$, there exist complex coefficients $(c_m^n)_{1\leq|m|\leq N_2(h_n)}$ such that

$$z_n = \begin{bmatrix} \phi_n \\ \psi_n \end{bmatrix} = \sum_{1\leq|m|\leq N_2(h_n)} c_m^n \Phi_{m,h_n}. \tag{19}$$

The normalization condition (13) reads then

$$\sum_{1\leq|m|\leq N_2(h_n)} |c_m^n|^2 = 1. \tag{20}$$

Step 2
In this step, we prove that the following relations holds true

$$\psi_n = \frac{1}{\sqrt{2}} \sum_{m=1}^{N_2(h_n)} \left(c_m^n + c_{-m}^n\right) \varphi_{m,h_n}, \tag{21}$$

$$\sum_{M(h_n)<m\leq N_2(h_n)} \left|c_m^n + c_{-m}^n\right|^2 \to 0, \tag{22}$$

$$\sum_{1\leq|m|\leq M(h_n)} |\omega_n - \Lambda_{m,h_n}|^2 |c_m^n|^2 \to 0. \tag{23}$$

Note that, roughly speaking, relations (21) and (22) show that the projection of ψ_n on the high frequencies tends to 0 as n tends to $+\infty$. Relation (21) follows directly by taking the second component in (19) and by using (12). On the other hand, by using (19) and the fact that $\Phi_{m,h}$ is an eigenvector of A_{1h} associated to the eigenvalue $i\Lambda_{m,h}$, we have

$$i\omega_n z_n - A_{1h_n} z_n = \sum_{1\leq|m|\leq N_2(h_n)} i\left(\omega_n - \Lambda_{m,h_n}\right) c_m^n \Phi_{m,h_n} \tag{24}$$

From (16) and (21) it follows that

$$h_n^2 \left\| A_0^{\frac{1}{2}} \psi_n \right\|^2 = \sum_{m=1}^{N_2(h_n)} h_n^2 \Lambda_{m,h_n}^2 \left|c_m^n + c_{-m}^n\right|^2 \to 0. \tag{25}$$

Using the expression (6) of $\lambda_{p,q,h}$ and that $\lambda_{p,q} = p^2 + q^2$, it can be easily checked that $\frac{4}{\pi^2}\lambda_{p,q} \leq \lambda_{p,q,h} \leq \lambda_{p,q}$, for all $1 \leq p,q \leq N_1$, or equivalently

$$\frac{4}{\pi^2}\Lambda_m \leq \Lambda_{m,h} \leq \Lambda_m \qquad \forall\, 1 \leq m \leq N_2(h). \tag{26}$$

Relations (25), (26) and (15) imply (22). On the other hand, relations (26) and (25) clearly imply that there exists a constant C independent of h such that

$$h_n^4 \sum_{m=1}^{M(h_n)} \Lambda_{m,h_n}^4 \left|c_m^n + c_{-m}^n\right|^2 \leq C\varepsilon \sum_{m=1}^{M(h_n)} h_n^2 \Lambda_{m,h_n}^2 \left|c_m^n + c_{-m}^n\right|^2 \to 0. \tag{27}$$

On the other hand, a simple calculation shows that

$$\left[\begin{array}{c} 0 \\ h_n^2 A_{0h_n} \psi_n \end{array}\right] = \sum_{1\leq |m|\leq N_2(h_n)} \frac{h_n^2}{2} \Lambda_{m,h_n}^2 \left(c_m^n + c_{-m}^n\right) \Phi_{m,h_n}, \tag{28}$$

Relations (27) and (28) imply that

$$\left[\begin{array}{c} 0 \\ h_n^2 A_{0h_n} \psi_n \end{array}\right] - \sum_{M(h_n)<|m|\leq N_2(h_n)} \frac{h_n^2}{2} \Lambda_{m,h_n}^2 \left(c_m^n + c_{-m}^n\right) \Phi_{m,h_n} \to 0 \tag{29}$$

By using (18), (7) and (29) it follows that

$$\sum_{1\leq |m|\leq N_2(h_n)} i\left(\omega_n - \Lambda_{m,h_n}\right) c_m^n \Phi_{m,h_n}$$
$$+ \sum_{M(h_n)<|m|\leq N_2(h_n)} \frac{h_n^2}{2} \Lambda_{m,h_n}^2 \left(c_m^n + c_{-m}^n\right) \Phi_{m,h_n} \to 0.$$

Since the family (Φ_{m,h_n}) is orthogonal, the above relation implies (23).

Step 3
Consider the set

$$\mathcal{F} = \left\{ n \in \mathbb{N} \mid \exists\, m(n) \in \mathbb{Z},\ 1 \leq |m(n)| \leq M(h_n),\ \text{and } |\omega_n - \Lambda_{m(n),h_n}| < \frac{1}{8} \right\}.$$

In other words, $\mathcal{F}$ is constituted by those integers n such that ω_n is located in the "low frequency band". We distinguish then two cases:

<u>First Case:</u> **The set $\mathcal{F}$ is finite.** Then, for the sake of simplicity, we can suppose, without loss of generality, that $\mathcal{F}$ is empty. In this case, all the elements of the sequence (ω_n) are located in the "high frequency band". By using relation (23) in Step 2 and the above relation, we obtain that $<$

$\sum_{1\leq|m|\leq M(h_n)} |c_m^n|^2 \to 0$, i.e. that the low-frequency part of ψ_n tends to 0. Thus, the above relation, (21) and (22) in Step 2 imply that

$$\psi_n \to 0 \text{ in } H,$$

which contradicts (17).

Second case: **The set $\mathcal{F}$ is infinite.** Then, for the sake of simplicity, we can suppose, without loss of generality, that $\mathcal{F} = \mathbb{N}$. In this case, all the sequence ω_n is located in the "low frequency band". For all $n \in \mathbb{N}$, we introduce the set $\mathcal{F}_n = \left\{ m \in \mathbb{Z} \mid 1 \leq |m| \leq M(h_n) \text{ and } |\omega_n - \Lambda_{m,h_n}| < \frac{1}{8} \right\}$. Note that $\mathcal{F}_n$ is never empty (since it always contains $m(n)$) and represents the collection of low frequency eigenvalues located near ω_n. Set then $\widetilde{\psi}_n = \frac{1}{\sqrt{2}} \sum_{m \in \mathcal{F}_n} c_m^n \varphi_{m,h_n}$. The definition of $\mathcal{F}_n$, together with relation (23) of Step 2 imply that

$$\sum_{m \in \{1,\ldots,N_2(h_n)\} \setminus \mathcal{F}_n} |c_m^n|^2 \to 0. \tag{30}$$

Using now relations (21) and (22) of Step 2, we see that (30) exactly states that

$$\|\psi_n - \widetilde{\psi}_n\| \to 0. \tag{31}$$

The above relation implies that $\|B_{0h_n}^*(\psi_n - \widetilde{\psi}_n)\| \to 0$. This relation together with relation (16) of Step 1 show that

$$\|B_{0h_n}^* \widetilde{\psi}_n\| \to 0. \tag{32}$$

But on the other hand, applying lemma 3.2 in [5] on the uniform observability of low frequency packets of eigenvectors (note that $I_{h_n}(\omega_n) = \mathcal{F}_n$) , we get the existence of $\delta > 0$ such that for all $n \in \mathbb{N}$, we have

$$\|B_{0h_n}^* \widetilde{\psi}_n\|^2 > \delta^2 \sum_{m \in \mathcal{F}_n} |c_m^n|^2. \tag{33}$$

Gathering (30), (32) and (33), we finally obtain that $\widetilde{\psi}_n \to 0$ in H. By using (31), we obtain that $\psi_n \to 0$ which contradicts (17). The proof of theorem 1 is now complete.

References

1. Glowinski, R., Li, C.H, Lions, J-L.: A numerical approach to the exact boundary controllability of the wave equation. I. Dirichlet controls: description of the numerical methods. Japan J. Appl. Math., **7**, 1–76 (1990)

2. Infante, J. A., Zuazua, E.: Boundary observability for the space semi-discretizations of the 1-D wave equation. M2AN Math. Model. Numer. Anal., **33**, 2, 407–438 (1999)
3. Jaffard, S.: Contrôle interne exact des vibrations d'une plaque rectangulaire. Port. Math., **47**, 4, 423–429 (1990)
4. Liu, Z., Zheng, S.: Semigroups associated with dissipative systems, Chapman & Hall/CRC, Boca Raton (1999)
5. Ramdani, K., Takahashi, T., Tucsnak, M.: Internal stabilization of the plate equation in a square : the continuous and the semi-discretized problems. J. Math. Pures Appl., To appear
6. Tcheugoué Tébou, L. R., Zuazua, E.: Uniform exponential long time decay for the space semi-discretization of a locally damped wave equation via an artificial numerical viscosity. Numer. Math., **95**, 3, 563–598 (2003)
7. Zuazua, E.: Propagation, observation and control of waves approximated by finite difference methods. SIAM. Rev., **47**, 197–243 (2005)

SINGULAR PERTURBATION

An ε-Uniform Hybrid Scheme for Singularly Perturbed 1-D Reaction-Diffusion Problems

S. Natesan[1], R.K. Bawa[2] and C. Clavero[3]

[1] Department of Mathematics, Indian Institute of Technology, Guwahati-781 039, India
natesan@iitg.ernet.in

[2] Department of Computer Science and Engineering, Punjabi University, Patiala-147 002, India
rajesh_k_bawa@yahoo.com

[3] Departamento de Matemática Aplicada, Universidad de Zaragoza, Zaragoza, Spain
clavero@unizar.es

Summary. An ε–uniform second–order numerical method for singularly perturbed reaction-diffusion problems is proposed in this article. The difference scheme is based on cubic spline and classical finite difference scheme, which is applied on layer resolving Shishkin meshes. Uniform stability and uniform convergence of the scheme in the maximum norm are studied. A numerical example is presented to support the theoretical results.

1 Introduction

In this article we consider the following singularly perturbed self–adjoint boundary value problem (BVP):

$$Lu(x) \equiv -\varepsilon u''(x) + b(x)u(x) = f(x), \quad x \in D = (0,1), \tag{1}$$
$$u(0) = A, \quad u(1) = B, \tag{2}$$

where $\varepsilon > 0$ is a small parameter and b, f are sufficiently smooth functions such that $b(x) \geq \beta > 0$ on $\overline{D} = [0,1]$. Under these assumptions the problem (1-2) has a unique solution $u(x) \in \mathcal{C}^2(D) \cap \mathcal{C}(\overline{D})$, which may exhibit two boundary layers of width $O(\sqrt{\varepsilon})$ of exponential type at both end points $x = 0, 1$.

Singular perturbation problems (SPPs) arise frequently in various branches of science and engineering. The solution of SPPs has multi-scale behaviour and therefore, to solve them numerically, it is necessary to use uniformly convergent methods, for which the error does not depend on the value of the singular perturbation parameter ε. Various methods having this property are proposed in the literature (see the books of Farrell et al. [2] and Roos et al. [7] and references therein). Natesan et al. [6] proposed a booster method for the boundary value problem (1-2). In [3] the authors have devised HODIE schemes for (1-2) giving higher order uniform convergent finite difference schemes on Shishkin meshes. Kadalbajoo and Bawa [4] used cubic spline on variable mesh for solving singularly perturbed problems. Recently, Bawa and Natesan solved SPPs of reaction–diffusion type (1-2) by quintic spline [1]. In [8] quadratic spline is used to solve a nonlinear reaction-diffusion problem, showing their uniform convergence; nevertheless, the proof of the results is incomplete and not clear.

Here, we devise a numerical scheme based on variable mesh cubic spline for self-adjoint BVPs, and we apply it on a layer resolving Shishkin mesh. The cubic spline scheme fails to satisfy the discrete maximum principle on the outer region where the Shishkin mesh is coarse. To overcome this difficulty we shall use the classical finite difference scheme in that portion of the domain. By this way, we retain the stability of the scheme, which will be crucial in the proof of the ε-uniform convergence of the scheme and give a different way to the one used in [8]. In this paper we only show the uniform convergence of the method at the mesh points. In [5] a detailed analysis of the uniform convergence of the global solution and the normalized flux in the continuous domain is given; these properties show the advantages and the conveniences of the numerical method defined here. The method proposed here is easily extendable for 1D linear system of equations, and 1D nonlinear BVPs. Shishkin meshes for two dimensional problems have been studied extensively, for example, in [2, 7]. Although the present method is not only limited to 1D problems, its extension to higher dimensional problems and the error estimates involve more critical analysis and it will be postponed to our future work.

The article is organized as follows. The hybrid scheme is derived in Section 2. Error estimates are provided in Section 3. Finally, a numerical example is given in Section 4. Throughout this article, C denotes a positive constant independent of the diffusion parameter ε and the discretization parameter N.

2 The cubic spline-cum-finite difference scheme

In this section, first we derive the cubic spline scheme on a general non uniform mesh, and then we propose the hybrid scheme. Let $x_0 = 0, x_i = \sum_{k=0}^{i-1} h_k, h_k = x_{k+1} - x_k, x_N = 1, i = 1, 2, \cdots, N - 1$, be the mesh; then, for given values $u(x_0), u(x_1), \cdots, u(x_N)$ of a function $u(x)$ at the nodal points $x_0, x_1, \cdots, x_N$, there exists an interpolating cubic spline $s(x)$ with the following properties: (i) $s(x)$ coincides with a polynomial of degree three on each subinterval $[x_i, x_{i+1}], i = 0, \cdots, N - 1$; (ii) $s(x) \in C^2[0, 1]$ and (iii) $s(x_i) = u(x_i), i = 0, \cdots, N$.

It is well known that the cubic spline can be written in the form

$$s(x) = \frac{(x_{i+1} - x)^3}{6h_i} M_i + \frac{(x - x_i)^3}{6h_i} M_{i+1} + \left(u(x_i) - \frac{h_i^2}{6} M_i\right) \left(\frac{x_{i+1} - x}{h_i}\right) +$$
$$+ \left(u(x_{i+1}) - \frac{h_i^2}{6} M_{i+1}\right) \left(\frac{x - x_i}{h_i}\right), \quad x_i \leq x \leq x_{i+1}, i = 0, \cdots, N-1, \quad (3)$$

where $M_i = s''(x_i)$, $i = 0, \cdots, N$. From the basic properties of splines, it should satisfy the following 'condition of continuity':

$$\frac{h_{i-1}}{6} M_{i-1} + \left(\frac{h_i + h_{i-1}}{3}\right) M_i + \frac{h_i}{6} M_{i+1} = \left(\frac{u(x_{i+1}) - u(x_i)}{h_i}\right) -$$
$$- \left(\frac{u(x_i) - u(x_{i-1})}{h_{i-1}}\right), \quad i = 1, \cdots, N-1. \quad (4)$$

The continuity condition given above ensures the continuity of the first order derivatives of the spline $s(x)$ at the interior nodes. Then, substituting $-\varepsilon M_j + b(x_j)u(x_j) = f(x_j)$, $j = i, i \pm 1$ in (4), we get the linear system

$$\begin{cases} \left[\frac{-3\varepsilon}{h_{i-1}(h_i + h_{i-1})} + \frac{h_{i-1}}{2(h_i + h_{i-1})} b_{i-1}\right] U_{i-1} + \left[\frac{3\varepsilon}{h_i h_{i-1}} + b_i\right] U_i + \\ + \left[\frac{-3\varepsilon}{h_i(h_i + h_{i-1})} + \frac{h_i}{2(h_i + h_{i-1})} b_{i+1}\right] U_{i+1} = \left[\frac{h_{i-1}}{2(h_i + h_{i-1})}\right] f_{i-1} + f_i + \\ + \left[\frac{h_i}{2(h_i + h_{i-1})}\right] f_{i+1}, \end{cases}$$
$$(5)$$

which gives the approximations $U_1, U_2, \cdots, U_{N-1}$ of the solution $u(x)$ at the mesh points $x_1, x_2, \cdots, x_{N-1}$ (note that $U_0 = A$, $U_N = B$ are the natural discretizations of the Dirichlet boundary conditions). This cubic spline scheme is analyzed for stability and it is observed that for the corresponding stiffness matrix to be an M-matrix, a very restrictive condition is needed on the step size only in the outer region where the mesh is coarse. To overcome this difficulty, the following hybrid scheme is proposed, in which the classical central difference scheme is taken in the outer region and the above cubic scheme is applied only in the boundary layer regions:

$$L^N U_i \equiv r_i^- U_{i-1} + r_i^c U_i + r_i^+ U_{i+1} = q_i^- f_{i-1} + q_i^c f_i + q_i^+ f_{i+1}, \quad (6)$$

along with boundary conditions $U_0 = A$ and $U_N = B$, where, for $i = 1, \cdots, N/4 - 1$ and $3N/4 + 1, \cdots, N-1$, the coefficients are given by

$$\begin{cases} r_i^- = \frac{-3\varepsilon}{h_{i-1}(h_i + h_{i-1})} + \frac{h_{i-1}}{2(h_i + h_{i-1})} b_{i-1}; \quad r_i^c = \frac{3\varepsilon}{h_i h_{i-1}} + b_i; \\ r_i^+ = \frac{-3\varepsilon}{h_i(h_i + h_{i-1})} + \frac{h_i}{2(h_i + h_{i-1})} b_{i+1}; \\ q_i^- = \frac{h_{i-1}}{2(h_i + h_{i-1})}; \quad q_i^c = 1; \quad q_i^+ = \frac{h_i}{2(h_i + h_{i-1})}, \end{cases} \quad (7)$$

and for $i = N/4, \cdots, 3N/4$, the coefficients are given by

$$\begin{cases} r_i^- = \dfrac{-2\varepsilon}{h_{i-1}(h_i + h_{i-1})}; & r_i^c = \dfrac{2\varepsilon}{h_i h_{i-1}} + b_i; & r_i^+ = \dfrac{-2\varepsilon}{h_i(h_i + h_{i-1})}, \\ q_i^- = 0; & q_i^c = 1; & q_i^+ = 0. \end{cases} \quad (8)$$

Note that the stiffness matrix of the newly modified finite difference scheme (6) is an M-matrix.

3 Uniform convergence analysis on a Shishkin mesh

Before analyzing the uniform convergence of the hybrid scheme, we define an appropriate Shishkin mesh for the boundary value problem (1-2). On $\overline{D}$ a piecewise uniform mesh of N mesh intervals is constructed as follows: the domain $\overline{D}$ is divided into three subintervals as $\overline{D} = [0, \sigma] \cup (\sigma, 1-\sigma) \cup [1-\sigma, 1]$, for some σ such that $0 < \sigma \leq 1/4$. On the subintervals $[0, \sigma]$ and $[1-\sigma, 1]$ a uniform mesh with $N/4$ mesh intervals is placed, where $[\sigma, 1-\sigma]$ has a uniform mesh with $N/2$ mesh intervals. It is obvious that the mesh is uniform when $\sigma = 1/4$ and it is fitted to the problem by choosing σ as the following function of N, ε and σ_0: $\sigma = \min\left\{\frac{1}{4}, \sigma_0\sqrt{\varepsilon}\ln N\right\}$, where σ_0 is a constant to be fixed later. Further, in our analysis we assume that $\sigma = \sigma_0\sqrt{\varepsilon}\ln N$, which is the interesting case, and we denote the mesh size in the region $[\sigma, 1-\sigma]$ by $H = 2(1-2\sigma)/N$, and in the regions $[0, \sigma], [1-\sigma, 1]$ by $h = 4\sigma/N$. On this mesh, denoted by D^N, it is straightforward to prove that the truncation error of the hybrid scheme, for $i = 1, \cdots, N/4 - 1$ and $3N/4 + 1, \cdots, N - 1$, satisfies

$$\begin{aligned} \tau_{i,u} = & \frac{\varepsilon h^2}{8} u^{(iv)}(x_i) + r_i^+ R_4(x_i, x_{i+1}, u) + r_i^- R_4(x_i, x_{i-1}, u) + \\ & + q_i^+ \varepsilon R_2(x_i, x_{i+1}, u) + q_i^- \varepsilon R_2(x_i, x_{i-1}, u) - \\ & - q_i^- b_{i-1} R_4(x_i, x_{i-1}, u) - q_i^+ b_{i+1} R_4(x_i, x_{i+1}, u), \end{aligned} \quad (9)$$

and for $i = N/4, \cdots, 3N/4$, it holds

$$\begin{aligned} \tau_{i,u} = & -\varepsilon \left(\frac{h_i - h_{i-1}}{3}\right) u'''(x_i) - \frac{2\varepsilon}{4!}\left(\frac{h_i^3 + h_{i-1}^3}{h_i + h_{i-1}}\right) u^{(iv)}(x_i) + \\ & + r_i^+ R_4(x_i, x_{i+1}, u) + r_i^- R_4(x_i, x_{i-1}, u), \end{aligned} \quad (10)$$

where $R_n(a, p, g)$ denotes the remainder Taylor expansion.

It is well known (see [2]) that the solution $u(x)$ of the BVP (1-2) and its derivatives satisfy the following bounds: $|u^{(k)}(x)| \leq C\left(1 + \varepsilon^{-k/2} e(x, x, \beta, \varepsilon)\right)$ for $0 \leq k \leq j+1$, where $e(\xi_1, \xi_2, \beta, \varepsilon) = \exp(-\sqrt{\beta}\xi_1/\sqrt{\varepsilon}) + \exp(-\sqrt{\beta}(1-\xi_2)/\sqrt{\varepsilon})$. Further, its smooth and singular components, respectively denoted by v and w, satisfy $= u = v + w$ and also the bounds $|v^{(k)}(x)| \leq C$, $|w^{(k)}(x)| \leq C\varepsilon^{-k/2} e(x, x, \beta, \varepsilon)$.

Lemma 1. *The local truncation error (9-10) satisfies*

$$|\tau_{i,u}| \leq \begin{cases} CN^{-2}\sigma_0^2 \ln^2 N, & \text{for } 1 \leq i \leq N/4 - 1 \text{ and } 3N/4 + 1 \leq i < N, \\ C(N^{-2}\varepsilon + N^{-\sqrt{\beta}\sigma_0}), & \text{for } N/4 < i < 3N/4, \\ C(N^{-1}\varepsilon + N^{-\sqrt{\beta}\sigma_0}), & i = N/4 \text{ or } i = 3N/4. \end{cases}$$

Proof. We distinguish several cases depending on the location of the mesh points. First, when $x_i \in (0, \sigma) \cup (1 - \sigma, 1)$, from (9) we get

$$|\tau_{i,u}| \leq C[h^2 \varepsilon + h^2 \varepsilon^{-1}(e(x_i, x_{i+1}, \beta, \varepsilon) + e(x_{i-1}, x_i, \beta, \varepsilon))].$$

Using that $h = 4N^{-1}\sigma_0\sqrt{\varepsilon} \ln N$ and bounding the exponential functions by constants, we deduce that $|\tau_{i,u}| \leq CN^{-2}\sigma_0^2 \ln^2 N$, for $1 \leq i \leq N/4 - 1$ and $3N/4 + 1 \leq i < N$.

In second place, when $x_i \in (\sigma, 1 - \sigma)$, we must distinguish two cases. First, if $H^2 < \varepsilon$, we obtain

$$|\tau_{i,u}| \leq C\left(H^2\varepsilon + \frac{H^2}{\varepsilon}(e(x_i, x_{i+1}, \beta, \varepsilon) + e(x_{i-1}, x_i, \beta, \varepsilon))\right) \leq N^{-2}\varepsilon + N^{-\sqrt{\beta}\sigma_0}.$$

On the other hand, if $H^2 \geq \varepsilon$, it can been shown that

$$|\tau_{i,u}| \leq C\left(H^2\varepsilon + \int_{x_i}^{x_{i+1}} (x_{i+1} - \xi)\varepsilon^{-1} e(\xi, \xi, \beta, \varepsilon) d\xi + \int_{x_{i-1}}^{x_i} (\xi - x_{i-1})\varepsilon^{-1} e(\xi, \xi, \beta, \varepsilon) d\xi\right).$$

Integrating the first integral by parts, we get

$$\int_{x_i}^{x_{i+1}} (x_{i+1} - \xi)\varepsilon^{-1} e(\xi, \xi, \beta, \varepsilon) d\xi \leq C\left(H\varepsilon^{-1/2} e(x_i, x_i, \beta, \varepsilon) + \int_{x_i}^{x_{i+1}} \varepsilon^{-1/2} e(\xi, \xi, \beta, \varepsilon) d\xi\right) \leq C[e(x_i, x_i, \beta, \varepsilon) + e(x_i, x_{i+1}, \beta, \varepsilon)]$$

$$\leq CN^{-\sqrt{\beta}\sigma_0}.$$

A bound for the second integral can be found in a similar way. Using the fact that $H < 2N^{-1}$, we get $|\tau_{i,u}| \leq C(N^{-2}\varepsilon + N^{-\sqrt{\beta}\sigma_0})$. Finally we study the case $x_i = \sigma$ (similarly we can find the bound for the case $x_i = 1 - \sigma$). In this case we write the local truncation error in the form

$$\tau_{i,u} = r_i^+ R_2(x_i, x_{i+1}, u) + r_i^- R_2(x_i, x_{i-1}, u),$$

and again we distinguish two cases depending on the relation between H^2 and ε. When $H^2 \geq \varepsilon$, it is easy to prove that $|\tau_{i,u}| \leq CN^{-3} + N^{-\sqrt{\beta}\sigma_0}$. On the other hand, when $H^2 < \varepsilon$, we obtain $|\tau_{i,u}| \leq CN^{-1}\varepsilon + N^{-\sqrt{\beta}\sigma_0}$, which is the required result. ∎

Lemma 2. *Let $N \geq N_0$ be, where N_0 is the smallest positive integer such that $16\beta N_0^{-2}\sigma_0^2 \ln^2 N_0 < 6$. Then, it holds*

$$r_i^- < 0, \ r_i^+ < 0, \ |r_i^c| - |r_i^-| - |r_i^+| \geq \min(1, \beta) > 0.$$

Therefore, the stiffness matrix of the method (6) is positive definite and it satisfies a discrete maximum principle. Further, it is ε-uniform stable in the maximum norm.

Proof. Clearly, $r_i^- < 0$ and $r_i^+ < 0$ for $i = N/4, \cdots, 3N/4$. To see that $r_i^- < 0$ for $i = 1, \cdots, N/4 - 1$ and $i = 3N/4 + 1, \cdots, N - 1$, we proceed as follows: we have $r_i^- = -3\varepsilon/(h_{i-1}(h_i + h_{i-1})) + h_{i-1}b_{i-1}/(2(h_i + h_{i-1})) = -3\varepsilon/(2h^2) + b_{i-1}/4$, and from $h = 4N^{-1}\sigma_0\sqrt{\varepsilon} \ln N$ and $\beta \leq b_{i-1}$, it follows that $r_i^- < 0$.

Similarly, it can be shown that $r_i^+ < 0$ for $i = 1, \cdots, N/4 - 1$ and $i = 3N/4 + 1, \cdots, N - 1$. On the other hand, for $i = N/4, \cdots, 3N/4$, clearly it holds $|r_i^c| - |r_i^-| - |r_i^+| \geq \beta$. Finally, for $i = 1, \cdots, N/4 - 1$ and $i = 3N/4 + 1, \cdots, N - 1$, we have

$$|r_i^c| - |r_i^-| - |r_i^+| = \frac{3\varepsilon}{h_i h_{i-1}} + b_i - \frac{3\varepsilon}{h_{i-1}(h_i + h_{i-1})} + \frac{h_{i-1}b_{i-1}}{2(h_i + h_{i-1})} - \frac{3\varepsilon}{h_i(h_i + h_{i-1})} + \frac{h_i b_{i+1}}{2(h_i + h_{i-1})} = b_i + \frac{h_{i-1}b_{i-1}}{2(h_i + h_{i-1})} + \frac{h_i b_{i+1}}{2(h_i + h_{i-1})} \geq$$

$$\geq \beta + \left[\frac{h_{i-1}}{2(h_i + h_{i-1})} + \frac{h_i}{2(h_i + h_{i-1})}\right]\beta = \frac{3}{2}\beta,$$

and the result follows. ∎

Theorem 1. *Let $u(x)$ be the solution of (1-2) and $U_i, i = 0, 1, \cdots, N$ be the numerical solution of the hybrid finite difference scheme (6) at the mesh points of the Shihskin mesh. Then, the error satisfies*

$$|u(x_i) - U_i| \leq C\left(N^{-2}\ln^2 N + N^{-\sqrt{\beta}\sigma_0}\right), \quad i = 0, 1, \cdots, N, \quad (11)$$

and therefore, if the constant σ_0 satisfies $\sigma_0 \geq 2/\sqrt{\beta}$, the method (6) is uniformly convergent of order almost two.

Proof. Defining the discrete barrier function

$$\phi_i = C\left(N^{-2}\ln^2 N + N^{-2}\varepsilon + N^{-\sqrt{\beta}\sigma_0} + \frac{\sigma^2}{\sqrt{\varepsilon}}N^{-2}\psi(x_i)\right), \quad i = 0, \cdots, N,$$

where
$$\psi(z) = \begin{cases} z/\sigma, & 0 \leq z \leq \sigma, \\ 1, & \sigma \leq z \leq 1-\sigma, \\ (1-z)/\sigma, & 1-\sigma \leq z \leq 1, \end{cases}$$

by choosing a sufficiently large C, using Lemma 1 and the discrete maximum principle, it is straightforward to obtain the required result. ∎

4 Numerical experiments

In this section, we show the numerical results obtained from the hybrid scheme. The results are given in terms of the maximum point-wise errors and the corresponding rates of convergence in a table, and we display the maximum errors in loglog plots.

We consider the self-adjoint boundary–value problem

$$-\varepsilon u''(x) + (1 + x^2 + \cos x)u(x) = x^{4.5} + \sin x, \ x \in (0,1), \ u(0) = 1, \ u(1) = 1,$$

for which the exact solution is unknown. The maximum point-wise errors and the rates of convergence are calculated by using a variant of the double mesh principle (see [8]). Let U^N be the numerical solution on D^N and $\widetilde{U}^N$ the numerical solution on the mesh $\widetilde{D}^N$, where the transition parameter is now given by $\widetilde{\sigma} = \min\{1/4, \sigma_0\sqrt{\varepsilon}\ln(N/2)\}$. Then, the maximum errors are obtained by

$$E^N_\varepsilon = \max_{x_j \in D^N} |U^N(x_j) - \widetilde{U}^{2N}(x_j)|, \ E^N = \max_\varepsilon E^N_\varepsilon.$$

In addition, in a standard way, the rates of convergence and the ε-uniform order of convergence are calculated by $p_N = \log_2\left(E^N_\varepsilon/E^{2N}_\varepsilon\right)$ and $p_{uni} = \log_2\left(E^N/E^{2N}\right)$ respectively.

Table 1 displays the results for some values of ε and N, taking $\sigma_0 = 2$, and also the maximum errors for the range of values $\varepsilon = 2^{-2}, 2^{-4}, 2^{-6}, \cdots, 2^{-32}$. Further, the maximum errors are presented in Figure 1 in loglog scale. This clearly reveals our claim of second order uniform convergence, as well as the role of the parameter σ_0 in the definition of the transition parameter σ.

5 Conclusions

An hybrid scheme to solve singularly perturbed reaction–diffusion problems is presented in this paper. The proposed numerical method is a combination of cubic spline and a classical finite difference scheme. Error estimates are derived for this method showing that the error is independent of the singular perturbation parameter ε. Numerical results also reveal the same fact. Global

Table 1. *Maximum pointwise errors and rates of uniform convergence*

ε	Number of mesh points N						
	16	32	64	128	256	512	1024
2^{-2}	8.0643E-5	1.5789E-5	3.4201E-6	7.9057E-7	1.9091E-7	4.8311E-8	1.2151E-8
	2.3526	2.2068	2.1131	2.0500	1.9825	1.9913	
2^{-8}	1.4380E-2	3.0695E-3	7.6719E-4	1.8964E-4	4.7339E-5	1.1830E-5	2.9572E-6
	2.2279	2.0004	2.0163	2.0022	2.0006	2.0001	
2^{-16}	2.5860E-2	1.0086E-2	3.2023E-3	1.0724E-3	3.4875E-4	1.0978E-4	3.3851E-5
	1.3583	1.6553	1.5782	1.6206	1.6676	1.6973	
2^{-24}	2.5853E-2	1.0048E-2	3.2023E-3	1.0688E-3	3.4762E-4	1.0942E-4	3.3742E-5
	1.3635	1.6497	1.5831	1.6205	1.6676	1.6973	
2^{-32}	2.5853E-2	1.0046E-2	3.2023E-3	1.0686E-3	3.4755E-4	1.0940E-4	3.3735E-5
	1.3638	1.6494	1.5833	1.6205	1.6676	1.6973	
E^N	2.5864E-2	1.0209E-2	3.2316E-3	1.0837E-3	3.5235E-4	1.1090E-4	3.4198E-5
p_{uni}	1.3411	1.6596	1.5763	1.6209	1.6677	1.6973	

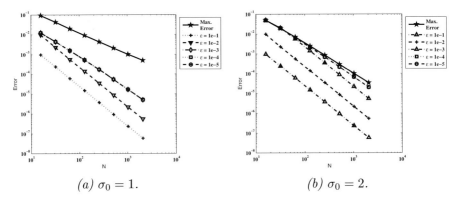

(a) $\sigma_0 = 1$. (b) $\sigma_0 = 2$.

Fig. 1. *Loglog plot of the error.*

error estimates for the numerical solution and for the numerical normalized flux are given in detail in [5].

References

1. Bawa, R.K., Natesan, S.: A computational method for self-adjoint singular pertrubation problems using quintic spline. Comput. Math. Appl., **50(8-9)**, 1371–1382 (2005)
2. Farrell, P.A., Hegarty, A.F., Miller, J.J.H., O'Riordan, E., Shishkin, G.I.: Robust Computational Techniques for Boundary Layers. Chapman & Hall/CRC Press, Boca Raton (2000)
3. Gracia, J.L., Lisbona, F., Clavero, C.: High order ε-uniform methods for singularly perturbed reaction-diffusion problems. Lecture Notes in Computer Science, **1988**, 350–358 (2001)

4. Kadalbajoo, M.K., Bawa, R.K.: Variable mesh difference scheme for singularly perturbed boundary value problems using splines. J. Optim. Theory and Appl., **9**, 405–416 (1996)
5. Natesan, S., Bawa, R.K., Clavero, C.: Uniformly convergent compact numerical scheme for the normalized flux and the global solution of singularly perturbed reaction-diffusion problems, (submitted for publication)
6. Natesan, S., Ramanujam, N.: Improvement of numerical solution of self-adjoint singular perturbation problems by incorporation of asymptotic approximations. Appl. Math. Comput., **98**, 199–137 (1999)
7. Roos, H.-G. , Stynes, M., Tobiska, L.: Numerical Methods for Singularly Perturbed Differential Equations. Springer, Berlin (1996)
8. Surla, K., Uzelac, Z.: A uniformly accurate spline collocation method for a normalized flux. J. Comp. Appl. Math., **166**, 291–305 (2004)

Solids

A Dynamic Frictional Contact Problem of a Viscoelastic Beam

M. Campo[1], J.R. Fernández[1], G.E. Stavroulakis[2] and J.M. Viaño[1]

[1] Departamento de Matemática Aplicada, Universidade de Santiago de Compostela. Facultade de Matemáticas, Campus Sur s/n, 15782 Santiago de Compostela, Spain
macampo@usc.es, jramon@usc.es, maviano@usc.es
[2] Department of Production Engineering and Management, Technical University of Crete, GR-73132, Chania, Greece and Department of Civil Engineering, Technical University of Braunschweig, Germany
gestavr@dpem.tuc.gr

Summary. We study the dynamic frictional contact of a viscoelastic beam with a deformable obstacle. The left end of the beam is rigidly attached and the horizontal movement of the right one is constrained because of the presence of a deformable obstacle. The effect of the friction is included in the vertical motion of the free end, by using Tresca's law or Coulomb's law. We recall an existence and uniqueness result. Then, by using the finite element method to approximate the spatial variable and an Euler scheme to discretize the time derivatives, a numerical scheme is proposed. Error estimates are derived on the approximative solutions. Finally, some numerical results are shown.

1 Introduction

Contact problems involving different types of materials appear in many structural and industrial problems and everyday life (see, e.g., the monographs [8, 10] and references therein). Recently, one-dimensional contact problems for beams and rods have been studied ([1, 3]), including the adhesion ([7]), the wear ([9]) or the damage ([2]).

In this paper, a model for the dynamic frictional contact of a viscoelastic beam with a deformable obstacle is numerically studied and solved. The model was introduced in [3], where the existence and uniqueness of solution, as well as its regularity, were studied. This work deals with the study of a numerical scheme, based on the finite element method to approximate the spatial variable and an Euler scheme to discretize the time derivatives. Moreover, in order to show the performance of the proposed algorithm, some numerical simulations are presented.

2 The variational formulation

In this section we briefly present the model and the variational formulation of the problem (see [3] for further details).

The beam is supposed to be rigidly attached at its left end, while the right one is free to come into frictional contact with a deformable obstacle. We will denote by $[0, L]$, $L > 0$, the reference configuration of the beam and by $[0, T]$, $T > 0$, the time interval of interest. The material is supposed to obey a constitutive law of Kelvin-Voigt type, and as contact conditions, a normal compliance condition for the horizontal displacement and a friction condition for the vertical one have been considered, that is,

$$-\sigma_H(L,t) = c_H(u(L,t) - g)_+, \quad t \in [0,T], \tag{1}$$

$$\left.\begin{array}{l} |\sigma_V(L,t)| \leq h(t), \\ \text{if } |\sigma_V(L,t)| = h(t) \Rightarrow \exists \lambda \geq 0\,; \widetilde{u}_t(L,t) = -\lambda \sigma_V(L,t), \\ \text{if } |\sigma_V(L,t)| < h(t) \Rightarrow \widetilde{u}_t(L,t) = 0, \end{array}\right\} t \in [0,T], \tag{2}$$

where σ_H and σ_V denote the horizontal and vertical stresses, respectively, u and $\widetilde{u}$ represent the respective horizontal and vertical displacements, g is the initial gap between the beam and the obstacle and $h(t)$ represents a friction bound. Two different cases are considered:

(i) $h(t) = c_V = \text{constant}$: it corresponds to the classical Tresca's conditions.
(ii) $h(t) = c_V(u(L,t) - g)_+$, for $c_V = \text{constant} > 0$: it leads to a particular case of Coulomb's conditions (it includes the previous case).

In order to derive a weak formulation of the problem, let us define the following variational spaces:

$$H = L^2(0,L), \quad E = \{w \in H^1(0,L)\,;\, w(0) = 0\},$$
$$V = \{z \in H^2(0,L)\,;\, z(0) = z_x(0) = 0\}.$$

Moreover, we denote by $(\cdot, \cdot)$ the classical inner product defined in $L^2(0, L)$ and, for a Hilbert space X, let $|\cdot|_X$ represent its norm.

Let $j_H(u, \cdot) : E \to \mathbb{R}$ and $j_V(u, \cdot) : V \to \mathbb{R}$ be the normal compliance and friction forms defined by

$$j_H(u,w) = c_H(u(L,t) - g)_+ w(L,t), \quad \forall w \in E,$$
$$j_V(u,z) = c_V(u(L,t) - g)_+ |z(L,t)|, \quad \forall z \in V.$$

Integrating by parts the equations of motion and using the previous boundary conditions, the variational formulation is then written as follows.

Variational problem VP. *Find the horizontal displacement* $u : [0,T] \to E$ *and the vertical displacement* $\widetilde{u} : [0,T] \to V$ *such that* $u(0) = u_0$, $u_t(0) = v_0$, $\widetilde{u}(0) = \widetilde{u}_0$, $\widetilde{u}_t(0) = \widetilde{v}_0$ *and for a.e.* $t \in (0,T)$,

$$(\rho u_{tt}(t), w) + a_1(u_x(t), w_x) + c_1(u_{xt}(t), w_x) + j_H(u(t), w)$$
$$= (f_H(t), w), \quad \forall w \in E, \qquad (3)$$

$$(\rho \widetilde{u}_{tt}(t), (z - \widetilde{u}_t(t))) + a_2(\widetilde{u}_{xx}(t), (z - \widetilde{u}_t(t))_{xx}) + c_2(\widetilde{u}_{xxt}(t), (z - \widetilde{u}_t(t))_{xx})$$
$$+ j_V(u(t), z) - j_V(u(t), \widetilde{u}_t(t)) \geq (f_V(t), (z - \widetilde{u}_t(t))), \quad \forall z \in V, \qquad (4)$$

where $a_i > 0$ and $c_i > 0$ ($i = 1, 2$) are the elastic and viscosity constants of the material, and f_H and f_V denote the density of body forces acting along the horizontal and vertical directions, respectively. The existence of a unique solution to problem VP has been proved in [3]. We recall the main result in the following.

Theorem 1. *Assume that $f_H \in W^{1,2}(0, T; H)$, $f_V \in \mathcal{C}([0, T]; H)$, $u_0 \in E$, $\widetilde{u}_0 \in V$, $v_0, \widetilde{v}_0 \in H$. Then, there exists a unique solution $\{u, \widetilde{u}\}$ to problem VP with $u \in W^{1,2}(0, T; E)$, $u_{tt} \in L^2(0, T; E')$, $\widetilde{u} \in W^{1,2}(0, T; V)$ and $\widetilde{u}_{tt} \in L^2(0, T; V')$.*

3 Numerical approximation

Now we describe a fully discrete scheme for the variational problem VP. For convenience, we will consider the variational problem in terms of the velocity fields $v(t) = u_t(t)$, $\widetilde{v}(t) = \widetilde{u}_t(t)$. In order to discretize the spatial variable, a uniform partition of $[0, L]$, denoted by $\{I_i\}_{i=1}^M$, is introduced, in such a way that $[0, L] = \cup_{i=1}^M I_i$. Let E^h and V^h be the following finite element spaces approximating E and V,

$$E^h = \{w^h \in E \; ; \; w^h_{|I_i} \in P_1(I_i), \quad 1 \leq i \leq M\},$$
$$V^h = \{z^h \in V \; ; \; z^h_{|I_i} \in P_3(I_i), \quad 1 \leq i \leq M\},$$

where $P_q(I_i)$ denotes the polynomial space of degree less or equal to q restricted to I_i. Moreover, we introduce a uniform partition of the time interval with the step size $k = T/N$ and the nodes $t_n = nk$, $n = 1, \ldots, N$. Finally, we use the notation $z_n = z(t_n)$ for a continuous function $z(t)$, and for a sequence $\{z_n\}_{n=0}^N$ we denote by $\delta z_n = (z_n - z_{n-1})/k$ the divided diferences. In this section no summation is assumed over a repeated index and c will denote a generic constant which does not depend on k, h or n.

Using an Euler scheme, we introduce the following fully discrete aproximation of problem VP.

Fully discrete problem VP^{hk}. *Find $v^{hk} = \{v_n^{hk}\}_{n=0}^N \subset E^h$ and $\widetilde{v}^{hk} = \{\widetilde{v}_n^{hk}\}_{n=0}^N \subset V^h$, such that $u_0^{hk} = u_0^h$, $v_0^{hk} = v_0^h$, $\widetilde{u}_0^{hk} = \widetilde{u}_0^h$, $\widetilde{v}_0^{hk} = \widetilde{v}_0^h$ and, for $n = 1, \ldots, N$,*

$$(\rho \delta v_n^{hk}, w^h) + a_1((u_n^{hk})_x, w_x^h) + c_1((v_n^{hk})_x, w_x^h) + j_H(u_{n-1}^{hk}, w^h)$$

$$= ((f_H)_n, w^h), \quad \forall w^h \in E^h, \qquad (5)$$

$$(\rho \delta \widetilde{v}_n^{hk}, z^h - \widetilde{v}_n^{hk}) + a_2((\widetilde{u}_n^{hk})_{xx}, (z^h - \widetilde{v}_n^{hk})_{xx}) + c_2((\widetilde{v}_n^{hk})_{xx}, (z^h - \widetilde{v}_n^{hk})_{xx})$$
$$+ j_V(u_n^{hk}, z^h) - j_V(u_n^{hk}, \widetilde{v}_n^{hk}) \geq ((f_V)_n, z^h - \widetilde{v}_n^{hk}), \quad \forall z^h \in V^h, \qquad (6)$$

where u_0^h, v_0^h, $\widetilde{u}_0^h$ and $\widetilde{v}_0^h$ are appropriate approximations of the initial conditions u_0, v_0, $\widetilde{u}_0$ and $\widetilde{v}_0$, respectively. Moreover, $u^{hk} = \{u_n^{hk}\}_{n=0}^N$ and $\widetilde{u}^{hk} = \{\widetilde{u}_n^{hk}\}_{n=0}^N$ denote the displacement fields defined by

$$u_n^{hk} = u_{n-1}^{hk} + k v_n^{hk}, \quad \widetilde{u}_n^{hk} = \widetilde{u}_{n-1}^{hk} + k \widetilde{v}_n^{hk}, \quad n = 1, \ldots, N.$$

The following error estimates result was proven in [5].

Theorem 2. *Let the assumptions of Theorem 1 hold and assume the additional regularity conditions*

$$u \in \mathcal{C}^1([0,T]; E), \quad \ddot{u} \in \mathcal{C}([0,T]; H), \qquad (7)$$

$$\widetilde{u} \in \mathcal{C}^1([0,T]; V), \quad \ddot{\widetilde{u}} \in \mathcal{C}([0,T]; H). \qquad (8)$$

Then, the following error estimates are obtained for all $\{w_j^h\}_{j=0}^N \subset E^h$ *and* $\{z_j^h\}_{j=0}^N \subset V^h$,

$$\max_{0 \leq n \leq N} \{|\widetilde{v}_n - \widetilde{v}_n^{hk}|_H^2 + |v_n - v_n^{hk}|_H^2\} + k \sum_{j=1}^N [|\widetilde{v}_j - \widetilde{v}_j^{hk}|_V^2 + |v_j - v_j^{hk}|_E^2]$$

$$\leq c \Big\{ \sum_{j=1}^N k \Big(|\dot{\widetilde{v}}_j - \delta \widetilde{v}_j|_H^2 + |\dot{v}_j - \delta v_j|_H^2 + |u_j - u_{j-1}|_E^2 + |\widetilde{v}_j - z_j^h|_V^2$$

$$+ |v_j - w_j^h|_E^2 + \widetilde{I}_j^2 + I_j^2 + |R(\widetilde{v}_j, z_j^h)| \Big) + |\widetilde{v}_0 - \widetilde{v}_0^h|_H^2 + |v_0 - v_0^h|_H^2$$

$$+ |u_0 - u_0^h|_E^2 + |\widetilde{u}_0 - \widetilde{u}_0^h|_V^2 + \max_{0 \leq n \leq N} |\widetilde{v}_n - z_n^h|_H^2$$

$$+ \max_{0 \leq n \leq N} |v_n - w_n^h|_H^2 + k^{-1} \sum_{j=1}^{N-1} |\widetilde{v}_j - z_j^h - (\widetilde{v}_{j+1} - z_{j+1}^h)|_H^2$$

$$+ k^{-1} \sum_{j=1}^{N-1} |v_j - w_j^h - (v_{j+1} - w_{j+1}^h)|_H^2 \Big\}, \qquad (9)$$

where

$$R(\widetilde{v}_n, z_n^h) = (\dot{\widetilde{v}}_n, z_n^h - \widetilde{v}_n) + c_2((\widetilde{v}_n)_{xx}, (z_n^h - \widetilde{v}_n)_{xx}) + a_2((\widetilde{u}_n)_{xx}, (z_n^h - \widetilde{v}_n)_{xx})$$
$$+ j_V(u_n, z_n^h) - j_V(u_n, \widetilde{v}_n) - (f_n, z_n^h - \widetilde{v}_n),$$

$$I_j = \Big| \int_0^{t_j} v(s) ds - \sum_{l=1}^j k v_l \Big|_E, \quad \widetilde{I}_j = \Big| \int_0^{t_j} \widetilde{v}(s) ds - \sum_{l=1}^j k \widetilde{v}_l \Big|_V.$$

Error estimates (9) are the basis for the convergence rate. Under adequate regularity conditions, we obtain the following corollary which states the linear convergence of the method.

Corollary 1. *Let the assumptions of Theorem 1 hold. Under the following additional regularity conditions*

$$u \in H^2(0,T;E) \cap \mathcal{C}^1([0,T];H^2(0,L)), \quad \ddot{u} \in L^2(0,T;H),$$

$$\widetilde{u} \in H^2(0,T;V) \cap \mathcal{C}^1([0,T];\mathcal{C}^3([0,L])), \quad \dddot{\widetilde{u}} \in L^2(0,T;H),$$

assume that the initial conditions satisfy

$$u_0 \in H^2(0,L), \quad v_0 \in H^1(0,L), \quad \widetilde{u}_0 \in H^3(0,L), \quad \widetilde{v}_0 \in H^1(0,L),$$

and define the discrete initial conditions by

$$u_0^h = \Pi^h u_0, \quad v_0^h = \Pi^h v_0, \quad \widetilde{u}_0^h = \widetilde{\Pi}^h \widetilde{u}_0, \quad \widetilde{v}_0^h = \Pi^h \widetilde{v}_0,$$

where Π^h and $\widetilde{\Pi}^h$ denote the standard Lagrange and Hermite interpolation operators over E^h and V^h, respectively. Then, there exists $c > 0$, independent of h and k, such that,

$$\max_{0 \leq n \leq N} \{|\widetilde{u}_n - \widetilde{u}_n^{hk}|_H + |u_n - u_n^{hk}|_H\} \leq c(h+k). \quad (10)$$

4 Numerical results

In order to show the behaviour of the numerical scheme presented in the above section, some numerical experiments have been done. In this section we describe the algorithm employed to solve the fully discrete problem VP^{hk}, and we resume some numerical results which demonstrate the performance of the method.

4.1 Numerical resolution

The variational equation (5) can be seen as a linear system and its resolution was done with Cholesky's method. In the case of the vertical problem, the term j_V is not differenciable. We introduce an efficient combination of a penalty-duality algorithm with a penalization of the friction condition as follows,

$$-\sigma_V(L,t) = \Phi_\mu(\widetilde{v}(L,t)), \text{ with } \Phi_\mu(r) = \begin{cases} -h(t) & \text{if } r < -\mu, \\ \dfrac{h(t)}{\mu} r & \text{if } r \in [-\mu, \mu], \\ h(t) & \text{if } r > \mu. \end{cases} \quad (11)$$

Using (11) instead of (2), another second kind variational inequality is derived for the vertical velocity,

$$(\rho\delta\widetilde{v}^{hk}_\mu, z^h - \widetilde{v}^{hk}_\mu) + a_2((\widetilde{v}^{hk}_\mu)_{xx}, (z^h - \widetilde{v}^{hk}_\mu)_{xx}) + c_2((\widetilde{u}^{hk}_\mu)_{xx}, (z^h - \widetilde{v}^{hk}_\mu)_{xx})$$
$$+ j_\mu(u^{hk}_n, z^h) - j_\mu(u^{hk}_n, \widetilde{v}^{hk}_\mu) \geq ((f_V)_n, z^h - \widetilde{v}^{hk}_\mu), \quad \forall z^h \in V^h, \qquad (12)$$

where $j_\mu(u^{hk}_n, \cdot) : V \to \mathbb{R}$ is a differentiable functional defined by

$$j_\mu(u^{hk}_n, v) = \begin{cases} -h(u^{hk}_n)v - h(u^{hk}_n)\dfrac{\mu}{2} & \text{if } v < -\mu, \\ \dfrac{h(u^{hk}_n)}{2\mu}v^2 & \text{if } v \in [-\mu, \mu], \\ h(u^{hk}_n)v - h(u^{hk}_n)\dfrac{\mu}{2} & \text{if } v > \mu, \end{cases}$$

where either $h(u^{hk}_n) = constant$ (Tresca's law) or $h(u^{hk}_n) = c_V(u^{hk}_n - g)_+$ (Coulomb's law). Problem (12) is solved using a penalty-duality algorithm introduced in [2]. Also, in [6] it was proved that

$$|\widetilde{v}^{hk}_n - \widetilde{v}^{hk}_\mu|_V \leq c\mu(h + k + |\widetilde{u}|_{C([0,T];V)}).$$

4.2 Numerical simulations

First example

As a first numerical example, and in order to see the behaviour of the algorithm, a sequence of numerical solutions based on uniform partitions of both the time interval and the spatial domain, have been computed. The spatial domain $[0,1]$ ($L=1$) is divided into n equal parts ($h = 1/n$). We start with $n = 100$, which is sucessively halved, and $k = 0.01$, taking as "exact" solution that obtained with $n = 12800$ and $k = 10^{-5}$. The following data were used in the simulations:

$$T = 1\,s, \quad a_1 = a_2 = 1000\,N, \quad c_1 = c_2 = 1\,N \cdot s,$$
$$\rho = 10^{-4}\,kg/m^3, \quad c_H = 10^3, \quad c_V = 5 \times 10^6, \quad \mu = 10^{-10},$$
$$f_H(x,t) = 400(e^t - 1)\,N/m, \quad f_V(x,t) = -10000(e^t - 1)\,N/m \text{ in } [0,1],$$
$$u_0(x) = \widetilde{u}_0(x) = 0\,m, \quad v_0(x) = \widetilde{v}_0(x) = 0\,m/s \text{ in } [0,1].$$

In this example the Coulomb's version of the friction law was employed, and the tangential stress does not achieve the friction bound, so the beam remains sticked to the obstacle in its right end. In Fig 1 the displacements are shown at different times, for the horizontal case on the left-hand side, and for the vertical one on the right-hand side. The numerical errors e^{hk} given by

$$e^{hk} = \max_{0 \leq n \leq N} \{|u_n - u^{hk}_n|_H + |\widetilde{u}_n - \widetilde{u}^{hk}_n|_H\},$$

are shown in Table 1 for different discretization parameters. The numerical convergence of the algorithm is clearly observed, although the linear convergence, stated in Corollary 1, was not achieved.

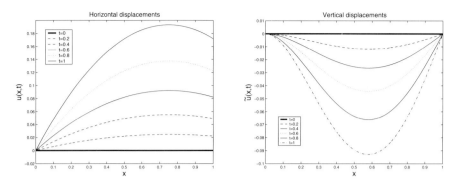

Fig. 1. Example 1: Horizontal and vertical displacements at different times.

Table 1. Numerical errors e^{hk} for some n and k

$n \downarrow k \rightarrow$	0.01	0.005	0.002	0.001	0.0005	0.0001
100	5.042e-3	5.042e-3	5.042e-3	5.042e-3	5.042e-3	5.042e-3
200	2.856e-3	2.856e-3	2.856e-3	2.856e-3	2.856e-3	2.856e-3
400	1.699e-3	1.699e-3	1.699e-3	1.699e-3	1.699e-3	1.699e-3
800	1.008e-3	1.008e-3	1.008e-3	1.008e-3	1.008e-3	1.008e-3
1600	5.639e-4	5.637e-4	5.637e-4	5.636e-4	5.636e-4	5.636e-4
3200	2.028e-4	2.015e-4	2.008e-4	2.007e-4	2.005e-4	2.003e-4

Second example

In this second example, Tresca's law has been considered for modelling the contact in a stick-slip case. Since in this example the problem decouples, the interest concerns only the vertical motion. The following data were used:

$$T = 1\,s, \quad a_2 = 1000\,N, \quad c_2 = 1\,N \cdot s, \quad \mu = 10^{-10}, \quad \rho = 10^{-4}\,kg/m^3,$$
$$f_V(x,t) = -10000t\,N/m \text{ for } (x,t) \in [0,1] \times [0,1], \quad h(t) = c_V = 3000\,N,$$
$$\widetilde{u}_0(x) = 0\,m, \quad \widetilde{v}_0(x) = 0\,m/s \quad \text{for } x \in [0,1].$$

In Fig 2 (left-hand side) the evolution in time of the tangential stress of the free end is plotted, while the vertical displacements at several times are shown on the right-hand side. We notice that the displacement of the contact node is produced when the absolute value of the stresses reaches the value $h(t) = 3000\,N$ at time $t = 0.67$.

Acknowledgements

This work was partially supported by the project "New Materials, Adaptive systems and their Nonlinearities; Modelling, Control and Numerical

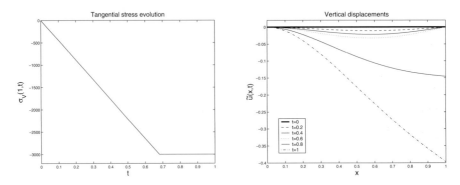

Fig. 2. Example **2**: Evolution of the tangential stress of the contact node and vertical displacements at several times.

Simulation" carried out in the framework of the european community program "Improving the Human Research Potential and the Socio-Economic Knowledge Base" (Contract n° HPRN-CT-2002-00284).

References

1. Andrews, K.T., Fernández, J.R., Shillor, M.: A thermoviscoelastic beam with a tip body. Comput. Mech., **33**, 225–234 (2004)
2. Andrews, K.T., Fernández, J.R., Shillor, M.: Numerical analysis of dynamic thermoviscoelastic contact with damage of a rod. IMA J. Appl. Math., **70**, no. 6, 768–795 (2005)
3. Andrews, K.T., Shillor, M., Wright, S.: On the dynamic vibrations of an elastic beam in frictional contact with a rigid obstacle. J. Elasticity, **42**, 1–30 (1996)
4. Bermúdez, A., Moreno, C.: Duality methods for solving variational inequalities. Comput. Math. Appl., **7**, 43–58 (1981)
5. Campo, M., Fernández, J.R., Stavroulakis, G.E., Viaño, J.M.: Dynamic frictional contact of a viscoelastic beam. Math. Model. Numer. Anal., to appear
6. Campo, M. Fernández, J.R., Viaño, J.M.: Numerical analysis and simulations of a quasistatic frictional contact problem with damage. J. Comput. Appl. Math., to appear
7. Han, W., Kuttler, K.L., Shillor, M., Sofonea, M.: Elastic beam in adhesive contact. Internat. J. Solids Structures, **39**, 1145–1164 (2002)
8. Laursen, T.A.: Computational contact and impact mechanics: fundamentals of modeling interfacial phenomena in nonlinear finite element analysis. Springer, Berlin (2002)
9. Sofonea, M., Shillor, M., Touzani, R.: Quasistatic frictional contact and wear of a beam. Dynam. Contin. Discrete Impuls. Systems, **8**(2), 201–218 (2000)
10. Wriggers, P.: Computational contact mechanics. Wiley, Berlin (2002)

Numerical Analysis of a Frictional Contact Problem for Viscoelastic Materials with Long-Term Memory

A. Rodríguez-Arós[1], M. Sofonea[2] and J. Viaño[1]

[1] Departamento de Matemática Aplicada, Universidade de Santiago de Compostela, Avda. Lope Gómez de Marzoa S/N, Facultade de Matemáticas, 15782 Santiago de Compostela, Spain, Fax: 34-981597054
`angelaros@usc.es, maviano@usc.es`

[2] Laboratoire de Mathématiques et Physique pour les Systèmes, Université de Perpignan, 52 Avenue Paul Alduy, 66860 Perpignan Cedex, France, Fax: 33-0468661760
`sofonea@univ-perp.fr`

Summary. We consider a mathematical model which describes the frictional contact between a viscoelastic body and an obstacle, the so-called foundation. The process is quasistatic and the behavior of the material is modeled with a constitutive law with memory. The contact is bilateral and the friction is modeled with Tresca's law. The existence of a unique weak solution to the model was proved in [15]. Here we describe a fully discrete scheme for the problem, implement it in a computer code and provide numerical results in the study of a two-dimensional test problem.

1 Introduction

Contact phenomena involving deformable bodies abound in industry and everyday life. For this reason, considerable progress has been made in their modelling and analysis, and the engineering literature concerning this topic is rather extensive. An early attempt to the study of frictional contact problems within the framework of variational inequalities was made in [3]. Comprehensive references on analysis and numerical approximation of contact problems include [5, 6] and, more recently, [4]. Mathematical, mechanical and numerical state of the art on Contact Mechanics can be found in the proceedings [7, 10] and in the special issue [14].

The present paper is devoted to numerical analysis of a problem of bilateral frictional contact. The process is quasistatic and the friction is modeled with the well known Tresca's law in which the friction bound is given. The behavior of the material is described with a linear viscoelastic constitutive law with long memory of the form

$$\sigma_{ij}(t) = \mathcal{A}_{ijkl}\varepsilon_{kl}(\boldsymbol{u}(t)) + \int_0^t \mathcal{B}_{ijkl}(t-s)\varepsilon_{kl}(\boldsymbol{u}(s))ds,$$

where $\boldsymbol{\sigma} = (\sigma_{ij})$ denotes the stress tensor, $\boldsymbol{u} = (u_i)$ is the displacement field, $\boldsymbol{\varepsilon}(\boldsymbol{u}) = (\varepsilon_{ij}(\boldsymbol{u}))$ denotes the linearized strain tensor and $\mathcal{A} = (\mathcal{A}_{ijkl})$, $\mathcal{B} = (\mathcal{B}_{ijkl})$ are given fourth order tensors. Details concerning such kind of constitutive laws can be found in [3, 9], for instance. The variational analysis of the problem was provided in [15]. There, its unique solvability was proved by using an abstract existence and uniqueness result for a class of evolutionary variational inequalities involving a Volterra-type integral term. In the present paper we describe a fully discrete scheme for the problem, involving finite difference discretization in time and finite element discretization in space, then we implement it in a computer code and provide numerical simulations.

The paper is organized as follows. In Section 2 we present the contact problem and recall the result obtained in [15] concerning its unique solvability. In Section 3 we describe the fully discrete approximations of the model and state error estimates results. Our main interest lies in Section 4 where we present numerical simulations in the study of a two-dimensional test problem.

2 The model and its well-posedness

The physical setting is the following. A viscoelastic body occupies a regular domain Ω of $\mathbb{R}^d$ $(d = 2, 3)$ with boundary Γ partitioned into three disjoint measurable parts Γ_1, Γ_2 and Γ_3 such that $meas\,(\Gamma_1) > 0$. We are interested in the evolution process of the mechanical state of the body in the time interval $[0, T]$ with $T > 0$. The body is clamped on Γ_1 and so the displacement field vanishes there. Surface tractions of density $\boldsymbol{f}_2$ act on Γ_2 and volume forces of density $\boldsymbol{f}_0$ act in Ω. We assume that the forces and tractions change slowly in time so that the acceleration of the system is negligible. On Γ_3 the body is in bilateral frictional contact with a rigid obstacle, the so-called foundation, and friction is modeled with Tresca's law. Under these assumptions, the classical formulation of the mechanical problem is the following.

Problem P. *Find a displacement field $\boldsymbol{u} : \Omega \times [0, T] \to \mathbb{R}^d$ and a stress field $\boldsymbol{\sigma} : \Omega \times [0, T] \to \mathbb{S}_d$ such that, for all $t \in [0, T]$,*

$$\boldsymbol{\sigma}(t) = \mathcal{A}\boldsymbol{\varepsilon}(\boldsymbol{u}(t)) + \int_0^t \mathcal{B}(t-s)\boldsymbol{\varepsilon}(\boldsymbol{u}(s))ds \text{ in } \Omega, \quad (1)$$

$$\text{Div}\,\boldsymbol{\sigma}(t) + \boldsymbol{f}_0(t) = \boldsymbol{0} \text{ in } \Omega, \quad (2)$$

$$\boldsymbol{u}(t) = \boldsymbol{0} \text{ on } \Gamma_1, \quad (3)$$

$$\boldsymbol{\sigma}(t)\boldsymbol{\nu} = \boldsymbol{f}_2(t) \text{ on } \Gamma_2, \quad (4)$$

$$\begin{cases} u_\nu(t) = 0, \ |\boldsymbol{\sigma}_\tau(t)| \le g, \\ |\boldsymbol{\sigma}_\tau(t)| < g \Rightarrow \dot{\boldsymbol{u}}_\tau(t) = \boldsymbol{0}, \\ |\boldsymbol{\sigma}_\tau(t)| = g \Rightarrow \exists \lambda \ge 0 \text{ s.t. } \boldsymbol{\sigma}_\tau(t) = -\lambda \dot{\boldsymbol{u}}_\tau(t) \end{cases} \quad \text{on} \quad \Gamma_3, \tag{5}$$

$$\boldsymbol{u}(0) = \boldsymbol{u}_0 \quad \text{on} \quad \Omega. \tag{6}$$

Here and below ν denote the unit outer normal on Γ, the subscripts ν and τ represent the *normal* and *tangential* components of vectors or tensors, respectively, and the dot above indicates the derivative with respect to the time; $\mathbb{S}_d$ is the space of second order symmetric tensors on $\mathbb{R}^d$, while "$\cdot$" and $|\cdot|$ represent the inner product and the Euclidean norm on $\mathbb{S}_d$ and $\mathbb{R}^d$, respectively; ε and Div are the *deformation* and *divergence* operators, defined by

$$\varepsilon(\boldsymbol{u}) = (\varepsilon_{ij}(\boldsymbol{u})), \quad \varepsilon_{ij}(\boldsymbol{u}) = \frac{1}{2}(u_{i,j} + u_{j,i}), \quad \text{Div } \boldsymbol{\sigma} = (\sigma_{ij,j}),$$

where the index that follows a comma indicates a partial derivative with respect to the corresponding component of the spatial variable; finally, the indices i, j, k and l run between 1 and d, and the summation convention over repeated indices is adopted.

Equation (1) is the viscoelastic constitutive law where $\mathcal{A} = (\mathcal{A}_{ijkl})$ represents the fourth order tensor of elastic coefficients and $\mathcal{B} = (\mathcal{B}_{ijkl})$ is the relaxation tensor. Equation (2) represents the equilibrium equation. Relations (3) and (4) are the displacement and traction boundary conditions, respectively, in which $\boldsymbol{\sigma}\boldsymbol{\nu}$ is the Cauchy stress vector. Conditions (5) are the frictional contact conditions, where u_ν denotes the normal displacement, $\boldsymbol{\sigma}_\tau$ represents the tangential stress and $\dot{\boldsymbol{u}}_\tau$ is the tangential velocity. Equality $u_\nu(t) = 0$ on Γ_3 shows that there is no loss of the contact during the process, that is, the contact is bilateral. The rest of conditions in (5) represent Tresca's law of dry friction where $g \ge 0$ is the friction bound function, i.e. the magnitude of the limiting friction traction at which slip begins. The inequality in (5) holds in the stick zone and the equality holds in the slip zone. Contact problems with Tresca's friction law can be found in [3, 8], and more recently in [1, 4] (see references therein for further details). Finally, (6) is the initial condition in which the initial displacement $\boldsymbol{u}_0$ is given.

We turn now to the variational formulation of Problem $\mathcal{P}$. To this end we use the spaces

$$Q = \{\,\boldsymbol{\sigma} = (\sigma_{ij}) \mid \sigma_{ij} = \sigma_{ji} \in L^2(\Omega)\,\},$$
$$V = \{\boldsymbol{v} \in H^1(\Omega)^d \mid \boldsymbol{v} = \boldsymbol{0} \text{ on } \Gamma_1,\ v_\nu = 0 \text{ on } \Gamma_3\},$$

which are real Hilbert spaces with the inner products

$$(\boldsymbol{\sigma}, \boldsymbol{\tau})_Q = \int_\Omega \sigma_{ij}\tau_{ij}\,dx, \quad (\boldsymbol{u}, \boldsymbol{v})_V = \int_\Omega \varepsilon_{ij}(\boldsymbol{u})\varepsilon_{ij}(\boldsymbol{u})\,dx$$

and the associated norms denoted $\|\cdot\|_Q$ and $\|\cdot\|_V$. We also use the space

$$\mathbf{Q}_\infty = \{\,\boldsymbol{\xi} = (\xi_{ijkl}) \mid \xi_{ijkl} = \xi_{jikl} = \xi_{klij} \in L^\infty(\Omega)\,\},$$

which is Banach with the norm

$$\|\boldsymbol{\xi}\|_{\mathbf{Q}_\infty} = \max_{0 \leq i,j,k,l \leq d} \|\xi_{ijkl}\|_{L^\infty(\Omega)}.$$

Also, for any real Banach space X we employ the usual notation for the spaces $C([0,T];X)$, $L^p(0,T;X)$ and $W^{k,p}(0,T;X)$, where $1 \leq p \leq \infty$ and $k = 1, 2, \ldots$.

In the study of the mechanical problem P we assume that

$$\mathcal{A} \in \mathbf{Q}_\infty, \tag{7}$$

$$\exists\, m > 0 \text{ such that } \mathcal{A}\boldsymbol{\xi} \cdot \boldsymbol{\xi} \geq m|\boldsymbol{\xi}|^2 \;\; \forall \boldsymbol{\xi} \in \mathbb{S}_d, \text{ a.e. in } \Omega, \tag{8}$$

$$\mathcal{B} \in W^{1,2}(0,T;\mathbf{Q}_\infty), \tag{9}$$

$$\boldsymbol{f}_0 \in W^{1,2}(0,T;L^2(\Omega)^d), \quad \boldsymbol{f}_2 \in W^{1,2}(0,T;L^2(\Gamma_2)^d), \tag{10}$$

$$g \in L^\infty(\Omega), \quad g \geq 0 \text{ a.e. on } \Gamma_3, \tag{11}$$

$$\boldsymbol{u}_0 \in V, \tag{12}$$

$$a(\boldsymbol{u}_0, \boldsymbol{v}) + j(\boldsymbol{v}) \geq (\boldsymbol{f}(0), \boldsymbol{v})_V \quad \forall \boldsymbol{v} \in V, \tag{13}$$

where the bilinear form $a : V \times V \to \mathbb{R}$, the function $\boldsymbol{f} : [0,T] \to V$ and the functional $j : V \to \mathbb{R}_+$ are defined by

$$a(\boldsymbol{v}, \boldsymbol{w}) = (\mathcal{A}\boldsymbol{\varepsilon}(\boldsymbol{v}), \boldsymbol{\varepsilon}(\boldsymbol{w}))_Q \quad \forall \boldsymbol{v}, \boldsymbol{w} \in V, \tag{14}$$

$$(\boldsymbol{f}(t), \boldsymbol{v})_V = \int_\Omega \boldsymbol{f}_0(t) \cdot \boldsymbol{v}\, dx + \int_{\Gamma_2} \boldsymbol{f}_2(t) \cdot \boldsymbol{v}\, da \quad \forall \boldsymbol{v} \in V,\, t \in [0,T], \tag{15}$$

$$j(\boldsymbol{v}) = \int_{\Gamma_3} g |\boldsymbol{v}_\tau|\, da \quad \forall \boldsymbol{v} \in V. \tag{16}$$

Proceeding in a standard way and using the notation (14)–(16) we obtain the following variational formulation of the contact problem (1)–(6), in terms of displacement.

Problem P_V. *Find the displacement field $\boldsymbol{u} : [0,T] \to V$ such that*

$$a(\boldsymbol{u}(t), \boldsymbol{v} - \dot{\boldsymbol{u}}(t)) + \Big(\int_0^t \mathcal{B}(t-s)\boldsymbol{\varepsilon}(\boldsymbol{u}(s))ds, \boldsymbol{\varepsilon}(\boldsymbol{v}) - \boldsymbol{\varepsilon}(\dot{\boldsymbol{u}}(t))\Big)_Q \tag{17}$$

$$+ j(\boldsymbol{v}) - j(\dot{\boldsymbol{u}}(t)) \geq (\boldsymbol{f}(t), \boldsymbol{v} - \dot{\boldsymbol{u}}(t))_V \quad \forall \boldsymbol{v} \in V, \text{ a.e. } t \in (0,T),$$

$$\boldsymbol{u}(0) = \boldsymbol{u}_0. \tag{18}$$

The well-posedness of the Problem P_V was proved in [15] and may be stated as follows.

Theorem 1. *Assume that* (7)–(13) *hold. Then Problem* P_V *has a unique solution* $\boldsymbol{u} \in W^{1,2}(0,T;V)$.

In the rest of the paper we assume that conditions stated in Theorem 1 hold.

3 Fully discrete approximation

We now consider a family of fully discrete schemes to approximate Problem P_V. We assume that Ω is a polyhedron. Let $\mathcal{T}_h$ be a finite element triangulation of $\overline{\Gamma}$ composed by d-simplex, compatible to the boundary decomposition $\Gamma = \overline{\Gamma}_1 \cup \overline{\Gamma}_2 \cup \overline{\Gamma}_3$, i.e., any point where the boundary condition type changes is a vertex of the triangulation. We denote by $h > 0$ the maximum diameter of triangles of $\mathcal{T}^h$ and we introduce the following finite element space:

$$Q^h = \{\boldsymbol{\tau}^h \in Q : \boldsymbol{\tau}^h_{|T^h} \in [P^0(T^h)]_s^{d \times d}, \forall T^h \in \mathcal{T}^h\},$$

$$V^h = \{\boldsymbol{v}^h = (v_i^h) \in [C(\overline{\Omega})]^d, \boldsymbol{v}^h_{|T^h} \in [P^1(T^h)]^d \ \forall T^h \in \mathcal{T}^h,$$

$$\boldsymbol{v}^h = \boldsymbol{0} \text{ on } \overline{\Gamma}_1, \ v_\nu^h = 0 \text{ on } \overline{\Gamma}_3\}.$$

Here $P^m(T^h)$ is the space of polynomials of degree less or equal to m on d variables. Also, we denote by $\mathcal{P}^h : V \to V^h$ the operator given by

$$(\mathcal{P}^h \boldsymbol{v}, \boldsymbol{v}^h)_V = a(\boldsymbol{v}, \boldsymbol{v}^h) \quad \forall \boldsymbol{v} \in V, \ \boldsymbol{v}^h \in V^h.$$

In addition to the finite-dimensional subspace V^h, we need a partition of the time interval: $[0,T] = \cup_{n=1}^N [t_{n-1}, t_n]$ with $0 = t_0 < t_1 < \cdots < t_N = T$. We denote by $k_n = t_n - t_{n-1}$ the length of the sub-interval $[t_{n-1}, t_n]$ and let $k = \max_n k_n$ be the maximal step-size. Since $\boldsymbol{u} \in W^{1,2}(0,T;V)$ and $\boldsymbol{f} \in W^{1,2}(0,T;V)$, the pointwise values $\boldsymbol{u}_n = \boldsymbol{u}(t_n)$ and $\boldsymbol{f}_n = \boldsymbol{f}(t_n)$ $(0 \leq n \leq N)$ are well-defined. Also, since $\mathcal{B} \in W^{1,2}(0,T;\mathbf{Q}_\infty)$, the pointwise values $\mathcal{B}^{n,j} = \mathcal{B}(t_n - t_j)$, $0 \leq j \leq n \leq N$, are well-defined. Note that, in particular, $\mathcal{B}^{n,n} = \mathcal{B}_0 = \mathcal{B}(0)$. Below, the symbol $\Delta \boldsymbol{u}_n$ represents the backward difference $\boldsymbol{u}_n - \boldsymbol{u}_{n-1}$, while $\delta_n \boldsymbol{u}_n = \Delta \boldsymbol{u}_n / k_n$ denotes the backward divided difference. And, for each time step t_n, the constants $\alpha_j^n > 0$ $(0 \leq j \leq n)$ denote the weights of the composed trapezoidal quadrature formula of $n+1$ points in $[0, t_n]$. Finally, note that in this section no summation is considered over the repeated indices n and j.

With the notation above, a family of fully discrete approximation schemes to Problem P_V is the following.

Problem P_V^{hk}. Find $\boldsymbol{u}^{hk} = \{\boldsymbol{u}_n^{hk}\}_{n=0}^N \subset V^h$ such that $\boldsymbol{u}_0^{hk} = \boldsymbol{u}_0^h = \mathcal{P}^h \boldsymbol{u}_0$ and

$$a(\boldsymbol{u}_n^{hk}, \boldsymbol{v}^h - \delta_n \boldsymbol{u}_n^{hk}) + \left(\sum_{j=0}^n \alpha_j^n \mathcal{B}^{n,j} \boldsymbol{\varepsilon}(\boldsymbol{u}_j^{hk}), \boldsymbol{\varepsilon}(\boldsymbol{v}^h - \delta_n \boldsymbol{u}_n^{hk})\right)_Q$$

$$+ j(\boldsymbol{v}^h) - j(\delta_n \boldsymbol{u}_n^{hk}) \geq (\boldsymbol{f}_n, \boldsymbol{v}^h - \delta_n \boldsymbol{u}_n^{hk})_V, \quad n = 1, \ldots, N.$$

Suppose that the assumptions in Theorem 1 hold and, in addition, assume that

$$k < \frac{2m}{\|\mathcal{B}_0\|_{\mathbf{Q}_\infty}}. \tag{19}$$

Then, by using arguments similar as those used in [11] we deduce that there exists a unique solution $\boldsymbol{u}^{hk} = \{\boldsymbol{u}_n^{hk}\}_{n=0}^N \subset V^h$ of the Problem P_V^{hk}.

Next, we assume that the solution of the problem P_V satisfies $\boldsymbol{u} \in W^{2,\infty}(0,T;V)$, the relaxation tensor verifies $\mathcal{B} \in W^{1,\infty}(0,T;\mathbf{Q}_\infty)$, $\mathcal{B}$ and $\dot{\mathcal{B}}$ are Lipschitz continuous functions on $[0,T]$ with values to $\mathbf{Q}_\infty$ and (19) holds. Under these assumptions, proceeding as in [4, 11] we can show that, for k sufficiently small, the following error estimate holds:

$$\max_{1 \le n \le N} \|\boldsymbol{u}_n - \boldsymbol{u}_n^{hk}\|_V^2 \le d_N \left(k^2 + Nk^3 + Nk^4 + Nk^5 + \|\boldsymbol{u}_0 - \boldsymbol{u}_0^{hk}\|_V^2 \right. \tag{20}$$
$$\left. + Nk \max_{1 \le n \le N} \left\{ \inf_{\boldsymbol{v}^h \in V^h} \{\|\dot{\boldsymbol{u}}(t_n) - \boldsymbol{v}^h\|_V + \|\dot{\boldsymbol{u}}(t_n) - \boldsymbol{v}^h\|_V^2 \} \right\} \right).$$

Here $d_N = c(1 + c(N+1)(N^2k^3 + k)e^{2c(N+1)(N^2k^3+k)})$ and c is a positive constant which depends on $\mathcal{A}, \mathcal{B}, \boldsymbol{f}, g$ and $\boldsymbol{u}$ but is independent on the discretization parameters h and k. Inequality (20) is the basis for error estimates. For example if we assume that

$$\dot{\boldsymbol{u}} \in C([0,T]; [H^2(\Omega)]^d), \quad \boldsymbol{u}_0 \in [H^2(\Omega)]^d,$$

and the partition of $[0,T]$ is uniform, then we obtain

$$\max_{1 \le n \le N} \|\boldsymbol{u}(t_n) - \boldsymbol{u}_n^{hk}\|_V \le c(k + h^{\frac{1}{2}}).$$

Moreover, if $\dot{\boldsymbol{u}}_\tau \in C([0,T]; [H^2(\Gamma_3)]^d)$, then it can be shown that

$$\max_{1 \le n \le N} \{\|\boldsymbol{u}(t_n) - \boldsymbol{u}_n^{hk}\|_V\} \le c(k + h),$$

which shows a linear convergence with respect to the parameters h and k. We refer to [12, 13] for the complete proof of this results.

4 Numerical simulations

To show the performance of the numerical method described in the previous section we have developed a FORTRAN-based software to solve variational inequalities of the second kind involving the functional $j : V \to \mathbb{R}$ given by (16). It combines a fixed point strategy with a method of duality-penalization based on the algorithm in [2]. We tested it in a number of problems and we present below two numerical simulations obtained in the study of a two-dimensional test problem. The physical setting is presented in Figure 1.

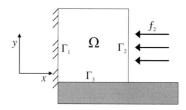

Fig. 1. Two-dimensional contact problem with friction.

For both simulations we considered a homogenous isotropic material with the modulus of Young $E = 10^6 \ N/m^2$ and the coefficient of Poisson $\kappa = 0.3$. A difficulty from the practical point of view is data storage when $\mathcal{B}$ is time dependent (which in fact is a more realistic case), thus constant values were taken here for the components of the relaxation tensor $\mathcal{B}$ (see [13] for details on this topic). Other data were: $T = 0.1 \ sec.$, $k = 1 \times 10^{-3} \ sec.$ (the time interval discretization being uniform), $\boldsymbol{f}_0 = (0,0) \ N/m^3$, $\boldsymbol{f}_2 = (-10^3, 0) \ N/m^2$ and $\boldsymbol{u}_0 = (0,0) \ m$.

For the first simulation, we considered a *high* value for the friction bound, $g = 1 \times 10^6 \ N/m^2$, while for the second one we took a *low* value, $g = 1 \ N/m^2$. The numerical results are shown in Figures 2 and 3, respectively, where the

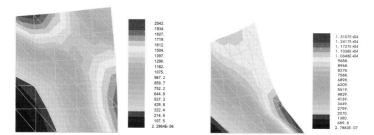

Fig. 2. Deformed configuration and stress distribution in Von Mises norm for $t = 1 \times 10^{-2} \ sec.$ (left) and $t = 0.1 \ sec.$ (right). Friction bound $g = 1 \times 10^6 \ N/m^2$.

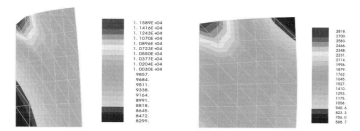

Fig. 3. Deformed configuration and stress distribution in Von Mises norm for $t = 1 \times 10^{-2} \ sec.$ (left) and $t = 0.1 \ sec.$ (right). Friction bound $g = 1 \ N/m^2$.

deformations are amplified for a better visual analysis. In Figure 2-left, we observe that, at $t = 1 \times 10^{-2}$ sec., the nodes close to contact zone are subjected to displacements which are smaller than those the nodes situated far from the contact zone are subjected to. The stresses on the nodes close to contact zone are important, as well. Nevertheless, in Figure 3-left we observe that the horizontal displacements are quite similar either the node is far or near to the contact zone. This is due to the low value of the friction bound g. On the other hand, in Figure 2-right and Figure 3-right, we observe that for $t = T = 0.1$ sec., the deformable body has recovered most of its original shape. This behavior represents a collateral effect of its memory.

Acknowledgments

This work is sponsored by the European Project "New Materials, Adaptive Systems and their Nonlinearities; Modelling Control and Numerical Simulation" (HPRN-CT-2002-00284) and Project "Contacto de materiales viscoplásticos y viscoelásticos; Formulación matemática y análisis numérico" of MCYT, Spain (BFM2003-05357)

References

1. Amassad, A., Shillor, M., Sofonea, M.: A quasistatic contact problem for an elastic perfectly plastic body with Tresca's friction, Nonlin. Anal. **35**, 95–109 (1999)
2. Bermúdez, A., Moreno, C.: Duality methods for solving variational inequalities. Comp. and Math. with Appl. **7**, 43–58 (1981)
3. Duvaut, D., Lions, J. L.: Inequalities in Mechanics and Physics, Springer-Verlag, Berlin (1976)
4. Han, W., Sofonea, M.: Quasistatic Contact Problems in Viscoelasticity and Viscoplasticity. In: Studies in Advanced Mathematics (30). Providence, R.I.: American Mathematical Society. Somerville, MA: Intl. Press (2002)
5. Hlaváček, I., Haslinger, J., Nečas, J., Lovíšek, J.: Solution of Variational Inequalities in Mechanics, Springer-Verlag, New York (1988)
6. Kikuchi, N., Oden, J. T.: Contact Problems in Elasticity: A Study of Variational Inequalities and Finite Element Methods. Philadelphia: SIAM, (1988)
7. Martins, J. A. C., Monteiro Marques, M. D. P.: (Eds.) Contact Mechanics. Dordrecht: Kluwer Academic Publishers, (2001)
8. Panagiotopoulos, P.D.: Inequality Problems in Mechanical and Applications, Birkhäuser, Basel (1985)
9. Pipkin, A. C.: Lectures in Viscoelasticity Theory. In: Applied Mathematical Sciences **7**. London: George Allen & Unwin Ltd. New York: Springer-Verlag, (1972)
10. Raous, M., Jean, M., Moreau, J.J.: (eds.), Contact Mechanics, Plenum Press, New York (1995)

11. Rodríguez–Arós, A. D., Sofonea, M., Viaño, J. M.: A Class of Evolutionary Variational Inequalities with Volterra-type Integral Term, *Mathematical Models and Methods in Applied Sciences* (M^3AS) 14, 555-577 (2004)
12. Rodríguez–Arós, A., Sofonea, M., Viaño, J. M.: Numerical Analysis of a Frictional Contact Problem for Viscoelastic Materials with Long-term Memory. Submitted.
13. Rodríguez–Arós, A. D.: Análisis variacional y numérico de problemas de contacto en viscoelasticidad con memoria larga. Doctoral Thesis. Departamento de Matemática Aplicada, USC. (2005)
14. Shillor, M.: Special Issue on Recent Advances in Contact Mechanics. Mathematical and Computer Modelling, **28,** 4–8 (1998)
15. Rodríguez–Arós, A., Sofonea, M., Viaño, J. M.: A class of integro-differential variational inequalities with applications to viscoelastic contact. Mathematical and Computer Modelling, **41,** 1355–1369 (2005)

A Suitable Numerical Algorithm for the Simulation of the Butt Curl Deformation of an Aluminium Slab

P. Barral and M.T. Sánchez

Department of Applied Mathematics. Universidade de Santiago de Compostela, 15782, Santiago de Compostela. Spain
patribr@usc.es, marua@usc.es

Summary. The aim of this work is to present several numerical strategies to adapt efficiently the numerical algorithm published in [2] for the calculus of the deformation of an aluminium slab during a semicontinuous casting.

1 Introduction

During casting processes, the liquid aluminium is poured onto a water-cooled mould which is called bottom block; when the aluminium begins to solidify the bottom block starts to descend leaving room for more liquid metal. The large thermal stresses due to the cooling jets cause the butt of the slab to bend and lose contact with the bottom block. To predict this deformation called butt curl, the aluminium alloy is treated as a thermo-viscoelastic material with a contact condition between the slab and the bottom block. Furthermore, during the casting, the slab grows with time far from the contact region, increasing the size of the problem and, consequently, the matrix dimension for its numerical simulation.

Several mathematical aspects of this mechanical problem have been studied by the authoresses: the weak formulation of the problem and its numerical simulation [2], the existence of a solution [4], the imposing of the metallostatic pressure on the liquidus-solidus interphase [3]. To solve the nonlinearities due to the contact condition and the constitutive law, the numerical solution in [2] was based on iterative algorithms involving two Lagrange multipliers which were approximated by a fixed point method. Nevertheless, although the algorithm was successfully tested with academic examples, its computational demands for the real problem led us to improve its efficiency.

In this work we present some numerical strategies to obtain a computationally efficient algorithm to simulate the butt curl deformation. We propose to approximate the contact multiplier using a generalized Newton's method and a penalization technique (see [7]). To approximate the viscoplastic multiplier, we use standard Newton's techniques without modification of the stiffness matrix at each iteration. This algorithm is fast, accurate and its convergence is independent of the algorithm's parameters; nevertheless, due to the contact condition, the stiffness matrix needs to be recalculated at each iteration. So, taking into account that usually the nonlinear boundary condition only involves a small part of the boundary, we propose to use a factorization of the stiffness matrix adapted to the problem's geometry. Furthermore, due to the large gradients of stresses with respect to time, it is necessary to employ an adimensionalization technique, an Armijo rule and an optimization of the time step to obtain a good convergence. Finally, numerical results are presented to show the applicability of this algorithm to casting processes.

2 Mathematical model

Due to casting symmetry, $\Omega(t)$ represents a quarter of the slab at the instant $t \in (0, t_f]$. The temperature field $T(x, t)$ at each point $x \in \Omega(t)$ is previously computed by using the mathematical model developed in [5]. The mechanical domain at each time instant t is the solidified part of the slab, denoted by $\Omega_s(t)$ (see Fig. 1).

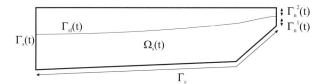

Fig. 1. Computational domain.

Under the small deformations assumption and in the quasistatic case, the slab behaviour is given by the equilibrium equations

$$-\text{Div}(\boldsymbol{\sigma}) = \mathbf{f} \text{ in } \Omega_s(t),$$

where $\mathbf{f}$ represents the volume forces due to the gravity. The boundary $\Gamma(t)$ of $\Omega_s(t)$ is split into five disjoint parts $\Gamma = \bar{\Gamma}_{sl}(t) \cup \bar{\Gamma}_C \cup \bar{\Gamma}_s(t) \cup \bar{\Gamma}_n^1(t) \cup \bar{\Gamma}_n^2(t)$. The upper boundary $\Gamma_{sl}(t)$, defined by the isotherm corresponding to the liquidus temperature, is subjected to the metallostatic pressure exerted by the overlying liquid metal

$$\boldsymbol{\sigma}\mathbf{n} = p_r\mathbf{n}, \ p_r(x,t) = \rho(T)g(x_3 - h(t)) \text{ on } \Gamma_{sl}(t),$$

where ρ is the density of the aluminium, g the gravitational acceleration, $h(t)$ the length of the slab at the instant t and $\mathbf{n}$ the unit outward normal vector to $\Gamma_{sl}(t)$. On Γ_C the body is in contact with the bottom block. On this boundary, to reproduce the butt curl deformation, we consider a Signorini unilateral frictionless contact condition

$$\boldsymbol{\sigma}_\tau = \mathbf{0},\ \sigma_n \leq 0,\ u_n \leq 0,\ \sigma_n u_n = 0 \text{ on } \Gamma_C,$$

where u_n and σ_n denote the normal components of displacements and stresses and $\boldsymbol{\sigma}_\tau$ is the tangential component of the stresses. The boundary Γ_n^1 denotes the part of the lateral face which has already solidified; so, it is free of forces. The boundary Γ_n^2 corresponds to the mushy region, which is confined by the mould; so, $\boldsymbol{\sigma}_\tau = \mathbf{0},\ u_n = 0$ on $\Gamma_n^2(t)$. Finally, on $\Gamma_s(t)$ we assume usual symmetry conditions.

The aluminium is considered a viscoelastic material with temperature dependent properties; so, the strain rate tensor is the sum of elastic, viscoplastic and thermal effects

$$\dot{\varepsilon}(\mathbf{u}) = \dot{\varepsilon}^e(\mathbf{u}) + \dot{\varepsilon}^{vp}(\mathbf{u}) + \dot{\varepsilon}^{th}(\mathbf{u}).$$

Elastic deformations are related to stresses by Hooke's law $\varepsilon^e(\mathbf{u}) = \Lambda_s(T)\boldsymbol{\sigma}$. The thermal expansion is related to the temperature by a generalized Arrhenius law

$$\dot{\varepsilon}^{th}(\mathbf{u}) = \alpha_s(T)\dot{T}\mathbf{I}$$

where α_s is the coefficient of thermal expansion. The relevant viscoplastic effects at high temperature $\dot{\varepsilon}^{vp}$ are described by Norton-Hoff law for secondary creep

$$\dot{\varepsilon}^{vp}(\mathbf{u}) = \nabla\Phi_q(\boldsymbol{\sigma}^D) = \theta(T)|\boldsymbol{\sigma}^D|^{q-2}\boldsymbol{\sigma}^D, \tag{1}$$

where $\boldsymbol{\sigma}^D$ is the deviatoric component of $\boldsymbol{\sigma}$, $\nabla\Phi_q$ denotes the gradient of the dissipation potential Φ_q and θ, q are material parameters (for aluminium slabs $q > 2$). Finally, to complete the model, we consider the initial conditions

$$\mathbf{u}(0) = \mathbf{u}_0(x),\ \boldsymbol{\sigma}(0) = \boldsymbol{\sigma}_0(x) \text{ in } \Omega_s(0). \tag{2}$$

3 Weak formulation

Let $2 \leq q < +\infty$ be the exponent of viscoplastic law (1) and p, $1 < p \leq 2$, its conjugate exponent. The usual weak formulation of this problem is given by: Find $\mathbf{u} \in W^{1,\infty}(0,t_f;\mathbf{U}_{ad}^p(t))$ and $\boldsymbol{\sigma} \in W^{1,\infty}(0,t_f;\mathbf{H}^q(t))$ such that

$$\int_{\Omega_s(t)} \boldsymbol{\sigma}(t) : \varepsilon(\mathbf{v}-\mathbf{u}(t))\,dx \geq \int_{\Omega_s(t)} \mathbf{f}(t)\cdot(\mathbf{v}-\mathbf{u}(t))\,dx +$$
$$\int_{\Gamma_{sl}(t)} p_r(t)\mathbf{n}\cdot(\mathbf{v}-\mathbf{u}(t))\,d\gamma,\ \forall \mathbf{v} \in \mathbf{U}_{ad}^p(t), \tag{3}$$

$$\varepsilon(\dot{\mathbf{u}})(t) = \overline{(\Lambda_s(T)\boldsymbol{\sigma})}(t) + (\nabla\Phi_q(\boldsymbol{\sigma}))(t) + \left(\alpha_s(T)\dot{T}\right)(t)\mathbf{I}, \tag{4}$$

a.e. $t \in (0, t_f]$, together with the initial conditions (14). Here, $\mathbf{U}_{ad}^p(t)$ denotes the admissible displacements field

$$\mathbf{U}_{ad}^p(t) = \{\mathbf{v} \in \mathbf{U}^p(t);\ v_n \leq 0 \text{ on } \Gamma_C\},$$
$$\mathbf{U}^p(t) = \{\mathbf{v} \in [W^{1,p}(\Omega_s(t))]^3;\ \text{Div}(\mathbf{v}) \in L^2(\Omega_s(t)),\ v_n = 0 \text{ on } \Gamma_n^2(t) \cup \Gamma_s(t)\},$$

and the corresponding spaces of stress fields are defined as

$$\mathbf{H}^q(t) = \{\boldsymbol{\xi} = (\xi_{ij});\ \xi_{ij} = \xi_{ji};\ \text{tr}(\boldsymbol{\xi}) \in L^2(\Omega_s(t));\ \xi_{ij}^D, (\text{Div}(\boldsymbol{\xi}))_i \in L^q(\Omega_s(t))\}.$$

4 Numerical solution

The main difficulties we must overcome in the numerical solution of this problem are the following:

- The solidified part of the slab, which is the computational domain of the mechanical simulation, changes with time. Furthermore, on the upper boundary, which is the isotherm of the liquidus temperature we must impose the metallostatic pressure due to the liquid metal.
- The computational domain grows with time and the zone with steep gradients changes, so the computational demands of the problem are considerable.
- The aluminium behaviour is non linear and depends strongly on the temperature field.
- To model the butt curl, we must solve a contact condition between the slab and the bottom block.

4.1 Imposing the metallostatic pressure

The upper boundary $\Gamma_{sl}(t)$ is the free boundary of the thermal problem and therefore it is obtained at each instant from the thermal simulation. In order to impose the metallostatic pressure on this boundary, we have used a fictitious domain method, extending the computational domain to the entire slab, but treating the liquid metal as a very weak elastic material confined by the mould. The elasticity tensor corresponding to the liquid metal is denoted by $\Lambda_l(T)$. The accurate of this method was proved by using asymptotic expansion techniques in [3]. With this technique the variational inequality (3) is defined over the complete slab and the integral over the thermal free boundary $\Gamma_{sl}(t)$ disappears. From now on, we will consider this new weak formulation:

$$\int_{\Omega(t)} \boldsymbol{\sigma}(t) : \varepsilon(\mathbf{v} - \mathbf{u}(t))\, dx \geq \int_{\Omega(t)} \mathbf{f}(t) \cdot (\mathbf{v} - \mathbf{u}(t))\, dx, \forall \mathbf{v} \in \mathbf{U}_{ad}^p(t), \quad (5)$$

$$\varepsilon(\dot{\mathbf{u}})(t) = \begin{cases} \overline{(\Lambda_s(T)\boldsymbol{\sigma})}(t) + (\nabla\Phi_q(\boldsymbol{\sigma}))(t) + \left(\alpha_s(T)\dot{T}\right)(t)\mathbf{I}, & \text{in } \Omega_s(t), \\ \overline{(\Lambda_l(T)\boldsymbol{\sigma})}(t), & \text{in } \Omega(t)\backslash\Omega_s(t), \end{cases} \qquad (6)$$

$$\mathbf{u}(0) = \mathbf{u}_0, \quad \boldsymbol{\sigma}(0) = \boldsymbol{\sigma}_0, \text{ in } \Omega(0). \qquad (7)$$

4.2 Mesh construction

Finite element approximation to problem (5)–(7) is considered in the usual way. We construct a family of finite-dimensional subspaces $\mathbf{U}_h^p(t)$ of $\mathbf{U}^p(t)$ by approximating the test functions by piecewise polynomials of degree one over a tetrahedral mesh $\mathcal{T}_h$ on the computational domain. The stresses are assumed constant within each element.

In order to design the mesh of the slab, we must take into account three essential factors:

- The contact zone corresponds with the part of the slab rested on the mould. So, the mesh is constructed in such a way that the first nodes correspond to the contact nodes.
- The slab deformation depends strongly on the thermal gradients, which are larger in the recently solidified zone, that varies with the time. So, the mesh is finest-grained where the thermal gradients are larger.
- The computational domain grows with time. Then, the mesh is structured in layers to model the filling process and it is reconstructed at each time step adding the amount corresponding to the metal poured during that step.

4.3 Nonlinear constitutive law

In order to avoid the nonlinearity due to the constitutive law, the numerical solution is based on maximal monotone operator techniques involving a viscoplastic multiplier, which is a fixed point of a nonlinear equation. In a first stage we solved the problem by using a fixed point method (see [2]). Although this algorithm is robust its convergence is very slow in real simulation. In order to improve the efficiency, we propose a new methodology which consists of applying a Newton's method to compute the multiplier. The viscoplastic law (6) is discretized in time by using an implicit Euler scheme at each time step t^{j+1}, $j = 0, \ldots, N-1$:

$$\varepsilon(\mathbf{u}^{j+1}) - \varepsilon(\mathbf{u}^j) - \left(\Lambda^{j+1}\boldsymbol{\sigma}^{j+1} - \Lambda^j\boldsymbol{\sigma}^j\right) - \alpha^{j+1}(T^{j+1} - T^j)\mathbf{I} = \Delta t\,\nabla\Phi_q\left((\boldsymbol{\sigma}^{j+1})^D\right), \qquad (8)$$

with $\Delta t = t_f/N$ and g^j an approximation of a given function $g(t)$ at time t^j. To compute the nonlinear term we define the viscoplastic multiplier $\mathbf{q}$

$$\mathbf{q}^{j+1} = \nabla \Phi_q \left((\boldsymbol{\sigma}^{j+1})^D \right) = (\nabla \Phi_q)_{\lambda_P} \left((\boldsymbol{\sigma}^{j+1})^D + \lambda_P \mathbf{q}^{j+1} \right), \qquad (9)$$

where $(\nabla \Phi_q)_{\lambda_P}$ denotes the Yosida approximation of $\nabla \Phi_q$. Therefore, from (8), the stress tensor is given by the relation

$$\boldsymbol{\sigma}^{j+1} = \left(\Lambda(T^{j+1}) \right)^{-1} \left(\varepsilon(\mathbf{u}^{j+1}) - \Delta t \mathbf{q}^{j+1} + \mathbf{F}^j \right), \qquad (10)$$

where

$$\mathbf{F}^j = \Lambda(T^j) \boldsymbol{\sigma}^j - \varepsilon(\mathbf{u}^j) - \alpha(T^{j+1})(T^{j+1} - T^j) \mathbf{I}. \qquad (11)$$

Then, the only problem to compute the stress tensor at each time step is to compute the viscoplastic multiplier. To overcome this difficulty, we approximate the multiplier by using a Newton method in the form

$$\mathbf{q}_k^{j+1} = (\nabla \Phi_q)_{\lambda_P} \left((\boldsymbol{\sigma}_k^{j+1})^D + \lambda_P \mathbf{q}_{k-1}^{j+1} \right) + \\ D(\nabla \Phi_q)_{\lambda_P} \left((\boldsymbol{\sigma}_k^{j+1})^D + \lambda_P \mathbf{q}_{k-1}^{j+1} \right) \left(\lambda_P \left(\mathbf{q}_k^{j+1} - \mathbf{q}_{k-1}^{j+1} \right) \right). \qquad (12)$$

The details of these computations can be found in [1].

Due to the large thermal gradients which appear in the slab, it is necessary to employ an adimensionalization technique on the stresses. This technique consists of choosing a reference stress and introducing new nondimensional unknowns in order to pass the magnitude of the stresses to the coefficients of the behaviour law to solve a similar problem for these new unknowns. Moreover, since the time interval of interest in the aluminium casting is $(0, 1000]$, we are interested in increasing the time step. If we do so, the method does not always converge. To solve this problem, we use an optimization technique on the time step in such a way that, given Δt, if convergence is not achieved, we reduce the time step until the algorithm converges. Furthermore, to stabilize the Newton method we employ an Armijo rule on the computation of the viscoplastic multiplier (see [1, 6]).

4.4 Contact condition

To solve the weak inequality (5), the numerical algorithm is based on maximal monotone operator techniques involving a contact multiplier, which will be denoted by p (see [1]). This multiplier is a fixed point of a nonlinear equation; in [1] we propose to solve it by using a generalized Newton's method which leads to the following algorithm:

At each time step t^{j+1}, given the starting values $(\mathbf{u}_0^{j+1}, \boldsymbol{\sigma}_0^{j+1}, p_0^{j+1})$, successive approximations $(\mathbf{u}_k^{j+1}, \boldsymbol{\sigma}_k^{j+1}, p_k^{j+1})$, $k \geq 1$, of the solution $(\mathbf{u}^{j+1}, \boldsymbol{\sigma}^{j+1}, p^{j+1})$ are computed by solving the equation

$$\int_{\Omega^{j+1}} (\Lambda^{j+1})^{-1} \varepsilon(\mathbf{u}_k^{j+1}) : \varepsilon(\mathbf{v}) dx + \frac{1}{\epsilon} \int_{(\Gamma_{C,k-1}^+)^{j+1}} (u_k^{j+1})_n\, v_n d\gamma =$$
$$\int_{\Omega^{j+1}} (\Lambda^{j+1})^{-1} \left(\Delta t \mathbf{q}_{k-1}^{j+1} \right) : \varepsilon(\mathbf{v}) dx - \int_{\Omega^{j+1}} (\Lambda^{j+1})^{-1} \mathbf{F}^j : \varepsilon(\mathbf{v}) dx +$$
$$\int_\Omega \mathbf{f}^{j+1} \cdot \mathbf{v}\, dx,\ \forall \mathbf{v} \in \mathbf{U}_h^p(t^{j+1}), \tag{13}$$

where $\boldsymbol{\sigma}_k^{j+1}$ is given by expression (10) and $\mathbf{q}_k^{j+1}$ by expression (12). In (13) the contact condition is imposed by the penalty term with parameter ϵ –small enough– on the faces with effective contact at each iteration

$$(\Gamma_{C,k-1}^+)^{j+1} = \{ C \in \mathcal{S}_h; (u_{k-1}^{j+1})_n + \lambda_C p_{k-1}^{j+1} > 0 \}, \tag{14}$$

$\mathcal{S}_h$ being the triangulation induced by the mesh on the boundary Γ_C. Once the displacement field is calculated the multiplier is updated by

$$p_k^{j+1} = \begin{cases} \frac{1}{\epsilon}(u_k^{j+1})_n & \text{on } (\Gamma_{C,k-1}^+)^{j+1}, \\ 0 & \text{on } \Gamma_C \backslash (\Gamma_{C,k-1}^+)^{j+1}. \end{cases}$$

Notice that at each iteration we must only compute the penalty term on the contact faces.

Matrix factorization

In casting processes the computational domain grows with time far from the contact region which is localized on a small part of the boundary. Then, we consider that the contact nodes correspond with the first ones of the mesh numbering. So, the classical factorization $\mathbf{K} = \mathbf{L}\mathbf{D}^*\mathbf{L}^T$ –where $\mathbf{D}^*$ is a diagonal matrix and $\mathbf{L}$ is a lower triangular matrix with unitary diagonal– is very expensive since it would be necessary to factorize the whole stiffness matrix at each iteration to compute the multipliers. Then, we propose to use a new factorization of the type $\mathbf{K} = \mathbf{U}\mathbf{D}^*\mathbf{U}^T$, where $\mathbf{U}$ is an upper triangular matrix with unitary diagonal. We employ a non standard storage by means of an upper skyline by rows; so, we storage the upper submatrix of $\mathbf{K}$ row by row from downwards to upwards and from the right to the left. At each iteration to achieve the convergence of the multipliers, we only compute and factorize the first rows of the matrix, corresponding to the contact nodes.

5 Numerical results

In this section we present the numerical results obtained in a real casting process. To show the efficiency of the new algorithm, we compare the cpu-time[1] and the number of iterations between the Newton and the fixed point algorithms. The mesh of the computational domain in different intants is shown in Figures 2 and 3. Let $(0, 170]$ be the time interval of interest. The constitutive law data are similar to those used in [2]. Since butt curl deformation is quick, we consider a small time step $\Delta t = 0.1$s and an initial mesh with 1320 elements and 732 nodes.

Fig. 2. Initial mesh. Fig. 3. Mesh after 170s of casting.

Fig. 4. Isotherms and butt curl deformation after 170s of casting.

Due to the optimization on the time step, the smaller time step paremeter used in the real simulation corresponds to $\Delta t = 7.8125 \times 10^{-4}$s. In Figure 4 we present the isotherms and the butt curl deformation after 170s.

The cpu-time was 7 hours and 28 minutes to solve the casting process using Newton's algorithm while the fixed point algorithm needed approximately 259 hours. Furthermore, the number of iterations is considerably reduced using the Newton's algorithm, since we obtained 10 iterations in contrast with 13623 iterations with the fixed point algorithm.

---

[1] The numerical solution was computed on a PC with Intel Pentium IV 3.00GHz processor running on LINUX.

References

1. Barral, P., Moreno, C., Quintela, P., Sánchez, M.T.: A numerical algorithm for a Signorini problem associated with Maxwell-Norton materials by using generalized Newton's methods. Comput. Methods Appl. Mech. Engrg., **195**, 880–904 (2006)
2. Barral, P., Quintela, P.: A numerical algorithm for prediction of thermomechanical deformation during the casting of aluminium alloy ingots. Finite Elem. Anal. Des., **34**, 125–143 (2000)
3. Barral, P., Quintela, P.: Asymptotic justification of the treatment of a metallostatic pressure type boundary condition in an aluminium casting. Math. Mod. Meth. Appl. Sci., **11**, 951–977 (2001)
4. Barral, P., Quintela, P.: Existence of solution for a contact problem of Signorini type in Maxwell-Norton materials. IMA J. Appl. Math., **67**, 525–549 (2002)
5. Bermúdez, A., Otero, M.V.: Numerical solution of a three-dimensional solidification problem in aluminium casting. Finite Elem. Anal. Des., **40**, 1885–1906 (2004)
6. Bertsekas, D.P.: Nonlinear Programming. Athena Scientific, Belmont (1995)
7. Ito, K., Kunisch, K.: Augmented Lagrangian methods for nonsmooth, convex optimization in Hilbert spaces. Nonlinear Anal., **41**, 591–616 (2000)

An Efficient Solution Algorithm for Elastoplasticity and its First Implementation Towards Uniform h- and p- Mesh Refinements

Johanna Kienesberger and Jan Valdman

Special Research Program SFB F013, 'Numerical and Symbolic Scientific Computing', Johannes Kepler University Linz
{johanna.kienesberger,jan.valdman}@sfb013.uni-linz.ac.at

Summary. The main subject of this paper is the detailed description of an algorithm solving elastoplastic deformations. Our concern is a one time-step problem, for which the minimization of a convex but non-smooth functional is required. We propose a minimization algorithm based on the reduction of the functional to a quadratic functional in the displacement and the plastic strain increment omitting a certain nonlinear dependency. The algorithm also allows for an easy extension to higher order finite elements. A numerical example in 2D reports on first results for uniform h- and p- mesh refinements.

1 Introduction

We consider the quasi-static initial-boundary value problem for small strain elastoplasticity with a linear hardening constitutive law, which can be abstractly formulated as a time-dependent variational inequality for unknown displacements and plastic strains fields. The question of the existence and uniqueness of the solution has been positively answered in [4] under the presence of hardening. It has been showed that a time-dependent variational inequality can be sufficiently approximated by a sequence of variational inequalities in given discrete times. Each of these variational inequalities contains a dissipative term coming from the plastic part of the model and represents an inequality of the second type according to [3]. Furthermore, the solution of each of these inequalities can be alternatively obtained as the minimizer of a certain convex energy functional, which is a functional depending on the unknown displacement smoothly and on the unknown plastic strain non-smoothly. The energy functional possesses a unique solution due to its strict convexity and coercivity.

Our main task here is the description of a new effective algorithm for finding such a solution. In addition to [6], where the basic parts of the algorithm have been explained, we concentrate on providing a more detailed description allowing straightforward implementation. The algorithm is based on the reduction of the functional to a quadratic functional in the displacement and the plastic strain omitting a certain nonlinear dependency. This can be understood as a linearization of the nonlinear elastoplastic problem. Then, the displacement field satisfies the linear Schur complement system after the elimination of plastic strains. The solution of this linear system can be efficiently computed by a multi-grid preconditioned conjugate gradient solver, whose convergence is already guaranteed [5].

The structure of the algorithm also allows for a direct generalization for higher degree polynomial finite elements. This is demonstrated in the numerical example, where the calculation for meshes of different sizes (h- uniform method) and polynomial degrees (p- uniform method) are presented.

2 The Model of Elastoplasticity

The elastoplastic body is assumed to occupy a bounded domain $\Omega \subset \mathbb{R}^d$ with a Lipschitz boundary $\Gamma = \partial \Omega$, where is the space-dimension. The local behavior is driven by the system of equalities and inequalities, see [4]:

$$\mathrm{div}\sigma + \mathrm{b} = 0 \tag{1}$$

$$\sigma = \sigma^T \tag{2}$$

$$\varepsilon(u) = \frac{1}{2}(\nabla u + (\nabla u)^T) \tag{3}$$

$$\sigma = \mathbb{C}(\varepsilon(u) - p) \tag{4}$$

$$\varphi(\sigma, \alpha) < \infty \tag{5}$$

$$\dot{p} : (\tau - \sigma) - \dot{\alpha}(\beta - \alpha) \leq \varphi(\tau, \beta) - \varphi(\sigma, \alpha) \tag{6}$$

Equation (1) describes the equilibrium of the stress tensor σ and outer volume body force b, equation (2) states the stress tensor's symmetry. The linearized elastic strain ε is defined in equation (3), whereas equation (4) represents the additive decomposition of the strain into its elastic part ε and its plastic part p. It also states the linear relation between the strain and the stress given by the elasticity tensor $\mathbb{C}$ which is defined for isotropic continua as

$$\mathbb{C}e = 2\mu e + \lambda(\mathrm{tr}e)\mathrm{i}, \tag{7}$$

where μ and λ are the Lamé coefficients, i is the identity matrix in $\mathbb{R}^{d \times d}$ and e a strain tensor. The trace operator $\mathrm{tr} : \mathbb{R}^{d \times d} \to \mathbb{R}^d$ of a matrix e is given by $\mathrm{tr}e := \sum_{i=1}^d e_{ii}$. The set of admissible stresses σ is steered by the the dissipational functional φ of equation (5), where the hardening parameter $\alpha \in \mathbb{R}_+$ represent memory (hysteresis) effects throughout the plastic deformations. The time development (the time derivative is denoted by $\dot{p} = \frac{\partial p}{\partial t}$) is given by the Prandtl-Reuσ normality law in equation (6). The scalar product of matrices $A, B \in \mathbb{R}^{d \times d}$ is defined such that $A : B = \sum_{i,j=1}^{d} A_{ij} B_{ij}$. Consequently, the induced matrix norm is the Frobenius norm $|A| := (\sum_{i,j=1}^{d} A_{ij}^2)^{\frac{1}{2}}$. For the local model above, the global initial value problem reads, see [4]:

Problem 1 (Variational formulation) *Let $b \in W^{1,2}(0, T; L^2(\Omega, \mathbb{R}^d))$ with $b(0) = 0$ be a given volume force. Find the displacement $u \in W^{1,2}(0, T; H_0^1(\Omega)^d)$, the plastic strain $p \in W^{1,2}(0, T; L^2(\Omega, \mathbb{R}^{d \times d}))$ such that $p(0) = 0$, the hardening parameter $\alpha \in W^{1,2}(0, T; L^2(\Omega, \mathbb{R}^m))$ such that $\alpha(0) = 0$, and the stress $\sigma \in W^{1,2}(0, T; L^2(\Omega, \mathbb{R}^{d \times d}))$ such that $\sigma(0) = 0$, and such that the system (1)–(6) is satisfied in a weak sense.*

It has been shown in [4] that Problem 1 can be reformulated as a single time-dependent variational inequality which possesses a unique solution under the presence of the positive hardening H given later. Using an implicit Euler time-discretization with an uniformly chosen Δt, we obtain a sequence of one time-step variational inequalities. The solution of each of these inequalities satisfies a minimization problem, which is obtained using the dual functional φ^* calculated by the Legendre-Fenchel transformation $\varphi^*(y) := \sup_x \{y : x - \varphi(x)\}$.

Problem 2 (One time step) *Find the minimizer $(u, p, \alpha) \in H_0^1(\Omega)^d \times L^2(\Omega, \mathbb{R}_{sym}^{d \times d}) \times L^2(\Omega, \mathbb{R}^m)$ of*

$$f(u, p, \alpha) := \frac{1}{2} \int_\Omega \mathbb{C}[\varepsilon(u) - p] : (\varepsilon(u) - p) dx + \frac{1}{2} \int_\Omega |\alpha|^2 dx$$
$$+ \Delta t \int_\Omega \varphi^*(\frac{p - p_0}{\Delta t}, \frac{\alpha_0 - \alpha}{\Delta t}) dx - \int_\Omega b\, u\, dx. \tag{8}$$

In comparison with Problem 1, σ has been replaced by $\mathbb{C}(\varepsilon(u) - p)$ and therefore it is no longer an unknown. $\mathbb{R}_{sym}^{d \times d}$ denotes real, symmetric $d \times d$ matrices. The values α_0 and p_0 are given from the previous time step t_0. The dissipational functional φ, its dual functional φ^*, as well as the hardening with parameter dimension m are specific for each hardening law such as isotropic hardening, kinematic hardening, its combination and the perfect plasticity as the limit case. For deriving an algorithm, the dual functional is calculated explicitely. There is $m = 1$ and the local minimization with respect to the hardening parameter α yields $\alpha = \alpha_0 + \sigma_y H |p - p_0|$ in the case of isotropic hardening. Problem 2 reduces to

Problem 3 (Isotropic hardening) *Find the minimizer $(u, p) \in H_0^1(\Omega)^d \times L^2(\Omega, \mathbb{R}_{sym}^{d \times d})$ of*

$$f(u, p) := \frac{1}{2} \int_\Omega \mathbb{C}[\varepsilon(u) - p] : (\varepsilon(u) - p) dx + \frac{1}{2} \int_\Omega (\alpha_0 + \sigma_y H |p - p_0|)^2 dx$$
$$+ \int_\Omega \sigma_y |p - p_0| dx - \int_\Omega b\, u\, dx \tag{9}$$

under the local constraint $\mathrm{tr}\,(p - p_0) = 0$.

$\sigma_y > 0$ is the initial yield stress and $H > 0$ the modulus of hardening. dev denotes the matrix deviator defined by $\text{dev} A := A - \frac{1}{d} \text{tr}(A) \cdot i$.

3 The algorithm

The solution algorithm is derived for Problem 3, i.e., for the isotropic hardening only. Modification to the kinematic hardening case is straightforward.

The objective functional in (9) contains the matrix norm term $|p - p_0|$, which is non-differentiable in the origin. Thus standard methods, e.g. Newton's method, do not apply. A remedy is the following regularization:

$$|\cdot|_\epsilon := \begin{cases} |\cdot| & \text{if } |\cdot| \geq \epsilon, \\ \frac{1}{2\epsilon} |\cdot|^2 + \frac{\epsilon}{2} & \text{if } |\cdot| < \epsilon, \end{cases} \qquad (10)$$

for some positive small ϵ. By this regularization we replace the original non-smooth objective $f(u,p)$ in (9) by an already smooth objective denoted as $\bar{f}(u,p)$. Thus, by introducing the plastic strain increment $\widetilde{p} = p - p_0$, the modified problem writes:

Problem 4 (Isotropic hardening regularized) *Find the minimizer* $(u, \widetilde{p}) \in H_0^1(\Omega)^d \times L^2(\Omega, \mathbb{R}^{d \times d}_{sym})$ *of*

$$\bar{f}(u, \widetilde{p}) := \frac{1}{2} \int_\Omega \mathbb{C}[\varepsilon(u) - \widetilde{p} - p_0] : (\varepsilon(u) - \widetilde{p} - p_0) \, dx - \int_\Omega b\, u \, dx \qquad (11)$$
$$+ \frac{1}{2} \int_\Omega \alpha_0^2 \, dx + \frac{1}{2} \int_\Omega \sigma_y^2 H^2 |\widetilde{p}|^2 \, dx + \int_\Omega \sigma_y (1 + \alpha_0 H) |\widetilde{p}|_\epsilon \, dx$$

under the constraint $\text{tr}\widetilde{p} = 0$.

The spatial discretization is carried out by the standard finite element method using finite elements of a fixed polynomial degree. For computational reasons the symmetric matrices, e.g. $\widetilde{p}$, are transformed into vectors $\widetilde{p} = (\widetilde{p}_{11}, \widetilde{p}_{22}, \widetilde{p}_{12})^T$ (in 2D) or $\widetilde{p} = (\widetilde{p}_{11}, \widetilde{p}_{22}, \widetilde{p}_{33}, \widetilde{p}_{12}, \widetilde{p}_{13}, \widetilde{p}_{23})^T$ (in 3D). The objective $\bar{f}(u, \widetilde{p})$ can now be discretized using the matrix and vector notation:

$$\frac{1}{2}(Bu - \widetilde{p})^T \mathbb{C}(Bu - \widetilde{p}) + \frac{1}{2}\widetilde{p}^T \mathbb{H}(|\widetilde{p}|_\epsilon)\widetilde{p} + (-B^T \mathbb{C} p_0 - b)^T u \longrightarrow \min! \qquad (12)$$

under the constraint $\text{tr } \widetilde{p} = 0$. Here, Bu denotes the discretized strain $\varepsilon(u)$. $\mathbb{H}$ is the Hessian of the discretized objective with respect to $\widetilde{p}$, it depends on $|\widetilde{p}|_\epsilon$ only and is computed as

$$\mathbb{H}(|\widetilde{p}|_\epsilon) = \left(\sigma_y^2 H^2 + \frac{2\sigma_y(1 + \alpha_0 H)}{|\widetilde{p}|_\epsilon} \right) \mathbf{N}, \qquad (13)$$

where the matrix $\mathbf{N}$ is defined, so that it holds $|p| = (p^T \mathbf{N} p)^{\frac{1}{2}}$. Thus

$$\text{2D: } \mathbf{N} = \begin{pmatrix} 2 & 1 & 0 \\ 1 & 2 & 0 \\ 0 & 0 & 2 \end{pmatrix}, \quad \text{3D: } \mathbf{N} = \begin{pmatrix} 1 & 0 & 0 \\ 0 & 1 & 0 \\ 0 & 0 & 1 \end{pmatrix} \oplus \begin{pmatrix} 2 & 0 & 0 \\ 0 & 2 & 0 \\ 0 & 0 & 2 \end{pmatrix},$$

where the symbol $\oplus$ denotes the direct sum of two matrices, so $\mathbf{N}$ is a 6×6 matrix in 3D. The trace-free plastic strain constraint is explicitely satisfied as follows: In 3D, it must hold that $\widetilde{p}_{33} = -\widetilde{p}_{11} - \widetilde{p}_{22}$. This linear condition is easily realized by projecting the arbitrary $\widetilde{p}$ onto a hyperplane, where the constraint $\mathrm{tr}\widetilde{p} = 0$ is satisfied. The projection matrix is denoted by P, the result of projection is then

$$\widetilde{p} = P\bar{p}. \tag{14}$$

In 2D plane strain model, which is of our interest, the "third" dimension components of the elastic strain are zeros, i.e., $\varepsilon_{13} = \varepsilon_{23} = \varepsilon_{33} = 0$. However, the plastic strain increment $\widetilde{p}$ as well as the stress σ have non-zero components

$$\widetilde{p}_{33} = -\widetilde{p}_{11} - \widetilde{p}_{22} \quad \text{and} \quad \sigma_{33} = \frac{\lambda}{2(\lambda+\mu)}(\sigma^e_{11} + \sigma^e_{22}) + 2\mu(\widetilde{p}_{11} + \widetilde{p}_{22}),$$

where $\sigma^e := \mathbb{C}Bu$ is the elastic part of the stress tensor. Therefore, $\widetilde{p}_{11}$ and $\widetilde{p}_{22}$ are arbitrary and no special projection as in 3D is required, i.e., $P = I$. The dependence of $\mathbb{H}$ on $|p|_\epsilon$ in (13) is "frozen" and the nonlinear functional (12) becomes a quadratic one. Its minimizer must fulfill the necessary condition

$$\begin{pmatrix} B^T\mathbb{C}B & -B^T\mathbb{C}P \\ -P^T\mathbb{C}B & P^T(\mathbb{C}+\mathbb{H})P \end{pmatrix} \begin{pmatrix} u \\ \bar{p} \end{pmatrix} + \begin{pmatrix} -b - B^T\mathbb{C}p_0 \\ P^T\mathbb{C}p_0 \end{pmatrix} = 0. \tag{15}$$

By eliminating $\bar{p}$ in (15) we obtain the Schur-Complement system in u:

$$B^T(\mathbb{C} - \mathbb{C}P(P^T(\mathbb{C}+\mathbb{H})P)^{-1}P^T\mathbb{C})Bu =$$
$$-b - B^T(\mathbb{C} + \mathbb{C}P(P^T(\mathbb{C}+\mathbb{H})P)^{-1}P^T\mathbb{C})p_0. \tag{16}$$

This linear system of equations for the vector of displacements u is solved by the conjugate gradient method with a geometrical multigrid preconditioner [2]. The spectral equivalence of the Schur-Complement matrix and the corresponding Schur-Complement matrix for the pure elasticity has been proved in [5]. Thus the convergence of the conjugate gradient method with a geometrical multigrid preconditioner is guaranteed. From the displacements u, the plastic strain increment $\widetilde{p}$ is calculated by the local minimization of (12). In the unregularized case ($\epsilon = 0$), the analytical solution from [1] states

$$\widetilde{p} = \frac{(\|\mathrm{dev}A\| - a)_+}{2\mu + \sigma_y^2 H^2} \frac{\mathrm{dev}A}{\|\mathrm{dev}A\|} \tag{17}$$

with the quantities $A = \mathbb{C}(Bu - p_0)$ and $a = \sigma_y(1 + \alpha_0 H)$. The operator $(\cdot)_+$ is defined by $(\cdot)_+ = \max(0, \cdot)$. In the regularized case ($\epsilon > 0$), $\widetilde{p}$ is solved by a local Newton's method, where the analytical solution (17) is used as an initial approximation.

An important practical aspect is the use of higher order finite elements, whose implementation may lead to an exponential convergence, see [7]. Since the trace-free condition of the plastic strain increment $\widetilde{p}$ is satisfied explicitly (cf. (14)), the polynomial ansatz functions for $\widetilde{p}$ can be taken one degree lower than the polynomial ansatz for the displacement without the loss of stability (inf − sup like condition). The convergence of the discrete solutions was proved for linear ansatz functions for u and piecewise constant ansatz functions for $\widetilde{p}$ in [1]. For our implementation, the displacement values u on an element are computed and stored in all integration points belonging to a given Gauss rule providing exact numerical integration of (16). The same integration points are taken to determine the plastic strain increment values by (17) and to reconstruct the finite element "shape" from them. This approach works nicely for linear ansatz functions for u, where only one Gauss point (center of a triangle) is required for the exact integration of the local matrices in (16). Then, the same point is used for computation of the locally constant $\widetilde{p}$. For the quadratic ansatz function for u, there are three Gauss integration points (centers of triangle edges) required per triangle and three basis functions corresponding to a locally linear $\widetilde{p}$. Nevertheless, in higher polynomial degree cases, the number of Gauss integration points for the exact integration of matrices in (16) is higher than the number of the finite element basis for one degree lower polynomial and thus a certain projection is applied. Summarizing our algorithm, the solution of Problem 4 is determined by

Algorithm 1 *Given initial displacement approximation u.*

1. *Calculate $\widetilde{p}$ locally using Newton's method with the initial approximation $\widetilde{p} = \frac{(||\mathrm{dev}A||-a)_+}{2\mu+\sigma_y^2 H^2} \frac{\mathrm{dev}A}{||\mathrm{dev}A||}$, where $A = \mathbb{C}(B(u) - p_0)$ and $a = \sigma_y(1 + \alpha_0 H)$.*
2. *Substitute $\widetilde{p}$ into $\mathbb{H}$ in (13) and assemble the global Schur-complement system (16).*
3. *Solve u from the global linear system (16) using multigrid PCG method.*
4. *Repeat steps (1)-(3) until convergence is reached.*
5. *Output displacement u and plastic strain increment $\widetilde{p}$.*

4 Numerical experiments

Algorithm 1 was implemented in the finite element solver NGSolve which is an extension package of the mesh generator Netgen [8] developed in our group. The testing geometry considered is the unit square depicted in Figure 1. The left edge is fixed in both x and y directions and the right edge is subjected to an outward acting force. The material parameters are $\lambda = 1000$ and $\mu = 1000$ as the Lamé parameters, the modulus of hardening is $H = 100$, and the initial yield stress is $\sigma_y = 6$. Several tests have been performed with different mesh

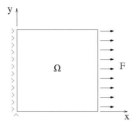

Fig. 1. Testing geometry

sizes and different orders of the polynomial ansatz functions for the strains and the stresses. Figure 2 shows the von-Mises stresses for uniformly refined h- and p- meshes. The columns represent results with different polynomial ansatz functions (p- method) and a fixed mesh. The rows show result for the same polynomial ansatz functions and different meshes (h- method). The stresses are scaled so that small values are darker shaded, the largest values are white colored and for most of the figures they also correspond to plasticity zones.

In Picture (a) both the mesh size and the order are chosen too coarse, the resulting stress is not reasonable from a physical point of view. From the Pictures (c), (f), and (i) it is obvious that for resolving singularities as they occur in the left corners, a finer local mesh-refinement (h- adaptive method) would be helpful. The great potential of higher order functions is approximating smooth functions on larger sized elements effectively, as demonstrated in Picture (g). Although there are only two elements in this calculation, a ninth degree polynomial functions for the stresses already provides the continuity on the common edge of these elements.

5 Conclusions and future work

In this paper, a new algorithm for the fast solution of the elastoplastic problems with hardening and its implementation together with the generalization to the higher-degree polynomial ansatz functions is presented. The future work will concentrate on theoretical analysis explaining convergence of the solution algorithm and on the implementation of the combined hp-adaptive method.

Acknowledgment

The authors are pleased to acknowledge support by the Austrian Science Fund 'Fonds zur Förderung der wissenschaftlichen Forschung (FWF)' under grant SFB F013/F1306 in Linz, Austria.

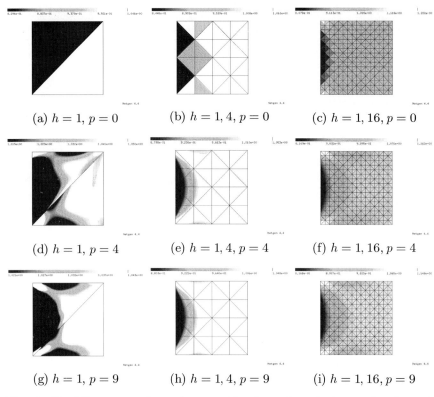

Fig. 2. Von-Mises stress for various combinations of mesh sizes and polynomial degrees of u.

References

1. Alberty, J., Carstensen, C., Zarrabi, D.: Adaptive numerical analysis in primal elastoplasticity with hardening. Computer Methods in Applied Mechanics and Engineering, **171, 3-4**, 175–204 (1999)
2. Braess, D.: Finite Elements: Theory, fast solvers, and applications in solid mechanics. Cambridge University Press (2001)
3. Glowinski, R. and Lions J. L. and Trémolières: Numerical analysis of Variational Inequalities. North-Holland, Amsterdam (1981)
4. Han, W., Reddy, B.D.: Plasticity: Mathematical theory and numerical analysis. Springer, Berlin Heidelberg New York (1999)
5. Kienesberger, J., Langer, U., Valdman, J.: On a robust multigrid-preconditioned solver for incremental plasticity problems, In: Proceedings of IMET 2004 - Iterative Methods, Preconditioning & Numerical PDEs, (2004)
6. Kienesberger, J., Valdman, J.: Multi-surface elastoplastic continuum - modelling and computations. Numerical mathematics and advanced applications, Proceedings of ENUMATH 2003, Springer, 539–548, (2004)

7. Nübel, V., Düster, A., Rank, E.: An rp-adaptive finite element method for elastoplastic problems. Submitted to Computational Mechanics (2004)
8. Schöberl, J.: NETGEN: An advancing front 2D/3D-mesh generator based on abstract rules. Comput. Vis. Sci., **1**, No. 1, 41–52 (1997)

UNBOUNDED DOMAINS

A LDG-BEM Coupling for a Class of Nonlinear Exterior Transmission Problems

Rommel Bustinza[1], Gabriel N. Gatica[2] and Francisco-Javier Sayas[3]

[1] Departamento de Ingeniería Matemática, Universidad de Concepción, Casilla 160-C, Concepción, Chile
    rbustinz@ing-mat.udec.cl
[2] Departamento de Ingeniería Matemática, Universidad de Concepción, Casilla 160-C, Concepción, Chile
    ggatica@ing-mat.udec.cl
[3] Departamento de Matemática Aplicada, Universidad de Zaragoza, Centro Politécnico Superior, María de Luna 3-50018, Zaragoza, Spain
    jsayas@unizar.es

Summary. We consider the coupling of local discontinuous Galerkin (LDG) and boundary element methods (BEM) for a class of nonlinear exterior transmission boundary value problems in the plane. We introduce suitable numerical fluxes in order to obtain the LDG formulation, and additional unknowns to couple it with the BEM formulation. Finally, we show that the rates of convergence are optimal with respect to the mesh size.

1 Introduction

The study of discontinuous Galerkin methods (see, for e.g., [1] for an overview) to deal with a large class of linear and nonlinear elliptic problems that comes from engineering and physics applications, have been increased lately. We refer, for example, to [14, 7, 8, 9], and [10], where linear problems such as the Poisson, Stokes, Maxwell and Oseen equations are treated with this approach. In addition, the study of these methods for solving a kind of nonlinear elliptic problems have been developed recently (see, e.g. [4, 5] and [13]). We remark here that the DG methods have the advantage of consider more general meshes (with hanging nodes, for e.g.), due to the fact that the inter-element continuity of the approximate solution is not strongly required. This property makes the method suitable for p and hp version, as well as for local adaptivity, subject that is still under development (see, e.g. [2, 3, 5], and [15]).

Recently, the analysis has turned to extend the applicability of this approach in combination with the boundary element method (BEM). Up to the authors' knowledge, there is only one work in this direction. In fact, in [12] a linear exterior elliptic problem is studied by coupling these both methods.

In this note, we report on the main results derived in [6] that extend the approach of [12] to the LDG-BEM coupling for a nonlinear exterior transmission problem, introducing further unknowns as well as suitable numerical fluxes. As a result, we obtain optimal rates of convergence, in the h-version context.

2 An exterior transmission problem

We begin by introducing Ω_0 as a simply connected and bounded domain in $\mathbb{R}^2$ with polygonal boundary Γ_0. Next, we let Ω_1 be an annular and simply connected domain surrounded by Γ_0 and another polygonal boundary Γ_1. Then, given $f \in L^2(\Omega_0)$, $g_0 \in H^{1/2}(\Gamma_0)$, $g_1 \in H^{1/2}(\Gamma_1)$, and $g_2 \in L^2(\Gamma_1)$, we consider the nonlinear exterior transmission problem:

$$-\operatorname{div} \mathbf{a}(\cdot, \nabla u_1) = f \quad \text{in} \quad \Omega_1, \quad u_1 = g_0 \quad \text{on} \quad \Gamma_0,$$

$$-\Delta u_2 = 0 \quad \text{in} \quad \mathbb{R}^2 \setminus (\bar{\Omega}_0 \cup \bar{\Omega}_1), \quad u_1 - u_2 = g_1 \quad \text{on} \quad \Gamma_1,$$

$$\mathbf{a}(\cdot, \nabla u_1) \cdot \boldsymbol{\nu} - \nabla u_2 \cdot \boldsymbol{\nu} = g_2 \quad \text{on} \quad \Gamma_1, \quad u_2(\boldsymbol{x}) = \mathcal{O}(1) \quad \text{as} \quad |\boldsymbol{x}| \to +\infty. \tag{1}$$

Hereafter, the nonlinear vector function $\mathbf{a}(\cdot, \cdot)$ is defined so that the corresponding operator becomes Lipschitz continuous and strongly monotone (see, e.g.[4]).

The next step here is to obtain the variational formulation of (1), applying the coupling of the LDG and BEM approaches. Then, following the ideas given in [12], we introduce another simple closed polygonal curve Γ such that its interior contains the support of f. In addition, we let Ω_2 be the open domain bounded by Γ_1 and Γ, and set $\Omega_e := \mathbb{R}^2 - (\bar{\Omega}_0 \cup \bar{\Omega}_1 \cup \bar{\Omega}_2)$ (see Figure 1). Besides, based on the behavior of u_2 at infinity, it is enough for us to seek $u_2 \in W^1(\mathbb{R}^2 \setminus \bar{\Omega}_0 \cup \bar{\Omega}_1)$ (see [11] for a definition of this Beppo–Levi space).

Then, (1) can be re-written equivalently, as the nonlinear boundary value problem:

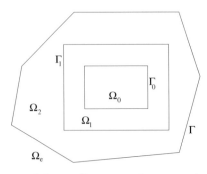

Fig. 1. Domain of the nonlinear exterior transmission problem.

$$-\operatorname{div} \mathbf{a}(\cdot, \nabla u_1) = f \quad \text{in} \quad \Omega_1, \quad u = g_0 \quad \text{on} \quad \Gamma_0, \tag{2}$$

the Laplace equation in the bounded region Ω_2:

$$-\Delta u_2 = 0 \quad \text{in} \quad \Omega_2, \quad u_1 - u_2 = g_1 \quad \text{on} \quad \Gamma_1,$$
$$\mathbf{a}(\cdot, \nabla u_1) \cdot \boldsymbol{\nu} - \nabla u_2 \cdot \boldsymbol{\nu} = g_2 \quad \text{on} \quad \Gamma_1, \tag{3}$$

and the Laplace equation in the unbounded region Ω_e:

$$-\Delta u_2 = 0 \quad \text{in} \quad \Omega_e, \quad u_2(\boldsymbol{x}) = \mathcal{O}(1) \quad \text{as} \quad |\boldsymbol{x}| \to +\infty, \tag{4}$$

coupled with the transmission conditions:

$$\lim_{\substack{\boldsymbol{x} \to \boldsymbol{x}_0 \\ \boldsymbol{x} \in \Omega_2}} u_2(\boldsymbol{x}) = \lim_{\substack{\boldsymbol{x} \to \boldsymbol{x}_0 \\ \boldsymbol{x} \in \Omega_e}} u_2(\boldsymbol{x})$$

$$\text{and} \quad \lim_{\substack{\boldsymbol{x} \to \boldsymbol{x}_0 \\ \boldsymbol{x} \in \Omega_2}} \nabla u_2(\boldsymbol{x}) \cdot \boldsymbol{\nu}(\boldsymbol{x}_0) = \lim_{\substack{\boldsymbol{x} \to \boldsymbol{x}_0 \\ \boldsymbol{x} \in \Omega_e}} \nabla u_2(\boldsymbol{x}) \cdot \boldsymbol{\nu}(\boldsymbol{x}_0) \tag{5}$$

for almost all $\boldsymbol{x}_0 \in \Gamma$, where $\boldsymbol{\nu}(\boldsymbol{x}_0)$ denotes the unit outward normal to $\boldsymbol{x}_0$.

Now, following the ideas given in [12], we introduce the gradients $\boldsymbol{\theta}_1 := \nabla u_1$ in Ω_1 and $\boldsymbol{\theta}_2 := \nabla u_2$ in Ω_2, as well as the fluxes $\boldsymbol{\sigma}_1 := \mathbf{a}(\cdot, \boldsymbol{\theta}_1)$ and $\boldsymbol{\sigma}_2 := \boldsymbol{\theta}_2$, as additional unknowns in Ω_1 and Ω_2, respectively. Next, we define the following auxiliary quantities that will later act as unknowns:

$$\lambda(\boldsymbol{x}_0) := \lim_{\substack{\boldsymbol{x} \to \boldsymbol{x}_0 \\ \boldsymbol{x} \in \Omega_2}} \nabla u_2(\boldsymbol{x}) \cdot \boldsymbol{\nu}(\boldsymbol{x}_0), \quad \gamma(\boldsymbol{x}_0) := \lim_{\substack{\boldsymbol{x} \to \boldsymbol{x}_0 \\ \boldsymbol{x} \in \Omega_e}} \nabla u_2(\boldsymbol{x}) \cdot \boldsymbol{\nu}(\boldsymbol{x}_0),$$

and

$$\varphi(\boldsymbol{x}_0) := \lim_{\substack{\boldsymbol{x} \to \boldsymbol{x}_0 \\ \boldsymbol{x} \in \Omega_e}} u_2(\boldsymbol{x}) - \kappa \quad \text{with} \quad \kappa := \frac{1}{|\Gamma|} \int_\Gamma u_2,$$

for almost all $\boldsymbol{x}_0 \in \Gamma$. In this way, (2)–(3) can be reformulated as:

$$\boldsymbol{\theta}_1 = \nabla u_1 \quad \text{in} \quad \Omega_1, \quad \boldsymbol{\sigma}_1 = \mathbf{a}(\cdot, \boldsymbol{\theta}_1) \quad \text{in} \quad \Omega_1, \quad -\operatorname{div} \boldsymbol{\sigma}_1 = f \quad \text{in} \quad \Omega_1,$$

$$\boldsymbol{\theta}_2 = \nabla u_2 \quad \text{in} \quad \Omega_2, \quad \boldsymbol{\sigma}_2 = \boldsymbol{\theta}_2 \quad \text{in} \quad \Omega_2, \quad -\operatorname{div} \boldsymbol{\sigma}_2 = 0 \quad \text{in} \quad \Omega_2,$$

$$u_1 = g_0 \quad \text{on} \quad \Gamma_0, \quad u_1 - u_2 = g_1 \quad \text{on} \quad \Gamma_1,$$

$$\boldsymbol{\sigma}_1 \cdot \boldsymbol{\nu}_1 - \boldsymbol{\sigma}_2 \cdot \boldsymbol{\nu}_1 = g_2 \quad \text{on} \quad \Gamma_1, \tag{6}$$

and the transmission conditions (5) become

$$\lim_{\substack{\boldsymbol{x}\to\boldsymbol{x}_0 \\ \boldsymbol{x}\in\Omega_e}} u_2(\boldsymbol{x}) = \varphi(\boldsymbol{x}_0) + \kappa \quad \forall\, (a.e.)\ \boldsymbol{x}_0 \in \Gamma \quad \text{and} \quad \lambda = \gamma \ \text{on} \ \Gamma. \tag{7}$$

We proceed now to obtain the discrete variational formulation that results by applying the BEM approach on the boundary Γ and the LDG method in the bounded domain $\Omega := \Omega_1 \cup \Gamma_1 \cup \Omega_2$.

3 LDG-BEM coupling

In this section, we introduce the coupling of local discontinuous Galerkin and boundary element methods for the nonlinear exterior transmission problem (6)–(7) and present their well-posedness, using the recent results in [12].

3.1 Meshes

We let $\{\mathcal{T}_{1,h}\}_{h>0}$ and $\{\mathcal{T}_{2,h}\}_{h>0}$ be families of shape-regular triangulations of $\bar{\Omega}_1$ and $\bar{\Omega}_2$ respectively, each made up of straight-side triangles K with diameter h_K and unit outward normal to ∂K given by $\boldsymbol{\nu}_K$. As usual, the index h also denotes $h := \max_{K\in\mathcal{T}_h} h_K$, where $\mathcal{T}_h := \mathcal{T}_{1,h} \cup \mathcal{T}_{2,h}$. Then, given $\mathcal{T}_h$, its edges are defined as follows. An *interior edge of* $\mathcal{T}_h$ is the (non-empty) interior of $\partial K \cap \partial K'$, where K and K' are two adjacent elements, both of them belonging to $\mathcal{T}_{1,h}$ or $\mathcal{T}_{2,h}$. An *interface edge* of $\mathcal{T}_h$ is the (non-empty) interior of $\partial K \cap \partial K'$, where K and K' are two adjacent elements belonging to different triangulations. Similarly, a *boundary edge* of $\mathcal{T}_h$ is the (non-empty) interior of $\partial K \cap \Gamma_0$ or $\partial K \cap \Gamma$, where K is a boundary element of $\mathcal{T}_h$. For each edge e, h_e represents its length. In addition, we define $\mathcal{E}(K) :=$ edges of K, $\mathcal{E}_{1,h}^{\text{int}}$: list of interior edges (counted only once) on Ω_1, $\mathcal{E}_{2,h}^{\text{int}}$: list of interior edges (counted only once) on Ω_2, $\mathcal{E}_h^{\text{int}} := \mathcal{E}_{1,h}^{\text{int}} \cup \mathcal{E}_{2,h}^{\text{int}}$, $\mathcal{E}_h^\Gamma$: list of edges on Γ, $\mathcal{E}_h^{\Gamma_0}$: list of edges on Γ_0, $\mathcal{E}_{1,h}^{\Gamma_1}$: list of edges (belonging to $\mathcal{T}_{1,h}$) on Γ_1, $\mathcal{E}_{2,h}^{\Gamma_1}$: list of edges (belonging to $\mathcal{T}_{2,h}$) on Γ_1, $\mathcal{E}_h^{\Gamma_1}$: list of interface edges, and i_h: interior grid generated by the triangulation, that is $\text{i}_h := \cup\{e\,:\ e \in \mathcal{E}_h^{\text{int}}\}$. Also, we let Γ_h^0 and Γ_h be the partition of Γ_0 and Γ, respectively, inherited by $\mathcal{T}_h$. In addition, we also assume that $\mathcal{T}_h$ is of *bounded variation*, which means that there exists $l > 1$, independent of the meshsize h, such that $l^{-1} \leq \frac{h_K}{h_{K'}} \leq l$ for each pair $K, K' \in \mathcal{T}_h$ sharing an interior/interface edge.

3.2 Averages and jumps

Next, we define average and jump operators. To this end, let K and K' be two adjacent elements of $\mathcal{T}_h$ and $\boldsymbol{x}$ be an arbitrary point on the interior edge $e = \partial K \cap \partial K' \subset \text{i}_h$. In addition, let q, and $\boldsymbol{v}$ be scalar-, and vector-valued functions, respectively, that are smooth inside each element $K \in \mathcal{T}_h$. We denote by $(q_{K,e}, \boldsymbol{v}_{K,e})$ the restriction of $(q_K, \boldsymbol{v}_K)$ to e. Then, we define the averages at $\boldsymbol{x} \in e$ by:

$$\{q\} := \frac{1}{2}\Big(q_{K,e} + q_{K',e}\Big), \qquad \{\boldsymbol{v}\} := \frac{1}{2}\Big(\boldsymbol{v}_{K,e} + \boldsymbol{v}_{K',e}\Big).$$

Similarly, the jumps at $\boldsymbol{x} \in e$ are given by

$$[\![q]\!] := q_{K,e}\boldsymbol{\nu}_K + q_{K',e}\boldsymbol{\nu}_{K'}, \qquad [\![\boldsymbol{v}]\!] := \boldsymbol{v}_{K,e} \cdot \boldsymbol{\nu}_K + \boldsymbol{v}_{K',e} \cdot \boldsymbol{\nu}_{K'}.$$

On boundary/interface edges e, we set $\{q\} = q$, $\{\boldsymbol{v}\} := \boldsymbol{v}$, as well as $[\![q]\!] := q\boldsymbol{\nu}$, and $[\![\boldsymbol{v}]\!] := \boldsymbol{v} \cdot \boldsymbol{\nu}$.

3.3 The LDG-BEM formulation

Given a mesh $\mathcal{T}_h$, we proceed as in [12] and introduce two new partitions $\Gamma_{\widetilde{h}}$ and $\Gamma_{\bar{h}}$ of Γ, independent of the partition Γ_h inherited from $\mathcal{T}_h$, with edges $\widetilde{e}$ and $\bar{e}$, respectively. Next, denoting by $\mathcal{E}_{\widetilde{h}}^{\Gamma}$ and $\mathcal{E}_{\bar{h}}^{\Gamma}$ the corresponding list of edges, and considering $\widetilde{k}, \bar{k} \in \mathbb{N}$, we introduce the following discrete spaces:

$$X_{\widetilde{h}} := \{\xi_{\widetilde{h}} \in L^2(\Gamma) : \ \xi_{\widetilde{h}}|_{\widetilde{e}} \in P_{\widetilde{k}}(\widetilde{e}) \ \ \forall \widetilde{e} \in \mathcal{E}_{\widetilde{h}}^{\Gamma}\},$$

$$Y_{\bar{h}} := \{\psi_{\bar{h}} \in \mathcal{C}(\Gamma) : \ \psi_{\bar{h}}|_{\bar{e}} \in P_{\bar{k}}(\bar{e}) \ \ \forall \bar{e} \in \mathcal{E}_{\bar{h}}^{\Gamma}\}, \qquad (8)$$

$$Z_{\bar{h}} := \{\mu_{\bar{h}} \in L^2(\Gamma) : \ \mu_{\bar{h}}|_{\bar{e}} \in P_{\bar{k}-1}(\bar{e}) \ \ \forall \bar{e} \in \mathcal{E}_{\bar{h}}^{\Gamma}\}.$$

From here on, the superscript 0 on any of these spaces will denote their restrictions to that functions having zero integral over Γ. Moreover, by $P_k(K)$ we denote the space of polynomials of total degree at most k on K.

Then, we wish to approximate the solution of (6)–(7) by $(\boldsymbol{\theta}_h, \boldsymbol{\sigma}_h, u_h, \varphi_{\bar{h}}, \lambda_{\bar{h}}, \gamma_{\widetilde{h}}) \in \boldsymbol{\Sigma}_h \times \boldsymbol{\Sigma}_h \times V_h \times Y_{\bar{h}}^0 \times X_{\widetilde{h}}^0 \times Z_{\bar{h}}^0$, with $\boldsymbol{\theta}_h := (\boldsymbol{\theta}_{1,h}, \boldsymbol{\theta}_{2,h}), \boldsymbol{\sigma}_h := (\boldsymbol{\sigma}_{1,h}, \boldsymbol{\sigma}_{2,h}) \in \boldsymbol{\Sigma}_h := \boldsymbol{\Sigma}_{1,h} \times \boldsymbol{\Sigma}_{2,h}$, and $u_h := (u_{1,h}, u_{2,h}) \in V_h := V_{1,h} \times V_{2,h}$, where for any $i \in \{1, 2\}$, we define

$$\boldsymbol{\Sigma}_{i,h} := \Big\{\boldsymbol{\tau} \in \boldsymbol{L}^2(\Omega_i) : \ \boldsymbol{\tau}\big|_K \in \boldsymbol{P}_r(K) \ \ \forall K \in \mathcal{T}_{i,h}\Big\}$$

$$V_{i,h} := \Big\{v \in L^2(\Omega_i) : \ v\big|_K \in P_m(K) \ \ \forall K \in \mathcal{T}_{i,h}\Big\},$$

with $m \in \mathbb{N}$ and $r = m$ or $r = m - 1$.

Finally, defining the so-called numerical fluxes as in [6], and coupling the LDG method with the BEM approach, we arise to the global discrete LDG-BEM formulation: *Find* $(\boldsymbol{\theta}_h, \boldsymbol{\sigma}_h, u_h, \varphi_{\bar{h}}, \lambda_{\widetilde{h}}, \gamma_{\bar{h}}) \in \boldsymbol{\Sigma}_h \times \boldsymbol{\Sigma}_h \times V_h \times Y_{\bar{h}}^0 \times X_{\widetilde{h}}^0 \times Z_{\bar{h}}^0$ *such that*

$$\begin{aligned}
\int_\Omega \bar{\mathbf{a}}(\cdot, \boldsymbol{\theta}_h) \cdot \boldsymbol{\zeta} - \int_\Omega \boldsymbol{\sigma}_h \cdot \boldsymbol{\zeta} &= 0 & \forall \, \boldsymbol{\zeta} \in \Sigma_h, \\
\int_\Omega \boldsymbol{\theta}_h \cdot \boldsymbol{\tau} - \left\{ \int_\Omega \nabla_h u_h \cdot \boldsymbol{\tau} - S(u_h, \boldsymbol{\tau}) \right\} &= G_h(\boldsymbol{\tau}) \; \forall \, \boldsymbol{\tau} \in \Sigma_h, \\
\left\{ \int_\Omega \nabla_h v \cdot \boldsymbol{\sigma}_h - S(v, \boldsymbol{\sigma}_h) \right\} + \boldsymbol{\alpha}(u_h, v) - \langle \lambda_{\tilde{h}}, v \rangle &= F_h(v) \; \forall \, v \in V_h, \\
\langle \xi_{\tilde{h}}, u_h \rangle - \langle \xi_{\tilde{h}}, \varphi_{\tilde{h}} \rangle &= 0 & \forall \, \xi_{\tilde{h}} \in X_{\tilde{h}}^0, \\
\langle \lambda_{\tilde{h}}, \psi_{\tilde{h}} \rangle + \langle W \varphi_{\tilde{h}}, \psi_{\tilde{h}} \rangle - \langle (\tfrac{1}{2} I - K') \gamma_{\tilde{h}}, \psi_{\tilde{h}} \rangle &= 0 & \forall \, \psi_{\tilde{h}} \in Y_{\tilde{h}}^0, \\
\langle \mu_{\tilde{h}}, (\tfrac{1}{2} I - K) \varphi_{\tilde{h}} \rangle + \langle \mu_{\tilde{h}}, V \gamma_{\tilde{h}} \rangle &= 0 & \forall \, \mu_{\tilde{h}} \in Z_{\tilde{h}}^0,
\end{aligned} \qquad (9)$$

where $\bar{\mathbf{a}}(\cdot, \boldsymbol{\theta}_h) := \begin{cases} \mathbf{a}(\cdot, \boldsymbol{\theta}_{1,h}) & \text{in } \Omega_1 \\ \boldsymbol{\theta}_{2,h} & \text{in } \Omega_2 \end{cases}$, ∇_h denotes the piecewise gradient, and $S : H^1(\mathcal{T}_h) \times \boldsymbol{L}^2(\Omega) \to \mathbb{R}$ and $\boldsymbol{\alpha} : H^1(\mathcal{T}_h) \times H^1(\mathcal{T}_h) \to \mathbb{R}$, stand for the bilinear forms given by:

$$S(w, \boldsymbol{\tau}) := \int_{I_h} [\![w]\!] \cdot (\{\boldsymbol{\tau}\} - [\![\boldsymbol{\tau}]\!] \boldsymbol{\beta}) + \int_{\Gamma_0} w \, (\boldsymbol{\tau}_1 \cdot \boldsymbol{\nu}_1) + \int_{\Gamma_1} (w_1 - w_2) \boldsymbol{\tau}_1 \cdot \boldsymbol{\nu}_1,$$

$$\boldsymbol{\alpha}(w, v) := \int_{I_h} \alpha \, [\![w]\!] \cdot [\![v]\!] + \int_{\Gamma_0} \alpha \, w \, v + \int_{\Gamma_1} \alpha \, (w_1 - w_2)(v_1 - v_2),$$

for all $w := (w_1, w_2), v := (v_1, v_2) \in H^1(\mathcal{T}_h) := H^1(\mathcal{T}_{1,h}) \times H^1(\mathcal{T}_{2,h})$ and $\boldsymbol{\tau} := (\boldsymbol{\tau}_1, \boldsymbol{\tau}_2) \in \boldsymbol{L}^2(\Omega) := \boldsymbol{L}^2(\Omega_1) \times \boldsymbol{L}^2(\Omega_2)$.

The linear operators $G_h : \boldsymbol{L}^2(\Omega) \to \mathbb{R}$ and $F_h : H^1(\mathcal{T}_h) \to \mathbb{R}$, which are defined by

$$G_h(\boldsymbol{\tau}) := \int_{\Gamma_0} g_0 \, \boldsymbol{\tau}_1 \cdot \boldsymbol{\nu}_1 + \int_{\Gamma_1} g_1 \, \boldsymbol{\tau}_1 \cdot \boldsymbol{\nu}_1,$$

$$F_h(v) := \int_{\Omega_1} f \, v_1 + \int_{\Gamma_0} \alpha \, g_0 \, v_1 + \int_{\Gamma_1} \alpha \, g_1 \, (v_1 - v_2) + \int_{\Gamma_1} g_2 \, v_2.$$

for all $\boldsymbol{\tau} \in \boldsymbol{L}^2(\Omega)$ and $v \in H^1(\mathcal{T}_h)$. In addition, by V, K, K', and W, we denote the boundary integral operators associated to the single, double, adjoint of the double, and hypersingular layer potentials, respectively, while I denotes the identity operator.

The stabilization parameters α and β are chosen so that the solvability of the discrete LDG-BEM formulation is guaranteed. Therefore, we require that $\alpha \in P_0(I_h \cup \Gamma_h^0 \cup \mathcal{E}_h^{\Gamma_1})$ and $\boldsymbol{\beta} \in \boldsymbol{P}_0(I_h)$. Indeed, $\boldsymbol{\beta}$ can be chosen as the null vector.

3.4 Well-posedness

Well-posedness of the discrete formulation system (9) is established in [6], by endowing V_h with the norm

$$\|v\|_h^2 := \|\nabla_h v\|_{[L^2(\Omega)]^2}^2 + |v|_h^2 \quad \forall v \in H^1(\mathcal{T}_h),$$

where

$$|v|_h^2 := \|\alpha^{1/2}[\![v]\!]\|_{[L^2(I_h^1 \cup I_h^2)]^2}^2 + \|v\|_{L^2(\Gamma_0)}^2 + \|\alpha^{1/2}(v_1 - v_2)\|_{L^2(\Gamma_1)}^2,$$

for all $v \in H^1(\mathcal{T}_h)$.

We also consider the standard L^2–norm for $\boldsymbol{\Sigma_h}$. In addition, as in [12], we introduce the seminorm:

$$|\xi|_{\tilde{h}} := \sup_{0 \neq \psi_{\tilde{h}} \in Y_{\tilde{h}}} \frac{\langle \xi, \psi_{\tilde{h}} \rangle}{\|\psi_{\tilde{h}}\|_{1/2,\Gamma}} \quad \forall \xi \in H^{-1/2}(\Gamma),$$

which becomes a norm, equivalent to $\|\cdot\|_{-1/2,\Gamma}$ on $X_{\tilde{h}}$, under additional requirements on the independent mesh $\Gamma_{\tilde{h}}$ (see [12] for details). Our strategy is to prove the unique solvability of a suitable reduced formulation, equivalent with (9), which is obtain by introducing some lifting operators (see [6] for further details). Therefore, assuming enough regularity on the exact soution, we can state our main result in the following theorem, whose proof can be found in [6].

Theorem 1. *The coupled LDG-BEM scheme* (9) *is uniquely solvable. In addition, there exists $C > 0$, independent of the meshsizes, such that*

$$\|u - u_h\|_h + |\lambda - \lambda_{\tilde{h}}|_{\tilde{h}} + \|\boldsymbol{\theta} - \boldsymbol{\theta}_h\|_{0,\Omega} + \|\boldsymbol{\sigma} - \boldsymbol{\sigma}_h\|_{0,\Omega} + \|\varphi - \varphi_{\tilde{h}}\|_{1/2,\Gamma}$$

$$+ \|\gamma - \gamma_{\tilde{h}}\|_{-1/2,\Gamma} \leq C \left\{ \left(h^{\min\{\delta_2, m\}} + \bar{h}^{\min\{\delta_2, \bar{k}+1/2\}} \right) \|u_2\|_{1+\delta_2,\Omega_2} \right.$$

$$\left. + \tilde{h}^{\min\{1, \tilde{k}\}} \|\lambda\|_{3/2,\Gamma} + h^{\min\{\delta_1, m\}} \|u_1\|_{1+\delta_1,\Omega_1} + h^{\min\{\delta_1, r+1\}} \|\boldsymbol{\sigma}_1\|_{\delta_1,\Omega_1} \right\}.$$

(10)

Acknowledgements

This work has been partially supported by CONICIYT-Chile through FONDECYT Projects No. 1050842 and 7050209, and the FONDAP Program in Applied Mathematics, by the Dirección de Investigación of the Universidad de Concepción, and by Spanish FEDER/MCYT Project MTM2004-019051 and by a grant of Programa Europa XXI (Gobierno Aragón + CAI).

References

1. Arnold, D.N., Brezzi, F., Cockburn, B., Marini, L.D.: Unified analysis of discontinuous Galerkin methods for elliptic problems. SIAM Journal on Numerical Analysis, **39**, 5, 1749-1779 (2001)
2. Becker, R., Hansbo, P., Larson, M.G.: Energy norm a posteriori error estimation for discontinuous Galerkin methods. Computer Methods in Applied Mechanics and Engineering, **192**, 723-733 (2003)
3. Bustinza, R., Cockburn, B., Gatica, G. N.: An a-posteriori error estimate for the local discontinuous Galerkin method applied to linear and nonlinear diffusion problems. Journal of Scientific Computing, **22**, 1, 147-185 (2005)
4. Bustinza, R., Gatica, G. N.: A local discontinuous Galerkin method for nonlinear diffusion problems with mixed boundary conditions. SIAM Journal on Scientific Computing, **26**, 1, 152-177 (2004)
5. Bustinza, R., Gatica, G. N.: A mixed local discontinuous Galerkin method for a class of nonlinear problems in fluid mechanics. Journal of Computational Physics, **207**, 427-456 (2005)
6. Bustinza, R., Gatica, G. N., Sayas, F. J.: On the coupling of local discontinuous Galerkin and boundary element methods for nonlinear exterior transmission problems. Preprint 06-09, Departamento de Ingeniería Matemática, Universidad de Concepción, (2006)
7. Castillo, P., Cockburn, B., Perugia, I., Schötzau, D.: An a priori error analysis of the local discontinuous Galerkin method for elliptic problems. SIAM Journal on Numerical Analysis, **38**, 5, 1676-1706 (2000)
8. Cockburn, B., Kanschat, G., Schötzau, D.: The local discontinuous Galerkin method for linear incompressible fluid flow: a review. Computer and Fluids, **34**, 491-506 (2005)
9. Cockburn, B., Kanschat, G., Schötzau, D.: The local discontinuous Galerkin method for the Oseen equations. Mathematics of Computation, **73**, 569-593 (2004)
10. Cockburn, B., Kanschat, G., Schötzau, D., Schwab, C.: Local discontinuous Galerkin methods for the Stokes system. SIAM Journal on Numerical Analysis, **40**, 1, 319-343 (2002)
11. Galdi, G.P.: An Introduction to the Mathematical Theory of the Navier-Stokes Equations. I: Linearised Steady Problems. Springer-Verlag (1994)
12. Gatica, G.N., Sayas, F.J.: An a-priori error analysis for the coupling of local discontinuous Galerkin and boundary element methods. Mathematics of Computation, to appear.
13. Houston, P., Robson, J., Süli, E.: Discontinuous Galerkin finite element approximation of quasilinear elliptic boundary value problems I: the scalar case. IMA Journal of Numerical Analysis, **25**, 726-749 (2005)
14. Perugia, I., Schötzau, D.: An hp-analysis of the local discontinuous Galerkin method for diffusion problems. Journal of Scientific Computing, **17**, 561-571 (2002)
15. Riviere, B., Wheeler, M. F.: A posteriori error estimates and mesh adaptation strategy for discontinuous Galerkin methods applied to diffusion problems. Preprint 00-10, TICAM, University of Texas at Austin, USA (2000)

High Order Boundary Integral Methods for Maxwell's Equations: Coupling of Microlocal Discretization and Fast Multipole Methods

L. Gatard[1,2], A. Bachelot[1] and K. Mer-Nkonga[2]

[1] MAB, Université Bordeaux 1, 351 cours de la libération, 33405 Talence cedex.
[2] CEA-CESTA, BP 2, 33114, Le Barp, France.
ludovic.gatard@math.u-bordeaux1.fr

Summary. An efficient method to solve time harmonic Maxwell's equations in exterior domain for high frequencies is obtained by using the integral formulation of Després combined with a coupling method based on the Microlocal Discretization method (MD) and the Multi-Level Fast Multipole Method (MLFMM) [1]. In this paper, we consider curved finite elements of higher order in the MLFMM and in the MD/MLFMM methods. Moreover, we give some improvements of the MD/FMM method.

1 Introduction

Let us consider the diffraction of an electromagnetic wave by a homogeneous 3D bounded obstacle Ω with boundary Γ. We solve time harmonic Maxwell's equations for high frequencies by using the integral equations of Després (EID) [3]. After discretization, we solve a dense linear system with a size N proportional to the wave number k. By using the MLFMM [5], [9] based on the reduction of the interactions generated by the Green kernel, we reduce the complexity of a matrix-vector product to $O(NlnN)$ [6].

Another strategy to solve Maxwell's equations is to reduce the size of the linear system by using the MD method [4]. The approximation of the phase function of the unknown leads to a number of degrees of freedom of order $k^{\frac{2}{3}}$ instead of k^2. However, it is necessary to approximate Γ by a mesh with $O(k^2)$ elements. In order to speed up the calculation of the system, the MLFMM is used [1], [2]. These papers show a lack of accuracy in the case of a Léontovitch type impedance boundary condition. In order to achieve a high level of accuracy, we consider the use of curved finite elements of higher order.

In this paper, we present in section 2 the EID and the MLFMM, in section 3 the interaction of finite elements of higher order with MLFMM and in section 4 the coupling method MD/MLFMM. We give in section 5 numerical results.

2 The integral equations of Després (EID) and MLFMM

We solve Maxwell's equations associated with Léontovitch Impedance Boundary Condition (IBC) by using the EID [3]. Let Z be the relative impedance of the surface Γ, we denote $R = (Id - Z)(Id + Z)^{-1}$ the reflexion coefficient characterizing the IBC. We have to solve the following system:

$$\begin{pmatrix} (1+\beta)Id + (\mathcal{A}^\infty)^*\mathcal{A}^\infty & -\mathcal{T}^\star + i\beta Id \\ \mathcal{T} - i\beta Id & \beta Id + (\mathcal{A}^\infty)^*\mathcal{A}^\infty \end{pmatrix} \begin{pmatrix} U \\ V \end{pmatrix} + \mathcal{N}_R U = \begin{pmatrix} G \\ 0 \end{pmatrix}, \quad (1)$$

where $U = (\sqrt{i}J, \sqrt{i}^{-1}M)^t$ and $V = (J', M')^t$; $J = n \wedge H_{|\Gamma}$ and $M = -n \wedge E_{|\Gamma}$ are the equivalent currents associated with the magnetic (resp. electric) field H (resp. E). In (1), $\beta \in]0,1[$, and the integral operators are defined by:

$$\mathcal{T} = \begin{pmatrix} T_r & K_r - \frac{1}{2}n\wedge \\ K_r - \frac{1}{2}n\wedge & T_r \end{pmatrix}, \quad (\mathcal{A}^\infty)^*\mathcal{A}^\infty = \begin{pmatrix} T_i & K_i \\ K_i & T_i \end{pmatrix},$$

where T_r, K_r (resp. T_i, K_i) are the real (resp. imaginary) part of the integral operators of the Stratton-Chu formulae. $\mathcal{N}_R$ is the coupling term ($\mathcal{N}_R = 0$ if $R = 0$); $G = (-\sqrt{i}n\wedge g, \sqrt{i}^{-1}g)^t$ where g is the right-hand side of the IBC. The discrete problem is obtained by using the finite element (FE) method: we use the Nédélec edge-based FE [7]. To solve the discrete problem, we can use an iterative method in which the MLFMM is used to speed up the matrix-vector products [6]. We describe succintly the MLFMM algorithm [5], [9].

We enclose Γ in a cube and we subdivide recursively this cube as an octree. Let $x_i \in \Gamma$, and let X_i be the center of the box C_i including x_i (i=1,2). Then, $x_1 - x_2 = r_{12} + r$, where $r = r_1 - r_2$ and $r_i = x_i - X_i$. When $|r_{12}| > |r|$, the variables of the Green kernel can be separated:

$$G(x_1, x_2) = \frac{e^{ik|x_1-x_2|}}{4\pi|x_1-x_2|} \simeq \frac{ik}{(4\pi)^2} \sum_{p=1}^{P} \omega_p e^{ik\hat{s}_p \cdot r_1} T_{M,r_{12}}(\hat{s}_p) e^{-ik\hat{s}_p \cdot r_2},$$

where $T_{M,r_{12}}$ is the transfer function, $\hat{s}_p$ a point on the unit sphere S^2, and $k = 2\pi/\lambda$ the wave number (λ is the wavelength). $M = kd + c_0(kd)^{\frac{1}{3}}$ is the truncation parameter, d the diameter of the multipole boxes, c_0 a constant, and $P = (M+1)(2M+1)$ the number of quadrature points on S^2.

Due to $|r_{12}| > |r|$, to compute a matrix-vector product AY with $A_{ij} = \overline{\alpha_i}\alpha_j G(x_i, x_j)$ ($\alpha_i, \alpha_j \in \mathbb{C}$), we write $A = A^{far} + A^{near}$ such that $(A^{far})_{ij} = 0$ if $\widetilde{C}$ is close to C and $(A^{far})_{ij} = A_{ij}$ if not, with $x_i \in C$, $x_j \in \widetilde{C}$, and $\widetilde{C}$ close to C means $\widetilde{C}$ and C have at least a common vertex. Let $\overline{V}(C) = \{\widetilde{C}|\widetilde{C}$ far from C and $\widetilde{C}_a$ close to $C_a\}$ if level = 3, and $\overline{V}(C) = \{\widetilde{C}|\widetilde{C}$ far from $C\}$ if level $>$ 3, where C_a is the antecedent of C in the octree.

The MLFMM algorithm to compute the multipole approximation of the far interactions, $A_{app}^{far}Y$, is given by (be careful, the level index l is not written):

- step 0: $T_{M,r_{C\widetilde{C}}}(\hat{s}_p)\ \forall l \in \{3,\cdots,L_f\}, \forall r_{C\widetilde{C}}, \widetilde{C} \in V(C), \forall p \in \{1,\cdots,P\}$.

- step 1: at level L_f, $\forall \widetilde{C}, \forall p \in \{1,\cdots,P\}$, $F_{\widetilde{C}}(\hat{s}_p) = \sum_{j|x_j\in\widetilde{C}} e^{ik\hat{s}_p\cdot(X_{\widetilde{C}}-x_j)}\alpha_j Y_j$.

- step 2: Interpolation of $F_{\widetilde{C}}(\hat{s}_p)$, $\forall l \in \{L_f-1,\cdots,2\}, \forall \widetilde{C}, \forall p \in \{1,\cdots,P\}$.

- step 3: $\forall l \in \{3,\cdots,L_f\}$, transfer from $\widetilde{C} \in V(C)$ to C $\forall C$, $\forall p \in \{1,\cdots,P\}$, $G_C(\hat{s}_p) = \sum_{\widetilde{C}\in V(C)} T_{M,r_{\widetilde{C}C}}(\hat{s}_p) F_{\widetilde{C}}(\hat{s}_p)$.

- step 4: Anterpolation of $G_C(\hat{s}_p)$, $\forall l \in \{3,\cdots,L_f\}, \forall C, \forall p \in \{1,\cdots,P\}$.

- step 5: Integration on S^2 at level L_f: $\forall i \in \{1,\cdots,N\}$, $x_i \in C$, $(A_{app}^{far}Y)_i = \frac{-k}{(4\pi)^2}\overline{\alpha_i}\sum_{p=1}^P \omega_p e^{ik\hat{s}_p\cdot(x_i-X_C)} G_C(\hat{s}_p)$.

3 Finite elements of higher order and MLFMM

We solve (1) by using an edge-based FE method [7]. The size of the system obtained is proportional to the number of elements of the triangulation $\mathcal{T}_h$. Because the use of FE of higher order gives results with a better accuracy than FE of lower order, we expect the use of curved FE of higher order allows us to reduce the number of unknowns. We are going to study the effects of FE of higher order and coarser mesh in MLFMM. We use the basis functions of Graglia [8] of order q, given by:

$$\Lambda_{ijk}^\beta(\xi) = \Gamma_{ijk}^\beta(\xi)\Lambda_\beta(\xi), \quad \beta \in \{1,2,3\}, \quad i+j+k=q+2,$$

where $\xi = (\xi_1,\xi_2,\xi_3)$ are the barycentric coordinates, Λ_β the RWG basis functions, and $\Gamma_{ijk}^\beta(\xi) = N_{ijk}^\beta \frac{(q+2)\xi_\beta}{i_\beta}\hat{L}_i(q+2,\xi_1)\hat{L}_j(q+2,\xi_2)\hat{L}_k(q+2,\xi_3)$ a scalar polynomial. N_{ijk}^β is a normalization constant, i_β is taken to be i, j, or k for $\beta = 1$, 2, or 3 and $\hat{L}_i(q+2,\xi_j) = \frac{1}{(i-1)!}\Pi_{k=1}^{i-1}(q\xi_j - k)$, if $2 \leq i \leq q+1$ and $\hat{L}_i(q+2,\xi_j) = 1$ if $i = 1$.

We develop in the following the modifications in MLFMM due to the use of higher order FE.

- **Number of levels in the octree:** The octree is obtained by subdividing a cube of size d_0 enclosing Γ into smaller cubes until the size of the cube's edge at the finest level d^{10} satisfies $kd^{10} \simeq 1$. This criterion is valid when we use a mesh Γ_h^{10} with ten points per wavelength ($\lambda/10$) to approximate Γ. We denote h_{max}^n the largest edge of Γ_h^n. As for Γ_h^{10}, $d^{10} \sim \lambda/4 = 2.5 h_{max}^{10}$ is equivalent to $kd^{10} \simeq 1$, we define a general criterion for Γ_h^n with $n \in \mathbb{N}^\star$:

$$d^n = Max(ch_{max}^n, \lambda/(2\pi)), \quad \text{with} \quad c > 1. \tag{2}$$

Due to (2), the number of MLFMM's levels $L_f^n = \frac{ln(d_0)-ln(d^n)}{ln 2}$ depends on h_{max}^n. For example, with a mesh Γ_h^5, we lose one level in MLFMM.

- **Cost of a matrix-vector product:** We compare the MLFMM complexity for a matrix-vector product when order 1 FE with a mesh Γ_h^{10} and 3 Gaussian quadrature points by patch of Γ_h^{10} (case 1) are used, and when curved order 2 FE with a mesh Γ_h^5 and 7 Gaussian quadrature points by patch of Γ_h^5 are used (case 2). In the MLFMM algorithm, steps 0, 2, 3 and 4 are less expansive in case 2 because case 2 has one level less than case 1. However, for case 2, to achieve a good accuracy the number of Gaussian points increases due to the order of FE. Also due to the larger size of boxes at fine level, the number of quadrature points on S^2 is larger. Thus, steps 1 and 5 of the MLFMM are more expansive for case 2.

The use of curved FE of higher order has also an impact on the computation of the near matrix which is computed at the finest level of the MLFMM one time for all the matrix-vector products of the iterative solution. The computation of the near matrix is more expansive for case 2 than for case 1 because of the increase of the near interactions due to the large patch size. For curved FE, we can choose two ways based on change of variables to compute the singular interactions. The first way lies on the use of a mesh refinement method (curved element is approximated by smaller plane elements), and the second way lies on direct calculus from curved elements.

To overcome these problems, Chew [9] considers multipole boxes with a distribution of quadrature points instead of patches of Γ_h^n in the octree. As a point has no spatial extent, the octree has the same number of levels as for order 1 FE. This approach reduces the memory requirements of MLFMM, but increases the CPU time for steps 2, 3 and 4 of MLFMM which represent the most expansive steps of MLFMM. We chose to emphasize on the CPU time rather than the memory requirements. We can see in the numerical results the efficiency of higher order FE and the cost of the transfers in MLFMM.

4 Microlocal discretization (MD) and MLFMM

We reduce the size of (1) by using the MD method [4]. In MD, when we discretize the unknown, we give some information on the oscillatory behaviour of the solution by the approximation of the phase function of the unknown. When Ω is a convex domain illuminated by an incident plane wave, a phase function approximation is given by $\phi(x) = k\phi_0(x) + O(k^{\frac{1}{3}}) = k\hat{k} \cdot x + O(k^{\frac{1}{3}})$, where $\hat{k}$ is the direction of the incident wave. We have a new unknown $\widetilde{J}$ such that $J = \widetilde{J}e^{ik\phi_0}$, where J and $\widetilde{J}$ are currents on Γ. Because $\widetilde{J}$ is less oscillatory than J, we can approximate $\widetilde{J}$ with a number of degrees of freedom of order $k^{\frac{2}{3}}$ instead of k^2. However, for MD, a high degree of approximation of Γ is needed and we use a double mesh: a coarser one Γ_c for the unknown with

$N_c \sim O(k^{\frac{2}{3}})$ elements, and a finest one Γ_f for the surface with $N_f \sim O(k)$ elements. Thus, if we consider the matrix A seen before, we have to compute:

$$A_{ij} = \sum_{i_0=1}^{N_i} \sum_{j_0=1}^{N_j} \alpha_{jj_0} \overline{\alpha_{ii_0}} G(x_{ii_0}, x_{jj_0}) e^{ik\phi(x_{jj_0})} \overline{e^{ik\phi(x_{ii_0})}},$$

where N_i (resp. N_j) is the number of elements $\overline{K_{ii_0}} \in \mathcal{T}_f$ (resp. $\overline{K_{jj_0}} \in \mathcal{T}_f$) of $K_i \in \mathcal{T}_c$ (resp. $K_j \in \mathcal{T}_c$), and x_{ii_0} (resp. x_{jj_0}) is a quadrature point on $\overline{K_{ii_0}}$ (resp. $\overline{K_{jj_0}}$). $\mathcal{T}_c$ (resp. $\mathcal{T}_f$) is a triangulation associated with Γ_c (resp. Γ_f). Because the computation cost of the matrix of (1) is of order $O(k^2)$, we speed up the matrix calculation of (1) by an original use of MLFMM [1], [2]. The algorithm of MD/MLFMM to compute A_{app}^{far} is given by:

- step 0: $\forall l \in \{L_0, \cdots, L_f\}, \forall r_{C\widetilde{C}}, \widetilde{C} \in \overline{V}(C), \forall p \in \{1, \cdots, P\}, T_{M, r_{C\widetilde{C}}}(\hat{s}_p)$.

- step 1: at level L_f, $\forall i \in \{1, \cdots, N_c\}$, $\forall C$ such that $\overline{K_i} \cap C \neq \emptyset$, $\forall p \in \{1, \cdots, P\}, F_{iC}(\hat{s}_p) = \sum_{i_0 | x_{ii_0} \in C} \alpha_{ii_0} e^{ik\hat{s}_p \cdot (X_C - x_{ii_0})}$.
The radiation functions at the upper levels are computed by interpolation

- step 2: $\forall l \in \{L_f, \cdots, L_0\}, \forall i, j \in \{1, \cdots, N_c\}$

$$(A_{app}^{far})_{ij} = \frac{-k}{(4\pi)^2} \sum_{p=1}^{P} \omega_p \sum_{C|\overline{K_i} \cap C \neq \emptyset} \overline{F_{iC}(\hat{s}_p)} \sum_{\substack{\widetilde{C}|\overline{K_j} \cap \widetilde{C} \neq \emptyset \\ \widetilde{C} \in \overline{V}(C)}} T_{M, r_{C\widetilde{C}}}(\hat{s}_p) F_{j\widetilde{C}}(\hat{s}_p).$$

L_0 is the level of MLFMM where the MD/MLFMM algorithm starts.

The memory requirements and the time consuming of the MD/MLFMM are respectively of order $O(N_f)$ and $O(N_f^{\frac{4}{3}} \ln N_f + N_{iter} N_f^{\frac{2}{3}})$. We speed up the solution and reduce the memory requirements. In (1), because Id and $\frac{1}{2} n \wedge$ are sparse operators, we compute them at each iteration of the solution. Thus, we just compute and store the lower part of the other operators in (1) due to the fact that T_r, K_r, T_i and K_i are hermitian operators. We reduce the CPU time to compute the matrix and the memory requirements by a factor 2.

We optimize the calculation of the singular interactions for the near matrix. This optimization has a bigger impact for MD than for MLFMM, because the computation of the matrix represents for MD the major part of the CPU time used. As for MLFMM, we use for MD change of variables to eliminate efficiently the singularity and to compute accurately the singular interactions.

To improve the accuracy for Léontovitch type impedance boundary condition, we implemented curved FE of higher order with respect to the error estimations [4]. We study as well the size of the coarse mesh Γ_c which provides accurate results with an impedance $Z = 1$ when Γ is a sphere. For a frequency of 0.6 GHz, Γ_c has five points per wavelength, and for a frequency of 1 GHz Γ_c has three points per wavelength, with order 1 FE. With order 2 FE, we can use a mesh Γ_c coarser as we can see in numerical results.

5 Numerical results

We give first results for FE of higher order and MLFMM. The iterative solver used is a flexible GMRES. For each test case, we compare the results obtained with plane order 1 FE associated with a mesh Γ_h^{10} (case 1 of Section 3) and curved order 2 FE associated with a mesh Γ_h^5 (case 2 of Section 3). For each test case, we give the bistatic Radar Cross Section (RCS) and the RCS error [9] defined by: $RCS_{err} = \sqrt{\frac{1}{Q}\sum_{i=1}^{Q}|\sigma_{ref}^i - \sigma_{app}^i|^2}$, where Q is the number of sampling points, σ_{ref} the RCS reference solution, and σ_{app} the RCS numerical solution. We give also the convergence of the GMRES residual, and a table with the CPU time. There are three columns in the CPU time table: the first one gives the CPU time of the near matrix, the second one gives the CPU time for one GMRES iteration of the MLFMM matrix-vector product, and the last one gives the CPU time of the iterative solution. For order 2 FE, because we can choose two ways to compute the singular interactions of the near matrix, there are two numbers in the first column. The first way lies on the use of a mesh refinement method, and the second way lies on direct calculus from curved elements. But, as the first way is more time consuming than the second way, we just consider in the total time the last way to compute singular interactions.

The test cases with the sphere allow us to compare numerical solutions with the exact solutions given by the Mie series.

NASA almond ($Z = 1$): This test case does not have an exact solution. So, the reference solution is obtained with order 1 FE associated with a fine mesh Γ_h^{20}. The number of degrees of freedom is about 21000. FE of order 2 provide a better accuracy than FE of order 1 with the same CPU time (table 1 and figure 1).

Perfectly absorbing sphere ($Z = 1$): The number of degrees of freedom is about 75000. The RCS results are good for both order 1 and order 2 FE, but the order 2 FE solution is less time consuming (table 2 and figure 2). The gains in CPU time increase with the size of the problem. We have same results for $Z = 0$.

Table 1. CPU time for the NASA almond with $Z = 1$ ($N = 21000$)

Computation Time (s)	Near Mat.	MLFMM Product	Total (21 iters GMRES)
Order 1 FE ($\lambda/10$)	15	56	1191
Order 2 FE ($\lambda/5$)	162 / 35	55	1190

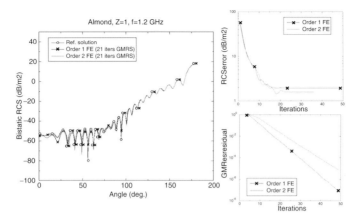

Fig. 1. MLFMM, NASA almond, $Z = 1$, $f = 1.2GHz$

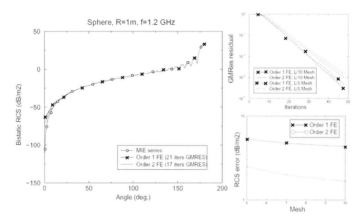

Fig. 2. MLFMM, unit sphere, $Z = 1$, $f = 1.2GHz$

Table 2. CPU time for the unit sphere with $Z = 1$ ($N = 75000$)

Computation Time (s)	Near Mat.	MLFMM Product	Total (33 iters GMRES)
Order 1 FE ($\lambda/10$)	50	182	6056
Order 2 FE ($\lambda/5$)	430 / 100	150	5050

Perfectly abosrbing sphere in the MD/MLFMM case: We can see the efficiency of higher order FE in the coupling method (figure 3). With order 2 FE, we have a good accuracy for small RCS (-50dB), and the size of the coarse mesh Γ_c is about $\lambda/1.5$ with a frequency $f = 1GHz$.

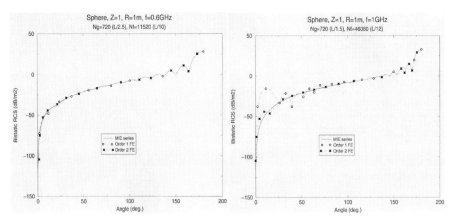

Fig. 3. MD/MLFMM, unit sphere, $Z = 1$

6 Conclusion

We have studied the use of FE of higher order for two kinds of methods in order to speed up the solution of integral equations and to achieve a good level of accuracy. In MLFMM, by using FE of higher order, the CPU time is reduced for a better accuracy. The EID/MD/MLFMM solution obtained with higher order FE allows us to have a great accuracy in the difficult case of perfectly absorbing objects. We plan to work on optimizations of the MD/MLFMM algorithm and on an improvement of the approximation of the phase function for non-convex objects.

References

1. Bachelot, A., Darrigrand, E., Mer-Nkonga, K.: Coupling of a multilevel fast multipole method and a microlocal discretization for the 3-D integral equations of electromagnetism. C. R. Acad. Sci. Paris, Ser, I **336**, 505-510 (2003)
2. Darrigrand, E.: Coupling of Fast Multipole Method and Microlocal Discretization for the 3-D Helmholtz Equation. J. Comput. Phys., **181**, 126-154 (2002)
3. Collino, F., Després, B.: Integral equations via saddle point problems for time-harmonic Maxwell's equations. J. Comput. and Appl. Math., **150**, 157-192 (2003)
4. Abboud, T., Nédélec, J.-C., Zhou, B.: Improvement of the Integral Equation Method for High Frequency Problems. Third international conference on mathematical aspects of wave propagation phenomena, SIAM, 178-187 (1995)
5. Coifman, R., Rokhlin, V., Wandzura, S.: The fast multipole method for the wave equation: A pedestrian prescription. IEEE Antennas Propagat. Mag., **35**, 7-12 (1993)
6. Mer-Nkonga, K., Collino, F.: The fast multipole method applied to a mixed integral system for time-harmonic Maxwell's equations. in: JEE 02: European Symposium on Numerical Methods in Electromagnetics, ONERA, 121-126 (2002)

7. Nédélec, J.-C.: Mixed Finite Elements in $\mathbb{R}^3$. Numer. Math., **35**, 315-341 (1980)
8. Graglia, R.D., Wilton, D.R., Peterson, A.F.: Higher Order Interpolary Vector Bases for Computational Electromagnetics. IEEE Trans. Antennas and Propagat., **45**, 329-342 (1997)
9. Donepudi, K.C., Song, J., Jin, J.-M., Kang, G., Chew, W.C.: A Novel Implementation of Multilevel Fast Multipole Algorithm for Higher Order Galerkin's Method. IEEE Trans. Antennas and Propagat.,**48**, 1192-1197 (2000)

Indirect Methods with Brakhage–Werner Potentials for Helmholtz Transmission Problems

María–Luisa Rapún[1] and Francisco–Javier Sayas[2]

[1] Dep. Matemática e Informática, Universidad Pública de Navarra
mluisa.rapun@unavarra.es
[2] Dep. Matemática Aplicada, Universidad de Zaragoza
jsayas@unizar.es

Summary. In this work we propose and analyse numerical methods for Helmholtz transmission problems in two and three dimensions. The methods we analyse use combined single and double layer potentials to represent the interior and exterior solution of the transmission problem. The corresponding boundary integral system includes weakly singular and hypersingular boundary integral operators on the interface. Its invertibility is equivalent to the unique solvability of the transmission problem, since the use of the above mentioned potentials does not introduce spurious eigenmodes in the formulation. We give necessary and sufficient conditions for the convergence of general Petrov–Galerkin schemes for solving the resulting system, providing some concrete methods for the two dimensional case. Some numerical experiments are shown.

1 Problem

Let Ω_{int} be a bounded simple connected open set in $\mathbb{R}^n$ for $n=2$ or 3 and $\Omega_{\mathrm{ext}} := \mathbb{R}^n \setminus \overline{\Omega}_{\mathrm{int}}$. The common interface Γ is assumed to be connected and smooth. For the integral formulation, Lipschitz regularity is sufficient, but some additional smoothness is required for the numerical analysis.

The problem we consider consists of a system of Helmholtz equations with different wave numbers,

$$\Delta u + \lambda^2 u = 0, \quad \text{in } \Omega_{\mathrm{ext}}, \tag{1a}$$

$$\Delta u + \mu^2 u = 0, \quad \text{in } \Omega_{\mathrm{int}}, \tag{1b}$$

coupled through continuity conditions on Γ,

$$u|_\Gamma^{\mathrm{int}} - u|_\Gamma^{\mathrm{ext}} = g_0, \tag{2a}$$

$$\alpha\, \partial_n u|_\Gamma^{\mathrm{int}} - \beta\, \partial_n u|_\Gamma^{\mathrm{ext}} = g_1. \tag{2b}$$

As usual in Helmholtz exterior problems, we also impose the Sommerfeld radiation condition at infinity

$$\lim_{r\to\infty} r^{\frac{n-1}{2}}(\partial_r u - \imath\lambda u) = 0, \qquad (3)$$

that has to be satisfied uniformly in all directions. Some relevant fields where problem (1–3) appears are the scattering of acoustic and thermal waves in time–harmonic situations in media with piecewise constant properties.

Throughout this work we will assume that λ, μ, α and β are non–zero parameters chosen such that problem (1–3) has a unique solution in $H^1(\Omega_{\text{int}})$ and $H^1_{\text{loc}}(\overline{\Omega}_{\text{ext}})$. In particular $\alpha \neq -\beta$. Conditions guaranteeing uniqueness can be found in [10] and references therein.

2 Boundary integral formulation

Traditional boundary integral formulations of problem (1–3) based on the representation formula (see [2]) or indirect formulations with single layer potentials (see [10]) fail when either $-\lambda^2$ or $-\mu^2$ are Dirichlet eigenvalues of the Laplace operator in Ω_{int}. Our proposal is an indirect method that provides integral equations that are uniquely solvable for all wave numbers.

We introduce the fundamental solution to the Helmholtz equation $\Delta u + \rho^2 u = 0$,

$$\phi_\rho(\mathbf{x},\mathbf{y}) := \begin{cases} \dfrac{\imath}{4} H_0^{(1)}(\rho|\mathbf{x}-\mathbf{y}|), & \text{if } n=2, \\[4pt] \dfrac{\exp(\imath\rho|\mathbf{x}-\mathbf{y}|)}{4\pi|\mathbf{x}-\mathbf{y}|}, & \text{if } n=3, \end{cases}$$

and the associated single and double layer potentials

$$\mathcal{S}^\rho \varphi := \int_\Gamma \phi_\rho(\,\cdot\,,\mathbf{y})\,\varphi(\mathbf{y})\,d\gamma_\mathbf{y} \;:\; \mathbb{R}^n \longrightarrow \mathbb{C},$$

$$\mathcal{D}^\rho \varphi := \int_\Gamma \partial_{n(\mathbf{y})}\phi_\rho(\,\cdot\,,\mathbf{y})\,\varphi(\mathbf{y})\,d\gamma_\mathbf{y} \;:\; \mathbb{R}^n \longrightarrow \mathbb{C}.$$

Then, we look for the solution to (1–3) as

$$u = \begin{vmatrix} (\mathcal{S}^\mu - \imath\eta\mathcal{D}^\mu)\varphi, & \text{in } \Omega_{\text{int}}, \\ (\mathcal{S}^\lambda - \imath\eta\mathcal{D}^\lambda)\psi, & \text{in } \Omega_{\text{ext}}. \end{vmatrix} \qquad (4)$$

The new unknowns will be the densities $\varphi, \psi \in H^{1/2}(\Gamma)$ and $\eta > 0$ is just a fixed parameter. In fact we could use different values of η for the interior and exterior potentials. We will keep a single one for the sake of simplicity. This combination of a single and a double layer potential, commonly referred as a Brakhage–Werner potential, was independently suggested in [1, 6, 9] for solving the exterior Dirichlet problem for the Helmholtz equation. The use of this kind of potentials for Helmholtz transmission problems is not new (see [3, 5]), but to the best of our knowledge, there is no previous work concerning numerical methods. The proof of the equivalence of the boundary formulation and the original problem that we propose here is also less involved.

If we take u as in (4), by definition it satisfies (1) and (3). Therefore, it only remains to impose the transmission conditions (2). With this purpose, we introduce the boundary integral operators

$$V^\rho \varphi := \int_\Gamma \phi_\rho(\,\cdot\,,\mathbf{y})\,\varphi(\mathbf{y})\,d\gamma_\mathbf{y} \;:\; \Gamma \longrightarrow \mathbb{C},$$

$$J^\rho \varphi := \int_\Gamma \partial_{n(\,\cdot\,)}\phi_\rho(\,\cdot\,,\mathbf{y})\,\varphi(\mathbf{y})\,d\gamma_\mathbf{y} \;:\; \Gamma \longrightarrow \mathbb{C},$$

$$K^\rho \varphi := \int_\Gamma \partial_{n(\mathbf{y})}\phi_\rho(\,\cdot\,,\mathbf{y})\,\varphi(\mathbf{y})\,d\gamma_\mathbf{y} \;:\; \Gamma \longrightarrow \mathbb{C},$$

$$W^\rho \varphi := -\partial_{n(\,\cdot\,)}\int_\Gamma \partial_{n(\mathbf{y})}\phi_\rho(\,\cdot\,,\mathbf{y})\,\varphi(\mathbf{y})\,d\gamma_\mathbf{y} \;:\; \Gamma \longrightarrow \mathbb{C}.$$

Then, by the jump relations of the single and double layer potentials it follows that

$$u|_\Gamma^{\text{int}} - u|_\Gamma^{\text{ext}} = \left(V^\mu + \imath\eta\left(\tfrac{1}{2}I - K^\mu\right)\right)\varphi - \left(V^\lambda - \imath\eta\left(\tfrac{1}{2}I + K^\lambda\right)\right)\psi,$$

$$\alpha\,\partial_n u|_\Gamma^{\text{int}} - \beta\,\partial_n u|_\Gamma^{\text{ext}} = \alpha\left(\tfrac{1}{2}I + J^\mu + \imath\eta\,W^\mu\right)\varphi - \beta\left(-\tfrac{1}{2}I + J^\lambda + \imath\eta\,W^\lambda\right)\psi.$$

Thus, (2) can be written as

$$\mathcal{H}\begin{bmatrix}\varphi\\\psi\end{bmatrix} := \begin{bmatrix} V^\mu + \imath\eta\left(\tfrac{1}{2}I - K^\mu\right) & -V^\lambda + \imath\eta\left(\tfrac{1}{2}I + K^\lambda\right) \\ \alpha\left(\tfrac{1}{2}I + J^\mu + \imath\eta\,W^\mu\right) & \beta\left(\tfrac{1}{2}I - J^\lambda - \imath\eta\,W^\lambda\right) \end{bmatrix}\begin{bmatrix}\varphi\\\psi\end{bmatrix} = \begin{bmatrix}g_0\\g_1\end{bmatrix}. \tag{5}$$

By the properties of the integral operators involved, $\mathcal{H} : H^{1/2}(\Gamma) \times H^{1/2}(\Gamma) \to H^{1/2}(\Gamma) \times H^{-1/2}(\Gamma)$ is bounded. To prove invertibility we will use the Fredholm alternative. With this aim we recall some well–known properties of the operators in $\mathcal{H}$ (see [7]):

- There exists an elliptic operator $W_0 : H^{1/2}(\Gamma) \to H^{-1/2}(\Gamma)$ such that $W^\lambda - W_0$, $W^\mu - W_0 : H^{1/2}(\Gamma) \to H^{-1/2}(\Gamma)$ are compact.
- When Γ is smooth, J^λ, $J^\mu : H^{-1/2}(\Gamma) \to H^{-1/2}(\Gamma)$ and K^λ, $K^\mu : H^{1/2}(\Gamma) \to H^{1/2}(\Gamma)$ are compact.
- For $\eta > 0$, the operators $V^\lambda - \imath\eta\left(\tfrac{1}{2}I + K^\lambda\right)$, $V^\mu + \imath\eta\left(\tfrac{1}{2}I - K^\mu\right) : H^{1/2}(\Gamma) \to H^{1/2}(\Gamma)$ are invertible.

Theorem 1. $\mathcal{H}$ *is Fredholm of index zero.*

Proof. We define

$$\mathcal{H}_0 := \begin{bmatrix} \frac{\imath\eta}{2}I & \frac{\imath\eta}{2}I \\ \imath\eta\, \alpha W_0 & -\imath\eta\beta W_0 \end{bmatrix}. \tag{6}$$

Then $\mathcal{H} - \mathcal{H}_0$ is compact. Moreover,

$$\mathcal{H}_0 = \imath\eta \begin{bmatrix} I & 0 \\ 2\alpha W_0 & -I \end{bmatrix} \begin{bmatrix} \frac{1}{2}I & \frac{1}{2}I \\ 0 & (\alpha+\beta)W_0 \end{bmatrix}$$

and therefore, it is invertible since $W_0 : H^{1/2}(\Gamma) \to H^{-1/2}(\Gamma)$ is invertible and $\alpha \neq -\beta$. ∎

Theorem 2. $\mathcal{H}$ *is injective.*

Proof. We decompose

$$\mathcal{H} = \begin{bmatrix} I & 0 \\ A_{\lambda\mu} & I \end{bmatrix} \begin{bmatrix} V^\mu + \imath\eta(\frac{1}{2}I - K^\mu) & -V^\lambda + \imath\eta(\frac{1}{2}I + K^\lambda) \\ 0 & H_{\lambda\mu} \end{bmatrix},$$

with

$$A_{\lambda\mu} := \alpha(\tfrac{1}{2}I + J^\mu + \imath\eta W^\mu)(V^\mu + \imath\eta(\tfrac{1}{2}I - K^\mu))^{-1}$$

and

$$H_{\lambda\mu} := A_{\lambda\mu}(V^\lambda - \imath\eta(\tfrac{1}{2}I + K^\lambda)) - \beta(-\tfrac{1}{2}I + J^\lambda + \imath\eta W^\lambda).$$

The proof is reduced now to showing inyectivity of $H_{\lambda\mu} : H^{1/2}(\Gamma) \to H^{-1/2}(\Gamma)$. Let us assume that $H_{\lambda\mu}\xi = 0$ and define

$$u := \begin{vmatrix} (\mathcal{S}^\lambda - \imath\eta\mathcal{D}^\lambda)\xi, & \text{in } \Omega_{\text{ext}}, \\ (\mathcal{S}^\mu - \imath\eta\mathcal{D}^\mu)(V^\mu + \imath\eta(\tfrac{1}{2}I - K^\mu))^{-1}(V^\lambda - \imath\eta(\tfrac{1}{2}I + K^\lambda))\xi, & \text{in } \Omega_{\text{int}}. \end{vmatrix}$$

Then, u satisfies (1) and (3). Furthermore, by the jump relations of the layer potentials, $u|_\Gamma^{\text{int}} = (V^\lambda - \imath\eta(\tfrac{1}{2}I + K^\lambda))\xi = u|_\Gamma^{\text{ext}}$ and also $\alpha\,\partial_n u|_\Gamma^{\text{int}} - \beta\,\partial_n u|_\Gamma^{\text{ext}} = H_{\lambda\mu}\xi = 0$. Thus, u is a solution to the homogeneous transmission problem, that is, to (1–3) with $g_0 = g_1 = 0$, and therefore $u = 0$. Taking the exterior trace, $(V^\lambda - \imath\eta(\tfrac{1}{2} + K^\lambda))\xi = 0$ and thus $\xi = 0$. ∎

As a direct consequence of Theorems 1 and 2, $\mathcal{H} : H^{1/2}(\Gamma) \times H^{1/2}(\Gamma) \to H^{1/2}(\Gamma) \times H^{-1/2}(\Gamma)$ is an isomorphism.

3 Petrov–Galerkin methods

For the numerical approximation we will use the same discrete space $X_N \subset H^{1/2}(\Gamma)$ for the unknown densities. We then test the hypersingular equation with X_N but use a simpler, i.e. less regular space $Y_N \subset H^{-1/2}(\Gamma)$, to test the other equation. For well–posedness we require that $\dim X_N = \dim Y_N$.

We consider discretizations of the form

$$\begin{vmatrix} \varphi_N, \psi_N \in X_N \\ \left(\mathcal{H}\begin{bmatrix} \varphi_N \\ \psi_N \end{bmatrix}, \begin{bmatrix} v_N \\ \phi_N \end{bmatrix}\right) = \left(\mathcal{H}\begin{bmatrix} \varphi \\ \psi \end{bmatrix}, \begin{bmatrix} v_N \\ \phi_N \end{bmatrix}\right), \quad \forall v_N \in Y_N, \ \phi_N \in X_N. \end{vmatrix} \quad (7)$$

For the sake of brevity, we will say that (7) is the Petrov–Galerkin $\{X_N \times X_N; Y_N \times X_N\}$ method for $\mathcal{H} : H^{1/2}(\Gamma) \times H^{1/2}(\Gamma) \to H^{1/2}(\Gamma) \times H^{-1/2}(\Gamma)$.

Theorem 3. *Method (7) is convergent in $H^{1/2}(\Gamma) \times H^{1/2}(\Gamma)$ if and only if the $\{X_N; Y_N\}$ method is convergent for $I : H^{1/2}(\Gamma) \to H^{1/2}(\Gamma)$.*

Proof. In both cases X_N satisfies the approximation property in $H^{1/2}(\Gamma)$:

$$\inf_{\phi_N \in X_N} \|\phi - \phi_N\|_{1/2,\Gamma} \to 0, \quad \forall \phi \in H^{1/2}(\Gamma).$$

Since $\mathcal{H} - \mathcal{H}_0$ is compact, convergence of the $\{X_N \times X_N; Y_N \times X_N\}$ method for $\mathcal{H}$ is equivalent to convergence of the same method for $\mathcal{H}_0$ (see [4, Theorem 3.7]). Moreover,

$$\mathcal{H}_0 = \imath\eta \begin{bmatrix} I & 0 \\ 0 & W_0 \end{bmatrix} \begin{bmatrix} \frac{1}{2}I & \frac{1}{2}I \\ \alpha I & -\beta I \end{bmatrix},$$

and the right–most operator in this decomposition is an isomophism in $X_N \times X_N$ with uniformly bounded inverse. Therefore, convergence is also equivalent to convergence of the Petrov–Galerkin $\{X_N \times X_N; Y_N \times X_N\}$ method for the diagonal operator in the decomposition above, or equivalently, to convergence of the $\{X_N; Y_N\}$ method for $I : H^{1/2}(\Gamma) \to H^{1/2}(\Gamma)$ since the Galerkin $\{X_N; X_N\}$ method is convergent for the elliptic operator $W_0 : H^{1/2}(\Gamma) \to H^{-1/2}(\Gamma)$. ∎

4 Numerical approximation in two dimensions

From now on we will assume that Γ is a $\mathcal{C}^\infty$–curve in $\mathbb{R}^2$ and that there exists a 1–periodic regular parameterization $\mathbf{x} : [0,1] \to \Gamma$ with $|\mathbf{x}'| > 0$, $\mathbf{x}(t) \neq \mathbf{x}(s)$, $s - t \notin \mathbb{Z}$. In the new setting, we will deal with the 1–periodic Sobolev spaces (see [4, Chapter 8] or [11, Chapter 5])

$$H^s := \{\phi \in \mathcal{D}' \mid |\widehat{\phi}(0)|^2 + \sum_{0 \neq k \in \mathbb{Z}} |k|^{2s}|\widehat{\phi}(k)|^2 < \infty\},$$

$\mathcal{D}'$ being the space of 1–periodic distributions at the real line and $\widehat{\phi}(k)$ the Fourier coefficients of ϕ.

We can define now parameterized versions of the layer potentials,

$$\mathcal{S}^\rho \varphi := \int_0^1 \phi_\rho(\,\cdot\,, \mathbf{x}(t))\, \varphi(t)\, dt \;:\; \mathbb{R}^2 \longrightarrow \mathbb{C},$$

$$\mathcal{D}^\rho \varphi := \int_0^1 |\mathbf{x}'(t)|\, \partial_{n(t)} \phi_\rho(\,\cdot\,, \mathbf{x}(t))\, \varphi(t)\, dt \;:\; \mathbb{R}^2 \setminus \Gamma \longrightarrow \mathbb{C},$$

and of their related operators. When we look for the solution to (1–3) using Brakhage–Werner potentials in parameterized form, we obtain an equivalent system of boundary integral equations with exactly the same structure as (5) with the parameterized versions of the boundary integral operators and with the functions $g_0 \circ \mathbf{x}$ and $|\mathbf{x}'|\, g_1 \circ \mathbf{x}$ in the right–hand side.

It is easy to prove, with due adaptations, that all the previous results at the continuous level remain valid when replacing the traditional Sobolev spaces $H^s(\Gamma)$ by the corresponding 1–periodic ones.

For the numerical approximation we consider families of subspaces of $H^{1/2}$ and $H^{-1/2}$ with the same finite dimension. The preceding analysis can also be adapted to show that convergence of the $\{X_N \times X_N; Y_N \times X_N\}$ method for $\mathcal{H}: H^{1/2} \times H^{1/2} \to H^{1/2} \times H^{-1/2}$ is equivalent to the convergence of the simpler $\{X_N; Y_N\}$ method for the identity operator in $H^{1/2}$.

From the results in [10] and with the same kind of techniques, it is rather simple to prove that the following couples provide convergent methods satisfying the indicated convergence estimates:

i) **Periodic smoothest splines.** Consider two staggered uniform grids formed by the nodes $x_i := i/N$, $x_{i-1/2} := (i - 1/2)/N$, $i = 1, \ldots, N$ to define for $m \geq 0$ the spaces $X_N := \{p_N \in H^{m+1} \,|\, p_N|_{[x_i, x_{i+1}]} \in \mathbb{P}_{m+1}\}$ and $Y_N := \{p_N \in H^m \,|\, p_N|_{[x_{i-1/2}, x_{i+1/2}]} \in \mathbb{P}_m\}$. With this choice, the method converges in the natural norm, and moreover, if $s < m + 3/2$, $-m - 1 \leq s \leq t$, $-m - 1/2 < t \leq m + 2$ and $\varphi, \psi \in H^t$, then

$$\|\varphi - \varphi_N\|_s + \|\psi - \psi_N\|_s \leq C_{s,t}(1/N)^{t-s}(\|\varphi\|_t + \|\psi\|_t). \tag{8}$$

In particular, for smooth data (and boundary) the optimal convergence rate is $2m + 3$.

ii) **Periodic smoothest splines on non–uniform grids.** We take $0 < x_0 < x_1 < \cdots < x_N = 1$ and $x_{i+1/2} := (x_i + x_{i+1})/2$ to define the spaces of piecewise linear and constant functions $X_N := \{p_N \in H^1 \,|\, p_N|_{[x_i, x_{i+1}]} \in \mathbb{P}_1\}$ and $Y_N := \{p_N \in H^0 \,|\, p_N|_{[x_{i-1/2}, x_{i+1/2}]} \in \mathbb{P}_0\}$. For the convergence analysis we require that

$$1/C_1 \leq h_{i-1}/h_i \leq C_1, \quad 6 - (h_{i-1/2}/h_{i+1/2} + h_{i+1/2}/h_{i-1/2}) \geq C_2 > 0,$$

where $h_i := x_i - x_{i-1}$ and $h_{i+1/2} := (h_i + h_{i+1})/2$. As in the uniform case, if $s < 3/2$, $-1 \leq s \leq t$, $-1/2 < t \leq 2$ and $\varphi, \psi \in H^t$, then (8) holds. Therefore the method has cubic convergence order in the weaker norm.

iii) **Trigonometric polynomials**. The last method we propose is indeed a pure Galerkin method defined by choosing $X_N = Y_N = \mathrm{span}\{\exp(2\pi k\imath \cdot)$, $-N/2 \leq k < N/2\}$ to obtain a convergent method. Assuming that the data and the boundary are smooth, the densities are approximated with superalgebraic convergence order.

We want to point out that for smooth but not $\mathcal{C}^\infty$-curves, it is more suitable to use spline approximations than spectral ones. We refer to [10] for a detailed discussion on this topic.

5 A numerical example

To conclude we present a numerical experiment. We use a fully implementable version of the method with piecewise linear and constant functions on uniform grids obtained by applying adequate quadrature rules to approximate the integrals involved. The method is an adaptation of one proposed in [10] using the ideas in [8] for the approximation of the integrals related with the hypersingular operators. To define a pointwise approximation of (4), we replace φ and ψ by the discrete ones and apply simple midpoint rules. Theoretically, when the functions on the right hand side are $\mathcal{C}^2$, this approximation has quadratic convergence order in $\mathbb{R}^2 \setminus \Gamma$, i.e.

$$|u(\mathbf{z}) - u_N(\mathbf{z})| = \mathcal{O}_\mathbf{z}(N^{-2}), \qquad \mathbf{z} \notin \Gamma.$$

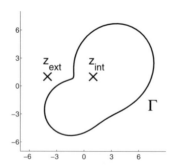

Fig. 1. Geometry

To test the method we consider the obstacle represented in Fig. 1, defined by the parameterization $\mathbf{x}(t) = (r(t)\cos(2\pi t), r(t)\sin(2\pi t))$ with $r(t) = 5 + 2(\cos(2\pi t) + \sin(2\pi t))$. We choose $\lambda = 1 + \imath$, $\mu = 1$, $\alpha = 2$, $\beta = 1$ and take

Table 1. Relative errors

N	$\mathbf{z}_{\text{ext}}$	e.c.r.	$\mathbf{z}_{\text{int}}$	e.c.r.
64	2.16(-4)		7.92(-4)	
96	8.34(-5)	2.34	3.56(-4)	1.96
144	3.44(-5)	2.18	1.59(-4)	1.97
216	1.47(-5)	2.08	7.14(-5)	1.98
324	6.49(-6)	2.03	3.18(-5)	1.99

g_0 and g_1 such that the solution to (1–3) is $u(\mathbf{z}) = \exp(\imath\mu(\sqrt{2}/2, -\sqrt{2}/2) \cdot \mathbf{z})$ in Ω_{int} and $u(\mathbf{z}) = H_0^{(1)}(\lambda|(3,5) - \mathbf{z}|)$ in Ω_{ext}. The relative errors and the estimated convergence rates written in Table 1 are computed at the points $\mathbf{z}_{\text{ext}} = (-4, 1)$ and $\mathbf{z}_{\text{int}} = (1, 1)$ taking N nodes for the discretization. The ratio between consecutive grids is $3/2$. We observe that, as expected, the convergence rate is quadratic and that the errors in both domains have the same size.

Acknowledgements

The authors are partially supported by Gobierno de Navarra project ref. 18/2005, MEC/FEDER project MTM–2004–01905 and by DGA–Grupo consolidado PDIE.

References

1. Brakhage, H., Werner, P.: Über das Dirichletsche Aussenraumproblem für die Helmholtzsche Schwingungsgleichung. Arch. Math., **16**, 325–329 (1965)
2. Costabel, M., Stephan, E.: A direct boundary integral equation method for transmission problems. J. Math. Anal. Appl. **106**, 367–413 (1985)
3. Kleinman, R.E., Martin, P.A.: On single integral equations for the transmission problem of acoustics. SIAM J. Appl. Math **48**, 307–325 (1988)
4. Kress, R.: Linear integral equations. Second edition. Springer–Verlag, New York, 1999.
5. Kress, R., Roach, G.F.: Transmission problems for the Helmholtz equation. J. Mathematical Phys. **19**, 1433–1437 (1978).
6. Leis, R.: Zur Dirichletschen Randwertaufgabe des Aussenraumes der Schwingungsgleichung. Math. Z., **90**, 205–211 (1965)
7. McLean, W.: Strongly elliptic systems and boundary integral equations. Cambridge University Press, Cambridge, 2000.
8. Meddahi, S., Sayas, F.–J.: Analysis of a new BEM–FEM coupling for two dimensional fluid–solid iteraction. Numer. Methods Partial Differential Equations **21**, 1017-1154 (2005)

9. Panich, O.I.: On the question of the solvability of the exterior boundary–value problems for the wave equation and Maxwell's equations. Russian Math. Surveys, **20**, 221–226 (1965)
10. Rapún, M.-L., Sayas, F.-J.: Boundary integral approximation of a heat–diffusion problem in time–harmonic regime. Numer. Algorithms, **41**, 127–160.
11. Saranen, J., Vainikko, G.: Periodic integral and pseudodifferential equations with numerical approximation. Springer-Verlag, Berlin, 2002.

A FEM–BEM Formulation for a Time–Dependent Eddy Current Problem

S. Meddahi[1] and V. Selgas[2]

Departamento de Matemáticas, Universidad de Oviedo
[1]`salim@orion.ciencias.uniovi.es`, [2]`selgas@orion.ciencias.uniovi.es`

Summary. We study in this paper a time–dependent eddy current problem posed in the whole space. We propose a weak formulation that can be rewritten as a well–posed saddle point problem when the constraint satisfied by the magnetic field in the dielectric medium is handled by means of a Lagrange multiplier. Furthermore, we provide a BEM–FEM formulation of the problem that leads to a semi–discrete Galerkin scheme based on Nedelec's and Raviart–Thomas finite elements. Finally, we analyze the asymptotic behavior of the error in terms of the mesh size parameter.

1 Introduction

The *eddy current* problem is a magneto–quasistatic sub–model of Maxwell's equations that is commonly used in electrical power engineering. Formally speaking, such sub–model is obtained by neglecting the displacement currents in Ampère's law.

Generally, the eddy current problem is posed in the whole space with decay conditions on the electromagnetic field at infinity. This fact leads to an additional difficulty for numerical schemes based on finite elements. In [2] and [3], such a drawback is overcome for *time–harmonic* problems by incorporating the far field effects through non-local boundary conditions and reducing the computational domain to the conductor Ω_c. In fact, both papers exploit the well–known symmetric method of Costabel for the coupling of finite elements (FEM) and boundary elements (BEM). However, in [2] the problem is formulated in terms of the electric field, while [3] extends the early work of Bossavit and uses a model based on the magnetic field **H**. In the two cases, the Galerkin discretization relays on the **curl**-conforming finite elements of Nédélec.

Nevertheless, in transient processes and even in some situations with a sinusoidal supply voltage, it is not possible to assume a time-harmonic behavior for the whole electromagnetic system. Thus, we undertake here the *time–dependent* eddy current problem, and introduce and analyze a weak BEM–FEM formulation of this problem.

One effort of our formulation is to make no restrictions on the topology of Ω_c or on the regularity of its boundary. Indeed, when Ω_c is non–simply connected, both [2] and [3] require the construction of cumbersome (and expensive) cutting surfaces in order to deal correctly with the discrete problem. Recently, in [1] the **H**–based formulation of the time–harmonic eddy current problem (posed in a bounded domain) is rewritten with a saddle point structure that is free from the above restrictions. Such a formulation is obtained by introducing a Lagrange multiplier associated to the constrain satisfied by **curl H** in the insulator. We adopt here the same strategy and we show that this technique can be extended to the case of a time–dependent eddy current problem posed in the whole space.

2 Model problem

We want to approximate the eddy currents induced in a passive conductor $\Omega_c \subset \mathbb{R}^3$ by a current source with density $\mathbf{J}(t, \mathbf{x})$. Without loss of generality, we assume that the bounded domain Ω_c is connected, and eventually multiply–connected, and we denote by Σ_i ($i = 1, \cdots, I$) the connected components of its boundary Σ. Supposing that Σ is Lipschitz continuous, we may introduce **n** as the unit outward normal to Ω_c at almost every point of Σ.

The electric and magnetic fields $\mathbf{E}(t, \mathbf{x})$ and $\mathbf{H}(t, \mathbf{x})$ solve the following eddy current model:

$$\mu \partial_t \mathbf{H} + \mathbf{curl}\,\mathbf{E} = 0 \quad \text{in } (0, T) \times \mathbb{R}^3, \tag{1}$$

$$\mathbf{curl}\,\mathbf{H} = \mathbf{J} + \sigma\,\mathbf{E} \quad \text{in } (0, T) \times \mathbb{R}^3, \tag{2}$$

$$\mathrm{div}(\varepsilon\,\mathbf{E}) = 0 \quad \text{in } (0, T) \times (\mathbb{R}^3 \setminus \overline{\Omega_c}), \tag{3}$$

$$\int_{\Sigma_i} \varepsilon\,\mathbf{E} \cdot \mathbf{n}\, dS = 0 \quad \text{in } (0, T),\ \forall i = 1, \cdots, I, \tag{4}$$

$$\mathbf{E}(0, \mathbf{x}) = \mathbf{E}_0(\mathbf{x}) \quad \text{and} \quad \mathbf{H}(0, \mathbf{x}) = \mathbf{H}_0(\mathbf{x}) \quad \text{in } \mathbb{R}^3, \tag{5}$$

$$\mathbf{E}(t, \mathbf{x}) = O(\tfrac{1}{|\mathbf{x}|}) \quad \text{and} \quad \mathbf{H}(t, \mathbf{x}) = O(\tfrac{1}{|\mathbf{x}|}) \quad \text{as } |\mathbf{x}| \to \infty. \tag{6}$$

The permeability $\mu \in L^\infty(\mathbb{R}^3)$ and the electric permittivity $\varepsilon \in L^\infty(\mathbb{R}^3)$ are real valued and bounded functions that satisfy the conditions

$$\mu_1 \geq \mu(\mathbf{x}) \geq \mu_0 > 0 \text{ in } \Omega_c \quad \text{and} \quad \mu(\mathbf{x}) = \mu_0 \text{ in } \mathbb{R}^3 \setminus \overline{\Omega_c},$$
$$\varepsilon_1 \geq \varepsilon(\mathbf{x}) \geq \varepsilon_0 > 0 \text{ in } \Omega_c \quad \text{and} \quad \varepsilon(\mathbf{x}) = \varepsilon_0 \text{ in } \mathbb{R}^3 \setminus \overline{\Omega_c},$$

where $\mu_1, \mu_0, \varepsilon_1$ and ε_0 are positive constants. In addition, the conductivity $\sigma \in L^\infty((0, T) \times \mathbb{R}^3)$ may depend on time and satisfies the following property for a.e. $t \in (0, T)$:

$\sigma_1 \geq \sigma(t, \mathbf{x}) \geq \sigma_0 > 0$ at a.e. $\mathbf{x} \in \Omega_c$ and $\sigma(t, \mathbf{x}) = 0$ at a.e. $\mathbf{x} \in \mathbb{R}^3 \setminus \overline{\Omega}_c$,

being σ_1, σ_0 two positive constants.

Let us assume that $\operatorname{div} \mathbf{H}_0$ and $\mathbf{J}$ have bounded supports. Then we can introduce a connected and simply–connected bounded domain $\Omega \subseteq \mathbb{R}^3$ with

$\overline{\Omega}_c \cup \operatorname{support}(\operatorname{div} \mathbf{H}_0) \subset \Omega$ and $\operatorname{support}(\mathbf{J}(t)) \subset \Omega$ for a.e. $t \in (0, T)$.

Moreover, we may take Ω so that its boundary $\Gamma := \partial \Omega$ is Lipschitz continuous, connected and simply–connected, and we denote by $\mathbf{n}$ the unit outward normal to Ω at almost every point of Γ. We represent by $\Omega_d := \Omega \setminus \overline{\Omega}_c$ and $\Omega_e := \mathbb{R}^3 \setminus \overline{\Omega}$ the complement of Ω_c in Ω and in $\mathbb{R}^3$, respectively.

In the sequel, we write $\mathbf{J}^c := \mathbf{J}|_{(0,T) \times \Omega_c}$ and $\mathbf{J}^d := \mathbf{J}|_{(0,T) \times \Omega_d}$. From the property (2) and the hypothesis $\operatorname{support}(\mathbf{J}) \subset [0, T] \times \Omega$, we deduce that $\mathbf{J}^d$ must satisfy the following compatibility conditions for a.e. $t \in (0, T)$:

$$\operatorname{div} \mathbf{J}^d(t) = 0 \text{ in } \Omega_d,$$
$$\int_{\Sigma_i} \mathbf{J}^d(t) \cdot \mathbf{n} \, dS = 0 \, \forall i = 1, \cdots, I, \qquad (7)$$
$$\mathbf{J}^d(t) \cdot \mathbf{n} = 0 \text{ on } \Gamma.$$

To introduce the functional frame of our model problem (1–6), we define the spaces

$\mathbf{U} := \{\mathbf{q} \in \mathbf{L}^2(\mathbb{R}^3); \ \operatorname{div} \mathbf{q} = 0, \ \operatorname{\mathbf{curl}} \mathbf{q} = \mathbf{0} \text{ in } \Omega_e\}$, $\mathbf{X} := \mathbf{U} \cap \mathbf{H}(\operatorname{\mathbf{curl}}, \mathbb{R}^3)$,

$\mathbf{U}_0 := \{\mathbf{q} \in \mathbf{U}; \ \operatorname{\mathbf{curl}} \mathbf{q} = \mathbf{0} \text{ in } \mathbb{R}^3 \setminus \overline{\Omega}_c\}$, $\mathbf{X}_0 := \mathbf{U}_0 \cap \mathbf{X}$.

Notice that $\mathbf{X}$ and $\mathbf{U}$ are closed subspaces of $\mathbf{H}(\operatorname{\mathbf{curl}}, \mathbb{R}^3)$ and $\mathbf{L}^2(\mathbb{R}^3)$, respectively. Similarly, $\mathbf{X}_0$ and $\mathbf{U}_0$ are closed subspaces of $\mathbf{X}$ and $\mathbf{U}$, respectively. Moreover, the space $\mathbf{X}_0$ is densely embedded in $\mathbf{U}_0$; cf. [4, Lemma 3.1]. Then we may define $\mathbf{X}'_0$ as the dual space of $\mathbf{X}_0$ pivotal to $\mathbf{U}_0$ and consider

$$W^1(0, T; \mathbf{X}_0, \mathbf{X}'_0) := \{\mathbf{q} \in L^2(0, T; \mathbf{X}_0); \ \partial_t \mathbf{q} \in L^2(0, T; \mathbf{X}'_0)\}.$$

Besides, we introduce the space

$$\mathbf{M}(\Omega_d) := \Big\{\mathbf{m} \in \mathbf{L}^2(\Omega_d); \ \operatorname{div} \mathbf{m} = 0 \text{ in } \Omega_d,$$
$$\int_{\Sigma_i} \mathbf{m} \cdot \mathbf{n} \, dS = 0 \, \forall i = 1, \cdots, I, \ \mathbf{m} \cdot \mathbf{n} = 0 \text{ on } \Gamma\Big\}.$$

For each $\mathbf{m} \in \mathbf{M}(\Omega_d)$, we define

$$(\mathcal{E}\mathbf{m})(\mathbf{x}) := \begin{cases} 0 & \text{if } \mathbf{x} \in \Omega_e, \\ \mathbf{m}(\mathbf{x}) & \text{if } \mathbf{x} \in \Omega_d, \\ \nabla\rho(\mathbf{x}) & \text{if } \mathbf{x} \in \Omega_c, \end{cases}$$

where $\rho \in H^1(\Omega_c)$ is a harmonic function uniquely determined (up to an additive constant) by the Neumann boundary condition

$$\frac{\partial \rho}{\partial \mathbf{n}} = \mathbf{m} \cdot \mathbf{n} \quad \text{on } \Sigma.$$

We also consider

$$(\mathcal{M}\mathbf{m})(\mathbf{x}) := \frac{1}{4\pi} \operatorname{curl} \int_{\mathbb{R}^3} \frac{(\mathcal{E}\mathbf{m})(\mathbf{y})}{|\mathbf{x}-\mathbf{y}|} \, d\mathbf{y} \quad \forall \mathbf{m} \in \mathbf{M}(\Omega_d).$$

Then, it turns out that operator

$$\mathcal{M}: \mathbf{M}(\Omega_d) \to \mathbf{X} \cap \mathbf{H}^1(\mathbb{R}^3),$$

is linear continuous and satisfies $\operatorname{curl}(\mathcal{M}\mathbf{m}) = \mathbf{m}$ in Ω_d for every $\mathbf{m} \in \mathbf{M}(\Omega_d)$; cf. [4, Lemma 3.3].

We propose the following global weak formulation of problem (1–6):

find $\mathbf{H} \in W^1(0,T;\mathbf{X}_0,\mathbf{X}_0') + \mathcal{M}(\mathbf{J}^d(t))$ such that

$$\frac{d}{dt}\int_{\mathbb{R}^3} \mu \mathbf{H}(t) \cdot \mathbf{q} \, dx + c(t, \mathbf{H}(t), \mathbf{q}) = L(t, \mathbf{q}) \quad \text{in } \mathcal{D}'(0,T), \, \forall \mathbf{q} \in \mathbf{X}_0, \quad (8)$$

$$\mathbf{H}(0) = \mathbf{H}_0,$$

where $L(t, \mathbf{q}) := \int_{\Omega_c} \sigma(t)^{-1} \mathbf{J}^c(t) \cdot \operatorname{curl} \mathbf{q} \, dx$ and

$$c(t, \mathbf{q}_1, \mathbf{q}_2) := \int_{\Omega_c} \sigma(t)^{-1} \operatorname{curl} \mathbf{q}_1 \cdot \operatorname{curl} \mathbf{q}_2 \, dx.$$

Notice that the second equation of (8) is meaningful, as $W^1(0,T;\mathbf{X}_0,\mathbf{X}_0')$ is continuously embedded in $\mathcal{C}^0([0,T];\mathbf{U}_0)$. The following result is a consequence of the classical Theorem of J.L. Lions; see [4, Theorem 3.4] for the details.

Theorem 1. *Assume that* $\mathbf{J}^d \in H^1(0,T;\mathbf{M}(\Omega_d))$ *and* $\mathbf{J}^c \in L^2((0,T) \times \Omega_c)$. *Assume also that* $\mathbf{H}_0 \in \mathbf{U}$ *satisfies the compatibility condition* $\operatorname{curl}\mathbf{H}_0 = \mathbf{J}^d(0)$ *in* Ω_d. *Then problem* (8) *has a unique solution* $\mathbf{H}$.

Next, we handle the constraint $\operatorname{curl}\mathbf{H} = \mathbf{J}^d$ in $(0,T) \times \Omega_d$ by means of a Lagrange multiplier. More precisely, we consider the following saddle–point structure corresponding to problem (8):

find $\mathbf{H} \in W^1(0,T;\mathbf{X},\mathbf{X}')$ and $\mathbf{r} \in L^2(0,T;\mathbf{M}(\Omega_d))$ such that

$$\frac{\mathrm{d}}{\mathrm{d}t}\left[\int_{\mathbb{R}^3} \mu\,\mathbf{H}(t)\cdot\mathbf{q}\,dx + d(\mathbf{q},\mathbf{r}(t))\right] + c(t,\mathbf{H}(t),\mathbf{q}) = L(t,\mathbf{q}) \quad \forall \mathbf{q}\in\mathbf{X},$$
$$d(\mathbf{H}(t),\mathbf{m}) = \int_{\Omega_d} \mathbf{J}^d(t)\cdot\mathbf{m}\,dx \quad \forall \mathbf{m}\in\mathbf{M}(\Omega_d),$$
$$\mathbf{H}(0) = \mathbf{H}_0,$$
(9)

where

$$d(\mathbf{q},\mathbf{m}) := \int_{\Omega_d} \mathbf{m}\cdot\operatorname{curl}\mathbf{q}\,dx \quad \forall \mathbf{q}\in\mathbf{X},\,\mathbf{m}\in\mathbf{M}(\Omega_d).$$

The following result is proven in [4, theorem 4.2].

Theorem 2. *Under the hypothesis of Theorem 1, problem (9) has a unique solution. Moreover, the field $\mathbf{H}$ is the same for problems (8) and (9).*

In order to deduce a BEM–FEM formulation of problem (9), we first notice that, given $\mathbf{q}\in\mathbf{X}$, there is a unique harmonic potential

$$\varphi \in W^1(\Omega_e) := \left\{\psi \in \mathcal{D}'(\Omega_e);\; \frac{\psi}{\sqrt{1+|\mathbf{x}|^2}} \in L^2(\Omega_e),\; \nabla\psi \in \mathbf{L}^2(\Omega_e)\right\},$$

such that $\nabla\varphi = \mathbf{q}|_{\Omega_e}$. In particular, as φ is harmonic in Ω_e, its trace and normal derivative $\dfrac{\partial\varphi}{\partial\mathbf{n}}$ on Γ are related through the boundary integral equations

$$\varphi = (\tfrac{1}{2}\mathcal{I}+\mathcal{K})\varphi - \mathcal{V}\frac{\partial\varphi}{\partial\mathbf{n}} \quad \text{and} \quad \frac{\partial\varphi}{\partial\mathbf{n}} = -\mathcal{H}\varphi + (\tfrac{1}{2}\mathcal{I}-\mathcal{K}^*)\frac{\partial\varphi}{\partial\mathbf{n}} \quad \text{on }\Gamma. \quad (10)$$

We have denoted by $\mathcal{V}$ and $\mathcal{K}$ the single and double layer potentials:

$$\mathcal{V}\lambda(\mathbf{x}) := \int_\Gamma E(\mathbf{x},\mathbf{y})\,\lambda(\mathbf{y})\,dS_\mathbf{y} \quad \text{and} \quad \mathcal{K}\lambda(\mathbf{x}) := \int_\Gamma \frac{\partial E(\mathbf{x},\mathbf{y})}{\partial\mathbf{n}(\mathbf{y})}\,\lambda(\mathbf{y})\,dS_\mathbf{y} \quad \forall\mathbf{x}\in\Gamma,$$

being $E(\mathbf{x},\mathbf{y}) := \dfrac{1}{4\pi|\mathbf{x}-\mathbf{y}|}$ the fundamental solution of Laplace equation in $\mathbb{R}^3$. We have also denoted by $\mathcal{K}^*$ and $\mathcal{H}$ the dual operator of $\mathcal{K}$ and the hypersingular operator, respectively. It is given by

$$\mathcal{H}\lambda(\mathbf{x}) := -\frac{\partial}{\partial\mathbf{n}(\mathbf{x})}\left[\int_\Gamma \frac{\partial E(\mathbf{x},\mathbf{y})}{\partial\mathbf{n}(\mathbf{y})}\,\lambda(\mathbf{y})\,dS_\mathbf{y}\right] \quad \forall\mathbf{x}\in\Gamma.$$

Let $\mathbf{H}_\times^{-1/2}(\Gamma)$ be the dual space of $\mathbf{H}_\times^{1/2}(\Gamma) := \{\mathbf{q}\times\mathbf{n};\,\mathbf{q}\in\mathbf{H}^1(\Omega)\}$ pivotal to $\mathbf{L}_t^2(\Gamma) := \{\mathbf{q}\in\mathbf{L}^2(\Gamma);\,\mathbf{q}\cdot\mathbf{n}=0\text{ on }\Gamma\}$. For any $\lambda \in H^{1/2}(\Gamma)$, we introduce

$$\operatorname{curl}_\Gamma(\lambda) := (\nabla\varphi)\times\mathbf{n} \quad \text{on }\Gamma,$$

where $\varphi \in H^1(\Omega_e)$ satisfies that $\varphi|_\Gamma = \lambda$ on Γ. Then, the differential operator

$$\mathbf{curl}_\Gamma \colon H^{1/2}(\Gamma) \to \mathbf{H}_\times^{-1/2}(\Gamma)$$

is continuous and is known as the vectorial *surface rotational* operator; moreover, the operator

$$\mathbf{curl}_\Gamma \colon H^{1/2}(\Gamma)/\mathbb{R} \to \{\mathbf{q} \times \mathbf{n};\ \mathbf{q} \in \mathbf{X}(\Omega)\}$$

defines an isomorphism; cf. [3, Proposition 2.5]. Thus, recalling that it holds the identity $\mathcal{K}2 = 1$, we may introduce the linear continuous operator

$$\mathcal{B} \colon \mathbf{X}(\Omega) \to H^{1/2}(\Gamma)/\mathbb{R}$$

$$\mathbf{q} \mapsto \mathcal{B}\mathbf{q} := (\tfrac{1}{2}\mathcal{I} - \mathcal{K}) \circ \mathbf{curl}_\Gamma^{-1}(\mathbf{q} \times \mathbf{n}).$$

We also define the following version of the vectorial single layer potential:

$$\mathcal{V}_\times \mathbf{u}(\mathbf{x}) := \left[\int_\Gamma E(\mathbf{x},\mathbf{y})\,\mathbf{u}(\mathbf{y})\,dS_\mathbf{y}\right] \times \mathbf{n}(\mathbf{x}) \quad \forall \mathbf{u} \in \mathbf{H}_\times^{-1/2}(\Gamma).$$

In [3, Lemma 3.4], it is shown that this operator is related to the hypersingular operator through

$$\int_\Gamma \mathcal{H}\psi\,\varphi\,dx = \int_\Gamma (\mathbf{curl}_\Gamma \varphi \times \mathbf{n}) \cdot \mathcal{V}_\times(\mathbf{curl}_\Gamma \psi)\,dS \quad \forall \psi, \varphi \in H^{1/2}(\Gamma). \tag{11}$$

We consider $\widehat{\mathbf{X}} := \mathbf{X}(\Omega) \times H_0^{-1/2}(\Gamma)$, with

$$H_0^{-1/2}(\Gamma) := \left\{\eta \in H^{-1/2}(\Gamma);\ \int_\Gamma \eta\,dS = 0\right\}.$$

For each $\widehat{\mathbf{q}}_1 := (\mathbf{q}_1, \eta_1) \in \widehat{\mathbf{X}}$ and $\widehat{\mathbf{q}}_2 := (\mathbf{q}_2, \eta_2) \in \widehat{\mathbf{X}}$, we define the bilinear bounded form

$$a(\widehat{\mathbf{q}}_1, \widehat{\mathbf{q}}_2) := \int_\Omega \mu\,\mathbf{q}_1 \cdot \mathbf{q}_2\,dx + \mu_0 \int_\Gamma ((\mathbf{q}_2 \times \mathbf{n}) \times \mathbf{n}) \cdot \mathcal{V}_\times(\mathbf{q}_1 \times \mathbf{n})\,dS$$

$$+ \mu_0 \int_\Gamma \eta_2 \mathcal{V}\eta_1\,dS + \mu_0 \int_\Gamma (\eta_2\,\mathcal{B}\mathbf{q}_1 - \eta_1\,\mathcal{B}\mathbf{q}_2)\,dS.$$

Let us combine the first equation of problem (9) and the integral equations (10). Using the property (11) to eliminate the hypersingular operator from the equations, we obtain the following BEM–FEM variational formulation of problem (1–6):

find $\widehat{\mathbf{h}} := (\mathbf{h}, \lambda) \in L^2(0, T; \widehat{\mathbf{X}}) \cap \mathcal{C}^0([0, T]; \mathbf{L}^2(\Omega) \times H_0^{-1/2}(\Gamma))$
and $\mathbf{r} \in L^2(0, T; \mathbf{M}(\Omega_d))$ such that

$$\frac{d}{dt}\left[a(\widehat{\mathbf{h}}(t), \widehat{\mathbf{q}}) + d(\mathbf{q}, \mathbf{r}(t))\right] + c(t, \mathbf{h}(t), \mathbf{q}) = L(t, \mathbf{q}) \quad \forall \widehat{\mathbf{q}} \in \widehat{\mathbf{X}}, \qquad (12)$$

$$d(\mathbf{h}(t), \mathbf{m}) = \int_{\Omega_d} \mathbf{J}^d(t) \cdot \mathbf{m}\, d\mathbf{x} \quad \forall \mathbf{m} \in \mathbf{M}(\Omega_d),$$

$$\widehat{\mathbf{h}}(0) = \widehat{\mathbf{h}}_0,$$

where $\widehat{\mathbf{h}}_0 := (\mathbf{H}_0|_\Omega, \mathbf{H}_0|_{\Omega_e} \cdot \mathbf{n})$, being $\mathbf{H}_0 \in \mathbf{U}$ the initial magnetic field.

Theorem 3. *Under the hypothesis of Theorem 1, problem (12) has a unique solution. Moreover, its solution is related to that of problem (9) by* $\mathbf{h} = \mathbf{H}|_\Omega$, $\lambda = \mathbf{H}|_{\Omega_e} \cdot \mathbf{n}$ *and the Lagrange multiplier* $\mathbf{r}$ *is the same for both problems.*

3 Semi–discrete problem

From now on, we assume that Ω and Ω_c are Lipschitz polyhedra. Let $\{\mathcal{T}_h\}_h$ be a regular family of meshes of Ω such that each $T \in \mathcal{T}_h$ is a tetrahedra completely contained in $\overline{\Omega}_c$ or in $\overline{\Omega}_d$. We denote by $\mathcal{T}_h(\Omega_d)$ the restriction of $\mathcal{T}_h$ to Ω_d. In addition, we represent by $\mathcal{T}_h(\Gamma)$ the triangular mesh induced by $\mathcal{T}_h$ on Γ. As usual, h denotes the longest diameter of the tetrahedra in $\mathcal{T}_h$.

We consider a conforming discretization of the space $\mathbf{H}(\mathbf{curl}, \Omega)$ by means of the Nédélec edge finite elements:

$$\mathcal{ND}_h(\Omega) := \left\{\mathbf{q} \in \mathbf{H}(\mathbf{curl}, \Omega);\ \mathbf{q}|_T \in \mathcal{ND}(T)\ \forall T \in \mathcal{T}_h\right\},$$

being $\mathcal{ND}(T) := \left\{\mathbf{a} \times \mathbf{x} + \mathbf{b};\ \mathbf{a}, \mathbf{b} \in \mathbb{R}^3\right\}$. We approximate the magnetic field by means of the space

$$\mathbf{X}_h(\Omega) := \mathcal{ND}_h(\Omega) \cap \mathbf{X}(\Omega).$$

We estimate the boundary unknown using elements that are piecewise constant on the triangulation $\mathcal{T}_h(\Gamma)$ of Γ:

$$\Lambda_h := \left\{\eta \in L^2(\Gamma);\ \int_\Gamma \eta\, dS_\mathbf{x} = 0,\ \eta|_F \in \mathbb{R}\ \forall F \text{ face of } \mathcal{T}_h(\Gamma)\right\}.$$

We introduce a conforming discretization of $\mathbf{H}(\mathrm{div}, \Omega_d)$ using the lowest order Raviart–Thomas finite elements:

$$\mathcal{RT}_h(\Omega_d) := \{\mathbf{q} \in \mathbf{H}(\mathrm{div}, \Omega);\ \mathbf{q}|_T \in \mathcal{RT}(T)\ \forall T \in \mathcal{T}_h(\Omega_d)\},$$

where $\mathcal{RT}(T) := \{a\mathbf{x} + \mathbf{b};\ a \in \mathbb{R},\ \mathbf{b} \in \mathbb{R}^3\}$. We approximate the Lagrange multiplier by means of the space

$$\mathbf{M}_h(\Omega_d) := \mathcal{R}\mathcal{T}_h(\Omega_d) \cap \mathbf{M}(\Omega_d).$$

Then the semi-discrete scheme corresponding to problem (12) reads find $\widehat{\mathbf{h}}_h(t) := (\mathbf{h}_h(t), \lambda_h(t)) \in \mathbf{X}_h(\Omega) \times \Lambda_h$ and $\mathbf{r}_h \in \mathbf{M}_h(\Omega_d)$ such that

$$\frac{d}{dt}\left[a(\widehat{\mathbf{h}}_h(t), \widehat{\mathbf{q}}) + d(\mathbf{q}, \mathbf{r}_h(t))\right] + c(t, \mathbf{h}_h(t), \mathbf{q}) = L(t, \mathbf{q}) \quad \forall \widehat{\mathbf{q}} \in \mathbf{X}_h(\Omega) \times \Lambda_h,$$
$$d(\mathbf{h}_h(t), \mathbf{m}) = \int_{\Omega_d} \mathbf{J}^d(t) \cdot \mathbf{m} \, dx \quad \forall \mathbf{m} \in \mathbf{M}_h(\Omega_d), \tag{13}$$

for each $0 < t \leq T$, and satisfying approximated initial conditions

$$\mathbf{h}_h(0) \approx \mathbf{h}_0 \quad \text{and} \quad \lambda_h(0) \approx \lambda_0. \tag{14}$$

It is important to remark that, as shown in [3, 4], the inverse operator of $\mathbf{curl}_\Gamma$ is involved in the definition of the bilinear form $a(\cdot, \cdot)$, but it is not used in the effective calculus of the associated matrix. Therefore, the numerical method turns out to be implementable and computationally efficient.

The following result is shown in [4, Theorem 7.3].

Theorem 4. *Problem (13–14) admits a unique solution.*

For the sake of simplicity, we assume that $\mathbf{h}_h(0) \in \mathbf{X}_h(\Omega)$ is the Nédélec interpolate of $\mathbf{h}_0$ and that $\lambda_h(0) \in \Lambda_h$ is characterized by

$$\int_\Gamma \eta \, \mathcal{V}\lambda_h \, dS = \int_\Gamma \eta \, \mathcal{B}\mathbf{h}_h(0) \, dS \quad \forall \eta \in \Lambda_h.$$

We conclude with two results on the convergence of our numerical method; cf. [4, Theorem 8.2, Corollaries 8.3 and 8.4].

Theorem 5. *Let us assume that the exact solution has the regularity*

$$\mathbf{h}(t) \in \mathcal{C}^1([0, T]; \mathbf{H}^s(\mathbf{curl}, \Omega) \cap \mathbf{X}(\Omega))$$

for some $1/2 < s \leq 1$. Then,

$$\sup_{t \in [0,T]} \|\mathbf{h}(t) - \mathbf{h}_h(t)\|^2_{\mathbf{L}^2(\Omega)} + \sup_{t \in [0,T]} \|\mathbf{h}(t) \times \mathbf{n} - \mathbf{h}_h(t) \times \mathbf{n}\|^2_{\mathbf{H}_\times^{-1/2}(\Gamma)}$$
$$+ \int_0^T \|\mathbf{h}(t) - \mathbf{h}_h(t)\|^2_{\mathbf{H}(\mathbf{curl}, \Omega_c)} \, dt + \sup_{t \in [0,T]} \|\lambda(t) - \lambda_h(t)\|^2_{H^{-1/2}(\Gamma)}$$
$$\leq C h \left[\sup_{t \in [0,T]} \|\mathbf{h}\|^2_{\mathbf{H}^s(\mathbf{curl}, \Omega)} + \sup_{t \in [0,T]} \|\partial_t \mathbf{h}\|^2_{\mathbf{H}^s(\mathbf{curl}, \Omega)} \right].$$

Corollary 1. *Let us assume that the Lagrange multiplier has the regularity*

$$\mathbf{r} \in L^2(0, T; \mathbf{H}^r(\Omega_d) \cap \mathbf{M}(\Omega_d))$$

for some $1/2 < r \leq 1$. Then, under the hypothesis of Theorem 5,

$$\int_0^T \|\mathbf{r}(t) - \mathbf{r}_h(t)\|^2_{\mathbf{L}^2(\Omega_d)} \, dt$$
$$\leq C h \left[\sup_{t \in [0,T]} \|\mathbf{h}\|^2_{\mathbf{H}^s(\mathbf{curl},\Omega)} + \sup_{t \in [0,T]} \|\partial_t \mathbf{h}\|^2_{\mathbf{H}^s(\mathbf{curl},\Omega)} + \int_0^T \|\mathbf{r}(t)\|^2_{\mathbf{H}^r(\Omega_d)} \, dt \right].$$

References

1. Alonso, A., Hitpmair, R., Valli, A.: Mixed finite element approximation of eddy current problems. IMA J. Numer. Anal., **24**, 255–271 (2004)
2. Hiptmair, R.: Symmetric coupling for eddy current problems. SIAM J. Numer. Anal., **40**(1), 41–65 (2002)
3. Meddahi, S., Selgas, V.: A mixed–FEM and BEM coupling for a three–dimensional eddy current problem. M2AN Math. Model. Numer. Anal., **37**(2), 291–318 (2003)
4. Meddahi, S., Selgas, V.: An **H**–based BEM–FEM formulation for a time dependent eddy current problem. Submitted.

Mixed Boundary Element–Finite Volume Methods for Thermohydrodynamic Lubrication Problems*

J. Durany[1], J. Pereira[1] and F. Varas[1]

Dpto. Matemática Aplicada II. Universidad de Vigo. Campus Marcosende.
36271-Vigo (Spain)
durany@dma.uvigo.es, curro@dma.uvigo.es

Summary. This work is focused on a steady coupled model for pressure and temperature computations in lubricated journal bearing devices, including the thermal exchange with the environment through the bush and the shaft. The thermohydrodynamic problem is decoupled through a fixed point procedure. In this way, a finite element method for the hydrodynamic Reynolds equation with a cavitation model of Elrod-Adams is applied. Then, the energy equation in the lubricating film is solved by a second–order cell–vertex volume method and the heat equation on the bush by using a P1 collocation boundary element method while the very simple model considered for the shaft is straightforwardly integrated. Finally, some remarks are made about the extension of the present algorithm to the corresponding transient problem.

1 Introduction

In journal bearings modelling, the isothermal theory has often been used as a simplification of many problems that lead to calculate the pressure distribution. However, the dissipated energy by viscous effects is very significant when the device operates under high rotation and/or considerable imposed loads. In these cases, a thermal problem must be solved in the lubricant film and a coupled system of partial differential equations is obtained. The coupling is given by the viscosity influence in the hydrodynamic equation, and the velocity field obtained from pressure gradients that it is introduced in the energy equation. Additionally, the bush and shaft thermal exchange with the external environment must be included in the model.

Usually, energy equation in the lubricant film is solved with first–order schemes (due to the simple upwinding used for convective terms) and finite

---

* This work has been supported by projects MTM2004-05796-C02-02 of the MEC of Spain and PGIDT05PXIC32202PN of Xunta Galicia.

difference or finite element methods are considered to solve the thermal problem in the bush (see, for instance, [1, 5, 7, 9] and [10] and references therein). Both types of approaches lead to a high computational cost if an accurate solution must be computed.

In this work the steady thermohydrodynamic problem is decoupled through a fixed point procedure. To solve the hydrodynamic subproblem a finite element method (FEM) for the Reynolds equation with the Elrod-Adams cavitation model is applied. The solution of this free boundary problem is obtained by means of a duality method applied to a maximal monotone operator (see [4]). Next, the energy equation in the lubricating film is solved by using a cell–vertex volume method (FVM) (see Morton–Stynes–Süli [8]). The main advantage of this method lies in the possibility of retrieving second–order convergence under some assumptions.

The analysis also takes into account the heat transfer by conduction within the bush and shaft. In the first case the bushing temperature distribution is computed by using a P1 collocation boundary element method (BEM). This approximation joined to the symmetry properties lead to a low computational cost.

Additionally, since the fixed point procedure is formulated through an artificial time-stepping, we briefly consider the extension of the present algorithm to the (real) evolution problem.

2 Thermohydrodynamic mathematical model

2.1 Hydrodynamic cavitation model

The non-dimensional fluid domain between the shaft and the bush is first transformed through a change of variable into $\Omega_1 = [0, 2\pi] \times [0, 1]$ for the $(\theta, \bar{z})$ coordinates and the heigh:

$$\bar{h} = 1 + \varepsilon \cos\theta, \tag{1}$$

where ε is the journal-bearing eccentricity coefficient which is assumed to be given (in practice it must be computed by solving an inverse problem related to the mechanical load acting on the shaft).

So, in the geometrical system $(\theta, \bar{y}, \bar{z})$ the Reynolds equation for newtonian fluids with constant density takes the expression (see [6], for example):

$$\frac{\partial}{\partial \theta}\left(\bar{h}^3 \bar{G} \frac{\partial \bar{p}}{\partial \theta}\right) + \eta^2 \frac{\partial}{\partial \bar{z}}\left(\bar{h}^3 \bar{G} \frac{\partial \bar{p}}{\partial \bar{z}}\right) = \frac{\partial}{\partial \theta}\left(\bar{h} - \bar{h}\frac{\bar{I}_2}{\bar{J}_2}\right), \tag{2}$$

where $\bar{p}$ is the fluid pressure, $\eta = R_s/L$, a coefficient relating radius and shaft length, and

$$\bar{G} = \int_0^1 \frac{\bar{y}}{\bar{\mu}} \left(\bar{y} - \frac{\bar{I}_2}{\bar{J}_2} \right) d\bar{y} \quad , \quad \bar{I}_2 = \int_0^1 \frac{\bar{y}}{\bar{\mu}} d\bar{y} \quad , \quad \bar{J}_2 = \int_0^1 \frac{d\bar{y}}{\bar{\mu}} \quad , \quad (3)$$

where the fluid viscosity, $\bar{\mu}$, depends on the temperature following the behaviour law:

$$\bar{\mu} = e^{-\beta T_0 (\bar{T}-1)}, \quad (4)$$

with β the thermoviscous coefficient and T_0 the reference temperature.

The mathematical formulation of the cavitation free boundary problem is posed by means of the so called Elrod-Adams model (see [3], for example). It introduces an additional unknown to the original problem: the saturation ϑ, which represents the lubricating fluid concentration (i.e. the volume fraction filled with the lubricant fluid, that takes the value 1 for the fluid part, Ω_1^+, and takes any other value between 0 and 1 for the cavitated one, Ω_1^0). In this way, the hydrodynamic problem can be written as follows:

To find (p, ϑ) such that:

$$\frac{\partial}{\partial \theta} \left(\bar{h}^3 \bar{G} \frac{\partial \bar{p}}{\partial \theta} \right) + \eta^2 \frac{\partial}{\partial z} \left(\bar{h}^3 \bar{G} \frac{\partial \bar{p}}{\partial z} \right) = \frac{\partial}{\partial \theta} \left(\bar{h} \left(1 - \frac{\bar{I}_2}{\bar{J}_2} \right) \right) , \quad (5)$$

$$p > 0 \text{ and } \vartheta = 1 \text{ in } \Omega_1^+ ,$$

$$\frac{\partial}{\partial \theta} \left(1 - \frac{\bar{I}_2}{\bar{J}_2} \right) = 0, \quad p = 0, \quad 0 \le \vartheta \le 1 \text{ in } \Omega_1^0 , \quad (6)$$

$$\bar{h}^3 \bar{G} \frac{\partial \bar{p}}{\partial n} = (1 - \vartheta) \bar{h} \left(1 - \frac{\bar{I}_2}{\bar{J}_2} \right) \cos(\mathbf{n}, \mathbf{i}), \quad p = 0 \text{ on } \Sigma , \quad (7)$$

$$p = 0 \text{ in } \partial \Omega_1 , \quad \vartheta = \vartheta_0 \text{ on } \Gamma_0 , \quad (8)$$

where Σ represents the free boundary between the lubricated region (Ω_1^+) and the cavitated one (Ω_1^0), $\mathbf{n}$ the normal vector to Σ, $\mathbf{i}$ the unitary normal vector pointing to θ-direction and Γ_0 the boundary fluid supply ($\theta = 0$).

2.2 Fluid thermal model

In many situations the axial gradients of temperature can be neglected. So, the dimensionless energy equation for temperature $\bar{T}_f$ is posed in terms of longitudinal and thickness coordinates $(\theta, \bar{y})$ (see [6], among others):

$$Pe \left[\bar{u} \frac{\partial \bar{T}_f}{\partial \theta} + \frac{1}{\bar{h}} \left(\bar{v} - \bar{u} \bar{y} \frac{\partial \bar{h}}{\partial \theta} \right) \frac{\partial \bar{T}_f}{\partial \bar{y}} \right] - \frac{1}{\bar{h}^2} \frac{\partial^2 \bar{T}_f}{\partial \bar{y}^2} = N_d \frac{\bar{\mu}}{\bar{h}^2} \left(\frac{\partial \bar{u}}{\partial \bar{y}} \right)^2 \text{ in } \Omega_f , \quad (9)$$

where Pe is the Peclet number, N_d a nondimensional parameter associated to viscous dissipation, $\Omega_f = [0, 2\pi] \times [0, 1]$ the nondimensional longitudinal section and the velocity components, $\mathbf{v} = (\bar{u}, \bar{v})$, depending on the fluid pressure:

$$\bar{u} = \bar{h}^2 \frac{\partial \bar{p}}{\partial \theta}\left(\bar{I} - \frac{\bar{I}_2}{\bar{J}_2}\bar{J}\right) + \frac{\bar{J}}{\bar{J}_2} \quad ; \quad \bar{v} = -\bar{h}\int_0^{\bar{y}}\left(\frac{\partial \bar{u}}{\partial \theta} - \frac{y}{\bar{h}}\frac{d\bar{h}}{d\theta}\frac{\partial \bar{u}}{\partial \bar{y}}\right)d\xi, \quad (10)$$

with $\bar{I}_2$, $\bar{J}_2$ defined in (3) and the new integrals, $\bar{I}$, $\bar{J}$, given by:

$$\bar{I} = \int_0^{\bar{y}} \frac{\xi}{\bar{\mu}} d\xi \quad ; \quad \bar{J} = \int_0^{\bar{y}} \frac{d\xi}{\bar{\mu}} . \quad (11)$$

The conservation energy equation (9) must be reformulated in the cavitated region to take into account the mixture vapour-lubricant. Assuming that the vapour velocity is equal to the lubricant one and the thermal generation in the vapour is null, the thermal problem can be rewritten in Ω_f following the expression (9), but now Pe and Nd depend on the lubricant concentration:

$$Pe = \frac{\omega C^2[\rho_f c_f \vartheta + \rho_a c_a(1-\vartheta)]}{[k_f \vartheta + k_a(1-\vartheta)]} \quad ; \quad Nd = \frac{\mu_0 \omega^2 R_s^2}{[k_f \vartheta + k_a(1-\vartheta)]T_0}, \quad (12)$$

with $\rho_f, c_f, k_f, \rho_a, c_a, k_a$ denoting the density, specific heat and thermal conductivity of the fluid and gas, respectively, ω the rotation speed, C the clearance of the journal-bearing, μ_0 the viscosity at the reference temperature, R_s the journal radius and T_0 the reference lubricant temperature.

Equation (9) is completed by adding Dirichlet conditions on the boundary supply, $\theta = 0$, and on the contact boundaries with shaft, ($\bar{y} = 1$), and bush, ($\bar{y} = 0$).

2.3 Bush and shaft thermal models

The bush thermal model can be formulated as a diffusion problem with boundary conditions related to the fluid temperature distribution on the internal boundary, and the environment temperature on the external one:

$$-div(k_b \nabla T_b) = 0 \quad in \ \Omega_b, \quad (13)$$

$$-k_b \frac{\partial T_b}{\partial n} = h_a(T_b - T_a) \quad on \ \Gamma_b^e, \quad (14)$$

$$i) \ T_b = T_0 \bar{T}_f \quad and \quad ii) -k_b \frac{\partial T_b}{\partial \mathbf{n}} = k\frac{T_0}{C\bar{h}}\frac{\partial \bar{T}_f}{\partial \bar{y}} \quad on \ \Gamma_b^i, \quad (15)$$

where T_b denotes the bush temperature, Ω_b the annular bush domain with the internal boundary Γ_b^i and the external one Γ_b^e, k_b and k the diffusion coeficients in the bush and the fluid, respectively, $\mathbf{n}$ the unitary normal vector to the boundaries, T_a the environment temperature and h_a the convective external flux coefficient.

On the other hand, an uniform temperature model is considered for the shaft due to the fact that it is highly rotating. So, the thermal equilibrium in the shaft is given by

$$\int_{\Gamma_s} k \frac{T_0}{C\bar{h}} \frac{d\bar{T}_f}{d\bar{y}} \, dl = h_s(T_s - T_a) \,, \tag{16}$$

where T_s is the shaft temperature, Γ_s represents the boundary in contact with the fluid, h_s is the environment convective coefficient and the normal derivative on the boundary is the corresponding vertical one.

3 Numerical solution of hydrodynamic and fluid thermal models

The numerical solution of the thermohydrodynamic problem in the fluid has been the subject of some of the previous works developed by the authors. This section presents an abstract of the outline explained in Durany–Pereira–Varas [4].

The hydrodynamic problem is solved by means of an algorithm based on the following elements:

- reformulation of the hydrodynamic problem as an artificial evolution one including the convective terms into the material derivative through the corresponding characteristic curves. Next, it will be integrated by means of an explicit scheme,
- space semidiscretization of the lubrication problem with cavitation by means of a finite element scheme applied to the Elrod–Adams model (5)–(8) for the pressure. The solution of the free boundary problem is obtained by using a duality algorithm applied to a monotone maximal operator (see [3]).

The discretization of the thermal model in the lubricant is based on:

- spatial semidiscretization by means of a second–order cell–vertex finite volume scheme (see [8]), that incorporates a fourth–order artificial dissipation term in order to equilibrate the number of equations and degrees of freedom and also to improve the front–capturing features. This scheme allows to reduce the oscillations at the fronts adding a minimal numerical dissipation,
- addition of an artificial thermal inertia to obtain a fictitious evolution problem to solve the nonlinearity as well as the coupling with the thermal models in the bush and shaft:

$$\omega_l \frac{d\mathbf{T}_l^h}{dt} = \mathbf{N}(\mathbf{T}_l^h), \tag{17}$$

where $\mathbf{T}_l^h$ represents the vector of nodal temperature approximations, $\mathbf{N}(\mathbf{T}_l^h)$ is the matrix with the assignations of the nodal residuals and ω_l denotes the artificial inertia.

4 Numerical solution in the bush and the shaft

The thermal problem (13), (14),(15ii) can be easily rewritten as an integral equation

$$k_b\, c(\bar{x})T_b(\bar{x}) + \int_{\Gamma_b} k_b\, T_b \frac{\partial G_{\bar{x}}}{\partial n}\, d\Gamma - \int_{\Gamma_b} G_{\bar{x}}\, k_b \frac{\partial T_b}{\partial n}\, d\Gamma = 0, \quad (18)$$

and then it is discretized by using the boundary element method (BEM) with P_1 continuous elements and a collocation scheme. This approximation can use the exact representation of the geometry by taking the mesh generated from the edges of the finite volume mesh used for the fluid thermal model. On the other hand, the boundary condition (15i) is forced by identifying the nodal values in both discretizations.

To solve the coupling with the lubricant thermal model a new artificial thermal inertia is introduced. This lead to compute the solution of the following pseudo–evolution problem

$$\omega_b \frac{d\mathbf{T}_b^h}{dt} = \mathbf{A}\mathbf{T}_b^h + \mathbf{B}\mathbf{\Phi}^h, \quad (19)$$

where $\mathbf{T}_b^h$ represents the vector of nodal temperature approximations, $\mathbf{\Phi}^h$ the discretized heat flux vector related to the boundary conditions, ω_b the artificial thermal inertia and $\mathbf{A}$, $\mathbf{B}$ are the (non–sparse) matrices associated to the boundary element discretization (the geometrical symmetry allows to compute only one arrow of each matrix).

Finally, the shaft thermal model (16) is also converted to a fictitious evolution problem by means of another artificial thermal inertia ω_s:

$$\omega_s \frac{dT_s}{dt} = \int_{\Gamma_s} k_f \frac{T_0}{C\bar{h}} \frac{d\bar{T}_f}{d\bar{y}} - h_s(T_s - T_a). \quad (20)$$

5 Numerical solution of the global problem

The global scheme of resolution of the thermohydrodynamic coupled problem consists of a fixed point procedure. In particular, we consider given an initial guess of the steady (discrete) temperatures in the lubricant, bush and shaft: $\mathbf{T}_{l,0}^{h,s}$, $\mathbf{T}_{b,0}^{h,s}$ and $\mathbf{T}_{s,0}^{h,s}$, and then proceed as follows:
For $k = 1, 2, 3, ..., k_{max}$

(a) calculate $\bar{G}^k$, $\bar{I}_2^k$ and $\bar{J}_2^k$, defined by (3), with $\mathbf{T}_{l,k-1}^{h,s}$ in (4), and solve the hydrodynamic problem (5)–(8) by using the techniques described in Section 3 to obtain $\mathbf{p}_k^h$ and $\boldsymbol{\vartheta}_k^h$.

(b) calculate $\bar{u}^k$ and $\bar{v}^k$, defined by (10), with $\mathbf{p}_k^h$ and $\mathbf{T}_{l,k-1}^{h,s}$, and solve the global thermal problem by a segregated temporal integration with fixed time step $\Delta t = 1$:

For $n = 1, 2, 3, ..., n_{max}$

(1) semi–implicit integration scheme for the temperature $\mathbf{T}_{l,k}^{h,n}$ in the fluid

$$(\omega_l \mathbf{I} + \mathbf{N}_l^k)\mathbf{T}_{l,k}^{h,n} = \omega_l \mathbf{T}_{l,k}^{h,n-1} + \mathbf{N}_{nl}^k(\mathbf{T}_{l,k}^{h,n-1}) + \mathbf{b}_k^{n-1}, \qquad (21)$$

where a splitting of $\mathbf{N}$ in the linear part $\mathbf{N}_l$ (where diffusion terms responsible for stiffness are included) and the non–linear one $\mathbf{N}_{nl}$ is used. Here, $\mathbf{b}_k^{n-1}$ incorporates the boundary Dirichlet conditions from $\mathbf{T}_{b,k}^{h,n-1}$ and $\mathbf{T}_{s,k}^{h,n-1}$,

(2) computation of fluid–bush and fluid–shaft fluxes: $\mathbf{q}_{b,k}^{h,n+1/2}$, $q_{s,k}^{n+1/2}$,

(3) explicit integration of the bush thermal scheme:

$$\omega_b \mathbf{T}_{b,k}^{h,n+1} = (\mathbf{A} + \omega_b \mathbf{I})\mathbf{T}_{b,k}^{h,,n} + \mathbf{B}\boldsymbol{\Phi}_{b,k}^{h,n+1/2}, \qquad (22)$$

(4) explicit integration of the shaft thermal equation:

$$\omega_s T_{s,k}^{n+1} = \omega_s T_{s,k}^n + q_{s,k}^{n+1/2} - h_s(T_{s,k}^n - T_a), \qquad (23)$$

(5) if the thermal convergence test holds, take $\mathbf{T}_{l,k}^{h,n+1}$, $\mathbf{T}_{b,k}^{h,n+1}$ and $T_{s,k}^{n+1}$ as $\mathbf{T}_{l,k+1}^{h,s}$ $\mathbf{T}_{b,k+1}^{h,s}$ and $T_{s,k+1}^s$, and go to (c)

(c) if the global convergence test holds stop

The global algorithm is quite robust (typically, convergence of the global fixed point scheme with relative tolerances in infinite norm of 10^{-3} takes no more than 10–15 global iterations) and easy to extend to fully three-dimensional thermal models (since segregation used in the integration of the thermal problem allows to decouple them).

At the same time, fictitious thermal inertia coefficients, ω_l, ω_b and ω_s, play the role of relaxation parameters controlling the convergence of the algorithm for the thermal problem. More precisely, convergence is attained, for ω_l fixed, if ω_b and ω_s are large enough.

6 Extension to the evolution problem

The extension of the present model to describe the evolution problem is quite straightforward, incorporating a transient term $\frac{\partial}{\partial t}(\vartheta h)$ in Elrod-Adams equation and true thermal inertia terms (in contrast with the fictitious ones previously considered) in energy equations. Additionally, a model describing the mouvement of the shaft must be provided (since eccentricity can no longer be considered as given); if the shaft is considered as a rigid body this model will consist of an ODE system (relating aceleration of the center of gravity and the external to pressure loads).

The schemes previously presented to solve the hydrodynamic and fluid thermal models (Section 3) can be easily extended to the evolution problem by modifying the characteristic curves (to incorporate the corresponding transient term) in the solution of the Elrod-Adams equations, and including the thermal inertia terms (in the fluid thermal model) in the cell residuals of the finite volume scheme. On the other hand the thermal and mechanical models in the shaft are straightforwardly integrated.

Nevertheless, the adaption of the boundary element scheme used for the thermal model in the bush is not so easy. First, the integral equation associated to the evolution problem involves time integration over the boundary (and then a high computational cost and a large storage, which are precisely the features to be avoided). Other formulations, as the use of the Laplace transform, make difficult the coupling with the other schemes. Alternatively, boundary element – dual reciprocity methods (DRM) can be used to obtain time–stepping algorithms.

Discretization by DRM has anyway some drawbacks (in addition to the lack of maximum principle preservation):

- convergence using radial basis functions is rather slow and thus demands computational costs similar to those needed by finite element methods,
- the system of functions used in DRM can be enriched with eigenfunctions (with better approximation properties) but the scheme is strongly sensitive to the (not obvious) choice of the collocation points and the collocation technique spoils completely the good approximation properties,
- the evolution problem can also be discretized in time before the integral equation is obtained (see [2]), and the resulting non-homogeneous Helmholtz problem treated with DRM, but serious consistency drawbacks for not-so-fine meshes arise.

References

1. Costa, L., Miranda, A.S., Fillon, M., Claro, J.C.P.:An analysis of influence of oil supply conditions on the thermohydrodynamic performance of a single-groove journal bearing. Proceedings of the I MECH E Part J,J. of Engrg. Tribol., **217**, 133–144 (2003)
2. Costabel, M.: Time-dependent problems with the boundary integral equation method. In: Encyclopaedia of Computational Mechanics, E.Stein-R.de Borst-T.J.R. Hughes (eds.), John Wiley and Sons, (2004)
3. Durany, J. , García, G., Vázquez, C.: Numerical simulation of a lubricated hertzian contact problem under imposed load. Fin. Elem. in Anal. and Design **38** , 645–658 (2002)
4. Durany, J. , Pereira, J., Varas, F.: A cell-vertex finite volume method for thermohydrodynamic problems in lubrication theory. Comput. Methods Appl. Mech. Engrg. (in press).
5. Fillon, M., Bouyer, J.: Thermohydrodynamic analysis of a worn plain journal bearing. Tribol. Int. **37**, 129–136 (2004)
6. Frene, J., Nicolas, D., Degueurce, B., Berthe, D., Godet, M.: Lubrification hydrodynamique. Eyrolles, Paris, (1990)

7. Kucinschi, B., Fillon, M., Frêne, J., Pascovici, M.: A transient thermoelastohydrodynamic study of steadily loaded plain journal bearings using finite element method analysis. ASME J. Tribol. **122**, 219–226 (2000)
8. Morton, K.W., Stynes, M., Suli, E. : Analysis of a cell-vertex finite volume method for convection-diffusion problems. Math. of Comput. **66** , 1389–1406 (1997)
9. Paranjpe, R.S., Han, T.: A Study of the thermohydrodynamic performance of steadily loaded journal bearings. ASME J. Tribol. **37**, 679–690 (1994)
10. Zhang, C., Yi, Z., Zhang, Z.: THD analysis of high speed heavily loaded journal bearings including thermal deformation, mass conserving cavitation and turbulent effects . ASME J. Tribol. **122**, 597–602 (2000)

Water Pollution

Numerical Modelling for Leaching of Pesticides in Soils Modified by a Cationic Surfactant

M.I. Asensio[1], L. Ferragut[1], S. Monedero[1], M.S. Rodríguez-Cruz[2] and M.J. Sánchez-Martín[2]

[1] Departamento de Matemática Aplicada, Universidad de Salamanca, Salamanca, España
mas@usal.es, ferragut@usal.es, smonedero@usal.es
[2] Departamento de Química y Geoquímica Ambiental, Instituto de Recursos Naturales y Agrobiología CSIC, Salamanca, España
sorocruz@usal.es, mjesussm@usal.es

Summary. We present two different one-dimensional models for transport of solutes in soils. We numerically solve the mathematical problems arising from these models using stabilization techniques, maintaining a low computational cost guaranteeing the stability of the schemes for any regimen to assure a right parameter adjustment. The numerical experiments are based on real data from laboratory experiments.

1 Introduction

The suitable use of organic pesticides in agriculture is profitable but the mobility of these substances in soils represents a potential threat to the environment, particularly to groundwater resources.

Physically-based environmental simulation models are less expensive and quicker than other experimental strategies, and their use can be considered as an essential tool in the decision making.

The problems that we are facing are: the development of appropriate models, the numerical solution of the mathematical problems arising and the adjustment of the parameters.

We present two different one-dimensional models: a classical linear equilibrium model for transport of non volatile solutes [8]; and a non-equilibrium sorption model to represent non-equilibrium processes during transport, which includes the chemical non-equilibrium [6], or two sites models, and the physical non-equilibrium, or two regions models [10].

The mathematical problem to be solved is basically an unsteady linear convection-diffusion-reaction one-dimensional problem. The main difficulty in the numerical approximation of this kind of partial differential equation is the accurate modelling of the interaction between convection, diffusion and reaction processes. This forces to work with small time and spaces steps, but if we want to fit the parameters with a low computational cost we need to work with coarse meshes, so the solution is to use stabilization techniques. Moreover, if we do not know a-priori the relative weight of the diffusive, convective and reactive terms, we need to guarantee a scheme stable for any regimen to assure a right parameter adjustment. The numerical scheme used here is based on the *Link Cutting Bubble* strategy (LCB) proposed by Brezzi et al. [4] for the steady problem, that authors have adapted for the non steady case in [3] where this scheme is compared to more classical stabilization techniques.

The data used for the numerical experiments are real data from laboratory experiments carried out by the authors.

2 Laboratory experiments

In this section we briefly describe the laboratory experiments that have provided the data for the parameter adjustment of the model proposed.

These laboratory experiments study the effect of the modification of two soils (S2 and S4), with low organic matter content ($< 2\%$) and with varying clay contents, with a cationic surfactant, octadecyltrimethylammonium (ODTMA) on the immobilization of different hydrophobic pesticides: Linuron, Atrazina and Metalaxyl. The adsorption and mobility of hydrophobic pesticides in the soil depends mainly on the content of soil organic matter, so these clayey soils are bad adsorbents of these compounds. The injection of cationic surfactant increases the organic matter content of soils and offers a tool as temporal adsorbent barriers.

For the experiments with no modified soils, glass leaching columns ($3cm \times 20cm$) were packed with $100g$ of natural soil previously sieved. Each column was saturated with water and allowed to dry. The pore volume of the packed columns was estimated. Then, $1mL$ of a solution of each of the three pesticides at $1000 \mu g m L^{-1}$ in methanol was added to the top of the columns. The columns were then washed by continuously applying $500mL$ of water with a peristaltic pump: i.e. under a saturated flow regime. Leach fractions of $15mL$ were successively collected using an automatic fraction collector in which the concentrations of pesticides were measured. To determine the leaching of pesticides in the soil modified with the surfactant, the soil was first loaded with ODTMA by pumping through the column an aqueous solution containing $6.4meq$ of ODTMA for soil S2 or $2.5meq$ ODTMA for soil S4. The Chloride was used as a non reactive trazer. For more details about the experiments and the properties of soils and pesticides see [14].

3 Models describing leaching of solutes in soils

In this section we summarize two different one-dimensional models describing the leaching of solutes in soils.

3.1 Linear equilibrium sorption model

The *linear equilibrium sorption model* describes the one-dimensional transport of non-volatile solute during steady state flow in homogeneous porous media and uniform soil moisture distribution. Taking into account degradation of both phases of solute (dissolved and adsorbed), and linear equilibrium sorption isotherms, the model can be expressed in dimensionless form as a convection-diffusion equation as

$$R\frac{\partial C}{\partial t} - \frac{1}{Pe}\frac{\partial^2 C}{\partial z^2} + \frac{\partial C}{\partial z} + \mu C = 0 \qquad (1)$$

where R is the retardation factor, C is the relative resident solute concentration in soil, t is the dimensionless time, z is the dimensionless space, Pe is the Peclet number and μ is the dimensionless first order degradation coefficient, including both phases.

Equation (1) deals with, in this order, the variation in time of the dissolved and adsorbed concentration, the hiydrodynamic diffusion flux, the convective flux and the degradation in both phases. A more detailed explanation of this model can be found in [8], and a brief resume in [1].

3.2 Non-equilibrium models

When bi-model porosity leads to a two-regions flow or in those situations where there are different sorption processes, *non-equilibrium models* must be considered. *Chemical non-equilibrium models* consider that adsorption on some of the sorption sites is instantaneous while adsorption on the remainder sites is governed by first-order kinetics. These non equilibrium models are called *two-site* models (see [6]) and consider steady flow and degradation in both phases. On the other hand, *physical non-equilibrium* is often modelled using a two-regions dual-porosity type formulation (see [10]). The *two-region* transport model assumes that the dissolved phase can be partitioned into mobile and immobile regions. In the mobile region, the solute transport is due to convection and diffusion, in the immobile region, the convective transport is null and diffusion is dominant. Solute exchange between the two dissolved regions is modelled as a first-order process.

Although the concepts are different for both chemical and physical non-equilibrium models, the processes can be described by the same dimensionless equations for linear adsorption and steady state conditions (see [9]). These dimensionless equations, corresponding to the non equilibrium models, chemical or physical, are,

$$\beta R \frac{\partial C_1}{\partial t} - \frac{1}{Pe} \frac{\partial^2 C_1}{\partial z^2} + \frac{\partial C_1}{\partial z} + w(C_1 - C_2) + \mu_1 C_1 = 0 \qquad (2)$$

$$(1-\beta) R \frac{\partial C_2}{\partial t} - w(C_1 - C_2) + \mu_2 C_2 = 0 \qquad (3)$$

where the subscripts 1 and 2 refer to the instantaneous and non-instantaneous adsorption sites for the chemical non equilibrium model (resp. mobile and immobile regions for the physical non equilibrium model); β is the partitioning coefficient between the two kind of sites (resp. regions); R, C, t, z, Pe and μ represent the same dimensionless heading as in the equilibrium model, and w is the dimensionless mass transfer coefficient.

3.3 Boundary and initial conditions

On the upper boundary we need to represent the application of a total amount of solute (C_0) at a fixed rate for a relative short period of time (t_0) that is much smaller than the time scale of interest. This can be mathematically represented as,

$$-\frac{1}{Pe}\frac{\partial C}{\partial z} + C\Big|_{z=0} = \delta(t) \quad \text{and} \quad -\frac{1}{Pe}\frac{\partial C_1}{\partial z} + C_1\Big|_{z=0} = \delta(t) \qquad (4)$$

where $\delta(t)$ is a step function (see [1]).

Assuming concentration continuity at the end of the column, we can consider a homogeneous Newmann condition on the lower boundary [12].

$$\frac{\partial C}{\partial z}\Big|_{z=1} = 0 \quad \text{and} \quad \frac{\partial C_1}{\partial z}\Big|_{z=1} = 0. \qquad (5)$$

Finally, we assume null initial solute concentration for both models.

4 Numerical methods

The two models described, from the mathematical point of view, are basically an unsteady linear convection-diffusion-reaction one-dimensional problem. The main difficulty in the numerical approximation of this kind of partial differential equation is the accurate modelling of the interaction between convection, diffusion and reaction processes. This forces to work with small time and spaces steps, but if we want to fit the parameters with a low computational cost we need to work with coarse meshes. In order to maintain a low computational cost and assure an accurate numerical solution we propose to use stabilization techniques. Moreover, if we do not know a-priori the relative weight of the diffusive, convective and reactive terms, we need to guarantee a scheme stable for any regimen to assure a right parameter adjustment.

Following an idea mentioned in section 14.3.2 of [13], authors proposed in [2] and [3] a simple and nonstandard approach: to adapt the stabilization techniques developed for the steady problem to deal with the unsteady problem. In particular, one of the approaches proposed by authors consists in first discretize the time derivative, and then discretize in space the corresponding family of steady problems (one per each time step), where the reactive term now depends on the inverse of the time step. The classical stabilization techniques as the *Stream-line Upwind Petrov-Galerkin* (SUPG) method by Brooks et al. (see [5]), the *Galerkin Least Squares* (GaLS) method by Hughes et al. (see [11]) and the *Douglas-Wang* (DWG) formulation by Douglas and Wang, (see [7]), have been compared with the more recent *Link Cutting bubbles* strategy proposed by Brezzi et al. (see [4]). This last scheme has good behaviour in all regimes, either convection or reaction dominated. Therefore, this is the more appropriate scheme for the problem treated here, as maintains a low computational cost and gives accurate numerical solutions independently of the regimen.

The idea behind the LCB strategy is to enrich the finite element space V_h by adding a space of *discrete* bubbles V_B. The space V_B is built, element by element, by taking piecewise linear bubbles on a suitable subgrid of two nodes. The key point is the location of the extra nodes (i.e., the bubble nodes): this is done so that the approximation to the solution of the local (bubble) problems is stable and accurate in all regimes, i.e., either convection or reaction dominated. We refer to [4] for an explanation of the steady case and other details.

We describe the numerical scheme for the equilibrium model, for the non-equilibrium model the scheme is similar, taking into account that the second equation of this model is an ordinary differential equation.

The problem to be solved for the equilibrium model, given by (1) and the corresponding initial and boundary conditions, can be written as,

$$\begin{cases} R\dfrac{\partial C}{\partial t} + \mathcal{L}C = 0 & \text{in} \quad [0,1] \times (0,T), \\ \left(-\dfrac{1}{Pe}\dfrac{\partial C}{\partial z} + C\right)\Big|_{z=0} = \delta(t) & \text{on} \quad (0,T), \\ \dfrac{\partial C}{\partial z}\Big|_{z=1} = 0 & \text{on} \quad (0,T), \\ C\big|_{t=0} = 0 & \text{on} \quad [0,1], \end{cases} \qquad (6)$$

where we denote by $\mathcal{L}$ the convection-diffusion-reaction operator,

$$\mathcal{L} := -\dfrac{1}{Pe}\dfrac{\partial^2}{\partial z^2} + \dfrac{\partial}{\partial z} + \mu I \qquad (7)$$

and I denotes the identity operator.

We first carry out the time discretization of (6), by using Crank-Nicholson scheme; if $N = T/k$ and k is the time step, we obtain

$$\frac{R}{k}C^{n+1} + \frac{1}{2}\mathcal{L}C^{n+1} = \frac{R}{k}C^n - \frac{1}{2}\mathcal{L}C^n, \quad n = 0, 1, \ldots, N-1, \quad C^0 = 0, \quad (8)$$

which yields a family of "steady" convection-diffusion-reaction problems. Observe that by proceeding in this way, for k small we are led to a family of reaction-dominated problems. To write (8) in a more compact form, we define the associated convection-diffusion-reaction operator

$$\widetilde{\mathcal{L}} := \frac{R}{k}I + \frac{1}{2}\mathcal{L} = -\frac{1}{2Pe}\frac{\partial^2}{\partial z^2} + \frac{1}{2}\frac{\partial}{\partial z} + \left(\frac{R}{k} + \frac{\mu}{2}\right)I, \quad (9)$$

In this way, (8) can be rewritten as

$$\widetilde{\mathcal{L}}C^{n+1} = \frac{R}{k}C^n - \frac{1}{2}\mathcal{L}C^n, \quad n = 0, 1, \ldots, N-1, \quad C^0 = 0. \quad (10)$$

The LCB discretization of (10) is obtained constructing the subgrid for the bubbles space V_B, but now with respect to the operator $\widetilde{\mathcal{L}}$, (see [3] for details). Then LCB can be viewed in two different ways: as a plain Galerkin method or as a stabilized scheme. As the former, the method reduces to,

For $n = 0, 1, \ldots, N-1$, find $C_E^{n+1} \in V_E$ such that, $\forall \psi_E \in V_E$
$$\widetilde{a}(C_E^{n+1}, \psi_E) = \frac{1}{2}(\delta^{n+1} + \delta^n)\psi_E(0) + \frac{R}{k}(C_E^n, \psi_E) - \frac{1}{2}a(C^n, \psi_E), \quad (11)$$

where $\widetilde{a}(\cdot, \cdot)$ (resp. $a(\cdot, \cdot)$) is the bilinear form associated to $\widetilde{\mathcal{L}}$ (resp. $\mathcal{L}$), and the enriched space $V_E = V_h \oplus V_B$ is regarded as a space of piecewise linear functions on a suitable refined grid.

As to the latter, in the spirit of the augmented spaces, one can directly compute the approximation on the original coarse grid, i.e., the approximate solution without the bubble part: by means of *static condensation* of the bubbles degrees of freedom, which entails the solution of the corresponding local problems, we are led to a scheme involving only the coarse approximation. In other words, the linear system remains the same size. It is worth pointing out that operating in the subspace of the coarse grid rather than in the enriched subspace, the computational cost is kept nearly the same as for a classical stabilization techniques.

5 Parameter adjustment

The data provided by the laboratory experiments measure the solute amount over leach fractions successively collected, but the model measures instantaneous relative concentrations of the solute. So we need to normalize the data in order to be able to compare the relative concentration collected until the instant t_n, $Exp(t_n)$; with the corresponding value computed with the approximate solution of the model, that is, we compute concentration at z=1 (column's end) during all the experimental time $C(z = 1, t)$ solving model problems, and then we compute the discrete function,

$$Sol(t_n) = \int_0^{t_n} C(z=1,t)dt.$$

So, the error function to minimize in the parameter adjustment is,

$$Error = \sum_n (Sol(t_n) - Exp(t_n))^2.$$

For the parameter adjustment we use a genetic type algorithm.

6 Examples

The parameter adjustment have been done solving the corresponding problems with the described numerical scheme for a time step $k = 0.005$ and mesh size $h = 0.3$. The experiments shown in Fig. 1 and Fig. 2 correspond to the metalaxyl and soils S4 natural (equilibrium model) and modified (non-equilibrium model). Left figures shown the approximate relative concentration at the end of the column experiment, and right figures compare the approximate and experimental collected solute.

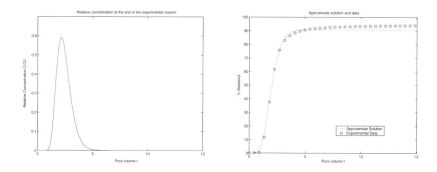

Fig. 1. Metalaxyl and natural soil S4.

Acknowledgement

Supported by the research projects: CGL2004-06171-c03-03/cl1 and CTM2004-00381/TECNO from the Education and Science Ministry (Spain) and FEDER funds (European Union); SA078A05 from the Castilla y Leon government.

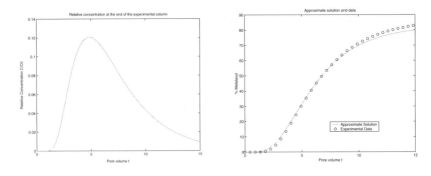

Fig. 2. Metalaxyl and modified soil S4.

References

1. Alvarez, J., Herguedas, A., Atienza, J., Bolado, S.: Modelización numérica y estimación de parámetros para la descripción del transporte de solutos en columnas de suelo en laboratorio. Monografías INIA, **91**, Madrid, (1995).
2. Asensio, M.I., Ayuso, B., Sangalli, G.: Bubbles for the stabilization of transient convection-diffusion-reaction problems. In: Proceedings of ECCOMAS 2004, Jyväskylä, 24-28 July 2004.
3. Asensio, M.I., Ayuso, B., Sangalli, G.: Coupling stabilized finite element methods with finite difference time integration for advection-diffusion problems. I.M.A.T.I. Technical report. (2005)
4. Brezzi, F., Hauke, G., Marini, L.D., Sangalli, G.: Link-cutting bubbles for the stabilization of convection-diffusion-reaction problems. Math. Models Methods Appl. Sci., **13**, 445–461 (2003). Dedicated to Jim Douglas, Jr. on the occasion of his 75th birthday.
5. Brooks, A.N., Hughes, T.J.R.: Streamline upwind/Petrov-Galerkin formulations for convection dominated flows with particular emphasis on the incompressible Navier-Stokes equations. Comput. Methods Appl. Mech. Engrg., **32**, 199–259 (1982). FENOMECH '81, Part I (Stuttgart, 1981).
6. Cameron, D.R., Klute, A.: Convective-dispersive solute transport with a combined equilibrium and kinetic adsorption model. Water Resour. Res., **13**(1), 183–188 (1977).
7. Douglas, J.Jr., Wang, J.P.: An absolutely stabilized finite element method for the Stokes problem. Math. Comp., **52**, 495–508, (1989).
8. van Genuchten, M.Th., Cleary, R.W.: Movement of solutes in soil: computer-simulated and laboratory results. G.H. Boltz (ed.), Soil Chemistry B: Phsysico-Chemical models. Elsevier, Amsterdam, 349–386 (1982).
9. van Genuchten, M.Th., Wagenet, R.J.: Two-Site/Two-Region models for pesticides transport and degradation: theoretical development and analytical solutions. Soil Sci. Soc. Am. J., **53**(5), 1303–1310 (1989).
10. van Genuchten, M.Th., Wierenga, P.J.: Mass transfer studies in sorbing porous media: I. Analytical solutions. Soil Sci. Soc. Am. J., **40**(4), 473–480 (1976).
11. Hughes, T.J.R., Franca, L.P., Hulbert, G.M.: A new finite element formulation for computational fluid dynamics. VIII. The Galerkin/least-squares method for

advective-diffusive equations. Comput. Methods Appl. Mech. Engrg., **73**, 173–189 (1989).
12. Kreft A., Zuber, A.: On the physical meaning of the dispersion equation and its solutions for differen initial and boundary conditions. Chem. Engn. Sci., **33**, 1471–1480, (1978).
13. Quarteroni, A., Valli, A.: Numerical approximation of partial differential equations. Springer Series in Computational Mathematics, **23**. Springer Verlag, Berlin, (1994).
14. Rodríguez-Cruz, M.S., Sánchez-Martín, M.J., Andrades, M.S., Sánchez-Camazano, M.: Immobilization of pesticides in soils modified by cationic surfactants: effect of octadecyltrimethylammonium cation. Proc. Int. Workshop. Saturated and Unsaturated zone: integration of process knowledge into effective models. Rome, Italy, (2004).

Formulation of Mixed-Hybrid FE Model of Flow in Fractured Porous Medium*

Jiřina Královcová, Jiří Maryška, Otto Severýn and Jan Šembera

Faculty of Mechatronics and Interdisciplinary Studies,
Technical University of Liberec,
Hálkova 6, 461 17 Liberec, Czech Republic
jirina.kralovcova@tul.cz, jiri.maryska@tul.cz,
otto.severyn@tul.cz, jan.sembera@tul.cz

Summary. We encounter many problems, within which we try to model the groundwater flow in the disrupted rock massifs using numerical models. So-called "standard approaches" such as replacement by porous medium or double-porosity models of discrete stochastic fracture networks appear to have constraints and limitations, which make them unsuitable for the large-scale long-time hydrogeological calculations. This article presents the mathematical formulation of model based on a new approach to the modelling of groundwater flow, which combines the two above-mentioned approaches. The approach considers three substantial types of objects within a structure of modelled massif important for the groundwater flow – small stochastic fractures, large deterministic fractures and lines of intersection of the large fractures. The systems of stochastic fractures are represented by blocks of porous medium with suitably set hydraulic conductivity. The large fractures are represented as polygons placed in 3D space and their intersections are represented by lines. Thus flow in 3D porous medium, flow in 2D and 1D fracture systems, and communication among these three systems are modelled together.

1 Rock Massif environment

Numerical modelling of the hydraulical, geochemical and transport processes in fractured rock attracts the attention of many scientists more than forty years. The first numerical models of such processes were created in late 60's of the last century. According to [2], there existed more than thirty software packages claimed to solve problem of the fluid flow in fractured rock in 1994.

Despite the fact, there is a lot of open and unresolved problems in the field of research. The reason for that lies in the nature of the problem – lack of input data, their uncertainty and often low accuracy, high computational

---

* This work was supported with the subvention from Ministry of Education of the Czech Republic, project code 1M4674788502.

cost are the main difficulties we encounter when we try to simulate processes in the fractured rock. Avoiding these difficulties is usually possible only at a price of simplification of the problem.

The hydrogeological research brought following empirical knowledge about the rock environment and groundwater flow in them:

- The rock matrix can be considered hydraulically impermeable.
- Even the most compact massifs are disrupted by numerous fractures.
- Most of the fractures are relatively small, with the characteristic length less than one meter.
- The groundwater flow in the small fractures has significant store capacity and play important role mainly in the transport processes.
- It is barely possible to obtain exact parameters of all the fractures. They are treated in statistical way.
- The most of the liquid is conducted by relatively small number of large fractures. The spatial position of these fractures is usually detectable.
- The fastest groundwater flux is observed on intersections of the large fractures. These intersections behave like "pipelines" in the compact rock massifs.

These facts lead us to the conclusion that there are in general three different objects involved in conduction of the groundwater through a compact rock: small fractures, large fractures and intersections of large fractures.

There are two possible approaches to the modelling of flow in environment of **small fractures**: employment of the *s*tochastic discrete fracture networks or the *h*omogenisation and replacement with porous media. The first one is more suitable for small problems, see [5]. On the other hand the second approach is much better applicable for large problems. Fractured rock disrupted only by small fractures can by relatively well homogenised and replaced by hydraulically equivalent porous media environment. The methods of homogenisation and setting the hydraulic parameters of the porous media can be found for example in [1].

The large fractures are relatively well known. The *d*iscrete fracture networks approach works well in this case. Then, the fractures are represented as 2D objects (polygons) placed in 3D space.

The intersections of large fractures are relatively rare in the rock massifs, but significant for the flow. The velocity of the flow on the intersections of fractures can be higher in order of magnitude than velocity on the fractures. They can be represented by 1D objects placed in 3D space. Similar way within the model can be represented for example possible borehole.

On the base of our experience in groundwater flow modelling we decided to build up a model that could treat all the above-mentioned objects. The model incorporates 3D blocks, considered as pourous medium, 2D fractures, standing for real rock fractures, and 1D lines, representing significant pipelines in the underground. Some of the practical aspects of the new model were already mentioned in [4].

2 Linear steady Darcy's flow

In general, the linear steady Darcy's flow, which we expect in a considered underground domain, is described by the equations

$$\mathbf{u} = -\mathsf{K}\,\nabla p\,, \tag{1}$$
$$\nabla \cdot \mathbf{u} = q\,, \tag{2}$$

where $\mathbf{u}$ is the vector of velocity of the flow, p is the hydraulic pressure, K is the second order tensor of hydraulic conductivity (symmetric, positive definite), and q is a function expressing the density of sources or sinks of the fluid.

For the particular task evaluation we consider Dirichlet's boundary condition on a part of boundary and Neumann's boundary condition on the rest of boundary of considered domain. Additionally, it is also possible to employ Newton's boundary condition, which combines the two mentioned.

3 Mathematical formulation of the problem

We consider three domains Ω_3, Ω_2 and Ω_1, $\Omega_3 \subset R^3$, $\Omega_2 \subset R^3$, $\Omega_1 \subset R^3$. Ω_3 is a simply connected three dimensional polyhedral domain. Ω_2 is a finite set of mutually connected polygons placed in 3D space and Ω_1 is a finite set of mutually connected line segments placed in 3D space.

The boundary of each domain is divided into two parts – Dirichlet's part and Neumann's part. Boundary of domain Ω_3: $\partial\Omega_3 = \Gamma_{3,D} \cup \Gamma_{3,N}$, $\Gamma_{3,D} \neq \emptyset$, $\Gamma_{3,D} \cap \Gamma_{3,N} = \emptyset$. Boundary of domain Ω_2: $\partial\Omega_2 = \Gamma_{2,D} \cup \Gamma_{2,N}$, $\Gamma_{2,D} \neq \emptyset$, $\Gamma_{2,D} \cap \Gamma_{2,N} = \emptyset$ and Boundary of domain Ω_1: $\partial\Omega_1 = \Gamma_{1,D} \cup \Gamma_{1,N}$, $\Gamma_{1,D} \neq \emptyset$, $\Gamma_{1,D} \cap \Gamma_{1,N} = \emptyset$. ($\partial\Omega_1$ is a set of discrete points).

Then, the potential driven flow in Ω_3 domain is described by the following system of equations:

$$\mathbf{u}_3 = -\mathsf{K}_3\,\nabla p_3 \quad \text{in} \quad \Omega_3\,, \tag{3}$$
$$\nabla \cdot \mathbf{u}_3 = q_3 + \widetilde{q}_{23} + \widetilde{q}_{13} \quad \text{in} \quad \Omega_3\,, \tag{4}$$
$$p_3 = p_{3,D} \quad \text{on} \quad \Gamma_{3,D}\,, \tag{5}$$
$$\mathbf{u}_3 \cdot \mathbf{n} = u_{3,N} \quad \text{on} \quad \Gamma_{3,N}\,. \tag{6}$$

Accordingly, we can write down the systems of equations for the other domains Ω_2 and Ω_1.

The flow between different domains is considered as a source of fluid. The quantity $\widetilde{q}_{ij}$ stands for the source of fluid due to interactions between different domains Ω_i and Ω_j and it is expressed as

$$\widetilde{q}_{ij} = (p_i - p_j)\sigma_{ij}\,, \quad i \neq j\,,$$
$$\sigma_{ij} = 0 \quad \text{in} \quad (\Omega_i \cup \Omega_j) \setminus (\Omega_i \cap \Omega_j)\,,$$

$$p_i = 0 \quad \text{in} \quad \Omega_j \setminus \Omega_i ,$$
$$p_j = 0 \quad \text{in} \quad \Omega_i \setminus \Omega_j .$$

The quantity σ_{ij} stands for the permeability between domains Ω_i and Ω_j. $\sigma_{ij} = \sigma_{ji}$ so $\widetilde{q}_{ij} = -\widetilde{q}_{ji}$.

The solution of the problem are values of physical quantities p_3, p_2, p_1, $\mathbf{u}_3$, $\mathbf{u}_2$, $\mathbf{u}_1$ over the considered domains. To be able to solve the task we need to know values of material parameters K_3, K_2, K_1, σ_{12}, σ_{13}, σ_{23} and values of boundary conditions $p_{3,D}$, $p_{2,D}$, $p_{1,D}$, $u_{3,N}$, $u_{2,N}$, $u_{1,N}$.

4 Mixed-hybrid formulation

We consider three domains with their boundaries as it was mentioned above. Next we define partition of each domain into a set of subdomains. Let denote $\tau_{3,h}$ the partition of the domain Ω_3, $\tau_{2,h}$ the partition of the domain Ω_2 and $\tau_{1,h}$ the partition of the domain Ω_1:

$$\tau_{3,h} = \{e; e \in \Omega_3, \cup_{e \in \tau_{3,h}} \bar{e} = \Omega_3, e_i \cap e_j = \emptyset \text{ for } i \neq j\} ,$$
$$\tau_{2,h} = \{e; e \in \Omega_2, \cup_{e \in \tau_{2,h}} \bar{e} = \Omega_2, e_i \cap e_j = \emptyset \text{ for } i \neq j\} ,$$
$$\tau_{1,h} = \{e; e \in \Omega_1, \cup_{e \in \tau_{1,h}} \bar{e} = \Omega_1, e_i \cap e_j = \emptyset \text{ for } i \neq j\} .$$

Next we denote the sets of points on all non-Dirichlet faces in each domain:

$$\Gamma_{3,h} = \cup_{e \in \tau_{3,h}} \partial e \setminus \Gamma_{3,D} ,$$
$$\Gamma_{2,h} = \cup_{e \in \tau_{2,h}} \partial e \setminus \Gamma_{2,D} ,$$
$$\Gamma_{1,h} = \cup_{e \in \tau_{1,h}} \partial e \setminus \Gamma_{1,D} .$$

We do not prescribe any special demands for the mutual position of subdomains of of different dimensions.

For each $\Omega \in \{\Omega_1, \Omega_2, \Omega_3\}$, $\Gamma_h \in \{\Gamma_{1,h}, \Gamma_{2,h}, \Gamma_{3,h}\}$, $\tau_h \in \{\tau_{1,h}, \tau_{2,h}, \tau_{3,h}\}$, $e \in \tau_h$ we use next function spaces: We use the standard space of square-integrable functions $L^2(\Omega)$ and the standard Sobolev space of scalar functions with square-integrable weak derivatives $H^1(\Omega)$. We then have the space $H^{\frac{1}{2}}(\Gamma_h)$ of functions, that are traces of some functions from the corresponding $H^1(\Omega)$. For each subdomain e, we denote by $\mathbf{H}(div, e)$ the space of vector functions with square-integrable weak divergences and we define $\mathbf{H}(div, \tau_h)$ as the space of the $\mathbf{L}^2(\Omega)$ functions whose restriction to each subdomain $e \in \tau_h$ is from $\mathbf{H}(div, e)$.

In the following expressions, β denotes the inverse of K i.e. hydraulic resistance, f^e denotes the restriction of function f on subdomain e, $(f, g)_e$ denotes the integral form $\int_e fgmDx$ and $< f, g >_{\partial e}$ denotes the integral form $\int_{\partial e} fgmDx$.

Next we can derive the mixed-hybrid formulation on particular domains. Considering a particular subdomain $e \in \tau_{3,h}$, we need to fulfil equations

$$\beta_3^e \mathbf{u}_3^e + \nabla p_3^e = 0 \,,$$
$$\nabla \cdot \mathbf{u}_3^e - \widetilde{q}_{23}^e - \widetilde{q}_{13}^e = q_3^e \,.$$

We use test functions $\mathbf{v}_3 \in \mathbf{H}(div, \tau_{3,h})$ and $\phi_3 \in L^2(\Omega_3)$ so way

$$(\beta_3^e \mathbf{u}_3^e, \mathbf{v}_3^e)_e + (\nabla p_3^e, \mathbf{v}_3^e)_e = 0 \,,$$
$$(\nabla \cdot \mathbf{u}_3^e, \phi_3^e)_e - ((p_2^e - p_3^e)\sigma_{23}^e, \phi_3^e)_e - ((p_1^e - p_3^e)\sigma_{13}^e, \phi_3^e)_e = (q_3^e, \phi_3^e)_e \,.$$

After application of Green's formula on the term $(\nabla p_3^e, \mathbf{v}_3^e)_e$ we get

$$(\beta_3^e \mathbf{u}_3^e, \mathbf{v}_3^e)_e - (p_3^e, \nabla \cdot \mathbf{v}_3^e)_e + <\psi_3^e, \mathbf{v}_3^e \cdot \mathbf{n}^e>_{\partial e} = 0 \,,$$
$$(\nabla \cdot \mathbf{u}_3^e, \phi_3^e)_e - ((p_2^e - p_3^e)\sigma_{23}^e, \phi_3^e)_e - ((p_1^e - p_3^e)\sigma_{13}^e, \phi_3^e)_e = (q_3^e, \phi_3^e)_e \,.$$

where ψ_3^e stands for value of p_3^e on the boundary of e, i.e. ∂e, and $\mathbf{n}$ stands for outer normal of ∂e.

Next, let us change over from particular subdomain to the whole domain and proceed to the summation over all subdomains of $\tau_{3,h}$

$$\sum_{e \in \tau_{3,h}} \{(\beta_3^e \mathbf{u}_3^e, \mathbf{v}_3^e)_e - (p_3^e, \nabla \cdot \mathbf{v}_3^e)_e + <\psi_3^e, \mathbf{v}_3^e \cdot \mathbf{n}^e>_{\partial e}\} = 0$$

$$\sum_{e \in \tau_{3,h}} \{(\nabla \cdot \mathbf{u}_3^e, \phi_3^e)_e - ((p_2^e - p_3^e)\sigma_{23}^e, \phi_3^e)_e - ((p_1^e - p_3^e)\sigma_{13}^e, \phi_3^e)_e\} =$$
$$= \sum_{e \in \tau_{3,h}} (q_3^e, \phi_3^e)_e \,.$$

Considering the whole domain, we need to ensure the balance on internal faces of the partition $\tau_{3,h}$. For balance of $\mathbf{u}_3$ on the internal face, which interconnects subdomains e_k and e_l, holds

$$\mathbf{u}_3^{e_k} \cdot \mathbf{n}^{e_k} + \mathbf{u}_3^{e_l} \cdot \mathbf{n}^{e_l} = 0 \,.$$

Accordingly, we test the equation by the function $\mu_3 \in H^{\frac{1}{2}}(\Gamma_{3,h})$ and after the summation over all internal faces of partition $\tau_{3,h}$ we get

$$\sum_{e \in \tau_{3,h}} <\mathbf{u}_3^e \cdot \mathbf{n}^e, \mu_3^e>_{\partial e \cap (\Gamma_{3,h} \setminus \Gamma_{3,N})} = 0 \,.$$

We employ the balance equation and boundary conditions and derive the following system of integral equations on partition $\tau_{3,h}$

$$\sum_{e \in \tau_{3,h}} \{(\beta_3^e \mathbf{u}_3^e, \mathbf{v}_3^e)_e - (p_3^e, \nabla \cdot \mathbf{v}_3^e)_e + <\psi_3^e, \mathbf{v}_3^e \cdot \mathbf{n}^e>_{\partial e \cap \Gamma_{3,h}}\} =$$
$$= -\sum_{e \in \tau_{3,h}} <p_{3,D}, \mathbf{v}_3^e \cdot \mathbf{n}^e>_{\partial e \cap \Gamma_{3,D}} \,, \quad (7)$$

$$\sum_{e \in \tau_{3,h}} \{-(\nabla \cdot \mathbf{u}_3^e, \phi_3^e)_e + ((p_{23}^e - p_3^e)\sigma_{23}^e, \phi_3^e)_e + ((p_{13}^e - p_3^e)\sigma_{13}^e, \phi_3^e)_e\} =$$

$$= -\sum_{e \in \tau_{3,h}} (q_3^e, \phi_3^e)_e, \quad (8)$$

$$\sum_{e \in \tau_{3,h}} <\mathbf{u}_3^e \cdot \mathbf{n}^e, \mu_3^e>_{\partial e \cap \Gamma_{3,h}} = \sum_{e \in \tau_{3,h}} <u_{3,N}^e, \mu_3^e>_{\partial e \cap \Gamma_{3,N}}. \quad (9)$$

The same way we derive the system of integral equations on partition $\tau_{2,h}$ and $\tau_{1,h}$:

$$\sum_{e \in \tau_{2,h}} \{(\beta_2^e \mathbf{u}_2^e, \mathbf{v}_2^e)_e - (p_2^e, \nabla \cdot \mathbf{v}_2^e)_e + <\psi_2^e, \mathbf{v}_2^e \cdot \mathbf{n}^e>_{\partial e \cap \Gamma_{2,h}}\} =$$

$$= -\sum_{e \in \tau_{2,h}} <p_{2,D}, \mathbf{v}_2^e \cdot \mathbf{n}^e>_{\partial e \cap \Gamma_{2,D}}, \quad (10)$$

$$\sum_{e \in \tau_{2,h}} \{-(\nabla \cdot \mathbf{u}_2^e, \phi_2^e)_e + ((p_{32}^e - p_2^e)\sigma_{32}^e, \phi_2^e)_e + ((p_{12}^e - p_2^e)\sigma_{12}^e, \phi_2^e)_e\} =$$

$$= -\sum_{e \in \tau_{2,h}} (q_2^e, \phi_2^e)_e \, ' \quad (11)$$

$$\sum_{e \in \tau_{2,h}} <\mathbf{u}_2^e \cdot \mathbf{n}^e, \mu_2^e>_{\partial e \cap \Gamma_{2,h}} = \sum_{e \in \tau_{2,h}} <u_{2,N}^e, \mu_2^e>_{\partial e \cap \Gamma_{2,N}}, \quad (12)$$

$$\sum_{e \in \tau_{1,h}} \{(\beta_1^e \mathbf{u}_1^e, \mathbf{v}_1^e)_e - (p_1^e, \nabla \cdot \mathbf{v}_1^e)_e + <\psi_1^e, \mathbf{v}_1^e \cdot \mathbf{n}^e>_{\partial e \cap \Gamma_{1,h}}\} =$$

$$= -\sum_{e \in \tau_{1,h}} <p_{1,D}, \mathbf{v}_1^e \cdot \mathbf{n}^e>_{\partial e \cap \Gamma_{1,D}}, \quad (13)$$

$$\sum_{e \in \tau_{1,h}} \{-(\nabla \cdot \mathbf{u}_1^e, \phi_1^e)_e + ((p_{31}^e - p_1^e)\sigma_{31}^e, \phi_1^e)_e + ((p_{21}^e - p_1^e)\sigma_{21}^e, \phi_1^e)_e\} =$$

$$= -\sum_{e \in \tau_{1,h}} (q_1^e, \phi_1^e)_e, \quad (14)$$

$$\sum_{e \in \tau_{1,h}} <\mathbf{u}_1^e \cdot \mathbf{n}^e, \mu_1^e>_{\partial e \cap \Gamma_{1,h}} = \sum_{e \in \tau_{1,h}} <u_{1,N}^e, \mu_1^e>_{\partial e \cap \Gamma_{1,N}}. \quad (15)$$

Some of the terms have slightly different meaning in 1D. The term $\nabla \cdot \mathbf{u}$ is simply derivative of $\mathbf{u}$ and the term $<\mathbf{u} \cdot \mathbf{n}, \mu> = (u_a n_x + u_b n_x)\mu$, where u_a, u_b are values of $\mathbf{u}$ on the boundary points and n_x is outer normal in the corresponding boundary point.

In the equations (8), (11), (14) the functions $p_{ij}^{e_j}$ for $e_j \in \tau_{j,h}$, $i \neq j$, $i, j \in \{1, 2, 3\}$ stand for

$$\sum_{e_i \in \tau_{i,h}} p_{ij}^{e_i, e_j}, \quad (16)$$

where $p_{ij}^{e_i,e_j}$ are any functions fulfilling the following conditions:

$$\text{for } i < j : p_{ij}^{e_i,e_j} \in L^2(e_j) : \int_{e_j} p_{ij}^{e_i,e_j} mDx = \int_{e_i \cap e_j} p_i mDx , \qquad (17)$$

$$\text{for } i > j : p_{ij}^{e_i,e_j} \in L^2(e_i \cap e_j) : \int_{e_i \cap e_j} p_{ij}^{e_i,e_j} mDx = \frac{|e_i \cap e_j|_j}{|e_i|_i} \int_{e_i} p_i mDx , \qquad (18)$$

where $|e|_i$ means the i-dimensional measure of e.

Next introduce the function space **Z**:

$$\begin{aligned}\mathbf{Z} = &\ \mathbf{H}(div, \tau_{3,h}) \times \mathbf{H}(div, \tau_{2,h}) \times \mathbf{H}(div, \tau_{1,h}) \times \\ & \times L^2(\tau_{3,h}) \times L^2(\tau_{2,h}) \times L^2(\tau_{1,h}) \times \\ & \times H^{\frac{1}{2}}(\Gamma_{3,h}) \times H^{\frac{1}{2}}(\Gamma_{2,h}) \times H^{\frac{1}{2}}(\Gamma_{2,h}) .\end{aligned}$$

Definition 1. *We call the function*

$$\bar{\mathbf{z}} = (\mathbf{u}_3, \mathbf{u}_2, \mathbf{u}_1, p_3, p_2, p_1, \psi_3, \psi_2, \psi_1) \in \mathbf{Z}$$

the weak solution of mixed hybrid formulation of the problem of flow in fracture porous medium, if for all functions

$$\mathbf{z} = (\mathbf{v}_3, \mathbf{v}_2, \mathbf{v}_1, \varphi_3, \varphi_2, \varphi_1, \mu_3, \mu_2, \mu_1) \in \mathbf{Z}$$

$\bar{\mathbf{z}}$ *satisfies the equations (7)–(15).*

5 Finite element approximation

Let $\tau_{3,h}$ be a partition of Ω_3 into simplex elements, $\tau_{2,h}$ be a partition of Ω_2 into triangle elements and $\tau_{1,h}$ be a partition of Ω_1 into line elements.

For the approximation of function spaces $\mathbf{H}(div, \tau_{3,h})$, $\mathbf{H}(div, \tau_{2,h})$ and $\mathbf{H}(div, \tau_{1,h})$ we use the Raviart-Thomas spaces of piecewise linear functions $\mathbf{RT}_{-1}^0(\tau_{3,h})$, $\mathbf{RT}_{-1}^0(\tau_{2,h})$, $\mathbf{RT}_{-1}^0(\tau_{1,h})$, for details about the approximation function spaces, see [3].

For the approximation of function spaces $L^2(\tau_{3,h})$, $L^2(\tau_{2,h})$, $L^2(\tau_{1,h})$ we use the multiplicator spaces of piecewise constant functions $M_{-1}^0(\tau_{3,h})$, $M_{-1}^0(\tau_{2,h})$, $M_{-1}^0(\tau_{1,h})$.

For the approximation of function spaces $H^{\frac{1}{2}}(\Gamma_{3,h})$, $H^{\frac{1}{2}}(\Gamma_{2,h})$, $H^{\frac{1}{2}}(\Gamma_{1,h})$ we use the multiplicator spaces of piecewise constant functions $M_{-1}^0(\Gamma_{3,h})$, $M_{-1}^0(\Gamma_{2,h})$, $M_{-1}^0(\Gamma_{1,h})$.

As a result of approximation we get a system of linear algebraic equations, which can be written in the following form:

$$\begin{aligned}
\mathbb{A}_3\mathbf{V}_3 \quad\quad\quad\quad +\mathbb{B}_3\mathbf{P}_3 \quad\quad\quad\quad +\mathbb{C}_3\mathbf{\Psi}_3 &= \mathbf{q}_{3D}\\
\mathbb{A}_2\mathbf{V}_2 \quad\quad\quad\quad +\mathbb{B}_2\mathbf{P}_2 \quad\quad\quad\quad +\mathbb{C}_2\mathbf{\Psi}_2 &= \mathbf{q}_{2D}\\
\mathbb{A}_1\mathbf{V}_1 \quad\quad\quad\quad +\mathbb{B}_1\mathbf{P}_1 \quad\quad\quad\quad +\mathbb{C}_1\mathbf{\Psi}_1 &= \mathbf{q}_{1D}\\
\mathbb{B}_3^T\mathbf{V}_3 \quad\quad +\mathbb{D}_3\mathbf{P}_3 +\mathbb{D}_{32}\mathbf{P}_2 +\mathbb{D}_{31}\mathbf{P}_1 &= \mathbf{q}_{3E}\\
\mathbb{B}_2^T\mathbf{V}_2 \quad\quad +\mathbb{D}_{32}^T\mathbf{P}_3 +\mathbb{D}_2\mathbf{P}_2 +\mathbb{D}_{21}\mathbf{P}_1 &= \mathbf{q}_{2E}\\
\mathbb{B}_3^T\mathbf{V}_1 \quad +\mathbb{D}_{13}^T\mathbf{P}_3 +\mathbb{D}_{21}^T\mathbf{P}_2 +\mathbb{D}_1\mathbf{P}_1 &= \mathbf{q}_{1E}\\
\mathbb{C}_3^T\mathbf{V}_3 \quad\quad\quad\quad\quad\quad\quad\quad\quad\quad\quad\quad\quad &= \mathbf{q}_{3N}\\
\mathbb{C}_2^T\mathbf{V}_2 \quad\quad\quad\quad\quad\quad\quad\quad\quad\quad\quad\quad\quad &= \mathbf{q}_{2N}\\
\mathbb{C}_1^T\mathbf{V}_1 \quad\quad\quad\quad\quad\quad\quad\quad\quad\quad\quad\quad\quad &= \mathbf{q}_{1N}
\end{aligned}$$

The resulting system matrix is symmetric, sparse of characteristic internal structure, indefinite. The blocks $\mathbb{A}_i$ are positive definite. The properties enable to use specialised solvers of linear equation systems to make the process of solving more effective.

6 Conclusions

The presented formulation of water flow is based on connection of 1D, 2D, and 3D porous medium systems via only the source terms. It allows to construct meshes of the three systems independently on each other. This feature is crucial for ability of modelling large-scale real-world problems.

Actually, we have implemented the model based on this formulation. It is a subject of testing at the time. The results of small-scale tests, which are not presented in this article, show qualitatively good behaviour of the model. The most recent open problems, we are solving, are the identification of σ_{ij} in real problems and behaviour of the model in large scale.

References

1. Bogdanov, I.I., Mourzenko, V.V., Thovert, J.F.: Effective permeability of fractured porous media in steady state flow. Water Res. Research, **39** (1993)
2. Diodato, D.M.: Compendium of fracture flow models. Centre for Environmental Restoration Syst., Energy Syst. Division, Argonne National Laboratory, USA (1994) Available on: http://www.thehydrogeologist.com/docs/cffm/cffmtoc.htm.
3. Kaasschieter, E.F., Huijben, A.J.M.: Mixed-hybrid finite elements and streamline computation for the potential flow problem. Numerical Methods for Partial Differential Equations, **8**, 221–266 (1992)
4. Maryška, J., Severýn, O., Tauchman, M., Tondr, D.: Modelling of the groundwater flow in fractured rock – a new approach. In: Proc. of Algoritmy 2005, Slovak University of Technology, Bratislava, **10** (2005)
5. Maryška, J., Severýn, O., Vohralík, M.: Mixed-hybrid finite elements and streamline computation for the potential flow problem. Computational Geoscience 8/3, **18**, 217–234 (2005)

Newton–Type Methods for the Mixed Finite Element Discretization of Some Degenerate Parabolic Equations

Florin A. Radu[1], Iuliu Sorin Pop[2] and Peter Knabner[1]

[1] Institute of Applied Mathematics, University Erlangen-Nürnberg, Martensstr. 3, D-91058 Erlangen, Germany
{raduf, knabner}@am.uni-erlangen.de
[2] CASA, Technische Universiteit Eindhoven, P. O. Box 513, 5600 MB Eindhoven, The Netherlands
ipop@win.tue.nl

Summary. In this paper we discuss some iterative approaches for solving the nonlinear algebraic systems encountered as fully discrete counterparts of some degenerate (fast diffusion) parabolic problems. After regularization, we combine a mixed finite element discretization with the Euler implicit scheme. For the resulting systems we discuss three iterative methods and give sufficient conditions for convergence.

1 Introduction

Degenerate parabolic equations appear as models for reactive flow in porous media, or for phase transitions. By degeneracy we mean nonlinearities in the diffusion or in the time derivative term, which may vanish for certain values of the unknown function. To be specific, we consider the Richards' equation in the mixed form (see, [9] and [10] for details)

$$\begin{aligned}\partial_t b(u) + \nabla \cdot \mathbf{q} &= f & \text{in } J \times \Omega, \\ \mathbf{q} &= -(\nabla u + k(b(u))\mathbf{l}) & \text{in } J \times \Omega,\end{aligned} \quad (1)$$

where $J = (0, T]$ denotes the time interval and $\Omega \subset \mathbb{R}^d$ ($d = 1, 2$, or 3) is a domain with smooth boundary Γ. To comply with the physical background of the problem, the partially saturated flow through a porous medium, we take $\mathbf{l}$ as a constant vector, but the discussion can be easily extended to more general cases. Further, $b(\cdot)$ is an increasing function whose derivative may vanish, leading to a possible change of type from parabolic to elliptic. The equation has to be endowed with initial and boundary data.

Due to the degeneracy, the solutions of (1) may lack regularity, restricting the possible choices of suitable discretization methods. A useful tool in the analysis of such kind of problems is the regularization of the nonlinearities, which can also be applied for developing numerical schemes. This is also the idea we are following here. Further, we discretize the above equation by combining a mixed finite element method (MFEM) with the backward Euler scheme. This leads to fully discrete problems that are nonlinear, requiring suitable iterative algorithms. In this paper we discuss different approaches related to the Newton method, and give general convergence conditions in terms of the discretization parameters.

The general framework is introduced in Section 2, as well as the fully discrete nonlinear scheme. In Section 3 we study the convergence of the Newton method, which is shown to be of order $(1+r)$, with r being the order of the Hölder continuity of the first derivative of the coefficient functions. Next we use the apriori estimates from [10] to show that the solution computed at the previous time step is a good starting point for the iterations. A similar approach is considered in Subsection 3.1, this in connection with an explicit discretization of the convection term. In this case we obtain only a linear convergence. Finally, we consider the scheme proposed by Jäger and Kačur [3, 4], for which a linear convergence is proved. The conclusions are presented in the last section.

2 The fully discrete problem

We consider the system (1) together with the initial condition $u = u^0$ at $t = 0$ and some boundary conditions on Γ. For the ease of presentation we only consider here homogeneous Dirichlet boundary data, but all the results can be extended to more general cases. For the nonlinearities and the data we assume the following:

(A1) $b(\cdot) \in C^1$ is nondecreasing and Lipschitz continuous.

(A2) $k(b(\cdot))$ is continuous, bounded and satisfies

$$|k(b(z_2)) - k(b(z_1))|^2 \leq C_k(b(z_2) - b(z_1))(z_2 - z_1), \text{ for all } z_1, z_2 \in \mathbb{R}.$$

(A3) $b(u_0)$ is essentially bounded, and $u_0 \in L^2(\Omega)$.

(A4) $|b'(x) - b'(y)| \leq \gamma_1 |x - y|^r$, for all $x, y \in \mathbb{R}$, where $r \in (0, 1]$.

(A5) $|(k \circ b)'(x) - (k \circ b)'(y)| \leq \gamma_2 |x - y|^r$ for all $x, y \in \mathbb{R}$.

Since b' vanishes for some arguments, the problem changes its type from parabolic to elliptic. To overcome this we replace b in (1) by

$$b_\epsilon(u) := b(u) + \epsilon u,$$

$\epsilon > 0$ being a small perturbation parameter. In this way the assumptions on b are applying also for b_ϵ. A straightforward consequence of the Hölder continuity of the first derivative of $b_\epsilon(\cdot)$ and $k \circ b(\cdot)$ is the following lemma (see [5, p. 350]):

Lemma 1. *Assuming (A4) and (A5), then for all reals x and y we have*

$$|b_\epsilon(x) - b_\epsilon(y) - b'_\epsilon(y)(x-y)| \leq \frac{\gamma_1}{1+r}|x-y|^{1+r}, \qquad (2)$$

$$|k \circ b(x) - k \circ b(y) - (k \circ b)'(y)(x-y)| \leq \frac{\gamma_2}{1+r}|x-y|^{1+r}. \qquad (3)$$

Let $N \geq 1$ be an integer giving a time step $\tau = T/N$, $t_n = n\tau$ ($n = 1,\ldots,N$) and $\mathcal{T}_h$ be a triangulation of the domain Ω into closed d-simplexes. Here h stands for the mesh-size. Further we denote by W the space $L^2(\Omega)$, while $V = H(\text{div}, \Omega)$ is the space of d-dimensional vector functions having all components and the divergence in $L^2(\Omega)$. The standard notations for the Sobolev spaces and the related norms will be used. By $(\cdot,\cdot)$ we mean the $L^2(\Omega)$-inner product and $\|\cdot\|$ denotes the norm in $L^2(\Omega)$. The discrete subspaces $W_h \times V_h \subset W \times V$ are then defined as

$$\begin{aligned}W_h &:= \{p \in W|\ p \text{ is constant on each element } T \in \mathcal{T}_h\}, \\ V_h &:= \{\mathbf{q} \in V|\ \mathbf{q}_{|T} = \mathbf{a} + b\mathbf{x}, \mathbf{a} \in \mathbb{R}^d, b \in \mathbb{R} \text{ for all } T \in \mathcal{T}_h\}.\end{aligned} \qquad (4)$$

Applying implicit Euler and MFEM to the regularized equations leads to the following fully discrete problems written in a mixed weak form

Problem $\mathbf{P}_n$. *Let $n \in \{1,\ldots,N\}$ and $p_h^{n-1} \in W_h$ be given. Find $(p_h^n, \mathbf{q_h^n}) \in W_h \times V_h$ such that for all $w_h \in W_h$ and $\mathbf{v_h} \in V_h$*

$$(b_\epsilon(p_h^n), w_h) + \tau(\nabla \cdot \mathbf{q_h^n}, w_h) = (b_\epsilon(p_h^{n-1}), w_h), \qquad (5)$$

$$(\mathbf{q_h^n}, \mathbf{v_h}) - (p_h^n, \nabla \cdot \mathbf{v_h}) + (k(b(p_h^n))\mathbf{l}, \mathbf{v_h}) = 0. \qquad (6)$$

Once this point is reached there are several possibilities to proceed. First, the resulting algebraic system can be enlarged by adding Lagrange multipliers on edges, this being the mixed hybrid finite element method (see [9]). Another way is to compute the mass matrix using an adequate quadrature formula, which allows an explicit and global elimination of the flux variable, leading to an equivalent finite volume scheme (see [1]). Alternatively, a robust linearization procedure is proposed in [8]. That scheme is converging linearly even in the non-regularized case. In the following we discuss some Newton–type methods for Problem $\mathbf{P}_n$, including convergence results for the Jäger-Kačur relaxation applied to our problem.

Below i is used to index the iteration, whereas n is indexing the time step. Accordingly, $\{p_h^n, \mathbf{q_h^n}\}$ denotes the solution pair at the nth time step and

$\{p_h^{n,i}, \mathbf{q_h^{n,i}}\}$ the solution pair at iteration i. Considering the pressure and the flux errors

$$e_p^i := p_h^{n,i} - p_h^n, \quad \mathbf{e_q^i} := \mathbf{q_h^{n,i}} - \mathbf{q_h^n}, \tag{7}$$

convergence means that both $\|e_p^i\|$ and $\|\mathbf{e_q^i}\|$ vanish as $i \to \infty$. Notice that proving convergence in the pressure immediately implies the flux convergence.

In what follows we make use of the following elementary lemma, which can be proved by mathematical induction:

Lemma 2. *Let $r > 0$ and $\{x_n\}_{n \geq 0}$ a sequence of real positive numbers satisfying*

$$x_n \leq \alpha x_{n-1}^{1+r} + \beta x_{n-1} \quad \forall n \geq 1. \tag{8}$$

Assuming that

$$\alpha x_0^r + \beta < 1, \tag{9}$$

the sequence $\{x_n\}_{n \geq 0}$ converges to zero.

3 The Newton scheme

The method developed by Newton (1642-1727) in his book "Method of Fluxions", written in 1671 and published in 1736, is also called "Newton–Raphson", because Raphson was the first who published it in 1690. Applying the guidelines of the general Newton method to Problem $\mathbf{P}_n$ gives:

Scheme $\mathbf{N}_{n,i}$. *Fix $n \in \{1, \ldots, N\}$ and $i > 0$ and assume p_h^{n-1}, and $p_h^{n,i-1}$ given in W_h. Find $(p_h^{n,i}, \mathbf{q_h^{n,i}}) \in W_h \times V_h$ such that for all $w_h \in W_h$ and $\mathbf{v_h} \in V_h$*

$$(b_\epsilon'(p_h^{n,i-1})(p_h^{n,i} - p_h^{n,i-1}), w_h) + \tau(\nabla \cdot \mathbf{q_h^{n,i}}, w_h) \\ = (b_\epsilon(p_h^{n-1}) - b_\epsilon(p_h^{n,i-1}), w_h), \tag{10}$$

$$(\mathbf{q_h^{n,i}}, \mathbf{v_h}) - (p_h^{n,i}, \nabla \cdot \mathbf{v_h}) + ((k \circ b)'(p_h^{n,i-1})(p_h^{n,i} - p_h^{n,i-1})\mathbf{1}, \mathbf{v_h}) \\ = -(k(b(p_h^{n,i-1}))\mathbf{1}, \mathbf{v_h}), \tag{11}$$

Remark 1. The choice of a starting point that is close to the solution is important for the convergence of the Newton scheme. Therefore we start here the iterative process with the solution at the previous time step, $p_h^{n,0} = p_h^{n-1}$. This choice ensures the convergence of the scheme, as we will show below.

Under certain restrictions on the time step τ with respect to the regularization parameter ϵ and the mesh size h, and assuming $b_\epsilon(\cdot)$ and $k \circ b(\cdot)$ of class $C^{1,r}$ leads to a order $(1+r)$ convergent scheme. In particular, if $b_\epsilon(\cdot)'$ and $k \circ b(\cdot)'$ are Lipschitz continuous, the scheme (10)–(11) is quadratically convergent. In what follows we make this sentence more precise, by showing first the following

Proposition 1. *Assume (A4) and (A5). For a small enough τ we have*

$$\|e_p^i\|^2 + \frac{\tau}{\epsilon}\|\mathbf{e_q^i}\|^2 \le C\frac{\frac{\gamma_1^2}{\epsilon} + \tau\gamma_2^2}{(1+r)^2}h^{-rd}\epsilon^{-1}\|e_p^{i-1}\|^{2+2r}, \quad (12)$$

where $C > 0$ does not depend on the discretization parameters.

Proof. Subtracting (5) from (10), respectively (6) from (11) gives

$$(b'_\epsilon(p_h^{n,i-1})(p_h^{n,i} - p_h^{n,i-1}), w_h) + \tau(\nabla \cdot (\mathbf{q_h^{n,i}} - \mathbf{q_h^n}), w_h)$$
$$= (b_\epsilon(p_h^n) - b_\epsilon(p_h^{n,i-1}), w_h) \quad (13)$$

and

$$(\mathbf{q_h^{n,i}} - \mathbf{q_h^n}, \mathbf{v_h}) - (p_h^{n,i} - p_h^n, \nabla \cdot \mathbf{v_h}) + ((k \circ b)'(p_h^{n,i-1})(p_h^{n,i} - p_h^{n,i-1})\mathbf{l}, \mathbf{v_h})$$
$$= ((k(b(p_h^n)) - k(b(p_h^{n,i-1})))\mathbf{l}, \mathbf{v_h}), \quad (14)$$

for all $w_h \in W_h$ and $\mathbf{v_h} \in V_h$. Taking $w_h = e_p^i$ and $\mathbf{v_h} = \tau\mathbf{e_q^i}$ in the above, adding the resulting two equalities and observing that $p_h^{n,i} - p_h^{n,i-1} = e_p^i - e_p^{i-1}$, $\mathbf{q_h^{n,i}} - \mathbf{q_h^n} = \mathbf{e_q^i}$ and $p_h^{n,i} - p_h^n = e_p^i$ we obtain

$$(b'_\epsilon(p_h^{n,i-1})(e_p^i - e_p^{i-1}), e_p^i) + \tau\|\mathbf{e_q^i}\|^2 = (b_\epsilon(p_h^n) - b_\epsilon(p_h^{n,i-1}), e_p^i)$$
$$-\tau((k \circ b)'(p_h^{n,i-1})(e_p^i - e_p^{i-1})\mathbf{l}, \mathbf{e_q^i}) + \tau((k(b(p_h^n)) - k(b(p_h^{n,i-1})))\mathbf{l}, \mathbf{e_q^i}).$$

This is further equivalent to

$$(b'_\epsilon(p_h^{n,i-1})e_p^i, e_p^i) + \tau\|\mathbf{e_q^i}\|^2 = (b_\epsilon(p_h^n) - b_\epsilon(p_h^{n,i-1}) + b'_\epsilon(p_h^{n,i-1})e_p^{i-1}, e_p^i)$$
$$+\tau((k(b(p_h^n)) - k(b(p_h^{n,i-1})) + (k \circ b)'(p_h^{n,i-1})e_p^{i-1})\mathbf{l}, \mathbf{e_q^i}) \quad (15)$$
$$-\tau((k \circ b)'(p_h^{n,i-1})e_p^i\mathbf{l}, \mathbf{e_q^i}).$$

Using now Lemma 1 and the inequality $|ab| \le \delta a^2 + b^2/(4\delta)$ (for all reals a and b, and $\delta > 0$) we estimate the first term on the right as follows

$$T_1 \le \int_\Omega |b_\epsilon(p_h^n) - b_\epsilon(p_h^{n,i-1}) + b'_\epsilon(p_h^{n,i-1})e_p^{i-1}| |e_p^i|\, d\mathbf{x}$$

$$\le \int_\Omega \frac{\gamma_1}{1+r}|e_p^{i-1}|^{1+r}|e_p^i|dx \le \frac{\gamma_1^2}{\epsilon(1+r)^2}\|e_p^{i-1}\|_{L^{2+2r}}^{2+2r} + \frac{\epsilon}{4}\|e_p^i\|^2. \quad (16)$$

Similarly, for the second term on the right in (15) we get

$$T_2 \leq \tau \frac{\gamma_2^2}{(1+r)^2}\|e_p^{i-1}\|_{L^{2+2r}}^{2+2r} + \frac{\tau}{4}\|\mathbf{e_q^i}\|^2. \tag{17}$$

Finally, for the last term in (15) we use the boundedness of $(k \circ b)'(\cdot)$ to obtain

$$T_3 \leq \tau C_1^2 \|e_p^i\|^2 + \frac{\tau}{4}\|\mathbf{e_q^i}\|^2. \tag{18}$$

Using (16)–(18) into (15), since $b'_\epsilon \geq \epsilon$ we obtain

$$\frac{3\epsilon}{4}\|e_p^i\|^2 + \frac{\tau}{2}\|\mathbf{e_q^i}\|^2 \leq \frac{\frac{\gamma_1^2}{\epsilon} + \tau\gamma_2^2}{(1+r)^2}\|e_p^{i-1}\|_{L^{2+2r}(\Omega)}^{2+2r} + \tau C_1^2 \|e_p^i\|^2. \tag{19}$$

For $\tau C_1^2 \leq \frac{\epsilon}{4}$, and using the inverse estimate for discrete polynomial spaces (see [2, p. 111] or [11, p. 705])

$$\|e_p^{i-1}\|_{L^{2+2r}(\Omega)} \leq Ch^{-\frac{rd}{2+2r}}\|e_p^{i-1}\|, \tag{20}$$

the estimate (19) becomes

$$\frac{\epsilon}{2}\|e_p^i\|^2 + \frac{\tau}{2}\|\mathbf{e_q^i}\|^2 \leq \frac{\frac{\gamma_1^2}{\epsilon} + \tau\gamma_2^2}{(1+r)^2} Ch^{-rd}\|e_p^{i-1}\|^{2+2r}. \tag{21}$$

Now (12) follows straightforwardly. ∎

Now we can proceed with the convergence of the Newton iteration.

Theorem 1. *Assuming (A1)–(A5), if $\tau = O(\epsilon^{\frac{r+2}{r}} h^d)$, then the Newton scheme (10)–(11) is $(1+r)$-order convergent.*

Proof. We first recall the stability estimate proved in [10], Proposition 3.5:

$$\|e_p^0\|^2 := \|p_h^n - p_h^{n-1}\|^2 \leq C\frac{\tau}{\epsilon}. \tag{22}$$

This shows that $p_h^{n,0} := p_h^{n-1}$ is a good starting point for the iteration. With $\tau = O(\epsilon^{\frac{r+2}{r}} h^d)$ we get

$$C(\frac{\gamma_1^2}{\epsilon} + \tau\gamma_2^2)h^{-rd}\frac{\tau^r}{\epsilon^{1+r}} < 1. \tag{23}$$

Using now the estimate given in Proposition 1, as well as Lemma 2 with $\beta = 0$ we obtain that Scheme $\mathbf{N}_{n,i}$ is convergent and of order $1+r$. ∎

Remark 2. In the non-degenerate case ϵ can be replaced by a constant. Then the $(1+r)$-order convergence is ensured if τ is of order h^d.

Remark 3. The Newton method for the mixed finite element discretization of nonlinear second-order elliptic problems is also studied in [6]. Assuming more regularity ($p \in H^{5/2+\epsilon_0}(\Omega)$), convergence is proved by duality techniques. In the non-degenerate case, our result is similar to the one in [6].

3.1 A Newton–like scheme

We present now a modification of the Newton scheme that can be applied whenever the assumption (A5) does not hold. When compared to Scheme $\mathbf{N}_{n,i}$, the only difference is in the discretization of the convection term.

Scheme $\mathbf{L}_{n,i}$. Fix $n \in \{1, \ldots, N\}$ and $i > 0$ and assume p_h^{n-1}, and $p_h^{n,i-1}$ given in W_h, with $p_h^{n,0} = p_h^{n-1}$. Find $(p_h^{n,i}, \mathbf{q_h^{n,i}}) \in W_h \times V_h$ such that for all $w_h \in W_h$ and $\mathbf{v_h} \in V_h$

$$(b'_\epsilon(p_h^{n,i-1})(p_h^{n,i} - p_h^{n,i-1}), w_h) + \tau(\nabla \cdot \mathbf{q_h^{n,i}}, w_h)$$
$$= (b_\epsilon(p_h^{n-1}) - b_\epsilon(p_h^{n,i-1}), w_h), \quad (24)$$

$$(\mathbf{q_h^{n,i}}, \mathbf{v_h}) - (p_h^{n,i}, \nabla \cdot \mathbf{v_h}) + (k(b(p_h^{n,i-1}))\mathbf{l}, \mathbf{v_h}) = 0. \quad (25)$$

Due to the discretization of the convection, the above scheme is simpler than the Newton method. This at the expense of a reduced convergence order, which becomes linear:

Theorem 2. *Assuming (A1)–(A4), if $\tau = O(\epsilon^{\frac{r+2}{r}} h^d)$, the Newton–like scheme (24)–(25) converges linearly.*

The proof goes along the lines of Theorem 1, see [9] for details.

3.2 The Jäger and Kačur scheme

In [3, 4], W. Jäger and J. Kačur have proposed a relaxation scheme for doubly degenerate parabolic problems. The scheme is semi-implicit, and for solving the emerging nonlinear problems an iterative scheme has been proposed. A similar scheme is discussed in [7], Section 2.3.2, where convergence is proved for the conformal discretization and in the absence of convection. In a simplified version, applied to our specific context, the Jäger–Kačur iteration reads

Scheme $\mathbf{JK}_{n,i}$. Fix $n \in \{1, \ldots, N\}$ and $i > 0$ and assume p_h^{n-1}, and $p_h^{n,i-1}$ given in W_h, with $p_h^{n,0} = p_h^{n-1}$. Find $(p_h^{n,i}, \mathbf{q_h^{n,i}}) \in W_h \times V_h$ such that for all $w_h \in W_h$ and $\mathbf{v_h} \in V_h$

$$(\lambda_{i-1}(p_h^{n,i} - p_h^{n-1}), w_h) + \tau(\nabla \cdot \mathbf{q_h^{n,i}}, w_h) = 0, \quad (26)$$

$$(\mathbf{q_h^{n,i}}, \mathbf{v_h}) - (p_h^{n,i}, \nabla \cdot \mathbf{v_h}) + (k(b(p_h^{n,i-1}))\mathbf{l}, \mathbf{v_h}) = 0, \quad (27)$$

with $\lambda_0 = b'_\epsilon(p_h^{n-1})$ and for $i \geq 1$

$$\lambda_i = \frac{b_\epsilon(p_h^{n,i}) - b_\epsilon(p_h^{n-1})}{p_h^{n,i} - p_h^{n-1}} = \int_0^1 b'_\epsilon(p_h^{n-1} + s(p_h^{n,i} - p_h^{n-1}))\, ds.$$

Theorem 3. *Assuming (A1)–(A4) with $r = 1$, if $\tau = O(\epsilon^3 h^d)$, the Jäger and Kačur scheme (26)–(27) is linearly convergent.*

The proof combines some ideas from [4] with the techniques used in the previous section, details being given in [9]. We only mention here the usage of the apriori estimate $\|e_p^0\|^2 \leq C\frac{\tau}{\epsilon}$. In some special cases, assuming more regularity for the solution of (1), the previous estimate becomes $\|e_p^0\|_{L^\infty(\Omega)} \leq C\frac{\tau^j}{\epsilon^k}$, with $j, k > 0$ (see [4, Theorem 5.1]). This leads to a substantial improvement in the restriction on τ, which can now be chosen of order ϵ^3. In other words, convergence is achieved independent of the mesh diameter h. Unfortunately, for the general case such an estimate cannot be obtained.

4 Conclusions

The paper deals with some iterative methods for solving the nonlinear systems resulting after a complete discretization of equation (1) by the MFEM. For overcoming the difficulties due to degeneracy we first perturb the original degenerate equation to obtain a regular parabolic one, and then apply an Euler time stepping. The order of the Newton method depends on the smoothness of the nonlinearities. In particular, if these are of class $C^{1,1}$, quadratic convergence has been obtained. Further, a simplified Newton–like scheme, as well as the Jäger and Kačur approach are considered. Both are converging linearly.

References

1. Baranger, J., Maitre, J-F., Oudin, F.: Connection between finite volume and mixed FE methods. RAIRO Model. Math. Anal. Numer., **30** , 445–465 (1996)
2. Brenner, S. C., Scott, L. R.: The mathematical theory of finite element methods, Springer-Verlag, New York (1991)
3. Jäger, W., Kačur, J.: Solution of doubly nonlinear and degenerate parabolic problems by relaxation schemes, Math. Model. Numer. Anal. **29**, 605–627 (1995)
4. Kačur, J.: Solution to strongly nonlinear parabolic problems by a linear approximation scheme, IMA J. Numer. Anal. **19**, 119–145 (1999)
5. Knabner, P., Angermann, L.: Numerical methods for elliptic and parabolic partial differential equations, Springer Verlag (2003).
6. Park, E.J.: *Mixed finite elements for nonlinear second-order elliptic problems*, SIAM J. Numer. Anal. **32**, 865–885 (1995).
7. Pop, I.S.: Regularization methods in the numerical analysis of some degenerate parabolic equations, Preprint 98-43 (SFB 359), IWR, Heidelberg (1998)
8. Pop, I.S., Radu, F.A., Knabner, P.: Mixed finite elements for the Richards' equation: linearization procedure, J. Comput. Appl. Math. **168**, 365-373 (2004)

9. Radu, F.A.: Mixed FE discretization of Richards' equation: error analysis and application to realistic infiltration problems, PhD-Thesis, Erlangen (2004)
10. Radu, F.A., Pop, I.S., Knabner, P.: Order of convergence estimates for an Euler implicit, mixed FE discretization of Richards' equation, SIAM J. Numer. Anal. **42**, 1452-1478 (2004)
11. Woodward, C., Dawson, C.: Analysis of expanded mixed finite element methods for a nonlinear parabolic equation modeling flow into variably saturated porous media, SIAM J. Numer. Anal. **37**, 701–724 (2000)

WAVES

Domain Decomposition Methods for Wave Propagation in Heterogeneous Media

R. Glowinski[1], S. Lapin[2], J. Periaux[3], P.M. Jacquart[4] and H.Q. Chen[5]

[1] University of Houston
   roland@math.uh.edu
[2] University of Houston
   slapin@math.uh.edu
[3] Pole Scientifique Dassault Aviation
   jperiaux@free.fr
[4] Dassault Aviation
[5] Nanjing University of Aeronautics and Astronautics

Summary. The main goal of this paper is to address the numerical solution of a wave equation with discontinuous coefficients by a finite element method using domain decomposition and semimatching grids. A wave equation with absorbing boundary conditions is considered, the coefficients in the equation essentially differ in the subdomains. The problem is approximated by an explicit in time finite difference scheme combined with a piecewise linear finite element method in the space variables on a semimatching grid. The matching condition on the interface is taken into account by means of Lagrange multipliers. The resulting system of linear equations of the saddle-point form is solved by a conjugate gradient method.

1 Formulation of the problem

Let $\Omega \subset \mathbb{R}^2$ be a rectangular domain with sides parallel to the coordinate axes and boundary Γ_{ext} (see Fig. 1). Now let $\Omega_2 \subset \Omega$ be a proper subdomain of Ω with a curvilinear boundary and $\Omega_1 = \Omega \setminus \bar{\Omega}_2$. We consider the following linear wave problem:

$$\begin{cases} \varepsilon \dfrac{\partial^2 u}{\partial t^2} - \nabla \cdot (\mu^{-1} \nabla u) = f \text{ in } \Omega \times (0, T), \\ \sqrt{\varepsilon \mu^{-1}} \dfrac{\partial u}{\partial t} + \mu^{-1} \dfrac{\partial u}{\partial \mathbf{n}} = 0 \text{ on } \Gamma_{ext} \times (0, T), \\ u(x, 0) = \dfrac{\partial u}{\partial t}(x, 0) = 0. \end{cases} \quad (1)$$

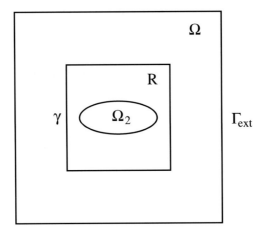

Fig. 1. *Computational domain.*

Here $\nabla u = (\frac{\partial u}{\partial x_1}, \frac{\partial u}{\partial x_2})$, **n** is the unit outward normal vector on Γ_{ext}; we suppose that $\mu_i = \mu|_{\Omega_i}$, $\varepsilon_i = \varepsilon|_{\Omega_i}$ are positive constants $\forall i = 1, 2$ and, $f_i = f|_{\Omega_i} \in C(\bar{\Omega}_i \times [0,T])$.

Let $\varepsilon(x) = \{\varepsilon_1 \text{ if } x \in \Omega_1, \varepsilon_2 \text{ if } x \in \Omega_2\}$ and $\mu(x) = \{\mu_1 \text{ if } x \in \Omega_1, \mu_2 \text{ if } x \in \Omega_2\}$. We define a weak solution of problem (1) as a function u such that

$$u \in L^\infty(0,T;H^1(\Omega)), \frac{\partial u}{\partial t} \in L^\infty(0,T;L^2(\Omega)), \frac{\partial u}{\partial t} \in L^2(0,T;L^2(\Gamma_{ext})) \quad (2)$$

for a.a. $t \in (0,T)$ and for all $w \in H^1(\Omega)$ satisfying the equation

$$\int_\Omega \varepsilon(x) \frac{\partial^2 u}{\partial t^2} w dx + \int_\Omega \mu^{-1}(x) \nabla u \cdot \nabla w dx + \sqrt{\varepsilon_1 \mu_1^{-1}} \int_{\Gamma_{ext}} \frac{\partial u}{\partial t} w d\Gamma = \int_\Omega f w dx \quad (3)$$

with the initial conditions

$$u(x,0) = \frac{\partial u}{\partial t}(x,0) = 0.$$

Note that the first term in (3) means the duality between $(H^1(\Omega))^*$ and $H^1(\Omega)$.

Now we formulate problem (3) variationally as follows:

Let

$$W_1 = \{v \in L^\infty(0,T;H^1(\tilde{\Omega})), \frac{\partial v}{\partial t} \in L^\infty(0,T;L^2(\tilde{\Omega})), \frac{\partial v}{\partial t} \in L^2(0,T;L^2(\Gamma_{ext}))\},$$

$$W_2 = \{v \in L^\infty(0,T;H^1(R)), \frac{\partial v}{\partial t} \in L^\infty(0,T;L^2(R)))\},$$

Find a pair $(u_1, u_2) \in W_1 \times W_2$, such that $u_1 = u_2$ on $\gamma \times (0,T)$ and for a.a. $t \in (0,T)$

$$\begin{cases} \int_{\widetilde{\Omega}} \varepsilon_1 \frac{\partial^2 u_1}{\partial t^2} w_1 dx + \int_{\widetilde{\Omega}} \mu_1^{-1} \nabla u_1 \cdot \nabla w_1 dx + \int_R \varepsilon(x) \frac{\partial^2 u_2}{\partial t^2} w_2 dx \\ + \int_R \mu^{-1}(x) \nabla u_2 \cdot \nabla w_2 dx + \sqrt{\varepsilon_1 \mu_1^{-1}} \int_{\Gamma_{ext}} \frac{\partial u_1}{\partial t} w_1 d\Gamma = \int_{\widetilde{\Omega}} f_1 w_1 dx + \int_R f_2 w_2 dx \\ \text{for all } (w_1, w_2) \in H^1(\widetilde{\Omega}) \times H^1(R) \text{ such that } w_1 = w_2 \text{ on } \gamma, \\ u(x,0) = \frac{\partial u}{\partial t}(x,0) = 0. \end{cases} \quad (4)$$

Now, introduce the interface supported Lagrange multiplier λ (a function defined over $\gamma \times (0,T)$), problem (4) can be written in the following way:

Find a triple $(u_1, u_2, \lambda) \in W_1 \times W_2 \times L^\infty(0, T; H^{-1/2}(\gamma))$, which for a.a. $t \in (0,T)$ satisfies

$$\int_{\widetilde{\Omega}} \varepsilon_1 \frac{\partial^2 u_1}{\partial t^2} w_1 dx + \int_{\widetilde{\Omega}} \mu_1^{-1} \nabla u_1 \cdot \nabla w_1 dx + \int_R \varepsilon(x) \frac{\partial^2 u_2}{\partial t^2} w_2 dx$$
$$+ \int_R \mu^{-1}(x) \nabla u_2 \cdot \nabla w_2 dx + \sqrt{\varepsilon_1 \mu_1^{-1}} \int_{\Gamma_{ext}} \frac{\partial u_1}{\partial t} w_1 d\Gamma + \int_\gamma \lambda(w_2 - w_1) d\gamma \quad (5)$$
$$= \int_{\widetilde{\Omega}} f_1 w_1 dx + \int_R f_2 w_2 dx \text{ for all } w_1 \in H^1(\widetilde{\Omega}), w_2 \in H^1(R);$$

$$\int_\gamma \zeta(u_2 - u_1) d\gamma = 0 \text{ for all } \zeta \in H^{-1/2}(\gamma), \quad (6)$$

and the initial conditions from (1).

Remark 1 *We selected the time dependent approach to capture harmonic solutions since it substantially simplifies the linear algebra of the solution process. Furthermore, there exist various techniques to speed up the convergence of transient solutions to periodic ones (see, e.g. [3]).*

2 Time discretization

In order to construct a finite difference approximation in time of problem (5), (6) we partition the segment $[0,T]$ into N intervals using a uniform discretization step $\Delta t = T/N$. Let $u_i^n \approx u_i(n \Delta t)$ for $i = 1, 2$, $\lambda^n \approx \lambda(n \Delta t)$. The explicit in time semidiscrete approximation to problem (5), (6) reads as follows:

$u_i^0 = u_i^1 = 0;$

for $n = 1, 2, \ldots, N-1$ find $u_1^{n+1} \in H^1(\widetilde{\Omega})$, $u_2^{n+1} \in H^1(R)$ and $\lambda^{n+1} \in H^{-1/2}(\gamma)$ such that

$$\int_{\widetilde{\Omega}} \varepsilon_1 \frac{u_1^{n+1} - 2u_1^n + u_1^{n-1}}{\Delta t^2} w_1 dx + \int_{\widetilde{\Omega}} \mu_1^{-1} \nabla u_1^n \cdot \nabla w_1 dx$$
$$+ \int_R \varepsilon(x) \frac{u_2^{n+1} - 2u_2^n + u_2^{n-1}}{\Delta t^2} w_2 dx + \int_R \mu^{-1}(x) \nabla u_2^n \cdot \nabla w_2 dx \quad (7)$$
$$+ \sqrt{\varepsilon_1 \mu_1^{-1}} \int_{\Gamma_{ext}} \frac{u_1^{n+1} - u_1^{n-1}}{2\Delta t} w_1 d\Gamma + \int_\gamma \lambda^{n+1}(w_2 - w_1) d\gamma =$$
$$\int_{\widetilde{\Omega}} f_1^n w_1 dx + \int_R f_2^n w_2 dx \text{ for all } w_1 \in H^1(\widetilde{\Omega}), w_2 \in H^1(R);$$

$$\int_\gamma \zeta(u_2^{n+1} - u_1^{n+1}) d\gamma = 0 \text{ for all } \zeta \in H^{-1/2}(\gamma). \quad (8)$$

Remark 2 *The integral over γ is written formally; the exact formulation requires the use of the duality pairing $\langle .,. \rangle$ between $H^{-1/2}(\gamma)$ and $H^{1/2}(\gamma)$.*

3 Fully discrete scheme

To construct a fully discrete space-time approximation to problem (5), (6) we will use a lowest order finite element method on two grids semimatching on γ (Fig. 2) for the space discretization. Namely, let $\mathcal{T}_{1h}$ and $\mathcal{T}_{2h}$ be triangulations of $\widetilde{\Omega}$ and R, respectively.

We denote by $\mathcal{T}_{1h}$ a coarse triangulation, and by $\mathcal{T}_{2h}$ a fine one. Every edge $\partial e \subset \gamma$ of a triangle $e \in \mathcal{T}_{1h}$ is supposed to consist of m_e edges of triangles from $\mathcal{T}_{2h}$, $1 \leq m_e \leq m$ for all $e \in \mathcal{T}_{1h}$.

Let $V_{1h} \subset H^1(\widetilde{\Omega})$ be the space of the functions globally continuous, and affine on each $e \in \mathcal{T}_{1h}$, i.e. $V_{1h} = \{u_h \in H^1(\widetilde{\Omega}) : u_h \in P_1(e) \ \forall e \in \mathcal{T}_{1h}\}$. Similarly, $V_{2h} \subset H^1(R)$ is the space of the functions globally continuous, and affine on each $e \in \mathcal{T}_{2h}$.

For approximating the Lagrange multiplier space $\Lambda = H^{-1/2}(\gamma)$ we proceed as follows. Assume that on γ, $\mathcal{T}_{1h}$ is twice coarser than $\mathcal{T}_{2h}$; then let us divide every edge ∂e of a triangle e from the coarse grid $\mathcal{T}_{1h}$, which is located on γ ($\partial e \subset \gamma$), into two parts using its midpoint. Now, we consider the space of the piecewise constant functions, which are constant on every union of half-edges with a common vertex (see Fig.2).

Further, we use quadrature formulas for approximating the integrals over the triangles from $\mathcal{T}_{1h}$ and $\mathcal{T}_{2h}$, as well as over Γ_{ext}. For a triangle e we set

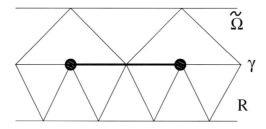

Fig. 2. Space Λ is the space of the piecewise constant functions defined on every union of half-edges with common vertex.

$$\int_e \phi(x)dx \approx \frac{1}{3}\text{meas}(e)\sum_{i=1}^{3}\phi(a_i) \equiv S_e(\phi)$$

where the a_i's are the vertices of e and $\phi(x)$ is a continuous function on e. Similarly,

$$\int_{\partial e} \phi(x)dx \approx \frac{1}{2}\text{meas}(\partial e)\sum_{i=1}^{2}\phi(a_i) \equiv S_{\partial e}(\phi),$$

where a_i's are the endpoints of the segment ∂e and $\phi(x)$ is a continuous function on this segment.

We use the notations:

$$S_i(\phi) = \sum_{e \in \mathcal{T}_{ih}} S_e(\phi), \; i=1,2, \text{ and } S_{\Gamma_{ext}}(\phi) = \sum_{\partial e \subset \Gamma_{ext}} S_{\partial e}(\phi).$$

Now, the fully discrete problem reads as follows:
Let $u_{ih}^0 = u_{ih}^1 = 0$, $i=1,2$;
for $n = 1, 2, \ldots, N-1$ find $(u_{1h}^{n+1}, u_{2h}^{n+1}, \lambda_h^{n+1}) \in V_{1h} \times V_{2h} \times \Lambda_h$ such that

$$\begin{cases} \dfrac{\varepsilon_1}{\Delta t^2} S_1((u_{1h}^{n+1} - 2u_{1h}^n + u_{1h}^{n-1})w_{1h}) + S_1(\mu_1^{-1}\nabla u_{1h}^n \cdot \nabla w_{1h}) \\ + \dfrac{1}{\Delta t^2} S_2(\varepsilon(x)(u_{2h}^{n+1} - 2u_{2h}^n + u_{2h}^{n-1})w_{2h}) + S_2(\mu^{-1}(x)\nabla u_{2h}^n \cdot \nabla w_{2h}) \\ + \dfrac{\sqrt{\varepsilon_1 \mu_1^{-1}}}{2\Delta t} S_{\Gamma_{ext}}((u_{1h}^{n+1} - u_{1h}^{n-1})w_{1h}) + \int_\gamma \lambda_h^{n+1}(w_{2h} - w_{1h})d\gamma = \\ S_1(f_1^n w_{1h}) + S_2(f_2^n w_{2h}) \text{ for all } w_{1h} \in V_{1h}, w_{2h} \in V_{2h}; \end{cases} \quad (9)$$

$$\int_\gamma \zeta_h(u_{2h}^{n+1} - u_{1h}^{n+1})d\gamma = 0 \text{ for all } \zeta_h \in \Lambda_h. \quad (10)$$

Note that in $S_2(\varepsilon(x)(u_{2h}^{n+1} - 2u_{2h}^n + u_{2h}^{n-1})w_{2h})$ we take $\varepsilon(x) = \varepsilon_2$ if a triangle $e \in \mathcal{T}_{2h}$ lies in Ω_2 and $\varepsilon(x) = \varepsilon_1$ if it lies in $R \setminus \Omega_2$, and similarly for $S_2(\mu^{-1}(x)\nabla u_{2h}^n \nabla w_{2h})$.

Denote by $\mathbf{u}_1$, $\mathbf{u}_2$ and $\boldsymbol{\lambda}$ the vectors of the nodal values of the corresponding functions u_{1h}, u_{2h} and λ_h. Then in order to find $\mathbf{u}_1^{n+1}$, $\mathbf{u}_2^{n+1}$ and $\boldsymbol{\lambda}^{n+1}$ for a fixed time t^{n+1} we have to solve a system of linear equations such as

$$\mathbf{Au} + \mathbf{B}^T \boldsymbol{\lambda} = \mathbf{F}, \qquad (11)$$

$$\mathbf{Bu} = 0, \qquad (12)$$

where matrix $\mathbf{A}$ is diagonal, positive definite and defined by

$$(\mathbf{Au}, \mathbf{w}) = \frac{\varepsilon_1}{\Delta t^2} S_1(u_{1h} w_{1h}) + \frac{1}{\Delta t^2} S_2(\varepsilon(x) u_{2h} w_{2h}) + \frac{\sqrt{\varepsilon_1 \mu_1^{-1}}}{2\Delta t} S_{\Gamma_{ext}}(u_{1h} w_{1h}),$$

and where the rectangular matrix B is defined by

$$(\mathbf{Bu}, \boldsymbol{\lambda}) = \int_\gamma \lambda_h (u_{2h} - u_{1h}) d\Gamma,$$

and vector F depends on the nodal values of the known functions u_{1h}^n, u_{2h}^n, u_{1h}^{n-1} and u_{2h}^{n-1}.

Eliminating u from equation (11) we obtain

$$\mathbf{BA}^{-1}\mathbf{B}^T \boldsymbol{\lambda} = \mathbf{BA}^{-1}\mathbf{F}, \qquad (13)$$

with a symmetric matrix $\mathbf{C} \equiv \mathbf{BA}^{-1}\mathbf{B}^T$.

Remark 3 *A closely related domain decomposition method applied to the solution of linear parabolic equations is discussed in [1].*

4 Energy inequality

Let h_{min} denote the minimal diameter of the triangles from $\mathcal{T}_{1h} \cup \mathcal{T}_{2h}$. There exists a positive number c such that the condition

$$\Delta t \leq c \min\{\sqrt{\varepsilon_1 \mu_1}, \sqrt{\varepsilon_2 \mu_2}\} h_{min} \qquad (14)$$

ensures the positive definiteness of the quadratic form

$$\mathcal{E}^{n+1} = \frac{1}{2}\varepsilon_1 S_1\left(\left(\frac{u_{1h}^{n+1} - u_{1h}^n}{\Delta t}\right)^2\right) + \frac{1}{2}S_2\left(\varepsilon\left(\frac{u_{2h}^{n+1} - u_{2h}^n}{\Delta t}\right)^2\right) +$$

$$\frac{1}{2}S_1\left(\mu_1^{-1}\left|\nabla\left(\frac{u_{1h}^{n+1} + u_{1h}^n}{2}\right)\right|^2\right) + \frac{1}{2}S_2\left(\mu^{-1}\left|\nabla\left(\frac{u_{2h}^{n+1} + u_{2h}^n}{2}\right)\right|^2\right)$$

$$-\frac{\Delta t^2}{8}S_1(\mu_1^{-1}|\nabla(\frac{u_{1h}^{n+1}-u_{1h}^n}{\Delta t})|^2)-\frac{\Delta t^2}{8}S_2(\mu^{-1}|\nabla(\frac{u_{2h}^{n+1}-u_{2h}^n}{\Delta t})|^2), \quad (15)$$

which we call the discrete energy.

System (9), (10) satisfies the energy identity

$$\mathcal{E}^{n+1}-\mathcal{E}^n+\frac{\sqrt{\varepsilon_1\mu_1^{-1}}}{4\Delta t}S_{\Gamma_{ext}}((u_{1h}^{n+1}-u_{1h}^{n-1})^2)=$$
$$\tfrac{1}{2}S_1(f_1^n(u_{1h}^{n+1}-u_{1h}^{n-1}))+\tfrac{1}{2}S_2(f_2^n(u_{2h}^{n+1}-u_{2h}^{n-1})) \quad (16)$$

and the numerical scheme is stable: there exists a positive number $M = M(T)$ such that

$$\mathcal{E}^n \leq M\Delta t \sum_{k=1}^{n-1}(S_1((f_1^k)^2)+S_2((f_2^k)^2)) \; \forall n \quad (17)$$

Remark 4 *For more details on the derivation of these energy relations (and their application to convergence and stability analysis) see [4].*

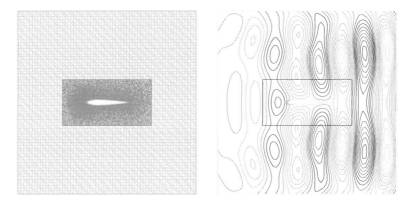

Fig. 3. *Finite element mesh (left) and contour plot of the real part of the solution for $f = 0.6$ (right). Incident wave is coming from the left.*

5 Numerical experiments

In order to solve the system of linear equations (11)–(12) at each time step we use a Conjugate Gradient Algorithm in the form given by Glowinski and LeTallec [2].

We consider problem (9)–(10) with a source term given by the harmonic planar wave:

$$u^{inc} = -e^{ik(t-\mathbf{a}\cdot\mathbf{x})}, \tag{18}$$

where $\{x_j\}_{j=1}^2$, $\{a_j\}_{j=1}^2$, k is the angular frequency and $|\mathbf{a}| = 1$.

For our numerical simulation we consider the following cases: the first with the frequency of the incident wave $f = 0.6$ GHz, the second with $f = 1.2$ GHz, and the third one with $f = 1.8$ GHz which gives us wavelengths $L = 0.5$ meters, $L = 0.25$ meters and $L = 0.16$ meters respectively.

We performed numerical computations for the case when the obstacle is an airfoil with a coating (Figure 3) and Ω is a 2 meter × 2 meter rectangle. The coating region Ω_2 is moon shaped and $\varepsilon_2 = 1$ and $\mu_2 = 9$.

We show in Figure 4 the contour plot of the real part of the solution for the incident frequency $f = 1.2$ GHz and $f = 1.8$ GHz for the case when the incident wave is coming from the left.

A crucial observation for the numerical experiments mentioned is that despite the fact that a mesh discontinuity takes place over γ together with a weak forcing of the matching conditions, we do not observe a discontinuity of the computed fields.

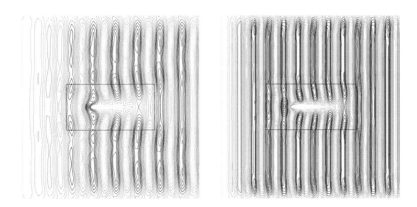

Fig. 4. Contour plot of the real part of the solution for $f = 1.2$ (left) and $f = 1.8$ (right). Incident wave is coming from the left.

References

1. Glowinski, R.: Finite Element Methods for Incompressible Viscous Flow Vol. IX of *Handbook of Numerical Analysis*, Eds. P.G. Ciarlet and J.L. Lions. North-Holland, Amsterdam (2003)
2. Glowinski, R., LeTallec, P.: Augmented Lagrangian and Operator Splitting Methods in Nonlinear Mechanics. SIAM,Phildelphia, PA (1989)
3. Bristau, M. O., Dean, E.J., Glowinski, R., Kwok, V., Periaux, J.: Exact Controllability and Domain Decomposition Methods with Non-matching Grids for

the Computation of Scattering Waves. In: Domain Decomposition Methods in Sciences and Engineering. John Wiley and Sons (1997)
4. Lapin, S.: Computational Methods in Biomechanins and Physics. PhD Thesis, University of Houston, Houston (2005)

Galbrun's Equation Solved by a First Order Characteristics Method

Rodolfo Rodríguez[1] and Duarte Santamarina[2]

[1] GI^2MA, Departamento de Ingeniería Matemática Universidad de Concepción
  Casilla 160-C, Concepción, Chile
  `rodolfo@ing-mat.udec.cl`
[2] Departamento de Matemática Aplicada, Universidade de Santiago de
  Compostela, 15706 Santiago de Compostela, Spain
  `duartesr@usc.es`

Summary. This paper deals with a time-domain mathematical model for a linearized acoustics problem in the presence of an uniform flow. First, the resulting initial-boundary value problem is rewritten in a suitable functional framework; then a time discretization is proposed. Finally stability and error estimates are stated.

1 Galbrun's equation

Propagation of acoustic disturbances in nonuniform flows is a subject of great interest in many practical problems, particularly in transport engineering with automotive exhaust systems, aeronautical turbofan engine inlet ducts, etc. The understanding of this phenomenon is a central feature for the prediction of noise and for designing components that efficiently attenuate sound.

Galbrun's equation is used to study sound propagation in flows. It exactly describes the same physical phenomenon as the linearized Euler's equations but is derived from an Eulerian-Lagrangian description and is written only in terms of the Lagrangian perturbation of the displacement (**u**).

Let $\Omega \subset \mathbb{R}^n, (n = 2, 3)$ be the (bounded) domain filled with a perfect fluid (dissipation and heat conduction are neglected) driven by a velocity field $\mathbf{v}_X$.

An acoustic disturbance in the fluid at time t_0 (which is a small isentropic perturbation), produces a displacement $(\mathbf{u}(\mathbf{x}, t_0))$ in the position of a particle **x**. By the effect of the fluid flow, that displaced particle follows another trajectory and occupies at time t_1 a position which may differ from the one that it should be occupying without perturbation in $\mathbf{u}(\mathbf{x}, t_1)$ (see Fig. 1). As we are in a regime of small disturbances, a linearization of **u** can be done and the result models the physical behaviour of the acoustic problem in flows. The equation governing these small perturbations is

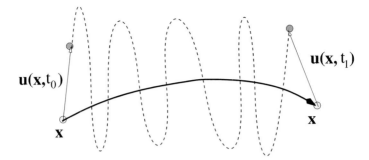

Fig. 1. Lagrangian perturbation of the displacement of a fluid particle at times t_0 and t_1.

$$\rho_X \frac{D^2 \mathbf{u}}{Dt^2} = \text{Div}\left[\rho_X c_X^2 \text{Div}\mathbf{u} I + \text{Div}\mathbf{u}\pi_X I - \pi_X \mathbf{u}^t\right] + \mathbf{f}, \quad (1)$$

which is the so-called Galbrun's equation. Where

- ρ_X stands for the density of the fluid (associated to the mean flow X),
- c_X is the sound speed (associated to the mean flow X),
- π_X is the static pressure (associated to the mean flow X),
- $\frac{D^2 \mathbf{u}}{Dt^2}$ represents the second total derivative of the displacement perturbation. Where the total derivative for a vector field is

$$\frac{D\mathbf{w}}{Dt} = \frac{\partial \mathbf{w}}{\partial t}(\mathbf{x}, t) + \text{grad}\mathbf{w}(\mathbf{x}, t)\mathbf{v}_X(\mathbf{x}, t),$$

which is composed of the partial derivative in time and a convective term which appears because $\mathbf{x}$ is also moving by the effect of the fluid flow. In the previous equation, $\mathbf{v}_X$ represents the velocity of the fluid flow (which is assumed to be time-independent),
- $\mathbf{f}(\mathbf{x}, t)$ is a volume force over Ω.

A detailed description of all the steps for the deduction of the above mentioned equation can be found, for example in [7], but there exist an extensive bibliography on this topic.

As a first step we assume that the static pressure π_X associated to the mean flow without perturbation is zero. This assumption is justified in [9]. Therefore, our equation is reduced to

$$\rho_X \frac{D^2 \mathbf{u}}{Dt^2} = \nabla\left[\rho_X c_X^2 \text{Div}\mathbf{u}\right] + \mathbf{f}. \quad (2)$$

To obtain a complete meaningful problem we will need to impose boundary and initial conditions:

$$\mathbf{u} \cdot \boldsymbol{\nu} = 0, \quad \text{on } \Gamma \times (0, T), \quad (3)$$

$$\mathbf{u}(\mathbf{x}, 0) = \mathbf{u}_0(\mathbf{x}), \quad \text{in } \Omega, \quad (4)$$

$$\frac{D\mathbf{u}}{Dt}(\mathbf{x},0) = \mathbf{v}_0(\mathbf{x}), \qquad \text{in } \Omega, \tag{5}$$

where $\Gamma = \partial\Omega$. Notice that in (3), viscosity effects have been neglected.

To obtain a weak formulation we multiply by a test function $\mathbf{w} \in H_0(\text{Div}, \Omega)$, where,

$$H_0(\text{Div}, \Omega) = \{\mathbf{w} \in H(\text{Div}, \Omega) \,:\, \mathbf{w} \cdot \boldsymbol{\nu} = 0 \text{ on } \Gamma\}.$$

We integrate in Ω and apply Green's formula to obtain:

$$\int_\Omega \frac{D^2}{Dt^2}\mathbf{u}\cdot\mathbf{w} + \int_\Omega c_X^2 \,\text{Div}\,\mathbf{u}\,\text{Div}\,\mathbf{w} = \int_\Omega \frac{1}{\rho_X}\mathbf{f}\cdot\mathbf{w} \qquad \forall \mathbf{w} \in H_0(\text{Div}, \Omega); \tag{6}$$

for the sake of simplicity we have considered ρ_X and c_X constant parameters.

In what follows we will introduce an abstract functional framework which will allow us to analyze the problem written above as well as its numerical approximation.

- Let $V = H_0(\text{Div}, \Omega)$ and $H = L^2(\Omega)^n$ be Hilbert spaces endowed with respective norms $\|\cdot\|$ and $|\cdot|$. Identifying H with its topological dual space and denoting by V' the dual space of V, we have the following dense inclusions:

$$V \hookrightarrow H \hookrightarrow V'.$$

We also denote by $(\cdot,\cdot)$ the corresponding inner product on H, by $\|\cdot\|_*$ the corresponding dual norm on V', and by $\langle\cdot,\cdot\rangle$ the duality pairing between V' and V.

- Let $a\colon V \times V \to \mathbb{R}$, be the continuous bilinear form:

$$a(\mathbf{u},\mathbf{w}) := \int_\Omega c_X^2 \,\text{Div}\,\mathbf{u}\,\text{Div}\,\mathbf{w}.$$

The bilinear form a is satisfies Garding's inequality in V, i.e., there exists $\lambda > 0$ and $\mu > 0$ such that

$$a(\mathbf{u},\mathbf{u}) + \lambda|\mathbf{u}|^2 \geq \mu\|\mathbf{u}\|^2 \qquad \forall \mathbf{u} \in V.$$

- $L(t) \in V'$ defined by

$$\langle L(t), \mathbf{v}\rangle := \int_\Omega \frac{1}{\rho_X}\mathbf{f}\cdot\mathbf{v} \qquad \mathbf{v} \in V.$$

The complete variational problem reads as follows (see [4]).

Problem 1. To find a vector field $\mathbf{u}\colon [0,T] \to V$ satisfying

$$\left\langle \frac{D^2\mathbf{u}}{Dt^2}(\cdot), \mathbf{w}\right\rangle + a(\mathbf{u}(\cdot),\mathbf{w}) = \langle L(\cdot),\mathbf{w}\rangle \qquad \forall \mathbf{w} \in V,$$

and the initial conditions

$$\mathbf{u}(\mathbf{x}, 0) = \mathbf{u}_0 \quad \text{and} \quad \frac{D\mathbf{u}}{Dt}(\mathbf{x}, 0) = \mathbf{v}_0.$$

2 Characteristic curves

In this section we define the characteristic lines associated to a time-dependent vector field $\mathbf{v}_X$ (velocity of the flow without perturbation), and study some properties satisfied by them. The aim is to numerically solve Problem 1 using characteristics for time-integration (see [6]). We propose a well posed scheme for its numerical solution, for which stability and consistency error are stated.

Characteristic lines/curves associated with vector field $\mathbf{v}_X$ through $(\mathbf{x}, t)$ in $\omega \times (0, T)$ are the trajectories of a vector function

$$\begin{aligned} X(\mathbf{x}, t; \cdot) : (0, T) &\longrightarrow \mathbb{R}^n, \\ \tau &\longrightarrow X(\mathbf{x}, t; \tau), \end{aligned}$$

which are obtained by solving the initial value problem

$$\begin{cases} \dfrac{\partial X}{\partial \tau}(\mathbf{x}, t; \tau) = \mathbf{v}_X(X(\mathbf{x}, t; \tau), \tau), \\ X(\mathbf{x}, t; t) = \mathbf{x}. \end{cases} \quad (7)$$

Remark 1. If $X(\mathbf{x}, t_0; \tau)$ is the movement of the fluid and $\mathbf{x}$ is the position of a particle at time t_0 (i.e. $\mathbf{x} = X(\mathbf{x}, t_0; t_0)$), at time t_1 the particle will be occupying position $X(\mathbf{x}, t_0; t_1)$.

The basic idea of the characteristics method is well known in convection-diffusion problems. There exists an extensive bibliography from the begining of the eighties ([6, 5]) up to now ([1, 2, 8]). For example, if we have a first order total time derivative, it is usually dicretized by the two point formula

$$\frac{\partial}{\partial \tau} \phi(X(\mathbf{x}, t_{n+1}; \tau), \tau) \xrightarrow{discret.} \frac{\phi(X(\mathbf{x}, t_{n+1}; t_{n+1}), t_{n+1}) - \phi(X(\mathbf{x}, t_{n+1}; t_n), t_n)}{\Delta t}$$

$$= \frac{\phi^{n+1}(\mathbf{x}) - \phi^n(X(\mathbf{x}, t_{n+1}; t_n))}{\Delta t}.$$

Remark 2. Commonly, an approximation is also done to compute the characteristic curves, for example:

$$X(\mathbf{x}, t_{n+1}; t_n) \approx \mathbf{x} - \mathbf{v}_X^{(n+1)}(\mathbf{x}) \Delta t.$$

which is an Euler approximation of the ODE (7). Higher order methods have also been used in order to obtain more accurate approximate solutions (see for example [1, 8]).

In this work we assume that the characteristic lines are computed exactly.

Proposition 1. Let the velocity field, $\mathbf{v}_X \in \mathcal{C}^0(0,T;\mathcal{C}(\Omega))^n$, be Lipschitz continuous with respect to the first variable and vanishing on Γ. Then,

$$X(\mathbf{x},t;\cdot) : \tau \in [0,T] \longrightarrow \overline{\Omega}.$$

Remark 3. As a consecuence of Proposition 1, our domain Ω is fixed, it is not time-dependent.

3 Numerical approximation

Let us introduce a time discretization scheme to solve Problem 1. We start with some notation.

- Let us denote by
$$X^{p,q}(\mathbf{x}) := X(\mathbf{x},t_q;t_p)$$
the characteristic lines.
- We divide the time interval $[0,T]$ in N equidistant subintervals and we denote $\Delta t = T/N$ the measure of each of those subintervals. We denote the mesh points $t_n = n\Delta t$ for $n = 0, 1, \ldots, N$.
- In what follows, we use the notation $\mathbf{u}^n(\mathbf{x}) := \mathbf{u}(\mathbf{x}, t_n)$ for a function $\mathbf{u}(\mathbf{x},t)$.
- We supose that $L \in \mathcal{C}(0,T;V')$ can be evaluated at the mesh points.

Lemma 1. Let $\mathbf{v}_X \in \mathcal{C}^0\left(0,T;\mathrm{W}^{1,\infty}(\Omega)\right)^n$ and $\mathbf{v}_X \equiv \mathbf{0}$ on Γ. If $\mathbf{u} \in \mathrm{L}^2(\Omega)^n$, then

$$|\mathbf{u} \circ X^{p,q}|^2 \leq e^{\|\mathbf{v}_X\|_{\mathcal{C}^0(0,T;\mathrm{W}^{1,\infty}(\Omega))^n}|t_q - t_p|}|\mathbf{u}|^2, \qquad (8)$$

for $p, q = 0, \ldots, N$.

Corollary 1. Under the same assumptions of previous Lemma, if $\mathbf{u} \in \mathrm{L}^2(\Omega)^n$ and $\|\mathbf{v}_X\|_{\mathcal{C}^0(0,T;\mathrm{W}^{1,\infty}(\Omega))^n}|t_q - t_p| < 1$, then there exists a positive constant C, such that

$$|\mathbf{u} \circ X^{p,q}|^2 \leq (1 + C|t_q - t_p|)|\mathbf{u}|^2, \qquad (9)$$

for $p, q = 0, \ldots, N$ and where C is related to $\|\mathbf{v}_X\|_{\mathcal{C}^0(0,T;\mathrm{W}^{1,\infty}(\Omega))^n}$ and some residual terms.

The proof of these two previous results can be found in [1].

Remark 4. In our particular case, as Galbrun's equation is only a second order equation in time, we will only have that $(t_q - t_p) = 2\Delta t$, i.e. $|q - p| = 2$, therefore the time discretization step will need to verify

$$\Delta t \leq \frac{1}{2\|\mathbf{v}_X\|_{\mathcal{C}^0(0,T;\mathrm{W}^{1,\infty}(\Omega))^n}}.$$

3.1 Time discretization

As a first approach we introduce the following algorithm.

Algorithm 1. *For* $\mathbf{u}_0, \mathbf{v}_0 \in V$ *and* $L^n \in V'$, $n = 0, \ldots, N$, *being given data, let:*

- $\mathbf{u}^0 = \mathbf{u}_0$,
- $\mathbf{u}^1 \in V$ *be the solution of*

$$\mathbf{u}^1 = \mathbf{u}_0 \circ X^{0,1} + \Delta t \, \mathbf{v}_0 \circ X^{0,1}, \tag{10}$$

- *and, for* $n = 1, \ldots, N-1$, $\mathbf{u}^{n+1} \in V$ *be the solution of*

$$\left(\frac{\mathbf{u}^{n+1} - 2\mathbf{u}^n \circ X^{n,n+1} + \mathbf{u}^{n-1} \circ X^{n-1,n+1}}{\Delta t^2}, \mathbf{w} \right) + a\left(\mathbf{u}^{n+1}, \mathbf{w}\right)$$
$$= \left\langle L^{n+1}, \mathbf{w} \right\rangle \quad \forall \mathbf{w} \in V. \tag{11}$$

Remark 5. As it can be seen in Section 3.3, it is obvious that this discretizacion will generate a solution of order Δt in time.

Remark 6. It is simple to show that this algorithm is well-posed for any $\Delta t > 0$. Consequently, we have existence and uniqueness of the semi-discrete solution.

Remark 7. Other more complicated schemes can be used by combining the Newmark method with a higher-order mehtod of characteristics to obtain second-order methods in Δt (see [1, 2, 8]).

3.2 Stability

The following results can be obtained:

Lemma 2. *If*

$$\Delta t < \frac{1}{2\|\mathbf{v}_X\|_{\mathcal{C}^0(0,T;W^{1,\infty}(\Omega))^n}},$$

then

$$\left|\mathbf{u}^1\right| + \left|\frac{\mathbf{u}^1 - \mathbf{u}^0 \circ X^{0,1}}{\Delta t}\right| \leq C\left[|\mathbf{u}_0| + |\mathbf{v}_0|\right].$$

PROOF.- The result follows by straightforward computations from (10). ∎

Theorem 1. *If* $L \in \mathcal{C}(0, T; V')$, *then there exists a* $C > 0$ *such that, for any time step* Δt, *satistfying*

$$\left[\Delta t + 3\lambda(\Delta t)^2\right] < 2 \quad \text{and} \quad \Delta t < \frac{1}{2\|\mathbf{v}_X\|_{\mathcal{C}^0(0,T;W^{1,\infty}(\Omega))^n}}, \tag{12}$$

we have

$$\|\mathbf{u}^{n+1}\| + \left|\frac{\mathbf{u}^{n+1} - \mathbf{u}^n \circ X^{n,n+1}}{\Delta t}\right|$$
$$\leq C\left[\|\mathbf{u}_0\| + |\mathbf{v}_0| + \Delta t \sum_{r=1}^{n} \|L^{r+1}\|_*\right]$$
$$n = 0, \ldots, N-1.$$

PROOF.- The proof of this result follows from combining stability theorems in [1] and [3]. ∎

3.3 Error estimate

The aim of the present section is to estimate the difference between the *discrete* solution $\mathbf{u}^n$ and the exact solution of the continuous problem $\mathbf{u}(\mathbf{x}, t_n)$. As usual, we need to impose some extra regularity assumptions.

We introduce the Banach space for a non-negative integer m

$$Z^m = \{\varphi \in \mathcal{C}^j(0,T; \mathrm{H}^{m-j}(\Omega)^n);\ 0 \leq j \leq m\}.$$

The following theorem states error estimates for the approximate solution obtained with Algorithm 1, provided that the solution of Problem 1 is smooth enough.

Lemma 3. *Let* $\mathbf{u}_0, \mathbf{u}_1 \in V$ *and* $L \in \mathcal{C}(0,T; V')$. *Let* $\mathbf{u}$ *be the solution of Problem 1. Let* $\mathbf{u}^1$ *be the solution of our Algorithm 1. If*

$$\mathbf{u} \in Z^2,$$

then there exists a positive constant C *such that*

$$\left|\mathbf{u}(t_1) - \mathbf{u}^1\right| \leq C\Delta t \|\mathbf{u}\|_{Z^2}.$$

Theorem 2. *Let* $\mathbf{u}_0, \mathbf{u}_1 \in V$, *and* $L \in \mathcal{C}(0,T; V')$. *Let* $\mathbf{u}$ *be the solution of Problem 1. Let* $\mathbf{u}^1$ *and* $\mathbf{u}^{n+1}$, $n = 1, \ldots, N-1$, *be obtained by means of Algorithm 1 with* $L^n = L(t_n)$. *If*

$$\mathbf{u} \in Z^3$$

then there exist a positive constant C *such that for every time-step* $\Delta t > 0$ *satisfying constraint (12), the following error estimates hold true for all* $r = 0, \ldots, N-1$:

$$\left|\frac{D\mathbf{u}}{Dt}(t_{r+1}) - \frac{\mathbf{u}^{r+1} - \mathbf{u}^r \circ X^{r,r+1}}{\Delta t}\right| \leq C\Delta t \|\mathbf{u}\|_{Z^3}, \tag{13}$$

$$\|\mathbf{u}(t_{r+1}) - \mathbf{u}^{r+1}\| \leq C\Delta t \|\mathbf{u}\|_{Z^3}. \tag{14}$$

Acknowledgment

R. Rodríguez was partially supported by FONDAP in Applied Mathematics, Chile. D. Santamarina is partially supported by MEC research project DPI2004-05504-C02-02, Spain.

References

1. Bermúdez, A., Nogueiras, M., Vázquez, C.: Numerical analysis of degenerate convection-diffusion-reaction problems with higher order characteristics/finite elements. Part I: Time discretization. (To appear in) SIAM J. Numer. Anal.
2. Bermúdez, A., Nogueiras, M., Vázquez, C.: Numerical analysis of degenerate convection-diffusion-reaction problems with higher order characteristics/finite elements. Part II: Fully discretized scheme and quadrature formulas. (To appear in) SIAM J. Numer. Anal.
3. Bermúdez, A., Rodríguez, R., Santamarina, D.: Finite element approximation of a displacement formulation for time-domain elastoacoustic vibrations. J. Comput. Appl. Math., **152**, 17–34 (2003).
4. Berriri, K., Bonnet-Bendhia, A. S., Joly, P.: Numerical analysis of time-dependent Galbrun Equation in an infinite duct. (To appear) Waves'05.
5. Douglas, J. Jr., Russell, T.F.: Numerical methods for convection-dominated diffusion problems based on combining the method of characteristics with finite element or finite difference procedures. SIAM J. Numer. Anal., **19**, 871–885 (1982).
6. Pironneau, O.: On the transport diffusion algorithm and its application to the Navier-Stokes equation. Numer. Math., **38**, 309–332 (1982).
7. Poirée, B.: Les équations de l'acoustique linéaire et non-linéaire dans un écoulement fluide parfait. Acustica, **57**, 5–25 (1985).
8. Rui, H., Tabata, M.: A second order characteristic finite element scheme for convection-diffusion problems. Numer. Math., **92**, 161–177 (2002).
9. Treyssède, F., Gabard G., Ben Tahar M.: A mixed f.e.m. for acoustic wave propagation in moving fluids based on an Eulerian-Lagrangian description. J. Acoust. Soc. Am., **113**(2), 705–716 (2003).

Open Subsystems of Conservative Systems

Alexander Figotin[1] and Stephen P. Shipman[2]

[1] University of Californina, Irvine, CA 92697
    afigotin@uci.edu
[2] Louisiana State University, Baton Rouge, LA 70803
    shipman@math.lsu.edu

Summary. The subject under study is an open subsystem of a larger linear and conservative system and the way in which it is coupled to the rest of system. Examples are a model of crystalline solid as a lattice of coupled oscillators with a finite piece constituting the subsystem, and an open system such as the Helmholtz resonator as a subsystem of a larger conservative oscillatory system. Taking the view of an observer accessing only the open subsystem we ask, in particular, what information about the entire system can be reconstructed having such limited access. Based on the unique minimal conservative extension of an open subsystem, we construct a canonical decomposition of the conservative system describing, in particular, its parts coupled to and completely decoupled from the open subsystem. The coupled one together with the open system constitute the unique minimal conservative extension. Combining this with an analysis of the spectral multiplicity, we show, for the lattice model in particular, that *only a very small part of all possible oscillatory motion of the entire crystal, described canonically by the minimal extension, is coupled to the finite subsystem.* **Keywords:** open system, subsystem, conservative extension, coupling, delayed response, reconstructible. ©A Figotin, SP Shipman

1 Overview

When one has to treat a complex evolutionary system involving a large number of, or infinitely many, variables, it is common to reduce it to a smaller system by eliminating certain "hidden" variables. The reduced system, involving only the "observable" variables, becomes a non-conservative, or *open system*, even if the underlying system is conservative, or closed. This is not surprising since generically any part of a conservative system interacts with the rest of it. In the reduced system, the interaction with the hidden variables is encoded in its dispersive dissipative (DD) properties. For classical material media, including dielectric, elastic, and acoustic, the interaction between proper fields and the matter, which constitutes the hidden part of the system, is encoded into the so-called material relations, making them frequency dependent and consequently making the open system dispersive and dissipative.

Often it is an open DD system, described by frequency-dependent material relations, that we are given to study, and the conservative system in which the open system is naturally embedded may be very complicated. A natural question is, how much information about the underlying conservative system remains in the reduced open one? The answer is provided by the construction of the minimal conservative extension of the given DD system [2], which is unique up to isomorphism. This minimal extension is a part of the entire conservative system—it is the part that is detectable by the open system through the coupling to the entire system. We ask, how big a part of the original conservative system is this minimal extension? This is a question we address in this paper. The answer is clearly related to the nature of the coupling between the observable and hidden variables. Although the term "coupling" is commonly used to describe interactions, its precise meaning must be defined in each concrete problem. We make an effort to provide a general constructive mathematical framework for the treatment of the coupling.

In this paper, we concentrate on the detection of one part of a system by another, or, equivalently, the extent of reconstructibility of a conservative system from the dynamics of an open subsystem. More detailed analysis of this problem as well as the study of the decomposition of open systems by means of their conservative extensions will be presented in another work.

Motivating examples

We have already mentioned the classical problems of electromagnetic, acoustic, and elastic waves in matter. Detailed accounts of the construction of the minimal conservative extension are given in [2, 8].

Another important example is of an object coupled to a heat bath through surface contact. It has been observed for crystalline solids that certain degrees of freedom do not contribute to the specific heat [3, Section 3.1], [4, Section 6.4]. It appears that some of the admissible motions of the solid cannot be excited by the heat bath through the combination of surface contact and internal dynamics. This can be explained though high multiplicity of eigenmodes arising from symmetries of the crystal.

A concrete toy model consists of an infinite three-dimensional lattice of point masses as the total system, each mass being coupled to its nearest neighbors by springs, and a finite cube thereof as the observable subsystem. The coupling of the cube to the rest of the lattice takes place only between the masses on the surface of the cube and their nearest neighbors outside the cube. We discuss this system in Example 1 below, in which we show that the cube is able to detect only a relatively small part of the entire lattice, the rest of which remains dynamically decoupled.

One more example is the phenomenon of anomalous acoustic or electromagnetic transmission through a material slab, or film, can also be viewed from the point of view of coupled systems. The governing equation is the wave equation or the Maxwell system in space. A leaky guided mode in a material slab interacts with plane wave sources from outside the slab, giving rise to anomalous scattering behavior [7, 6]. A single mode of the slab constitutes a one-dimensional subsystem, which, under weak coupling to the ambient medium, say air, interacts with a portion of the entire system in space, decoupled from the rest. We do not analyze this problem in this paper, but attempt to develop a framework for studying like problems.

List of symbols

H_1, H_2, $\mathcal{H}$: Hilbert spaces
v_1, v_2, $\mathcal{V}$, f_1, f_2, $\mathcal{F}$: Hilbert space-valued functions of time
Γ, Ω_1, Ω_2, Ω, $\mathring{\Omega}$, $\mathring{\Gamma}$: operators in Hilbert space
a_1, a_2: operator-valued functions of time
$\mathcal{O}$: orbit
Ran: range
dim: dimension
$\mathbb{C}$: the complex number field
$\mathbb{Z}$: the ring of integers
$\mathcal{Q}$: a cube in $\mathbb{Z}^3$
Δ_j: finite difference operators

2 Open systems within conservative extensions

Often an observable open system in a Hilbert space H_1 of the form

$$\partial_t v_1(t) = -i\Omega_1 v_1(t) - \int_0^\infty a_1(\tau) v_1(t-\tau) \, mD\tau + f_1(t) \quad \text{in } H_1, \qquad (1)$$

in which $a_1(t)$ is the operator-valued delayed response, or retarded friction, function, is known to be a subsystem of a linear conservative system in a larger Hilbert space $\mathcal{H}$, in which the dynamics are given by

$$\partial_t \mathcal{V}(t) = -i\Omega \mathcal{V}(t) + \mathcal{F}(t), \quad \mathcal{V}(t), \mathcal{F}(t) \in \mathcal{H}, \qquad (2)$$

where $\Omega : \mathcal{H} \to \mathcal{H}$ is the self-adjoint frequency operator. The structure of the open system within the conservative one can be seen by introducing the space H_2 of hidden variables, defined to be the orthogonal complement of H_1 in $\mathcal{H}$: $H_2 = \mathcal{H} \ominus H_1$. With respect to the decomposition $\mathcal{H} = H_1 \oplus H_2$, Ω has the form

$$\Omega = \begin{bmatrix} \Omega_1 & \Gamma \\ \Gamma^\dagger & \Omega_2 \end{bmatrix}, \qquad (3)$$

in which Ω_1 and Ω_2 are the self-adjoint frequency operators for the internal dynamics in H_1 and H_2, and $\Gamma : H_2 \to H_1$ is the coupling operator. In this paper, we assume for simplicity that Γ is bounded. The results hold, essentially unchanged, for unbounded coupling; details of how to treat this case are handled in [2]. The dynamics (2) with respect to the decomposition into observable and hidden variables become

$$\partial_t v_1(t) = -i\Omega_1 v_1(t) - i\Gamma v_2(t) + f_1(t), \quad v_1(t), f_1(t) \in H_1, \quad (4)$$
$$\partial_t v_2(t) = -i\Gamma^\dagger v_1(t) - i\Omega_2 v_2(t) + f_2(t), \quad v_2(t), f_2(t) \in H_2.$$

Solving for $v(t)$ gives

$$\partial_t v_1(t) = -i\Omega_1 v_1(t) - \int_0^\infty \Gamma m E^{-i\Omega_2 \tau} \Gamma^\dagger v_1(t-\tau) \, mD\tau + f_1(t) \quad \text{in } H_1, \quad (5)$$

from which we see that the delayed response function $a_1(t)$ is related to the dynamics of the hidden variables and the coupling operator by

$$a_1(t) = \Gamma m E^{-i\Omega_2 t} \Gamma^\dagger, \quad (6)$$

and it is straightforward to show that $a_1(t)$ satisfies the no-gain dissipation condition

$$\operatorname{Re} \int_0^\infty \int_0^\infty \overline{v(t)} a(\tau) v(t-\tau) \, mDt \, mD\tau \geq 0 \quad \text{for all } v(t) \text{ with compact support.} \quad (7)$$

A natural question to ask is whether every system of the form (1) whose friction function $a_1(t)$ satisfies the condition (7) is a subsystem of a conservative system. The answer is positive, and there exists in fact a unique minimal extension up to isomorphism [2]. This extension, or, equivalently, the form (6), is canonically constructible through the Fourier-Laplace transform $\hat{a}_1(\zeta)$ of $a_1(t)$. It follows that all open systems of this type can be studied as a subsystem of a larger closed one.

This minimal conservative extension should be viewed as the space H_1 of observable variables coupled to the subspace of the original space of hidden variables H_2 that is detectable by the observable system; we denote this coupled subspace by H_{2c}. The influence on H_1 of this subsystem of hidden variables is manifest by $a_1(t)$ and reconstructible by $a_1(t)$, up to isomorphism. The decoupled part of H_2, denoted by $H_{2d} = H_2 \ominus H_{2c}$, is not detectable by the reduced open system (5) in H_1.

The detectable part of the hidden variables may be a very restricted subspace of the naturally given space of hidden variables. We will show that, if the coupling is of finite rank, in particular, if the observable system is finite dimensional, then the spectral multiplicity of the conservative extension is finite. This leads to the following observation: Suppose our system of hidden variables is modeled by nearest-neighbor interactions in an infinite multidimensional lattice or the Laplace operator in continuous space, both of which have infinite multiplicity, and suppose that our observable system is a finite-dimensional resonator (perhaps very large, but finite). Then there is a huge subspace of the hidden variables that is not detected by the resonator, in other words, there are many hidden degrees of freedom that are not detected by the resonator, and which, in turn, do not influence its dynamics.

In this discussion, the roles of H_1 and H_2 may just as well be switched. One may solve for $v_2(t)$ and obtain an analogous expression to (1) with delayed response function $a_2(t) = \Gamma^\dagger m E^{-i\Omega_1 t}\Gamma$. H_1 is then decomposed into its coupled and decoupled parts: $H_1 = H_{1c} \oplus H_{1d}$.

With respect to the decomposition of $\mathcal{H}$ into the coupled and decoupled parts of the observable and hidden variables,

$$\mathcal{H} = H_{1d} \oplus H_{1c} \oplus H_{2c} \oplus H_{2d}, \tag{8}$$

the frequency operator Ω for the closed system in $\mathcal{H}$ has the matrix form

$$\Omega = \begin{bmatrix} \Omega_{1d} & 0 & 0 & 0 \\ 0 & \Omega_{1c} & \Gamma_c & 0 \\ 0 & \Gamma_c^\dagger & \Omega_{2c} & 0 \\ 0 & 0 & 0 & \Omega_{2d} \end{bmatrix}. \tag{9}$$

The minimal conservative extension of the system (5) in H_1 within the given system $(\mathcal{H}, \Omega)$ is the space generated by H_1 through Ω, or the *orbit* of H_1 under Ω, denoted by $\mathcal{O}_\Omega(H_1)$. Similar reasoning can be applied to H_2. We therefore obtain

$$\mathcal{O}_\Omega(H_1) = H_1 \oplus H_{2c}, \tag{10}$$
$$\mathcal{O}_\Omega(H_2) = H_{1c} \oplus H_2. \tag{11}$$

The orbit of a subset S of $\mathcal{H}$ is

$$\mathcal{O}_\Omega(S) = \text{closure of } \{f(\Omega)v \mid f \in C_0^\infty(\mathbb{R}), v \in S\}.$$

If Ω is bounded, $\mathcal{O}_\Omega(S)$ is equal to the smallest subspace of $\mathcal{H}$ containing S that is invariant, or closed, under Ω. Equivalently, it is the smallest subspace of $\mathcal{H}$ containing S that is invariant under $(\Omega - i)^{-1}$; this latter formulation is also valid for unbounded operators. The relevant theory can be found, for example, in [1] or [5].

The closed subsystem $(H_{1c} \oplus H_{2c}, \Omega_c)$ with frequency operator

$$\Omega_c = \begin{bmatrix} \Omega_{1c} & \Gamma_c \\ \Gamma_c^\dagger & \Omega_{2c} \end{bmatrix},$$

is in fact *reconstructible* by either of the open subsystems $(H_{1c}, \Omega_{1c}, a_1(t))$ or $(H_{2c}, \Omega_{2c}, a_2(t))$. Equivalently, $(H_{1c} \oplus H_{2c}, \Omega_c)$ is the unique minimal conservative extension, realized as a subsystem of $(\mathcal{H}, \Omega)$, of each of its open components separately. This motivates the following definition.

Definition 1 (reconstructibility). *We call a system $(\mathcal{H}, \Omega)$ together with the decomposition $\mathcal{H} = H_1 \oplus H_2$ reconstructible if $H_{1d} = 0$ and $H_{2d} = 0$, that is, $(\mathcal{H}, \Omega)$ is the minimal conservative extension of each of its parts.*

The next theorem asserts the existence of a unique reconstructible subsystem of $(\mathcal{H}, \Omega)$ that contains the images of Γ and $\Gamma^\dagger$ and gives a bound on the multiplicity of Ω, as we have discussed above. Define

$$\mathring{\Omega} = \begin{bmatrix} \Omega_1 & 0 \\ 0 & \Omega_2 \end{bmatrix}, \quad \mathring{\Gamma} = \begin{bmatrix} 0 & \Gamma \\ \Gamma^\dagger & 0 \end{bmatrix}. \tag{12}$$

Theorem 1 (system reconstruction). *Define*

$$H_{2c} = \mathcal{O}_\Omega(H_1) \ominus H_1, \qquad H_{2d} = H_2 \ominus H_{2c}, \tag{13}$$
$$H_{1c} = \mathcal{O}_\Omega(H_2) \ominus H_2, \qquad H_{1d} = H_1 \ominus H_{1c}. \tag{14}$$

Then

$$H_{1c} \oplus H_{2c} = \mathcal{O}_\Omega(H_{1c}) = \mathcal{O}_\Omega(H_{2c}) = \mathcal{O}_\Omega(\operatorname{Ran}\mathring{\Gamma}). \tag{15}$$

In particular, $H_{1c} \oplus H_{2c}$ is reconstructible and

$$\text{multiplicity}\,(\Omega_c) \leq \min\left(2\operatorname{rank}(\Gamma), \dim(H_{1c}), \dim(H_{2c})\right), \tag{16}$$

in which Ω_c denotes the restriction of Ω to $H_{1c} \oplus H_{2c}$.

Proof. That $H_{1c} \oplus H_{2c}$ is invariant under $(\Omega - i)^{-1}$ (or Ω, if Ω is bounded) and contains the range of $\mathring{\Gamma}$ is evident from the decomposition (9) of Ω. To prove the first equality in (15), let

$$\mathcal{O}_\Omega(H_{1c}) = H_{1c} \oplus H'_{2c},$$

in which $H_{2c} = H'_{2c} \oplus H''_{2c}$. We see that $H_1 \oplus H'_{2c}$ is closed under $(\Omega - i)^{-1}$, and since $H_1 \oplus H_{2c}$ is the smallest subspace of $\mathcal{H}$ that is closed under $(\Omega - i)^{-1}$, we have $H''_{2c} = 0$. The second equality in (15) is proved similarly.

Since $\operatorname{Ran}\mathring{\Gamma} \subseteq H_{1c} \oplus H_{2c}$, we have $\mathcal{O}_\Omega(\operatorname{Ran}\mathring{\Gamma}) \subseteq H_{1c} \oplus H_{2c}$. It remains to be proved that $H_{1c} \oplus H_{2c} \subseteq \mathcal{O}_\Omega(\operatorname{Ran}\mathring{\Gamma})$. First,

$$\mathcal{O}_\Omega(\operatorname{Ran}\mathring{\Gamma}) = \mathcal{O}_{\Omega,\mathring{\Gamma}}(\operatorname{Ran}\mathring{\Gamma}) \tag{17}$$
$$[\text{because } \mathring{\Gamma}(\mathcal{O}_\Omega(\operatorname{Ran}\mathring{\Gamma})) = \operatorname{Ran}\mathring{\Gamma} \subseteq \mathcal{O}_\Omega(\operatorname{Ran}\mathring{\Gamma})]$$
$$= \mathcal{O}_{\mathring{\Omega},\mathring{\Gamma}}(\operatorname{Ran}\mathring{\Gamma}) \quad [\text{because } \mathring{\Omega} = \Omega - \mathring{\Gamma}] \tag{18}$$
$$= \mathcal{O}_{\mathring{\Omega}}(\operatorname{Ran}\mathring{\Gamma}) \tag{19}$$
$$= \mathcal{O}_{\mathring{\Omega}}(\operatorname{Ran}\Gamma) \oplus \mathcal{O}_{\mathring{\Omega}}(\operatorname{Ran}\Gamma^\dagger) \tag{20}$$
$$[\text{because } \operatorname{Ran}\mathring{\Gamma} = \operatorname{Ran}\Gamma \oplus \operatorname{Ran}\Gamma^\dagger]$$
$$= \mathcal{O}_{\Omega_1}(\operatorname{Ran}\Gamma) \oplus \mathcal{O}_{\Omega_2}(\operatorname{Ran}\Gamma^\dagger). \tag{21}$$

From this we see that $\mathcal{O}_{\Omega_2}(\operatorname{Ran}\Gamma^\dagger) \subseteq H_{2c}$. Since $\left(H_2 \ominus \mathcal{O}_{\Omega_2}(\operatorname{Ran}\Gamma^\dagger)\right) \perp \operatorname{Ran}(\Gamma^\dagger)$, we have that $H_1 \oplus \mathcal{O}_{\Omega_2}(\operatorname{Ran}\Gamma^\dagger)$ is $(\Omega-i)^{-1}$-invariant, so that $H_{2c} \subseteq \mathcal{O}_{\Omega_2}(\operatorname{Ran}\Gamma^\dagger)$ by the minimality of $H_1 \oplus H_{2c}$ with respect to closure under $(\Omega-i)^{-1}$. Therefore, $H_{2c} = \mathcal{O}_{\Omega_2}(\operatorname{Ran}\Gamma^\dagger)$; similarly, $H_{1c} = \mathcal{O}_{\Omega_1}(\operatorname{Ran}\Gamma)$. We conclude that $H_{1c} \oplus H_{2c} = \mathcal{O}_\Omega(\operatorname{Ran}\mathring{\Gamma})$, and this finishes the proof of (15).

To prove (16), note that the multiplicity of Ω_c is the minimal number (which could be infinity) of generating vectors needed to generate $H_{1c} \oplus H_{2c}$ by Ω_c, or, equivalently, by Ω. Thus, by (15), the multiplicity of Ω_c is bounded by the dimension of H_{1c}, the dimension of H_{2c}, and the dimension of the range of $\mathring{\Gamma}$. Since the dimensions of the ranges of Γ and $\Gamma^\dagger$ are equal and $\operatorname{Ran}\mathring{\Gamma} = \operatorname{Ran}\Gamma \oplus \operatorname{Ran}\Gamma^\dagger$, we see that the range of $\mathring{\Gamma}$ has twice the dimension of the range of Γ. This completes the proof of the Theorem. ∎

Example 1 (lattice). Let $\mathcal{H}$ be the Hilbert space of square-summable complex-valued functions on the integer lattice $\mathbb{Z}^3 = \{n = (n_1, n_2, n_3) \mid n_1, n_2, n_3 \in \mathbb{Z}\}$,

$$\mathcal{H} = \left\{ f : \mathbb{Z}^3 \to \mathbb{C} \,\Big|\, \sum_{n \in \mathbb{Z}} |f(n)|^2 < \infty \right\},$$

and let Ω be the discrete Laplace operator:

$$\Omega f = \sum_{j=1}^{3} \Delta_j f,$$

in which $(\Delta_j f)(n) = f(n+e_j) - 2f(n) + f(n-e_j)$ and e_j is the j-th elementary vector in $\mathbb{Z}^3$ (e.g., $e_1 = (1, 0, 0)$).

Let H_1 be the finite-dimensional subspace of $\mathcal{H}$ consisting of complex-valued functions on the lattice cube

$$\mathcal{Q} = \{n = (n_1, n_2, n_3) \mid 0 \le n_j < N, j = 1, 2, 3\},$$

which is isomorphic to $\mathbb{C}^{N^3}$. Since Ω involves only nearest-neighbor interactions, the range of Γ is the space of complex-valued functions on the surface of $\mathcal{Q}$, which has dimension $6N^2 - 12N + 8$. Therefore, by Theorem 1, the multiplicity of the restriction of Ω to the minimal conservative extension of H_1 in $\mathcal{H}$ is no greater than $12N^2 - 24N + 16$. However, the multiplicity of Ω in $\mathcal{H}$ is infinite, showing that the restriction of Ω to H_{2d}, the decoupled part of $H_2 = \mathcal{H} \ominus H_1$ has infinite multiplicity. This is a very large space of degrees of freedom that are not detected by the cube $\mathcal{Q}$ and therefore do not influence its dynamics.

3 Discussion

Based on the minimal conservative extension of an open system, we develop a clear mathematical framework for the widely used concept of coupling. In this paper, we have focused on the amount of information about a conservative system that is encoded in a given open subsystem and the reconstruction of that part of the conservative system that is equivalent to the abstract minimal extension. The efficiency of the construction is demonstrated by a concrete statement showing, by analysis of spectral multiplicity, that very often this extension is a very small part of the conservative system. In ongoing work, we analyze the interaction between the spectral theories of the internal dynamics of two systems and a coupling operator between them and its bearing on the decomposition of an open system into dynamically independenty parts.

To give a sense of the potential of the approach, we mention as problems that are naturally addressed in the framework of conservative extensions (1) the classification and analysis of eigenmodes and resonances, (2) applications to the construction of dynamical models for thermodynamics, and (3) transmission of excitations in complex inhomogeneous media.

An interesting conclusion of our studies of coupling of open systems and the spectral multiplicity is that there can be degrees of freedom which are completely decoupled from the rest of the system. Since the spectral multiplicity is a consequence of a system's natural symmetries, one can consider such a decoupling as an explanation for so-called "frozen" degrees of freedom observed in the treatment of the specific heat for crystalline solids (Dulong-Petit law) [3, Section 3.1]. The analysis of the specific heat involves the law of equipartition of energy and the number of degrees of freedom, and in order to agree with the experiment one has to leave out some degrees as if they were not excited and can be "frozen", [3, Section 3.1], [4, Section 6.4].

References

1. Akhiezer, N. I., Glazman, I. M.: Theory of Linear Operators in Hilbert Space. Dover, New York (1993)
2. Figotin, A., Schenker, J.: Spectral Theory of Time Dispersive and Dissipative Systems. J. Stat. Phys. **118** (1), 199–263 (2005)
3. Gallavotti, G.: Statistical Mechanics, A Short Treatise. Springer, Berlin (1999)
4. Huang, K.: Statistical Mechanics. Wiley (1987)
5. Reed, M., Simon, B.: Functional Analysis, Vol. I. Academic Press, New York (1972)
6. Shipman, S.P., Venakides, S.: Resonant transmission near nonrobust periodic slab modes. Phys. Rev. E., **71**, 026611-1–10 (2005)
7. Tikhodeev, S.G., Yablonskii, A.L., Muljarov, E.A., Gippius, N.A., Ishihara, T.: Quasiguided modes and optical properties of photonic crystals slabs. Phys. Rev. B, **66**, 045102 (2002)
8. Tip, A.: Linear absorptive dielectrics. Phys. Rev. A, **57**, 4818–4841 (1998)

Author Index

Abdulle, A., 743
Abgrall, R., 861
Alaa, N. E., 926
Alvarez-Dios, J.A., 937
Alvarez-Vázquez, L.J., 1040
Amari, T., 917
Amat, S., 629
Ambroso, A., 852
Andallah, L. S., 217
Andrianov, N., 814
Antonietti, P. F., 423
Apel, T., 299
Aràndiga, F., 654
Araya, R., 752
Aregba, A. W., 638
Aregba-Driollet, D., 373, 638
Arregui, I., 319
Asensio, M.I., 328, 601, 1175
Audebert, B., 646
Audusse, E., 181
Ayuso, B., 328, 423

Babovsky, H., 217
Bachelot, A., 609, 1137
Baeza, A., 198, 654
Bañas, L., 531
Banasiak, J., 618
Barral, P., 1108
Barrenechea, G. R., 752
Bawa, R.K., 1079
Beirão da Veiga, L., 1051
Bendahmane, M., 381
Bentaleb, L., 761

Berezovski, A., 703
Bermejo, R., 880
Berthon, C., 843
Boffi, D., 575
Bösing, P. R., 457
Bonaventura, L., 207
Bonito, A., 487
Boulbe, C., 917
Boulmezaoud, T.Z., 917
Braack, M., 495, 770
Braconnier, B., 843
Bristeau, M.O., 181
Brusche, J.H., 585
Buffa, A., 3
Bujanda, B., 987
Bürger, R., 387
Burman, E., 487, 504, 512, 779
Busquier, S., 629
Bustinza, R., 1129

Cabaleiro, J.C., 937
Cakoni, F., 119
Campo, M., 1091
Carbou, G., 539
Carlini, E., 723, 732
Casal, G., 937
Castro, M. J., 288, 662
Castro-Díaz, M. J., 190
Cavalli, F., 404
Cea, M., 151
Cendán, J. J., 319
Chacón, T., 279
Chalons, C., 852

Cheggour, A., 926
Chen, H.Q., 1203
Ciarlet, P. Jr, 547
Claudel, J., 843
Clavero, C., 1079
Clemens, M., 996
Clopeau, T., 362
Codina, R., 21
Coquel, F., 646, 814, 852
Cristiani, E., 723

Dahmen, W., 39
Decoene, A., 181
Devigne, V.M., 362
Dolejší, V., 432
Donat, R., 654
Dostál, Z., 62
Dumas, L., 1007
Durand, O., 1007
Durany, J., 1164

El Alaoui, L., 512
Ern, A., 79, 504, 512
Escudero, A., 629

Falcone, M., 732
Farge, M., 822
Feistauer, M., 440
Felcman, J., 225
Fernández, J.R., 1091
Fernández, E. D., 279
Fernández-Nieto, E. D., 190
Ferragut, L., 601, 1175
Ferreiro, A. M., 190
Ferretti, R., 732
Fierro, F., 269
Figotin, A., 1220
Forcadel, N., 723
Frochte, J., 871

Gaevskaya, A., 308
Galán del Sastre, P., 880
García, J. A. , 288
Gardini, F., 243
Gastaldi, L., 575
Gatard, L., 1137
Gatica, G. N., 1129
Glowinski, R., 1203
Godlewski, E., 852

Gómez, M., 279
González, J. M., 288
Gorshkova, E., 252
Gravemeier, V., 788
Grimm, V., 557
Gritzki, R., 805
Guermond, J-L., 79, 796

Hallo, L., 861
Heinrich, B., 467
Heinrichs, W., 871
Heltai, L., 423, 575
Herrero, H., 889
Hoppe, H.W. H., 308
Horák, D., 62
Horáček, j., 897
Hoyas, S., 889

Ignat, L. I., 593
Iserles, A., 97
Ivorra, B., 1007

Jacquart, P.M., 1203
Jamelot, E., 547
Janovská, D., 945, 954
Janovský, V., 954
Javierre, E., 712
John, V., 336
Jorge, J.C., 987
Jung, B., 467

Karasözen, B., 1031
Karlsen, K. H., 381
Kienesberger, J., 1117
Knabner, P., 1192
Knobloch, P., 336
Knopp, T., 805
Kozakevicius, A., 387
Kozakiewicz, J.M., 618
Královcová, J., 1184
Kubera, P., 225
Kunoth, A., 39
Kuzmin, D., 233, 345

Labbé, G., 539
Lagoutière, F., 852
Lamby, P., 831
Lapin, S., 1203
Lattanzio, C., 396

Leentvaar, C.C.W., 975
Liska, R., 671
Lube, G., 805
Lukáš, D., 1015
Lyly, M., 1059

Mancho, A. M., 889
Marazzina, D., 448
Martínez, A., 1040
Maryška, J., 1184
Massjung, R., 831
Maugin, G.A., 703
Meddahi, S., 1155
Mer-Nkonga, K., 609, 1137
Miglio, E., 207
Missirlis, N. M., 354
Mohammadi, B., 1007
Möller, M., 233
Monedero, S., 601, 1175
Monk, P., 119
Morice, J., 609
Moritzen, K., 1023
Mozolevski, I., 457
Muñoz, M.L., 662
Mulet, P., 198
Müller, S., 831

Naldi, G., 404
Natesan, S., 1079
Neittaanmäki, P., 252
Niiranen, J., 1051, 1059
Nkonga, B., 843
Nørsett, S.P., 97

Olver, S., 97
Oosterlee, C.W., 975
Opfer, G., 945
Ostermann, A., 564
Ouazzi, A., 520

Parés, C., 288, 319, 662
Parumasur, N., 618
Pennacchio, M., 475
Pereira, J., 1164
Periaux, J., 1203
Perrier, V., 861
Pop, I. S., 362, 1192
Postel, M., 814
Prudhomme, S., 796

Rösch, A., 299
Radu, F. A., 1192
Ramdani, K., 1068
Rapún, M. L., 1146
Raviart, P.-A., 852
Repin, S., 135, 252, 308
Richter, T., 770
Roche, J.R., 926
Rodríguez, R., 1212
Rodríguez-Arós, A., 1099
Rodríguez-Cruz, M.S., 1175
Rommes, J., 963
Roussel, O., 761

Saleri, F., 207
Sánchez, M.T., 1108
Sánchez-Martín, M.J., 1175
Sangalli, G., 328
Santamarina, D., 1212
Sayas, F.-J., 1129, 1146
Schäfer, M., 1031
Schneider, K., 822
Segal, A., 585, 712
Seguin, N., 852
Selgas, V., 1155
Šembera, J., 1184
Semplice, M., 404
Severýn, O., 1184
Shipman, S. P., 1220
Simoncini, V., 475
Sofonea, M., 1099
Solín, P., 683, 691
Stavroulakis, G.E. , 1091
Stefanica, D., 62
Steinmetz, T., 996
Stenberg, R., 1051, 1059
Stiemer, M., 260
Stiriba, Y., 831
Süli, E., 457
Sváček, P., 897

Takahashi, T., 1068
Tanabe, K., 954
Tenaud, C., 761
Thalhammer, M., 564
Tran, Q. H., 814
Trillo, J.C., 629
Tucsnak, M., 1068
Turek, S., 520, 906

Tzaferis, F. I., 354

Uğur, Ö., 1031
Urbach, H.P., 585

Váchal, P., 671
Valdman, J., 1117
Valentin, F., 752
Vallet, G., 412
Van der Zwaag, S., 712
Van Duijn, C.J., 362
Van Noorden, T., 963
Varas, F., 1164
Vázquez, C., 319
Vázquez-Méndez, M.E., 1040
Veeser, A., 269

Vejchodský, T., 683, 691
Vermolen, F., 712
Viaño, J., 1091, 1099
Vilar, M., 1040
Vorloeper, J., 39
Vuik, C., 585, 712

Wan, D., 906
Wimmer, G., 996
Winkler, G., 299

Yapıcı, K., 1031

Zítka, M., 683, 691
Zuazua, E., 151

Printing: Krips bv, Meppel
Binding: Stürtz, Würzburg